海岸海洋科学
研究与实践

Coastal Ocean Science *Study and Practice*

（上 册）

王 颖 著

 南京大学出版社

图书在版编目(CIP)数据

海岸海洋科学研究与实践:上下册 / 王颖著.
— 南京:南京大学出版社,2021.4
ISBN 978 - 7 - 305 - 24300 - 4

Ⅰ.①海… Ⅱ.①王… Ⅲ.①海岸－海洋学－文集
Ⅳ.①P7 - 53

中国版本图书馆 CIP 数据核字(2021)第 051072 号

出版发行　南京大学出版社
社　　址　南京市汉口路 22 号　　　　邮　编　210093
出 版 人　金鑫荣

书　　名　**海岸海洋科学研究与实践(上下册)**
著　　者　王　颖
责任编辑　蔡文彬　　　　　　　编辑热线　025 - 83686531

照　　排　南京南琳图文制作有限公司
印　　刷　徐州绪权印刷有限公司
开　　本　787×1092　1/16　印张 125.75　字数 3050 千
版　　次　2021 年 4 月第 1 版　2021 年 4 月第 1 次印刷
ISBN 978 - 7 - 305 - 24300 - 4
总 定 价　498.00 元

网址:http://www.njupco.com
官方微博:http://weibo.com/njupco
官方微信号:njupress
销售咨询热线:(025) 83594756

王颖院士

序

　　整理文集出版的过程中,引发了对生平的回顾。执笔作序,浮想联翩:深深感谢父母给予我生命、抚育我成长与就学受教育;青年时代成长于新中国,共和国的建立为国人开辟了一条康庄大道,我深深感谢党和国家的关怀与培养,获得在科学、文化、道德与品质的健康成长,明确为祖国服务的终身志愿;感谢学校师长们无私的多年辛勤教导,使我长知识、获能力,坚定地在科技兴国的岗位上努力工作;感谢同事、朋友与家人的理解、支持与热心相助……这些发自肺腑的真心感谢的同时,使我领悟到:个人的成长与发展离不开时代的特点与感召。诚挚实意地响应时代发展的需要,扎扎实实地立足于本岗位工作,会获得与时代前进步伐一致的进步。

　　自己所经历的时代,具有新旧交替与变革的特点:父母生长于清末民初,封建帝制崩溃,民众共和萌芽的初始阶段,得以入学受教育,20世纪30年代国共合作,父亲王奇峰(峙亭)* 得以与刘伯承、徐向前在朱德将军领导下,三位师长指挥129师与骑兵第四师共同抗日,艰苦卓绝、转战华北与河南一带。"1938年11月,日军决心铲除骑四师,在装甲部队配合下大举来攻,王率部且战且走,一夜大雨中受寒高烧、经月不退,12月底病逝西安,得年仅41岁。"20世纪40~50年代,我和姐弟得以系统完成小学与中学教育,均考入高校,却因能否心无旁贷坚持努力学习而学有所成或半途成家肄业;或因专业不符心愿彷徨于校门退出而失去完成高等教育之区别,人生的道路自有不同。我的体会是适应国家与社会需求,安心于本岗位,不斤斤计较,毋需盲目地追求很重

---

　　* 王奇峰见《青天白日勋章》一书第78页,台湾知兵堂出版社,2011年11月。

要。新中国的建设发展赋予了青年一代成长的康庄大道。

本文集汇集之文始于 1956 年及 60 年代初，我大学毕业后的研究生学习阶段，当年号召"教育为无产阶级服务、与生产劳动相结合"，在进行海岸地貌与沉积学基础理论学习之同时，参加生产实践而从事海港选址与海岸带砂矿勘探工作。坚持近三十年的工作实践，学习领悟到调查研究海岸带动力——波浪、潮汐、海流作用，河流入海水沙效应，与海岸带地质基础、岩层结构特性，两者间的相互作用过程、其效应与所形成的地貌特点，论证海岸带发育的成因与发展动态，勘定了胶辽半岛海滨锆英石、金红石等稀有元素砂矿分布与 C1 级矿量；更多的实践是海港选址，通过海岸带动力地貌调查与剖面定位观测，选定"深水、浪稳、泥沙回淤少"的港址。历经夏日酷热、冬春寒风与晕船之苦，成功地完成了渤海、黄海与南海若干港口的选建：秦皇岛新港区，南李庄海军码头，山海关船厂港，广西北海地角海军码头，广东调顺与海南铁炉海军港以及扩建三亚港等中小型港口；发展完成了海南洋浦港，江苏洋口港与唐山曹妃甸等重要深水港址的选建与防止航道骤淤等问题。其中，利用渤海潮流通道深槽与古滦河河口沙坝曹妃甸岛组合建成的 10 万～30 万吨级的曹妃甸深水大港，利用在黄海的古长江河谷与辐射强潮流动力结合组建的 15 万吨级洋口深水港，以及在海南利用火山喷溢玄武岩流阻断的构造下沉古河道与强劲潮流往复冲刷作用选建的洋浦港，是应用海岸动力地貌原理成功选建深水港的例证。

海岸调查与选港建设的实践，使我提高了认识，理解了中国海岸带以季风波浪、潮汐作用与大河的交互作用为特点的自然发展过程与人类活动效应，总结发表了有关"潮滩分带性""贝壳堤建造与海岸发育蚀积""潮汐汊道港湾海岸"以及"火山海岸"发育等论文以及出版《海岸地貌学》《海洋地貌学》与《中国海洋地理》等理论与区域专著，并进一步应用于教学与海岸带建设发展实践中，完成一系列有关文章。

1979 年到 1982 年，作为党和国家培养的首批派赴北美留学的中青年之一,我被选拔赴加拿大研修三年。在 Dalhousie 大学海洋系与地质系学习,提高了海洋科学理论;参加了 Bedford 海洋研究所大西洋地质中心的新斯科舍海岸与大西洋考察工作;获得海洋科学领域理论知识扩展与海洋调研能力的进步,结合在祖国海岸研究工作的坚实基础,使我对高纬冰川作用海岸海洋特性体会深入,总结完成了鼓丘海岸发育的成因类型与阶段性特点,大西洋深海平原海洋地质环境与核废物埋藏稳定性分析,以及应用电子显微镜图相总结了"石英砂表面结构与作用营力特点的模式图集"等,深获美国、加拿大海洋科学家的肯定和赞誉。在加拿大获得参加 Pandor Ⅱ 号在圣劳伦斯湾深潜海底调研项目,观察到海洋悬浮体的絮凝过程,下沉基岩与砾石海岸的现代沉积环境与附着沉积特点,直视海底,其感受与研究效应是深刻的。

在加拿大的深造学习与实践过程中,认识到海-陆过渡带的海岸海洋(Coastal Ocean),是介于陆地与海洋间的一个独立的环境地带,而地球表面是由大陆、海陆过渡带及海洋三部分组成的。这是我对海陆交互作用带自然组体认识的飞跃:研究领域从早年的沿海陆地工作,明确延伸到海岸带(Coastal Zone),进而认识到海岸海洋是一个独立的环境体系,它介于大陆与大洋之间,范围从海岸带延伸至大陆架、大陆坡至坡麓的大陆隆(Continental Rise),是海陆之间的过渡体系,地壳结构兼有陆壳的硅铝层与洋壳的硅镁层,即 Si-Al/Si-Ma,构成地球壳体表面的三大组成地带,互有联系与影响。研究对象明确了,工作领域扩展了,我认识到中国海的绝大部分为海岸海洋,仅台湾岛以东濒临大洋。90 年代以后我写的文章主要在海岸海洋,研究工作不仅限于自然过程与海岸带建设,而且涉及到疆域归属,工作区域逐渐转入南海,出任了由教育部批准,在南京大学建立的中国南海研究协同创新中心主任,致力于南海岛礁与海疆权益工作。21 世纪以来,由于体力限制,难以下海,工作主要集中于南海岛礁与海疆权益研究。我所撰写的文章与对研究生培养,中心内容主要是这一领域的调查研究。这项工作密切结合国家利益,研究内容将海域的资源环境与疆域历史沿革相结合,为政治、为国防服务。开始查询第二次世界大战后，中美英苏同盟国协定及民国史的文档，对比分析民国以来出版的不同时期与版本的地图，获得充分与一致的证据，故而明确论证了南海断续线是我国的海疆国界线，而且，国际上对海洋疆界线均以断续线表示。这项工作使我深深体会到：海洋科学乃至地球科学的资源环境研究必须落实到为国家建设服务，必须与人文科学结合，服务于人类生存的持续发展。文集中还有一部分是纳入了教学工作体会与指导研究生所写的文章。我基本不参加研究生文章中的署名，除非是我的的确确总结与写作了研究心得才列名于文中。本文集中特列出与研究生合作的文章，是希望承前启后，为与新生一代的合作，留下"雪中鸿迹"，实为教学工作的一个反映。

年逾 82 岁，记忆减退，记下岁月长河中的阶段回忆，以志文集之序言，供参考指正。

王颖 2017.6.2 于

南京大学仙林校区昆山楼 B226 办公室

致感谢：王敏京系统打印文稿，付印前对全文进行了两次全面认真校对与相应的更正。葛晨东与张永战组织文稿录入。

目　录

第一篇　中国海岸海洋地质地貌与环境资源

第二篇　中国典型区域海岸海洋研究

第一篇
中国海岸海洋地质地貌与环境资源

位居亚洲与太平洋之交,中国边缘海独具特色。气、水、岩、生交互作用、季风波浪与潮汐作用特色,大河贯通陆地、移山填海,人类开发历史悠久,赋予中国具有海岸海洋环境、向海递次发展成陆与资源特点。探索陆海交互作用效应与和谐发展规律,实为海岸海洋科学(Coastal Ocean Science)研究之要旨。

潮汐作用与潮流动力、大河供沙与滩涂蚀、积变化,构成滩涂资源与成陆过程之发展。潮滩沉积相反映着自然演变历史。

发源于西部山地的大河,源远流长,贯穿大陆、汇入边缘海海洋,填积海盆,影响海平面变化,是区域以至全球变化对人类生存环境的直接效应。人类活动效应与世纪性的海平面变化,对中国海岸海洋作用过程与趋势性发展之影响突出。以区域海岸海洋环境变化的深入剖析为例证,探索其演变因由与变化趋势。

渤海海底地貌[*]

渤海是深入我国大陆的内海,它东北—西南向长约 470 km,东西向宽约 295 km,平均水深 18 m,最大水深 70 m。

渤海的轮廓受构造断裂所控制。它的东侧是一条北北东向的大断裂,从辽河口起沿辽东半岛西岸经庙岛群岛西侧到莱州湾与郯城庐江大断裂相联;它的西侧也是一条北北东向的断裂,它从辽西沿岸延伸到沙垒田东侧,至渤海湾与黄河口一带,构造线转为东西向。这两条大断裂之间为沉降带,即渤海盆地。

渤海的古老岩石基础是前寒武纪的变质岩^①。古生代沉积了下古生界的海相碳酸盐层,上古生界极薄甚至缺失。中生代主要为陆相的凝灰质砂岩、凝灰岩,并有泥岩、油页岩、砂砾岩以及玄武岩、石膏等夹层,这些反映了当时构造运动和火山活动均较强烈。中生代渤海四周大部分上升隆起,而渤海相对下沉。渤海主要是新生代的沉降盆地,新第三纪以来的大规模下沉一直持续到现代。渤海在第三纪堆积了厚层的河湖相地层,其中夹有火山堆积与局部的海相沉积。第四系下部主要为河流相,上部为三角洲相与海相沉积。第四系皆未成岩,与上第三系逐渐过渡。第四系上面普遍盖着现代海相沉积。

渤海海底地貌就是在这个背景上发展形成的。

下面简要介绍一下渤海海底的多种类型的地貌情况。

一、渤海海峡与冲刷槽

渤海海峡位于辽东老铁山与山东蓬莱之间,宽 117 km(图 1)。庙岛群岛横亘其间,使海峡分割成若干水道,其中北部的老铁山水道为黄海与渤海的主要通道。从黄海进入渤海的潮流,经过海峡时,由于水道狭窄,束水流急,使该处潮流流速高达 5 kn,这强劲的潮流,沿海底冲刷,日久形成了规模巨大的冲刷槽。

潮流通过老铁山水道分成两支,使海底冲刷槽亦有两条,其中向西北的一条长约 100 km,可达金州湾外,深槽呈 U 形,水深超过 30 m,槽底局部涡流所造成的深潭,最大水深 78 m。另一条冲刷槽自老铁山水道向西,穿过渤海中部,呈现为伸向渤海湾北部的舌形深槽,这条冲刷槽深度较小,一般深 30 m,最大深度 60 m。老铁山水道内有的基岩出露,有的地方堆积着海底冲刷形成的砾石。

---

* 王颖:《海洋战线》,1977 年第 6 期,第 5—8 页。
① 前寒武纪、古生代、中生代与新生代(又分第三纪与第四纪)均为地质时期,其距今年龄大体上是:前寒武纪为 6 亿年前;古生代为 2.8 亿~6.0 亿年;中生代为 0.8 亿~2.8 亿年;新生代为 0.01 亿~0.8 亿年。

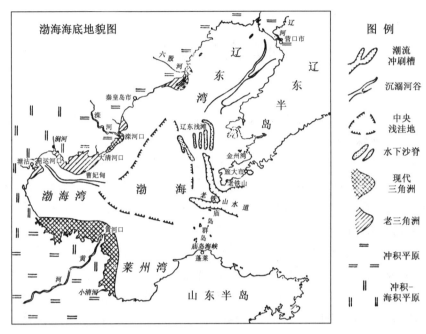

图 1　渤海海底地貌图

南部庙岛海峡水深 30 m,冲刷槽规模小,底质为粗大砾石,有些砾石表面已附生瓣鳃类生物,表明该处水流速度不大,砾石已停积海底,很少活动。

南北水道之间,有南北砣矶岛水道与长山水道,其水深超过 20 m,水道底部有基岩出露,也有砾石堆积,砾石中夹有砂礓结核。这几条水道以西的大片海底岩礁上,盖着一层砂砾层,砂砾是潮流冲刷海底形成深槽时的产物,砂砾被潮流携带到海峡以西,受底部岩礁阻滞,堆积为海底砂砾滩。

二、潮流三角洲

在老铁山水道西北,冲刷槽末端的海底,分布着呈指状排列的六道沙脊,地名为"辽东浅滩"。沙脊西北走向与潮流方向平行,顶部水深 8～18 m,东高西低,沙脊之间的水道深度 25～30 m。"辽东浅滩"表层是细砂与粉砂淤泥,含有大量贝壳砂,浅滩内有砂砾质,基础部分可能有古辽河河口沉积。辽东浅滩是个"潮流三角洲",是潮流出老铁山西北冲刷槽以后,水流分散,流速减低,于冲刷槽末端形成的指状沙堆积。

"辽东浅滩"的规模大,地貌格局特殊,它的成因不同于一般的潮流三角洲。"辽东浅滩"南端与冲刷槽相连,而北端接着古辽河沉溺谷地,浅滩的基础可能是古辽河三角洲,古三角洲受到改造,其上叠加了老铁山水道深槽冲刷产物。可以认为,它是古辽河三角洲与现代潮流三角洲的综合体。

这类巨大的海底砂质堆积体,往往成为蕴藏丰富的石油、天然气的有利条件,因此对它进行调查研究,是一件很有意义的工作。

三、海底河谷

从"辽东浅滩"向北,可以看到平坦的海底上有一条狭长的谷地。它呈西南—东北向,蜿蜒伸展到辽河口,又折向西北与大凌河口相接,全长超过 180 km,河谷深 5~7 m,两坡不对称,东侧谷坡陡峻,西侧平缓。这条水下谷地是沉溺海底的古辽河河谷,这段古辽河河谷是沿渤海东侧大断裂(即郯庐大断裂的北延部分)构造软弱地带发育的。目前,水下谷地仍是辽河入海径流及潮流的通道,谷地内潮流流速 1~2 kn,现代辽河口的泥沙仍沿谷地输送到东侧沿岸及"辽东浅滩"一带。

另一条海底河谷在渤海湾北部,西北—东南走向,上段与蓟运河口相接,下段渐转为东西向,与老铁山水道的冲刷槽相连,这也是一条沿断裂构造发育的河谷,沉溺于海底后受潮流冲刷改造,成为潮流进入渤海湾的主要通道,潮水冲刷还使蓟运河口成为喇叭状。

在平坦的粉砂淤泥质海底上,能保持一条完整的水下谷地,这主要是潮流作用的结果,它们是渤海海底上现存的巨大侵蚀地貌。由于有潮流不断把泥沙沿水下谷地向外输送,因此,在有沉溺河谷的河口,不发育三角洲或者三角洲规模不大,这时,河流的泥沙主要通过水下谷地堆积于谷地末端。

四、河流三角洲

河口的水下三角洲,是渤海海底另一引人注目的地貌。

辽东湾西部六股河口,分布着规模不同的两个三角洲。在河口以东,水深 5 m 以内的是现代三角洲,它是向海突出的弧形浅滩,浅滩上有两列沙堤,高差 3~5 m,由砂砾质组成,不含泥质。在六股河口的西南,水深 5~20 m 之间有三列水下沙脊,与海岸斜交向西南方向伸展,当地称为"三道沙干",沙脊与海底的高差为 9~13 m,向海坡陡,向陆坡缓。沙脊由砂砾质组成,成分与六股河泥沙相同,砂砾呈次棱角状,表面干净无生物附着,是受到海底水流作用尚在运动。沙脊之间凹地为砂质淤泥,无砾石沉积,是河流带入的悬浮物。这就是古六股河三角洲,因河口改道,海底泥沙供应减少,老三角洲受近岸带潮流冲刷,改造成与潮流流向一致的几道沙脊。

滦河有新老三个三角洲。位于滦河口外的现代三角洲是 1915 年滦河迁至此出口后开始发育的,该海岸带盛行南风和东南风,河流泥沙入海后搬运不远即被风浪掀动带向岸边,堆积成沙坝,形成了由沙坝环绕的三角洲。三角洲的外围至水深 10 m。滦河泥沙主要是长石、石英质中细砂,含较多的磁铁矿、绿帘石及石榴石等重矿。中细砂入海后,主要被堆积成海岸沙坝,在沙坝环绕的河口湾、潟湖内,沉积了粉砂与淤泥。

在现代滦河三角洲东北侧大蒲河以东的海底,有一个由中细砂、粉砂组成的三角洲,其下界延伸到水深 15 m,它是滦河在七里海、大蒲河一带出口时堆积的老三角洲。目前,岸边还堆积着起源于滦河泥沙的大型海岸沙丘。

规模最大的滦河三角洲,是在大清河口与涧河之间,它是滦河在 1453—1813 年自大清河一带出口时所形成的三角洲。石臼坨、蛤坨、曹妃甸等沙岛是当时环绕河口的沙坝,其成分是已磨圆的长石、石英质细砂,含磁铁矿、绿帘石等重矿砂,并含有大量具河口特点的贝壳(近江牡蛎、文蛤、青蛤等)。自 1813 年滦河改道后,原大清河一带三角洲泥沙供应减少,在风浪作用下,原河口沙坝受到冲蚀,如曹妃甸原有庙宇僧人,现今庙已沦没水下。沙坝冲蚀后,其后侧的平原海岸亦遭到潮水浸淹。

由于滦河的摆荡改道,致使东起大蒲河,西到涧河,南到水深 15～20 m,这一广大范围内,形成一系列三角洲堆积体,但无论新老三角洲,它们的特点都是有沙坝环绕河口的、具双重岸线的"封闭式"三角洲。

在渤海,规模最大的三角洲还是黄河三角洲。黄河三角洲界于渤海湾与莱州湾之间,是一个巨大的扇形三角洲,前缘到达水深 15 m。

黄河入海径流量平均每年约 485 亿 m³,每年入海泥沙约 12 亿 t,利津以下黄河分流多,尾闾河道经常变化,加之黄河口沿海潮差小,浪流作用不十分强烈,因而易于形成规模巨大的三角洲。在黄河的历史中曾多次注入渤海,巨量的泥沙充填海底,对渤海海岸与海底地貌的形成与发展有重要的作用。目前的三角洲是 1855 年黄河自黄海北归后所形成的。据测计,黄河的泥沙其中约 1/5 落在口门附近,堆积成河口沙嘴以及沙嘴两侧的烂泥湾。河口沙嘴生长很快,如 1964 年元月—1965 年夏,一年多的时间内,河口沙嘴向海伸长达 10 km,沉积为粉砂,厚度约 2.5 m,它直接盖在原来的淤泥岸滩上。河口两侧是淤泥质的"烂泥湾",亦在淤积伸展。当河口改道,原河口沙嘴及"烂泥湾"均受风浪冲刷,迅速后退。随着入海河道的改道变动,三角洲的堆积与冲刷岸段也随之变化。自 1855 年以来,黄河三角洲向海推进的速度,在河口附近平均每年生长 250～320 m,河口外围两翼每年约 100～200 m。

黄河水下三角洲的发展,使渤海深水区逐渐向北推移。

黄河入海泥沙除在三角洲沿岸堆积外,大部分悬浮物质被水流携带,形成淤泥流向三个方向扩散。主要一支是随北东—东向的余流,向东走,后又向南到莱州湾沉积下来;另一支随河口射水冲入渤海中部;较少的泥沙随较弱余流,并顺海底斜坡向西北方向流去,成为渤海湾淤泥沉积的主要来源。

五、中央盆地及海底沉积

渤海中央部分,位于辽东湾、渤海湾、莱州湾与渤海海峡之间,水深 25 m,是一个北窄南宽近似三角形的浅洼地,洼地中部低下,东北部稍高。中央盆地沉积着分选良好的黄色细砂,其周围沉积着粉砂。

渤海海底沉积物有一定的分布规律:基岩海底与砾石沉积,主要分布于渤海海峡及辽东湾两侧靠近山地的地段(如旅大及山海关、秦皇岛沿岸)。砾砂及中细砂,分布于辽东湾近岸带受山地河流供沙处以及"辽东浅滩"。细砂沉积在山东半岛北岸及渤海中央盆地。粉砂在辽东湾湾顶,莱州湾西部,渤海湾北部以及中央盆地四周皆有广泛分布。淤泥分布于黄河三角洲外围及渤海湾南部。粉砂及淤泥是渤海海底现代沉积的主要组成。

通常,近岸带是粗质沉积,离岸远的海底是细的沉积。在以波浪作用为主的辽东、冀东、山东北部沿岸带,就是自岸向海沉积物粒径逐渐变小。但在以潮流作用为主的渤海湾,却是低潮线附近沉积最粗,自低潮线向陆及向海,随着波浪、潮流作用减弱,而泥沙粒径均逐渐减小。渤海中央盆地是砂质沉积,这是自海峡进入渤海的潮流通过中央盆地,细粒泥沙被水流掀动,难以停积。另外,中央盆地的细砂分布的深度、位置与辽东浅滩相邻近,细砂分选良好,标志着该处为一沉溺的古海滨。

　　渤海海底三角洲系与水下河谷相伴分布的现象,以及近岸带普遍分布的海底埋藏风化壳,标志着渤海是陆地沉溺而成。在渤海中部海底曾发现一披毛犀的牙齿,披毛犀是耐寒的草原动物,生长时代是距今 1 万年前至 5 万年,说明当时渤海是一寒冷气候的草原环境。渤海的三角洲相与海相主要出现于上第四系与现代沉积,整个渤海水深不超过 80 m以及考虑到渤海海峡的形成时代(如:北西向断裂与北东向断裂交切;庙岛群岛各岛屿在相同高度的地方分布着黄土;海峡水道中及西部海底沉积中有砂礓结核等情况),可以认为现代渤海是更新世末全新世初形成的。地球气候在冰后期(距今约 11 000 年)以来逐渐变暖,世界洋面普遍上升,海水从黄海经渤海海峡进来,淹没了平原、洼地,形成了今天的渤海。

　　渤海海底有着类型众多、沉积深厚的三角洲系与古海岸堆积体,其中蕴藏着丰富的石油和天然气。我国广大石油工人坚决贯彻党的战略决策,发扬大庆精神,正在奋力开发渤海海底的石油资源。让林立的石油井架与渤海的滚滚波涛竞相辉映吧!

南海的海底[*]

辽阔的南海,位于我国大陆的南部,是太平洋的边缘海,其西界为中南半岛与马来半岛,东界菲律宾与加里曼丹,南界苏门答腊岛和加里曼丹岛之间的隆起地带(邦加岛一带),面积约为 340 万 km²。

在万顷碧波的海面上,散布着我国的大大小小岛屿:在北部有台湾、澎湖和海南等岛屿环抱,中部点缀着东沙、西沙、中沙群岛和黄岩岛;在南部,分布着繁星般的南沙群岛。其中于 4°N 附近的曾母暗沙等礁滩,是我国最南部的领土。

深邃的南海,它是我国毗连的最深海域。南海的平均深度超过 1 000 m,最大深度为 5 567 m。

雄伟的南海,如果拉开厚厚的海水"帷幕",就可以见到它的底部的地貌,气势宏伟而壮丽。

南海海底的总轮廓,近似一个长轴为东北—西南向的菱形海盆,海底地势是西北高、东南低,并且自海盆边缘向中心部分呈阶梯状下降。在海盆的四周边缘,分布着大陆架,其水深在 200 m 以内,外缘转折点平均深度约 150 m;在大陆架以外分布着呈阶梯状下降的大陆坡,其水深介于 150～3 600 m,我国的东沙、西沙、中沙和南沙群岛,都是分布于大陆坡山脊上的礁岛;在大陆坡终止处,是南海海盆的中央部分,该处是深度超过 3 600 m 的一片坦荡的深海平原,平原的中部突立着几座数千米高的火山峰,有的已接近海面,成为水下暗礁……这些,就是南海海底的基本轮廓。

南海海底的轮廓型式不是偶然的,是受基底构造所控制,地质构造成为其地形轮廓的骨架。

在南海,东北向的构造断裂居主导地位,它可能与南海的构造成因有关,由于断裂,南海可划分为三个主要的东北向构造带。

南海中央海盆,表现为拉裂出现的深部构造带,基底岩层属大洋型地壳的玄武岩类物质,纵波速度 6.5～6.6 km/s,上覆较薄的 4.3～4.5 km/s 的第二层(可能是火山喷发物、块状珊瑚以及固结的沉积层)和纵波速度为 2.1 km/s 的顶部松散沉积层,在这个构造带内还有一系列东北向裂隙与火山喷发活动。

中央海盆的南北两侧为南沙块断构造带与西沙块断构造带,它们均呈东北—西南向分布,相对于中央海盆其下沉较小。每个构造带的两侧皆有大断裂,而块断本身又为一系列小的断裂所切割。

在块断构造带的向大陆一侧,是两个具有北东向构造脊环绕的新生代沉降盆地。大体上以红河口到海南岛之间的西北向联线为界,在此线的东北侧,即我国两广沿海沉降盆地,主要为东北向构造;而在联线的西南,一系列沉降盆地与隆起则渐以西北向构造为主。

---

[*]　王颖:《海洋战线》,1975 年第 3 期,第 25 - 28 页。

块断构造带东侧时代较近的构造系列为南北向。

南海东部,从我国的台湾岛到菲律宾群岛以及巴拉望岛,出现了一系列岛弧与海沟相伴分布的格局。这种现象不是偶然的,它与南海海盆东部是大洋型基底的现象一致,反映着其形成发展密切受太平洋板块活动的影响,而南海其他部分与亚洲大陆关系比较密切,受大陆构造活动影响较大。可以说,海底地貌基本上是海底构造的反映。下面,让我们进一步看看南海的海底地貌吧!

一、大陆架

南海大陆架主要分布于海区的北、西、南三面,是亚洲大陆向海缓缓延伸的地带。大体上是西北与西南部大陆架宽大,宽度一般超过 200 km,而西部陆架较窄,一般约数十千米。东部是岛缘陆架,范围窄狭。南海大陆架外缘明显转折点的深度大约为 150 m(80浔)。如:我国沿海大陆架,大体上在中国台湾与海南岛以南这一带,宽度自 190 km 到280 km 不等,一般宽度超过 250 km,坡度约 0.6‰;而南海西南大陆架则为巽他大陆架的一部分,其宽度超过 300 km,我国的南安礁、南康暗沙和曾母暗沙位于这个大陆架的北部;西部越南沿海大陆架较窄,一般仅数十千米,但坡度较大,自 1‰到 6‰不等。

南海大陆架的基底岩层与相邻大陆一致,主要是花岗岩与片麻岩的老剥蚀面,剥蚀面上覆盖了老第三纪的沉积层,它们已固结成岩,并与基底岩层一起经褶皱变形,形成一列列东北向的构造脊,构造脊在大陆架边缘地带,像水下堤坝一样拦截了河流从亚洲大陆冲刷下来的碎屑物质,在脊后浅海盆地中沉积,形成了现在南海的堆积型大陆架。大陆架中的新生代沉积厚度达 2 000 m,在北部湾有些地方沉积厚度已超过 3 000 m,大陆架堆积始于老第三纪,但主要堆积时期是新第三纪与第四纪。这些由沉降盆地的厚层新生代浅海沉积物所形成的大陆架中,蕴藏着丰富的石油与天然气矿藏。

南海东部,沿菲律宾群岛以及巴拉望等岛的外缘分布的是岛缘陆架,其宽度小,坡度陡。如:北吕宋与马尼拉一带岛缘陆架宽 5～10 km,坡度为 29.2‰～14.6‰,南部巴拉望与沙巴一带,陆架宽度约 43～90 km,坡度为 3‰～1.8‰。南海东部岛缘陆架是侵蚀-堆积型的,有的地方受冲刷而基岩裸露,有的地方沉积着自各岛冲刷下来的泥沙。岛缘陆架上的沉积层不厚。

南海大陆架表层沉积,主要是从大陆冲下来的泥沙。陆架内侧(靠近陆地部分)受现代河流的影响,沉积着粉砂、淤泥或黏土(再向岸,在现代波浪作用带,泥沙被波浪分选,愈近岸边,颗粒愈粗,或为细砂,或为砂砾不等,亦有一些地方基岩裸露)。而外侧,在淤泥或黏土带以外,普遍分布着一个粗砂沉积带,粗砂的成分,主要是石英,表面已染有黄色,夹杂有贝壳及海绿石,粗砂带为冰期低海面时的海滨沉积,由于冰后期以来的海浸被淹没于海底,因其位置于现代海岸泥沙扩散带以外,故尚未被掩埋。

二、大陆坡

南海大陆坡分布于大陆架的外缘,水深介于 150～3 600 m 之间,呈阶梯状下降。大致在水深约 150 m 处,由平坦的大陆架变为陡坡,水深急剧加大,陡坡下,隔以深沟,大约至水深 1 000～1 800 m 的地方,地形又转缓,成为一断续相连的平坦面,宽度可达数百千米,平坦面外侧,又是一个急陡坡,直降到水深 3 600 m 处,达到了南海深海平原,大陆坡就终止了。

南海阶梯状大陆坡,是受块断构造控制形成,并为次要的断裂所分割。阶梯状大陆坡的宽平的台阶面,相对于南海深海平原而言,是一个"海底高原"。海底高原的基底岩性、上覆沉积等与邻近的大陆架一致,也是花岗岩或片麻岩的基底,在白垩纪末形成一剥蚀面,上覆老第三纪沉积与基底岩层经褶皱变形,形成一列列东北向的构造脊,这些构造脊与一些火山峰组成了盘亘在海底高原上的海岭,在海岭的脊部常常发育了珊瑚礁,我国的东沙群岛、西沙群岛、中沙群岛与南沙群岛等,就是分布于阶梯状大陆坡的海底高原面上的珊瑚礁群岛。

西沙群岛与中沙群岛是由珊瑚环礁所组成,有的出露于海面成为珊瑚岛,有的隐伏于海面下成为暗礁。如中沙群岛,主要就是暗礁。西沙、中沙所依据的海底高原面是属南海北、西部的阶梯状大陆坡(简称西沙—中沙大陆坡),这个高原面位于水深 1 000～2 000 m 之间,由于受断裂分割,所以,其断续相间分布,绵延达 540 km。其内侧以 23‰的陡坡与海南岛南部大陆架相接,外侧是一个 37‰～10‰的陡壁,以 1 000～2 000 m 的高差直插深海平原。如:中沙群岛外侧有一个 51°的陡坡,形成了巨大的海底悬崖。沿着海底高原外侧的陡斜坡,常形成局部的海底上升流,它带来南海深部的营养盐,加上高原而上珊瑚礁内的溶解矿物,因此,在海底高原上的珊瑚礁岛地带,成为丰产的渔场。

南沙群岛所在是南海南部-东南部大陆坡(简称为南沙大陆坡),与西沙—中沙大陆坡隔以深海平原遥遥相望,它也是阶梯状的大陆坡。在大约 1 400～2 000 m 水深之间,是一个呈东北—西南向伸展的海底高原,高原面纵长 925 km,横宽约 335 km。高原内缘:西南坡以 7‰的单斜坡与巽他大陆架北部相接;南坡以 18‰的坡度与曾母暗沙一带的大陆架相接;东南坡以 13‰的坡度与巴拉望海槽相接,海槽深达 2 560 m,它的内缘以 35‰的坡度与沙巴一带的大陆架相连。南沙海底高原外缘以 36‰～73‰的坡度急剧下降至深海平原。

南沙海底高原平均水深约 1 700 m,虽然这个高原面较西沙—中沙高原面略为完整,但高原上地形起伏亦极不规则,有一些大的谷地切割了高原,而在东北向的山脊上建造了南沙群岛珊瑚礁。南沙群岛也多为环礁,我国古代,曾称郑和群礁为团沙群岛,即指其形若圆环。由于珊瑚礁沿东北向海岭发育,并且受南海盛行的东北与西南季风的影响,活珊瑚在迎风浪一侧生长的较礁湖一侧更为旺盛,所以,南沙群岛多为呈东北—西南向延伸的椭圆形环礁,如太平岛就像个梭子。这些珊瑚礁岛,位于海面上,虽然面积不大,但位于浩瀚的南海之中,位置重要,在军事、航行、渔业与海上气象观测上,皆具有独特的作用。例如:南威岛环礁面积约 15 km²,但它的礁湖有一个缺口,水深达 14 m,可使海轮自此深水通道进入礁湖内避风,因而这些岛屿起了南海航线上的重要中途岛作用。但是,也有不少环礁、礁滩是隐伏于海面下数米到数十米深处的暗沙和暗礁。南沙海底高原上,由于岛、礁众多,水道纵横,水深变化大,不利船只通行,所以,被称为航行上的"危险地带"。

南沙大陆坡的基底岩层,上覆沉积与构造亦与西沙—中沙大陆坡一致。老的花岗岩与片麻岩的基底与老第三纪沉积被褶皱隆起,形成了一列列东北向构造脊——它们组成了高原面上的海岭,在东北向海岭与火山峰上建造了南沙群岛珊瑚礁。目前珊瑚礁基已位于 1 100 m 深处。这种情况表明:新第三纪(东北向构造山脊形成后)珊瑚礁生长发育以来,其基底地块是缓缓下沉的,所以,珊瑚礁能随着下沉的地块逐次向上增长而堆积了巨厚的珊瑚礁体。通常,活珊瑚繁殖的下限是 60 m,南海有的地方,由于下沉剧烈,珊瑚礁礁顶已位于 100 m 深处,礁体已不会再向上增长了。

根据西沙-中沙海底高原与南沙海底高原的基底岩层、上覆构造(东北向构造脊普遍分布于大陆架边缘和大陆坡上)以及地貌结构等均与相邻的大陆架一致,但此处深度大,有巨大的东北向断裂分割以及海岭上有上千米厚的珊瑚礁等情况,表明了:南海中呈块断下沉的海底高原是沉没、断折的古大陆架,断裂主要是东北向的,与下沉相伴发生,其发生时间始于新第三纪,但多次活动,第四纪时新构造活动仍剧烈,因而使一些珊瑚礁灰岩成明显的背斜,并有近东西向的断层使高原及构造脊分割错动。这种由块断下沉的古大陆架所构成的阶梯状大陆坡是南海大陆坡的重要特色。由于南海大陆坡的这样一种形成过程,也决定了在海底高原面上,可能还保留着昔日大陆架上所形成的油气藏。

南海东部的大陆坡,位于菲律宾群岛外缘,该处大陆坡:① 狭窄而陡峻,大部分宽度不超过 70 km,南部沙巴一带坡度约为 27‰,而西吕宋海沟、马尼拉海沟处坡度为 109‰～164‰;② 大陆坡呈狭窄的阶梯状下降,坡麓分布着深海沟,如:西吕宋海槽与马尼拉海沟;③ 大陆坡遭受到较多的水下峡谷的切割。一些大的水下峡谷穿越了大陆坡形成海峡通道。如巴拉巴克海峡、民都洛海峡、巴布延海峡……,除巴拉巴克海峡外,这些谷地规模较大,多数下达深海平原,并于大陆坡坡麓峡谷出口处,由浑浊流堆积了海底扇。也有一些水下峡谷切穿大陆坡而进入到深海沟,于海沟内形成浑浊流沉积。

三、深海平原

这是南海海盆的中央部分,大部分水深在 3 600 m 以上,范围介于西沙、中沙和南沙大陆坡之间,地形平坦,是一个深海平原,平原呈东北—西南向延伸,纵长 1 600 km,其中部最宽处(相当于珠江口到巴拉望岛一线所经部分)为 725 km,西南部较窄为 200 km,其东北部最窄处为 158 km。平原自北向南倾斜,北部水深约 3 400 m,而南部水深为 4 200 m左右,超过 4 400 m 的深度分布于深海平原南部。在深海平原的中部矗立着一些海山孤峰——海底火山喷发堆积的火山锥。这些海山高出深海平原 500～900 m,或者高 1 500～3 500 m 不等,其中最高的山峰为 3 904.15 m,已接近海面。

深海平原的东北端(位于台湾岛以南)、西南端的深水谷地充填着厚层的浑浊流沉积,在谷地的出口与大陆坡坡麓各水下峡谷的末端,都堆积着海底扇。这些堆积体有的已受到褶皱,形成一些东北向的小型山脊。在深水平原的这一部分中,蕴藏着潜在的石油资源。

大部分深海平原的底质为含有抱球虫软泥的黏土,同时,沉积中普遍含有火山灰。

在平原的中部与东部,沿东北向的大裂隙尚有一系列火山岩流喷发活动。

南海海底地貌的特征和结构,反映着南海的发展过程,研究南海海底地貌,有助于阐明南海的成因与动力机制,有利于人们有目的地去开发海底矿藏。

中国海海底地质地貌[*]

渤海、黄海、东海、南海界于亚洲大陆与太平洋之间,台湾东岸濒临太平洋,欧亚板块与太平洋板块的相互作用,形成一系列北东—南西向的隆起与沉降等构造带。它们自西向东,时代由老至新,构成了中国近海海底地貌的基本骨架。由于长江、黄河、珠江等大河向海输送了巨量泥沙,填充了沉降地带,形成了辽阔的、有隆脊围绕的堆积型大陆架,孕育了富饶的海底矿藏(图1和图2)。又因多次地壳运动与海面变化的影响,形成了宏伟的

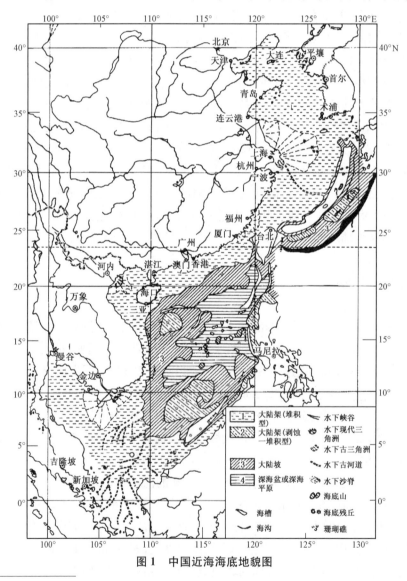

图1　中国近海海底地貌图

---

　　* 王颖,朱大奎,金翔龙:中国科学院《中国自然地理》编辑委员会编,《中国自然地理——海洋地理》,北京:科学出版社,1979年,第5-52页。

阶梯状大陆坡与深海平原。从而构成了特色显著的中国近海海底地貌——具有单一大陆架的黄、渤海，主要是大陆架但有部分大陆坡和海槽的东海，以及类型丰富的南海及台湾以东海底。

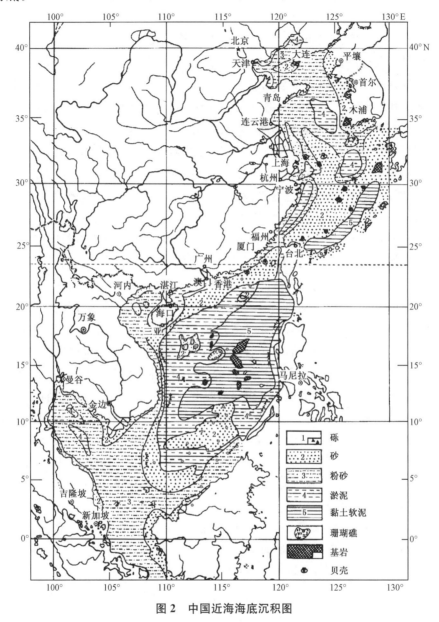

图 2 中国近海海底沉积图

一、渤海海底

渤海，由山东半岛、辽东半岛所环抱，是深入我国大陆的内海，东北—西南纵长约470 km，东西向宽约 295 km，其面积约为 77 000 km²，平均深度 18 m，最大深度 70 m。渤海可分为辽东湾、渤海湾、莱州湾、中央浅海盆和渤海海峡五部分。海底坡度平缓，海底表层为现代沉积物所覆盖，仅渤海海峡有前寒武纪变质岩及中生代花岗岩出露。

（一）海底地质

渤海是一个中、新生代沉降盆地,中、新生代的隆起与拗陷都明显地受到基底古构造与古地貌的控制。

渤海的基底是前寒武纪变质岩。在古生代,其构造发展与华北拗陷相似,沉积了以下古生界为主的海相碳酸盐层,中、上石炭统、二叠系均极薄甚至缺失。中生界、侏罗系为陆相的厚层的凝灰质砂岩及轻度变质的石英砾岩,砂岩中夹有数层薄煤层;白垩系为厚达200 m的杂色凝灰质砂岩、凝灰岩,夹有泥岩和油页岩的砂砾岩,以及玄武岩、石膏等夹层。中生代渤海四周大部分地层上升隆起,而渤海相对下沉。

新生代,早第三纪时,渤海地区受老地形差异影响,由于断陷作用形成分割性凹陷,各凹陷内均有沉积中心。由于断陷的多旋迴性,所以下第三系特别厚。在凹陷内,沉积了厚约2 000～4 000 m的灰绿、灰白色砂岩、砂砾岩与灰绿、深灰、紫红、紫褐色泥岩以及鲕状灰岩、生物灰岩与油页岩。砂岩、泥岩分选层次皆好,具微层理,主要是湖相沉积。早第三纪中期,由于构造活动强烈,有火山活动,所以沉积层很不稳定,并夹有数层玄武岩与凝灰岩。早第三纪沉积与其下部基底为不整合接触,有的覆盖在侏罗、白垩系上,有的却直接覆盖在前寒武纪变质岩上;它与上部的晚第三纪沉积亦非连续接触。

晚第三纪时,渤海全区急剧地拗陷式下沉,与四周地区明显地区分开。沉积中心由渤海边缘向渤海湾与中央部分转移,上第三系厚达2 000 m以上。主要是灰绿和棕红杂色的泥岩与砂岩或粉砂岩,其粗细韵律交替明显,具有良好微层理,为湖相沉积。上部有透镜体分布的棕黄或灰黄色砂岩,并具有一定分选性及不同磨圆度的砂、泥岩的河流相沉积,还有含海相介形虫的海相沉积。沉积层的厚度与颜色向海区中部加厚加深,大部分沉积层中含有钙质结核和石膏夹层,有些沉积层具有网纹花斑与铁锰质结核,表明沉积层曾暴露于空气中,遭受到风化以后由于不断下沉而埋藏。晚第三纪沉积厚度大,分布广泛,表明当时渤海经历了统一的、大规模的下沉运动,沉降幅度大,延续时间长,沉积环境稳定。总之,在第三纪,本区为遍布着河流和湖泊的下沉拗陷环境,沉积了厚层的河湖相堆积物,其中夹有海相与火山堆积物。

第四纪的沉积厚度达300～500 m。其下部130 m为棕黄色、灰黄色的砂质黏土与灰—灰黄色粉砂、细砂互层,夹有薄层砂砾层及钙质结核,底部有砾石层,为河流相堆积;其上部为土黄—棕黄色、灰色的黏土,夹有粉砂质黏土及粉砂,在灰色的含砂较多的黏土中富含瓣鳃类与腹足类壳体,这一层似为河-海交互作用的三角洲堆积与海相堆积。第四系皆未成岩,与第三系逐渐过渡,两者沉积稳定,巨厚的沉积层遍及整个渤海区。第四系上面均覆盖有现代海相沉积。上述情况,反映出自晚第三纪以来的下沉运动,经过第四纪而持续至今。

渤海新生代的构造线,主要是北东至北北东向,在渤海湾及黄河口附近为近东西向。第三纪玄武岩沿此两组构造线相交处喷发溢流。渤海东部有一条北北东走向的大断裂,大致从辽河口开始,沿辽东半岛西岸经庙岛群岛西侧到莱州湾,与郯城一庐江大断裂相连。渤海西部为北北东向的沙垒田东侧大断裂,它向东北延伸到辽西海岸。这两条大断裂之间即为沉降拗陷带。渤海在第三纪时沿此构造带形成一系列狭长形湖泊,它进一步下沉,并遭受海侵,即为目前的沉降拗陷盆地,而大断裂的外侧是沿岸隆起(如,东侧的胶辽隆起,西侧的山海关隆起等)。

渤海海底表层沉积物的分布特点是四周海湾颗粒较细,而向中央浅海海盆颗粒逐渐变粗。如渤海的辽东湾沉积以粗粉砂、细砂为主,渤海湾以粉砂淤泥和黏土质淤泥为主,莱州湾则以粉砂质沉积物占优势,而中央海盆分布着粉砂和砂,其中尤以广布的分选良好的细砂为特征。渤海海峡的沉积物呈斑状分布,变化较大,分选性差,有砾石及贝壳碎屑出现。沉积物中的碳酸钙含量由四周向中心减低,呈环状分布,有机质的含量则中部比四周高。渤海各海湾的底质分布明显地受沿岸河流作用的影响,而海盆中央部分是古海滨沉积,但某些地方后来受潮流冲刷再塑造。

(二) 海底地貌

在地貌上渤海是一个大陆架上的浅海盆地,地貌类型单一(图3),可分为五区,现将各区的构造、地貌与表层沉积作进一步综合分析。

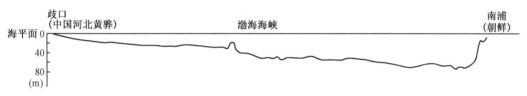

图3 渤海湾—朝鲜半岛西岸海底地形剖面图

1. 渤海海峡

在辽东老铁山与山东蓬莱之间,宽105 km,庙岛群岛罗列其中,使海峡分割为若干水道,以北部的老铁山水道为主。该处最大深度78 m,是黄海海水进入渤海的主要通道。从黄海进入渤海的潮流,经海峡时,由于过水断面变窄,束水流急,致使该处潮流速度高达5 kn。海流通过老铁山水道,分成两支,一支向西北,沿海底冲刷出一条长达80 km,宽30 km的U形深槽,局部涡流壶穴水深达70 m,深槽的北端分布着指状排列的六道水下沙脊(图4),通称"辽东浅滩",表层为细砂、粉砂淤泥及贝壳砂,浅滩内有砂、砾石,基底部分可能有古辽河河口沉积。现代辽东浅滩是个"潮流三角洲",是潮流流出老铁山西北冲刷槽以后,水流分散,流速减低,于冲刷槽末端形成指状沙脊堆积。沙脊西北走向与潮流方向平行,顶部水深8~18m,东高西低。沙脊之间的水道,深度达25~30 m。通过老铁山水道的另一股潮流向西流动,越过整个中央盆地,形成伸向渤海湾北部的舌状深槽。老铁山水道内堆积着长轴为2~7 cm的砾石,砾石成分为石英岩、花岗岩以及硅质灰岩、千枚岩等,是沿岸及海底的冲刷产物,有些地方出露着基岩。

南部的庙岛海峡水深18~23 m,底质为粗大的砾石,成分为石英岩和硅质灰岩,砾石表面附有瓣鳃类底栖生物,表明砾石已停积海底,活动性不大。

在南、北水道之间,还有几条水道,水深都超过20 m。水道底部,有基岩出露,也有砾石堆积,砾石中夹有砂礓结核。这几条水道的西部有大片砂砾滩,覆盖在海底岩礁上。

2. 辽东湾

位于渤海北部,在长兴岛与秦皇岛联线以北,海底地形自湾顶及东西两侧向中央倾

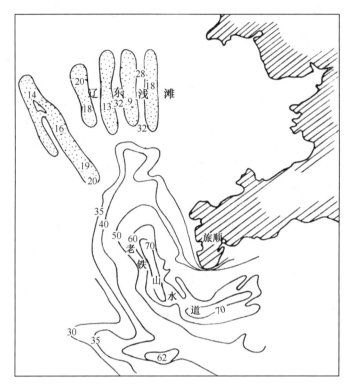

图 4　老铁山水道及辽东浅滩平面图

斜,且湾东侧较西侧为深,最大水深 32 m,位于湾口的中央部分。辽东湾的构造是处于两条大断裂之间的地堑型拗陷,拗陷内尚有一系列呈长条状相间排列的北东向的凹陷与凸起构造。整个辽东湾内均被晚第三纪以来的厚层沉积物所覆盖。上第三系从湾顶向渤海中央逐渐加厚。

辽东湾湾顶与辽河下游平原相连,水下地形平缓,沉积了由辽河带入海中的泥沙,湾顶为淤泥,其外侧为细粉砂,而东西两岸则分别与千山山地及燕山山地相邻,水下地形坡度较大,通常两侧近岸部分的坡度达 5‰,底质为中细砂或粗砂与砾石,砂粒表面光洁,无杂质沾染,重矿物含量较高,砂层中淤泥含量很少,这表明砂砾均系新近沉积的,并在风浪掀动下搬运移动。离岸较远处,在水深 8 m 以下,坡度减为 1‰,沉积物以粉砂黏土为主,并含有少量砂粒,砂粒表面无光泽,而有锈斑污点。底质明显地分为两层,上层为无臭的灰黄色淤泥或砂质淤泥,厚 2～4 cm,为现代沉积;下层为灰褐色—黑色粉砂淤泥,质地滑腻黏稠,有臭味,表明沉积较久,且稳定而不易被掀起移动。辽

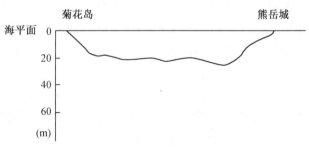

图 5　辽东湾海底地形剖面图

东湾中央地势平坦,几乎全为黑色微臭的淤泥沉积(图 5)。

辽东湾东西两侧水下地貌较复杂,在砂质海滩外围,常分布有与海岸平行的水下沙

堤,通常有1～2列,由砂组成,向海坡较缓,向陆坡较陡,是风浪作用下泥沙向岸移动过程中形成的。河口大多有水下三角洲,由砂砾及粉砂黏土混杂而成,通常有明显的堆积平台与较陡的前坡。有的河口三角洲被改造为水下沙脊,如六股河口有水下浅滩,其西侧为两列沙脊,高差3～5 m,由砂砾组成,不含泥质。再向西有三列与岸斜交的水下沙脊(图6),高差9～13 m,向陆坡缓(为2.3‰),向海坡陡(为3.6‰),由细砂、砾石、淤泥混杂构成。沙脊上砂粒纯净,次棱角状,砾石表面干净无生物附着,表明砾石尚在运动中;而沙脊之间凹地为砂质淤泥,未发现砾石,是当前河流带入的悬浮物,沙脊原为古六股河口水下三角洲,后经潮流冲刷改造而成。

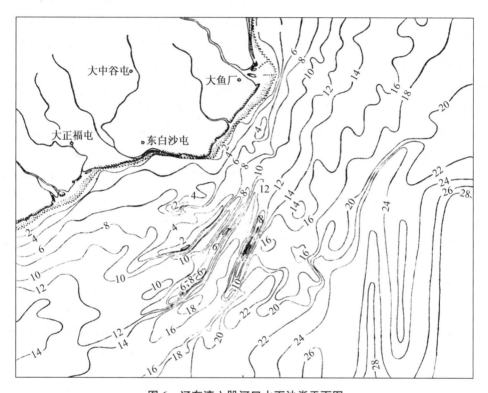

图6 辽东湾六股河口水下沙脊平面图

在辽东湾中部海底(杨屯河口外-8 m)亦分布着较多的砾石,其磨圆度好,砾石表面呈灰黑色,生长着苔藓虫,砾石已经风化,核心呈水湿状的同心圆圈,这表明砾石在海底沉积已久,为古河口堆积物,辽东湾近岸海底有二级水下阶地,分布在-2 m与-8 m,在基岩海岸外围为二级水下磨蚀阶地,岩滩宽500 m,坡度为5‰,表面参差起伏,礁石丛生,而在河口与砂质海岸外围则为二级水下堆积阶地。

辽东湾东部有一狭长的水下谷地,它沿大凌河口向东与辽河口相连,再折向东南。与辽东半岛西海岸平行向西南延伸,可达复州湾外,长约180 km,谷形明显,谷底相对低下5～7 m,至水深25 m处,谷的宽度加大,逐渐消失。谷坡不对称,西坡平缓,东坡因邻近辽东山地较陡。这一带潮流流速1～2 kn,不足以形成较强的海底冲刷。从其形态与分布位置来看,这条水下谷地是沉溺于海底的古辽河河谷,而古辽河河谷是沿郯城—庐江大断裂构造软弱地带发育的。目前,这条水下河谷仍为辽河入海径流及潮流的通道,因而未被沉积物所填满,保持了明显的谷地形态。同时,水下河谷也是现代辽河口泥沙输送的渠道。它将辽河的粉砂淤泥物质沿谷地输送到东侧沿岸及"辽东浅滩"一带。由于潮流不断把泥沙沿水下谷地向外输送,故现代辽河口三角洲规模不大,并且辽东半岛西侧沿岸有粉砂淤泥沉积。

3. 渤海湾

渤海湾是一向西凹入的弧形浅水海湾,构造上与沿岸地区同为一拗陷区,构造线为东西向,湾内凹陷与凸起呈雁行排列,整个渤海湾有厚达 3 000 m 以上的新生代沉积层,目前仍处于下沉堆积过程中,故水下地形平缓单调,等深线与海岸线平行,海湾水深一般小于 20 m。

渤海湾北部 20 m 的深水区紧贴岸边,这里有一条呈西北—东南走向的水下谷地,上段与蓟运河河口相接,下段渐渐转为东西向,与老铁山冲刷槽相连,这也是一条沿断裂构造发育的河谷,沉溺于海底,以后受潮流冲刷改造,成为潮流进入渤海湾的主要通道。由于河口段输沙量少,潮水冲刷,致使蓟运河河口成为喇叭状。在平坦的粉砂淤泥质海底上能保持一条完整的水下谷地,主要是潮流作用的结果。在水下河谷以北的曹妃甸一带,分布着数列水下沙脊,呈北东走向,高出海底 11~18 m,其间有较深的沟槽。沙脊由磨圆良好的中细砂及大量贝壳碎屑组成,有较多的近江牡蛎、刀蛏、镜蛤及扁玉螺等河口浅滩生物遗骸。据其物质组成,其泥沙系来自滦河及附近海底,是老的滦河水下三角洲受波浪、潮流冲刷改造而成,因滦河改道北移,泥沙供应不足,沙脊因受冲刷而逐渐缩小,被冲刷下来的泥沙却散布于附近海底。因此,渤海湾北部海底泥沙颗粒较粗。

黄河于渤海湾南侧入海,其大量泥沙入海后,一部分泥沙在潮流与东南向浪、流作用下,向北部运移扩散,多年辗转运移,加上历史时期黄河曾自海河一带入海,以及现代海河水系泥沙的影响,因而,渤海湾海水浑黄,堆积作用盛行。沉积物主要是粉砂与淤泥,且以淤泥为主。海底地形十分单调平缓,平均坡度小于 0.25‰。海底大致自南向北,自岸向海倾斜,海湾中部与北部水深较大,这也是黄河入海的泥沙,有一部分向北运移的原因之一(图 7)。

图 7　黄河口—辽东湾海底地形剖面图

4. 莱州湾

莱州湾以黄河三角洲与渤海湾相隔开,海湾开阔,水下地形平缓单调,并向中央盆地缓缓倾斜(图 8)。莱州湾的水深,大部分在 10 m 以内,最深处 18 m,位于海湾的西部。莱州湾在构造上为一个凹陷区,新生代沉积厚度达 8 000 m。辽河口—庙岛西大断裂①在莱州湾东部通过,该断裂的东侧为上升区,即鲁北沿岸山地,近岸海底有礁石突露,断裂之西

---

①　即郯—庐大断裂通过渤海部分。

即莱州湾现代沉积区，有较厚的沉积；莱州湾底质以粉砂占优势，东部为细粉砂，向黄河口方向黏粒逐渐加多；南部为分选良好的粗粉砂，向西逐渐变细；西部是含黏粒较多的粉砂。沿岸沉积颗粒较粗，东部沿岸为细砂，是

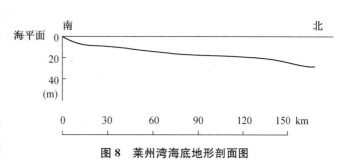

图 8　莱州湾海底地形剖面图

海峡与蓬莱一带入海的泥沙，在东北向强、常风浪与涨潮流推动下，缓缓向西运移，在蓬莱以西形成大片砂质浅滩与沿岸沙嘴，随着沙嘴的发展使屺姆岛等岛屿和陆地相连。它们拦阻了自东而西的沿岸泥沙纵向运移。但在浅平的细砂质滩底上，激浪作用活跃，于浅滩上部形成一列列与岸线平行的水下沙堤。西部沿岸受黄河入海泥沙的影响，沉积颗粒较东部细小，主要是细粉砂。

界于渤海湾与莱州湾之间的黄河三角洲，是一个巨大扇形的三角洲。黄河口附近在地质构造上为东西向凸起，向东倾伏。黄河入海径流量，平均每年约为 480 亿 m^3，平均年输沙量约为 12 亿 t。据测计，其五分之一在口门附近落淤形成河口沙嘴（粗粉砂）和沙嘴两侧的烂泥湾（粉砂质淤泥）。黄河主流入海处，河口沙嘴以每年 2～3 km 的惊人速度向海淤进。但入海口经常改变，其两侧淤进速度较慢。河口泥沙直接影响口门外两侧约 20 km 的范围，在此范围之外，是过去河流入海的地段，目前由于泥沙供应不足，沿岸遭受冲刷后退，后退速度每年从数米到 300 m 不等。冲刷岸段范围约 130 km，其海底为粉砂，而细颗粒的淤泥被带入海中，冲刷岸段以外的两侧，岸线相对稳定，冲淤变化较小。由于黄河河口经常改道，三角洲大致以利津为中心的扇形面向海推进，其推移速度多年平均大体上是：中段平均每年外伸 290～370 m；而两翼平均每年 120～190 m。这样巨大的扇形三角洲，不仅形成沿岸的广阔平原，而且在渤海湾南部与莱州湾北部平坦海底上，造成一个巨大的圆弧形水下三角洲，其范围北起大口河，南到小清河，水下三角洲前缘延伸到 −15 m 左右深度。

黄河入海的泥沙，除在口门堆积外，大部分呈悬浮状态，分三个方向扩散，黄河口外主要余流方向是北东—东，黄河的大部分泥沙随流东去，向南转入莱州湾沉积；另一部分泥沙随河口射水直接冲入渤海深水区；较少部分泥沙随较弱余流向西北方向流去，成为渤海湾的重要泥沙来源。黄河口外水下三角洲的发展，使渤海深水区逐渐向北推移。

5. 渤海中央盆地

位于渤海三个海湾与渤海海峡之间，水深 20～25 m，是一个北窄南宽，近似三角形的盆地，盆地中部低洼，东北部稍高，在构造上，这里是渤海东西两条大断裂之间的一个最大的地堑型凹陷。凹陷连续，分割性甚小，新生代沉积厚度达 5 000 m，是上第三系沉积的中心。渤海四周三个海湾的上第三系底板均向中央盆地倾伏。中央盆地的底质，中心部分为分选良好的黄褐色细砂，其深度、位置与辽东浅滩相邻近，说明该处为沉溺的古海滨，细砂区的周围为粉砂，并向各海湾延伸。渤海中央盆地中心分布着细砂，而周围却分布着粉砂，这有异于一般规律。这是因为有一支自海峡进入渤海的潮流，经中心部分贯通向西，

使古海滨沉积物中的细粒部分被潮流冲刷带走,而留下细砂。

总言之,渤海为胶辽两半岛环抱的大陆架内海,深度小,海底地形平坦,在平缓海底上的地貌类型却各有特色,如,渤海南部巨大的黄河水下三角洲,沙坝围封的滦河水下三角洲,以及被潮流改造成为三条水下沙脊的古六股河水下三角洲。沉溺于海底的长大的古辽河谷地,与延伸至渤海中央盆地的蓟运河—海河水下河谷,渤海海峡老铁山水道西北侧的潮流冲刷槽,以及深槽北端出口处巨大的“潮流三角洲”。这些海底地貌都反映了渤海海底发育的历史及现代动态过程。

新生代以来,渤海的形成过程大体上是,受郯城—庐江大断裂与沙垒田东断裂的控制,早第三纪时,渤海地区断裂下沉,形成一系列狭长的湖泊与洼地。

晚第三纪,渤海大规模普遍下沉,一直持续到第四纪。当时,渤海主要是呈拗陷下沉的湖泊,因而渤海海区有巨厚的晚第三纪、第四纪的河-湖相堆积。

新生代渤海海峡已初具雏形。第四纪时,由于水动型的海面变化,渤海可能有数次海侵,在渤海 20～25 m 深处,还保留一些古海滨遗迹。

由于,在渤海中部海底采集到一个披毛犀(*Rhinoceros antiquitatis*)的牙齿(左上侧的第二个上臼齿)。披毛犀系耐寒草原动物,时代距今 1 万～5 万年前,这枚化石表明了在晚更新世大理冰期低海面时,渤海中部曾为陆地草原;河海交互作用的三角洲相沉积与海相沉积主要出现于上第四系与现代沉积中;整个渤海水深不超过 80 m;以及考虑到渤海海峡的形成时代,如,北西向断裂与北东向断裂交切,庙岛群岛各岛屿在相同高度的地方分布着黄土,海峡水道中及其西部口门外海底沉积中有砂礓结核等种种情况分析,现代渤海主要形成于晚更新世末与全新世初期,由于气候转暖,世界洋面普遍上升,海水从黄海经渤海海峡进来,淹没了平原、洼地、河、湖,形成了现代的渤海。

二、黄海和东海海底[*]

黄海和东海在地质、地貌有一定联系,故以一并叙述。

(一)地质构造基础

黄海、东海在地质构造上位于新生代环太平洋构造带的西部边缘岛弧内侧,海域内的主体构造走向为北北东,由一系列中、新生代的大致平行相间排列的隆起带和拗陷带所组成,在同期或晚后的北西向构造的影响下,构成了黄、东海的海底构造骨架,成为黄、东海宽广的堆积型大陆架得以发育的基础和边界条件。

因北北东向构造对古老构造的改造,故黄、东海的基底极为复杂,且各海区曾经历了不同的地质发育阶段,而今日的黄海和东海则是全新世初冰后期海面上升的结果。

根据相邻陆地和海区资料,现将海区内主要地质构造单元自北而南依次简述如下:

1. 胶辽隆起

大体上从我国的庐江—郯城—苏北燕尾港至朝鲜的海州一线以北的黄海海区,在构

---

* 《化石》1974 年 1 期,此部分主要由金翔龙执笔。

造上统属于中朝准地台的胶辽隆起。其基底由前寒武系的结晶片岩、片麻岩、大理岩、石英岩等变质岩系组成。古生代地层的发育,和我国华北地区相类似。中生代时基底迭遭断裂破坏,有侏罗、白垩系陆相碎屑岩和火山岩系的堆积,并受酸性火成岩的侵入。新生代地层在周缘陆地不甚发育,在朝鲜半岛西海岸的安州附近等地有陆相第三纪地层的分布,推测向海底延伸。根据海区物探资料,结合现代沿岸岛屿众多,基岩港湾曲折,有的地方海底基岩裸露等地貌现象,可以说明,本区在第三纪时期,基本上仍处于一个隆起的背景,并为渤海盆地创造了封闭和巨厚沉积的有利条件。

2. 南黄海—苏北拗陷带

本带东南大体沿浙江的江山、绍兴经九段沙至朝鲜的沃川一线,与闽浙隆起为界,基底亦由前寒武系的变质岩组成。从寒武系直至中、下三叠统以海相碳酸盐为主的建造,地层走向总体呈北东东至近东西向,与基底基本一致。中生界的上三叠统至上白垩统,在江苏地区为一套碎屑岩夹中酸性火山岩,总厚超 5 000 m。新生代地层的最大厚度大于3 000 m。下第三系为一套湖相的砂、泥岩建造,沉积时受古地理条件的限制,盆地分割性较强。上第三系含有海相化石。地震剖面揭示存在有三个明显的不整合面,分别代表三次构造运动。上新世至第四纪地层几近水平产状。由此可见,本区在新生代时经受了大规模的断陷,接受了巨厚的第三纪地层的沉积,为勘探海上油气田提供了有利的物质基础。

3. 浙闽隆起带

主体在我国浙、闽东部的陆地上,向东北延伸入黄海、东海海底,越济州岛至朝鲜半岛的东南部。隆起带大致位于黄、东海的交接处,济州岛、苏岩礁、虎皮礁、花鸟山和舟山群岛等组成一系列岛礁线,是隆起带在海底地形上的表现。隆起带的基底岩系由两套组成,下部为前寒武纪的变质岩系,上部为中生代的火山碎屑岩系。古老的变质岩系大致分布在绍兴—江山、邵武—河沉断裂与上虞—丽水—华安—大埔断裂之间,由同位素年龄测得为 10 亿年(建瓯群)及 16.5 亿年(陈蔡群)的晚元古代基底组成。在隆起区的东部有一系列燕山运动形成的北北东、北东向断裂,大量中酸性火山岩系和花岗岩沿此分布,即著名的东南沿海火山岩带。需要指出的是在浙闽东部(福建福鼎、浙江石浦)发现了晚古生代浅海相碎屑岩为主的沉积岩,其中有硅质岩、结晶灰岩、火山碎屑岩等,遭受不同的区域变质作用。因此,在广泛覆盖的中生代火山岩系之下可能还有海西褶皱带的存在。基底岩系之上覆盖着厚约 800 m 的新生代地层,一般为上第三系和第四系,个别地段可能有下第三系,它们与下伏岩系均呈不整合接触。

4. 东海陆架拗陷带

此带几乎囊括整个东海大陆架,向北延至对马海峡,往南达台湾海峡中部的北港隆起,总的呈北北东走向,是东海新生代沉积的主体。它的基底岩系性质不很清楚,大体是中生代(和古生代)复杂变质的沉积岩和火山岩。基底岩系之上是巨厚的第三纪、第四纪地层,最厚可达 9 000 m 左右。我国著名地质学家李四光很早就指出了东海是新华夏构造体系第一沉降带的组成部分,具有优越的含油气远景。这一套新生代地层的下段是经过构造变动的晚第三纪地层,上段是晚第三纪和第四纪地层,上、下段之间不整合,中新统与上新—更新统之间也有局部不整合。上段地层以晚第三系为主,一般向北变薄(对马海峡处厚度仅达 200 m),向南变厚,多在 2 000 m 以上,在台湾省,石油钻井钻进了 5 000 m以上的晚第三纪地层,已知海相中新统为主要产油层。

5. 东海陆架边缘隆褶带

它沿着东海大陆架的外缘而分布,为一条状隆起褶皱带,总体走向为北北东,两侧断裂发育。该带向西南经钓鱼岛等岛屿可与台湾褶皱的山脉相连,是为南段。钓鱼岛等岛屿的露头由下第三系砂砾岩组成,夹有煤线,并有中酸性侵入岩,此外,还有晚第三纪的火山岩分布。向东北方向经鸟岛、五岛列岛可追索到日本九州的北部,是为北段。五岛列岛主要由中新世绿色凝灰岩组成,并有花岗岩侵入。在中段,中新世地层减薄,尖灭或缺失。折射层速度也证实缺失相当中新统的 3.6~4.4 km/s 的速度层,其下主要为 4.7~5.3 km/s 高速的下第三系。磁力资料亦揭示为一基底小于 1 000 m 的隆起,并有沿断裂分布的次级异常。据此,本隆褶带可能主要为早第三纪或早第三纪以前地层组成,并有中新世及后期火山岩分布的褶皱带。在其形成过程中,两侧相对长期持续沉降,隆褶带作为一天然堤坝,拦截了来自中国大陆的泥沙补给,使之在东海陆架拗陷带堆积了巨厚的第三纪沉积,奠定了目前大陆架的基础。

6. 冲绳海槽张裂带

冲绳海槽张裂带地处地形急剧变陡的大陆坡和海槽区,在构造上为走向北北东的张裂构造带。海槽北与日本强张裂的、下第三系组成的天草褶皱带相联,在晚第三纪和第四纪时有大量的基性火山岩喷发;南与台湾省东部的苏澳构造带相联,该带由包括下第三系在内的结晶杂岩组成。海槽本身可能由复杂变质的下第三系岩层组成(纵波速度 3.5~5.5 km/s),在其深而平的海底,发育一套充填式沉积,厚度超过 1 200 m,其时代可能属第三纪晚期至第四纪,并有少量近期砂质混浊岩和火山碎屑岩。冲绳海槽内的地壳热流量较高,大于 2 μcal/cm²,海槽纵槽一带热流量达到 3~5 μcal/cm²,年轻断裂相当发育,海槽的东北部还见到一些高耸于海底的火山岩潜山,凡此均显示出它是一个扩张初期的裂谷构造,张裂活动可能始于中新世晚期。

7. 琉球岛弧-海沟系

它属太平洋西部的岛弧-海沟构造带,顺琉球群岛而分布。琉球岛弧是双列岛弧,它的内弧在东海(位琉球群岛与冲绳海槽之间),主要由晚第三纪火山岩(中-上新世安山岩)组成,称之为琉球古火山带,它是在冲绳海槽裂开时与东海陆架边缘隆褶带上的火山同时形成的,在地形上表现为一些水下山脊或岛屿(如吐噶喇群岛等)。外弧即琉球群岛,由古生代、中生代变质岩和褶皱的晚、早第三纪地层组成,有花岗岩、辉长岩和现代火山活动。琉球岛弧之东便是琉球海沟,其水深超过 6 000 m,是太平洋的菲律宾板块与欧亚板块相碰撞的消亡带,海沟以东的地壳变为典型的大洋型结构地壳。

(二) 海底地貌

黄海和东海的海底地貌以宽阔的大陆架为其特点。这个大陆架与中国大陆一片相连,是世界最宽的大陆架之一。根据黄、东海大陆架的地形变化规律、地貌特征与地质基础,充分说明,它是中国大陆向海的自然延伸部分。它自黄、东海我国海岸带的低潮线开始,以平缓的坡度(1′10″~1′20″)向海倾斜,直到水深为 140~180 m 处坡度发生突然转折,坡度倍增 60 倍以上(平均坡度 1°10′),成为大陆坡而进入冲绳海槽为止。

黄海是一个浅海,全部在大陆架上(图 9),海面宽度最大达 378 nmile,面积约 38 万 km²。黄海海底地势向反 S 状的中轴线倾斜,平均坡度 1′21″,平均深度 44 m(北部平均深度 38 m,南部平均深度 46 m),最大水深 140 m,出现于朝鲜济州岛的北侧。

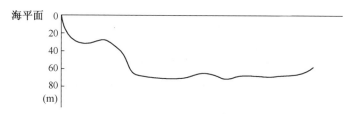

图 9 黄海大陆架(山东荣成以东)海底地形剖面图

东海开阔,略呈扇形,扇面撒向西太平洋。它东北—西南向长约 700 nmile,东西最宽约 400 nmile,总面积约 77 万多 km²,平均水深 370 m。西部为宽阔的大陆架,占东海总面积的 2/3 强;东部为大陆坡,它以冲绳海槽与琉球岛架相隔,范围占东海总面积的 1/3 弱(图 10 和图 11)。我国的台湾岛位于东海大陆架的东南边缘,它与福建陆地挟持着台湾

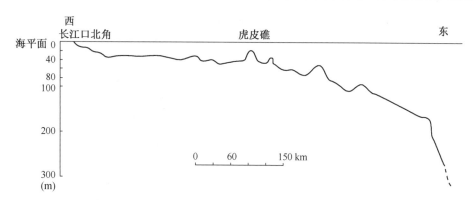

图 10 东海大陆架(长江口北角以东)海底地形剖面图

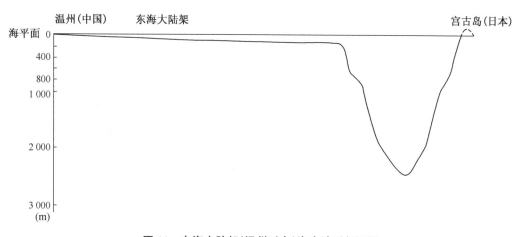

图 11 东海大陆架(温州以东)海底地形剖面图

海峡。台湾海峡北部宽约 93 nmile,南部宽约 200 nmile,长 150 nmile,大部分水深小于 60 m,多位于大陆架上(图 12 和表 1)。

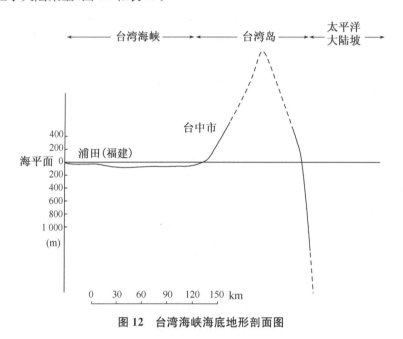

图 12　台湾海峡海底地形剖面图

表 1　黄、东海几个代表断面的宽度

断　面	宽　度
成山角—朝鲜半岛长山串	114 nmile
35°N 断面	378 nmile
长江口北—日本男女群岛	346 nmile
瓯江口—黄尾屿	205 nmile
三都澳—富贵角	105 nmile
湄洲岛—台中海岸	86 nmile

　　地质构造是构成海底地貌的骨架,陆地来沙和海水动力条件的变化对海底地貌的塑造,也有深刻的影响。其中,特别是中国大陆的一些大河携带来的极其丰富的泥沙,对黄、东海海底地貌影响尤大,只要从现代海底地势起伏的情况,就可看出长江、黄河在海底地貌形成过程中所起的巨大作用。陆源碎屑物质填没了一系列构造拗陷和水下谷地,漫过了一系列构造隆起和水下丘岭,河流把泥沙一直输送到大陆坡上,甚至到冲绳海槽中。从而形成了现在的大陆架,只有少数的峰顶,突露在沉积面上,成为岩岛和暗礁。如立于大陆架前缘的钓鱼岛等岛屿,以及大陆架上的苏岩礁、虎皮礁等。因此,黄、东海大陆架应属于堆积型的大陆架。第三纪,特别是第四纪以来冰期、间冰期更迭交替所引起的海面变化,使大陆架多次暴露成为陆地和多次受到海侵。最后一次海侵是距今 20 000～15 000 年间开始的。海面从大理冰期末的低海面逐渐上升到现在的位置。所以,黄海是一个冰后期的浅海盆,台湾海峡再度形成浅海峡,东海则从狭窄的海域发展为现代东海。

　　根据海底地貌特征,现分为黄海北部区、黄海南部和东海大陆架区、大陆坡及海槽区几个部分叙述。

1. 黄海北部的海底地貌

虽然一般以成山角至朝鲜半岛长山串以北为黄海北部海区,但在海州湾以北,海底地貌无显著差别。所以,这里所指的黄海北部是指海州湾以北的黄海部分。

黄海南宽北狭。沿 35°N,宽度为 378 nmile。由于山东半岛突出,海面向北变狭,至成山角—长山串断面,海面宽仅 114 nmile,自此向北,又渐拓宽,西与渤海相接。

黄海海底平缓开阔,深水轴线偏近朝鲜半岛(图 13 和图 14),其深度大多为 60～80 m,东部的坡较陡,一般为 0.7‰;西部坡较缓,一般为 0.4‰左右。两侧不对称的斜坡交汇处,为一条轴向近南北的水下洼地。该处是自东海进入黄海的暖流通道。水下洼地偏于黄海东侧,洼地以东的海底沉积主要为来自朝鲜山地的砾质物质,而洼地以西的广大海底,沉积着被黄河、长江自中国大陆携运下来的粉砂、淤泥以及黏土物质。

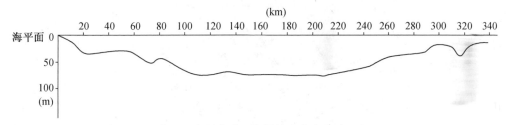

图 13　中国山东桑沟湾—朝鲜德积岛海底地形剖面图

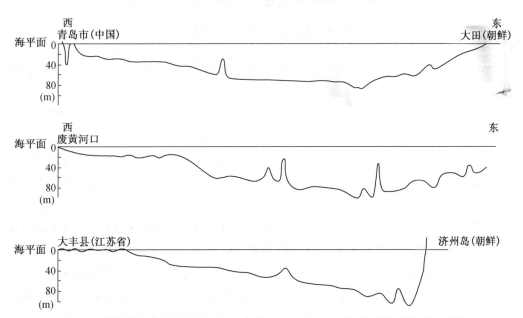

图 14　黄海南部大陆架(青岛以东、废黄河口以东、大丰县以东)海底地形剖面图

在黄海北端西朝鲜湾一带,由于朝鲜半岛为一隆起构造,地势高峻,不断经受剥蚀,半岛和西部山地河流水量大,流势急,携来大量粗粒物质沉积于朝鲜半岛近岸处,形成砂质沉积带。冰后期海面上升,在潮流与河流作用下,形成多条平行的水下沙脊,呈东北走向,与潮流方向一致(图 15)。沙脊规模较大,脊顶高出海底 7～30 m,平均高约 20 m,两脊之间隔约 0.8～4 nmile,为潮流通道,落潮时亦宣泄河流淡水。此水下沙脊的形成与基岩构

造无关,而是由于此处潮差大(超过3 m)、潮流急(1~2 kn,两脊之间为2 kn),致使海底沙滩在潮流的冲刷改造作用下,逐渐形成了与流平行的"潮流脊"。分布于鸭绿江口与大同江口之间的大片海底潮流脊,构成黄海北部海底地貌的一个重要特色。

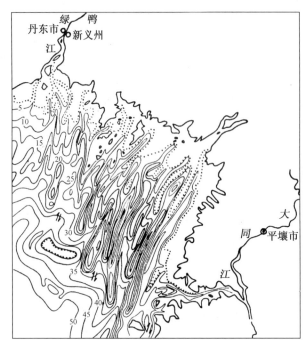

图15 黄海西朝鲜湾潮流脊平面图

辽东半岛东侧,河流较少,砂质来源不多,潮流作用较弱,潮流脊不发育,仅有由沉溺的山丘形成环陆罗列的岛屿,如长山列岛。

黄海以渤海海峡与渤海相接。由于海面束狭,岛屿丛崎、潮流迅急,海底受到冲刷,形成较深的水道。老铁山水道是一个涨潮流冲刷槽,庙岛水道则以落潮流占优势。沿岸流沿着山东半岛北岸流出,受到山脉与岬角的影响,形成局部的涡动,产生局部深潭,如威海遥远嘴附近深达61 m;成山角附近,水流逼转,流速很大,涡动强烈,沿岸海底冲刷出一个规模很大的深槽,最大水深达80 m。

在38°N以南的黄海两侧,多分布着宽广的水下阶地,西侧阶地比较完整,东侧阶地则受到强烈的切割(图16和图17),水下地形非常复杂,此与潮流作用强烈有关。潮波从南向北传播,受到地球偏转力影响,使朝鲜半岛西岸潮差增大,有的地方潮差达8.2 m,从而潮流也非常迅急,最大流速达9.5 kn,在这样强劲的潮流作用下,有些水下谷地被冲刷,深度超过50 m。

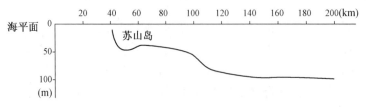

图16 苏山岛向东南—20 m水下阶地剖面图

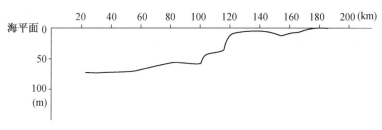

图 17　黄海江华湾－10 m，－40 m 水下阶地剖面图

山东半岛南北两岸都有水下阶地分布，北岸－20 m 阶地甚为宽广，南岸则有深度为 20～25 m 以及 25～30 m 的阶地局部保存完好。阶地面多平坦，前缘受到侵蚀，有许多沟谷，有的地方阶地以陡坡直插到水深 50 m 以下的平缓海底。

黄海东侧水下阶地分布的深度与黄海西侧并不一致，朝鲜半岛沿岸两级水下阶地，为水深 15 m 左右和水深 40 m 左右，这反映出构造运动在地区上的差异。

黄河虽然目前不在本区入海，但通过渤海海峡或历史时期在苏北入海扩展而来的泥沙，使本区海水相对混浊。水中泥沙沉积下来，覆盖了黄海北部大部分海底，以粉砂为主，间夹有淤泥。在海区的东侧，由于距离大陆泥沙来源较远，且潮流较强，细物质难以停积，使古海滨砂带暴露出来。局部地区，也有沙质淤泥带分布，这是朝鲜半岛河流所带来的物质在这里沉积的结果。

2．黄海南部与东海大陆架

黄、东海大陆架东南前缘以弧形突出，面临冲绳海槽，此大陆架以宽度大、坡度小，并受大河深刻影响为特点。

黄海南部的东侧与东海相通，地势向东南倾斜；西侧地势平坦，平均坡度只有 15° 左右，水深约 30 m，那里存在着一些水下三角洲。黄海南部有一系列小岩礁，如苏岩礁、鸭礁、虎皮礁等，它们与济州岛联成一条北东方向分布的岛礁线，是黄海与东海的天然分界线。

东海大陆架北宽南窄，海底向东南缓倾，平均坡度 1′17″，平均水深 72 m，但大部分海域的水深为 60～140 m，大陆架外缘转折在 140～180 m 深处。东海大陆架约以水深 50～60 m 分为东、西两部，西部岛屿林立，如舟山群岛等，水下地形复杂，坡度稍陡；东部开阔平缓，仅在其东南边缘有岛屿和岩礁，如我国的台湾与钓鱼岛等岛屿（图 18）。

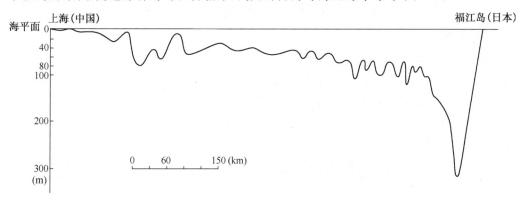

图 18　东海大陆架(中国上海—日本福江岛)海底地形剖面图

黄海南部与东海大陆架的地貌,虽然经受现代的水动力作用与泥沙运动的塑造影响,但是,却以大量地、清晰地保存着昔日的沿海平原与古海滨地貌为特征。它又可分为复式的规模巨大的古三角洲、宽广的古海滨以及长江古河道等地貌单元。

(1) 复式的古三角洲

从海州湾向南到杭州湾以北,有一个规模巨大的水下三角洲平原。它在平面上呈扇形展开,组成物质主要为石英、长石质的粉砂、淤泥粉砂,含有贝壳碎屑,具有河流沉积特点的交错层理,说明它是古代大河形成的三角洲平原,后来为海水淹没,潜伏于海面之下。古三角洲的东端约在 125°15′E,即苏岩礁与虎皮礁一线,这些岩礁是被三角洲沉积物掩埋的岛屿。沿着古三角洲的前缘斜坡,发育了一系列水下沟谷,是斜坡水流切割与粉砂质沉积物液化滑塌而成的前坡谷。近年,在水下三角洲平原的前缘地带,水深 50~60 m 处采集到棕色的极细砂[①],该砂层厚度超过 1m,大体上沿等深线方向分布,但已被前坡谷切断,间断分布于前坡谷两岸。砂的成分主要是石英与长石,含少量贝壳,其性质与两广沿海以及秦皇岛海滨的红棕色的古海岸阶地沉积类似,很可能也是更新世期间的一个海岸沉积。对它的研究将有助于确定古三角洲平原的时代。

长江和黄河都是以输沙量巨大著称。长江平均年输沙量 4.7 亿 t,黄河入海的泥沙高达 12 亿 t。更新世时,长江曾有相当长的时间在苏北入海,在河口堆积成三角洲,并不断向海推展,逐渐发展成为巨大的三角洲平原。黄河在更新世和历史时期也曾有相当长的时间从苏北汇入黄海,它所输送的泥沙也成为建造这个大陆架的物质基础。

位于苏北沿海,叠置在三角洲北部的古黄河三角洲,前缘约与 20~25 m 等深线相当,组成物质以细砂和粉砂为主,局部有淤泥沉积。它可能形成于晚更新世末期,并受历史时期黄河从苏北入海的影响。历史时期黄河曾有 727 年(公元 1128—1855 年)从苏北入海,它的巨量的泥沙使苏北海岸迅速向海推进。同时,也以大量的物质堆积在它的口外古三角洲上。现在这个水下三角洲的物质,受到潮流作用,游移不定,形成一系列的暗沙,散布在苏北沿岸。

上述的两个交互重叠的三角洲,组成了复式的古三角洲。此外,在长江口,还分布着面积较小的长江现代水下三角洲。它的顶部位于河口拦门沙的外面,三角洲前缘水深为 15~20 m。这组规模巨大的三角洲体,分布在黄海西南部与东海西北缘的广大范围内,除大三角洲体的南部遭受后来水流破坏外,其他大部分多保存完好,构成了本区引人注目的海底地貌。

(2) 古海滨

东海大陆架的东侧,深度大于 50~60 m 处,地势坦荡,起伏和缓,为宽广的砂质沉积带。它从台湾海峡向北延伸到朝鲜海峡,向南延伸到南海大陆架。砂带沉积物为分选极好到好,负偏态或接近于零的正偏态的细砂与中细砂(贝壳砂),有的地方还有砾石和粗砂。自西向东,物质由细变粗。细砂为石英、长石质,砂粒浑圆,细砂中重砂组合为角闪石、绿帘石、石榴石、榍石、锆石和电气石等,并有海相自生矿物黄铁矿充填有孔虫壳体,呈黑色的鲕状集合体。砂质沉积中含有丰富的软体动物残体、有孔虫、少量珊瑚及其他钙质

---

① 据国家海洋局第二海洋研究所。

有机质。愈向陆架边缘,贝壳碎片,有孔虫壳体含量愈高,达 20% 以上,可称贝壳砂。贝壳破碎,表面已经污染,壳色发暗,成为黄棕色、黄褐色和灰黑色,并普遍遭到磨损与溶蚀,有些壳体已胶结成砾石,有孔虫为浅水型的毕克卷转虫、小仿轮虫和中华丽花介等。外壳破碎,有些被磷灰石、海绿石等海底自生矿物所交代,壳体呈圆球状。上述现象表明,沉积物形成后,水深加大,距岸渐远,环境安定下来,导致生成这些自生矿物;细砂沉积由于在缺氧与扰动较小的海底停积较久,故砂色较暗,多为黑灰色。

这种细砂是滨海和浅海环境下的沉积,它是在更新世冰期低海面时,河流带来了寒冻风化所形成的粗粒物质。当时因大陆架狭窄,古海滨属高能海滨环境,泥沙在波浪作用下进一步粗化,堆积于现在大陆架外缘地带。冰期末,海面上升,河流流量加大,更搬运了大量陆地机械风化产物入海,形成大量砂质沉积。它们部分地掩覆着冰期的砂质沉积,并且随着海面上升,海浸范围加大,砂带逐渐向西展宽。因此,大陆架砂带既有冰期低海面时的沉积,亦有冰期后的沉积,但以海进的海滨沉积为主。

近年来,在虎皮礁附近采集到北方原始牛的下颚骨,在日本男女群岛一带采集到猛犸象的牙齿,在大陆架东部水深 100～150 m 处,采集到大量牡蛎、文蛤、红螺、扁玉螺、杂色蛤仔以及楔形条纹蛤等海岸带及河口区贝类的外壳,经用放射性碳法测定年代为 15 000～20 000 年。这些资料表明,在晚更新世大理冰期时,海面降低,黄、东海大陆架曾为一陆地平原。栖息着大型的寒冷气候的动物群;至更新世末,即大约距今 20 000～15 000 年间,海面从水深 150～160 m 的古海岸线逐渐上升,海面淹没了平原东部,形成了有大量海岸带浅水贝壳的砂带,即今大陆架水深 100～150 m 的砂质区,曾是当时的古海滨。冰后期海浸规模加大,砂带向西展宽,平原、河谷与古海滨皆沉于海底。大陆架的西侧,由于受到现代河流泥沙的覆盖,沉积物较细,为粉砂或淤泥。现代长江泥沙基本上仅影响到 120°20′E 左右。黄海水团带来的现代悬浮物明显影响到 30°N 以北、127°E 以西的区域。加之,黑潮主流沿水深 200～1 000 m 的冲绳海槽斜坡北上,其主轴流路稳定,表面流速 2～4 kn。黑潮水流影响大陆架东侧,使现代河流入海泥沙难以至此扩散停积,古海滨砂带未被掩埋,只是其沉积结构经过了一些改造①,但基本特性,与现代所处环境仍不适称,是所谓的大陆架上"残留沉积"。

(3) 古河谷

东海大陆架上,从长江口水下三角洲向外,延伸一条水下谷地,谷底沉积为灰色细砂贝壳。在马鞍列岛与嵊泗列岛之间,为一深谷,至浪岗山列岛一带逐渐向东南扩展,谷形宽浅,至水深 100 m 附近(29°N,125°E 附近)稍转向东,至大陆架前缘以急坡峡谷形式进入冲绳海槽。这条水下谷地是因海面上升而淹没的长江古河道。古河道呈西北—东南向延伸,向陆与长江口相接应,继之,似与呈西北向分布的高宝湖、洪泽湖以及南四湖等一脉相承;向海与宫古洼地遥相对应。分析这条古河道是沿这组北西向的大断裂构造发育而成。虽然在海面上升的过程中,古河道沉溺海底,但仍有水流作用(潮流流速自 0.3～1.0 kn 不等),谷形保存完好,并沿谷地向外输送泥沙。古河道的下段,接近冲绳海槽部分,受浑浊流作用,沿海槽斜坡下切形成水下峡谷,并于峡谷末端的海槽底部,堆积了海底

---

① 受现代水流作用,细颗粒被带走,并有生物性沉积加入。

扇。这种沿断裂带发育并沉溺于海底的长大古河道,亦构成中国近海大陆架的一个具普遍性的主要特色。

在31°15′N,124°E,沿地形低洼处向东南至30°30′N,127°45′E,分布着分选极好的厚砂层,其中亦夹有薄层的粉砂与黏土,局部有植物碎屑富集层和牡蛎层。有些夹有小砾石,富含片状矿物。这层沉积具有河流相特征,估计亦为一古河道遗迹,但尚需进一步查明河谷之分布。

长江古河道所流经的浙闽沿岸大陆架,一般都较平缓,深水距岸较近。而近岸带的水下地貌则较为复杂。由于岸线曲折,港湾深邃,岛屿罗列,在强潮作用下,有些海底受到强烈冲刷,有些海峡通道,基岩裸露,如舟山群岛某些水道水深超过100 m。河流带来的化学风化的细物质,以及被沿岸漂流运来的泥沙,在近岸海底沉积为黏土质淤泥,覆盖在陆架砂带之上。而距岸较远的陆架上仍然保持以细砂为主的底质。

3. 台湾海峡

台湾海峡位于东海大陆架南部,与南海陆架相连。它的北界从我国福建省的平潭岛到台湾省的富贵角,相距约93 nmile。南界从福建省东山岛到台湾省最南端鹅銮鼻,宽约200 nmile。台湾海峡水深较小,地形复杂(图19)。海峡大部分水深小于60 m,平均深度为80 m。海峡东南部水深在140~150 m发生坡折,以40‰~50‰的急坡进入南海海盆。大陆坡被许多水下峡谷切割,基岩裸露。由64个小岛组成的澎湖列岛位于海峡中部,其中以澎湖岛最大,高出海面50~60 m,由更新世玄武岩组成。台湾浅滩位于东山岛与澎湖列岛之间,东西长约110 nmile,南北宽约50 nmile。呈椭圆形近东西向分布,是台湾海峡最浅的地方。平均水深在20 m左右,最浅处约10 m左右。台湾浅滩由若干个东西向的、连续排列的水下沙堤所组成,这些砂堤大都是北陡南缓,紧密排列,沙堤最大宽度达1 000 m。台中浅滩是个小浅滩,位于澎湖列岛东北,台中以西,水深约40 m,最浅处约10 m,近东西向分布,浅滩上分布着平缓的沙丘。

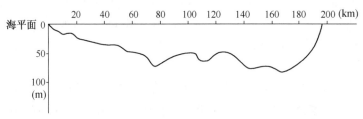

图19　台湾海峡25°N(湄州湾以东)海底地形剖面图

澎湖水道,位于台湾海峡的东南部,界于澎湖列岛与台湾南部西岸之间,是沿北东和北西两组断裂发育的三角形水下谷地。该处构成了台湾海峡中最深的地段。基隆水道,始于台湾海峡东北,其深度在观音岸外约为80 m,沿台湾北部海岸向东北深度增大。

台湾海峡沉积物除福建省沿海和台湾岛西海岸有泥质沉积外,主要由砂质组成。海峡中部存在着宽阔的粉砂、淤泥沉积,宽达85 km,从台湾岛西海岸24N°~24°45′N,穿过台湾海峡中部往西北延伸,与福建省平潭岛外的泥质沉积相连。泥、粉砂沉积的颜色大多是中灰到深灰色,其大陆坡上为青灰色泥质沉积。

砂质沉积分布面积最广,褐灰色或褐色的粗砂、中砂分布在台湾浅滩,灰色或深灰色的细砂分布在台湾海峡东北部与台中浅滩。一般说来,砂质沉积物分选极好到好,多呈负偏态。

台湾海峡砂质沉积中含有大量的软体动物残体和完整的贝壳,并含有孔虫、介形虫、苔藓虫和少量的珊瑚碎屑,含量大于10%。一般说来,近海岸含量较低,愈向海含量逐渐增高,以台湾浅滩含量最高达50%以上。上述特征表明,台湾海峡的砂质沉积亦为一古海滨沉积。海峡底部局部有砂砾和基岩,主要分布在澎湖列岛的西南面、台湾南北两端以及福建沿海某些岛屿周围。

4. 东海大陆坡与冲绳海槽

东海大陆坡位于东海大陆架的东南侧外缘,分布于东海大陆架的明显转折点之下。东海大陆架在水深140～180 m处坡度发生转折,然后在宽约30～60 nmile的范围内,以大于1°的陡坡急转直下进入冲绳海槽,至坡麓附近坡度又减缓(图20～图24)。所以,冲绳海槽的西坡,也即东海大陆坡。

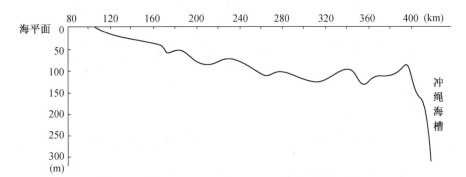

图20　飞云江口南—钓鱼岛南海底地形剖面图

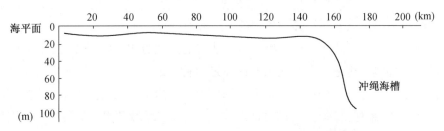

图21　东海30°N,126°E至29°N,128°E海底地形剖面图

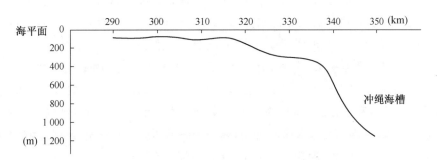

图22　东海30°N,124°E至28°N,127°E海底地形剖面图

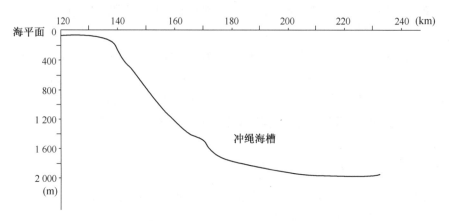

图 23 东海 27°N,123°E 至 23°N,124°E 海底地形剖面图

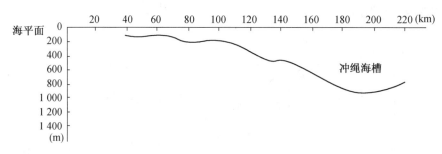

图 24 东海 31°N,127°E 至 30°N,130°E 海底地形剖面图

东海大陆坡的水深,大体上界于 140 m(或 180 m)到 1 000 m 之间。东北段稍浅,大陆坡脚水深小于 1 000 m;西南段陡深,水深大于 1 000 m。但各段皆表现为单一的陡斜坡。大陆坡的上端与大陆架转折带衔接,各段转折带皆表现出明显的构造地貌——沿陆架边缘分布的隆脊或高地以及环绕隆脊外围的裂隙谷地(照片 1)。

冲绳海槽是东海大陆架与琉球群岛岛缘陆架的天然分界。海槽东坡,即琉球群岛的西北侧陆坡。冲绳海槽是一个北北东向的弧形海槽。北浅南深,北部水深一般为 600~800 m,南部水深则为 2 500 m 左右,最大水深出现在台湾东北,超过 2 719 m。海槽两坡陡峭,沿坡发育了一系列由浑浊流切割的水下峡谷,在大陆坡坡麓水下峡谷出口处堆积了浑浊流相的海底扇。由于海槽内堆积了厚度超过 1 200 m 的沉积物,故谷底平缓,使海槽呈 U 形剖面。沿海槽的轴心部分,大体上在 26°N 以北的范围内,分布着一系列海山,海山相对高度约 800 m,最高的海山顶部水深为 92 m,大部分海山是沿北东—南西向张裂构造喷发的火山峰,有几个尚系近代活动的火山(见照片 2、照片 3)。

东海大陆架外缘至水深 400 m,有一条与等深线大体平行的细砂带。它与大陆架细砂带相比,介形虫与贝壳碎片含量显著减少,而辉石与云母类片状矿物却有所增加;含有丰富的浮游有孔虫和少量的珊瑚。细砂分选好到极好,负偏态。细砂带以下为灰白色有孔虫—粉砂—泥,自南至北呈条带状分布,遍布于海槽西坡,至 29°30′N 以北,宽度明显增大。此种沉积中,砂、粉砂、泥含量均大于 20%。砂级沉积中,绝大多数是微体生物壳体,主要是有孔虫,有些地方有孔虫含量达 80% 以上。海槽底部为黏土软泥,仍含不少的微体生物壳体,$CaCO_3$ 含量为 20%～30%,中心部分达 30%以上。海槽泥质沉积中受火山沉积影响明显。不少地方的黏土软泥中有浮石,其个体大小不等,大者大于 1 cm。在海山及海槽底部隆起的附近,黏土软泥沉积中夹有灰白色的火山灰层。铁、锰含量急剧增高。海槽沉积中陆源矿物组合虽然有所降低,但仍占相当比重,出现有较多的紫苏辉石。

在晚更新世大理冰期最盛时,海岸线位于水深 150～160 m,低海面时中国大陆的河流在今大陆架外缘直接注入冲绳海槽,河流将大陆泥沙带到海槽沉积。冰后期,海面上升,黑潮进入东海后沿冲绳海槽西坡北上。高温高盐的黑潮水体,有利于浮游有孔虫繁殖;加上火山活动,故冲绳海槽沉积中,以陆源碎屑物质为主,同时接受了较多的生物沉积与火山沉积。现代的底流作用还带来了河流的、大陆架的细粒物质。

自东海大陆架转入冲绳海槽有一系列差异。地质上是由稳定的大陆地壳,转为构造活动的过渡性地壳;地形坡度由 1′增大到 1°以上,深度增至 2 000 m;地貌上,大陆架主要是陆地与古海岸的地貌经浅海环境改造的,如古河谷、残丘、古海岸阶地、古沿岸堤、与三角洲等,而冲绳海槽主要是构造地貌及海洋环境形成的;在沉积上,大陆架以陆源碎屑沉积占绝大部分,海槽中生物沉积、火山沉积亦占相当比例。因此,东海大陆架是中国大陆的自然延伸,而冲绳海槽构成了东海大陆架与琉球岛弧的天然分界。

5. 琉球群岛岛架

冲绳海槽之东为露出海面的琉球群岛和日本九州,以及各岛屿在水下的岛架。岛架宽度不大,在九州处约为 30～50 nmile,在琉球群岛附近为 2～20 nmile。岛架地形复杂,沙滩、岩滩密布,并且岛架分布不连续,各岛间已为水下峡谷隔断,并有三个深度超过 1 000 m 的海峡,使冲绳海槽与太平洋相通,其中以石垣岛西北的海峡最深,水深达 2 000 m 以上。

琉球群岛的外坡,急剧加深,坡度达 10°左右。但在 1 800～4 000 m 深处有一由火山质沉积组成的阶地。阶地以下,坡度更大达 13°左右,降入 6 000 m 水深的琉球海沟中。

综上所述,黄、东海自第三纪陆架边缘隆褶带产生以来,奠定了大陆架的基底,该隆褶带成为大陆架的边缘堤坝,它阻拦了黄河、长江从中国大陆侵蚀搬运来的大量泥沙,使它们沉积在西侧的拗陷带内,形成了黄、东海堆积型大陆架。第四纪海面的升降变化,使黄、东海大陆架几经沧海桑田之变,大量化石资料表明,晚更新世大理冰期时海面降低,黄、东海大陆架出露为平原,一些大型的寒冷动物群得以越过平原到达边缘诸岛。嗣后,海面上升,又淹为大陆架,所以,现在的黄海、东海大陆架及台湾海峡,主要是全新世以来海侵的结果。

三、台湾以东太平洋区

我国台湾的东海岸直接濒临浩瀚的太平洋,这一海区的北界大致相当于日本琉球群岛的先岛群岛,南侧以巴士海峡与菲律宾的巴坦群岛相隔。

台湾及其附近岛屿(如兰屿、火烧岛及钓鱼岛等岛屿)是我国通向太平洋的最前哨。这一海域构造、海底地貌主要特点为:地形起伏变异大,地貌类型齐全,有大陆架、大陆坡与大洋盆;地质构造复杂,数个构造带相交汇,是亚洲大陆与太平洋的复杂接触带。

（一）海域地质构造

沿西太平洋边缘地区，发育着一系列岛屿-海沟构造带，台湾岛弧也是其中之一，台湾在构造上地位重要，因为它是东北方的琉球岛弧构造带及南方的吕宋岛弧构造带的交汇枢纽。各构造带情况如下：

1. 琉球岛弧构造带

该构造带呈弧形向东南突出，即所谓"正向岛弧构造带"。

2. 台湾构造带

它是向西突出的弧形构造带，即向亚洲大陆凸出的"反向弧构造带"。从西向东包括四个互相平行的次一级构造带，即台湾西部新第三纪冒地槽，台湾山脉双变质带，台东纵谷断裂带，台湾东部新第三纪优地槽。

台湾海岸山脉以东的火烧岛及兰屿，岩性为第四纪的火山岩。根据折射地震及重力观察的推算，地壳厚度在福建沿岸为 35 km 左右；台湾海峡、澎湖列岛、台湾西部平原和丘陵区为 30～35 km；台湾山脉为 35～40 km；台东海岸山脉为 20～27 km；菲律宾西部为 6～8 km。从重力资料看，台东纵谷以东，为正异常，以西为负异常。地壳厚度的变化及正负重力异常的出现，以及台湾山脉存在一低温高压与高温低压的双变质带，都说明了台湾东部包括火烧岛和兰屿等地区，太平洋与大陆间的相互作用是很强烈的。

3. 菲律宾吕宋岛弧构造带

该构造带是指北部巴坦群岛、巴布延群岛及吕宋岛。吕宋岛是一个很复杂的岛屿，它的北半部构造线向西突出，即凸向南海；而其南半部则向东北方突出，凸向太平洋。它们从西向东包括三个次一级大致平行的构造带，即西海岸科迪勒拉中央山脉变质岩带，卡加延（嘉牙鄢）河谷第三系沉积拗陷，东部马德里火山岩带。

以上诸构造带在台湾以东的海区交汇，它们均成线状延伸，且多为拗陷—隆起—拗陷类型，可以相互对比。

在上述构造带的交汇带以东，可见一些海岭，深度在 6 000 m 以上，已属太平洋洋盆。深海底为波速 6.0 km/s 的地层，为大洋层，其上覆沉积物厚约 200 m，地形平坦。

（二）海底地貌

台湾以东海底地貌的特征是，狭窄的岛缘陆架，其上或出露剥蚀的基岩，或者堆积着由陆上剥蚀而来的砾砂或中细砂。陆架外侧是陡窄的大陆坡直插入海沟或洋底。按其地貌特征，可分三段。

1. 北段

从三貂角至苏澳南面的乌石鼻。这一段海深 600～1 000 m，海底坡度较缓，大陆架较其他两段稍宽，大体上从北到南，约 5.4 nmile～8.6 nmile～4.3 nmile，坡度为 20‰～12.5‰～25‰。海底地貌以大陆坡为主，三貂角以东即接近冲绳海槽的西端（图 25）。而龟山岛与苏澳湾紧邻琉球海岭西端的先岛群岛（图 26）。冲绳海槽与琉球海岭在这一段皆为近东西向伸延，海底沉积在近海槽部分为灰色深海软泥，而邻近群岛处为白色细砂。

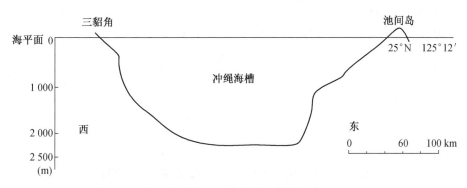

图25 中国三貂角—日本官吉列岛海底地形剖面图

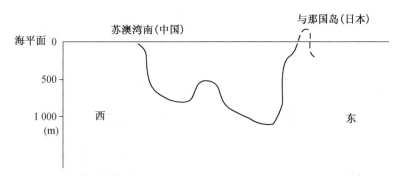

图26 中国苏澳湾南(24°30′N,121°47′E)至日本与那国岛海底地形剖面图

2. 中段

从乌石鼻至三仙台,海岸是断崖峭壁,崖下即临深海,陆架狭窄,宽度约 1~2 nmile;大陆坡坡度更为窄陡。花莲岸外以东 19 nmile 处,水深达 3 700 m(图 27),其东南 16 nmile 处,已临 4 420 m 深的洋底。

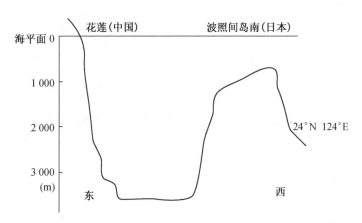

图27 中国台湾省花莲至日本波照间岛南(24°N,124°E)海底地形剖面图

中段海深一般超过 3 000 m,海底坡度约为 125‰~150‰。海底北部与先岛群岛相接,其岛缘陆架与大陆坡均十分陡峭(约 10°)。海底南部横亘着东西向的琉球海沟。在两者之间为宽广阶地,其中有厚数百米至两千米的沉积物,表层为深海钙质软泥,或火山灰黏土,近岛屿处有砾、砂。阶地以南为 13°的陡坡,并急剧下降至超过 6 000 m 深度的琉球海底。海沟纵向坡度约 6‰（图 28）。海沟底部覆盖着薄层的浑浊流沉积，海沟南坡较缓，坡度约为 25‰。琉球海沟的南面是宽广平坦的菲律宾海盆（图 29）。

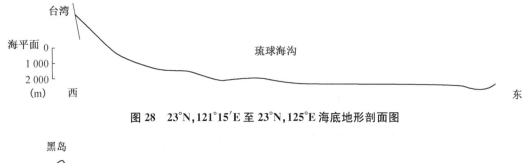

图 28　23°N，121°15′E 至 23°N，125°E 海底地形剖面图

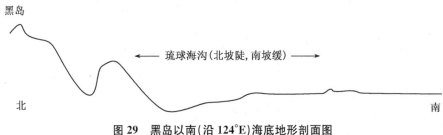

图 29　黑岛以南（沿 124°E）海底地形剖面图

3. 南段

南段为台湾东南海域，有两列受南北向隆褶带控制的水下岛链。西部的水下岛链是台湾岛台湾山脉向海延伸部分，向南可达吕宋岛以西的南北向海岭；东部的水下岛链是由我国的火烧岛、兰屿等向南延伸到吕宋岛东部的马德里山。后一条岛链东坡最陡，急转直下菲律宾海盆（图 30）。菲律宾海盆水深约 5 000 m，地形平坦开阔，普遍沉积了薄层的深海软泥。软泥下，即为大洋型玄武岩。在东西两岛链之间，为一深度超过 4 000 m 的南北向深海槽，谷坡陡峭，谷底深度大于 5 000 m。

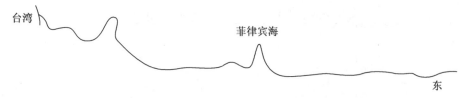

图 30　22°N，120°52′E 至 22°N，125°E 海底地形剖面图

四、南海海底

南海海域辽阔，其轮廓大体上为一边呈北东—南西向、而另一边呈北—南向的斜菱形。

南海海底也是长轴为北东—南西方向的菱形盆地。盆地处于西沙—中沙群岛与南沙群岛之间，是一个深度在 4 000 m 以上的深海平原。

盆地的西北边沿，地形平坦，从两广沿海向东南倾斜，在台湾与海南岛联线内侧，水深在 200 m 以内，其中北部湾是一个半封闭的浅海海盆，大部分水深在 20～50 m，最大水深 80 m，湾内海底平坦，自西北向东南倾斜。从东沙群岛向南水深加大，为 1 000～2 000 m，东沙群岛与中沙—西沙群岛间水深为 1 000~3 000 m,中沙—西沙群岛所依附的基底为一海底高原,高原的东南侧深度急剧加大,下降到中央海盆。

盆地的东侧,即南海的东部边缘地区,由于分布着巴拉望海槽、马尼拉海沟与西吕宋海槽,所以深度大、坡度陡,水深超过 5 000 m。

西沙、中沙与南沙群岛分布于中央海盆两侧海底高原上,其依据是,海山顶部发育的珊瑚岛有的尚隐伏于海面之下。岛礁数目众多,大小不一,其间水道纵横,深度变化多端(图 31)。南海的西南部,大致从湄公河口到马来西亚(沙捞越)一线以西,海底地形平坦,为深度在 200 m 以内的浅海。

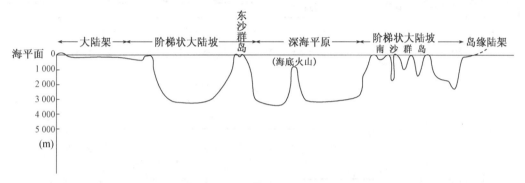

图 31　南海珠江口高润岛—巴拉巴克海峡邦立岛海底地形剖面

(一) 海底地质构造

南海的轮廓型式不是偶然的,是受基底构造所控制,其特点如下:

(1) 北东向的断裂与隆起成为南海构造的主干,尤以北东向断裂居主导作用,它可能与南海的构造成因有关,北东向断裂不止一次发生,其时代始于晚第三纪。由于断裂,南海可划分为三个主要的北东向构造带。

南海中央海盆,位于中沙与南沙群岛之间是北东走向的狭长海盆。构造上是居于北东向拉开断裂地带之间,由于地壳的均衡补偿作用而出现的地幔物质上隆地带。基底岩层属大洋型地壳的玄武岩类物质,为厚约 2 000~4 000 m 的纵波速度 6.6 km/s 的第三层(层 3);上覆 1 000~2 000 m 厚的纵波速度为 4.3~4.5 km/s 或 3.7~3.9 km/s 的第二层(层 2),可能属火山喷发物、块状珊瑚和固结的沉积岩;顶部为厚约 1 000 m,纵波速度为 2.1 km/s 的松散沉积层。第三层下即为莫霍面(图 32)。南海中央海盆的第三层较正常洋壳的第三层约薄 1/2,其位置较太平洋洋壳层 3 的正常位置浅 1.5 km(有的地方浅 2~3 km)(图 33 和图 34),这种情况既反映着南海洋壳的区域性特征,亦反映着海盆底部洋壳补偿性的上升。目前,在这个构造带内还有一系列北东向裂隙与火山喷发活动。

中央海盆的南北两侧是沉降的块断构造带。西北侧的是西沙—中沙块断构造带,走向北东,断续延伸到东沙、台湾浅滩与澎湖列岛一带。其基底是中生代或者更老的岩层,因隆起上升,曾经遭受风化剥蚀,新生代早第三纪地层也隆起成为一系列北东向的构造脊,在上第三系上发育了厚达千米的珊瑚礁层,根据珊瑚礁的厚度、块断带目前所处的水深位置和基底风化壳被埋藏的深度,反映着自晚第三纪以来,块断的各部分有不同幅度的

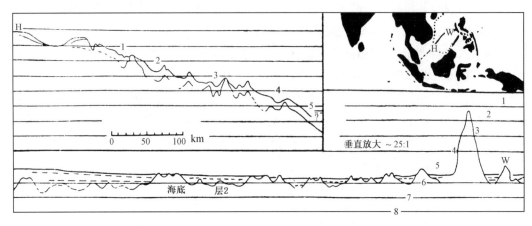

图 32 南海盆地地震剖面(据 Ludwig W. J., Hayes D. E., Ewing J. I., 1967)

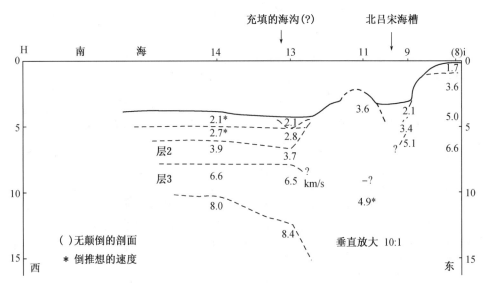

图 33 北吕宋以西地震剖面(据 Ludwig W. J., 1970)

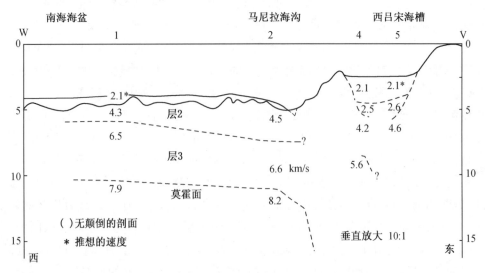

图 34 中吕宋以西地震剖面(据 Ludwig W. J., 1970)

沉降。在中央盆地的东南侧是南沙块断构造带,基本情况与西沙—中沙块断构造带相似,但隔以中央海盆遥相对峙。每个块断构造带的两侧都有北东向的大断裂,使块断带与相邻构造单元分隔开来。而块断带本身,又为次一级的,时代较晚的北西向断裂所分割。

(2) 在上述北东向构造带的大陆侧,为一系列新生代沉降盆地。大体上以红河为界,以东为我国两广沿海沉降盆地。这里,在陆地所发育的一系列北北东向的深大断裂都直接延伸入海,通常表现为一些大的重力阶梯带和线性磁力异常带,但是在海区这些断裂的方向大都偏转为北东至北东东方向。由于这些断裂的存在,构成了数列同走向的隆起的构造脊,控制着盆地的分布和地层的发育,并在现代海底地形上有明显的反映。东西向的构造在海区也有重要影响,上述的许多陆地上北北东向断裂到了海区发生偏转的现象,可能即与受东西向的构造干扰有关。在红河大断裂以西,是一系列北西向构造的沉降盆地与隆起,如湄公—南沙西南盆地、暹罗湾盆地以及其间的隆起地带。

(3) 南海东部,从我国的台湾岛到菲律宾群岛以及巴拉望岛附近海底伴生着一系列的海槽与海沟,如:吕宋海槽、马尼拉海沟与巴拉望海槽等。在海槽中,新生代沉积在晚更新世至全新世时褶皱隆起,形成数列近于平行的南北向构造脊,它们向北延伸到台湾岛台湾山脉的南端。这种岛弧与海沟相伴分布的格局不是偶然的,它与南海中央海盆有大洋型基底的现象一致,反映了南海东部密切受太平洋底部的构造活动影响,而南海其他部分则与亚洲大陆关系密切,受大陆构造活动影响较大。

(二) 海底地貌

南海海底地貌的总特征是:海底地势是西北高、东南低,自海盆边缘向中心部分呈阶梯状下降。在菱形盆地的四周边缘分布着大陆架;在大陆架的外缘分布着呈阶梯状下降的大陆坡;在大陆坡终止处,是南海海盆的中央部分,该处是一片坦荡的深海平原,平原中部有几座数千米高的海山,有的已接近海面,成为水下暗礁。

1. 大陆架

南海大陆架主要分布于海区的北、西、南三面,是亚洲大陆向海缓缓延伸的地带,大陆架外缘明显转折点水深约为150~200 m,有些剖面,转折点深度超过200 m,有的甚至达到375 m。它环绕着陆地,大体上是西北部与西南部宽度大,西部宽度减小;而东部狭窄,是岛缘陆架。具体情况大约如下:

(1) 南海北—西北部

我国华南沿海大陆架宽度见表2。

表 2 南海北—西北部大陆架宽度

地 名	大陆架宽度	坡 度
台湾南部(高雄)	7.5 nmile	14.3‰
汕头(南澳岛)	106 nmile	0.77‰
珠江口	137 nmile	0.54‰
电白	148 nmile	0.53‰
海南岛南部	50 nmile	2‰

北部湾由岸边向中央部分逐渐加深,最大深度 80 m,所以整个海湾皆位于大陆架上。自湾顶至湾口,大陆架宽 261 nmile,坡度为 0.3‰(见照片 4、照片 5 和照片 6)。

（2）南海西部

大陆架宽度不大（表3）。

表 3　南海西部相邻各国沿海大陆架宽度

地　名	大陆架宽度	坡　度
岘　港	65 nmile	1.22‰
归　仁	15 nmile	5.3‰
金兰湾	26 nmile	3.0‰
湄公河口	与巽他陆架连成一片	

（3）南海南部巽他大陆架

包括马来西亚、印度尼西亚诸岛与中南半岛之间的区域,暹罗湾全部为大陆架,亦属巽他大陆架的一部分,这个大陆架是世界上最宽的陆架区之一。我国南沙群岛的南屏礁、南康暗沙、立地暗沙、八仙暗沙和曾母暗沙等,是位于水深 8～9 m 与 20～30 m 之内的珊瑚礁,它们位于这个大陆架的北部,该处陆架宽约 150 nmile。

（4）南海东部岛缘陆架

陆架范围窄狭（表4）。南海大陆架有两种类型,一种是堆积型的,另一种是侵蚀-堆积型的,以堆积型为主。南海东部岛缘陆架是侵蚀-堆积型的;我国沿海大陆架是堆积型的。南海大陆架的基底岩层与相邻大陆一致,主要是老的花岗岩与片麻岩,经长期上升剥蚀成一片起伏不平的地面。新生代沉降为盆地,其中覆盖着早第三纪地层,后经褶皱隆起,形成一系列北东向的构造脊,像水下堤坝般地拦阻了河流从亚洲大陆冲刷下来的碎屑物质,于堤后堆积下来,形成了堆积型大陆架。大陆架中的新生代沉积厚度,一般超过 2 000 m。当泥沙于堤脊内侧堆满后,会超过堤脊,于外面另一列堤脊的内侧再行堆积,如此前进,陆架逐渐扩大。如香港附近的大陆架上,发现一个构造脊被掩埋于 1 400 m 厚的沉积层下（照片7、照片8）。

表 4　南海东部岛缘陆架宽度

地　名	大陆架宽度	坡　度
沙　巴	37～43 nmile	2.12‰～1.63‰
巴拉望岛	25～31 nmile	3.97‰～3.17‰
马尼拉	5.0 nmile	15.9‰
北吕宋	2.4 nmile	31.8‰

南海大陆架地形平缓,大多自陆地向深海倾斜,坡度逐渐加大。我国沿海的大陆架在 50 m 水深处有一地形坡折,使大陆架呈台阶状。南部巽他大陆架上,自邦加岛一带向东北沿坡而下,有一条深 80～90 m 的水下谷系,河口部分水深超过 100 m,它是冰期低海面时的河流侵蚀成的,冰后期海面上升被海水淹没,又为潮流冲刷加深,成为海底大陆架上

的溺谷系。南海南部大陆架上,分布着一些珊瑚礁、滩与水道,海底地形起伏变化较大。南海东部大陆架,冲刷切割较多,例如,沙巴北部巴拉巴克海峡附近的大陆架外缘深度为180 m,其上有深达 40～60 m 的沟谷,谷中有些地方填充了沉积物,有些地方出露了石质海底。

几条穿越南海大陆架的深水道,将南海与相邻海域相互贯通。大陆架北部与东海的台湾海峡相连;南部经过卡里马塔海峡、加斯帕海峡与爪哇海相通,这两个海峡局部冲刷深度超过 40 m,已接近基底岩床;西部通过深度为 30 m 的马六甲海峡与印度洋相通;东部通过巴士海峡深水道与太平洋相通。

南海大陆架表层底质,在我国沿海呈现为与海岸平行的北东—西南向带状分布。

在两广沿海的韩江口与珠江口外,受河流冲积物的补给,在沿岸 20～30 m 等深线内分布着粉砂质黏土;在现代河流泥沙扩散带以外(在河口附近相当于 50 m 水深以外的地区)分布着砂质沉积。砂质(包括粉砂)沉积在南海大陆架上,分布较广,在汕头沿海大致从 30 m 等深线向外即为砂质沉积带,沉积颗粒从岸向海由细砂—粉砂—粗砂递变。珠江口外是在 50 m 等深线处交错分布着细砂与粉砂,砂的成分主要是石英长石,含较多的自生海绿石(占重矿总量的 20％～25％以上)和较多的碳酸钙,细砂中含有钛、磁铁矿等重矿物,矿物成分向外海减少,而贝壳与有孔虫成分却逐渐递增。砂层为青灰色,略具臭味。种种特征表明,其沉积海底为时已久。砂带的外界大致与大陆架的外缘水深相当,但颗粒组成变细。砂带是更新世末期低海面或全新世初海面开始上升时的古海滨沉积,当时由于气候尚冷,机械风化作用强,故海滨沉积物的颗粒较粗,而冰后期以来,气候转暖,尤其华南沿海高温多雨,以化学风化为主,颗粒较细,砂带内侧的粉砂质黏土是被河流携运覆于砂上的现代沉积,如:珠江是华南最大的河流,年径流量 3 700 亿 m^3,每年输沙 8 544 万～1 亿 t,珠江口的淡水入海后与海水混合,形成一低盆的冲淡水团,分布在雷州半岛与汕头附近水深约 40 m 以内的浅水区。每年 5—8 月西南季风期间,珠江冲淡水随季风沿岸向东北漂移,5—6 月末珠江径流最大期间,这支漂流可伸展到 117°E 以东;8 月为此漂流的鼎盛时期;而夏季时,珠江口出现一个低盐水舌,其轴线直向东南;但粤西沿岸冲淡水是沿海岸向西移,一直到达海南岛东北角。每年 10 月至翌年 5 月,南海为东北季风,广东沿岸冲淡水团随季风向西南移动,盐度较夏季为大,但其扩散影响范围很宽。珠江冲淡水影响到达处,海底沉积了黄褐色粉砂质黏土,其影响不到的地方古海滨砂仍暴露于海底。

琼州海峡内,由于:① 峡窄流急,海峡宽 14 nmile,涨潮流流速最大可达 5～6 kn,冲刷力强,海峡内局部水深达 100 m,峡底难以停积细颗粒泥沙。② 海南岛最大河流南渡江,是一条多沙性河流,它在琼州海峡处入海,于河口堆积了大片砂质浅滩,浅滩泥沙受往复潮流作用,被带到河口东西两侧堆积。基于上述两个原因,造成琼州海峡内碎石、砂砾与中、粗砂堆积带,也有些地方出露了基岩。

琼州海峡西侧,由于过水断面开阔,潮流扩散,流速减低,形成了与渤海海峡西侧相似的潮流三角洲。数条指状的水下沙脊,自海峡西口向西和西北方向延伸倾伏,沙脊间与两侧较为低下,是水流通道。潮流三角洲堆积物主要是砂,通道内为砾砂,颗粒较海峡内沉积为细。根据潮流三角洲所在位置及其延伸形态与物质组成,表明它是涨潮流形成的。

北部湾沉积以粉砂为主。在北部,沿岸表层沉积是细砂,向海依次递变为粉砂和淤泥,再向外又逐渐变粗,大约在水深 5 m 处,即出现粗砂沉积。成分主要是石英砂、夹贝壳碎屑,砂粒初经磨圆,表面已染黄,沉积物略具臭味,表明沉积海底较久,很少受到现代水流的扰动,粗砂或砾砂围绕着北部湾沿岸陆地呈带状分布,其外界水深为 10~25 m,它是古海滨堆积。由于北部湾沿岸河流不多,带入海湾中的泥沙较少,所以古海滨砂带在较浅的海底即出现。粗砂带以外海底沉积是粉砂,在一些沉溺的谷地中有粉砂质黏土沉积。北部湾西部与西南部皆为粉砂底质,成分和颜色都受河流泥沙影响,北部湾中的各个岩石岛屿,如海南岛西侧,肥猪龙岛和白龙尾岛的周围,底质是中砂与细砂。海南岛周围细砂中富含角闪石、绿帘石、蓝闪石和黝帘石等重矿物,细砂带可延伸到 50 m 深处,细砂沉积既有邻近基岩岛屿的风化剥蚀产物,亦可能有古海滨沉积。北部湾东南侧水深较大,分布着粉砂质黏土,并呈现为北东—南西向的带状分布。

自暹罗湾至南海西南大陆架,沿岸底质为砂,陆架轴心部分为黏土质软泥,其他绝大部分底质为粉砂,但在大陆架外侧,即朝向南海深水区方向,沉积了粗砂,亦是更新世末及全新世初海侵早期海面较低的古海滨沉积。

南海东部岛缘陆架为侵蚀-堆积型的,底质分布不甚规则,或为基岩裸露,或沉积着自陆地侵蚀下来的碎屑物质,成分主要是砂。

南海大陆架表层沉积以陆源为主,大陆架内侧沉积受现代河流泥沙影响较大,而外部的砂质沉积属陆源的残留堆积物,由于冰后期以来,长期停积于海底,沉积物已经染色,并产生了海绿石与碳酸钙等物质。

概括来说,南海大陆架亦主要是由陆源物质于构造脊内侧储留堆积成的,堆积始于早第三纪,但主要堆积时期是晚第三纪与第四纪。更新世末与全新世初,大陆架基本形成,嗣后,大陆架内侧叠加堆积了现代细颗粒泥沙。

2. 大陆坡

南海大陆坡分布于大陆架的外缘,水深界于 150 m 到 3 600 m 之间,多呈阶梯状。通常在 −150 m 的地方,海底地形由平坦的大陆架明显地转为一陡坡,水深急剧加大,陡坡下有条深沟,大约在 1 000~1 800 m 之间,地形又转缓,为一宽达数百海里断续相连的平台面,其上海岭横亘、岛礁众多,平台面外又为一急陡坡,直降至水深 3 600 m 处,地形又转平缓,而到达中央海盆底部——南海深海平原(照片 5 和照片 6)。南海阶梯状的大陆坡是受块断构造控制形成,并为次要的断裂所分割,阶梯状大陆坡的宽阔平台面,相对于深海平原,也可以认为是耸立于南海深海平原之上的海底高原。在海底高原外缘,常形成局部海底上升流,致使高原顶部的岛礁地带成为丰产的南海渔场。

南海大陆坡可划分为:① 中央海盆的北坡;② 海南岛南部大陆坡;③ 中央海盆南—东南坡;④ 中央海盆东坡。

(1) 中央海盆的北坡

位于我国台湾与珠江口之间。大陆坡特点是以东沙群岛为中心呈现为阶梯状的凸形坡而向中央海盆倾斜。

台湾南端一带的大陆坡,坡度陡,为 26.5‰;台湾浅滩以南,大陆坡是上部陡(介于 −150 m 与 −2 560 m 之间,坡度为 28.4‰),而下部缓(在 −1 828 m 到 −3 600 m,坡度为 4.3‰)。

东沙群岛一带是凸形缓坡,大陆坡上部(约相当于 700 m 水深以内)坡度平缓(3.4‰),而在 700 m 水深之外,坡度急剧加大,以 27.7‰的坡度倾入深海平原。在这一带的大陆坡上有一些岩礁与珊瑚岛,突立于坡面上;大陆坡下部还有一些小的水道切割着斜坡。在珠江口外的大陆坡上,有一条北西—东南向的深沟(最大深度是 3 300 m),将北坡与海南岛南部大陆坡区分开(图 35 和图 36)。

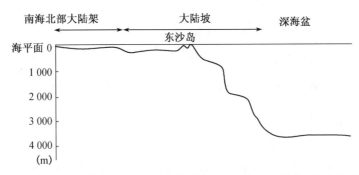

图 35 广东沿海(22°15′N,116°E)至南海东北部(19°10′N,119°10′E)海底地形剖面图

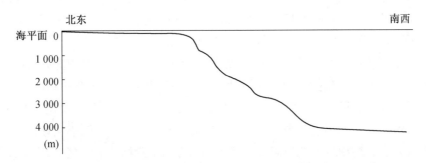

图 36 珠江口外(22°20′N,114°03′E 至 18°30′N,115°15′E)海底地形剖面图

(2) 海南岛南部大陆坡

范围介于珠江口外深沟与越南南部之间,其内侧在水深 150～1 000 m 之间是以 27.5‰的坡度与海南岛大陆架外缘相接。1 000 m 水深以下,为一宽度约 267 nmile 的海底高原;其外侧以 28.4‰～116.2‰的陡壁直接降到 3 600 m 的深海平原。例如,中沙群岛外侧有个 51°的陡坡,这种陡坡—海底高原—陡坡与深海平原相接的形式,使得这一带的大陆坡具有明显的阶梯状特征(图 37 和图 38)。

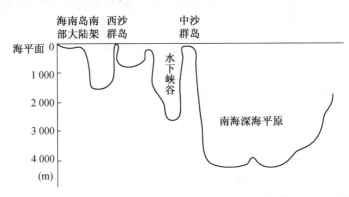

图 37 海南岛南部(18°11′N,109°48′E)至南海东南部(12°32′N,118°40′E)深海盆地海底地形剖面图

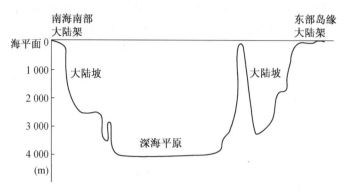

图 38 12°N,109°15′E 至 12°N,120°E 海底地形剖面图

海底高原与基底岩层与其内侧大陆架一致,在高原东部的一系列岭脊上,发育了众多的珊瑚岛与浅滩,前者主要是环礁。如中沙群岛与西沙群岛都是环礁,也是沿着海底高原上的北东向构造脊而发育的。构造脊的基底岩层与邻接的大陆架基底岩层一致,是花岗岩或片麻岩。基底岩层深度为 1 100～1 300 m,超过珊瑚礁生长下限达 1 000 m 以上,表明海底是下沉的。下沉总幅度超过 1 000 m,主要是地壳下沉,其次为海面上升。伴随着地壳的缓慢下沉,而珊瑚礁建造不断发育,其时代主要是上新世以来。

珊瑚群岛内的各环礁之间水深不大,界于数十米到 200 m,但中沙群岛与西沙群岛之间的海底高原面下,却有近 1 100 m 的深水道。在西沙群岛的西北部和北部,有一条近东西向的狭窄海槽。海槽长约 227 nmile,西宽东窄(西部宽 7 nmile,东部宽 3.2 nmile),西部水深为 1 870 m,而东部水深为 3 170 m,已与海底平原沟通。磁测表明该处分布着磁性较强的基性岩类。海槽似为拉裂而成。这条海槽分割了阶梯状的大陆坡,成为几个块段,各块段内亦有小的水道切过大陆坡面,注入深海平原。

海底高原上的山脊向西延伸,高度渐低,并被隔裂为一个个孤立的小高地。深海盆地的西南坡是一个单一的倾斜坡,自大陆架外缘起,坡度从 16.6‰(水深 150～1 800 m)到 8.0‰(水深 1 800～3 600 m 处的大陆坡麓),然后过渡到深海平原。

因此,海南岛南部大陆坡呈阶梯状(照片 2),其海底高原的基底岩层,上覆构造(东北向构造脊普遍分布于大陆架和海底高原上)均与相邻的大陆架一致,但此处深度大,有巨大的东北向断裂分割,而高原的海岭上有千余米厚的珊瑚礁,表明这个呈块断下沉的海底高原是沉没、断折的古大陆架,断裂主要是东北向的,它与下沉相伴发生,时代始于晚第三纪,但经多次活动。第四纪时新构造活动仍剧烈,有些珊瑚礁抬升至海面上,并有近东西向的断层,使海底高原及其构造脊均有分割错动。这种由块断下沉的古大陆架所构成的阶梯状大陆坡是南海大陆坡成因上的重要特色。正是因为这样,致使在阶梯状大陆坡的海底高原上,还保留着当时形成于大陆架上的石油天然气矿藏。

北部的凸形大陆坡,在形态上与这一区稍有差异,但从基底岩层与地貌结构来看,也是下沉的古大陆架。

(3)中央海盆南—东南坡

它与海南岛南部大陆架隔以深海平原,遥相对峙,也是阶梯状的大陆坡。在 1 800 m深处为一个呈北东—南西方向延展的海底高原,它的北东方向长约 500 nmile,北西方向

宽 181 nmile。在高原的内外两侧皆为陡坡,并以海槽与大陆架相接,其具体情况是:

海底高原的西南端上接北巽他陆架(大纳吐纳岛东北方的延长线上),自陆架外缘至海底高原顶部,大陆坡坡度是 7.5‰,是单一的倾斜面(图 39)。

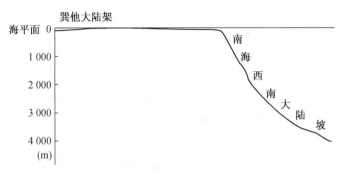

图 39　马来半岛—南海深海平原(10°40′N,112°25′E)海底地形剖面图

自曾母暗沙一带的南沙西南陆架的外缘,至海底高原的大陆坡坡度较陡为 16.3‰。

海底高原的东南部与巴拉望海槽相衔接,巴拉望海槽大致从沙巴外部延伸到巴拉望岛,其北东向长 314 nmile,北西向宽 25～37 nmile,槽底为一平底谷地,其深度在 2 560 m 以上,最深为 3 475 m(位于巴拉巴克岛以西)。巴拉望海槽东南坡内接沙巴陆架,自水深 150 m 至 2 560 m 处的大陆坡坡度为 35.6‰,而巴拉望海槽的西北坡,外接海底高原,其坡度为 13‰。

海底高原外侧,大陆坡陡,普遍以 39.8‰～79.5‰坡度急剧下降至深海平原(图 40)。

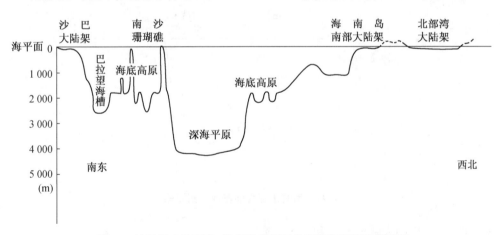

图 40　沙巴经南沙群岛、海南岛南部至北部湾海底地形剖面图

海底高原边缘水深介于 1 500～2 000 m,一般水深超过 1 800 m,高原面上地形起伏极不规则,依附其上建造了众多珊瑚礁,并且有一些大的谷地切割于高原内。我国的南沙群岛即立于这个高原上,它们有的是出露于海面的礁岛,有的是隐伏于水面下的浅滩与暗礁,后者往往是沉没于水下的环礁,并多呈东北—南西向拉长的椭圆形,这种形态反映着珊瑚礁的发育受南海盛行的东北与西南季风的影响,因为活的珊瑚在迎风浪一侧较向礁湖一侧生长得更为旺盛,同时亦反映珊瑚礁岛的基底多半依据高原上的北东向岭脊分布。目前礁基位于 1 100 m 的深度,这反映了珊瑚礁生长始于晚第三纪(北东向山脊形成后),

并保持着一定的速度伴随巨大地块的缓慢下沉而生长。珊瑚礁体的厚度，基本上反映了下沉的幅度。根据地貌与地质构造特点看，这个断块下沉的深海高原也是沉没、断折的古陆架。由这种生长着大量珊瑚礁的沉没古陆架所构成的阶梯状大陆坡，是南海大陆坡的特点之一。而位于深海平原两侧的阶梯状大陆坡在岩性、构造上的一致性，反映着它们是被拉开分裂出来的古陆块。

（4）中央海盆东坡

位于菲律宾群岛的外缘，大陆坡的特点有三：

① 大陆坡范围狭窄而坡度陡峻。大部分大陆坡的宽度很少超过 38 nmile；其中较缓的坡度，如巴拉望岛屿外缘的陆坡坡度是 17.7‰，而其他地方如西吕宋海槽与马尼拉海沟处坡度陡，达 170.8‰～119.3‰。

② 大陆坡呈狭窄的阶梯状下降，坡麓分布着海槽或深海沟。例如，吕宋岛陆架外缘分布着一系列海槽与海沟。以仁牙因湾为界，其北为深约 3 200 m 的北吕宋海槽；其南是长 112.8 nmile、宽 29 nmile（南北两端变窄）的西吕宋海槽。西吕宋海槽槽底深度为 2 600 m，海槽两侧有三列近于平行的南北向构造脊，是由晚更新世至全新世时褶皱隆起的新生代沉积所构成。其向北延伸可达我国台湾的南部，在构造脊内拦积了厚达 1 000～1 500 m 的沉积层，致使槽底平坦，并呈现为向西倾斜的平台下降至大陆坡坡麓的马尼拉海沟。马尼拉海沟以与吕宋岛近于平行的方向从 16°40′N 向南分布到 13°N，海沟深度超过 4 800 m（图 41）。

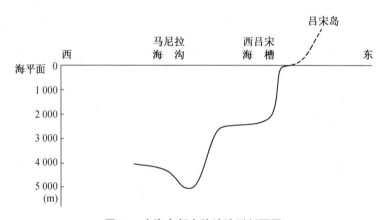

图 41　南海东部大陆坡地形剖面图

其他，在沙巴的大陆坡上也有一些小阶地。

③ 大陆坡遭受到较多的水下峡谷的切割，一些大的水下峡谷皆穿越大陆坡而形成海峡通道，如巴拉巴克海峡、民都洛海峡、巴布延海峡以及巴士海峡等。除巴拉巴克海峡外，这些谷地规模较大，大多下达深海平原，并于大陆坡坡麓峡谷出口处由浑浊流堆积了海底扇。还有些水下谷道，是切穿大陆坡而下达深海槽中。

总起来看，南海的大陆坡是块断下沉的古陆架所组成的阶梯状大陆坡，这是南海大陆坡的第一个特点，巨大的块断阶梯与构造活动密切有关，块断下沉可能始于中新世或上新世，但第四纪时有急剧的断裂活动。南海东部大陆坡呈狭窄的阶梯下降，但坡麓分布着海槽与海沟，这是南海大陆坡的第二个特点。大陆坡的这些特点皆与南海海盆的形成有密切的关系。

3. 南海深海平原

此为南海中央最深部分,即南海中央海盆,范围界于中沙与南沙群岛大陆坡之间,大部分水深在3 600 m以上。由于底部平坦,可视为一深海平原。南海深海平原亦作北东—南西向延伸,纵长795 nmile,其中最宽处(相当于自珠江口外到巴拉望岛一线上)为360 nmile;西南部较窄为99 nmile;东北部最窄为78 nmile。平原自北向南倾斜,北部水深约为3 400 m,南部水深为4 200 m左右,其中有不少地方深度超过4 400 m。

深海平原的中部分布着一些孤立的海山,它们是由海底火山喷发形成的,表现为孤立的锥体或几个火山锥的结合体。这些海山高出深海平原500～900 m,甚至1 500～3 500 m,其中最高的突起在深海平原上3 904.15 m,大部分海山尚未达到海面。

深海平原的东北端与西南端是两个充填着沉积物的深水谷地,谷地出口与前述各水下峡谷的末端一样,堆积着海底扇,这些堆积体有的已隆起成为北东向的小型山脊。在大陆边缘的海底扇隆起中可能蕴藏着潜在的石油资源。

深海平原的沉积为含球房虫、放射虫与火山灰的黏土质软泥,近期在深海沉积中,发现有锰结核。

从地貌结构看,南海深海平原是亚洲大陆边缘经拉开分裂引起深部的玄武岩流补偿性上升而出现的部分,在平原的中部与东部沿着北东向大裂隙尚有一系列火山岩流的喷发活动。

五、结语

综上所述,中国近海海底地质地貌特点可归纳如下:

(1)中国近海系沿西太平洋并由一系列边缘海组成,它们的内侧是亚洲大陆,外侧是太平洋西部的岛弧-海沟系。岛弧-海沟系将边缘海与太平洋分隔开来,在边缘海中围堵了从亚洲大陆冲来的大量陆源物质;岛弧-海沟系又是太平洋西部剧烈的构造活动带,现代火山与地震非常活跃,它们是新生代环太平洋构造带的一部分。台湾岛以北的岛弧呈外凸状(朝太平洋方向突出),台湾岛及菲律宾北部则作内凸状(朝亚洲大陆突出),是为"反弧";菲律宾以南又为外凸弧。南北两列在台湾地区交汇,台湾反弧是个引人注目的构造现象,其成因还需要进一步研究。

在构造上,整个边缘海及其邻区都是由几条相间排列的隆褶带和拗陷带构成的,它们的走向为北北东至北东,形成时代是由西向东逐渐变新。最东侧为北北东向的现代岛弧-海沟系和新生代的东海陆架边缘隆褶带(或称台湾褶皱带),向西为新生代的东海陆架拗陷带、中生代的浙闽隆起带、中新生代黄海南部拗陷与前古生代胶辽隆起。这种北北东—北东向的构造体系可能是太平洋板块和亚洲板块多次相互作用的结果。

在南海,居主导地位的北东向断裂可能与南海的构造成因有关,大体上南海可划分为三个北东向的主要构造带:南海海盆中央表现为拉开断裂的古陆块之间的地幔物质上隆带;其两侧为块断沉降的刚性古陆块。在三个北东向构造带的西北侧与西南侧为新生代的沉降盆地;在东侧,为复杂的岛弧-海沟系。

南海的西半部及渤、黄、东海与亚洲大陆构造关系密切,而南海东半部及台湾以东海域受太平洋构造活动影响较大。

(2)大陆架是大陆边缘倾斜平缓的海底地带,是陆地向海的自然延伸。它的宽度从低潮线起向深海方向倾斜,直到坡度显著增大的转折点为止。

中国近海大陆架是世界上最宽的大陆架区之一。黄海和渤海整个位于大陆架上;东海大陆架的宽度从北向南为350～130 nmile,其外缘转折点水深约为140～180 m;南海两广沿岸大陆架宽度为100～140 nmile,转折点水深约150～200 m;台湾以东大陆架狭窄仅数海里,其外转折点水深约150 m。

中国近海大陆架的基底,主要是中生代白垩纪末期的剥蚀面,其岩层与相邻大陆一致。在这个基础上,由新生代沉积构成了堆积型的大陆架。

中国近海大陆架有两种成因类型,一种是堆积的,另一种是侵蚀-堆积的,以堆积型为主。大体上呈北东向的构造脊,把从中国大陆侵蚀搬运下来的泥沙,拦截堆积在沉降盆地中,泥沙填充了沉降盆地而成为大陆架浅海。当内侧的盆地被泥沙填满后,盆地外缘的构造脊失去了堤坝作用,泥沙则越过构造脊向前堆积,因此大陆架范围不断向海发展。中国沿海大陆架主要是由黄河、长江、珠江等大河入海泥沙堆积而成。目前,黄海、东海盆地正在沉降中,冲绳海槽尚未受到大量填充。两广沿海盆地亦处在沉降充填中。因此,由构造脊围封的、被大河泥沙填充堆平的大陆架,是中国近海大陆架的主要成因特色。在大陆架的巨厚新生代沉积层中,形成和圈闭着丰富的石油与天然气矿藏。

(3)大陆坡是大陆与大洋交接活动带,分布在大陆架外缘,是向深海过渡下降的地带,它像一条窄带一样围绕着各大陆,至大陆坡坡麓即为大洋底部。

中国近海的大陆坡,在东海、台湾以东太平洋海域与南海东部,表现为陡窄的阶梯与海槽、海沟相伴分布的特点,它们是西太平洋新生代的构造活动带,火山、地震活动频繁。南海西部大陆坡的特征,是由巨大的海底高原组成的宽广阶梯状的大陆坡,高原上的岭脊上分布着许多珊瑚礁岛,这种大陆坡是由晚第三纪以来沉降、断折的古大陆架所形成的。根据珊瑚礁基部目前所在的水深,可见自晚第三纪以来,该处断陷、下沉超过1 000 m。

深海盆分布于南海与台湾以东海域,基底具有非典型的大洋型玄武岩,海盆的成因需进一步研究。

总之,从中国近海海底各地貌单元的分布的广度以及各地貌单元的成因关系来看,大陆架是一个关键部分。

(4)中国近海海底沉积特征:

1)大陆架沉积主要是黄河、长江、珠江等大河由中国大陆冲刷搬运入海的泥沙,颗粒组成与矿物成分各具有其相邻陆地的区域特征;其次为岛屿的冲刷与生物成因的堆积。第四纪冰期、间冰期气候和海平面的变化,对大陆架沉积物的特性和分布有深刻的影响。

渤、黄海大陆架沉积以粉砂为主,但边缘受河流影响,沉积有变异。西部由于黄河作用,有细颗粒的淤泥沉积;而黄海东部由于朝鲜半岛山地河流来沙,并受潮流冲刷改造,因而有粗颗粒的砂质沉积。

东海与南海大陆架沉积有一致性,近岸部分为现代沉积,从岸向海颗粒由粗至细,在东海是粉砂—淤泥,在南海是细砂—粉砂—淤泥。大致在水深50 m以外,海底广泛出露着砂质沉积,即古海滨砂带,由于未被其后沉积所覆盖,仍然出露海底,是"残留沉积"。其中含较多的贝壳、有孔虫、钙质砂与一些海绿石等自生矿物。

南海、东海大陆架砂带宽度除受底部地形控制而有不同外,在东海,由于内侧有长江古河道汇集水沙,外侧有黑潮流经扰动,故砂带出露宽,在长江口外水深30 m处底质即为暗色细砂,向外延伸可达水深400 m处的冲绳海槽西侧边沿。南海限于沿岸来沙的特征与规模,砂带较窄,砂中的贝壳与自生矿物含量的比例相对地更高些。

海峡与岛屿周围多为沿岸与海底冲刷产物,由于浪流作用强,海底沉积为砂、砾或基岩出露。

2)大陆坡沉积较复杂。特点是:

① 海底高原沉积变化大,其上大量发育着珊瑚礁,在礁体周围有珊瑚砂与粉砂沉积。在高原面与深水道中沉积着青灰色淤泥、球房虫淤泥与红黏土。

② 狭窄的阶梯与陡坡上,或者基岩裸露,或为灰白色的有孔虫砂,以及细砂与砂泥沉积。沿着大陆坡上的岛屿或海岭周围,也有珊瑚礁发育。中国近海的大陆坡上,普遍有珊瑚礁或珊瑚碎屑沉积,这是海区自然条件特点的一个显著反映。

③ 沿大陆坡的水下峡谷,浑浊流冲刷搬运了泥沙,至大陆坡坡麓或海槽底部堆积,其颗粒大小混杂,有灰、暗色的细砂与淤泥。海槽中心部分,主要是泥质沉积,此外尚有浮石、火山碎屑、火山灰以及珊瑚碎屑等生物沉积。

3)深海盆沉积物主要是球房虫与火山灰所形成的黏土质软泥。南海深海平原有锰结核沉积,在海底火山或裂隙喷发处有玄武岩流及浮石等物质。

(5)中国近海大陆架形成以来,曾几经海陆变化,除构造因素外,第四纪气候变迁所造成的海面升降有深刻的影响,据现有资料可做三点推论。

1)从渤、黄海到南海,大陆架上普遍分布着沉溺的古河道、古三角洲、阶地、残留的砂带沉积以及埋藏的风化壳、古海滩砂岩等。其分布位置从水深数米到150 m不等,根据地貌综合分析,目前可划出四条古海滨带,即水深:① 20~25 m;② 50~60 m;③ 100 m左右;④ 140~160 m。它们是第四纪以来,海面变化不同阶段的产物。

2)据化石定年资料:

① 渤海中部海底采到披毛犀牙齿;东海虎皮礁附近采到北方原始牛下颚骨;钓鱼岛附近及日本男女列岛一带发现有猛犸象牙齿等,它们均属更新世晚期至全新世初期的寒冷草原动物群。

② 据东海大陆架砂带上的软体动物化石:水深60 m海底采到蚬贝,水深100~150 m的海底采集到红螺、文蛤、牡蛎、杂色蛤仔等海岸带贝壳,[14]C定为15 000~20 000年,表明:晚更新世大理冰期低海面时,曾为辽阔平原,大约到距今20 000年更新世末,沿海平原逐渐被淹成为海岸带与浅海。冰后期以来,海侵规模加大,浅海与海岸带进一步受淹,海水加深,遂成今日现状。因此,现代中国大陆架浅海主要是全新世形成的。

3)考古与有关测量资料表明:更新世末到全新世初以来,海面上升幅度约150 m。海面上升具有振荡性与间歇性的特点。大约距今5 000~7 000年时,海面上升已接近现代海岸位置。嗣后,海面变化幅度较小,大体上相对稳定。但若缩小时间幅度,按世纪性趋势看,目前海面处于微小上升趋势中。

参考文献

[1] 李四光. 1972. 天文、地质、古生物. 北京：科学出版社.

[2] 黄汲清，任纪舜等. 1974. 对中国大地构造若干特点的新认识. 地质学报，48(1)：36 - 52.

[3] 秦蕴珊. 1963. 中国陆棚海的地形及沉积类型的初步研究. 海洋与湖沼，5(1)：71 - 85.

[4] 陈吉余. 1957. 长江三角洲江口段的地形发育. 地理学报，23(3)：241 - 253.

[5] Ph. H. 奎年著，梁元博译. 1963. 海洋地质学. 北京：中国工业出版社，第七章.

[6] Wageman, J. M., Hilde, J. W. C., Emery, K. O. 1970. Structural framework of East China Sea and Yellow Sea. *American Association of Petroleum Geologists Bulletin*., 54(9)：1611 - 1643.

[7] Parke, M. L. Jr., Emery, K. O., Szymankiewiczr, Reynolds, L. M. 1971. Structural framework of continental margin in South China Sea. *American Association of Petroleum Geologists Bulletin*, 55(5)：723 - 751.

[8] Emery, K. O. 1968. Relict sediments on continental shelves of the world. *AAPG Buletin*., 52(3)：445 - 464.

[9] Niino, H., Emery, K. O. 1961. The sediments of shallow portion of South and East China Sea. *Geological Society of America Bulletin*, 72(5)：731 - 762.

[10] Shepard, F. P., Emery, K. O., Gould, H. R. 1949. Distribution of sediments of East Asiatic Continental shelf. Allan Hancock Found. Occ. Paper, 9：64.

[11] Fairbrige, R. W. (Editor) 1966. The Encyclopedia of Oceanography. New York, Reinhold Publishing Corporation. 994 - 998, 238 - 243, 829 - 837.

[12] GLLG, J. G. 1970. The bathymetric chart of the South China Sea. The Kuroshio. 21 - 28.

[13] Ludwig, W. J., Hayes, D. E., Ewing, J. I. 1967. The Manila trench and West Luzon trough——Ⅰ4, bathymetry and sediment distribution. *Deep Sea Research*, 14：533 - 544.

[14] Ludwig, W. J. 1970. The Manila trench and west Luzon trough——Ⅲ, seismic-refraction measurements. *Deep Sea research*, 17：553 - 571.

[15] Ludwig, W. J., Murauchi, S., Den. N, Buhl, P., Hotta, H., Ewing, M., Asanuma, T., Yoshii, T., and Sakajiri, N. 1973. Structure of East China Sea-West Philippine Sea margin off Southern Kyushu, Japan. *Journal of Geophysical Research*, 78(14)：2526 - 2536.

[16] Leyden, R., Ewing, M. and Murauchi, S. 1973. Sonobuoy refraction measurements in East China Sea. *American Association of Petroleum Geologists Bulletin*, 57(12)：2396 - 2403.

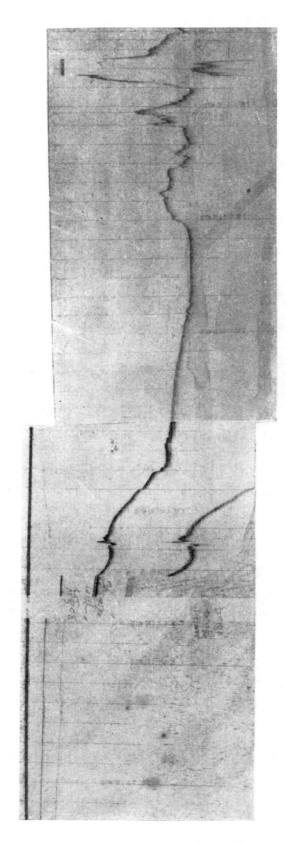

照片1 东海大陆架转折带及东海大陆坡地形测深剖面

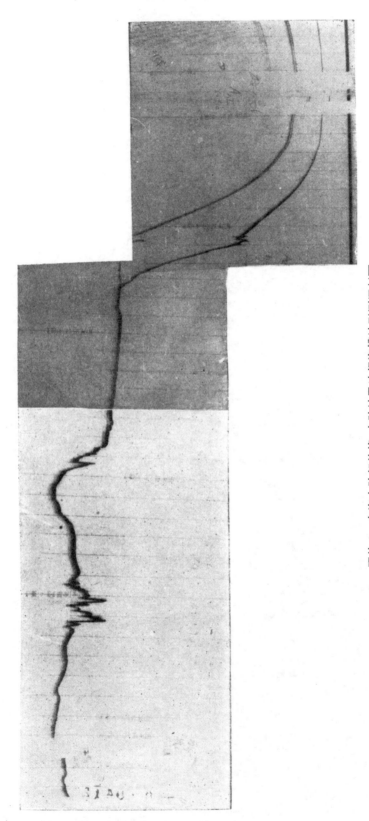

照片 2　东海大陆架外缘、大陆坡及冲绳海槽地形测深剖面

照片 3 东海大陆架地形测深剖面

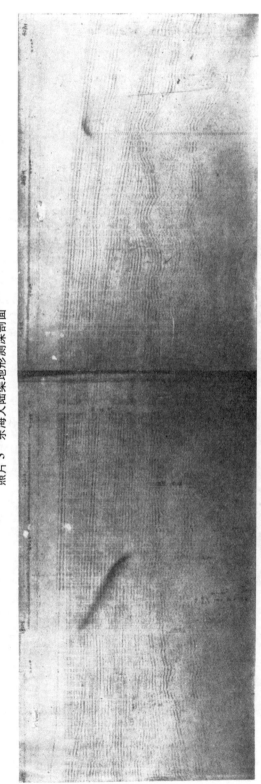

照片 4 海南岛以东地震剖面(从剖面中可看出陆架以平缓的坡度逐渐过渡为陆坡,在陆架边缘基底褶皱隆起,并可看出地层向陆一侧明显加厚)

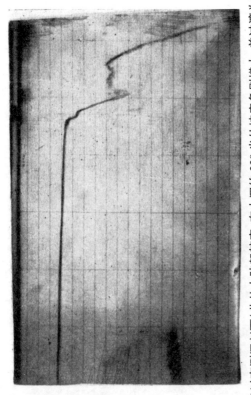

照片 5　海南岛以南测深剖面（此处大陆架较窄，在水深约 220 米处坡度急剧增大，并过渡为阶梯状大陆坡）

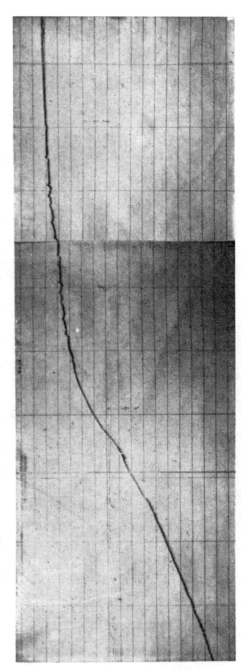

照片 6　位于照片 4 剖面以西的测深剖面（表示了大陆架外缘及大陆坡）

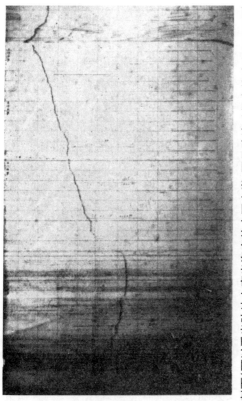

照片 7　海南岛以东测深剖面（可以看出陆坡上高出海底的水下山脊，而且陆坡以很大的坡度迅速降至 3 200 m 左右的深海谷地）

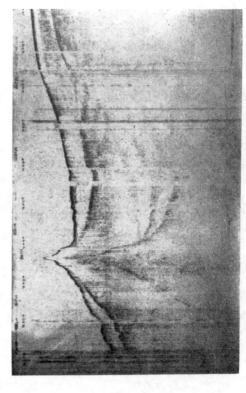

照片 8　与照片 7 同一测线的地震剖面（水下山脊向陆一侧堆积有厚超 1 000 m 的沉积，向海一侧地层明显减薄）

中国海洋地质专题调查[*]

一

中国海地处亚洲板块、太平洋板块与印度洋板块之交,基底构造复杂,构造运动活跃,加之受季风气候与含沙量大的大河作用影响,因此,海域环境独具特色:活动的大陆边缘,扩展的陆地,边缘海经受近期的张裂变动但主要被河流的泥沙所充填。调查研究中国海,不仅在石油天然气资源、金属矿产资源、地震预报和自然环境开发保护方面具有重要的应用价值,而且具有全球性的重要理论意义。

地球表面70%是被海洋所覆盖,而60%的海底深度超过2 000 m,平均深度超过4 500 m。直到20世纪50年代,大海隔在洋底与人类之间的帷幕还未曾被掀开过一个角。以前关于地球演变的理论主要建立在对29%出露的地表部分的研究基础上。60年代,板块构造学说与新的地球动力模式使人们逐渐明确:洋底是研究地球变动的关键地区。大陆是在洋底的更新变动与运动中发生着变化,不研究洋底最关键的活动部分就不可能对陆地地质有深刻的了解。地球最活动的部分,即沟通地球内外之间的动力通道是在洋中脊的大裂谷与大陆边缘的深海沟。这两处是研究洋壳的形成、发展、消亡以及向陆壳过渡与演化的关键地带,是当代地球科学研究的焦点之一。由于研究手段的困难,当前研究侧重于海陆过渡带的大陆边缘。

东亚大陆边缘是以一系列边缘海环以弧形列岛以及岛外侧的深海沟为特点。即构造体系上的海沟、岛弧与弧后盆地复合体系。从全球角度着眼,这个体系主要分布于亚洲大陆东部与西太平洋边缘部分。它是地球上地形起伏高差最大,地震、火山与岩浆活动极为活跃;构造变动、沉积、变质与成矿作用极为复杂的地带。我们应据此有利条件,开展有计划地调查研究,以期在这一项具全球性重大意义的理论课题上做出贡献。并为我国近海油气资源钻探开发积累资料和提供服务。

中国海的调查范围包括海岸、大陆架浅海以及大陆边缘岛弧海沟系深海。调查研究的成果深度因基础工作不同而有所区别。海岸带是在详细调查的基础上总结规律,加强对海岸资源与海岸环境的开发利用与规划管理。大陆架浅海是摸清资源与探明环境,进行有计划的开发与利用。深海区宜抓住关键地区进行调查,为进行重大的地学理论总结而占领阵地,积累资料,并在一两项关键课题上有所建树。到2000年,本专题应具有下列成果:

1. 中国海区海岸与海底地质的一系列基本资料与图件(如:表面地质——地貌与沉

---

* 王颖:《我国海洋开发战略研究论文集》,北京:海洋出版社,1985年,第292-294页。

1984年载于:(南京大学海洋地貌与沉积研究室编)《海洋开发战略课题(55):中国海的专题调查》1-6页,1984年7月。内容有少许改动。

积、基底构造、地层、重、磁等）。

2. 黄海、渤海、东海、南海区域专著以及为解决某些应用问题的专门著述。

3. 在重大的地学理论（如：东亚大陆边缘地壳构造、板块运动及边缘海形成演化等问题）的某些方面有所建树，并为进行重大的理论著述提供关键地区的基本资料。

西方诸国自二次大战以来，舰艇活动、海底通讯的发展以及战后大量舰艇转入民用加强了调查研究手段，海底油气藏的开发利用以及领土与海洋资源的开发等，大大促进了海洋地质科学的发展，板块学说的兴起带动了地球科学的革命。因此，在北美、西欧各国，海洋地质学受到重视，经费、设备与研究成员方面具有强大的实力。国外研究工作具有以下特点：

（1）调查船设备先进，有关的单位可充分利用设备与航次，开展多学科的调查，如：音响测深、测地层剖面、海底摄影、重磁测量，取表层及柱状沉积样品、CTD观测以及热流量测定等，这些已成为每个航次的基本工作项目，并应用深海钻探成果。国外注意把地貌与沉积等表面地质工作和深部构造与岩石学研究结合起来进行综合分析。

（2）基础理论的研究与应用项目密切结合。如：美洲与大西洋边缘、北非与地中海边缘的研究已为石油天然气资源开发做出了贡献；北美大陆西部与太平洋东北边缘的研究工作是把地震、火山的监测预报与板块边界的活动特点研究有机地结合起来；对夏威夷群岛火山活动的调查亦结合探索板块活动的动力机制；印度河与恒河深水扇的研究是结合了油气资源的开发。国际研究着重于板块深部结构及海底扩张的动力机制，研究地区集中于洋中脊大裂谷与板块迎缘俯冲带，特别是深海沟洋壳消亡带。八十年代以来，研究焦点集中于大陆边缘，如：日本对邻近的日本海沟以及最深的马利亚那海沟与热液矿床的研究；美苏对太平洋海沟系的调查活动等。

我国海洋地质工作起步较迟（20世纪50年代中期），装备不足并受限于当时的形势条件，主要在海岸与浅海区工作，侧重于地貌与沉积等表面地质工作，海底构造与地球物理工作力量薄弱。70年代以来，随着开发我国东部海底石油资源，石油部、地质部、海洋局与中国科学院等单位开展了黄、渤、东、南诸海区大陆架地质调查，积累了地球物理勘测资料，进行了区域海底地质构造研究，有些工作已扩及到海槽与深海盆。我国地学界的著名学者不仅传播介绍了国外研究的成果与动向，并著文探讨亚洲东部与西太平洋边缘地区的地质构造与板块运动问题，对我国海洋地质学发展起了重要的推动作用。但是，在我国由于：① 基本手段不足。船只与设备不齐全配套，或有少量高精仪器，但缺乏齐备的、先进的基本工作手段。调查活动花费巨大，但有限的航次中不能充分开展多学科或一个学科内多项目的系统工作。② 人员与现有装备分散，门户之限而不能充分发挥现有人力与设备力量。如：高校中有一批科研力量，但缺乏从事海洋地质调查的基本装备与经费；生产部门包括新建的海洋研究单位，具有一定的装备与较充足的经费，但缺少科研力量尚未形成梯队，忙于生产任务而对长远规划和理论课题考虑不足。③ 远洋调查的活动范围与涉外问题等，仍使我们的调查研究工作与国际合作交流受到限制。因此，我国对大陆边缘构造体系尚未着手于系统的专题研究。④ 我国目前从事研究的单位多，分属于科学院、海洋局、教育部、地质部与石油部五大系统。一方面海区工作研究不够，资料不足，另一方面工作又多有重复。从现阶段状况看，我国海洋地质工作，可能需要建立由有关单位专家组成的指导委员会,统筹、协调与指导有关全局性的调查研究。

二

建议本专题的调查项目如下:

1. 中国海岸的特点、成因、发展过程以及自然资源与海岸带开发管理原则。

本课题是在已进行的海岸带普查的基础上,对重点海岸段落进行深入地调查研究与规划。

拟包括下列分课题:

(1) 区域海岸特征、资源与开发管理原则。

(2) 主要海岸类型的成因发展过程研究*。

(*具中国特色的海岸类型:如河口湾与大河三角洲;淤泥质平原海岸;潮汐汊道港湾岸以及南海珊瑚礁岛等)。

2. 中国各边缘海的构造、沉积与地貌特点、成因发展机制与油气藏分布规律的调查研究。

拟包括下列课题:

(1) 南海的成因与构造演化。

(2) 东海(包括冲绳海槽)的形成演化、大地构造特征及成矿作用。

(3) 黄、渤海的形成、发展与油气藏分布规律。

(4) 边缘海沉积作用过程与模式。

进行本课题工作时,宜有意识地争取参加日、美、法所组织的有关西太平洋与日本海的调查研究工作,并搜集鄂霍次克海的调查研究资料,以兹研究西太平洋边缘海的成因过程。

3. 东亚大陆边缘岛弧海沟系构造与板块活动机制的调查研究。

此项包括三个分课题的调查研究工作:

(1) 台湾岛反弧的成因。

(2) 菲律宾板块与东亚板块的现代活动特点与构造影响。

(3) 大陆边缘浊流形成机制与沉积相。

上述课题应包括下列海域在内的调查研究:

东海:

① 菲律宾板块边缘——琉球海沟——岛弧——冲绳海槽东海大陆架——浙江海岸,与陆地衔接。

② 菲律宾板块边缘——台湾岛弧*——台湾海峡——福建海岸。

(*台湾海域目前虽难以前往工作,但作为中国海的规划应该包括它)。

南海:

③ 菲律宾板块——菲律宾海沟——菲律宾岛弧——马尼拉海槽——南海盆地——南海大陆坡——大陆架——广东海岸——通过横断山脉与内陆东西向构造带(即印度洋板块与亚洲板块的作用带)相接。

　　以上三题概括了各海域、不同部分、不同研究程度的主要中心问题,应在中国海专题调查中予以适当地安排。三题宜在现在就着手进行工作,其中第一个课题宜在 1990 年前完成总结,第三课题是在 2000 年前拿出成果,而第二题的部分成果应在 1990 年完成,部分成果稍延迟。我想,开展各项专题调查时,要有相应的单位从研究的角度来承担主要调查任务并组织协作。全国宜有高一级的学术指导委员会进行统一协调的安排,使达到集中人员与物力之优势进行有成效的调查研究,避免发生过多的重复浪费与遗漏。在开始进行调查时,即注意到积极参加有关的国际合作调查工作,及早应用深海钻探项目的调查成果。在技术项目引进、保密条例、对外交流活动经费与外文图书资料订阅办法等方面应作切实的改进。

中国海岸海洋地貌科学研究[*]

一、海岸带动力过程与地貌研究

(一) 20 世纪 60 年代以来海岸带科学的蓬勃发展

我国现代海岸科学的蓬勃发展始于 1957—1958 年间,缘于长江口上海港的治理和天津新港泥沙来源与回淤研究。交通部与中国科学院地理研究所以及北京大学先后邀请了 И. В. 萨莫伊洛夫讲授"河口学",В. П. 曾科维奇系统地讲授"海岸科学",О. К. 列昂杰夫讲授"海岸与海底地貌",Е. Н. 涅维斯基介绍"海岸水文学研究"等。参加听讲的成员后来多成为现代海岸科学的实践者。

20 世纪 60 年代以来,海岸带科学蓬勃发展,陆续翻译出版一系列欧美学者的论著,如 H. Kuenen 的《海洋地质学》(1950),苏联科学院院士 В. П. 曾科维奇的《海岸基本原理》(1967),О. К. 列昂杰夫的《海岸与海底地貌学》(1965),英国 C. A. M. King 的《海滩和海岸》(1972),及美国 F. P. Shepard 的《海底地质》(1978)。欧美海岸与海底研究是相互交流而推动着学科进步,20 世纪后期的海岸研究是将海陆交互作用地带结合在一起。海岸不仅是指沿海的陆地部分,而是涵盖海陆作用最活跃的地带,包括陆上与水下两部分,实为海岸带;海岸带的发育受海、气动力作用控制,也受陆地结构与陆源泥沙量多少的影响。

在中国海岸带研究工作中,始终贯彻为海港建设、围涂造地等海岸工程建设和海岸发展规划服务,采纳应用欧美海岸科学新理论,同时又总结发展中国在海岸研究中的理论与技术成果,推动着海岸科学的进步。

1. 海岸带的定义及认识发展

20 世纪中期以后,海岸研究最重要的进步是明确了海岸是海洋与陆地交互作用的过渡地带,其范围既包括沿岸陆地,又包括水下岸坡,这个地带是浅海波浪、潮汐与水流对沿岸的积极作用而变化活跃,同时受到入海河流及人类活动极为深刻的影响。具有代表性的海岸带定义与范围可归纳为以下四种。

(1) 苏联海岸学家 В. П. 曾科维奇及 О. К. 列昂杰夫(1955)把海岸带定义为现代海洋与陆地交互作用的带,上部为出露的现代水上阶地及其后缘,中部为海陆相互作用的海滩,下部海底地区称为水下岸坡,其界限一方面是海洋的平均水面线(或海岸线),另一方面是海浪作用开始扰动海底泥沙的深度。

海陆交互作用在地貌上的表现不只可以在现代海岸带的范围内保存在地形中,并且

　　* 王颖:中国科协主编,中国海洋学会编著(苏纪兰主编,王颖副主编),《中国海洋学学科史》,第八章第二节,北京:中国科学技术出版社,2015 年,第 232 - 258 页。

亦能在离海岸很远的地方保存下来。使海岸线变动的因素如海洋中水量变化,海与洋盆容积的变化,海岸带内岩石圈构造变动等,都使海岸地貌的分布面积要比现代的海岸和水下岸坡为广。

(2) C. A. M. King(1972)基本遵循着 D. W. 约翰逊的海岸概念,但对不同类型海岸剖面又作了进一步的分类,将砂质海岸分出滨外带、滨面带、后滨带,并解释了各带主要的动力条件与地貌构成。后滨是指大潮高潮时进流作用上限以上的地带,它很少直接受到波浪作用。前滨是随着潮水涨落而有规律地为海水淹没或疏干的部分,在无潮海,前滨很狭窄,只包括了大浪进流与退流之间的距离。滨外带的范围是从海水经常覆盖的最高点起,到正常情况下水底泥沙基本不动的那个深度为止。

(3) F. P. 谢帕德(Shepard,1978)将海岸概念作了修正并提出海岸(Coast)是海滨直接向陆上的宽广地带,海岸包括了海蚀岸和上升阶地以及海滨向内的低地。海滨线是陆地与海水相交处的水边线,海滨从平均低潮线(或较低低潮线)至海浪携运泥沙的内缘地带。Shepard 把海滨与海滩结合起来,而海岸具有向内陆伸展的概念。应当说,Shepard 的海岸概念是可取的。

(4) 王颖、朱大奎(1994)总结出海岸的概念:现代海岸是指海陆交界处相互作用变化活跃的地带,包括沿岸陆地与水下岸坡(图1)。其上界是现代波浪作用的上限,在陡峻海岸是海蚀崖的顶部,在平缓的砂质海岸是海滩的顶部,但由于风浪、风暴越流的作用,亦应将海滩、海岸沙丘及其后侧的潟湖低地列入现代海岸的范围。海岸的下界是波浪开始扰动海底泥沙之处,这个界线随波浪作用的强度而变动,一般来说,是在水深相当于波浪长

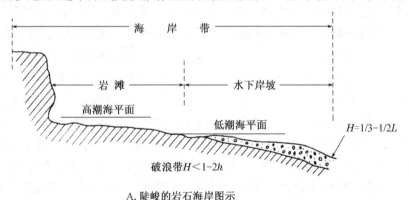

A. 陡峻的岩石海岸图示

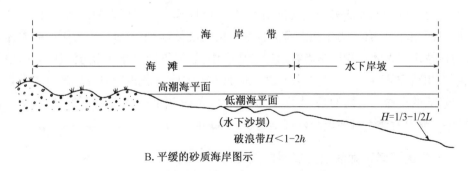

B. 平缓的砂质海岸图示

图 1　王颖、朱大奎总结的海岸带图式(海岸地貌学,1994)

度的 1/2 或 1/3 处。基岩、砂质与淤泥质海岸的不同类型见图 1。所以,海岸包括三部分:沿岸陆地(longshore land),包括海蚀崖、海岸沙丘、潟湖洼地、港湾等;潮间带(inter-tidal zone)包括海滩、岩滩、潮滩(tidal flat);水下岸坡(submarine coastal slope)。海岸的三部分是一个整体,它们相互之间有着成因上的联系,其发展变化是相互影响、相互制约的。

2. 天津新港泥沙来源与回淤研究

20 世纪中期,中国大陆经济复苏,对外经贸的发展对海港的需求,尤其是对首都门户天津新港的需求十分迫切。在新港的治理扩建中,提出对"天津新港回淤与泥沙来源研究",建立了"回淤研究工作站"(天津海洋科学研究所前身)。交通部与中国科学院、教育部联合聘请了一批苏联海岸与河口专家来华讲学:1957 年萨莫伊洛夫来华讲授河口学,1958 年邀请苏联科学院院士 B. Π. 曾科维奇、莫斯科大学地貌学教授 O. K. 列昂杰夫及苏联海洋研究所海洋水文研究员 E. H. 涅维斯基在天津讲学。

为落实天津新港回淤研究,1958 年中国科学院海洋研究所组织了一支相当规模的"渤海湾动力地貌调查队",队长为中国科学院海洋研究所的尤芳湖助研(竺可桢院长的前任秘书),陆上有南、北两支队伍:北队调查从滦河至新港以北沿岸河流对港口回淤的影响,队长为北京大学研究生王颖,副队长为北师大教师张如意,成员有华南师大地理系教师刘南威、周祜生,南京大学教师俞锦标等。南队负责调查黄河泥沙对新港的影响,队长为华东师大地理系讲师陈吉余,学生王宝灿、虞志英、恽才兴及沈焕庭等。海上调查主要采取底质样品与震动活塞沉积孔柱状样品,由中国科学院海洋研究所蔡爱智、李成治、高明德及张宏文等负责。

经过两个季度的调查研究与实验分析,调查队于年底完成总结。获知滦河的入海泥沙以细砂为主,主要分布于河口地区形成沙坝环绕的潟湖—三角洲海岸;古滦河曾从曹妃甸一带入海,但泥沙未越过南堡;滦河因陆地构造掀升,河道自西南向东北逐步迁移,曾迁徙至大清河入海,形成打网岗、月坨、石臼坨等一系列沙坝;后逐渐迁移至流经姜各庄以北的现代河口,泥沙主要堆积在河口三角洲,不会越过浪窝口。因为滦河上游建坝拦水,入海水沙量减少,所以滦河泥沙对新港无影响。南队的研究认为:黄河的淤泥质主要输入莱州湾与渤海湾交界的河口区,会辗转向北输送,但主要影响到歧口。以后的逐次调查,结果相同。渤海动力地貌调查明确了天津新港回淤的泥沙主要源于海河口大沽坝浅滩及渤海湾本区因风浪掀扬泥沙,而被潮水带入。通过天津新港等海港调查研究实践,培养了多学科交叉的人才,推动了我国海岸科学的发展。

促进海岸科学发展的另一重要推动力,是 20 世纪 60 年代初以周恩来总理提出"三年改变港口面貌"的号召。当年在渤海湾首次考察的成员,后来大多成为我国海岸海洋研究院校的科技领军人才。我国学者在海岸带研究中,始终贯彻海洋动力与泥沙运移,地貌与沉积结构相结合的科学思维与技术路线,通过调查与实验,获得所研究地区海岸发育历史与发展趋势的真谛。遗憾的是,十多年的"文化大革命"中断了海岸带开发,摧白了两代人的少年头,海岸海洋学科建设与发展推迟了近 30 年。

(二) 中国海岸带地貌研究发展的贡献

中国海岸海洋地貌是以季风波浪、潮汐与潮流作用以及大河泥沙的补给效应为特征的动力组合,加上人类活动的悠久历史,促成了自然过程的进一步变化效应。经历半个多世纪的海岸调查研究实践,中国海岸海洋科学研究学者阐明了中国海岸的成因-形态分类;在以潮流作用控制发育的潮滩海岸、潮汐汊道港湾海岸,以及河海交互作用与平原海岸发育演变方面做出了系统的科学理论贡献。

1. 中国海岸成因与形态分类

基于海岸选港动力地貌研究及20世纪80年代第一次全国海岸带与滩涂普查的调研成果,王颖(1980)从海岸成因、地貌特点及发育动态相结合的原则,将中国海岸归纳为四大类型10个亚类(图2)。

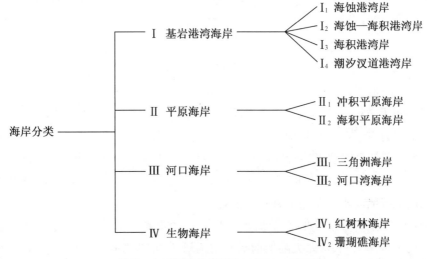

图2 中国海岸分类(Wang Ying, 1980)

(1)基岩港湾海岸:发育于基岩山地与海洋交界带,具有突出的岬角和凹入的海湾,海岸线曲折,水下岸坡很陡,波能量高,海岸沉积为粗砂和砾石(Wang 和 Aubrey,1987)。代表性的海蚀港湾海岸为辽东半岛南端;海蚀-海积型港湾岸为山东半岛南岸;海积型港湾岸为厦门的曾厝黄厝、冀东秦皇岛沿岸;潮汐汊道港湾为浙、粤、琼等沿岸。

(2)平原海岸:岸坡平缓,坡度小于1/1 000,多处为1/4 000,海岸沉积物由细粒泥沙组成——细砂、粉砂与淤泥,潮流作用主导,在海岸平原外缘发育宽广的潮滩与水下岸坡。冲积平原海岸主要分布于渤海湾与莱州湾沿岸、松辽平原外缘和江苏黄海沿岸,以及闽、粤沿海(Wang 和 Zhu,1994)。渤海湾平原海岸与江苏黄海沿岸是典型的海积平原海岸。

(3)河口海岸:主要分布于大河入海处,河流与海洋相互作用,受海岸轮廓、海岸坡度等不同因素影响发育为三角洲或河口湾(三角港),如黄河三角洲海岸、长江三角洲海岸及杭州湾喇叭型河口湾。

(4)生物海岸:主要分布于华南沿海及南海海域的珊瑚礁海岸和红树林海岸。在台湾、北部湾、海南岛及近岸的小岛有岸礁、堡礁型珊瑚礁海岸断续分布。南海诸岛中除西沙群岛东岛的高尖石是由凝灰熔岩构成外,其余基本上由珊瑚环礁组成(黄金森,1982;曾昭璇,曾宪中,1989)。中沙群岛黄岩岛亦是以基岩为主体。

中国海岸带大体上可以杭州湾为界,杭州湾以北的海岸,因构造差异,而具有稳定隆起的基岩港湾海岸与断续沉降的平原海岸相间分布;杭州湾以南的海岸,基本是因海水浸淹隆起的山地所构成的基岩港湾海岸,仅在海湾内有局部的平原、沙滩或泥涂。在全球海平面上升的背景下,海岸的升降反映地体构造升降与海平面上升双重作用的结果(王颖、金翔龙等,1979)。

2. 中国海岛普查与海岛成因分类

1989—1995 年全国海岛资源综合调查,是我国在 20 世纪海岸海洋研究工作中的又一具有里程碑意义的重要成果。我国海岛近万个,岛屿岸线长达 14 000 km,总面积约 7.54 万 km² ,约占陆域面积的 0.8%(杨文鹤,2000)。海岛分布南北跨越 38 个纬度,东西跨越 17 个经度。中国海岛根据成因分为大陆岛和海洋岛。大陆岛(占 95%)又可分为基岩岛与冲积岛;海洋岛(约占 5%)又可分为珊瑚岛和火山岛。

(1)基岩岛原为山地丘陵,由于海平面上升而被浸淹为岛屿。以台湾岛最大,海南岛次之。杭州湾外的舟山群岛是我国最大的群岛,含大小岛屿 1 390 多个。

(2)冲积岛是由大陆河流带来的泥沙冲积而成,地势比较低,平坦宽广,地形起伏不大。主要分布于淤积剧烈的大河口近岸海域,以长江河口段和江苏沿岸的沙岛最多,如崇明岛与辐射沙脊群(朱大奎,1984)。

(3)珊瑚岛是由珊瑚虫的骨骼所构成,主要分布于南海。包括东沙群岛、西沙群岛、中沙群岛与南沙群岛。岛屿的特点是地势低,一般海拔 4~5 m,面积小,以平方米计算,往往是在以沉降的海山为基底的背景上发育的。

(4)火山岛在我国分布亦多。澎湖列岛有 97 个岛屿,全部为火山喷发熔岩所组成。其他还有位于台湾东北、东海大陆架前缘的钓鱼列岛,北部湾的涠洲岛(火山口)、斜阳岛,台湾以东太平洋海域的绿岛、兰屿与龟山岛等。

中国大多数基岩岛与火山岛的轮廓均受构造控制,海岛的排列方向或单个海岛的长轴方向均与构造线方向有良好的相关。火山岛与珊瑚岛在平面形态上大都以圆锥形、环形为主;冲积岛在平面展布上以扇形、指状、长条状或椭圆形较多,与海域的水流动态相一致。

3. 淤泥质平原—潮滩海岸研究

(1)潮滩海岸

中国海陆交互带特性之一是潮汐与潮流作用显著,加上诸多河流源自西部高山、高原地带,东流入海,入海泥沙集聚、堆积成沿海平原,平原海岸岸坡缓,于平原海岸外缘潮间带及潮下带上部发育了广阔的潮滩。中国的潮滩规模大,岸线总长度达 4 000 km,渤海湾潮滩宽 3~4 km,南黄海苏北海岸潮滩宽 10~13 km(Wang,1983)。潮滩是人类活动频繁的地区,很多工程建设在潮滩环境或古代潮滩沉积上,所以研究潮滩沉积在理论上和实践上有很大的意义。

①潮滩沉积。其一,平原潮滩研究。1961—1963 年,南京大学海岸研究组在渤海湾与莱州湾进一步调查天津新港泥沙来源与岸滩变化时,从岸陆向海布置了多条固定的动力、地貌与沉积断面,进行了为期两年的冬、春、夏、秋季节测量。认识到渤海湾潮滩自高潮位向低潮位可分出四个沉积带:龟裂带、内淤积带、滩面冲刷带及外淤积带(表 1,王颖、

朱大奎,1990)。这四个带春秋季节分带明显,夏季风浪大,冲刷带受破坏物质粗化,SE向海风时有浮泥在滩面上堆积。冬季滩上多冰丘,解冻后形成众多的浮泥堆。

表1 渤海湾潮滩沉积—地貌分带特征

沉积带	龟裂带	内淤积带	滩面冲刷带	外淤积带
位置	大潮高潮位—平均高潮位以上	小潮高潮位	中潮位	中潮—大潮低潮位
动力	大潮潮流浸淹,平时蒸发,日晒	潮流,主要是涨潮流的作用	潮流往复作用,落潮流作用突出	潮流、波浪
地貌	多龟裂的泥滩,盐沼湿地,滩面坡度0.5‰~0.6‰	泥沼潮滩,多浅凹地;坡度0.67‰,通行沉陷	长条形冲刷体及潮水沟;冲刷体长轴与潮流向平行	极细砂与粗粉砂砂质潮滩,波痕及流水波痕
沉积作用	基本稳定,沉积作用弱,仅在大潮高潮位时沉积,而风暴潮时侵蚀	悬移质(泥)主要在此沉积,一年中滩面曾淤高5 cm	涨潮流沉积作用,落潮流沿潮水沟冲刷,形成冲刷体	潮流从外海向潮滩输沙,波浪从潮下带向低潮位输沙
沉积特征	黏土与黏土质淤泥水平纹层,具龟裂纹、虫穴	黏土质淤泥,水平纹层,多虫穴	黏土与细粉砂互层,水平页状纹层	细砂及粗粉砂,各种交错层理,透镜体

潮滩分带性是一项重大发现,充分反映了平缓海岸潮流塑造岸坡的动力与沉积过程。

苏北平原海岸因岸外有南黄海辐射沙脊群隐蔽而潮滩更加宽广,最宽处超过13 km,平均坡度0.2‰。沉积物主要是源于古长江的粉砂质,物质普遍较渤海湾粗。江苏沿岸潮流以东台市弶港为中心,是南北两个潮流系统的汇合区,弶港潮差最大。潮位决定了潮滩的高程,亦影响到潮滩的宽度,涨潮流速明显大于落潮流速,在一个潮周期中泥沙做向岸净运动。在平均高潮位至小潮高潮位间含沙浓度最大,是沉积作用最强的部位(朱大奎、许廷官,1982)。

波浪对江苏潮滩也有较明显的影响,波浪扰动使潮滩水层含沙浓度增加,形成各种波痕和斜层理,而风暴天气波浪对潮滩侵蚀和沉积物的分选作用就更明显了,形成各种侵蚀形态及风暴沙层(任美锷、张忍顺、杨巨海等,1983)。

中国对平原潮滩海岸研究的成果,如潮流动力特征、大河泥沙来源与自海向陆的泥沙横向再搬运、潮滩沉积—地貌分带性及潮滩发育动态等,为外国学者所注目并引入沉积相专著中(Dalrymple R W,1992)。

其二,港湾潮滩研究。对港湾潮滩海岸的全面调查研究始于1980—1988年全国海岸带和海涂资源综合调查。港湾内潮滩主要是由于长江入海泥沙沿海岸南下,最南达福鼎,汇合浙闽沿海河流输入的悬移质,形成沿岸的一支稳定的淤泥流。淤泥物质又被涨潮流携运至狭长海湾内落淤为厚层泥质滩涂,其分带性不如平原海岸潮滩显著(王颖、吴小根,1991)。

② 潮滩动力环境。潮滩分布于平缓海岸的潮间带及潮下带上部,是海陆交互作用的堆积地貌,发育有大量细颗粒泥沙供应、海岸坡度平缓(<1/1 000)、波浪作用衰减、潮

流动力活跃的地区(Wang,1983;Wang et al.,1990)。潮流是潮滩形成演变的主要动力,近岸潮波变形产生的时间不对称导致了流速不对称,涨潮流速大于落潮流速,导致沿海的细颗粒泥沙向岸的再搬运,并在潮间带堆积(Postma H,1967;朱大奎、许廷官,1982),但对潮间带中下部砂质推移质的影响很小(朱大奎、高抒,1985)。潮间带流速分布的不对称性,使最大含沙量与最大沉积量出现于潮滩中部(朱大奎等,1986)。由于波浪参与的程度强弱与频率不同,使潮滩整体上又可分为细砂质潮滩、粉砂质潮滩及淤泥质潮滩三种主要类型(王颖、朱大奎,1990)。

③ 现代潮滩地貌分带性与沉积相。潮滩地貌具有分带性(Wang,1983;Dalrymple,1992),从海向陆有相应的四个沉积相带(王颖,2000;Evans,1965)。潮滩分带性是普遍性的规律,是潮流动力在滩涂上递变的反映。分带性的名称在一些文献中各有不同,但是,潮滩底部是砂质粉砂滩,上部是泥质滩,潮滩中部为泥沙混合滩,却是潮滩沉积相共同的特征(王颖,2000;Evans,1965)。当潮滩遭受侵蚀后退,或在风暴潮期间受波浪冲刷,常会形成贝壳质海滩或贝壳沙堤,因此,潮滩沉积中常伴有贝壳层。贝壳多为粉砂质或淤泥滩的种属。上下两层在剖面上表现为泥层与砂层的交替沉积。潮滩沉积是在低潮滩沉积基底的基础上,通过"横向沉积"的拓宽和"垂向沉积"的增厚发展起来的,一个潮滩环境的旋回层厚度约为20~30 m(朱大奎、许廷官,1982)。

(2) 生物作用叠加形成的潮滩堆积地貌

潮滩海岸贝壳堤与河口湾牡蛎礁是淤泥质平原海岸由生物作用叠加形成的次一级堆积地貌,十分特殊,分别代表着一定的成因环境。

其一,贝壳堤。贝壳堤是发育于岸坡平缓(1‰~2‰),具有中等强度的激浪作用,粉砂质或黏土粉砂质平原海岸的沿岸堤(滩脊)。组成此类沿岸堤的物质主要是软体动物的贝壳或壳屑,其中掺杂着少量细砂、粉砂或黏土,它代表着特定的、遭受海侵冲刷之潮滩海岸环境(Wang and Ke,1989)。中国在河海交互作用海岸带的研究最为深入,居当代前列,贝壳堤是其中之一。例如在渤海湾西部海滨平原上的5列贝壳堤,是自公元前5000年以来,由大河泥沙汇入浅海,被浪潮再堆积而成,以其沉积组成与结构"记录"了当时海岸环境信息。

通过1961—1963年在渤海湾的调查,王颖(1963,1964)讨论了沿海两列贝壳堤与古海岸发育的关系。天津地质矿产研究所自20世纪90年代至今对渤海湾平原贝壳堤进行了深入调查研究,多层次采样定年与校正(王宏,1995、2001、2002、2003;王强,1991),印证了李世瑜(1962)、赵希涛(1980)和徐家声(1994)等关于渤海湾西岸平原尚存有2~3列更老的贝壳堤的看法。这5列贝壳堤从海向陆为:

① Ⅰ贝壳堤,沿海滨特大潮线以上分布,历史上以它为基础建成海堤。它始形成于元至元年间。

② Ⅱ贝壳堤,分布于Ⅰ贝壳堤的向陆侧,主要在海河以南分布。它始形成于东汉王景治黄河成功后(公元70年)。

③ Ⅲ贝壳堤,仍保存较完整,南起天津市东南约90 km的大港区沙井子,向北至田庄坨。平原上的村庄仍落在贝壳堤上。

④ Ⅳ贝壳堤,分布于黄骅县前苗庄南面的大坑河及大港区的沈清庄、大苏庄和翟庄

子一带。多埋藏地下1～2 m深处,沉积组成中砂质成分增多,底部为黑色淤泥,仍为潮滩带沉积。

⑤ Ⅴ贝壳堤,分布于Ⅳ堤以西的东孙村前苗庄与翟庄一带。

其二,河口湾—平原牡蛎礁。牡蛎礁与贝壳堤成因相类似,均为由生物壳体组成的潮滩平原海岸堆积体,形成于海平面缓缓上升过程中。但区别是:贝壳堤形成于高潮岸边陆地上,贝类经激浪流击打、搬运已处于死亡状态,而牡蛎礁是牡蛎生长于海岸浅水域坚硬的基底上,是活体在不断生长加积的,礁体的厚度是海面升高的标志。

渤海湾西北部平原海岸牡蛎礁分布于蓟运河与潮白新河间的倒三角形平原地带及蓟运河口内(图3),该处相当于蓟运河古河口湾,其范围远较现代蓟运河河口湾的规模宏大,可称为河口湾平原(王强,1991;王宏等,1996、2006、2010;范昌福等,2005、2008、2010)。

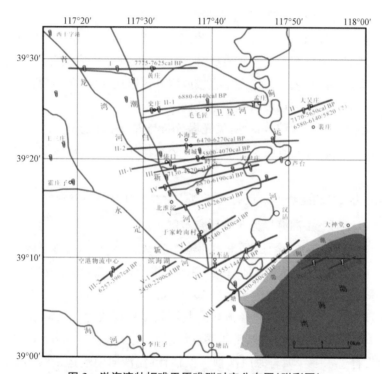

图3　渤海湾牡蛎礁平原礁群时空分布图(附彩图)

资料来源:据商志文等(2013)最新报告绘制。

河口湾平原突出的标志性沉积是分布着多列牡蛎礁,堆积发育,形成最具代表性的河口湾—牡蛎礁平原,在全新世海岸沉积地貌与环境演化方面,具有重要的科学价值。

渤海湾西北岸牡蛎礁曾引起较广泛的科研关注,先后有30多篇论著发表:翟乾祥和李世瑜(1962)阐述了牡蛎礁分布及其古海面意义;翟乾祥提出组成牡蛎礁的生物主要是长牡蛎(Ostrea gigans),是温暖气候时海水影响之生物,并做了^{14}C定年(1976);嗣后,中国科学院的赵希涛等(1979、1996)、彭贵等(1980)、李元芳等(1988)、李秀文等(1990)先后发表了研究成果,加深了对牡蛎礁的认识。提高其科学理论意义的是天津地质调查中心海岸带与第四纪地质研究室的王强、王宏、苏盛伟、王海峰、商志文、李建芬、范昌福等。

2010—2013 年来,天津地质调查中心海岸带与第四纪地质研究室承担天津市古海岸与湿地国家级自然保护区"贝壳堤与牡蛎礁新发现与新研究"项目①,连续发表 5 篇最新研究成果,其中 4 篇成为系列论文发表于《地质通报》,集中反映了新发现的鱼岺子贝壳堤和空港物流中心、滨海湖两处牡蛎礁,新增建了 72 个 AMS ^{14}C 年龄数据,进一步研究了巨葛庄、鱼岺子、板桥农场三分场、上古林和青坨子贝壳堤,罾口河、空港物流中心、大吴庄和岭头牡蛎礁。这些研究成果阐明了组成渤海湾西北岸牡蛎礁的主要牡蛎种属其生态特征与适应环境,牡蛎礁的沉积结构与分布、年代及其衰亡原因等(苏盛伟等,2011;王海峰等,2011;王宏等,2011;王海峰等,2012)。总结出具有里程碑意义的科学进步,使人们得以加深与提高研究渤海湾海岸平原发展过程科学意义的重视。

研究表明,牡蛎礁分布是自陆向海,时代由老到新,反映着该平原由浅海湾而逐渐淤填过程;在近 7000 年的时间内,该处始终是牡蛎礁繁殖发展的场所,直至公元前 1000 年左右时(相当历史上的西周时期)明显地衰亡。黄河于 3000BC—1128AD 堆积了以天津为中心的古黄河三角洲(高善明等,1980),黄河自天津一带入海输入大量淤泥黏土物质,改变了渤海湾北部的海水浑浊度与底质成分,估计是牡蛎礁停滞发育的原因。渤海湾西北部牡蛎礁具有独特的代表性。该海岸宜按成因并冠以地名,定名为"蓟运河河口湾—牡蛎礁平原海岸",为今后海岸环境研究奠定范例。

潮滩海岸动力、沉积、地貌分带性,贝壳堤分布与古海岸变迁记录,以及牡蛎礁河口湾遗证,是我国地貌与第四纪地质学家在河-海交互作用与平原海岸发现的系统性辉煌成就。

4. 受季风波浪与构造掀升形成的迁徙型沙质三角洲体系海岸研究

这类海岸以滦河三角洲海岸为典型研究实例。

滦河是一条多沙性河流,全长 877 km,年输沙量 24.08×10^6 t,含沙量曾高达 3.9 kg/m^3 (高善明等,1980;程天文,赵楚年,1984)。滦河是季节性河流,年径流量 3.89×10^9 m^3, 6—9 月夏秋季降雨期,径流量高达 34.3×10^8 m^3,占全年径流量的 73%,夏季 7—8 月降水占总径流量的 56%,季节性暴雨径流导致其尾闾多变迁改道。滦河沿岸潮差较小,平均为 1~1.5 m,而沿岸波浪作用强,常年盛行偏东风,有效波高达 3.8 m,形成泥沙自海向岸横向搬运,发育了水下沙坝与海岸沙坝。据估计,泥沙向南输运量约为 384×10^4 t/a, 向北输移为 76.7×10^4 t/a(钱春林,1994)。滦河平原沿岸均有沙坝环绕,即内侧为原始的平原岸线,外侧为沙坝海岸,两者之间成为潟湖,是外动力以波浪作用为主的河口三角洲沉积模式(王颖等,2007)。

5. 潮汐汊道港湾海岸研究

中国沿海普遍受潮汐涨落及潮流影响,尤其是苏、浙沿海强潮区以及海南岛因山地抬升而周边形成港湾海岸,受潮汐作用发育的潮汐汊道港湾海岸特色显著。我国对潮汐汊道港湾的关注与研究,始自 20 世纪 60 年代对海南岛海岸之研究,通过开发实践,逐渐认

---

① 商志文,王宏,苏盛伟,等。天津古海岸与湿地国家级自然保护区贝壳堤、牡蛎礁新发现与新问题研究项目成果报告,中国地质调查局天津地质调查中心(天津地质矿产研究所),2013。

知其发育规律(Wang et al.,1987)。

潮汐汊道的稳定性按自然发展规律,潮汐汊道将日趋淤积消亡。由潮流作用越过的汊道较为稳定,一般能维持较大的水深、通常发育在落潮流速大于涨潮流速的汊道,但口门下游海岸因泥沙不足而受侵蚀(任美锷、张忍顺,1984)。

海南洋浦湾与新英湾是很典型的潮汐汊道港湾。南京大学海岸与海岛实验室自1983—1984年开始进行动力、地貌与沉积研究,继之于1984—1985年及1987—1991年中加联合调查研究,掌握了系统的基础资料与动态变化趋势成果,阐明了洋浦湾的稳定性,明确可建深水港(王颖、朱大奎,1990)。

二、海岸海洋地貌研究——21 世纪中国海陆交互作用体系研究进展

上述具体的科学研究事例反映出海洋地貌研究的前两个阶段:第一个阶段(20 世纪30 年代前后),研究沿海陆地是分辨海岸类型的初始阶段;第二个阶段(50—60 年代),认识到海岸带实为海陆交互作用的过渡地带,研究范围包括沿岸陆地与水下岸坡。波浪变形、季风波浪与沿岸流,潮汐变化与潮流作用,河流输沙以及人类活动使海岸带具有不同的类型与发展变化过程。至 20 世纪 90 年代,人们意识到:海岸带的发展变化与海陆相互作用带密切相关,并备受人类生存活动的直接与间接影响。海陆交互作用带是晚更新世末以来,全新世地质历史发展中的一个最新篇章,是对海岸海洋地貌认识的第三阶段,研究海陆过渡带全貌。

中国海陆交互作用过渡地带具有一系列地质构造与地貌发育特征,研究成果主要体现在以下方面。

(一) 海陆作用过程与中国边缘海

我国除台湾岛以东直临太平洋外,大陆所濒临的黄(渤)海、东海与南海,均为欧亚大陆与太平洋之间的边缘海,大洋与大陆两大板块相互碰撞作用所形成的岛弧-海沟系,将边缘海与太平洋分隔开来,并围堵了河流自亚洲大陆携运的陆源泥沙,发育了堆积型大陆架。岛弧-海沟系属于新生代西太平洋构造活动带,现代火山与地震活动频繁。边缘海主体轮廓受几条相间排列的呈 NNE-NE 向的隆褶带与拗陷带所控制,其时代由西向东逐渐变新。最东侧为现代岛弧-海沟系,其内侧是冲绳海槽与东海陆架边缘隆褶带,台湾褶皱带属此隆褶带的南延部分;再向西依次为新生代的东海陆架拗陷带;中生代的浙闽隆起带;中新生代的黄海南部拗陷带;以及古生代的胶辽隆起带。在拗陷带发育了新生代的大陆架,而隆褶带在上升过程中遭受侵蚀与剥蚀作用,表现为基岩裸露的半岛或岛屿。

南海可划分出三个 NE 向的主体构造带:南海中央为拉张断裂开的海盆,有地幔物质上涌为洋壳,以及喷溢堆积的海底火山;中央海盆两侧是经多次拉张,成块断沉降的西沙与南沙古陆块,构成海底高原-阶梯状大陆坡。在上述三个 NE 向构造带的西北侧与西南侧为新生代的南海陆架与巽他陆架沉降盆地;东侧相邻马尼拉海沟、南沙海槽与吕宋岛弧。

中国海经历了中新生代以来的海-陆相互作用与演变过程,其地质构造与巨量的陆源泥沙汇入控制着海底地貌的发育。

(二) 河海交互作用与中国海岸海洋地貌发展研究

1. 现代河海交互作用与效应研究

王颖等选择 5 个不同类型的河流展示其不同的泥沙运动与河口沉积的特性以及对相邻陆架之影响(王颖等,2007),包括:① 强潮型动力的鸭绿江河口湾,形成从陆向海与从海向陆的双向水流交汇沉积的潮流沙脊体系;② 季风、波浪为主导动力的滦河口,以泥沙的横向运动为主,形成沙坝环绕的双重海岸;③ 弱潮型、多沙的黄河口,径流于两侧堆积指状沙咀,沿岸流自黄河口外携运泥沙向渤海湾延伸为淤泥舌;④ 径流与沿岸流组合作用的沉积模式,以长江口为代表,泥沙沿岸向南输运为主导,形成浙江沿海之淤泥滩的港湾海岸;⑤ 充填河口湾的三角洲,以珠江为代表,在河流分汊与会潮点处泥沙堆积。

2. 古海岸平原发育与大陆架堆积研究

现代中国边缘海大陆架在晚更新世时期曾是海岸平原,海平面的大幅度变化导致岸线的大范围退进,河流的长距离伸缩和大尺度迁移,河-海交互作用是形成平原海岸与浅海泥沙的主要动力过程。王颖等研究表明(2012):东海大陆架与南黄海大陆架实由"古扬子大三角洲沉积体系"(Paleo-Yangtze Grand Delta System)所组成。它包括:基底的古扬子大三角洲(面积可能达 38 万 km^2),是中、晚更新世由古长江、古黄河两条世界级大河输水供沙,受季风波浪和潮流作用形成的,其发育时代应在长江贯通下游汇入黄、东海以后;其上叠置发育了规模逐次减小的古江河三角洲(65 330 km^2)、南黄海辐射沙脊群(22 470 km^2)、全新世—现代长江三角洲(约 10 000 km^2)和历史时期的废黄河三角洲(约 4 100 km^2),组成巨型的复合三角洲体系。其表层经全新世以来海侵改造发育了波浪与潮流共同作用的沙脊地貌。

3. 边缘海大型古三角洲体系研究进展

前人虽未谈及黄、东海大三角洲体系,但在相关论著中曾多次涉及堆积型大陆架上河海交互作用堆积,并提出"扬子浅滩"或"扬子大沙滩"(刘振夏 1996;叶银灿等,2004)、"长江口大浅滩"或"长江大沙滩"(Chen Jiyu et al,1985;陈吉余,2007)的概念。

陈吉余等(1957、1959)最早在对长江口外水下地形分析中肯定地提出"长江口外具有显著的呈扇形分布的古代水下三角洲,面积达 7 000 km^2,前缘水深在海面下 50 m,它的平面中心在 $32°18'N$,与现代长江三角洲主泓—南泓道的出口显然不符合,说明长江的水下三角洲在发展过程中是向南移动的"。任美锷(1986)在《江苏省海岸带和海涂资源综合调查报告》中明确提出"辐射状沙脊群叠置发育在古黄河和古长江三角洲上"、"古长江三角洲在玉木冰期末—冰后期初形成和后退中,遭受了古滨岸海滩化动力改造"。国家海洋局第一海洋研究所和中国科学院地理科学与资源研究所 1984 年出版的《渤海黄海地势图》,展示出古长江中期三角洲和古黄河、古长江早期三角洲。古长江中期三角洲保存尚属完整,其外界范围在北部为 -20 m,在中部辐射沙脊群外围为 -45 m,在东南部古三角洲远端的外围水深为 50 m。秦蕴珊(1989)在《黄海地质》研究中,明确指出"旧黄河—古

长江复合三角洲地貌位于长江口以北,射阳至弶港以东海域,其外缘界限北面至－30 m
等深线,在东及东南面为－60～－65 m 等深线"。Liu et al.(2010)通过浅地层剖面和 4
个钻孔推测了南黄海有一个大型古三角洲的存在,时代为 MIS3,向海分布可达－50 m 等
深线。李全兴(1990)认为位于长江口外东侧的扬子浅滩与北部的废黄河三角洲这片不规
则的扇形隆起(－50 m 等深线)是古长江—古黄河复合三角洲。金翔龙(1992)在《东海海
洋地质》中明确界定了长江口从北向南迁移过程与现代三角洲之发育。刘振夏指出扬子
浅滩水下外界是水深 25～55 m(1997)。

国际上对大河的研究,反映出不同的河海相互作用效果:密西西比河汇入沉降中的墨
西哥湾,泥沙滑落入深海,只在沿河流口门两侧堆积了鸟足状三角洲(Fisk et al,1954;
Roberts,1997;Coleman,1998;Shepard,1973),且由于河流输沙量的减少及海平面上
升,该三角洲进一步沉溺(Blum & Roberts,2009)。亚马逊河流量大,但向海输送的泥沙
量未及长江、黄河丰富,其泥沙主要输入大西洋,并沿海底堆积为深海平原(Milliman,
1979;Damuth & Flood,1984、1988;Figueiredo et al,2011)。只有中国既有发育于岛
弧-海沟系后侧的边缘海,又有来自亚洲内陆大河输入的泥沙,才能形成由大三角洲体系
组成的宽阔的堆积型陆架以及孕育于沉积层中的油气矿藏等。

南京大学海岸地貌与沉积研究室成员在江苏省海岸带资源调查时(1980—1984)认识
到:"江苏的潮滩淤涨显著的岸段,位于南黄海辐射沙脊群波影区——射阳河口向南至吕
四段,与沙脊群的蔽浪作用有关。"1993—1996 年,王颖等在从事南黄海辐射沙脊群调查
时,发现"南黄海辐射沙脊群是全新世海侵由潮流改造的古长江三角洲堆积体""沙脊群北
部的废黄河水下三角洲受蚀,细颗粒物质沿岸向南输运;沙脊群中部枢纽区主潮流通道是
承袭古长江河谷发育的;伴随着现代海平面上升,在波浪作用下,沙脊展宽的同时,沙脊群
形成沿岸潮滩淤积的重要泥沙来源"(王颖,2002)。但是,在海平面持续上升的背景下,
2006—2010 年"中国近海海洋综合调查与评价"调查发现:"辐射沙脊群有冲有淤,或时冲
时淤,总体上却趋于蚀淤动态平衡,并未出现整体规模的显著缩小或向岸迁移"。显然,近
海海域存在巨大的沙源供给。

先期研究启迪我们向外海探索,找寻沙源地。在从事"中国近海海洋综合调查与评
价"专项研究中,对辐射沙脊群进行了多孔钻探,进一步了解到:古长江在江苏海岸的出口
较已知的黄沙洋—烂沙洋(辐射沙脊群中部枢纽区)更北;而黄河不仅在历史时期曾直流
入黄海,在辐射沙脊群北部钻孔揭示的晚更新世沉积中,也多次出现较厚的黄色粉砂与黏
土沉积。结合遥感与测量的海底制图中所提供的海底地形地貌特征信息,以及《江苏省地
图集》(史照良,2004)的遥感影像图,清晰展现出在辐射沙脊群外侧的南黄海内陆架上还
有一个更大的三角洲,分布范围覆盖整个苏北海域,北部达到海州湾,宽度小,外围水深约
30 m;中部宽大,位于辐射沙脊群外围,水深 30～50 m,陡坡界限明显,向南宽度变小,延
伸至长江口外,呈现受蚀残缺形态;该三角洲外缘－30 m 与－50 m 等深线形成地形转折,
显然为古长江与古黄河加积而成,故命名为古江河三角洲。其范围远比现代的长江三角
洲、废黄河三角洲大,比南黄海辐射沙脊群的面积大出 3 倍。该海域底质主要是细砂、粉
砂、砂质粉砂、粉砂质砂。北部多黏土质粉砂,富含结核(粒径长约数毫米至数厘米,钙质
结核和软体动物遗壳),表明系黄河携运之泥沙。细砂之重矿组成表明沉积物源主要来自

长江(王颖,2002)。

最近研究发现,古江河三角洲体是叠置在外围的一个更大的三角洲上(王颖等,2012)。这也是辐射沙脊群现今在陆地江、河泥沙补给已急剧减少,而海平面日益上升的背景上,仍能保持蚀、积动态平衡的重要原因,即海域外围有大三角洲的砂质补给。从遥感影像与海图分析,该大三角洲范围北起灌河口(34°30′N),南达马祖列岛南部(26°N),东部海域界限相当于−100～−150 m等深线,接近东海陆架边缘。大体上,以现代长江口与杭州湾为轴心,呈230°弧形向海扩展,覆盖黄、东海大陆架,面积约38万 km²,基本组成物质是细砂与粉砂,是全球罕有尚保留于大陆架的浅海地貌,可定名为"古扬子大三角洲体系"(图4,王颖等,2012)。这一完整的大三角洲地貌在南黄海—东海海底地形图中得到印证(图5)(Berne S et al,2002;李家彪,2008)。以它为基底,上面叠置着晚更新世古江河三角洲体、南黄海辐射沙脊群、全新世—现代长江三角洲,以及历史时期的废黄河三角洲,五部分组成巨大的三角洲复合体系,泥沙与水动力互有关联,至今尚未进行过系统研究。古扬子大三角洲体系是东亚—太平洋边缘海河海相互作用的最大地质载体,保留着中、晚第四纪以来河海交互作用、海陆变迁之环境变化的重大信息,探索其形成发展,将是对地球科学理论的重大贡献。大三角洲体系的组成物质,充分反映了其是来自我国内陆的泥沙在地质历史长河中的积累,这将是大陆架划界归属的有力证据。

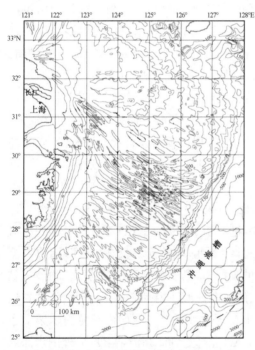

图 4　古扬子大三角洲体系(王颖等,2012)

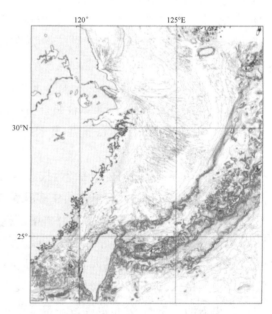

图 5　南黄海与东海海底地形(李家彪,2008)

现代海岸带与大陆架古海岸环境关系密切,既具有成因联系,又有着水、沙动力与生物活动的交换效应。古扬子大三角洲体系反映出南黄海—东海大陆架上部的沉积结构,蕴藏了中、晚更新世以来海陆交互作用带地貌形成的动力环境,陆源泥沙输入量,以及河-海作用过程的效应与发展历史。

三、新构造运动活跃的海陆交互作用过渡带——台湾岛东侧太平洋海域与南海

台湾东侧太平洋海域是指琉球群岛以南,巴士海峡东北的太平洋水域,绝大部分水深大于 4 000 m,海底地形直接由岛坡向深海平原过渡,无海沟存在(郑彦鹏,2012)。

南海是濒临中国大陆最大的边缘海。海底地貌类型齐全,包括大陆架与岛架,张裂沉降的阶梯状大陆坡与单斜坡,以及有海底火山分布的深海平原。

南海与台湾以东海域的系统性海洋研究工作均始于第二次世界大战后,台湾岛以海岸带研究深入,而南海因油气资源开发而海域研究渐进。

(一)台湾岛群研究进展

台湾岛地质发育的历史,可以概述为新生代形成而构造活动显著,这凝聚着台湾地质与海洋学家陈汝勤、林斐然、陈民本、俞何兴、何春荪、郑伟力等多年研究成果的概括。

1. 台湾海岸地貌研究

以台湾师范大学石再添教授的成果具有系统性与初始里程碑的地位。他研究了台湾岛四周海岸:北部海岸的计量研究,南端海岸的珊瑚礁,台湾东部海岸系列如:花东沿岸陆棚边缘峡谷、台湾东部苏花及礁溪断层海岸域的地形学计量研究、台湾东部花东海岸的地形学计量研究、台湾东部东棚海岸域的地形学计量研究、台湾西部海岸线的演变及海埔地的开发、台湾西南部嘉南洲潟湖海岸地形及其演变、台湾西部的剖面地形,这些均在台湾师范大学研究报告中发表(石再添,1970、1975、1976、1977、1978、1979、1980、1981)。石再添教授与他的学生和继任者如张瑞津、林雪美等,对台湾西南部、东部的河口区地形学研究、石再添与张瑞津、许民阳、沈淑敏关于琉球屿的海阶及珊瑚礁定年研究,以及石再添、邓国雄、许民阳、杨贵三关于台湾花东海岸海阶的地形学研究等(石再添等,1995、1991、1998),研究涉及台湾岛各边海岸地貌,尤其在海岸阶地与地貌计量方面具有特色。石再添多年关注海峡两岸的学术交流,推动地貌学研究与教学进步。此外,1980 年华南师范学院曾昭璇著《台湾海岸地貌》(华南师院科研丛刊),对台湾西部平原海岸地貌、西北部台地海岸地貌、南端山地海岸地貌,以及北部和东部海岸地貌进行了比较系统的研究(曾昭璇,1980)。

对台湾海岸海洋地貌进行深入的成因与发育趋势研究,具有第二个里程碑地位的是台湾大学地理环境资源学系王鑫和台大师生团队的工作。王鑫以地球圈层系统——气、水、岩、生相互作用的观点研究台湾本岛与周边诸岛的地貌形成与发育演变,获得该区海岸与海岛成因发育的科学真谛。他十分注重将地貌研究成果应用于公众教育并促进自然保护工作,撰写出版的关于台湾群岛自然环境、资源专著与丛书约 27 部。

2. 台湾以东太平洋海域

1979 年由科学出版社出版的《中国自然地理·海洋地理篇》,在"中国海底地质地貌"章节中,由王颖执笔,撰写了"台湾以东太平洋海底"。1996 年科学出版社出版由王颖主

编的《中国海洋地理》专著,在第一篇"海洋环境资源"中第三节是"台湾以东太平洋海底",其中包括"海底地质构造、海底地貌与海底沉积"。

2012年《中国区域海洋学——海洋地貌学》(王颖主编,2012)由海洋出版社出版,是反映我国"十二五"期间"近海海洋综合调查与评价专项"的成果总结。海洋地貌专辑中设置了5篇,第1篇渤海,第2篇黄海,第3篇东海,第4篇南海,第5篇即台湾以东太平洋海域。第5篇中的第16章为台湾以东太平洋海域环境特点,内容包括:(16.1)海洋地理特征;(16.2)海底地质构造。由国家海洋局第一海洋研究所郑彦鹏执笔,内容包括:台湾东部碰撞造山带,琉球沟-弧-盆体系,西菲律宾海盆,基底深大断裂。系统地反映了该海域的海底地质构造与动力的基本特征;海洋动力(16.3)由南京大学刘绍文执笔,内容包括风和降水、潮汐、海流和潮流。第17章是对该海域海底地貌所作的首次系统总结,包括(17.1)海岸类型(东北部基岩港湾海岸、东部断层悬崖海岸与南部珊瑚礁海岸),(17.2)海底地貌,包括海岸带水下岸坡、岛架、岛坡、海沟与大洋底部;(17.3)为海底地貌成因分析;(17.4)总结海岛、台湾本岛、绿岛、兰屿与龟山屿。第18章阐述该海域的海岸海洋地质灾害(海岸侵蚀、地震、海啸、火山和泥火山以及热带风暴灾害),第17、第18两章均由刘绍文执笔完成。

2013年1月,科学出版社出版的由王颖主编的《中国海洋地理》是中国自然地理系列专著之一(王颖等,2013),反映着学科研究的进步,内容均有增展。在第二篇区域海洋地理篇章中,第12章是台湾以东太平洋海域,其内容涉及该海域的气、水、岩、生各个方面。这一版《中国海洋地理》的最大特色是由刘瑞玉负责的中国海海洋生物一章,是首次在海洋地理学中增写了海洋生物篇章,空前丰富了内容,大大提高了该书的学术与应用价值。该篇中第2章,海底地质地貌的第3节为台湾以东太平洋海域,包括海底地质构造、海底地貌与海底沉积,由王颖执笔;在第二篇区域海洋地理的东海一章中之第4节由刘绍文、于堃执笔,介绍台湾海峡的地质地貌、气象水文、动植物区系,并撰写了第5节台湾岛的地质地貌、气候、河流水文、土壤与自然资源等。第二篇的第12章由刘绍文、孙祝友、傅命佐、郑彦鹏专门撰写台湾以东太平洋海域,共5节。第1节内容为海底地质构造,增加了地球物理场特征与断裂构造;第2节充实了海底地貌特征、底质与沉积及海底地貌成因分析;其余三节分别为气象水文、海洋灾害与海洋资源。

（二）南海海洋地貌研究进展

南海海岸与海底地貌类型齐全,学者对南海的研究程度不一致,海岸研究深入,尤以沿我国大陆海岸研究深入,而海底研究不足,对南海中部岛屿的调查研究更待深入。2012年由海洋出版社出版的《中国区域海洋学——海洋地貌学》,系统总结了已有的南海研究工作。

南海海岸带研究涉及海域总论的专著不多,具代表性的是赵焕廷、张乔民、宋朝景等著的《华南海岸和南海诸岛》(1999),李建生著《华南沿海地区海相地层与全新世地层划分》(1998),王文介等著《中国南海海岸地貌沉积研究》(2007)。总体进展尤其是关于海平面的研究有明显的进步:刘以宣等(1993)讨论了南海晚近海平面变化与构造升降的初步研究;王颖与吴小根(1995)探讨关于海平面上升与海岸侵蚀,多以海南岛沿岸海平面变化为实例。后有白鸿叶等(2004)的《中国闽粤沿海现代海平面上升的海岸地貌响应》;黄镇国、张伟强(2004)的《南海现代海平面变化研究的进展》;颜梅等(2008)发表了《全球及中国海海平面变化研究进展》。上述研究表明,南海海区海平面变化与地貌效应的研究逐步深入,且始终与全球海平面变化相联系。

1. 南海海岸研究进展

其进展突出地表现在潮汐汊道港湾海岸、珊瑚礁海岸与红树林海岸的研究中,同时,在海岸研究的方法与开发利用均有反映。

(1) 潮汐汊道港湾海岸

潮汐汊道港湾在华南多沿断裂走向或山地河谷下游发育。任美锷、张忍顺(1984)介绍了有关潮汐汊道的理论,指出纳潮量与口门断面是维持潮汐汊道动力的重要因素。王颖、陈万里(1982)提出三亚港、榆林港及铁炉港口处潮汐汊道港湾的特性,多为海湾内沙坝潟湖型。王文介(1984)研究了华南沿海潮汐通道类型特征;王文介、李绍宁(1988)探讨了清澜潟湖沙坝——潮汐通道体系的沉积环境和沉积作用;王文介等(1991)总结潮汐汊道的动力与沉积结合,完成《潟湖和潮汐水道沉积》。张乔民等(1985)认为湛江湾溺谷型潮汐水道发育是与清澜港有区别的潮汐汊道,主要为沉溺的基岩谷地与海湾,张乔民等(1995)综述了华南海岸沙坝—潟湖型潮汐汊道演变。

南京大学海岸与海岛开发国家试点实验室与加拿大贝德福德海洋研究所(Bedford Institute of Oceanography)在海南三亚、洋浦地区进行了一系列潮汐汊道港湾的水文动力、海底地形、底质与沉积层等断面定点重复测试,发表了系列文章,包括王颖等(1987)讨论海南潮汐汊道港湾特性与深水港建设;王颖、朱大奎(1990)利用沉溺谷地型潮汐汊道建设洋浦深水港的总结;朱大奎等(1992)论证泥沙不致形成港口回淤之动力机制;王颖等(1992)就三亚湾的海洋地质环境论证利用白排可建深水大港,是将潮汐汊道港湾研究应用于建港的进步。中山大学罗章仁总结出版了《华南港湾》(1992);邵全琴(1996)研究了海南潮汐汊道表层沉积特征。其后,王颖等出版了《海南潮汐汊道港湾海岸》专著,总结了海南三种成因的潮汐汊道港湾类型:沉溺谷地体系——水深条件好,潮汐控制的海洋环境,如榆林港、清澜港;构造断裂带或软硬岩层交接地带体系,如东寨——铺前港;沙坝潟湖体系,如三亚港、铁炉港、新村港、黎安港、坡头港、港北港、博鳌港等。论述潮汐汊道港湾的动力、泥沙作用环境,并分别论述了海南岛南部、西部、北部与东部的潮汐汊道港湾,最后阐明潮汐汊道港湾发育演变。

(2) 生物作用参与的珊瑚礁与红树林海岸

珊瑚礁海岸研究始于海南岛,专文论述的主要是岸礁研究,而且主要作为岸线年代与变迁之证据。王国忠等(1979)首先发表了关于海南岛鹿回头珊瑚礁的沉积相带研究,赵希涛等(1979)发表了《关于鹿回头珊瑚礁的形成年代及其对岸线变迁的反映》,涉及较多类型的礁体与分布。沙庆安、潘正蒲(1981)发表了《小东海全新世—现代礁岩的成岩作用》。王颖等在《海南潮汐汊道港湾》一书中,系统分析了三亚湾大东海、鹿回头珊瑚礁平台,抬升礁、海滩岩及落笔洞等处的地貌特征,认识到大部分礁平台为全新世的沉积,因小铲低潮线已并陆,洋浦湾大铲是海南岛西岸唯一的离岸岛礁。

涠洲岛是位于北部湾中的火山口顶部而发育的海岛,周边基岩岸有珊瑚礁分布。黄金森(1997)论述了《北部湾涠洲岛珊瑚海岸沉积》;亓发庆等(2003)详细论述了涠洲岛火山口地貌结构、火山岩与火山碎屑岩产状与分布,以及海岸背叠式海岸沙坝的定年结果,主要是全新世及近代沉积;王国忠等(1991)详细阐明了涠洲岛火山口岛地貌与沉积相。

大陆沿岸珊瑚礁研究以雷州半岛居多。包砺彦(1989)论述了雷州半岛青安湾海滩沉积与地形发育;余克服等(2002)阐述了《雷州半岛珊瑚礁生物地貌与全新世多期高海面》,阐明了角孔珊瑚与块状珊瑚之分布,反映着激浪冲刷强烈的位置所在,同时给出了校正后的年代,全新世中期至现代;据赵焕庭等(2002、2009)给出的数据,雷州半岛灯楼角珊瑚礁最老的年龄为 7 120±165 aB. P.。雷州半岛珊瑚礁年代与海南岛南岸大致相同,海南三亚市郊红塘一带珊瑚礁与海滩岩年龄大体相当,起始年代可能约自 8 000 aB. P.,但以 6 000~5 000 aB. P. 繁殖居多。

南海海岸地貌研究的论文,尚有吴正、吴克刚(1987)关于《海南岛东北部海岸沙丘的沉积特征及其发育模式》;王颖、周旅复(1990)关于《海南岛西北部火山海岸研究》;殷勇等(2002、2006)关于探地雷达对海南岛博鳌海岸沙坝的研究、对博鳌地区沙坝—潟湖沉积之研究和在海南岛东北部海岸调查中的应用。

红树林海岸自然分布于闽、粤、琼、台沿海。1963 年王颖发表于《地理》上的《红树林海岸》,始于 1956 年访问印度时,在孟买象岛一带穿越生长于玄武岩上的红树林及 20 世纪 60 年代初在北海见到红树林而完成该文;毛树珍(1991)从生物地理与海岸沉积系统研究综合论述红树林海岸现代沉积;陶思明(1999)从生态学角度研究红树林生态服务功能及其保护;张乔民等(2001)从海岸地理科学角度深入研究了"红树林宜林海洋环境指标",为促淤防冲及综合利用红树林资源提供科学依据;郑德璋等(1997)讨论了广东红树林及其保护重要性,麦少芝、徐颂军(2005)论述广东红树林资源的保护与开发;连军毫(2005)论述广东红树林现状及发展对策,都反映出地区对红树林自然资源之关注,提出开发与保护并重之建议。广西山口红树林保护区建设促进了公众的重视,李春干(2004)论述广西红树林的现状及发展对策;范航清等(2005)论述山口红树林滨海湿地与管理,是有针对性地关注到资源开发利用与环境保护,均需关注到可持续发展;兰竹虹、陈桂珠(2007)和傅秀梅等(2009)分别讨论了南中国海地区红树林的利用和保护、中国红树林资源状况及其药用调查的资源现状、保护与管理,反映出研究视野扩展至全海域及国家,以及与人类健康的关系。

2. 海底与岛礁地貌研究

目前对南海海底的研究成果,远不及对海岸带的研究,已有的研究仍是以毗连大陆海岸带的居多,反映着我国现有海洋研究的特点:重视海岸与部分大陆架,重视极地海洋,在远海部分工作不足,对群岛岛礁也需加大工作力度。

南海海底调查研究工作最多的是中国科学院南海海洋研究所与国土资源部南海部门的工作,中科院系统重视调查工作及研究成果的出版。

(1)南海海底地貌综合性研究以冯文科、鲍才旺(1982)的《南海地形地貌特征》为代表,继之中国科学院南海海洋研究所于 1982 年,以海区整体概念,由科学出版社发表了《南海海区综合调查研究报告(一)(二)》,以及《南海地质构造与陆缘扩张》。

(2)中国科学院南海海洋研究所 1987 年由科学出版社出版了《珍贵的曾母暗沙——中国南疆综合调查研究报告》,中国科学院南沙综合考察队 1989 年出版了《南沙群岛及其邻近海区综合调查研究报告（一）（二）》。继之，姚伯初（1991、1996、2001、2004）相继论述了南海海盆在新生代的构造演化、南沙海槽的构造特征及其构造演化史、南海的天然气水合物矿藏，以及大南海地区新生代板块构造活动。这一系列成果表达了对南海海域释理是从构造格局到新生代运动之进展，并应用于新能源矿藏之分析。郭令智（1948）的英文论文《中国南沙群岛郑和群礁地貌学》由钟晋梁 1988 年译为中文发表，可能是关于南沙群岛地貌研究最早的中、英文本的文章；谢以萱（1991）发表《南沙群岛海区地形基本特征》，共同组成了南沙群岛的一组论文。

（3）《南海——边缘海的几个地形地貌问题》（曾成开、王小波，1986），《南海陆缘扩张地貌》（谢以萱，1986），《南海新生代沉积盆地类型与演化系列》（钟建强，1997），是一组早期关于南海地貌的文章；刘昭蜀、赵焕庭等 2002 年出版了专著《南海地质》。

（4）南海北部包括西沙海槽在内的文章有：陈俊仁等（1983、1985）《南海北部－50 m 古海岸线的初步研究》及《南海北部－20 m 古海岸线研究》；冯文科等（1988）《南海北部晚第四纪地质环境》；丁巍伟等（2010）《南海北部陆坡海底峡谷形成机制探讨》；王海荣等（2008）《南海北部陆坡的地貌形态及其控制因素》；丁巍伟等（2009）《南海北部陆架－陆坡区新生代构造－沉积演化》，以及刘方兰、吴庐山（2006）《西沙海槽海域地形地貌特征及成因》，反映出对近大陆区研究工作的丰富。

（5）关于南海西部研究的论文两篇：鲍才旺、吴庐山 1999 年文章《南海西部的陆坡地貌类型及其特征》，及 1999 年鲍才旺文章《南海西部陆坡的海岭及其特征》，均发表于《南海西部海域地质构造特征和新生代沉积》专著中。

（6）黄岩岛是中沙群岛中唯一出露海面以上的岛屿，其余均在水下。黄金森（1980）《南海黄岩岛的地貌特征》和谢以萱（1980）《中沙群岛水下地形概况》，构成对中沙群岛的总体研究，是我国关注拥有主权海域进行深入了解的证据；王叶剑等（2009）《晚中新世南海珍贝——黄岩海山岩浆活动及其演化》，更说明我国对中沙群岛坚定不移地一贯重视。

（7）涉及南海其他部分的代表文章：邱燕等（2008）《南海西南海盆花岗岩的发现及其构造意义》；阎贫、刘海龄（2005）《南海及其周缘中新生代火山活动时空特征与南海形成模式》，填补了对南海深海平原缺少研究之不足；颜佳新（2005）《加里曼丹岛和马来半岛中生代岩相古地理特征及其构造意义》，填补了对南海南部及巽他陆架的研究空白；王颖、马劲松（2003）《南海海底特征、疆界与数字南海》系统概括了南海海底地貌（大陆架、岛缘陆架、大陆坡、岛坡、深海平原）、东沙、西沙、中沙与南沙群岛及周边海疆，以体现对传统海疆权益之维护。

以上概括表达了对南海海洋地貌研究的进展过程，需进一步加强对巽他陆架、南海东部岛弧海沟系所反映的地貌结构以及南海东北部的研究。

四、中国海岸海洋地貌研究小结

综上所述，可将本节内容总结为以下几点：

（1）中国海岸海洋地貌研究特点

在中国海岸海洋地貌研究历史过程中，反映出我国海岸科学研究，始终抓住边缘海海陆交互作用过渡带这一特定环境为研究客体,研究成果反映出亚太边缘海的区域构造结构,季风波浪与潮汐、潮流动力作用,以及大河水流泥沙之影响是控制区域海岸带发育过程及地貌特点的主导成因机制。全新世海平面上升与人类活动对海岸与海底地貌发育有着显著的影响效应。

(2)中国海岸海洋研究发展的重要里程碑阶段

① 现代中国海岸带研究主要起始于1957—1958年间长江口—上海港的治理,及天津新港泥沙来源与回淤研究。苏联海岸科学家曾科维奇、列昂杰夫及涅维斯基来华的系统讲学,中国学者以现代海岸科学的观点与方法,开始了对中国海岸带的深入研究,动力、地貌与沉积相结合系统地研究分析海陆过渡带的发生、发展与变化趋势,获得了我国海岸发育的真谛。

② 20世纪60年代初期,国家提出"三年改变港口面貌"的号召,促进了海岸动力、地貌与沉积研究的规模化发展,密切结合港口选建生产实践,促进海岸带科学研究的纵深化,建港的成败,检验了研究成果的科学水平。

③ 20世纪80年代全国海岸带和海涂资源综合调查,90年代全国海岛资源综合调查为中国海岸研究的第三个里程碑,是全国规模的系统化研究,并积累出版了首批全国海岸滩涂与海岛调查的系统成果。

④ 2000年以来,"908"专项开展的全国海岸带调查及海岛调查,是中国海岸带研究的第四个里程碑:系统调查了自1990年以来沿海大规模开发的效应,比较海岸带与海岛在自然与人类活动效应下的变化与发展趋势,有益于可持续性地开发利用海岸海洋资源,建立开发与生态环境和谐发展的途径。

当代21世纪,我国正在从海陆一体化研究,将海岸海洋与深海海洋结合研究,深海大洋是发展中的研究重点。极地海洋始终被我国研究关注,从南极到北极,并且还关注到世界屋脊的巅峰。

(3)海岸海洋研究中的亮点

① 从地质构造基础、季风波浪、潮汐与潮流与大河泥沙活动的动力作用以及人类活动效应,对全国海岸进行了简明、扼要的成因分类,便于了解与应用。

② 对河海交互作用的平原海岸环境特点、地貌发生、发展变化趋势以及人类活动做出系统的总结,区别出渤海湾与黄海西部大型平原海岸之异同,以及两者与华南海湾平原海岸的区别。在潮滩地貌、沉积分带性与动力相关以及沉积相于石油地质方面之应用;在以潮汐动力为主的平原海岸,因波浪侵蚀海岸,发育了贝壳堤,以贝壳堤为古海岸标志分析海岸冲淤变化历史过程;以及平原海岸牡蛎礁发育所反映的粉砂质河口湾环境方面,均有世界前沿水平的系统成果。

③ 河海交互作用与中国大陆架堆积发展以及长江—黄河大三角洲体系的开创性研究,具国际前沿水平,同时标志着海岸科学研究衔接陆架海底,成为中国海岸海洋科学发展的一个重要里程碑。

参考文献

[1] 李庆远.1935.中国海岸线的升沉问题[J].地理学报,2(2):129-166,19.

[2] Johnson D W. 1919. Shore Processes and Shoreline Development [M]. New York：John Wiley & Sons.

[3] Von Heidenstam H. 1922. The Growth of the Yangtze Delta [J]. Journal of the North China Branch of the Royal Asiatic Society，Vol. 53：21 - 36.

[4] Ting V K. 1919. Geology of the Yangtze Valley below Wuhu，Shanghai Harbour Investigation，Ser. Ⅰ. Report 1[R]. Shanghai：Whangpoo Conservancy Board，1 - 83.

[5] 陈国达. 1950. 中国岸线问题[J]. 中国科学,1(2～4):351 - 373.

[6] Richthofen F V. 1912. China. Ergebnisse eigener Reisen und darauf gegründeter Studien[J]. BJ，Ⅲ：403.

[7] Arnold Heim. 1929. Fragmentary Observation in the Region of Hongkong，Compared with Canton [R]. Annual Report，Geological Survey Kwangtung and Kwangsi，3(2)：1 - 32.

[8] W. Panzer. 香港海岸有何失序？（罗开富译）[N]. 国立中山大学日报,1934 - 05 - 04.

[9] Lin K T. 1937. Movement of the Strandline near Fuchow[J]. Bull. Geol. Soc. China，17：343 - 348.

[10] Hou T F，Wang Y L，and Chang C C. 1935. Geological Reconnaissance between Lungyan and Amoy，Fukien[J]. Bull. Geol. Surv. China，25.

[11] 马廷英. 1942. 闽海岸线之变动[J]. 中国地理研究所,海洋集刊.

[12] 高振西. 1942. 福建之山脉水系及海岸[J]. 福建省地质土壤所年报,2:19 - 28.

[13] 何春荪. 1947. 过去五十年内台湾地质之研究[J]. 地质论评,12(5):397 - 424.

[14] Chang L. S.，Chow M. C. 1949. Chen P. Y.. The Tainan Earthquake of Dec. 5，1946[J]. Bull. Geol. Survey，Taiwan，1：11 - 20.

[15] B. Ⅱ曾柯维奇（Zenkovich V. P.）. 1967. Processes of Coastal Development[M]. Edinburgh：Liver and Boyd.

[16] O. K 列昂杰夫著. 1965. 海岸与海底地貌学[M]. 王乃樑等译. 中国工业出版社.

[17] King C A M. 1972. Beaches and Coasts [M]. London：Edward Arnald.

[18] Shepard F P. 1973. Submarine Geology [M]. Third Edition. Harper & Row，Publishers，New York，Evanston，San Francisco，London.

[19] Philip H. Kuenen. 1950. Marine Geology[M]. New York，London：Wiley，Chapman & Hall.

[20] 王颖，朱大奎编著. 1994. 海岸地貌学[M]. 北京:高等教育出版社,3.

[21] Wang Ying. 1980. The Coast of China[J]. Geoscience Canada，7(3):109 - 113.

[22] Wang Y，Aubrey D. 1987. The Characteristics of the China Coastline[J]. Continental Shelf Research，7(4)：329 - 349.

[23] Wang Ying，Zhu Dakui. 1994. Tidal Flats in China[J]. Oceanography of China Seas，2：445 - 456.

[24] 黄金森. 1982. 中国珊瑚礁的岩溶特征[J]. 热带海洋,1(1):12 - 20.

[25] 曾昭璇,曾宪中. 1989. 海南岛自然地理[M]. 北京:科学出版社.

[26] 王颖,金翔龙等. 1979. 中国自然地理·海洋地理. 北京:科学出版社.

[27] 杨文鹤. 2000. 中国海岛[M]. 北京:海洋出版社.

[28] 朱大奎. 1984. 江苏海岛的初步研究[J]. 海洋通报,3(2):34 - 36.

[29] Wang Ying. 1983. The Mudflat Coast of China[J]. Canadian Journal of Fisheriesand Aquatic Sciences，40(Suppl. 1)：160 - 171.

[30] 王颖,朱大奎. 1990. 中国的潮滩[J]. 第四纪研究,4:291 - 300.

［31］朱大奎，许廷官.1982.江苏中部海岸发育和开发利用问题［J］.南京大学学报,18(3):799-818.

［32］任美锷,张忍顺,杨巨海等.1983.风暴潮对淤泥质海岸的影响［J］.海洋地质与第四纪地质,3(4):1-24.

［33］Dalrymple R W. 1992. Tidal Deposition System, Facies Models Response to Sea Level Changes ［M］. Geological Association of Canada, 195-218.

［34］王颖,吴小根.1991.浙闽港湾潮滩与沉积的组合特征［J］.南京大学学报(地理版),12:1-9.

［35］Wang Ying, Collins M B, Zhu Dakui. 1990. A Comparison Study of Open Coast Tidal Flat: the Wash(U. K.), Bohai Bay and West Yellow Sea(Mainland China)［C］. Proceedings of International Symposium on the Coastal Zone. Beijing: China Ocean Press, 120-134.

［36］Postma H. 1967. Sediment Transport and Sedimentation in the Environment［M］. In: Lauff G H ed. Estuaries, American Association for Advancement of Science Publication, 83: 158-199.

［37］朱大奎,高抒.1985.潮滩地貌与沉积的数字模型［J］.海洋通报,4(5):15-21.

［38］朱大奎,柯贤坤,高抒.1986.江苏海岸潮滩沉积的研究［J］.黄渤海洋,4(3):19-27.

［39］王颖.2000.潮滩沉积动力过程与沉积相［M］.见:苏纪兰,秦蕴珊主编.当代海洋科学学科前沿.北京:学苑出版社,177-182.

［40］Evans G. 1965. Intertidal Flat Sediments and their Environments of Deposition in the Wash ［M］. Jul Geol Soc, London, 121: 209-240.

［41］王颖,朱大奎, 曹桂云.2003.潮滩沉积环境与岩相对比研究［J］.沉积学报,21(4):539-545.

［42］Wang Ying, Ke Xiankun. 1989. Cheniers on the East Coastal Plain of China［J］. Marine Geology, 90(4): 321-335.

［43］李世瑜.1962.古代渤海湾西部海岸遗址及地下文物的初步调查研究［J］.考古,(12):652-657.

［44］王颖.1962.中国粉砂淤泥质平原海岸发育因素及贝壳堤形成条件［C］.中国地理学会1961年地貌学术讨论会,论文摘要,北京:科学出版社.

［45］王颖.1964.渤海湾西部岸滩特征;渤海湾西部贝壳堤与古海岸问题［C］.中国海洋湖沼学会1963年学术论文摘要汇编,北京:科学出版社,55-57.

［46］王颖.1964.渤海湾西部贝壳堤与古海岸线问题［J］.南京大学学报(自然科学版),8(3):424-443.

［47］Wang Hong, Keppens E, Nielsen P, et al. 1995. Oxygen and Carbon Isotope Study of the Holocene Oyster Reefs and Paleo-environment Reconstruction on the Northwest Coast of Bohai Bay, China ［J］. Marine Geology, 124: 289-302.

［48］王宏.2001.渤海湾牡蛎礁与新构造运动:几个基本问题的讨论［M］.见:卢演俦,高维明,陈国星等主编.新构造与环境.北京:地震出版社.

［49］王宏.2002.渤海湾贝壳堤与近代地质环境变化［G］.见:前寒武纪第四纪文集.北京:地质出版社,183-192.

［50］王宏.2003.渤海湾泥质海岸带近现代地质环境变化研究(Ⅱ):成果与讨论［J］. 第四纪研究,23(4):283-407.

［51］王强,李秀文,张志良等.1991.天津地区全新世牡蛎滩的古海洋学意义［J］.海洋学报,13(3):371-382.

［52］赵希涛,张景文,焦文强等.1980.渤海湾西岸的贝壳堤［J］.科学通报,25(6):279-281.

［53］徐家声.1994.渤海湾黄骅沿海贝壳堤与海平面变化［J］.海洋学报,16(1):68-77.

［54］王宏,李凤林,范昌福等.2004.环渤海湾海岸带^{14}C数据集(Ⅰ)［J］.第四纪研究,24(6):601-613.

［55］王宏.1996.渤海湾全新世贝壳堤和牡蛎礁的古环境［J］.第四纪研究,1:71-79.

[56] 范昌福,李建芬,王宏等.2005.渤海湾西北岸大吴庄牡蛎礁测年与古环境变化[J].地质调查与研究,28(2):124-129.

[57] 王宏,范昌福,李建芬等.2006.渤海湾西北岸全新世牡蛎礁研究概述[J].地质通报,25(3):315-331.

[58] 范昌福,王宏,裴艳东等.2008.渤海湾西北岸滨海湖埋藏牡蛎礁古生态环境[J].海洋地质与第四纪地质,28(1):33-41.

[59] 王宏,商志文,李建芬等.2010.渤海湾西侧泥质海岸带全新世岸线的变化与海洋的影响[J].地质通报,29(5):627-640.

[60] 范昌福,王宏,裴艳东等.2010.牡蛎壳体的同位素贝壳年轮研究[J].地球科学进展,25(2):163-173.

[61] 翟乾祥,丁云鹏.1962.渤海西岸古遗址分布及海岸变迁图(1:200 000).

[62] 赵希涛,耿秀山,张景文.1979.中国东部20000年来的海平面变化[J].海洋学报,1(2):269-281.

[63] 赵希涛,韩有松,李平日等.1996.区域海岸演化与海面变化及其地质记录[M].施雅风主编.中国海面变化.济南:山东科学技术出版社.

[64] 彭贵,张景文,焦文强等.1980.渤海湾沿岸晚第四纪地层^{14}C年代研究[J].地震地质,2(2):71-78.

[65] 李元芳,安凤桐.1985.天津平原第四纪微体化石群及其古地理意义[J].地理学报,40(2):155-168.

[66] 李秀文,赵福利.1990.^{14}C年代测定报告(TD)I[G].第四纪冰川与第四纪地质论文集(第六集:^{14}C专辑).北京:地质出版社.

[67] 苏盛伟,商志文,王福等.2011.渤海湾全新世贝壳堤:时空分布和海面变化标志点[J].地质通报,30(9):1382-1395.

[68] 王海峰,裴艳东,刘会敏等.2011.渤海湾全新世牡蛎礁:时空分布和海面变化标志点[J].地质通报,30(9):1396-1404.

[69] 王宏,陈永胜,田立柱等.2011.渤海湾全新世贝壳堤与牡蛎礁:古气候与海面变化[J].地质通报,30(9):105-141.

[70] 王海峰,王宏,范昌福等.2012.天津空港牡蛎礁:中全新世环境恶化与新构造控礁作用[J].地质通报,31(9):1387-1393.

[71] 高善明,李元芳,安凤桐等.1980.滦河三角洲滨岸沙体的形成和海岸线变迁[J].海洋学报,2(4):102-114.

[72] 程天文,赵楚年.1984.我国沿岸入海河川径流量与输沙量的估算[J].地理学报,39(4):418-427.

[73] 王颖,傅光翙,张永战.河海交互作用沉积与平原地貌发育[J].第四纪研究,27(5):674-689.

[74] 钱春林.2007.引滦工程对滦河三角洲的影响[J].地理学报,1994,49(2):158-166.

[75] Wang Ying, Schafer C T, Smith J N. 1987. Characteristics of Tidal Inlets Designated for Deep Water Harbour Development, Hainan Island[J]. China Proceedings of Coastal & Port Engineering in Developing Countries, 1(1):363-369.

[76] 任美锷,张忍顺.1984.潮汐汊道的若干问题.海洋学报,6(3):352-360.

[77] 王颖,朱大奎.1990.洋浦港海岸地貌与海岸工程问题[J].南京大学学报[自然科学(地理学)],11:1-11.

[78] Milliman J D, Syvitski J P M. 1992. Geomorphic/tectonic Control of Sediment Discharge to the Ocean:The Important of Small Mountainous Rivers[J]. Journal of Geology, 100(5):525-554.

[79] Wang Y, Ren ME, Syvitski J. 1998, Sediment Transport and Terrigenous Fluxes[M]. In:Brink K

H，Robinson A R eds. The Sea，Vol. 10：The Global Coastal Ocean Processes and Methods. New York，Toronto：John Wiley & Sons，253 - 292.

[80] Wang Ying，Ren Mei'e，Zhu DK. 1986. Sediment Supply to the Continental Shelf by the Major Rivers of China[J]. Journal of the Geological Society，London，143(6)：935 - 944.

[81] Curray J R，Moore D G. 1971. Growth of the Bengal Deep Sea Fan and Denudation in the Himalayas[J]. Geological Society of American Bulletin，82(3)：563 - 572.

[82] 王颖，邹欣庆，殷勇等. 2012. 河海交互作用与黄东海海域古扬子大三角洲体系研究[J]. 第四纪研究，32(6)：1055 - 1064.

[83] 刘振夏. 1996. 对东海扬子浅滩成因的再认识[J]. 海洋学报，18(2)：85 - 92.

[84] 叶银灿，庄振业，来向华等. 2004. 东海扬子浅滩砂质底形研究[J]. 中国海洋大学学报，34(6)：1057 - 1062.

[85] Chen J Y，Zhu H F，Don YF，et al. 1985. Development of the Changjiang Estuary and its Submerged Delta[J]. Continental Shelf Research，4(1~2)：47 - 56.

[86] 陈吉余. 2007. 中国河口海岸研究与实践[M]. 北京：高等教育出版社，156 - 199.

[87] 陈吉余. 1957. 长江三角洲江口段的地形发育[J]. 地理学报，23(3)：241 - 253.

[88] 陈吉余，虞志英，恽才兴. 1959. 长江三角洲的地貌发育[J]. 地理学报，25(3)：201 - 220.

[89] 任美锷. 1986. 江苏省海岸带与海涂资源综合调查报告[M]. 北京：海洋出版社，120 - 133.

[90] 国家海洋局第一海洋研究所，中国科学院地理科学与资源研究所. 1984. 渤海黄海地势图[M]. 北京：地图出版社.

[91] 秦蕴珊. 1989. 黄海地质[M]. 北京：海洋出版社，24 - 30.

[92] Liu J，Saito Yoshiki，Kong XH，et al. 2010. Delta Development and Channel Incision during Marine Isotope Stage 3 and 2 in the Western South Yellow Sea[J]. Marine Geology，278(1~4)：54 - 76.

[93] 李全兴. 1990. 渤海黄海东海地质地球物理图集[M]. 北京：海洋出版社.

[94] 金翔龙. 1992. 东海海洋地质[M]. 北京：海洋出版社.

[95] Liu Z X. 1997. Yangtze Shoal—A Modern Tidal Sand Sheet in the Northwestern Part of the East China Sea[J]. Marine Geology，137(3~4)：321 - 330.

[96] Fisk H N，McFarlan E，Kolb C R，et al. 1954. Sedimentary Framework of the Modern Mississippi Delta[J]. Journal of Sedimentary Petrology，24(2)：76 - 99.

[97] Roberts H H. 1997. Dynamic Changes of the Holocene Mississippi River Delta Plain：The Delta Cycle[J]. Journal of Coastal Research，13(3)：605 - 627.

[98] Coleman J M，Roberts H H，Stone G W. 1998. Mississippi River Delta：An Overview[J]. Journal of Coastal Research，14(3)：698 - 716.

[99] Blum M D，Roberts H H. 2009. Drowing of the Mississippi Delta due to Insufficient Sediment Supply and Global Sea-level Rise[J]. Nature Geoscience，2：488 - 491.

[100] Milliman J D. 1979. Morphology and Structure of Amazon Upper Continental Margin[J]. American Association of Petroleum Geologists Bulletin，63(6)：934 - 950.

[101] Damuth J E，Flood R D. 1984. Morphology，Sedimentation Processes and Growth Pattern of the Amazon Deep-sea Fan[J]. Geo-Marine Letters，3(2~4)：109 - 117.

[102] Damuth J E，Flood R D，Kowsmann R O，et al. 1988. Anatomy and Growth Pattern of Amazon Deep-sea Fan as Revealed by Long-range Side-scan Sonar(GLORIA) and High-resolution Seismic

Studies[J]. American Association of Petroleum Geologists Bulletin，72(8)：885－911.

[103] Figueiredo J，Hoorn C，van der Ven P，et al. 2011. Late Miocene Onset of the Amazon River and the Amazon Deep-sea Fan：Evidence from the Foz do Amazonas Basin[J]. Geology，37(7)：619－622.

[104] 王颖.2002.黄海陆架辐射沙脊群[M].北京:中国环境科学出版社.

[105] 史照良.2004.江苏省地图集[M].北京:中国地图出版社.

[106] Berne S，Vagner P，Guichard F et al. 2002. Pleistocene Forced Regressions and Tidal Sand Ridges in the East China Sea[J]. Marine Geology，188(3～4)：293－325.

[107] 李家彪.2008.东海区域地质[M].北京:海洋出版社.

[108] 郑彦鹏.2012.台湾以东太平洋海域环境特点[M].王颖主编.中国区域海洋学——海洋地貌学,第17章.北京:海洋出版社,647－656.

[109] 戴昌鳳.2003.台湾的海洋[M].台北:远足文化事业股份有限公司.

[110] 俞何兴,陈汝勤.1994.台湾海域之沉积盆地[M].台北:渤海堂文化公司.

[111] 陈民本,俞何兴,郑伟力等.1998.台湾四周海域之海底地形沉积[C].庆祝台湾大学海洋研究所成立三十周年学术研讨会论文摘要集,60－70.

[112] 石再添.台湾海岸地形学计量研究系列.台湾师范大学地理系研究报告,1970,52;1975a,3;1975b,1;1976,2;1977,3;1978,4;1979,5;1980a,6;1980b,4;1981,7.

[113] 石再添,张瑞津,林雪美等.台湾河口海岸研究系列报告:台湾师范大学地理系研究报告,1995a,23;1995b,24;1991,17;1988,14.

[114] 曾昭璇.1980.台湾海岸地貌[J].华南师院科研丛刊.

[115] 王鑫.台湾海岸研究专著系列:
　　① 台湾的地形景观[M].1980.台北:度假出版社.
　　② 北海岸风景特定区生态资源调查报告[R].台湾大学地理学系,1982,131.
　　③ 台湾的海岸地形[J].大自然,1985a,6;12－26.
　　④ 台湾沿海地区自然环境保护计划研究[M].台北:台湾大学地理学系,1985b.
　　⑤ 火炎山自然保留区生态之研究报告[R].台北:台大地理系,1987.
　　⑥ 泥岩恶地地景保留区之研究报告[R].台北:台大地理系,1988a.

[116] 王鑫.2004a.台湾的特殊地景——北台湾(台湾地理百科)[M].台北:远足文化事业有限公司.

[117] 王鑫.2004b.台湾的特殊地景——南台湾(台湾地理百科)[M].台北:远足文化事业有限公司.

[118] 中国科学院《中国自然地理》编辑委员会.1979.中国自然地理海洋地理篇[M].北京:科学出版社.

[119] 王颖.1996.中国海洋地理[M].北京:科学出版社.

[120] 王颖.2012.中国区域海洋学——海洋地貌学[M].北京:海洋出版社.

[121] 王颖,刘瑞玉,苏纪兰.2013.中国海洋地理[M].北京:科学出版社.

[122] 赵焕庭,张乔民,宋朝景等.1999.华南海岸和南海诸岛地貌与环境[M].北京:科学出版社.

[123] 李建生.1988.华南沿海地区海相地层与全新世地层划分[J].地理学报,43(1):19－34.

[124] 王文介等.2007.中国南海海岸地貌沉积研究[M].广州:广东经济出版社.

[125] 刘以宣,詹文秋,陈欣树等.1993.南海邻近海平面变化与构造升降初步研究[J].热带海洋,12(3):24－26.

[126] 王颖,吴小根.1995.海平面上升与海滩侵蚀[J].地理学报,50(2):118－127.

[127] 白鸿叶,王晓岚,邱维理等.2004.中国闽粤沿海现代海平面上升的海岸地貌响应[J].北京师范大学学报,40(3):404－410.

［128］黄镇国,张伟强.2004.南海现代海平面变化研究的进展［J］.台湾海峡,23(4):530-535.

［129］颜梅,左军成,傅深波等.2008.全球及中国海海平面变化研究进展［J］.海洋环境科学,27(2):197-200.

［130］王颖,陈万生.1982.三亚湾海岸地貌的几个问题［J］.海洋通报,1(3):37-45.

［131］王文介.1984.华南沿海潮汐通道类型特征的初步研究［G］.中国科学院南海海洋研究所.南海海洋科学集刊(第五集).北京:科学出版社,19-30.

［132］王文介,李绍宁.1988.清澜潟湖—沙坝—潮汐通道体系的沉积环境和沉积作用［J］.热带海洋,7(3):27-38.

［133］王文介.1991.潟湖和潮汐水道现代沉积［M］.北京:科学出版社,142-150.

［134］张乔民,宋朝景,赵焕庭.1991.湛江湾溺谷型潮汐水道的发育［J］.热带海洋,4(1):48-57.

［135］张乔民,陈欣树,王文介等.1995.华南海岸沙坝潟湖型潮汐汊道口门地貌演变［J］.海洋学报,17(2):69-77.

［136］Zhu D K,Wang Y,Smith J N,et al.1992. Sediment Transport Processes in Yangpu Bay Hainan Island［M］. Island Environment and Coast Development,Nanjing University Press,157-182.

［137］Wang Y,Zhu D K,Schafer C T.1992. Marine Geology and Environment of Sanya Bay Hainan Island,China［M］. 南京: Nanjing University Press,125-156.

［138］罗章仁等.1992.华南港湾［M］.广州:中山大学出版社.

［139］邵全琴,王颖,赵振家.1996.海南潮汐汊道的现代沉积特征研究［J］.地理研究,15(2):84-91.

［140］王颖等著.1998.海南潮汐汊道港湾海岸［M］.北京:中国环境科学出版社.

［141］王国忠,周福根,吕炳全等.1979.海南岛鹿回头珊瑚岸礁的沉积相带［J］.同济大学学报,2:70-89.

［142］赵希涛.1979.海南岛鹿回头珊瑚礁的形成年代及其对海岸线变迁的反映［J］.科学通报,24(21):995-999.

［143］沙庆安,潘正莆.1981.海南岛小东海全新世—现代礁岩的成岩作用［J］.石油与天然气地质,2(4):312-327.

［144］黄金森.1997.北部湾涠洲岛珊瑚海岸沉积［J］.热带地貌,2:1-3.

［145］亓发庆,黎广钊,孙永福等.2003.北部湾涠洲岛地貌的基本特征［J］.海洋科学进展,21(1):41-50.

［146］王国忠,全松青,吕炳全等.1991.南海涠洲岛区现代沉积环境和沉积作用演化［J］.海洋地质与第四纪地质,11(1):69-81.

［147］包砺彦.1989.雷州半岛南部青安湾海滩的沉积特征和地形发育［J］.热带海洋学报,8(2):75-83.

［148］余克服,钟晋梁,赵建新等.2002.雷州半岛珊瑚礁生物地貌带与全新世多期相对高海平面［J］.海洋地质与第四纪地质,22(2):27-33.

［149］王丽荣,赵焕庭,宋朝景等.2002.雷州半岛灯楼角海岸地貌演变［J］.海洋学报,24(6):135-144.

［150］赵焕庭,王丽荣,宋朝景等.2009.广东徐闻西岸珊瑚礁［M］.广州:广东科技出版社.

［151］吴正,吴克刚.1987.海南岛东北部海岸沙丘的沉积构造特征及其发育模式［J］.地理学报,42(2):129-141.

［152］王颖,周旅复.1990.海南岛西北部火山海岸的研究［J］.地理学报,45(3):321-330.

［153］殷勇,朱大奎,关洪军等.2002.应用探地雷达方法对海南岛博鳌海岸沙坝的研究［J］.海洋地质与第四纪地质,22(3):119-123.

［154］殷勇,朱大奎,王颖等.2002.海南岛博鳌地区沙坝—潟湖沉积及探地雷达(GPR)的应用［J］.地理

学报,57(3):301-309.

[155] 殷勇,朱大奎,I. P. Martini. 2006. 探地雷达(GPR)在海南岛东北部海岸带调查中的应用[J]. 第四纪研究,26(3):462-469.

[156] 王颖. 1963. 红树林海岸[J]. 地理,3:110-112.

[157] 毛树珍. 1991. 红树林海岸现代沉积[M]. 见:王文介等编著. 华南沿海和近海现代沉积. 北京:科学出版社,142-150.

[158] 陶思明. 1999. 红树林生态系统服务功能及其保护[J]. 上海环境科学,18(10):439-441.

[159] 张乔民,隋淑珍,张叶春等. 2001. 红树林宜林海洋环境指标研究[J]. 生态学报,21(9):1427-1437.

[160] 郑德璋等. 1997. 广东省红树林及其保护的重要性[J]. 广东林业科技,13(1):8-14.

[161] 麦少芝,徐颂军. 2005. 广东红树林资源的保护与开发[J]. 海洋开发与管理,1:44-48.

[162] 连军豪. 2005. 广东红树林的现状及发展对策[J]. 广东科技,11:37-38.

[163] 李春干. 2004. 广西红树林的数量分布[J]. 北京林业大学学报,26(1):47-52.

[164] 范航清,陈光华,何斌原等. 2005. 山口红树林滨海湿地与管理[M]. 北京:海洋出版社.

[165] 兰竹虹,陈桂珠. 2007. 南中国海地区红树林的利用和保护[J]. 海洋环境科学,26(4):355-360.

[166] 傅秀梅,王亚楠,邵长伦等. 2009. 中国红树林资源状况及其药用研究调查Ⅱ——资源现状、保护与管理[J]. 中国海洋大学学报,39(4):705-711.

[167] 冯文科,鲍才旺. 1982. 南海地形地貌特征[J]. 海洋地质研究,2(4):80-93.

[168] 中国科学院南海海洋研究所. 1982. 南海海区综合调查研究报告(一)(二)[R]. 北京:科学出版社.

[169] 中国科学院南海海洋研究所海洋地质构造研究室. 1988. 南海地质构造与陆缘扩张[M]. 北京:科学出版社.

[170] 中国科学院南海海洋研究所. 1987. 曾母暗沙——中国南疆综合调查研究报告[R]. 北京:科学出版社.

[171] 中国科学院南沙综合科学考察队. 1989. 南沙群岛及其邻近海区综合调查研究报告(一)(二)[R]. 北京:科学出版社.

[172] 姚伯初. 1991. 南海海盆在新生代的构造演化[J]. 南海地质研究,3:9-23.

[173] 姚伯初. 1996. 南沙海槽的构造特征及其构造演化史[J]. 南海地质研究,8:1-13.

[174] 姚伯初. 2001. 南海的天然气水合物矿藏[J]. 热带海洋学报,20(2):20-28.

[175] 姚伯初,万玲,吴能友. 2004. 大南海地区新生代板块构造运动[J]. 中国地质,31(2):113-122.

[176] 郭令智. 1988. 中国南沙群岛郑和群礁的地貌学[J]. 1948. 原载:Acta Geologica Taiwanica,钟晋梁译. 南海研究与开发,1:56-61.

[177] 谢以萱. 1991. 南沙群岛海区地形基本特征[G]. 见:南沙群岛及其邻近海区地质地球物理及岛礁研究论文集(一). 北京:海洋出版社.

[178] 曾成开,王小波. 1986. 南海——边缘海的几个地形地貌问题[J]. 东海海洋,4(4):32-40.

[179] 谢以萱. 1986. 南海的陆缘扩张地貌[J]. 热带海洋,5(2):12-19.

[180] 钟建强. 1997. 南海新生代沉积盆地的类型与演化系列[J]. 南海研究与开发,3:15-18.

[181] 刘昭蜀,赵焕庭等. 2002. 南海地质[M]. 北京:科学出版社.

[182] 陈俊仁,冯文科,赵希涛. 1983. 南海北部-50米古海岸线的初步研究[J]. 地理学报,38(2):176-187.

[183] 陈俊仁等. 1985. 南海北部-20米古海岸线研究[C]. 见:中国第四纪海岸线学术讨论文集,230-240.

［184］冯文科,薛万俊,杨达源.1988.南海北部晚第四纪地质环境［M］.广州:广东科技出版社.

［185］丁巍伟,李家彪,李军.2010.南海北部陆坡海底峡谷形成机制探讨［J］.海洋学研究,28(1): 26－31.

［186］王海荣,王英民,邱燕等.2008.南海北部陆坡的地貌形态及其控制因素［J］.海洋学报,30(2): 70－79.

［187］丁巍伟,黎明碧,何敏等.2009.南海中北部陆架—陆坡区新生代构造—沉积演化［J］.高校地质学报,15(3):339－350.

［188］刘方兰,吴庐山.2006.西沙海槽海域地形地貌特征及成因［J］.海洋地质与第四纪地质,26(3): 7－14.

［189］鲍才旺,吴庐山.1999.南海西部陆坡地貌类型及其特征［M］.见:姚伯初等.南海西部海域地质构造特征和新生代沉积.北京:地质出版社.

［190］鲍才旺.1999.南海西部陆坡的海岭及其特征［M］.见:姚伯初等.南海西部海域地质构造特征和新生代沉积.北京:地质出版社.

［191］黄金森.1980.南海黄岩岛的一些地质特征［J］.海洋学报,2(2):112－123.

［192］谢以萱.1980.中沙群岛水下地形概况［J］.海洋科技资料,1:39－45.

［193］王叶剑,韩喜球,罗昭华等.2009.晚中新世南海珍贝——黄岩海山岩浆活动及其演化:岩石地球化学和年代学证据［J］.海洋学报,31(4):93－102.

［194］邱燕,陈国能,刘方兰等.2008.南海西南海盆花岗岩的发现及其构造意义［J］.海洋通报,27(12): 2104－2107.

［195］阎贫,刘海龄.2005.南海及其周缘中新生代火山活动时空特征与南海的形成模式［J］.热带海洋学报,24(2):33－41.

［196］颜佳新.2005.加里曼丹岛和马来半岛中生代岩相古地理特征及其构造意义［J］.热带海洋学报,24 (2):197－200.

［197］王颖,马劲松.2003.南海海底特征、疆界与数字南海［J］.南京大学学报,11:56－64.

The Mudflat System of China[*][①]

Much of the coast of China consists of mudflats that extend along ~200 km of the coastline. This coastal type is most prominent along the Yellow Sea coast to the north of Hangzhou Bay and along the Bohai Sea coast of the Great Plain of North China (Fig. 1). This paper describes the turbid water environment of the mudflat coast. This entire coastal area has subsided over 300 m since the beginning of the Cenozoic, and has been built by the huge volumes of sediment brought down by several large rivers, notably the Yellow River and Chang Jiang (Yangtze River). The great plain built by these sediments extended onto the continental shelf at times of low sea level and was partly submerged during postglacial transgression. At present, the Yellow River and the Yangtze River continue to deliver sediment to the coast and continental shelf of North China, forming extensive coastal mudflats in Bohai Bay and to the north of the Yangtze River.

The intertidal flat is the major constituent of the mudflat coast and is also the active zone of the coast. Two contrasting types of intertidal mudflats have been studied by a coastal dynamic research group from the University of Nanking. Between 1963 and 1965, sediment dynamics were monitored on the mudflats along the west coast of Bohai Bay, including topographic and hydrological measurements during different seasons and tidal periods, together with observations on changes in morphology, sediment, and material movement along several fixed profiles (sediment size nomenclature is shown in Table 1). Since 1980, similar systematic work has been started on the north Jiangsu coast.

---

  * Ying Wang: *Canadian Journal of Fisheries and Aquatic Sciences*, 1983, Vol. 40, Supplement No. 1: pp. 160－171.

  ① This paper forms part of the Proceedings of the Dynamics of Turbid Coastal Environments Symposium convened at the Bedford Institute of Oceanography, Dartmouth, N. S. , Canada, September 29—October 1, 1981.

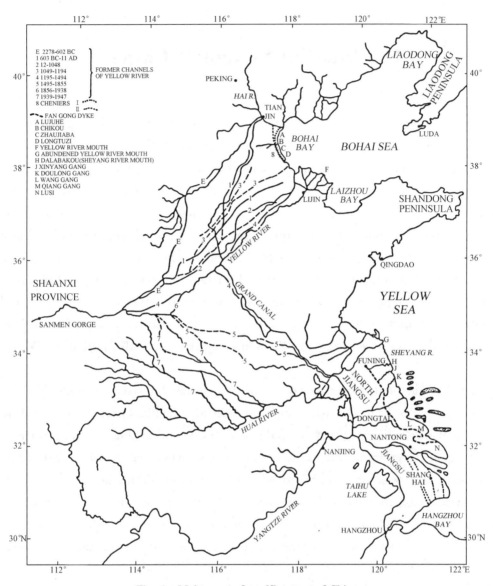

Fig. 1 Major part of mudflat coast of China

Table 1 Sediment classification used during the 1960s and 1970s by Chinese marine geologists

	Mean diameter (mm)
Gravel	>1
Coarse sand	1~0.5
Medium sand	0.5~0.25
Fine sand	0.25~0.1
Coarse silt	0.1~0.05
Fine silt	0.05~0.01
Clay	<0.01

Physical Environment

In Bohai Bay the water is less than 20 m deep and the coastal slopes are gentle. Tidal currents are the major agents of sediment transport and, with plenty of sediment supplied by the Yellow River, they develop the typical intertidal mudflats. Along the north Jiangsu plain the coast is exposed to the large waves of the Yellow Sea, and as a result waves are important in addition to the currents in sediment movement. The coastal sediment here is also coarser than in Bohai Bay. The extensive mudflats along the coastline of the north Jiangsu plain develop because of the protection that is provided by the offshore bars of the former Yellow River delta.

BOHAI BAY

The Yellow River delta has been located to the south of Bohai Bay since 1855, the eighth major recorded change in its course since 2278 B. C. (Fig. 1). The apex of the fan-shaped delta is at Lijin. The coastline of the delta is 170 km long and the total area of the delta is 5 860 km^2 (Pang and Si 1980). The annual water discharge of the Yellow River is 43% of the total water discharge to the Bohai Sea, and its annual sediment discharge is the largest in the world. The mean suspended sediment concentration near the mouth is approx. 24. 7 kg/m^3.

The coarser sediments are deposited close to the river mouth to form the river-mouth bars, and the finer fraction (mean diameter=0. 006 mm) is always moved by longshore currents as a distinct sediment plume in longshore drift, often called "fluid mud".

Nearly 400 million t of sediment per year is deposited in the lower reaches of the river. As a result, the river bed becomes progressively higher and its course longer, as it progrades seaward (Table 2). Every 6~8 yr the prograding river breaks the banks and changes to a new course because of siltation (Wang 1980). As a rule, the course changes from north to northeast, to east and southeast, and then to the north again. This process of clockwise migration has an approximate 50-yr cycle (Pang and Si 1979).

Table 2　The rate of progradation of the Yellow River delta

Time period	Area of delta	Progradation rate	
		River	Delta
1855—1980	5 860 km^2	1. 5 km/yr	50 km^2/yr
1964—1971	400	1. 97~3. 45 km/yr	57 km^2/yr

The tides in Bohai Bay are mainly semidiurnal. The average tidal range is about 3 m with a minimum (0. 8 m) around the mouth of the Yellow River. Along the coast the currents are rectilinear. Flood currents dominate towards the west (210°~310°), and ebb currents towards the east（60°～95°or 135°），except at the Yellow River mouth. Here tidal periods are unequal; the ebb tide lasts 7 h and the flood 4 h. As aresult, the velocity of the flood is higher than the ebb, especially on the intertidal flat. The highest concentrations of suspended particulate matter（SPM）（Table 3）are found at the beginning of the flood tide when the water is shallow and turbulent. The SPM concentrations decrease to a minimum at high tide as the water becomes less turbulent. During the ebb tide, the suspended sediment concentration is low because of the long-lasting ebb period with lower current speeds. As a result, the principal transport of sediment is by the flood tide in a landward direction. Very fine sand and coarse silt （mean diameter＝0.077 mm）are transported mainly as traction load. Fine silt （mean diameter＝0.002 6 mm）is transported in suspension in the bottom 20 cm of the water column, and clay（mean diameter＝0.006～0.008 mm）in the upper layer of water.

Table 3　Hydrographic measurement along the middle part profile of Bohai Bay, July 5—6, 1965

Station (on profile)		+2.0 m	±0 m	−2 m	−4 m
Water depth (m)		0.2～1.8	0.2～3.8	1.5～5.4	3.5～6.8
Flood	Velocity (m/s)	0.3	0.3～0.25	0.34～0.52	0.2～0.57
	Direction	265°		55°～40°	60°～70°
Ebb	Velocity (m/s)	0.2	0.2～0.3	0.3～0.45	0.2～0.4
	Direction	35°		250°～260°	280°
Wave	λ(m)	0.1	0.2	0.2～0.45	0.29～0.4
	T(s)	1～2.2	2.8～4.0	1.6～3.0	3.0～4.0
Sediment concentration ‰		165.4[a] 5.2 3.7	4～6	2～3	

[a] NOTE: The maximum sediment concentration（164.5‰）appears at the first hour of the flood. It decreases by one-half during the second hour, by one-tenth at the third hour, and only one-thirtieth, i. e., 5.2‰, at the fourth hour, then all between 5.2‰ and 3.7‰.

In Bohai Bay the waves are mainly wind-induced with very little swell. The wave climate reflects the directionand speed of the wind. There are strong waves during winter and spring, when northeast winds predominate. Extreme wave height is over 4 m and average maximum wave height is 1.9 m. During the summer and autumn, southeast and east waves predominate with the average height in the order of 0.4～0.5 m. The effect of wave action on the mudflat is weak, because the water is shallow during the period of flood tide. The waves stir up the sediment which is then transported by tidal currents along the intertidal flat. In addition to tidal currents, there also are wind-induced currents with velocities between 10 and 30 cm/s. These currents are important in carrying suspended sediment as a longshore drift, but not in transverse movement on the flat.

NORTH JIANGSU COAST

Much of the North Jiangsu coast is developed on the pre-1855 delta of the Yellow River (Fig. 1), and lies immediately north of the Yangtze River.

The annual sediment discharge of the Yangtze River is only about one-third of that of the Yellow River (Table 4), and at the present time most is deposited in the channels at the river mouth. Sediment deposited outside these channels is remobilized as longshore drift towards the south, where it forms the mudflats in the bays of Zhejiang and Fujian Provinces to the south. Longshore drift towards the north from the Yangtze River mouth contributes very little to the development of the North Jiangsu coast.

Table 4　Hydrographic data for the Yellow and Yangtze rivers

	Yellow River	Yangtze River
Total drainage area (km^2)	752 443	1 807 199
Total length of river (km)	5 464	6 380
Average water discharge (m^3/s) Maximum water discharge (m^3/s) Minimum water discharge (m^3/s)	1 480 10 400 5.2	29 200
Average sediment concentration (kg/m^3) Maximum sediment concentration (kg/m^3)	24.7 184.0	0.616
Average annual sediment discharge (10^8 t) Maximum annual sediment discharge (10^8 t) Minimum annual sediment discharge (10^8 t)	12.0 21.0 9.0	4.78
Percent of suspended sediment with size <0.025 mm Percent of suspended sediment with size >0.025 mm	1950—1959　60.7% 1974—1977[a]　40.6% 1950—1959　39.3% 1974—1977[a]　59.4%	

[a] After completion of Sanmen dam.

Along the coast of the Yellow Sea semidiurnal tides are predominant with different tidal ranges (Table 5). The average tidal range is 1.5~1.7 m around the abandoned Yellow River mouth; and 2.6~3.7 m at the Yangtze River mouth and Dongtai. The maximum tidal range at the Yangtze River is 6.02~6.8 m. To the north of this area, average tide is 3.0~3.4 m, and maximum is 4.5~6.44 m. Off Jiangsu Province there are rotary currents along the sea shore and tidal current speeds range between 50 and 100 cm/s. Flood currents are faster than ebb currents, which last nearly 8~10 h. The wave heights are smaller (0.6~1.2 m) because the coast is protected by offshore sandy ridges. Outside of the sand ridges, wave height can reach 4 m. Along the North Jiangsu coast, waves are also controlled by the wind. There are northeast/north-northeast waves during winter with maximum heights of 2.9~4.1 m, and southeast waves predominantly during summer, with maximum heights in the range 1.7～3.2 m. In autumn, easterly swell prevails.

Table 5　Some topographic and hydrographic data of the North Jiangsu coast

Location	Altitude (m)	Rate of coastline retreat (−) or progress (+) (m/yr)	Average tidal range (m)	Flood period (h)	Ebb period (h)	Width of tidal flat (km)	Slope of tidal flat (‰)
Abundant Yellow River mouth	1.40	−230	1.58	4:00	8:25	3.3	1.26
Sheyang River mouth	1.70	+330	2.12	3:17	8:17	7.7	0.30
Xinyang Gang	2.0	+300	2.08	3:42	8:38	13.0	0.23
Dou-Long Gang	2.3	+500	2.01	3:09	9:16	18.5	0.21
Wang Gang	4.6	+400		2:17	10:08	18.0	
Qiang Gang	3.0	+500	2.01	3:13	9:12	14.5	0.91
Lüsi	4.1		1.90	3:08	9:17	3.7	1.0

SEDIMENTS OF THE NORTH CHINA MUDFLAT COAST

The coastal mudflats of the North China plain consist of three zones. From land to sea these are the salt marsh plain, intertidal mudflats, and submarine coastal slope (Fig. 2).

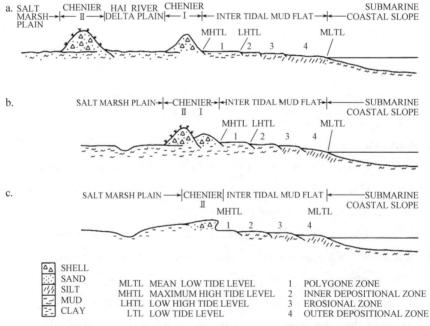

Fig. 2　Coastal profile of (a) prograding part of west coast of Bohai Bay, (b) stable part of west coast of Bohai Bay, and (c) slightly eroded part of west coast of Bohai Bay.

Salt marsh plain—The marsh plains are located above the maximum high tidal level, are normally flat, and may contain several lagoonal depressions. The vegetation is sparse.

Along the northwest coast of Bohai Bay, salt marsh plains are the main marine depositional features. These salt marsh plains contain shell ridges (cheniers) (Fig. 1 and Table 6) (Wang 1964). The shell ridges were formed by breaking waves as the principal shore-forming agent. The sediment was mainly silt or sandy silt and the coastal slope in the range of 1~2 m/km (Wang 1964; Ren and Cheakai 1980). At present the shore of Bohai Bay consists of mud with tidal currents as the major shore-forming processes, and under the present condition, the shell ridges are not formed.

Along the coast of North Jiangsu, there is an extensive alluvial plain containing at least one old shell ridge, with a C^{14} date of $4\,800 \pm 150$ yr B. P. (Zhu 1980). At the seaward margin of the plain is a salt marsh, termed the grass flat zone by Zhu (1980, Fig. 3). It is about 2~5 km wide, with a slope between 0.1 and 0.2 m/km. This zone is only occasionally flooded with water. The flat consists of silty clay, which was carried in suspension by extreme flood tides. The sediment is about 2~3 m thick with micro laminations. The salt content of the sediment is less than 0.3‰ and organic matter is 1%~2%. The zone is covered with a thick growth of various grasses, mainly *Seluropus liuttoralis* var. *sinensis*, *Suaeda salsa*, *Phragmites ausfralis*, and *Imperate cylindrica* var. *major*.

Intertidal mud flat—In Bohai Bay, the width of the intertidal mudflat is generally 3~7 km, increasing toward the south to 16~18 km at the Yellow River delta. The intertidal flat can be divided into four geomorphological and sedimentary zones (Wang et al. 1964), which, from land to sea, are the polygon zone, inner depositional zone, erosional zone, and outer depositional zone (Table 7, Fig. 4).

The *polygon zone* is located between lower and maximum high tides, with its landward limit marked by a chenier. The zone is 70~870 m wide with slopes of 0.5~0.6 m/km. Sediment is greyish-yellow clay, with gray silty mud 1.5 m below the surface. Only during spring high tide is this zone covered by water. Normally, high water does not cover the polygon zone and under the dry conditions polygonal desiccation cracks develop in the clay. These average 10~50 cm in diameter with 1~2 cm-wide cracks. Where there is freshwater influence some vegetation may grow, including several grass species such as *Suaeda glance* Bunge, *Limonium bicolor* Kuntie, and *Phragmites communis* Irin.

The *inner depositional zone* consists of soft mud in the form of pools from 500 to 1 000 m in diameter with an average slope of 0.67 m/km. Mud and clayey mud is deposited here during the flood tide. The soft sediment makes walking through this area difficult.

Table 6　Shell beach ridges of the west coast of Bohai Bay

Name of chenier	Position	Relative height (m)	Width (m)	Form	Sediment	Note	Age (yr)
I	Along shoreline above maximum high tide	0.5~1.0 (northward of Hai River) 2.0 (southward of Hai River)	20~30	Discontinuous arc beaches and branching ridges along advanced river mouth area. Continuous ridges along advanced and stable coast	Shells and shell fragments mainly *Arca (Anadara) subcrenata* Lischkei, with several silt layers. Also there are a few china fragments. All in the natural deposited laminations. Unconsolidated.	Almost bare vegetation	300~600
II	On land, distance from coast range from 40km at Hai River mouth to zero at Chikou and ZhauJiabo	5	100~200	Basically continuous ridge along the coast except in the southern eroded coast.	Shells and shell sand. *Arca (Anadara) subcrenata* Lischkei, *Meretrix* Linné, *Umbonium toxensi* (Croose), *Tympanotomus cingulatus* (Gmelin), *Rapana rhomasian* Crosse and Papna, *Pechilionsis* Graban and King, *Dosinia japonica* Reeve, etc., with little silt and net weight all in natural deposit state and little consolidated.	With grass, bushes, and fresh water	1 300~1 900

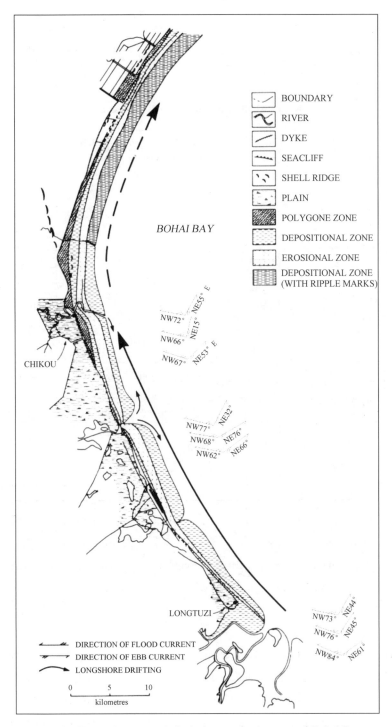

Fig. 3 Dynamic geomorphological map of west coast of Bohai Bay

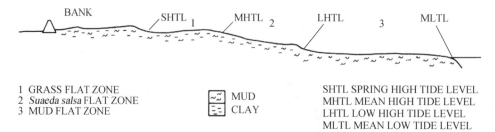

1 GRASS FLAT ZONE
2 *Suaeda salsa* FLAT ZONE
3 MUD FLAT ZONE

MUD
CLAY

SHTL SPRING HIGH TIDE LEVEL
MHTL MEAN HIGH TIDE LEVEL
LHTL LOW HIGH TIDE LEVEL
MLTL MEAN LOW TIDE LEVEL

Fig. 4　Mudflat profile of North Jiangsu coast

Table 7　The intertidal zones of the west coast of Bohai Bay

Zone	Outer depositional zone	Erosional zone	Inner depositional zone	Polygon zone
Location	Immediately shoreward of the low tide level	Around mid-tide level	Immediately below the lower high tide level	Between lower high and maximum tide level
Morphology	Wave or current ripple	Overlapping fish scales structure or tidal channels	Mudflat of mud pools	Polygon fields or marsh (near river mouth)
Sediment	Very fine sand or coarse silt	Alternating laminae of fine silt and mud	Mud	Clay or clayey-mud
Dynamics	Tidal currents and microwaves	Tidal current, mainly ebb currents	Tidal currents, mainly flood currents	Tidal currents and evaporation due to solar heating
Processes	Deposition of thin layers over large area	Deposition during flood, erosion during ebb current	Extensive deposition	Stable (during maximum tide deposition, erosion during storm)

The *erosional zone* occurs around the midtide level. In northern Bohai Bay, the zone is 500 m wide with a slope of 0. 57 m/km, and exhibits alternating laminae of fine silt and clay. The surface is eroded into an overlapping series of scours called fish-scale structures, 3~18 m long and 0. 5 m wide which dip seaward. Scarps between each layer of scales are 2 to 10 cm high and face the coast. These are erosional forms produced by ebb flow.

In the middle part of Bohai Bay, the erosional zone is about 1 000 m wide, with a slope of 0. 42 m/km, and consists of mud. Small tidal channels may be eroded in the mud. This zone is eroded mainly by the ebb tide. The sediments are carried to margins of the zone, mainly toward the sea where the silt settles in the outer depositional zone.

The *outer depositional zone* is present between the erosional zone and the low tide level. It is approximately 1 000 m wide and has a 0. 5 m/km slope. The zone consists of coarse silt with current ripple marks on the surface. The ripples have a wavelength of

3~7 cm, with the steep slope facing seawards along the outer part of this zone and shorewards along the inner part. In the middle part of the zone they are symmetrical. At the low tide level, the ripples are overlapped by micro wave marks. Wave action always influences the zone of low tide level, where coarser material (coarse silt and very fine sand) is deposited. The sediments become finer both offshore and onshore.

On the North Jiangsu coast, only three zones are recognized (Zhu 1980).

The *Suaeda salsa flat zone* corresponds approximately to the polygon zone and partly to the inner depositional zone. It is 4~5 km wide with a slope of 0.1~0.2 m/km, and extends from maximum high tide level to mean high tide level. Tidal channels extending to this zone are parallel or branched, normally 10 m wide, and 0.2~0.5 m deep. Between channels, the sediment is laminated silty clay. There are occasional polygons, impregnated with extensive burrows of invertebrates such as *Arenicola* and *liyoplax dentinierosa*. Over one-third of the zone is covered by *Suaeda salsa*.

The *mud flat zone* extends between mean high tide level and lower high tide level. The slope of the zone increases to 2~3 m/km. It consists of alternating laminae of clay and silt. Tidal channels are concentrated in this zone. Silt is deposited mainly in the channels, but a thin layer of the silt may also be deposited on the flat during extreme high tides. Mainly clay is deposited on the flat. This zone may compare with the outer part of the inner depositional zone and the erosional zone of Bohai Bay. Big rice grass (*Spartina anglica*) is being planted on the inner part of this zone. This grass traps mud and is thus 10 cm higher than the surface of the surrounding mudflat.

The *silt zone* between mean high tide level (or low high tide) and mean low tide level occupies nearly half the width of the whole flat. The slope here is between 0.3~ 0.4 m/km; the sediment is 60%~70% silt (0.1~0.01 mm) and 24%~30% fine sand (>0.1 mm). The sediment is transported mainly as traction load and ripple marks are common on the flat.

Two sediment units are distinguished in vertical section through the mudflats of North Jiangsu. The upper unit is principally well-laminated clay and occurs mainly above mean high tide level, i. e. in the high flat. It is 2~3 m thick, but its total thickness is not over half of the maximum tidal range. The lower unit is sand or silt. It is found mainly along a zone from the mean tide level to the bottom of the submarine coastal slopes at a water depth of 25 m. The sandy unit is 20 m thick along much of the North Jiangsu coast.

Submarine coastal slope—In Bohai Bay the submarine coastal slope is very gentle. The sediment becomes finer offshore, from sandy silt at the low tidal level to very fine mud in depths of more than 10 m (Table 8, Fig. 5, profile of 5a, b, c, d).

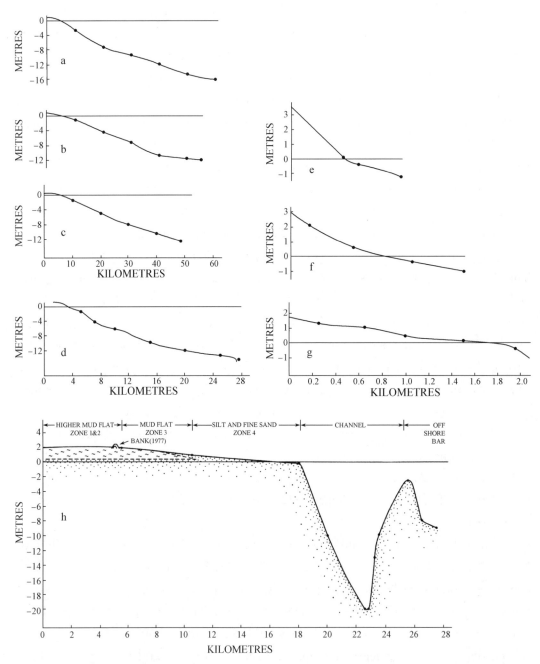

Fig. 5　Submarine coastal slope profile of （a） Lüjühe, （b） Chikou, （c） Longtuzi, （d） Diaokou （1965 river mouth of Yellow River）, （e） northern part of abundant Yellow River mouth （Location G, Fig. 1.）, （f） southern part of abundant Yellow River mouth, （g） DalaBakou （river mouth of Sheyang River）, and （h） Wang gang profile.

Table 8　Submarine coastal slope data of Bohai Bay

Locality	Range (m)	Slope (‰)	Notes
Lüjühe	0 to −16	0.26	Deposited and advanced part (a)
Chiko	0 to −12	0.21	Stable part (b)
Zhau Jia Bo	0 to −14	0.23	
Longtuzi	0 to −12.5	0.24	Slightly eroded part (c)
Diaoko (Yellow River mouth)	0 to −14	0.51	Along sand bars of 1965 river mouth

The submarine coastal slope of North Jiangsu also shows gentle slopes and concave profiles in the inner part of the offshore (Table 9 and Fig. 5e, f, g). However, the slopes are steeper here than in the Bohai Bay and the profiles are complex toward the sea (Fig. 5h) because of submarine sand ridges developed at the former Yellow River mouth. The sediment of the submarine coastal slope of North Jiangsu is mainly fine sand.

Table 9　Submarine coastal slope data of North Jiangsu

Locality	Range(m)	Slope(‰)	Notes
Abandoned Yellow River mouth	0~5 −5~−10	1.66 1.25	Rapid eroded part
Dala Ba Kou	0~−10	1.66	Stable part
Xinyang Gang	0~−10	2.27	Deposited part
Wang Gang	0~−20	4.39	
Chuang Dong Gang	0~−10	3.9	Rapid deposit part

CONTEMPORARY COASTAL DYNAMICS

In Bohai Bay, the principal source of sediment is the Yellow River, which plays a predominant role in the development of the mudflat coast of west Bohai Bay. The plume of suspended river sediment is the major source of coastal sediment. The "fluid mud" is deposited only in the upper part of the tidal flats or in deep water.

Three years of monitoring indicate that the flats are areas of rapid sediment deposition. The net depositional rate varies from 0.2 to 0.6 m annually and is concentrated in the outer and inner depositional zones. The higher rates of deposition are in the southern part of the Bay, closer to the Yellow River, where mainly mud is deposited. More silty sediment is deposited in the northern part of the Bay.

The tidal flat zonation is the result of the varying importance of different tidal processes acting on the flat, and is summarized in Table 7. Sediment type on the flat is controlled by the velocity of the tides. On the very low gradient, frictional drag is of major importance. The deposition of very fine sand and coarse silt on the lower flat, alternating laminae of fine silt and mud on the middle flat, and clay on the upper flat is a result of settling lag (discussed by Postma 1967; Wang et al. 1964) and the decrease in current velocity landwards as water depth decreases. Although the ebb current is slower than the flood current, gravitational acceleration on the steeper middle flat leads to slight erosion. The alternating mud and silt laminae of this zone accentuate the erosional features.

Seasonal changes in depositional processes are also important. In Bohai Bay, freezing starts in late November. Coastal ice to a thickness of 0.5 m accumulates mainly between high and middle tide level. Normally there are alternating layers of ice and mud. During March, ice melts and leaves patches of mud on the flat. During May and June the wind is predominantly northeasterly. The mud flat suffers from erosion by the strong wind waves, but sediment is redeposited during the summer. With southeasterly and easterly winds, the flat is covered with yellowish mud that was carried from offshore and originated in the Yellow River.

Tidal currents are the major agents of transport along the North Jiangsu coast, but wave action is also important. The tidal flat here is wider than along the west coast of Bohai Bay. The Sheyang River separates two different coastal sedimentary regimes. To the north the coast isretreating, especially around the former mouth of the Yellow River, with the development of shelly beach ridges (cheniers) (Fig. 6). To the south a 200 km length of coastline is prograding, protected by a field of submarine sand ridges (Ren and Cheakai 1980). These sand ridges are about $5 \sim 10$ m high and extend for about 200 km from north to south, and 90 km from west to east. The ridges represent an old Yellow River delta formed between 1194 and 1855 A. D. Since the return of the Yellow River to Bohai Bay, strong tidal currents have reshaped the deltaic features to the field of sandy shoals. Further to the south, from Qianggang southwards, the coast is retreating behind a narrow tidal flat (Table 5).

Fig. 6 Modern cheniers on coast of Jiangsu Province(附彩图)

HISTORICAL EVOLUTION

During historic times，there have been eight major shifts in the course of the Yellow River（Wang 1964；Pang and Si 1979；Shen 1979），which at times flowed into Bohai Bay and at other times to the North Jiangsu coast of the Yellow Sea. The Yellow Sea is named after the turbid mud plume from the former Yellow River delta in North Jiangsu in historic times.

Coastal progradation occurs in the vicinity of the active delta of the Yellow River，as illustrated by the Bohai Sea coast at present. In contrast，erosion occurs along abandoned deltaic coasts（Fig. 7）.

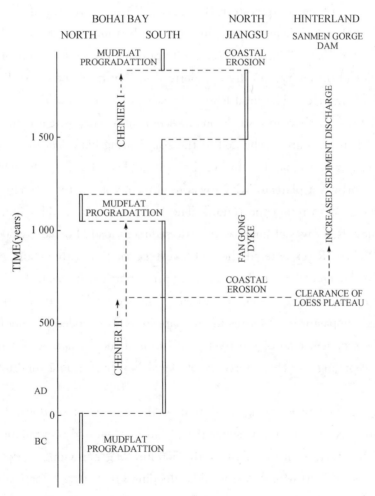

Fig. 7 Summary of changes in the course of the lower Yellow River in the last 2000 yr and their effects on coastal development

The Yellow River changed its course from the Yellow Sea back to Bohai Sea in 1855. Since then，the North Jiangsu coastline has retreated by as much as 30 m per year. The most rapidly eroding sections are along 150 km of the coast where the Yellow

River mouth used to be (Wang 1961). In this area nearly 1 400 km^2 of land area has been lost since 1855. During the last 50 yr the coast has retreated 230 m (Zhu 1980), and the 1957 coastline is now located under 1 m of water. No retreat has occurred where the coastline is protected by offshore bars, such as the middle part of the North Jiangsu coast.

The coarse and shelly sediment of the cheniers that occur in the salt marsh plain of both the Bohai and North Jiangsu coasts must have been formed under erosional and wave-dominated conditions, rather than under a depositional tidal current regime.

In early historical times (prior to 11 A. D.) the Yellow River entered northwestern Bohai Bay near the present Hai River. In 12 A. D. , the river shifted to close to its present course, entering southwestern Bohai Bay. Erosional retreat of the former delta formed Chenier II, which began to take its shape at the time of the Eastern Han Dynasty (70 A. D.), after Wang Ching had successfully put the Yellow River under control.

The sediment of the Yellow River is derived mainly from the Loess Plateau in the middle reach of the river. The total load is 16×10^8 tons after the river passes through the Loess Plateau. Each year 4×10^8 tons of sediment are deposited in the lower reach plain, and 12×10^8 tons are discharged to the sea. During historical times the sediment load of the Yellow River has fluctuated. Before the Tang Dynasty the population was lower on the northwest plateau, which was used mainly as pasture. There were plenty of forests and grass to retain the runoff. The loess was more stable, and during that time the Yellow River was cleaner, with little sediment load. For nearly 500 yr after 71 A. D. , the Yellow River entered the southwest Bohai Sea without any major course shifting or floods, and Chenier II developed on the northwest coast. During the Tang Dynasty (618—906 A. D.), people migrated to the west. The settlers cut the forests and cultivated the pastures. As a result the bare loess was eroded by runoff and wind, increasing the sediment load of the river. This caused the formation of a large Yellow River delta, changing the characteristics of Bohai Bay by forming mudflats along the coast.

In 1049 the Yellow River again shifted to the northwest of Bohai Bay, thereby terminating the accumulation of Chenier II and causing renewed progradation.

In the 9th to 11th centuries A. D. , the 580-km-long Fan Gong Dyke was built to protect the coast of North Jiangsu from tidal flooding (Ji 1980). After 1194 the Yellow River changed its course from the Bohai Sea to the Yellow Sea. This caused a rapid progradation of the North Jiangsu shoreline because of the increased supply of sediment. By 1453 a strip of land 15 km wide had been deposited seaward of the Fan Gong Dyke. Subsequently, during the next 7 centuries, the Yellow River delta prograded 90 km into the Yellow Sea creating a new area of 15 700 km^2. Since then the Fan Gong Dyke has been a road a long distance from the shore.

In Bohai Bay, Chenier I started to form when the Yellow River changed its course to the Yellow Sea (Wang 1964; Zhau et al. 1979). During this time Bohai Bay did not receive sediment, the mudflat suffered from wave erosion, and shell ridges were also formed along the coast.

In 1855 the Yellow River returned to the Bohai Sea again, forming mudflats adjacent to the shell ridges along the coast. It is possible to trace the development of the coast since that time by using the chenier as a marker. Thus, prograding areas (Fig. 2A), stable areas (Fig. 2B), and slightly eroded coast (Fig. 2C) can be distinguished.

This shift in the Yellow River also caused the retreat of the North Jiangsu coast due to wave erosion. The submarine delta of the Yellow River was reshaped by strong tidal currents to a system of Submarine sandy ridges and channels.

The dynamic processes of the natural environment are in a precarious equilibrium. Human activities such as cutting forests, cultivating grassland, and building dams may suddenly change this equilibrium and cause serious results. In China there are good examples of this kind of human interference. The historical examples of the control of the Yellow River by Wang Ching and the erosion of the Loess Plateau started during the Tang Dynasty have already been discussed.

There are also more recent examples of environmental modifications. In the early 1950s a series of dams was built in the estuaries of the North Jiangsu plain. The purpose of the project was to harness the tidal currents, and to use fresh river water for irrigation. However, the results were disastrous. All the river channels in front of the dams were silted up because of the reduced tidal circulation. Eventually, the dams had to be opened to wash away the silt and to maintain the depth of the river channel (Wang 1961).

At Sanmen Gorge a dam was built to separate the middle and lower reaches of the Yellow River. The Sanmen Reservoir rapidly accumulated silt, and changed the base level of the Wei River, which is a tributary to the Yellow River upstream of Sanmen Gorge and very important for irrigation of Shaanxi Province. As a result, the fertile soil of the Wei River plain and the lower reach plain of the Yellow River became salty. The water quality at the mouth of the Yellow River was also changed, influencing the migration route of prawns. Difficulties such as siltation and erosion were also encountered when a harbor was built on a mudflat (Wang, 1980).

There are also examples of successful environmental modification such as the planting of the big rice grass on mudflats to enhance the deposition of sediment. It is therefore very important to understand the nature of the environment before starting any construction along the coast. Human activity should not interfere with the natural processes.

Acknowledgements

I am indebted to Drs D. J. M. Piper and G. Vilks who critically reviewed the original manuscript and contributed many useful suggestions incorporated in the revised version. I wish particularly to thank Prof. Meie Ren and Da-Kuei Zhu for their encouragement and continual support of this study. Thanks are extended to my other Chinese colleagues for their contribution in field and laboratory work. I am grateful to Dr D. Forbes for critical comments on the manuscript.

References

JI, C. 1980. The building of the Fangong Embankment and its uses. *Fudan J.* (*Social Science*), Suppl. Ed. , August 1980: 56 - 61 (In Chinese with English abstract).

POSTMA, H. 1967. Sediment transport and sedimentation in the environment. *In* Lauff [ed.], Estuaries. AAAS Publication 83: 158 - 179.

PANG, J. , AND SI. SHUHENG. 1979. The estuary changes of Huanghe River l. Changes in modern time. *Oceanol. Limnol. Sinica*, 10(2): 136 - 141 (In Chinese with English abstract).

PANG, J. , AND SI. SHUHENG. 1980. Fluvial process of the Huanghe River Estuary. *Oceanol. Limnol. Sinica*, 10(4): 295 - 305 (In Chinese with English abstract).

REN, M. , AND C. ZSNG. 1980. Late Quaternary continental shelf of East China Sea. *Acta Oceanol. Sinica*, 2(2): 106 - 111.

SHEN, H. W. 1979. Some notes on the Yellow River. EOS 60: 545 - 546.

WANG, Y. 1961. The properties of China's silted plain coast and problems of seaport's construction. Postgraduate thesis, Departmem of Geology and Geography, Peking University. 1961: 2 (In Chinese).

WANG, Y. 1964. The Shell coast and ridges and the old coastlines of the west coast of the Bohai Bay. *Acta Scientiarum Naturalium universitati Nankinesis*, 8(3): 424 - 442 (In Chinese with English abscract).

WANG, Y. 1980. The coast of China. *Geosci. Can.* , 7(3): 109 - 113.

WANG, Y. , D. -K. ZHU, X. -H. GU, AND C. -C. CHUEI. 1964. The characteristics of mudflat of the west coast of Bohai Bay. Abstracts of theses in the Annual Symposium of Oceanographical and Limnological Society of China in 1963, Wahan, Scicnce Press. The Research on Siltation of Newport, (2): 49 - 63 (In Chinese).

ZHU, D. -K. 1980. The problems concerning the development and utilization of Jiangsu coast. Science Symposium of Nanking University, Nanking, May 20, 1980. (In Chinese).

ZHAU, X. , S. -S. GUANG, AND J. -W. ZHANG. 1979. The sea level fluctuation during last 20 000 years in East China. *Acta Oceanol. Sinica*, 1(2): 269 - 280 (In Chinese).

中国的潮滩*

一、引言

潮滩(tidal flat)是指淤泥质海岸潮间带浅滩。在国外主要分布在荷兰、法国等大西洋沿岸,英国沃什湾,北美芬地湾、加利福尼亚湾,南美的亚马逊河口、圭亚那沿岸等处。中国潮滩规模大,在世界上具有特殊地位,潮滩岸线总长约 4 000 km,主要可分为平原型和港湾型两类[1]。前者在大河入海平原沿岸,如辽东湾、渤海湾、江苏沿岸、杭州湾;后者分布在浙、闽、粤沿岸一些港湾内(图 1),是注入海湾的河流及涨潮带来的海域泥沙沉积形成的。

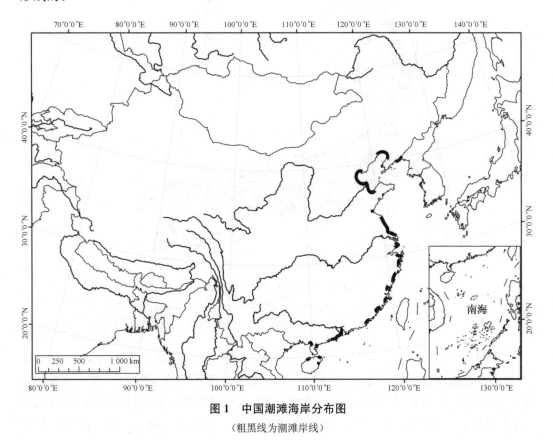

图 1　中国潮滩海岸分布图

(粗黑线为潮滩岸线)

潮滩是个重要的沉积带,是海陆相地层分界的重要标志层,它本身也是个生油环境,在石油地质上有重要意义。潮滩是人类活动频繁的地区,很多工程建设在潮滩环境或古

* 　王颖,朱大奎:《第四纪研究》,1990 年 12 月第 4 期,第 291 - 300 页。
国家自然科学基金资助项目。

代潮滩沉积上,所以研究潮滩沉积在理论上和实践上有很大的意义。

二、平原型潮滩

1. 渤海湾潮滩

渤海湾沿岸是黄河三角洲冲积平原和滦河三角洲冲积平原,湾顶在歧口,湾顶部分海岸坡度最为平缓,向两侧海河口及黄河三角洲坡度逐渐增大(图 2)。渤海湾波浪主要是风浪,常风向为 SW - NE,强风向为 N、NE,而强浪为 NE,7 级 NE 风时浑水带宽达 30 km。沿岸是不正规半日潮,平均潮差 1.65 m。涨潮流向 W、NW,落潮流向 NE、NEE。

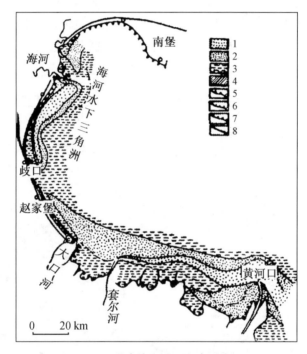

图 2 渤海湾潮滩沉积类型图

1. 粗粉砂 2. 细粉砂 3. 粉砂质黏土 4. 黏土质软泥

5. 淤进的岸线 6. 稳定的岸线 7. 蚀退的岸线 8. 贝壳堤

据渤海湾多次水文测验研究[2],潮滩上的水文特征是:① 涨潮流速大于落潮流速。② 最大流速出现在高潮与低潮之间的半潮面,余流主要受风的作用。③ 含沙量涨潮大于落潮;外海含沙量低,进入潮滩水边线(0 m 线)含沙量突然增高,且向高潮位继续增大(表 1 和表 2)。潮滩上沉积物搬运过程是:潮滩上刚上水时水层薄,流速较快,近底层紊动强,使滩上沉积物被扰动。大风天时滩上水呈泥浆状,所以风浪起着扰动掀沙作用,而潮流则将悬移质搬运,滩上的沉积物不仅从外海带来,还有很大一部分是滩面沉积物被风浪水流掀动,而随涨潮流向高潮位搬运。当潮位增高,水层增厚,底层水扰动减弱,含沙浓度也逐渐降低,至高潮时为最低值。落潮时因流速小于涨潮,历时长,含沙浓度较涨潮时小。这样,在高潮滩上发生沉积,使潮滩形成泥质沉积层。

表1　渤海湾驴驹河断面

测点(m)	涨潮		落潮		余流		输沙	
	流速(cm/s)	流向(°)	流速(cm/s)	流向(°)	流速(cm/s)	流向(°)	输沙率(t/m·24h)	方向(°)
+1.5	31	338	29	140			3.6	305
+0.2	35.4	279	29.3	97			6.0	267
-2	29	305	36	65	9	140	7.8	31
-5	41	268	47	82	5	71	3.0	60
-7	40	297	33	81				

表2　渤海湾歧口断面

测点(m)	涨潮		落潮		余流		含沙量(10^{-4})		输沙	
	流速(cm/s)	流向(°)	流速(cm/s)	流向(°)	流速(cm/s)	流向(°)	涨潮	落潮	输沙率(t/m·24h)	方向(°)
+2	40	271	21	95	0~1	7	10	8.04	1.2	238
±0	50	290	27	96	0~3	7	9.89	6.71	0.7	261
-2	58	268	33	84	3	7	2.99	2.64	1.6	261
-4	57	252	36	118	12	247	2.37	2.29	1.5	346
-5	57	260	35	108		247	0.79	0.71		

渤海湾潮滩自高潮位向低潮位可分出四个沉积带[2,3]：龟裂带、内淤积带、滩面冲刷带及外淤积带(表3)。该四个带春秋季节分带明显,夏季风浪大冲刷带受破坏物质粗化,SE风时有浮泥在滩面上堆积。冬季滩上多冰丘,解冻后成众多的浮泥堆。

表3　渤海湾潮滩沉积分带

沉积带	龟裂带	内淤积带	滩面冲刷带	外淤积带
位置	大潮高潮位—平均高潮位	小潮高潮位	中潮位	中潮—大潮低潮位
动力	潮流,蒸发,日晒	潮流,主要是涨潮流的作用	潮流,落潮流作用明显	潮流,波浪
地貌	龟裂的泥滩,盐沼湿地滩面坡度0.5‰~0.6‰	泥质潮滩,多浅凹地坡度0.67‰	长条形冲刷体及潮水沟冲刷体长轴与潮流流向平行	砂质潮滩波痕及流水波痕
沉积作用	基本稳定的沉积作用弱,仅在大潮高潮位时沉积,风暴潮时侵蚀	悬移质(泥)主要在此沉积一年滩面淤高5cm	涨潮流沉积作用落潮流沿潮水沟冲刷,形成冲刷体	潮流从外海向潮滩输沙,波浪从潮下带向低潮位输沙
沉积特征	黏土、黏土质淤泥水平纹层具龟裂纹、虫穴	黏土质淤泥水平纹层多虫穴	黏土与细粉砂互层水平纹层	细砂及粗粉砂各种交错层理

渤海湾潮滩沉积物主要来自黄河及海河,湾顶的歧口附近是它们的分界处。按粒度(见图2)黄河入海的泥沙成舌状向北伸到歧口附近,而海河口也有一舌状伸向歧口。黄河口沉积物为粗粉砂,套尔河以北为细粉砂、粉砂质黏土和黏土质软泥。海河物质较粗,主要是粉砂、极细砂。从矿物组成亦可分出黄河的和海河的两组,黄河泥沙重矿物含量低(占1%),磁铁矿少,赤铁矿多,稳定矿物(石榴石、锆石、电气石)及次生矿物(绿帘石、赤铁矿)增多,而不稳定矿物消失,外形上失去原有晶形,表面模糊,反映黄河物质经过长距离搬运,而海河物质来自燕山山地火山岩及变质岩系,来源较近,矿物特征相反。

根据渤海湾潮滩浅层钻探相连的剖面,潮滩沉积主要有两层:上层是泥质沉积,黏土与黏土质粉砂互层,呈水平层理,厚度约2 m;下层是粉砂与极细砂,厚数米。这反映了平面上从高潮位泥质龟裂带到低潮位粉砂、极细砂外淤积带的变化。

2. 江苏的潮滩

江苏淤泥质海岸线长883.6 km,历史上长江、黄河曾在苏北入海,带来巨量泥沙,江苏沿岸潮差较大(2~4 m),潮汐作用强,潮滩宽10~13 km,最宽36 km,平均坡度0.2‰(图3)。

沿岸受季风影响,冬季N、NE向波浪,夏季SE向,年平均波高0.6~1.2 m,台风时出现强浪。潮流以东台县弶港为中心,是南北两个潮流系统的汇合区,故潮差亦以弶港为最大。潮位决定了潮滩的高程,亦影响到潮滩的宽度。

在江苏潮滩,涨潮流速明显大于落潮流速,最大涨潮流速与最大含沙量同时出现,即在涨潮后1 h两者同时出现,而落潮最大含沙量则在落潮最大流速出现以后1h。潮流进入潮滩时大体与岸线呈20°~30°夹角,近岸边潮流与岸线夹角变大,至高潮位潮流方向大致与岸呈直角,即成为向岸与离岸的往复流。这样,在一个潮周期中泥沙做向岸净运动。在平行高潮位至小潮高潮位间含沙浓度最大,是沉积作用最强的部位[4]。

波浪在江苏潮滩也有较明显的影响,波浪扰动使潮滩水层含沙浓度增加,形成各种波痕和斜层理,而风暴天气波浪对潮滩侵蚀和沉积物的分选作用就更明显了,形成各种侵蚀形态及风暴沙层[5]。

江苏潮滩有四个沉积相带:① 草滩(湿地)沉积带;② 泥滩沉积带;③ 泥-粉砂滩沉积带;④ 粉砂-细砂滩沉积带。随着潮滩的发育,这四个带向海推进,向陆一侧的沉积就会依次覆盖在它向海一侧的沉积带上,当潮滩发育趋于成熟时就形成一个完整的潮滩沉积相序,其下部是砂质沉积层,上部为泥质沉积。下部粉砂、细砂是低潮位高能流态下的底移质沉积,位于低潮位至波浪作用的下界(海岸斜坡下界),沉积层厚15~20 m,这层是潮滩沉积的基础。砂质沉积以底移质运动,从水下向岸推进,使潮滩向海增加其宽度。上部泥质沉积是潮流搬运的悬移质,它使滩面增高。当滩面达到大潮高潮位以上就很少再被海水淹没,一般不再堆积加高,所以泥质沉积的最大厚度相当于中潮位至大潮高潮位(2~3 m)。这两者之间既有悬浮的泥质沉积又有底移的砂质沉积,在剖面上表现为泥层与砂层的交替沉积。

江苏潮滩的重矿物可分两个区:南部长江矿物组合为普通角闪石、磁铁矿、磷灰石、褐铁矿、胶磷石、锆石及岩屑,其中前三种是特征矿物;北部古黄河矿物组合为普通角闪石、绿帘石、褐铁矿、锆石、电气石、钛铁矿、石榴石,其中特征矿物有锆石、电气石、钛铁矿、石

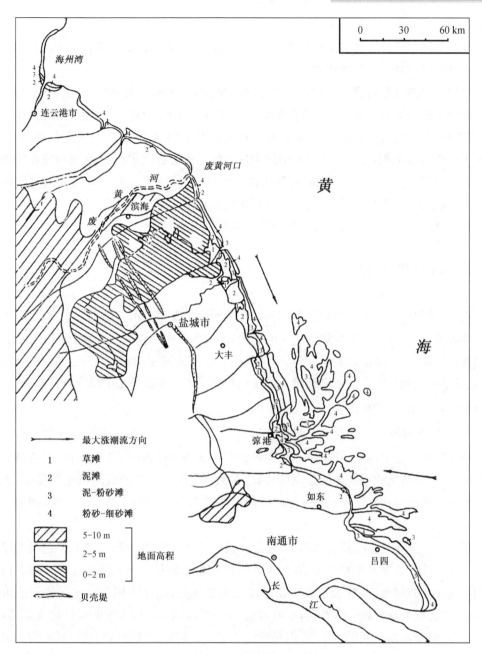

图 3　江苏潮滩图

榴石。重矿物含量由高潮位(含量 0.84%)向低潮位(5.1%)逐渐增高。

江苏潮滩中高潮位(草滩、泥滩)多草本花粉,向海方向草本花粉逐渐减少,至低潮位无草本花粉而出现木本花粉。这表明,草滩是当地的花粉,低潮位木本花粉是沿岸流从河流(长江)带来的。近岸(草滩、泥滩)物质未向海搬运,沉积物主要从海向岸搬运。微体化石分布也说明潮滩物质的向岸搬运过程。在泥滩带主要浅水相有有孔虫——卷转虫 *Ammonia*、希望虫 *Elphidium*、九字虫 *Nonion*,同时有浮游的、生活于水深>200 m 的抱球虫 *Globigerina*、圆球虫 *Orbulina*,占 10.3%,生活于水深>50 m 的箭头虫 *Bolivina*、五块虫 *Quinguloculina*,占 7.2%。深水种属与滨岸的混在一起,说明岸外海底沉积物向潮滩输送[6]。

江苏潮滩的成分反映了沉积物来源有三:长江泥沙、废黄河三角洲侵蚀产物以及岸外海底(辐射状沙洲区)的侵蚀物质。

平原型潮滩沉积特征 根据中国主要的平原潮滩(渤海湾、江苏海岸)的研究,可得出平原潮滩沉积的基本特征:① 潮滩在平面上有高潮位泥滩、中潮位泥砂混合滩、低潮位粉砂、细粉砂滩三个沉积带,相应地在剖面上具潮滩二元相结构:上层为泥质沉积,厚 2~3 m,下层为砂质沉积,厚 10~15 m,中间为过渡的泥砂混合沉积。② 潮滩组成比较均一稳定,包括粒度成分、化学组成、矿物及微体生物等,这反映平原型潮滩由大河供应沉积物,影响范围大,沉积持续时间长。③ 潮滩沉积体大,面积数十至数百千米。大多位于地质构造沉降区,可连续几个旋回,厚度为几十米至百多米。

三、港湾型潮滩

浙江、福建海岸沿着一些构造断陷盆地构造断裂带发育了不同类型的港湾。浙、闽沿岸大部分地区潮差较大,平均潮差在 4 m 以上,港湾区大多为强潮区,流速在 1 m 以上,而港湾深入内陆,湾内波浪微弱,均在 0.5 m 以下。长江冲淡水沿海岸南下会同浙、闽沿岸流,在冬季可达闽北海岸。强潮环境及适量细粒物质供应使港湾中潮滩发育。浙、闽沿岸潮滩岸线长约 1 500 km。

1. 乐清湾潮滩

浙江乐清湾是个构造断陷盆地,从湾口到湾顶长约 40 km,平均水深 10 m,港湾面积 250 km²。乐清湾是我国著名的强潮区,湾内最大潮差可达 8.34 m,潮流呈往复流,涨潮流速大于落潮流速;含沙量 0.4 kg/m³,湾内潮滩宽 1.8 km,坡度 1‰~1.4‰,潮滩面积 134.8 km²①。

乐清湾潮滩主要由细粒物质组成,表层灰黄色淤泥 20~40 cm,下部灰黑色淤泥厚数米至 10 m,但底部为河流相的亚黏土砂砾沉积。潮滩的粒度组成非常一致,中值粒径 8.73φ,属黏土粒级范围,从表层到下部粒度无明显变化,层理不明显。重矿物组成可分两类:湾顶区域是铁矿物—帘石类,前者含量 88.9%,后者 5.3%,该区重矿物含量高,但粒径较小,是受湾顶小河供砂影响,但影响范围不大。其他区域为角闪石—铁矿物—帘石类,其组成相同,反映乐清湾大部分区域有统一的物质来源,是潮流搬运的海域来砂。

2. 沙埕港潮滩

福建省福鼎县的沙埕港是一狭长形的港湾,纵深 55.5 km,宽约 1.85 km,水深大多在 15 m 以上,面积 111.23 km²。湾内岛屿面积 6.05 km²,潮滩面积 53.81 km²(图 4)。该处平均潮差 4.10 m,最大潮差 6.90 m,潮流在湾内为往复流,流速 1~1.5 m/s,最大 2 m/s,落潮流速大于涨潮流速。湾内地形封闭狭长,风浪很小,8~10 级台风过境湾内波浪也仅 0.5 m。

---

① 吴小根,1988,浙闽港湾潮滩沉积。南京大学理学硕士论文。

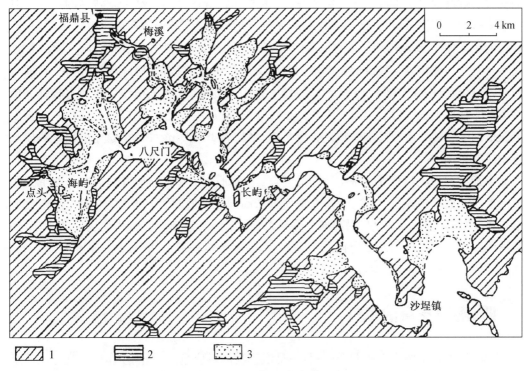

图 4　福建沙埕港潮滩图

1. 基岩山地　2. 冲积平原　3. 潮滩

沙埕港潮滩分布于湾顶及两侧小湾汊,不连续,零散分布,一般宽度不足 1 km,坡度较大(5‰～10‰),潮滩表层为 30 cm 厚的黄色淤泥,下部为厚数米的青灰色淤泥,底部为河流相砂砾沉积。湾内各处潮滩的沉积物组成有含砂砾质的和淤泥质的两类。

含砂砾质的潮滩,分布于湾顶及港湾两侧沟口,砂含量 4.6%～65%,黏土 22.5%～66%,细砾 9.3%,重矿物含量亦高(236 mg/100 g)。沿岸短小的溪沟注入,在沟口堆积为扇形地,再有潮流供应的细粒物质,组成面积较小、零散分布的含砂砾潮滩、淤泥质潮滩,黏土粒级 57.5%～65%,粉砂 35%～42.5%,重矿物含量低(7.3 mg/100 g),一般不受沿岸溪沟供沙的影响。从沙埕港湾口、湾外潮滩到湾顶,物质有变细的趋势。两组潮滩的重矿物组成相似,含铁矿物(62%)、帘石类(22.8%)、白钛石(11%)和锆石等,均有自生黄铁矿。这些反映沙埕港潮滩细粒物质是潮流搬运的海域来砂,影响遍及整个港湾,但量不大,目前港湾潮滩堆积作用甚微。

3. 厦门湾潮滩

厦门湾是断陷沉溺的港湾,地形比较开阔,湾内风浪作用显著。该处为正规半日潮,平均潮差 4 m,最大潮差 6.96 m,落潮流速(0.89 m/s)大于涨潮流速(0.77 m/s)。湾内潮滩面积 87.3 km²[①]。厦门湾潮滩宽度数百米至 1～2 km,坡度 1‰～2‰,滩面多潮水沟。潮滩沉积在垂向上有明显的韵律变化,出现薄层的泥砂互层,层理明显,而波痕流痕形成

———————————

① 系指浔江、西港目前潮滩的面积。

的交错层理尚少见。潮滩的粒度组成复杂,主要有砂、粉砂质砂、黏土质砂、砂质黏土、黏土质粉砂,在北部侵蚀岸的潮滩上还有砾质砂。但砂分布广,砂质在潮滩中占重要组分。粒度的频率曲线都呈多峰态,反映了物质多种来源,有北部台地侵蚀产物、湾底浅滩侵蚀改造及沿岸河流供砂。粒度的概率累积曲线上悬移组分和跃移组分所占比例相当,成分较粗,悬移组分中含有细砂成分,跃移组分中以中砂为主,反映沉积环境中动力较强。厦门湾潮滩剖面上部(高潮位)受波浪作用,下部(低潮位)主要受潮流作用,形成横向上粒度上粗下细的特征。厦门湾潮滩重矿物含量高,平均为 151.2 mg/100 g,这是受沿岸物质就近供应的影响。

港湾型潮滩沉积特征 基本特征有三点:① 物质组成单一,全是淤泥,在剖面上无明显的上层泥质沉积、下层砂质沉积;在平面上从高潮位到低潮位无明显的相带(泥滩、泥砂混合滩及粉砂-细砂滩)。② 局部成分分异大,受当地河流、海岸侵蚀供应物质的影响,使潮滩沉积物在短距离内发生变化。③ 沉积体较小,厚度一般为几米,少数厚度大于 10 m,面积一般为几平方千米,少数为几十平方千米。

四、潮滩发育的条件

潮滩的发育主要决定于两个因素:一是沉积物的供应量,二是潮汐作用。凡是泥沙供应量丰富、潮差大的区域,潮滩发育就宽广;当沉积物供应量减少或消失,波浪和潮流的侵蚀将使潮滩缩小。因此,沉积物供应量成为潮滩发育的重要因素,潮汐作用的差异构成了潮滩与潮滩沉积结构上的差异。

海岸带沉积物来源主要有河流的、海底与海岸侵蚀的产物,而海底的泥沙大多是过去的河流供应物质。

黄河年输沙量 11×10^8 t,入海泥沙 64% 在河口堆积,使三角洲岸线以平均每年 50 m 的速度淤长,其余 36% 属沿岸运动,使渤海湾、莱州湾泥沙供应丰富,潮滩发育。随着黄河多次注入,渤海湾沿岸普遍受到黄河泥沙的供应,成为我国主要的潮滩分布区。历史时期中,黄河在山东半岛两侧摆动,1128—1855 年在江苏注入黄海,黄河泥沙使江苏岸线迅速淤涨,当时废黄河口向海伸展 90 km,陆地面积增长 16 000 km²。1855 年黄河北归,注入渤海,江苏海岸失去巨量泥沙供应,废黄河三角洲潮滩受到侵蚀,岸线迅速后退。因此,渤海湾和江苏潮滩都是受河流泥沙供应量影响的典型地区。

江苏潮滩还受到海底泥沙供应的影响,江苏岸外分布着一片辐射状沙洲区,南北长 200 km,东西宽 90 km,水深 0~25 m,由 70 多条沙脊与深槽相间形成,它是晚更新世时长江在江苏入海的三角洲于海面上升后受潮流改造而成。目前受沙洲区掩护的江苏潮滩每年淤高 2~10 cm,增宽 20~100 m,按 1980—1984 年三次测验计算[1],潮滩的堆积量为 7.7×10^8 t/a,而现代长江入海泥沙沿岸北上为 0.35×10^8 t/a,废黄河三角洲侵蚀物质沿岸南下为 1.09×10^8 t/a,外海进入沙洲区的泥沙为 0.32×10^8 t/a。因此,江苏潮滩每年淤积量中有 76%(5.84×10^8 t)来自岸外沙洲区海底侵蚀产物。

---

① 傅命佐,1984,江苏岸外辐射沙脊群的形成、演变与发展趋势。南京大学理学硕士论文。

　　浙、闽港湾潮滩的沉积特征也反映了河流及海域来沙都是潮滩发育的重要因素。长江悬移质泥沙及浙、闽沿岸入海河流的悬移质沿岸向南运移,成为浙、闽港湾沉积物的主要来源。按物质组成,长江入海泥沙影响到福建北部沙埕港,而厦门湾的潮滩主要受当地河流及风化侵蚀物质影响。

　　潮汐作用影响到潮滩规模和沉积特征。

　　在物质供应充分的条件下,潮滩宽度与潮差大小相适应。潮汐大小、潮汛的潮位变化将影响到潮滩沉积的结构[7]。如图5,A点为大潮汛时砂质沉积的上界,B点为小潮汛时砂质沉积的上界,在一次从小潮—大潮—小潮的周期中,将在潮滩中留下一个向岸尖灭的砂质薄层。大小潮周期的多次重复,在A、B之间潮滩剖面上形成泥-砂混合沉积层。在渤海湾大潮汛与小潮汛的水位差为1.91 m,这将使A与B的水平距离为1 500 m,整个潮滩相带发生位移。江苏潮滩大小潮的水位差,将使沉积相带水平位移1 900 m,潮滩沉积剖面复杂化。大潮的流速是小潮的2倍,在江苏潮滩大潮汛沉积物

$$Md=5.4\phi$$

小潮汛为6.3ϕ,这使泥滩带出现黏土与细粉砂的纹层[8]。

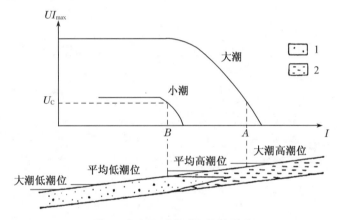

图5　大小潮对潮滩沉积的影响

1. 砂质沉积　2. 泥质沉积

UI_{max}. 大潮涨潮流速　U_c. 小潮涨潮流速　I. 与低潮位的距离

　　在渤海湾及江苏潮滩上,涨潮流向岸过程中潮流方向变化,在低潮位与岸交角20°～30°,中潮位40°,至泥滩带70°～80°,以致大体与岸垂直。在英国沃什湾潮滩也有相似现象[9]。潮流方向及流速的变化影响到含沙浓度等变化,使潮滩产生分带出现泥质的、泥砂混合的及砂质的三个沉积带[7]。

　　潮滩的发育是受到沿岸陆地及海洋环境条件的影响,潮滩沉积的结构特征也就是海岸环境变动的反映,对潮滩环境与沉积的综合研究将有助于对现代潮滩的利用及古潮滩沉积的识别。

参考文献

[1] 朱大奎. 1986. 中国海涂资源的开发利用问题. 地理科学,6(1):34-40.

［2］Wang Ying. 1983. The Mudflat System of China. *Can. J. Fish. Aguat. Sci.*，40(1)：160-171.

［3］Wang Ying and Ke Xiankun. 1989. Cheniers on the East Coastal Plain of China. *Marine Geology*，9 (4)：321-335.

［4］朱大奎,许廷官. 1982. 江苏中部海岸发育和开发利用问题. 南京大学学报(自然科学版),(3)：799-818.

［5］任美锷,杨巨海,章大初. 1983. 风暴潮对淤泥质海岸的影响. 海洋地质与第四纪地质,3(4)：1-24.

［6］朱大奎,柯贤坤,高抒. 1987. 江苏潮滩沉积研究. 黄渤海海洋,4(3)：19-28.

［7］朱大奎,高抒. 1985. 潮滩地貌与沉积的数学模型. 海洋学报,4(5)：15-21.

［8］任美锷,张忍顺,杨巨海. 1984. 江苏王港地区淤泥质潮滩的沉积作用. 海洋学报,3(1)：40-54.

［9］Evans，G. 1965. Intertidal Flat Sediments and Their Environments of Deposition in the Wash. *Quart. J. Geol. Soc.*，121(1～4)：209-245.

浙闽港湾潮滩与沉积的组合特征*

中国潮滩主要可分为平原型与港湾型两大类[1][2]。平原型分布于大河三角洲平原外围,如渤海湾及江苏沿海等。港湾型潮滩相对地规模小,类型复杂,主要分布于华东、华南基岩丘陵地区的港湾内。本文以浙江、福建沿海五个港湾的实地考察研究,来讨论港湾潮滩的发育条件与沉积特点,并按动力条件与沉积类型划分为三个亚类型,作进一步的叙述讨论。

一、港湾型潮滩发育条件

浙闽港湾大多属潮汐汊道型,其潮滩发育受湾内动力条件控制,同时亦受湾外海域强潮环境和沿岸输沙的影响。

(一) 湾外海域的强潮环境

浙闽岸外的东海海域是强潮海区,多属正规半日潮,潮差大,潮流速亦强,据实测,象山港至崇武岸外海域潮流流速都在 100 cm/s 以上,近底层潮流速亦达 30 cm/s[3],该流速值可以搬运粉砂级物质(表 1)。

表 1 浙闽沿岸潮差变化

	测　站						
潮差	海门	璇门港	龙湾	瑞安	沙埕	崇武	厦门
平均潮差(m)	4.01	5.15	4.50	4.44	4.13	4.27	3.98
最大潮差(m)	6.30	8.43	7.21	6.39	5.28	5.12	4.95

(二) 岸外沿岸流输沙

浙闽港湾岸外东海海域有三大流系,即长江冲淡水、浙闽沿岸流和台湾暖流[4],它们消长的季节变化直接控制了浙闽沿岸泥沙运移的数量和方向。

紧邻浙闽沿岸北部入海的长江,其年径流量为 9 320×10⁸ m³,年输沙量 4.61×10⁸ t,此外有 2×10⁸ t 溶解物质注入东海,长江入海泥沙有 20%～30% 被浙闽沿岸流挟带南下[5],受台湾暖流和浙闽沿岸流季节性消长影响,长江泥沙向南,夏季一般至象山港,冬季可达福建北部沿海[6]。

* 王颖,吴小根:《南京大学学报(地理学专辑)》,1991 年第 12 期,第 1-9 页。
高等学校博士学科点专项科研基金资助课题。

（三）湾内低能环境

潮流、波浪和注入湾内的河流是浙闽港湾三种动力要素。受 X 型断裂构造影响，浙闽港湾岸线曲折，港湾深入陆地，岸外岛屿罗列。这种地形限制了湾外海域强潮动力和波浪的入侵，加之缺乏大河注入，使湾内水动力较弱。

潮流是浙闽港湾主要的动力，外海潮波传入港湾，受港湾和浅滩地形影响，使流速从湾口向湾顶减小，湾内一般是涨潮历时大于落潮历时，涨潮平均流速小于落潮平均流速。港湾潮汐周期受外海潮波控制，属胁迫振动，湾内潮波的反映使潮差自湾口向顶逐渐增大（表 2 和表 3）。

表 2　乐清湾的潮流特征

潮流特征				
位置	$\bar{V}_f(\text{m/s})$	$\bar{V}_e(\text{m/s})$	$t_f(\text{h:m})$	$t_e(\text{h:m})$
湾口	0.59	0.63	7:00	5:12
湾顶	0.36	0.69	6:30	5:24

$\bar{V}_f$—涨潮平均流速；$\bar{V}_e$—落潮平均流速；t_f—涨潮历时；t_e—落潮历时。

表 3　乐清湾潮差

站　名	湾　口	华秋洞	清　江	东门村
离湾口距离（km）	0	23	30	40
平均潮差（m）	4.20	4.54	4.59	5.0
同湾口潮差比	1.00	1.09	1.10	1.19

影响浙闽港湾的波浪主要是风浪。受亚热带季风气候影响，浙闽沿海风向的季节性变化明显，春夏多东南风，秋冬偏北风。台风和寒潮形成本区主要的大风。沿海大陈站实测多年平均波高 1.2 m，最大浪高可达 14.4 m，当外海波浪传入港湾时，受地形屏蔽，使湾内波浪明显衰减。而港湾内水域吹程小，难以生成大的风浪。因此港湾内波浪作用小，平均波高都在 0.5 m 以下。

浙闽港湾沿岸注入的河流均属中小型山溪性河流，源短流急，集水面积小，径流量小，含沙量低，年内枯洪变化大。如乐清湾有 30 多条小河注入，集水面积 1 280 km²，注入的年径流量 10.3×10^8 m³，而乐清湾一个潮的纳潮量可达 19.3×10^8 m³。故注入湾内河流水量与潮流量相比，不足 1%。

（四）港湾型潮滩的类型

潮流是浙闽港湾主要的动力因素，波浪及河流为次要因素。由于潮流、波浪及河流的组合关系在不同的港湾具有差异性，使相应的潮滩沉积也具有差异的特征，据此，本文将港湾潮滩分出三种亚类型：

（1）潮流型港湾潮滩，潮流作用占绝对优势，波浪和河流作用都很微弱。

（2）河流型港湾潮滩，潮流作用仍占主导，但河流作用相对较强，其沉积中反映潮流与河流交互作用的特点。

（3）波浪型港湾潮滩，潮滩作用为主，而季节性波浪作用亦强，其沉积具有潮流与波浪作用的特征。

二、潮流型港湾潮滩沉积

浙江的乐清湾瓯江南岸及福建泉州湾属于潮流型港湾潮滩。

乐清湾三面是丘陵低山，冰后期海面上升所形成的海湾伸入陆地 40 多 km，口外有大门、小门、鹿西诸岛为屏障。泉州湾口外有大坠、小坠等岛礁，湾口有拦门沙浅滩。这两个港湾的封闭性较好，湾内潮流占绝对优势。如乐清湾，河流径流量仅占纳潮量的 0.15%（图 1 和图 2），湾内平均潮流流速为 0.5 m/s。

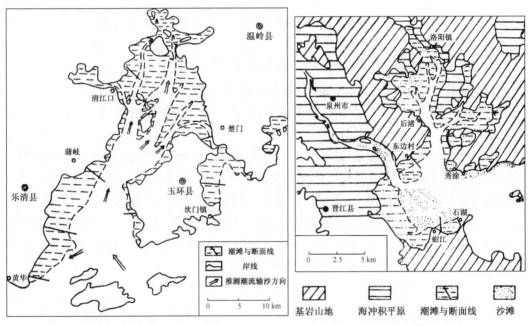

图 1　乐清湾略图　　　　　　　　　　图 2　泉州湾潮滩分布图

乐清湾潮滩沉积物黏土级含量 42%～75%，大多数在 60% 以上，粉砂级含量 25%～56.8%，绝大多数是 6～8Φ 的细粉砂，以及少量极细砂（图 3），其粒度频率曲线呈双峰型。中值粒径（MD）是反映沉积物介质平均动能情况，乐清湾沉积物中值粒径为 8.7Φ（7.8～9.5Φ），处黏土级，属低能环境。潮滩沉积物中的重矿物含量普遍很低，每 100 g 干样中含量 17 mg，但多处发现有自生黄铁矿，这些皆表明为低能环境的特征。

泉州湾潮滩沉积物以黏土质粉砂为主，中值粒径为 6.2～7.1Φ，粉砂含量 49%～61%，黏土含量比较稳定，为 33.5%～36%，重矿物含量亦低，为 62.7 mg（0.63%），且普遍含有自生黄铁矿。

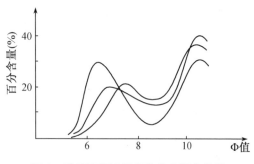

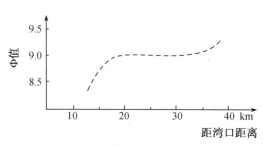

图 3 乐清湾潮滩沉积物粒度频率曲线　　　　图 4 乐清湾潮滩沉积物

这类细颗粒潮滩沉积物与当地港湾海岸所产生的泥沙迥异,主要来自湾外海域,两湾的潮滩沉积粒度,均具有自湾口向湾顶递减的特征(图4)。在乐清湾的玉环岛,1977年修建了促淤蓄淡的璇门港堵口工程,至 1987 年,海域来泥沙在璇门拦海堤外淤积厚度达 36 m,使堤外水道淤浅消失,原水道两侧的潮滩已连成一片,而堤内侧水道却仍存在,淤积甚微,这是湾外海域来沙的例证。

由于单一潮流作用,动力较弱,物质细,故这类潮滩沉积构造单一,粒径变化不明显,缺乏明显的层理构造。乐清湾潮滩沉积层可分两层:上层为灰黄色淤泥,层厚自湾口向湾内变薄(20～10 cm),下层为青灰色淤泥,厚度超过 5 m,二层的粒度组成一致,为长期稳定的堆积环境。

温州湾南岸的潮滩属潮流型,由于位置朝向开阔海域,潮流作用与波浪作用叠加的影响作用均稍有增强,但潮滩仍处于缓慢的淤长状态,大约每年向海伸展 10 m。潮滩沉积具有轻微的分异特征:高潮位与低潮位的潮滩沉积物细,是粉砂质黏土或黏土质粉砂(中值粒径 8.3Φ、7.5Φ);而中潮位潮滩沉积物稍粗,为粗粉砂或砂质粉砂(4.3Φ),中潮位流速大,是最大含砂量分布带。沉积剖面表现为砂质黏土,夹有分散的粉砂质透镜层,表层偶有黏土质浮泥(图5)[7]。

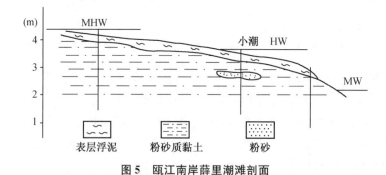

图 5 瓯江南岸薛里潮滩剖面

三、河流型港湾潮滩沉积

福建省沙埕港是狭长弯曲的溺谷型港湾,港湾长度超过 55 km,水域面积 111.2 km²。口门宽 1.5 km,湾口有岛屿为屏障(图6),湾内平均潮差 4.0 m,纳潮量 4.6×10⁸ m³,落潮流速(1.5 m/s)大于涨潮流速(0.5 m/s)。沙埕港沿岸多山溪性小河注入港湾,据水北溪高滩站水文资料,该站集水面积341 km²,1970—1979年,年平均径流深度为1 245.77 mm,年侵蚀模数为196.5 t/km²。注入沙埕港水系总集水面积1 128.8 km²,按此推算,沙埕港年平均径流为140×10⁸ m³,年平均输沙量为22.1×10⁴ t。

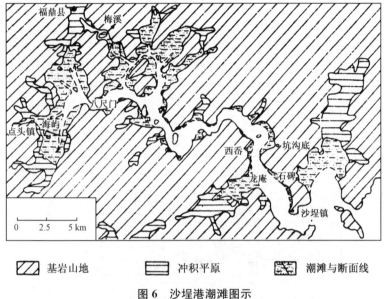

图6　沙埕港潮滩图示

沙埕港口门狭窄,湾长而曲折,似为海水拓宽的河谷,故港湾内作用微弱,即使8～10级台风影响,湾内波高也仅0.5 m。特殊港湾地形也限制了外海潮波向湾内的传播,造成了湾内潮流强度的显著衰弱,因而河流作用相对地突出,使沙埕港具有河流型港湾潮滩的发育条件。

沙埕港潮滩沉积具有淤泥质和砂砾质双重特性,淤泥质主要分布在低潮水边线附近,黏土含量为57.5%～65%,粉砂35%～43%。中值粒径8.5～9.4Φ,重矿物含量7.3 mg/100 g,从湾口向湾顶淤泥质潮滩的粒度逐渐变细,反映了其物质来自湾外海域,是由潮流向湾内搬运而堆积的,不受注入海湾的小河的影响。砂砾质分布在各小河河口,呈现为山溪河流沉积物叠加混合在潮滩沉积物上,其砂砾成分46%～65%,黏土含量22%～62%,重矿物含量236 mg/100 g,其粒度组成具有河流相的特征(图7)。

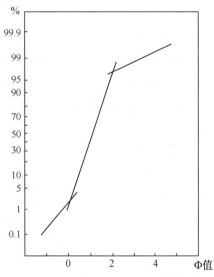

图7　梅溪断面潮滩沉积概率累积曲线

受山溪河流季节性变化影响,枯水季节,山溪河流作用微弱或停顿,潮滩主要接受湾外海域细粒悬浮物质,洪水季节则接受河流输沙而改变上部潮滩沉积特征,结果形成低潮位潮滩是潮流作用的黏土沉积,向中潮位及高潮位渐过渡为河流沉积与潮流沉积混合的砂砾质黏土。

四、波浪型港湾潮滩沉积

厦门湾由九龙江河口湾、厦门西港和浔江三部分海域组成。杏集、高集海堤使厦门西港和浔江海域成为单口式的港湾(图8)[8]。

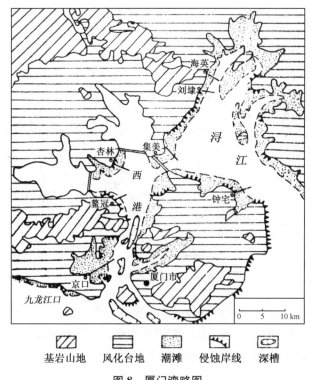

基岩山地　风化台地　潮滩　侵蚀岸线　深槽

图8　厦门湾略图

厦门港为半日潮,平均潮差 3.98 m,进入港湾的潮波受地形影响及海岸的反射波作用而形成驻波,湾内涨落潮流均较弱。如西港口门涨潮平均流速,大潮汛时为 29～44 cm/s,小潮汛时 13～14 cm/s。落潮平均流速,大潮汛时 22～28 cm/s,小潮汛时 13～18 cm/s,但湾内风浪作用显著,出现频率88%,平均波高 0.2～0.4 m,平均最大波高 1 m,台风和寒潮影响下波高可达 1.6 m。当4～5级的风与高潮叠加时,使沿岸红土台地遭受显著冲刷与蚀退,湾内-2 m 水深以内的海底浅滩亦受到波浪的扰动与侵蚀[9]注入西港及浔江湾海域的河流较小,主要的河流是浔江湾的西溪,其年径流量 $3.97×10^8$ m^3,年输沙量 $6×10^4$ t,河流对厦门湾的作用甚小。

厦门湾潮滩沉积物有砂、粉砂质砂、黏土质砂、黏土质粉砂和粉砂黏土五类。粒级范围宽,含有-1Φ 至 9Φ 各粒级组分,粒度频率的曲线上具多峰的特点(图9),沉积物的分选与组成具有局部差异的特征,重矿物含量高,为 151.2 mg/100 g。各粒级组成中,砂分布最广,在调查的 7 条断面中均有砂分布,其他粒级沉积物因地而异。这些特征表明,厦门湾潮滩沉积物的多源性,来自湾外海域,沿岸小河及海蚀产物。

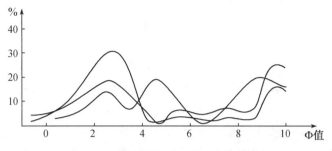

图9 厦门湾潮滩沉积物粒度频率曲线

厦门湾潮滩坡度 1‰～2‰,滩面多潮水沟及侵蚀洼地。在垂直剖面上,物质成分和颜色有明显的韵律变化,泥与砂质沉积成薄层互层,泥质沉积物(粉砂黏土、黏土质粉砂)是潮流沉积,在风浪平静季节,整个厦门湾外处于潮汐作用环境,接受湾外海域细粒物质,台风及冬季风暴季节,海岸及近岸浅滩受侵蚀改造,向潮滩供应砂质沉积,构成沉积物粒级比较宽,形成砂质与泥质互层的特征。

五、结论

(1)港湾型潮滩是在低能的潮汐港湾环境下形成的,但湾外的强潮动力与大河悬移质泥沙原是港湾型潮滩发育的必要条件。长江悬移质泥沙以及浙闽沿岸入海河流悬移质泥沙受沿岸流搬运向南运移扩散,又为潮流再搬运到湾内沉积,是浙闽潮滩发育的重要因素。它改变了原始的基岩港湾岩滩的特征,而叠置发育了宽阔的淤泥质潮间带滩涂。

(2)各海湾潮滩沉积物粒度与成分表明,源于长江的悬移质泥沙,被潮流搬运的间接影响,可达福建的沙埕港,虽然该湾已具有当地沙源的表现。沙埕港以南,各海湾以当地海域的泥沙来源为主,长江悬移质泥沙影响反映甚微。

(3)动力条件对沉积环境具有决定性影响,浙闽港湾主要受潮流作用控制,同时也受到波浪与河流作用的影响。根据潮流、波浪与河流三者关系的差异,浙闽港湾潮滩可分为三种亚类型:① 潮流型;② 河流型;③ 波浪型。潮流型港湾潮滩是正常潮流沉积环境产物,为淤泥质,其沉积物与潮流动力条件相适应,根据港湾开敞程度,又可分为隐蔽的与开敞的两种,前者沉积物组成一致,沉积构造单一,以块状层理为主,沉积环境长期稳定。后者沉积构造主要是水平层理和粉砂透镜体,滩面有分带性,垂直层序是下粗上细的"双向"结构。

河流型港湾潮滩与波浪型港湾潮滩,这两类沉积环境具有季节性变化特点。前者是正常的潮流沉积叠加了山溪性河流作用的产物,由于洪水期河流作用的影响,局部港湾岸段近岸潮滩沉积物粗化,甚至保留了河流相沉积特征,这与其所处潮流的沉积环境不相适应,可能反映古地貌的局部制约影响。后者是正常的潮流沉积叠加了季节性波浪作用的产物,在波浪作用有效范围的近岸段,潮滩沉积物粗而粒级分配宽,并形成自岸向海粒度变细的滩面结构,具潮滩(Tidal flat)与海滩(Beach)混合特征或过渡型的结构。

参考文献

［1］朱大奎.1986.中国海涂资源的开发利用问题.地理科学,(1):34-39.

［2］Wang Ying. 1983. The Mudflat System of China. *Canada J. F. Aguat Sci.*, 40(1):160-171.

［3］孙湘平.1981.中国沿海海洋水文气象概况.北京:科学出版社.

［4］袁耀初等.1982.中国海洋陆架环流的单层模式.海洋学报,4(1):1-15.

［5］Wang Ying. 1986. Sediment Supply to the Continental Shelf by the Major River of China. *Journal of the Geological Society*, London, 143:935-944.

［6］蔡爱智.1982.长江入海泥沙的扩散.海洋学报,4(1):78-88.

［7］王宝灿等.1981.温州海区海岸带和潮间带基本特征.见:海岸带温州试点区报告.上海:华东师大出版社.161-190.

［8］王寿景.1987.厦门港湾海洋环境综合调查报告(1).台湾海峡,6(4):349-357.

［9］刘维坤等.1984.厦门港海底地貌及其冲淤变化.台湾海峡,3(2):179-188.

Tidal Flats in China [*]

Ⅰ. INTRODUCTION

Tidal flats in China occur on a larger scale and are associated with two different types of coast. The total length of coastline represented by tidal flat is about 4 000 km long. The tidal flat coast is dominated by tidal processes and is often associated with a large sediment supply.

Major tidal flats in China are developed along the fringing zone of the North China plain where tidal range is about 3 m and the velocity of tidal currents is less than that found in the estuaries. The tidal flats are between 3 to 18 km in width and are much more extensive than in estuarine areas. Because the slope of the coast zone is very gentle (the gradient is less than 1/1 000), wave action is restricted to offshore areas and tidal currents are the dominant dynamic processes on the flats. The tidal flats along the East China Sea and the South China Sea are developed in narrow embayments or other types of sheltered coasts. Fetch within such environments limits the wind/wave interaction. Hence, tidal currents are, once again, the major agents controlling sedimentary processes. Thus, major factors which control the development of tidal flats are tidal dynamics, the gentle coastal slope, and the abundance of fine-grained sedimentary material.

Silt is the major constituent of the deposits, ranging from fine to coarse silt. Very fine sands and clays are also present on most tidal flats. All such sediments are carried onto the flats mainly by the flood tidal current which brings material eroded from the adjacent seabed; this process infers a landward transport. Sediments on tidal flats in China, being fluvial in origin, have been supplied to the coastline during times of lowered sea level by large rivers, such as the Yellow River and the Changjiang (Yangtze) River. In contrast, most of sediments on the tidal flats along the Atlantic coast were laid down originally as offshore glacial deposits during lower sea level stands of the last glaciation.

Tidal currents carry sediments in traction, saltation, and in suspension towards the land. Velocities decrease because of friction from the intertidal zone seabed (Wang, 1963, 1983; Evans, 1965). Thus, sediment grain size decreases on intertidal flats from

---

[*] Ying Wang, Dakui Zhu; Zhou Di et al. (eds), *Oceanology of China Seas*, Vol. 2, pp. 445 – 456, Kluwer Academic Publishers, 1994.

low water level towards high water level, and in the vicinity of deep waters. Generally, flood tides attain higher velocities and have shorter duration than the corresponding ebb. Furthermore, the extensive accretion on flats suggests that flood-tide deposition has been dominant over ebb-tide erosion. Even though the flood tide may be of longer duration, given the additional energy from waves, the flood tide is still dominant. Micro-morphological features, such as ripples and small creeks, exposed on intertidal flats during the passing of the ebb tide, reflect the last phase of the ebb current.

Morphological, sedimentary, and benthic faunal zonations are typical features on intertidal flats everywhere; which are a reflection of changes in tidal dynamics. Zonations have been recorded in the sediment as ripple bedding, polygon cracks, a mosaic of sandy coarse bedding,or as lenticular and other structures that are intervened with primary structures of fine lamination and delicate bedding. Sedimentary facies can be used to identify sedimentary environment of ancient deposits.

II. CLASSIFICATION AND REGIONAL SETTING

Tidal flats in China can be classified into two major types: the plain type and the embayment type, depending on their coastal setting. The plain type is developed on a large scale and is distributed in lower reaches of large rivers and river deltas or along the edge of coastal plains. Tidal flats associated with plains in the Bohai Sea are the head of the Liaodong Bay near the mouth of the Liaohe River, and the Bohai Bay and the Laizhou Bay in the vicinity of the modern Yellow River delta. The coastal plain along the Yellow Sea is near the North Jiangsu coast where the ancient deltas of the Yellow and Changjiang rivers occurred. In the East China Sea, the coastal plain of the Changjiang River delta and the estuary of the Hangzhou Bay are under the influence of fluvial sediment influx. The embayment type is developed in long, narrow, and rocky tidal inlets along the coasts of the East China Sea and South China Sea. For example, the embayments in Zhejiang, Fujian, Guangdong, Hainan provinces (Fig. 1). These embayments may have local sediments from small rivers or from coastal erosion, but main agents of sediment delivery are flood tides which erode the offshore sea-bed. In South China, under tropical and subtropical climate conditions, most bay head environments and shallow waters formed behind sand bars are dotted by mangrove forests. The mangroves have trapped particles of silt and clay-size through the filtering action of their well-developed root systems. The associated muddy flats have special features and unique sedimentary dynamics.

Tidal flats can be classified also into three types according to their characteristic morphology and sedimentary features. This approach is especially important for coastal plain flats that face the open sea,because the classification shows the different levels of

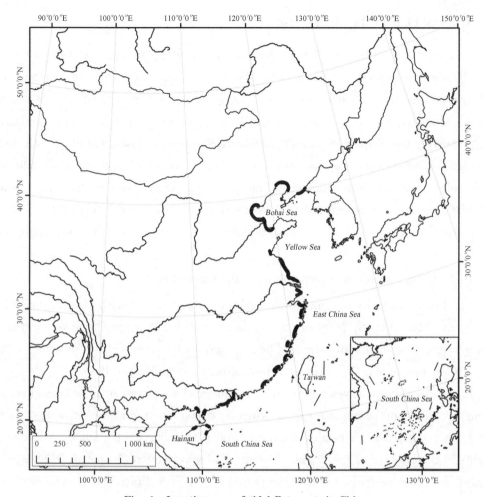

Fig. 1 Location map of tidal flat coasts in China

wave energy input in addition to the predominating effects of tidal dynamics (Wang, Collins, and Zhu 1990). Coastal plain tidal flats are divided into: i) Sand flat. The Wash of East England is an example of an intertidal flat with high wave energy influences. In China, a similar type of flat appears in the Xiamen Bay, Fujian Province, ii) Silt flat. This well developed flat occurs along the northern part of Jiangsu plain coastline facing the Yellow Sea. A medium level of wave energy stirs up sand and silt material offshore, and then transports the material onshore through the action of flood currents, iii) Mudflat. This type is exemplified by the tidal flats of the Bohai Bay. Tidal currents are the only predominant transport agents with minimal wave influences, but with a large source of clay and fine silt size material supplied by the Yellow River. Only this kind of flat has a "Fluid mud" zone that marks the mean high tide level. An abnormal type of boulder flat is developed in the Arctic environment because of frequent of floating ice that is carried onto the flats by storms and tidal currents (Dale, 1985).

A. Tidal Flats of the Plain Coast

This type of tidal flats is the most important one along the Yellow Sea coast north of the Hangzhou Bay and along the Bohai Bay coast of the Great Plain of North China. This coastal area has subsided over 300 m since the beginning of the Cenozoic and has been built out by huge volume of sediment brought down to the coast by several large rivers, notably the Yellow River and the Changjiang rivers. The great plain built up by these sediments extend onto the continental shelf at times of low sea level, and was partly submerged during the postglacial transgression. At the present, the Yellow River and the Changjiang River continue to deliver sediment to the coast and continental shelf, forming extensive coastal tidal flats in the Bohai Bay and to the north of the Changjiang River.

1. Tidal Flats Along the Coastal Plain of the Bohai Bay

The Bohai Bay is located in the west of the Bohai Sea and with a water depth less than 20 m. The center tip point of the arc-shaped gulf is located in Qikou where it has a very gentle submarine coastal slope. The slope gradually increases on both sides to the Haihe River mouth and towards the Yellow River mouth (Wang, 1983).

In the Bohai Bay, waves are mainly wind-induced with very little swell. There are strong waves during winter and spring, when northeast maximum wave height is 1.9 m. During the summer and autumn, southeast and east waves predominate with average heights in the order of 0.4~0.5 m. The effect of wave action on the tidal flat is weak because the water is shallow during the period of the flood tide. The waves stir up the sediment offshore which is then transported by tidal currents along the intertidal flat. When NE wind waves reach a Beaufort Scale of 7, the turbid coastal waters can be 30 km wide. In addition to tidal currents, there are also wind-induced currents with velocities between 10 and 30 cm/s. These currents are important in carrying suspended sediment as longshore drift, but cannot move sediment in a transverse direction across the flat. Tides in the Bohai Bay are of the irregular semidiurnal type. The tidal range is 1.65 m in average and more than 3 m in maximum, and the minimum is 0.8 m around the mouth of the Yellow River. Currents are rectilinear along the coast. Flood currents dominate towards the east (65°~95° or 135°), except at the Yellow River mouth. Tidal periods are unequal; the ebb tide lasts 7 h and the flood 4 h. As a result, the velocity of the flood is higher than the ebb, especially on the intertidal flat. Maximum current velocities appear during the middle period between high tide and low tide. The highest concentration of suspended particulate matter (SPM) is found at the beginning of the flood tide when the water is shallow and turbulent. The SPM concentration decreases to a minimum at high tide as the water becomes deeper and less turbulent. During the ebb tide, the suspended sediment concentration is low because of the long-lasting ebb period with lower current speeds. As a result, the principal transport of sediment is by the flood tide in a landward direction. Very fine sand and coarse silt are transported mainly as traction load. Fine silt is transported in suspension in the bottom 20 cm of the water column, and clay in the upper part of the column.

The coastal plain profile of North China consists of three parts from land to sea. They are: the salt marsh plain, intertidal flats, and the submarine coastal slope.

1) Salt marsh plain. The marsh plain is located above the maximum high tidal level, landward of shell beach ridges (cheniers) or artificial banks. The plains are flat and contain several lagoon depressions. Vegetation is sparse.

2) Intertidal flat. In the Bohai Bay, the width of the intertidal flat is generally 3~ 7 km, increasing toward the south to 16~18 km at the Yellow River delta. The intertidal flat can be divided into 4 geomorphological and sedimentary zones (Wang el al., 1964). From land to sea, these are: the polygon zone, the inner depositional zone, the erosional zone, and outer depositional zone (Table 1).

Table 1　The Zonation of the Tidal Flats of the Bohai Bay

Zone	Outer depositional zone	Erosional zone	Inner depositional zone	Polygon zone
Location	Immediately shoreward of the low tidal level	Around mid-tide level	Immediately below the lower high tide level	Between lower high and maximum tidal level
Morphology	Sandy silt flat with wave or current ripple	Tidal channels follow ebb current's direction, with overlapping fish scale erosional structures between channels or elongated depressions	Mud pools	Polygon field with wormhole; Marsh (near river mouth)
Sediment	Very fine sand or coarse silt	Alternating laminate of fine silt and clay	Clayey silt and clay	Clay or silt clay
Dynamics	Tidal currents and micro-waves	Tidal currents, mainly ebb currents	Tidal currents, mainly flood current	Tidal currents and evaporation due to solar heating
Process	Deposition of thin layers over large area	Deposition during flood, erosion during ebb flow	Extensive deposition, mainly by SPM	Stable (deposition during maximum tide, erosion during storm)

The polygon zone is located between lower high and maximum high tides with its landward limit marked by a chenier. The zone is 70~80 m wide with slopes of 0.5~ 0.6 m/km. The characteristic sediment is greyish-yellow clay, with grey silt clay 1.5 m bellow the surface. The zone is only covered by water during the spring high tide, and under dry conditions, polygonal desiccation cracks develop in the clay. These average 10~50 cm in diameter with cracks that are 1~2 cm wide, where there is a freshwater influence, several grass species may grow such as *Suacda salsa* Bunge, *S. glauac*, *Limonium sinensis* Kumtie, *Phragmites australis* and so on.

The inner depositional zone consists of soft mud in the form of pools with an average slope of 0. 67 in/km. Clayey silt and clay deposit here during the flood tide.

The erosional zone occurs at about mid-tide level. In the northern Bohai Bay, this zone is 500 m wide with a slope of 0. 57 m/km. It exhibits alternating laminae of fine silt and clay, and the surface is eroded into an overlapping series of scours called fish scale structures. These structures are 3~18 m long and 0. 5 m wide and dip seaward. Scarps between each layer of scales are 2 to 10 cm high and face the coast. These are erosional forms produced by the ebb flow. In the middle part of the Bohai Bay, the erosional zone is about 1 000 m wide, with a slope of 0. 42 m/km and consists of silty clay. Tidal creeks may be eroded into the mud by ebb tide currents. The sediments are carried to the margin of the zone, mainly toward the sea where silt settles in the outer depositional zone.

The outer depositional zone is present between the erosional and the low tide level. It is approximately 1 000 m wide and has a 0. 5 m/km slope. The zone consists of coarse silt and very fine sand with current ripple marks on the surface. The ripples have a wavelength of 3~7 cm, with their steep slope facing seaward along the outer part of this zone and shoreward along the inner part. In the middle part of the zone they are symmetrical. At the low tide level, the ripples are overlapped by microwave marks. Wave action always influences the low tide zone, where coarser material (coarse silt and very fine sand) is deposited. The sediments become finer in both offshore and onshore directions.

3) Submarine coastal slope. This slope is very gentle in the Bohai Bay. The sediment becomes finer offshore, changing from a sandy silt at the low tidal level to a very fine silt and clay in water depths of more than 10 m.

In the Bohai Bay, the principal source of sediment is the Yellow River, which plays a predominant role in the development of the tidal flat coast. The plume of suspended river sediment is the major source of coastal sediment. "Fluid mud" is deposited only in the upper part or tidal flats or in deep water. Several years of monitoring indicated that the flats are areas of rapid sediment deposition with relatively high deposition in the outer and inner depositional zones.

The tidal flat zonation is the result of the spatially-varying nature of different tidal processes acting on the flat. Sediment type on the flat itself is controlled by the velocity of tidal currents. On very low gradients, frictional drag is of importance. The deposition of very fine sand and coarsesilt on the lower flat, alternating laminae of fine silt and mud on the middle flat and clay on the upper flat, reflects a settling lag process (Postma, 1967; Wang et al, 1964) and a landward decrease in current velocity as water depth decreases. Although the ebb current has a lower velocity than the flood current, gravitational acceleration on the steeper middle flat promotes slight erosion that is recorded by alternating clayey and silt laminae.

Seasonal changes in depositional processes are also important. In the Bohai Bay, freezing starts in late November. Coastal ice up to a thickness of 0.5 m accumulates mainly between the high and middle tide levels. Normally, there are alternating layers of ice and mud. During March, the ice melts, leaving patches of mud on the flat. During May and June, winds are predominantly northeasterly. The tidal flat experiences erosion by strong wind waves, but sediment is redeposited during the low energy conditions of the summer. With southeasterly and easterly winds, the flat is covered by a yellowish fluid mud. Tidal flat zonation is apparent during spring and autumn. Sediments on tidal flats of the Bohai Bay can be traced to two sources: a tongue of fine materials extends from the Yellow River and reaches Qikou, another tongue from the Haihe River consists of silt and very fine sand; it also reaches Qikou. Thus the bay head is the boundary of two sources of sediment. Heavy minerals also indicate two sources. The Yellow River sediments have a low heavy mineral content of 1%, consisting of more stable minerals (garnet, zircon, tourmaline) and an increasing proportion of secondary minerals (allochite, hematite). All of the heavy minerals have lost their original crystal shapes and most unstable minerals have disappeared, indicating that the sediments have been transported over a long distance. Sediments from the Haihe River indicate an igneous rock and metamorphic rock source from the nearby Yanshan Mountains. The mineral composition of the Haihe sediment is distinctly different to that of the Yellow River.

2. Tidal Flats Along the Coastal Plain of North Jiangsu

Much of the North Jiangsu coast is developed on the pre-1855 delta of the Yellow River that lies immediately north of the Changjiang River. The total coastline of tidal flat in north Jiangsu is 883.6 km long. During historical times, the Changjiang River was also present in north Jiangsu, entering the Yellow Sea from Qianggang. The two rivers brought in large quantities of sediment which has influenced the geomorphology of the coast. Tidal processes are powerful along this coast. Thus, 10~13 km wide tidal flats have been developed. Their averageslope is 2/10 000 and their maximum width is 36 km.

The annual sediment discharge of the Changjiang River is 478 million tons, which is only one-third that of the Yellow River; at the present time most of this sediment load is deposited in the channel at the river mouth. Sediment deposited outside the river mouth channel is remobilized as longshore drift and carried towards the south where it forms the tidal flats in the embayments of Zhejiang and Fujian provinces. Longshore drift towards the north from the Changjiang River mouth contributes little to the development of the North Jiangsu coast.

Along the coast of the Yellow Sea, semidiurnal tides are predominant with varying tidal ranges. The average tidal range is 1.5～1.7 m around the abandoned Yellow River mouth and along the southern part of the coast. The average tidal range is 3.0～3.4 m in the middle part of the north Jiangsu coastline, with the maximum tidal range of 6～9.28 m in the area of the Huangsha Yang Channel. Tidal elevation determines the altitude of tidal flats and also has an effect on their width. Flood currents, which last nearly 8～10 h, are faster than ebb currents. The wave heights in this area are smaller (0.6～1.2 m) because the coast is protected by offshore sandy ridges. Outside sandridges, wave heights can reach 4 m. Along the North Jiangsu coast, waves are also controlled by the wind. Northeast/north-northeast waves prevail during the winter season with maximum height of 2.9～4.1 m. Southeast waves predominate during the summer, with maximum heights in the range of 1.7～3.2 m. In the autumn, an easterly swell prevails.

On the tidal flats, maximum flood tidal current appears to be associated with maximum sediment concentration which lags the tide by about one hour. Maximum sediment concentration during the ebb also appears about one hour after the maximum ebb flow. Tidal currents approach the coast at an angle of 20°～30°. The angle increases to about 90° on the high tidal flat level, thereby giving rise to rectilinear currents that flow onshore or offshore. During one tidal cycle, sediment movement is mainly landward, and the area of highest deposition is in the middle part of the flat between high tidal level and low high tidal level. The sediment depositional pattern is parallel to the HHTL and LHTL boundaries and is characterized by the highest sediment concentration seen on the flat (Zhu and Xu, 1982).

Wave action influences tidal flats through winnowing which increase sediment concentration. These high concentrations promote the formation of sedimentary structures such as ripples, inclined bedding, and storm records of scores and sandy layers (Ren et al., 1983).

Basically, they are 3 zones on the intertidal flat. A mud zone forms the upper part of the flat near the high tidal level. This zone is characterized mainly by fine-grained material which forms silty-clay or clayey-silt thin laminae. A silt zone dominated by ripples is located in the lower part of the intertidal flat, immediately above the lower tidal level. This zone forms the widest part of the intertidal flat and is composed of sediment deposited mainly by traction processes. These sediments fall within the coarse silt or very fine sand size range. It is basically a homogeneous silt deposit, with relatively thick laminae structures. A silt or silty-mud zone occurs over the middle part of the intertidal zone. It is composed of transitional deposits derived mainly from fine silt-sized material moving in saltation and clay-sized material which is deposited from suspension. These two depositional mechanisms give rise to alternating thin laminae. Erosional or micro erosional features are quite common in this zone. They may indicate that this part of the flat has a steeper slope which tends to focus the stronger ebb current activity. The zonation of the North Jiangsu coastline is a typical example of an accretionary intertidal flat, with silt as the main textural component over the entire range of the flat. This textural characteristic reflects the dominance of tidal current processes. Even though the flat occasionally suffers from storm wave erosion, it soon returns to a normal profile. Such a flat may be referred as a "silt flat". Including the supratidal salt marshes or grass flat, four zones of sedimentary facies can be recognized. As the tidal flat extends and progrades towards the sea, these zones gradually supersede one another. After a tidal flat reaches a mature stage, an entire series of tidal flat sedimentary facies can be seen in the stratigraphic cross-section. The section begins with a lower layer of sandy deposits that is overlain by an upper layer of clayey deposits. The lower silt and fine sand deposits represent bed-load sediments carried by higher energy currents that are produced near low tidal level. Sediments carried by these currents have been deposited on the lower tidal flat between low tidal level and outer boundary of submarine coastal slope. These sediments form base of the tidal flat deposit; their thickness is about $15 \sim 20$ m. Sandy material moves as traction load from the submarine slope advancing towards land, gradually expanding the size of the tidal flat. The upper layer clayey sediments are carried in suspension by tidal currents and accumulate on the highest part of the tidal flat. Although the surface of tidal flat reaches the spring high tidal level, it is only occasionally flooded by sea water. As a result, the surface of the tidal flat cannot be built up further by material carried in suspension. The maximum thickness of clayey deposits is usually $2 \sim 3$ m and equivalent to the height difference between the levels of middle tide and spring high tide. There are intervals of suspended clayey deposits and of tractional bed-load sandy deposits, giving rise to the alternating clay layer and sand layer deposits observed in the section.

Heavy minerals that occur in the tidal flats of North Jiangsu can be classified into two provinces. One is the Changjiang mineral complex, which consists of hornblende, magnetite, apatite, hyposiderite, collophane, zircon and rock fragments. The first three minerals are characteristic. The second province represents ancient Yellow River mineral complex. Its characteristic minerals include hornblende, epidote, hyposiderite, zircon, tourmaline, titanioferrite and garnet. Among these, zircon, tourmaline, titanioferrite and garnet are the major and characteristic ones. Heavy mineral content increases from the high tidal level (0.84%) to the low tidal level (5.1%).

Grass pollen species occur in the sediment of the high tidal level of tidal flats (grass flat and mud flat). They decrease in number and species from landward to seaward and are absent in the low tidal level. The pollen of tree species are characteristic of low tidal level tidal flat deposits. This distribution implies that the grass pollen is derived from a local source on the upper flat while the tree pollen represents material carried by longshore drift from river sources such as the Changjiang. This distribution indicates that sediment movement is from sea towards land and that sediment transport in a seaward direction is rare. This pattern of sediment movement is also evidenced by the distribution of microfossils. Tidal flat sediments contain shallow water foraminifera species such as *Ammonia*, *Elphidium* and *Nonion*, which are associated with 10.3% of planktonic marine species of the genera *Globigerina* and *Orbulina*. In this area, these planktonic species typically live in water depths of more than 200 m. There are also benthic species such as *Bolivina* (7.2%) and *Quinqueloculina* which are typically associated with water depths of more than 50 m in this area. The mixed species of deep water foraminifera with shallow water forms also demonstrates that sediments move from deep water to the shallow water environment of the flats (Zhu et al., 1986).

The sedimentary composition of tidal flats along north Jiangsu coast indicates three sediment provinces: Changjiang River sediment, sediments derived from erosion of the abandoned Yellow River delta, and offshore submarine sand ridges.

In summary, the tidal flat of the plain coast of China has certain basic characteristics of sedimentation conditions:

1) There are three zones developed on the tidal flat. Mud flat occurs at the high tidal level; muddy silt mixed flat at the middle tidal level; and silt flat at the low tidal level. The stratigraphic section consists of 2～3 m thick muddy deposits of the upper layer and 10～15 m thick sandy deposits of the lower layer; a mixed muddy and sandy deposit comprises a transitional middle layer.

2) Tidal flat deposits are homogeneous and rather stable with respect to their grain size, chemical, mineral, and microfossil compositions.

3) Tidal flat deposits represent a large scale geologic feature; their total area can range from tens to several hundreds of square kilometers. They occur mainly in areas of subsidence and can be formed over several sedimentary cycles reaching a thickness of several tens to hundreds of meters.

B. Tidal Flats of Embayment Coast

Different types of coastal embayments have developed along the coastal fault zone of Zhejiang and Fujian provinces. These coastal areas are associated with large tidal ranges, more than 4 m in the average. Powerful tidal currents are found in most embayments; current speeds can exceed 1 m/s, but wave energy is weak because the bays extend inland and are sheltered. The average wave height in these bays is less than

0.5 m. Diluted fresh water from the Changjiang River flows to the south and mixes with the long shore drift of the Zhejiang-Fujian coast; it can travel to the northern part of Fujian coast during winter season. Thus, the powerful tidal environment combined with fine-grained sediment supplied by diluted fresh water provides requisite conditions for developing tidal flats in these embayments. This kind of tidal flat coastline is about 1 500 km long.

1. Tidal Flat in the Leqing Bay

The Leqing Bay is 40 km long from bay mouth to bay head with an average water depth of 10 m. Total area of the embayment is 250 km². The maximum tidal range is up to 8.34 m. Rectilinear currents flow towards the bay and backwards to the East China Sea. Flood current velocity is higher than that of the ebb flow. Sediment concentration is 0.4 kg/m³. The average width of tidal flats in the Leqing Bay is 1.8 km, coastal slopes are $1/1\,000 \sim 1.4/1\,000$, and the total area of tidal flats is 134.8 km².

Tidal flats in this area consist of fine grain materials. The surface layer is a $20 \sim 40$ cm thick deposit of greyish yellow mud; the lower layer is a few meters to 10 m thick deposit of greyish dark mud. The bottom sediment layer of the flats consists of fluvial deposits of silt-clayey sands and shingles. The tidal flat deposits is of a homogenous clay size with a medium diameter (Md) of 8.73φ. There is no variation in grain size and no apparent stratification. Heavy minerals in this tidal flat consist of two suites: i) In the bay head area, there is a magnetite-epidote assemblage (88.9% and 5.3%). The heavy mineral content there is relatively high but consists of relatively finer grains supplied by local rivers whose influence is restricted to the bay head. ii) Other parts of the bay are characterized by a hornblende-magnetite-epidote assemblage of heavy minerals. This assemblage indicates that sediments in the large part of the Leqing Bay are mainly from the sea and carried into the bay by tidal currents.

2. Tidal Flats of the Shacheng Bay

The Shacheng Bay is a narrow embayment 55.5 km in length and 1.85 km wide, located in Fuding County of Fujian Province. Its total area is 11.23 km² with a water depth of at least 15 m in most parts of the Bay. There are 6.05 km² of islands in the bay, and the tidal area is 53.81 km². The average tidal range is 4.10 m and the maximum tidal range is 6.90 m. Rectilinear flood and ebb currents typically flow through the bay with a velocity $1 \sim 1.5$ m/s and a maximum velocity 2 m/s; the ebb current is faster than flood. Wind waves in the bay are suppressed. Even during a typhoon with wind speed of $8 \sim 10$ Beaufort scale, the wave height is only 0.5 m.

Tidal flats in the Shacheng Bay are distributed in the bay head and along both sides of the inlets, forming a discontinuous pattern. The width of tidal flats is less than 1 km with a relatively steep slope in the order of $5/1\,000 \sim 10/1\,000$. Surface sediments consist of 30 cm thick yellowish mud. The lower layer is greenish-grey mud of several meters thick, and the bottom layer consists of fluvial sand and gravel. The tidal flat deposits in the bay can be divided into two types, gravel-sandy and muddy.

The gravel-sandy type is distributed in the head of the bay and in mouths along the sides of the bay. Sand content is 4.6%~65%; clay 22.5%~66%; small gravel 9.3%; heavy mineral about 236 mg/100 g. This kind of flat is the result of coastal streams that discharge to the bay. They form small alluvial fans at stream mouths, that combines with fine material carried into the bay by flood tidal currents.

The muddy tidal flat type is not influenced by stream discharges. Muddy tidal flat deposits contain up to 57.5%~65% clay sizes sediment, 35%~42.5% of silt, and heavy mineral content of about 7.3 mg/100g. The grain size of tidal flat sediments decreases from the bay mouth to the bay head. Heavy mineral contents of both types of tidal flats in the Shacheng Bay are similar, iron minerals 62%, epidote 22.8%, leucoxene 11%, and zircon. All associated with secondary pyrite. These sedimentary features indicate that fine materials of these flats were carried by tidal currents from seaward areas. Mud is distributed all over the bay but is limited in quantity. The present sedimentation rate of tidal flats in the bay is small.

3. Tidal Flats of the Xiamen Bay

The Xiamen Bay is a fault-subsided and submerged embayment with more open water areas. Consequently, wind wave action predominates. Tides in the bay are semidiurnal; the average tidal range is 4 meters. Maximum tidal range is 6.9 in and the ebb current speed (0.89 m/s) is larger than that of the flood current (0.77 m/s). The tidal flat area is 873 km² in the bay; its width is from several hundred meters to 1 or 2 km. The flat slope is 1/1 000 ~ 2/1 000 and is dissected tidal creeks. Apparent rhythmic changes of sediment strata can be seen in stratigraphic sections. There are intervals composed of thin layers of mud and sand with rare occurrence of cross bedding, reflecting wave or current-formed ripples. Tidal flat sediments are a mixture of sands, silty sands, clayey sands, sandy clay, and clayey silt. There are gravelly sands appear on tidal flat located on the erosional coast in the north. Sand is distributed over wide areas and represent the major component of tidal flat deposits observed in the bay. Sediments are derived from northern erosional coastal terraces, submarine banks and coastal river discharges. Waves act on the deposits found on the high tidal level of tidal flats while tidal currents have a major influence at the low tidal level. These relationships give rise to an upper coarser and a lower finer facies of the depositional profile. Heavy mineral content is relatively high in the sediments, the average content is 151.2 mg/100 g reflecting a local sediment supply from the erosion of bedrock embayment.

Sedimentary characteristics of embayment type of tidal flat can be summarized as follows:

1) Sediments on the tidal flats are mostly muddy deposits. There is neither apparent stratification of upper muddy or lower sandy layers, nor clearly surface zonation between the high tidal and low tidal levels.

2) The multi-sediment sources of some tidal flats cause the changes of particle size and composition over short distances.

3) Tidal flat deposits are small-scale formations. They are typically several meters thick and occasionally reach thickness of more than 10 m. The area of sedimentation is usually several square kilometers and up to several tens square kilometers.

Ⅲ. CONCLUSIONS

Tidal flat development is controlled mainly by the quantity of sediment supply and by tidal action. Wherever there is an abundant sediment supply in association with a large tidal range, a tidal flat will develop. Whenever the sediment supply stops or decreases, the tidal flat will succumb to wave and tidal current erosion. Thus sediment supply is a major controlling factor, and different kinds of tidal action shape different type of tidal flats with their own characteristic sedimentary structures.

1) River sediment discharge, coast erosion and offshore submarine erosion are the processes that supply sediment to the tidal flat. Most of the offshore-source sediments have a fluvial origin, especially in China.

Sedimentary characteristics of the Zhejiang and Fujian coast tidal flats also demonstrate that river discharge and offshore sediment supply are major factors for developing these features. The fine sediments discharged by the Changjiang River are transported as far as to the Shacheng Bay in the northern Fujian Province. In contrast, the Xiamen Bay tidal flats are formed from local sediments supplied by coastal rivers and weathered coastal bedrocks.

2) Tidal action affects the spatial scale and sedimentary characteristics of tidal flats. If the sediment supply is sufficiently large, the width of the tidal flat is dependent primarily on tidal range. The varying characteristics of the tide (e. g., spring tide or neap tide) and the accompanying tidal level influence the nature of sedimentary structures (Zhu and Gao, 1985). In a sequence of neap tide to spring tide, a thin sand wedge is deposited from seaward to landward. After several such cycles, a mixed sediment layer of clay and sand form between the tidal flats of A and B as shown in Fig. 2. In the Bohai Bay, there is a 1. 91 m difference between the levels of spring high tide and neap high tide; the horizontal distance of the two levels is 1 500 m. Consequently, the entire tidal flat facies shifts in the distance. In the Jiangsu coastal zone, the tidal level difference between spring high tide and neap high tide causes a shift of 1 900 m horizontally in the sedimentary facies of tidal flats. As a result, a complex sedimentary

structure is produced on the tidal flat. In Jiangsu, the tidal current velocity of the spring tide is twice that of the neap tide. Sediments deposited on the tidal flat during the spring hish tide has a medium diameter of $Md=5.3\phi$, and a medium diameter of 6.5ϕ during the neap tide. This process forms laminae of clay and fine silt (Ren et al., 1984).

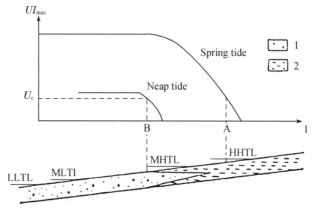

1　sandy deposits　2　mud deposits

UI_{max}　Flood current velocity of spring tide

U_c　Ebb current velocity of neep

I　Distance to low tidal level

LLTL　Low low tidal level

MLTL　Mean low tidal level

MHTL　Mean high tidal level

HHTL　High high tidal level

Fig. 2　Sediment function to the tidal levels

A is the upper limit of sand deposition during spring high tide.

B is the upper limit of sand deposition during neap high tide.

In the Bohai Bay and Jiangsu, the direction of flood tidal current changes as it approaches coast flats. At low tide, it approaches the coastline at an angle of $20°\sim30°$ relative to the tidal current direction. This angle increases to $40°$ at the middle tidal level, and to $70°\sim80°$ during the upper tidal level, i. e., almost perpendicular to the coast. Similar results have been observed on the tidal flats in the Wash (Evans, 1965). Changes in the direction and velocity of tidal currents influence sediment content, that promotes the zonation of three textural types, i. e., a muddy flat, a muddy and sandy mixed flat and a sandy flat in the Jiangsu area (Zhu and Gao, 1985).

3) Tidal flat sedimentary structures reflect the range of environmental variation in the coastal zone. The study of tidal flat will yield new understandings for the interpretation of ancient tidal flat facies and for the management and utilization of modern tidal flat environments.

This project was supported by National Science Foundation of China. Project coding: SCIEL21192105.

References

Dale, E. J. 1985. Physical and Biological Zonation of Intertidal Flat at Frobisher Bay. In: N. W. T. 14th Arctic Workshop, Nov. 6—8, 1986, Halifax. 180 – 181.

Evans, G. 1965. Intertidal flat sediments and their environments of deposition in the wash. *Quart. J. Geo. Soc.*, 121: 209 – 245.

Postma, H. 1967. Sediment transport and sedimentation in the environment. In: LAUFF (ed.). Estuaries, AAAS Publication 83. 158 – 199.

Ren, Mei-e, Zhang, Ren-shun, and Yang, Ju-hai. 1983. The influence of storm tide on mud plain coast. *Marine Geol. and Quaternary Geol.*, 3(4): 1 – 24.

Ren, Mei-e, Zhang, Ren-shun, and Yang, Ju-hai. 1984. Sedimentation on tidal mudflat in Wanggang area, Jiangsu Province, China. *Marine Science Bulletin*, 3(1): 40 – 54.

Wang, Ying. 1983. The Mudflat coast of China. *Canadian J. of Fisheries and Aquatic Sciences*, 40 (Supplement 1): 160 – 171.

Wang, Ying, Collins, M. R, and Zhu, Da-kui. 1990. A comparative study of open coast tidal flats: the Wash (U. K.), Bohai Bay and West Yellow Sea (Mainland China). In: *Proceedings of International Symposium on the Coast Zone*. 120 – 130.

Wang, Ying and Ke, Xian-kun. 1990. Chenniers on the east coastal plain of China. *Marine Geology*, 90: 321 – 335.

Wang, Ying and Zhu, Da-kui. 1990. Tidal flats of China. *Quaternary Sciences*, 12: 291 – 300.

Wang, Ying and Zhu, Da-kui, Gu, Xi-he and Chuai, Cheng-qi. 1964. The characteristics of mudflat of the west coast of Bohai Bay. In: Abstracts of Theses in the Annual Symposium of Oceanographical and Limnological Society of China, 1963, Wuhan. 55 – 56.

Zhu, Da-kui, and Gao, Shu. 1985. A mathematical model for the geomorphic evolution and sedimentation of tidal flats. *Marine Science Bulletin*, 4(5): 15 – 21 (in Chinese).

Zhu, Da-kui, Ke, Xian-kun, and Gao, Shu. 1986. Tidal flat sedimentation of Jiangsu coast. *J of Oceanography of Huanghai and Bohai Seas*, 4(3): 19 – 27 (in Chinese).

Zhu, Da-kui and Xu, Tin-guan. 1982. The coast development and exploitation of middle part of Jiangsu coast. *Acta Scentiarum University Nankinesis*, 3: 799 – 818 (in Chinese).

中国的海洋海涂资源[*]

海洋海涂(Tidal Flat)通称海涂、滩涂,主要指淤泥质海岸的潮间带浅滩,广义的海涂还包括部分未被开发的生长着一些低等植物的潮上带及低潮时仍难以出露的水下浅滩。海涂在国外主要分布在大西洋沿岸的荷兰、法国等国及北美的芬地湾、加利福尼亚湾,南美的亚马逊河口、圭亚那沿岸等处。中国的海涂规模巨大,在世界上具有特殊地位,在渤海、黄海、东海、南海四大海区的沿岸都有分布,其岸线总长可达 4 000 km。随着我国耕地面积的迅速减少,海涂作为潜在的土地资源,将成为我国经济发展新的增长点[1]。

一、中国海涂资源的概况

(一) 我国海涂资源的分布特点

我国海涂资源分布在北起辽宁,南至广西的沿海,主要分布于大河三角洲平原海岸及华南海岸的一些港湾内[2](图 1)。我国海涂的总面积约 235 万 hm²,现每年仍以 2 万～

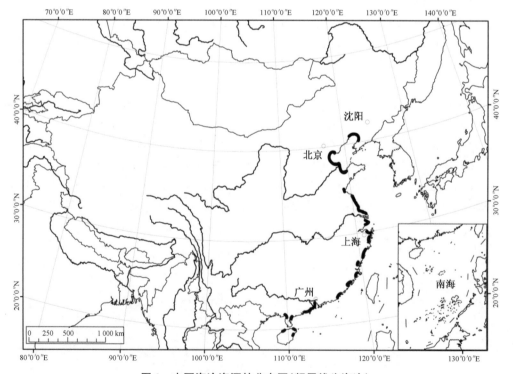

图 1　中国海涂资源的分布图(粗黑线为海涂)

---

* 杨宝国,王颖,朱大奎:《自然资源学报》,1997 年 12 卷 4 期,第 307 - 316 页。

3 万 hm² 的速度淤涨,其中,江苏省的海涂面积最大,约占全国的 28%(表 1)。

表 1 中国海涂资源的分布

海区	地区	海涂面积(hm²)	占全国总量百分比	占本海区百分比
渤海		57.19	24.3	100
	辽东湾	14.58	6.2	25.5
	河北	11.07	4.7	19.4
	天津	5.87	2.5	10.3
	莱州湾	25.67	10.92	44.9
北黄海		16.48	7.0	100
	辽东沿岸	9.60	4.1	58.3
	山东半岛	6.88	2.9	41.8
南黄海		65.33	27.8	100
	江苏	65.33	27.8	100
长江口		9.04	3.8	100
	上海	9.04	3.8	100
东海		49.34	21	100
	浙江	28.86	12.3	58.5
	福建	20.48	8.7	41.5
南海北岸		32.85	14	100
	广东	22.8	10	69.4
	广西	10.05	4.3	31.6
海南岛		4.87	2.1	100
	海南	4.87	2.1	100

注:数据来源于全国海岸带和海涂资源综合调查报告、1991~1994 年中国统计年鉴、1994 年江苏省统计年鉴等。

(二)中国海涂的土地资源地位

由于城镇道路建设、结构调整、退耕还牧及自然灾害等因素的影响,我国的耕地面积从 1958 年开始不断减少,到 1990 年累计减少 1 630 万 hm²,平均每年减少 50.42 万 hm²,到 1990 年全国的耕地面积降为 9 600 万 hm²,广东、福建、江苏、浙江人均耕地不足 0.067 hm²(1 亩)。根据国家土地部门的估计,今后我国耕地将以每年 40 万 hm² 左右的速度递减,即使每年垦荒 20 万 hm²,到 20 世纪末耕地仍将减少 100 万 hm²,而届时我国人口将增加到 13 亿,人均耕地将由目前的 0.084 hm²(1.26 亩)减少为 0.072 hm²(1.08 亩)。因而,目前我国以占世界 7% 的耕地,养活着占世界 22% 的总人口,人地矛盾越来越突出。海涂作为后备的土地资源,其开发利用对我国未来经济的发展会有重要的作用。按 1990 年全国土地资源调查,我国海涂资源为 235 万 hm²,占全国土地面积的 0.23%,而当时全国所有城镇工矿用地 187 万 hm² (占全国土地面积的 0.2%),海涂面积比全国城镇工矿用地还大 48 万 hm²,且海涂多分布在经济发达、土地资源紧缺的东部沿海地区,因此,海涂资源的开发在全国土地利用中具有重要地位[3]。

二、中国海涂资源的成因、类型

(一) 中国海涂的成因

影响海涂发育的因素很多,如海岸带泥沙供给的数量、海岸带的原始坡度、海岸发育的时间、波浪作用的强度、潮差等,其中最为重要的因素是泥沙供应量及潮差的大小。

我国每年由河流入海的泥沙量为 25 亿 t 左右[4],为海涂的发育提供了丰富的物源,此外海域来沙也部分贡献于海涂的发育,如苏北岸外的辐射沙洲和闽、浙的港湾海涂都有海域来沙。我国入海的年平均径流量为 1 815.4 亿 m³[5],占全国河流径流量的 69.8%,径流将大量的淡水、泥沙、溶解矿物及腐殖质带入海洋,使沿岸海涂发育并富含各种营养成分,适宜于发展海涂围垦、水产养殖与捕捞,成为沿海农、林、渔、牧业及轻工业的原料基地(表 2)。

表 2　中国海涂的主要特征指标

海区	底质	径流量 (10⁹ m³)	输沙量 (10⁹ t)	底栖生物量 (g/m²)	表层水温 (℃)	降水量 (mm)
渤海	泥质粉砂	538	8.47	111	13.0	632
北黄海	粉砂质泥	315	0.02	432	12.1	703
南黄海	粉砂	436	0.49	57	15.5	986
长江口	泥	10 573	5.32	38	17.5	1 124
东海	泥质粉砂	1 988	0.28	63	19.0	1 207
南海北岸	泥质粉砂	37 856	0.89	327	23.5	1 794
海南岛	细粉砂	163	0.01	507	25.6	1 696

潮汐作用影响到海涂的规模与沉积特征,潮流对淤泥质海涂的宽度有重要影响,海涂宽度与潮差大小相适应,在美国的路易斯安那州,潮差为 0.5 m,海涂宽度仅为 0.1~0.3 km,而在韩国西岸的潮差为 5~6 m,海涂宽度可达 5~30 km。在我国的江苏沿岸的许多中潮差岸段,物源丰富,海岸的坡度小,因此许多岸段的海涂宽达 10 km 以上。此外,海涂上潮流的变化特征决定了海涂上的沉积相带,沉积物的分布特征和海岸动力的空间分布一致,近岸质细,向海逐渐变粗[6,7]。

由于中国大河向海岸带输送大量泥沙并对海岸塑造产生巨大影响的独特性,我国大河尾闾的变化对海涂的演变也有巨大影响,其中最为典型的是黄河入海口的改变。例如,1128 年,黄河从江苏注入黄海,使苏北海涂迅速向海推进,至 1855 年,海涂增长面积达 15 700 km²,现为江苏沿海平原,占江苏面积的 1/7。1855 年,黄河北徙从渤海入海,苏北的废黄河三角洲开始遭受侵蚀,共冲掉土地约 1 400 km²,由于缺乏泥沙供应,至今海涂仍在不断后退。现在黄河入海口几乎每 10 年就会变化一次，对海涂的演变也有较大影响。例如，1956 年黄河从刁口入海，黄河泥沙向西北方向输送，渤海的大部分海涂向海延伸；而 1972 年后黄河改由小清河入海，黄河泥沙改为向南输送，因此小清河以南的地区海涂产生淤积，而小清河以北的海涂则开始遭受侵蚀而后退。像黄河口海涂在短时期内发生这样巨大的变化，在世界上是极为罕见的，原因是黄河口供沙量巨大，海涂随泥沙供应量而变化。

（二）中国海涂资源的类型

我国海涂主要分为平原型与港湾型两类。渤海湾、苏北沿岸是平原型海涂，而浙、闽及广东、海南沿海则分布着许多港湾型海涂。

平原型海涂主要分布在辽东湾、渤海湾、江苏沿岸、长江口、杭州湾、珠江口等地，其特征是：① 在平原上具分带性，自陆向海可分为 3 个平行的相带，即高潮位泥滩、中潮位的泥砂混合滩、低潮位粉砂或细粉砂滩 3 个沉积相带，相应地在垂直剖面上也具有海涂的二元相结构，上层为泥质沉积，厚 2～3 m，下层为砂质沉积，厚 10～15 m，中间为过渡的砂泥混合沉积；② 海涂组成均一稳定，包括粒度、化学组成、矿物及微体生物等，这反映了平原型海涂主要由大河提供沉积物，其影响规模大、持续时间长；③ 海涂的表面形态上为凸形，在平均高潮位及平均低潮位各有一个上凸点，滩面的坡度为 0.002%；④ 海涂的沉积规模大，面积大，沉积层深厚，面积可达数十至数百平方千米，且因大多位于地质构造沉降区，经过多旋回的沉积过程，沉积层厚度可达百米。

港湾型海涂主要分布于浙、闽、粤的一些港湾内，大部分地区潮差较大，平均潮差大于 4 m，是强潮环境，由于有适量的细颗粒泥沙的供应，使港湾中海涂发育，虽然分布零星，但总长度也很长，浙、闽沿岸的海涂岸线长度就可达 1 500 km，其主要特点是：① 物质组成单一，全是淤泥，因而在平面和垂向上没有明显的分带现象；② 在港湾的不同部位成分发生分异，主要是受物源的影响，一些部位以河流供沙为主，而大部分地貌部位则以海域侵蚀的沉积物为主，这使得海涂沉积物在短距离内发生一些变化；③ 沉积体规模小，厚度一般仅为几米，面积一般为几平方千米，反映了缺乏大量长期物源的特征；④ 潮差大，潮流作用强，物质较细，易被冲刷，冲淤变化频繁[7,8,9]。

三、海涂的利用现状及存在问题

（一）我国海涂的利用现状

我国的海涂利用目前主要是高潮滩围垦发展农业、水产养殖业，中、低潮滩发展贝类水产的捕捞采集，其利用率较低。近 50 年来，全国累计围海面积达 120 万 hm²，为我国沿海地区发展农业、盐业、养殖业提供了大量的用地，也为沿海港口建设、城市工业等提供了土地。但各省的开发很不平衡，浙江、福建、广东的开发程度最高，江苏其次，而天津、河北、广西等开发程度较低（表 3）。

表3　中国海涂的利用现状

海区 资源量	渤海	北黄海	南黄海	长江口区	东海西岸	南海北岸	海南岛
1984年利用面积2 244(km²)	307.4	439.8	444.3	4.5	549.8	480.2	18
1984年养殖产量247.3(10³ t)	6.4	22	9.4	0.5	190	19	—
潮间带资源量3 480(10³ t)	261	1 148	150	2	651	1 218	—

(二) 存在的问题

当前,中国海涂的开发在沿海经济发展中起了重要作用,特别是近10年来海涂种植业与养殖业的迅速发展,使沿海人民脱贫致富。但是,由于海涂开发技术力量仍较落后,经济效益较低,还存在着如下问题:① 由于全球变暖,海面上升等因素的影响,近几年来海岸的自然灾害有加剧的趋势,风暴潮、台风、巨浪等突发性自然灾害会直接影响海涂的开发,如1994年和1995年我国沿海的风暴潮过程分别造成了191.7亿元和87亿元的直接经济损失①。② 缺乏科学理论的指导,有许多盲目开发的现象存在。华南沿海的一些地区在港湾内盲目围垦,导致港湾内纳潮量减少,潮汐汊道发生淤积,航道变浅,造成巨大损失;在海涂开发的过程中,不了解海涂冲淤变化规律就进行海涂养殖的情况在全国范围内很多见,往往是刚养殖不久,海涂就开始侵蚀后退,造成了投资的巨大损失[10]。③ 目前,我国已围垦的海涂面积达115万hm²,而其中有部分是围而未垦,除了经济及技术上的原因外,缺乏水资源是限制开垦的重要原因。1994年,中国被列为全世界人均淡水资源13个贫水国之一,人均占有水量仅2 400 m³[11],只相当于世界人均的1/4,居世界第109位。黄河三角洲淡水资源紧缺,因此,黄河三角洲沿岸虽有大量的海涂资源,但其开发会受到水资源的制约。长江三角洲和珠江三角洲淡水资源丰富,但由于近15年来经济的迅速发展,水质污染严重,同时,沿海低地过量开采浅层与深层地下水,导致地下水位下降,淡水层逐渐枯竭,水环境不断恶化,并引起地面沉降及海水入侵,这些都将制约长江三角洲和珠江三角洲沿岸海涂的开发利用。④ 海涂开发的面积较少,已开发的海涂多属分散、小规模的,仍以出售初级产品为主,效益很低。大部分地区仍以海涂养殖业为主,缺乏全方位的开发指导思想,海涂的利用率较低。此外,在养殖过程中,缺乏科学技术的指导,育种、增养殖等问题一直没有得到很好的解决,许多地区多年在同一片海涂上养殖同一品种的海产品(如对虾),由于病害及营养物质的消耗,产量起伏很大。

四、海涂开发利用的战略与对策

(一) 开发利用的战略

针对我国海涂资源的自然特点、各地区的经济及技术条件,提出如下开发战略。

---

① 国家海洋局,中国海洋环境年报,1994、1995。

1. 改变产业结构与方向

我国的海涂开发目前仍集中在渔业及农业上，忽视了海涂作为工业、城市、港口基地的功能。今后应走海涂综合开发之路，在自然与经济条件合适的岸段，将海涂开发为工业、港口、城镇用地，以提高海涂开发的经济效益；海涂城镇可为海涂的深度和广度开发提供产、供、销的场所，并可充分利用海涂的旅游资源、海涂水产品、沿海港口等优势，使海涂各种资源优势都得以发挥，与城市的发展相得益彰，最终实现海涂的社会可持续发展。海涂作为城市用地的功能已日益重要，上海金山石油化工公司就围筑海涂 1 100 hm²，1995年其企业总产值达 58 亿元，利税超过了 12 亿元。随着浦东的经济发展，仍需围筑大片海涂，为上海提供城镇工业用地及粮、棉、油等农副产品；天津开发区的建设也是利用海涂土地资源的很好范例。我国淤泥质海岸缺乏深水大港，随着港口建设的加快，海涂作为港口用地的功能将不断增强。江苏岸外的辐射沙洲区经系统研究后，其中地形稳定的潮流通道可开发为深水航道；南通如东的洋口港即是利用潮流通道来建设 10 万 t 级深水港，沿海海涂开发为大型燃气轮机电厂及相关的临港工业；渤海湾沿岸的曹妃甸亦将利用曹妃甸深槽建设大型石油中转港沿岸海涂随之开发。

海涂农业应变单一投资为多角度、多元化、多方位的投资结构，增加对农业发展中先进技术和设施等现代生长要素应用的投资，积极采用现代化的生产手段和与之相适应的经营方式；海涂渔业以传统的鱼、虾、贝、藻等成品为主；在产品形态上，由原料商品、低值产品为主逐步转为深度加工、多次增值的最终产品为主；实行商品生产及加工业、交通能源和港口运输、对外贸易和旅游的综合开发同步发展，突破走围垦—养垦—开垦的老路子和以种植业为主的传统发展模式，综合开发利用沿海海涂的土地、生物、盐业、旅游资源以及市场、技术、人力资源和港口、电厂、运河等，实行层次式开发和区域式开发相结合。

2. 改善投资环境

我国目前海涂开发的投资环境缺乏凝聚力，在发展横向经济联合、引进外资方面，各地虽有相应的优惠政策，但能真正付之实行的或能实行但最终兑现的并不多见，加上海涂开发中自然、文化教育、交通运输、基础设施等其他方面因素的制约，使得海涂开发获得的投资较少。明确投资方向、投资重点，增加资金的利用率，调动开发经营者的积极性，增加海涂开发的物质投入，保持海涂开发中物质、能量输入与输出的平衡，创造优良的投资环境，吸引大量投资是当务之急。

3. 发展海涂生态经济，实现海涂持续发展

海涂开发的实现是在不断改进海涂自然资源的利用效率的过程中，满足人民日益增长的物质要求，而海涂可持续、稳定、协调的发展要求人类在满足当前自身需要的同时，不危及下一代人满足其自身需要的能力，这就促使我们在海涂开发的过程中要以增加永久性社会财富和福利为最终开发和管理的目的，而不是只顾眼前利益，耗竭资源，破坏海涂环境。

海涂开发是经济再生产与自然再生产的矛盾统一体。经济再生产的人为因素及调控机制（政策、法规、技术、人口、社会）呈不断发展的增长型，因而对海涂资源及其环境的需求也是无限的，而一定时期内海涂生态系统的资源和环境条件的供给则是有限度的，其系统生产力和资源更新、增殖能力也不是无限的，一旦经济再生产对海涂生态系统的资源和环境需求超过海涂生态系统承载量，海涂生态系统则会不堪重负而逐步退化。因此，海涂经济的生态发展是一条可供选择的持续发展之路[12]，它意味着人类—生物—环境的相互作用在多层次、多尺度下，通过生态适应化、内在化和整体化的自组织和自进化作用，逐步协同耦合成生态系统质量提高的过程。

在此理论支持下，海涂开发必须进行海涂资源的全面调查，在此基础上制订海涂开发规划，对海涂资源做出合理的经济评价，划分海涂功能区，从而借助于经济杠杆（资源价格、租税等），实现海涂资源的优化配置。海涂开发的步骤应是：首先应根据海涂的自然适应性，确定海涂开发的方式，分出适合集中开发的地区、不适于开发的地区、需控制开发的地区、需进行自然保护的地区；然后根据人口极限承载数量、水资源承载力等环境最大承载量因子确定海涂开发的强度、规模，以海涂能承受而又不致受到严重损害为原则；最后在此基础上选择适合的生态经济类型。目前已有学者针对海涂开发利用提出了一些生态经济类型[13]，如：综合开发生态经济类型（立体式开发，一体化发展）、潮间带养殖生态经济类型、盐业生态经济类型、森林生态经济类型、芦苇生态经济类型、大米草红树林生态经济类型、海滨旅游生态经济类型、野生动植物资源自然保护区生态经济类型等。根据海涂演变特点及利用方式来划分生态经济类型亦是海涂开发中一项有意义的工作，需进一步研讨。

4. 实现海涂开发的一体化管理

一体化管理（integrated management）是当前海洋开发管理中一个全新的概念。海涂开发的一体化管理是指在海涂开发过程中，在开发计划制订、资源利用、环境保护等方面进行综合考虑和协调，在行政管理上实行统一的调度和实施，以确保海涂开发的程度与其所依赖的自然体系的永续利用能力和存活能力之间保持平衡。海涂的一体化管理将以一定的法律依据，协调海涂开发的经济效益、环境效益、社会效益，使海涂发展走上更合理的良性循环，从而保证海涂的可持续发展。

（二）我国海涂资源开发利用的区划

海涂资源区划的目的是确定根据资源利用潜力的同一性原则，确定今后的主攻方向，避免利用的盲目性。区划的原则是根据海涂的自然条件、海涂的类型等，在考虑经济、社会、技术等因素的基础上进行实用区划[14]。海涂利用方式是在一定的自然和社会经济条件下，经过不断适应和改造自然环境而发展起来的，因此，海涂的利用区划必须从现状出发。根据我国海涂的地理位置、气候条件、海涂类型、生产特点和开发利用方向，把我国的海涂资源大致划分为 5 个利用区。

1. 辽东区

指辽东半岛东、西海岸和辽西走廊沿岸，年平均气温为 9.4 ℃，海涂底质以粉砂质泥为主，底栖生物量极为丰富，可达 366 g/m²，但目前的海涂养殖产量仅为 0.5 t/hm²，有较大潜力，存在问题是养殖苗种没有彻底解决，养殖技术落后，单产低。今后海涂利用以渔业与农业为主，应采取增殖与养殖并举，单品种放养与多品种套养、轮养相结合等方针，为逐步过渡到以增殖对虾资源为主的对虾生产阶段准备条件;配合辽河三角洲油田、辽东湾油气田的油气开发、锦州及葫芦岛港的建设,发展为石油天然气开发基地和加工工业区。

2. 黄河三角洲区

现代黄河三角洲区从天津至莱州湾,海涂以淤泥质粉砂为主,本区主要的气候特征是冬夏冷热极差大,冬季最低气温可达-20℃以下,沿海海涂在冬季普遍出现海冰现象,因此高潮位生物量少,在目前的经济技术条件下,开发利用困难较多。开发的有利条件是台风频率小,年均值不足一,因而风暴潮对海涂开发的破坏较小。由于受水资源的限制,今后的发展方向仍应以海涂渔业为主并配合海涂农业的开发,同时加速黄骅港等深水港的建设。本区内深水港的建设可使黄河海港与内河水道相联结,对促进黄河流域的经济发展有重要意义;此外,深水港可承担胜利油田、大港油田部分原油的外运,大幅度降低运输成本。如位于黄河三角洲西北角的滨洲港是良好的深水港址,河口内航道保持了较大的水深,一些地方水深大于10 m,可建设成为35 000 t级的深水港,并可以港口为依托,体现海涂作为油气开发基地的功能[15]。海涂渔业今后应以苗种放养、资源增殖为主,近期可以中国对虾、魁蚶、文蛤为重点,随着育苗种类增多,可增加梭子蟹、金乌贼等品种,综合利用,提高效益。但海涂农业需根据海涂的自然分带,因地制宜。

3. 长江三角洲区

本文将其界定为从江苏的连云港至浙江温州,其自然条件相近,海涂形成主要受长江沉积物的影响。该区海涂面积巨大,海涂底质以砂质粉砂、粉砂及泥质粉砂分布最广,气温的年平均值为14.2℃,底栖生物的平均生物量为70 g/m²。区内的江苏及浙江对围垦海涂发展农业已有成功的经验,上海围垦海涂发展海涂大农业及工业城镇均已取得巨大效益。今后本区海涂的开发方向将是海涂大农业、海涂及近海渔业,以大型深水港为依托的临港工业与海岸工业以及城镇建设等。

值得一提的是江苏省海涂外有大面积的辐射沙洲,总面积为2万 km²,其中0 m以上的巨型沙脊8条,面积为2 125 km²,是潜在的土地资源;其余为水下沙脊及水深150~30 m的潮流深槽,是珍贵的港口资源,可开发深水港及临港工业区,这些将为海涂开发提供一个新的类型和模式。长江三角洲地区是我国经济最为发达的地区,随着本区经济的进一步发展与上海国际航运中心的建设,本区内将形成以上海港为中心,北仑港、太仓港、洋口港、连云港等港口为次中心港的港口群①,港口群的建设将与区内大城市、大企业的发展相互促进,创造本区经济再次腾飞的契机。

4. 华南港湾区

本区主要包括福建、广东的港湾型海涂资源。本区属亚热带海洋性季风气候,气温年均值在16℃~20℃之间,海涂底栖生物量处于全国的较低水平,海涂的利用程度为全国最高,利用面积达549 km²,占全国已利用面积的1/4,闽南一带的海涂养殖率已高达40%以上。存在的问题是放养密度大,且同一种类长期放养,生产力降低。为了配合本区经济的迅速发展,今后应把本区海涂开发利用方向引向围垦水域的开发利用,积极发展海

---

① 任美锷.改革开放形势下我国港口发展条件分析兼论建设上海国际航运中心问题.1997.

涂工业,利用海涂推动经济发展。海涂渔业应降低水域的负载,提高单产,避免病害,应积极发展港湾养鱼、虾或多种混养。

5. 珠江三角洲区

本区包括广东及广西,属热带、亚热带季风气候,气温年平均值为 22 ℃～23 ℃,台风出现的频率高,海涂底栖生物量高,为 327 g/m²,生物资源丰富,自然环境优越已利用海涂的面积达 481 km²,占全国已利用面积的 21％,但其中广西因土壤肥力低,仅占南海北岸区已利用面积的 6％左右,而广东则存在着围垦的盲目性,因而不垦的现象严重,海涂养殖经营单一,经济效益不大,养殖的品种少,产量低。考虑到两广地区台风多、风暴潮危害大的特点,今后海涂的发展方向主要为海涂大农业,发展以精细蔬菜、花卉、果品为主的创汇农业及观光农业,并利用珠江河道水网开发港口及临港加工区、城镇,同时还应积极发展海涂旅游业,珠江三角洲海涂将成为综合开发、经济高密度发展的示范区域。

参考文献

[1] Wang Ying. 1983. The mudflat of China. *Canadian Journal of Fisheries and Aquatic Sciences*, 40 (Supplement 1): 160 – 171.

[2] Wang Ying and Zhu Dakui. 1994. Tidal Flats in China. In: Oceanology of China Seas, (2). The Netherlands: Kluwer Academic Publishers. 445 – 456.

[3] 吴传钧,郭焕成. 1994. 中国土地利用. 北京:科学出版社. 130 – 131.

[4] 史运良. 1989. 黄河入海水沙变化与河口治理. 南京大学学报(地理版),(10):24 – 34.

[5] 程天文,赵楚年. 1984. 主我国要河流入海径流量、输沙量及对沿岸的影响. 海洋学报,(4):461 – 471.

[6] 王颖. 1990. 中国的海涂. 第四纪研究,(4):295 – 330.

[7] Ying Wang, Mei-e Ren, Dakui Zhu. 1986. Sediment supply to the continental shelf by the major rivers of China. *Journal of the Geological Society*, London, 143(6): 935 – 944.

[8] Ying Wang and Xiankun Ke. 1989. Cheniers on the East Coastal Plain of China. *Marine Geology*, 90(4): 321 – 335.

[9] 朱大奎. 1984. 中国海涂资源的开发利用问题. 地理科学,6(1):34 – 40.

[10] 任美锷. 1985. 中国的淤泥质海涂沉积研究的若干问题. 热带海洋,4(2):6 – 14.

[11] 钮茂生. 1995. 中国水利报,1995.12.2.

[12] 胡聃. 1996. 实现可持续性——生态发展模式探讨. 自然资源学报,11(2):101 – 106.

[13] 章成逸. 1990. 海涂开发生态经济类型的形成、划分及其综合评价. 海洋与海岸带开发,7(2):2 – 7.

[14] 王中元. 1991. 中国浅海海涂渔业区划. 杭州:浙江科学技术出版社. 1 – 60.

[15] 任美锷. 1994. 中国的三大三角洲. 北京:高等教育出版社. 97 – 98.

Environmental Characteristics and Related Sedimentary Facies of Tidal Flat Example from China [*]

1. The distribution of tidal flat coasts

Tidal flats and associated muddy sedimentation are especially well developed along lowland shorelines in a wide range of environments from the Arctic to the tropics. It is distributed along the Atlantic Ocean coast and in distal settings, such as Frobisher Bay in the Canadian Arctic and the Bay of Fundy, along the North Sea coasts of the Netherlands, Britain, Germany and France, and along the Atlantic trailing edge coasts of North and South America, such as Amazon River estuary and the Guyana coasts. In the Pacific region, tidal flats are represented in a variety of morphodynamic environments, such as those of the shallow shelves of northern Australia; the Pacific Islands, where they may be associated with coral reefs and lagoonal mangrove forests; in the indented embayments of steepland island arc complexes with weathered volcanic sediment supplies such as the muddy coasts of Japan, Indonesia, New Guinea and New Zealand, and those of the west Korean coast, which are associated with river mouth settings and a huge fine sediment supply from offshore. Tidal flats in China are extensive covering about 2 M ha of intertidal area, and extending along some 4 000 km of coastline.

Major tidal flats in China are developed along the fringing zone of the North China plain and are related to the large Yellow and Changjiang Rivers, along the Bohai Sea and the Yellow Sea where tidal range is about 3 m and the velocity of tidal currents is less than 1 m/sec, which is less than that found in the estuaries. The tidal flats are between 3 to 18 km wide and are much more extensive than they are in estuarine areas. Because the coastal slope is very gentle (the gradient is less than 1/1 000), wave action is restricted to offshore areas and tidal currents drive the dominant dynamic processes on the flats. The tidal flats along the East China Sea and the South China Sea are developed in long-narrow embayments or other types of sheltered coasts (Wang & Zhu, 1990). Short fetches within such environments limit the wind/wave interaction. Hence, tidal

* Ying Wang, Dakui Zhu, Guiyun Cao: *Proceedings of Tidalite 2000*—"*Dynamics, Ecology and Evolution of the Tidal Flats*"-*Fifth International Conference on Tidal Environments ABSTRACTS*, pp. 158 - 161, edited by Park, Y. A., Chun, S. S., Choi, K. S., June 12—14, 2000, Hoam Convention Center, Seoul National University, Seoul Korea; *Proceedings of Tidalite 2000 Special Publication*, pp. 1 - 9, edited by Yong A. Park and Richard A. Davis, Jr. (-Guest editors-), The Korean Society of Oceanography, 2001.

currents are, once again, the major agents controlling sedimentary processes (Fig. 1).

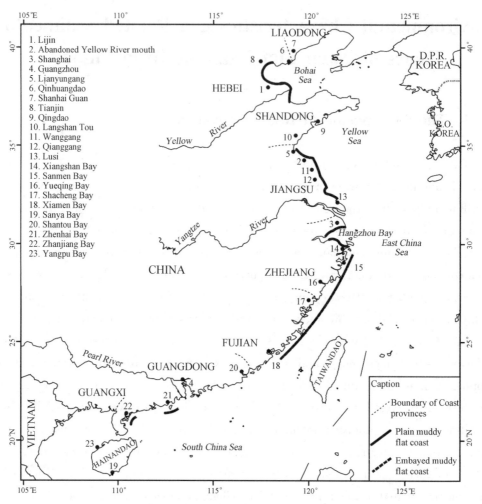

Fig. 1　Distribution of muddy coasts & associated tidal flats in China

1.1　Sedimentary dynamics and geomorphology of tidal flats

The major factors which control the development of tidal flats are tidal dynamics, gentle coastal slopes, and abundant fine-grained sedimentary materials.

Silt is the major constituent of the deposits, ranging from fine to coarse. Very fine sands and clays are also present on most tidal flats. All such sediments are carried onto the flats mainly by the flood tidal current which has brought material eroded from the adjacent seabed since late Holocene transgression; this process infers a landward transport. Sediments on offshore seabed, being fluvial in origin, have been supplied to the coast area during times of lowered sea level by large rivers, such as the Yellow River and the Changjiang (Yangtze) River, and river sediments are still the major provenance today for tidal flat development in China.

The processes of rhythmitic tides preserve in the sediment record as repetitive thick-thin graded two to four laminal or ripple beddings, indicate daily or monthly tidal periods, and tides of diurnal, semi diurnal or mixed patterns (Allen. et al 1995, ERIK P. Kvale 1999).

Tidal currents carry sediment in traction, saltation, and in suspension towards the land. Velocities decrease because of friction from the intertidal zone seabed (Wang, et al. , 1963; Wang, 1983; Evans, 1965). Thus, sediment grain size decreases on intertidal flats from low water level towards high water level, and again in the vicinity of deep water. Fluid mud, which appears on tidal flats after storms, is carried by longshore drift from adjacent river mouths (Wang et al. , 1990). Generally, flood tides attain higher velocities and have shorter duration than the corresponding ebb. Furthermore, the extensive accretion on flats suggests that flood-tide deposition has been dominant over ebb-tide erosion. Even though somewhere the flood tide may be of longer duration, given the additional energy from waves, the flood tide is still dominant. Micro-morphological features, such as ripples and small creeks, exposed on intertidal flats during the passing of the ebb tide, reflect the last phrase of the ebb current. Tidal action affects the spatial scale and sedimentary characteristics of tidal flats. If the sediment supply is sufficiently large, the width of the tidal flat is dependent primarily on tidal ranges. The varying characteristics of the tides (e. g. , spring tide or neap tide) and the accompanying tidal level influence on the nature of sedimentary structures (Zhu, et al. , 1985). In a sequence of neap tide to spring tide, a thin sandy wedge is deposited from seaward to landward. After several such cycles, a mixed sediment layer of clay and sand forms between the tidal flats (Fig. 2). In Bohai Bay, there is a 1. 91 m difference between the levels of spring high tide and neap high tide;

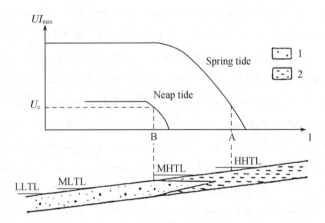

Fig. 2 Tidal level variation and sediment distribution of inter-tidal flat (after Zhu & Gao, 1985)

1. Sandy silt, very fine sands   2. Clayey silt, clay

UI_{max}: Velocity of spring high tidal current

U_c: Velocity of neap high tidal current

I: Distant to low tidal current

the horizontal distance of the two levels is 1 500 m. Consequently, the entire tidal flat facies shifts with distance.

In the Jiangsu coastal zone, the tidal level difference between spring high tide and neap high tide of 3m causes a shift of about 1 900 m horizontally in the sedimentary facies of tidal flats. As a result, a complex sedimentary structure is produced on the tidal flat (Fig. 2). In Jiangsu, the tidal current velocity of the spring tide is twice that of the neap tide, sediments deposited on the tidal flat during the spring high tide have a medium diameter of $Md=5.3\phi$, and a medium diameter of 6.5ϕ during the neap tide. This process forms laminae of clay and fine silt (Ren, et al., 1984).

Tidal flats can be classified further into three types according to their characteristic morphology and sedimentary features, the classification shows the different levels of wave energy input in addition to the predominating effects of tidal dynamics (Wang, et al., 1990). Coastal plain tidal flats are divided into: (1) Sand flat. The Wash of East England is an example of an intertidal flat with high wave energy influences. In China, a similar type of flat appears in the Xiamen Bay, Fujian Province. (2) Silt flat. This well developed flat occurs along the northern part of Jiangsu plain coast facing the Yellow Sea. A medium level of wave energy stirs up sand and silt material offshore, and then transports the material onshore through the action of flood currents, this may be the typical and major tidal flat. (3) Mud flat. This type is exemplified by the tidal flats of the Bohai Bay. Tidal currents are the only predominant transport agents with minimal wave influences, but with a large source of clay and fine silt size material supplied by the Yellow River. Only this kind of flat has a Fluid mud zone (0.5~1.0 km wide in the Bohai Bay and 2~4 m thick of fluid and soft mud), it marks the mean high tide level. An abnormal type of boulder flat is developed in the Arctic environment because of frequent of floating ice that is carried onto flats by storms and tidal currents (Dale, 1985).

Morphological, sedimentary, and benthic faunal zonations are typical features on intertidal flats everywhere (Table 1). There are salt marsh or clay polygon zone between superior high tide level and high tide level, muddy flat below high tide level, mixed mud-silt flat with microerosional features developed along half tide level. And sandy silt flat with ripples on the lower flat. The zonation features reflect variation in tidal dynamics. Zonations have been recorded in the sediment strata as typical fine lamination and delicate bedding of silt, silty clay and fine sandy silt. With microstructures such as polygon cracks, wormholes or microtunnels, plant roots, ripples etc, and the fine silt or clayey laminal beddings have mosaic structures of sandy coarse bedding or lenticular or other erosional structures intervened. Sedimentary structures and the thickness or scale of these beddings reflect also the range of environmental variation in the coastal zone. Thus study of modern sedimentary facies can be used to identify sedimentary environment of ancient deposits especially for tracing the land-sea interacted zone.

Table 1 Comparison of various tidal flat zonations (after Wang, Collins and Zhu 1990)

Zonation	The Wash, North Sea, U.K. (after Evans 1965)	North Jiangsu Coast, Yellow Sea, China (Zhu and Xu 1982)	Bohai Bay, Bohai Sea, China (Wang et al. 1964)	Bay of Fundy, Atlantic, Canada (Wang & Zhu, 1990)	Frobisher bay, Arctic, Canada (Dale, 1985)
Upper flat (Clayey)	A. Salt Marsh Inundated only during Spring High Tide. Flat with muddy depressions. Well-laminated silt-clays, clays and clayey silt. Very few sands. B. Higher Mud Flats Located between Spring High Tidal level and High tide level with forms of elongate depressions, shallow channel system, ripples and mud cracks. Laminated silty sands and sand silt.	I. Supratidal Zone(grass flat) Located above Spring High Tidal level. Only flooded during Storm High Tide. Very thin laminae of clayey silt with organic layers. II. Suasda salsa Mud Flat Located between Spring to Mean High Tidal level. Laminae of clayey silt with thin layers of silt. Worm holes and polygons (mud cracks) on the tidal flat surface.	Salt Marsh Plain and Lagoon Depression. I. Polygon Zone Clay or silty deposits with mud cracks. Only flooded by Spring High Tide. II. Inner Depositional Zone Muddy flat located below Mean High Tidal level.	I. Beach II. Salt Marsh III. Mud Flat	I. Beach II. Finer Sediment Flat (sandy silt)
Middle flat (clayey-silt)	C. Inner sandy Flats Less well laminated sands and silty sand, with muddy layers. Ripples and scour and fill structures. D. Arenicola. Sand Flats Poorly satisfied sands and silty sand, with minor fine-graind sediments mixed by the worm Arenicola marina E. Lower Mud Flat Poorly sorted laminated sands and silty sands intermixed with muddy layers.	III. Mud-Silt Flat Located between Mean Tidal level to Neap High Tidal level. Alternating laminae of clayey silt and silt, medium laminae with cross-bedding.	III. Erosional Zone of mud-silt flat Around mid-tide level, alternating laminate of fine silt and silty clay. Overlapping fish scale sedimentary structures or tidal creeks. IV. Outer Depositional Zone of Sandy Silt Flat	IV. Sandy-silt Flat With granular sands in the tidal creek	III. Boulder Flat IV. Boulder Ridges V. Very Bouldery Flat

(Continued)

Location Author		The Wash, North Sea, U.K. (after Evans 1965)	North Jiangsu Coast, Yellow Sea, China (Zhu and Xu 1982)	Bohai Bay, Bohai Sea, China (Wang et al, 1964)	Bay of Fundy, Atlantic, Canada (Wang & Zhu, 1990)	Frobisher bay, Arctic, Canada (Dale, 1985)
Lower flat (sandy-silt)		F. Lower Sand Flats Fine. Medium sands and silty sands with thicker laminae, transverse larger ripple marks with steep lee slopes facing ebb direction.	IV. Sandy Silt Flat Fine sands and silt flat with wave and current ripples.	V. Very fine sands and coarse silt with ripple forms.	V. Silty Ripple Flat VI. Tidal Channel	VI. Graded Flat
Classification		Sand Flat Tidal dominant flat with strong influence by wave action especially during storm periods deposit sandy materials.	Silt Flat Tidal current dominated flat with slight or occasional wave influences. Silt deposits originate mainly from offshore submarine delta of old Changjiang River	Mud Flat Tidal currents dominate intertidal zone flat with very fine sediment originally from the Yellow River. Very gentle coastal slope to form classic intertidal zonation of four units along 400 km long coastline.	Sandy-silt Flat Mixed type developed in the clayey-sandy bedrock embayment with high magnitude, current dynamic regime.	Boulder-muddy Flat Developed in the Arctic Ocean region, where coastal zone characterized by frequent floating ice, clay and boulders drop onto flat during storms.

1. 2　Comparative study on sediment facies of tidal flat-Present and Past

A comparative study has been carried out between modern tidal flats of North Jiangsu Coast and the siltstone of lower section of the Huangmaqing Formation of middle Triassic series, which presence is located in land near Nanjing City of the Jiangsu Province. It has also extended the study on the Zhanjiang Formation of late Pliocene to early Pleistocene Period in the northern Hainan Island of the South China Sea.

The result represents an excellent sequence of land-sea interaction with sediment facies of tidal flats.

2. The tidal flats of North Jiangsu coast

The tidal flats of North Jiangsu coast are 3～18 km wide developed along the inter-tidal zone of 700 km long plain coasts. Semidiurnal tide with an average tidal range of about 3 m, but the maximum tidal range is 9. 28 m in the Xiaoyangkou tidal channel located in the offshore area between Wanggang and Qianggang (Fig. 1). Sediment supply is plentiful and mainly from Changjiang River in the south, the abandoned Yellow River mouth in the north, and the eroding sandy ridges offshore. It is a typical type of tidal flat with silt as its sediment (Table 1).

The superficial zonation features of tidal flat shown as following:

(1) Upper flat: salt marsh with clay cracks and worm holes exposed during the dry period of ebb tides. (Fig. 3)

(2) Muddy flat is located between spring to mean high tide level. (Fig. 4)

Fig. 3　Salt marsh with clay cracks and worm holes exposed during ebb tide, upper tidal flat of north Jiangsu coast, China

Fig. 4　Muddy flat of north Jiangsu coast, China

(3) Mud-silt flat is located between mean high tide level to neap high tide level, with erosional morphology as dominant characteristics. (Fig. 5)

(4) Lower flat of sandy silt and silt deposits with ripples as it's dominant features.

(Fig. 6，Fig. 7)

Fig. 5 Erosional features and laminae bedding in the mixed mud-silt flat, located around the middle of tidal flat, North Jiangsu coast, China

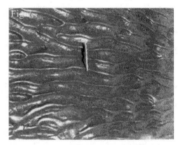

Fig. 6 Lower flat of sandy silt deposits with ripples as it's dominant features, North Jiangsu Coast, China

Fig. 7 Ripples on the lower tidal flat, North Jiangsu Coast, China

To combine the surficial features and section structures, that sedimentary facies of modern tidal flat can be summarized as (Fig. 8).

(1) Fine lamination and delicate bedding of silt, or very fine sandy silt and clayey silt or silty clay are the typical structures of tidal flat in the muddy coasts. The laminae reflect the changes of periodic tide dynamics.

(2) Coarser sediments such as coarse silt, sandy silt, even very fine sands, predominate in the lower section of sediment strata, ripples, lenticular interaction of flaser beddings, dune-shaped cross bedding, herringbone cross bedding, rhyotaxitic or fine streamline structures are common in the section.

(3) In the middle flat, lamination of silt and mud interval beddings predominate and have more erosional features, eddies, and lenticular or herringbone shaped cross-stratification, these structures are normally in a mosaic pattern.

(4) Clayey laminations are predominant as the characteristics of upper flat sedimentation, sedimentary facies are with structures of polygon or desiccated clay cracks' clay pebbles' wormholes' microtunnels' plant roots' bioturbation or filled with calcareous or pyrites materials, worm pillars, lime nodules etc.

Tickness (m)	Profile	Sub-facies	Sediments	Sedimentary Structures
0.5~1.0	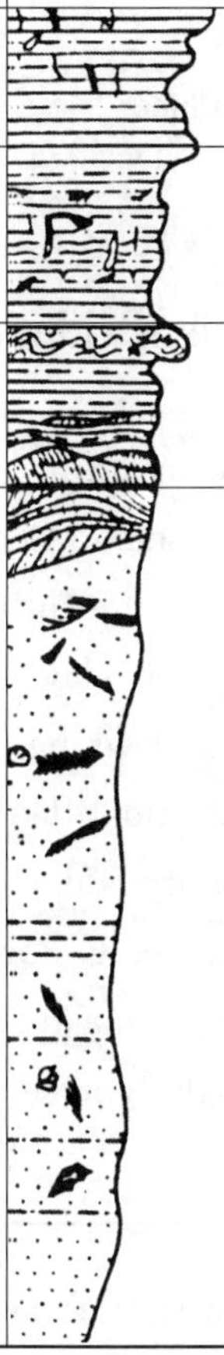	Subratidal flats of salt grass or marsh	Silty caly or clayey sit	Laminae bedding with roots, worms burrows, clay cracks
1.0~3.0		Mudflats of high tidal level	Clayey silt	Laminae bedding with polygons or desiccation cracks, ripples, lenticular beddings, roots and burrows structures
1.0~2.0		Mixed flats of mean T. L.	Silty clay/silt	Clayey and sandy interlaminations, ripples and flaser beddings, bioturbation, roots burrows and tubes, and lateral accumulation structures.
15.0~20.0		Silt flats of Low T. L.	Silt or fine sand	Dune-shaped cross bedding and herringbone beddings, grain rhytmic stratification of coarser and fine sediments

Fig. 8　Facies sequence of tidal flat sedimentation of North Jiangsu coast, China

(5) Storm deposits on the tidal flat appear as thinner layers of clayey sands, or dune-shaped structures with cross beddings.

(6) Shell beach ridges (cheniers) developed along the landward boundary of inter-tidal flat, as well as the salt marshes. Thus, mollosca shells' plants remnants' microfossils' or pollens of inter-tidal zone are combined evidence to indicate the sedimentary environment.

The sediment facies are commonly shown on the natural profile or in bore holes. It reflects a normal sedimentation sequence of accumulated tidal flat in a lower energy environment. The sedimentation pattern can be recognized as a vertical accumulation in the upper flat, and a lateral deposit along the lower tidal flat (Fig. 9). The large area of well developed tidal flat along north Jiangsu coast can be used as a classic sedimentation pattern of tidal flats in the study of ancient sedimentary rocks.

Fig. 9　Sedimentation patterns of tidal flat, North Jiangsu coast, China

A. vertical deposit; B. lateral transversal deposit

3. The comparative study on siltstone for tracing ancient tidal flats

The sedimentary characteristics of ancient tidal flats can not be found in a single profile or location, even with the variations within modern tidal flat deposition. Generally, there are several major similarities determined by the lower energy environment of tidal dynamics. The tidal bedding (including thin laminations, inter layered clay and silt bedding, lenticular bedding, wave bedding and flaser bedding), and fining-upward sequence of sediments both make up the key elements to research the tidal-flat facies. Under the view point, that the lower part of Huangmaqing Formation of middle Triassic series of Purple Mountain in Nanjing area (Fig. 10) and the Zhanjiang Formation (Fig. 11) of late Pliocene to early Pleistocene strata, Macun, Haikou, China are ancient tidal flat deposits similar to the sedimentary pattern of present north Jiangsu coastal tidal-flats.

Tickness (m)	Profile	Facies	Sediment	Characteristic Sedimentary Structures
>0.3		High tidal flats subfacies	clavish siltation	laminae bedding, microripples, vertical worm tubes (burrow) filled up by calcareous material
1.8		Middle tidal flats subfacies	medium-coarse siltation and fine silt stone interval	Ripple-cross limination, lenticular bedding, wave-shapled bedding, sandy/clay interval thinner layers
>0.4		Low tidal flats subfacies	coarse silt stone	Interfelted ripples, wave ripples, ripple cross bedding, herringbone cross bedding

Fig. 10　Sedimentary sequence of lower section Huangmaqing Formation middle

Triassic series Purple Mountain, Nanjing,China

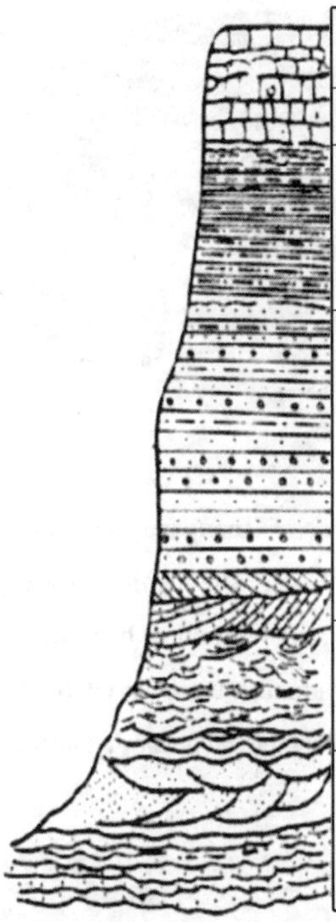

Thick-ness (m)	Facies	Sediment	Characteristic Sediment Structures
1.7	Basalt		
1.1	high micro-ripples, tidal flats	brownish grey colour muddy silt	laminae bedding with partly contain clayer concretion
1.8	mosic Sand-granule layer	medium coarse sands contain granule &. bedding thinner layers of fine sands	parallel lamination, inter-stratified bedding, herringbone cross
0.4	middle tidal flats	silty fine sands of silt	flaser bedding, ripple cross bedding, lenticular bedding sandy/clay intervals
>2.0	low tidal flats	silty fine sands	herringbone cross bedding, wave ripples, sandy/clay interval thinner bedding

Fig. 11　Zhanjiang Formation of late Pliocene to early Pleistocene period, Macun, Haikou, Hainan Island, China

(1) The natural profiles of the silt stone of Huangmaqing and the half consolidated silty and sandy strata of Zhanjiang Formation are with well preserved fining-upward grain sequence, and structures of rhythm thick-thin graded laminae 'ripples' lenticular' and herringbone cross bedding 'polygon cracks' worm tubes and calcareous fillers. Figure 12~15 show the sedimentary facies of Zhanjiang formation and figs. 16~18 show the Huangmaqing formation features.

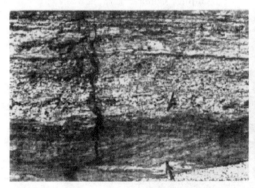 

Fig. 12　Lower section of Zhanjiang formation Macun profile. Herringbone cross bedding in the sandy granule layer, and a erosional surface between the layer and upper lamination layer.

Fig. 13　Wave ripples, lenticular bedding streamline laminae indicated the condition with powerful currents, and gradually transferred to a sandy granule layer above. Zhanjiang Formation, Macun profile, Hainan, China

Fig. 14　Ripples on the surface, upper layer has polygon cracks. Lower Zhanjiang Formation

Fig. 15　Middle-lower sections of Zhanjiang Formation, Macun profile shown the fining-upward sequence, ripples and lenticular bedding in the lower part, gradually, transferred to the thinner laminae bedding of sandy-clayey intervals

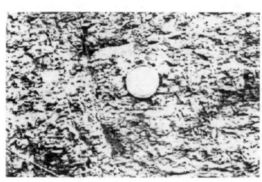

Fig. 16　Vertical worm burrows, filled up by hard calcareous are iron material, preserved well in the laminated silt stone, Huangmaqing Formation, Purple mountain profile, Nanjing, China

Fig. 17　Ripples on the surface of lower Huangmaqing Formation, Purple mountain profile, Nanjing, China

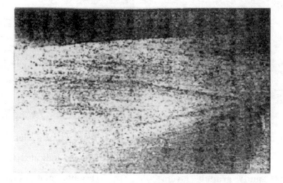

Fig. 18　Herring bone cross bedding and lenticular bedding in the lower section of Huangmaqing Profile

(2) The comparison of other elements of three sediment strata.

① Paleontology analyses indicate that Zhanjiang Formation of Quaternary strata and the lower section of Huangmaqing Formation of middle Triassic strata Period were deposited in the shallow water and brackish environment, especially that the wormtubes ($d=5\sim7$ mm, and $2\sim3$ mm) were the burrows of skolithos, which lived in the sea shore environment. (Table 2).

Table 2　Comparison of Tidal Flat: Paleontology Group

Region / Species	Water Environment	Normal Sea Water	Brackish (or Euryhaline)	Fresh Water
Tidal Flat of North Jiangsu		Globigerina, Orbulina, Lagena, Quinqueloculina	Ammonia beccarii, Elphidium, Ammonia convexidorsa, Nonion, Sinocytheridea, Albileberis sinensis, Neomonoceratina dongtaiensis	
Zhanjiang Formation, Hainan, Macun		Globigerina, Quinqueloculina, Asterorotalia, Pseudorotalia	Metacypris, Cythere	
Lower Section of Huangmaqing, Nanjing, Jiangsu			Skolithos Xiangxiella, Yangzicaris, Bakevellia	Stellatochara, stenochara, Darwinula

② Salinity comparison

• Present salinity of tidal-flats along North Jiangsu coast are 24. 6‰~32‰, and the average salinity is 28. 5 ‰. In the area with the river influence, the salinity is 18‰ to 24‰, the average value is 21‰, these values belong to polyhaline of mixed salty water (maxohaline).

• Paleosalinity of Zhanjiang Formation is 6. 4‰ to 8. 4‰, which is the mesohaline of mixed salty water, and the value of 2. 5‰ belongs the Oligohaline of mixed salt water (Table 3).

Table 3　Paleosalinity of Zhanjiang Formation, Macun profile

Sample	Location	Boron B_k (ppm)	Paleosalinity Sp‰
M03	High Tidal Flats	19. 6	8. 4
M04		13. 6	6. 4
M07	Sand Granular Layer	4. 2	2. 5
M08	Middle Tidal Flats	15. 2	6. 9
M11	Low Tidal Flats	14. 7	6. 5

• Paleosalinity of lower Huangmaqing Formation is 12. 3‰ to 17. 6‰, which belongs the mesohaline of mixed salt water (Table 4).

Table 4　Paleosalinity of Lower Huangmaqing Formation

Sample	B_k(ppm) In Kaolinite	Paleosalinity Sp‰	Average Salinity ‰
Z55・≤	30. 2	17. 6	
Z58・±	20. 0	12. 3	14. 6
Z38・∞	24. 1	14. 3	

③ Size analysis on the three sedimentary strata have shown the well related similarities of fine sediment component with the silty material as the major types(Figs. 19，20，21).

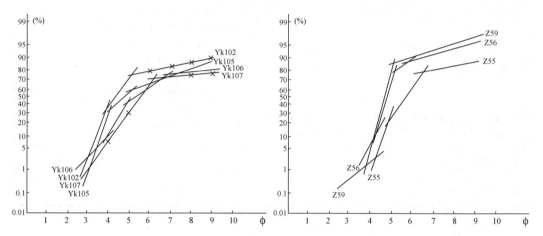

Fig. 19　Size probability diagram of north Jiangsu tidal flat(Rudong area)

Fig. 20　Size probability diagram of lower Huangmaqing Formation

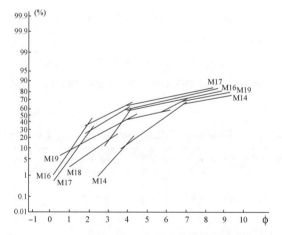

Fig. 21　Size probability diagram of Zhanjiang Formation

4. Conclusion

(1) Sedimentary facies of low wave energy and gentle-sloped coastal flats is

characterized by predominant tidal dynamics, finer sediments of silt and clay, and an upward fining of deposits in the form of rhythmic delicate lamination of coarser and finer grain materials intervals. Flood and ebb tidal periods associated with dynamic processes have formed the sedimentary structures with both subaerial and submarine features, and related biogenetic remnant.

(2) Tidal flats are not only developed in the wide range of climate region, but also developed during different periods of geological time. The study on the present sedimentary processes and related sedimentary facies of tidal flats is important and can be used as key indicaters to identify the dynamic processes and environmental natures, even the land-sea interacted history of ancient fine grained sedimentary strata. The study can be extended to estimate the porosity of the sedimentary rocks for water and petroleum reservoir analysis.

Acknowledgements

Acknowledge to Professor H. Jesse Walker from Louieiana State University for his help to review the manuscript, correction the English version and the valuable suggestions, Mr. Zhansheng Niu from Nanjing University to typewrite the manuscripts and to prepare the photos and figures.

References

Archer, A. W., Kuecher G. J., and Kvale E. P. 1995. The role of tidal-velocity asymmetries in the deposition of silty tidal rhythmites (CarBoNIFERous, Eastern Interior Coal Basin, U. S. A). *Journal of Sedimentary Research*, A65(2): 408 – 416.

Dale, E. J. 1985. Physical and biological zonation of intertidal flat at Frobisher Bay. In: N. W. T. 14th Arctic Workshop, Nov. 6—8, 1986, Halifax. 180 – 181.

Evans, G. 1965. Intertidal flat sediments and their environments of deposition in the Wash. *Quart. J. Geol. Soc.*, 121: 209 – 245.

Kvale, E. P., Johnson H. W., Sonett, C. P., Archer, A. W. and Zawistoshi A. 1999. Calculating lunar retreat rates using tidal rhythmites. *Journal of Sedimentary Research*, 69(6): 1154 – 1168.

Ren, Mei-e, Zhang, Ren-shun and Yang, Ju-hai. 1984. Sedimentation on tidal mudflat in Wanggang area, Jiangsu Province, China. *Marine Science Bulletin*, 3(1): 40 – 54.

Wang, Ying, Zhu, Dakui and Gu, Xihe. 1963. The characteristics of southwest coast of Bohai Bay. In: Proceedings of 1963 Symposium of Chinese Society for Oceanography and liminology, Science Press. 55 – 57.

Wang, Ying. 1983. The mudflat coast of China. *Canadian J. of Fisheries and Aquatic Sciences*, 40 (supplement 1): 160 – 171.

Wang, Ying, Collins, M. B. and Zhu, Dakui. 1990. A comparative study of open coast tidal flats: the Wash (U. K.), Bohai Bay and West Yellow Sea (Mainland China). In: Proceedings of International Symosium on the Coast Zone 1988. Beijing: China Ocean Press. 120 – 130.

Wang Ying and Zhu Dakui. 1990. Tidal flat of China. *Quaternary Science*，(4):291 - 300.

Zhu，Dakui and Gao，Shu. 1985. A mathematical model for the geomorphic evolution and sedimentation of tidal flats. *Marine Science Bulletin*，4(5):15 - 21 (in Chinese).

Zhu，D. and Xu，T. 1982. The coast development and exploitation of middle part of Jiangsu coast. *Actascentiarum University Nankinesis*，3:799 - 818 (in Chinese).

Tidal Flat and Associated Muddy Coast of China [*]

1 SUMMARY

Tidal flats are a major geomorphic feature along the coasts of China and cover an area of 2 M ha. Two major physiographic types may be identified, namely, (i) those fringing the shorelines of major plains which occur extensively in the north of China, and (ii) those occurring along embayed steepland coasts, which are found mainly along the south east coast. The former are associated with the reworked alluvial sediments deposited within the coastal nearshore and shelf environments by the two major rivers, while the latter reflect more localized catchment and tidal embayment depositional processes and exist mainly within elongated embayments along steepland coasts. The tidal flats developed off the North China Plain are between 3 to 18 km in width and are much more extensive than those associated with estuarine embayed areas.

Typical geomorphic features of the tidal flats include a relatively steeper slope within the high tide zone, a broad near flat mid-tidal zone, and a steeper low tidal to sub-tidal zone. The average slope across the entire system is less than 1 m/km. A striking characteristic of the tidal flat deposits is the very high proportion of mud-sized sediment. Generally, muddy clay-rich sediments predominate within the high and supra-tidal zones. The broad and extensive mid-tidal flat is generally dominated by silt with muddy lenses, while the low-to-sub-tidal zone features both sand and muds.

Tidal processes play an important role in the evolution of the tidal flats. For the extensive fringing tidal flats of the North China Plain, tidal range is about 3 m and the velocity of tidal currents is less than that found in the embayment tidal flat environments. The tidal flats along the East China Sea and the South China Sea are developed in narrow embayments or along other types of sheltered coasts. Fetch within such environments limits wave generation, and tidal currents likewise play important roles. Thus, major factors influencing evolution of the tidal flats as the dominant morphology of muddy coast are tidal dynamics, the gentle coastal slope, and the abundance of available fine-grained detrital sediments.

---

[*] Ying Wang, Dakui Zhu, Xiaogen Wu: T. Healy, Y. Wang and J. -A. Healy (Editors), *Muddy Coasts of the World: Processes, Deposits and Function*, chapter 13, pp. 319 - 345, Elsevier Science B. V., 2002.

2　INTRODUCTION

Tidal flat, and associated muddy sedimentation, is especially well developed along lowland shorelines in a widerange of environments from the Arctic to the tropics. It is distributed along the Atlantic Ocean coast and in distal settings, such as Frobisher Bay in the Canadian Arctic, in the Bay of Fundy, along the North Sea coasts of The Netherlands, Britain, Germany and France, the Amazon River estuary and the Guyana coasts, and along the Atlantic "trailing edge" coasts of North and South America. In the Pacific region, tidal flats are represented in a variety of morphodynamic environments, such as those of the shallow shelves of northern Australia, in the Pacific Islands, where they may be associated with coral reefs and lagoonal mangrove forest, in the indented embayments of steepland island arc complexes of Japan, Indonesia, New Guinea and New Zealand, and those of the west Korean coast, which are associated with river mouth settings.

The tidal flats in China occur on a large scale—covering about 2 M ha of intertidal area, and extending along some 4 000 km of coastline (Fig. 1). They are associated with two different physiographic types of coast, namely tidal flats fringing the north China plain, and those associated with embayments, often estuarine in nature, which occur mainly along the southeastern China coastline.

Tidal flats are a potential new land resource that have evolved gradually during the Holocene. They are of great importance to China because of the dense population distribution along the coastal zone. Human utilization of the tidalflat can be successful, or unsuccessful, depending upon the degree of understanding of the natural processes and state of evolution of the different types of tidal flats.

In this chapter we present the major geomorphic, environmental and sedimentation characteristics of the various types of tidal flat formedalong the coasts of China, especially those characteristics contributing to the development of 'muddy coasts'. The major formative processes and influences, which in our interpretation lead to the formation of the various tidal flat types, and the properties of muddy coastal deposits associated with them, are also presented.

3　CLASSIFICATION AND REGIONAL SETTING

Tidal flats in China can be classified into two major types: the *coastal plain type* and the *embayment type* depending upon the coastal setting.

The *coastal plain type* is developed on a large scale and is distributed in the lower reaches of large rivers and river deltas or along the edge of the North China coastal

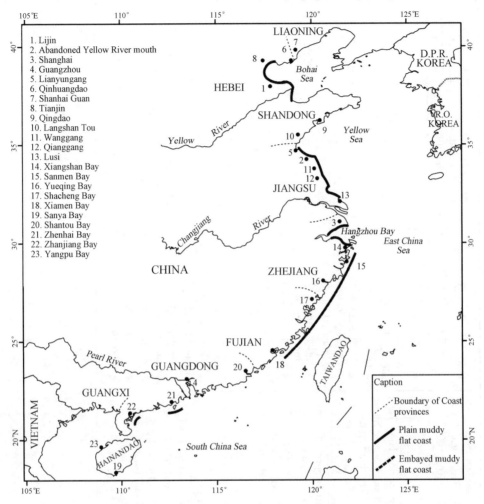

Fig. 1 Location of tidal flat associated with muddy coasts in China

plain. Tidal flats associated with coastal plains in the Bohai Sea include the head of Liaodong Bay near the mouth of the Liaohe River, and in Bohai and Laizhou bays in the vicinity of the modern Yellow River delta. Additionally, coastal plain type tidal flats occur along the Yellow Sea North Jiangsu coast near the ancient deltas of the Yellow and Changjiang rivers, while in the East China Sea tidal flats are found along the coastal plain of the Changjiang River delta and within Hangzhou Bay which is under the influence of high fluvial sediment influx.

The *embayment type* of tidal flat is developed along the elongated narrow rocky tidal inlets along the coasts of Zhejiang, Fujian, Guangdong, and Hainan provinces facing the East China Sea and South China Sea (Fig. 1). The heads of these embayments receive local sediment input from surrounding catchments, coastal erosion, and headward sediment transport by tidal currents.

Tidal flats can also be classified according to their characteristic morphological and sedimentary features (Table 1, from Wang, Collins and Zhu 1990). This approach is

Table 1　Comparison of various tidal flat zonations with approximate morphological linkages (after Wang, Collins and Zhu 1990)

Location Author / Zonation	The Wash, North Sea, U.K. (after Evans 1965)	North Jiangsu Coast, Yellow Sea, China (Zhu and Xu 1982)	Bohai Bay, Bohai Sea, China (Wang et al, 1964)	Bay of Fundy, Atlantic, Canada (Wang & Zhu, 1990)	Frobisher bay, Arctic, Canada (Dale, 1985)
	A. Salt Marsh Inundated only during Spring High Tide. Flat with muddy depressions. Well-laminated silt-clays, clays and clayey silt. Very few sands. B. Higher Mud Flats Located between Spring High Tidal level and High Tidal level with forms of elongate depressions, shallow channel systems, ripples and mud cracks. Laminated silty sands and sandy silt.	I. Supratidal Zone (grass flat) Located above Spring High Tidal level. Only flooded during Storm High Tide. Very thin laminae of clayey silt with organic layers. II. Suasda Salsa Mud Flat Located between Spring to Mean High Tidal level. Laminae of clayey silt with thin layers of silt. Worm holes and polygons (mud cracks) on the tidal flat surface.	Salt Marsh Plain and Lagoon Depression. I. Polygon Zone Clay or silty deposits with mud cracks. Only flooded by Spring High Tide. II. Inner Depositional Zone Muddy flat located below Mean High Tidal level.	I. Beach II. Salt Marsh III. Mud Flat	I. Beach II. Finer Sediment Flat (sandy silt)
	C. Inner Sandy Flats Less well laminated sands and silty sand, with muddy layers. Ripples and scour and fill structures.	III. Mud-Silt Flat Located between Mean Tidal level to Neap High Tidal level. Alternating laminae of clayey silt and silt, medium laminae with cross-bedding.	III. Erosional Mud-Silt Flat Around mid-tide level, alternating laminae of fine silt and silty clay. Overlapping fish scale sedimentary structures or tidal creeks.	IV. Sandy-Silt Flat With granular sands in the tidal creek.	III. Boulder Flat
	D. Arenicola Sand Flats Poorly stratified sands and silty sand, with minor fine-grained sediments mixed by the worm *Arenicola marina*.				IV. Boulder Ridges

(Continued)

Location Author	The Wash, North Sea, U.K. (after Evans 1965)	North Jiangsu Coast, Yellow Sea, China (Zhu and Xu 1982)	Bohai Bay, Bohai Sea, China (Wang et al 1964)	Bay of Fundy, Atlantic, Canada (Wang & Zhu, 1990)	Frobisher bay, Arctic, Canada (Dale, 1985)
Zonation	E. Lower Mud Flat — Poorly sorted laminated sands and silty sands intermixed with muddy layers. F. Lower Sand Flats — Fine, medium sands and silty sands with thicker laminae, transverse larger ripple marks with steep lee slopes facing ebb direction.	IV. Sandy Silt Flat — Fine sands and silt flat with wave and current ripples.	IV. Outer Depositional Zone of Sandy Silt Flat — Very fine sands and coarse silt with ripple forms.	V. Silty Ripple Flat VI. Tidal Channel	V. Very Bouldery Flat VI. Graded Flat
Classification	**Sand Flat** — Tidal dominant flat with strong influence by wave action especially during storm periods to deposit sandy materials.	**Silt Flat** — Tidal current dominated flat with slight or occasional wave influences. Silt deposits originate mainly from offshore submarine delta of old Changjiang River.	**Mud Flat** — Tidal currents dominate intertidal zone flat with very fine sediment originally from the Yellow River. Very gentle coastal slope to form classic intertidal zonation of four units along 400 km long coastline.	**Sandy-silt Flat** — Mixed type developed in the clayey-sandy bedrock embayment with high magnitude, current dynamic regime.	**Boulder-muddy flat** — Developed in the Arctic Ocean region, where coastal zone characterized by frequent floating ice, clay and boulders drop onto flat during storms.

particularly appropriate for coastal plain fringing tidal flats that face the open sea, and the details are discussed below.

3. 1　Tidal flats of the plains coast

Tidal flats fronting coastal plain predominate along the Yellow Sea coast north of Hangzhou Bay and along the Bohai Bay coast of the Great Plain of North China. This entire coastal area has undergone isostatic sinking in excess of 300 m since the beginning of the Cenozoic, and has been built out by the huge volume of sediment brought down to the coast by several large rivers, notably the Yellow River and the Changjiang. The great plain built up by these sediments extended onto the continental shelf at times of low sea level, and was partly submerged during the postglacial transgression. At present, the Yellow and Changjiang rivers continue to deliver sediment to the coast and continental shelf, forming extensive coastal tidal flats in Bohai Bay and to the north of the Changjiang River mouth.

The intertidal flat is the major component of the tidal flat coast and is presently undergoing active morphodynamic change. Two contrasting subtypes of tidal flats are described here as examples.

3. 2　Tidal flats along the coastal plain of Bohai Bay

Bohai Bay is located in the western sector of the Bohai Sea. It faces the Bohai Strait which has water depths <20 m. Coastal plains in this area include the Haihe River delta plain in the north, and the Yellow River delta plain in the south. The fulcrum point of the arc-shaped bay is located at Qikou where there is a very gentle submarine coastal slope. The slope gradually increases on both sides towards the Haihe and Yellow river mouths (Wang 1983) (Fig. 2).

In Bohai Bay, the waves are mainly wind-induced but with very little swell. The local wave climate reflects the direction and speed of the seasonal wind fields. There are strong waves from the northeast during winter and spring, which generate significant wave heights of 1. 9 m. During summer and autumn, southeast and easterly waves predominate with average heights in the order of 0. 4~0. 5 m (Wang et al. 1964; Wang 1983). The wave orbital motion stirs up the sediment, which is then transported by shore-parallel tidal currents along the intertidal flat. When northeast winds reach Beaufort Force 7, the "turbid fringe" of the coastal waters can be 30 km wide. In addition to tidal currents, waters covering the intertidal flats are subject to wind stress induced currents, which attain speeds of between 10 and 30 cm/s. These currents appear to be important for carrying suspended sediment in shore-parallel directions. The tides in Bohai Bay are irregular semidiurnal, with an average range of 1. 65 m and maximum range exceeding 3 m. However, the minimum range is only 0. 8 m around the

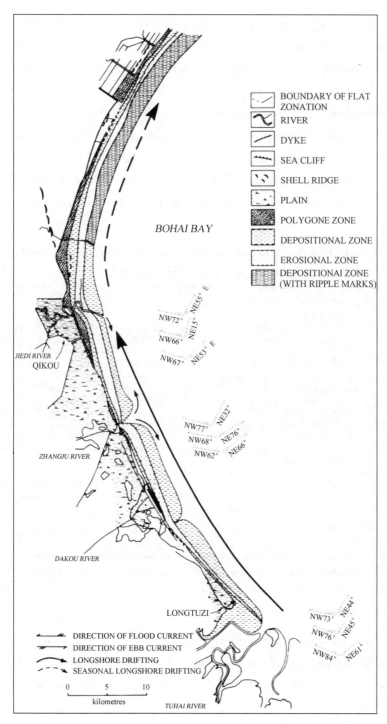

Fig. 2 Tidal flat coast in Bohai Bay (from Wang 1983) showing the morphological sedimentary zones and direction of flood and ebb tidal current and suspended transport, and longshore drift directions

mouth of the Yellow River. Flood currents dominate towards the east except at the Yellow River mouth. Tidal periods are unequal, the ebb lasting 7 h and the flood 4 h. As a result, the velocity of the flood is higher than the ebb, and this phenomenon has a particular impact on the intertidal flat. Maximum current velocities occur at mid-tide, but the highest concentrations of suspended particulate matter (SPM) and turbidity are found at the beginning of the flood tide when the water is shallow and turbulence is maximized throughout the water column.

The SPM concentration decreases to a minimum at high tide as the water becomes deeper and less turbulent. During the ebb tide, the suspended sediment concentration is lower because of the longer-lasting ebb flow with lower current speeds (Table 2, Fig. 3). As a result, the principal transport of sediment is by the flood tide in a landward direction.

Table 2 Hydrological measurement across the coastal tidal flat of Bohai Bay
July 5th—6th, 1965 (after Wang 1983)

Location of survey stations relative to MSL across the tidal flat		+2.0 m	±0 m	−2 m	−4 m
Water depth (m)		0.2~1.8	0.2~3.8	1.5~5.4	3.5~6.8
Flood	Velocity (m/s)	0.3	0.25~0.3	0.34~0.52	0.2~0.57
	Flow direction	265°		55°~40°	60°~70°
Ebb	Velocity (m/s)	0.2	0.2~0.3	0.3~0.45	0.2~0.4
	Flow direction	35°		250°~260°	280°
Wave	Wave length: λ (m)	0.1	0.2	0.2~0.45	0.29~0.4
	Period: T (s)	1~2.2	2.8~4.0	1.6~3.0	3.0~4.0
Sediment concentration (ppt)*		165.4 5.2 3.7	4~6	2~3	

* Note: the maximum sediment concentration (165.4 ppt) appears after the first hour of the flood tide. It decreases to 50% during the second hour, 10% in the third hour, and is only 1/30th, i.e. 5.2 ppt in the fourth hour, after which values range between 5.2~3.7 ppt.

Current velocities were measured with Ekman Current Meters. Water samples were collected using horizontally aligned sampling bottles on a vertical pole. Suspended sediment concentration was obtained by settling method, over a full 24 hours tidal cycle.

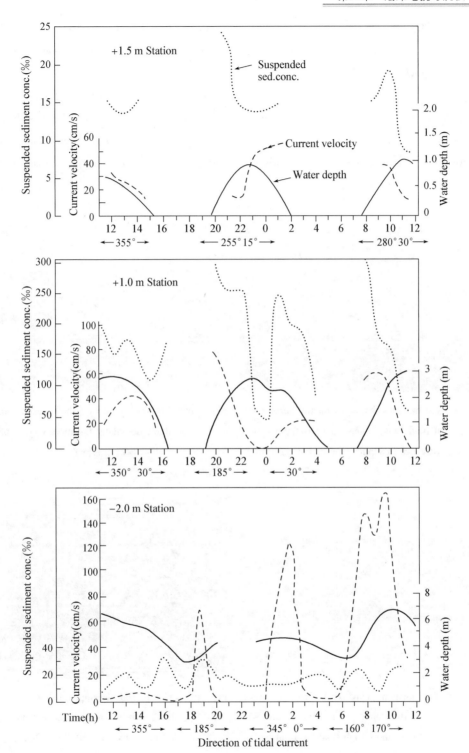

Fig. 3 (facing page ▶) **Tidal currents and suspended sediment concentration at 3 stations across the tidal flat at Wanggang in northern Jiangsu province (from Zhu and Xu 1982).**

- The +1. 5 m station was positioned at the mean high tidal level, and 1 km from the coastal dyke in the mud zone;
- The +1. 0 m station was positioned at the mid tidal level, which was 5 km away from the coastal dyke in the silty-mud zone; and
- The −2. 0 m station was located at the spring, low tidal level, 15 km away from the coastal dyke in the silt zone.

The North China coastal plain profile consists of three morphological units from land to sea. They are: salt marsh plain, intertidal flats, and submarine coastal slope.

Salt marsh plain

The marsh plain is located above the maximum high tidal level, landward of shell beach ridges (cheniers) or artificial stopbanks (dykes). The plains are of low relief and contain several lagoon-like depressions (Fig. 4a). Vegetation is sparse. Sediment of the salt marsh plains is mainly silt or sandy silt, and the terrain slopes are in the range of 0. 9~3 m/km (Wang and Ke 1990).

Fig. 4a Salt marshes developed on the upper part of mud flats, where there is fresh water influence to support the marsh vegetation

Fig. 4b Modern cheniers (shell beach ridges) developed along the muddy coast of the Yellow Sea(附彩图)

Fig. 4c The polygon (clay crack) zone on mud flat of the Bohai Sea and the Yellow Sea occurring above high tidal level but below maximum high tidal level

Fig. 4d Muddy flat developed below high tidal level or below polygon zone (clay crack zone) in the area of the Bohai Sea and the Yellow Sea coasts of China

Fig. 4e **Erosional creeks developed on the middle part of tidal flat in the muddy coast of the Bohai Sea**

Fig. 4f **Silty ripples. The surface ripple zone developed on the lower part of tidal flat along the west Bohai Sea coast of China**

For the Bohai Bay coast, salt marsh plains are the main depositional features, but two large cheniers, Chenier Ⅰ and Chenier Ⅱ (Wang 1964; Wang and Ke 1990; Zhao 1986) are identified along the coastline (Fig. 4b). Under present conditions, chenier shell ridges are not being formed on the prograding coasts, and thus the geomorphic form of the cheniers indicates a change in the evolutionary development process, perhaps an erosional phase.

Intertidal flat

In Bohai Bay, the width of the intertidal flat is generally $3\sim7$ km, increasing toward the south to $16\sim18$ km at the Yellow River delta. The intertidal flat can be divided into 4 geomorphological and sedimentary zones (Table 1; Wang et al. 1964). Moving seawards, these are: the polygon zone, an inner depositional zone, the erosional zone, and an outer depositional zone (Table 3).

Table 3　The zonation of the tidal flats of Bohai Bay (after Wang 1983)

Zone	Outer Depositional Zone	Erosional Zone	Inner Depositional Zone	Polygon Zone
Location	Immediately shoreward of the low tidal level	Around mid-tide level	Immediately below the lower high tide level	Between lower high and maximum high tidal level
Morphology	Sandy silt flat with current ripples	Tidal channels follow ebb current direction, with overlapping fish scale erosional structures between channels or elongated depressions	Mud pool field below mean high tidal level	Polygon fields with wormholes; marsh (near river mouth)
Sediment	Very fine sand to coarse silt	Alternating laminae of fine silt and clayey silt	Clayey silt and clay	Clay or silty clay

(**Continued**)

Zone	Outer Depositional Zone	Erosional Zone	Inner Depositional Zone	Polygon Zone
Dynamics	Tidal currents and small waves	Mainly ebb tidal currents	Mainly flood tidal current	Tidal currents and evaporation
Processes	Deposition of thin layers over large area	Deposition during flood, erosion during ebb flow	Extensive deposition, mainly of suspended particulate matter	Stable (deposition during maximum tides, erosion during storms)

The *polygon zone* (Fig. 4c) is located between the lower high and maximum high tide level with its landward limit marked by a chenier. The zone is 70~80 m wide with slopes of 0.5~0.6 m/km. Its characteristic sediment is greyish-yellow clay, with grey silty clay 1.5 m below the surface. The zone is covered by water only during spring high high tides. Because high tidal water does not normally cover the polygon zone, under dry conditions polygonal desiccation cracks develop in the clay. These average 10~50 cm in diameter with cracks that are 1~2 cm wide, where there is a freshwater influence, some grass species such as *Suaeda salsa* Bunge, *S. glauac*, *Limonium sinensis* Kumtie, and *Phragmites australis* may grow.

The *inner depositional zone* (Fig. 4d) which is 500 to 1 000 m wide, consists of >1 m thickness of soft mud, while average land slope is about 0.67 m/km. Soft clayey silt and clay are deposited here during the flood tide. A variation of the typical inner depositional zone is seen in Bohai Bay where mud percentage is significantly reduced.

The *erosional zone* (Fig. 4e) occurs at about mid-tide level. In northern Bohai Bay, this zone is 500 m wide with a slope of about 0.57 m/km. It exhibits sediments of alternating laminae of fine silt and clay, and the surface is eroded into a series of overlapping scours termed "fish scale" structures that are produced by the ebb flow. These structures are 3~18 m long and 0.5 m wide and dip seaward. Scarps between each layer of scales are 2 to 10 cm high and face the coast. In the middle part of Bohai Bay, the erosional zone is about 1 000 m wide, with a slope of 0.42 m/km and consists of silty clay. Small tidal creeks may be eroded into the tidal flat surface mud due to the draining ebb tidal waters. The fine sediments eroded from this zone are carried mainly toward the sea where they settle in the outer depositional zone.

The *outer depositional zone* is present between the erosional zone and the spring low tide level. It is approximately 1 000 m wide and has a 0.5 m/km slope. The zone consists of coarse silt and very fine sand with ripple marks on the surface (Fig. 4f). The ripples have a wavelength of 3~7 cm, with their steep slope facing seawards along the outer part of this zone and shorewards along the inner part. In the middle part of the zone they are symmetrical, suggesting a multiple wave origin. The sediments become finer in both offshore and onshore directions.

The *submarine coastal slope* is found seaward of the spring low tide level. This slope is very gentle in Bohai Bay. The sediment becomes finer offshore, changing from a sandy silt at the low tidal level to a very fine silt and clay in water depths exceeding 10 m (Fig. 5).

Principal source of sediment in Bohai Bay is theYellow River, which has played a predominant role in the evolution of the tidal flat coast. The plume of suspended river sediment is the major source of coastal sediment. "Fluid mud" is deposited only in the upper part of the tidal flats or in deep water. Several years of monitoring (during 1961 to 1964, 1985, 1987 and 1993) by the Coast Laboratory of Nanjing University indicate that these flats are areas of rapid sediment deposition with relatively higher deposition in the outer and inner depositional zones (Wang et al. 1964; Zhu and Xu 1982; Wang 1983; Wang and Zhu 1994). Spatially, relatively higher rates of sedimentation occur in the southern part of the bay closer to the Yellow River, where mainly silty clay is deposited, demonstrating the importance of sediment supply in the evolution of these broad muddy tidal flats. A great wedge of silty sediment is deposited in the northern part of the bay even though the Yellow River migrated to the southern part of the present delta in 1976, demonstrating the importance of northward littoral drift of the muddy sediment.

The pattern of *zonation* across the tidal flat reflects the spatially-varying nature of different tidal processes acting on the flat. Sediment texture on the flat itself is influenced by the strength of the tidal currents. The deposition of very fine sand and coarse silt on the lower flat, alternating laminae of fine silt and mud on the middle flat, and clay on the upper flat reflects a settling lag process (Postma 1967; Wang et al. 1964) that is controlled by a landward decrease in tidal current velocity and water depth.

Seasonal changes in depositional processes are also important. In Bohai Bay, freezing starts in late November. Coastal ice, attaining a thickness of 0.5 m becomes stranded between the high and mid tide levels. Normally one can observe alternating layers of ice and mud. During March, the ice melts leaving patches of mud on the flat. During May and June, winds are predominantly northeasterly, and at this time the tidal flat is subjected to strong wind waves inducing erosion of the surface sediments. However during the low wind and wave energy conditions of summer the high suspended sediment load waters allow deposition of sediments. With southeasterly and easterly winds, the flat becomes covered by a yellowish fluid mud originating from the Yellow River.

Sediment origin on the Bohai Bay tidal flats can be traced to two sources. Coarse silt, fine silt, silt-clay and clayey soft mud originate from the Yellow River in the south. The sediment distribution pattern delineates a tongue of fine materials extending from the Yellow River and reaching to Qikou. Another tongue from the Haihe River consists

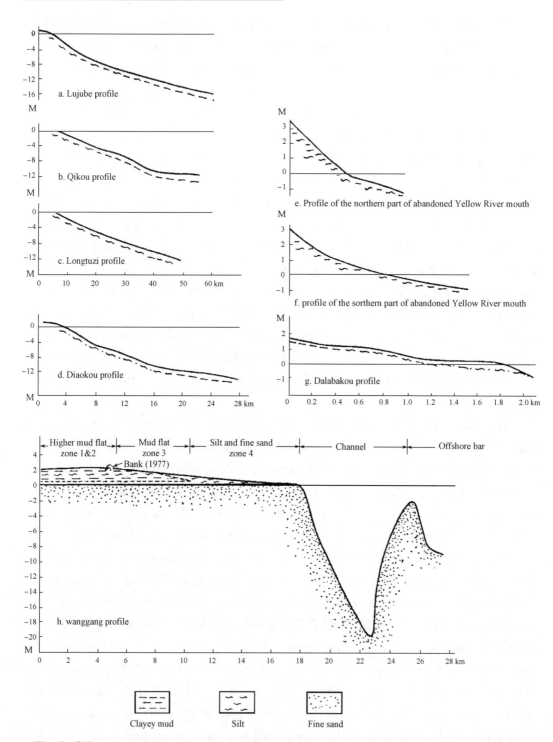

Fig. 5 Submarine coastal profiles of Bohai Bay and North Jiangsu (from Wang 1983): (a) Lujuhe profile, (b) Qikou profile, (c) Longtuzi profile. (d) Diaokou profile (1965 river mouth of Yellow River), (e) Profile of the northern part of abandoned Yellow River mouth, (f) Profile of the southern part of abandoned Yellow River mouth, (g) Dalabakou profile (river mouth of Sheyang River), and (h) Wanggang profile.

of silt and very fine sand, which also reaches to Qikou. Heavy minerals likewise reflect two sources. Yellow River sediments have a low heavy mineral content (1%) consisting of more stable minerals (garnet, zircon, tourmaline) and an increasing proportion of secondary minerals (allochite, hematite). All of the heavy minerals have lost their original crystal shapes and most unstable minerals have disappeared, indicating that the sediments have been transported over a long distance. In contrast sediments from the Haihe River suggest an igneous and metamorphic provenance from the nearby Yanshan Mountains.

3.3　Tidal flats along the coastal plain of North Jiangsu

Much of the North Jiangsu coast is developed on the pre-1855 delta of theYellow River that lies north of the Changjiang River. The tidal flat coastline in North Jiangsu is about 884 km long. In historical times the Changjiang River also flowed through North Jiangsu, entering the Yellow Sea at Qianggang. The two rivers brought in vast quantities of sediment which has strongly influenced the geomorphic evolution of the coast and resulted in formation of 10~13 km wide tidal flats of average slope 1 : 5 000 and maximum width 36 km (Fu and Zhu 1986; Wang and Zhu 1991) (Figs. 6 and 8).

Along the Yellow Sea coast tides are semidiurnal. Average tidal range is 1.5~1.7 m around the abandoned Yellow River mouth and along the southern sector. In the middle sector, average tidal range is 3.0~3.4 m, with the maximum tidal range of 6~9.3 m occurring in the area of the Huangsha Yang Channel. The Pacific tidal wave progresses from the southeast, while a local tide propagates from the Shandong Peninsula in the north, both converging in the Qianggang area to produce the largest tidal ranges along China's coastal zone (Wang et al. 1998). Tidal range influences the elevation of the tidal flats and also has an effect on their width. Tidal currents-typically between 50 and 100 cm/s over the tidal flats-are asymmetrical, with the flood lasting 8~10 hours, and maintaining higher speeds than the ebb.

On the tidal flats, maximum flood tidal currents appear to precede the maximum suspended sediment concentration by about one hour. Maximum sediment concentration during the ebb also appears to lag about one hour after the maximum ebb flow. Tidal currents approach the shelf at an angle of $20°~30°$, increasing to about $90°$ on the high tidal flat. Thus currents over the tidal flats are essentially diabathic. During a full tidal cycle, residual sediment transport is onshore to an area of deposition in the middle reaches of the tidal flats (Zhu and Xu 1982).

Wave heights over the tidal flats are relatively small ($H_{1/3} = 0.6~1.2$ m), but seaward of the sand ridges they can reach 4 m. Waves from the northeast prevail during the winter season with maximum significant heights of 2.9~4.1 m. Waves from the southeast are predominant during the summer, with maximum significant heights in the

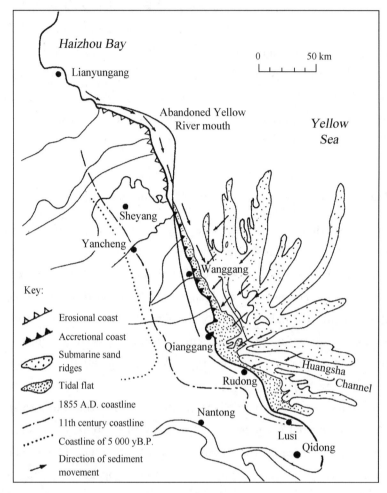

Fig. 6　Tidal flat coast of North Jiangsu showing the extent of tidal flat relative to the submarine sand ridges, and erosional and accretional coastline (after Fu and Zhu 1986)

range 1. 7~3. 2 m. In the autumn an easterly swell prevails. Wave action influences the tidal flats through agitation of the bottom fine sediments which increases fine suspended particulate material in the water column, and by promoting the formation of sedimentary structures such as ripples, inclined bedding, and storm records of scours and lenses of sandy deposits (Ren et al. 1983).

In geomorphic terms, three zones may be identified on the North Jiangsu intertidal flat (Wang, Collins and Zhu 1990):

- A *mud zone* consisting of thin laminae of silty clay and clayey-silt comprises the upper part of the flat near the high tidal level.

- A *silty-mud zone* occurs over the middle reaches of the intertidal zone. It is comprised of transitional deposits derived mainly from fine silt and clay-sized material re-deposited from suspension.

- A *coarse size zone* forms the widest part of the intertidal flat and extends out to the spring low tidal level. The sediments fall within the coarse silt or very fine sand size range and the surface is dominated by small ripples.

The zonation of the North Jiangsu coastline is believed typical of an accretionary intertidal flat, with silt as the main textural component over the entire reaches of the flat. Four sedimentary facies may be recognized in stratigraphic section:

- Supratidal salt marshes or grass flat comprise the highest part of the section.
- An upper zone of clayey sediments has accumulated on the highest part of the tidal flat surface. These sediments are usually of $2 \sim 3$ m maximum thickness, and have been deposited from suspension by tidal currents.
- Alternating intercollated clays and silty deposits are observed in the middle part of the section, and represent storm deposits intermixed with fair weather clay deposition from suspension.
- The lower silt and fine sand deposits, about $15 \sim 20$ m thick, have been deposited on the lower tidal flat between low tidal level and the outer boundary of the coastal submarine coastal slope. They represent sediment that has been transported onshore from the seaward lying submarine ridges.

Heavy minerals that occur in the tidal flats of North Jiangsu can be classified into two provinces. One is the Changjiang mineral complex, which consists of hornblende, magnetite and apatite as primary minerals, with hyposiderite, collopane, zircon and rock fragments as secondary minerals. The second province represents an ancient Yellow River mineral complex characterized by zircon, tourmaline, titanoferrite and garnet. Heavy mineral content increases from the high tidal level (0.84%) to the low tidal level (5.1%).

In terms of microfossils, the tidal flat sediments contain shallow water foraminifera species such as *Ammonia*, *Elphidium* and *Nonion*, of which about 10% are associated with planktonic genera such as *Globigerina* and *Orbulina*. In this area these planktonic species typically live in water depths of more than 200 m. There are also benthic species such as *Bolivina* (7.2%) and *Quinqueloculina* which are usually associated with water depths of more than 50 m in this area. The mixture of deep and shallow water foraminifera implies that sediments move from deeper offshore environments to the shallow water environment of the flats (Zhu et al. 1986).

In summarizing, the tidal flats of the north China plain coast are characterized by three zones: (i) mud flat occurs at the high tidal level; (ii) muddy silt mixed flat occurs at the mid-tidal level; and (iii) silt-to-fine sand flat is found at the low tidal level. The coastal plain with extensive tidal flat development is related closely to the continuity of sediment supply from the Yellow River.

4 TIDAL FLATS OF THE EMBAYED COAST

Coastal embayments have developed along the 1 500 km of tectonically block-faulted coast of Zhejiang and Fujian provinces. These coastal areas are associated with large tidal ranges, exceeding 4 m on average. Powerful tidal currents are found in most of the embayments with current speeds exceeding 1 m/s, but wave energy here is weak ($H_{1/3}$ < 0. 5 m) because the bays extend inland and are sheltered. Diluted fresh water containing suspended sediment from the Changjiang River tends to flow to the south and mixes with the parabathic flow along the Zhejiang-Fujian coast. Thus, the fine-grained sediment supplied by the Changjiang offshore deposits provides one source of fine grained material for development of muddy tidal flats in these embayments.

4. 1 Tidal flat in Yueqing Bay

Yueqing Bay is located north of the Oujiang River mouth in Zhejiang Province (Fig. 1). It occupies a tectonic down-faulted basin that was originally 40 km long from its mouth to bayhead and is aligned in a NE to SW direction parallel to the main trend of coastline. Total area of the embayment is 250 km². With an average water depth of 10 m, Yueqing Bay is renowned in China for its powerful tides featuring tidal ranges of up to 4. 2 m at the bay mouth and up to 5. 0 m at the bay head, with maximum tidal range ≥ 8. 3 m. Tidal dynamics decrease the velocity and increase the tidal period duration from the bay mouth to the bay head, with ebb tidal currents stronger than the flood currents (Table 4; Wang and Wu 1991). Typical suspended sediment concentrations are very high at 400 mg/l. The average width of tidal flats in Yueqing Bay is 1. 8 km, coastal slopes are 1~1. 4 m/km, and the total area of tidal flats in the bay is 134. 8 km².

Table 4　Tidal currents of the Yueqing Bay (after Wang and Wu 1991)

Location	Flood Current (m/s)	Ebb Current (m/s)	Flood Duration (h:m)	Ebb Duration (h:m)
Bay Mouth	0. 59	0. 63	7:00	5:12
Bay Head	0. 36	0. 69	6:30	5:24

The tidal flats consist of a surface layer of 20~40 cm thick greyish-yellow uniform clay (*Md* of 8. 73ϕ) overlying greyish dark mud a few to 10 m thick. The bottom sediment layer of the flats consists of fluvial deposits of silty-clayey sands and gravels. Heavy minerals in the sands of the bayhead tidal flats comprise a magnetite-epidote complex (88. 9% and 5. 3%) related to supply from surrounding catchments. Other parts of the bay are characterized by a hornblende-magnetite-epidote complex of heavy minerals indicating a provenance linked to Oujiang River discharge.

4.2 Tidal flats of Shacheng Bay

Shacheng Bay is a narrow embayment 55.5 km in length and 1.85 km wide located at the northern boundary of Fujian Province. The total area is 11 km² with a water depth of at least 15 m in most parts of the bay. There are 6 km² of islands in the bay, and the tidal area is 54 km² (Fig. 7). Tidal flats occur as a discontinuous pattern in the bay head and in inlets along both sides of the major arms. Width of the tidal flats is normally <1 km with a relatively steep slope in the order of 5~10 m/km. Average tidal range is 4.1 m with a maximum of 6.9 m. Tidal currents are typically 1~1.5 m/s, with maximum velocity of 2 m/s; ebb currents are faster than the flood. Because the outline of the bay is long narrow and curved, wind waves in the bay are suppressed. Even during a typhoon with wind force of 8~10 Beaufort scale, the wave heights reach only 0.5 m.

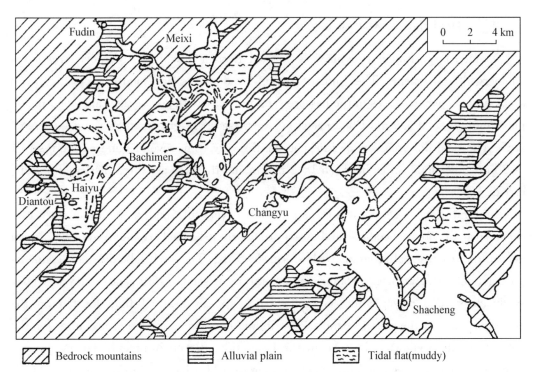

Fig. 7 Shacheng Bay as an example of an estuarine embayed tidal flat system and showing extent of tidal flat and muddy deposits

Broadly, two types of tidal flat are evident, namely *gravel-sandy* and *muddy*.

- The *gravel-sandy type* occurs at the head of the bay and in inlet mouths along the sides of the bay. The sediments comprise sand (4.6%~65%), clay (22.5%~ 66%), and gravel (10%), with heavy mineral content about 236 mg/100 g. This type of tidal flat results from coastal streams discharging sediment to the coast forming small alluvial fans at stream mouths which build out as a series of small gravel-sandy tidal flats.

- The *muddy tidal flat type* result from estuarine depositional processes and an abundant mud supply. Sediments consist of 30 cm of surficial yellowish mud grading into a several metre thick deposit of greenish grey mud, which overlies a bottom layer of sand-gravel fluvial sediments. Muddy tidal flat deposits contain up to 57.5%~65% of clay, 35%~42.5% of silt, and a heavy mineral content of about 7.3 mg/100 g.

Heavy mineral contents of the tidal flats in Shacheng Bay consist of 62% iron minerals, epidote (22.8%), leucoxene (11%), and zircon, all associated with secondary pyrite, suggesting that the fine-sized materials of these flats were transported into the bay by tidal currents, ultimately originating from Changjiang River discharges. The present vertical sedimentation rate on the tidal flats in the bay is small, suggesting that the present suspended sediment supply is also small.

4.3 Tidal flats of Xiamen Bay

Xiamen Bay, Fujian Province, is formed from a downfaulted graben creating a submerged embayment. The Jinmen Islands straddle the bay mouth enclosing broad areas of open water. Tides in the bay are semi-diurnal, with an average range of 4 m and maximum range of 6.9 m. Ebb speeds (0.89 m/sec) are greater than flood (0.77 m/sec). The tidal flat area is ~87 km^2 and widths vary from several hundred metres to 1 or 2 km. The slope varies between 1~2 m/km and the flats are often dissected by tidal creeks.

In stratigraphic section, alternating thin layers of mud and sand, with rare occurrence of cross bedding and wave or current-formed ripples, reflect normal deposition processes alternating with storm deposits. The surficial tidal flat sediments are relatively poorly sorted, comprising a mixture of sands, silty sands, clayey sands, sandy clays, and clayey silts. Gravelly sands appear on tidal flats located on the erosional coast in the north. Sand is distributed over wide areas and accounts for the majority of the sediment observed in the bay's tidal flat environment. Heavy mineral content is relatively high at 151.2 mg/100 g, reflecting a local sediment supply from the surrounding catchment.

4.4 Mangrove forests in estuarine embayment environments

Mangrove forests are distributed south of latitude 27°N, but only a few species survive north of 24°N, such as those found in the estuarine embayed environments of Shacheng Bay in Fujian province. Mangrove jungle environment occurs in the tropical zone of Guangxi and Hainan provinces where 38 species have been found. The major species are *Bruguiera conjugata*, *B. sexangula*, *Ceriops tagal*, *Kandelia candel*,

Rhizophora mucronata, *Lumnitzera racemosa*, *Aegiceras corniculatum*, *Avicennia marina* and *Exvoecaria agallocha*. Rhizophoraceae are among the predominant species.

Mangrove growth is confined to the bayhead or to lagoonal environments that occur behind sand barriers. As mangrove colonization proceeds, it creates mud flat areas, either expanding over sand beaches or onto bedrock and coral reef surfaces, where fine silt or clayey sediments can be mixed with sandy sediment, or can accumulate a thickness $> 40 \sim 50$ cm under tropical mangrove forests. Consequently, mangrove stands represent a distinctive subtype of tidal flat in the coastal embayment environment of the tropical and subtropic climatic zones where tidal creeks and dense infaunal worm populations are common.

4.5 Summary of embayed tidal flat characteristics

(i) Sediments on the tidal flats are predominantly muddy deposits.

(ii) Multi-sediment sources result in changes of particle size and composition over short distances on some tidal flats.

(iii) Tidal flat deposits are relatively small scale formations. They are typically several metres thick and occasionally reach a thickness of more than 10 m. Usually, the total area of sedimentation is of order several km².

5 INFLUENCE OF THE YELLOW RIVER IN TIDAL FLAT EVOLUTION

The annual sediment discharge of the Yellow River is 1.1 billion tonnes (Ren 1994); and about 64% of that discharge is deposited near the river mouth. Consequently the coastline of the Yellow River delta has been prograding at a rate of ~ 50 m/year. The other 36% of the total sediment discharge becomes associated with the littoral sediment transport which provides sediment for tidal flat formation in Bohai and Laizhou bays (Wang, Ren and Syvitski 1998).

During historical times, the lower reaches of the Yellow River have migrated to discharge on both sides of the Shandong Peninsula. Between 1128—1855 A. D., the Yellow River discharged into the Yellow Sea from the Jiangsu coast providing large quantities of sediment for the coastline to prograde some 90 km. Since 1855, the Yellow River has flowed northward to the Bohai Bay, so that the tidal flats along the abandoned Yellow River delta are now eroding, and the coastline is retreating rapidly ($15 \sim 100$ m/y) (Wang and Aubrey 1987). This phenomenon will likely recur in the future (Fig. 8).

Human intervention has resulted in large impacts on the Yellow River. These include diverting water volumes of order 200×10^8 m³/a for agricultural and urban water supply since the 1970's (Wang and Zhang 1998). As a result, the river channel for more

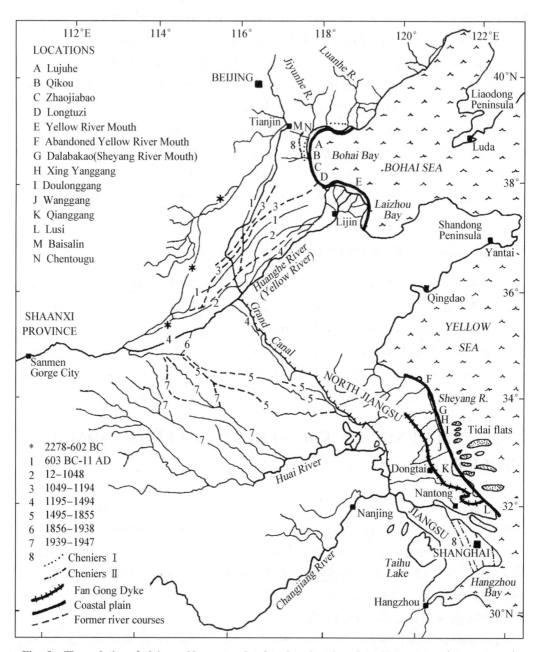

Fig. 8 The evolution of plain muddy coasts related to the migration of the Yellow River (Yellow River) (Wang, Collins and Zhu 1990)

than 683 km upstream of the mouth (87% of its lower reaches) dried out for 122 days in 1995. In comparison, the channel dried for only 41 days in the 1960s, 71 days in the 1970s, and 103 days in the 1980s. But in 1997 the channel was dry for 226 days and the length of dried channel increased to 700 km from the river mouth. It can be expected that the river mouth and the delta coastline of the Yellow River will be eroded again, and this would change drastically the characteristics of the local muddy coastal environment.

6　INFLUENCE OF THE CHANGJIANG RIVER AND SUBMARINE RIDGES

There are some 70 sand ridges located off the Jiangsu coast in water depths up to 25 m. This sand ridge field is more than 200 km long from north to south (32°00′N to 33°48′N), and 140 km wide from east to west (120°40′E to 122°10′E). It is the remnant of the Changjiang River delta which was deposited when the river entered the Yellow Sea from North Jiangsu during the late Pleistocene. The delta has been reworked by tidal currents into a radial sand ridge field during the rise of Holocene sea level. These sand ridges now shelter the tidal flat of the central Jiangsu coast so that it can continue to accumulate sediments. The flats increase between 2~10 cm in vertical elevation and 20~100 m in width each year. From three years of measurements during 1980—1984 the mean annual rate of sediment accumulation on the tidal flats was ascertained as 0.77 billion tonnes (Ren 1986). Present-day fluvial sediments from the northern branch of the Changjiang River are discharged at a rate of 35 M tonnes yearly. Sediments derived from the eroded coast of the abandoned Yellow River delta contribute 109 M tonnes per year. The sediment supply from the offshore sand ridge fields is 32 M tonnes per year. Thus, some 18% of the sediments reaching the Jiangsu tidal flats originate from offshore sand ridge fields.

Sedimentation characteristics of the Zhejiang and Fujian coast tidal flats also demonstrate that river discharge and offshore sediment supply are major factors in the development of these features. Suspended sediments from the Changjiang River and fluvial sediments from coastal rivers of Zhejiang Province constitute the major sediment supply to the embayments of this area. According to the textural evidence, fine sediments discharged by the Changjiang River are transported as far as Shacheng Bay in the northern part of Fujian province. In contrast, Xiamen Bay tidal flats are formed from local sediments that are supplied by adjacent rivers, and by the weathering and erosion of coast bedrock. However, the situation will change following the completion of the Three Gorges dam when the sediment discharges of the Changjiang River will be significantly reduced.

7　INFLUENCE OF TIDES

Tidal action affects the spatial scale and sedimentation characteristics of tidal flats. If sediment supply is sufficiently large, then the width of the tidal flat is dependent primarily on tidal range. The varying characteristics of the tide (e. g. spring tide or neap tide) and the accompanying change in water level, will influence the nature of sedimentary structures (Zhu and Gao 1985) (Fig. 9). In Figure 9, "A" is the upper limit of sand deposition during the spring high tide, and "B" is the upper limit of sand deposition during the neap high tide. In a sequence of neap tide to spring tide to neap tide, a thin sand wedge may be deposited from seaward to landward. After several cycles, alternating layers of clay and sand form on the tidal flats between A and B. In Bohai Bay, there is a 1.91 m difference between the spring and neap high tidal levels, and the horizontal distance between them is 1 500 m. Consequently, the entire tidal flat facies shifts laterally during spring-neap cycles. In the Jiangsu tidal flats, the difference between spring and neap high tide causes a shift in the sedimentary facies by 1 900 m horizontally. As a result, a complex sedimentary structure is produced on the tidal flat. In Jiangsu, the current velocity of the spring tide is twice that of the neap tide. Sediments deposited on the tidal flat during the spring high tide have a medium diameter of $Md=5.3\phi$, and a medium diameter of only 6.5ϕ during the neap tide. This process is reflected in deposition of laminae of clay and fine silt (Ren et al. 1984).

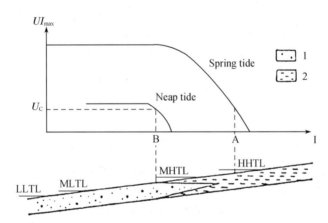

Fig. 9 Sedimentation processes relative to spring and neap tide variation
(from Zhu and Gao 1985)

1. represents sandy deposits, 2. muddy deposits. UI_{max} is the spring tide flood current velocity and U_c is the ebb current velocity of the neap tide. LLTL is low low tide level, MLTL is mean low tide level, MHTL is mean high tide level and HHTL is high high tide level.

8 CONCLUSION

From research in China, tidal flat development is perceived to be controlled mainly by the quantity of sediment supply and by tidal dynamics. Wherever there is an abundant sediment supply in association with a large tidal range, a tidal flat will develop. Whenever the sediment supply stops or decreases, the tidal flat will succumb to wave and tidal current erosion. Consequently, sediment supply is a major controlling factor. Different kinds of tidal action may shape different types of tidal flats, each with their own characteristic sedimentary structures.

River sediment discharge, coastal erosion and offshore submarine erosion are three processes that supply sediment for the tidal flat construction. Most of the offshore source sediments have a fluvial origin, especially in China's case. The tidal flats of Bohai Bay and North Jiangsu are typical examples of flats that are dominated by fluvial sediment supply, and are now impacted by human intervention on the river system as well as by developments on the muddy coast.

Acknowledgements

We would like to express our thanks to Dr. C. T. Schafer and to Professor Terry Healy and Judy-Ann Healy for reviewing the manuscript and carefully editing the English version. Thanks are extended to our Chinese colleagues for their contribution to field and laboratory work.

References

Dale, E. J. 1985. Physical and Biological Zonation of Intertidal Flat at Frobisher Bay. *North West Territories 14th Arctic Workshop*, November 6—8, 1986, Halifax. 180 – 181.

Evans, G. 1965. Intertidal flat sediments and their environments of deposition in the Wash. *Quarterly Journal of the Geological Society*, 121: 209 – 245.

Fu, M. and D. Zhu. 1986. The sediment sources of the offshore submarine sand ridge field of the coast of Jiangsu province. *Journal of Nanjing University (Natural Sciences)*, 22(3): 536 – 544.

Postma, H. 1967. Sediment transport and sedimentation in the environment. In: G. H. Lauff (editor), *Estuaries*. American Association for Advancement of Science Publication 83. 158 – 199.

Ren, M. 1986. The Report of the Comprehensive Survey on the Coasts and Tidal Flats of Jiangsu Province, China. Ocean Press, Beijing. 19 – 134.

Ren, M. 1994. Impact of climate change and human activity on flow and sediment of rivers: the Yellow River (China) example. *GeoJournal*, 33(4): 443 – 447.

Ren, M., R. Zhang and J. Yang. 1983. The influence of storm tide on mud plain coast. *Marine Geology and Quaternary Geology*, 3(4): 1 – 24.

Ren, M., R. Zhang and J. Yang. 1984. Sedimentation on tidal mudflat in Wanggang area, Jiangsu Province, China. *Marine Science Bulletin*, 3(1): 40 – 54.

Wang, Y. 1964. The shell beach ridges and the old coastline of the west coast of Bohai Bay. *Acta Scientiarium (Naturalium) Universitatis Nankinensis*, 8(3): 424 – 442.

Wang, Y. 1983. The mudflat coast of China. *Canadian Journal of Fisheries and Aquatic Sciences*, 40(1): 160 – 171.

Wang, Y. and D. G. Aubrey. 1987. The characteristics of the China coastline. *Continental Shelf Research*, 7(4): 329 – 349.

Wang, Y., M. B. Collins and D. Zhu. 1990. A comparative study of open coast tidal flats: the Wash (U. K.), Bohai Bay and West Yellow Sea (Mainland China). In: *Proceedings of the International*

Symposium on the Coastal Zone. China Ocean Press，Beijing. 120 – 134.

Wang，Y. and X. Ke. 1990. Cheniers on the east coastal plain of China. *Marine Geology*，90：321 – 335.

Wang，Y.，M. Ren and J. Syvitski. 1998. Sediment Transport and Terrigenous Fluxes. In：*The Sea*. Volume 10. John Wiley and Sons Inc.，New York. 253 – 292.

Wang，Y. and X. Wu. 1991. The characteristics of mudflats and sedimentation in the embayments of Zhejiang and Fujian province，China. *Journal of Nanjing University* (*Geography*)，12：1 – 9.

Wang，Y. and Y. Zhang. 1998. Influences on coastal environment by human activity and break off water discharge of the Yellow River. *Journal of Nanjing University* (*Natural Sciences*)，34 (3)：257 – 270.

Wang，Y. and D. Zhu. 1990. Tidal Flats of China. *Quaternary Sciences*，12：291 – 300.

Wang，Y. and D. Zhu. 1991. Submarine Sand Ridges：Unique Marine Environment. *Coastal Zone ′91*，American Society of Civil Engineers. 3377 – 3380.

Wang，Y. and D. Zhu. 1994. Tidal Flat in China. *In*：D. Zhou，Y. Liang and C. K. Tseng (editors)，*Oceanography of China Volume* 2. Kluwer Academic Publishers，Dordrecht. 445 – 456.

Wang，Y.，D. Zhu，X. Gu and C. Chual. 1964. The characteristics of mudflat of the west coast of Bohai Bay. In：*Abstracts of Theses in the Annual Symposium of the Oceanographical and Limnological Society of China*，1963，Wuhan. 55 – 57.

Wang，Y.，D. Zhu，L. Zhou，X. Wang，S. Jiang，H. Li，B. Shi and Y. Zhang. 1998. Evolution of radiative sand ridge field of the South Yellow Sea and its sedimentary characteristics. *Science in China* (*Series D*)，28(5)：385 – 393.

Zhao，X. 1986. Development of cheniers in China and their reflection on the coastline shift. *Scientia Geographica Sinica*，6(4)：293 – 304.

Zhu，D. and S. Gao. 1985. A mathematical model for the geomorphic evolution and sedimentation of tidal flats. *Marine Science Bulletin*，4(5)：15 – 21 (in Chinese).

Zhu，D.，X. Ke and S. Gao. 1986. Tidal flat sedimentation of Jiangsu coast. *Journal of Oceanography of Huanghai and Bohai Seas*，4(3)：19 – 27 (in Chinese).

Zhu，D. and T. Xu. 1982. The coast development and exploitation of middle part of Jiangsu coast. *Acta Scentiarum University Nankinesis*，3：799 – 818 (in Chinese).

潮滩沉积环境与岩相对比研究[*]

一、潮滩动力环境

潮滩(Tidal flat)分布于平缓海岸的潮间带及潮下带上部,是海陆交互作用的沉积地貌,发育于有大量细颗粒泥沙(以 4Φ 至 8Φ 的粉砂为主)供应,海岸坡度平缓(<1/1 000),波浪作用衰减,潮流动力活跃的地区[1,2]。潮流是潮滩形成演变的主要动力,近岸潮波变形产生的时间不对称导致了流速不对称,涨潮流速大于落潮流速,导致沿海的细颗粒泥沙向岸地再搬运,并在潮间带堆积。在一个潮周期中,流速分布的特点符合滞后效应,是悬移质堆积的基本条件[3,4],但对潮间带中下部砂质推移质的影响很小[5]。潮间带流速分布的不对称性,使最大含沙量与最大沉积量出现于潮滩中部[6,7]。

根据潮流动力及沉积物运动特点,潮间带具有三个动力沉积带:潮间带上部是滞后效应起作用的泥质沉积带;潮间带下部是受流速——时间不对称效应控制的砂质堆积带;在上述两大沉积带之间的潮间带的中部,存在着同时受滞后效应与不对称效应的过渡带,发育了泥-砂混合带。

波浪在潮滩地带,动能衰减,其作用是掀动浅水区泥沙,使再悬浮而增加潮滩水层含沙浓度,并影响泥沙沉积形成波痕和斜层理。风暴天气时,波浪对潮滩的侵蚀和沉积物的再分选作用明显,形成各种侵蚀形态与风暴砂层沉积。由于波浪参与的程度强弱与频率不同,潮滩整体又可分为细砂质潮滩、粉砂质潮滩和淤泥质潮滩三种主要类型[11]。在极地高纬区的浮冰融解在砂、泥滩上还可混杂有砾石[12]。

二、现代潮滩地貌分带性与沉积相

潮滩海岸沉积物的分布的规律,反映了潮滩潮流的动力作用特点,在低潮水边线处物质最"粗",为粗粉砂与极细砂;向岸逐渐变细为粗粉砂与细粉砂、细粉砂及黏土质粉砂;潮滩最上部有黏土质沉积。潮滩地貌具有分带性[1,8],由此,从海向陆有四个带,同样有相应的四个沉积相带[7,9,10]。低潮滩为粉砂——细砂波痕沉积带,分布于低潮位以上[1,2,8]。中潮滩为泥质——粉砂质叠层沉积带,具有落潮水流于潮滩中部流速加大形成的滩面冲刷体、冲刷洼地与潮水沟。高潮滩为泥质沉积带,沉积颗粒细,经常为水淹没,形成泥沼。潮上带,特大高潮才被淹没,为盐沼湿地黏土质沉积带,具虫孔、虫塔等生物扰动构造及日晒的龟裂,降水或有淡水自岸堤渗出而发育湿地草滩。

潮滩分带性是普遍性的规律,是潮流动力在滩涂上递变的反映。分带性的名称在一些文献中各有不同,但是,潮滩底部的砂质粉砂滩,上部是泥质滩,潮滩中部为泥沙混合

* 王颖,朱大奎,曹桂云:《沉积学报》,2003 年 12 月第 21 卷 4 期,第 539-546 页。

滩,却是潮滩沉积相共同的特征[8,9],当潮滩遭受侵蚀后退,或在风暴潮期间受波浪冲刷,常会形成贝壳质海滩或贝壳沙堤,因此,潮滩沉积中常伴有贝壳层。贝壳多为粉砂质或淤泥滩的种属。

随着潮滩的淤积发展,这四个带向海推进,原来在近陆侧的沉积带会依次覆盖在近海侧的沉积带上。当潮滩发育趋于成熟时,就形成了一个完整的从下向上逐渐变细的潮滩沉积相序:下部为粉砂、细砂质沉积,上部为泥质沉积。下部的粉砂、细砂是低潮位高能流态下的推移质沉积,界于低潮位与波浪作用的下界(海岸斜坡的下限)间,沉积层厚度约15～20 m,这是潮滩沉积的基底。上部泥质沉积是潮流搬运的悬移质,它使滩面增高。当滩面达到大潮高潮位以上,就很少被海水淹没,一般不再堆积加高,所以泥质沉积层的最大厚度相当于中潮位至大潮高潮位的间距,约为2～3 m。上下两层在剖面上表现为泥层与砂层的交替沉积。潮滩沉积是在低潮滩的沉积基底的基础上,通过"横向沉积"的拓宽和"垂向沉积"的增厚发展起来的,一个潮滩环境的旋回层厚度约为20～30 m(图1)。

图1　潮滩沉积模式(A. 垂向沉积　B. 侧向沉积)

潮滩沉积的特征结构是水平纹层或"百页"状的砂-泥薄互层理,因往复的潮汐作用所成。大潮时,纹层的厚度大,小潮时的层薄;砂质层带色浅,泥质的层带色深,同一组叠层中,泥质纹层厚度小于砂质纹层。在半日潮地区,每14.77 d为一大、小潮循回,因此可形成28个潮流薄层;全日潮区每半个月可形成14个或更少些的薄层[8]。潮滩微地貌与沉积分带性亦反映于沉积层结构中,如:上部沉积层中的泥裂、龟裂纹及其内的次生填充物(砂质、钙质或黄铁矿质沉积)、虫穴、虫塔、根系空隙及其内填充物;下部沉积层中的砂质与粉砂质波痕、流痕、豆荚状或鲕状体、脉状层理、斜层理或鱼刺状交错层理,潮水沟摆动所遗留的镶嵌堆积砂层、透镜体、风暴潮的沉积出现于不同层位中,其沉积质地较粗,具有侵蚀痕迹、透镜状体与丘状砂层等结构,或含泥块、泥砾等上部潮滩的蚀余堆积物。这些微结构与泥层、砂层、泥沙交互层相结合,构成了潮滩环境所赋有的沉积相。

潮滩沉积的各相带可多次反复交替,一个大层序的厚度大多为20 m。潮滩沉积中会嵌有河口沉积,后者具有二元相结构及咸淡水交汇的贝类残体。

三、潮滩沉积岩相对比分析

潮滩环境的沉积相被应用于古沉积岩的形成环境分析中,如David B Mackenzie(1972)对科罗拉多下白垩统的Dakota组古潮滩研究(Gimsburg R. N. 1975);夏文杰对云南思茅中侏罗统的和平乡组古潮滩研究,提出潮滩沉积具有自下向上变细的沉积层序[13];A. J. Tankard综合南非有关资料提出潮滩沉积岩相的理论层序[14],其相序与现代潮滩沉积结构特征,与江苏潮滩剖面的相序极为相似。加拿大Dalhousie大学地质系教授 P. Schenk 于 1982—1983 年介绍他所研究的渤海地区岩层的"竹叶状构造"及蛇纹大理岩中的多边形、块体结构，系古潮滩的龟裂带遗证，该项研究被应用于油储孔隙度分析。

本文以现代潮滩沉积相与南京钟山北坡的黄马青组及海南澄迈县马村的湛江组两组沉积岩进行对比分析，拟解释两地层所反映的古沉积环境。

(一) 黄马青组

位于南京钟山北麓的下五旗村，地层倾向与山坡向相反，出露中三叠统黄马青组。

本文研究的是黄马青组下段，由紫色、浅紫色、夹少数灰绿色薄层泥质粉砂岩，粉砂岩和下部夹灰色石英砂岩组成。常现轮藻化石，以 *Stellatochara*（星轮藻属）和 *Stenochara*（直立轮藻属、狭孢藻属）为主；介形类 *Darwinula*（达尔文介）丰富；小型半咸水-海水双壳类 *Bakevellia*（贝荚蛤属）出现。古植物 *Aunalepis*（具沟三缝孢属）、*Zeilleri*（葵勒蕨）的出现是本段的特色。本段岩石的微量元素 $Sr：Ba=0.6$，显示下段是滨岸相沉积环境[15]。上段是紫红色碎屑岩，钙质结核和虫管构造发育，轮藻化石丰富，仍以 *Stellatochara*、*Stenochara* 为特征，半咸水双壳类和陆生植物 *Aunalepis* 不再出现。$Sr：Ba=0.4$。

下五旗村黄马青组下段岩相剖面由下而上为：

下部为灰紫色粗粉岩段，厚 0.4 m，未见底，灰色砂质粗粉砂岩为主，偶夹由 3 mm 厚的紫色粗粉砂与 3 mm 左右的黄色粗粉砂组成的韵律层。有浪成对称与不对称波痕（波长 $L=6\sim8$ cm，波高 $H\approx1$ cm）（图版Ⅰ-1）。

中部由薄层粗粉砂层与紫色的中细粉砂层组成，厚 1.8 m，粗粉与细粉砂互层组成不间类型的层理构造，其中以透镜层理，砂/泥薄互层理最为发育（图版Ⅰ-2），下部偶尔有波状层理。

上部为紫红色泥质粉砂岩段，上覆第四纪坡积物，厚度大于 0.3 m，水平层理发育，其形态如页岩，具钙质填充之虫管（图版Ⅰ-3），有时偶夹微波状层理。

在南京钟山北麓黄马村的黄马青组下段的剖面与下五旗剖面中的Ⅱ段上部以及Ⅰ段相似，Ⅲ段则不发育。主要有紫红色的粉砂岩组成，夹有薄层的灰色细砂质粉砂岩，两者相互组成砂/泥薄互层理、具有虫管鱼刺状交错层理（图版Ⅰ-4）与透镜层理，单个纹层厚约 0.4 cm。

下五旗村的黄马青组下段的沉积相序，可综合成图 2，自下而上为：

Ⅰ段：相当于低潮滩的上部，厚度大于 0.4 m，未见底，灰紫色的砂质粗粉砂岩，有浪成波痕，鱼刺状交错层理，波痕交错层理发育。与江苏潮滩相比，缺少"修饰"波痕。结合黄马村剖面具有干涉波痕。此段相当于江苏潮滩的粉砂-细砂滩亚相。

Ⅱ段：中潮滩。是黄马青组下段古潮滩的"内核"，由灰白色的粗粉砂层与灰紫色的中细粉砂层组成，透镜层理相当发育，砂/泥薄互层理相当于江苏潮滩中上部的薄互层理，砂层比泥层要少，波痕交错层理发育，波状层理主要在下部。总体上此段潮滩层理仅相当于江苏潮滩的砂-泥混合滩亚相的上部，无压扁层理。物质组分以泥质＞砂质为特征，厚 1.8 m。

Ⅲ段：相当于高潮滩的下部，未见顶，被第四纪坡积物覆盖。厚度＞0.3 m，由薄层的水平纹层状的紫红色泥质粉砂岩组成，其形态如页岩，常称其为"薄页状结构"。水平层理

厚度 M	剖面	位相（相序）	沉积物	特征沉积结构
>0.3	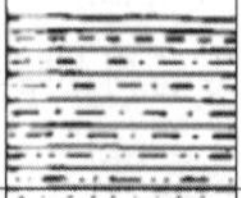	高潮滩亚相（Ⅲ段）	泥质粉砂岩	水平层理、微波状层理，垂直虫管
1.8	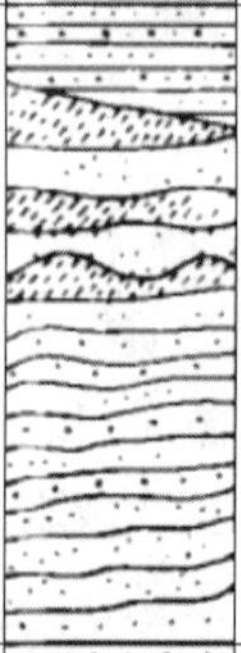	中潮滩亚相（Ⅱ段）	中粗粉砂岩与细粉砂岩互层	波痕交错层理、透镜层理、波状层理、砂/泥薄互层理
>0.4	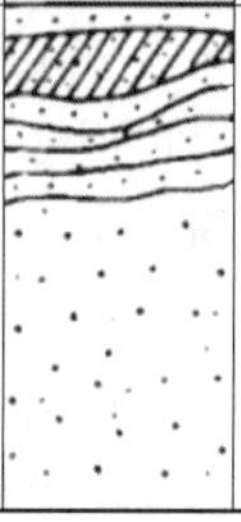	低潮滩亚相（Ⅰ段）	粗粉砂岩	干涉波痕、浪成波痕波浪交错层理鱼刺状交错层理

图 2　南京钟山北坡中三叠统黄马青组下段剖面

发育，有时略显波状起伏，此段相当于江苏潮滩的盐蒿泥滩亚相。

综合黄马青组下段相序特征：① 具有向上变细的层序；② 潮汐层理（包括透镜层理、波状层理以及薄互层理）发育；③ 悬移质堆积的水平层理与推移质堆积的波痕交错层理并存；④ 细粒沉积物（粉砂）为黄马青组下段形成的物质基础，由此认为下五旗剖面的黄马青组下段沉积岩源于潮滩沉积。

（二）海南省澄迈县马村下更新统湛江组

位于澄迈北港海湾，因海蚀使下更新统湛江组出露组成沿岸高 3～5 m 的海蚀崖。马村剖面可作为现代潮滩与三叠纪黄马青潮滩岩对比的中间例证。剖面自上而下为：

顶层，玄武岩，厚 1.7 m，柱状节理发育，有气孔构造。

第 2 层：灰色泥层，厚 30～60 cm，具水平层理，水平纹层厚约 2 mm，局部夹有微波状层理。

第 3 层：黏土与粉砂质泥层互层，棕色、蓝灰色，厚约 50 cm，每层厚约 5～6 cm，上部棕色为主，向下蓝灰色增多，具水平层理，棕灰相间为其特征，粒度总体表现为上细下粗（图版Ⅰ-5）。

第 4 层：灰黄色粗粉砂、细砂以及棕色砂砾层互层，厚约 1.8 m，本段铁锰质成分较多，有火山碎屑物，具平行层理，单个灰黄色的粉砂层厚 2～3 mm，棕色砂砾层单层厚 2～15 cm。

第 5 层：棕色的中细砂层，厚约 15 cm，鱼刺状交错层理发育，两斜层系方向相反，两斜层系之间为侵蚀界面。

第6层:黄色的砂质粉砂与棕色砂层,波状层理,透镜层理特别发育,波状层理的沙波波长为7～8 cm,波高1～2 cm(图版Ⅰ-6),层厚20 cm。

第7层:黄色的粗粉砂与黄绿色的中细砂互层,其表面波痕发育。一种是波长约3 cm、波高约0.3～0.5 cm,波脊线不连续,此种波痕属水流成因的,另一种是浪成波痕,波长7 cm、波高2 cm。

该处湛江组为一套未固结、或部分为弱钙质胶结的白色、灰白色砂砾、细砂、粉砂和黏土互层,时代为早更新世(Q_1),其上覆地层为北海组地层或早期火山岩。

海南马村湛江组的沉积相序(图3)可分为三个部分组成(即三个亚相),按沉积顺序自下向上依次为:

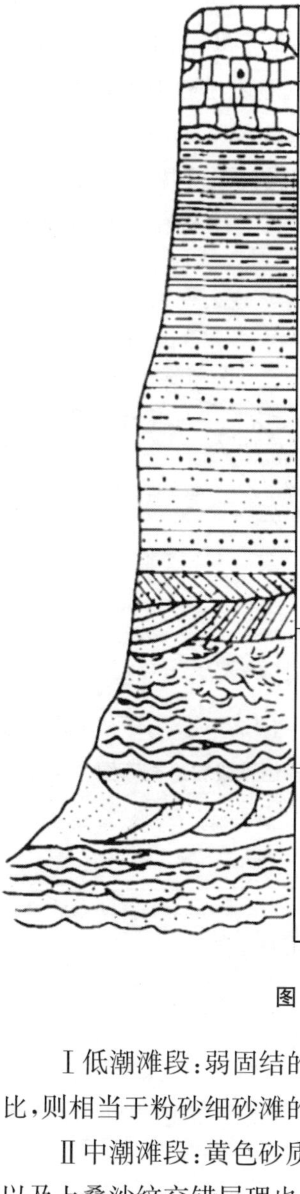

	厚度 m	位相相序	沉积物	特征沉积结构
	1.7	玄武岩		
	1.1	高潮滩（Ⅲ段）	棕灰色泥质粉砂岩	水平层理夹微波状层理局部黏土质结核
	1.8	砂砾层	含极细粒中粗砂夹少量薄层细砂	平行层理 互层层理 鱼刺状交错层理
	0.4	中潮滩（Ⅱ段）	粉砂质细砂或粉砂	波状层理、波痕交错层理透镜层理、砂泥薄互层理
	>2.0	低潮滩（Ⅰ段）	粉砂质细砂	鱼刺状交错层理 再作用面,浪成波痕砂/泥薄互层理

图3　海南马村下更新统湛江组古潮滩沉积相序

Ⅰ低潮滩段:弱固结的黄绿色粉砂质细砂层,具有浪成波痕、水流波痕,与江苏潮滩相比,则相当于粉砂细砂滩的上部。未见底,厚度大于2 m。

Ⅱ中潮滩段:黄色砂质粉砂与棕色的细砂组成,波状层理特别发育,鱼刺状交错层理以及上叠沙纹交错层理也发育,其上部发育砂/泥薄互层理,局部有透镜层理,这段相当于江苏潮滩的砂-泥混合滩。与江苏潮滩的砂-泥混合滩不同的是出现上叠沙纹层理,这种层理代表沉积环境的推移质与悬移质均丰富,沉积速率较大,发育在沉积物周期性快速堆积的环境。可能与河流泥沙汇入有关,结合上覆黄棕色砂砾层的特征,代表原潮滩环境发生突变。这段厚30～40 cm。

Ⅲ高潮滩段:由棕色与蓝灰色的黏土质粉砂与粉砂质黏土互层,砂-泥薄互层理发育,表现为棕色与蓝灰色相间,其上部为水平层理夹微波状层理,这相当于江苏潮滩的盐蒿泥滩亚相。但未见大量龟裂纹。厚1.1 m。

分析表明,无论是现代潮滩还是古潮滩都难以保存完整的理论层序,不同地区和不同时代的潮滩各有差异,但主要特点相似,宜做组合特征比较研究。

为阐明钟山黄马青组与马村湛江组的沉积环境,进行了泥沙粒度组成、古盐度与微生物的对比。

潮滩为低能环境,沉积以粉砂为主[2]。矿物组分为石英、长石、云母、黏土矿物及极少量的重矿物。沉积物粒度概率曲线表明:潮滩下部以推移质为主形成粉砂波痕及交错层理;潮滩中部以跃移质为主,夹杂少量悬移质形成砂泥交互层理;潮滩上部以悬移质为主形成水平纹层之泥滩带。江苏潮滩沉积物分析,黄马青组进行显微镜下薄片分析,以及对湛江组半胶结沉积层进行轻压与脱钙分析后,结果表示出相似的三段状粒度概率曲线(图4～图6)。

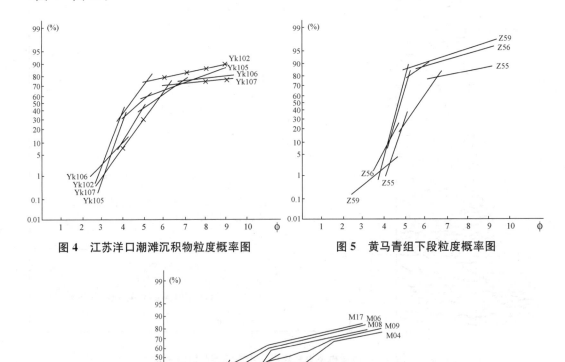

图4　江苏洋口潮滩沉积物粒度概率图　　　图5　黄马青组下段粒度概率图

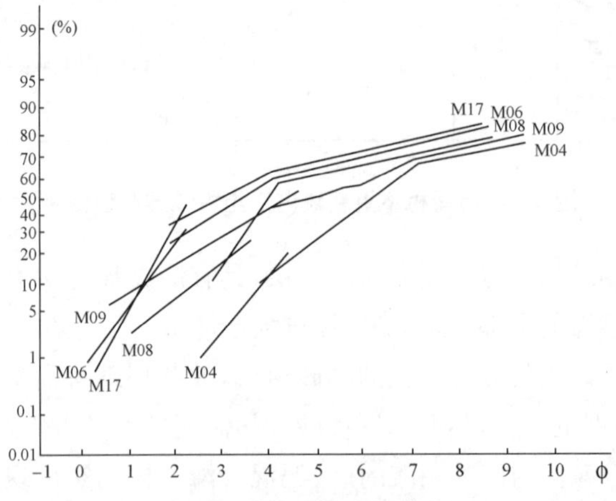

图6　湛江组粒度概率图

潮滩位于陆地环境和海洋环境之间,滩面上的水介质为海陆过渡型的半咸水,潮滩的生物种群具有海陆混生、广盐性或特殊的半咸水种属特性。盐度分析表明:

(1) 江苏潮滩水介质的含盐度都低于海水正常盐度,吕四潮滩含盐度为 24.6‰~32‰,平均值为 28.5‰;射阳港以北潮滩滩面水介质盐度为 18‰~24.9‰,平均值 21‰,(据 1990 年 5 月实测)。因陆地径流及降水影响,潮滩形成的水介质盐度值往往偏离正常海水盐度值。在干燥气候区与潟湖共生的潮滩或潮上带的盐沼的水介质会大大地咸化,但此类潮滩沉积压实固结后常伴有膏盐沉积,更易于识别。

(2) 古盐度计算是根据黏土矿物从溶液中吸收硼并将其固定,在海水体系溶液中硼的浓度是盐度的线性函数[1],用 X 射线衍射仪分析测出潮滩沉积中黏土矿物的百分含量,用提纯的黏土矿物经等离子光谱(ICP 法)测定微量元素 B 含量。分析结果(表 1 和表 2)与半咸水盐度分类表(表 3)[15,16] 比较表明,该两组岩层均属中盐水环境沉积,其中湛江组砂砾层古盐度偏低,反映出有淡水之影响。

表 1　马村湛江组古盐度

样号	位置	硼含量/×10⁻⁶	古盐度/‰
M03	高潮滩	19.6	8.4
M04		13.6	6.4
M07	砂砾层	4.2	2.5
M08	中潮滩	15.2	6.9
M11	低潮滩	14.7	6.5

表 2　钟山黄马青组下段古盐度

样号	高岭石中硼含量/×10⁻⁶	古盐度/‰	平均古盐度/‰
Z55°≤	30.2	17.6	
Z58°±	20.0	12.3	14.6
Z38°∞	24.1	14.3	

表 3　1958 威尼斯半咸水盐度分类方案[17,18]

	类别	盐度/‰	含氯度/‰
混盐度	淡水	0~0.5	<0.3
	少盐水(oligohaline)	0.5~5.0	0.3~3.0
	中盐水(mesohaline)	5.0~18.0	3.0~10.0
	多盐水(polyhaline)	18.0~30.0	10.0~16.5
	真盐水(euhaline)	30.0~40.0	16.5~22.0
	超盐水(hyperhaline)	>40.0	22.0

---

① 中国科学院南海海洋研究所. 华南沿海第四纪地质调查报告,1976.

江苏潮滩以有孔虫组合最能反映海陆过渡相性质,毕克卷转虫、凸背卷转虫—希望虫—九字虫,属于半咸水环境的产物,他们主要出现在盐蒿滩上。介形虫往往表现为异地生物,经潮流带到潮滩上沉积下来。在废黄河口、灌河口和射阳河口的径流流出处有大量的广盐性介形虫发育,有中华丽花介、中国洁面介及东台新单角介等种属①。潮滩的微体生物以广盐性或特殊的半咸水种属为主,也有正常海水种的抱球虫、瓶虫、五块虫、圈球虫及滩面上的陆生植物,构成江苏潮滩生物组合特征。

南京黄马青组下段化石丰富,具有滨岸相—海陆交互作用沉积环境的生物组合特征[17]。其中轮藻种的数量多,个体大,以 *Stellatochara-Stenochara*（星轮藻—直立轮藻）组合的特征,现代轮藻可在各种水域中出现,可视为广盐性植物。另外有双壳类的 *Bakevellia* 贝菜蛤和 *Nytilu6* 克菜蛤,克菜蛤一类属广盐性半咸水软体动物化石,常常是海陆过渡相的良好标志[18]。介形虫类单调,全为达尔文介,它基本上为淡水,有的可在弱咸水（最大盐度 7‰）生存,因此主要代表陆生,若在潮滩中出现一般为异地生物。陆地植物以石松类出现。总之,黄马青组下段的古生物同样具有海陆过渡相特征。

湛江组的生物化石以海相为主,其中的原筛藻主要在河口湾环境,浪花介为半咸水的。

综合比较三者的生物特征,其共同特点是广盐性生物普遍发育,黄马青组下段出现陆生植物,且正常海相化石缺乏（表4）。

表4　微体古生物组合特征表

水介质 / 种属 / 区域	正常海水	半盐水（或广盐性）	淡水
江苏潮滩	抱球虫、圆球虫、瓶虫、五块虫	毕克卷转虫、凸背卷转虫、希望虫、九字虫、中华丽花介、中国洁面介、东台新单角介	
海南马村湛江组	抱球虫、五块虫、美丽星轮虫、假轮虫	圆筛藻、浪花介	
南京黄马青组下段		壳菜蛤、偏顶蛤形贝菜蛤	达尔文介、石松类轮藻类

四、结语

应用现代潮滩沉积相特征,沉积物粒度级配、盐度、微体生物组合及沉积相序可以较客观地辨别中三叠统黄马青组下段与下更新统湛江组(马村)为古潮滩沉积,马村湛江组尚具有河口湾环境的淡水成分影响。

古今潮滩环境沉积相主要的特征为:

物质组分为粉砂、泥质粉砂,显示低能的水体环境。具潮汐层理特征:砂-泥薄互层

---

① 中国科学院南海海洋研究所.南海海岸地貌论文集,1975.

理,透镜层理以及波状层理,局部伴有以水流的鱼刺状交错层理。波痕、虫管、草丛根系发育显示滨岸浅水环境,钙质填充物反映着周期性海水淹没与露干的滩涂特点。为半咸水水介质的微古组合与古盐度。向上变细的潮滩相层序。潮滩沉积结构相的基本结构可综合为图 7[19]。

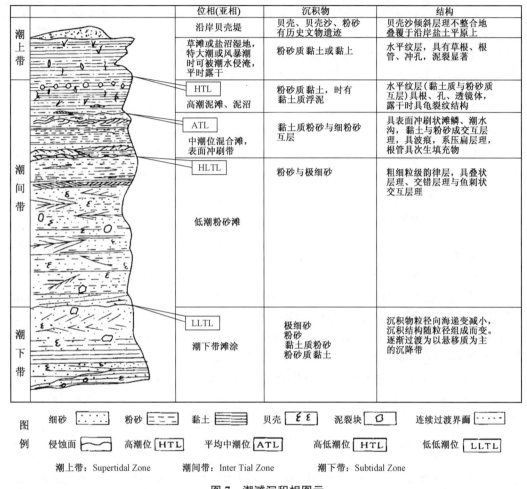

		位相(亚相)	沉积物	结构
潮上带		沿岸贝壳堤	贝壳、贝壳沙、粉砂有历史文物遗迹	贝壳沙倾斜层理不整合地叠覆于沿岸盐土平原上
		草滩或盐沼湿地,特大潮或风暴潮时可被潮水侵淹,平时露干	粉砂质黏土或黏土	水平层理,具有草根、根管、冲孔,泥裂显著
潮间带		HTL 高潮泥滩、泥沼	粉砂质黏土,时有黏土质浮泥	水平纹层(黏土质与粉砂质互层)具根、孔、透镜体,露干时具龟裂纹结构
		ATL 中潮位混合滩,表面冲刷带	黏土质粉砂与细粉砂互层	具表面冲刷状滩鳞、潮水沟,黏土与粉砂成交互层理,具波痕,系压扁层理,根管具次生填充物
		HLTL	粉砂与极细砂	粗细粒级韵律层,具叠状层理、交错层理与鱼刺状交互层理
		低潮粉砂滩		
潮下带		LLTL 潮下带滩涂	极细砂粉砂黏土质粉砂粉砂质黏土	沉积物粒径向海递变减小,沉积结构随粒径组成而变。逐渐过渡为以悬移质为主的沉降带

图例

细砂	粉砂	黏土	贝壳	泥裂块	连续过渡界面
侵蚀面	高潮位 HTL	平均中潮位 ATL	高低潮位 HTL	低低潮位 LLTL	

潮上带:Supertidal Zone　　潮间带:Inter Tial Zone　　潮下带:Subtidal Zone

图 7　潮滩沉积相图示

潮滩发育广泛于从极地到热带的气候环境,以及从地质时期的古生代至现代,具有长期延续发育的特点。其沉积质地、层序与结构在分析油气藏的储存方面具有重要的应用价值。

参考文献

[1] Wang Ying. 1983. The mud flat coast of China [J]. *Canadian Journal of Fisheries and Aquatic Sciences*, 40(Supp.): 160-171.

[2] Wang Ying, Collins M B, Zhu Dakui. 1990. A comparision study of open coast tidal flat, the Wash (U.K.), Bohai Bay and West Yellow Sea (China) [A]. In: Proceedings of International Symposium on the Coastal Zone. Beijing: China Ocean Press. 120-134.

[3] Postma H. 1967. Sediment transport and sedimentation in the environment [A]. In: Lauff G H ed.

Estuaries，American Association for Advancement of Science Publication，83．158 - 199．

［4］朱大奎,许廷官. 1982. 江苏中部海岸发育［J］. 南京大学学报,18(3)：799 - 818 ［Zhu D K, Xu T G. 1982. The coast development of Jiangsu Coast ［J］. *Acta Scentiarun University Nankinesis*，18(3)：799 - 818］.

［5］朱大奎,高抒. 1985. 潮滩地貌与沉积的数字模型［J］. 海洋通报,4(5)：15 - 21 ［Zhu D K, Gao S. 1985. A math matical model for the geomorphic evolution and sedimentation of tidal flats［J］. *Marine Science Bulletin*，4(5)：15 - 21］.

［6］朱大奎,柯贤坤,高抒. 1986. 江苏潮滩沉积的研究［J］. 黄渤海海洋,(3)：19 - 27 ［Zhu D K, Ke X K, Gao S. 1986. Research on tidal flat sediment of Jiangsu Coast ［J］. *The Yellow and the Bo Sea*，(3)：19 - 27］.

［7］王颖. 2000. 潮滩沉积动力过程与沉积相［A］. 见：苏纪兰,秦蕴珊主编. 当代海洋科学学科前沿［C］. 北京：学苑出版社. 177 - 182 ［Wang Y. 2000. Sediment dynamic and sedimentation facies［A］. In：Su J L, Qin Y S, eds. The Front of Ocean Science ［C］. Beijing：Culture Park Press. 177 - 182］.

［8］Dalrymple R W. 1992. Tidal Deposition System，Facies Models Response to Sea Level Changes ［M］. Geological Association of Canada，195 - 218．

［9］Evans G. 1965. Intertidal flat sediments and their environments of deposition in the Wash ［M］. Jul Geol Soc. London，121 (1 - 4)：209 - 240．

［10］Collins C M B，Amos G，Evans L. 1981. Observation of some sediment transport processes over intertidal flats ［M］. The Wash, U. K. Spec Publ. Int. Assoc. Sediment，5：81 - 98．

［11］王颖,朱大奎. 1990. 中国的潮滩［J］. 第四纪研究,10(4)：291 - 299［Wang Y, Zhu D K. 1990. Tidal flat of China ［J］. *Quaternary Science*，10(4)：291 - 299］.

［12］Dale E J. 1985. Physical and biological zonation of intertidal flat at Frobisher Bay. In：N. W. T. 14th Arctic Workshop, Nov. 6 - 8，Halifax. 180 - 181．

［13］夏文杰. 1982. 陆源碎屑潮汐沉积的判别标志［J］. 石油实验地质,4(4)：285 - 293［Xia W J. 1982. Indicators of continental detrital sediment ［J］. *Petroleum Geology* & *Experiment*，4(4)：285 - 293］.

［14］刘宝珺. 1980. 沉积岩石学［M］. 北京：地质出版社［Liu B J. 1980. Sedimentology［M］. Beijing：Geological Publishing House］.

［15］陈楚震. 1988. 苏南地区三叠纪生物地层,江苏地区下扬子准地台,震旦纪—三叠纪生物地层［M］. 南京：南京大学出版社. 315 - 363［Chen Chuzhen. 1988. Bisotratigraphy from Sinian to Triassic in Jiangsu region, Yangtse platform ［M］. Nanjing：Nanjing University Press. 315 - 363］.

［16］Couch E L. 1971. Calculation of paleosalinites from Boron and clay mineral data ［J］. *AAPG Bulletin*，155：1829 - 1837．

［17］汪品先等. 1980. 海洋微体古生物论文集［C］. 北京：海洋出版社［Wang P X, et al. 1980. Proceedings on Ocean Micopaleontology［C］. Beijing：Ocean Press］.

［18］同济大学海洋地质系. 1980. 海、陆地层辨别标志［M］. 北京：科学出版社［Department of Oceanal geology in Tongji University. 1980. Identified Symbol of Ocean and Land Geosphere ［M］. Beijing：Science Press］.

［19］Wang Y, Zhu D, Cao G. 2001. Environmental characteristics and related sedimentary facies of tidal flat example from China ［A］. In：Yong A Park and Richard A Davis Jr，eds. Proceedings of Tiadalites 2000. The Korea Society of Oceanography. 1 - 9．

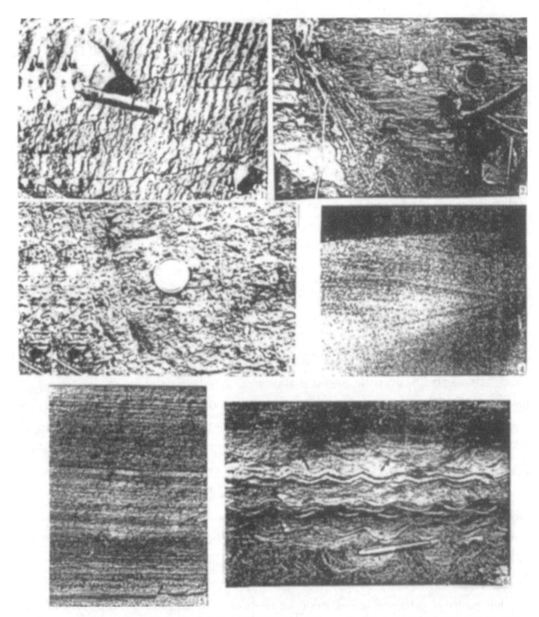

图版Ⅰ　说明 1. 钟山北坡下五旗黄马青组下段的波状层理; 2. 钟山北坡下五旗黄马青组下段的透
　　　　镜层理及砂-泥薄互层理; 3. 钟山北坡下黄马村剖面紫红色粉砂岩水平层理及垂直的虫管
　　　　已为钙质填充; 4. 钟山北坡黄马村剖面,紫红色粉砂岩中的鱼刺状交错层理; 5. 海南澄迈
　　　　马村湛江组剖面水平层理及向上粒级变细; 6. 海南澄迈马村湛江组剖面中层波状层理并逐
　　　　渐过渡到上部砂砾层,反映水动力增强。

The Natural Formation of Coastal Tidal Flats*

Coastal tidal flats are developing in a wide range of climatical environments from the Arctic to the tropic. The main factors to control the development of tidal flats are tidal dynamics, very gentle coastal slope and the abundance or fine-grained sediment.

Typical tidal flats in China are developed in the mesotidal Bohai Sea, Where the Yellow river discharge, and on its pre – 1855 delta 600 km to the south on the Yellow Sea coast. Such flats are normally 3km wide, but up to 18km wide at the delta. Because of the gradient of gentle slope is less than 1/1000, wave action is restricted offshore, thus, tidal currents are the dominant dynamic process on the flat.

Tidal flats along the East and South China Seas are associated with narrow embayments or other types of sheltered coast. Fetch within such enviroments limits the wind/wave interaction, hence, tidal currents are once again, the major agents controlling sedimentary processes. Silt is the major of sediments, ranging from fine to coarse silt, deposited over the largest part or areas or tidal flats. Fine sands and clay are also present on most tidal flats. The original source or such sediments may be terrigenous, however, with sediments on the chinese tidal flats being fluvial in origin, having been supplied to the coastline at lowered sea level by the large rivers of Yellow River and ChangJiang.

Three types of tidal flats have been recognized and the differences are caused by the slight changes in the dynamics regimes or the regions. A silt flat is a typical type of tidal flat with three zonations: these are mudflat on the upper flat, muddy silt flat on the middle part, and sandy silt flat on the lower part of the intertidal zone. A medium level of wave energy stirs up material from offshore submarine sandy deposit of old Changjiang River delta and transports it on to shore by flood-tidal currents; a sandy flat is an example of tidal flat with high wave energy influences; mud flats are with only minimal wave influences but with large quantities of fine grained material of clay and silt. There are four distinguishable zonations with typical features are: polygon zone located above mean high tidal level, muddy pool zone located below, an erosional zone with interbeded mud and silt deposits beyond the muddy pools, and an outer deposited zone consists of very fine sands and coarse silt with ripple forms located above low tidal level.

---

　　* Ying Wang: *The Fifth MICE Symposium for Asia and the Pacific*, Abstracts, E15, Nanjing, China, 1988. 8. 2 – 9.

To recognize and understand the natural processes of tidal flat will be of benefit in the exploitation of the coastal zone，as the tidal flats represent the natural growth of new land from the sea.

The Nature of Tidal Flat[*]

Tidal flats are developed along low land coastline in a wide range of climate environments. The development is mainly controlled by quantity of sediment supply in association with tidal processes. Silt is the major constituent of the deposits, very fine sands and clays are also present on most tidal flats. Flood tidal current brings material eroded from adjacent seabed and transports it in traction, siltation and suspension towards the land. Velocities decrease to landwards in response to fraction from the intertidal zone seabed. Ebb tidal current erodes the middle part of flat and redeposits sediments on the lower tidal flats. Thus, the zonation features of morphology and sediment are typical nature of intertidal flat everywhere. Basically muddy flat is on the upper part of intertidal zone, sandy flat on the lower part and a laminae intervals of silty and muddy flat on the middle part of the intertidal zone. Zonation recorded in the strata including fine lamination and delicate bedding with ripples, cracks and mosaic of cross bedding and scour marks to form a special facies, which can be used to identify the sedimentary environment of ancient deposits.

Three types of tidal flat: silt flat, muddy flat and sandy flat have been recognized as they reflect the different level of additional wave actions.

There are 2 million hectares of tidal flat in China. As an important land resource, it can be used for aquaculture, pastoral, cotton and rice fields, forest or industry utilization depending upon the zonation, i. e. the evolution stage of different part of tidal flat.

---

* Ying Wang, Dakui Zhu: American Geophysical Union, *Western Pacific Geophysics Meeting*, O12B, CC: 406, Mon 1510h, The Marine Intertidal Environment (Presiding: D J Reed, LUMCON), Hong Kong, July 25 - 29, 1994 (Published as a supplement to EOS, June 21, 1994).

The Coast of China[*]

Introduction

The coast of China extends in a 18 400 kilometre long arc from the mouth of the Yalu River on the China-Korea border in the north to the mouth of the Beilun River on the China-Viet Nam border in the south（Fig. 1）. The total length of the coastline，including the more than 6 000 islands，is approximately 32 000 km. The general outline

Fig. 1　The Coastline of China

＊　Ying Wang：*Geoscience Canada*，1980，Vol. 7，No. 3：pp. 109 - 113.

of the coast is controlled mainly by geological structure and deposition from a number of large rivers. The coast is acted upon by monsoon wind waves and tidal currents.

Geological Structure

The Continental Shelf and adjacent Mainland of China is composed of a series of NE or NNE trending uplifted and depressed belts that intersect obliquely with the coastline. From west to east, these are the Bohai Basin (Fig. 2), the Shandong-Liaodong uplift, the southern Yellow Sea depressed belt, the Zhejiang-Fujian uplift, the East China Sea depressed belt and Taiwan fold belt. A bedrock-embayed coast with rock islands is formed along the uplift belts, while a plains coast, usually with a large river, is always formed along the depressed belts. In the South China Sea, NE trending block faults define the main trend of the bedrock embayed coast, with inlets developed along NW trending faults.

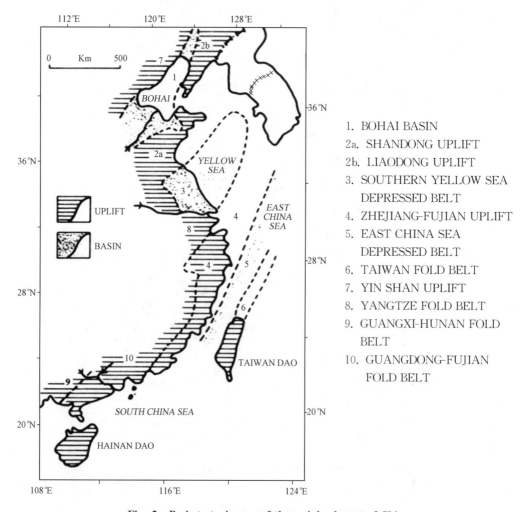

1. BOHAI BASIN
2a. SHANDONG UPLIFT
2b. LIAODONG UPLIFT
3. SOUTHERN YELLOW SEA DEPRESSED BELT
4. ZHEJIANG-FUJIAN UPLIFT
5. EAST CHINA SEA DEPRESSED BELT
6. TAIWAN FOLD BELT
7. YIN SHAN UPLIFT
8. YANGTZE FOLD BELT
9. GUANGXI-HUNAN FOLD BELT
10. GUANGDONG-FUJIAN FOLD BELT

Fig. 2　Basic tectonic map of the mainland coast of China

Rivers

Three major rivers and many minor rivers carry large volumes of fresh water and sediment to the coast of China (Table Ⅰ). River influence on the coast is not limited to the vicinity of the mouth; river sediment is redistributed along the plains coastline, where it is a major factor in coastal development.

Table Ⅰ　Annual water and sediment discharge of the three major rivers of China

River	Annual Water Discharge 10^8 m^3	Annual Sediment Discharge 10^8 Tonne
Yellow River	485. 65	12. 0
Chang-Jiang (Yangtze River)	9 793. 5	4. 0~5. 0
Pearl River	3 700. 00	0. 85~1. 0

Coastal Processes

The coastal tides of China are formed as a result of the tides of the Pacific Ocean interacting with the coastline and submarine topography of the continental shelf. They therefore have great regional variations. The tides of South China Sea are predominantly diurnal, with a small range (1 or 2 m). The Bohai, Yellow and East China Seas have a dominant semidiurnal tide with a large range (3 m or more), producing strong tidal currents. Tidal bores occur in many estuaries. Over one-third of the coastline of northern China is made up of tidal flats.

The action of surface waves, principally wind waves, is the main mechanism in shaping most of the coast. Wave patterns are dominated by the monsoon. Northerly waves prevail in the winter, moving southwards to reach the north of Taiwan Province in September, 10°N in October, and covering the entire seaboard in November. During January the wave directions change clockwise along the edge of the Mongolian high pressure system. This change is related to the monsoon, with northwesterly wave direction in Bohai Sea gradually swinging toward northerly to northeasterly in the southern East China Sea and South China Sea.

Southerly waves prevail in the summer, moving northward from the South China Sea. They appear first in February, and by May they dominate a wide area to the south of 5°N. In July southerly waves prevail along the entire seaboard. The wave directions change anti-clockwise along the edge of the Indian Ocean low pressure system, being southwesterly in the South China Sea to the south of 15°N. These change northwards through southerly to southeasterly waves in the East China, the northern Yellow and Bohai Seas.

There is a transitional period between the summer and the winter, during which the wind directions fluctuate, and there are no prevailing waves.

The Major Coastal Types and Their Characteristics

The sea coast of China can be classified in two major types: the bedrock embayed coast and the plains coast. Both may contain river mouth coasts, which are important because of the large populations that they support. In the south, under the tropical and subtropical climate, the coastline is also modified by the growth of coral reefs and mangrove swamps.

Bedrock-Embayed Coast. The coast develops where the mountains meet the sea, and is characterized by irregular headlands, small bays and islands. The submarine coastal slopes are very steep, and as a result, wave energy is high, influencing both erosion and deposition. Coastal sediment is coarse sand and cobbles. Detailed coastal morphology is related to wave action and is dependent on coastline exposure, submarine slopes, the type of bedrock and sediment supply. This type of coast is found along the Liaodong Peninsula, Shandong Peninsula, the Zhejiang, Fujian, Guangdong and Guangxi province.

The bedrock embayed coast can be subdivided in four subtypes according to the stage of development.

1) Marine-erosional embayed coast: This is developed on hard crystalline rock where the rate of erosion is slow, and there has been little coastal modification since the post glacial rise of sea level. Little sediment is supplied by either rivers or coastal erosion, so sediment along this type of coast is deposited only in a few areas in the form of small bay head beaches and sand spits. The main coastal features are thus erosional, rather than depositional. This type of coast is favourable for the building of sea ports, such as the Luda ports on the Liaodong Peninsula.

2) The Marine erosional-depositional type: This is most commonly developed where there are Mesozoic granites covered with a thick layer of weathered deposits, such as those on the Shandong Peninsula. This material is readily eroded by wave action and by the many small rivers that transport large amounts of sandy material to the sea. Therefore, erosional features such as rock benches and terraces and depositional features, bay-bars, sand spits and tombolos are developed. The sediments locally contain valuable placer deposits, including zircon, rutile, monazite and gold.

3) Marine depositional type: The bedrock of this type of coast is relatively soft. Marine erosion has supplied large amounts of sediment, that forms a straight prograding

barrier shoreline normal to the prevailing wave direction. A narrow coastal plain is built in front of an irregular line that represents the original bedrock coastline. Thus the embayed coast matures to the plains coast. This type of coast is very common in south China.

4) The tidal inlet-embayed coast: This type of coast is also common in south China. Here the major NE-SW trend of the coastline is cut by long, narrow bays following NW trending faults. Small islands or sand spits protect the mouths of the bays, forming tidal inlets, which are swept clean by the powerful ebb tidal currents. Sediment is deposited in the form of ebb deltas outside the tidal inlets. This type of inlet is suitable for harbours.

Plains Coast. On the basis of genesis, the plains coast can be subdivided into two types.

1) The alluvial plains coast: Most of this type of coast is located seaward of mountain ranges. The plains are built by fluvial sediments from the mountain rivers. The plains continue seaward and the submarine coastal slopes are low with gradients of 1:100 to 1:1 000. As a result, there is a very wide breaker zone, with active sediment movement both longshore and onshore. Barrier bars, sand spits, submarine bars and extensive sandy beaches are the typical features of this coast. Many beaches are prograding, with several beach ridges, and lagoons behind the bars.

2) The marine depositional plains coast: This type of coast is flat and very extensive, and is located on the lower parts of large rivers in areas of subsiding basement, such as the north Jiangsu plain, North China plain and the Liaohe plain. The coastal slopes are very gentle, with gradients in the order of 1:1 000 to 1:5 000. As a result, effective wave action is well off shore; thus tidal currents play a major role in shaping the coast. The sediment consists mainly of suspended silt and mud transported by rivers. This fine sediment cannot be deposited in the shallow nearshore zone, but accumulates either offshore in deeper water or on the upper part of tidal flats.

A typical coast profile passes from salt flats through intertidal mudflats to the submarine coastal slope. The salt flat extends inland from a shell ridge (Chenier) above the intertidal zone. The shell ridges are preserved for a considerable time and are useful indicators of the location of ancient coastlines. Occasionally relict lagoons are found behind the shell ridges. The intertidal mud flats can be extensive as a result of the three metre tidal ranges. For example, along the west coast of Bohai Bay, tidal flats are 4 to 6 kilometres wide (see Fig. 3), with slopes of 1 : 1 000 to 1 : 3 000. In the southern part of the North Jiangsu Plain the intertidal mud flats are wider still. On the basis of several years of research it was discovered that the intertidal flat can be divided in four geomorphological and sedimentary zones which are the result of tidal currents acting on the flat (Table Ⅱ). The submarine coastal slope is below the intertidal zone. The seaward slope from the intertidal flat is very gentle, in the order of 1:500 to 1 : 1 500, with a concave profile. Over 10 kilometres offshore, the water is only 3 to 5 metres deep. Occasionally there are 2 or 3 metre high shell banks and depressions of similar dimensions. The sediment becomes finer off shore, from sandy silt at the low tide level to very fine mud in depths of more than 10 metres.

Fig. 3　Yellow River Delta.

Table Ⅱ　The intertidal zones

Zone	1. Outer Depositional Zone	2. Erosional Zone	3. Inner Depositional Zone	4. Polygon Zone
Location	immediately shoreward of the low tide level	around mid tide level	immediately below the lower high tide level	between lower high and maximum tide level
Morphology	wave or current ripples	overlapping 'fish scale' structure or tidal channels	mud flat or mud pools	polygon fields or marsh (near river mouth)
Sediment	very fine sand or coarse silt	alternating laminae of mud fine silt and mud	mud	clay or clayey-mud
Dynamics	tidal currents and microwaves	tidal currents, mainly ebb currents	tidal currents, mainly flood currents	tidal currents and evaporation due to solar heating

Zone	1. Outer Depositional Zone	2. Erosional Zone	3. Inner Depositional Zone	4. Polygon Zone
Processes	deposition of thin layers over large area	deposition during flood, erosion during ebb current	extensive deposition	stable (during maximum tide deposition, erosion during storm)

The plains coast represents a balance between wave erosion and deposition of unconsolidated sediment that is dependent on sediment supply, principally from rivers. If the amount of the sediment supplied is larger than the amount eroded, the coastline progrades; if the sediment supply is reduced or stopped, wave erosion causes the coastline to retreat. Relative changes in sea level do not effect to any great extent the development of the plains coastline. For example, the west coast of Bohai Bay, which receives the sediments of the Yellow River, is prograding despite a steady regional subsidence since the late Miocene. During the time when the Yellow River changed its course to the south and entered the Yellow Sea, the coastline retreated, due to the lack of sediment.

The River Mouth Coast. The Yellow River has the highest rate of sediment discharge of any Chinese river, and carries the sediments into Bohai Bay (Fig. 3). The bay is shallow with a maximum depth of 40 metres; the tidal range is only about one metre at the river mouth; and wave action is also minimal. The sediment of the Yellow River is deposited along both sides of the river mouth in the form of 'finger bars'. The growth of the finger bars is rapid; we measured 10 km progradation in a single year. To either side, the bars protect bays in which fine sediment collects, forming mud flats. Beyond the mud-bays the older delta shoreline is being eroded, because the modern river sediment does not reach these areas. The lower Yellow River changes its course every 6 to 8 years, forming a new set of finger bars. The abandoned channel finger bars are slowly eroded by waves and the sediment is redeposited along the shoreline, so that with time a large and symmetrical fan delta is formed around the distributaries. Within a period of 120 years a fan-shaped delta prograded seawards from Lijin with a distal margin in 15 metres of water.

The Chang Jiang (Yangtze) River mouth faces the open East China Sea (Fig. 4). Wave action at the mouth is stronger than at the Yellow River. Most of the sediment accumulates in the inner part of the river mouth, and as a result, a series of sand islands and river-mouth banks are formed. Depending on the wave direction, some sediments are transported to the north or south by the longshore currents forming extensive mudflats, some just seaward of the cliffs of the bedrock coast to the south. There is a submerged ancient Chang Jiang river channel crossing the continental shelf. It starts at some distance from the modern river mouth and extends to the Okinawa Trough. At the present time tidal currents in this channel reach 50 cm/s and thus may transport sediment to the Continental Shelf, and possibly to the trough. Thus the modern Chang Jiang river delta is relatively small, and is growing towards the southeast in the shape of the bird beak-like protrusions.

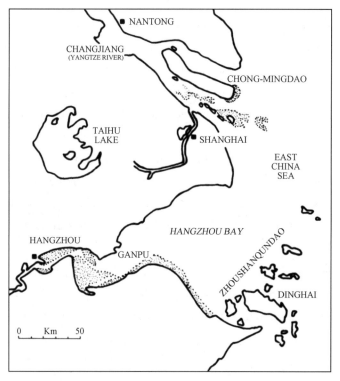

Fig. 4　Changjiang Delta and Hangzhou Estuary

The Pearl River delta is transitional between the estuarine type and the delta type of coastline, with sediment accumulating in a bedrock bay sheltered by islands (Fig 5).

The estuaries, such as Hangzhou Bay (just south of Shanghai, Fig 4), carry very little sediment. The tidal range is high up to 9 metres, depending on the shape of the coastline, and there is a distinct tidal bore. During the historical period, the Hangzhou estuary has tended to shoal and as a result, the tidal ranges have decreased.

Coral Reef and Mangrove Coast (Biogenic Coast). Most of the South China Sea Islands are atolls which are still growing. The basement of the atolls consists of Tertiary fold ridges. Block faulting and subsidence have resulted in reefs over 1 000 metres thick (Fig 6). Fringing reefs are found surrounding Hainan Island, Taiwan and many other small islands. At present, the coral growth is poor.

In China, the mangrove can grow south of latitude 27°N. Along the mainland coast the mangrove swamps grow intermittently at river mouths, bays or lagoons, but the growth is stunted with only small bushes (mainly *Avicennia*). Mangrove jungles grow along the northeast coast of Hainan island.

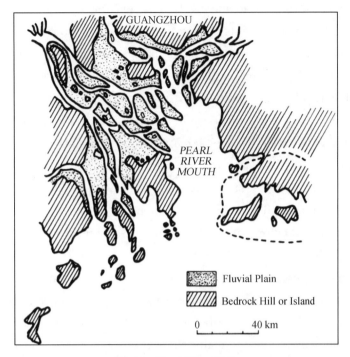

Fig. 5　Pearl River Delta

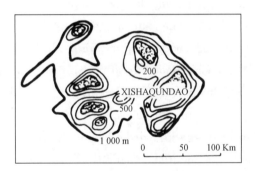

Fig. 6　The atolls of Xishaqundao

Coastal Research in China

Systematic study of coastal geomorphology in China started in the nineteen fifties. The term 'sea coast' in China has a wide meaning. It includes the land along the coast, the beaches and the submarine coastal slope, i. e., the entire coastal zone. As a result, we investigate the whole coast from backbeach to shore and offshore, out to depths where waves start to move sediment. The work is therefore carried out on land over the intertidal zone and at sea. We analyze the sea coast from the point of view of morphology, dynamics and sedimentology. This comprises the 'coastal dynamic geomorphology'. The aim of this research is to define the source and movement of the sediments and predict the developmental trends. On the basis of these analyses, we determine the character and genesis of the coast. Through such research, we choose the ideal sites for new harbours, new docks or navigation routes in order to avoid the problem of siltation. For example, Tianjin New Harbour was built on mud flats of Bohai Bay. Shortly after its completion, the harbour had to be deepened due to extensive siltation, and a total of six million m^3 of silt was removed annually. This example triggered extensive coastal research programs and subsequently, many harbour sites were chosen on the basis of thorough studies of coastal geomorphology. Unfortunately, some harbours still have been built without proper study, and have serious siltation problems. To avoid these, extensive coastal dynamic geomorphologic studies are now required before building any harbours. Thus practical requirements have advanced coastal science.

For this type of research, there is a team of coastal geomorphologists, physical oceanographers and engineers working together on the same project. Depending on the research aims, each participant contributes with plans for the project. The work of the coastal geomorphologists, who normally come from universities and institutes, is the basis of most of these studies. The Department of Transport supports this work with technicians, money, ships, vehicles and drilling equipment. When the project is finished a report is submitted to them. The data are also used for scientific research. For example, in the Tianjin New Harbour project, we studied the mud flat coast and the Yellow River delta; with the Chinhuangtau oil port project, we investigated the barrier bar coast; with the South China port projects we investigated tidal inlets and coral reefs, and with the Chang Jiang River mouth project, we studied the estuary. This research accelerated the advancement of the knowledge about the coast of China. For example, we were able to establish the mudflat zonation, the proper significance of the shell ridges, and the dynamics of estuarine channels, among others.

We suffer from a lack of contacts between ourselves and scientists outside China. We would like to establish and maintain close cooperation with scientists of other countries and participate in international meetings and publish in international journals.

Acknowledgements

I would like to thank Drs. G. Vilks, D. J. W. Piper and C. Tang for assistance in writing this paper. D. Buckley, H. B. S. Cooke and D. Frobel provided critical comments.

Annoted Bibliography

Chai, I-Chi and Xian-Yan Lee. 1964. Some characters of coral reef of southern coast of Hainan Island.

Oceanology and Limnology，6（2）：205－218.

Chao，Qiong-Ying，Hao-Ming He and Xien-Luan Cheng. 1977. The Huang He Delta and shoreline development. In：Geomorphological Symposium of the Geographical Soc. China，Tiejin.

Jin，Xiang-Lung. 1977. Submarine geology of the Yellow Sea and the East China Sea. Ocean，5.

Wang，Ying，Da-Kuei Zhu and Xi-He Gu. 1964. The characteristics of mud flat of the west coast of Bohai Bay. In：Abstracts of theses in the Annual Symposium of Oceanographical and Limnological Society of China in 1963，Wuhan. Beijing：Science Press. 55－56.

Wang，Ying，Da-Kuei Zhu，Xi-He Gu，Cheng-Qi Cui. 1964. The characteristics of mud flat of the west coast of Bohai Bay. *The Research on Siltation of Newport*，(2)：49－64.

Wang，Ying. 1976. The Sea Coast of China. Beijing：Science Press.

Wang，Ying. 1976. The Sea Bottom of the South China Sea. *Ocean*，2.

Wang，Ying. 1977. The submarine geomorphology of Bohai Sea. *Ocean*，6：5－8.

You，Bao-Sung. 1977. The offshore wind of the China Sea. In Press.

关于海岸升降标志问题[*]

近年来,国内在研究岸线升降与海面变化问题方面有了长足的进展。相应地,沉积物的多项分析,植物孢粉与微体古生物资料的应用,以及定年方法与技术装备等方面都有了明显的提高。有数据、有分析力的文章如雨后春笋般地出现在各类刊物上,这种现象反映出古老而又年轻的地球科学在我国具有强大的生命力。

发展着的概念与传统观念的对立是存在于科学发展的整个过程中的。人们在科学实践活动中逐渐认识到事物本质的某个方面,同时,也还需要通过理论总结来提高这些认识,以期应用它去认识更多的现象,去揭示更多的实质,如此发展,推向深广。因此,进行实践,进行总结,展开讨论,进一步提高认识并指导实践,这是科学工作的正常过程。任何事物都是从不知到知的,如果我们一开始就知道清楚了,那么也就没有科学研究了。

关于中国海岸升降问题的研究中,曾有着一些传统的观念,从现代海岸地貌学的观点看,值得商榷。但至今仍被一些书籍所采用,具有一定的影响。因此,有必要对一些问题开展讨论,以期得到较为符合实际的认识。

一

中国海岸的"南降北升"论,可能受 D. W. 约翰逊海岸分类的影响,主要从海岸外形,提出以钱塘江、长江口为界,北方平坦的沙岸是由于"海退"而从海底生长出来,属上升海岸;而南方为下沉海岸,多溺谷港湾。

现代海岸地貌学的研究成果表明,有沙坝环绕的潟湖平原海岸正是"下沉"的海岸地段所特有的。三角洲平原海岸亦是发育于下沉区。地质资料充分说明,广大的平原海岸地区正处于下沉凹陷带,渤海湾西岸与苏北平原海岸,都是位于新生代的沉降凹陷带。而杭州湾以南基岩港湾岸,却位于浙闽隆起构造带,应属于构造上升的海岸地段。但是,冰后期的海面上升又使我国海岸线普遍具有海侵的影响。区域构造升降运动对现代海岸发育过程,是通过改变海岸的高差、坡度或物质结构状况而发生影响。这种影响通常较之活跃于海岸带的波浪、水流与泥沙运动过程的影响要缓慢、微小得多。因此,构造升降运动对现代海岸过程可以看作是间接因素。例如,松散沉积组成的平原海岸,其岸线进退充分反映着波浪冲蚀与泥沙供给(主要是大河泥沙的供给)之间的动态平衡关系。如果泥沙补给量与波浪冲蚀量相当,则岸线表现为稳定的;若泥沙供应量小,甚至根本断绝了供应,则海岸受冲蚀而岸线迅速后退,如果泥沙供应量大,则海岸线不断淤积向海推进。例如今日的黄河三角洲地区,虽然地处沉降凹陷与海面上升的背景上,但岸线仍不断地向海淤进[1]。把这种平坦的新出露的海岸认为是"上升的沙岸"是错误的。既使在杭州湾以南的

---

　*　王颖:《南京大学学报》(自然科学版),1983 年第 4 期,第 745 - 752 页。

基岩海岸,由于有丰富的细颗粒泥沙供应,以及潮流作用在我国中部海岸发育中具有重要的影响,因此,在基岩港湾海岸的岩滩上,发育了背叠式的淤泥质岸滩。

其次,位于不同构造背景上的海岸与邻接的大陆架普遍发现有沉溺的古河谷(古辽河、古海河、古长江、古珠江等)、古三角洲(古六股河三角洲、古辽河三角洲、古黄河三角洲、古长江三角洲系、古闽江三角洲、古珠江三角洲等)以及古海岸带或海岸带平原(分布于水深 8 m;25 m±;50～60 m;100;120 m 以及 150 m±等[2,3,4,5])。各沉溺地貌分布的水深几乎可一一对应,而其中以－25 m 及－50 m 者为普遍,这种现象不是偶然的。位于现代海岸带的溺谷河口湾,不仅分布于浙、闽、粤沿海,同时也是我国中部海岸的特征现象,如苏北的几条天然河口:灌河口、射阳河口、新洋河口、斗龙港口等皆表现为明显的喇叭型。位于北方海岸的蓟运河、大口河河口亦有类似的现象,山东半岛某些段落表现为明显的"里亚式"海岸型式。这些地貌现象的分布遍及我国南北,表明由冰后期海面上升所引起的岸线相对下降,要比区域构造运动的影响更为直接与显著。

但是,各海域沉溺地貌分布的水深有差异,除了由于发展历史阶段不同外,可能有由于区域构造运动的差异所引起的。

因此,在研究海岸升降问题时,须注意确定那些真正反映岸线升降的标志或现象,确定其升降的时代——是过去升降的结果,还是反映现代的升降或者是今后升降的趋势。在研究岸线升降幅度时,须注意在海面普遍上升的背景上,其净效果如何。

二

海岸是有升降变化的,海岸带的海蚀地貌、堆积地貌或人为地貌,出现于现代海岸动力作用范围之外的,可用来作为判别岸线变迁的标志。通常,用堆积地貌来判断岸线升降,其可靠程度较高。只有能长期保存的发育于基岩地区的显著海蚀地貌才是判断海岸升降难得的宝贵标志。例如,北美纽芬兰东北岸、圣约翰(St. Johns)附近是沉积变质岩组成的海岸,冰后期以来已抬升了数十米,但仍清晰地保存着原来海岸岬湾曲折的形态。该处海岸年龄较短,是大冰盖融化后,地壳弹性回升的地段。又如,北美悉尼(Sydney)沿岸,完整地保存着桑加蒙间冰期的海蚀阶地,阶地面覆盖着威斯康辛冰期的冰碛物。这些海岸形态为分析岸线升降提供了可靠的标志。在我国,由于地处温湿地带,风化剥蚀作用较强,加之我国海岸带开发历史悠久,古海岸遗迹常常遭受较多的改造破坏,当然仍有不少有价值的遗迹保存下来,关键在于正确地识别岸线升降的标志。

(一) 海蚀穴

发育于海蚀型的基岩海岸段落的高潮线附近,由于波浪的冲刷掏蚀、夹带砂砾的研磨以及海水的溶蚀与风化作用,因而沿高潮水位线附近的岩层构造软弱处常形成龛状凹穴。海蚀穴的形态多样,但主要特点有三:① 沿岩层层面发育的海蚀穴,削切岩层层面。在花岗岩类岩体上发育的海蚀穴,其"上颚"适应水流的冲刷反卷作用多呈现为卷浪形式,或者形似鸟啄;"下颚"部分由于遭受较为频繁的砂砾研磨作用而蚀退快,多形成低平的水漫坡。② 海蚀穴的规模不一,有的可成为一列卷浪式的冲刷槽,有的仅是一龛穴。大的海蚀中还有一次一级的水流掏蚀凹穴以及浪花引起风化造成的蜂窝状孔洞。

上述特征可帮助我们鉴定海蚀穴。其次,由于海蚀作用使海蚀不断扩大,进而使上部岩石崩塌而形成海蚀崖。海蚀崖的高度随岸线向陆地推进而加高。伴随着海蚀崖的后退而于崖前方发育了岩滩(或称海蚀平台),岩滩的规模形态不一,取决于岩性与海蚀作用久暂,坚硬岩石形成崎岖不平的岩滩,其上部还有残留的海蚀柱。因此,各种海蚀地貌是成组分布的。如果位于现代海面以上较高的地方还保存着古海蚀穴,那么,附近还可能有海蚀崖或岩滩的遗迹相伴分布。由于海蚀柱或岩礁为坚岩组成,所以孤立的岩体上也能长期保留古海蚀遗迹。

在花岗岩、片麻岩或其他非均质岩石组成的海岸岩壁的不同高度上,发育着一些规模不大、密集成蜂窝状的孔洞,这是溅浪及饱含盐水和水汽的海风吹蚀崖面,使结晶岩类崖壁上产生差别风化与进一步吹蚀的结果[6]。在陡岸,溅浪常可达十数米高。笔者在旅大、冀东、山东半岛、连云港地区、潮汕地区及海南岛等处陡峻岩岸带皆见到这种崖面浪花风化所形成的凹痕,虽然它们高出现代海面十数米高,但它们是现代海岸的产物,不是古海蚀穴。在古海蚀崖面上,亦可能有古浪花风化的遗迹,它可以作为岸线变化的辅助标志。但其具体高程,不能作为高海面或地壳抬升的高度。

海湾由于淤积作用形成平原,原湾顶或湾侧的海蚀崖成为湾顶平原的死海蚀崖,其基部的海蚀穴可能仍出露。该处的标高,虽较目前海面略高,海水已不能到达,但这是正常的海岸堆积过程的结果,不一定反映海岸上升。

抬升的古海蚀穴确实是存在的,主要于海岛或临海的山地(可能是古海岛)人类开发活动稀少的地方保存完好。如厦门岛曾山,于海拔 120 m 处保存着完好的卷浪型海蚀穴、岩滩和海蚀柱的组合。连云港地区北云台山保存着数级完好的古海蚀穴[7](表 1),古海蚀岩滩和海蚀崖遗迹相伴分布,几为一岸线变迁遗证的天然博物馆。

表 1 江苏云台山古海蚀地貌分布状况

海拔高度	地 点	地貌特征
600 m	北云台山大桅尖	海蚀穴,岩滩,浪花风化穴
450 m	北云台山上竹园后六场	海蚀穴海蚀崖,岩滩,浪花风化穴
320 m	北云台山上竹园	海蚀穴,岩滩(长 300 m,宽 120 m),海蚀崖与浪花风化穴
200 m	北云台山宿城南	海蚀穴,岩滩,海蚀崖,浪花风化穴
120 m	北云台牛屁股山	海蚀崖残迹与岩滩,浪花风化穴

在上述几级古海蚀穴中,又以 120 m 与 320 m 的保存较完好并分布较广。

历史时期云台山尚系海岛[8],中云台山与南北云台之间以海峡相隔,正如今日连云港市与东西连岛之间隔以海峡相类似。到明末清初之际,即 1494 年黄河全流夺淮入海,黄河泥沙大量汇入,云台山间海峡才最后淤积成陆。由于人类对该岛开发迟,因而古海蚀穴等尚保存完好,并成为高海蚀面的见证。

(二) 海岸阶地

海岸阶地有两类:海蚀形成的与堆积形成的,两类阶地常相伴分布。海蚀阶地发育于岬角地区,是昔日的岩滩抬升而成。因此,它是一崎岖的向海倾斜的基岩面(沿厚层风化壳发育的海蚀阶地是平坦的),其上保存着少量砾石与砂的松散堆积物,或者剥蚀殆尽而无堆积,阶地的后缘有一明显的坡折,即古海蚀崖的遗迹。分布于相邻海湾的海积阶地,系昔日的湾顶海滩,由具有一定分选的砂质沉积所组成,砂层具水平层理或小角度的斜层理,夹大量贝壳屑。经长期风化的老海积阶地,贝壳碎屑可能经风化而淋失。海积阶地的构成有三种类型:① 背叠海滩型;② 具双斜坡的自由海滩,其上有若干列沿岸堤(滩脊);③ 沿岸沙坝并陆而成。海积阶地与海蚀阶地的高度,即使是同一时期形成的,也不是一样的。正如现代的岬角岩滩上界位于高潮线附近,而邻近的海湾海滩,甚至袋状海滩的顶部可高出高潮水位 4 m 之多。尚未固结成岩的松散堆积组成的古海积阶地,由于后日的侵蚀变化,其高度常低于相应时期的海蚀阶地。

海岸阶地在我国分布普遍,并具有多种海岸地貌组合分布之特征。如:① 秦皇岛港东侧有 5 m 海蚀阶地,其背侧有死海蚀穴与海蚀崖,海蚀崖顶部为 15 m 高的南山海蚀阶地。阶地由花岗岩组成,上覆厚度约 0.6 m 的砂层,为均匀的石英,长石质中砂夹有少量 2～6 cm 的扁平砾石。砾石包括花岗岩与火山岩类,系当地的海蚀产物。该海蚀阶地的东侧分布着一列 5 m 高的堆积阶地,系棕色均质中砂组成的老沿岸堤,向东延伸被沙河口的砂砾堤系列所覆盖成不整合交接。这两组沿岸堤所形成的海积阶地是不同时期高海面产物[①]。② 辽东钲锚湾 20 m 高的黑山海蚀阶地,具有海蚀崖残迹与海成砂砾质堆积以及古文化遗迹。低于 20 m 阶地的东西两侧死海蚀崖前方皆叠加着 5 m 高的海积阶地。③ 广东汕头广沃湾一带古岛屿有 40～60 m 高的海蚀阶地,其上散布着扁平状砾石。附近地区分布着一级 5 m 高的红砂阶地,阶地主干部分是一连岛沙坝。北海半岛与海南岛南部有类似的情况。

山东半岛南部,沿岸有四级阶梯状地形,其高度依次为 5 m、15～25 m、40 m 以及 60～80 m。这几级阶地在岚山头一带及连云港地区的东西连岛等小岛亦很清楚。浙闽沿岸及一些岛屿上如大鹿山岛处亦有保存。古海蚀地貌遗迹在岛屿上保存完好不是偶然的,这些岛屿系屹立于海底的坚岩山地,四周环海而海蚀地貌发育,加之人迹稀少,后日的破坏影响少。古海积阶地在舟山群岛等处亦有保存。由于未胶结的砂砾质堆积易为风化剥蚀所破坏,其次,在海蚀作用强盛的岬角、岩礁与小岛上,海积地貌在类型与规模上远不及海蚀地貌广泛。旅大地区现代海蚀地貌远较海积地貌发育、广泛,可为一例证。成山头的 5 m 阶地[9]大浪时尚可作用到,但在连云港海峡一侧所分布的则非大浪阶地。而北云台山、锦屏山、灌云县的大伊山、芦伊山等已位于陆上,尚保存着 20 m 与 40 m 的海蚀阶地。北云台山还有前述的高阶地。由于这些阶地背侧有死海蚀崖的坡折陡坎或海蚀遗迹,故可以与海侵淹没的剥蚀面区别开来。目前的问题是这些阶地形成的时代与造成阶地抬升的原因,是由于陆地上升,还是由于海面下降,或者是两者兼有尚难肯定。600 m 与 300 m 的高海面遗迹在英国等处有报道[10],被认为是第三纪的高海面,并提出如果是在稳定地区具有类似的效果,那么,可能是由于全球性构造——水动型的海面变化所引

---

① 据[14]C 定年资料。沙河口砂砾堤时代距今 195a±213a(海洋局二所,HL83016);棕砂层老沿岸堤下部的海积平原中砂、泥样与贝壳时代为 7352a+318a(海洋局二所 HL83017)。

起。第三纪时构造活动频繁，一些高山与构造海盆在那时形成，会影响到海面有较大的变化[11]。连云港到岚山头一带在大构造单元上也属"稳定"地区，阶地上已发育了厚层红色风化壳，似乎形成的时代较老。但云台山一带在中生代末期亦有断块上升活动，红色风化壳在该区分布到较低的阶地面上。所以，老海面下降或古岸线抬升的原因尚需进一步研究。5 m 阶地可能与冰后期的高海面有关。15～20 m 阶地在北美系桑加蒙间冰期产物，我国的如何？需进一步工作。总之，这是个有趣而复杂的问题。确定了古海岸阶地的分布后，需进一步确定其时代；研究岸线升降变化的原因；还要结合水下沉溺阶地与沉溺古风化壳[2]等现象分析升降变化的顺序以及现代海岸升降的动态。这是几个不同的问题，需要分别地予以阐明。

（三）沿岸堤

沿岸堤是由激浪流携带沉积物堆积在海滩上部的堆积体，它是沿着岸线分布的。根据物质来源与海岸动力条件不同，可以发育为砾石堤、砂砾堤与沙堤等。沙堤的上部还可叠加风力吹积的沙丘。贝壳堤也是沿岸堤的一种，发育于粉砂与淤泥质平原海岸，是潮间带泥滩遭受波浪冲刷改造时所发育的。

在沿岸堤的发育过程中，由于沉积物的粒径不同，泥沙补给量的多寡、岸坡的坡度与海岸暴露程度不同，入射波浪性质与强度不同等，使得沿岸堤的高度、宽度与延伸的距离都会随之不同[12]。例如秦皇岛附近的砂砾堤、荣成湾的砾石堤、三亚湾的沙堤等，各个堤之间以及同一条堤都会随着上述因素的不同而形态要素有变化。不仅沿岸堤，即使如大连附近小岛上的袋状海滩，由于海岸位置与朝向不同及暴风浪作用之结果，现代海滩发育成阶梯状，上部海滩高出临近海面的海滩 4 m。在河口地区由于海岸淤积发展较快，沿岸堤不断适应前进的岸线，而使同一列岸堤可在河口地区分叉，成一组复式沿岸堤。如，渤海湾蓟运河口的贝壳堤[13]、潮汕平原的西溪河口分叉状沙堤等[14]。又如海南岛三亚湾因海湾平原逐渐淤积展宽，来自山地的河流在穿越平原时，比降减小而减低了输沙能力，继而影响了现代岸堤的规模，较老的岸堤规模为小[15]。

海岸动力与泥沙供应条件的变化对岸堤规模的影响（包括高度的变化），远比地壳运动或者海面升降的影响更为显著，如渤海湾西岸的Ⅰ贝堤，受到黄河泥沙汇入的影响，海水变浑，岸坡变缓。因而贝壳种属改变（适合于细砂底质的贝类减少，适合于泥类的贝类增多），贝壳堤的成分与Ⅱ贝堤相比，发生了变化（粉砂与黏土含量增多），同时贝壳堤的规模（高度等）也相应发生变化[16]。

沿岸堤是指示岸线位置的好标志，它的形态敏感地反映着当地海岸动力与泥沙组成条件。用它的高度来分析海面变化或岸线升降时，要区别原始的成因因素与日后的高程变化。

堆积地貌及沉积物标志，是确定岸线升降与环境变化的有力证据，缺点是暴露于地面上的标志易遭日后的破坏，或次生变化较大。

（四）珊瑚礁

珊瑚礁是判断海岸环境与岸线升降变化的敏感标志。值得注意的是：

(1) 为什么变位的珊瑚礁存在于小岛上，这是否意味着它灵敏地反映了区域性的构造变动。

(2) 处于相同时代、相同地点、具有相同海面环境的珊瑚礁，与由珊瑚碎屑、砂砾胶结而成的海滩砂岩，它们分布高度是不同的。同一时期同一地区不同海岸段落的海滩砂岩，由于海岸朝向不同，其分布高度亦不同。因此，要注意区分珊瑚礁与海滩岩，要十分小心地根据海滩岩分布高度的差异来论证岸线升降变化。

(五) 盐沼湿地

多半发育于沉溺的冲积平原海岸，或河口湾内。国外有人根据湿地内侵来分析海面变化[17]。这在高纬度曾受冰川强烈刨蚀，而现代河流泥沙量不大的地区，是一种较好的标志。在我国由于冲积平原不断向海淤进，而湿地效果不明显。在某些特定的地段，如苏北废黄河口，水沙来源断绝又无人工堵口，湿地沼泽向内陆延伸变化，是研究海面变化的良好标志。在潮差较大的海区，盐沼湿地沉积层的高程变化可达 $4\sim5$ m，大潮高潮位以上湿地，多为半盐生植物群落，而平均高潮位为盐生植物群落。

潟湖接受来自海岸沙坝与周缘陆地的物质。沉积中掺杂着砂、粉砂与黏土等物质，而以富含贝壳、植物根系等有机质及呈灰黑色为特征。

在潮汕地区，沿达濠山地南侧现代潟湖平原上有一级抬升的潟湖阶地。向海侧分布着红色砂质的海成阶地。海南岛三亚地区亦有古潟湖沉积与红砂阶地伴生的现象。

沉积物是分析岸线变化比较明确的标志。如海底发现风化壳与淡水泥炭层；山麓地区出现由潟湖沉积所组成的阶地，平原上出现老的海岸沙堤或沙坝等。困难之处在于有些特征不明显的沉积层，如：均匀砂层，无贝壳，无明显的层理结构，无植物残体等有机质。总之，缺乏足以进行判断分析的剖面。在这种情况下，可以根据它的分布特点以及附近相伴生的沉积与地貌作为分析依据。同时，研究沉积物本身，如砾石的形态量计，砂粒的分选程度与表面结构以及黏土质的含量与矿物组成等亦有助于以分析沉积物的成因、沉积环境与变化历史。同样，对整个沉积层的结构变化进行沉积相分析是很重要的。

总之，有关的地貌标志，进行全面地对形态、沉积与动力的综合分析，则会找出有力的依据而作出较为客观的判断。

岸线升降与海面变化，涉及广泛的研究工作。这项研究，应建立在大量实地调查的基础上，根据海岸发展的自然规律与第四纪以来自然环境的变化历史并结合现代的水准测量资料，进行扎扎实实的工作。可以期望，我们会在研究新生代中国岸线的升降与西北太平洋边缘海的海面变化问题上作出重要贡献。

参考文献

[1] WANG Ying. 1980. The Coast of China. *Geoscience Canada*, 7(3): 109-113.

[2] 王颖,朱大奎,金翔龙. 1979. 中国海海底地质地貌. 见:中国自然地理(海洋地理). 北京:科学出版社.

[3] 任美锷,曾成开. 1980. 论现实主义原则在海洋地质学中的应用. 海洋学报,2(2):94-111.

[4] 赵焕庭.1981.珠江河口湾伶仃洋的地形.海洋学报,3(2):255-274.

[5] 冯文科,鲍才旺,陈俊仁,赵希涛.南海北部海底地貌初步研究.海洋学报,4(4):462-472.

[6] 任美锷.1965.第四纪海面变化及其在海岸地貌上的反映.海洋与湖沼,7(3):295-305.

[7] 朱大奎.1984.江苏海岛的初步研究.海洋通报,3(2):36-46.

[8] 张传藻.1980.云台山的海陆变迁.海洋科学,4(2):36-38.

[9] 陈国达.1950.中国岸线问题.中国科学,1(2-4):351-373.

[10] King, C. A. M. 1975. Introduction to Marine Geology and Geomorphology. Edward Arnold.

[11] 杨怀仁,徐馨.1980.中国东部第四纪自然环境演变.南京大学学报(自然科学版),(1):121-144,附表4页.

[12] WANG Ying, Piper D. J. W. 1982. Dynamic Geomorphology of Drumlin Coast of southeast Cape Breton Island. *Maritime Sediments and Atlantic Geology*, 18(1): 1-27.

[13] 王颖.1965.渤海湾北部海岸动力地貌.海洋文集,(3):25-35.

[14] 朱大奎.1981.汕头湾海岸地貌与港口建设问题.见:1977地貌学术讨论会文集.北京:科学出版社.

[15] 王颖,陈万里.1982.三亚湾海岸地貌几个问题.海洋通报,1(3):37-45.

[16] 王颖.1964.渤海湾西部贝壳堤与古海岸线问题.南京大学学报(自然科学),8(3):424-440,照片3页.

[17] Grant, D. R. 1970. Recent Coastal submergence of the Maritime, Province Proc, N. S. Inst 27 supplement 383-102.

Sediment Supply to the Continental Shelf by the Major Rivers of China[*]

The modern continental shelf of the China Seas was a coastal plain during the Late Pleistocene. River processes were the major factors in both forming the plain and distributing sediment. Thus, understanding the sedimentary processes of different types of rivers and the characteristics of the sediment that they carry is very important for the study of continental shelf sedimentation.

The coastal rivers of China carry enormous amounts of sediment into the sea which has greatly influenced the evolution of the coast and adjacent continental shelf. According to modern data (Tong & Cheng 1981) a total of 2×10^9 t/yr of fluvial sediment is discharged into the China Seas. Of this, approximately $1\ 210 \times 10^6$ t/yr is deposited in the Bohai Sea, 15×10^6 in the Yellow Sea, 631×10^6 in the E China Sea and 144×10^6 in the S China Sea and the Pacific. Each year, a total of 1 km^3 of sediment is carried by rivers to the continental shelf. The highest rate of deposition is in the Bohai Sea, which with a total area of 78 000 km^2 and an annual sediment discharge of 16.8 kg/m^2, produces a sedimentation rate of 8 mm per year; at this rate with an average water depth of 18 m it would be completely filled in 2 250 years, if basin subsidence was not active. Of course, past rates of sedimentation cannot be accurately estimated from present rates. However, the continental shelf of the China Seas has been influenced greatly by river processes.

The ancient Yellow River and the Changjiang River (Yangtze) crossed the full width of the continental shelf (Ren & Zang 1980). Detritus on the continental shelf was supplied by rivers during geological time. In this paper we present five different types of rivers showing different characteristics of sediment movement and distribution in the estuaries, and their influence on the adjacent continental shelf (Fig. 1).

Yalu River

The Yalu River flows from the Changbai Mountains in NE China to the north end of the Yellow Sea (data in Table 1 from Xue & Pan 1983); 80% of the total water and sediment discharge occurs during summer flood, from June to September.

---

* Ying Wang, Mei-e Ren, Dakui Zhu: *Journal of the Geological Society*, *London*, 1986, Vol. 143: pp. 935 – 944.

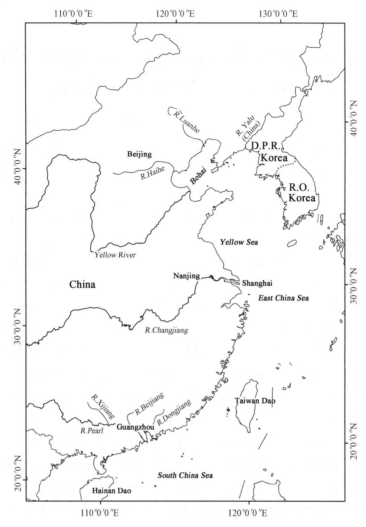

Fig. 1　Location map of the major rivers entering the China Seas

Table 1.　Basic data of five major rivers of China

River	Drainage area (km)	River length (km)	Annual water discharge (m³)	Annual sediment discharge*	Average sediment concentration†	Tidal range (average) m
Yalu	64 000	859	27.8×10⁹	4.75×10⁶	0.33~0.42 kg/m (ebb-flood)	4.48
Luan He	44 900	870	38.9×10⁸	24.08×10⁶	3.94	1.50
Yellow	752 443	5 464	48.5×10⁹	11.9×10⁸	37.7 (Shanxian)	0.80
Changjiang	1 807 199	6 380	9.25×10¹¹	4.86×10⁸	0.544	2.77
Pearl	452 616	2 197	3.7×10¹¹	(0.85~1.0)×10⁸	0.12~0.334	0.86~1.63

* tonne，†kg/m³

The river channel cuts through granite and Precambrian metaquartzite and has a relatively steep gradient along its lower reaches. Here, the meander belt has a width of only 1. 5 to 3 km. The sediment of the narrow fluvial plain consists of a thin layer of medium and fine sand. Because of these conditions only a small amount of sandy sediment is deposited in the river channel. Because of macrotidal processes, the river mouth has formed a wide funnel-shaped estuary (Fig. 2). The powerful tides of the estuary are irregularly semi-diurnal with an average range of 4. 48 m, and a maximum range of 6. 92 m. Outside the estuary, tidal currents are perpendicular to the sea coast, with an average velocity of 1. 25~1. 50 m/s. In the channel of the estuary the velocity of the tidal currents is about 0. 65 m/s on average. The flood tide period is longer at the river mouth, decreasing up-stream; the ebb tide period is correspondingly shorter and increases upstream. The difference between the periods of flood and ebb is 15~18 minutes at the river mouth. At the upper limit of the tidal currents in the Dandong the difference is 2 h 28 min. It is evident that the estuary is a zone of powerful flood tides.

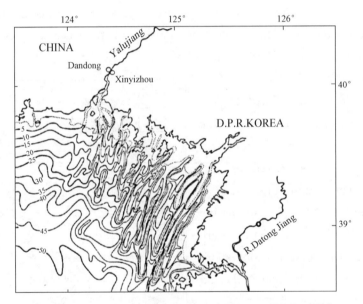

Fig. 2　Sketch map of Yalu Estuary (Water depth in metres)

During the summer, SW and S winds and waves strengthen the flood tides; in the winter, NE and N winds predominate. Normally there is an average wave height of 1 m in the bay mouth of the estuary, with a maximum wave height of 3. 3 m at Dalu Island outside the estuary.

The sediment in the estuary of the Yalu River consists mainly of coarse and sandy material. There are pebbles and shingle in the channels of the upper estuary, and medium to fine sands covered with a thin layer (less than 1 m thick) of silty mud in the areas of sandy shoal. Outside the estuary, an extensive area of fine to medium sands is distributed on the inner shelf. The outer boundary of this area, where the water is 10 m deep, is 20 to 30 km away from the river mouth.

The sediment in the estuary, including both suspension and traction load, is supplied mainly by the Yalu River and is carried far out to sea; even the traction load can be transported 20~30 km from the river mouth. The strong tidal currents outside the estuary have prevented the development of a submarine delta. However, these currents have transported the sediments towards the sea, creating a series of linear sand ridges and parallel to the direction of the tidal currents. The relief between ridges and troughs measures about 15 m over a distance of 1 km (Fig. 2). The estuary has increased the tidal prism of W Korea Bay. The convergence and divergence of the tidal prism during each tidal cycle aids in developing the sandy ridge fields. The sandy ridges consist mainly of fine sand (0.16~0.18 mm) with some coarse silt. These mobile fine materials are winnowed and redeposited by flood tidal current into zones. There is a progression of medium sand to fine sand to silty fine sand to clayey silt in zones towards the estuary. According to observed data in a channel of the estuary during the period of a tidal cycle, the mean sediment concentration is 0.42 kg/m^3 during the flood tide, and 0.33 kg/m^3 during the ebb. The net input of sediment is 2 100 t per cycle. Sedimentation is occurring on the tidal flats, located along both side of the estuary at a height of 0.2 to 0.5 m above mean high sea level. The sediments are mainly reworked by the sea. Several sandy shoals remain on the inner shelf in water depths of about 1 m and their fine sand is stirred up and winnowed by wave action (Fig. 3).

The sediments of the submarine tidal ridges off the Yalu Estuary are moved towards the west by long-shore drift, caused by the dominant SE and ESE wind-induced waves.

In summary, the estuary of the Yalu River and adjacent continental shelf are characterized by macrotides and an abundance of sandy sediments. There is a bidirectionally distributed zone of bottom sediments from the river down to the continental shelf. The principal source of pebbles and sand is the metamorphic rocks in the mountains. The sediments, transported by run-off, are deposited according to grain size from coarse to fine-grain downstream.

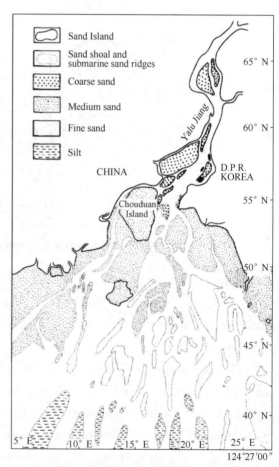

Fig. 3 Map of submarine topography and sediment distribution in the Yalu Estuary

The sandy shoals on the inner shelf off the estuary have been reworked by tidal currents, especially the flood current, winnowing the sediment and transporting it back towards the estuary. As current velocity decreases landward, the sediments from inner shelf become finer.

The transition zone, acted upon both by run-off and tidal processes, forms the estuary tidal bank. This is characterized by two layers, the river gravels deposited on the bottom and the muddy layer deposited by tidal currents on the top. The muddy materials cannot be deposited in the deep channels because the velocity of currents there rises above 0. 60 m/s. However, through progradation, the channel is becoming shallower and is filling up with the muddy sediments. The tidal bank sediments are surrounded by finer material.

On the sand shoals off the estuary the muds have been washed out by wave action and powerful currents. As a result, residual fine sands are the main type of sediment in the area of the sand shoals.

Under the dynamic activity of tidal currents, the sand shoals have been formed into a series of sandy ridges. These are oriented parallel to the tidal current, and have a tendency to migrate towards the west. The modern sediments supplied by the Yalu River influence the continental shelf to a water depth of 15 m.

Luan He River

The delta of the Luan He River is located in the transition zone between the Ianshan Mountain uplift belt and the subsiding basin of the Bohai Sea. Subsidence is therefore the major tendency (data in Table 1 from Tong & Cheng 1981). The mean sediment concentration at 3. 94 kg/m³ is the fourth largest among China's rivers. The abundant sediments enter the sea forming a prograded delta. The Holocene Delta of the Luan He River consists of four lobes (Fig. 4): (1) the east lobe formed during early to middle Holocene; (2) the west lobe, during the middle and

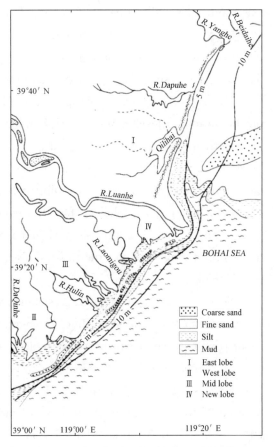

Fig. 4　Map of sediment distribution of the Luan He River delta and adjacent continental shelf.

late Holocene; (3) the middle lobe, during the historical period; and (4) the newer lobe formed since 1915 (Wang1963; Zhu 1980).

Wave patterns in the area of the inner shelf off the mouth of the Luan He River are dominated by the monsoons. In the winter, N and NE waves prevail, and E and SE waves in the summer. Wave dynamics are the major factor in the transport of sediment. The greatest waves occur with the dominant wind which is mainly ENE.

The significant wave height($H^{1/3}$) is 3. 8 m and the period is 7. 8 s. The maximum water depth of the breakers is 5. 5 m in the area.

The tides are different on each side of the river mouth. North of the Luan He River mouth there is an irregular semidiurnal tide with a NE flood current and a SW ebb current. South of the river mouth the tide is irregularly diurnal with a SW flood current and a ebb. The velocity of the ebb current is greater than that of the flood; the tidal range of the area is about 1. 5 m.

The sediment distribution of the inner shelf is also divided into two parts at the river mouth. To the south, there are narrow sediment belts parallel to the shorelineof fine sand, coarse silt, fine silt, and muddy clay. The belt of fine yellowish sand is well sorted ($S_0 = 1. 5 \sim 1. 7$). It meets the coarse silt belt at the breaker zone where water depth is 5 m. The belt of dark grey muddy clay is located in waters more than 9 m deep. There are some small authigenic grains of pyrite in the muddy clay. These grains form 0. 05% ~ 2% of the total heavy minerals. This indicates a quiet, reducing environment on the bottom. Eleven cores 1. 5 ~ 1. 8 m long were taken from sea bed, at water depths greater than 9 m. Analysis revealed at the top of the core a thin, dark-grey, muddy layer containing inter-stratified silty beds; the lower part of the core, is a yellowish, silty, fine-sand bed. The latter sediment is cross-stratified, well-sorted, and contains organic material. It is believed that the sediments were deposited on a mud-flat or the spit of an old river mouth, and then covered with the muddy clay of inner shelf deposition during the modern transgression. The present shelf deposit is only 0. 4 ~ 0. 8 m thick.

North of the Luan He River, sedimentation is entirely different. Sandy deposits extend 20 km from the shore line to a water depth of 13 m. Fine and medium sands occur which are yellow to brownish-yellow. Grain size varies, with 60% ~ 90% being 0. 1 ~ 0. 2 mm and 10% ~ 30% being 0. 25 ~ 0. 5 mm, with few grains of silt or clay. The fine sand is well-sorted with a coefficient of 1. 3 to 1. 8. The medium sands are distributed from a water depth of over 10 m to 13 m. Grain size is as follows: 60% ~ 70% are 0. 25 ~ 0. 5 mm; and 10% ~ 38% are 0. 1 ~ 0. 25 mm; less than 2% is silty material. The medium sands are also well sorted, $S_0 = 1. 5 \sim 1. 8$. The log-probability curves of the fine and medium sands are similar to the curves of the beach dunes (Fig. 5), which have a higher population of saltation and are well sorted. The surface of the rounded quartz grains have density-of-impact pits causing frosted surfaces. There are solution holes and silica precipitation in the concavities. These are all characteristic features of quartz sands in the environment of a submerged beach dune.

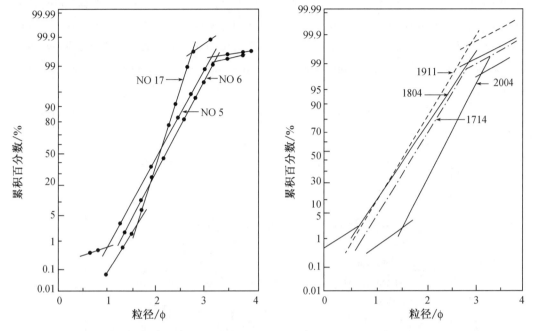

Fig. 5 Grain size distribution of (a) beach dune sands and (b) inner shelf sands of the Luan He

Thus the submarine sand deposits offshore were coastal beach and beach dunes during a period of lower sea level. They were submerged during the Holocene sea level rise.

The sediment of the whole inner shelf of the area is mainly supplied by the Luan He River, the delta of which has many channels and inlets. Some of the inlets were ancient river mouths formed as the river channel migrated from south to north during historic time. Between A. D. 1453—1813, the river channel flowed from the Daqing River, and migrated to the Laomi Channel in 1813. Since 1915 the water has flowed from the modern river mouth (Wang 1963). Many sandy bodies such as point bars, river mouth bars, and sand shoals were left in the area of ancient river mouths. When the river moved away, the sediment supply was cut off and erosion of these sandy bodies began. Sediment distribution in the area has shown a dynamic balance with wave action. Sand has been deposited in the wave breaking zone; transitional deposits of silt mixed with sand and clay have been deposited in the shallow water zone of wave transformation; and muddy clay has been deposited in the deeper water environment of the inner shelf.

Since 1915 the modern delta has prograded, and there is a series of distributary channels across the fan-shaped delta. The abundant river sediments enter the sea nearshore and then are pushed offshore by E and SE waves, especially during summer flood season. E and SE waves advance towards the coast transversely. They form offshore sand barriers enclosing the river mouth and the coastal zone nearby. Some of the sand barriers are the top parts of sand shoals, which rest unconformably upon the bottom sediments. The fore slopes of the barriers are steeper than the lee slopes. The river sediments are mainly deposited in the shallow water environment protected by the sand barriers. Well-laminated and inter-stratified silt and fine sand are developed leeward of the sand barriers. The fine sandy materials come from the sea, either by overwash of sand barriers, or by tidal currents through the inlets between the barriers. Along the coast in a zone of decreased wave action behind offshore sand barriers, tidal flats develop which overlap the delta plain of the original coastline (Fig. 6).

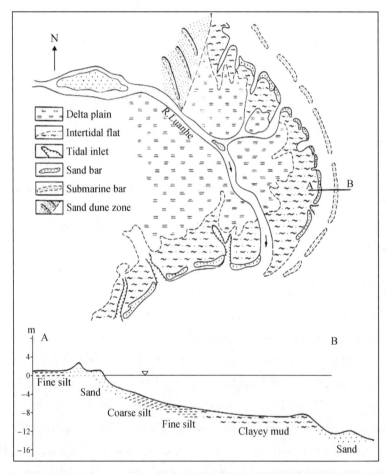

Fig. 6　Geomorphology and sediment distribution of the Luan He River delta with a cross-section profile (Distance A－B＝1 000 m)

The modern river sediment is deposited mainly in the area of new delta. The longshore drift, encouraged by tidal currents, extends 15 km to Chilihai to the north, and 20 km to Hulinko to the south. The river effluents fall eastward and reach the foot of the delta at a water depth of 7 m, 2.5 to 5.0 km from the river mouth.

In summary, the modern river delta of the Luan He River is under the control of monsoon-produced waves. The sediment movement along the coast is mainly transverse with limited long shore drifting of 20 km. The sedimentation model of the modem coast shows the coarse-grained sediments forming sand barriers offshore, the transitional deposits of sand and silt forming laminae in the lagoon behind the sand barriers, and the fine grains of muddy silt and clay depositing in the coastal tidal flats on land. As the lower river channel migrated to form new deltas, the old deposits at the abandoned river mouth were winnowed by wave action and reformed into narrow belts of sediment parallel to the coastline. Beyond the modern submarine coastal slope there are relict sand deposits on the surface of the inner shelf.

Yellow River

The Yellow River is noted for two characteristics: the huge volumes of silt it carries and the shifting courses of its lower reaches. During the past 4 000 years, there have been eight major shifts in the course of the Yellow River (Pang & Si 1979; Wang 1983) (Fig. 7). The northern most course flowed north of Tian Jin and passed through the Hai River into northwestern Bohai Bay, the southern most course passed through the Huai River into the Yellow Sea. In 1855, the Yellow River migrated from the Yellow Sea back to the Bohai Sea where it remains today. The annual sediment discharge of the river is 11.9×10^8 t (Tong & Cheng 1981). Sixty four percent of the sediment is deposited on the delta and mudflat and the remaining 36% is passed farther out to sea (Pang & Si 1980; Wang 1983) (Fig. 8).

The wave action is minimal at the river mouth, stirring up the sediments only during heavy storms. Tidal currents and residual currents are the major agents in transporting the sediment. The tidal range is $0.5 \sim 0.8$ m at the river mouth increasing gradually to 2.0 m on both sides of the delta. Maximum velocity of the tidal current is 1.5 m/s at the river mouth. The direction of maximum flood current is WNW, that of maximum ebb current ESE.

The fine-grained deposits off the Yellow River mouth are stirred up by shallow water waves and transported by longshore drift, either to the NW, or to the SE, the former being dominant.

The coarser sediment of the Yellow River, which is coarse silt, is deposited along both sides of the river mouth in the form of 'finger bars'. The growth of the finger bars is rapid, especially in the early stages of a new channel; 10 km of progradation in a single year occurred in 1964 (Wang 1980). Here, the submarine coastal slope is steep. In a water depth from $1 \sim 15$ m, the slope is approximately 1/760, caused by rapid deposition.

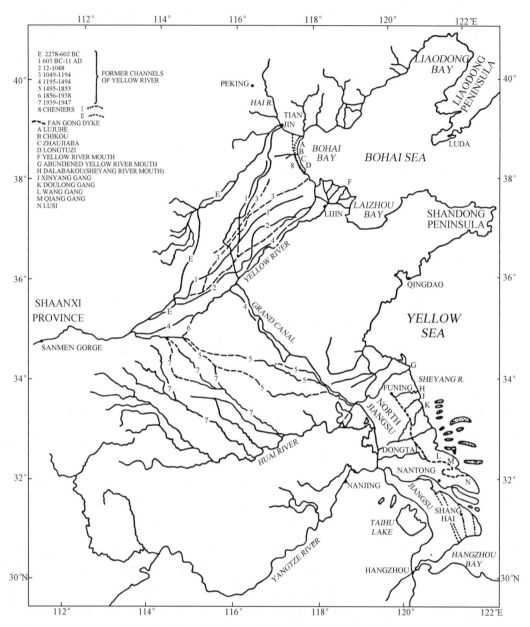

Fig. 7 Map of abandoned channels of the Yellow River

On either side of the river mouth, bars protect bays in which fine silt and silty clay collect forming mud flats. Therefore, the two areas of modern delta deposition are the river channel inlet and the muddy bays.

Beyond the muddy bays the old delta shoreline is being eroded, as the modern river sediment does not reach the coast. The Lower Yellow River changes its course every six to eight years, forming a new set of finger bars. The abandoned channel finger bars are eroded slowly by waves and the sediment is redeposited along the shoreline. Thus, coarse silt covers the fine sediment in the muddy bays while muddy deposits appear on

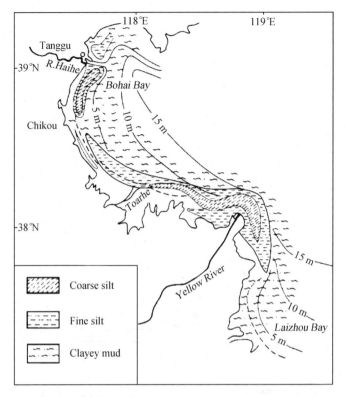

Fig. 8　Sediment distribution of Bohai Bay and the Yellow River delta

the eroded finger bars.

In cross-section, there are alternate beds of subaerial silt and marine mud in the lower layer and marine mud and shell-fragment sands in the upper layer. During a period of more than 100 years a fan-shaped delta prograded seawards from Lijin with its distal margin in 15 m of water. The total coastline of the delta is 160 km long from the Toarhe River in the north to the Xiaoqing River in the south (Fig. 9). The total progradation of coast has been 20. 5～27. 9 km since 1955. With an average progradation of 0. 2～0. 27 km per year, an average of 23. 5 km² of land has accumulated each year.

The map of surface sediment distribution in Bohai Bay shows the divergence zone of the Yellow River silt (Fig. 8). There are two sediment sources, the Yellow River in the south and the Hai River in the north. The Yellow River silt tongues run NW to Bohai Bay, whilst that of the Hai River trends towards the south. They join at Chikou in the Middle Bay.

The sediments derived from each river have slightly differing heavy mineral assemblages. Heavy minerals such as garnet, zircon and tourmaline are common in the Yellow River sediments. Some hematite, a little magnetite but very little biotite are present. Most of the grains of these minerals are oxidized, have frosted surfaces, and low transparency. The heavy mineral suite of the Hai River is more unstable with well developed crystals, clean surfaces, and better transparency. There is a higher

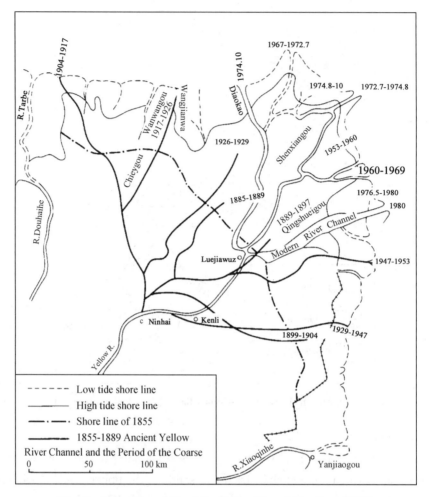

Fig. 9　The modern delta of the Yellow River (after Pang 1979)

percentage of biotite and magnetite than in the Yellow River sediment.

The clay minerals in Yellow River sediments are rich in illite, montmorillonite, kaolinite and some pelhamite. Montmorillonite has a distinct diffraction peak.

In summary, the Yellow River material is mainly deposited in the delta, with the remainder being carried by tidal and residual currents mainly to the NW. In 1980 the Yellow River changed its course to the south of the 1964 channel. Sediment was now carried into Laizhou Bay on the seaward side of which, river-derived sediments enter 15 m of water, where the foreslope of the submarine delta has been built. The distribution area of sediments supplied by the Yellow River has now advanced into the deeper part of the Bohai Sea.

Changjiang River

The Changjiang, the largest river in China, is the main source of sediments to the continental shelf of the East China Sea (data in Table 1 from Wang *et al.* 1983). The semidiurnal tidal range varies from 4. 6 m to 0. 17 m (average 2. 77 m). The velocity of tidal currents in the estuary is affected by discharge and the water level of the estuary. Normally, the ebb current velocity during the flood season is larger than during the dry season. The average velocity of flood tidal currents is 0. 95 m/s (maximum 2. 08 m/s), and of ebb currents is 1. 11 m/s (maximum 2. 5 m/s). The tidal prism has a linear relation with tidal ranges. The spring tidal prism is 5.3×10^9 m³ during the flood season and 3.9×10^9 m³ during the dry season. The observed maximum wave is $H = 2.3$ m, $L = 40$ m, $T = 5.6$ s. Waves are caused mainly by wind which blows SE during the summer and autumn, and NE in spring (Huang 1981).

There is a system of two currents off the Changjiang Estuary. One is a low temperature, high salinity, longshore current from the north which passes between $122° \sim 123°$E towards the south; the other is a high temperature, high salinity branched tongue, mainly from the surface water of the Kuroshio(Yu *et al.* 1983). The sediment output of the Changjiang River is carried mainly south by the longshore current. However, part of the sediment, carried by low salinity water, passes between the two saline tongues to the east. Normally, diluted water flows out towards the SE, but when the runoff is above average the low salinity water body flows NNE. When the runoff is below average, no tongue of diluted water is apparent.

According to observations, the modern effluents of the Changjiang River reach 122° 40′E at their distal margin(Yu *et al.* 1983). Sediment concentration is 1. 06 kg/m³ in the area between $121°50′ \sim 122°20′$E; maximum concentration in the bottom layer is 8 kg/m³. This is twice the concentration of the Changjiang Estuary. A submarine delta system has been developed in water between 5 and 40 m deep, which consists of clayey silt, with a 20 to 90 mm thick layer of 'fluid mud' on the top. The sedimentation rate is 5. 333 m per thousand years (You *et al.* 1983), and is estimated at 9. 0 mm/yr as determined by Pb²¹⁰. In the fore margin of the submarine delta the sediments are silty sand, silty clay, and discontinuous fluid mud, which is less than 40 mm thick. Beyond the margin, where the warm current flows, the fluid mud disappears, exposing relict fine sands in a weakly oxidized environment.

The sedimentation of the Changjiang River Estuary is characterized by rapid deposition of a submarine delta, sand shoals, and sand bars at the river mouth. The lower channel of the Changjiang River has shifted gradually to the south since the time of high post-glacial sea level. With the shifting of the river mouths the sand shoals and sand bars of the old river mouth merged into the northern fluvial plain. There have been five huge sand bodies left by the Changjiang River since 7 000 yr BP (Li 1979). The sixth sand body is developing as the channel of the modern river is moving south (Fig. 10). Since the river mouth has not shifted for a long time, large sand shoals and a submarine delta are forming.

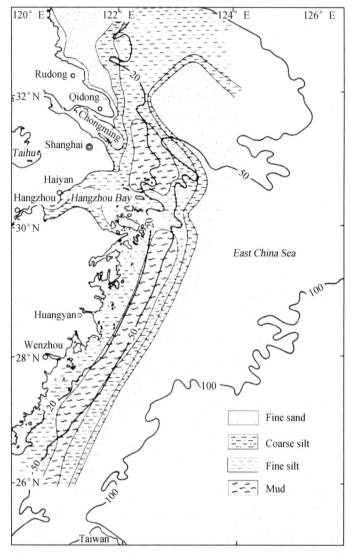

Fig. 10 Sediment distribution of the Changjiang Estuary and adjacent inner shelf (Water depth in metres)

There are two submarine deltas formed by the Changjiang River on the inner shelf of the East China Sea. The larger one 40 000 km² in area formed about 36 000 yr BP, according to C^{14} dating; it extends to a water depth of 60 m at its outer boundary. The smaller modem delta extends to a water depth of $25\sim30$ m; it formed since 7 000 yr BP, and is superimposed on the older delta.

One ancient Yellow-Yangtze River delta remains to the north on the inner shelf of the Yellow Sea, off the N Jiangsu coast. It was formed by the ancient Yellow River and its sediments can be divided into two parts. The northern part consists of greyish yellow silty sand and clayey silt. It contains homogeneous, smooth, fine grain (Median grain size$= 1.5 \sim 5.0$ μm); organic matter 0.88%; phosphorus 0.33%; and calcium carbonate 0.78%. The southern part, occurring at the junction of the Pacific and Yellow Sea tidal waves is a sedimentation field of radiating sand ridges (Ren & Zang 1980; Ren *et al.* 1983; Zhu & Xu 1982). These radiating sand ridges have been formed partly from the ancient delta of the Yellow River and partly from the Changjiang by the two systems of tidal currents.

The sand ridge field is 160 km long from north to south and contains more than 70 sand ridges. The top of the highest is 5 m above low tide. The individual ridges are 10 to 100 km long and 5 to 10 km wide; they are an average of 10 m thick and composed of well-sorted shell sand. There is poorly-sorted silty clay and clayey silt in the deep channels between the ridges and outside the sand ridge fields. In section, the ridges show fine grains in laminated layers in the lower part and coarse grains and cross-bedding in the upper part, reflecting a change from low energy to a high energy environment. When sediment supply was cut off as the Yellow River returned to the Bohai Sea the sand ridges suffered erosion. The eroded materials separated, and either were carried back to the sheltered coastal zone developing into a vast tidal flat, or were transported by currents along the north Jiangsu coast to the south.

In summary, the sediment distribution off the Changjiang River Estuary is influenced by the Changjiang runoff and the longshore current running from N to S. Since the Holocene sediments have been mainly diffused to the southeast. Sand bars, sand shoals, and submarine deltas are the major forms. As the river channels shifted the sand bodies were eroded and reformed by currents. Some sand bars and shoals were redeposited on the river plain, but most of the submarine sand bodies were reformed into sand ridges. The distal margin of the Changjiang effluents is near Jizhou Island of Korea (Zhou 1983) but most of the sediments are deposited in an 50 km wide area which extends from the estuary to the adjacent inner shelf.

Pearl River

The Pearl River is located in S China in a tropical climate. Because of the abundant precipitation and vegetation, discharge of water along the course of the river is larger, but the sediment discharge into the estuary is smaller, than from previously discussed rivers. The average water discharge rate is 11 000 m/s (other data in Table 1).

At the Pearl River mouth, tides are irregularly semidiurnal with an average tidal range of $0.86 \sim 1.63$ m. The original Pearl River mouth was a funnel-shaped bedrock embayment, developed along a NW trending fault. The bay is sheltered by a series of islands, which follow the major NE-SW trend of the coastline (Wang 1980). Three tributaries of the Pearl River, the Sijiang, Dongjiang, and Beijiang, empty into the bay.

The Pearl River sediments are deposited in a sheltered environment and fill the embayment, forming a delta; a transition between the estuarine type and the delta type of coastline is produced.

There are eight inlets of Pearl River tributaries in the delta. River sediments have accumulated as point bars, river mouth banks or sand shoals, overwash fans, and tidal deltas around each river mouth. With these sand bodies developed, the suspended materials, both from the sea and from the tributaries, can be deposited as a mud cap. Thus, binary structured tidal flats have been formed (Ren 1964; Li 1983).

The sediments on the continental shelf adjacent to the Pearl River mouth can be divided into two belts (Fig. 11). The inner shelf deposit of river-produced sediments is 35 km wide with its outer boundary at a water depth of $20\sim30$ m. From the coast seaward, sedimentation occurs in the order of sand, silt and clayey mud, showing a modern sedimentary series of wave dynamics. Sediments in this belt (Li 1980) contain little carbonate ($CaCO_3$ $5\%\sim10\%$), few heavy minerals ($<5\%$) and a large amount of organic material ($1\%\sim2\%$).

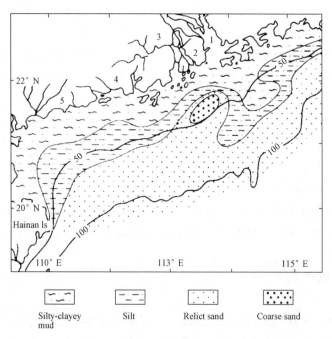

Silty-clayey mud   Silt   Relict sand   Coarse sand

Fig. 11 Sediment distribution of the Pearl River delta and adjacent inner shelf; water depth in metres. 1, Pearl River; 2, Zhongshan; 3, Gao Yao; 4, Yang Jiang; 5, Dian Bai.

The outer shelf deposit which is separated from the inner by a narrow transition zone is characterized by relict well-sorted fine and medium sands containing more than 2% heavy minerals. Shell fragments comprise $25\%\sim50\%$ of the material in this belt, with many foraminifera which reach 90% concentration at the continental shelf break.

The organic matter is lower than 1‰, and there is no organic matter in the fine sand belt (Li 1980).

There is a longshore current from NE to SW during all seasons of the year which carries the muddy clay westwards (Zhao 1983). This is a major source of mud for the tidal inlets along the coast west of the Pearl River Estuary. During summer, some muddy materials are also diffused to the outer shelf by NE drift.

Conclusion

The five types of river mouths and nearby continental shelves discussed in this paper can be divided into three dynamic types:

(1) The high energy wave regime.

This is seen on the inner shelf off the Luan He River. Sediment is distributed as a series of narrow belts oriented parallel to the coast. Mainly, there is a transverse sediment movement by wave action.

(2) Tidal current regimes.

In the macrotidal estuary of the Yalu River, sediment distribution follows the direction of tidal currents which are oriented either perpendicularly or obliquely to the coast (Fig. 12). The microtidal environment of the Yellow River mouth and nearby continental shelf shows sediment either deposited as a finger pointing to the sea or distributed as sediment tongues oriented parallel to the coast.

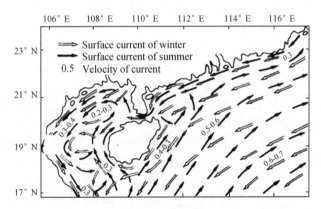

Fig. 12　Current system of the northern S China Sea (velocity in m/sec)

(3) River runoff and longshore current regimes.

In the Changjiang and Pearl River Estuaries and adjacent shelves sediment accumulates as sand shoals or sand banks. The sediment diffuses to the sea following the direction of the joint forces of the two types of currents.

Geologically, each river examined occurs in an environment where sediment supply is so great that regardless of marine energy, all five areas are sites of sedimentation.

The amount of deposition is directly related to the geographic stability of the river mouth. When the river mouth shifts, sediment supply is reduced or stopped at that site. The previously deposited sediments undergo erosion and the coarser grained sediment is reformed as a series of sand ridges along the direction of the current.

Most of the sediment from the rivers is deposited on the inner shelf near the river mouth. The maximum measured distance of longshore sediment movement is 105 km (Yellow River, Bohai Bay). The farthest distance sediment is transported into the deeper water area of the continental shelf is 50 km (Changjiang River, E China Sea). The distal margin of the Pearl River deposits is approximately 35 km from the shore at a water depth of $20 \sim 30$ m. Sand beyond these margins is relict after the last sea level low of the Pleistocene Period. Modern sediment distribution by the Luan He River is at a water depth of 10 m, 4 km from the coast. Beyond this are the relict sands of lower sea level. It seems that river driven sediment is mainly deposited near the river mouth and on the nearby inner shelf except when sea level changes occur.

River deltas advance gradually onto the continental shelf. During the lower sea levels of the late Pleistocene, the coastal plain which form today's continental shelf was crossed by many rivers. The ancient Yellow and Changjiang Rivers built a series of deltas during that time. These delta systems provided the framework which built the continental shelf of the China seas.

(M. E. Schenk assisted the authors in the revision of the English translation of the original manuscript.)

References

HUANG, SHEN. 1981. General description and deformation of the river bed of the Yangtze estuary. [In Chinese]. *Research on Navigation of Yangtze Estuary*, 1: 2 - 30.

LI, CHONG XIAN. 1979. The characteristics and distribution of Holocene sand bodies in Changjiang River delta area. [In Chinese]. *Acta Oceanologia Sinica*, 1: 252 - 68.

LI, CHUEI ZHUNG. 1980. The characteristic of the surface sediments of the northern Shelf of the South China Sea. [In Chinese], *Nanhai Studia Marina Sinica*, 1: 35 - 50.

LI, CHUN CHU. 1983. Dynamics and sedimentations of Madao-Men Estuary. *Tropical Geography*, 1: 27 - 34.

PANG, J. & SI, SHUHENG. 1979. The estuary changes of Huang He River. *Oceanologia et Limnologia Sinica*, 10: 136 - 41.

PANG, J. & SI, SHUHENG. 1980. Eluvial processes of the Huang He River estuary. *Oceanologia et Limnologia Sinica*, 11: 293 - 305.

REN, MEI-E. 1964. Geomorphological characteristics of the Pearl River estuary and vicinity. [In Russian]. *Acta Scientiarum Naturalium Universitatis Nankinensis*, Ⅷ: 136 - 46.

REN, MEI-E. 1974. The subaqueous delta of Hai He River and silting problem of Tiantsin new port. *Acta Scientiarum Naturalium Universitatis Nankinensis*, 1: 80 – 90.

REN, MEI-E. 1981. Sedimentation on tidal mudflat in Wanggang Area, Jiangsu Province, China. *Marine Science Bulletin*, 3: 40 – 50.

REN, MEI-E &. ZANG, CHENKAI. 1980. Quaternary Continental Shelf of East China Sea. *Acta Oceanologica Sinica*, 2: 94 – 111.

REN, MEI-E, ZHANG, REN-SHUEN &. YANG, GYU-HAI. 1983. Sedimentation on tidal mudflats of China. [In Chinese]. *Proceedings of International Symposium on Sedimentation on the Continental Shelf, with Special Reference to the East China Sea*, 1: 1 – 9.

TONG, QICHENG &. CHENG, TIAN-WEN. 1981. Runoff, *In*: HUANG, BINGWEI (ed.) *Physical Geography of China* [In Chinese]. Science Press. 6 – 121.

WANG, KANGSHAN, SU, JILAN, &. DONG LIXIAN. 1983. Hydrographic features of the Changjiang Estuary. [In Chinese]. *Proceedings of International Symposium on Sedimentation on the Continental Shelf, with Special Reference to the East China Sea*, 1: 137 – 47.

WANG, YING. 1963. The Coastal dynamic Geomorphology of the northern Bohai Bay. *In*: *Collected Oceanic Works*. [In Chinese]. 3: 25 – 35.

WANG, YING. 1980. The coast *of* China. *Geoscience*, 7: 109 – 13.

WANG, YING. 1983. The mudflat coast of China. *Canadian Journal of Fisheries and Aquatic Sciences*, 40 (Suppl. 1): 160 – 71.

YOU, KUN YUAN, SUI, LIANGREN &. QIAN, JIANG-CHU. 1983. Modern sedimentation rate in the vicinity of the Changjiang Estuary and adjacent continental shelf. [In Chinese]. *Proceedings of International Symposium on Sedimentation on the Continental Shelf, with Special Reference to the East China Sea*, 1: 590 – 605.

YU, HONGHUA, ZHENG, DACHENG &. JIAN, JINGZHENG. 1983. Basic hydrographic characteristics of the studied area. [In Chinese]. *Proceedings of International Symposium on Sedimentation on the Continental Shelf, with Special Reference to the East China Sea*, 1: 295 – 305.

ZHAO, HUANTING. 1983. Hydrological characteristics of the Zhujiang (Pearl River) delta. *Tropic Oceanology*, 2: 108 – 17.

ZHOU, FUGEN, 1983. Automorphous calcite crystal in seawater of the northeastern East China Sea. [In Chinese]. *Proceedings of International Symposium on Sedimentation on the Continental Shelf, with Special Reference to the East China Sea*, 1: 447 – 61.

ZHU, DA KUE. 1980. The coastal evolution of the east Hebwi Province. [In Chinese]. Symposium of China's Coast and Estuary, Shanghai.

ZHU, DA KUE &. XU, TINGGUAN. 1982. The coast development and exploitation of middle Jiangsu Province. [In Chinese]. *Acta Scientiarum Nafuralium Universitatis Nankinesis*, 3: 743 – 818.

The Characteristics of the China Coastline[*]

INTRODUCTION

CHINA'S coastline is approximately 32 000 km long, 18 400 km of which encompass the mainland from the Yalu River at the China-Korea border to the China-Vietnam border. The remaining 13 600 km or so of coastline belong to China's offshore islands, of which there are more than 6 000. Both the sediment and water discharge of the major rivers dominate the evolution of China's coastline. Seasonal monsoon winds control the wave climate for the mainland China coast, while the marginal seas control the tidal characteristics.

In the recent historical past the coastline of China has undergone rapid change. Most notably, the two major rivers, the Yellow River and the Chang Jiang (Yangtze River), have migrated great distances during the past 4 000 years. The present position of the Yellow River, centered near Lijin (Figs 1 and 2), was established in 1855, when the river migrated approximately 420 km to the north from the north Jiangsu coast. Since 2278 B. C. , the Yellow River has changed position eight times, the last major change occurring in 1855 (Fig. 3; PANG and SI, 1979; WANG, 1983). At the present location of the Yellow River, in southern Bohai Bay, major shoreline accretion has occurred during the past 130 years, the delta building seaward 20~28 km in this time period (WANG et al. , 1984).

Along the north Jiangsu coast, where the Yellow River emptied from 1128 to 1855, erosion has been extensive. The coast here has retreated about 17 km since 1855 (REN et al. , 1983a), with coastal retreat presently exceeding approximately 30 m y^{-1} along a 150 km coastal reach (WANG, 1961).

The Chang Jiang (Yangtze River) has steadily migrated southward and deposited five large sand bodies during the past 7 000 years (LI, 1979). This southerly migration has exerted a major influence on coastal development in south Jiangsu and north Zhejiang province, with Shanghai itself built upon the raised former river delta. Clearly not only the sediment supply, but also the position of the major rivers of China influence coastal development to a significant degree.

Besides these natural changes in river course, anthropogenic influence has impacted coastal evolution. For example, the sediment load of the Yellow River increased

＊　Ying Wang, David G. Aubrey: *Continental Shelf Research* ,1987, Vol. 7, No. 4: pp. 329－349.

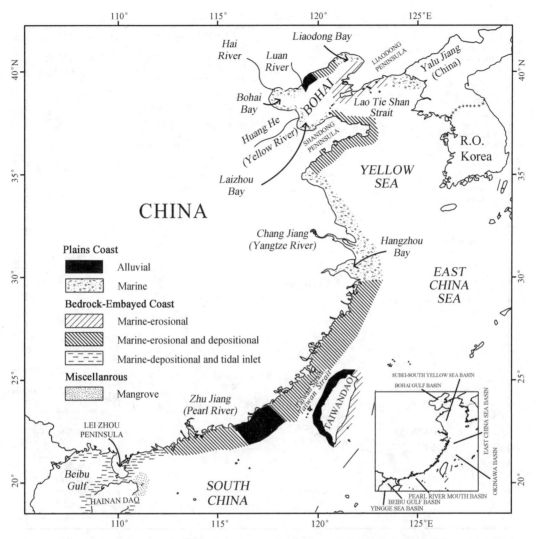

Fig. 1 Location map for China, incorporating major geographical names and features

Stippling indicates coastal classification, from this study. Insert: distribution of Cenozoic basins according to LI (1984) (after EMERY and AUBREY, 1986).

dramatically as a result of human activities (WANG, 1983). In early historical times (prior to 11A. D.) the Yellow River entered northwestern Bohai Bay near the present Hai River. In 12A. D. the river shifted to near its present position. Prior to and during the Western Zhou Dynasty (1100—771 B. C.) population was lower in the pasture-dominated Loess Plateau of northwestern China, the major source of sediment to the Yellow River. The abundant forested land and grass (32 million hectares, 53% of the land) stabilized the soil, thus the sediment load in the Yellow River was lower than present. Since then, especially during the Tang Dynasty (618—907 A. D.), settlers migrated to the west, cutting forests and cultivating the land, and causing rapid soil erosion. With only 3% of the land now forested, the sediment load has risen

Fig. 2 Map of China with political boundaries and town locations referred to in text. Insert: massifs and foldbelts as interpreted by EMERY (1983) (after EMERY and AUBREY, 1986)

dramatically, with an average sediment concentration of 25 kg m^{-3} in the Yellow River at present (WANG, 1983).

Also in the past, man has tried to tame the vagaries of the annual floods in the major rivers. As far back as the Song Dynasty (960—1127 A. D.), river bank dikes and other engineering methods of river control were constructed (WANG, 1983). More recently (during the 1950s) a series of dams was constructed in the estuaries of the North Jiangsu Plain, in an effort to harness the tidal currents and provide fresh river water for irrigation. The lower part of the estuaries silted rapidly because of reduced flood and tidal flow, requiring that all dams be opened to flush the silt and maintain the proper river channel depth (WANG, 1961). These and other human activities have influenced shoreline development in China and will continue to do so in the future.

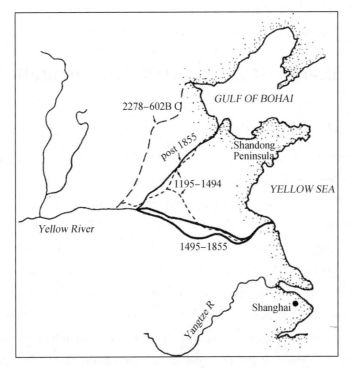

Fig. 3　Locations of the Yellow River in the historical past

The present paper is a classification and description of the China coastline, extending from the coastal plains or foothills out to the inner continental shelf. It is a synthesis of past Chinese and international research on China's coastline, with an emphasis on Chinese research during the past 20 years. In particular, it draws heavily on research performed in the Department of Geography at Nanjing University, Jiangsu Province,especially by the Marine Geomorphology and Sedimentology Laboratory. To the extent possible the work of many other Chinese coastal geoscientists is incorporated. This synthesis is more complete than those by SCHWARTZ (1982) and BIRD and SCHWARTZ(1985).

For consistency, local nomenclature is used where possible, but anglicized names that have become standard in the literature are retained. For instance, Bohai Sea has gained common acceptance even though it is redundant (hai in Chinese means sea). This redundancy is retained to distinguish the smaller Bohai Bay from the larger Bohai Sea that encompasses Bohai, Laizhou, and Liaodong bays (Fig. 1). Major rivers are represented by either their full Chinese names (Huang He, Chang Jiang, Zhu Jiang) or by their anglicized names (Yellow, Yangtze, Pearl rivers), but never by a mixture of the two usages. However, for the lesser known rivers where anglicized names have not gained general acceptance, the Chinese name is used with the English noun (Yalu River, Guan River, etc.). Redundancy in the Chinese and English words for river is avoided. City names are presented in their current Chinese form (Pinyin) with clarification provided where needed. The authors believe these conventions yield more consistency than is apparent in past Chinese and international literature.

MAJOR INFLUENCES ON COASTLINE MORPHOLOGY

Geology

The two primary geological controls on coastal morphology are the positions of the massifs and foldbelts, and the position of the major sedimentary basins (WANG, 1980; EMERY and AUBREY, 1986). Primary basins of the China region are the Bohai Sea, the Subei-South Yellow Sea, the East China Sea, the Okinawa, the Pearl River delta, the Beibu Gulf, and the Yingge Sea basins (Fig. 1, insert). Of these, the Bohai Sea, the Yellow Sea, and the Beibu Gulf basins exert the most direct influence on coastal morphology. Tectonics of the region can be simplified into a series of massifs, basin fills, and foldbelts (Fig. 2, insert; EMERY and AUBREY, 1986). Precambrian massifs occur along the western margin of Guangxi Province, the Shandong Peninsula, and the Liaodong Peninsula. Much of the rest of the coast has bedrock of Mesozoic or Neogene basin fills and foldbelts. Combined, the basins and foldbelts form a series of northeast or north-northeast trending belts that intersect the coastline obliquely (WANG, 1980).

Basins contain $2.5 \sim 9.5$ km of Cenozoic sediments that are dominated by continental (42%), transitional (22%), neritic (31%), and marine (5%) sediments (EMERY and AUBREY, 1986). These sediment thicknesses, deposited in regions of restricted elevation range, exceed those expected from compaction alone; tectonic subsidence must be responsible for much of this accumulation.

The massifs are dominated by Precambrian rocks. The alternation of massifs and foldbelts (with ophiolites along many boundaries) is interpreted by EMERY (1983) as repeated breakup of continental crust by plate divergence and translation, followed by deposition of thick sediment sequences in the new seaways, and ending with plate convergence that produced foldbelts. These same distributions of massifs have also been interpreted as massifs from other continents that were added to Asia (BEN-AVRAHAM, 1979). Regardless of the interpretation, the intervening foldbelts represent thick sediment accumulations that subsided after plate convergence.

Rivers

Five rivers, the Yalu, Luan, Yellow, Yangtze and Pearl, discharge 90% of China's total annual contribution of 2.0×10^9 tons of sediment to the marginal seas (Table 1). The Yellow, Yangtze, and Pearl rivers supply most of the total freshwater discharge to the marginal seas (Table 1). Estimates of annual freshwater discharge range from $1.2 \times$

10^{12} m^3（GUAN，1983）to more than 1. 4×10^{12} m^3（WANG，1980）. This discharge controls the character of the Continental Coastal Water（CCW），which in turn influences sediment dispersal from the major rivers. Sediment supply and freshwater influx distinguish the riverine from other coastal environments of China.

Table 1　Annual water-sediment discharge for rivers in China

	Water（m^3）	Sediment（tons）
Yellow River（Huang He）	4. 9×10^9	1. 2×10^9
Yangtze River（Chang Jiang）	930×10^9	4. 9×10^8
Pearl River（Zhu Jiang）	370×10^9	1. 0×10^8
Luan River	4. 6×10^9	2. 4×10^3
Yalu River	28×10^9	4. 8×10^6
Liao River	6×10^9	41×10^6
Daling River	1×10^9	36×10^6
Hai River	2×10^9	81×10^6
Huai River	—	14×10^6
Total	1. $2 \times 10^{12} \sim 1. 4 \times 10^{12}$	2. 0×10^9

Sources：MILLIMAN and MEADE（1983），GUAN（1983），TONG and CHENG（1981），WANG（1980），WANG et al.（1984），QIAN and DAI（1980）.

Yellow River. The Yellow River carries the largest sediment load of any river in the world. Although its freshwater discharge is modest（Table 1），reaching an average of 49×10^9 m^3 y^{-1}，its sediment discharge is immense，averaging 1. 2×10^9 tons y^{-1}（TONG and CHENG，1981）. Sixty-four percent of the sediment is deposited on the delta and mudflat，while the remaining 36% is transported into the Bohai Sea（PANG and SI，1980；WANG，1983）. The Yellow River sediment，composed primarily of coarse silt，is deposited as finger bars along the river mouth. Progradation of these finger bars is rapid，depositing as much as 10 km in 1964 alone（WANG et al.，1964）. To either side of the river mouth fine silt and clay collect in bays that are protected by coarser bars. Seaward of these muddy bays erosion of former deltas occurs，as the river changes its position every 6~8 years，although not in as drastic a fashion as a few centuries earlier. Since 1855，a delta has covered a distance of 160 km alongshore near Lijin，prograding seaward 20~28 km（WANG et al.，1984）. This progradation represents an average annual accumulation of 23. 5 km^2 in plan view.

Yangtze River（Chang Jiang）. The Yangtze River with its large estuary（CHEN et al.，1979，1985；MILLIMAN et al.，1984）is the largest river in China，having the fourth largest water discharge in the world（MILLIMAN and MEADE，1983）. It's annual freshwater runoff is 9. 3×10^{11} m^3，70% of which occurs in the months from May to October. It is the main source of sediments to the East China Sea, with an average annual sediment discharge of 4.9×10^8 tons. Although it discharges more water than the Yellow River (by a factor of 18), its concentration of sediment is much less (0. 5 kg m³; WANG et al., 1984), and its total sediment load also is lower by a factor of 2. 5. The sediments of the Yangtze reach 122°40′E longitude at its distal margin. The spring tidal prism is 5.3×10^9 m³ during the wet season, and 3.0×10^9 m³ during the dry season (WANG et al., 1984). The large freshwater discharge contributes to the dynamics of the CCW (see later section on low-frequency currents).

Sedimentation offshore of the Yangtze River is rapid, taking the form of a submarine delta, sand shoals, and sand bars at the mouth of the river. Five major sand bodies have been left by the Yangtze River since 7 000 a B. P. (LI, 1979) that either have merged into the north fluvial plain or form extensive sand ridges offshore of Jiangsu Province. Since the river mouth shifted continuously to the south, all of these former sand bodies are north of the present Yangtze River mouth. The sand ridge field is 160km long, from north to south, encompassing more than 70 minor sand ridges. Individual ridges are $10 \sim 100$ km long and $5 \sim 10$ km wide. The top of the highest ridge is 5 m above low tide (ZHU and XU, 1982). Sediments of the Yangtze River are largely silts, but include a significant fraction of fine sand, in contrast with the sediments of the Yellow River. The Yangtze has extensive offshore bedforms, particularly along it former river mouth positions, whereas the Yellow River lacks these extensive shoals (they are also missing in the eroding former Yellow River mouth in north Jiangsu). Present Yangtze River sediments are advected towards the south by coastal currents in the inner shelf and only a smaller part is carried to the north (WANG et al., 1984).

Pearl River (ZhuJiang). The Pearl River is in southern China, where a tropical climate influences its water and sediment discharge. With its abundant precipitation and lush vegetation, the water discharge is high, while sediment discharge is relatively lower. The annual runoff of the Pearl River is about 3.7×10^{11} m³, while its annual sediment discharge is about 1.0×10^8 tons (WANG, 1980).

Three tributaries enter the estuary (the Si, Dong, and Bei rivers). Most sediments are deposited within the embayment, forming the Pearl River delta. Eight inlets from the Pearl River tributaries connect the estuary to the South China Sea. Offshore, two belts mirror the sediment influx. The inner shelf deposit (35 km wide) continues to derive its sediments from the Pearl River, grading from sand nearshore to mud offshore. The outer shelf has relict sands. A coastal current carries river sediments to the southwest throughout the year, serving as a source of mud for tidal inlets along the downdrift coastline (WANG et al., 1984). Zhongshan University (Guangzhou) has studied the geology and sedimentation of the Pearl River and its tributaries for the past 20 years. Most publications are in Chinese only, with a few exceptions (see YING and

CHEN, 1984; INSTITUTE OF COAST AND ESTUARY, 1984; LI and WANG, 1985).

Many other smaller rivers locally influence sedimentation and coastal morphology. WANG et al. (1984) discussed the Yalu and Luan rivers that have smaller freshwater and sediment discharges. Approximately a thousand smaller rivers contribute both freshwater and sediments to the nearshore. As an example of the many small rivers, Hai Nan Island in southern China alone has 154 rivers, 38 of which contribute significantly to the coastal evolution of the area. Altogether, the small rivers of the mainland of China contribute about 2.1×10^8 tons of sediment per year to the coast and marginal seas (WANG et al., 1984).

Climate

A major influence on China's coastal processes is the monsoon wind pattern of Asia. In the winter, the Mongolian high-pressure system dominates the geostrophic winds. The anticyclonic gyre, beginning in January, yields northwesterly winds in the north, and northerly or northeasterly winds in the South China and East China seas. During the summer, the Indian Ocean low-pressure system dominates, creating a cyclonic (anticlockwise) gyre. These winds are southwesterly in the South China Sea (south of latitude 15°N), changing northwards to southeasterly winds in the East China, Yellow, and Bohai seas. A transitional season exists during which winds fluctuate between the two dominant weather systems (WANG, 1980).

Winds at many coastal locations are summarized by the U. S. DEPARTMENT OF COMMERCE (1981) and more recently by SUN et al. (1981). Wind roses, monthly wind summaries, and dominant winds are provided by SUN et al. (1981) for 17 coastal sites.

Typhoons

Generated primarily from July to October, typhoons are destructive because of their large waves as well as their associated storm surges (reaching 5 m or so above MWL). Genetically, typhoons are tropical cyclones exceeding 64 kn (32 m s^{-1}) in windspeed. These typhoons normally develop in the northwestern Pacific, moving west or westnorth-west across the Philippines and South China Sea, dissipating rapidly when they reach the Asian continent. Typhoon frequency is higher for the South China Sea than for the other marginal seas of China (SUN et al., 1981).

In the South China Sea, most tropical cyclones occur during August, September and October (Table 2). The annual average number of tropical cyclones exceeding winds of 17 m s^{-1} is 1.1, however the standard deviation about this average value exceeds the mean. Four typhoons were experienced in 1894 within 17 days, three within 10 days in

1887，but none for almost four years between 1932 and 1936.

Table 2　Number of tropical cyclones per month during 1884—1961 that caused winds of 17 m s^{-1}
and above.　Data for Hong Kong (WILLIAMSON, 1970)

Jan.	Feb.	Mar.	Apr.	May	June	July	Aug.	Sept.	Oct.	Nov.	Dec.
0	0	0	0	1	6	20	20	27	10	4	0

Annual number of typhoons generated in the northwest Pacific is 28，with considerable year-to-year variability (all typhoon statistics are from SUN *et al.*，1981). These data are monthly values，including maximum and minimum number of typhoons per month over the 21 year period of observation (1949—1969；Table 3). Of the total number of cyclones generated，only a small number reach the South China Sea. A compilation of monthly typhoon occurrence shows that on the average nine typhoons reach the South China Sea yearly，most of these in July-September，whereas the peak for the entire western Pacific is earlier. This difference in typhoon timing reflects the seasonal dependence of typhoon tracks in the Pacific. Of the typhoons reaching the South China Sea，there are four major tracklines. In September and October，40% of all typhoons travel over Hai Nan Island on into Vietnam. In March-June，22% of the annual typhoons travel north into Guangdong Province or Fujian Province. In spring and summer，another 20% of the total annual number of typhoons passes to the north-northeast，over Taiwan，or towards Japan. Finally，in winter，10% of the typhoons pass to the west into Vietnam，missing the China coast.

Table 3　Number of typhoons reaching South China Sea (includes storms
with prolonged winds exceeding 11 m s^{-1})

	Month												Yearly total
	1	2	3	4	5	6	7	8	9	10	11	12	
A	—	—	1	4	10	15	29	30	45	24	27	11	196
B	—	—	0.05	0.19	0.48	0.71	1.38	1.43	2.14	1.14	1.29	0.52	9.33
C	—	—	0.51	2.04	5.10	7.65	14.80	15.31	22.96	12.24	13.88	5.57	100
D	—	—	1	2	2	2	3	3	6	4	4	2	

A，Total number from 1949 to 1969；B，yearly average；C，% in month；D，greatest number in single month over period of observation.

From SUN *et al*. (1981).

In the period 1949—1969，the China coast experienced 77 strong typhoons (winds exceeding 32 m s^{-1})，61 weak typhoons (winds of 17~32 m s^{-1})，and 65 tropical storms (winds from 11 to 17 m s^{-1}). This yields an average of 9.7 major storms per year reaching land，6.6 of which are typhoons. Most of these storms occur in the period July-October. Typhoon landfalls are separated on a province-by-province basis，for months May-November. Most typhoons impact Guangdong Province; Taiwan is the next hardest hit, followed distantly by Fujian and Zhejiang Province.

Wind waves

Waves rarely have been measured during typhoons, and most accounts are anecdotal, quoting nearshore waves exceeding 20 m in height. ZHANG and LI (1980) compared calculations of typhoon waves with observations. They presented tables of wave height ($H_{1/10}$) and period for different fetches, wind speeds, water depths, and durations. For instance, a typhoon with wind speed of 32 m s^{-1}, blowing for 10 h over a fetch of 500 km with average water depth of 40 m, can generate a wave of 12 s period, and 13 m height. Most typhoons will generate waves less than this height, but some evidently can generate much higher waves, to 20 m. The recent failure of a drilling rig in the South China Sea during a storm accentuates the need to carefully consider typhoon waves.

Although there are exceptions, wave patterns along most of China are dominated by monsoon winds. Northerly waves prevail in the winter, their influence moving progressively southwards from the Yellow Sea, reaching Taiwan in September, latitude 10°N in October, and spanning the entire coast by November. In January, the wave directions rotate clockwise, with northwesterly waves in Bohai gradually swinging northerly to northeasterly in the East China and South China seas (WANG, 1980).

Southerly waves prevail in the summer, moving northward from the South China Sea. They appear first in February, and by May they dominate a wide area to the south of latitude 5°N (WANG, 1980). In June, southwesterly and southerly waves prevail in the South China Sea. In the Strait of Taiwan, northeasterly and southwesterly waves become equally frequent, while southerly waves prevail in the East China Sea and southern Yellow Sea. The northern Yellow and Bohai seas are dominated by southeasterly waves in June. In July, southerly waves prevail along the entire seaboard. Wave directions change anticlockwise, from southerly in the South China Sea (south of latitude 15°N), through southerly to southeasterly in the East China, Yellow and Bohai seas. During the periods of transition for the monsoons, wind directions fluctuate and there are no prevailing waves.

Locally, large variations occur in both wave height and direction. Commonly a mix of waves having different sources occurs at any time, reflecting local fetches and local wind patterns, as well as distant swell. As an example, eastern Hai Nan Island is affected not only by monsoon winds, but also by distant South Pacific swell. The north Jiangsu coast is dominated by waves generated by monsoon winds acting on the Yellow Sea. In southern Jiangsu, by contrast, the wind waves are modified strongly by the offshore sand ridges, making its wave climate much more moderate than the relatively unprotected northerly counterpart.

Waves in the South China Sea are largest of those in all China marginal seas, because of the large fetch and greater water depths (SUN *et al*. , 1981). The shallow, extensive shelf of the East China and Yellow seas limits the size of the waves in this region. The waves of the Bohai Sea are the smallest, as the fetch is restricted. Wave periods are under 5 s, and since winter monsoons are northwesterly here the waves tend to be largest near the Lao Tie Shan Strait separating Bohai Sea from the Yellow Sea. Water depths in the Bohai are also relatively shallow (the mean water depth is 18 m), so large waves are rare.

SUN *et al*. (1981) presented a wave compilation from various sources around China, including wind wave conditions (sea) for each of 14 locations around China, for the months January, April, July, and October, in the form of percent occurrence of waves from particular directions, and for swell conditions, again for the months of January, April, July and October. Monthly average wave height for 13 of these same coastal stations are presented for the same months, and an annual average given. Annual average wave height is largest off Fujian Province, and the north part of the South China Sea (1. 1 and 1. 4 m, respectively; Table 4).

Table 4　Average monthly wave height (m) for various China coastal stations

	N. Jiangsu	Zhejiang	Fujian (Pingtan)	Fujian (N)	Guangdong	W. Hai Nan (Basue)	Guangxi (in Gulf)	(North) South China Sea
A	0. 9	1. 0	1. 1	1. 2	0. 9	0. 9	0. 4	1. 6
B	0. 9	1. 0	0. 8	0. 9	0. 8	0. 8	0. 4	1. 0
C	0. 9	1. 0	0. 7	0. 9	0. 8	0. 9	1. 0	1. 4
D	0. 9	1. 0	1. 2	1. 3	1. 0	0. 7	0. 5	1. 4
E	0. 9	1. 0	0. 9	1. 1	0. 9	0. 8	0. 6	1. 4

For 4 months (A, January; B, April; C, July; D, October) and yearly average (E). Data from SUN *et al*. (1981).

Wave period data for 14 Chinese stations document the shortest periods are in the north (from 1 to 4s), with longer periods in Fujian Province and in the north part of South China Sea. According to these data, waves with periods exceeding 7 s are rare. Since typhoons occur for a small percentage of total time, they have little impact on the climatological summaries. These summaries do not reflect open ocean or open sea conditions, since waves have been measured only in shallow coastal waters where frictional attenuation is significant.

Average periods and average directions of wave propagation are available for 13 Chinese coastal stations, for the same four months of the year (eight of which are presented as Table 5). These data show the average wave period varies from 2.5 s up to 7 s, with maximum monthly wave periods ranging up to 14 s in the south. Maximum monthly averages reach 12.8 s in the north part of the South China Sea.

Table 5 Average wave periods (s) and directions for 4 months

			N. Jiangsu	Zhejiang	Fujian (Pingtan)	Fujian (N)	Guangdong	W. Hai Nan (Basue)	Guangxi (in Gulf)	(North) South China Sea
A I	II		4.4	6.0	5.8	4.0	3.8	3.7	4.5	5.1
	III		E	NW	E	SE	ENE ESE	SSW	SW	NW
	IV		8.6	11.2	8.7	9.6	5.8	6.8	5.2	8.9
B I	II		4.8	6.0	5.5	5.1	3.9	4.0	4.9	3.8
	III		ENE	WSW	NE	SE	ENE	SSE	SW	SSW
	IV		9.7	10.0	7.7	6.5	5.9	6.9	7.0	8.1
C I	II		5.0	7.1	5.8	5.7	7.9	3.8	4.3	3.9
	III		E	SE	E	SE	WNW	SSW WSW	SSW	WSW
	IV		13.7	19.8	8.8	11.5	7.9	7.0	8.8	10.9
D I	II		4.7	5.6	5.8	4.9	4.0	3.3	3.8	3.9
	III		ENE	NE ENE	E	SE	SSE		SSW	NE
	IV		10.9	10.8	9.1	9.1	6.8	9.5	5.9	12.8

I, Monthly averages; II, average period (s); III, average direction; IV, maximum average period over years of observation.

Data from SUN *et al.* (1981).

A, January; B, April; C, July; D, October.

Tides

Tides in the Chinese marginal seas are variable in type and amplitude. Tides in the Bohai, Yellow and East China seas are semidiurnal, while the tides in the South China Sea are either diurnal or semidiurnal, having considerable geographic variability. Within the Bohai Sea, tides are predominatly semidiurnal although they can be irregularly semidiurnal or irregularly diurnal. Here tidal range averages about 3 m except near the mouth of the Yellow River, where it is only 0.5~0.8 m (WANG *et al.*, 1984).

Tides of the Yellow and East China seas are semidiurnal, with a large variability in range. Near the Yalu River mouth, tidal range averages 4.5 m. Near the abandoned Yellow River mouth, tidal range is 1.6 m, while near Shanghai the range is 1.9 m. LARSEN and CANNON (1983), SHAN *et al.* (1983), CHOI (1984) and LARSEN *et al*. (1985) discussed the tides of this region in more detail.

Tides in the South China Sea are more complex, varying from diurnal to semidiurnal over its extent. The diurnal tides are locally amplified, particularly in the Gulf of Tonkin (Beibu Gulf) and Gulf of Thailand, where they are near resonance. Otherwise, tides are largely semidiurnal, with ranges of generally less than 2 m. YE and ROBINSON (1983) modeled the M_2 and K_1 tides in the area, both in sea surface and in velocity. Tidal currents in the South China Sea exceed 50 cm s^{-1} in restricted locations, including the Strait of Hai Nan.

SUN *et al*. (1981) discussed the coastal tides of China, though showing no cotidal or corange charts. WRYTKI (1961) published corange charts for the dominant semidiurnal and diurnal tides, as well as a chart of type of tide (semidiurnal, diurnal, or mixed), for the South China Sea. LARSEN *et al*. (1985) provided cotidal charts for the Yellow, Bohai, and East China seas. YIN (1984) discussed tides near Taiwan.

MAJOR COASTAL TYPES OF CHINA

In this summary, the coastal classification of WANG (1980) has been adopted. Two major coastal classes have been identified: bedrock-embayed coasts and plains coast (Fig. 1).

Bedrock-embayed coast

Characterized by irregular headlands, bays, and islands, these develop where mountains meet the sea. Four subtypes are identified.

Marine erosional-embayed coast. Developed on hard crystalline rock, erosion is slow and the coast is modified slowly. Sparse sediment supply limits coastal deposition, leaving a dominance of erosional geomorphic features.

Marine erosional-deposition type. Most commonly developed where Miocene granites are overlain by weathered deposits, these coasts are easily eroded. Erosional features (rock benches and terraces) and depositional features (bays bars, tombolos, sand spits) develop here.

Marine depositional type. With relatively erodable bedrock, marine erosion supplies large quantities of sediment for prograding shorelines. This bedrock-embayed coast eventually matures to a plains coast, and is common in South China.

Tidal inlet-embayed coast. Common again in South China, the northeast-southwest trend of the coast is interrupted by northwest-trending inlets along faults with the same orientation. Tidal inlets are formed with sand spits, creating lagoons and estuaries in many cases suitable for harbors.

Plains coast

Based on genesis, these are classified into two types.

The alluvial plains coast. Located generally seaward of mountain ranges, plains are built of fluvial sediments. Plains continue seaward with low gradients, resulting in a wide breaker zone and active sediment movement. Barrier bars, sand spits, submarine bars, and extensive beaches are typical features of this coast.

Marine depositional plains coast. Flat and very extensive, these are located on the lower parts of large rivers in areas of subsiding basement. Coastal slopes are extremely gentle (1:1 000 to 1:5 000), leading to wave breaking well offshore; the nearshore commonly is dominated by tidal processes. Sediment from rivers is deposited either offshore, or high on the extensive tidal flats.

Both the bedrock-embayed coast and the plains coast may incorporate two minor types of coast. The first, the river mouth coast, is limited to the major rivers (the Yellow, the Yangtze, the Pearl), containing both delta and estuarine features. The second type is biogenic, found in the south of China, incorporating coral reef and mangrove coasts. Mangroves are limited south of latitude 27°N, while coral reefs occur as far north as Taiwan.

REGIONAL SUMMARY OF CHINA COASTAL CHARACTERISTICS

Following is a brief summary description of China's coast, separated into four regions. The Bohai Sea (sector 1) is a river-dominated environment. The Yellow Sea (sector 2) is tide-dominated, with waves playing a secondary role, along with the former and present river processes. The third sector (the East China Sea coast) is wave-dominated, having large tidal range in some tidal inlets and estuaries. The Yangtze River (Chang Jiang) contributes to coastal development. The fourth sector (the South China Sea coast) is largely wave-dominated.

Bohai Sea

The Bohai Sea coastline extends from the southern tip of the Liaodong Peninsula to the northern tip of the Shandong Peninsula, having its eastern border at the straits of LaoTe Shan. The tides in the Bohai Sea, though generally semidiurnal, are complex, varying from irregularly semidiurnal to irregularly diurnal, having a tidal range of 3 m except near the Yellow River mouth. Because of its geometry, Bohai Bay has its greatest fetch to the northeast; the observed value of $H_{1/10}$ is 4.8 m, with a period of 7.8 s. Average maximum wave height is 1.9 m. Mean and low-frequency flows are complicated by the geometry of the bay, which has an average water depth of 18 m and a total area of

78 000 km^2. Much low-frequency water motion is due to storms, with the resultant flow similar to that of the tides. As discussed by CHOI (1984), the low-frequency flow generally is cellular, with one eddy moving clockwise to the north, the second moving anticlockwise to the south.

The dominant control of sedimentation in the Bohai Sea is river influx, contributing 1.21×10^9 tons of sediment each year. The two major rivers are the Yellow River (discharging 1.2×10^9 tons) and the Luan River (2.4×10^3 tons). Other rivers, including the Liao and the Hai, discharge the remainder of the sediment to the sea. As discussed by WANG (1980) and EMERY and AUBREY (1986), most of the Bohai Sea is a subsidence basin, with up to 9 km of sediment. This subsidence must be tectonic, since sediment accumulation exceeds that expected from sediment compaction alone. The Liaodong and Shandong peninsulas are regions of uplift, as is the northwest corner of Bohai Sea. These patterns have resulted in most of the Bohai being classified as plains coast, with part forming bedrock-embayed coasts.

The west coast of the Liaodong Peninsula is a bedrock-embayed coast of marine erosional type. Locally in the heads of the bays, modern mud from the Bohai Sea overlies the sparse material from erosion of the peninsular mountains. The Liao River, adjacent to Liaodong Peninsula, carries mainly fine sediment (silts and fine sands) to the shore. Stretching to the west of the Liaodong Peninsula about 100 km is a plains coast, dominated by delta processes.

The west Shandong coast is also a bedrock-embayed coast, of mixed erosional depositional type. The highly weathered granitic material provides sediment to the coast, forming sand spits and other accumulating forms. This section of coast begins in the east part of Laizhou Bay, beyond the influence of the modern-day Yellow River.

In the Luan River region, south of Qinhuangdao, the shore is a plains coast, backed by extensive sand dunes. The Luan River also provides sediments to the inner shelf. To the south of the river, there are narrow sediment belts distributed parallel to the shoreline: fine sand, coarse silt, fine silt, then muddy clay. Pyrite in the muddy offshore area in water depths exceeding 9 m suggests quiet reducing bottom conditions. North of the Luan River sandy deposits dominate the inner shelf, persisting in the 20 km wide band out to water depths of 13 m. The fine-and-medium sands have textures of former beach dunes, now submerged (WANG et al., 1984).

Most of the inner shelf sediments of this region are supplied by the Luan River, which has migrated in the past. Many sand bodies such as point bars, river mouth bars, and sand shoals, left in the area of ancient river mouths, have undergone wave erosion. The influence of the modern river delta extends both to the north (15 km) and south (20 km), responding to waves and tides.

North of the Luan River the coast is an alluvial plain coast, from Qinhuangdao to approximately Shan Hai Guan. From Shan Hai Guan north to the plains coast near the Liao River, the coast is classified as bedrock-embayed of mixed erosional-depositional type.

Near the former Yellow River mouths the coast is eroding and cheniers have formed, while near the present river mouth the beach is prograding rapidly. Much of this coastal region is classic mudflat coast, up to 6 m wide. The mudflat zonation here is the typical salt marsh plain, intertidal flats (with four subzones), and submarine coastal slope (WANG, 1963). The salt marsh plains are the main marine depositional features. There are four old coastlines of chenier formed during the erosional periods of the mudflat when waves dominated. Presently, tidal currents dominate this region so cheniers are not forming actively. On the inner shelf, sediments change from sandy silt at the low tide level, to very fine mud at depths exceeding 10 m. The submarine coastal slope is gentle, ranging from 2 : 10 000 in the stable or slightly eroding coastal sections, up to 5 : 10 000 in the sand bar region of the 1965 river mouth position (WANG, 1983).

Yellow Sea Coast

This sector has two parts: a northern sector extending from the east Liaodong Peninsula to the Yalu River on the China-North Korea border, and a sector from the southern Shandong Peninsula extending south, including north Jiangsu Province. Waves in this area are generally small, since they are generated locally in the shallow Yellow Sea (Tables 4 and 5). Recently, the Yellow Sea tidal behavior has been clarified by both theoretical and observational work (LARSEN and CANNON, 1983; CHOI, 1984; LARSEN et al., 1985). There are few observations of low-frequency and mean flows in this region, and theory has not advanced far in describing this frequency band. The extensive freshwater inflow dominates the coastal circulation in a gravitational sense, coupled at least indirectly to the Kuroshio. A few current meter and hydrographic observations of this area are described by BEARDSLEY et al. (1983, 1985), and LIMEBURNER et al. (1983).

The bedrock-embayed coast near the Yalu River mouth differs from the remainder of the Liaodong Peninsula. The Yalu River drains 64 000 km² of the Changbai Mountains in northeast China. Eighty percent of the annual discharge reaches the Yellow Sea during summer flood, from June to September. Tides here are irregularly semidiurnal, with an average range of 4. 5 m, and a maximum range of 6. 9 m. Tidal currents both within and outside the estuary are strong, and exert strong control over sediment distribution. Sediments within the estuary are coarse sands, with occasional pebbles and shingles (WANG et al., 1984). Outside the estuary, an extensive area of fine-to-medium sands is distributed on the inner shelf, out to a water depth of 10 m (a distance of 20~30 km away from the river mouth). Strong tidal currents transport the sediment seaward from the estuary, where linear sand ridges are formed parallel to the direction of the estuarine tidal currents. Relief of these sand ridges is 15 m about 1 km from the river mouth. Sediments of these sand ridges are mainly fine sand with some coarse silt. These sand ridges migrate slowly to the west, under the influence of tides and waves (WANG *et al.*, 1984).

The second bedrock-embayed coast of this sector extends from the east Shandong Peninsula southwards towards Lianyungang, including the region near Qingdao. This extensive bedrock-embayed coast is of marine erosional-depositional type, with waves eroding the coast and small rivers transporting sediment to the coast. Erosional features, such as sea cliffs and rock benches, and depositional features, such as sand spits and baymouth bars, coexist.

The mudflat coast of the north Jiangsu Province is the most extensive in China. Extending south from near Lianyungang to the Yangtze River, this fluvial-marine depositional plains coast has undergone considerable change in recent historic time. The north Jiangsu mudflat coast is located along an area which has been receding in recent years. EMERY and AUBREY (1986) show the region from the Yangtze northward to about the former Yellow River mouth to be submergent, at rates of 2 mm y^{-1} near the south, to near zero in the north. North of the former Yellow River mouth the coast is emergent, as suggested by EMERY and AUBREY (1986) from recent tide-gauge records, and supported by observations of raised seacliffs near Lianyungang (found at elevations of 5, 15~20, 40, 60~80, 120, 200, 320, 450, and 600 m; WANG, 1983). These raised terraces suggest relative sea level here has been lowering not only recently but also in the geological past, a result of eustasy and tectonism. Unfortunately it is not possible to compare quantitatively recent trends in relative sea levels by using tide-gauge records in that area to those of the past combining seacliff elevation with radiocarbon dating. Thus, although this section of coast from Gang Shan Tou south to the Yangtze is a plains coast, the northern part is emergent, while the southern part is submergent.

This coastal sector is strongly influenced by tides, while to the north near the former Yellow River mouth waves are also important. Sediment eroded from mudflats by waves is transported primarily to the south, having some minor transport to the north towards Lianyungang. The erosional zone extends approximately 118 km, with intertidal mudflats only 0.5 to 1 km in width (ZHU and XU, 1982), consisting of very fine sand and silt. In the 700 years during which the Yellow River discharged sediment here, approximately 15 000 km^2 of delta was built (WANG, 1983). Since 1855, the old delta has eroded 1 400 km^2 due to lack of sediment supply; present-day land loss is 2.4 km^2 y^{-1} (WANG, 1983). The counties of north Jiangsu are building dikes along the coast to protect valuable farmland from erosion as the tidal flats diminish under wave action.

In the region from Sheyang to Dueng Zhao Guang, the mudflats are rapidly prograding. Sediment comes directly from the eroding former Yellow River mouth through longshore transport in the nearshore, and indirectly through onshore transport of eroded sediment, which has been transported alongshore by the coastal currents, and subsequently moved onshore primarily by tidal action. Mudflats reach $10\sim13$ km in width, having a slope of 1 : 5 000, and are divided into four zones: supratidal zone (grass flat), mudflats, mud-sand flats, and silt-sand flats (ZHU and XU, 1982). These mudflats are protected from wave action by the extensive offshore sand ridges, many of which are intertidal.

From Dueng Zhao Guang south to approximately the north bank of the Yangtze River, the mudflat coast is eroding (REN et al., 1983b). Near Lusi, the mudflat is narrow, eroding under the combination of larger waves and swift tidal currents. Mudflats represent the result of a conflict between tidal sedimentation and wave erosion, with sediment supply serving as a third factor. When waves are large, mud flats cannot accumulate. When waves are small and there is an adequate source of sediment, mudflats can prograde under the influence of tides.

East China Sea

The third sector extends from the Yangtze River mouth south to Nan Ao Island, near the border between Fujian and Guangdong provinces. Coastal evolution here is strongly influenced by the Yangtze River, both from its modem sedimentation and reworking of its ancient deposits. Tides are generally strong in this region, as are waves that are generated across large fetches of the deeper East China Sea. The lee (west) side of Taiwan differs in having only low waves.

There are three major coastal areas within this sector. The northern region extends from the Yangtze River mouth south to Hangzhou Bay. It is a delta plain coast, prograding steadily under the continued sediment supply from the Yangtze River. Coastal morphology is characterized by two major esturaries: the Yangtze estuary in the north, and the Hangzhou estuary in the south. The Hangzhou estuary, with its 9 m tidal range, is perhaps best known for its tidal bore, which has considerable historical significance. Predictions of the time and magnitude of the tidal bore date back more than 1000 years. Several cheniers testify to the progradation of the shoreline.

The second coastal region is a bedrock-embayed coast extending from Hangzhou Estuary south to Nan Ao Island, just south of the northern border of Guangdong Province. This bedrock-embayed coast is of mixed erosional-depositional type, having erosion near the major headlands and marine deposition within the bays. The many coastal embayments here trend northwesterly, in the interior of which mud deposits overlie rock platforms. This mud is supplied primarily by the Yangtze River, as the

Yangtze material is moved seaward, then transported to the south under the influence of inner shelf currents, and finally moved shorewards by tidal currents. Many islands protect the bay mouths, the islands also following the general northeasterly trend of the Zhejiang-Fujian fault zone. These islands serve as sources for sediment to the bays, but more importantly shelter the bays from waves so fine sediments can be deposited within the upper reaches of the bays.

This region also has some tidal inlet bedrock-embayed sections, which are more typical of the southernmost sector (sector 4). For example, Xiangshan Bay, Sanmen Bay and Leqing Bay (Zhejing Province), and Quanzhou Bay (Fujian Province) are all tidal inlet coasts. Even these tidal inlet coasts accumulate mud inside, through tidal processes. Whereas the bays in this section tend to be depositional (mud in the upper reaches, sand in the lower reaches), most headlands are erosional.

The third coastal region in sector 3 is the Taiwan coast. The east coast is an erosional fault coast, where steep seacliffs drop off sharply into deep water of the Pacific Ocean. The west coast is more complex, consisting of an alluvial plain coast with marine deposition forming sand bars and lagoons. Marine processes are responsible for reworking the alluvial material into depositional features. In the south part of Taiwan Strait are fringing coral reefs, both along the main island and along the many small islands immediately adjacent.

The geological history of sector 3 shorelines is recorded in ancient deposits both onshore and offshore. The onshore evidence consists of raised terraces at elevations of 5, 10, 20, 40, 60, and 80 m above sea level (WANG, 1983). The age of these terraces has not been established. Offshore evidence for coastal submergence is extensive. Along the Zhejiang-Fujian coast, there are two submerged bedrock terraces (at $20\sim25$ and $50\sim60$ m depths). Paralleling the coast up to Hangzhou Bay, these terraces also have not been dated. They are in the region which EMERY and AUBERY (1986) determined from tide gauge records to be subsiding, consistent with submergence of former terraces. North of Hangzhou Bay evidence for former coastlines is slightly different. Here accumulated coastal forms (such as deltas, sandy beaches) testify to lower relative sea levels. These must be dated to verify the chronology of coastal development. Perhaps the largest of these offshore accumulated features is the old Yangtze delta, known as the Great Yangtze Bank. In addition to these features on the inner shelf, there is also evidence of former north-south trending coastlines at depths of 100, 120, and 150 m (the latter dating 18 000 y B. P. ; REN and TSENG, 1980).

South China Sea

Of all the coastal sectors, the South China Sea sector is perhaps the most diverse. It encompasses a large area of the China coast, from Nan Ao Island south to the Beilun

River (the border between China and North Vietnam), as well as the many islands offshore, the largest of which is Hai Nan Island. The diversity of coastal types makes it difficult to summarize the coastline in a systematic fashion; instead this section provides specific examples of the major coastal types.

Tides are important locally, but since the South China Sea tides are generally diurnal, their associated currents are smaller. The high rainfall and higher temperatures of the tropics results in significant sediment supply. While the Pearl River contributes sediment to the south, many smaller rivers exert local control over sedimentation, separating the coast into a series of interacting coastal segments.

North Guangdong Province is characterized by granite Bedrock-embayed coasts of mixed erosional-depositional type. There are many small islands at the mouth of the bays, the latter which generally trend northwesterly. Besides eroding headlands, there are also large sand beaches. Heavy weathering which is typical of tropical climates allows higher erosion rates and greater supply of sediment to the beaches. An example of the poorly consolidated sedimentary material is the Zhan-Jiang formation, exposed along the mainland of Lei Zhou peninsula and in Hai Nan Island, composed of sand material with minor clay content. The combination of weathered igneous and metamorphic rocks and the easily eroded sandstone results in large accumulating forms on original bedrock-embayed coasts. In southern Hai Nan Island, which is genetically a bedrock-embayed coast, sections of the coast have built out extensively through a series of sand bars and lagoons. Near Sanya, there is a series of eight bars and lagoons which have built the coast out approximately 9 km. To the east of Sanya, the bedrock-embayed coast still has headlands actively eroding by wave action, accompanied by sedimentation in the bays. These forms are also common to the Guangxi Province, in southwest China.

Tidal inlet coasts are common in Guangxi Province and Hai Nan Island, as these embayments follow secondary east-west and northwest structural trends (superimposed on the dominant northeast fabric). On the mainland, the embayments follow this secondary structural fabric. Barriers generally trend northeasterly, crossing the mouths of the bays. There are abundant mudflats along the upper reaches of the embayments, the mud coming both from local sources and from the Pearl River to the northeast. Sandy beaches stretch along the barriers and the lower reaches of the embayments. Examples of tidal inlet coasts are Shantou, Zhenhai, and Zhanjiang bays. These tidal inlet coasts form good harbors, as they have good ebb flushing properties (WANG, 1980).

Submerged coastlines exist off this sector, with 8, 20, and 50 m deep terraces documented. They are generally of the accumulated form, rather than the bedrock platforms characteristic of the Zhejiang-Fujian area. Emerged terraces reflect smaller scale tectonics. An example is near the part of San Ya, Hai Nan Island, where marine terraces cut into bedrock.

Mangroves extend up to 27° north latitude along the South China Sea. They are mainly mangrove bushes (*Bruguiera conjugata*, *Aegiceras corniculatum*, *Rhizophora mucronata*, *Avicennia marina*) with a few jungles (*Bruguiera conjugata*, *Acanthus ilicifolius*) on Hai Nan Island. The mangroves are generally located in protected areas, such as lagoons and in the protected embayments behind sand barriers. Northeast Hai Nan Island has extensive mangrove development.

Fringing reefs are common on many of the islands as well as along part of the mainland. These fringing reefs are found in the embayments near the headlands, facing into the dominant waves, afforded some protection by the embayments.

Coral atolls are the dominant forms of most South China Sea islands whose basement is generally near 1 200 m below sea level (CHEN, 1978). These atolls provide biogenic sediment for coastline morphology, and thus differ from the remainder of the beaches of China.

CONCLUSION

The development of the China coastline results from a complex interaction of tectonics, local geology, river processes, waves, tides and storms. With its broad north-south extent, ranging from latitude 42°N well into the tropics, varying climate dominated by monsoon conditions and typhoons impacts the coast to different degrees. With the active tectonism following a general northeast structural trend, coastal sectors can be divided into emergent and subsiding regions, reflected in the classification of bedrock and plains coasts. Net erosion in any of these sectors is dependent not only on vertical movement but also on sediment supply, primarily from rivers. For instance, the plains coast within Bohai Sea progrades rapidly as the Yellow River empties into this relatively shallow embayment. Conversely, the plains coast of North Jiangsu Province is rapidly eroding, losing the silt deposited here when the Yellow River occupied this coastal sector prior to 1855. Waves and tides combine to erode rapidly this unstable coast.

The sediment input of the historically unstable Yangtze, Pearl and many other smaller rivers, and the varying exposure of the coast to both locally generated seas and distantly generated swells, create considerable variability in coastal evolution in China. Superimposed on this natural variability is the strong influence of man, dating back more than 1 000 years to the Tang Dynasty, as he settled and de-stabilized the extensive loess plains of China, accelerating the rapid local erosion and increasing heavy silt loads of the Yellow River. In more recent times this human impact continues as China dams many of its rivers for flood control, irrigation, and navigation. Offsetting these benefits have been the conflicting detrimental effects of increased siltation in the rivers (requiring periodic opening of the dams for navigation improvement) and reduced sediment supply to the rivers. Since riverine sedimentation is linked strongly to flood events which the dams are designed to tame, sediment input to the shore is reduced with consequent decline in coastal stability. With it population still growing and an increasing need for irrigation to sustain the food supply, man can be expected to continue to divert water from the rivers of China; diligent intervention by coastal scientists will be required to reduce the adverse impacts from this, and other, human activities.

Acknowledgements

This article was written while D. Aubrey was a guest of Y. Wang in China. Support for the visit was provided by the Coastal Research Center of Woods Hole Oceanographic Institution, and by the Chinese Department of Transportation through a grant to Y. Wang. Partial support for completion of this paper was provided by the Department of Commerce, NOAA National Office of Sea Grant, under Grant No. NA83-AA-D - 00049. Improvements to this paper were suggested by K. O. Emery, E. Uchupi, and an anonymous reviewer. Contribution No. 6026 of the Woods Hole Oceanographic Institution.

REFERENCES

BEARDSLEY R. C. , R. LIMEBURNER, K. LE, D. HU, G. A. CANNON and D. J. PASHINSKI. 1983. Structure of the Changjiang River plume in the East China Sea during June 1980. In: *Sedimentation on the continental shelf with special reference to the East China Sea*, China Ocean Press, Beijing, China. 265 - 284.

BEARDSLEY R. C. , R. LIMEBURNER, H. YU and G. A. CANNON. 1985. Discharge of the Changjiang (Yangtze River) into the East China Sea. *Continental Shelf Research*, 4: 57 - 76.

BEN-AVRAHAM Z. 1979. The evolution of marginal basins and adjacent shelves in east and southeast Asia. *Tectonophysics*, 45: 269 - 288.

BIRD E. C. F. and M. L. SCHWARTZ. 1985. *The World's coastline*. Van Nostrand Reinhold Company, NY.

CHEN J.-Y. 1978. A preliminary discussion on Quaternary geology of Xisha Qundao Islands of South China. *Scientia Geoligica Sinica*, 1: 45 - 56.

CHEN J.-Y., C.-X. YAN, H.-G. XU and Y.-G. DONG. 1979. The development model of the Chang Jiang River estuary during the last 2000 years. *Acta Oceanologia Sinica*, 1: 103 - 111.

CHEN J.-Y. , H.-F. ZHU, Y.-F. DONG and J.-M. SUN. 1985. Development of the Changjiang estuary and its submerged delta. *Continental Shelf Research*, 4: 47 - 56.

CHEN T. and Y. HUANG. 1982. Climatic oscillation of the South China Sea typhoon activities during the recent 100 years. *Tropic Oceanology*, 1: 139 - 142.

CHOI B. H. 1984. A three-dimensional model of the East China Sea. In: *Ocean hydrodynamics of the Japan and East China Sea*, Elsevier, Amsterdam, Netherlands. 209 – 224.

CHOI B. H. 1985. Computation of meteorologically-induced circulation in the East China Sea. In: *Marine geology and physical processes of the Yellow Sea*, Y. A. PARK, O. H. PILKEY and S. W. KIM, editors, Korea Institute of Energy and Resources, Seoul, Korea. 36 – 52.

EMERY K. O. 1983. Tectonic evolution of East China Sea. In: *Sedimentation on the continental shelf with special reference to the East China Sea*, China Ocean Press, Beijing, China. 80 – 90.

EMERY K. O. and F. -H. YOU. 1981. Sea-level changes in the western Pacific with emphasis on China. *Oceanologia et Limnologia Sinica*, 12: 297 – 310.

EMERY K. O. and D. G. AUBREY. 1986. Relative sea-level changes from tide-gauge records of eastern Asia Mainland. *Marine Geology*, 72: 33 – 45.

GUAN B. -X. 1979. Some results from the study of the variation of the Kuroshio in the East China Sea. *Oceanologia et Limnologia Sinica*, 10: 297 – 306.

GUAN B. -X. 1983. A sketch of the current structures and eddy characteristics in the East China Sea. In: *Sedimentation on the continental shelf with special reference to the East China Sea*, China Ocean Press, Beijing, China. 56 – 79.

INMAN D. L. 1980. Shore processes and marine archaeology. In: *Oceanography in China*, CSCPRC Report No. 9, (in China, ch. 6, 47 – 65). National Academy of Science, Washington, DC. 106.

INSTITUTE OF COAST AND ESTUARY. 1984. Twenty years of research on the coastal and estuarine science (1964—1984). Department of Geography, Zhongshan University, Guangzhou, PRC. 103.

LARSEN L. H. and G. A. CANNON. 1983. Tides in the East China Sea. In: *Sedimentation on the continental shelf with special reference to the East China Sea*, China Ocean Press, Beijing, China. 337 – 350.

LARSEN L. H. , G. A. CANNON and B. H. CHOI. 1985. East China Sea tide currents. *Continental Shelf Research*, 4: 77 – 103.

LI C. -C. and W. -J. WANG. 1985. Sedimentation on the Zhujiang River mouth region. In: *Modern sedimentation in coastal and nearshore zone of China*, M. -E. REN, editor, China Ocean Press, Beijing, China. 230 – 251.

LI C. -X. 1979. The characteristics and distribution of Holocene sand bodies in Changjiang river delta area. *Acta Oceanologia Sinica*, 1: 252 – 268.

LI D. -S. 1984. Geologic evolution of petroliferous basins on continental shelf of China. *American Association Petroleum Geologist Bulletin*, 68: 993 – 1003.

LIMEBURNER R. , R. C. BEARDSLEY and J. ZHAO. 1983. Water masses end circulation in the East China Sea. In: *Sedimentation on the continental shelf with special reference to the East China Sea*, China Ocean Press, Beijing, China. 285 – 294.

MILLIMAN J. D. and R. H. MEADE. 1983. World-wide delivery of river sediment to the oceans. *Journal of Geology*, 92: 1 – 21.

MILLIMAN J. D. and Q. -M. JIN, editors. 1985. Sediment dynamics of the Changjiang Estuary and the adjacent East China Sea. *Continental Shelf Research*, 4: 1 – 4.

MILLIMAN J. D. , Y. HSUEH, D. -X. HU, D. J. PASHINSKI, H. -T. SHEN, Z. -S. YANG and P. HACKER. 1984. Tidal phase control of sediment discharge from the Yangtze River. *Estuaine*,

Coastal and Shelf Science, 19: 119 - 128.

PANG J. and S. -H. SI. 1979. The estuary changes of Huang He River. *Oceanologia et Limnologia Sinica*, 10: 136 - 141.

PANG J. and S. SI. 1980. Fluvial processes of the Huang He River estuary. *Oceanologia et Limnologia Sinica*, 11: 293 - 305.

PUTNAM W. C., D. I. AXELROD, H. P. BAILEY and J. T. MCGILL. 1960. Natural coastal environments of the world. Technical report, University of California at Los Angeles. 140+maps.

QIAN N. and D. -Z. DAI. 1980. Problems of river sedimentation and the present status of its research in China. In: *Chinese Society Hydraulic Engineers*, *Proceedings of the International Symposium on River Sedimentation*, 1: 1 - 39.

REN M. -E. and C. -S. TSENG. 1980. Late Quatemary continental shelf of east China. *Acta Oceanologica Sinica*, 2: 1 - 9.

REN M. -E., R. -S. ZHANG and J. -H. YANG. 1983a. Sedimentation on tidal mudflat of China-with special reference to Wanggang Area, Jiangsu Province. In: *Sedimentation on the continental shelf with special reference to the East China Sea*, Vol. 1, China Ocean Press, Beijing, China. 1 - 19.

REN M. -E., R. -S. ZHANG, J. -H. YANG and D. -C. ZHANG. 1983b. The influence of storm-tide on mud plain coast. *Marine Geology and Quaternary Geology*, 3: 1 - 24.

SCHWARTZ M. L. 1982. *The encyclopedia of beaches and coastal environments*. Hutchinson Ross Publishing Co., Stroudsbourg, PA. 940.

SHAN G. -L., Z. -P. LIU, Z. -J. WANG, H. -D. XU and G. -Y. LEI. 1983. Numerical simulation of the tidal mixing in the Bohai Sea, I. The numerical simulation of the principal semi-diurnal constituent in the Bohai Sea. *Oceanologia et Limnologia Sinica*, 14: 419 - 431.

SUN H., J. YAO, Y. HUANG, B. LONG, B. XU and H. TENG. 1981. *Summary of hydrography and meteorology of the China coastline*. Science Press, Beijing. 159.

TONG Q. -C. and T. -W. CHENG. 1981. *Physical geography of China*. Science Press. 6 - 121.

U. S. DEPARTMENT OF COMMERCE. 1981. *Climatological and oceanographic atlas for mariners*. Washington, DC.

WANG Y. 1961. The properties of China's silted plain coast and problems of seaport's construction. Postgraduate thesis, Department of Geology and Geography, Beijing University.

WANG Y. 1963. The coastal dynamic geomorphology of the northern Bohai Bay. *Collected Oceanic Works*, Nanjing University, 3: 25 - 35.

WANG Y. 1980. The Coast of China. *Geoscience Canada*, 7: 109 - 113.

WANG Y. 1983. The mudflat coast of China. *Canadian Journal of Fisheries and Aquatic Sciences*, 40: 160 - 171.

WANG Y., D. -K. ZHU, X. -H. GU and C. -C. CHUEI. 1964. The characteristics of mudflat of the west coast of Bohai Bay. Abstracts of these in the Annual Symposium of Oceanographical and Limnological Society of China in 1963. Wuhan, Science Press. 49 - 63.

WANG Y., M. -E. REN and D. -K. ZHU. 1984. Sediment supply by the major rivers of China to the Continental shelf. Marine Geomorphology and Sedimentology Laboratory, Nanjing University, Nanjing, PRC. 32.

WILLIAMSON G. R. 1970. Hydrography and weather of the Hong Kong fishing grounds. *Hong Kong*

Fisheries Bulletin, 43 – 49, plus figures.

WYRTKI K. 1961. *Physical oceanography of the southeast Asian waters*. Scripps Institution of Oceanography, La Jolla, CA. 195.

XUE H. -G. and J. -H. PAN. 1983. The problems of navigating the west channel of the Yalu River. *Symposium on Harbour Navigation*, Dandong, China.

YE A. L. and I. S. ROBINSON. 1983. Tidal dynamics in the South China Sea. *Geophysical Journal of the Royal Astronomical Society*, 72: 691 – 707.

YIN F. 1984. Tides around Taiwan. In: *Ocean hydrodynamics of the Japan and East China Seas*, T. ICHIYE, editor, Elsevier, Amsterdam. 301.

YING Z. -F. and S. -G. CHEN. 1984. Notable features of the mixing in the Lingdingyang Bay. *Acta Oceanologia Sinica*, 3: 1 – 12.

ZHANG J. and S. LI. 1980. A scheme for typhoon wave calculation in the northern part of the South China Sea. *Nanhai Studia Marina Sinica*, 1: 135 – 164.

ZHU D. -K. 1980. The coastal evolution of the east Hebei Province. Symposium on China's coasts and estuaries, Shanghai, China.

ZHU D. -K. and T. -G. XU. 1982. The coastal development and exploitation of middle Jiangsu Province. *Acta Scientiarum Naturalium Universitis Nankinensis*, 3: 799 – 818.

Coast Research in China *

BASIC CHARACTERISTICS OF THE CHINA COAST

The coast of China lies between the Eastern Continent of Asia and the Marginal Seas of the West Pacific. The coastline of the mainland extends from the mouth of the Yalu River in the north to the mouth of the Beilun River in the south, and is over 18 400 kilometres long. The total length of the coastline including the 6 500 islands, in approximately 32 000 km.

The coastal processes and evolution of China are mainly controlled by geological structures, large river influences, monsoon wind waves, and the tidal process of the Pacific Marginal Seas. River sediment and fresh water inputs to the sea have shown a great influence upon the evolution of the coast via monsoon waves and tidal currents. It has characteristics unique in the world for example: A total of more than two billion tons of fluvial sediment per year is discharged into the China seas (Wang et al. , 1986). This means a total of $1km^3$ of sediment is annually carried by river to the coastal zone and continental shelf of the China seas. The Yellow River (Huanghe) has the largest sediment discharge with 11×10^8 tons as the annual average and the maximum is 21×10^8 ton; although its freshwater discharge is modest, only 411×10^8 m^3/year. Thus per cubic metre of Yellow River water has an average value of 24. 7 kg of sediment (the maximum value is about 37. 7 kg/L). A great quantity of sediment is continuously transported to the Bohai Sea. Most of the 2 000 km mudflat coast is developed in the Bohai Sea where the present Yellow River discharges, and on its pre-1855 delta 600 km to the south on the Yellow Sea coast. Besides this, the two major rivers, the Yellow River, and the Changjiang River (Yangtze) have migrated great distances during the past 4 000 years. The present North Jiangsu coastal plain was formed by two rivers; even the continental shelf of the East China Sea was a huge alluvial plain during the period of last low-sea level (Ren and Zeng, 1980).

Two major types of coast have been identified (Wang, 1980): Bedrock—embayed coasts and plain coast (Fig. 1. , after Wang & Aubrey, 1987).

---

* Ying Wang: The Geographical Society of China ed. , *Recent Development of Geographical Science in China* , pp. 183 - 196, Beijing: Science Press, 1990.

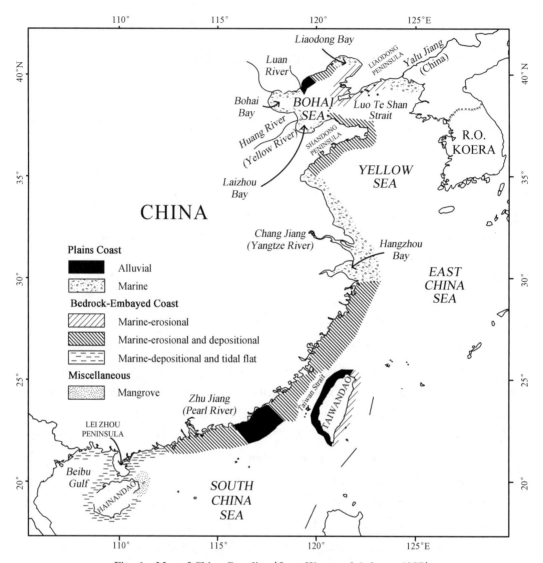

Fig. 1 Map of China Coastline (from Wang and Aubrey, 1987)

BEDROCK—EMBAYED COAST

Characterized by irregular headlands, bays and islands, it develops at the place where mountain meets the sea. The coastal slopes are steep with high energy wave action. Four subtypes are identified.

Marine erosional-embayed coast: it develops on hard crystalline rock, its erosion rate is small and the coast is modified slowly. A spare sediment supply limits coastal deposition, leaving a dominance of erosional geomorphic features.

Marine erosional-depositional type: It mostly develops at the place where mesozoic granite, are overland by weathered deposits and the coast are easily eroded. Erosional features (rock benches and terraces) and depositional features (bay bars, sand spits, tomboloes) develop here.

Marine depositional type: With relatively erodible bedrock, marine erosion supplies large quantities of sediment for prograding shoreline. The coast eventually matures to a plain coast, commonly in south China.

Tidal inlet-embayed coast: Common again in south China, the northeast-southwest trend of the coast is interrupted by northwest-trending inlets along faults with the same orientation. Small islands or sand spits project into the mouth of the bays, forming tidal inlets which are swept clean by the powerful ebb tidal currents. However, sediments are deposited as tidal deltas at the ends of the deep pass channel, connecting the open sea and the inner bays.

PLAIN COAST

Based on genesis, these are classified into two types:

1) The alluvial plain coast: located generally seaward of mountain ranges; the plains are built on fluvial sediments, with low gradients towards the sea, resulting in a wide breaker zone and active sediment movement. Barrier bars, sand spits, submarine bars, and extensive beaches are typical features of this coast.

2) Marine depositional plain coast: Flat and very extensive, these are located on the lower parts of large rivers in areas of subsiding bedrock. Coastal slopes are extremely gentle (1:1 000 to 1:5 000), leading to wave breaking offshore; the nearshore is commonly dominated by tidal processes. Sediment from the rivers is deposited either offshore or on the extensive tidal flats. The coast profile passes from salt low-land through intertidal mudflats to the submarine coastal slope. The salt low-land extends inland from a shell ridge (chenier) above the intertidal zone. The shell ridges usually form along the tidal flat coast and are useful indicators of the location of ancient coastlines.

Both the bedrock—embayed coast and the plain coast may incorporate two minor types of coast. The first, the river mouth coast, is limited to the mouth of major rivers, containing both delta and estuarine features. The second type is biogenic, found in south China, incorporating coral reef and mangrove coasts. Mangroves are limited to south of latitude 27 degrees north, while coral reefs occur as far north as Taiwan.

Four sectors can be recognized according to regional dynamic characteristics (Wang and Aubrey, 1987): As a whole the Bohai Sea (sector 1) is a river-dominated environment. The Yellow Sea (sector 2) is tide-dominated, with waves playing a secondary role, along with the former and present river processes. The East China Sea Coast (sector 3) is wave dominated, having a large tidal range in some tidal inlets and estuaries. The Chang Jiang contributes to coastal development. The South China Sea Coast (sector 4) is largely wave-dominated.

COASTAL RESEARCH IN CHINA

The systematic study of sea coast in China started in the late 1950's under the influence of soviet scientific trends. A group of coast experts, including the famous coast authority V. P. Zenkovich, who joined the Chinese siltation studies of Tianjin New-Port in 1958, held a series of lectures in Beijing University and conducted reconnaissance work along the Chinese coastline from Liaodong, Shandong Peninsulas to the Pearl River delta and Hainan Island. We adopt the definition of Sea coast as a coastal zone, which includes the land along the sea shore, the beaches or intertidal flat and the submarine coastal slope. Thus we investigate the sea coast from back beach to shore and offshore, out to the water depth where waves start to move sediment. The work is therefore carried out on land over the intertidal zone and at the sea, and to study the coast from the point of view of geomorphology, dynamics and sedimentology. This includes coastal dynamic Geomorphology, and the work is carried out by a team of coastal geomorphologists, sedimentologists, physical oceanographers and coastal engineers. The aim of the research is to define the source and movement of sediments, the coastal characteristics of erosion and deposition, the genesis of the coast, and to predict the future evolution of the coastline.

Much of the work is directed towards the location of sea harbours, quays and navigation channels. Which means that coast research in China is connected closely with applied projects. Sea port construction is an important step for Chinese coastal zone development, as a limited number of sea harbours can hardly meet the increasing volume of transportation. The average is one harbour for each 500 km coastline, with very little number of sea ports scattered along the quite longer coastline. Up to 1985, we have a total of 400 berths, of which 199 are deepwater berths with a depth of 10 m or more. (Yen, 1987). The total volume of traffic has increased fromsome 10 million tons in 1985. However, the majority of larger sea ports are located in the north of China. In the 1 000 km long coastline of north Jiangsu, there is only Lianyun Gang Harbour at the north end of the province. Bedrock embayments are distributed widely in the south of China, nearly one hundred coastal bays can be developed into middle or large sea harbours. For harbour construction, there are the contradictory factors of "deepwater and calm sea" in the natural coastal environment especially along China's coast zone with the great sediment supplied by rivers, even mud flats have developed inside longer bays in the areas of Zhejiang and the northern part of Fujian Province owing to long shore drifting from the mouth of Changjiang River. Thus, the siltation of harbours has become a serious problem. A typical example is the siltation of Tianjin New Port. (Wang, 1980, Yen, 1987). The port was first constructed by the Japanese from 1939 to 1945. It is situated on the mudflat of the Bohai Sea with the Haihe River at the south and the Jiyunhe River at the north, emptying into the vicinity of the bay. The harbour is embraced by two breakwaters, enclosing a water area of 18 km^2. Not all the water area was utilized, only about 2.5 km^2 were occupied by ship berths and the approach channel was maintained at the required depth by dredging. (Fig. 2. from Yen 1987.) The two breakwaters have not yet been extended to the designed water depth and openings were left here and there along the breakwaters (when the Japanese left the harbour), the longest one, named the north opening, is 1 300 m in length. The conditions of the harbour grew rapidly worse and the water depth in the approach channel was reduced to only 3 m during 1949 (Yen, 1987). Extensive dredging has been carried out since 1949, and by the end of 1952, the water depth was increased to 9 m. But the main trouble was still the siltation of the harbour and the water depth could be maintained only by powerful dredging. Each time the harbour area was extended to meet the demands of increasing traffic, the amount of siltation was increased accordingly. A total of 6 000 000 m^3 of silt was removed annually. This example triggered extensive coastal research programs. As a result, it was ascertained that silt from the Yellow River was the main source of material forming the entire mudflat of the Bohai Sea, but the Haihe River was also responsible. To reduce siltation, all the openings were closed, part of the redundant water area was refilled with dredged soil and the extension of the two breakwaters was also contemplated. In 1958, sluice gates was built at the mouth of Haihe River. This has a prominent effect in reducing the harbour siltation.

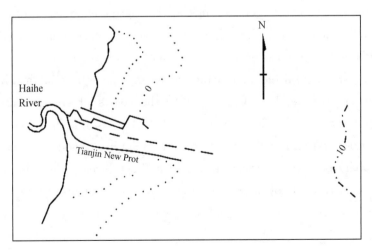

Fig. 2   Status of Tianjin New Port (from Yen, 1987)

At present, the 24 km approach channel has been widened from 60 m to 120 m and 150 m with a water depth of 11 m (Yen. 1987). Subsequently, many harbour sites were chosen on the basis of thorough studies of coastal dynamic geomorphology. Unfortunately, many harbours still have been built without proper study, and have serious siltation problems. To avoid these, extensive coastal dynamic geomorphologic studies are now required before the building of any new harbours. Thus practical requirements have advanced coastal science. For this type of research, there is a team of Geoscientists and engineers working on the same project. The work of the coastal geomorphologists, who normally come from universities and institutes, forms the basis of most of these studies. The Department of Transport supports this work with technicians, money, ships, vehicles and drilling equipment. When the project is finished, a report is submitted to them. Scientific research was promoted by these projects, for example, mudflat coast and Yellow River delta in the Tianjin New Port Project; Investigation of the sand barrier coast with the Qinhuangdao oil port Project; study of the estuaries with Chang Jiang River mouth project, study of the tidal inlets and coral reefs with the South China Sea port project. This research accelerated the advancement of knowledge about the coast of China. For example, we were able to establish the mudflat zonation, the proper significance of shell beach ridges, and the sedimentary dynamics of Tidal inlets.

Another type of applied study is the exploitation of the valuable placer deposits. These are zircon, rutile, titanic, iron, monazite and gold, which are mostly deposited in the fine sediments of sand beaches, sand spits, bay mouth bars, sand barriers and tomboloes, but are limited in sand dunes and river mouth deltas. The placers were formed during late Pleistocene and Holocene eras (Ruan, 1988). In areas such as Liaodong, Shandong peninsula, Fujian, Guangdong, and Hainan Provinces there are mesozoic granites. The coast there belongs to the Marine erosional— depositional type of bedrock embayment coast. This kind of project has also improved the study on coasts, sand-body formations and submarine coastal slope sediments.

Specific coasts of coral reef and mangrove swamps have been studied in detail from the southern coast of the East China Sea to the coasts of Guangdong and Hainan provinces and the south China Sea Islands. The study has achieved greater understanding of coral reef evolution during the Holocene period and has provided the evidence for analysis of the tectonic movement and sea level changes in different coast areas. Six natural reserves of mangrove forest have been set up in Fujian, Guangdong and Hainan Provinces to provide examples showing the natural beauty and benefit of the ecosystem.

From 1980—1985, China carried out a comprehensive survey along the whole coast zone of the mainland. The main purpose was to understand natural resources of the coast and tidal flat as well as its quantity and qualities; to get the basic data of coast environment and social economics; to make a comprehensive review on the environment and resources of China's coast; and to work out the plans for full scale exploitation of coastal resources. Large amount of scientific data including: Coastal lithology and structure, geomorphology, sedimentology, oceanography, climate, soil, fishery and aquatic production, botany and zoology, agriculture, coastal engineering etc. have been collected. It encompassed a great deal of work and was the first time in ten coastal provinces and cities of China that systematic coast data has been ascertained. It is also a great help to the development of China's coastal zones. As a continuation of the comprehensive survey of coasts, China now carries out survey work on sea islands. Attention is mostly paid to the medium size islands. The work is not only a comprehensive survey of planmaking but also an application of the results of studies on each island, to develop each island to be specialized according to its environmental and natural resource. All of these measures will increase interest in coastal zone development, and improve the exploitation of coastal resources, especially in: energy (wind and marine dynamic), placer minerals, tidal flat and wild land use, sport and tourism industry, and harbour and coast bay resources.

MAJOR PROGRESS OF CHINA COAST RESEARCH

The speed of coastal research in China has increased since the 1980's. Because of the limited space here, I can only summarize a few of the major topics, according to the point of view of the authors.

1. The Plain Coasts are still hot spots of research and with apparent progress both theoretically and practically. The plain Coast is formed along the depressed belts, usually in the lower reaches of a large river. The Liaohe River Plain, the Yellow River delta Plain and the North Jiangsu Plain, which were formed by the Yellow and Changjiang Rivers, are the major types of plains coast discussed here. Three components have advanced recently, even though the results of each part are not equal.

1) The intertidal flat is a predominant feature of this extensive plain coast. The main factors which determine the development of tidal flats are the very gentle slopes of plain coasts, tidal dynamics, and the abundance of fine-grained sedimentary material. (Wang et al., 1988). The width of any particular tidal flat is governed mainly by the tidal range; similarly morphological characteristics and their associated sedimentary structures are controlled by the tidal energy, (current velocities), large tidal ranges are always accompanied by well developed flats, such as Hangzhou Bay in China, Bay of Fundy, and the Severn Estuary abroad. Typical tidal flats in China are developed where the tidal range is about 3 m and the velocity of tidal currents is not as large as in the estuaries referred to previously. Such tidal flats are between 6 to 18 km in width and are much more extensive than in the estuarine areas. The major types of tidal flats in China

are developed along the fringe zone of the North China Plain, where the slope of the coast zone is less than 1 : 1 000. As a result of this, the wave action is restricted to offshore areas and tidal currents are the dominant dynamic processes on the flats. In comparison, some of the tidal flats in South China, along the East China and South China Seas are associated with long narrow embayments or other types of sheltered coast. Fetch within such environments limits the wind/wave interaction, hence tidal currents are, once again, the major agents controlling sedimentary processes. Silt is the major sediment, ranging from fine to coarse silt, deposited over the largest part of tidal flats. Very fine sands and clays are also present on most tidal flats. All such sediments deposited on tidal flats are carried mainly onto the flats by flood tidal currents, bringing material from the adjacent seabed: this infers to transport from seaward to landward. The original source of such sediments may be terrigenous. Actually, the sediments on the Chinese tidal flats are fluvial in origin and have been supplied to the coastal zone at lowered sea level by the Yellow and Changjiang Rivers.

The sediment, morphology and benthic zonation features of intertidal flats are distinct phenomenon in a wide range of climate environments (Dale 1985, Evans 1965, Thompson 1986), and are also a reflection of changes in the tidal dynamics. Three types of coastal tidal flats have been recognised in China and the U. K. by comparison (Wang et al. , 1988).

A silt flat well developed along the North Jiangsu coast of the Yellow Sea (China) is a typical type of tidal flat with three zones, namely, mud zone on the upper flat, muddy silt zone on the middle part and sandy silt zone on the lower part of the intertidal zone. A medium level of wave energy stirs up material from the submarine sandy ridges of the old Changjiang River delta and transports it onto the shore by the flood tidal currents.

A mud flat has developed along Bohai Bay in the China inland Sea of Bohai, with only minimal wave influences but with large quantities of fine grained material of clay and silt which is supplied mainly by Yellow River. There are four distinguishable zonations along the 400 km long coastline: a polygon zone consisting of clay cracks is distributed on the upper part of the intertidal zone which is only flooded during the spring high tides. The classic feature of the mudflat is a 0. 5~1. 0 km wide belt of semi-fluid mud located below the mean high tidal level as an inner depositional zone. Only in the Bohai Sea, where there is a large quantity of sediment supplied by the Yellow River plume to support the coast, is the mudflat well developed with a wide zone of semi-fluid mud. Outside of this zone, there appears to be an erosional zone with interbedded silt, clay and silt deposits, and a silt ripple zone, known as the outer depositional zone, is located on the lower part of the intertidal zone.

In comparison, a sandy flat developed in the Wash, along the coastline of eastern

England, U. K. is an example of a tidal flat with high wave energy influences. Coarse sand and shingles are present offshore as glacial deposits during a period of lower sea level. They are now reworked by wave to be deposited by tidal currents as a sand flat.

Differences between the three types of tidal flat are considered to be in response to the varying levels of wave energy input. A sedimentation model of the typical tidal flat is shown as two-layers (Zhu and Xu, 1982):

a) The top part consists of vertical deposits of mud suspension (A.), carried by the flood tide. The thickness of the layer is limited by the tidal range between mid and spring tidal levels.

b) Overlapped by A. It is a transverse deposit of silt or fine sand (B.) moving as a traction load. The maximum thickness of the layer may by the same as the vertical distance between the low tide level and the coastal wave action base.

However, during the typhoon season, a strong tropical cyclone wind (Beaufort Scale $10 \sim 11$) causes high waves which are 5 times greater than the average wave height. (Ren et al. , 1983). Strong waves coupled with swift tidal current cause widespread surface erosion of the tidal flats, especially scouring in the lower intertidal zone. But in other parts of the tidal flat, the early phase scouring is followed by aggradation and only accretion occurs in the upper part of the supratidal zone. (Ren et al. , 1983). The sediments deposited during the typhoon season are distinctly different from the sediments deposited during ordinary weather. In the upper intertidal zone, typhoon deposits are coarse silt (Md 4.93ϕ) with good sorting ie: only with very little fine portion, while the underlying sediments are mainly medium and fine silt (Md 6ϕ) with poor sorting. Typhoon deposits are also characterized by its sediment structures: thin horizontal beddings in their middle and upper parts (Ren et al. , 1983).

Tidal creeks extend across the whole tidal flat zone, as a series of meandering creeks, which in some parts, are bordered by levees. Tidal current dynamics in creeks show there is a series of asymmetrical phenomena (Ding and Zhu, 1985), as 10% of the complete tidal prism which comes into the creek on the flood results in a residual down-channel flow on the flood. Within the creeks, the ebb discharge is greater than the flood discharge throughout a complete tidal cycle. Maximum ebb currents are higher than on the flood and the ebb duration is longer. Thus, the creeks are ebb-dominated tidal controls. The tidal creeks are not only dissecting the intertidal flat, but also acting as a pass transporting the sediment and creek deposits form a mosaic formation within the intertidal flat deposits. Sediments in the creeks are coarser than those deposited on the flats and sometimes appeared as silty cross-beddings or sandy lens in the sediment profiles.

In summary, tidal flat sediment have the following structures:

a) Thin bedding layers, especially the alternating laminate of fine silt and silty clay, which is the main component and structure of tidal flats, and indicates the systematic sedimentary processes of tidal cycles.

b) Coarse silt or sandy silt is the lower part deposit and has pod, ripple beddings or horizontal beddings, pod features are common in the coarse silt bedding as micro ripples shown in the section.

c) Sandy silt or very fine sand beddings with cross or herringbone shaped beddings, are normally a mosaic formation in the laminate groups.

d) Silty clay laminate as upper part of tidal flat deposits have clay cracks, worm burrows and root mottlings, sometimes the burrows are filled with lots of small pills or iron pyrite materials, and also calcareous materials fill up the root fissures.

e) Storm deposits of thinly interlayered silt, with cross beddings.

2) Shell beach ridges (cheniers) are distributed in the upper part of silty tidal flats, and are a distinct land-form associated with the extended plain coast. The cheniers consist mainly of shells and shell fragments with a little proportion of silicate materials of silts and fine sands, which is different to normal beach ridges. Due to fresh water precipitation remaining in the shell deposits, the ridges are the green island on the salty lowland. Cheniers are interesting topics to archaeologists, coastal geomorphologists and geologists. Since the early sixties (Chen et al., 1959, Li, 1962, Wang, 1964), scientists have carried out investigations on a large scale and dated the shell ages by using ^{14}C method. According to recent publications (Zhao 1987) four series of cheniers are distributed in the coastal plains of the west part of Bohai Sea and the West Yellow Sea, more than five cheniers are in the vicinity of Shanghai, and also four cheniers have been dated on the outer continental shelf of the East China Sea.

Most of the ^{14}C data of molluse shells are reliable and can be identified as the old coastline environment. However, it is normal for beach ridges to be formed with shell fragments, but not all the ridges are cheniers, some deposited forms might be old sand barriers as shown in the North Jiangsu coastal plain, some might be shallow water deposits with shell fragments as seen in the West Bohai Coast Plain. Attention should be paid to the under water features, there needs to be sufficient evidence of ridge forms and sediment structures to identify the chenier series. Cheniers indicate a special coast environment when tidal flat of the plain coast is retreating due to insufficient sediment supply. The molluse shells can be dug out by break wave action, and carried landwards by swash to accumulate shell beach ridges on the upper part of the inter tidal flat. The evolution of both cheniers I and Ⅱ (counted from sea to landward) at the west side of the Bohai Bay, as well as along the Jiangsu coast, appears to be related to the history of the shifting lower course of the Yellow River. Coastal progradation occurred in the vicinity of the active delta of the Yellow River. On the contrary, erosion took place along the abandoned deltaic coast and it's in this erosional and wave dominated tidal flat environment that coarse shell cheniers are formed. According to the history of the shifting Yellow River Mouth, this occurred alternatively on the coastal plain of Bohai Sea and Jiangsu. As an illustration, the behaviour of the lower course of the Yellow River and the related coastal development in the last 2 000 years can be briefly summarized (Fig. 3, Wang, 1983.)

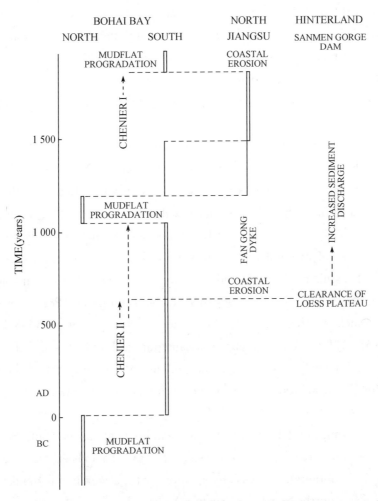

Fig. 3 Summary of changes in the course of the lower Yellow River in the last 2 000 years and its effects on coastal development (From Wang, 1983)

3) The plain coast represents a balance between wave erosion and deposition of sediment that is dependent on sediment supply, principally from rivers. If the amount of sediment supply is larger than the amount eroded, the coastline progrades; if the sediment supply is reduced or stopped, wave erosion causes the coastline to retreat. Relative changes in sea level do not effect to any great extent the development of the plain coastline. For example, at the present location of the Yellow River, in Southern Bohai Sea, major shoreline accretion has occurred during the past 130 years, the delta forming seaward at 20 ～ 28 km (Wang et al. , 1986), despite a steady regional

subsidence since the late Miocene period. Along the abandoned Yellow River mouth in the North Jiangsu coast, where the Yellow River emptied from 1128 to 1855, erosion has been extensive. The coast has retreated about 17 km since 1855. (Ren et al., 1983). Thus, efforts have been made to construct detached breakwater parallel to the coast, and to construct groins at the end of the longitudinal breakwaters in a combined project to protect the tidal flat coast erosion (Yu et al., 1987).

2. Tidal inlet embayments are distributed widely along the coast zone of the East China and South China seas, eg. Xiangshan Bay, Sanmen Bay, Leqing Bay, (Zhejiang province), Quanzhou Bay, Xiamen Bay (Fujian province), Shantou Zhan Jiang Bay, (Guangdong province), Yulin Bay and Yangpu Bay (Hainan province) are all large tidal inlets.

Tidal inlets should be distinguished from tidal channels which connect open sea at both ends. Such as the Jintang channel, which is near the Xiangshan Bay of Zhejiang Province, where the deep water port of Beilun is situated (Ren and Zhang, 1985). Three types of tidal inlets have been summarized in China as 1) the embayments-lagoon type on sandy or rocky coasts such as at Baihai port and Shantou Bay; 2) estuarine inlets on mouths of small of medium rivers which may be on mud plain coasts, such as the west channel of Yalujiang Estuary and the Huang Pu Jiang Estuary in Shanghai; 3) artificial inlets enclosed by breakwaters such as the Tianjin New Port. Improvement of navigation channels of these inlets follows the principle of the O'Brian P-A formula (O'Brian, 1967, 1980). Where accurate oceanographical and littoral drift data are not available, a careful analysis of coastal morphology and sedimentology may provide a useful clue for the evaluation of the value of inlets in navigation (Ren and Zhang, 1985).

Recently, the mean tidal prism (P) of 32 tidal inlets along the South Sea Coasts and mouth cross-sections area below mean sea level (A) have been calculated (Zhang, 1987). The P-A optimum regression equation with a minimum standard deviation (solved through using the regression analysis) is in the power function form with the power approaching unity, however the coefficient C doubles the sandy coasts in the United States, and is only 1/10 as the rocky bay coasts in Japan. The P-A relation here shows either a very high correlation, or a very obvious unstability.

Not only in P-A relation formula is the coefficient e changeable, P and A points deviate apparently from the regression line, frequency distribution of the P/A ratio is discrete, but also the evolutionary directions of the P/A ration are not consistant during the developmental process of the same inlet, when tidal prism reduces dramatically under the impact of reclaimation. P-A relation is a quite complex problem which should be further studied and improved. Tidal prism is an important factor, and so are the geologico-geomorphologic conditions and hydrodynamics of the different inlets. (Zhang,

1987).

Tidal inlets are the major components among 64 embayments in Hainan Island; and the inlets can be divided into three types according to genetic patterns. (Wang et al., 1987): 1) Tidal inlets developed along structural fault zone, where there is also a boundary weak zone of volcanic rock and sandy deposits of old terraces, as the Yangpu inlet, the largest inlet located in the northwest part of the island. 2) Sandy barriers enclosed embayment, such as Sanya, Xincun, and Qinlan Inlets. 3) The Holocene transgression flooded mountain or River Valleys. Yulin and Yalong Bays belong to this type. All three types of tidal inlets are suitable for building sea harbours and can serve multipurposely. There are two factors in the maintenance of the inlet bay and the water depth of navigation route. a) The factor of the tidal prism, which is decided by tidal range and the duration ratio of flood and ebb tides. The ebb current is the natural dredge washing out sandy sediments to benefit the navigation channel. It is important to keep or extend the larger area of the inlet bay for retaining larger quantities of flood sea water. b) Determining the sediment sources along coastline and inlet, and decreasing the coast sediment supply are both important steps producing natural flushing out of the sediment or sediment bypassing by long shore drifting. Sedimentation of the larger tidal inlets of Hainan Island consists of three dynamic phases according to our recent study on Lehigh core data.

Inner bay phase; River influences at the bay head, sediments are coarse material with sand or gravel on the form of sand banks or shoals. Deposits in the outer part of Inner bay are mainly found on flood tidal deltas, and are well sorted fine sediments.

Deep channel phase. Coarse sand or gravel deposited in the narrow pass, and sometimes mixed with fluid mud, fine materials of silt deposited in the deep channels extended to both sides of the narrow pass.

Outer bay phase. Mainly ebb tidal deposits of mixed finer sedimentation, but with storm deposits of sandy lens as its typical features. In the ebb tidal delta at the end of the deep channel in the outer bay, sediments are coarser and may be mixed with local shells or coral fragments.

Outside of the outer bay. Seabed sediments are more homogenous but with storm lenses and less animal trace marks than in the bays.

3. Many other interesting topics can not be introduced in detail because of the limited space. The publication of Modern Sedimentation in coastal and Nearshore zones of China edited by Professor Ren Mei'e, has summarized the major results of Chinese coast research, which is valuable for detailed references (Ren et al., 1985). Given here are only a few supplements for further studies.

1) Sand barriers and coastal dunes are developed along alluvial plain coasts such as the Luanhe River delta and the Hanjiang River delta. The sand deposits are controlled

by monsoon-produced waves, E and SE waves advance towards the coast transversely with limited longshore drifting from the river mouths. They form offshore sand barriers enclosing the river mouth and the coast. Some of the sand barriers are the top parts of sand shoals, which rest uncomfortably upon the bottom sediments. The foreslopes of the barriers are steeper than the lee slopes, and the river sediments are mainly deposited in the shallow water environment protected by the sand barriers. Well-laminated and inter-stratified silt and fine sand are developed leeward of the sand barriers. The fine sandy materials come from the sea, either by the overwash or by tidal currents through the inlets between the barriers. Along the coast in a zone of decreased wave action behind offshore sand barriers, tidal flats develop which overlap the delta plain of the original coastline. With delta progradation, several series of sand barriers can be formed parallel to the coastline. These features reflect the constant directions of monsoon wave dynamics.

Several larger coast dunes, 25～42 m high, are very close to the present seashore, with a beach zone of less than 100 metres wide, they are presently being undermined by wave attack, such as the dune coast along the northern part of Luanhe River delta plain. This refers to different circumstances of air, sea, and coastland interactions. The larger sand dune coast is developed along the seashore with abundant sands; Dry, cold or semi-arid climates with a strong prevailing wind on land, and a gentle submarine coastal slope, i. e. 1 : 1 000 or a medium tide range which allows wide enough exposure of beach zone during ebb tides for wind blowing sands to form the huge sand dunes. A comparative study on the sediments among sand dunes, on the nearshore and offshore seabed and the^{14}C data of the foundation of coast dunes (Gao et al. , 1980), have proved that the huge sand dunes were formed during 8 015±105 a B. P. , the first cold phase and the lower sea level of the Holocene period (Yang and Xie, 1984, Wang and Zhu 1987). This result is also proved by the dune coast along the Mauritanian coast of the Atlantic Ocean.

2) The evidence of coastal tectonic regimes.

There are a number of terraces along the bedrock embayed coasts of China's seas, especially along the coasts of small island.

Due to lack of sedimentary evidence it is often hard to decide whether they are elevated wave cut platforms or surfaces of denudation.

By using a combination of coastal features, such as sea notches, sea benches, sea stacks and sea cliffs it is possible to identify an elevated coast terrace. For example, a series of elevated coast terraces with heights of 5 metres, 15～20 m, 40 m, 60～80 m, 120 m, 200 m, 320 m, 450 m and 600 m around the ancient island of Yuntai Mountain in Jiangsu Province can be distinguished via geomorphologic evidence.

The 120 m high coast terraces can be identified at Zhenshan mountain, Xiamen of

Fujian Province by the combined features of trough-shaped sea notches, rocky bench and sea stacks (Wang, 1983).

Sixty sea notches single or trough-shaped with a few pebbles have been found at heights of 120 m, 200 m and 250 m in the vicinity of Dalian area of the Liaodong Peninsula (Han, 1986).

Elevated sand barriers and lagoon systems have been found in the Southern part of Hainan Island at heights of 10 m, 15 m and 30 m. Three periods of volcanism are documented in the North West Coast of Hainan Island. The first eruptions occurred during late Pliocene to early Pleistocene, forming a basaltic hinterland with some lava covering the lower sand barriers and forming a veneer of contact metamorphic alteration. The second eruption was in the Pleistocene, depositing volcaniclasties with well-preserved ripple beddings. The latest eruptions were in the Holocene with many small breached craters forming volcano coasts and islands in the area of Beibu Bay and the Yingge Sea. The last eruption elevated adjacent shingle beaches by 15 metres and raised the tidal flat and shallow bay deposits to form a 20 m high terrace. The volcanic belts, with tholeiite basalt lithologies, migrate through time from the east to the west. These tectonic events may have developed the taphrogenic fault zone in the Beibu Bay and Yingge Sea. Present barrier coral reefs have developed only on the northwest side of Hainan Island. This perhaps reflects recent seabed volcanic activity (Wang, 1986).

4. "Man Is a Geological Agent" Human impact on morphology and sedimentation of the coastal zone of north China gives sufficient evidence that river and sea form an intergrated system (Ren, 1989). Both conclusions are drawn from a specific style of China coast research on the larger river influences in the natural environment, and the long history of human activity, which directly or indirectly affected the major coastal landforms and sedimentation (Ren, 1989).

1) The coastal plain of North Jiangsu and the associated subaqueous delta of the abandoned Yellow River were the direct result of human diversion of the Yellow River southward for 726 years between 1124—1854 AD. During that time the coastal plain prograded seaward by 30~50 km.

2) The present Yellow River delta is river-dominated and microtidal, but has an accurate form due indirectly to man's cultivation of the loess plateau and the huge sediment load of 1.1 billion tons per year resulting from accelerated erosion.

3) The accumulation rate of the Bohai Sea is also greatly affected by human activity. Detailed analysis of historical data shows that during the last 5 000 years, the Yellow River did not flow into the Bohai Sea for 758 years, and annual sediment input into the Bohai sea reached 1.1 billion tons for only 840 years. This data together with coring and seismic profiling data seems to indicate that the accumulation rate of the Bohai Sea averages about 0.6~0.7 m/ka in the Holocene period (Ren. 1989).

References

[1] Chen Jiyu, Yu Zhiying and Caixing. 1959. Development of landform of the Chang Jiang Delta. *Acta Geographica Sinica*, 25: 201 – 220, in Chinese (with Russian abstract).

[2] Dale Janis, E. 1985. Physical and biological zonation of intertidal flats at Frobisher Bay. N. W. T. In: Abstracts of 14th Arctic Workshop, Dartmouth, Canada.

[3] Ding Xianrong and Zhu Dakui. 1985. Characteristics of tidal flow and development of Geomorphology in tidal creeks. Southern Coast of Jiangsu Province. In: Proceeding of China-West Germany Symposium on engineering.

[4] Evans, G. 1965. Intertidal flat sediments and their environments of deposition in The Wash. *Quat. Jnl. Geol. Soc.* London, 121: 209 – 245.

[5] Gao Shanming, Li Yuanfong, An Feng tong and Li Fengzin. 1980. The formation of sand bars on the Luanhe River Delta and the change of the coastline. *Acta Oceanologica Sinica*, 2(4): 102 – 114.

[6] Han Mukang. 1986. The discovery of High Sea level remnants in the Dalian area. *Acta Oceanologica Sinica*, 3(6): 793 – 795.

[7] Li Shiyu. 1962. Reconnaissances along the west Coast of the Bohai Gulf. *Archeology*, 12: 652 – 657 (in Chinese).

[8] O'Brian, M. P. 1967. Equilibrium flow area of tidal inlets on sandy coasts. In: Proceedings of the 10th Coastal Engineering Conferences, 1. 676 – 686.

[9] O'Brian, M. P. 1980. Comments on Tidal Entrance on sandy coasts. In: Proceedings of the 17th Coastal Engineering Conference Ⅲ. 2504 – 2016.

[10] Ren Mei'e and Zeng Chenkai. 1980. Late Quaternary continental Shelf of East China. *Acta Oceanologica Sinica*, 2(2): 106 – 111.

[11] Ren Mei'e, Cai Aichi, Reng Huaichen, Zhan Zhizheng, Qin Ming, Li Chunchu, Li Zhongsi, Hsia Dongshin and Huang Jinshen. 1985. Modern sediment in coastal and nearshore zone of China. China Ocean Press, Springer-Verlag.

[12] Li Chunchu, Li Zhongsi, Hsia Dongshin and Huang Jinshen. 1985. Modern sedimentation in coastal and nearshore zone of China. China Ocean Press, Springer-Verlag.

[13] Ren Mei'e. 1989. Human impact on the coastal morphology and sedimentation of North China. *Scientia Geographica Sinica*, 9(1): 1 – 7.

[14] Ren Mei'e and Zhang Renshun. 1985. On tidal inlet of China. *Acta Oceanologica Sinica*, 4(3): 423 – 432.

[15] Ren Mei'e, Zhang Renshun and Yang Juhai. 1983. Sedimentation on Tidal Flat of China-with special reference to Wanggang Area, Jiangsu Province. In: Proceedings of International Symposium on Sedimentation on the Continental Shelf, with special reference to the East China Sea, V. 1. 1 – 19.

[16] Ruan Ting. 1988. Types of the marine place deposits in China. *Journal on Ningbo University*, 1 (2): 49 – 62.

[17] Wang Ying, Ren Mei'e and Zhu Dakui. 1986. Sediment supply to the continental shelf by the major river of China. *Journal of the Geological Society*, London, 143: 935 – 944.

[18] Wang Ying. 1980. The coast of China. *Geoscience.* Canada，7(3)：109－113.

[19] Wang Ying and David G. Aubrey. 1987. The characteristics of the China Coastline. *Continental Shelf Research*，7(4)：329－394.

[20] Wang Ying, Schafer C. T. , Smith J. N. 1987. Characteristics of tidal inlets designated for deep water harbour development, Hainan Island. In：Proceedings of Coastal and Port Engineering in Developing countries, V. 1. China Ocean Press. 363－369.

[21] Wang Ying. 1983. The mudflat coast of China. *Canadian Journal of Fishery and Aquatic Science*，40(Supplement 1)：160－171.

[22] Wang Ying, Collins M. B. and Zhu Dakui. A Comparative Study of Open Coast Tidal Flat：The Wash (U. K), Bohai Bay and West Yellow Sea (Mainland China). In：Proceedings of International symposium on the coastal zone. (in press).

[23] Wang Ying. 1964. The Shell Coast ridges and the old coastlines of the West Coast of the Bohai Bay. *Journal of Nanjing University*，8：420－440 (in Chinese with English abstract)

[24] Wang Ying and Zhu Dakui. 1987. An approach on the formation causes of coastal sand dunes. *Journal of Desert Research*，7(3)：29－40.

[25] Wang Ying. 1983. Some problems on the coast morphologic analysis of the sea level changing. *Journal of Nanjing University* (Natural Sciences)，4：745－752.

[26] Wang Ying. 1986. Sedimentary Characteristics of The Coast of Hainan Island，Effects of A Tropical climate and Active Tectonism. In：Abstract of 12th IAS meeting, Canberra.

[27] Yang Huairen and Xie Zhiren. 1984. Sea level changes along the East Coast of China over the last 20 000 years. *Oceanologica ET Limnologia Sinica*，15(15)：1－12.

[28] Yen Kai. 1987. Coastal and Port Engineering in China In：Proceedings of Coastal and Port Engineering in Developing Countries, V. 1. China Ocean Press. 3－31.

[29] Yu Guohua，Bao Shudong and Li Zegang. 1987. Deposition, erosion and control on the silt-muddy beach. In：Proceedings of Coast and Port Engineering in Developing Countries，V. 1. China Ocean Press. 397－403.

[30] Zhang Qiaoming. 1987. Analyses of P-A correlationship of Tidal Inlet along the Coasts of South China，V. 1. China Ocean Press. 412－422.

[31] Zhao Xitao. 1986. Development of Cheniers in China and their Reflection on the coastline shift. *Scientia Geographica Sinica*，4.

[32] Zhu Dakui and Xu Tienguan. 1982. The coast development and exploitation of middle part of Jiangsu Coast. *Acta Sceniarum Naturalium Universitata Nankinesis*，3：799－818 (in Chinese).

The Coastal Environments of China[*]

INTRODUCTION

China coastline extent 18 400 km along mainland from the Yalu River mouth at the China-Korea border to the Beilun River mouth at China-Vietnam border. The total length of coastline, including the more than 6 000 islands, is approximately 32 000 km (Fig. 1).

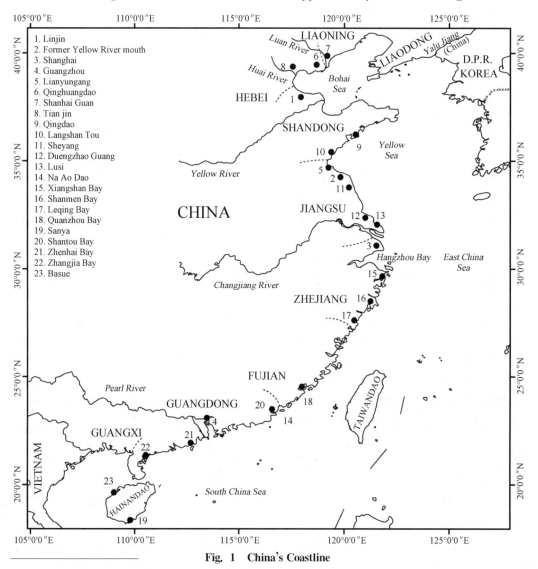

1. Linjin
2. Former Yellow River mouth
3. Shanghai
4. Guangzhou
5. Lianyungang
6. Qinghuangdao
7. Shanhai Guan
8. Tian jin
9. Qingdao
10. Langshan Tou
11. Sheyang
12. Duengzhao Guang
13. Lusi
14. Na Ao Dao
15. Xiangshan Bay
16. Shanmen Bay
17. Leqing Bay
18. Quanzhou Bay
19. Sanya
20. Shantou Bay
21. Zhenhai Bay
22. Zhangjia Bay
23. Basue

Fig. 1   China's Coastline

* Ying Wang：王颖主编,《第五届国际海岸带学术讨论会议论文集(Proceedings of the Fifth MICE Symposium for Asia and the Pacific：Ecosystem and Environment of Tidal Flat Coast Effected by Human Being's Activities)》1－7 页,南京大学出版社,1990 年 10 月。

The coastal processes and evolution of China is mainly controlled by geological structure, larger river influences, monsoon wind waves, and the tidal process of the Pacific margin seas.

GEOLOGICAL STRUCTURE

A series of NE or NNE trending uplifted and depressed belts that intersect obliquely from continental shelf to the coastline. From northwest to the southeast, these are the Bohai Basin, the Shandong-Liaodong uplift, the Southern Yellow Sea depressed belt, the Zhejiang-Fujian uplift, the East China Sea depressed belt and Taiwan fold belt (Fig. 2). The distribution of coast morphological types is controlled by geological and tectonic setting, particularly zones of uplift and subsidence. Such as, a bedrock-embayed coast with rock islands is formed along the uplift belts, while a plain coast, usually with a large river, is always formed along the depressed belts. In the South China Sea, NE trending block faults define the main trend of the bedrock embayed coast, with inlets developed along NW trending fault.

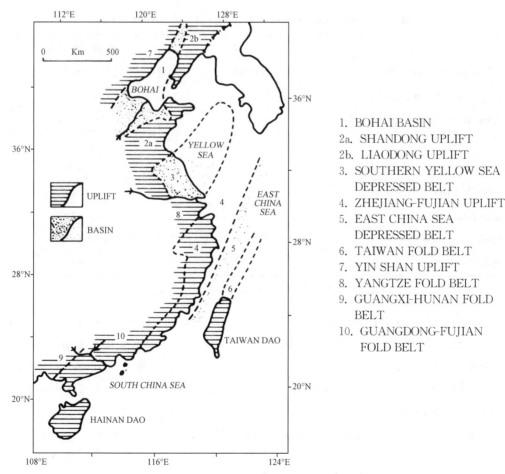

1. BOHAI BASIN
2a. SHANDONG UPLIFT
2b. LIAODONG UPLIFT
3. SOUTHERN YELLOW SEA DEPRESSED BELT
4. ZHEJIANG-FUJIAN UPLIFT
5. EAST CHINA SEA DEPRESSED BELT
6. TAIWAN FOLD BELT
7. YIN SHAN UPLIFT
8. YANGTZE FOLD BELT
9. GUANGXI-HUNAN FOLD BELT
10. GUANGDONG-FUJIAN FOLD BELT

Fig. 2　Basic tectonic map of the mainland coast of China

RIVERS

There are more than 1 500 rivers enter the China Seas, river carry large volumes of fresh water and sediment to the coast area of China. According to modern data, a total of more than two billion tons of fluvial sediment per year are discharged into China Seas (Table. 1). That means a total of 1 km^3 of sediment is carried by rivers to the adjacent continental shelf of China Seas annually. The Yellow River has largest sediment discharge of 11.9×10^8 ton as the annual average and the maximum is 21×10^8 ton. Per cubic meter of water with 24.7 kg of sediment as an average value, the maximum value is about 37.7 kg/l. A great quantity of sediment is transported to the Bohai Sea quantity of sediment is transported to the Bohai Sea continuously. The total area of the Bohai Sea is 78 000 km^2, with an average water depth of 18 m, and the annual sediment discharge in the sea is 16.8 kg/m^2, which produces a sedimentation rate of 8 mm per year. At this rate, Bohai Sea would be completely filled in 2 250 years, if the basin subsidence is considerably little.

Table 1　Annual water-sediment discharges of major rivers in China

River	Water（m^3）	Sediment（tons）
Yellow River（Huang He）	49×10^9	1.2×10^9
Yangtze River（Chang Jiang）	930×10^9	4.9×10^8
Pearl River（Zhu Jiang）	370×10^9	1.0×10^8
Yalu River	28×10^9	4.8×10^6
Liao River	9×10^9	20×10^6
Luan He River	4.9×10^9	24×10^6
Qian-tang Jiang River	35×10^9	5×10^6
Mien Jiang River	62×10^9	8×10^6
Han Jiang River	26×10^9	8×10^6

（Sources: Cheng and Zhao, 1985; Wang and Aubrey, 1987）

River influence on the coast is not limited to the vicinity of the mouth, alluvial sediment is also redistributed along the plains coastline, where it is a major factor in coastal development. For an example: most of 2 000 km mudflat coast is developed in the Bohai Sea, where modern Yellow River discharges, and on its pre‑1855 delta 600 km to the south on the Yellow Sea coast. Besides, the two major rivers, Yellow River and Chang Jiang River (Yangtze) have migrated great distances during the past 4 000 years. The present position of north Jiangsu coastal plain was formed by two rivers, even the continental shelf of East China Sea was a huge alluvial plain during the period of last low sea level.

COASTAL PROCESSES

The monsoon wind pattern of Asia is a major influence on China's coastal processes. In the winter, the Mongolian high-pressure system dominates the geostrophic winds. The anticyclonic gyre, beginning in January, yields northwesterly winds in the north, and northerly or northeasterly winds in the South China and East China Seas. During the summer, the Indian Ocean low-pressure system dominate, creating a cyclonic gyre. These winds are southwesterly in the South China Sea (south of latitude 15°N), changing northward to southeastly winds in the East China, Yellow, and Bohai Seas. A transitional season exists during which winds fluctuate between the two dominant systems (Wang and Aubrey 1987).

Typhoons are tropical cyclones exceeding 32 m/s in wind speed and creating large waves as well as their associated storm surges. Typhoons normally develop in the northwestern Pacific, moving west or westnorthwest across the Philippines and South China Sea, dissipating rapidly when they reach the Asian continental during the period of July to October. Typhoon frequency is higher for the South China Sea than for the other marginal seas of China (Sun et al. 1981). Annual number of typhoons generated in the northwest Pacific is 28, with considerable year-to-year variability. Of the total number of cyclones generated, only a small number reach the South China Sea, and there are four major tracklines. In September and October, 40% of all typhoons travel over Hainan Island on into Vietnam. In March-June, 22% of the annual typhoons travel north into Guangdong Province or Fujian Province. In spring and summer, another 20% of the total annual number of typhoons passes to the north-northeast, over Taiwan, or towards Japan. Finally, in winter, 10% of the typhoons pass to the west into Vietnam, missing in the China coast (Sun et al. 1981, Wang and Aubrey 1987). In the period of 1949—1969, the China coast experienced 77 strong typhoons (winds exceeding 32 m/s), 61 weak typhoons (winds of $17 \sim 32$ m/s) and 65 tropical storms (winds from 11 to 17 m/s).

Waves rarely have been measured during typhoons, quoting nearshore waves exceeding 20 m in high, and 5 m high wave recorded during typhoon storm in the area of Sanya harbour, Hainan Island. The recent failure of a drilling rig in the South China Sea during a storm accentuates the need to carefully consider typhoon waves.

Wave patterns along most of China are dominated by monsoon although there are exceptions. Northerly waves prevail in the winter, their influence moving progressively southwards from the Yellow Sea, reaching Taiwan in September, latitude 10°N in October, and spanning the entire coast by November. In January, the wave directions rotate clockwise, with northwesterly wave in Bohai gradually swinging northerly to northeasterly in the East China and South China Seas (Wang, 1980). Southerly waves prevail in the summer, moving northward from the South China Sea. They appear first in February, and by May they dominate a wide area to the south of latitude 5°N (Wang, 1980). In June, southwesterly and southerly waves prevail in the South China Sea. In the strait of Taiwan, northeasterly and southwesterly waves become equally frequent, while southerly waves prevail in the East China Sea and southern Yellow Sea. The northern Yellow and Bohai Seas are dominated by southeastly waves in June. In July, southerly waves prevail along the entire seaboard. Wave directions change anticlockwise, from southerly in the South China Sea (south of latitude 15°N), through southerly to southeasterly in the East China, Yellow and Bohai Seas. During the periods of transition for the monsoons, wind directions fluctuate and there are no prevailing waves.

Tides in the China Seas are variable in type and amplitude as a result of the tides of the Pacific Ocean interacting with the coastline and submarine topography of the continental shelf. In the Bohai, Yellow and East China Seas, tides are semidiurnal, while the tides in the South China Sea are either diurnal or semidiurnal, having considerable geographic variability. Tidal ranges in the Bohai Sea averages about 3 m except near the Yellow River mouth and Qinhuangdao, where tidal ranges are less than 1 m. Tides of the Yellow and East China Seas are with a large variability in range. Near the Yalu River mouth, tidal range averages 4.5 m; near the abandoned Yellow River mouth, tidal range is 1.6 m, and 2 m for most of Jiangsu Province coast, while near Shanghai the range is 1.9 m. Largest tide range appears in the East China Sea. It is 6 to 9 m in the Hangzhou Bay, and more than 8 m in the Leqing Bay and other embayments along the coast of Zhejiang province. Tidal bores occur in many estuaries with strong tidal currents. Over one-third of the coastline of north China is made up of tidal flats. Tides in the South China Sea vary from diurnal to semidiurnal, with ranges is smaller, normally 1~2 m.

The major coastal types of China:

1. Bedrock-embayed coast

The coast is formed along the uplift belts where the mountains meet the sea, thus the coastal slopes are steep, wave energy is high and the coast lines are irregular with many headlands, small bays and islands. The coastal morphology is related to wave action, lithology of coast rocks, and sediment supply of the coastal zone. This type of coast is found along the Liaodong Peninsula, Shandong Peninsula in the north China. The Zhejiang, Fujian, Taiwan, Guangdong and Guangxi Provinces in the South China.

The bedrock embayed coast can be subdivided in four subtypes according to the stage of development (Wang 1980).

（1）Marine erosional-embayed coast

Developed on hard crystalline rock where the rate of erosion is slow, and the coast features are modified slowly since the postglacial rise of sea level. Sparse sediment supply limits coastal deposition, leaving a dominance of erosional geomorphic features. This type of coast is suitable for setting up of sea harbour, such as Dalian harbour on the Liaodong Peninsula.

（2）Marine erosional-depositional type. Most commonly developed where mesozoic granites are overlain by weathered deposits. Thus there are plenty sediment supply along the coast zone either, by wave erosion or by river input, as a result, both erosional features （cliff, beach etc.） and depositional features （sand spit, bay bars and tombolo） are developed well along the coast, such as the coast on Shandong Peninsula.

（3）Marine depositional type, with relatively erodable bedrock, marine erosional supplies large quantities of sediment for prograding barrier shoreline which enclosed the original irregular bedrock embayed coastline. This type of coast is common in South China.

（4）Tidal inlet-embayed coast. Common in South China. The major NE-SW trend of the coastline is cut by long, narrow bays following NW trending fault, small islands or sand barrier protect the mouths of bays forming tidal inlets. Because of the larger water area of the bay and powerful tidal currents, these tidal inlets are idea set for sea harbours.

2. Plains coast

Based on genesis, the plains coast can be subdivided into two types.

（1）The alluvial plains coast. Located generally seaward of mountain ranges, plains are built of fluvial sediments with a seaward gradients of $1:100$ to $1:1\,000$, resulting in a wide break zone and active sediment movement on shore or long shore. Barrier bars, sand spits and extensive beaches are the typical features of this coast.

（2）Marine depositional plains coast. Flat and very extensive, the plains are located on the lower parts of large rivers in the area of subsiding basement. Coast slopes are extremely gentle （$1:1\,000$ to $1:5\,000$）, leading to wave breaking well offshore; the nearshore is dominant by tidal processes. Sediment from rivers is deposited either offshore or landward to build up the extensive tidal flats.

The plains coast represents a dynamic balance between wave erosion and coastal sediment supply, which is mainly from large rivers. If the amount of the sediment supplied is larger than the amount erode, the coastline progrades; if the sediment supply is reduced or stopped, wave erosion causes the coastline to retreat. For example, the west coast of Bohai Bay, which receive the sediment at Yellow River, is prograding despite a steady regional subsidence since the late mesozoic. During the time of 1128 to 1855 A. D. when the Yellow River entered the Yellow Sea in the South, the coastline of

Bohai Bay retreated, due to the lack of sediment but the north Jiangsu coast along the Yellow Sea prograded fast. However, since 1855, the Yellow River back to Bohai Sea, the coastline of abandoned Yellow River Mouth in north Jiangsu has retreated about 17 km (Ren et al 1983), with coastal retreat presently exceeding approximately 30 m/y along the 150 km coast reach.

3. The river mouth coast

The Yellow River carries the largest sediment load of any river in the world. Sixty-four percent of the sediment is deposited on the delta and mudflat, while the remaining 36% is transported into Bohai Sea. A fan shaped of delta has been formed since 1855, the delta has covered a distance of 160km along shore near Lijing, pograding seaward 20~28 km. This progradation represents an average annual accumulation of 23.5 km^2 in plain view (Wang and Aubrey 1987).

The Changjiang River (Yangtze) mouth faces the Open East China Sea, wave action at the river mouth is stronger than that at the Yellow River. Most of the sediment accumulates in the miner part of the river mouth to form a series of sand banks and islands. There is a submerged ancient river channel which starts at some distance from the modern river mouth and extends the continental shelf reaching to the Okinawa Trough. Thus, the modern Changjiang River delta is rather small, and is growing towards the southeast in a bird beak-shaped protrusions.

The Pearl River delta in the South China Sea is a transitional pattern from an estuarine to a delta, sediment accumulated in the original bedrock bay sheltered by islands.

The Hangzhou Bay is an estuary with tidal range of 6 to 9 metres, depending on the shape of the coastline and the submarine topography, there is a distinct tidal bore appeared in the Ganpu.

4. Biogenic coast-coral reef and mangrove coast

Most of the South China Sea islands (Xisha and Nansha islands) are atolls which are still growing, even though, the bedrock basements of reefs have been subsidence more than 1 000 metres since Tertiary. Fringing reef are found surrounding Hainan Island, Taiwan and other small island, and at the Beihai Peninsula. The reefs start growing during Holocene time.

Mangroves are mainly distributed in the south of latitude 27°N, but only few species of mangrove bushes grow up in the north to 24°N. There are 20 species of mangrove in China, and the mangroves belong to orient assemblage. Mostly, the mangrove swamps grow intermittently at river mouths, bay head areas and in the lagoons. Because of the local people cuts off the mangrove woods, which is widely to destroy the forests and natural ecological balance. Recently, four of the mangrove protection parks have been founded in China. One is in the Xiamen area on the mainland, three are in the Hainan Island. It hopes to set up good examples for showing the natural beauties of mangrove swamps and its profits.

China had carried out the comprehensive surveys along whole coast zone during 1980 to 1986, and has worked out plans on full scale exploitation of marine resources and the coastal environments. Attention has been paid especially on the island development for recent three years. All the steps will also improve the scientific research on the coast and adjacent continental shelf of China.

REFERENCES

[1] Cheng T. W. and C. N. Zhao. 1985. Water and Sediment Discharge of Chinese Major Rivers and Its Influence to Coast Zone. *Acta Oceanologica Sinica*, 7(4): 460 – 471.

[2] Ren M-E, R. S. Zhang and Yang J. H. 1983. Sedimentation on Tidal Mudflat of China-With Special Reference to Wanggang Area, Jiangsu Province. In: Proceedings of SSCS, V. 1. China Ocean Press. 1 – 19.

[3] Sun H. , J. Yao, Y. Huang, B. Long, B. Xu and H. Teng. 1981. Summary of Hydrography and Meteorology of the China Coastline. Science Press, Beijing.

[4] Wang Y. 1980. The Coast of China. *Geoscience*, *Canada*, 7:109 – 113.

[5] Wang Y, M. -E. Ren and D. K. Zhu. 1986. Sediment Supply to the Continental Shelf by the Major River of China. *Journal of the Geological Society*, *London*, 143: 935 – 944.

[6] Wang Y. and D. G. Aubrey. 1987. The Characteristics of the China Coastline. *Continental Shelf Research Shelf Research*, 7(4): 329 – 349.

Several Aspects of Human Impact on
the Coastal Environment in China
Examples: Through the River-Sea System[*]

32 000 km long China's coastline is located in the east side of Asia and adjacent the Pacific Ocean and its margin seas. The coastal zone extend across three climate zones of temperate, subtropical and tropical zone, with the effects of monsoon winds, tidal waves and larger river sediment supply. The coasts of China vary in the environmental natures, resources abundance and evolutional history.

Coastal zone in China was forbidden and wild area during mostly historical time as "the blocked country" policy by the feudal dynasty. Even though the fishery and salt production were also controlled by feudal government. Only during this century, coastal zone has been populated, and speeds up the development since late 1970's as a result of present government open policy. It becomes China's golden coast in 1990's.

Along China coastal area, there are nine provinces and two cities controlled by Central Government. The coastal land is 14% of China's total area, and it has centralized 41% of China's population, 56% of total output value and 70% of larger cities. Recently, four large city groups are formed as: around Bohai Sea cities with Tianjing as centre, the Changjiang River delta group with Shanghai as its centre, Southern Mingjiang River delta group with Xiamen as its centre, and the Zhujiang(Pearl River) delta group with Guangzhou as its centre. The four group includes 14 open harbour cities, three economic special zones and five seashore open zones (Fig. 1, from China's Daily Overseas Press). The four groups absorb more than 80% of total foreign investment in China. The east side of China along sea side is characterized by rapid developing economy and heavy population with 386 persons/km² as its average. However, the wild land, tidal flats and embayments of coastal zone are valuable resources for China developing.

In the past, human impact on the coast environment mainly through river-sea system, as several larger rivers across larger area of China, from the west to the east, which brings great influences on the China's coastal zone. Such as the Yellow River

* Ying Wang, Chendong Ge: *Proceedings of 1993 PACON CHINA SYMPOSIUM: Estuarine and Coastal Processes*, pp. 596 – 603, James Cook University Publisher, North Queensland, Townsville, Australia, 1994.

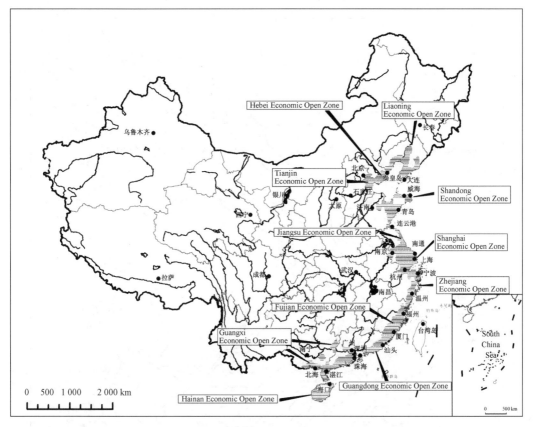

Fig. 1　Coastal Economic Open Zone of China

(Huang He) is unique for its extremely heavy sediment load (11.9×10^8 tons y^{-1}) inputed to sea, rather small water discharge (485×10^8 $m^3 y^{-1}$), and migrated river channel in the lower reaches (Wang et al, 1986). Silts are the main component of sediment, and have been carried out by the Yellow River from the loess plateau mainly since west Han Dynasty (B.C. 2 century, Shi, 1981) as immigrant cultivation which destroyed forests and pastures, which caused serious soil erosion, and huge quantity of sediments have been carried out to sea to form great North China plain. As a result, it changes the coastal water environment both in the Yellow Sea and the Bohai Sea. Because of the Yellow River has experienced eight major changes in its lower course since 2278 B.C., discharging either into the Bohai Sea, or into the Yellow Sea via the Huai River(淮河)*. The shifted distances between north and south were more than 600 km long (Wang, 1983, Fig. 2). Two major changes of the river channel migration were created artificially as a result of human being's activities for military purpose.

---

*　River name in Chinese, added by author in Nov. 2016.

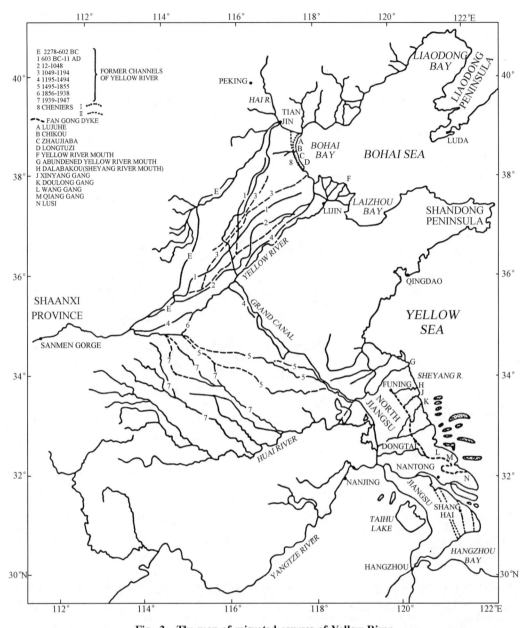

Fig. 2 The map of migrated courses of Yellow River

（1）In November 15th of A. D. 1128，Du Chun，the official of Capital City（Kai Feng 开封）of the Southern Song Dynasty，excavated the Yellow River bank at "Li Gu Ferry"（Hua County of Henan Province）in order to use flood water as a weapon to prevent an invasion of Jin Dynasty soldiers from the North（Song history，Jin history，Ren，1989）. Even though this effort was unsuccessful，it initiated a change that caused the Yellow River to flow through the Si River（泗河），one of the distributions of the Huai River. Since then，the Yellow River shifted often in the Huai River area as a disaster to local people，and built up the North Jiangsu coast plain，more than 40 km wide, and its associated subaqueous delta. Until 1855, the Yellow River shifted back to the present location in the Southern Bohai Bay, major shore line accretion has occurred during the past 130 years, the delta building seaward 20~28 km in this period and with larger area of mud tidal flat developed. The North Jiangsu coastline around abandoned Yellow River Delta area has retreated about 17 km since 1855 as lost the huge quantity of river sediment supply, the coast feature is different with silt flats and shell beach ridges developing along high tidal level shoreline.

(2) May 1938, the military committee of nationalist ordered the Chinese Army to excavate the Yellow River bank between Zhenzhou City and Zhuangmu County to force the river flow to the southern, in an effort to stop the Japanese Army from advancing into the western and southern parts of China. After twice explosion during June 2nd to 9th, 1938, the river rushed out of Channel in the place named Hua Yuan Kou(花园口) of Henan Province, and resulting in flood developed rapidly. It flooded 44 counties in the three province (Fig. 3). 12.5 million people died as a consequence of that operation (The Water Conservancy Committee of the Yellow River, 1982). The Yellow River returned to its original channel in 1947, the flood area was 54 000 km² and completely changed from one of fertilized farm lands to wild bedland with sand dune field and saline alkali soil depressions. The natural river system of the Huai River was destroyed and

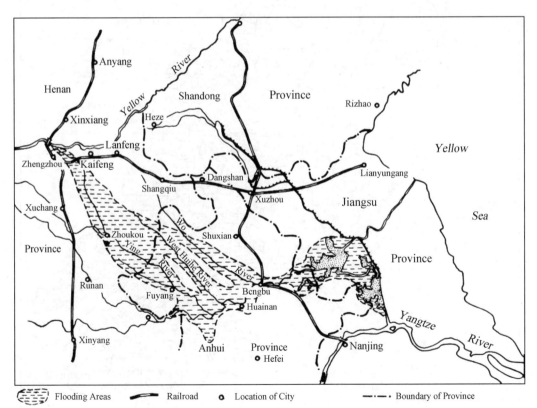

Fig. 3 **The flooded area of Yellow River in 1938 diversion**

the whole area suffered from disasters of sand storm, mobile dunes, drought and water logging often. Local people lost their houses and farm land to be refugees to the neighbour provinces for several decades. It has taken 40 years to re-establish conditions suitable for farming, such as, to set up drainage system and irrigation network, to plant forest and grasses for fixing the dune sands etc. Present time, the abandoned Yellow River area becomes one of most productive wine fields and fruit farms in the North China. However, the natural environment has been changed completely since the military operation in 1938. Using flooded Yellow River to repel an invasion had been repeated twice in Chinese history, and also failed twice, but has had influences on environment lasting for a long time plus the loss of huge amount of properties and lives of the native. It is better to be smart to learn from the past mistakes for avoiding the future ones.

North Jiangsu plain had been previously divided by numerous canals and channel networks in the early 1950's as a strategy for frontier defense. This project changed the natural environment (Fig. 4) with the positive effect of bringing more farmland under

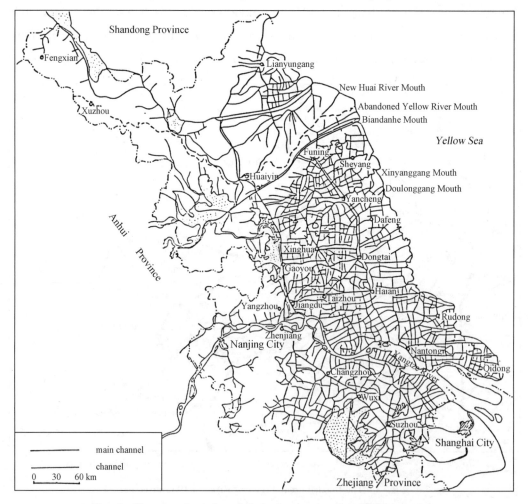

Fig. 4 The map of canals and channel networks of north Jiangsu plain

irrigation. Today, the North Jiangsu plain is one of the major grain producing area in China as a result of more than 40 years peaceful period since then. However, the human activity related to this event occurred over a very short period, but the environment impact of this activity would still last for a long time.

In present, human impact on the coast zone of China has been more and more effective as the large scale of economic development on coastal zone. Still, human impacts on the coastal environment through the river-sea system appear more and more often and seriously, especially in the China's coastal zone with great influences from numerous rivers.

1. To dam up river mouth for preventing salt water intruding and preserving fresh water resources along plain coast area. Then, the construction causes serious under-dam siltation in the lower part of river mouth. The dams also have blocked fish, prawn and crabs breeding up stream from river mouth. As a result, local people lose economic income from the decreasing precious sea food, such as silver fish in river mouth Jiyunhe (蓟运河) of Bohai Bay, and crabs along North Jiangsu coasts of west Yellow Sea. The silted river mouth channels become smaller and shallower day by day, gradually to lose navigable water depth, and suffering flood disasters often such as the situation in the Changjiang River delta during flooded event of 1991 Summer in China.

2. To divert river discharge for urban water supply, because decreasing sediment discharge at same time, as a result, to cause the coast erosion at the delta area as lost the sedimentary dynamic balance. Such as in the Luanhe River, original water discharge was 4.19×10^9 m^3 with total sediment supply of 1.066×10^7 tons. After diverting the water discharge of 3.55×10^8 m^3, the delta coastline has been retreated in a rate of 20 m/yr since 1980 (Qian, 1991, theses).

3. Using river as a natural channel to flush out chemical wastes of industry, through irrigation of polluted water which damaged corps. Such as in the area of Jiyunhe River during Spring of 1974, total of 2 700 hectares wheat fields came to harm by toxic river. Besides, the rich nutritious water and sediment entering the sea, it causes serious red tide in the coast zone. For an example in the summer of 1978, red tide happened in the Dagukou(大沽口) area of Bohai Sea few days after a heavy rain thus rapid run off brought huge quantity of heavy metal and chemical waste to the sea. The red tide almost killed aquiculture of prawn. Even though, there are still many cities in China using coast land as garbage field. Such as Shanghai in the summer season, the daily garbage reaches 10 000 tonnes, which is a new pollution source of COD to the coastal area.

4. Over exploitation of under ground water in the coastal zone. Such as in the Laizhou(莱州) and Longkou(龙口) county of Shangdong Province, under ground water table subsidence $15 \sim 30$ m in the late 1980's compare with 1977, thus destroyed the natural water balance between fresh water and salt one, then salt water intrusion are

more than 100 km. Thus, people lack of good quality water for drinking, for irrigation and industry using, as a result, it causes larger area of farm land to be abandoned.

Over exploitation of under ground water and heavy load of construction on the coastal low land and delta area has caused rapid ground subsidence to accelerate local sea level rising, such as: 0. 5 m subsidence has been recorded during 1966 to 1985 in the Tanggu(塘沽) Harbour of the Haihe River(海河) delta according to the State Survey and Mapping Bureau data in 1992. Relative sea level rising during the same period was 2 mm/y in the Pearl River delta; 11. 5 mm/y at Changjiang River delta, and 24. 7 mm/y at abandoned Yellow River delta according to recent study by Ren, Mei-e et al (Ren, 1993).

Sea level rising and the decreasing of river sediment supply, plus the human being's impacts directly through the beach sand mining. It has speed up the sandy beach erosion in the rate of 1～3 m/yr as its average in the most parts of China's coastline.

5. Reclamation of tidal inlet embayment has decreased tidal prims and changed the natural flushing pattern. Thus, increasing siltation on the navigation channel of the deep water harbour. For an example, Sanya Harbour in the south of Hainan Island, because human being's reclamation, the water area of tidal inlet embayment has been decreased 2/3 from 4 140 000 m^2 to 1 393 300 m^2, and the tidal prism decreased 51. 6% from 4 906 800 m^3 to 2 531 340 m^3. The harbour suffers siltation from almost no siltation in the channel ten years before to present: the channel needs to be deepened annually.

River-sea system is a natural cycle to keep the environment balance between air and water, land and ocean. Even though human impacts on land, it is still directly or indirectly to add the influences on the coastal water through the river connection.

It is urgent to set up and improve a systematic regulation and management of coast zone preserving natural environment balance for longer period of fast development in the coast area.

REFERENCE

Shi, Nianhai. 1981. The distribution and migration of farm land, pasture and forests of the loess plateau. *Historical Geography*, 1: 21 – 31 (All in Chinese).

Wang, Ying. 1983. The mudflat coast of China. *Canadian Journal of Fisheries and Aquatic Science*, 40: 160 – 171.

Song History. Original Records of Gao Zhong. (in Chinese).

Jin History. Geographical Records. (in Chinese).

Ren, Mei-e. 1989. Human Impact on the coastal morphology and sedimentation of North China. *Scientia Geograpgica Sinica*, 9(1): 1 – 7 (in Chinese).

The Water Conservancy Committee of Yellow River. 1982. The summary of water conservancy history of

Yellow River. Water Conservancy Press.

Qian, Chunlin. 1991. Effects of Water Conservancy projects to the Luanhe River Delta in the Luanhe River Basin (theses in Chinese).

Ren, Mei-e. 1993. Coastal lowland vulnerable to sea level rise in China. In: Proceedings of PACON, 93.

Sea Level Rising and Coastal Response[*]

1. Sea Level Changes in the Past 100 Years

The early systematic treatment of recent sea level changes in China was by Emery and You (Emery, K. O. and You, Fanghu, 1981), their study was based on tide-gauge records of only 8 stations. A later paper by Emery and Aubrey (Emery, K. O- and Aubrey, D. G. , 1986) discussed relative sea-level changes in China by using both simple regression analysis and eigen analysis. But insufficient tidal gauge records, and incomplete analysis on local environmental conditions. The same holds true for works of some Chinese scientists (e. g. Wang, Z. H. , 1986), and detailed studies in recent years have improved our knowledge of sea-level changes in Shanghai (Chen, X. Q. , 1991, Wang, B. C. et al, 1991). Since 1990's more comprehensive and more reliable accounts of relative sea level changes for whole China have been published one after others (Ren, Mei-e, 1993; Shi, Yafeng et al, 1990).

It is generally estimated that global sea level rise over the past 100 years is 1～2 mm/yr (UNESCO, IGBP etc). In China, according to State Bureau of Geodetic Survey, the rate of sea level rise in the last several decades is 2～3 mm/yr. The mean rate of sea level rise is 1. 4 mm/yr, and it is continuously to rise in the near future (State Bureau of Geodetic Survey of China, 7th July, 1992). Scientists have studied survey data from 9 stations along China Seas, it shows that coastal sea level has risen 19 cm in the East China Sea and 20 cm in the South China Sea during recent hundred years. The result of calculating 102 station data of tide gauge in the world are as following: the average sea level rise is 15 cm during last of sea level in the Pacific, and 39. 6 cm rising in the Indian Ocean as the maximum one.

Analysis of tide gauge data of 32 stations in China shows that over the last 30～80 years relative sea level in 20 stations is rising but that in 12 stations is falling (Ren, Mei-e, 1993) (Fig. 1, Table 1). Stations with large rate of relative sea level rise are located in the lowland plain or delta area, such as Lusi (吕泗)[**] in the north Jiangsu

---

　*　Ying Wang and Xiaogen Wu: *Canada Coastal Zone'94 "Cooperation in the Coastal Zone"*, *Conference Proceeding*, 1994, Vol. 4, pp. 1831 - 1840.

　修订后的摘要载于 1995 年 12 月 23～25 日在北京举办的 IGBP 科学研讨会和第四届科学咨询会联合会议摘要集——*Natural and Anthropogenic Changes: Impacts on Global Biogeochemical Cycles——GLOBAL CHANGE*, *Book of Abstracts*, pp. 37。

　**　place name in Chinese added by author in Nov. 2016.

coastal plain along the Yellow Sea，Tanggu(塘沽) of Tianjin in the Haihe river mouth where was the Old Yellow River Delta，Wusong(吴淞) Station of Shanghai in the Yangtze River Delta. Here high rate of relative sea level rise is chiefly due to local land subsidence. Stations with falling sea level are located in uplifting bedrock hilly coasts such as Qinhuangdao，Yantai，Qingdao，Lianyungang and especially Fugang on behalf of the east coast of Taiwan Island (Liu，C. C.，1989).

Fig. 1　The location of tidal gauge station in China (After Ren，Mei-e，1993)

TABLE 1　Relative sea-level changes at major tide-gauge station in China during the last 20～80 years

No.	Station	Period	Rise(＋) or Fall(－)	Rate(mm/a)
1	Dalian(大连)	1970—1989	＋	3.07
2	Huludao(葫芦岛)	1960—1989	＋	1.86
3	Qinhuandao(秦皇岛)	1960—1989	－	1.95
4	Tanggu(塘沽)			

（Continued）

No.	Station	Period	Rise(+) or Fall(−)	Rate(mm/a)
4(1)	Beipaotai(北炮台)	1910—1963	+	1.77
4(2)	Beipaotai	1937—1953	−	0.43
4(3)	6 meter(6 米)	1959—1985	+	1.20
5	Yangjiaogou(羊角沟)	1952—1986	−	1.20
6	Longkou(龙口)	1961—1989	−	1.31
7	Yantai(烟台)	1960—1989	−	3.39
8	Qingdao(青岛)	1952—1985	−	1.07
9	Shijiusuo(石臼所)	1968—1989	−	0.85
10	Lianyungang(连云港)	1963—1985	−	1.86
11	Lusi(吕四)	1969—1989	+	8.72
12	Wusong(吴淞)	1912—1936	+	2.50
		1957—1983	+	3.00
13	Changyu(长屿)	1960—1989	+	2.42
14	Kanmen(坎门)	1960—1961	+	2.10
		1963—1989	+	2.10
15	Sansha(三沙)	1973—1989	+	1.10
16	Pintan(平潭)	1967—1989	+	1.77
17	Xiamen(厦门)	1958—1989	+	2.31
18	Dongshan(东山)	1960—1989	+	0.24
19	North Point, Hong Kong（香港北角）	1962—1988	+	2.60
20	Taipokau, Hong Kong	1970—1988	−	5.00
21	Chiwan(赤湾)	1971—1985	−	2.00
22	Sanpanzhou(汕板洲)	1967—1984	+	0.39
23	Huangpu(黄浦)	1967—1985	−	0.69
24	Macao(澳门)	1925—1973	−	0.09
25	Sanzao(三灶)	1966—1985	+	3.28
26	Zhaipo(乍浦)	1959—1989	+	2.60
27	Weizhou(涠洲)	1960—1989	+	6.09
28	Haikou(海口)	1976—1989	+	3.85
29	Keelung(基隆)		+	6.00
30	Taixi(太溪)		+	27.80
31	Hengchun(恒春)		+	0.20
32	Fugung(福岗)		−	24.50

（After Ren, Mei-e, 1993）

As a whole, the eustatic sea level rise along China coastal zone is in the rate of 1.4 to. 1.7 mm/yr rising, after correction on the rate of vertical movement, and continuously rising from south to north recently with the general trend that sea level is gradually higher from north to south (Zhou, Tian Hua et al, 1992)(Fig. 2).

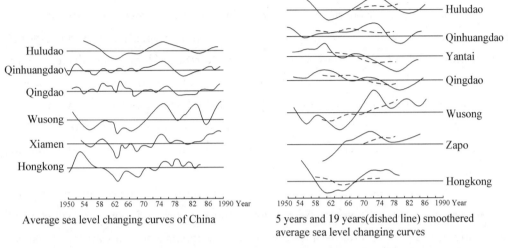

Average sea level changing curves of China

5 years and 19 years(dished line) smoothered average sea level changing curves

Fig. 2　Average sea level changing curves of China(After Zhou, Tianhua et al, 1992)

2. Relative Sea-level Rise Over the Next Century

In estimating global sea level rise over the next century, the best estimate from Intergovernmental Panel on climate change (IPCC,1992) under Scenario A. Business as usual is adopted which gives a projected sea level of 18 cm (4.5 mm/yr) in 2030 and 66 cm (6.0 mm/yr) in 2100 with high estimate at 110 cm and low estimate at 31 cm. However, there are considerable uncertainties in these estimates. First, future sea level rise is strongly dependant on future rise of global mean air temperature which, in turn, is a factor of the magnitude of future increase of concentrations of greenhouse gases, particularly CO_2 from the energy sector span a broad range of futures. According to 1992 IPCC estimate, the difference between the highest and lowest estimate for 2100 is about 7 times (3 billion tons carbon versus 5 billion TC) and that for 2030 is about 2 times (14 billion TC versus 7 billion TC). As CO_2 emissions are chiefly of anthropogenic origin, their magnitude depends strongly on socio-economic factors, such as population and economic growth, changes in economic structures, energy prices government policies etc. which change greatly through time and are difficult to predict with precision, even more difficult for modelling. Another major uncertainty is stability of the west Antarctic Ice Sheet, the world's only remaining marine based ice sheet which survived the last deglaciation. Owing to its large volume, if disintegrated and melted, it would raise the global sea level about 6 m. However, the relative stability/instability of the west Antarctic Ice Sheet in the last interglacial and in the future global warming is still poorly known and there is yet no conclusive answer on this important issue.

Third, recent improved knowledge of the emission and behaviour of greenhouse gases may modify previous estimate of rate of global warming. For example, global emissions of methane from rice paddies may be less than previously estimated. The cooling effect of aerosols (airborne particles) from sulphur emissions may have effect a significant part of the greenhouse warming in the Northern Hemisphere during the past several decades. Therefore, if sulphur emissions continue to increase, the previous global warming rate is likely to be reduced and with this, also global sea level rise rate.

Under these circumstances, it seems prudent to adopt a short-term (2030) prediction on global sea level rise. The reason of using the best estimate under "business as usual" scenario is:

(1) "Business as usual" scenario will be essentially true to the condition in the near future (say next $20 \sim 40$ yrs). As the greenhouse gases emissions are substantially increasing in the past several decades and many developing countries are most likely to continue to rely coal as their major source of energy, there is no reason to predict that the emission of CO_2 would not be continuously increasing in the near future.

(2) The best estimate is generally regarded as the most likely to occur. It has already been used in the construction of some major coastal defense works. For example, in the Netherlands, a projected sea level rise of 60 cm over the next century has been added in the design and construction of coastal defense works of the Delta Project. Thus the best estimate is also a more realistic estimate.

In predicting relative sea level rise in China in 2030, we use 1992 IPCC best estimate, 4.5 mm/yr, as the background value. Estimates of local subsidence in 2030 are from the latest figures provided by the local authorities and also from a careful consideration of local environmental conditions and current subsidence rates. Owing to rapidly changing socio-economic conditions, the predictions of local subsidence as well as local relative sea level rise must be regarded as tentative. Projected relative sea level rise in 2030 for major Chinese deltas is shown in Table 2 (Ren, 1993).

The future relative sea level rise should be discussed in some detail because it is the most important parameter in any model for assessing coastal risks. From the Table 2, it can be readily seen that rate of the projected relative sea level rise in 2030 for Chinese deltas and coastal lowlands is much higher than global sea level rise, being 3 times higher in Old Yellow River Delta and nearly 2 times higher in Yangtze River Delta. Therefore, their potential coastal risks from future sea level rise are great.

TABLE 2　Relative Sea Level Rise in Old Yellow River Delta, Modern Yellow River Delta, Yangtze River Delta and Pearl River Delta (mm/yr)

A. 1956—1985			
locality	Eustatic sea level	Land subsidence	Relative sea level
Old Yellow River Delta (Tanggu)	1.5	23.2*	24.7
Modern Yellow River Delta	1.5	3~4	4.5~5.5
Yangtze River Delta (Wusong)	1.5	10**	11.5
Pearl River Delta	1.5		2.0
B. Estimate for 2030			
Old Yellow River Delta	4.5	10	14.5
Modern Yellow River Delta	4.5	3~4	7.5~8.5
Yangtze River Delta	4.5	3~5	7.5~9.5
Pearl River Delta	4.5	1~1.5	5.5~6.0

* average rate of land subsidence (1966—1985) from precise levelling from benchmark at Tanggu Station to a master benchmark on the rocky foundation in Baodi(宝砥) (north of Tianjin City) is 24.5 mm/yr but Baodi benchmark is rising at a rate of 1.3 mm/yr. Therefore, net land subsidence is 23.2 mm/yr.

** total amount of correction for land subsidence at Wusong station is 1 110 mm between 1923—1966, averaging 25.2 mm/yr. Between 1977 and 1990, land subsidence at Wusong Station is 12.1 mm/yr, 10 mm/yr is a lower estimate. (After Ren, 1993).

3. Impact of Sea Level Rise on Coastal Regions

Impact of future sea level rise on Chinese coastal regions will be profound. The most important are:

(1) Inundation　Large area of Chinese Deltas and coastal lowland, with high rate of relative sea level rise, will be vulnerable to sea inundation. Total area liable to inundation is at least 35 000 km² with a population about 45 million, including a great part of downtown area of important coastal cities of Shanghai, Tianjin and Guangzhou if deltas and coastal plain are without coastal defense or are protected by low dikes. Actually, almost all deltas and coastal plain are currently protected by dikes. Some dikes are very solid, having a safety frequency of 1/1 000 yrs, as anti-flood dikes along the Huangpu River, Shanghai (206 km long) under construction. But a great part of dikes are of low standard, unable to stand against the present severe storm surges. It is important that these dikes are upgraded to reduce the risk of inundation.

It should also be mentioned that a part of the area liable to inundation is not directly due to sea invasion but is resulted from the increase of probability of flooding due to difficulty in drainage caused by sea level rise. A notable example is the lowlands in

Jiangsu coast and in the east of Taihu Lake. Appropriate water conservancy measures are necessary to mitigate flood risks in these lowlying plains.

(2) With, rising sea level, tidal current will penetrate a greater distance upstream in coastal rivers with the result that polluted water from the upstream will be more difficult to discharge into the sea and tends to remain almost stationary at the interface between the stream flow and tidal current. The situation is particularly serious in Shanghai and Pearl River Delta. With future sea level rise, water pollution, and sea water intrusion will cause the problems of fresh water supply to the larger cities such as Shanghai, Tianjin and Guangzhou. More frequent and stronger storm surge will damage the harbour wharves and cause even to inland water way transportation, such as in the area of Yangtze River delta.

(3) Coastal and beach erosion As a result of sea level rising it happens in the sandy coast plain, even though there are sediment supply from small rivers, such as: 1.5～2 m/y retreat in the Shandong and Liaodong Peninsula in the Bohai Sea and Yellow Sea; 2～5 m/y in Zhejiang and Fujian Province along the East China Sea. The reason of coast retreat is not only by the rising of sea level, but also by the artificial mining. Human effects give significant influence to the coastline evolution in China, even to Yellow River migration. The net result of coastline erosion by sea level rising along reddish sandy terrace coasts, with the retreat rate at 0.7～1.5 m/y along Fujian and Guangxi coasts in the East and South China Sea. Actually the reddish sandy terraces were the old coast dune and beach ridge, and mostly lack of sediment supply from modern coast.

It is limited erosion along bedrock coast as most bedrock coasts are granite and gneiss, which are hard to resist the coast erosion. Several parts show that the 0.07～0.1 m/y retreating as the serious example along the East China Sea (Table 3) (Ren, 1993).

TABLE 3 Rate of Coastline Retreating Caused by Sea Level Rising in China (After Ren, 1993)

Type	Location	Annual Retreating Rate(m/y)
Sandy Beaches	Shandong & Liaodong Peninsula	1.5～2.0
	Zhejiang & Fujian Provinces	2.0～5.0
Consolidated reddish sandy terraces	Fujian & Guangxi Provinces	0.7～1.5
	Granite & Gneiss rock coasts	Limited erosion
	Along East China Sea	0.07～0.1 as maximun

Following the rise of sea level, it increases the submarine coastal slope, has decreased gradually wave winnowing on submerged coastal sediments. At the same time, erosion on the upper beach by break waves has been enhanced. On the other

TABLE 4　The net result of beach erosion in China by 0.5 m of sea level rising during next century

| Location | | Modern Beach | | | | Estimated Beach Response of 0.5 m Sea Level Rising | | | | | | | | |
| | | | | | | Natural Flooding | | | Beach Erosion | | | Sum Value | | |
		Length (m)	Average Width (m)	Relative Height (m)	Areas (m²)	Beach retreating (m)	Lose of the area (m²)	Rate of lose(%)	Beach retreating (m)	Lose of the area (m²)	Rate of lose(%)	Beach retreating (m)	Lose of the area (m²)	Rate of lose(%)
Dalian	Xing Hai Park	2 125	264.7	6.1	562 510	76.6	130 950	23.3	26.5	56 314	10.0	86.6	187 264	33.3
	Dongshan Hotel	510	62.4	3.8	31 824	18.1	9 236	29.0	15.8	8 078	25.4	33.9	17 314	54.4
	Grand Beach	→756	122.4	3.8	92 560	23.0	20 853	22.5	24.7	22 144	23.9	47.7	42 997	46.4
	Summary	3 391	202.6		686 894	18.1~76.6	161 039	23.4	15.8~26.5	86 536	12.6	33.9~86.6	247 575	36.0
Qinghuangdao	Bei Dai He	7 850	199.0	5.9	1 562 460	28~155.6	548 354	35.1	48.8	383 080	24.5	76.8~204.4	931 434	59.6
	West Xiang He Zhai	3 124	433.3	6.4	1 353 540	34~127.8	304 986	22.5	41.5	129 650	9.6	75.5~169.3	434 636	32.1
	Shandong Bao	756	89.5	3.5	74 652	5.8	4 385	5.9	25.4	19 202	25.7	31.2	23 587	31.6
	Summary	11 730	255.0		2 990 652	5.8~155.6	857 725	28.7	25.4~48.8	531 932	17.8	31.2~204.4	1 389 657	46.5
Qingdao	Qingdao Bay	1 356	314.8	6.0	348 520	46.3	62 762	18.0	37.9	51 455	14.8	84.2	114 217	32.8
	Hui Quan Wan	1 124	219.9	6.0	247 150	37.2	41 841	16.9	38.6	43 386	17.6	75.8	85 227	34.5
	Fu Shan Sou Mouth	1 625	423.5	5.4	688 250	48.2	78 325	11.4	26.4	42 932	6.2	74.6	121 257	17.6
	Summary	4 105	312.8	5.0~9.2	1 283 920	37.2~46.3	182 928	14.3	26.4~38.6	137 773	10.7	74.6~84.2	320 701	25.0
Beihai	Wai Sha	2 530	179.3	6.2	453 750	9.4~33.3	61 686	13.6	27.9	70 587	15.6	37.3~61.2	132 273	29.2
	Da Dun Hai	5 516	1 115.4	5.0~9.2	6 152 470	92~250.0	926 820	15.1	48.1	265 335	4.3	140.1~298.1	1 192 155	19.4
	Bai Hu Tou	5 165	917.3	5.0~7.2	4 738 090	88.7~320.5	743 476	15.7	45.2	233 458	4.9	133.9~365.7	976 934	20.6
	Summary	13 211	858.7		11 344 310	33.3~320.5	1 731 982	15.3	27.9~48.1	569 380	5.0	37.3~365.7	2 301 362	20.3
Sanya	Da Dong Hai	2 650	241.6	5.9	640 260	89	235 755	36.8	12.2	32 330	5.0	101.2	268 085	41.8
	Ya Long Bay	8 880	412.7	5.4~13.4	3 665 060	27.1~75.4	429 750	11.7	12.7	112 776	3.1	39.8~88.1	542 526	14.8
	San Ya Sand Bar	16 360	440.0	3.3~11.6	7 198 430	33.4~70.2	863 467	12.0	43.6	713 296	9.9	77~113.8	1 576 763	21.9
	Summary	27 890	412.5		11 503 750	27.1~89	1 528 972	13.3	12.2~43.6	858 402	7.5	39.8~113.8	2 387 374	20.8
Total		60 327	461.0	3.3~13.4	27 809 526	5.8~320.5	4 462 646	16.0	12.2~48.8	2 184 023	7.9	31.2~365.7	6 646 665	23.9

hand, following the rise of sea level, the slopes of the river beds have reduced, decreasing the fluvial sediment discharges. The lack of coastal sediment supply is a world-wide phenomenon, combined with the frequency of storm surges and EL NINO events. Beach erosion and sand barriers retreating to landward are the comprehensive result of stronger hydrodynamic processes and a smaller volume of sediment supply to the coast. It is estimated, by using the Bruun's rule (Bruun, P. , 1988), that the most favourable sand beaches will lose around $15\% \sim 60\%$ of their present area, while sea level is continually rising to 50 cm higher by the year of 2100 (Table 4). Although present erosion of mud flat coast in China is largely triggered by sediment starvation due to human activities rather than by sea level rise, it will certainly be aggravated by future accelerated sea level rise. However, because of difference in size of sediment grains, coastal erosion of Chinese mud coast may not exactly follow "Bruun's Model" (Ren, Mei-e, 1993). Coast protection and beach nourishment are the major methods used in such circumstance.

REFERENCES

Bruun, P. 1988. The Bruun rule of erosion by sea level rise: a discussion of large-scale two and three dimensional usages. *Journal of Coast Research*, 4: 627 - 648.

Chen, Xiqing. 1991. Sea level changes since the early 1920's from the long records of two tidal gauges in Shanghai, China. *Journal Coastal Research*, 7(3): 787 - 799.

Emery, K. O. and You Fanghu. 1981. Sea-level changes in the Western Pacific with special emphasis on China. *Oceanologica et Limnologica Sinica*, 12(4): 297 - 310.

Emery, K. O. and Aubrey, D. G. 1986. Relative sea level changes from tide-gauge records of Eastern Asia Mainland. *Marine Geology*, 72: 33 - 45.

Emery, K. O. and Aubrey, D. G. 1988. Coastal neo-tectonic of the Mediterranean from tide gauge records. *Marine Geology*, 81: 41 - 52.

Huang, Liren, et al. 1990. An isostatic datum for sea level change study along the coastal area in China. In: Studies on climatic and sea level changes in China. China Ocean Press. 62.

Liu, Chi-Ching. 1989. Impact of crustal deformation on tide gauge records. In: Proceedings Geological Society of China, 32(4): 321 - 338.

Ren, Mei-e. 1993. Sea level changes in China over the last 80 years. *Journal of Coastal Research*, 9(1): 229 - 241.

SCOR WG89. 1990. The response of beaches to sea level changes: A review of predictive models. *Journal of Coastal Rescarch*, 7(3): 895 - 921.

Shi, Yafeng and Fan, Jianhua. 1990. Progress in the study of climate and sea level changes in China. In: *Studies on climatic and sea level changes in China*. China Ocean Press. 1 - 6.

Wang, Baochan, et al. 1991. An analysis of seasonal variations in sea level along the Changjiang estuary, China. Shanghai: East China Normal University, manuscript.

Wang, Zhihao. 1986. The sea level changes in 20th century. In: China Sea Level Changes. Beijing:

China Ocean Press. 237 - 245.

Yim，W. W. S. 1991. An analysis of tide gauge and storm surges data in Hong Kong. *Hong Kong Meteorological Society Bulletin*，1(1)：16 - 22.

Zhou，Tianhua，et al. 1992. Study on recent sea level changing trend along coastal zone of China. *Acta Oceanologica Sinica*，14(2)：1 - 8.

海平面上升与海滩侵蚀[*]

一、世纪性的海平面上升

海平面变化由不同的作用过程形成，具有长、短周期的不同变化。扼要综述海平面变化的研究成果，有助于阐明海岸过程背景，主导变化作用与海岸效应的内在机制。

长周期海平面变化被概括为水动型(eustatic)与均衡型(isostatic)两类。全球规模的变化影响到海水的总量或海盆的体积为水动型海平面变化，是由于构造运动、洋盆被沉积物充填、冰川作用或水体密度的变化所产生。均衡型表现为地方性的变化，由于陆地相对于静态海面的挠曲活动所形成，或伴随冰川后退由于均衡作用形成的区域上升，或地区沉陷而成。最重要的长周期海平面变化是构造-水动型的(tectono-eustatic)。冰川-水动型(glacio-eustatic)的变化，虽在地质历史时期发生较少，但在过去的300万年期间海平面变化与陆地上冰盖的生长与消融有关。形成于晚第三纪的南极冰盖是水动型海平面变化的最重要的因素，其生长消退导致第四纪海平面变化迅速的响应[①]。

应用稳定同位素定年法，对采自陆架不同深处的泥炭层、潮间带有机体与化石、海滩岩及海成阶地的年代测定，已获得有关距今50 000～40 000年时期海平面变化的局部资料。了解到距今30 000年至25 000年前的间冰期时，海平面与现代海面高度相当。由于最后一次冰期开始，冰川生长而海平面下降，下降的最大值估计为75～130 m(多数人采用低于现代海面100 m的数值)，发生于18 000年前。全新世海侵约始于17 000～15 000年前，海平面上升迅速，速率可达8～10 mm/a。此上升持续到7 000年前，该时的海面约相当于现在海面的10 m深处。距今5 000年时，海面上升率剧减为1 mm/a，此速率一直保持到近期的200年期间。对以海成阶地或沿岸堤保存下来的高海面，大多数观察者认为是地方性的构造抬升效应而非起源于水动型，并认为现代海平面是全新世海侵以来最高的位置[①]。

海平面继续在变化，证据来自验潮站水位记录分析，展现出一个世界范围的海平面上升，源于冰川的进一步融化与海洋水体热膨胀。近200年的验潮资料(瑞典Brest站位记录始自1704年，荷兰Amsterdam站位始自1682年，以Brest站自1807年的记录最为标

---

　　[*]　王颖，吴小根：《地理学报》，1995年第50卷第2期，第118-127页。

　　同年载于：《海平面变化与海岸侵蚀专辑，海岸与海岛开发国家试点实验室年报(1991—1994)》，119-128页，南京大学出版社，文章题目为"海平面上升与海滩效应"。

　　1993年曾载于：包浩生主编，《任美锷教授八十华诞地理论文集》，28-38页，南京大学出版社，1993年，文章题目为"海平面上升与海滩效应"。

　　本文内容有少许修订。

　　[①]　Scientific Committee on Ocean Research，Working Group 89：The Changing Level of The Sea and Models of Beach Responses，1993.

准)反映出海平面上升趋势与大气温度、海水表层温度变化趋势呈良好的相关,并且自 1930 年以后,海平面上升速率增加。构造活动与人类影响使陆地水准发生变化,使海平面上升值形成明显的地区差异,纽约验潮站代表美国东海岸状况,近百年海平面上升速率为 3 mm/a,是海平面上升与相当数量陆地下沉的综合效应。南部的得克萨斯站位资料表明海平面上升速率的平均值达 6 mm/a,原因在于抽取地下水与原油而引起的地面沉降。美国西海岸俄勒冈站几乎未表示出相对的海平面上升,因为水动型的海平面上升与陆地抬升量相当。D. Aubrey 与 K. O. Emery 的工作试图将新构造运动上升值与全球性的水动型上升区别开来。虽然验潮站分布在南半球稀少,但从全球范围的验潮记录进行相近比较与趋势性分析,在过去 50 年到 100 年间,水动型的海平面上升值变化为 1～2 mm/a。尽管测算方法不同,但结果相近(表 1)。

表 1　据验潮资料所确定的全球水动型海平面变化*

研究者	上升速率(mm/a)	研究者	上升速率(mm/a)
Gutenberg (1941)	1.1±0.8	Barnett (1982)	1.51±0.15
Kuenen (1950)	1.2～1.4	Barnett (1984)	2.3±0.2
Lisitzin (1958)	1.1±0.4	Gornitz and Lebedeff (1987)	1.2±0.3
Wexier (1961)	1.2	Braate and Aubrey (1987)	1.1±0.1
Fairbridge and Krebe (1962)	1.2	Peltier and Tushingham (1989)	2.4±0.9
Hicks (1978)	1.5±0.3	Douglas (1991)	1.8±0.1
Emery (1980)	3.0	谢志仁 (1992)	0.7～1.2
Gornitz et al. (1982)	1.2		

＊据国际海洋研究科学委员会第 89 工作组

近数十年来海面加速上升与地球的温室效应有关。人们预测,由于全球变暖使冰川融溶与海水热膨胀,可使海面上升的数值如下:政府间气候变化专门委员会(IPCC-WG1, 1990)估计至 2050 年海平面上升 30～50 cm,至 2100 年海平面可能上 1 m。美国环境保护局预测到 2100 年海平面将上升 50～340 cm,相当于 5～30 mm/a 上升速率[1]。全美研究委员会的二氧化碳评估组(The committee of the National Research Council on Carbon Dioxide Assessment)提出,至 2100 年,海平面上升速率为 7 mm/a[2]。Van Der Veen (1988)估计,至 2085 年海平面上升率为 2.8～6.6 mm/a[3]。这些数据比过去 100 年来海平面 1～2 mm/a 的上升速率速高出 2～4 倍。

国家海洋局 1990 年公布了据 44 个站位的验潮资料分析结果:到 1989 年为止的近 30 年来,中国沿岸海平面平均上升速率为 0.14 cm/a[①]。国家测绘局于 1992 年 7 月发布根据 9 个观测站的资料分析结果:在过去 100 年中,中国东海与南海沿岸海平面分别上升 19 cm 与 20 cm,中国海平面的年上升率为 2～3 mm,未来海平面仍呈上升趋势[②]。同时,

①　国家海洋局. 1989 年中国海洋环境年报. 1990 年 3 月。
②　国家测绘局. 中国海平面每年上升二至三毫米,人民日报海外版,1992 年 7 月 8 日第 3 版。

发表了对世界的 102 年验潮站海平面记录的计算分析结果:在过去 100 年中,全球海平面平均上升 15 cm,太平洋海平面上升 10 cm,大西洋海平面上升 29 cm,印度洋海平面上升 39.6 cm。上述资料表明,海平面变化存在着海区差异与时段的差异,但过去 100 年的海平面上升数值是相近的,未来海平面上升趋势与速率增加是为大多数学者所肯定的。

短周期海平面变化,是由于大气与海洋作用过程的变化,如海水温度的地区性变化、海岸水流强度的改变、气压与风作用力与方向的改变等所造成的海平面年度变化、季节变化或日变化等。最突出的短周期海平面变化与太平洋的厄尔尼诺(EL Nino)的发生有关。在太平洋东岸的赤道附近的岛屿验潮站重复记录到,在不到一年的时间内,海平面变化达到 40~50 cm。在美国西岸,由于厄尔尼诺形成的海平面高达 10~20 cm。1982—1983 年间,俄勒冈州海岸由于厄尔尼诺与海平面季节变化造成海面在 12 个月内抬升达 60 cm。风暴潮所形成的增减水在孟加拉湾形成年海平面差异达 100 cm 的记录[4]。人类活动的影响,如过度抽取地下水或建筑物重载,使河口三角洲地区大面积沉降,加大海平面上升值,如天津新港码头自 1966—1985 年下沉达 0.5 m(国家测绘总局,1992 年)。从某种意义上讲,这类变化可归为短周期变化,通过人工措施可控制这类变化。短周期海平面变化对海岸带会形成灾害性破坏。而对海岸潜在效应的推究,研究工作应致力于世纪性的全球范围的水动型海平面上升。这种世纪性的、全球性范围的变化促进了风暴潮与厄尔尼诺现象发生频率的增加。

二、海岸侵蚀效应

全球海平面上升在海岸带的主要反应是海滩侵蚀和海岸沙坝向岸位移。组成海滩与沙坝的沉积物主要是砂级的,属波场中的沉积物,是由波浪自水下岸坡海底掀带,并被浪、流进一步搬运(以横向运动为主)至岸坡上部堆积的。泥沙或来源于河流供给,或来自海蚀岸段以及由近岸海底供沙。后者主要是古海岸堆积(如中国沿岸),或为冰期低海面时的冰川作用沉积(如欧、美沿岸)。由于海平面持续上升,加大的水深,使波浪对古海岸带的扰动作用逐渐减小而形成海底的横向供沙减少,却加强了激浪对上部海滩的冲刷。同时,逐渐升高的海平面,降低了河流的坡降,减小了河流向海的输沙量。因此,世界上大部分海滩普遍出现沙量补给匮乏。海平面上升伴随厄尔尼诺现象与风暴潮频率的增加,使水动力作用加强,加上泥沙量匮乏的综合效应,使海滩普遍遭受冲蚀,而沙坝向海坡受冲刷,与越流扇(overwash fan)的形成过程,综合表现为沙坝的向陆迁移。

Bruun P. 以图式表明了海平面上升与海滩变化效应[5,6]。Bruun 定律的大意是:随着海平面上升,海滩与外滨浅水区的均衡剖面呈现向上部与向陆的移动,海滨线的后退速率(R)与海平面增高(S)有关,即

$$R = \frac{L}{B+h}S \tag{1}$$

式中:h 是近滨沉积物堆积的水深;L 是海滩至水深 h 间的横向距离;B 代表滩肩的高度。关系式(1)亦可表示为

$$R = \frac{1}{\tan\theta}S \tag{2}$$

$\tan\theta \approx (B+h)/L$，是指沿着横距 L 的近滨平均坡度。砂质与砂砾质海岸的坡度大部分为 $1/100 \sim 1/200$ 间，即 $\tan\theta \approx 0.01$ 至 0.02。因此据公式（2）可得到 $R=50S$ 至 $100S$，表明微小的海平面上升可形成较大的海滨线后退。图 1 是对 Bruun 定律的图解。

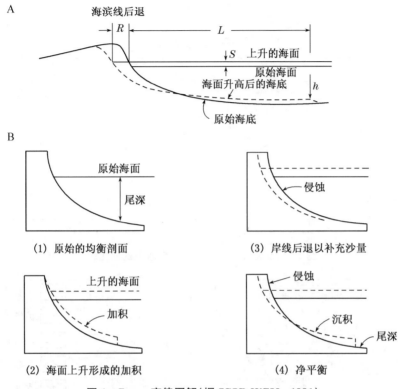

图1　Bruun 定律图解（据 SCOR WG89, 1991）

A. 由于海面上升引起海滩剖面的净变化，根据 Bruun 定律，海面上升（S）将引起外滨带堆积及海滩上部侵蚀，总的侵蚀后退率（R）。

B. 根据 Bruun 定律分析公式（1），由于海面上升（S），因此海滩向陆侵蚀后退（R）。

图 2 系秦皇岛海岸剖面重复测量记录[7]，其中南山灯塔岸段系海蚀基岩岸。1973 年 8 月较 1964 年 7 月所测剖面，显示海蚀崖与岩滩蚀退变低，仅岩滩外侧有砾石堆积，而石河口堆积海岸则显示上部海滩侵蚀与下部堆积。山东半岛平直砂岸海滩剖面重复测量资料也反映出类似的特点（图 3）。说明 Bruun 图式具代表性。Bruun 图式表明达到均衡剖面的海岸在海平面上升过程中海滩再造的情况，而不适宜于非堆积型的海岸与海蚀（如秦皇岛南山）或海积变化剧烈的岸段。

砂质海滩的侵蚀在中国是普遍的。如，1989 年 5 月至 1990 年 5 月辽东半岛盖县开敞砂质海滩侵蚀速率最大达 6.8 m/a，1989 年至 1993 年 4 年平均侵蚀后退速率约 2 m/a；1989 年到 1993 年 4 年间辽西六股河一带海滩蚀退率约 1 m/a[①]。近 20 年来山东半岛砂质海滩蚀退速率约 $1 \sim 2$ m/a，造成海滩砂亏损约 2×10^7 t/a[8]。海岸蚀退在河口段尤为严重，

---

① 庄振业，常瑞芬，苗丰民等. 鲁、辽砂质海岸蚀退研究. 中国海平面变化和海岸侵蚀工作组 1994 会议.

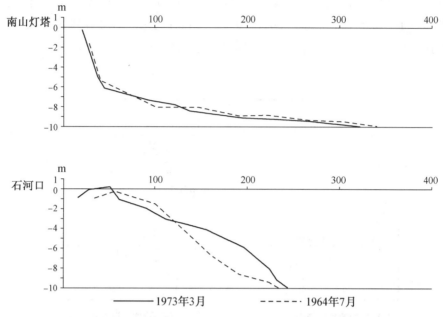

图2 秦皇岛1973年与1964年水下剖面比较

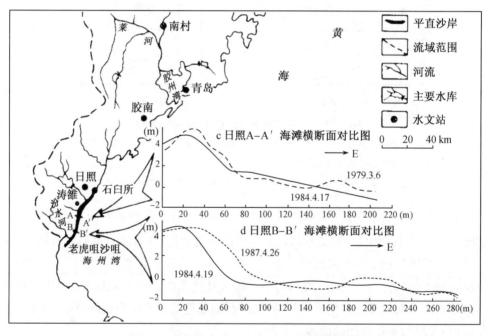

图3 山东半岛两段平直沙岸及其蚀退情况[5]

海浪冲毁了海滩防护林,威胁农田与建筑,咸化了滨海地下水。海滩侵蚀后退与世纪性的全球海平面上升有关,也受到人为的影响:河流中下游水库拦沙,减少了海滩砂之补给。人工采沙做建筑材料销售,使海滩砂益加亏损,失去海浪作用与泥沙补给之平衡,使海滩遭受侵蚀。如,滦河自引滦输水工程后,上游泥沙主要淤积于潘家口水库与大黑汀水库内,多年平均入海水量由 $41.9 \times 10^8 \ m^3$ 减为 $3.55 \times 10^8 \ m^3$,而多年平均输沙量由工程前的 $2\,219 \times 10^4 \ t$ 减至 $103 \times 10^4 \ t$。海岸泥沙补给骤减,滦河三角洲砂质海岸由加积而转为蚀退,口门岸滩蚀退率达 300 m/a,岸外沙坝蚀退率 25 m/a,海岸蚀退速率较工程前约增加 6 倍,潟湖淤泥层普遍于沙坝外缘出露[9]。自 20 世纪 70 年代以来,浙闽沿岸砂质海滩或沙丘冲蚀后退约 1~4 m/a,老岸堤组成的红砂台地蚀退速率高达 0.4~1 m/a,而基岩岬角岸段蚀退速率约为 0.1 m/a①。海滩侵蚀主要发生于台风或寒潮大浪期间,而后逐渐加积成平缓剖面,由于泥沙亏损与世纪性的海平面持续上升,净效果表现为海岸的后退。在构造上升的丘陵或岛屿海岸段,海平面上升的效果不甚明显。但是,由于人工采沙而导致海滩侵蚀使滩肩消失的现象却是普遍的。如江苏赣榆县九里沙滩,水下取沙做建材出售。海滩受蚀几尽,现已禁止采沙并修建水泥堤防冲,但沙滩风光已消失。海南岛三亚湾由于采沙加速海滩冲蚀,海滩剖面降低,木麻黄林亦遭受损坏。

砂质海滩自低潮线向下至激浪带外缘,宽度大,脊槽起伏,粗细砂夹杂,激浪带外缘有陡坡坡折。低潮水边线附近为 1°~2° 平坦坡的细粒沙滩。高低潮间的海滩宽度不大一般不超过 50 m,相对高度小于 1 m,海面间或有脊湾交错的滩尖嘴微地形。高潮线附近坡度增大至 4°~7°,砂粒增大并夹杂贝壳或海藻残体,部分陡滩坡度约 12°。特大高潮线以上多为长草的沙丘或沿岸沙堤,其向海坡可增大至 20°,沙堤高度 2~5 m 不等,沙丘叠加处高度可达 10 m 或更高。沿岸堤系全新世的海滩脊或晚更新世的古海滩,大部分已发展为海滩上部的沙丘带,不经常受到海浪冲刷,可视为一天然的海滨屏障。上述各带系海滩的整体结构,由于人工采沙或其他原因,破坏了海滩水下部分的动态平衡——海浪动力与泥沙供应间的平衡,会招致上部海滩遭受冲刷破坏。

海平面上升与海滩侵蚀是全球性现象,为此,海洋研究科学委员会(Scientific Committee on Ocean Research,SCOR)已成立专门工作组进行研究,作为工作组的成员之一,作者结合中国海岸实际介绍了 SCOR 的主要结论。中国科学院地球科学部提出:预估当 21 世纪海平面上升 50 cm 时,中国主要海滨旅游海滩的变化数据。本文选择了大连、秦皇岛、青岛、北海、三亚等 5 处著名海滨旅游区内的若干海滩进行了分析计算(表2)。上述各地均有我们多年考察的实测剖面数据,海岸段近滨带外界水深,除北海为 -2 m 外,其余均采用 -5 m,再结合大比例尺地形图与海图,可以获得有关参数。作者经过对多处海岸剖面重复测量结果对比研究后,认为 Bruun 公式基本上反映海滩变化的自然规律,接受了为国际海洋界所肯定的 SCOR89 工作组研究成果。作者结合对我国海滩研究的认识,并对 Bruun 图加以修正(图4)。图中滨线采用低潮海滨线,$\triangle y$ 为预定的海平面上升幅度;$\triangle x_1$ 表示因海平面上升使部分海滩受淹没而产生的后退量;$\triangle x_2$ 为海平面上升而产生的海滩侵蚀后退量,由于海平面上升而形成的海滩总后退量为 $\triangle x_1$ 与 $\triangle x_2$ 之和,计算的基本依据是海滩趋向于在海平面变动情况下形成新的均衡剖面。表2 总结了各海滨沙滩在 21 世纪海平面上升 0.5 m 后的淹没与冲蚀后退数值。各海滨沙滩面积损失的最小值为 12.7%(亚龙湾),因为该处海岸坡度较大,最大值达 66%(北戴河海滨),上述海滩面积总损失量可达 266×10^4 m²。实际损失值可能要大于上述预算数值,因为激浪与风暴潮作用将更加频繁,其影响的范围更大,大部分海滩均会遭受海水淹侵冲蚀。

---

① 据福建省地震局姚庆元实测资料。

表 2 海平面上升 0.5 m 对我国重要海滨旅游区海滩的影响

海滨位置		长 (m)	平均宽 (m)	相对高 (m)	面积 (m²)	海平面上升 0.5 m 之海滩响应预测								
		现代海平面之海滩				海滩淹没			海滩侵蚀			综合效应		
						滨线后退 (m)	损失面积 (m²)	损失率 (%)	滨线后退 (m)	损失面积 (m²)	损失率 (%)	滨线后退 (m)	损失面积 (m²)	损失率 (%)
大连	星沙公园	2 125	68.5	6.1	145 613	6.8	14 450	9.9	26.5	56 314	38.7	33.3	70 764	48.6
	东山宾馆	510	42.2	3.8	21 645	7.4	3 774	17.4	15.8	8 078	37.3	23.2	11 852	54.7
	大沙滩	756	56.3	3.8	42 560	7.9	5 972	14.0	24.7	18 674	43.9	32.6	24 646	57.9
	小 计	3 391	61.9		209 818	6.8~7.9	24 196	11.5	15.8~26.5	83 066	39.6	23.2~33.3	107 262	51.1
秦皇岛	北戴河	7 850	87.1	5.9	683 456	8.7	68 295	10.1	48.8	383 080	56.1	57.5	451 375	66.1
	西向河寨	3 124	223.6	6.4	698 466	6.7	20 930	3.0	41.5	129 650	18.6	48.2	150 580	21.6
	山东堡	756	88.2	3.5	66 672	7.5	5 670	8.5	25.4	19 202	28.8	32.9	24 872	37.3
	小 计	11 730	123.5		1 448 594	6.7~8.7	94 895	6.6	25.4~48.8	531 932	36.7	32.9~57.5	626 827	43.3
青岛	青岛湾	1 356	72.8	6.0	98 650	8.5	11 526	11.7	37.9	51 455	52.2	46.4	62 981	63.9
	汇泉湾	1 124	70.6	6.0	79 356	7.0	7 868	9.9	38.6	43 386	54.7	45.6	51 254	64.6
	浮山所口	1 625	193.1	5.4	313 857	8.9	14 462	4.6	26.4	42 932	13.7	35.3	57 394	18.3
	小 计	4 105	119.8		491 863	7.0~8.9	33 856	6.9	26.4~38.6	137 773	28.0	35.3~46.4	171 625	34.9
北海	外 沙	2 530	60.8	6.2	153 750	5.8~9.5	17 254	11.2	27.9	70 587	45.9	33.7~37.4	87 841	57.1
	大墩海至电白寨	5 516	258.4	5.0~9.2	1 425 588	5.4~9.8	41 926	2.9	48.1	265 335	18.6	53.5~57.9	307 261	21.5
	电白寨至白虎头	5 165	183.2	5.0~7.2	946 363	5.4~8.7	36 457	3.9	45.2	233 458	24.7	50.6~53.9	269 915	28.6
	小 计	13 211	191.2		2 525 701	5.4~9.8	95 637	3.8	27.9~48.1	569 380	22.5	33.7~57.9	665 017	26.3
三亚	大东海	2 650	81.5	5.9	215 905	7.9	20 935	9.7	12.2	32 330	15.0	20.1	53 265	24.7
	亚龙湾	8 880	166.1	5.4~13.4	1 475 184	6.8~9.8	74 592	5.1	12.7	112 776	7.6	19.5~22.5	187 368	12.7
	三亚湾	16 360	296.2	3.3~11.6	4 846 024	5.6~10.2	137 654	2.8	43.6	713 296	14.7	49.2~53.8	850 950	17.5
	小 计	27 890	234.4		6 537 113	5.6~10.2	233 181	3.6	12.2~43.6	858 402	13.1	20.1~53.8	1 091 583	16.7
总 计		60 327	185.9	3.3~13.4	11 213 085	5.4~10.2	481 765	4.3	12.2~48.8	2 180 553	19.4	20.1~57.9	2 662 314	23.7

注：表中海滩面积指低潮低潮位以上包括沿岸沙坝在内的海滨沙滩面积。

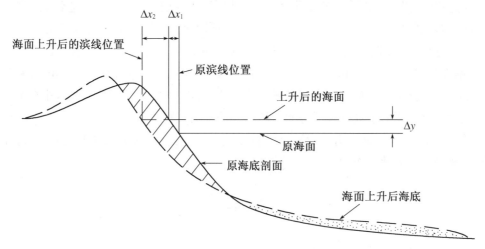

图 4 海面上升使海滩遭受淹没与侵蚀

（$\triangle y$ 为海面上升幅度，$\triangle x_1$、$\triangle x_2$ 分别为海滨线因海滩遭受淹没和侵蚀而产生的后退量；假定海面上升前后的海滩剖面均已达到平衡）

三、海滩侵蚀预测与对策

海滨是旅游胜地，以阳光、沙滩与海鲜三 S 著称，供人们增进健康、陶冶心情开展体育、研究与经营活动。20 世纪 80 年代末以来，国际旅游业已超过石油工业与汽车制造业，成为国际最大的产业，发达国家的海滨旅游业产值约占旅游业总值的 2/3。由海平面上升造成的海滨沙滩的冲蚀破坏，不仅丧失了旅游休憩之场所，而且还会危及滩后沙丘带、潟湖水域、沿岸建筑，蚕食岸陆土地与破坏陆地环境，所造成的经济损失与社会影响是不容忽视的。

日益发展的海滩侵蚀已引起各界人士的关注，并成为海岸工程研究的热点课题。当前防护海滩侵蚀最有效的措施是海滩喂养（Beach Nourishment），并辅以导堤促淤或外防波堤掩护，视海岸环境的特点而定。这种措施已为欧、美、日等国广泛应用。采用海滩砂人工补给法，必须对目标海岸段充分调查研究，包括海岸与海底地形、波浪折射、激浪带的横向与纵向泥沙运动、风力运沙与沙丘带活动状况、沉积物粒径与分布、海岸冲刷与堆积特点、海岸演变与地质过程以及航片与海图的重复测量等。通过调查确定沙源、泥沙粒径、人工海滩型式、防浪掩护的方式以及人工海滩可维持的期限等，然后进行供设计与施工所需的数学模拟。例如，在有一定潮差的海岸段落人工补充的沙量（m³/a）需增加 40％的耗损量[10]，再求出按需要与经费条件所能达到的维持年限（5 年、10 年、12 年），最后计算出应补充的总砂量。同时，计算确定人工海滩的长宽比、铺设部位、预定的高度、海滩坡度以及选用砂的粒径等。如选用的砂较原海滩砂细，则均衡剖面的坡度较平缓，可能招致较大的失砂量。目前多开采外滨古海岸砂补充现代海滩，该处水深已超过海岸泥沙活动带，有限量地采沙不会形成对现代海岸过程的破坏。人工堆沙部位以沙丘带坡麓与低潮水边线以下 −1 m 水深处为宜，该处为海滩活跃地带，最需补充沙量。虽然铺沙后改变不了海滩过程性,仍会发生季节性变化,但是,在相当长的期限内,为该海滨造就了一条美丽的沙滩。若配以少量防波堤建筑,则人工海滩可预期保持滩体的基本稳定。比如,由南京大学海岸与海岛开发国家试点实验室设计的三亚小东海人工海滩,其海滩长宽比为 2:1～4:1,铺设范围介于 $-1.0\,m～2.5\,m$,人工海滩填砂选用三亚市以西 30 多 km 处粒径为 $1～2\,mm$ 的天然老沙坝砂。为尽可能减少今后人工海滩的维护性回填砂量,小东海人工海滩还设计有必要的丁坝及潜堤等起保滩作用的辅助工程措施①。

总之,在全球海平面上升、环境变化以及采取有效对策的研究与实施过程中,地学工作具有相当重要的独到作用。

参考文献

[1] Hoffman,J,Keyes,D. and Titus,J. G. 1983. Projecting Future sea level Rise:Methodology,Estimates to the Year 2100,Research Needs U. S. Environment Protection Agency. Washington D. C.

[2] Revelle,R. R. 1983. Probable Future Changes in sea level Resulting From Increased Atmospheric Carbon Dioxide,Changing Climate. National Research Council Report Washington,D. C. National Academy Press. 433－448.

[3] Van Der Veen,C. J. 1988. Projecting Future sea level. *Survey in Geophysics*,9:389－418.

[4] SCOR Working Group 89. 1991. The Response of Beaches to sea level Changes,A Review of Predictive Models. *Journal of Coastal Researeh*,7(3):895－921.

[5] Bruun,P. 1962. Sea-level rise as a cause of shore erosion. Journal Waterways and Harbours Division,American Society Civil Engineers,88(WWI):117－130.

[6] Bruun,P. 1988. The Bruun Rule of erosion by sea level rise:A discussion of large-scale two-and-three-dimensional us-ages. *Journal of Coastal Research*,4:627－648.

[7] 南京大学海洋科学研究中心. 1988. 秦皇岛海岸研究. 南京:南京大学出版社.

[8] 庄振业,陈卫栋,许卫东. 1989. 山东半岛若干直平砂岸近期强烈蚀退及其后果. 青岛海洋大学学报,19(1):90－98.

[9] 钱春林. 1994. 引滦工程对滦河三角洲的影响. 地理学报,49(2):158－166.

[10] Hendrzk. J. 1992. Verhagen Method for Artificial Beach Nourishment. In:23rd International Conference on Coastal Engineering,Book of Abstracts. 593－594.

---

① 南京大学海岸与海岛开发国家重点实验室. 海南鹿回头及小东海海滩改造利用可行性研究报告,1993.

Plain Coast Changes: Human Impacts and River-Sea System Control ——Examples from China[*]

INTRODUCTION

Plain coasts are developed more than 2 000 km long mainly marginal to the major river deltas in China. On the basis of genesis, the plains coast can be divided into two types (Wang, 1980):

(1) The Alluvial Plain Coast

Most of these plains are located between the mountain ranges and sea shore, such as the alluvial plains located along the west side of Taiwan mountains, and the Luanhe River Plain located on the east side of Hebei province along the Bohai Sea. The plains consist of fluvial sediments from the mountain rivers. As the plain reaches the sea, it makes up the submarine coastal slope, with the gradients of 1/100 to 1/1 000. This results in a very wide breaker zone. The breakers, the swash and the backwash are the main types of wave action on the coast. The material movement is active in both the longshore and onshore directions. The barrier bars, sand spits, submarine bars and extensive sandy beaches are the typical features along this coast. The beaches may contain several beach ridges, with lagoons behind the bars.

(2) The Marine Deposited Coast

This type of coast is very extensive, flat and located at the lower parts of large rivers. Regionally as a rule, the basement is subsiding such as the North Jiangsu Plain, North China Plain and the Liaohe Plain. The coast slopes are very gentle, with the gradients in the order of 1/1 000~1/5 000. As a result, the effect of wave action is well offshore. Tidal currents play a major role in shaping the coast. The sediment consists mainly of silt and mud brought in suspension by rivers. This fine sediment cannot be deposited in the shallow nearshore zone, but is either deposited offshore in the deeper

---

* Ying Wang: *BORDOMER 95-COASTAL CHANGE ACTES/ PROCEEDINGS*. International Conference organized by BORDOMER & COI/IOC-UNESCO. Feb. ,1995, Bordomer, France. Volume I, Session I-V, pp. 172 – 179.

water or carried by the flood tide to the upper part of the coast, where it may remain. In these plains, then, the sediments could have been deposited either by marine action or mixed river and marine action (delta). The geomorphological subdivisions of this type of coastline are (Wang, 1983):

1) Salt marsh. These extend inland from a shell beach ridge (chenier) above the intertidal zone. The shell beach ridges are preserved for a considerable time and are useful indicators of the location of ancient coastlines. Occasionally relict lagoons are found behind the shell beach ridges.

2) Intertidal mudflats. These can be extensive as a result of a 3 m tidal range. For example, along the west coast of the Bohai Bay, tidal flats are $4\sim6$ km wide, with slope $0.3\%\sim1.0\%$. In the middle part of the North Jiangsu Plain, the intertidal mud flat is even wider still. Zonation features of morphology and sedimentation are typically the result of tidal currents operating on the flat: sandy flat with ripples on the lowerflat; silty and mud mixed flat with erosional microfeatures around mid-tidal level; muddy flat or mud pools located below low high tidal level; and clayey flat with polygon features or salt marshes located between high and maximum high tidal level (Wang et al., 1990).

3) Submarine coastal slope. This extends below the intertidal zone. It is gentle in the order of $0.7\%\sim2\%$ with a concave profile. The water is shallow, only about $3\sim5$ m at a distance of over 10 km offshore. The sediment becomes finer offshore; it is sandy silt at the lower tidal level and very fine mud in the deeper parts. Thus, the marine deposited plains are with tidal currents as it's leading dynamic action. Silt is the major component of sediment deposited on the tidal flat. Sediment becomes finer from low tidal level both landwards and offshore. These features are completely different to these of the alluvial plain coast.

The plain coasts represent a balance between wave erosion and deposition of unconsolidated sediment that is dependent on sediment supply, principally from rivers. If the amount of the sediment supplied is larger than the amount of eroded, the coastline progrades; if the sediment supply is reduced or stopped, wave erosion causes the coastline to retreat (Wang, 1980). Relative changes in sea level do not effect to any great extent the development of the plains coastline. For example, the west coast of the Bohai Bay, which receives the sediment of the Yellow River, is prograding despite a steady regional subsidence since the late Mesozoic (Wang, 1980).

River sediment discharges to the ocean were and are the major sediment supply to the plain coastal formation and the dynamic processes of coastal zone in China. It is estimated that total of 20.14×10^8 tons of river sediment discharged to the ocean annually (Chen and Zhao, 1985; Table1), which is about 20% of annual sediment inputs to the ocean by the world rivers(Milliman and Meade, 1983).

Table 1 Average water and sediment discharge of the major rivers of China

Sea area	River	Area of drainage		Average water discharge			Average sediment discharge		
		km²	(%)	10^9 m³/yr	%	mm	10^4 t/yr	%	t/km²/yr (yield)
Bohai Sea	Liao	164 101	12.3	86.98	10.9	53	1 849	1.5	113
	Luan	44 945	3.4	48.69	6.1	108	2 268	1.9	505
	Yellow	752 443	56.3	430.78	53.7	57	111 490	92.2	1 482
	$\sum$**	961 492	72.0	566.45	71.7	59	115 607	95.6	1 202
	$\sum$All	1 335 910	100.0	801.46	100.0	60	120 881	100.0	905
Yellow Sea	Yalu	63 788	19.1	251.34	44.8	394	195	13.3	31
	$\sum$All	334 132	100.0	561.45	100.0	168	1467	100.0	44
East China Sea	Changtze	1 807 199	88.4	9 322.67	79.7	516	46 144	73.1	255
	Qiantang	41 461	2.0	342.39	2.9	826	437	0.7	105
	Min	60 992	3.0	615.87	5.3	1 010	768	1.2	126
	$\sum$**	1 909 652	93.4	10 280.93	87.9	538	47 349	75.0	248
	$\sum$All	2 044 093	100.0	11 699.32	100.0	572	63 060	100.0	308
South China Sea	Hanjiang	30 112	5.0	258.78	5.4	859	719	7.5	239
	Pearl	452 616	77.3	3 550.32	73.6	784	8 053	84.2	178
	$\sum$**	482 728	82.4	3 809.10	79.0	789	8 772	91.7	182
	$\sum$All	585 637	100.0	4 821.81	100.0	824	9 592	100.0	164
Pacific Ocean		11 760	100.0	268.37		2 282	6 375		5 421
Total	$\sum$All China rivers	4 311 562		18 152.44		421	201 375		467

HUMAN IMPACTS AND RIVER SEA SYSTEM RESPONSE

In the past, human impact on the coastal evolution has been mainly through the river-sea system, because of several large rivers across the larger area of China, from the west, Qinghai-Tibet Plateau, to the east, the China Seas, which bring great influences to the China's coastal zone. For example, the Yellow River is unique for it's extremely heavy sediment load (11.9×10^8 ton/yr), rather small water discharge(485× 10^8 m³/yr), and migrated river channel in the lower reaches(Wang et al, 1986). Silts are the main component of sediment, and have been carried out by the Yellow River from the Loess Plateau mainly since the west Han dynasty (2nd Century B. C. ; Shi, 1981). Because of immigrant cultivation, forests and pastures were destroyed, which caused serious soil erosion, and huge quantity of sediments have been carried out to sea to form the great North China Plain. As a result, it changed the coastal water environment both in the Yellow Sea and the Bohai Sea.

The Yellow River has experienced eight major changes in its lower course since 2278 B. C. (Fig. 1.), discharging either into the Bohai Sea, or into the Yellow Sea via the Huai River. The shifted distances between north and south were more than 600 km long (Wang, 1983). Two major changes of the river channel migration were created artificially as a result of human activities for military purpose.

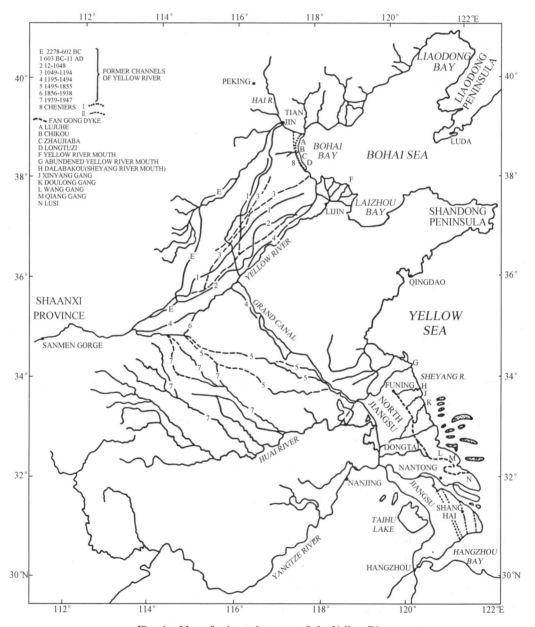

Fig. 1　Map of migrated courses of the Yellow River

One major event was in A. D. 1128. On November 15th, the capital official of Southern Sung Dynasty excavated the Yellow River bank at Kaifeng in order to use flood water as a weapon to prevent an invasion of Jin Dynasty soldiers from the north. Even though this effort was unsuccessful, it initiated a change that caused the Yellow River to flow through the Si River, one of the distributaries of the Huai River, to the south. Since then, the river shifted often in the Huai River area and entered the Yellow Sea for more than 720 years. With the enormous sediment supply from the Yellow River, the North Jiangsu coast along the Yellow Sea were created mudflat plain coast instead of the original sand barrier and lagoon system, and the plain prograded for 40 km wide, and its associated Yellow River delta, prograded 90 km into the Yellow Sea. The total area of accumulated coastal land was 157 000 km^2. It was 1/6 of present land area of Jiangsu province. However, the coastal zone along the Bohai Sea, at the same time, suffered from erosion, with the active coastal wave action, mudflats retreated and coastal shell beach ridges were formed instead of the original mudflat coast with tidal currents as coastal dynamics.

In 1855, the Yellow River returned back to the Bohai Sea. Since then, coastal erosion has been extensive along the North Jiangsu coast as a consequence of the cut in the huge volume of sediment supply, 15～30 m retreating annually along 150 km long abandoned Yellow River delta. A total of 1 400 km^2 of land has been lost completely, and a new rank of shell beach ridge has been formed by present day breaking wave patterns. It is still the major project of coastal defence along the delta. On the other hand, the total 160 km long coastline of modem Yellow River delta in the Bohai Sea has been prograded for 20. 5～27. 5km to the sea since the Yellow River shifted back in 1855. The net and average progradation is 0. 2～0. 27 km per year, i. e. , an average of 23. 5 km^2 land has accumulated each year. The new river mouth grows fast, 6～10 km in a single year, even though the sea level is rising and the tectonic subsidence in that area continuous since Tertiary time. Mudflats develop widely along the Bohai Bay and Laizhou Bay off the original chenier shoreline, with clayey silt sediment supplied by the Yellow River, and tidal currents as the major dynamic agent along the whole coastal zone. The example has shown clearly the human effects on coastal evolution through a river-sea system.

Future development of the modern Yellow River delta may be slower, because of the annual sediment discharge of the Yellow River has been decreased to 9. 5×10^8 tons annually as the annual water discharge is 37. 9 × 10^9 m^3, a consequence of water diversion for irrigation along middle and lower reaches. It is estimated that more than 150×10^8 m^3 of water/yr has been taken from the Yellow River for irrigation over the last decade, and with the volume of water about 1. 74×10^8 t of sediment per year. The increasing diversion of water for irrigation and other purpose over the next decade will withdraw considerable amounts of sediment from the Yellow River. The construction of the large Xiaolangde Reservoir, for example, scheduled to be completed by 2000A. D. could trap 3.3×10^8 t of sediment per year. Recently, sediment has been pumped from the Yellow River channel to widen and strengthen the main dikes along it slower reaches (total length about 1 400 km). In this way, large amounts of sediment were taken away from the river. Progress in soil conservation in the Loess Plateau, about 34. 8 % of total eroded area has been under control, then, 300 million tons of sediment flux have been decreased annually during recent 20 years (People's Daily. Overseas Edition, Jan. 4th, 1995). Thus, the discharge of Yellow River sediment to the sea maybe reduced $6 \times 10^8 \sim 7 \times 10^8$ t/yr in the next 20 years.

At present, human impact on the coastal zone of China has been more and more effective because of the large scale of economic development in the coast area. Still, human impacts on the coastal environment through the river-sea system appear more and more often and seriously, because there are more than 1 500 rivers entering the coastal zone of China.

1. To divert river discharge for urban water supply, then decreasing sediment discharge at same time, mentioned in example of the Yellow River, as a result, to cause the coast erosion at the delta area because a reduction in the sedimentary dynamic balance. Such as the Luanhe River, original water discharge was 4.19×10^9 m³ with total sediment supply of $2\ 219 \times 10^4$ tons. After diverting the water discharge of 3.55×10^8 m³ for water supply to Tianjing city, the sediment supply is only 103×10^4 tons, the delta coastline of Luanhe River has been retreated in a rate of 17. 4 m/yr since 1988 (Qian, 1991), and the salt water intrusion changing the salinity from 27. 3% to 32. 6% in the delta area. Solving one problem has caused other serious problems as the river-sea is a chose related system.

2. North Jiangsu Plain had been previously divided by numerous canals and channel networks in the early 1950's as a strategy for frontier defence. This project changed the natural environment (Fig. 2, Wang, 1994) with the positive effect of bringing more farmland under irrigation. For preventing salt water intruding and preserving fresh water resources along plain coast area, local people dammed up river mouth for most of rivers in North Jiangsu coast plain. Title constructions have caused serious under-dam siltation in the lower part of river mouth. The dams also have blocked fish, prawn and crabs breeding up stream from river mouth. As a result, local people not only lose economic income from the decreasing precious sea food, but also, the silted river mouth channels become smaller and smaller day by day, gradually losing navigable water depth, and often suffering flood disaster.

3. Using a river as a natural channel to flush out chemical wastes of industry, through irrigation of polluted water which damaged crops, and the rich nutritious water

Fig. 2　Artificial channel networks of the North Jiangsu Coastal Plain

and sediment entering the sea, it causes serious red tide in the coastal zone. The red tide almost "killed" aquiculture of prawn in the plain coast area.

4. Over exploitation of underground water in the plain coastal zone. Such as in the Laizhou and Longkou county of the plain coast along the Bohai Sea, underground water table subsided 15～30 m in the late 1980's compared with 1977. Thus it destroyed the natural water balance between fresh and saline water. As a result, salt water intruded more than 100 km and caused large area of farmland to be abandoned.

Over exploitation of underground water and heavy load construction on the coastal lowlandand delta area have caused rapid ground subsidence to accelerate local sea level rise, for example 0. 5 m subsidence has been recorded during 1966 to1985 in the Tianjing Tanggu Harbour of the Haihe River delta according to the State Survey and Mapping Bureau data in 1992. Relative sea level rise during the same period was 2 mm/yr in the Pearl River delta; 11. 5 mm/yr at Changjiang River delta, and 24. 7 mm/yr at abandoned Yellow River delta (Ren, 1993).

Sea level rising and the decreasing of river sediment supply, plus the human impacts directly through the beach sand mining. This has speeded up the sandy beach erosion to the rate of $1{\sim}3$ m/yr as its average in the most parts of China's coast line.

As the conclusion, human activity is one of the most dynamic agents in the coast formation, especially in the plain coast evolution through the river-sea dynamic system, where the impacts become more and more effective by using the advanced scientific techniques. The river-sea system is a natural cycle keeping the environmental balance between air and water, land and ocean. Even though the human activities are on land, they still directly or indirectly add to the influences on the coastal waters through the river connection. People should pay great attention to setting up a harmonious relationship between human activity and the natural environment. It is urgent to set up and improve systematic regulation and management of the coastal zone to preserve natural environmental balance for along period sustainable development in the coast area.

REFERENCE

Chen, Tian-wen & Chu-nian Zhao. 1985. Water and Sediment Discharges to the Ocean and Affected to the Coastal Zone by the Major Rivers of China (In Chinese). *Acta Oceanologica Sinica*, 7(4): 460 – 471.

Milliman, J. D. & R. H. Meade, 1983. World-wide Delivery of River Sediment to the Oceans. *The Journal of Geology*, 91: 1 – 21.

Shi, Nianhai. 1981. The Distribution and Migration of Farm Land, Pasture and Forests of the Loess Plateau (In Chinese). *Historical Geography*, 1: 21 – 31.

Qian, Chunlin. 1991. Effects of Water Conservancy Projects to the Luanhe River Delta in the Luanhe River Basin (Ph. D. Theses).

Wang, Ying. 1994. Effects of Military Activities on Environment in Eastern and Southeastern China. *Annual Science Report—Supplement of Journal of Nanjing University*, 30(English Series 2): 43 – 46.

Ren, Mei-e. 1993. Coastal Lowland Vulnerable to Sea Level Rise in China. In: Proceedings of PACON China Symposium. 3 – 12.

Wang, Ying. 1980. The Coast of China. *Geoscience Canada*, 7(3): 109 – 113.

Wang, Ying. 1983. The Mudflat Coast of China. *Canadian Journal of Fisheries and Aquatic Sciences*, 40(Supplement 1): 160 – 171.

Wang, Ying, M. B. Collins & Dakui Zhu. 1990. A Comparative Study of Open Coast Tidal Flat: The Wash (U. K), Bohai Bay and West Yellow Sea (China). In: Proceedings of International Symposium on the Coastal Zone 1988. China Ocean Press. 120 – 134.

Wang, Ying, Mei-e Ren & Dakui Zhu. 1986. Sediment Supply to the Continental Shelf by the Major Rivers of China. *Journal of the Geological Society*, London, 143: 935 – 940.

Sea-level Changes, Human Impacts and Coastal Responses in China[*]

SEA-LEVEL CHANGES IN THE PAST 100 YEARS

The earliest systematic analysis of recent sea-level changes in China was by EMERY and YOU (1981) based on tide-gauge records of eight stations. A later paper by EMERY and AUBREY (1986) discussed relative sea-level changes in China in greater detail using both simple regression analysis and eigen analysis of the records from thirteen stations together with analysis of local environment conditions. Similar work by Chinese scientists (e. g. WANG, 1986), and detailed studies in recent years have improved the knowledge of sea-level changes in several regions of China (ZHAO et al., 1990; CHEN, 1991; WANG et al., 1991). Since then, more comprehensive and more reliable accounts of relative sea-level changes for the whole of China have been published (REN, 1993, 1994; SHI and YAN, 1994).

It is generally estimated that global sea level rise over the past 100 years has been 1~2 mm/year (SCOR WORKING GROUP 89, 1991); that is, a total 10~25 cm rise estimated by IPCC "Climate Change" 1995 (1996). The State Oceanic Administration of China summarized tide-gauge data from 44 stations along the China coast over the last 30 years (1959—1989); the mean rate of sea-level rise was 1. 4 mm/year. According to the geodetic survey data from nine stations along China coast by the Station Survey Bureau of China in 1992, the sea-level rise during last hundred years has totaled 19 cm in the East China Sea, and 20 cm in the South China Sea; the rate of sea-level rise has been 2~ 3 mm/year, which is predicted to continue in the future. The result of calculating data from 102 tide-gauge stations in the world are as following: the average sea-level rise has been 15 cm during last hundred years; 29 cm in the Atlantic Ocean, 10 cm in the Pacific Ocean and 39. 6 cm in the Indian Ocean.

Neotectonic movements are variable along the coastal zone of China. Generally uplift has occurred along the bedrock-embayed coasts and the hilly coasts such as those around the Liaodong and the Shandong peninsulas, the hilly coasts of eastern Hebei province, eastern Taiwan Island, most parts of the coasts of Zhejiang, Fujian, Guangdong, Guangxi and Hainan provinces. Subsidence has occurred in the area of several large river deltas, and the area of sedimentary basins since the Pliocene

---

*　Ying Wang: *Journal of Coastal Research*, 1998, Vol. 14, No. 1: pp. 31 - 36.

(Fig. 1). Based on the different sedimentary evidence, it is clear that there is considerable variability in the rates of tectonic change, even within the same coastal zone. The summary table (Table 1) is adopted from the authoritative study by LU and DING (1994) and can be used as a reference to the geologic background of the regional sea-level variation of China's coasts.

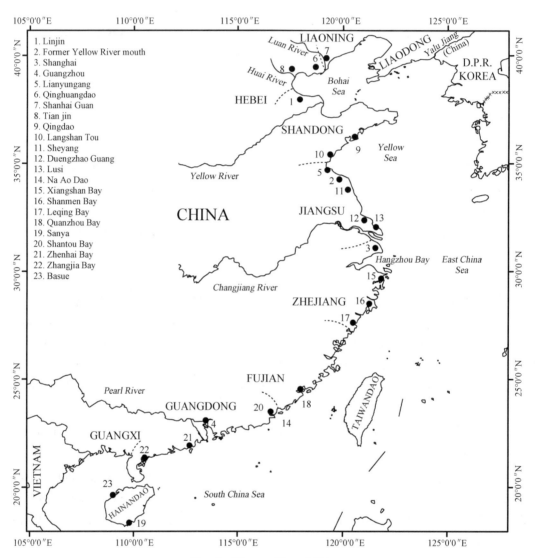

Fig. 1 The coastline of China

Table 1　Neotectonic movements along coast zone of China (from LU and DING, 1994)

Coast location	Type and value	Total Sum of Neotectonic Movement since Pliocene			Vertical Variation during Holocene		
		Uplift (m)	Subsidence (m)	Annual Rate (mm/a)	Uplift (m)	Subsidence (m)	Annual Rate (mm/a)
Liaodong peninsula		50~100		0.02~0.03	4~6		0.5~1.0
Lower Liao River delta			500~3 000	0.12~0.25		5~10, 25	0.3~1.0
Eastern Hebei & western Liaoning		100~200		0.03~0.06	4,5~10		0.3~0.5
West Bohai Bay & Laizhou Bay			500~700	0.18~0.25		10~20 (Bohai Bay)	1.5~2.5 5~10 (Tianjin)
Yellow River delta						0~10 (Laizou Bay)	0.5
Shandong peninsula		50~200		0.02~0.06	+	Stable	±1.0 <0.5 (Qingdao, Penglai)
North Jiangsu			100~300	0.03~0.1		0~5	±0.5
Changjiang River delta			20~500	0~0.2		<10	0.5
Zhejiang & northern Fujian		300~500		0.1~0.2			0.3~1.0
Eastern Guangdong		150~250		0.05~0.1			
(1) Lingjiang & Oujiang River			+	(1) 0.5~1.5			
(2) Mingjiang River				(2) 0.1~0.2			(1)(3) 0.5~1.0
(3) Jiulongjiang River			+	(3) 0.1~0.3		+	(5) 0.5
(4) Hanjiang River				(4) 1.0~2.0			
(5) Pearl River			+	(5) 0.5~1.5			
Guangxi Beihai & Hepu			+	0.2~0.3			
Hainan Island, North			400~600	0.15~0.2		−2~10	0.2~0.5
Hainan Island, South		+		0.05~0.2			
Taiwan Island (East)		1 500		3~5,5~7	40~50		6.2

After correcting for vertical movements, the annual average rate of mean sea-level rise along the coast zone of China is 2. 0 mm/year (CHEN *et al.* , 1994).

RELATIVE SEA-LEVEL RISE OVER THE NEXT CENTURY

The best estimate of global sea-level rise over the next century is from the Intergovernmental Panel on Climate Change (IPCC, 1992) under scenario A (Business as usual) which involves a projected sea level of 18 cm (4. 5 mm/year) in 2030, and 66 cm (6. 0 mm/year in 2100 with a high estimate at 110 cm and a low estimate at 31 cm). There are considerable uncertainties in these estimates. First, future sea-level rise is strongly dependent on the future rise of global mean air temperature which, in turn, is affected by future increases in concentrations of greenhouse gases, particularly CO_2, from the energy sector which span a broad range of futures. As CO_2 emissions are chiefly of anthropogenic origin, their magnitude depends strongly on socioeconomic factors, such as population and economic growth, changes in economic structures, energy prices, and government policies which change greatly through time, and are difficult to predict with precision. A second uncertainty is the stability of the West Antarctic Ice Sheet, the world's only marine based ice sheet to survive the last glaciation. Owing to its large volume it would, if disintegrated and melted, raise global sea level about 6 m. However, the relative stability/instability of the West Antarctic Ice Sheet in the last inter glacial and in the future under conditions of continued global warming is still poorly known, and there is no answer yet on this issue. Third, recent improved knowledge of the emission and behaviour of greenhouse gases may modify previous estimates of the rate of global warming. For example, global emissions of methane from rice paddies may prove less than previously estimated. The cooling effect of air-borne particles from sulphur emissions may have offset a significant part of the greenhouse warming in the northern hemisphere during the past several decades. It is possible therefore, that the previous global warming rate may be reduced, as may the rate of global sea-level rise.

HUMAN IMPACTS AND COASTAL RESPONSES

Sea-level rise will increase the water depth of the submarine coastal slope, and gradually decrease the winnowing action of waves on submerged coastal sediments, but erosion on upper beaches by breaking waves is likely to be enhanced. At the same time, the slopes of river beds will be reduced decreasing fluvial sediment discharges. On the other hand, human impacts involving diversion of river discharge for urban water supply, and the construction of dams for irrigation, have decreased sediment discharges enormously. In the Luanhe River for example, the original water discharge was 4.19×10^9 m^3 with a total sediment supply of 2.22×10^7 tonne. After diversion, the water discharge of 3.55×10^8 m^3 for water supply to Tianjin city reduced discharge and sediment supply, which fell to only 1.03×10^6 tonne. As a consequence, the delta coastline of the Luanhe River has been retreating at a rate of 17.4 m/year since 1988 (QIAN, 1994). The same situation has happened in the Yellow River; the annual sediment discharge of the Yellow River has decreased from 11×10^8 tonne to be 9.5×10^8 tonne as the annual water discharge now is only 3.79×10^8 m^3, a reduction associated with water diversion for irrigation along its middle and lower reaches. It is estimated that more than 1.5×10^{10} m^3 of water per year has been taken from the Yellow River for irrigation over the last decade, and with the volume of water about 1.74×10^8 tonnes of sediment per year. The increasing diversion of water for irrigation and other purposes over the next decade will withdraw considerable amounts of sediment from the Yellow River. As a result of both natural and human influences, a reduction in sediment supply to the coastal zone is a world-wide phenomenon, particularly when combined with the increasing frequency of storm surges and El-Niño events accompanying the sea-level rise. Beach erosion and sand-barrier retreat are the result of stronger hydrodynamic processes and a smaller volume than previously of sediment supply to the coast. Directly-monitored data indicate that sandy-coast retreat is occurring at a rate of $1.5 \sim 2$ m/year on average along the Shandong and Liaodong peninsula coasts even with sediment supply from many small rivers along the Bohai Sea and the Yellow Sea; the $2 \sim 5$ m/year average value in the Zhejiang and Fujian coasts along the East China Sea is exacerbated by the mining of sands along this coast. Along the red-sand terrace coast in Fujian and Guangxi in the East and South China Seas, the net rate of coastline erosion is $0.7 \sim 1.5$ m/year. Even though most bed-rock (granite and gneiss) coasts experience limited erosion, several parts exhibit as much as $0.07 \sim 0.1$ m/year retreat.

By using the Bruun role, the response of major tourist beaches in Dalian, Qinhuangdao along the Bohai Sea, Qingdao along the Yellow Sea, Beihai and Sanya along the South China Sea, has been estimated. It is predicted that they will lose $13\% \sim 66\%$ of their present area while sea level is continually rising to 50 cm higher by the year 2100 (Table 2). Beach protection and beachnourishment provide the principal management solutions to these problems.

The response in the muddy tidal flat coasts is more complex and potentially more serious, as the major types of muddy flat coasts are distributed along the lower reaches or delta plains of larger rivers in the areas of tectonic subsidence. Even within tectonically uplifted regions, such as the Zhejiang, Fujian and Guangdong coast zone, there are embayed tidal flats which experience significant subsidence (WANG, 1983, 1994).

Table 2　The predicted net results of beach erosion in, China by 0.5 m of sea-level rise during the next century (from WANG and Wu, 1995).

Location	Estimated Beach Response of 0.5 m Sea Level Rising												
	Modern Beach				Natural Flooding			Beach Erosion			Sum Value		
	Length (m)	Average Width (m)	Relative Height (m)	Areas (m²)	Beach Retreating (m)	Loss of the Area (m²)	Rate of Loss (%)	Beach Retreating (m)	Loss of the Area (m²)	Rate of Loss (%)	Beach Retreating (m)	Loss of the Area (m²)	Rate of Loss (%)
Dalian													
XingHaiPark	2 125	68.5	6.1	145 613	6.8	14 450	9.9	26.5	56 314	38.7	33.3	70 764	48.6
Dongshan Hotel	510	42.4	3.8	21 645	7.4	3 774	17.4	15.8	8 078	37.3	23.2	11 852	54.7
Grand Beach	756	56.3	3.8	42 560	7.9	5 972	14.0	24.7	18 674	43.9	32.6	24 646	57.9
Summary	3 391	61.9		209 818	6.8~7.9	24 196	11.5	15.8~26.5	83 066	39.6	23.2~33.3	107 262	51.1
Qinghuangdao													
Bei Dai He	7 850	87.1	5.9	683 456	8.7	68 295	10.0	48.8	383 080	56.1	57.5	451 375	66.1
West Xiang He Zhai	3 124	223.6	6.4	698 466	6.7	20 930	3.0	41.5	129 650	18.6	48.2	150 580	21.6
Shandong Bao	756	88.2	3.5	66 672	7.5	5 670	8.5	25.4	19 202	28.8	32.9	24 872	37.3
Summary	11 730	123.5		1 448 594	6.7~8.7	94 895	6.6	25.4~48.8	531 932	36.7	32.9~57.5	626 827	43.3
Qingdao													
Qingdao Bay	1 356	72.8	6.0	98 650	8.5	11 526	11.7	37.9	51 455	52.2	46.4	62 981	63.9
Hui Quan Wan	1 124	70.6	6.0	79 356	7.0	7 868	9.9	38.6	43 386	54.7	45.6	51 254	64.6
Fu Shan Sou Mouth	1 625	193.1	5.4	313 857	8.9	14 462	4.6	26.4	42 932	13.7	35.3	57 394	18.3
Summary	4 105	119.8		491 863	7.0~8.9	33 856	6.9	26.4~38.6	137 773	28.0	35.3~46.4	171 625	34.9

(Continued)

Estimated Beach Response of 0.5 m Sea Level Rising

Location	Modern Beach				Natural Flooding			Beach Erosion			Sum Value		
	Length (m)	Average Width (m)	Relative Height (m)	Areas (m²)	Beach Retreating (m)	Loss of the Area (m²)	Rate of Loss (%)	Beach Retreating (m)	Loss of the Area (m²)	Rate of Loss (%)	Beach Retreating (m)	Loss of the Area (m²)	Rate of Loss (%)
Beihai													
Wai Sha	2 530	60.8	6.2	153 750	5.8~9.5	17 254	11.2	27.9	70 587	45.9	33.7~37.4	87 841	57.1
Da Dun Hai	5 516	258.4	5.0~9.2	1 425 588	5.4~9.8	41 926	2.9	48.1	265 335	18.6	53.5~57.9	307 261	21.5
Bai Hu Tou	5 165	183.2	5.0~7.2	946 363	5.4~8.7	36 457	3.9	45.2	233 458	24.7	50.6~53.9	269 915	28.6
Summary	13 211	191.2		2 525 701	5.4~9.8	95 637	3.8	27.9~48.1	569 380	22.5	33.7~57.9	665 017	26.3
Sanya													
Da Dong Hai	2 650	81.5	5.9	215 905	7.9	20 935	9.7	12.2	32 330	15.0	20.1	53 265	24.7
Ya Long Bay	8 880	166.1	5.4~13.4	1 475 184	6.8~9.8	74 592	5.1	12.7	112 776	7.6	19.5~22.5	187 368	12.7
SanYa Sand Bar	16 360	296.2	3.3~11.6	4 846 024	5.6~10.2	137 654	2.8	43.6	713 296	14.7	49.2~53.8	850 950	17.5
Summary	27 890	234.4		6 537 113	5.6~10.2	233 181	3.6	12.2~43.6	858 402	13.1	20.1~53.8	1 091 583	16.7
Total	60 327	185.9	3.3~13.4	11 213 085	5.4~10.2	481 765	4.3	12.2~48.8	2 180 553	19.4	20.1~57.9	2 662 314	23.7

Background subsidence of muddy coasts has been occurring since the Tertiary and the total subsidence is quite high. Additional factors of over pumping groundwater for fresh water supply and irrigation, and overloading by constructions (the larger cities in China are located mostly in river mouth areas along the coastal zone) exacerbate modern subsidence of these areas. These processes have compacted sediments and caused rapid relative sea-level rise, especially in the delta plain coasts, where the major muddy flat coasts in China are located. Four examples are given below.

(1) Tianjin city and harbour are located in the northern part of the North China plain. They can be taken as an example of sites on the old Yellow River delta in west Bohai Bay. The rate of land subsidence is shown in Table 3 (from HAN 1994). Recently, expansion of the city area has been controlled in order to decrease the rate of subsidence, but in the coastal area to which much of the growth has been redirected, particularly around Tanggu and Hangu, the average rate of subsidence during 1989—1991 was 29 mm/year and 64 mm/year (HAN, 1994). Including land subsidence, the relative sea-level rise was 24.5 mm/year between 1956 and 1985 (REN, 1994), but reached 50 mm/year between 1983 and 1988 with additional land subsidence. The 1983—1988 average rate is far greater than the average rate of sea-level rise either of global or for the whole coast of China. At present sea levels, the frequent high water can inundate the area below the 2 m contour and can flood the area below 3 m contour line where high tide coincides with a storm surge which may reach 1 m in height (HAN, 1994). As a comparison, the rate of land subsidence in the modern Yellow River delta was 3～4 mm/year between 1956 and 1985, a reflection of the limited economic and urban development in that area at the time; the average rate of relative sea-level rise was 4.5～5.5 mm/year during the same period (REN, 1994). The rapid increase of relative sea-level rise here has brought disaster to the rapid development of Shengli oil field and the surrounding agriculture.

Table 3 Land subsidence (mm) 1985—1991 in the Tianjin, area (from HAN, 1994)

	Tianjin City	Tanggu Harbour	Hangu, NE Coast	Dagong, SE Coast
1975	130	144		
1980	89	36		
1981	119	140		
1982	94	188		
1983	71	116		
1984	58	137		
1985	86	100		
1986	62	54	40	41

（**Continued**）

	Tianjin City	Tanggu Harbour	Hangu, NE Coast	Dagong, SE Coast
1987	43	46	46	41
1988	24	29	53	25
1989	18	44	68	56
1990	15	19	66	18
1991	17	24	38	45

（2）The North Jiangsu coastal plain was formed by the coalescence of the ancient Yellow and Changjiang River deltas. It is the most extensive mud flat coast in China. The coast is in an early stage of economic development and may this represent the natural condition of such environments. Recent sea-level rise is $2\sim3$ mm/year here according to tide-gauge records. This is particularly worrying because any rise of sea level here will have severe results on account of the extensive low-lying nature of this coast. For example, the strong tropical storm from August 29 to September 1, 1992, which was superimposed on a high astronomical tide, caused the water level to rise $2\sim3$ m above the normal maximum flood tide level. This disaster resulted in the death of 300 people and an economic loss of about 9.2 billion yuan.

（3）Shanghai is representative of the situation at the Changjiang River mouth. This delta is the largest urban area in China with a large population and developing economy. Both over pumping of groundwater and overloading by constructions has caused serious land subsidence and rapid relative sea-level rise, even though artificial recharge of aquifers has been regarded as successful (Table 4, from LIU 1994). But the lowland suffers from storm surges and floods, phenomena which are increasing in both frequency and magnitude as a result of recent relative sea-level rise. The relative sea-level rise is $6.5\sim11.5$ mm/year, which had caused saline water intrusion, floods which have extended 170 km upstream from river mouth, and harmed industry, agriculture and the daily life of Shanghai residents.

Table 4   Land subsidence of Shanghai and its vicinity (from LIU, 1994)

Time Period	Subsidence Rate (mm)	
	Annual Average	Annual Maximum Average
1921—1948	24.0	42
1949—1956	40.0	96
1957—1961	110.0	287
1962—1965	69.0	164
1966—1992	2.5	19.3

（4）Pearl River (Zhujiang) delta. The ground subsidence of $0.5\sim1.5$ mm/year in the Pearl River delta is less than for other deltas, but increased human influences, such as the reclamation project to unite embankments and the damming of distributaries, have affected river-bed and water-level characteristics. Thus, there are two types of seasonal change in tidal level: estuary type and coastal type. The amplitudes of annual sea-level change has varied by 20~30 cm over the past 30~40 years. Relative sea-level rise has occurred at a rate of 1~2 mm/year since 1955 (CHENG and YANG, 1994). Rising sea-level effects on the estuary environment have caused the submergence of a large area of lowland, water logging of construction foundations, saline-water intrusion harming irrigation and drainage, threatening the town, cities, airports, and harbours along the coast.

The above examples indicate the effects and spatial variations of sea-level rise in the muddy coasts, particularly those affected greatly by human activity. Coastal responses differ from sand to rocky coasts. Coastal erosion is very variable depending on the coastal sediment supply, which is mainly from larger river inputs. Under conditions of continued rise of sea level, the frequency and amplitude of storm surges has been enhanced. This has caused coastal erosion, lowland flooding over coastal marshes, flats, and fishing villages; it has extended saltwater intrusion inland to harm coastal aquifers, agricultural irrigation and freshwater supply, and has also prevented the effective dispersal of flood waters and urban pollutants. The risk to China's coastal zone is manifest, as a total of 11 provinces and more than 60% of larger cities in China are located in the coastal zone, about one half of total annual agricultural products are from the coastal region, and the height to which most areas of the coastal plain rise is less than 2 metres above present sea level. A systematic study of the effects of sea-level rise along the whole coastal zone of China, especially the river mouth area and the lowland coasts, is required to understand properly the challenge of this situation and to solve the problems step by step.

LITERATURE CITED

CHEN, T. and YANG, Q. 1994. Studies on Secular Trends of Sea Level in Zhujiang River Estuary in the Past Decades. Beijing: Science Press. 53 - 62.

CHEN, X. 1991. Sea level changes since early 1920s from the long records of two tidal gauges in Shanghai, China. *Journal of Coastal Research*, 7: 787 - 799.

CHEN, Z., ZHOU, T., Yu, Y., and TIAN H. 1994. Mean Sea Level Changes along the Coast of China. In: *The Yellow Book of Geo No. 1 Chinese Academy of Science*. Beijing: Science Press. 40 - 44.

EMERY, K. O. and AUBREY, D. G. 1986. Relative Sea Level Changes from Tide-gauge Records of Eastern Asia Mainland. *Marine Geology*, 72: 33 - 45.

EMERY, K. O. and AUBREY, D. G. 1988. Coastal Neotectonics of the Mediterranean from Tide Gauge

Records. *Marine Geology*, 81: 41 – 52.

EMERY, K. O. and You, F. 1981. Sea-Level Changes in the West Pacific with Special Emphasia on China. *Oceanologica Liminologica Sinica*, 12: 297 – 310.

HAN, M. 1994. Impacts of Sea-Level Rise on the North China Coastal Plain and the Coast-Benefit Analysis of the Prevention Measures Concerned. In: *The effects and responses to China's delta area by the sea level rising*. Beijing: Science Press. 339 – 356.

IPCC. 1996. *Climate Change* 1995. Cambridge: Cambridge University Press. 359 – 406.

Liu, T. 1994. Analysis on Mechanism of Ground Subsidence in the Shanghai Area. 100 – 110.

Lu, Y. and DING, G. 1994. Neotectonic Movements in the Coastal Zone of China. In: *The Yellow Book of Geo No.* 1 *Chinese Academy of Science*. Beijing: Science Press. 63 – 74.

QIAN, C. 1994. Effects of Water Conservancy Projects in the Luanhe River Basin on Luanhe River Basin on Luanhe River Delta, Hebei Province. *ACTA Geographica Sinica*, 49: 158 – 166.

REN, M. 1993. Sea-Level Changes in China. *Journal of Coastal Research*, 9: 229 – 241.

REN, M. 1994. The Trends of Sea Level Rising in Huanghe, Changjiang and Zhujiang Deltas and a Prediction of Sea Level Rising in 2050. In: *The Yellow Book Geo No.* 1. *Chinese Academy of Science*. Beijing: Science Press. 18 – 26.

SCOR WORKING GROUP 89. 1991. The response of beaches to sea level changes, a review of predicative models. *Journal of Coastal Research*, 7: 895 – 921.

SHI, Y. and YAN, G. 1994. Sea Level Rise and Its Impacts in China. *The Yellow Book of Geo No.* 1. *Chinese Academy of Science*. Beijing: Science Press. 163 – 173.

VAN DER VEEN C. J. 1988. Projecting Future Sea Level. *Survey in Geophysics*, 9: 389 – 418.

WANG, B. , et al. 1991. An Analysis of Seasonal Variation in Sea Level along the Changjiang Estuary. China, Shanghai, East China Normal University Manuscript.

WANG, Y. 1983. The Mudflat Coast of China. *Canadian Journal of Fisheries and Aquatic Sciences*, 40 (supplement 1): 160 – 171.

WANG, Y. and Wu, X. 1995. Sea Level Rising and Beach Response. *ACTA Geographica Sinica*, 50: 118 – 127.

WANG, Y. and ZHU, D. 1994. Tidal Flats in China. *In*: *Oceanology of China Seas*. Dordrecht: Kluwer Academic, Volume 2. 445 – 456.

WANG, Z. 1986. The Sea Level Changes in 20th Century. In: *Sea Level Changes of China*. Beijing: China Ocean Press. 237 – 2245.

ZHAO, X. and WANG, S. 1992. The Relationship Between the Changes of Holocene Sea Level, Climate and Coast Evolution in China. In: *The Holocene Megathermal Climate and Environment of China*. Beijing Ocean Press. 11 – 120.

人类活动与黄河断流及海岸环境影响[*]

　　黄河是我国第二大河,源头海拔 4 368 m,全长 5 464 km,流域面积 $75.2×10^4$ km²,总落差 4 830 m(图 1)[1]。流域年平均气温-4 ℃～14 ℃[2],年平均降水量为 464 mm(1956—1979),流域内蒸发量大,有效蒸发量 350 mm。流域内径流系数变化大,上游兰州地区径流系数为 47.4%,河套平原地区约为 23.2%,中游黄土地区和下游平原仅为 15%左右[3]。

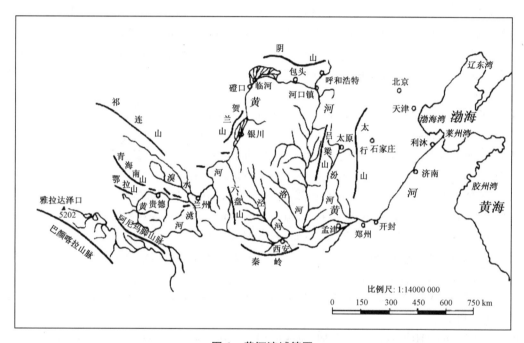

图 1　黄河流域简图

一、河流特点

　　黄河特点可概括为:水少沙多,水沙异源、时空分布不均匀,下游(三角洲)河道迁徙频繁。

(一) 水

　　黄河径流水量源自冰川、积雪融化与大气降水。黄河流域水资源总量[①]平均(1956—

---

　　[*]　王颖,张永战:《南京大学学报(自然科学)》,1998 年 34 卷 3 期,第 257 - 271 页。
　　①　水资源总量是指地表水与地下水净量之总和。

1979)为 744×10^8 m$^{3[4,5]}$,多年实测天然径流量的平均值为 580×10^8 m$^{3[1,6]}$,实测多年平均径流量在陕县(1919—1989)为 464×10^8 m$^{3[1]}$,在花园口(1919—1979)为 563×10^8 m$^{3[1,4]}$,在黄河尾闾三角洲始点利津站(1855—1988)为 493×10^8 m$^{3[1]}$。黄河径流水量主要来自兰州以上的上游地区,该区流域面积占全流域总面积的29.6%,径流量则占全流域多年平均天然径流量的55.7%,雨水补给占上游来水的70%;兰州至河口镇流域面积占全流域的21.7%,但基本无水量补入;河口镇至龙门流域面积占14.8%,来水量占12%;龙门至三门峡之间流域面积占25.4%,来水量占14%;三门峡至花园口流域面积占5.5%,来水量占17%;花园口以下(即下游地区)流域面积占3.0%,来水量占3.0%(表1)。可见,水量的空间分布是不均匀。

表 1　黄河流域各段来水来沙情况

站名	控制流域面积(×10⁴ km²)	区间流域面积(×10⁴ km²)	占流域总面积的%	多年平均径流量(×10⁸ m³)	区间来水(×10⁸ m³)	占流域天然径流量的%	多年平均输沙量(×10⁸ t)	区间来沙(×10⁸ t)	占平均输沙量的%
↓		22.26	29.6		323	55.7		1.15	7.2
兰州	22.26			323			1.15		
↓		16.33	21.7		−10	−1.7	(1935—1972)	0.44	2.8
河口镇	38.59			·313			1.59		
↓		11.17	14.8		72	12.4	(1954—1972)	9.61	60.1
龙门	49.76			385			11.2		
↓		19.08	25.4		79	13.6	(1934—1972)	4.50	28.1
三门峡	68.84			464			15.7		
↓		4.16	5.5	(1919—1989)	99	17.1	(1919—1958)		
花园口	73.00			563			13.1	0.30	1.9
↓		2.24	3.0	(1919—1979)	17	3.0	(1949—1972)		
利津	75.24			493			11.5		
				(1855—1988)			(1950—1972)		

据[1][2][7]有关数据计算。

黄河径流季节变化明显,每年有凌汛(3月)、伏汛(7、8月)与秋汛(9、10月),其中以伏汛水量大,凌汛危害明显,上中游高寒区3月融冰随水下泻,至平原河段,因河道狭窄,冰块叠积成坝,后来之水被堵而破堤泛滥。黄河径流年际变化突出,具丰水与枯水交替变化和连年枯水的特点(图2),连续枯水时段为3年、4年、5年、6年、7年、8年及11年。如1922—1932年的连续11年枯水,1969—1974年连续6年枯水,1977—1980年、1986—1989年连续4年枯水,以及1991—1997年连续7年枯水[1,6]。以利津站实测资料为例,1964年径流量高达 937×10^8 m^3,最小值仅为 91.5×10^8 m^3(1960年),两者比值达10.3(表2)[1,8]。显然,径流量时间分配极不均匀。

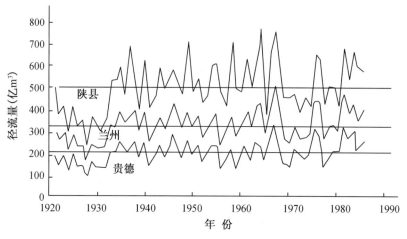

图2　陕县、兰州、贵德1919—1989年天然径流过程曲线

表2　黄河三门峡站和利津多年平均径流量与输沙量特征值

水沙量	测站	多年平均	实测时间	最大值	出现时间	最小值	出现时间	最大值/最小值
径流量 ($\times 10^8$ m³/a)	三门峡	464*	1919—1958	660★	1937	201★	1928	3.3
	利津	493*	1950—1987	937★	1964	91.5★	1960	10.2
			1855—1988	904*	1964	104*	1987	8.7
输沙量 ($\times 10^8$ t/a)	三门峡	15.6*	1919—1958	39.1**★	1933	4.88***★	1928	8.0
	利津	13.4*	1950—1987	21.08**★	1958	0.96★	1987	22.0
			1855—1988			0.66*	1987	31.9

据[1](＊)、[8](★)、[9](＊＊)计算。

黄河上、中游地区年平均降水量自20世纪50年代至80年代略有降低。90年代以来,气温升高,降水偏少12％,上游地区来水持续偏枯,1990—1995年6年中少来了一年的水量。受气候影响,黄河上游来水量已降至历史最低点,黄河兰州段1997年1至4月水量比多年平均值减少了三成(黄河水利委员会上游水文水资源局,1997)。黄河上中游水量调度委员会估计:1997年第一季度黄河上游来水比多年同期平均来水偏少1/4以上,以龙羊峡水库的入库量为指标,1997年1~3月,流量比多年平均流量小33 m³/s,达历史最小值。刘家峡水库流量也降低,两库蓄水量比1995年同期少24.9×10⁸ m³,比严重枯水年1996年少14.3×10⁸ m³,达到建库以来的最低值①。

(二) 沙

黄河含沙量大,多年平均含沙量高达37.6 kg/m³[1,7,8,10],洪汛期龙门站最大含沙量

---

① "黄河之水天上来"气势难现——上游来水量降至历史最低点.信息日报,1997年4月2日。

达 933 kg/m³（1966 年 7 月 18 日 9 时～7 月 20 日 19 时）与 826 kg/m³（1970 年 8 月 2 日 0 时～8 日 5 日 0 时）[9]。黄河多年平均输沙量为 16×10⁸ t[6]，在陕县（1919—1989）为 15.6×10⁸ t[1]，以利津站（1855—1988）为代表的入海输沙量为 13.4×10⁸ t[1]。进入下游河道的泥沙，约有 1/4 淤积在利津以上地段，3/4 入海（1/2 淤积在利津以下的河口三角洲及滨海地区，1/4 被输送至深海）。黄河泥沙 90% 来自中游黄土高原，河口镇至龙门河段黄土高原面积 11×10⁴ km²，区间输沙达 9.6×10⁸ t，占全流域输沙量的 60%；龙门至三门峡河段输沙量占 28%；上游来沙占 10%；三门峡以下仅占 2%（表 1）。显然，黄河泥沙空间分布不均匀。同时，水沙异源特点突出。

黄河泥沙季节变化明显，年内分配不均匀，水沙量均主要集中于汛期（7～10 月），汛期水量占年水量的 60% 左右，沙量更加集中，占年沙量的 85% 以上，且集中于几场暴雨洪水[2]。如三门峡站 1919—1960 年统计，其多年平均径流量为 424×10⁸ m³，年输沙量 15.9×10⁸ t，年内水沙集中于汛期 7～10 月，分别占全年的 60.8% 和 89.7%[1]。泥沙亦有丰、枯相间的周期性年际变化，且输沙量年际变幅远远大于径流量。三门峡站年最大输沙量达 39.1×10⁸ t（1933 年），最小仅为 4.88×10⁸ t（1928 年），两者比值达 8，利津站两者比值更是高达 32（表 2）。可见，黄河泥沙时间分配极不均匀。

黄河水沙异源，上游是主要的来水区，泥沙则集中产自中游黄土高原。年内，水沙集中于汛期（7～9 月），往往来自几次暴雨洪水。年际，水沙变化大，洪水泥沙的搭配视暴雨降落区域的不同而出现丰水多沙年、丰水少沙年、枯水多沙年或枯水少沙年的变化。同时，年沙量变幅大于年水量的变幅。

近年来，由于引水引沙量增加，黄土高原区水土保持工作不断推进，使 34.8% 的水土流失区获得控制，黄河入海年径流量已减少为 270×10⁸ m³，年输沙量约为 6.4×10⁸ t～9.0×10⁸ t（表 3）。

表 3   黄河流域降水、径流、输沙与引水引沙情况表

区、站		年代	1950—1959	1960—1969	1970—1979	1980—1989
中上游多年平均汛期降水量（mm）			332.00	329.00	326.00	324.00
花园口多年平均汛期径流量（×10⁸ m³）			371.00	378.00	318.00	360.00
利津站	径流量（×10⁸ m³）		464.00	513.00	304.00	270.00
	输沙量（×10⁸ t）		13.20	11.00	8.90	6.40
全流域	年引水量（×10⁸ m³）		125.00	175.00	233.00	274.00
	年引沙量（×10⁸ t）		1.63	1.36	2.54	1.61

据[1]、[11]、[12]计算。

（三）下游（三角洲）河道迁徙频繁

黄河自桃花峪出峡谷进入下游河段，平原河谷宽展，河流坡降骤低（平均比降 0.12‰），使水流速度突然降低，导致大量泥沙堆积。华北大平原的发育过程就是黄河搬移黄土高原的泥沙、移山填海的造陆——堆积三角洲、山麓冲积扇逐渐发育为泛滥平原的

过程。巨量泥沙于下游不断加积,形成高出平原地面 13 m(开封)与 21 m(新乡地区)的地上悬河。下游平原以千里长堤夹峙来维护黄河河道。在洪水期,易发生溃堤、循低而流的自然改道与泛滥。

自 1855 年黄河北归入渤海以来,140 年间,尾闾河道决口 50 多次,较大改道 10 次。尾闾河道变迁具有沿顺时针方向迁徙的特点:大致自北向转北东向,再转东向,继而转南东向,再返回至北向,自然改道周期约 6～10 年。当黄河入海年输沙量为 $9 \times 10^8 \sim 12 \times 10^8$ t 时,行水 10 年左右的尾闾河道已淤高而易发生改道[10]。华北平原的古黄河三角洲与古冲积扇,河道变迁的原因亦大致如此。

据历史记录,自公元前 602 年至 1938 年的 2540 年中,黄河下游决口 1 591 次,改道 26 次,大体上以 2 年左右一次的频率决口,以 100 年为周期改道[1]。自公元前 2278 年以来,黄河下游经历过 8 次大改道(图 3),北部沿永定河水系经天津入渤海,南部夺淮河水系入黄海,南北距离达数百公里。黄河改道,对下游与河口海岸生态环境带来一系列影响[10,13]。下游改道迁徙变动大是水少沙多流经平原区的河流的共同特点。西周战国时期(3000—2200 aB. P.),黄土高原地区草原与森林植被保存尚好[14,15],河流泛滥少;秦与西汉时(公元前 221—公元 8 年)移民开垦,肇使水土流失与河流泛滥;东汉—魏晋南北朝

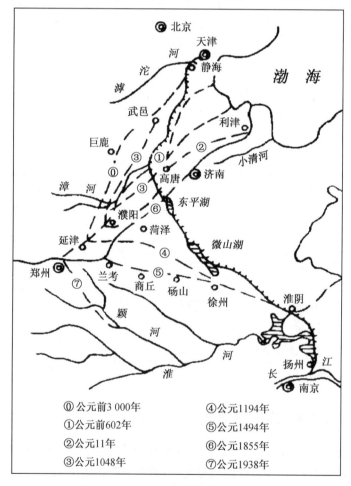

⓪ 公元前3 000年   ④ 公元1194年
① 公元前602年   ⑤ 公元1494年
② 公元11年   ⑥ 公元1855年
③ 公元1048年   ⑦ 公元1938年

图 3　黄河下游改道

(8—581年)废农还牧,黄河稳定流动达500多年[16,17,18]。唐时(公元900年),大规模移民边寨,屯垦黄土高原,砍伐林木与破坏草场,变牧畜为农耕,又召致严重的水土流失,黄河泥沙增多,嗣后,下游河道淤积加速,改变频频[19]。

二、断流——20世纪后期发生发展中的变化

(一) 从泛滥到断流

历史上,黄河下游决口泛滥频繁曾称谓为"黄患"(Yellow Sorrow),水灾范围北起天津,南达江淮,纵横25×10^4 km²,波及冀、鲁、豫、皖、苏五省。人民流离失所,良田变沙荒。新中国成立后,加固大堤、建库发电、引黄灌溉、束水排沙,减轻了黄河的水患。但是,黄河流域沿途开发,不断扩大灌溉与用水量,黄河"入不敷出",出现断流现象。

1960年,黄河下游出现断流。据利津站记录,自1972—1996年有19年断流(表4),累计断流57次①,共641 d,河床干涸可行车!

表4 黄河断流情况统计

时间段	断流年数	累计天数(d)	年断流最长时间(出现年份)	断流始/终日期	断流出现频率(%)	平均断流距离(km)
1960—1969	1	41	41(1960)	4月/6月	10	
1970—1979	6	71	19(1979)	4月/6月	60	242*
1980—1989	7	103	37(1981)	4月/6月	70	256*
1990—1996	6	467	136(1996)	2月/7月	86	392*

据黄委会资料与水文年鉴资料计算,*据《人民政协报》,1997.10.18.(第一版)

1991—1997年,黄河连年断流(图4),断流始期提前,断流时间延长,断流距离加大。1992年为一次断流高峰,利津站从4月28日开始断流,5月22日至7月22日连续断流

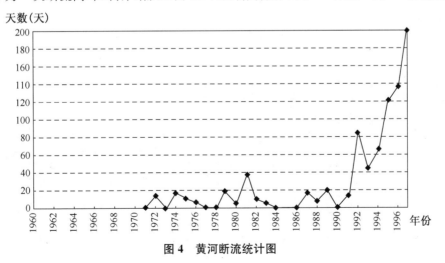

图4 黄河断流统计图

① 拯救黄河.科技日报,1997年5月10日,第一版。

61 天,全年累计断流 84 天,河口区达 128 天[20],断流河道长达 339 km;1995 年断流河道从河口延伸至开封陈桥附近,长达 683 km,为下游河长的 87%;1996 年断流时间更是累计长达 136 天,接近全年的 2/5。1997 年断流情况更为严重,2 月 7 日开始断流,利津站断流 10 次,经中上游人工调水至下游河道才结束断流,累计断流时间达 226 天,全年一半以上的时间黄河无水,甚至绵延至 9 月黄河秋汛期,干涸河道已扩展到开封柳园口,长约 700 km(表 5)。

表 5 1990 年代黄河断流情况(利津站)统计

年份	断流始/终日期(月、日)	断流天数	断流河道长度(km)
1991	5.16/6.1	14	178
1992	4.28/7.22	84	339
1993	2.13/6.	45	299
1994	4.4/6.	66	339
1995	3.4/7.14	122	683
1996	2.24/7.14	136	340
1997	2.7/10.	226	700

据黄委会资料统计。

(二)黄河断流与降水、径流及人工引水比较结果

(1)黄河下游断流现象始自 20 世纪 60 年代,至 90 年代愈益严重。60 年代,仅 1960 年断流;70 年代,6 年发生断流,4 年(1970 年、1971 年、1973 年、1977 年)未曾断流;80 年代则有 7 年发生断流,仅 3 年(1984 年、1985 年、1986 年)未发生断流;90 年代以来,除 1990 年 1 年未断流,至今已连续 7 年,年年断流,断流时段不断延长,形成峰值。

(2)断流是人为因素——用水量增加过量、滥伐滥垦使水土流失严重与自然因素——中上游汛期降水量变化使径流量递减双重作用的结果。如图 5 和图 6 所示,多年

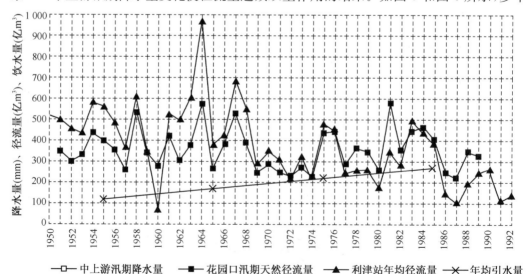

图 5 1950—1990 年黄河降水量与径流量及流域引水量变化曲线

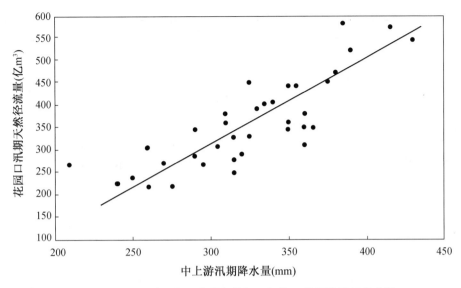

图6　黄河中上游汛期降水量与花园口汛期天然径流量相关曲线

统计资料表明：黄河径流量（R）的丰枯与中上游降水量（P）呈线性相关（$R=1.54P-147.60$），相关系数达0.83。中国5 000年来气候变化具有寒冷期与温暖期交替的规律，20世纪进入温暖期，气温逐渐升高，近期降雨减少，干旱现象上升[21,22]。自然因素（干旱年份）加上人工引水过多，致使黄河断流加剧（表6）。

表6　历年黄河断流概况

年份	中上游汛期降水量 （mm）	花园口汛期径流量 （×10⁸ m³）	下游断流天数 （天）	备注
1952	260	310	0	年引水量约125×10^8 m³
1957	270	275	0	
1958	430	540	0	大炼钢铁，滥伐林木，大办公
1959	355	345	0	社，开荒种地，召致严重的水土 流失，降水量不低，但水量难以 转化为径流
1960	290	297	41	
1965	210	280	0	年引水量约175×10^8 m³
1969	300	250	0	
1970	320	290	0	
1971	315	250	0	年引水量$175\times10^8\sim200\times$ 10^8 m³
1972	240	220	15	
1973	355	305	0	
1974	262	220	18	前一年干旱，春季断流
1975	355	440	11	
1976	375	450	7	年引水量约230×10^8 m³，前两 年丰水
1977	315	280	0	

(续表)

年份	中上游汛期降水量 （mm）	花园口汛期径流量 （×10⁸ m³）	下游断流天数 （天）	备注
1978	350	360	1	前一年径流量偏低
1979	365	350	19	经济复苏
1980	300	270	6	经济开始腾飞
1981	385	500	37	降水径流过于集中在汛期，春季断流 年引水量约 250×10⁸ m³
1982	310	360	10	
1983	325	450	5	
1984	380	470	0	
1985	340	405	0	
1986	250	240	0	
1987	275	220	17	年引水量超过 274×10⁸ m³
1988	360	350	8	
1989	315	330	20	前一年枯水，当年引水量大

上述可知：黄河断流是自然因素与人类活动影响的结果。20 世纪 60 年代黄河下游始发断流即是自然因素和人为影响双重作用的结果；70 年代早、中期，断流与降水丰枯密切相关；70 年代末经济复苏，80 年代初经济开始腾飞，工农业发展使用水量急剧增加，80 年代黄河全流域年均引水量达 274×10⁸ m³，引水过多，使黄河自身调节能力降低，除非连续数年降水丰沛，否则，降水量与径流量稍有变化即可能发生断流；80 年代中、后期，断流现象明显反映出上游汛期降水丰枯与下游断流发生间在时间上的滞后效应；90 年代以来，持续数年气候普遍升温、干燥，平均降水量减少，黄河流域引水量已达到 300×10⁸ m³，使 1991—1997 年下游连续 7 年断流。1997 年自 2 月 7 日以来，黄河已先后断流 10 次，当年最后一次来水是 9 月 15 日，25 天后，至 10 月 14 日起再度断流，虽经人工调水调节，全年累计断水亦达 220 天，汛期的 9 月亦发生断流，这是前所未有的严重事件。

（三）黄河断流临界值分析

20 世纪 60 年代，黄河流域年均引水量 125×10⁸ m³，这一时期既使遇到干旱年份，汛期降水量小于 260 mm，汛期径流量低于 330×10⁸ m³，但当年径流量净余值高于 150×10⁸ m³，黄河下游不会出现断流。当引水量为 175×10⁸ m³ 时，既使汛期降水量为 210 mm 低值，或汛期径流量仅为 280×10⁸ m³，但径流量净余值高于 100×10⁸ m³，仍有一定数量径流入海，黄河下游很少出现断流，机遇约为 1/10。所以，200×10⁸ m³ 的人工引水量，或 100×10⁸ m³ 的净余径流量可能是黄河下游断流与否的"临界值"，260 mm 的汛期降水量与 300×10⁸ m³ 的汛期径流量可能是相关数值的低限。

连续多年降水丰沛，即使有一年汛期降水量少于 260 mm，当年春季不一定出现断流。反之,连续多年干旱或有一年异常干旱,随后的1~2年即使降水量高,或降水过量集中于汛期,当年或来年甚至第三年春季仍会出现断流,且旱年后的第一年断流现象非常严重,随后断流天数可能减少。所以,春旱断流常发生于汛期降水量与径流量偏低年之后。

当人工引水量超过 200×10^8 m³,达到 250×10^8 m³ 后,已接近或超过黄河天然年径流的一半时,黄河的"自然调节能力"降低,"抗灾性"脆弱,即使汛期降水量超过 300 mm,径流量为 420×10^8 m³ 时,断流仍会发生。

(四)断流的人为因素分析

黄河流域是干旱与半干旱区,流域面积占国土面积的8%,多年平均径流量为全国径流量的2%,流域内人均水量为全国平均水平的25%,耕地亩均水量为全国平均数的17%[①],是水资源贫乏区。一方面,自然环境中因全球气候变暖与降水量减少的影响在干旱、半干旱地带尤其明显;另一方面,人类活动耗水量日益增加,黄河干流建成与在建的大型水利枢纽9座,蓄水淤沙量大(表7),加之流域内水库众多,其库容已达黄河年均径流量的84%[20]。规划发展的用水量亦是有增无减,大幅度增加用水量更导致水资源供求矛盾激化,成为黄河下游断水的主导因素。

表7 黄河干流水利枢纽概况

工程名称	控制面积($\times 10^4$ km²)	总库容($\times 10^8$ m³)	调节库容($\times 10^8$ m³)	用途	调节性能	开工日期/蓄水日期	已淤积量($\times 10^8$ m³)
龙羊峡	13.14	247.00	193.60	发电,防洪,灌溉	多年	1976/1986.10	
刘家峡	18.18	57.00	42.00	发电,防洪,灌溉	年	1958.9/1968.10	11.848 8
盐锅峡	18.27	2.20	0.10	发电,灌溉	日	1958.9/1961.3	1.646 5
八盘峡	21.59	0.49	0.09	发电	日	1969.11/1975.6	0.159 6
青铜峡	27.50	6.06	0.30	灌溉,发电	日、周	1958.8/1967.4	5.903 2
三盛公	31.40	0.80		灌溉,防洪		1959.6/1961.4	0.544 0
天桥	40.39	0.67	0.25~0.30	发电	日	1970.4/1977.2	0.418 3
三门峡	68.84	354.00		防洪,灌溉,发电		1957/1960.9	54.100 0
小浪底	69.45	126.50	51.00	防洪,灌溉,发电	季节	1991.9/?	

据[1]补充。

统计表明:20世纪90年代,黄河上游地区来水持续偏枯,1990—1996年间年平均降水量每年偏少约12%;1997年第一季度,上游来水比多年同期平均来水偏少1/4以上。90年代,黄河下游非汛期来水比50年代减少 24.5×10^8 m³,用水增加 81.5×10^8 m³。同期,黄河流域灌溉面积比50年代增长了1.6倍;上游地区耗水量比50年代增长了0.8倍,中游地区增长了1.0倍,下游地区增长了4.6倍[②]。花园口以上年耗水量从1919年

---

① 如何解除黄河下游断流之患. 光明日报,1997年5月11日。

② 拯救黄河. 科技日报,1997年5月10日,第一版.

的 39×10^8 m³,增加到 1949 年的 74×10^8 m³,1990 年则激增至 190×10^8 m³,相当于花园口多年平均径流量的 34%[6]。下游引黄灌溉面积于 1995 年达 35×10^4 ha,远远超过黄河供水能力,尤其是下游为防止春枯引不到水而在冬季蓄水入平原水库,使非灌溉期用水紧张。促使黄河断流时间逐年提早,断流历时和断流河长度逐年加长。每年枯季,沿黄河各地都从各自需要出发争引水、蓄水,甚至抢水,加剧了断流局面的形成。黄委会设计院预测:2000 年全流域总用水量可能达到 368×10^8 m³,2010 年将达 392×10^8 m³[1,6]。缺水状况将更加严峻。

80 年代,政府已制定了黄河可供水量分配方案,但未制定适用于枯水年的具体方案、监督与管理措施,对黄河水资源尚未建立统一调度、分级管理的体制和运行机制。所以,各段分水方案未具体落实。如,根据国家计委和水利部对沿黄各省(区)实行水量分配政策所确定的近中期引水量,80 年代以来均有超过[8]。

此外,节水意识淡薄,管理粗放,水资源利用率低,浪费严重。流域内农业灌溉用水占全河用水量的 90%,用水规模超前;流域内工业生产技术落后,管理水平低,用水重复利用率低(平均仅为 29%),流域内 1×10^4 元生产值平均耗水 802 m³[23],下游沿黄大中城市工业用水平均定额为 300~500 m³/万元(均为 1980 年不变价),比先进国家高 3~4倍[24]。地下水超采,局部受到污染;流域内废污水年排放量 80 年代初为 21.7×10^8 t,目前已近 33×10^8 t[24],地面水污染日趋严重,加重了黄河流域水资源紧张状况。水价过低也是一个原因,灌区农业用水水费及下游引黄渠首工程业供水水费不足 1 分/m³,远远低于供水成本,使水利设施年年因缺少经费而失修,降低了调蓄、引水功效,加重了水资源浪费状况。

三、断流对海岸环境的影响

(一) 对海岸带经济发展的影响

黄河断流,直接影响到黄河三角洲和下游平原地区鲁、豫两省及胜利与中原两大油田的生产与城乡居民生活,影响范围达 5.4×10^4 km²。河口地区 90 万城镇居民严重缺水,下游两岸约 1.4 亿居民吃水困难,饮用水质量差,水价高达 1.8 元/m³。由于断流和供水不足,1972—1996 年,农田直接受旱面积 70.42×10^4 ha,粮食减产 98.6×10^8 kg,工农业累计经济损失约 268 亿元,90 年代年均工农业损失已达 36 亿元。

水资源紧张已成为制约整个黄河下游乃至中游部分地区城市发展、工农业生产的关键因素。为缓解下游用水紧张状况,减少了中、上游农灌区农业用水,沿河各水库加大下泄量,影响到中、上游地区的农业生产,也使中、上游水利枢纽和水电站利用率降低,造成了潜在的巨大经济损失。

(二) 对海岸带自然环境的影响

黄河三角洲海岸及邻接的渤海湾与莱州湾,海岸类型、海岸演变与生态环境是河-海体系相互作用达到动态均衡的结果。当黄河断流时,淡水径流与泥沙输送量骤减,造成海岸自然环境天然动态平衡被打破,海岸发生一系列变化,其影响比经济损失更为深远。

（1）长达半年的黄河断流,减少了河流行水维护河道的自然能力,造成河道萎缩、恶化;减少了河水泛滥淤泥压制沙碱的能力,导致土壤性状恶化;同时,多年持续性的河流间断,将使地下水水面降低,淡水量下降,海水与地下盐水必沿河道及地层孔隙向岸、向陆入侵;造成三角洲及下游地区水质恶化,加速土壤盐渍化。加之,旱季气候干燥,东北风盛行,将加重下游灌区土壤的沙化过程,使农田恶化,水资源更加紧张。

（2）减少了入海径流量,季节性中断淡水向海携运有机质与营养盐的过程,断绝了河口区高生产力之源,使鱼、虾产卵、栖息环境与回游路线发生改变,降低鱼虾生产量。当年三门峡水库建成后,黄河径流递减,渤海湾的对虾、梭鱼与白虾均减产,渔民讲:"少了渤海湾那股甜水"——少了河川径流与随水而下之营养盐。淡水中断使河口三角洲区的苇滩、湿地生态环境变化,会对多种野生植物、水生生物及鸟类的繁衍造成损害,危及生物多样性,进而影响渤海海洋生物链与生态系统。

（3）黄河每年入海的巨量泥沙是河口三角洲土地得以稳定存在并向海淤进的重要物质来源。河水是携运泥沙的动力,黄河断流,入海泥沙中断与锐减,必然使三角洲陆地的冲、淤动态均衡发生变化,海岸带因丧失泥沙补给而发生侵蚀后退。加之,全球海平面上升,风暴潮增多,由松散泥沙堆积的三角洲海岸抗蚀力低,海岸侵蚀加剧,其变化影响更为严重。

1. 黄河改道对海岸环境影响的历史实例

自公元前 2278 年以来,黄河下游曾经历过 8 次大改道,其间或入渤海不同地点,或迁徙入黄海,移距达数百公里。在 8 次大改道中,有两次是人为的（1128 年与 1938 年）。第一次人为改道（即黄河的第五次大改道）是南宋建炎二年（1128 年）11 月 15 日,开封府尹杜充掘黄河堤以阻金兵南下,结果使黄河南迁,夺淮河,经泗水、颍水入黄海,历时达 727 年,其间 400 多年（1128—1546）,黄河下游分成数股在今黄河以南、淮河以北、颍河以东、泗河以西东西宽约 250 km 的大平原上经常泛滥,决口和改道,至 1578—1579 年筑堤治河以后,下游河道才被基本固定,直到 1855 年黄河才北归渤海。自 1128 年起的黄河夺淮入海,巨量的黄河泥沙使苏北海岸从以波浪作用为主的沙坝-潟湖岸,转变为淤泥质潮滩海岸。黄河泥沙不仅在江苏北部海岸迅速淤积,形成向海淤进达 90 km、面积达 7 160 km^2 的废黄河三角洲,而且使该三角洲南部海岸自此长期接受黄河或废黄河三角洲的泥沙补给,淤进向海超过 40km（以范公堤为基准）,共形成岸陆平原面积达 157 000 km^2,相当于 1/6 的江苏土地面积[19,25,26,27,28]。更为严重的是,这次人为决口彻底改变了黄淮平原的自然水系,不仅促使形成了中国最大的淡水湖之一——洪泽湖（面积 2 200 km^2）,而且改变了淮河水系,使淮河入江,甚至黄河一度也成为长江的支流。深刻地改变了整个华北平原和苏北平原的地理环境[18]。

1855 年,黄河北归至渤海湾。由于突然失去巨量泥沙供应,在岸线长达 150 km 的废黄河三角洲地区,海岸蚀退,年速率 15～30 m 不等,迄今约 1 400 km^2 的土地已入海,作者曾于 1961 年去灌溉总渠口门工作的土地现已入海数百米之远! 海岸防护仍是废黄河三角洲区的主要问题。海岸又渐转为细砂-粗粉砂与贝壳堤类型[13,25,29]。与之相应,废黄河三角洲侵蚀下来的泥沙则继续在苏北海岸淤积,形成广阔的新生土地资源。

另一方面,在渤海 160 km 长的黄河三角洲海岸带,1855 年以来向海淤进达 20.5～27.5 km,新生河口淤积快速,废弃河口遭受侵蚀,泥沙又向河口两侧堆积,大体上每年淤积约 23.5 km² 的土地,一个夏季淤进 6～10 km。这段渤海海岸又从 1855 年前的贝壳堤转为淤泥潮滩岸[13]。

海岸演化取决于海陆两组动力,虽然渤海地处下沉地带,海面上升速率 4.5～5.5 mm/a[30],但因每年 $8×10^8～12×10^8$ t 黄河泥沙汇入,所以,河流力量与泥沙压倒海潮与浪流,不断促使海岸向海推进。一旦泥沙供应中断,即使海洋动力不变,海岸亦逐渐侵蚀,从水下岸坡变陡至岸线后退。

黄河河口尾闾改道的自然周期为 6～10 年或 12 年,因东营市、油田发展需要,拟控制河口使之不再迁移。近期研究表明,将来河口将进一步偏转向东南,最终会封闭河口成为潟湖,但是随着泥沙量减少至 $6×10^8$ t,河口淤长速度会变慢,若河口泥沙断绝,或每年有 1/3 时期断绝,则海岸要塑造新的平衡,盐水入侵,土地(下游)因失去泛滥压碱,必进一步盐碱化,海岸受蚀后退必会逐渐发生而成为三角洲区的一项灾害。

2. 其他实例

(1) 滦河入海年径流量原为 $41.9×10^8$ m³,泥沙为 $2\,219×10^4$ t。"引滦济津"工程调拨了 $3.55×10^8$ m³ 淡水量,河流入海泥沙减少为 $103×10^4$ t,自 1988 年滦河三角洲在滦河口门段侵蚀后退年率为 17.4 m/a,海水入侵使下游河水盐分从 27.3‰ 升高至 32.6‰[31],水质恶化。

(2) 尼罗河于 1881 年在三角洲顶点修建滚水坝,1902 年修建阿斯旺水库,坝下来水来沙日益减少,导致三角洲强烈遭受侵蚀,沙咀平均每年后退 29 m,沙坝平均每年后退达 31 m[32]。1964 年阿斯旺高坝建成后,使下游丧失由于尼罗河泛滥使土地增肥的自然过程,而且由于泥沙不入海,加速了海岸退缩过程。流入地中海的泥沙和有机质锐减,使尼罗河口的沙丁鱼产量减少。

历史的实例说明:人类活动必须适应自然环境的发展规律,必须将人类活动纳入自然环境(水、土、气、生物)所能容忍的承载量内。

四、基于可持续发展观点的对策

第一,提高全民节水意识,保护水资源与水环境,有计划地利用有限的淡水资源。黄河流域的水资源总量是有限的,与流域内人口和面积相比,水资源是短缺的。在确定区域发展方向、重点、速度与产业结构等的过程中,应充分考虑水资源这一影响甚至制约流域内社会经济发展的关键因素。必须首先保证居民生活用水,其次保障中上游水利发电以及下游与河口有一定的行水输沙,必须充分认识到河口区缺少水沙的输出,必将带来严重的海岸侵蚀、海水、咸水入侵等一系列环境负效应。在此基础上,因地制宜地发展节水型的工农业。当前,首先要改变中上游农田传统落后的耕作方式,尤其是大水漫灌,提高水资源有效利用率;适当控制下游和三角洲地区农田面积的进一步盲目扩大,提高工业用水重复利用率,减少耗水型工业,严格控制污水排放,减少水污染;适当提高水价,制定合理的水价体系,有偿用水;制定与执行有效的用水方案,优化用水结构。

第二,协调好人地关系,研究黄河及流域环境特性与水资源承载力。一体化地考虑全流域发展:上游梯级水力开发、中游水土保持、水土资源涵养与下游防洪、治淤工程。既要使蓄洪与分段用水计划落实,又要保证一定数量的水、沙全年性入海,维持三角洲、渤海湾与莱州湾环境的动态平衡。

第三,系统的对全流域全年用水进行适宜的时空控制管理,科学合理地分配不同河段、不同地区在不同季节的用水量,并制定出旱情发生时有效的应急措施。要控制、管理黄河流域水资源,建立一体化的用水规划、管理体系。全流域引水量应控制在 200×10^8 m³,保持下游至少有 200×10^8 m³ 的水量供河道行水排沙,使年输沙量保持在 6×10^8 t 左右,以维持三角洲海岸的冲刷与堆积的动态均衡。当汛期降水量低于 260 mm,或汛期径流量小于 300×10^8 m³ 的警戒值时,次年引水量必须严格控制,重视发挥沿途水利枢纽的季节性与多年性的调蓄功能。下游与三角洲地区,宜发展牧业和养殖业,限制发展灌溉农业,限制盲目大量抽取地下水,控制高层建筑重载量,有效防止地面沉降、盐水入侵与海岸侵蚀。

第四,需慎重对待南水北调的设想,必须对长江流域、淮河流域、黄河流域进行一体化的系统研究,进行预后模拟实验,在充分有依据的情况下再进行规划、设计,以免形成新问题。在一定的时期之内,"引江济河"可能不是解决黄河流域缺水的有效办法。目前,最切实可行的出路就是优化流域内部用水结构,提高农业用水利用率(争取达到 50%～75%)和工业用水重复率(60%～80%),有效地使现阶段的引水量控制在 200×10^8 m³ 以下,这样,在小浪底水利枢纽建成后,随沿途调节能力的加强,还可保证流域内仍有 50×10^8 m³ 的可供调节引水量,保障流域内 5～10 年的经济进一步发展的用水要求,降低断流频率。

参考文献

[1] 钱意颖,叶青超,周文浩.1993.黄河干流水沙变化与河床演变.北京:中国建材工业出版,1-230.

[2] 崔宗培主编.1991.中国水利百科全书.北京:水利水电出版社,856-2473.

[3] 刘东生编著.1964.黄河中游黄土.北京:科学出版社,2-5.

[4] 水利电力部水文局.1987.中国水资源评价.北京:水利电力出版社,1-194.

[5] 水利电力部水利水电规划设计院.1989.中国水资源利用.北京:水利电力出版社,1-226.

[6] 赵业安,潘贤娣,李勇.1994.黄河水沙变化与下游河道发展趋势.人民黄河,(2):31-41.

[7] 张天曾.1990.中国水利与环境.北京:科学出版社,143-178.

[8] 任美锷主编.1994.中国的三大三角洲.北京:高等教育出版社,8-128.

[9] 任美锷,史运良.1986.黄河输沙及其对渤海、黄海沉积作用的影响.地理科学,6(1):1-12.

[10] Wang Ying, Mei-e Ren, Dakui Zhu. 1986. Sediment Supply to the Continental Shelf by the Major Rivers of China. *The Journal of Geological Society*, London, 143(6): 935-944.

[11] 王云璋.1991.黄河中上游 40 年降水变化与水沙关系初探.见:黄河流域环境演变与水沙运行规律研究文集(第一集).北京:地质出版社,16-26.

[12] 朱振源,贾绍凤.1993.黄河流域自然环境变化和人类活动对水沙的影响预测.黄河流域环境演变与水沙运行规律研究文集(第二集).北京:气象出版社,134-155.

[13] Wang Ying and Ke Xiankun. 1989. Cheniers on the east coastal plain of China. *Marine Geology*, 90：321 - 335.

[14] 吴祥定,钮促勋,王守春等. 1994. 历史时期黄河流域环境变迁与水沙变化. 北京:气象出版社, 65 - 121.

[15] 史念海,曹尔琴,朱士光. 1985. 黄土高原森林与草原的变迁. 西安:陕西人民出版社.

[16] 谭其骧. 1986. 何以黄河在东汉以后会出现一个长期安流的局面. 原载学术月刊,1962(2). 收录于 谭其骧主编,黄河史论丛. 上海:复旦大学出版社,72 - 101.

[17] 邹逸麟. 1989. 东汉以后黄河下游出现长期安流局面问题的再认识. 人民黄河,(2):60 - 65.

[18] 任美锷. 1997. 黄河与人生. 见:区域可持续发展理论,方法与应用研究. 开封:河南大学出版社, 78 - 95.

[19] Wang Ying. 1994. Effect of Military Activities on Environment in Eastern and Southeastern China. *Annual Science Report-Supplement of Journal of Nanjing University*,30(English Series2)：43 - 46.

[20] 黄定武,许克钜. 1994. 关于黄河三角洲的灌溉水源问题. 人民黄河,(5):43 - 46.

[21] 竺可桢. 1973. 我国近5000年来气候变迁的初步研究. 中国科学,(2):291 - 296.

[22] 南京大学气象系气候组. 1977. 关于我国东部地区公元1401—1900年五百年内的旱涝概况. 见:气候变迁和超长期预报文集. 北京:科学出版社,53 - 58.

[23] 张挺. 1993. 黄河水资源紧缺问题及对策. 人民黄河,(1):4 - 6.

[24] 高传德. 1993. 黄河流域水环境现状及治理措施. 人民黄河,(12):4 - 7.

[25] Wang Ying and David Aubrey. 1987. The Characteristics of the China Coastline. *Continental Shelf Research*,1(4)：329 - 349.

[26] 任美锷. 1988. 黄河三角洲整治的若干问题. 科技导报,(6):23 - 27.

[27] Ren Mei-e. 1992. Human impact on coastal landform and sedimentation-the Yellow River Example. *Geo Journal*,28(4)：443 - 448.

[28] 张忍顺. 1984. 苏北黄河三角洲及滨海平原的成陆过程. 地理学报,39(2):173 - 184.

[29] 王颖,朱大奎. 1990. 中国的潮滩. 第四纪研究,(4):291 - 299.

[30] Wang Ying. 1998. Sea-level changes, human impacts and coastal responses in China. *Journal of Coastal Research*,14(1)：31 - 36.

[31] 钱春林. 1994. 引滦工程对滦河三角洲的影响. 地理学报,49(2):158 - 166.

[32] 陈吉余. 1987. 三峡工程对长江河口影响初步分析. 见:王俨主编,长江三峡工程争鸣集专论. 成都: 成都科技大学出版社,230 - 233.

黄河断流与海岸环境反馈[*]

一、概况

黄河是我国的第二大河,源自青海雅拉达泽山东麓、海拔 4 368 m 的约古宗列渠,流经青、川、甘、宁、内蒙、晋、陕、豫、鲁九省(自治区),汇入渤海,全长 5 464 km。黄河具有水少沙多,水沙异源与时空分布不均,以及下游河道迁徙频繁的特点。历史上 4000 多年来,黄河泛滥与下游决口达 1 591 次,水灾范围北起天津,南达江淮,波及冀、鲁、豫、皖、苏五省,纵横范围达 25×10^4 km²,故有黄患之称。

黄河水资源总量为 744×10^8 m³,平均径流量为 580×10^8 m³,枯水年与丰水年径流量之比可达 1∶10.3。黄河平均含水量达 37.6 kg/m³,三门峡站年最大输沙量 39.1×10^8 t,最小仅 4.88×10^8 t,年入海泥沙量变动为 16 亿~11 亿~9 亿 t。

二、断流现象

1960 年,黄河下游出现断流。自 1972—1997 年有 17 年断流,累计发生断流达 867 天。自 1991 年以来黄河连年断流始期提前,断流时间延长,断流距离加大(表 1 和表 2)。

表 1　黄河断流统计表

时段	断流年数	累计天数(天)	年断流最长时间(出现年份)	断流始/终日期	断流频率(%)	断流距离(km)
1960—1969	1	41	41(1960)	4 月/6 月	10	
1970—1979	6	71	19(1979)	4 月/6 月	60	242
1980—1989	7	103	37(1981)	4 月/6 月	70	256
1990—1996	6	467	136(1996)	2 月/7 月	60	392

表 2　20 世纪 90 年代利津站黄河断流记录

年份	断流始/终日期(月、日)	断流天数	断流河道长度(km)
1991	5.16/6.1	14	178
1992	4.28/7.22	84	339
1993	2.13/6.	45	299
1994	4.4/6	66	339

* 王颖,张永战:刘荣川主编,《首届两岸尖端科学研讨会论文(摘要)集》,262 - 265 页,1999 年 10 月,南京大学编印。

（续表）

年份	断流始/终日期（月、日）	断流天数	断流河道长度（km）
1995	3.4/7.14	122	683
1996	2.24/7.14	136	340
1997	2.7/10	226	700

三、断流原因

断流是人为因素——用水过量、滥伐滥垦使水土严重流失与自然因素——中上游降水量变化使径流量递减，双重作用之结果。

黄河径流量（R）的丰枯与中上游降水量（P）呈线性相关，相关系数达 0.83。中国 5000 年来气候变化具有寒冷期与温暖期交替的规律，20 世纪进入温暖期，气温逐渐升高，近期降雨减少，干旱现象上升，而引水量过多导致断流。例证如下：

（1）50 年代降水正常，其中 1958 年与 1959 年偏高。1952 年与 1957 年中上游汛期降水量偏低，仅为 260 mm 和 270 mm，花园口汛期径流量分别为 310×10^8 m³ 和 275×10^8 m³，年平均引水量为 125×10^8 m³，50 年代未发生过断流。

（2）60 年代出现一次断流，却反映出人为作用的影响。1965 年，中上游汛期降水量达到极低值 210 mm，花园口汛期径流量约 280×10^8 m³，当年平均引水量 175×10^8 m³，下游净雨水量 105×10^8 m³，未发生断流。1960 年花园口汛期径流量 297×10^8 m³，1958 年与 1959 年中上游降水量高达 430 mm 和 355 mm，但 1960 年发生断流 41 天，人为影响系 1958 年大炼钢铁，砍伐树木；1960 年大办公社，开荒种地，招致水土流失严重，降水量减少，但水量又难以转化为径流，终使下游河床干涸。

（3）70 年代早、中期，降水量普遍偏低，从 1969 年至 1974 年黄河连续 6 年枯水，年平均引水量达 233×10^8 m³，断流发生与降水丰枯有关，如表 3。

表 3　1970 年代花园口黄河断流记录

年份	中上游降水量（mm）	花园口汛期径流量（$\times10^8$ m³）	人工引水量（$\times10^8$ m³）	断流天数（天）
1972	240	220	175～200	15
1973	355	305	200	0
1974	262	220	200	18
1975	355	440	200	11
1976	375	450	230	7
1977	315	280	230	0
1978	350	360	230	1

所以，当上游汛期降水量小于 260 mm，花园口汛期径流量小于 300×10^8 m³，年引水量大于 200×10^8 m³ 时，净余径流量小于 100×10^8 m³，下游往往发生断流。当头一年汛

期降水量偏低时,第二年春季往往发生断流,即使以后两年汛期降水量高,若水量未高到补足前一年水量的亏空,第三年春季下游仍将断流,但断流持续时间比前一年缩短。

(4) 70年代末至80年代初期,汛期降水量正常,未低于300 mm,但连续5年发生断流。当时经济开始腾飞,工农业发展,使用水量增加,全流域引水量已达274×10^8 m^3,过量引水,使黄河自身调节能力降低,易断流,虽时间不长,但频频发生断流,已难于分析断流延时长短的具体原因。

(5) 80年代中、后期,断流现象明显反映中、上游汛期降水丰枯与下游断流发生间的时间滞后效应(表4)。

表4　20世纪80年代黄河花园口段断流记录

年份	中上游汛期降水 (mm)	花园口汛期径流量 (×10^8 m^3)	年引水量 (×10^8 m^3)	断流天数 (天)
1979	365	350	250±	19
1980	300	270	250	6
1981	385	500	274	37
1982	310	360	274	10
1983	325	450	274	5
1984	380	470	274	0
1985	340	405	274	0
1986	250	240	274	0
1987	275	220	274	17
1988	360	350	274	8
1989	315	330	274	20

(6) 90年代,黄河上游地区来水持续降低。

1990—1996年间降水量每年减少约12%,1997年第一季度上游来水较多年平均来水偏少1/4以上。90年代黄河下游非汛期来水比50年代减少24.5×10^8 m^3,用水量增加81.5×10^8 m^3。黄河流域灌溉面积比50年代增长1.6倍,上游耗水量比50年代增长0.8倍,中游地区增长1.0倍,下游增长4.6倍。花园口以上耗水量在1990年激增至190×10^8 m^3,相当于多年平均径流量的34%。下游引黄灌溉面积于1995年达35×10^4 ha,黄河流域引水量已达300×10^8 m^3,远远超过黄河供水能力。自1991—1997年下游连续7年断流,尤其是下游为防止春枯而在冬季蓄水入平原水库,使非灌溉期用水也紧张,结果,使缓和断流时间逐年提早,断流历时和断流河段逐渐加长,1997年,下游断流达226天,断流河段长度达700 km,人工引水调剂后才缓解。

综上所述,可窥见黄河下游断流发生的机制:

(1) 当引水量为125×10^8 m^{3}时,即使汛期降水量小于260 mm,或汛期径流量低于300×10^8 m^3,但径流量净余值高于150×10^8 m^3,下游不会出现断流。

(2) 当引水量为175×10^8 m^{3}时,即使汛期降水量为210 mm低值,汛期径流量为

280×10^8 m^3，但当年径流量净余值高于 100×10^8 m^3，仍有一定径流入海，黄河下游很少出现断流，断流发生概率为 1/10。所以，200×10^8 m^3 的人工引水量，或 100×10^8 m^3 的净余径流量是黄河下游断流与否的"临界值"，260 mm 的汛期降水量与 300×10^8 m^3 的汛期径流量可能是相关数值的低限。

（3）连续多年降水丰沛，即使有一年汛期降水量少于 260 mm，当年春季不一定出现断流。反之，连续多年干旱或有一年异常干旱，随后 1～2 年即使降水量高，或降水过量集中于汛期，当年或来年春季仍会出现断流，且旱年后的第一年断流现象非常严重，随后断流天数可能减少。所以，春旱断流常发生与汛期降水量与径流量偏低年之后。

（4）当人工引水量超过 200×10^8 m^3，达到 350×10^8 m^3 后，已接近或超过黄河天然年净流量的一半时，黄河的自然调节能力降低，"抗灾性"脆弱，即使汛期降水量超过 300 mm，径流量为 420×10^8 m^3，断流仍会发生。

四、影响与对策

黄河断流直接影响的三角洲平原面积达 5.4×10^4 km^2，断流的直接经济损失累计达 268 亿元。断流招致河流行水维护河道的自然能力减弱，使河道萎缩恶化；减少与季节性中断入海径流，改变河口生态环境，湿地退化，生物种类减少；减少与季节性中断入海泥沙量，促使三角洲岸线侵蚀后退，断流犹如历史上黄河改道，经淮河入黄海引起渤海湾与苏北黄海两地河口海岸环境发生剧烈变化。黄河断流令人深思：正确认识河流自然过程与环境的发展变化规律，协调人类活动，应在适应自然规律的条件下开发应用水资源与水环境，重视与自然环境的和谐关系，解决黄河断流的对策是提高全民节水意识，全流域规划与计划用水，将引水量控制在 200×10^8 m^3 以内，设法保证下游 200×10^8 m^3 径流供河道行水与排沙。

黄河断流与海岸反馈[*]

李白诗云"君不见黄河之水天上来,奔流到海不复回",仅仅两句就将黄河奔腾磅礴的气势体现得淋漓尽致。黄河是我国第二大河,发源于青海巴颜喀拉山脉北部的雅拉达泽山(海拔 5 202 m)东麓的约古宗列渠,由积雪与冰川融水等涓涓细流汇集而成咆哮的大河。

由于受地质构造与地貌控制,使得黄河河道蜿蜒曲折。首先在甘肃省的积石山,形成黄河的第一大转折——东南折向西北流,后再次转折向东北流,在贵德一带形成第三次转折向东行,并形成一系列峡谷(龙羊峡、松巴峡、积石峡、刘家峡),这一段河水清澈,水流湍急,峡谷夹峙,气势磅礴;自兰州河流进入黄土高原,并转折向东北,在阴山南麓复又东流,形成富庶的河套平原,所谓"黄河百害,唯富一套",河套确实是塞北江南,黄河流域是中华民族文化之摇篮,其作用不可低估。黄河在山西、陕西黄土高原间自北向南流,水流急、水势大(龙门峡可见一斑),汇集了大量黄土质泥沙,含沙量高达 37.6 kg/m³。河流在陕西和河南之间,再转折向东,完成了第六次大拐弯,形成黄河的第一大特点——河道多次大转折,最后一马平川奔向东北而汇入渤海。黄河干流自西向东,切穿不同的地质构造与岩层,途经青海、四川、甘肃、宁夏、内蒙古、山西、陕西、河南、山东 9 个省(自治区),全长 5 464 km,最终汇入大海,河势雄伟。

黄河的第二大特点是水少、沙多。黄河流域面积为 752 443 km²。水源来自于融化的冰川和积雪以及大气降水,水资源总量平均为 744×10⁸ m³,多年平均天然径流量约560×10⁸ m³,入海径流量为 493×10⁸ m³[1]。黄河入海径流量是长江入海径流量(9 250×10⁸ m³)的 1/19(6%),是密西西比河流量的 1/15,是亚马逊河流量的 1/183,尼罗河流量的 1/2。黄河径流水量主要来自兰州以西的上游地区,该区流域面积占全流域总面积的29.6%,径流量约占 55.7%,雨水补给占上游来水的 70%[2]。黄河径流年际变化突出,具丰水与枯水交替变化和连年枯水的特点。黄河每年有凌汛(3 月,亦称桃汛)、伏汛(7、8月)与秋汛(9、10 月),其中以伏汛水量大,凌汛或桃花水季节危害明显,上游河水解冻,而河套平原地区因纬度向北高出 5°,河水仍冰封,加之河道浅平,上游来的融冰水难以排泄,使冰凌拥塞推挤,冰块叠积成坝,后来的水流被堵而破堤泛滥,曾数次动用空军炸冰坝以解危急。

黄河平均含沙量 37.6 kg/m³[1],洪汛期曾达 933 kg/m³[3],素有"斗水升沙"之说。黄河 90% 的泥沙来自中游黄土高原,河口镇至龙门段黄土高原面积 11×10⁴ km²,区间输沙达 9.6×10⁸ t,占全流域输沙量的 60%[2]。黄河出黄土高原后的多年平均输沙量高达16×10⁸ t,其中约 4×10⁸ t 堆积于下游泛滥平原区,约 12×10⁸ t 出海(其中 1/2 淤积在河

* 王颖、张永战:《科学》,1999 年第 51 卷第 5 期,第 40-44 页。

口三角洲和滨海地区,1/4 被输送至深海)。其入海输沙量是长江输沙量(4.86×10^8 t)的 2 倍多,是亚马逊河年输沙量的 2 倍,密西西比河输沙量的 3 倍,尼罗河输沙量的 9 倍。黄河沙量往往集中于汛期的几场暴雨洪水,亦有丰、枯相间的周期性年际变化,入海年输沙量最大达 21.08×10^8 t(1958 年),最小仅为 0.66×10^8 t(1987 年)[1],两者比值高达 32。近 20 年来,由于入海水量减少(人为与自然造成的),同时,黄土高原地区已有 34.8% 的水土流失区获得控制,每年减少泥沙量达 3.0×10^8 t,黄河入海输沙量已减少到约 9.0×10^8 t/a。

黄河的第三大特点是下游改道频繁。黄河自河南桃花峪出峡谷后,进入下游平原地区,河谷突然展宽,河床坡度降低(平均比降 0.012%),使水流速度突然减缓,导致大量泥沙迅速堆积,从中游黄土高原带来的泥沙约有 1/4 至 1/3 堆积在下游,黄河下游是平原河道,尾闾是三角洲水系。华北大平原的发育过程,就是黄河搬移黄土高原的泥沙、移山填海的造陆过程,即堆积三角洲、山麓冲积扇逐渐发育为泛滥平原的过程。巨量泥沙在下游不断淤积,使河床加高。例如,在开封和新乡地区,黄河河床高出地面 13～21 m,成为地上悬河。当河道淤高到一定程度后,多易发生循低而流的自然改道,下游平原则以千里长堤夹峙来维护黄河河道。历史上,黄河下游曾有八次大改道。自 1855 年,黄河北归入渤海以来,130 年间,尾闾河段较大的改道有 10 次,大约每条流路的平均行水期约 12 年左右[4]。

从历史上看,西汉时,黄土高原水土侵蚀较重;东汉期间,水土流失减轻,黄河曾稳定流动达 500 年;而后至唐时,由于大规模移民边寨屯垦黄土高原,变牧为农,砍伐林木与破坏草场,又导致水土严重流失。黄河的河性变化、河道变化始终是自然因素与人为影响交互作用的结果。

一、从泛滥到断流

黄河曾因其下游泛滥频繁而称谓为"黄患"(Yellow Sorrow),自公元前 602 年至 1938 年的 2540 年中,黄河下游决口 1591 次[1],水灾范围北起天津,南达江淮,纵横 25 万 km^2,波及冀、鲁、豫、皖、苏五省,人民流离失所、良田变沙荒。新中国成立后,经过加固大堤、建库发电、引黄灌溉、束水排沙,减轻了黄河的水患,但灾害仍未解决。治水主要以加筑堤防、修坝建库等"堵"的一系列办法为主,未将黄河流域水沙源与供需作系统研究。近年来,沿途不断开发,迅速扩大灌溉与用水规模,"入不敷出",黄河出现断流现象。

黄河下游断流始于 1960 年,嗣后断流相继发生,现在是断流时间延长,断流日期提前,断流距离加大。70 年代,黄河下游 6 年出现断流,共计断流日 71 天,最长为 19 天(1979 年);80 年代,黄河下游 7 年出现断流,共计断流日 103 天,最长为 37 天(1981 年)。

90 年代以来,黄河下游连年断流,愈演愈烈。1995 年,下游从 3 月 4 日开始断流,至 7 月 14 日累计断流 122 天,断流河道长达 683 km,为下游河长的 87%;1997 年断流情况更为严重,2 月 7 日开始断流,全年利津站先后断流 10 次,后经中上游人工调水至下游河道才结束断流,累计断流时间达 226 天,全年一半以上的时间黄河无水,甚至延续至 9 月黄河秋汛期,干涸河道扩展到开封柳园口,长约 700 km(图 1,表 1)。

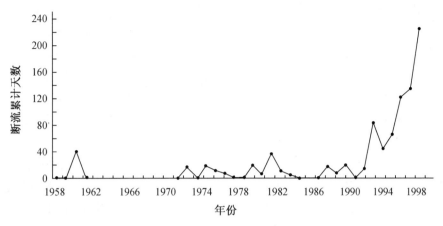

图1 黄河断流统计图

表1 90年代黄河断流情况(利津站)统计[2]

年份	断流日～终止日(月、日)	累计断流天数	断流河道长度(km)
1991	5.16～6.1	14	178
1992	4.28～7.22	84	339
1993	2.13～6.	45	299
1994	4.4～6.9	66	339
1995	3.4～7.14	122	683
1996	2.24～7.14	136	340
1997	2.7～10.	226*	700

注:表中的"＊"代表人工调水,才结束断流。

二、断流的原因

黄河流域属于干旱与半干旱区,流域面积占国土面积的8％,多年平均径流量仅为全国径流量的2％,流域内人均水量为全国平均水平的25％,耕地亩均水量为全国平均数的17％,是水资源贫乏区。由于气候变暖,气温升高,降水减少,花园口以上的中、上游年平均降水在1990—1995年间,平均每年大约偏少12％。1997年第一季度黄河上游来水比同期平均来水偏少约1/4以上。同期龙羊峡水库水的入库流量,比多年平均流量小33 m^3/s,达历史最小值。刘家峡水库流量也降低,两库蓄水量比1995年同期少24.9×10^8 m^3,比严重枯水年1996年少14.3×10^8 m^3,达到建库以来的最低值[2]。

60年代,黄河流域年均引水量125×10^8 m^3,这一时期即使遇到干旱年份,汛期降水量小于260 mm,汛期径流量低于300×10^8 m^3,但当年径流量净余值高于150×10^8 m^3,黄河下游不会出现断流。当引水量为175×10^8 m^3时,即使汛期降水量为210 mm低值,或汛期径流量仅为280×10^8 m^3,但径流量净余值高于100×10^8 m^3,仍有一定数量径流入海,黄河下游很少出现断流,断流概率约为1/10。

90 年代,黄河下游非汛期来水比 50 年代减少 24.5×10^8 m³,用水增加了 81.5×10^8 m³。同期黄河流域的灌溉面积,比 50 年代增长了 1.6 倍;上游地区耗水量比 50 年代增长了 0.8 倍,中游地区增长了 1.0 倍,下游地区增长了 4.6 倍[2],全流域年均总引水量已达 274×10^8 m³[1]。花园口以上年耗水量从 1919 年的 39×10^8 m³,增加到 1949 年的 74×10^8 m³,1990 年则激增至 190×10^8 m³,相当于花园口多年平均径流量的 34%[5]。若 2000 年全流域总用水量达到 368×10^8 m³,那么 2010 年将达 392×10^8 m³[1],缺水状况将更加严峻。

黄河流域农业灌溉用水,占全河用水量的 90%。用水规模超前,大水漫灌,水资源有效利用率只有 30%~50%,局部地区甚至产生严重的负效应,如宁夏、内蒙古河套地区,人口不足全流域的 10%。用水量却占全河用水量的 34%,其每亩灌溉水量分别为全国平均数的 3.2 倍和 1.5 倍,大水漫灌浪费水,冬季蒸发使地下水位抬升,导致土壤盐碱化,再用水压碱洗盐,造成用水恶性循环。工业生产技术落后,管理水平低,水重复利用率平均仅为 29%,生产平均耗水比先进国家高 6~7 倍。地下水超采,局部受到污染;流域内废污水年排放量 80 年代初为 21.7×10^8 吨,目前已近 33×10^8 吨[6],地面水污染日趋严重,加重了黄河流域水资源紧张状况。

通过分析表明,200×10^8 m³ 的人工引水量,或 100×10^8 m³ 的净余径流量,可能是黄河下游断流与否的"临界值";260 mm 的汛期降水量与 300×10^8 m³ 的汛期径流量,可能是相关数值的低限。当人工引水量超过 200×10^8 m³,达到 250×10^8 m³ 后,已接近或超过黄河天然年径流的一半时,黄河的"自然调节能力"降低,"抗灾性"脆弱,即使汛期降水量超过 300 mm,径流量为 420×10^8 m³ 时,断流仍会发生。

同时断流有一定的滞后效应。连续多年降水丰沛,即使有一年汛期降水量少于 260 mm,当年春季不一定出现断流。反之,连续多年干旱或有一年异常干旱,随后的 1~2 年即使降水量高,或降水过量集中于汛期,当年或来年甚至第三年春季仍会出现断流。干旱年后的第一年断流现象非常严重,春旱断流常发生于汛期降水量与径流量偏低年之后。

对于黄河断流的成因,一方面,自然环境中因全球气候变暖与降水量减少的影响,在干旱、半干旱地带尤其明显;另一方面,人类活动耗水量日益增加,黄河干流建成与在建的大型水利枢纽 9 座,中小水库众多,其库容已达黄河年平均径流量的 84%。规划发展的用水量尚有增无减,大幅度增加用水量更导致水资源供求矛盾激化。因此,黄河断流是自然因素与人类活动影响双重作用的结果,人类活动加剧了自然环境变化的影响。

三、断流对海岸环境的影响

黄河三角洲海岸及邻接的渤海湾与莱州湾,海岸生态环境与海岸演变,是河海体系相互作用达到动态均衡的结果。当黄河断流时,供水输沙量骤减,海岸动态平衡被打破,进一步调整到新平衡的过程中,带来一系列的变化,其影响比经济损失更为深远。

首先,长达半年的黄河断流,会减少河流行水维护河道的自然能力,导致河道萎缩、恶化,降低河水泛滥淤泥压制沙碱的能力。并且多年持续性的断流,会使地下水水面降低,海水与地下盐水向岸、向陆入侵,造成三角洲及下游地区水质与土壤性状恶化。

其次,断流会季节性中断淡水向海携运有机质与营养盐的过程,断绝河口区高生产力之源,使鱼、虾产卵、栖息环境与洄游路线发生改变,降低鱼虾生产量。也会导致河口三角洲区的苇滩、湿地生态环境变化,对多种野生植物、水生生物及鸟类的繁衍造成损害,危及生物多样性。

再有,由于黄河每年入海的巨量泥沙,是河口三角洲土地得以稳定存在,并向海淤进的重要物质来源,断流会使三角洲地区的冲、淤动态均衡发生变化,海岸带因丧失泥沙补给而发生侵蚀后退。加之全球海平面上升,风暴潮增多,由松散泥沙堆积的三角洲海岸抗蚀力低,海岸侵蚀加剧,其变化影响更为严重。

黄河改道对海岸环境的影响有许多的历史实例。自公元前 2278 年以来,黄河下游曾经历过 8 次大改道,其间或入渤海不同地点,或迁徙入黄海,移距达数百千米。8 次大改道中,有两次是人为的(1128 年与 1938 年)。其中,公元 1128 年黄河改道始于南宋建炎二年 11 月 15 日,开封府尹杜充掘黄河堤以阻金兵南下,结果使黄河南迁,夺淮河,经泗水、颍水入黄海,历时达 727 年,直到 1855 年黄河才北归渤海。这次人为决口彻底改变了黄淮平原的自然水系,不仅促使形成中国最大的淡水湖之一——洪泽湖(面积 2 200 km²),而且改变了淮河水系,使淮河入江,甚至黄河一度也成为长江的支流。巨量的黄河泥沙进入黄海,使苏北海岸从以波浪作用为主的沙坝-潟湖岸,转变为淤泥质潮滩海岸。黄河泥沙不仅在江苏北部海岸迅速淤积,形成向海淤进达 90 km、面积达 7 160 km² 的废黄河三角洲,并使该三角洲南部海岸接受黄河与三角洲的泥沙补给,岸线向海淤进超过 40 km(以范公堤为基准),共形成岸陆平原面积达 15.7×10⁴ km²,相当于 1/6 的江苏省土地面积。

在 1855 年,黄河北归至渤海湾,在岸线长达 150 km 的苏北废黄河三角洲地区,因突然失去巨量泥沙供应,海岸蚀退,年速率达 15～30 m。迄今已有 1 400 km² 的土地被大海吞没[7],1961 年笔者在苏北灌溉总渠口门工作时的土地,至 80 年代后期位于海中距岸线达数百米之遥! 海岸防护成为废黄河三角洲地区的主要问题。

自 1855 年黄河返归渤海入海后,现代黄河三角洲海岸向海淤进达 20.5～27.5 km,新生河口淤积迅速,大体上每年淤积约 23.5 km² 的土地,一个夏季淤进 6～10 km[7],而现代三角洲上的废弃河口海岸仍遭受侵蚀后退。

海岸演化取决于海陆两组动力,虽然渤海地处下沉地带,海面上升速率 4.5～5.5 mm/a[8],但因每年 8×10⁸～12×10⁸ t 黄河泥沙汇入,所以,河流力量与泥沙压倒海潮与浪流,不断促使现代黄河三角洲海岸向海推进。一旦泥沙供应中断,即使海洋动力不变,海岸亦逐渐侵蚀,从水下岸坡变陡至岸线后退,这是必然的趋势。历史时期,黄河河口尾闾改道的自然周期平均约为 12 年。现因东营市与油田的发展,人工控制河口使不再向北或向东北迁徙,但河口会进一步南偏与转向东南,河口沙咀会封闭河口成为潟湖。随着小浪底水库的建成,泥沙量减少至 6×10⁸ t,河口淤长速度会变慢。若河口泥沙断绝,或每年有 1/3 时期断绝,则海岸侵蚀后退,盐水沿河口入侵,下游会进一步盐碱化,逐渐成为三角洲地区的一项新灾害。

河流入海水量、沙量减少的海岸效应不乏实例。如滦河入海年径流量原为 41.9×

10^8 m³，泥沙为 $2\,219\times10^4$ t。"引滦济津"工程从滦河调拨了 3.55×10^8 m³ 淡水量，使得滦河入海泥沙减少为 103×10^4 t，造成自 1988 年以来，滦河三角洲在滦河口门段，侵蚀后退年率为 17.4 m/a，海水入侵使下游河水盐分从 2.73‰ 升高至 3.26‰，水质恶化[9]。尼罗河于 1881 年在三角洲顶点修建滚水坝，1902 年修建阿斯旺水库，坝下来水来沙日益减少，导致三角洲强烈遭受侵蚀，沙咀平均每年后退 29 m，沙坝平均每年后退达 31 m[10]。1964 年阿斯旺高坝建成后，使下游丧失由于尼罗河泛滥使土地增肥的自然过程，而且由于泥沙不入海，加速了海岸退缩过程。流入地中海的泥沙和有机质锐减，使尼罗河口的沙丁鱼产量减少。

历史的事件说明，黄河断流对工农业、人民生活所产生的影响与环境反馈是可以预见的。人类活动必须适应自然环境的发展规律，必须将人类活动纳入自然环境（水、土、气、生物）所能容忍的承载量内。

四、建议和对策

为了避免黄河断流及其带来的影响，应确保全流域引水量控制在 200×10^8 m³，保持下游至少有 200×10^8 m³ 的水量供河道行水排沙，使年输沙量保持在 6×10^8 t 左右，以维持三角洲海岸的冲刷与堆积的动态均衡。当汛期降水量低于 260 mm，或汛期径流量小于 300×10^8 m³ 的警戒值时，次年引水量就必须严格控制。当前的关键是节水，提高全民族节水惜水意识，保护水资源与水环境，有计划地利用有限的淡水资源。应用高新技术，多次、多层开发水资源利用方式，努力提高农业用水利用率（争取达到 50%～75%）和工业用水重复率（60%～80%），并切实有效地控制与减轻水污染。

对全流域环境、水资源要作系统深入的研究，掌握其变化发展的规律，确定容量与承载力，协调好人地关系。对流域全年用水进行系统的适宜的时空控制管理，科学合理地分配不同河段、不同地区在不同季节的用水量，制定出旱情发生时的有效应急措施。需要一体化地考虑全流域的发展：上游梯级水力开发，中游水土保持，水土资源涵养与下游防洪、治淤工程。既要蓄洪与分段用水计划落实，又要保持一定数量的水、沙的全年性入海，维持黄河三角洲、渤海湾与莱州湾的环境动态平衡。建立一体化的黄河流域水资源规划与管理体系（图2）。

同时，要植树造林，涵养水源，减少泥沙，降低下游河道改道的频率。重视发挥沿途水利枢纽的季节性与多年性的调蓄功能，并有计划地在流域内建设小水库群，发挥其蓄水调节能力，有效冲刷下游河道，增强河道自然泄水排沙能力。这是改变断流局面的根本措施。

跨流域引水工程，"引江济河"利用京杭大运河进行南水北调的东线方案切实可行，收效快。需进一步研究引水期、引水量以及对长江口生态环境的影响，在充分研究的基础上，再规划、设计与实施。此外，"朔天运河"计划，即从西藏雅鲁藏布江朔玛滩至天津长达 7 000 km 的南水北调、西水东用、联通南北东西几大水系的跨流域巨型调水工程，是富有创见的宏伟设想，以丰补欠、解决黄河断流、并调节水资源的流域分布、改变流域水环境特性。但是，其技术可行性、经济可能性及预后效果，需加以深入地分析。一项浩大的工程，不仅施工面大，工期漫长，费用昂贵，而且引水流域的地质地貌、沉积特性及气候条件复杂多变,有一系列重大问题需要认真研究。例如,这项工程所赋予自然界的巨大干涉与重大改变,会导致自然界怎样的反馈,在上游拦截水源后,如何妥善解决中、下游水资源利用的矛盾等,这些都值得进行深入的调查研究。

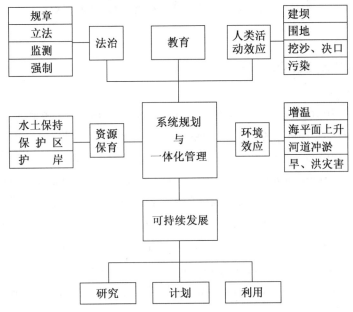

图 2　一体管理系统与生存环境可持续发展(河流流域)

参考文献

[1] 钱意颖,叶青超,周文浩.1993.黄河干流水沙变化与河床演变.北京:中国建材工业出版,1 - 230.

[2] 王颖,张永战.1998.人类活动与黄河断流及海岸环境影响.南京大学学报(自然科学),34(3):257.

[3] 任美锷,史运良.1986.黄河输沙及其对渤海、黄海沉积作用的影响.地理科学,6(1):1 - 12.

[4] Wang Ying, Mei-e Ren, Dakui Zhu. 1986. Sediment Supply to the Continental Shelf by the Major Rivers of China. *The Journal of Geological Society*, London, 143(6): 935 - 944.

[5] 赵业安,潘贤娣,李勇.1994.黄河水沙变化与下游河道发展趋势.人民黄河,(2):31 - 41.

[6] 高传德.1993.黄河流域水环境现状及治理措施.人民黄河,(12):4 - 7.

[7] Wang Ying and Ke Xiankun. 1989. Cheniers on the east coastal plain of China. *Marine Geology*, 90: 321 - 335.

[8] Wang Ying. 1998. Sea-level changes, human impacts and coastal responses in China. *Journal of Coastal Research*, 14(1): 31 - 36.

[9] 钱春林.1994.引滦工程对滦河三角洲的影响.地理学报,49(2):158 - 166.

[10] 陈吉余.1987.三峡工程对长江河口影响初步分析.见:王俨主编.长江三峡工程争鸣集专论.成都:成都科技大学出版社,230 - 233.

Drought in the Yellow River—an Environmental Threat to the Coastal Zone[*]

INTRODUCTION

The Yellow River, the second largest river in China, originates from the Yueguzonglie Creek on the eastern side of the Yagradagzê mountain (alt. 5 202 m) in the Bayan Kala Range, Qinghai province. The stream head is located at an altitude of 4 368 m. Runoff, mainly from direct precipitation but also from melting snow and glacier ice, forms small creeks, which eventually merge and turn into a torrential river passing through the high mountain passes of the upstream area. Downstream it has formed a series of canyons along the mountain section, such as the Longyang-, Sumba-, Jishi- and Liujia canyons. The sediment load is low in this region, due to the erosion resistance of the bedrock. The middle course of the river runs through the Loess Plateau where the semi-arid climate and heavy erosion of the unconsolidated loess give rise to a huge sediment load. As a result, the river here changes its nature completely to yellowish muddy water with an average sediment concentration in the range of 38 kg/m³.

Blocked by the Qinling Mountain Ranges while approaching the Tongguan district, the river turns eastward, and completes its sixth large turn. After passage through the canyon at the border of the loess plateau and entering the Xiaolangdi reservoir, the river flows northeastward through the great North China Plain, and finally enters the Bohai Sea in the Dongying area of Shandong province. The main stream of the Yellow River passes through nine provinces and autonomous regions: Qinghai, Sichuan, Gansu, Ningxia, Inner Mongolia, Shaanxi, Shaanxi, Henan and Shandong. The total length is 5 464 km, and the drainage basin area 752×10^3 km³ (Fig. 1, Table 1) (QIAN *et al.*, 1993).

The river passes through the arid and semi-arid climate zones with annual temperatures in the range $-4 \sim 14$ ℃ (Cui, 1991), lower in the northwestern—and higher in the southeastern part of the drainage basin. The average annual precipitation over the whole basin is 464 mm (1956—1979); on the Loess Plateau it is $200 \sim 300$ mm, rising to $500 \sim 700$ mm further south such as in the Wei River area. Precipitation falls mainly in August and September, often in thunderstorms with hail. The net evaporation

＊　Ying Wang, Gustaf Arrhenius and Yongzhan Zhang: *Journal of Coastal Research*, 2001, Special issue 34 (ICS 2000 New Zealand): pp. 503 – 515.

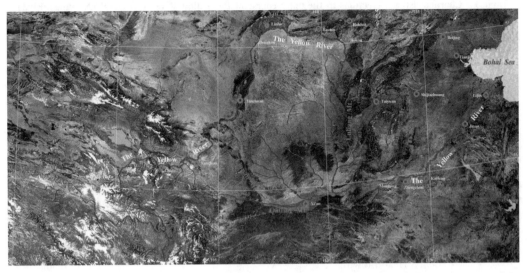

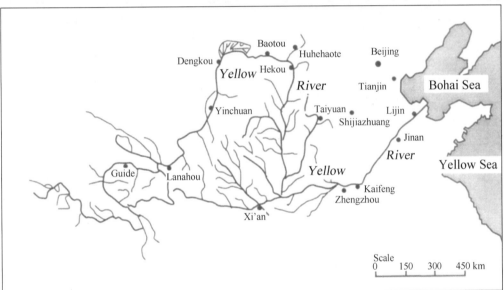

Fig. 1 **The Yellow River drainage basin. Above one is a satellite image (according to Landsat Image Map of China, Institute of Remote Sensing Application, Chinese Academy of Sciences (ed.), Science Press, 1991)(附彩图), below one is a sketch map.**

rate is as high as 350 mm per year culminating with rising temperature in the dry spring season. The runoff coefficient in the Yellow River basin area changes from 47. 4% in Lanzhou area of the upper stream to 23. 2% in the Hetao Plain, and to 15% in the middle area of the Loess Plateau and the lower reaches of alluvial plain (LIU *et al.*, 1964).

Table 1　Characteristics of sections in the main course of the Yellow River（After QIAN et al.，1993）

Section	From/to	Length (km)	Altitude difference (m)	Average gradient (‰)	Area (km²)	Larger branches entered	Land forms
Whole river	Headwater/river mouth	5 464	4 480	0.82	752 443	76	
Upper stream	Headwater/Hekou town	3 472	3 496	1.00	385 966	43	Qinghai and Nei Mongol Plateaus
	1. Headwater/Longyang Canyon	1 687	2 030	1.10	131 420	25	Qinghai Plateau
	2. Longyang Canyon/Xiaheyan county	794	1 220	1.54	122 722	8	Junction between Qinghai and Loess Plateaus
	3. Xiaheyan county/Hekou town	991	246	0.25	131 824	10	Yingchuan and Hetao Plains
Middle reach	Hekou town/Taohuayu county	1 206	890	0.74	343 751	30	Loess Plateau
	1. Hekou town/Yumenkou county	725	607	0.84	111 591	21	Jing-Shaan canyon
	2. Yumenkou county/Tongguan town	126	52	0.42	184 584	4	Feng-wei basin
	3. Tongguan town/Taohuayu county	355	231	0.65	47 576	5	Jing-Yu canyon
Lower reach	Taohuayu county/river mouth	786	94	0.12	22 726	3	Great alluvial North China Plain with two-step hung river
	1. Taohuayu county/Gao county	206	38	0.18	4429	1	
	2. Gao county/Taochengfu county	165	20	0.12	4 668	1	
	3. Taochengfu county/Lijin town	301	29	0.09	13 055	1	
	4. Lijin town/river mouth	104	7	0.07	574	0	Delta fan

FLOW CHARACTERISTICS OF THE YELLOW RIVER

The Yellow River is characterized by huge sediment loads，carried by relatively small volumes of water. Water and sediment to a large extent originate from different areas，and vary in distribution with the seasonal changes. Furthermore channels often migrate in the lower reaches，especially in the delta area.

Water sources, runoff and discharge

The sources of water for the Yellow River are rain and melt water from snow and glacier ice. The average (1956—1976) resident volume of water in the Yellow River drainage basin was 74×10^9 m³. The average annual discharge, monitored from 1919 to 1989, was 58×10^9 m³ (The hydrological data quoted here and below are from Qian *et al.*, 1993; WANG *et al.*, 1986; ZHANG, 1990; REN, 1994 and ZHAO *et al.*, 1994).

Ultimately about 49.0×10^9 m³ are discharged into the sea annually. As a comparison, the discharge from the Yellow River is about 1/19 of that from the Changjiang (Yangtze) River, 1/15 of the Mississippi River, 1/183 of the Amazon River, and 1/2 of the Nile.

The water in the Yellow River originates mainly from the upper stream drainage area ahead of the Lanzhou region. Even though this drainage area is only 29.6% of the entire watershed, the average annual volume corresponds to 55.7% of the total, with 70% is supplied by precipitation. The area between Lanzhou and Hekou with 21.7% of the total river drainage area is practically without additional water supply. The drainage area from Hekou to Longmen is 14.8% of the total and receives 12% of the water supply. The region between Longmen and Sanmenxia occupies 25.4%, and contributes 14% of the water supply. The region between Sanmenxia and Huayuankou occupies 5.5% of the total, with 17% of the total water supply. The 3% area below the Huayuankou district provides 3% of the total discharge volume (Table 2). Thus, varying quantities of water supply come from different areas and vary noticeably with season.

Table 2　Water and sediment discharge in individual sections of the Yellow River drainage basin (After ZHANG, 1990; CUI, 1991; QIAN et al., 1993)

Station	Cumulative drainage area (10³ km²)	Section drainage area (10³ km²)	Fraction of entire drainage area (%)	Average annual runoff volume (10⁹ km³)	Annual water supply in section (10⁹ km³)	Fraction of total supply (%)	Annual cumulative sediment load (10⁹ t)	Annual sediment load in section (10⁹ t)	Fraction of total sediment load (%)
↓		223	29.6		32.3	55.7		0.115	7.2
Lanzhou	223			32.3			0.115 (1935—1972)		
↓		163	21.7		−1.0	−1.7		0.044	2.8
Hekou town	386			31.3			0.159 (1954—1972)		
↓		112	14.8		7.2	12.4		0.961	60.1
Longmen	498			38.5			1.120 (1934—1972)		
↓		191	25.4		7.9	13.6		0.450	28.1

(Continued)

Station	Cumulative drainage area (10^3 km²)	Section drainage area (10^3 km²)	Fraction of entire drainage area (%)	Average annual runoff volume (10^9 km³)	Annual water supply in section (10^9 km³)	Fraction of total supply (%)	Annual cumulative sediment load (10^9 t)	Annual sediment load in section (10^9 t)	Fraction of total sediment load (%)
Sanmenxia	688			46.4 (1919—1989)			1.570 (1919—1958)		
↓		41.6	5.5		9.9	17.1			
Huayuankou	730			56.3 (1919—1979)			1.310 (1949—1972)	0.030	1.9
↓		22.4	3.0		1.7	3.0			
Lijin	752			49.3 (1855—1988)			1.150 (1950—1972)		

There are three flood periods in the Yellow River drainage basin in a year: melting-ice flood in March ("peach blossom flood"), summer flood in July and August, and autumn flood in September and October. Among them, the March flood is most dangerous. In the upper stream, fed by melt water from snow and glaciers, the narrow meandering river channels are during this season often choked by ice, temporarily blocking river flow. The accumulated water eventually breaks through the ice dams, resulting in catastrophic flooding downstream. The annual changes between high and low runoff prevail regularly but the total runoff periodically remains below the annual average for 3 - 11 years in a row (Fig. 2). Examples of these droughts are the 11 - year low-water period from 1922 to 1932, the 6 - year period from 1986 to 1989, and a 7 - year period from 1991 to 1997. The extremes in runoff rates, exemplified by the

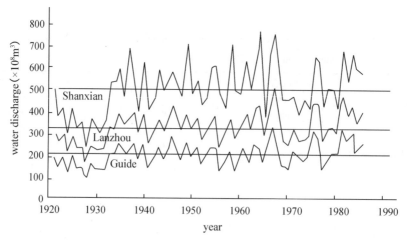

Fig. 2　**Water flow Guide Station, Lanzhou Station (upper reach) and Shanxian Station (middle reach) of the Yellow River, 1919—1989 (After Hydrological Bureau of Water Resources and Electric Power of China, 1987; Water Conservation and Hydroelectric Water Design and Plan Institute, Water Conservation and Electric Power Department, 1989).**

measurements at the Lijin hydrometric station were 93.7×10⁹ m³/year in the high water period of 1964, in contrast to the lowest of 9.2×10⁹ m³/a in the low water period of 1960; a ratio between high and low of 10.2 (Table 3, Fig. 3). These data illustrate the large seasonal and regional variations in the water flow regime in the Yellow River.

Table 3 Extreme and average annual water and sediment transport at Sanmenxia and Lijin hydrographic stations (After REN et al., 1986 (a); QIAN et al., 1993 (b); REN, 1994 (c) and CAI and WANG, 1999(d)).

	Station	Average	Monitoring period	Maximum	Year of occurrence	Minimum	Year of occurrence	Max/Min
Runoff volume (10⁹ m³)	Sanmenxia	46.4[b]	1919—1958	66.0[c]	1937	20.1[c]	1928	3.3
	Lijin	49.3[b]	1950—1987	93.7[c]	1964	9.15[c]	1960	10.2
			1855—1988	90.4[b]	1964	10.4[b]	1987	8.7
Sediment load (10⁹ t)	Sanmenxia	1.56[b]	1919—1958	3.91[abc]	1933	0.488[abc]	1928	8.0
	Lijin	1.34[b]	1950—1987	2.11[bc]	1958			
			1855—1988			0.066[bd]	1987	31.9

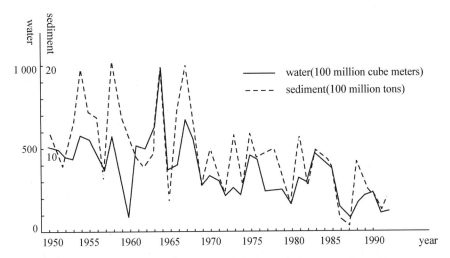

Fig. 3 Discharge of water and sediment Lijin station, 1950—1992.

The average annual precipitation in the upper and middle stream regions decreased slightly from the 1950's to the 1980's. Because of an increase in air temperature since 1990, the precipitation has been reduced by 12%, and the water supply from the upper stream has remained abnormally low during the entire decade. As a result what corresponds to almost an entire year of precipitation has been lost during the six-year period 1990 through 1995 and the water discharge from the upper Yellow River has dropped to the lowest level in the historical record.

The flow rate during the period January to April 1997 in the Lanzhou section fell more than 30% below normal level. The Water Control Committee of the Upper and

Middle Yellow River, reported a decrease in flow rate of 33 m³/sec during January to March of 1997 into the reservoirs of Longyang Canyon and Liujia Canyon. The quantity of water stored in these two reservoirs was reduced by 2.49×10^9 m³ in 1997 compared with the storage in 1995. The 1997 volume is even 1.43×10^9 m³ lower than during the seriously water deficient year of 1996, and represents the lowest value since the construction of the reservoirs.

Sediment discharge

The Yellow River is known for its huge sediment load. The average sediment concentration reaches 37.6 kg/m³. The highest load (933 kg/m³) was observed at the Longmen hydrometric station during the flood period (9AM on July 18th to 7PM on July 20th, 1966), and the second largest 826 kg/m³ from 0 AM on August 2nd to 0 AM on August 5th, 1970 (REN and SHI, 1986). The average annual sediment discharge is estimated at 1.60×10^9 t (ZHAO *et al.*, 1994). The amount entering the lower reaches of the river is represented by the load measurements at the Sanmenxia station with the average for the period 1919—1989 estimated at 1.56×10^9 t/year. Further downstream at the Lijin station, the load had decreased by 0.22×10^9 t/year.

About 1/4 of the sediment transported to the lower reaches of the river from Sanmenxia is deposited in the fluvial plain down stream of the Lijin station; the remaining 3/4, is carried to the sea with about 1/2 deposited in the river delta and the nearshore area, and the rest transferred to deeper water in the Bohai Sea.

90% of the Yellow River sediment comes from the Loess Plateau, transected by the middle reaches of the river between Hekou and Sanmenxia. In this area the river carries 60% of its total annual load between Hekou and Longmen and another 28% between Longmen and Sanmenxia at the lower edge of the Loess Plateau. In contrast to the Loess Plateau the upper stream area thus contributes only 10% of the total sediment load and the lower stream downwards from Sanmenxia supplies merely 2% (Table 2). As seen from these data, among the characteristic features of the Yellow River are the highly dissimilar sediment loads carried in different regions and the difference in sources of both runoff and suspended solids along the course of the river through physiographically different areas of its drainage basin.

Seasonal and monthly variations in sediment discharges are also pronounced. The highest rates of runoff and erosion naturally occur in the rainy season July to October when, on the average 60% of the water and 85% of the suspended load are supplied. In some years a series of rainstorms accounts for a major part of the annual sediment flux (CUI, 1991). Statistics from the Sanmenxia station between 1919 and 1960 indicate an average annual water runoff of 42.4×10^9 m³ and a sediment discharge of 1.59×10^9 t; both culminating during the flood period July-October with 60.8% and 89.7% of the annual totals respectively.

The variations in sediment load are even more pronounced than those in runoff. As examples, the maximum sediment discharge at Sanmenxia 3. 914×10^9 t in 1933 amounts to eight times the minimum, 0. 488×10^9 t in 1928; an even more extreme ratio of 32 is reached at Lijin with 2. 108×10^9 t in 1958 and 0. 066×10^9 t in 1987 (Table 3, Fig. 3).

The runoff from precipitation and the introduction of sediment load by erosion are not proportional to each other regionally. Most of the runoff comes from the upper reaches of the river while the major sediment load derives from the easily eroded Loess Plateau. However this distribution is also subject to substantial variations caused by changes in the regional distribution of rainstorms. This leads to all possible permutations of high and low runoff with high and low sediment load but under all conditions the annual change in sediment discharge is larger than the change in runoff. With about 35% of the Loess Plateau having recently been put under water conservation control, the annual water discharge from this area has now been reduced by 47% to 27. 0×10^9 m^3, down from 51. 3 × 10^9 in the 1960's. The sediment discharge was correspondingly reduced by 48% from 1. 32×10^9 t in the 1950's with a possibility for further reduction to about 0. 6×10^9 t by the year 2000 after completion of the Xiaolangdi reservoir (Table 4).

Table 4 Precipitation, runoff, sediment load, and diversion of water with suspended sediment in the river drainage basin (Calculated from WANG, 1991; QIAN et al. , 1993 and ZHU and JIA, 1993).

Period			1950—59	1960—69	1970—79	1980—89
Average precipitation during flood season in upper and middle reaches (mm/year)			332	329	326	324
Average discharge volume in flood season at Huayuankou station (10^9 m^3/year)			37. 1	37. 8	31. 8	36. 0
Lijin station	Annual runoff volume(10^9 m^3)		46. 4	51. 3	30. 4	27. 0
	Sediment load (10^9 t/year)		1. 32	1. 10	0. 890	0. 640
Whole drainage basin	Artificial water diversion (10^9 m^3/year)		12. 5	17. 5	23. 3	27. 4
	Associated diversion of suspended sediment (10^9 t/year)		0. 163	0. 136	0. 254	0. 161

RIVER DROUGHT IN RECENT TIME

From river flow to drought

The frequent flooding disasters in historical time have given the name "Yellow

Sorrow" to the Yellow River. The inundated areas have extended as far as to Tianjin in the north and North Jiangsu plain in the south, covering an area of 25×10^4 km^2 and affecting the five provinces of Shandong, Hebei, Henan, Anhui and Jiangsu. The flooding has led to displacement of the population and has turned fertile farmland into barren plains. Remedies sought in the 1950's included strengthening of the dikes, construction of dams for hydroelectric power generation, diversion of water for irrigation and control of the water flow to stem flash floods and associated siltation. These measures all helped to relieve the burden on the population from the vagaries of the river. However, with the expansion of economic development and the ensuing demands on water supply everywhere along the river without regional planning of water budgets, complete river drought became increasingly frequent in its lower reaches.

The first of these drought events occurred in 1960 and according to Lijin records were repeated in 19 of the 24 years between 1972 and 1996. During this period the flow of water ceased in the lower reaches of the river channel at 57[①] occasions and for a total of 641 days. Since 1991 river droughts have become regular annual events (Table 5, Fig. 4), happening ever earlier, lasting longer and extending over increasing distances. 1992 was a year with particular profusion of drought events with dry river bed extending over as much as 339 km. In that year the first drought, according to Lijin records, appeared on April 28th, and from May 22nd to July 22nd the drought was permanent. In the lower river plains the water flow ceased during a total of 84 days and at the river mouth druing 128 days (HUANG and XU, 1994). In 1995 the river channel turned dry from the mouth to Kaifeng, a distance of 638 km or about 87% of the entire channel over the lower river plain. In 1996 the lower reaches of the river were dried out for 136

Table 5　Interruption of water flow (river drought) in lower Yellow River (Calculated from the data given by Water Conservancy Committee of the Yellow River according to CPPCC Newspaper Oct. 18th, 1997)

Period	Number of years with incidence of river drought	Number of days with river drought in decade	Longest duration of river drought in specific year days (year)	Season with occurrences of river drought	Frequency of river drought (%) **	Average length of dry river bed (km)
1960—1969	1	41	41(1960)	April/June	10	
1970—1979	6	71	19(1979)	April/June	60	242*
1980—1989	7	103	37(1981)	April/June	70	256*
1990—1999	9	909	226(1997)	February/Oct.	90	700

* According to CPPCC Newspaper, Oct. 18th 1997.

** Frequency of river drought happened in that decade.

---

① Saving the Yellow River. Science and Technology Daily, May 10th, 1997.

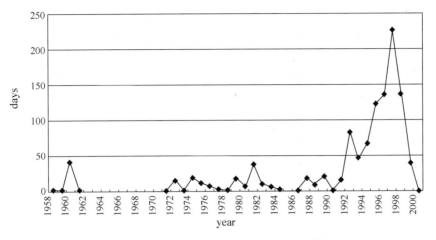

Fig. 4　Interruption of water flow 1958—2000

days; in 1997 for 226 days with ten separate occurrences beginning Feb. 7th as observed at the Lijin station. September marks the beginning of the normal fall flood season, however the drought still persisted and the dry channel extended over 700 km in 1997 (Table 6). The 1997 drought was terminated in October by artificial diversion of water from sources upstream.

Table 6　Drought events in lower reaches of the Yellow River in the 1990's

(Data from Water Conservancy Committee of the Yellow River)

Year	Drought period beginning/ending	Duration(days)	Length of dry river bed (km)
1991	16 May/June	14	178
1992	28 April/22nd July	84	339
1993	13th Feb. /June	45	299
1994	4th April/June	66	339
1995	4th March/14th July	122	683
1996	24th Feb. /14th July	136	340
1997	7th Feb. /Oct.	226	700
1998	Feb.	137	
1999	Feb.	42	

In 1998 there were 137 days of complete drought in the lower reaches of the river, and even one of the two lakes in the river head area dried out. As a countermeasure the Yellow River Conservation Committee of China that year established a general water diversion systems for the entire 2 500 km course of the river. This resulted in a marked reduction in the incidence of drought so that in 1999 the period of flow interruption began in February and lasted only for 42 days.

During the flood season sediment accumulates rapidly at the river mouth, but due to insufficient consolidation, some of this deposit is eroded away during the rest of the year.

With normal, high discharge of sediment the river delta has continued building up seaward to add about 600 km^2 of new land since 1976 when the river migrated and began discharging into the sea through the Qingshuigou channel.

However with the sharp drop in water—and sediment transport during the 1990's the picture changed dramatically. During this decade the average annual water discharge was reduced to only 27.8% of the average value for the last 50 - year period of 20th century. As a result the sediment transport also decreased precipitously with a drop of 27.2% from the value for the comparison period. Under the extreme conditions in 1997 the water—and sediment discharges at the Lijin station were as low as 18×10^9 m^3 and 0.18×10^9 t respectively.

As a consequence of the decreased sediment discharge into the Bohai Sea, the sediment—starved delta now suffers from net erosion even during flood season and is currently retreating at an annual rate of 0.5~1.5 km. During the flood season in the period Nov. 1990~Oct. 1991 the sediment transport at the Lijin station was 0.8×10^9 t. During this time the -2 m depth contour in the delta retreated 2 km landward and in the following years the erosional loss was even larger as the annual water—and sediment discharge continually decreased. In 1999 the annual water—and sediment discharge was only 1/2 of the average volume in the 1990's. Since the year 2000 the Xiaolangdi reservoir has been used for flood control leading to further decrease of the river water and sediment discharges. Thus there is a direct relationship between water flow and sediment discharge on one hand, and on the other the balance between erosion and deposition at the mouth of the river. This natural balance has been upset by excessive diversion of water and the Yellow River delta has developed from an accumulative, constructive pattern to one of erosion and destruction.

Relation of river drought to precipitation and runoff

As indicated above, the seasonal river drought is caused by a natural trend of climatic warming associated with decreasing precipitation in combination with human impact (see section below). The correlation between upstream precipitation and water discharge at the mouth of the river (r=0.84) is shown in Fig. 5 and 6. During the Pleistocene—Holocene, including the last 5 000 years, the climate in Northern China has oscillated between warmer and colder periods with related variations in humidity. The 20th century has been characterized by increasing temperature, reduced precipitation and more frequent droughts (ZHU, 1973). These natural factors together with increased human usage of water have heightened the severity of drought in the

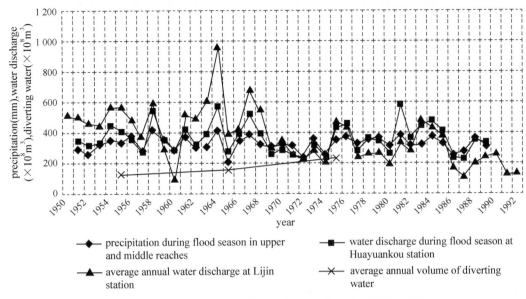

Fig. 5   Precipitation, water flow and water diversion 1950—1992.

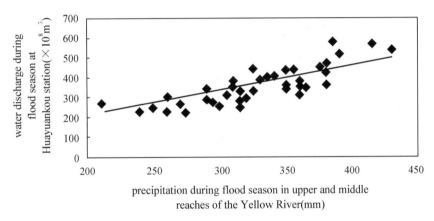

Fig. 6   Relationship between precipitation in flood season in upper and middle reaches on one hand and river water discharge on the other.

Yellow River (Table 7).

Unless the annual precipitation stays above a critical level, already small variations in precipitation and runoff may cause river drought. The fact that the precipitation upstream was relatively low during some years in the 1980's (e. g. in 1986) did not always cause river drought in those years but a delayed drought in those cases appeared in the following year (Table 7).

Table 7 Evolution of water supply in upper and middle reaches of the Yellow River and interruption of flow (river drought) in lower reaches.

Year	Precipitation in upper and middle reaches during flood season (mm)	Runoff at Huayuankou during flood season(10^8 m³)	Days without water in lower channel	Note
1952	260	310	0	Water diversion about 125×10^8 m³/a
1957	270	275	0	
1958	430	540	0	Deforestation and conversion of land to agricultural and industrial use.
1959	355	345	0	Numerous communities established for smelting of iron
1960	290	297	41	
1965	210	280	0	Water diversion about 175×10^8 m³/a
1969	300	250	0	
1970	320	290	0	
1971	315	250	0	
1972	240	220	15	Water diversion $175 \times 10^8 \sim 200 \times 10^8$ m³/a
1973	355	305	0	
1974	262	220	18	
1975	355	440	11	Drought last year，interruption of flow during spring
1976	375	450	7	
1977	315	280	0	Water diversion about 230×10^8 m³/a，abundant precipitation in last two years
1978	350	360	1	Water discharge was lower last year
1979	365	350	19	Economic recovery beginning
1980	300	270	6	Economy surges
1981	385	500	37	Precipitation largely concentrated to flood season
1982	310	360	10	Water diversion about 250×10^8 m³
1983	325	450	5	
1984	380	470	0	
1985	340	405	0	
1986	250	240	0	
1987	275	220	17	Water diversion exceeding 274×10^8 m³
1988	360	350	8	Enhanced drought，extensive water diversion
1989	315	330	20	

In the 1990's the average annual precipitation decreased; this together with climatic warming led to increased dryness. At the same time diversion of river water accelerated, reaching 30.0×10^9 m³/year in this decade. As a result of combination of these effects river drought occurred each year from 1991 to 1999; in 1997 the riverbed dried out ten times after a first occurrence on Feb. 7. On Sept. 15 diverted water was returned to the river but on Oct. 14 partial drought occurred again. In spite of contributions from precipitation and conservation control measures water ceased flowing in the lower reaches of the Yellow River during 226 days in 1997, something that had never happened before.

Analysis of critical runoff volume

During the 1960's there were years with precipitation as low as 260 mm, runoff limited to 30.0×10^9 m³ and diversion kept at 15.0×10^9 m³. Even under these relatively severe conditions continuous flow could be maintained along the entire river channel. When the water usage rose above 17.5×10^9 m³ in combination with precipitation as low as 210 mm and runoff during flood season down to 28.0×10^9 m³, river drought was still relatively infrequent (about 10 percent of the time) and some water was discharged into the sea. These observations indicate that a water diversion of 20.0×10^9 m³, leaving 10.0×10^9 m³ as discharge into the lower reaches of the Yellow River, represents a critical limit for maintaining a continuous flow of water through the river channel. Another expression of this critical limit is a precipitation of 260 mm combined with 30.0×10^9 m³ runoff during the flood season.

If several years of abundant rain are followed by a year with precipitation less than 260 mm, river drought may not necessarily occur in the following year during spring. Conversely if one or several years of serious drought are followed by one or two years of heavy precipitation, and even if this is concentrated into the flood season, flow interruption will occur in the lower reaches of the river channel during that year or the river drought may be delayed to the spring of the next one or two years. Furthermore, serious water deficiency in the river will follow a year of drought while in the subsequent year the situation improves.

When water usage exceeds 20.0×10^9 m³ and reaches 25.0×10^9 m³ it approaches or exceeds half of the average natural runoff of the Yellow River. Under these conditions the capability of the river for adjustment and protection against disaster is impaired. With water diversion at such high levels, river drought will continue to appear in the lower reaches of the river even if the rainfall under these conditions would be as high as 300 mm and the runoff would be at least 42.0×10^9 m³.

The human impact

The human factors include diversion and overuse of river water, irresponsible deforestation and agricultural exposure contributing to soil erosion, and furthermore increased loss of water by evaporation and reduced runoff, in large part associated with economic and social developments. Recovery of China's economy began at the end of the 70's and accelerated in the early 80's, when the industrial and agricultural boom led to a dramatic increase in water usage. In the 1980's the average annual volume of water diverted from the river reached 27.4×10^9 m³. The resulting water shortage in the lower reaches of the channel has impaired the natural ability of the river to adjust itself.

The Yellow River drainage basin is characterized by a semiarid climate. It occupies 8 percent of the total area of China, but the runoff represents only 2 percent of the country's total. The water distribution rate per person is only 1/4 of the national average. The water distribution per acre of farmland is 17 percent of the national average[1]. Thus the drainage region has a comparative lack of water resources. The climatic trend toward warmer and drier conditions is manifest in lower precipitation and is accompanied by a steady increase in human water consumption.

A total of nine major water control projects have been completed along the main course of the Yellow River, withdrawing large amounts of water and suspended sediment from the river (Table 8). The total capacity of the numerous reservoirs had already in 1994 reached 84 percent of the river's average annual flow (HUANG and XU, 1994) and the planned water usage is increasing. At the same time as the precipitation decreased in the 90's the irrigated area was increased 1.6 times resulting in an increase in water consumption of 8.15×10^9 m³. Compared to the 50's the water usage in upper reaches increased by a factor 0.8, in the middle reaches it doubled, while in the lower reaches the water diversion for irrigation increased to 4.6 times the value in the 50's[2]. The water usage in the section upstream from Huayuankou, which was 3.9×10^9 m³ in 1919, rose to 7.4×10^9 m³ in 1949 and to 19.0×10^9 m³ in 1990, corresponding to 34 percent of the average annual runoff at Huayuankou (ZHAO et al., 1994). The irrigation area diverted water from the downstream region of the river did already in 1995 reach 35×10^6 mu[3], which far exceeds the river's capacity for water supply. In the wintertime the reservoirs in lower reaches of river are used for storage in order to meet demands in the spring season, and this results in water deficiency also during time periods without need for irrigation. Consequently over the years river drought has appeared increasingly early, with increasing severity and duration. Each year during the drought season the districts along the river strive to divert and store water to meet their needs, competing between each other, and thereby worsening the river drought

---

① Sunny Daily, May 11th, 1997.

② Science and Technology Daily, May 10th, 1997.

③ Mu is the Chinese unit of area. 1 acre = 4 047 m² = 6.07 mu

problem. Water usage after the year of 2001 in the entire drainage area is estimated to reach 36.8×10^9 m^3 and in 2010 39.2×10^9 m^3 (QIAN *et al.*, 1993; ZHAO *et al.*, 1994), developments that are bound to further increase river drought.

Table 8　The main water control projects in the main course of the Yellow River

(After QIAN *et al.*, 1993)

Control project	Controlled area (10^3 km^2)	Total storage (10^9 m^3)	Regulated storage (10^9 m^3)	Function	Capability for adjustment	Inception of construction/ beginning of water storage	Storage loss by silt deposition (10^9 m^3)
Longyang Canyon	131.4	24.70	19.36	Power, flood control and irrigation	Many years	1976/Oct. 1986	
Liujia Canyon	181.8	5.700	4.200	Power, flood control and irrigation	Year	Sept. 1958/ Oct. 1968	1.184
Yanguo Canyon	182.7	0.220	0.010	Power and irrigation	Day	Sept. 1958/ March. 1961	0.164 7
Bapan Canyon	215.9	0.490	0.009	Power	Day	Nov. 1969/ June 1975	0.016 0
Qingtong Canyon	275.0	0.606	0.030	Irrigation and power	Day & week	Aug. 1958/ April 1967	0.590 3
Sanshenggong	314.0	0.080		Irrigation and flood control		June 1959/ April 1961	0.054 4
Tianqiao	403.9	0.067	0.025~ 0.030	Power	Day	April 1970/ Feb. 1977	0.041 8
Sanmen Canyon	688.4	35.40		Power, flood control and irrigation		1957/Sept. 1960	5.410
Xiaolangdi	694.5	12.65	5.100	Power, flood control and irrigation	Season	Sept. 1991/?	

Already in the 1980's the Chinese government prepared a water distribution plan for the Yellow River tailored to the water supply capacity. However, the plan did not include provisions for monitoring and management of water usage under drought conditions. Aside from those limitations the Government also failed to establish a hierarchical system prescribing the order in which the regulations should apply. As a result these could not be practically implemented. As an example the water usage for the near-term regional water distribution stipulated by the State Planning Committee and Ministry of Water Resources and Electrical Power was exceeded already in the 1980's (REN, 1994).

There is little public awareness of the need for water conservation and many loopholes exist in the rules for water conservation and regulation management. Backward technology and inefficient management prevent effective recycling of water, which now amounts to an average of only 29 percent (ZHANG, 1993). The industries in cities along the lower reaches of the river consume $300 \sim 500$ m³ of water for the production value of 10 000 yuan, exceeding the use rate in developed countries by a factor 3 to 4 (GAO, 1993). In some regions the ground water is being deplenished and water pollution occurs. The volume of contaminated waste water amounted to 2.17×10^9 t in the 1980's and reached 3.3×10^9 t in the early 90's (GAO, 1993), intensifying the shortage of usable water resources. Another contributing factor is the underpricing of irrigation water which in the lower reaches of the river is made available at only 1 fen /m³, far below the actual cost of distribution. The revenue from river water sales is insufficient for maintenance and improvement of water conservation facilities. The ability of those facilities to store and distribute water has therefore steadily decreased, intensifying the waste.

IMPACT ON THE COASTAL ENVIRONMENT

Effects on economic development of the coastal area

The river drought has had an immediate impact on agriculture and industrial productivity and on the quality of life in the river plain and delta, particularly in Shandong and Henan provinces, including the two oil fields Shengli and Zhongyuan. The affected area comprises 54×10^3 km². The 900 000 inhabitants of the region around the estuary and the 140 million living adjacent to the lower reaches of the river suffer a serious shortage of drinking water, the quality of available water is extremely poor and the price as high as 1.8 yuan/m³. The area of farmland affected by river drought from 1972 to 1996 exceeds 7×10^6 mu and grain production there has been reduced by 9.86×10^9 kg. The industrial and agricultural losses in the same time period are estimated at about 27×10^9 CNY. In the 1990's this loss exceeded 4×10^9 CNY.

The scarcity of water resources has inhibited municipal, industrial and agricultural development along the lower- and parts of the middle reaches of the Yellow River. In order to mitigate the problem, the amount of water has been reduced that is used for irrigation in the middle and upper reaches, and also the amount retained in the reservoirs. These measures, however, have affected agricultural productivity, water conservation facilities and hydroelectric power generation in those regions, creating large hidden economic losses.

Impact of river drought on the natural environment in the coastal zone

The earlier steady-state interaction between the marine and fluvial systems led to a dynamic equilibrium that established the coastal physiographic and ecological features and evolutionary trends in the Yellow River delta and the adjacent coastal regions in the Bohai and Laizhou Bays. With the loss of fresh water and sediment discharge the balance in the natural environment has been destabilized.

The absence of flow in the river during half of the year has lowered the ability of the river to maintain its course resulting in deformation and eventual disappearance of the shallow water channel. The lack of overflow into the surrounding plains during the flood season is also depriving these of fresh mud, burying the alkaline soil and thus restoring its fertility. The effects of many years with river drought are bound to include also a lowering of the ground water level and the availability of fresh water. Seawater will as a result invade the river channel and the wedge of underground salt water likewise progress further inland, penetrating the pore space in the sediment strata abandoned by the fresh water. This is turn will degrade the water quality and lead to salinization of the soil in the delta and in the lower reaches of the river. In addition, with the strong northeastern winds during the increasingly pronounced dry season desertification is becoming more severe in this region, exacerbating the scarcity of water and deteriorating the farmland.

The diminished discharge of fresh water carrying dissolved mineral—and organic nutrients into the river estuary has lowered the production of plankton, and consequently of fish and shrimp. The environmental change has, in addition to their grazing needs, negatively affected their migration routes and spawning grounds. Following completion of the Sanmenxia dam construction in the 1950's the decreasing river runoff was accompanied by lower catches of mullet and white shrimp in Bohai Bay. Fishermen ascribe this to "a lack of sweet fresh water in the bay" which is understandable in terms of the environmental factors mentioned above. The lack of fresh water is also bound to affect biodiversity and the biological food chain in the reed marshes and mud flats along the coast of the delta, threatening the survival of many wild plants, water organisms and birds along the Bohai Bay.

The huge mass of sediment, which was normally discharged by the river into the sea, maintained the outward growth of the delta and its addition to the land area. The decrease and often complete interruption flow in the river has strongly affected the sediment transport into the sea. This causes net erosion of the uncompacted delta deposits and regression of the coastline, further enhanced by storm surges acting in combination with the rise in sea level all with serious consequences for the coastal region.

Coastal effects of river channel migration

Upon entering the plains through Taohuayu from the canyon lands upstream the bed of the Yellow River broadens, the gradient decreases dramatically to an average of 0.012% and the flow velocity drops. As a result the sedimentation rate is markedly increased. In the past the sediment load picked up on the Loess Plateau by the river was deposited as a fluvial delta fan in the sea that occupied what is now the North China Plain, gradually enlarging the plain to its present size.

The large volume of sediment that is continually deposited in the lower reaches of the river channel causes the riverbed to rise above the surrounding land. As a result the riverbed is elevated 13 m above ground in Kaifeng and 21 m in Xinxiang in Henan province. Dikes, commonly of kilometer length are constructed to keep the river in its course. During the flood season the dikes are occasionally breached with flooding as result. Thus there is a natural tendency for the river to overflow its banks and to seek a new course through the lowest terrain.

Since 1855 when the Yellow River turned northward and flowed into the Bohai Sea, it has breached the dikes fifty times, at ten of these occasions with significant effects on its course. The changes tend to follow a clockwise pattern, starting from a northerly direction, turning northeast, east and southeast and then suddenly north again. Each such meandering cycle is completed in about ten years, corresponding to a total sediment discharge to the ocean of $0.9 \sim 1.2 \times 10^9$ t. This phenomenon extends into the past and explains the migrations of the river channel in the ancient delta and fluvial fan of the Yellow River.

Historical records of the course of the Yellow River extend back over 2607 years to 605 B.C. In this time period the banks in the lower reaches were breached 1591 times and the river changed its course 26 times, hence with a breach occurring about every two years and a change of course every hundred years. Historical records indicate that major changes in course of the river have occurred eight times since 2278 B.C. (Fig. 7). During this time the river migrated far north and flowed through the present Yongding River channel from Tianjin into the Bohai Sea. In its migration to the south the Yellow River took over the Huaihe River bed debouching into the Yellow Sea. The range of migration from north to south covers more than six hundred kilometers. The frequent changes in course of the river have severely impacted the ecosystems in the lower reaches of the river and in the delta (WANG *et al.*, 1986; WANG and KE, 1989).

The frequency of overflow and migration has been directly influenced by the erosion in the Loess Plateau upstream, providing a major part of the sediment load. During the period of Western Zhou Dynasty and the Warring States (771~221B. C.) the grass— and forest vegetation on the Loess Plateau remained relatively undisturbed (SHI *et al.*,

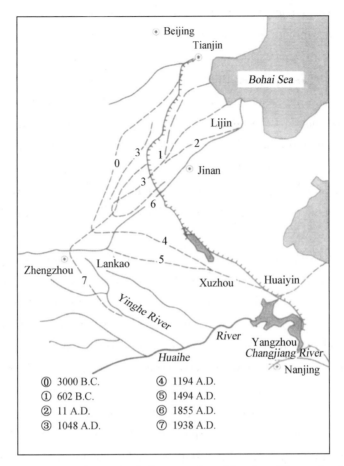

Fig. 7　Migration of the channel in the lower reaches of the river. The number from 0 to 7 indicates the main channels of the lower Yellow River at different times.

1985; WU *et al.*, 1994) and as a result overflows rarely occurred. From the time (221B. C. ～8A. D.) of the Qin Dynasty to the Western Han Dynasty extensive settlement took place in the area and the expanded cultivation caused extensive soil erosion and river outbreaks.

In the time period 8～589A. D. encompassing the Eastern Han Dynasty and the Northern and Southern Dynasties the region was transformed from farmland to pasture and the course of the river was stabilized for this entire period (TAN, 1962; ZOU, 1989; REN, 1997). However during the Tang Dynasty (618～907A. D.) a large scale migration to the west out of the capital city of Changan (near today's Xi'an) took place and settlements on the Loess Plateau destroyed the forest and opened the grassland to farming. This led to extensive soil erosion, which in turn increased the sediment load in the river, the rate of deposition in its lower reaches and the frequency of river migration events (WANG, 1994).

Two major migrations were induced artificially (1128 and 1938 A. D.). The first of these events, i. e. the fifth major migration of the Yellow River, took place on

November 15 in the second year of the Southern Song dynasty The capital (Kaifeng) official, named Du Chong, excavated the riverbank as a defense, by flooding, against the invasion by a Jin Dynasty army moving in from the north.

This strategy was unsuccessful but as a result the Yellow River flowed south through Si River and Yingshui River, and via the Huaihe River to the Yellow Sea; this course of the river lasted 727 years. During more than 400 years of this period (1128 – 1546A. D.) the lower reaches of the river broke up into a number of small branches. These river branches frequently broke through containing dams, overflowed the surrounding plains and migrated within the region south of the Yellow River, north of the Huai River, east of the Ying River, and west of Si River, extending more than 250 km from east to west.

From 1578 to 1579, following the construction of a major flood control dam, the lower river channel remained largely stabilized in its course until 1855A. D. when it returned to Bohai Bay. Since 1128, the river frequently sought its way to the Yellow Sea through the Huaihe River. Large quantities of Yellow River sediments caused the North Jiangsu coast to change from mainly wave process formed barrier-lagoon coast to tidal mudflat. The Yellow River sediments were rapidly deposited in the North Jiangsu coast region, creating the now abandoned Yellow River delta extending 90 km into the sea with an area of 7 160 km^2. Also, the coast south of this delta has continuously received sediment either from the river or from the abandoned delta since that time. This coastal plain extends 40 km into the sea with a flat area of 157 000 km^2, about 1/6 of the area of the Jiangsu province (ZHANG, 1984; WANG and AUBREY, 1987; REN, 1988 and 1992; WANG, 1994).

The change in the course of the river has also affected the natural water system of the Huang-Huai Plain. It led to the creation of Hongze Lake, one of China's largest fresh water lakes with an area of 2 200 km^2. It also altered the Huaihe River causing it to enter the Changjiang River channel, as a result converting the Yellow River to a tributary of the Changjiang River. These relocations fundamentally changed the physiographic features of the entire North China—and North Jiangsu plains (REN, 1997).

In 1855 the Yellow River returned north and entered the Bohai Bay. Due to the sudden loss of its sediment source the abandoned delta with a shore length of 150 km started to undergo erosion and began receding back towards land. The rate of erosion amounts to 15～30 m/year and about 1 400 km^2 that used to be land has now been inundated by the sea. What used to be the mouth of the main irrigation channel in 1961 is today located several hundred meters off shore. Protection of the coastal area is consequently a major problem in the abandoned Yellow River delta. The coastal sediment has changed to fine and coarse sand with cheniers (WANG and AUBREY,

1987；WANG and KE，1989；WANG and ZHU，1994）. Along with this, the eroded material from the abandoned delta is continually carried into the North Jiangsu coastal region, forming new land areas.

The 160 km long Yellow River delta coast has from 1855 progressed 20. 5~27. 5 km seaward. While the abandoned estuary is being eroded, the newly created estuary is rapidly accumulating sediment. Deposits are forming on both sides of it, adding approximately 23. 5 km² of new land per year. In a single summer the shoreline may extend 6~10 km. This part of the Bohai coast turned from cheniers which existed before 1855 to mud tidal flats (WANG and KE, 1989).

The coastal evolution is thus controlled by the interplay of marine and fluvial processes. The Bohai Sea is considered a region of tectonic subsidence, and sea level is currently rising by 4. 5~5. 5 mm/year (WANG, 1998). However, with the yearly addition of 0. 8~1. 2×10⁹ t of river-born deposits the river flow and sediment pressure resist the erosive action by waves and tidal currents, and as a result the coastline now gradually progresses outward. Once the sediment supply ceases, the erosive forces of the sea take over, the coast will be eroded, the nearshore slope of the bottom will gradually become steeper and the shoreline will recede.

The natural period of lateral migration of the Yellow River mouth is 6~10 years or 12 years. Because of the requirements for development of the local oil field and of the city of Dongying it is necessary to halt the migration. Recent investigations indicate that the river mouth will next move south, then southeast and eventually closing off, forming a lagoon. However, with the current decrease in sediment supply by 0. 6×10⁹ t per year the extension of the river mouth into the sea will slow down. If the sediment supply ceases completely for more than one third of the year, the coast will seek a new equilibrium, and the salt-water wedge will progress inland, pushing back the fresh ground water. Without addition of new river-born soil, burying the alkaline surface deposits, the ground will become barren and unsuitable for farming. The associated erosion and recession of the coastline will have disastrous effects on the river delta.

Comparative examples

The Luanhe River in the Northern part of Hebei Province used annually to discharge 4. 19×10⁹ m³ of water and 22. 19×10⁶ t of sediment into the sea. In a project for provision of water for the city of Tianjin 0. 355×10⁹ m³ of water was diverted from the river annually. This had the effect of diminishing the transport of sediment to the ocean to 1. 03 × 10⁶ t/year. From 1988, since the inception of water diversion, the shoreline in the delta estuary has been eroded back by 17. 4 m per year. Seawater invaded the lower river channel with deterioration of the water quality in the form of a salinity increase from 27. 3‰ to 32. 6‰ (QIAN, 1994).

The Gunshui dam was constructed in the Nile River delta in Egypt in 1881, and the Aswan reservoir upstream was completed in 1902. The flow of water and sediment below the reservoir decreased steadily, causing heavy erosion in the delta. The sand spit and sand bar receded at average rates of 29 m and 31 m/year respectively (CHEN, 1987). As a result of the completion of the Aswan dam in 1964 the natural process ceased in which the river annually used to overflow its banks, inundating and fertilizing the surrounding fields. The deficiency of sediment transported into the sea further exacerbated the erosion of the delta with recession of the coastline. The decreased transport of organic and mineral nutrients to the Mediterranean led to diminishing sardine catches in the Nile estuary.

These and numerous other examples from history illustrate the dangers arising when human activities do not conform to the regular processes in the natural environment. Human actions must not extend beyond the limits of flexibility of the natural and artificial processes that preserve the balance in nature between water, air, soil, flora and fauna.

RESOLUTIONS BASED ON CONSIDERATION OF A SUSTAINABLE DEVELOPMENT

First of all, awareness must be raised in people's minds of the need for water conservation and for protection of the water resources and the water environment. A well-considered plan should be developed for the use of the fresh water resources of the Yellow River. Considering the large drainage area, the water resources of the river are limited and deficient. In planning the direction and rate of development of industry, agriculture and society in the region, the water resources must be researched and carefully considered since water is a key factor in these developments. The efficiency of water usage must be increased and the backward traditional irrigation method of overflowing the fields must be changed in the upper and middle reaches of the river.

The heedless expansion of land under cultivation around the lower reaches of the river and in the delta must be brought under control. The water needs for human consumption and for the petroleum industry should be given priority. Water recycling capacity must be increased and industries with excessive and inefficient use of water should be curtailed. Strict measures must be taken for pollution control in order to improve water quality and to eliminate dumping of industrial waste. A realistic pricing system should be introduced, raising the price of water to cover actual costs and discourage waste. In general, practical water usage methods must be stipulated, and applied in order to improve the system.

Second, to harmonize the relationship between humans and land, research needs to be carried out on the particular characteristics of the river and its drainage basin, and on the maximum tolerance of the water resource for human impact. Developments in the drainage area must be considered on an integrated basis. Exploitation of water resources for hydroelectric power in the upper reaches of the river should be carried out step by step, and in conjunction with the necessary maintenance and conservation of water and soil in the middle reaches of the river. Measures should be taken to prevent flooding and siltation in its lower reaches. On one hand storage of water in dammed reservoirs need to be implemented during flood periods for later use, on the other a year-round minimum effective flow of water and sediment to the sea must be maintained. Only in that way can the environmental stability of the delta and the Bohai and Laizhou bays be ensured.

Third, a management system needs to be established for the entire drainage basin, allocating the water resources in time and space. Appropriate technical planning must take into account the water usage requirements along different sections of the river system and their seasonal variations. Emergency plans must be developed to deal with drought and flood situations. The same attention must be given to these problems as to the Chinese population problem. The amount of water diverted from the system must remain below 20.0×10^9 m^3/year in order to maintain uninterrupted flow of water and removal of excess sediment in the lower reaches. The average annual sediment discharge should be kept around 0.6×10^9 t so that the natural equilibrium state of the delta can be sustained. The warning signals are when precipitation in the flood season falls below 260 mm or when the runoff in the same season is less than 30.0×10^9 m^3. To achieve the necessary restrictions, serious attention must in the immediate future be paid to preventing excess usage of water and to permanently empower water conservation facilities along the river to regulate the seasonal water flow. In the lower river and delta areas development of animal husbandry is preferable over irrigation-intensive agriculture. Careless overdraft of ground water and construction of high rise buildings should be prohibited. These precautions will prevent land subsidence, salt-water invasion and coastal erosion.

Lastly strategies must be laid for eventually moving water from the south to the north. Simulation studies for such transport projects should be undertaken in the drainage basins of the Changjiang, Huaihe and Yellow Rivers. Such studies should provide sufficient basis for planning and design without encountering additional problems. However, on a short-term basis it would be unrealistic to rely on the plan for water transfer from Changjiang River to the Yellow River to solve the existing problem of water shortage in the latter. The most practical immediate solution is to improve the water usage structure by optimizing the efficiency of agricultural water usage with a goal of $50\% \sim 75\%$ and industrial water recycling with $60\% \sim 80\%$. The volume of water diversion should be kept below 20. 0×10^9 m³/year. By using the Xiaolangdi reservoir together with the natural water regulation in the river, 0.50×10^9 m³ of water can be permanently maintained in the drainage basin. This will satisfy the needs for economic development arising in the next five to ten years and diminish the occurrence of river drought in the Yellow River.

LITERATURE CITED

Cai, M. L. and WANG, Y. 1999. Evolution of the Yellow River delta and its impacts to the Bohai and the Yellow Seas. Hehai University Press, Nanjing. 1 – 213 (in Chinese).

Chen, J. Y. 1987. Primary analysis of the impact of the San Xia construction on the Changjiang river mouth. In: Wang, Y., (ed.) *Collected Works of Debate of Changjiang Sanxia Construction*. Chengdu Science and Technology University Press, Chengdu. 230 – 233 (in Chinese).

Cui, Z. P. (ed.) 1991. *China Hydrological Encyclopedia*. Water Conservation and Electric Power Press, Beijing. 856 – 2473 (in Chinese).

Gao, C. D. 1993. Present water environment and coping methods in the Yellow River drainage area. *People's Yellow River*, 12: 4 – 7 (in Chinese).

Huang, D. W. and Xu, K. J. 1994. Debate on source of water irrigation problem in the Yellow River delta. *People's Yellow River*, 5: 43 – 46 (in Chinese).

Hydrological Bureau of Water Conservation and Electric Power Department. 1986. China Water Resources Evaluation. Water Conservation and Electric Power Press, Beijing. 1 – 194 (in Chinese).

Liu, D. T. 1964. The Loess in the Region around the Middle Reach of the Yellow River. Science Press, Beijing. 2 – 5 (in Chinese).

Qian, C. L. 1994. Effects of the water conservancy projects in the Luanhe River basin on Luanhe River delta, Hebei province. *Acta Geographic Sinica*, 49(2): 158 – 166 (in Chinese).

Qian, Y. Y., Ye, Q. C. and Zhou, W. H. 1993. Changes of the Water and Sediment Discharge and Evolution of Riverbed in the Main Yellow River Channel. China Industrial Construction Material Press, Beijing. 1 – 230 (in Chinese).

Ren, M. E. 1988. Several problems in the Yellow River delta dredging. *Scientific and Technological Review*, 6: 23 – 27 (in Chinese).

Ren, M. E. 1992. Human impact on coastal landform and sedimentation-the Yellow River example. *GeoJournal*, 28(4): 443 – 448.

Ren, M. E. (ed.) 1994. The Three Major Deltas of China. Higher Education Press, Beijing. 8 – 128 (in Chinese).

Ren, M. E. 1997. The Yellow River and people's life. In: QIN, Y. C. and Li, X. J. (eds.), *Study on the Regional Sustainable Development Theory, Methods, and Practical Application*. Henan University Press, Kaifeng. 78 – 95 (in Chinese).

Ren, M. E. and Shi, Y. L. 1986. Sediment discharge of the Yellow River (China) and its effect on the sedimentation of the Bohai and the Yellow Sea. *Continental Shelf Research*, 6(6): 785 – 810.

Shi, N. H., Cao, E. Q. and Zhu, S. G. 1985. Forest and Grass Change in the Loess Plateau. Shanxi People Press, Xi'an. 1 – 200 (in Chinese)

Tan, Q. X. 1986. New perspective on the long period stable flow in the lower reach of the Yellow River after East Han Dynasty. First published in *Scientific Monthly* 1962 (2), Compiled in tan Q. X. (ed.), *Papers on Yellow River History*. Fudan University Press, Shanghai, 1986. 72 – 101 (in Chinese).

Wang, Y. 1994. Effect of military activities on environment in eastern and southeastern China. *Annual Science Report—Supplement of Journal of Nanjing University*, 30 (English series 2): 43 – 46.

Wang, Y. and Aubrey, D. 1987. The characteristics of the China coastline. *Continental Shelf Research*, 1(4): 329 – 349.

Wang, Y. and Ke, X. K. 1989. Cheniers on the east coastal plain of China. *Marine Geology*, 90: 321 – 335.

Wang, Y., Ren, M. E. and Zhu, D. K. 1986. Sediment supply to the continental shelf by the major rivers of China. *Journal of the Geological Society of London*, 143(6): 935 – 944.

Wang, Y. and Zhu, D. K. 1994. Tidal flats in China. *In*: Zhou D.; Liang, Y. B. and Tseng C. K. (eds.), *Oceanography of China*. Kluwer Academic Publishers, 2. 445 – 456.

Wang, Y. 1998. Sea-level changes, human impacts and coastal responses in China. *Journal of Coastal Research*, 14(1): 31 – 36.

Wang, Y. Z. 1991. Initial Investigation of the relationship of the changes in the precipitation and water and sediment discharge in the upper and middle reaches of the Yellow River. *In*: QIAN Y. Y. (ed.) *The Collected Works of the Yellow River Drainage Basin Environment Evolution and Regulation of Water and Sediment Discharge* (Vol. 1). Geological Press, Beijing. 28 – 35 (in Chinese).

Water Conservation and Hydroelectric Water Design and Plan. Institute of Water Conservation and Electric Power Department, 1989. China Water Resources Utilization. Water Conservation and Electric Power Press, Beijing. 1 – 226 (in Chinese).

Wu, X. D., Niu, C. X. and Wang, S. C., et al. 1994. Historical Climate, Water and Sediment Change in the Yellow River Drainage Basin. Atmosphere Press, Beijing. 65 – 121 (in Chinese).

Zhang, R. S. 1984. Land-forming history of the Huanghe River delta and coastal plain of North Jiangsu. *Acta Geographic Sinica*, 39(2): 173 – 184 (in Chinese).

Zhang, T. Z. 1990. China Water Conservation and Environment. Science Press, Beijing. 143 – 178 (in Chinese).

Zhang, T. 1993. Water resources scarcity and its solution in the Yellow River. *People's Yellow River*, 1: 4 – 6 (in Chinese).

Zhao, Y. A., Pan, X. D. and Li, Y. 1994. Changes of the volumes of water and sediment and evolution trend of its lower reaches of the Yellow River. *People's Yellow River*, 2: 31 – 41 (in Chinese).

Zhu, K. Z. 1973. Primary investigation of the climate change in the past 5000 years in China. *China Science*, 2: 291 – 296 (in Chinese).

Zhu, Z. Y. and Jia, Z. S. 1993. Prediction of the natural environment change and human activity impact on the water and sediment in the Yellow River. *In*: Yang, Q. Y. (ed.), *Collected Works on the Evolution of the Yellow River Drainage Basin Environment and Regulation of Water and Sediment Discharges* (Vol. 6). Atmosphere Press, Beijing. 134 – 155 (in Chinese).

Zou, Y. L. 1989. New perspective on the long period stable flow in the lower reach of the Yellow River after East Han Dynasty. *People's Yellow River*, 2: 60 – 65 (in Chinese).

河海交互作用与苏北平原成因[*]

一、引言

苏北平原位于江苏省长江北岸与连云港市的岚山头之间（32°10′～35°05′N,118°40′～120°30′E）。平原西界为一系列低山丘陵,大体上自岚山头向西延伸,至洪泽、高邮湖泊群后,呈弧形弯曲向南、再向东南延伸至太湖内侧,平原东部滨临南黄海。平原东侧滨海分布着两列保存较好的古海岸贝壳堤,西侧一列分布于盐城龙冈、大冈,并延伸至东台及梅里、太仓一带,高约 6 m,宽度超过 10 m。沉积层厚,底部多贝壳与细砂层,上部为黏土质粉砂与粉砂层,实为一列长大的海岸沙坝。其形成时代相当于全新世高海面时（6 500～5 600aB. P.）[1],当时的海岸坡度较大,激浪带宽,泥沙粒径较粗。东侧另一列贝壳堤分布于阜宁沟墩、上冈至东台,时代相当于新石器晚期（3 800±70aB. P.）的古海岸[1,2]。两列贝壳堤中段相距约 10～20 km,但向北逐渐合并,反映着古海岸各段的淤涨与稳定情况不同[3]。北宋期间（960—1127）修建的挡海堤（后世称为范公堤）是建在古贝壳堤基础上,说明当时该处尚濒临黄海。目前,两列贝壳堤相距现代海岸达 50～60 km,堤段多被辟建为公路。古贝壳堤以东盐土质平原上亦有一些长度不大的贝壳滩、脊残留,反映着海滨逐渐堆积为陆地及广阔滩涂的过程[4~6]。鉴于苏北东部滨海平原的成因已明确,本文研究的地区是位于滨海平原的内陆侧,即古贝壳堤以西与洪泽、高邮湖泊群之间的平原成因（图 1）。

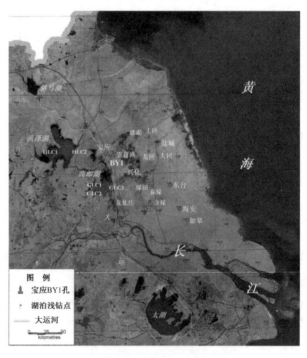

图 1 研究区位置图

* 王颖,张振克,朱大奎,杨競红,毛龙江,李书恒:《第四纪研究》,2006 年第 26 卷第 3 期,第 301 - 320 页。

同年载于:《第七届海峡两岸地貌学研讨会》,1 - 20 页. 南京大学;2006.8.21 - 27;以"河海交互作用与苏北平原建造"为题的中文长摘要载于《"中国海洋资源环境与南海问题"学术研讨会会议日程与论文(摘要)汇编》3 - 98～3 - 99 页,广西北海,2006 年 2 月 26 日—3 月 5 日。

国家自然科学基金项目(批准号:40271004)资助。

　　地质发展过程中,苏北地区是在古扬子地台上于白垩纪晚期受燕山运动影响而发育的苏北—南黄海断陷盆地,具有华夏式与新华夏式构造特征,基底为盐阜坳陷,建湖隆起与东台坳陷的结构格局。侏罗纪晚期火山喷发活动形成安山岩、粗面岩与凝灰岩层。一系列 NE、NNE 及 NW 向的断裂活动使坳陷与隆起分割,影响着各时代的沉积厚度与分布差异。第三纪时,坳陷中又形成次一级的地堑或凸起,并有早、中期火山喷发的玄武岩流堆积。整体上,苏北盆地堆积着达数千米的厚层新生代沉积[4,5,7]。第四纪是苏北由盆地发展为平原时期[7]。

　　前人在苏北地区的工作,着重于第四纪沉积地层与构造活动[5],第四纪气候变化、海平面变化及海岸线变化[1~3],或滨海平原及海域的动力、地貌、沉积与开发应用[6,8,9],或黄淮平原水系环境与灾害[10~12],尚未系统地阐述苏北平原之成因。本文期望在前人工作的基础上,探求苏北内平原的成因机制。

二、平原地貌特征与湖泊群探索

　　苏北平原河、湖、塘、渠纵横,平原地势起伏多变。大体上,呈现自西向东,即向海倾斜的趋势。但沿海因有古贝壳堤而地势增高,平原的最低洼处却在中部兴化与里下河流域,是高程在 0~2 m 的湖滩地及 5 m 以下的圩田平原(图 2)。这种地势分布,是否受基底构

图 2　苏北平原地势图[13](附彩图)

造与原始地貌影响？平坦的原野较难判别其成因与发育过程,尤其是苏北平原受长江、黄河及淮河水系迁徙多变和历史时期为防灾治水或为防御目的而多次开挖河渠的影响[12],难以判断其自然面貌。但是,若进一步分析地貌类型分布的组合特征,以及沿自然环境而分布的村庄聚落,仍能获得有关平原发育成因的一些启示。

(一) 基岩古海湾

苏北平原西侧止于一系列呈弧形环绕的基岩低山、丘陵与岗地($32°00'\sim35°05'$N),形似巨型的古海湾与岬角的遗留(图3)。

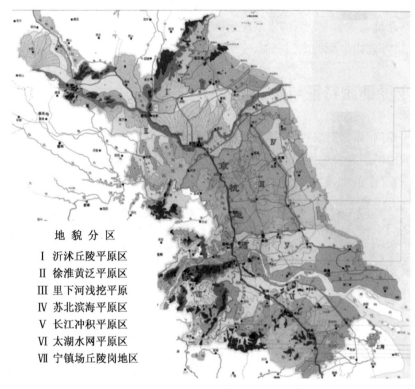

地 貌 分 区
Ⅰ 沂沭丘陵平原区
Ⅱ 徐淮黄泛平原区
Ⅲ 里下河浅挖平原
Ⅳ 苏北滨海平原区
Ⅴ 长江冲积平原区
Ⅵ 太湖水网平原区
Ⅶ 宁镇场丘陵岗地区

图3 江苏地貌图[13](附彩图)

环绕平原西北侧的山地,如岚山头、马陵山为剥蚀丘陵,高度约$50\sim100$ m,个别山峰,如大吴山,高达364 m。岚山头构成海州湾北岬角,该处为燕山期的花岗岩类组成并出露太古界与下元古界的变质岩;马陵山海拔50.8 m,中生代赤褐色砂砾岩组成,岩层水平具厚层风化壳,山顶平缓,向陆坡(西北坡)缓倾,向海坡有数层陡坎,陡坎高度约5 m,坡上覆盖含砂礓之黄土,表明马陵山陡坎坡形成于砂礓黄土层(Q_3)沉积以前。自马陵山向南至洪泽湖、高邮湖西侧,山地零散、低缓,其中徐州相山高程$50\sim100$ m,大洞山峰高达361 m;泗县与泗洪间低岗丘陵,高程$50\sim100$ m;盱眙南部的张八岭山地高$100\sim200$ m,其中裂山峰顶高194 m,狮子岭高231 m;天长与六合间为低山岗地,冶山峰顶高231 m。长江以南的宁镇山脉与宜溧山脉较高,多为$100\sim200$ m,并有超过400 m的山峰:九华山434 m,茅山山脉的鬐山410 m,宜溧山地的铜官山521 m,黄塔顶611 m;湖州与太湖地区山地多为200 m。南部山地地层完整,有太古代变质岩、古生代沉积岩、中生

代花岗岩,第三纪的砂砾岩、玄武岩以及第四纪的下蜀黄土层。虽然,山脉的走向多为
NE - SW 向,沿江分布的为近东西向,但其东侧,与湖泊群、与平原交接的一段仍呈弧形
环绕(见图 3)。这种组合反映着基岩山地与平原之间的成因联系——海湾与海积平原接
壤。在连云港、东海、新沂一带,尚保留着基岩山麓与海积平原面呈现为水平直接相交的
特点,向海侧岩壁尚有浪花风化、海蚀穴及抬升岩滩等地貌遗迹(图 4 照片 1,2,3 和 4)。

图 4 苏北平原地貌遗迹（附彩图）

1. 南云台山东南部山麓与海积平原呈水平面直接相交；2. 山麓海蚀崖与平原面成水平交接（沿连云港—灌云高速公路）；3. 大伊山古海蚀穴及海蚀阶地（高程 80 m）；4. 张宝山古海蚀穴（2.5 m 高）及海蚀崖壁上的浪花风化孔穴遗迹（距海 20 km，高程 8 m）；5. 河荡环绕的垛田与村舍；6. 苏北水网洼地垛田；7. 上冈贝壳堤剖面；8. 上冈堤基贝壳层

上述地貌组合的大势，反映着该平原内侧的山地为"基岩古海湾"岸的遗迹，则兴化地区低洼是因位居"古海湾"中部，原始基底低。

（二）水网区垛、圩、人居分布特点

苏北平原西侧有低山丘陵环绕，地势高；东部因大冈、上冈贝壳堤，其地势亦略高；北部因有废黄河决口扇堆积而地势增高；中部里下河平原地势低，其中以建湖与兴化间更为低洼，水网密布。村庄、人居或墓地多沿堤坝状、陆岛状高地而建，当地称为垛、圩、墩，如甘垛、垛田、三垛、曹垛、获垛、俞垛、大垛、新垛；下圩、中圩、老圩；塾墩、九里墩等。垛、墩系堆土筑高、防淹，四周有河荡、洼地环绕，如兴化的垛田镇。垛、圩多呈南北走向延续分布，大体上与现代海岸平行，一些垛、坝的基部尚残存天然沙基，田野土层中尚可拣到已经风化的青蛤、牡蛎壳片（见图 4 照片 5 和 6）。实质上，垛、墩主要是先民利用滨海沙堤或沙坝，添土增高，防潮侵而居，适应海滨低地环境的建筑遗迹。

以兴化南部之俞垛为例，该处海拔 5.9 m，已无岗、垛形态，但在田野地面挖渠处（32°38′59″N，120°00′40″E）出露自然剖面，查找到沉积证据，为海积沙堤或沙坝，人居沿天然高地而聚落。共分两层：上层为 0～3 m 厚灰黑色粉砂质黏土，层中（1.5～2.0 m 处）有密集的贝壳层，贝壳多已风化，主要为青蛤、白蛤。此层采样为俞垛 1 号，黏土层采取俞垛 2 号沉积样。下层 3～6 m 为灰黑色粉砂层及粉砂质黏土层，采样 3 号。在 20 m 深处出露棕黄色砂层。

将俞垛沉积层与平原北部上冈贝壳堤西侧平原沉积对比，在建新机械场（高程 8.5 m）以东砖瓦场 1.3 厚的上覆土层之下（33°33′30″N，119°59′12″E）观察到自然剖面（见图 4 照片 7 和 8），由上而下分别为：

（1）0～2.2 m，黑灰色粉砂质黏土。表层粉砂具有 1～2 mm 厚页状层理（取样上冈 1 号），中层为厚度 4～13 cm 的黏土质粉砂夹层（取样上冈 2 号），下部为粉砂层（取样上冈 3 号）。

（2）2.20～4.35 m，灰黄色粉砂质黏土（取样上冈 6 号）。

（3）4.35～5.80 m，灰黑色粉砂质黏土层（取样上冈 7 号）。

（4）5.8～7.4 m，灰黄色粉砂质黏土层，含贝壳（取样上冈 8 号）。

（5）出露 0.8 m，黑灰色粉砂质层（取样上冈 9 号）。

上冈样品采自贝壳堤内侧海积型平原。俞垛沉积层颜色、质地与有孔虫种属均与上冈处类同，所以，也是海相沉积（表 1）。其中有孔虫种属由加拿大学者 Dr. J.-P. Guilbault 鉴定。

上述分析证明俞垛处曾为海陆交互作用之浅海，垛、圩、墩等具有地域特色的民居村落点，是沿天然海岸高处集聚而发展的。

（三）湖泊群特点

沿弧形山地的东侧与里下河平原之间，分布着一系列湖泊，如：骆马湖、洪泽湖、白马湖、高邮湖、邵伯湖及北部赣榆县与东海县西山麓的小湖泊。长江以南，沿宜溧山麓东北侧分布着长荡湖、滆湖和太湖。一系列湖泊断续相延呈湖泊群组合形式，反映该处原始地势低洼；湖泊群西侧均有河流汇入，淮河水系流经处，基岩丘陵已被侵蚀间断，非现代河流一日之功；一些较大湖泊，如洪泽湖、高邮湖、骆马湖等可能曾是古河流（古淮河水系、古黄河水系）汇水基面；洪泽湖的西侧，高邮湖的西、北侧均有河流入湖形成的伸出型三角洲。前人研究认为："洪泽湖是于明、清时治水形成，湖中三角洲是 1850 年后黄河夺淮后形成"[11,14]。但是，洪泽湖北部以及洪泽湖与成子湖间的三角洲状堆积体上，覆盖着更新统沉积，其时代必早于 1850 年。文献中亦指出："洪泽湖形成以前，在淮河右岸散布有阜陵、万家、泥墩、破釜、白水等小型湖塘"[14]，"洪泽湖洼地前身是淮河流出低丘后形成的宽阔河谷洼地，全新世海侵曾否使该洼地沦为海域或潟湖……尚未得到直接证据，但全新世海侵曾引起洪泽湖区河谷洼地发育松散沉积"[11]。

上述表明洪泽湖的原始地面是低洼的。

关于高邮湖的形成发展，曾有详细的论述："高邮湖地区古为潟湖洼地平原"，"全新世 8000aB. P. 前后，海平面上升到高邮附近"，"冰后期最高海面时，长江口上溯至镇江与扬州之间……黄河、淮河入海口在淮阴市到叶云闸之间。高邮附近是被长江三角洲和淮河三角洲所夹的一个浅海湾"[15]。该论点明确，与本项研究所获结果较为一致。在淮阴南郊，地名为"海口"处的钵池山，观察到古淮河出海口之堆积层（图 5），共有两层：① 1.91 m 厚棕黄色夹粉土质粗砂层，已硬结，未显层理。内含多个砂质筒状结壳，内有淤泥充填，筒长 20 cm，筒径约 1 cm，直立或斜插状，系虫管之遗迹。底部为侵蚀界面，具浅谷形态。② 出露 1.5 m，棕红色粗砂层，已硬结，有较多虫管构造。

表 1　俞垛、上冈剖面有孔虫比较表

Locality 采样地点 / Sample 样品	Shanggang 上冈								Yuduo 俞垛	
	1f	2f	3f	7f	8f	8s	9f	9s	1f	2f
Oxides 氧化物						✓	■			
Plant matter（fresh to carbonized，not charcoal）植物屑	✓		✓	✓	✓	✓	✓		■	✓
Charcoal 碳屑	✓		✓	✓					g	✓
Seeds 种子										✓
Pollen/spore cases 孢粉			✓						✓	
Insect fgs 昆虫屑										✓
"shieldform phytolith" in WPX 盾状植物化石				✓						
Charophyte oogonia 轮藻植物										✓
Charophyte oogonia linings 轮藻卵原包裹									✓	✓
Candonida 玻璃介										?
Loxoconcha cf. *ocellata* 眼点弯贝介比较种										✓
Propontocypris euryhalina 广盐始海星介										✓
Sinocytheridea 中华丽华介	✓		✓	✓					✓	✓
Spinileberis pulchra 美丽刺面介				✓					✓	✓
Spinileberis furuyaensis 古屋刺面介										
Neomonoceratina crispate 皱新单角介				✓	✓					
Cytheropteron cf. *miurense* 三浦翼花介比较种					✓					
ostracod 介形虫					✓					
Acanthocythereis sp. 刺艳花介未定种					✓					
Hemicytheridea reticulata 半尾花介									✓	✓
Foraminifera（total，undifferentiated）有孔虫	■	■	■	■	■	■	✓	✓	■	■
Coscinodiscus like diatoms 园筛藻属			■							
other centrales（diatoms）硅藻属	✓									
urchin spines 海胆刺					✓					
sponge spicules 海绵针	✓									
"marine tubes"海相圆管		✓								
pyritized fossils 黄铁矿化石			✓							
authigenic pyrite（earthy of framboid）自生黄铁矿					■		✓			

说明：右侧竖排标注 介形虫 ostracods

□ 淡水　　■ 半咸水　　■ 咸水　　丰富　　✓ 出现

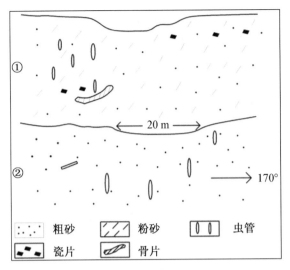

图 5　钵池山沉积剖面图

钵池山沉积剖面具有河流二元相结构及河口沉积的特点,底部河流粗砂层,渐变为漫滩相含粉砂、黏土质砂层。因沉积环境具有周期水位涨落之变化,故而软体动物筑管通气,管筒长度说明淹没时水层不厚,属潮上带沉积环境。硬结土层非现代沉积。

历史记载:2 500 年前江淮之间仍多水湿洼地;唐朝之后,有"三十六陂"之说;宋元时期,高邮湖区曾有 5 处湖泊;1194 年黄河决堤南流经泗水故道入淮河,历时 661 年,经黄河迁徙泛滥,苏北湖泊群不断扩大,有"五荡十二湖"之说;1565—1592 年间,明朝潘季驯治河,"束水攻沙",使黄河归为统一河道,1600 年时,高邮统汇为一大型湖泊[14,15]。苏北湖泊群经历了漫长的自然发展变化阶段与人为改造影响。成为淡水湖泊是在历史时期,但一系列湖泊分布的位置,却是因承袭"古海湾"。苏北大运河实为利用一系列湖泊、洼地与凿通湖间高地而成。

(四) 现代湖泊探索

现代洪泽湖、高邮湖是经过治淮兴修水利,与大运河浚深工程后所形成,规模范围与深度均有很大之变化。为探索洪泽湖、高邮湖及骆马湖的基底构成,本项目应用地球物理探测仪器,如:Geopulse 及 Chiper1 探测湖底地层,以 GeoAcustic 仪扫描湖底表面,然后选点进行湖底沉积柱采取,共设立 7 条断面(表 2 和图 6)。

表 2　苏北湖泊群测量断面位置

湖区	断面号与位置	方向	沉积柱样长度
高邮湖	GL Ⅰ (32°45′N,119°15′E→32°50′N,119°15′E)	SN	89 cm
	GL Ⅱ (32°49′N,119°15′E→32°46.023′N,119°15.016′E)	NS	118 cm
	GL Ⅲ (32°46.56′N,119°16.77′E→32°47.19′N,119°21.55′E)	SN	80 cm
洪泽湖	HL Ⅰ (32°16.22′N,118°39′E→32°16.27′N,118°49.6′E)	EW	112 cm
	HL Ⅱ (33°16.52′N,118°39.08′E→33°16.233′N,118°45.226′E)	WE	112 cm
骆马湖	LM Ⅰ (34°3.85′N,118°7′E→34°6.6′N,118°16.5′E)	SW-NE	砂层脱落
	LM Ⅱ (34°5.29′N,118°9.1′E→34°5.32′N,118°7.05′E)	NE-SW	砂层脱落

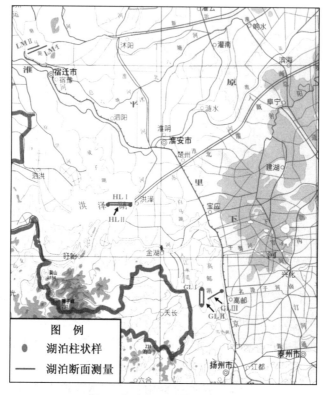

图6 湖泊断面测量与柱状样

但是,湖区探测工作十分困难。时逢初夏汛期前,各湖水浅不能成直线探测;骆马湖中多岛屿、多采砂船、多渔网,难以通行,湖底地形经采砂变动大,测量剖面不具代表性,湖底砂层难于采集沉积柱;洪泽湖及高邮湖多经人工疏浚扰动,淤泥层厚,反射信号弱,不显层次,不代表自然沉积环境。结果仅在洪泽湖 HL Ⅰ 断面获得穿透 40 m 深处的反射剖面,显示出在湖底 5~8 m 深处有 2 m 厚的水平沉积层,以下为 10 m 厚之砂层,砂层下为淤泥沉积。从 HL Ⅰ 剖面所获资料,说明洪泽湖基底埋藏深,该湖泊产生应早于历史时期。

1. 洪泽湖沉积物的研究

现代洪泽湖位于 33°06′~33°40′N 及 118°10′~118°52′E 之间,水域面积 1 576.9 km²,最大水深 4.37 m,平均水深 1.77 m,蓄水量 27.9×10⁸ m³。湖水补给主要为地表径流与大气降水,入湖河流主要在西部的淮河、濉河等[16]。

现代洪泽湖湖底沉积粒度、分选系数及峰态分析结果(图 7 和图 8)反映了洪泽湖现代沉积环境特点与沉积速率的变化(图 9)。图 7 是洪泽湖东岸近高良闸处 HL Ⅰ 钻孔,该处湖底水深 1.77 m,从表层至 1.1 m 深处,沉积物以砂质、粉砂为主,物源单一,分选较差,与该处经常受闸的启闭,船只活动的沉积动力环境相适应。对沉积柱进行²¹⁰Pb 测定,获知 0~40 cm 深处为近百年的沉积,其沉积速率为 0.4 cm/a;40 cm 以下为 100 年前的沉积物,依据 HL Ⅰ 沉积剖面的层位深度与其²¹⁰Pb 年龄间的相关系数 $r=0.975$(见图 9),用²¹⁰Pb 回归法计算并辅以¹³⁷Cs 时标验证法,获得该沉积柱底部 112 cm 深处的对应年代为 1720 年。HL I 沉积柱记录着 283 年来,洪泽湖三个方面的沉积信息:① 底层(112～83 cm),粗粉砂及细砂层,沉积时代约自 1720 至 1795 年。其中 102 cm 深处黏土质增加,可能反映着 1738 年期间洪泽湖水位超过 5 m 时之细粒沉降多。② 中层(83～45 cm),相当于 1795—1890 年沉积以粗粉砂和细砂为主,可能与 1881 年夏秋两季河水流量大,水流强相关。湖面水位虽上涨,但粗粒组分增长较多。③ 上层(45～0 cm),相当于 1890—2003 年时段之沉积,砂质粉砂,但黏土含量增加,可能为疏浚航道之影响所致。

图 8 为洪泽湖 HL II 钻孔沉积物粒度分析之综合结果,其物质组成表层为粉砂,中部为黏土质粉砂与少量粉砂夹层,下部又为黏土质粉砂。该孔位置接近湖中部,沉积物较岸边为细,含较多悬浮质黏土成分,并具有季节性粗、细变化,人为影响较少。

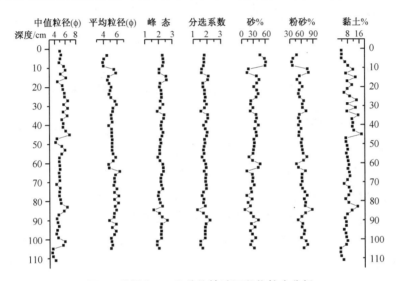

图 7　洪泽湖 HL I 站位钻孔沉积物粒度分析

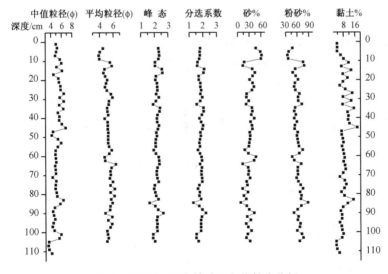

图 8　洪泽湖 HL II 钻孔沉积物粒度分析

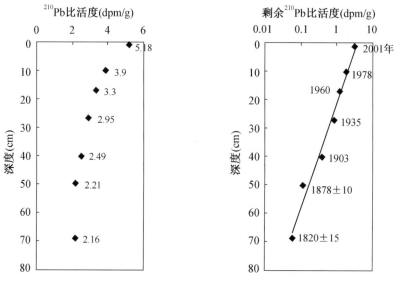

图9　洪泽湖 HLⅠ钻孔沉积²¹⁰Pb 比活度与深度关系

2. 高邮湖沉积物的研究

现代高邮湖位于洪泽湖东南方,界于金湖、高邮与天长之间($32°42'\sim33°04'$N 和 $119°06'\sim119°25'$E),湖区面积 674.7 km²,最大水深 2.4 m,平均水深 1.44 m,蓄水量 9.76×10^8 m³。目前主要是淮河通过三河入江水道汇入高邮湖,再经六道漫水闸泄入邵伯河后,汇入长江。高邮湖盆浅碟形,湖底平坦,平均高程 4.4 m,最低点高程 3.3 m,湖盆高出东部里下河平原 1.0~2.5 m[16]。在高邮湖取得 3 个钻孔(见表 2 和图 6),获得现代湖盆沉积特征与沉积速率。

GLⅠ孔位于湖中部狭颈段北部,GLⅡ孔位于南部,GLⅢ孔位于东部。分析结果(图 10、图 11 和图 12)表明:沉积物来源于入湖河流,以粉砂为主成分;北部沉积物较粗以粉砂质砂与砂质粉砂交互沉积,南部则以黏土质粉砂为主,间夹一些粉砂层,东部主要是黏土质粉砂,表层含粉砂层较多,沉积层分选程度较差。

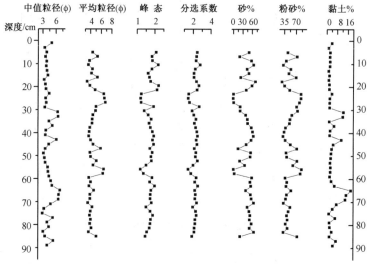

图 10　高邮湖 GLⅠ钻孔沉积物粒度分析

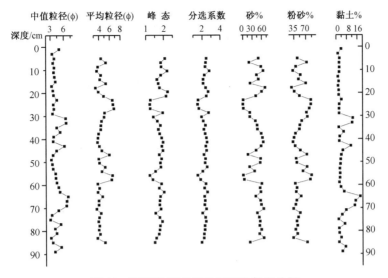

图 11　高邮湖 GL Ⅱ 钻孔沉积物粒度分析

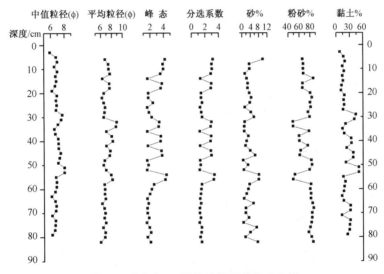

图 12　高邮湖 GL Ⅲ 钻孔沉积物粒度分析

选取 GL Ⅱ 孔进行了 ^{210}Pb 定年,获得各层与深度关系数(图 13)。据此获得高邮湖近百年来 3 个沉积阶段的沉积速率。

第 1 阶段　湖底 59～36 cm 深处,相当于 1928—1972 年期间,沉积速率 0.5 cm/a,表明当时湖泊环境较稳定,河流输入沙量变化不大。

第 2 阶段　36～16 cm 深处,相当于 1972—1982 年,沉积速率为 2.52 cm/a,此阶段沉积速率最大。这与 1969 年修建淮河入江水道、淮河水沙直接入湖,以及"大跃进垦荒",水土流失加剧,增加入湖泥沙有关。

第 3 阶段　16～0 cm(湖底表面),相当于 1982—2003 年,沉积速率为 0.7 cm/a,比前期沉积速率下降。由于 1980—1999 年间一系列水利工程,共治理 2.5×10^4 km^2 的水土流失区,渐发生良好效益。

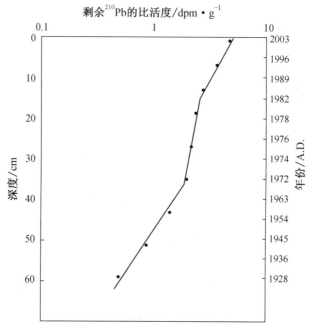

图 13 高邮湖沉积^{210}Pb 比活度与深度关系

总之,结合分析湖泊群地貌组合的特点与湖泊的沉积速率,可以了解苏北湖泊群在发生发展过程中,古海湾遗留,自然环境低洼是主导因素,现代湖泊形态与范围是经受历史时期与现代人类改造之结果。当前各湖泊仍受人类活动影响而淤积变浅。

三、沉积钻孔研究

平原西侧为"海湾型"的低山丘陵环绕,湖泊洼地遗迹以及低地平原上人居田野的分布型式,所构成之平原地貌特点,促发了本项目探索平原成因的沉积地层证据。

选择在洪泽湖、白马湖及高邮湖以东的平原低洼处,距离现代海岸 180 km 的宝应县望直港小学的操场中钻取宝应 1 号沉积孔(BY1,33°14′21″N,119°22′41″E),钻孔进深 146 m,取得沉积柱 97 m,不整合层面处的砂层有脱落。

BY1 孔沉积物中,有孔虫由加拿大学者 Dr. J.-P. Guilbault 鉴定(表3);沉积层粒度(图14)由南京大学海岸与海岛开发教育部重点实验室完成;中国科学院地质与地球物理研究所朱日祥进行了古地磁年代测定,南京大学鹿化煜测制曲线(图15);北京大学考古系^{14}C 测年实验室完成^{14}C 年代测定。

表3 宝应1号孔微体古生物鉴定*

		1	2	3	4	5	A1	6	7	8	9	10	11	12	13
	paleomagnetic chronology 古地磁年代														
咸水	sponge spicules 海棉针														
	marine and brackish water molluscs 海水与半咸水贝类							=							=
	echinoderm ossicles 棘皮类动物							=							
	bryozoan fragments 苔藓虫							=							
	barnacle fragments 藤壶属														
	foraminifera (total, undiff.) 有孔虫						1	1	16			A	A	A	A
	undifferentiated marine ostracods 未定种海洋介形虫													=	
	Bicornucythere bisanensis 美山双角花介														
	Neomonoceratina crispata 皱新单角介														
半咸水	*Spinileberis furuyaensis* 古屋刺面介														
	Spinileberis pulchra 美丽刺面介														
	Sinocytheridea spp. 中华美花介														=
淡水	undifferentiated ostracods, probably freshwater 未定种介形虫														
	Ilyocypris sp. 土星介属	=											=		
	Candonids 玻璃介	=													
	freshwater mollusks 淡水贝类	=													
	Arcellaceans 中文名待定														
	charophyte oogonia incl. linings 轮藻植物	=						=							
	insect and mite fragments 昆虫与螨屑		=					=							
	pollen/spore cases 孢粉		=	=											
	Seeds 种子	=			=			=							
	Charcoal 碳屑	=	=	=				=	=					=	
	plant matter (fresh to partly carbonized) 植物屑	=	=	=	=		=	=	=	=				=	
	Core depth (m) 孔深	0.98~1.03	1.25~1.30	1.45~1.50	1.95~2.00	3.00~3.07	4.97~5.02	7.45~7.55	12.25~12.30	14.05~14.15	14.55~14.64	15.00~15.29	15.77~15.87	16.55~16.65	16.75~16.80
	Sample number 样品号	1	2	3	4	5	A1	6	7	8	9	10	11	12	13

（续表）

分类	项目	A19	14	15	16	A27+17		A4	18	A5	A6	A20	A21+19	A22+20	21
	paleomagnetic chronology 古地磁年代														
咸水	sponge spicules 海棉针														=
咸水	marine and brackish water molluscs 海水与半咸水贝类					=									=
咸水	echinoderm ossicles 棘皮类动物	=													
咸水	bryozoan fragments 苔藓虫					?									
咸水	barnacle fragments 藤壶属					=		=							
咸水	foraminifera（total，undiff.）有孔虫	A	A	A	A	A	A	A	A	A	7		19	18	6
咸水	undifferentiated marine ostracods 未定种海洋介形虫		=			=	=								
咸水	*Bicornucythere bisanensis* 美山双角花介					=									
咸水	*Neomonoceratina crispata* 皱新单角介					=		=							
咸水	*Spinileberis furuyaensis* 古屋刺面介							=							
咸水	*Spinileberis pulchra* 美丽刺面介						=	=							
半咸水	*Sinocytheridea* spp. 中华美花介	=		=	=	=	=	=		A	=				
淡水	undifferentiated ostracods, probably freshwater 未定种介形虫													=	=
淡水	*Ilyocypris* sp. 土星介属														=
淡水	Candonids 玻璃介					=								=	=
淡水	freshwater mollusks 淡水贝类													=	=
淡水	Arcellaceans 中文名待定														
淡水	charophyte oogonia incl. linings 轮藻植物													=	
淡水	insect and mite fragments 昆虫与螨屑	=				=								=	=
淡水	pollen/spore cases 孢粉								=						
淡水	Seeds 种子	=												=	
淡水	Charcoal 碳屑	=	=	A	=		=	=	=				=		
淡水	plant matter（fresh to partly carbonized）植物屑	A		A	A	A	A	=	A			A	=	=	
	Core depth（m）孔深	17.20~17.25	17.25~17.30	17.48~17.52	17.96~18.00	18.71~18.81	19.25~19.35	19.80~19.90	20.40~20.44	20.60~20.65	20.75~20.80	21.20~21.30	22.35~22.40	23.80~23.90	24.15~24.23
	Sample number 样品号	A19	14	15	16	A27+17		A4	18	A5	A6	A20	A21+19	A22+20	21

（续表）

		22	23	24	25	26	27	28	29	30	31	32	A23	A8	A9
	paleomagnetic chronology 古地磁年代										B	M			
咸水	sponge spicules 海棉针	≡													
	marine and brackish water molluscs 海水与半咸水贝类														
	echinoderm ossicles 棘皮类动物														
	bryozoan fragments 苔藓虫												≡		
	barnacle fragments 藤壶属														
	foraminifera (total, undiff.) 有孔虫	8		1		1						13	1		A
	undifferentiated marine ostracods 未定种海洋介形虫														
	Bicornucythere bisanensis 美山双角花介														
	Neomonoceratina crispata 皱新单角介														
	Spinileberis furuyaensis 古屋刺面介														
	Spinileberis pulchra 美丽刺面介														
半咸水	*Sinocytheridea* spp. 中华美花介														≡
	undifferentiated ostracods, probably freshwater 未定种介形虫		≡		?	≡									
淡水	*Ilyocypris* sp. 土星介属			≡											
	Candonids 玻璃介	≡		≡											
	freshwater mollusks 淡水贝类														
	Arcellaceans 中文名待定														
	charophyte oogonia incl. linings 轮藻植物	≡													
	insect and mite fragments 昆虫与螨屑			≡										≡	≡
	pollen/spore cases 孢粉														
	Seeds 种子			≡											
	Charcoal 碳屑	≡		≡				≡	≡						
	plant matter (fresh to partly carbonized) 植物屑	≡	≡					≡	≡			≡			≡
	Core depth (m) 孔深	25.10~25.15	26.92~27.00	27.16~27.30	29.05~29.15	30.55~30.60	31.75~31.80	33.00~33.05	34.64~34.74	35.57~35.63	37.25~37.30	39.45~39.50	40.60~40.70	43.75~43.80	46.50~46.60
	Sample number 样品号	22	23	24	25	26	27	28	29	30	31	32	A23	A8	A9

（续表）

分类			33	34	35	36	37	38	39	40	41	42	43	44	45	46
	paleomagnetic chronology 古地磁年代	J														
咸水	sponge spicules 海棉针															
	marine and brackish water molluscs 海水与半咸水贝类															
	echinoderm ossicles 棘皮类动物															
	bryozoan fragments 苔藓虫															
	barnacle fragments 藤壶属															
	foraminifera（total，undiff.）有孔虫										1		1			
	undifferentiated marine ostracods 未定种海洋介形虫															
	Bicornucythere bisanensis 美山双角花介															
	Neomonoceratina crispata 皱新单角介															
	Spinileberis furuyaensis 古屋刺面介															
	Spinileberis pulchra 美丽刺面介															
半咸水	*Sinocytheridea* spp. 中华美花介															
淡水	undifferentiated ostracods，probably freshwater 未定种介形虫										=					
	Ilyocypris sp. 土星介属															
	Candonids 玻璃介		=	?.												
	freshwater mollusks 淡水贝类															
	Arcellaceans 中文名待定															
	charophyte oogonia incl. linings 轮藻植物															
	insect and mite fragments 昆虫与螨屑															
	pollen/spore cases 孢粉															
	Seeds 种子															
	Charcoal 碳屑						=				=			=	
	plant matter（fresh to partly carbonized）植物屑		=	=							=	=			=
	Core depth（m）孔深		47.30~47.40	48.55~48.60	49.40~49.50	51.00~51.10	52.25~52.35	53.40~53.50	55.75~55.85	56.30~56.40	57.10~57.15	58.15~58.25	59.00~59.10	59.30~59.35	60.00~60.10	60.80~60.85
	Sample number 样品号		33	34	35	36	37	38	39	40	41	42	43	44	45	46

（续表）

类别	项目	47	48	49	50	51	52	53	54	55	56	57	A10	A12	A24
	paleomagnetic chronology 古地磁年代														
咸水	sponge spicules 海棉针													=	
	marine and brackish water molluscs 海水与半咸水贝类														
	echinoderm ossicles 棘皮类动物														
	bryozoan fragments 苔藓虫														
	barnacle fragments 藤壶属														
	foraminifera（total, undiff.）有孔虫		1	2		1	1							3	2
	undifferentiated marine ostracods 未定种海洋介形虫														
	Bicornucythere bisanensis 美山双角花介														
	Neomonoceratina crispata 皱新单角介														
	Spinileberis furuyaensis 古屋刺面介														
	Spinileberis pulchra 美丽刺面介														
半咸水	*Sinocytheridea* spp. 中华美花介														?
淡水	undifferentiated ostracods, probably freshwater 未定种介形虫														
	Ilyocypris sp. 土星介属														
	Candonids 玻璃介														
	freshwater mollusks 淡水贝类														
	Arcellaceans 中文名待定													=	
	charophyte oogonia incl. linings 轮藻植物													=	=
	insect and mite fragments 昆虫与螨屑		=	=									=	=	=
	pollen/spore cases 孢粉														
	Seeds 种子														
	Charcoal 碳屑	=	A	=	=	=	=							=	
	plant matter（fresh to partly carbonized）植物屑	=	=	=	=	=	=		=		=	=		=	=
	Core depth（m）孔深	61.75~61.80	62.30~62.40	62.54~62.60	63.10~63.20	64.81~64.91	65.56~65.66	67.62~67.67	69.80~69.90	70.50~70.60	72.40~72.50	73.75~73.85	74.20~74.30	75.95~76.00	76.50~76.55
	Sample number 样品号	47	48	49	50	51	52	53	54	55	56	57	A10	A12	A24

（续表）

	A13	A25	A26	58	A15	A16	59	A17	A18
paleomagnetic chronology 古地磁年代	O				R			M	G
sponge spicules 海棉针									
marine and brackish water molluscs 海水与半咸水贝类									
echinoderm ossicles 棘皮类动物									
bryozoan fragments 苔藓虫									
barnacle fragments 藤壶属									
foraminifera (total, undiff.) 有孔虫	1	6			2	4			
undifferentiated marine ostracods 未定种海洋介形虫									
Bicornucythere bisanensis 美山双角花介									
Neomomoceratina crispata 皱新单角介									
Spinileberis furuyaensis 古屋刺面介									
Spinileberis pulchra 美丽刺面介									
Sinocytheridea spp. 中华美花介									
undifferentiated ostracods, probably freshwater 未定种介形虫									
Ilyocypris sp. 土星介属									
Candonids 玻璃介									
freshwater mollusks 淡水贝类									
Arcellaceans 中文名待定									
charophyte oogonia incl. linings 轮藻植物									
insect and mite fragments 昆虫与螨屑									〃
pollen/spore cases 孢粉									
Seeds 种子									
Charcoal 碳屑					〃				
plant matter (fresh to partly carbonized)植物屑	〃	〃	〃	〃	〃	〃		〃	〃
Core depth (m)孔深	79.85~79.90	80.60~80.70	83.70~83.80	86.00~86.10	87.90~87.95	90.00~90.10	91.65~91.70	93.50~93.60	96.65~96.75
Sample number 样品号	A13	A25	A26	58	A15	A16	59	A17	A18

咸水　半咸水　淡水

* 〃:代表微体古生物的出现,A代表微体古生物丰富,?代表微体古生物不定;B代表布容时(Brunhes),M代表松山时(Matuyama),G代表高斯时(Gauss),J代表加拉米洛时(Jaramillo),O代表奥尔都维亚时(Olduvai),R代表留尼旺尼亚时(Reunion);图15古地磁代号同

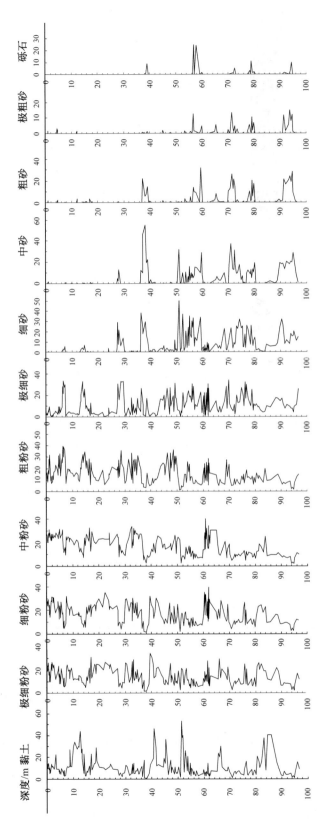

图 14 宝应 1 号孔（BY1）沉积物粒度分布曲线（%）

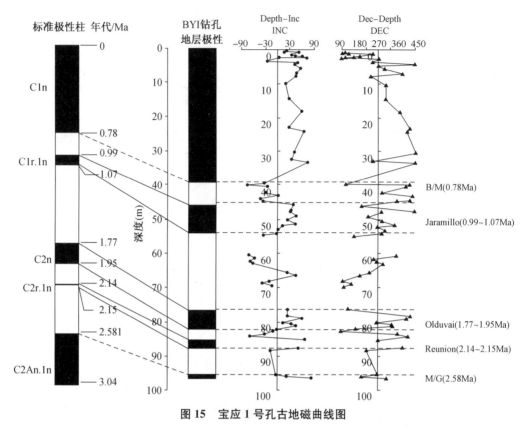

图 15　宝应 1 号孔古地磁曲线图

注：由于数据为钻孔数据，故地层极性柱主要根据磁倾角，参考磁偏角数据。标准极性柱数据引用自文献[17]

通过对沉积层粒度结构与沉积相、古地磁分析、微体古生物分析及地球化学元素分析等数据的综合研究，获得了第四纪近 2.58 Ma 以来基本连续的沉积剖面，了解该平原由河-海交互作用的形成过程，并将 97 m 长的 BY1 沉积柱划分为 10 个沉积相段（图 16）从下而上分别描述如下。

第 1 段（陆相）

（1）平原基底位于 95.22 m，即 M/G（松山/高斯）界面向下至 97 m 深处，是 2.58 Ma 前受火山喷发影响的风化剥蚀面。

（2）基底面上 95.6～81.0 m 深处为 15 m 厚的具有干湿季节变化的平原河流与湖泊沉积，含钙质结核、钙板，受火山喷发影响，岩层棕黑，底部含角砾岩与燧石。

第 2 段（海侵相）

从 81 m 至 79.66 m，浅灰色粉砂质砂层及黏土—砂—粉砂，已硬结，此层含有孔虫，系海水浸淹之平原堆积，时代于 1.95 Ma 以后，Olduvai（奥尔都维）正极性亚时期间。

第 3 段（陆相沉积）

（1）从 79.66～76.39 m 处，受火山喷发影响的混杂沉积。暗棕色、棕黑色及灰黄棕①，受火山喷发影响的混杂沉积，粉砂、砂及砾石，含小燧石岩块。顶部有 40 cm 厚棕黑

---

①　颜色按美国地质局岩层色谱标定（Rock-color chart，U. S. Geological Survey. The Geological Society of America，1979）。

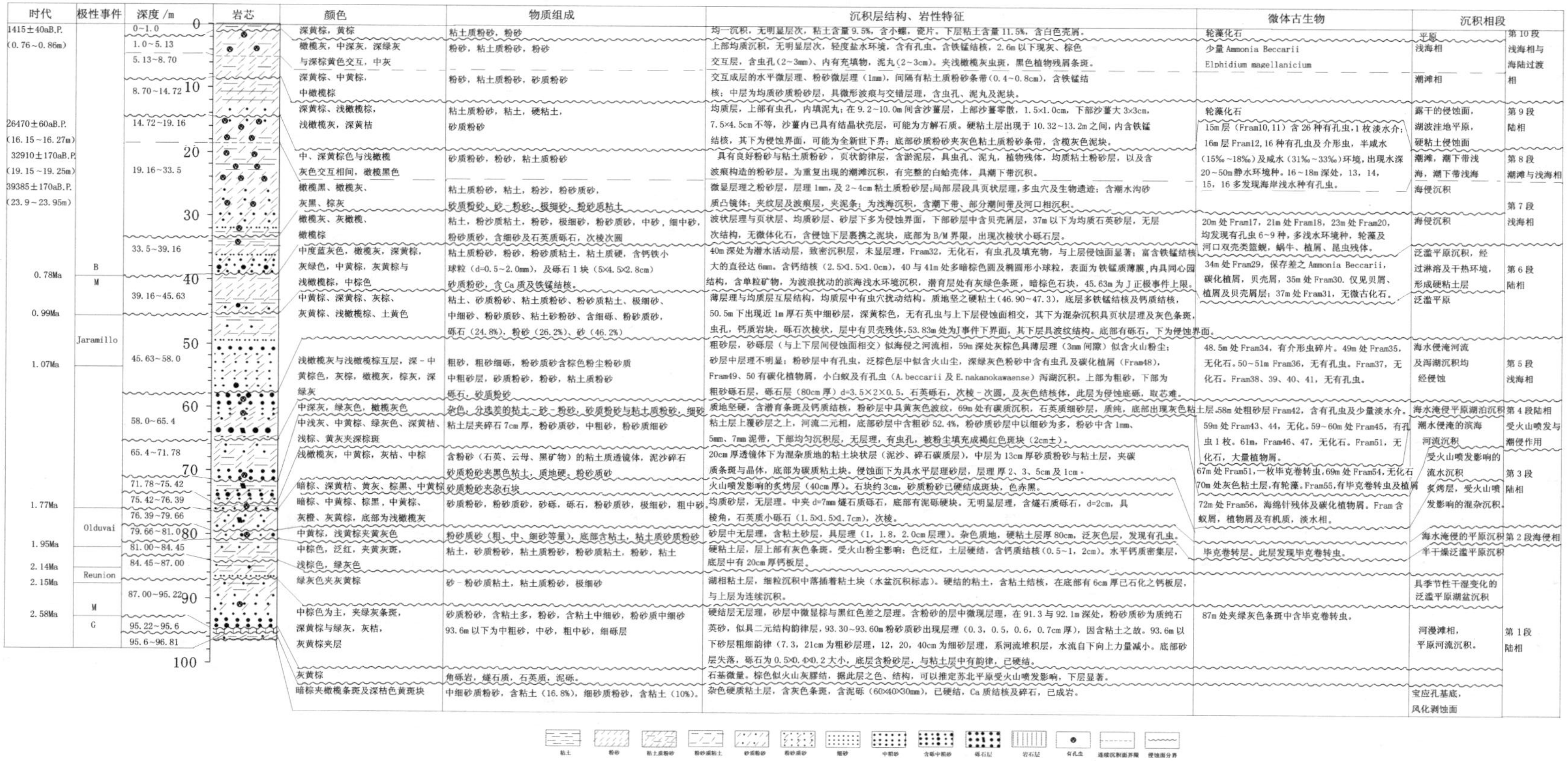

图16　苏北平原宝应1号孔综合剖面图

色炙烤层,砂质粉砂与石块均胶结成斑块,炙烤层约发生在 Olduvai 亚时刚刚结束的 1.77 Ma 之后。

(2) 76.39~75.42 m 深处,深黄棕及灰橙色,黏土与砂质粉砂硬结层,含碎石及炭质晶体。下部为具水平层理的粉砂层,受火山影响的流水沉积。

第 4 段(陆相)

受火山喷发影响及海潮浸淹的滨海平原沉积。

(1) 75.42~71.18 m,浅橄榄灰、中黄棕色,河流二元结构的黏土及砂、砾层,有虫孔,孔内被火山尘填充成褐红斑块。此层中含有海绵骨针残体,碳化植物碎屑及一只白蚁。此层为受到潮水浸淹的滨海河流沉积。

(2) 71.18~65.40 m,浅灰、绿灰与黄灰色夹杂,砂质粉砂、黏土质粉砂及细砂沉积,已硬结。含潜育条斑,钙质结核,碳粒碎屑,在灰色层中发现毕克卷转虫(*Ammonia beccarii*)及植物碎屑,73 m 处发现轮藻化石。此层是受到海水浸淹的平原湖泊沉积。

第 5 段(浅海相)

从 65.4~58.0 m 深处。浅橄榄灰与浅橄榄互层、棕灰与深绿灰色互层,上部为粉砂层及黏土粉砂层,具层理,含有孔虫、淡水介及碳化植屑,底部粗砂及砾石层,砾石经过搬运磨蚀为次棱—次圆形,此为海水浸淹河流及潟湖的滨海沉积及浅海沉积。

第 6 段(陆相)泛滥平原沉积

(1) 58.00~45.63 m,下层为浅橄榄棕、灰黄棕色含黏土、粉砂、砂、细砾及钙质岩块的混杂沉积,底部有硬黏土,层下为侵蚀界面,此层仍受海水影响而颜色泛灰。上层是黄棕色中细砂层,薄层理与均质层互层,顶部为侵蚀界面。本层为 1.07~0.99 Ma 的 Jaramillo 正磁极性亚时之沉积。

(2) 45.63~39.16 m,经过多次海水浸淹的泛滥平原沉积,蓝灰、橄榄灰、黄棕色黏土与粉砂层。40 m 处为具灰色条斑之潜育层,其下富含铁锰结核及钙结核,在 41 m 与 40 m 之间多暗棕色圆形与椭圆形小球粒(粒径为 0.5~2.0 mm),表面为铁锰质薄膜,具同心圆状结构,中含微粒单个矿物,如长石、锆石、金红石、钛铁矿等。这种鲕粒物质为海陆交互作用环境,具波浪扰动的滨海浅水沉积。鲕粒层底部有一砾石。顶部 39.16 m 处为 B/M (布容/松山)界面,时为 0.78 Ma。

第 7 段(海侵-浅海相,晚更新世沉积)

33.50~19.16 m 以橄榄黑、橄榄灰、灰黑、棕灰色为主,沉积以粉砂、黏土、极细砂为主。底层含石英质砾石(1.5 cm×1.5 cm×1.0 cm),次棱—次圆级磨圆度;中部为石英砂层,含贝壳屑层;上部粉砂层中具层理、具波痕与砂质透镜体。含毕克卷转虫、霜粒希望虫(*Elphidium nakanokawaense*)及光滑先希望虫(*Protelphidium glabrum*)等多种有孔虫,系浅海环境种(一般生活在水深 20 m 或 30~50 m,盐度为 31‰~32‰)与半咸水种(15‰盐度),含少量河口相篮蚬(*Carbicula* sp.)(中国科学院南京地质古生物研究所鉴定,样号为 1720~1725),一个蜗牛壳,植屑与昆虫残片。

[14]C 测年结果表明:在 23.90~23.95 m 为 39 385±170aB. P. ,19.15~19.25 m 为 32 910±170aB. P. 。该段为晚更新世沉积。

第 8 段(潮滩及潮下带浅海相,晚更新世沉积)

19.16～14.72 m,黄棕色与浅橄榄灰色交互沉积。粉砂、砂质粉砂与黏土质粉砂,重复出现具页状韵律层,虫孔、泥丸、波痕等标志的潮滩沉积[18],与潮下带均质黏土粉砂层。富含有孔虫及介形虫等微体古生物达 27 种,它们一般生活在水深 16～18 m 及 20～50 m,盐度 15‰～18‰的半咸水或海水(31‰～32‰)的环境;亦有河口环境的黑龙江河篮蛤(*Potamocorbula amurensie*)、篮蚬和篮蛤(*Corbula sp.*)及海岸海洋浅水有孔虫。^{14}C测年在 16.15～16.27 m 处为 26 470±60aB.P.。

第 9 段(陆相)

14.72～8.70 m,深黄棕、浅橄榄棕、灰、深黄橘色,黏土质粉砂、砂质粉砂与硬黏土,均质沉积,不显层理。9～10 m 处含沙薹层。上部沙薹 1.5 cm×1.0 cm,下部为 3 cm×3 cm 和 7.5 cm×4.5 cm,沙薹内部具方解石质结晶壳层。硬黏土出现于 10～13 m 处,含铁锰结核、轮藻化石。底部有裹携第 8 段的橄榄灰色泥块。此层为露干的湖泊洼地平原沉积,因暴露风化,具硬黏土层及侵蚀面。

第 10 段(浅海相与海陆过渡相)

底部 8.70～5.13 m 为潮滩相沉积。黄棕、橄榄棕色粉砂、黏土质粉砂、砂质粉砂交互成层的页状层理(1 mm)及黏土质条带(0.4～0.8 cm)。含虫孔,内充填泥丸。与第 9 段侵蚀面接触。

中部 5.13～1.00 m 以橄榄灰、中深灰、绿灰色为特征的粉砂层,轻度盐水环境,含有孔虫的毕克卷转虫及缝裂希望虫(*E. magellanicum*)。2.6 m 以下为灰色与棕黄色交互黏土质粉砂层,含虫孔及黑色植物碎屑。此层是海水淹没的潮滩沉积层。与上层陆地以侵蚀面交接。

上部居于侵蚀界面以上,从 1～0 m 层,黄棕色粉砂与黏土质粉砂层,含轮藻化石,淡水小螺,壳屑及碎瓷片。^{14}C 测定在 0.78～0.86 m 处为 1 415±40aB.P.。本层上部可能因剥蚀及人工平建操场而厚度减少。

总上述,宝应 1 号孔综合分析中,反映出沉积层以橄榄灰、黑灰与黄棕为主的颜色,质地、结构与海、陆环境、沉积动力间存在的良好相关性;古地磁极性时或亚时的正、反事件在相关沉积层中也有所反映,如炙烤层、火山灰影响、磁铁矿小球、燧石碎块等。其原因未解。

微体古生物鉴定的优势种数量与分布进一步证实了海相与陆相环境的客观区别。沉积层常量元素分析结果表明[19],在海相沉积层中 Mg、Ca、P、S 等海洋中的特征元素含量高;Si 和 Al 元素在陆相层中含量高。物质源于陆地丘陵残积层的 Fe 和 Mn 元素接近中国浅海区的含量;Ti 元素含量近似于海岸砂体沉积的含量。元素分析亦反映出多次海相的峰值[19],但其层段往往与沉积层表相分析所确定的海相层有上、下层位与次数之差。可能元素分析更反映内在联系,启示我们需要对本区域再做进一步的探索和研究。

宝应钻孔反映出苏北平原在第四纪时,经历过 4 次主要的海水作用环境,并有着多次海水侵泛陆地的短暂停留。这 4 个时期多发生于地球正向磁场时期。其中,最早的海相层位于 81.00～79.46 m 深处,时间相当于 1.95～1.77 Ma 的 Olduvai 正向磁极亚时期;第 2 层位于 65.4～58.0 m 处,于 Jaramillo 正向磁极亚时,即 1.07 Ma 前夕;第 3 层位于 33.5～14.72 m,形成于 0.78 Ma 的 Brunhes 正向磁极时发生以后,时间主要在晚更新世,约距今 39 385～26 470 年间,该次海洋环境的沉积过程显著;第 4 次发生于全新世,海侵影响较第 3 次小,并具有淤浅为陆地冲积平原的特征。分析沉积层结构与微体古生物表明:4 次均为浅海环境,深度不超过 100 m,多为 20～30 m 水深,部分水深为 50～60 m。盐度介于 31‰～32‰,但是,具有河口湾半咸水(盐度 15‰或小于 15‰)影响,并有着由浅海淤浅为潮滩的变化。本次研究成果,尤其是第 3 次海侵时代,较前人研究[20]的海侵时代更老,而且以 3 万年左右的浅海相与潮滩相沉积最厚。其原因可能是当时的海水入侵不仅来自东部海区,而且也来自南部的古长江河口湾,因为当时的河口、海岸位置与现代不同,海岸位于西部低山丘陵区,古长江河口湾处于扬州与镇江之间[13,15]。钻孔沉积以细粒物质(粉砂、细砂及黏土)为主,反映着平原堆积的泥沙源主要来自江、河,而粗粒沉积物(粗中砂、砂砾、砾石及纯净的石英质细砂)则起源于基岩海岸侵蚀产物,这在常量元素的分析中也有反映[19]。入海的泥沙、黏土质颗粒会搬运较远至深水处直接沉降淀积;粉砂、砂及细砾级泥沙,则被浪、潮、水流多次掀带、搬运与分选,最后沉积于相适应的动力减弱地带。陆相沉积层主要由源自古江、河水系的泥沙组成,堆积为其二元结构的河流泛滥平原,或杂色沉积的湖泊洼地平原。曾多次出露为陆地,经降水淋溶、日晒蒸发形成钙质或黏土质结核与钙板层。沙姜层的多次出现反映着当时是干湿季节交替变化显著的气候环境。平原在第四纪早、中期时,曾多次受到邻近区火山活动的影响,有多量漂浮的火山灰及火山碎屑物沉落,形成红棕色与黑棕色沉积层,甚至热融重结晶,或形成硬结炙烤层。但第四纪后期平原内动力构造活动渐趋平稳,而外动力的河海交替活动作用显著。

四、结论

(1) 环绕苏北平原的低山丘陵曾构成古海湾,时代约相当于新生代第三纪至第四纪早、中期,即燕山期花岗岩侵入体形成以后,经过长期剥蚀,但在晚更新世黄土覆盖于低山丘陵以前。古海滨环境曾经受过多次火山与构造活动影响;据第 2 段海侵相的层位深度推论,古海湾的深度不超过 100 m。

(2) 苏北平原是第四纪期间由江、河与浅海的交互作用堆积发育的。平原形成过程中经历了 4 次海水作用,及多次海水短期、小范围的浸淹活动。在 4 次海侵中,以晚更新世 3.9 万～2.6 万年前的浅海环境沉积更为显著。但是,古长江及大河水系的泥沙补给与平原建造更为主要。河、海作用不仅仅是携积泥沙,也有冲刷侵蚀活动,但总体上,沉积作用大于侵蚀效应,不仅由于有江、河的巨量泥沙汇入,也有来自冰期低海面平原沉积之再搬运与补给。第四纪后期苏北地区断裂下沉构造活动减缓,亦有利于泥沙堆积和加速平原成陆。

(3) 苏北平原形成过程中,海水入侵来自两方面:东部的黄海与南侧的古长江河口湾,当时的河口湾在现今的扬州与镇江之间,与古海湾相邻分布(见图 3)。河-海交互作用中,古长江泥沙供应与堆积作用重要,海相沉积层中均伴随着河口湾微体生物的信息。海水入侵是否还有另一通道? 需进一步研究古海湾西北侧骆马湖及微山湖洼地,探索古黄河和渤黄海通道以及与苏北平原的发育相关。

参考文献

[1] 陈万里,顾洪群,张道政.1998.江苏第四纪海侵及近代海岸变迁研究.江苏地质,22(增刊):45-50.
Chen Wanli, Gu Hongqun, Zhang Daozheng. 1998. Study on the Quaternary marine transgression and coastline change. *Jiangsu Geology*, 22(Supp. 1): 45-50.

[2] 朱诚,程鹏,卢春成等.1996.长江三角洲及苏北沿海地区7000年以来海岸线演变规律分析.地理科学,16(3):207-213.
Zhu Cheng, Cheng Peng, Lu Chuncheng et al. 1996. Palaeocoastline evolution law of Yangtze River Delta and coastal area of North Jiangsu Province since 7000 a B. P. *Scientia Geographica Sinica*, 16(3): 207-213.

[3] Wang Ying. 1983. The mudflat coast of China. *Canadian Journal of Fisheries and Aquatic Science*, 40 (Supp. 117): 160-171.

[4] 江苏省地质矿产局.1984.江苏省及上海市区域地质志.北京:地质出版社.357-375.
Jiangsu Bureau of Geology and Mineral Resources. 1984. Regional Geology of Jiangsu Province and Shanghai Municipality. Beijing: Geological Publishing House. 357-375.

[5] 张宗祜,周慕林,邵时雄主编.1990.1:2 500 000中华人民共和国及其毗邻海区第四纪地质图及说明书.北京:中国地图出版社.
Zhang Zonghu, Zhou Mulin, Shao Shixiong eds. 1990. 12 500 000 Atlas of Quaternary Geology of China and Vicinity Ocean Area. Beijing: Sinomaps Press.

[6] 王颖主编.2002.黄海陆架辐射沙脊群.北京:中国环境科学出版社.8-26,219-368.
Wang Ying ed. 2002. Radiactive Sandy Ridge Field on Continental Shelf of the Yellow Sea. Beijing: China Environment Press. 8-26, 219-368.

[7] 陈友飞,严钦尚,许世远.1993.苏北盆地沉积环境演变及其构造背景.地质科学,28(2):151-160.
Chen Youfei, Yan Qinshang, Xu Shiyuan. 1993. Evolution of the sedimentary environments in North Jiangsu Basin and its tectonic setting. *Scientia Geologica Sinica*, 28(2): 151-160.

[8] 任美锷主编.1986.江苏省海岸带和海涂资源综合调查报告.北京:海洋出版社.1-15,101-131.
Ren Mei'e ed. 1986. Comprehensive Survey Report on Coastal and Tidal Flat Resources in Jiangsu Province. Beijing: China Ocean Press. 1-15, 101-131.

[9] 中国科学院南京地理与湖泊研究所,江苏省海岸带和海涂资料源综合考察队主编.1988.江苏省海岸带自然资源地图集.北京:科学出版社.3-18.
Nanjing Institute of Geography and Limnology of Chinese Academy of Science, Comprehensive Survey Team of Coastal and Tidal Flat Resources in Jiangsu Province eds. 1988. Atlas of Coastal Natural Resources in Jiangsu Province. Beijing: Science Press. 3-18.

[10] 徐近之.1955.黄淮平原气候历史记载的初步整理.地理学报,21(2):181-190.
Xu Jinzhi. 1955. Study on climatic record during historic period of HuangHuai River plain. *Acta Geographica Sinica*, 21(2): 181-190.

[11] 杨达源,王云飞.1995.近2000年淮河流域地理环境的变化与洪灾——淮河中游的洪灾与洪泽湖的变化.湖泊科学,7(1):1-7.
Yang Dayuan, Wang Yunfei. 1995. On change of geographic environment and flood damage along the Huaihe River Basin during the last 2000 years. *Journal of Lake Sciences*, 7(1): 1-7.

[12] Wang Ying. 1994. Effect of military activities on environment in eastern and southeastern China. *Annual Science Report-Supplement of Journal of Nanjing University*, 30 (English Series2)：43 - 46.

[13] 史照良主编. 2004. 江苏省地图集. 北京：中国地图出版社. 5,6,9,13,25.
Shi Zhaoliang ed. 2004. Atlas of Jiangsu Province. Beijing：Sinomaps Press. 5，6，9，13，25.

[14] 王庆,陈吉余. 1999. 洪泽湖和淮河入洪泽湖河口的形成与演化. 湖泊科学,11(3):237 - 244.
Wang Qing, Chen Jiyu. 1999. Formation and evolution of Hongze Lake and the Huaihe River mouth along the lake. *Journal of Lake Sciences*, 11 (3)：237 - 244.

[15] 廖高明. 1992. 高邮湖的形成和发展. 地理学报,47(2):139 - 145.
Liao Gaoming. 1992. Formation and evolution of the Gaoyou Lake. *Acta Geographica Sinica*, 47 (2)：139 - 145.

[16] 王苏民,窦鸿身主编. 1998. 中国湖泊志. 北京：科学出版社. 261 - 297.
Wang Sumin, Dou Hongshen eds. 1998. Limnological Annals of China. Beijing：Science Press. 261 - 297.

[17] Cande S C, Kent D V. 1995. Revised calibration of the geomagnetic polarity time scale for the Late Cretaceous and Cenozoic. *Journal of Geophysical Research*, 100：6093 - 6095.

[18] 王颖,朱大奎,曹桂云. 2003. 潮滩沉积环境与岩相对比研究. 沉积学报,21(4):539 - 546.
Wang Ying, Zhu Dakui, Cao Guiyun. 2003. Study on tidal flat environment and it's sedimentary facies comparative. *Acta Sedimentologica Sinica*, 21(4)：539 - 546.

[19] 杨竞红,王颖,张振克等. 2006. 苏北平原 2.58Ma 以来的海陆环境演变历——宝应钻孔沉积物的常量元素记录. 第四纪研究,26(3):340 - 352.
Yang Jinghong, Wang Ying, Zhang Zhenke *et al*. 2006. Major element records of land-sea interaction and evolution in the past 2.58 Ma from the Baoying Borehole sediments, Northern Jiangsu Plain, China. *Quaternary Sciences*, 26(3)：340 - 352.

[20] 孙顺才. 1985. 苏北平原第四纪沉积及海面变化. 见：中国第四纪研究委员会,中国海洋学会编. 中国第四纪海岸线学术讨论会论文集. 北京：海洋出版社. 157 - 161.
Sun Shuncai. 1985. Quaternary deposits and changes of sea level in North Jiangsu Plain. In：Chinese Association for Quaternary Research, Chinese Society of Oceanography eds. Proceedings of Quaternary Coastline Symposium, China. Beijing：China Ocean Press. 157 - 161.

苏北平原 2.58 Ma 以来的海陆环境演变历史
——宝应钻孔沉积物的常量元素记录[*]

苏北平原是邻近南黄海盆地的陆上部分,位于江苏省的东北部,总面积约 32 000 km²。该区水网密布,有淮河、沂河等众多河流流经其间并东入黄海,其西部有京杭大运河穿其而过。苏北平原北邻废黄河故道,南近长江,东侧有两列古贝壳砂堤,它们大体上平行于现今海岸方向分布,可能形成于全新世高海面(4 200a～6 000aB. P.)[1],海拔 3～5 m,是海退海岸的象征。北宋时(960—1127)修建的防潮堤——范公堤就建造在天然的贝壳堤基础上,现距海约 60 km,其外侧已发育为海成平原和广阔的潮滩与海涂。平原西侧为一系列呈弧形分布的低山丘陵所环绕,有洪泽湖、高邮湖等众多湖泊形成的湖泊群。本文研究的是介于古贝壳堤西侧与湖泊群东侧之间的平原成因。

近年来,有关苏北平原古地理环境演变的研究日益增多[2～11],取得了一系列有意义的认识,但作为反映区域沉积特征的沉积物主元素的地球化学研究却较薄弱。已有研究表明,海岸带沉积盆地中沉积物的地球化学记录提供了过去环境变化的资料,代表区域沉积特征的沉积物主量元素组成可以反映不同沉积环境的信息[12～16]。因此本文试图通过对苏北平原宝应钻孔(BY1)2.58 Ma 以来不同沉积相沉积物的主量元素地球化学特征分析,来揭示河-海交互作用下海相沉积环境与陆相沉积环境的元素分布特点,探讨平原成因及其演化历史。

一、研究区地质概况

在大地构造位置上,苏北平原位于华北陆块和扬子陆块拼贴的接壤部位,晚白垩世以来该区由于受到太平洋板块和印度板块的影响,形成了以箕状断陷为特点的复合型沉积盆地[17]。第四纪以来受频繁的气候波动和海侵作用的影响,该区海陆交互作用十分明显[2,3,5,8,10,11,18,19],不仅有陆地泛滥平原的河湖相沉积,而且曾多次为海水所淹覆,发育了海相与滨海潮滩相砂和粉砂层,在地貌上经河-海交互堆积逐渐由盆地发展成平原。

苏北盆地中广泛分布着巨厚的中、新生代地层,是一个受深大断裂或断裂带控制的断陷型沉积盆地(图 1)。盆地始形成于白垩纪,在干旱的热带-亚热带气候下形成陆内红层沉积。白垩纪末至早第三纪,气候普遍温暖,形成了一套杂色碎屑岩含油建造。晚第三纪气候转冷,在开阔的盆地上沉积了以河流相为主的砂砾岩和杂色黏土岩。盆地及其周边

　* 杨竞红,王颖,张振克,J. -P. Guilbault,毛龙江,魏灵,郭伟,李书恒,徐军,季小梅:《第四纪研究》,2006 年第 26 卷第 3 期,第 340 - 352 页。

　摘要于同年载于:《"中国海洋资源环境与南海问题"学术研讨会 会议日程与论文(摘要)汇编》,3 - 125～3 - 126 页. 广西北海,2006. 2. 26～3. 5.

　国家自然科学基金项目(批准号:40271004)资助。

地区新生代玄武岩类以碱性系列为主,常以互层形式赋存于下第三纪沉积地层中,多为致密熔岩,局部夹火山碎屑岩;埋藏于盆地的玄武岩呈大面积广泛分布,在盆地中部、西部出露有第三纪钙碱性至碱性玄武岩,第四纪玄武岩出露较少[20,21]。盆地中断裂构造十分发育,NE 走向的响水-淮阴-盱眙断裂控制了盆地的北部边界,并受郯庐深大断裂带演化的极大影响。EW 走向的如东-扬州断裂和 NE 走向的金坛-如皋断裂发育于盆地南部,NW走向的南京-嘉山断裂控制了盆地西南部。这些深大断裂控制了第三纪晚期以后的沉积作用,也是该区新构造与地貌分区的界线。第四纪时全球气温普遍下降,冷暖的频繁变化造成海平面变化和海岸变迁,同样对苏北盆地的沉积环境有着深刻的影响。第四纪时该区沉积厚度为 100～340 m,主要有东台-海安和吕四-栟茶两个沉积区域。

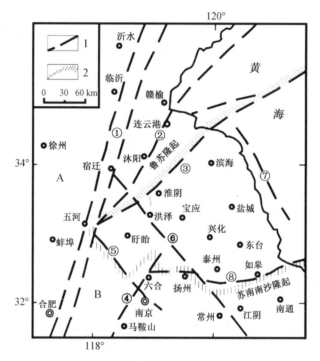

图1　苏北平原构造区划示意图[5,17]

1. 主要断裂(带)　2. 苏北-南黄海盆地范围

A. 华北陆块　B. 扬子陆块

① 郯庐断裂带　② 赣榆-泗阳断裂　③ 淮阴-响水断裂　④ 长江断裂带　⑤ 南京-嘉山断裂

⑥ 无锡-宿迁断裂　⑦ 苏北东海岸断裂　⑧ 扬州-如皋断裂

二、样品及分析方法

宝应位于苏北盆地的中西部低洼处($33°14'21''$N,$119°22'41''$E),东距现代海岸 120 km,西北有洪泽湖,向南有高邮湖,京杭大运河擦其西侧而过,第四系沉积厚度约 100 m。2004 年夏选择在宝应望直港镇中心小学操场钻孔取样(BY1),钻孔进深 146 m,实取沉积柱 97 m。根据沉积物颜色、组成和结构特点进行了层位划分,在北京大学考古系[14]C 测年实验室进行了沉积物的[14]C 定年,中国科学院地质与地球物理研究所进行了沉积物古地磁年代测定。在各层位选取有代表性的样品共 124 个进行了常量元素含量分析。沉积物的电子探针分析及 X 射线衍射分析在南京大学地球科学系成矿作用研究国家重点实验室完成。

常量元素分析是在南京大学现代分析中心采用 X 射线荧光光谱法（XRF）。样品自然风干后，用玛瑙研钵磨细至 200 目，在恒温箱中 60 ℃下干燥 24 小时，取 3 g 沉积物用熔融片法制片，在瑞士 ARL - 9800 型 X 射线荧光光谱仪上测定各种氧化物 SiO_2，Al_2O_3，TiO_2，Fe_2O_3，MgO，MnO，CaO，K_2O，Na_2O，P_2O_5 和 SO_2 的质量分数，然后根据质量百分数将主元素氧化物质量分数换算成各元素的质量分数。分析过程采用国家标准沉积物样 GSD - 1 全程监控，主元素质量分数的绝对偏差为 1.0%～1.5%，最大分析误差小于 0.5%。

三、BY1 钻孔的岩性特征

对 BY1 钻孔岩芯进行的古地磁年代测定表明[22]，B/M 界线位于 39.16 m 深度，45.63～53.83 m 处为 Jaramillo 正极性亚时，76～82 m 处为 Olduvai 正极性亚时，85～87 m 处为 Reunion 正极性亚时，Gauss 正极性时期的顶界位于剖面 95.22 m 深度。整个岩芯剖面揭示了约 2.58 Ma 的沉积历史。有孔虫鉴定结果发现，钻孔中多处层位出现一定量的不同种属的海相有孔虫组合[23～25]，从而进一步证实苏北平原曾经历了海陆交互作用。

本区 2.58 Ma 以前（95.22～97.00 m）为平原基底，以泛红棕色的杂色硬质黏土层为主，95.35～95.45 m 见较大泥砾岩块（7 cm×7 cm），以石英、绢云母为主，含少量方解石。基底上部 95.22 m 处为 M/G 界面，是 2.58 Ma 前的风化剥蚀面。

据沉积相可将 BY1 钻孔沉积剖面自下而上划分为 10 个沉积段。

（1）95.22～83.95 m（2.58 Ma～1.95 Ma）为半干燥泛滥平原的河湖相沉积。95.22～87.00 m（M/G 界面之上，2.58 Ma～2.15 Ma）以中黄棕色夹绿灰色的粉砂质砂与砂质粉砂为主，夹中-粗砂层；91 m 处见咸水种有孔虫（毕克卷转虫，*Ammonia beccarii*），似有海水的一定影响，具有典型的河流二元相结构和平原摆动的河流沉积特征；87.00～84.45 m（Reunion 正极性亚时，2.15 Ma～2.14 Ma）为绿灰色夹有灰黄棕色含细砂的粉砂质黏土，下层黏土硬结，含结核和钙板层，具干湿季节性变化的湖盆沉积特征；84.45～83.95 m 以浅黄棕色为主，夹黄灰色的硬质黏土的砂质粉砂，半干燥气候为主。

（2）83.95～79.66 m（约 1.9 Ma～1.8 Ma）为海水淹侵的海侵相沉积。以黄棕色为主的粉砂质砂，含黏土质粉砂，含毕克卷转虫。

（3）79.66～75.42 m（约 1.80 Ma～1.77 Ma）为泛滥平原的河湖相沉积。以深黄棕色为主的粉砂质砂，夹有砾石及粗中砂；76.61～76.64 m 处见块状细粒石英闪长岩，暗色矿物均已发生较强蚀变；76.5～75.9 m 见零星毕克卷转虫，而 76.7 m 为 Olduvai 正极性亚时结束，可能在 1.77 Ma 有一次潮侵的海陆交互作用。

（4）75.42～65.40 m 为受潮侵影响的滨海平原河湖相沉积。75.42～71.78 m 以浅橄榄灰、灰橘色粉砂质砂为主，含细砂及水平层理，具有河流二元相结构；71.78～65.40 m

具潮侵影响的平原湖泊沉积,绿灰色与黄棕色的杂色粉砂质砂和粗砂层与黏土、黏土质粉砂交杂沉积,夹泥质条斑和粗砂层,见毕克卷转虫及少量植物碎屑,可能为通海之湖;顶部的硬质黏土层经暴露、压实,为剥蚀面堆积。

(5) 65.4～58.0 m为受海水浸淹的浅海相沉积。厚约7 m的砂与粉砂层,重复出现的毕克卷转虫残壳表明该处为水深小于20 m的浅水环境,与侵蚀面相交出现的粗砂或砂砾层似经侵蚀的河流及潟湖相沉积,反映气候转暖下的一次明显的海进。

(6) 58.00～39.16 m(约1.00 Ma～0.78 Ma)为泛滥平原河湖相沉积。58.00～45.63 m以黄棕色为主,砂质粉砂夹浅橄榄灰色黏土层或泥质条带,见贝壳残体及植物碎屑,为河湖相沉积。在53.83～45.63 m为Jaramillo正极性亚时(1.07 Ma～0.99 Ma),以黄棕色的砂质粉砂、黏土质粉砂为主,夹有硬黏土层,局部具微层理且有虫穴扰动;53.3～53.6 m有致密块状泥岩,蚀变较强,主要为绢云母,含少量方解石及石英。46.5 m附近普遍发现毕克卷转虫和霜粒希望虫(*Elphidium nakanokawaense*)。45.63～39.16 m(0.99 Ma～0.78 Ma)多为黄棕色砂质粉砂、黏土质粉砂和粉砂质黏土,局部具微层理,含钙质结核、石英砂砾及蓝灰色黏土质斑块,顶部为B/M界线。40.78～41.03 m深处异常富集铁锰鲡粒,延至底部均有零星出现。该潜育层表明水位曾经常升降,在干湿交替环境下铁锰氧化物随氧化-还原条件变化而发生移动和局部沉淀,有着相对暖湿的气候条件;40.6 m附近发现多种源于大陆架中、外部的有孔虫,数量极少,可能反映在0.78 Ma地磁极性由正转负时,海平面曾有显著上升,或经历了一次较大的风暴潮[26～28]。

(7) 39.16～19.16 m(0.78 Ma～0.03 Ma)为海水浸淹的浅海相沉积。以橄榄灰-橄榄黑色为特征,黏土、黏土质粉砂、粉砂质黏土及细砂沉积,具波状和页状层理,侵蚀面发育,局部有波痕、纹层结构;底部B/M界线处有次棱状小砾石层。本段自下而上有孔虫含量逐渐增多,在27.3～23.8 m伴随淡水昆虫及植物碎屑,出现零星海绵骨针,24～21 m有少量来自含有海水和淡水混合的瓣鳃动物*Corbicula*及淡水介类,21.2 m附近植物碎屑开始大量富集,直至顶部;而20.70～19.16 m不仅富含毕克卷转虫和霜粒希望虫,而且普遍含有缝裂希望虫(*Elphidium magellanicum*),粗糙希望虫(*Elphidium hispidulum*),易变希望虫(*Elphidium subincertum*),光滑先希望虫(*Protelphidium glabrum*)和具瘤先希望虫(*Protelphidium tuberculatum*),零星发现的其他种类有孔虫达10种之多,海陆交互作用明显。采得^{14}C测年样品,结果表明19.15～19.25 m为32 910±170aB.P.,而23.90～23.95 m处为39 385±170aB.P.,说明在晚更新世30 kaB.P.左右本区有较大规模的海侵沉积(Q_3)。

(8) 19.16～14.72 m为潮下带浅海相沉积。黄棕色与浅橄榄灰色相间的砂质粉砂、黏土质粉砂和粉砂,具良好的粉砂与黏土质粉砂的页状韵律层。鉴定出有孔虫种类多达20种,其中以毕克卷转虫和霜粒希望虫为代表的优势种多指示水深小于20 m的浅水区及半咸水(盐度小于20‰)的温和气候环境,局部亦见盐度为31‰～32‰,水深在30～50 m甚至60 m的有孔虫出现,如缘刺纺轮虫(水深20～50 m,盐度31‰～32‰)、中国假元旋虫(20～50 m水深)、秀丽脐塞虫(一般出现在17～15 m处)及精美直小希望虫(水深<50 m,盐度<32‰,出现在19 m处),可能有较强的海水影响。此外,19.16～17.2 m发现大量植物残片。

(9) 14.72～8.70 m 为湖泊洼地相沉积。以深黄棕色为主夹橄榄灰的黏土质粉砂，含多个侵蚀间断面，局部可见虫孔、砂姜和铁锰结核；10.3～13.2 m 为与上、下层不整合接触的黄棕色硬黏土层，可能为全新世下界。

(10) 8.7～0 m 为全新世的浅海与海陆过渡相沉积。8.70～5.13 m 以深黄棕色为主夹橄榄棕的粉砂、黏土质粉砂，含虫孔，具水平层理和交错层理，属潮滩相沉积；5.13～1.00 m 为橄榄灰色的浅海相粉砂与黏土质粉砂，发现毕克卷转虫等 4 种有孔虫和一些淡水介类，表明本区在全新世仍有海水淹覆。已有研究资料表明[9,29]，早全新世（10 000～7 500aB. P. ）是我国东部海面的主要上升期，晚更新世以来苏北发生过 3 次海侵，且呈不断扩大之势，7 000aB. P. 时海侵达到极盛。洪泽湖东钻探资料也显示[4]，沭阳、刘集、泗洪、洪泽等地均有全新世海侵地层分布，在盱眙县东的维桥附近的全新世沉积中发现海陆过渡相的盾形生物化石（*Sinocytheriderlonga*）；早全新世时古淮河大约在盱眙县附近入海，高海面时盱眙县附近曾是古淮河口[9]。其东部发育的古贝壳砂堤是淤泥质、粉砂质海岸特有的滩脊，代表着约 5 000aB. P. 的古滨线，为海退海岸的重要标志。BY1 该段资料显示了本区在全新世也曾受到海水浸淹影响。1～0 m 是钻孔顶部层，深黄棕色为主的粉砂与黏土质粉砂，为陆相平原沉积。

以上岩相及岩性特征表明，2.58 Ma 以来本区沉积环境发生过较大变化，沉积物的颜色、组成、结构和微体生物化石均表现出河湖相、潮滩相、淡水湖相与滨浅海相交替演变的环境。第四纪以来主要经历了 4 次较大海侵，即 1.95 Ma 左右（83.95～79.66 m），1 Ma 左右（65.4～58.0 m），0.78 Ma～3 Ma（39.16～14.72 m）及全新世海侵（5.13～1.00 m），而以 0.78 Ma～3 Ma 间的海侵为最大，且晚期较早期规模大，与苏北平原上更新世（Q_3）发育巨厚海相层一致。杨怀仁等对黄河三角洲地区的第四纪海陆变迁的研究也表明[30]，早第四纪海进较弱，晚第四纪海进影响在范围和频率上有明显加强，即平原形成时曾遭受多次海水淹覆，海陆交互作用明显。地貌与沉积相综合研究证明，苏北平原内侧经历了浅海→潟湖→湖泊→潮滩→洼地，最后为河流冲积物覆盖成陆的过程，苏北平原整体上是在第四纪经河-海交互堆积形成。

四、常量元素分析结果与讨论

（一）常量元素变化特征

宝应钻孔沉积物中常量元素变化范围和平均质量分数见表 1。主元素质量分数分布模式为 Si＞Al＞Fe＞Ca＞K＞Mg＞Na＞Ti＞Mn＞P＞S。与中国东部上地壳的平均化学组成[31]比较，宝应钻孔沉积物具有高 Si，Ti，Mn，P，S，低 Na，Mg，Al，K，Ca，Fe 的特征（图 2a）。主元素在沉积物中的质量分数受区域地质背景、源岩组成、所处地理位置及相应气候条件、沉积环境所控制。苏北盆地新生代发育的巨厚的砂岩、泥岩沉积及碱性玄武岩夹层，早第三纪玄武岩起源于未经混染的、富集型岩石圈地幔，具有明显的 P 的正异常[34]，这些为后期的地质改造提供了丰富的物质来源。苏北盆地紧邻黄海南陆架海域，

表1 宝应钻孔沉积物中常量元素变化范围和平均质量分数

沉积相	沉积相段	深度/m	样品个数	Na/%	Mg/%	Al/%	Si/%	K/%	Ca/%	Ti/%	Fe/%	Mn/($\mu g \cdot g^{-1}$)	P/($\mu g \cdot g^{-1}$)	S/($\mu g \cdot g^{-1}$)
全钻孔		96.7~0	Aver(124)	0.55	0.81	6.45	31.62	1.98	2.08	0.36	3.02	966	713	348
			min	0.03	0.15	3.99	20.47	0.58	0.39	0.06	0.36	77	175	40
			max	1.05	1.69	8.65	39.37	2.90	10.58	0.55	19.91	26 719	15 753	1 562
浅海相与海陆过渡相	10	8.70~0	Aver(27)	0.52	0.85	6.97	30.94	1.74	1.53	0.45	3.54	674	457	334
			min	0.38	0.47	5.49	23.82	1.52	0.61	0.37	1.96	232	218	80
			max	0.76	1.52	8.10	35.38	2.03	6.46	0.50	5.32	1 394	742	601
湖泊洼地沉积相	9	14.72~8.70	Aver(8)	0.39	0.81	7.76	29.80	1.76	1.26	0.47	4.05	494	404	340
			min	0.24	0.63	6.50	25.56	1.41	0.63	0.39	2.75	387	175	240
			max	0.61	1.01	8.65	33.55	1.91	3.93	0.54	5.81	620	611	441
潮滩浅海相	8	19.16~14.72	Aver(9)	0.49	1.32	7.06	27.77	1.96	4.01	0.40	3.52	714	548	458
			min	0.33	0.92	5.83	23.21	1.73	2.99	0.37	2.15	387	480	200
			max	0.68	1.69	8.33	32.74	2.20	5.37	0.44	5.05	1 317	655	1 001
浅海相	7	39.16~19.16	Aver(32)	0.63	0.91	6.35	31.26	2.03	2.63	0.36	2.76	692	719	562
			min	0.32	0.31	5.13	23.50	1.25	0.76	0.11	0.76	232	262	120
			max	1.05	1.53	8.26	37.36	2.70	8.70	0.49	4.34	3 872	2 793	1 562
泛滥平原河湖相	6	58.0~39.16	Aver(15)	0.54	0.72	6.35	31.29	2.01	2.95	0.36	2.69	1 326	578	160
			min	0.19	0.19	4.34	24.75	1.62	0.56	0.07	0.52	155	218	40
			max	1.05	1.36	8.40	38.70	2.81	10.58	0.54	4.71	12 624	960	360

（续表）

沉积相	沉积相段	深度/m	样品个数	Na/%	Mg/%	Al/%	Si/%	K/%	Ca/%	Ti/%	Fe/%	Mn/$\mu g \cdot g^{-1}$	P/$\mu g \cdot g^{-1}$	S/$\mu g \cdot g^{-1}$
浅海相	5	65.4~58.0	Aver (7)	0.66	0.68	5.94	31.80	2.13	3.51	0.30	2.22	591	623	146
			min	0.19	0.19	4.34	24.75	1.80	0.56	0.07	0.52	155	218	40
			max	1.05	1.36	8.40	38.70	2.81	10.58	0.47	4.32	1781	960	200
滨海平原河湖相	4	75.42~65.40	Aver (7)	0.56	0.41	5.63	35.71	2.06	0.61	0.24	2.00	199	449	206
			min	0.03	0.19	4.21	33.33	0.58	0.45	0.11	0.77	155	218	160
			max	0.84	0.57	6.74	38.76	2.67	0.94	0.52	2.87	310	1571	320
泛滥平原河湖相	3	79.66~75.42	Aver (6)	0.56	0.47	4.92	33.66	2.26	0.87	0.17	4.86	6557	1658	120
			min	0.21	0.15	3.99	20.47	1.47	0.39	0.06	0.36	77	305	40
			max	0.85	0.92	6.08	39.37	2.73	1.54	0.34	19.91	26719	5367	240
海侵相	2	83.95~79.66	Aver (2)	0.39	1.15	6.63	31.20	1.78	2.12	0.41	3.15	658	633	120
			min	0.32	1.13	6.49	29.58	1.78	0.64	0.39	2.87	310	480	120
			max	0.45	1.16	6.77	32.83	1.78	3.60	0.43	3.43	1007	785	120
半干燥河湖相	1	95.22~83.95	Aver (9)	0.55	0.54	5.68	34.90	2.24	1.24	0.25	1.78	250	2003	125
			min	0.16	0.17	4.05	30.87	1.64	0.45	0.07	0.54	77	262	40
			max	0.88	0.87	7.01	39.24	2.76	4.50	0.41	2.82	542	15753	200
基底风化剥蚀面		96.70~95.22	Aver (2)	0.36	0.76	6.81	29.74	1.66	3.14	0.34	3.01	542	415	160
			min	0.24	0.67	6.66	27.28	1.43	0.98	0.29	3.01	465	393	160
			max	0.48	0.85	6.96	32.19	1.88	5.30	0.40	3.01	620	436	160

极易与海域的海水和物质进行交换,这很可能是导致 P,S 等元素质量分数明显高于中国东部上地壳平均化学组成的主要原因。

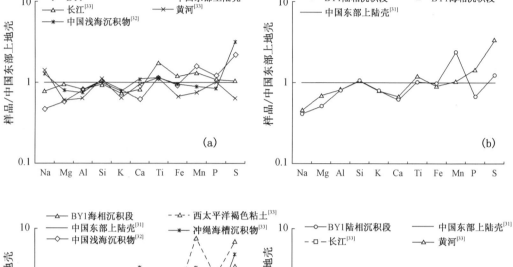

图 2 宝应钻孔沉积物与其他典型海、陆相沉积物相对于中国东部上地壳的常量元素比值对比图

(a)整个宝应钻孔沉积与其他典型海、陆相沉积物比较 (b)宝应钻孔海相沉积与陆相沉积比较

(c)宝应钻孔海相沉积与其他海相沉积物比较 (d)宝应钻孔陆相沉积与长江、黄河沉积比较

BY1 钻孔中,海相沉积物明显地较陆相沉积物含有较高的 Mg,Ca,P,S(图 2b),其中 S 平均含量是陆相沉积物的 2.7 倍,P 则是陆相沉积物的 2 倍,这些元素恰恰是海洋中的特征元素,与海洋生物及海水组成有着密切关系。海洋沉积物中的碎屑组分主要是通过各种途径从大陆搬运而来的陆壳风化产物[35],除 SiO_2,CaO,MgO 由于硅质和钙质生物壳体堆积而有明显的富集,MnO 一定程度上可通过化学沉积而富集外,其他主要成分,如 Al_2O_3,TiO_2,K_2O,Na_2O 等均主要赋存于碎屑组分中[36]。BY1 沉积物中 Si,K,Fe,Mn 在陆相组分中的平均含量高于海相,显示出陆源风化产物供给的特征。新生代海陆相互作用的研究表明[37],海陆相互作用体系对于构造运动或者洋面升降的反应十分灵敏,由于边缘海面积不大、通道浅窄,可以对海洋的气候信号产生一种放大效应;又由于边缘海紧靠大陆,可以对陆地气候产生比大洋更强的影响。边缘海的很大部分由于水浅在低海面时出露成陆。已有沉积记录证实冰期时生源沉积物减少,而陆源碎屑沉积物的比例增高[38]。

BY1 中海相沉积物常量元素组成与中国浅海沉积物[32]大体相近(图 2c),高 Si,Al,

Mn，P，S，且低 Na，Ca，Mg。其中 P 含量（880 $\mu g/g$）与长江（650 $\mu g/g$）、黄河（600 $\mu g/g$）及中国东部上陆壳沉积物中的 P 含量（600 $\mu g/g$）相距较远，且远远高于黄海表层沉积物中的总 P（411 $\mu g/g$）[39]，显然不具备"亲陆"性。P 是一种重要的生源要素，研究区海相沉积物的 P 含量介于半深海的冲绳海槽沉积物及属深海的西太平洋褐色黏土之间[33]，显示了明显的海源（海洋生物及海水）特征。异常高的 S 含量则介于中国浅海沉积物与冲绳海槽沉积物之间[33]，可能主要来自海盐及海洋生物。

BY1 钻孔中陆相沉积物 Si，Al 和 Fe 含量则介于黄河、长江沉积物之间[33]（图 2d），意味着两河流域的碎屑物质对本区陆相沉积有着一定的影响。在岩石风化过程中，比较活泼的 K，Na，Ca，Mg 等碱金属和碱土金属元素很容易被淋滤出来，而 Si，Al 等稳定元素则在残余相（风化产物）中富集[40]，被淋滤出来的 K 和 Mg 又很容易被风化产生的黏土矿物吸附或结合起来，从而相对富集于风化产物中，Na 和 Ca 则比较容易流失而致使在风化产物中含量较低[41,42]。对花岗岩风化壳的研究也表明，在风化产物中，Al，K，Mg 等元素含量较母岩富集，而 Na 和 Ca 含量则偏低，Ti 的含量没有明显差别[43]。

Si 元素高的原因有三个：① 沉积物中砂含量较高，矿物成分以石英、长石为主，表现为碎屑态硅的富集；② 可能预示周边源区尤其是从基岩丘陵侵蚀来源的物质中 Si 含量较高；③ 与海洋硅质生源的含量较高有关。因此，三种来源硅的叠加，导致了本区沉积物中具有较高含量的硅。

Al 是沉积物中仅次于 Si 和 O 的造岩元素，主要以各种铝硅酸盐矿物及其风化产物存在，广泛分布于沉积物中。BY1 海相沉积物的 Al 含量为 4.14%～8.33%，平均为 6.55%，与加拿大海盆沉积物中 Al 的含量（6.1%～8.6%）[44]相近，与研究区高的 Si 含量是相符的，亦反映着陆源物质对海相沉积的影响。

Fe 和 Mn 是典型的变价元素，具强烈的亲氧性，其迁移富集过程与环境关系密切。Fe 在 BY1 海相沉积物中的含量为 0.63%～5.32%，平均为 3.04%，低于加拿大海盆中的含量（3.68%～6.05%，平均为 5.13%）[44]，接近中国浅海沉积物平均值（3.10%）[32]。BY1 陆相沉积物 Fe 的平均含量介于长江与黄河沉积物[33]之间，反映着研究区地理环境介于南北气候交绥带及海陆交互作用的过渡带特征，物理风化较黄河强，而化学风化较长江弱。Darby 等[44]对加拿大海盆沉积物中 Fe 的形态分析结果表明，Fe 在加拿大海盆主要以碎屑态存在，可提取态 20% 左右，表明沉积物通过吸附获得的 Fe 量较少。BY1 沉积物 Mn 的含量平均为 966 $\mu g/g$，其中海相平均值为 634 $\mu g/g$，陆相平均值为 1 445 $\mu g/g$，均高于中国东部上地壳的平均化学组成（600 $\mu g/g$）[31]，这与研究区沉积物颗粒较细，对 Mn 的吸附量较大以及海侵阶段的暖湿环境、化学风化较强相一致。苏北盆地第三纪碱性玄武岩 Mn 含量达 1 440 $\mu g/g$[21]，为本区的高 Mn 提供了物质来源，含 Mn 岩石在风化过程中被水解，同时被氧化形成锰的高价氧化物 MnO_2。BY1 钻孔中铁锰结核的电子探针分析显示，Mn 含量变化在 9%～34%，在氧化-还原条件变化下而发生明显迁移富集。

BY1 钻孔海相沉积物中 Na 的平均值为 0.53%，陆相沉积物 Na 平均值为 0.48%，均明显低于中国东部上陆壳（1.16%）[31]；K 在海相层中平均为 1.96%，与中国浅海一致（1.93%）[32]。沉积物的 X 射线衍射分析表明，Na 主要赋存在钠长石、正长石和蒙脱石中；K 的赋存矿物主要为正长石和伊利石。在风化过程中 K 和 Na 极易淋滤析出，其淋滤程度与化学风化程度密切相关。苏北盆地地处暖温带,化学风化较强,Na 的淋滤与 K 的吸附均较大,因此研究区海相沉积物中 Na 的含量明显低于中国浅海,而 K 的含量近于中国浅海[32]。

Ca 和 Mg 为碱土造岩元素,在风化过程中容易进入溶液,并随溶液一起发生搬运;另一方面水体中的 Ca 和 Mg 易为生物吸收,形成碳酸盐骨骼。此外,当溶液中的 Ca 和 Mg 离子与碳酸根离子的浓度达到溶度积时,形成碳酸盐沉积。因此 Ca 和 Mg 在沉积物中的含量受控于物源、沉积环境和生物活动三个方面因素。Ca 在 BY1 钻孔海相沉积物中含量为 0.52%～8.70%,平均为 2.30%,其中 Q₃ 海侵沉积段平均值达 4.6%,高于中国浅海沉积物平均值(3.79%)[32],明显有着海洋沉积物的生物来源参与;陆相沉积物中 Ca 含量为 0.39%～10.58%,平均为 2.17%,低于中国东部上陆壳(3.42%)[31],与长江沉积物平均值(2.86%)近似[33],显示了陆源风化特征。Mg 在海相沉积物中含量变化在 0.19%～1.69%,平均为 0.91%,Q₃ 海侵沉积段平均值达 1.27%,高于中国浅海沉积物平均值(1.11%)[32],与钙的高异常相对应,是海源生物活动的体现;陆相沉积物中 Mg 含量为 0.15%～1.36%,平均为 0.67%,虽低于中国东部上陆壳(1.39%)[31],但接近于具有陆源特征的黄河沉积物平均值(0.84%)[33]。两元素在海相沉积物中的高含量反映着与陆源沉积的环境差异,表明它们主要与海洋生物相联系。

Ti 在地壳中虽然广泛分布,但在表生作用中比较稳定,属惰性元素,风化后难以形成可溶性化合物,基本上以碎屑矿物的形式(多为细砂或极细砂粒级的金红石与含钛磁铁矿)被搬运,不大受化学风化强度的影响,具有平面分布的"均匀性",在中国沿海多富集于海岸沙体中。BY1 沉积物中 Ti 含量变化在 0.06%～0.55%,平均为 0.36%,近于中国东部上陆壳平均值(0.31%)[31]。

综上所述,BY1 钻孔常量元素所揭示的平原沉积具有明显受物源控制、海陆交互过渡带的沉积特征。

(二) 常量元素垂直分布特征及其反映的环境演变过程

宝应 BY1 钻孔沉积物常量元素含量和元素对比值在剖面中不同沉积段呈明显的波动(图 3 和图 4)。95.22～97.00 m 为 2.58 Ma 以前的风化基底面,其沉积物相对于中国东部上地壳具有高的 Si 和 Ti,低 Na,Mg,Al,K,Ca,Fe,Mn 和 P 的特点;较中国浅海沉积物明显含有较高的 Si 和 Al。Si/Ti(90)高于中国大陆沉积物平均值(71),Si/Al 为 4.36,高于中国东部上陆壳平均值(3.79),反映风化剥蚀作用较强,意味着暖干气候。在 95.2～96.0 m 处 Ca 和 Mg 含量突然增高,与该段含钙质结核的硬黏土有关。钻孔常量元素的垂向变化记录了苏北盆地 2.58 Ma 以来海陆交互作用下的 10 个不同阶段的环境演化特征。

(1) 95.22～83.95 m(2.58 Ma～1.95 Ma)

相对于下覆层风化剥蚀面,该段 Si,Na,K 含量明显升高,Al,Mg,Ca,Fe,Mn 含量降低;Si/Ti 平均近 3 倍高于下覆剥蚀面,为整个沉积剖面最高,Si/Al 也同步上升到 6.47,反映了本段沉积环境与基底的明显不同。91 m 附近曲线有一较大波动:Ca,Mg,Fe,Mn,P 和 S 均出现峰值,与本段河湖相平均值相比,Mg 和 Fe 含量高出近 1.5 倍,Ca 含量高出

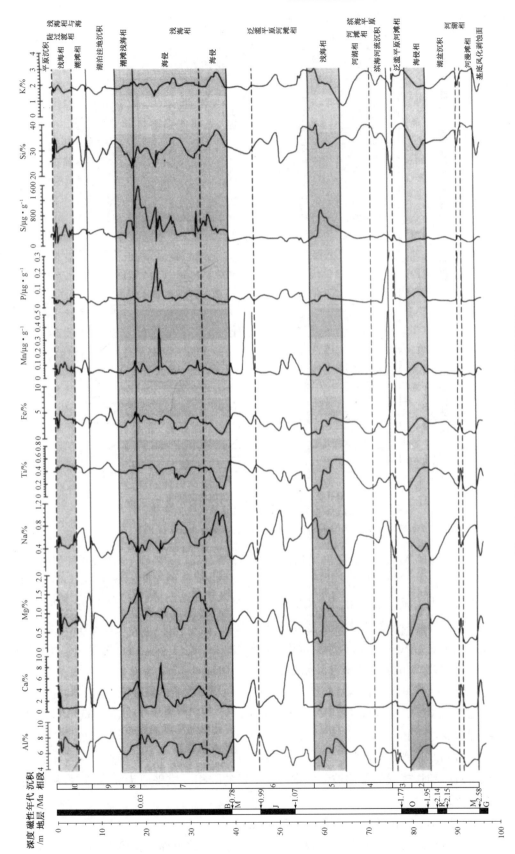

图3 宝应钻孔沉积物常量元素含量随深度变化图

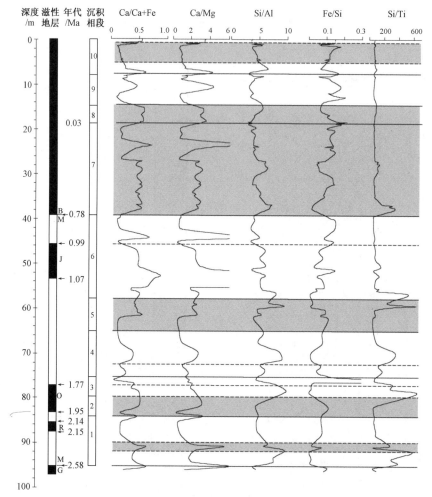

图4 宝应钻孔沉积物常量元素对比值的垂向变化

阴影部分为海侵或海水淹覆段

3.6倍,P的曲线波动最大,含量高出平均值的8倍。Si,Na,K与Si/Ti和Si/Al均呈相应低谷,所在层位中发现零星淡水植物屑及有孔虫,故很可能发生过一次海水淹覆,有海洋元素的加入。中国海岸带晚第四纪风暴沉积的研究发现[27,45],风暴以浸浪方式进入陆地时,可形成瞬时海洋环境,冲击和改造异地和原地沉积物,物质组分具有多元性和远源性。这段沉积是否与风暴潮有关,还有待进一步研究。上部84.7 m附近海源元素含量呈上升趋势,恰在Olduvai正极性事件开始之前。

(2) 83.95~79.66 m(约1.9 Ma~1.8 Ma)

海源元素Ca(2.12%),Mg(1.15%),P(633 μg/g),Fe(3.15%)和Mn(658 μg/g)平均含量继续增加,在剖面图上呈较宽的波峰,而Si(31.2%),Na(0.39%)和K(1.78%)含量及Si/Ti(77)和Si/Al均有降低,呈相应的波谷,反映了海侵的物源特点。本段正处在Olduvai正极性亚时,气候的偏湿波动使本区半干燥的陆相泛滥平原为海水淹侵,从而使物源发生变化。

(3) 79.66~75.42 m(约1.80 Ma~1.77 Ma)

平均含量Ca(0.87%),Mg(0.47%)和P(371 μg/g)显著降低而呈波谷,Si(33.66%),Na

（0.56％），K（2.26％）及 Si/Ti（77）和 Si/Al 均有升高而呈波峰，Fe（4.86％）和 Mn（6557 μg/g）仍继续增加；Olduvai 正极性亚时结束之处的 76.7 m（1.77 Ma）附近，各元素曲线波动剧烈，Ca，Mg，Fe，Mn，P 和 S 均呈峰值出现，尤其 Fe，Mn 和 P 具异常高值，有孔虫与海源元素相伴出现，很可能有一次海水淹覆或风暴潮。

（4）75.42～65.40 m

与下覆沉积段相比，各元素平均含量及元素对比值基本变化不大，72 m 附近有一明显波动，除 Fe/Si 呈波谷变化外，Si/Ti（77），Si/Al 和 Ca/（Ca+Fe）及 Ca/Mg 均表现出舒缓波峰，此时沉积环境从河流转为湖泊。

（5）65.4～58.0 m

Mg，Ca，P，S 等与海洋生物及海水有关的元素含量明显升高，Ca 含量虽低于中国浅海，在全区低 Ca 背景下，仍呈明显峰值，Mn 与 Fe 在剖面中亦呈现出波峰；Al，Na，K 和 Si 呈相应的低谷，而 Ca/（Ca+Fe）（0.43）及 Mg/Al（0.12）比值较下伏段明显增高，具有海相沉积物的高值特征[32]，沉积环境较之前的河湖沉积继续偏湿润波动，出现的毕克卷转虫，显示了一次海侵事件。

（6）58.00～39.16 m（约 1.00 Ma～0.78 Ma）

以 Jaramillo 正极性亚时结束（45.63 m）为界可分为上下两亚段。在 58.00～45.63 m（1.07 Ma～0.99 Ma）上亚段中，Si/Ti（142），Si/Al（5.51）与 Ca/Mg（5.48）值截然不同于下覆沉积段，前两者明显较低，其中 Si/Ti 值降低近 50％，而 Ca/Mg 值增加 2 倍多。P 和 S 出现全剖面最低值。P（623 μg/g）和 S（146 μg/g）平均含量具有中国东部上陆壳的低值特征[31]，在垂向分布图中呈低谷。在 53.83 m 的 Jaramillo 正极性亚时开始附近，元素含量曲线波动较大，Ca 和 Mg 出现异常高值，Fe，Mn，P 和 S 均出现小波峰，Si 和 Al 为异常低值，有可能反映泛滥平原的一次较弱的海水侵覆。Si/Al 大于东部上地壳（3.79）[31]，介于长江（4.42）与黄河（6.63）之间[33]，似有较强的亲源性，即陆源碎屑物质构成本段沉积的主体，总体上反映了较为干燥气候环境下的陆相泛滥平原沉积。在 45.63～39.16 m（0.99 Ma～0.78 Ma）下亚段，Si/Ti（63）和 Si/Al（4.21）继续走低，并为整个剖面最低段；Ca/Mg（2.26）有明显回落；Fe/Si（0.12）则大幅增高，Fe 与 Mn 的平均值不仅高于下覆层段，而且高于整个剖面平均值，与本段含大量铁锰结核一致，反映此段曾经历了反复的干湿变化。Jaramillo 正极性亚时结束时 Ca，Na，Mn 和 P 出现不同程度的波峰，Si，Al，Ti，Mg 和 K 相应出现波谷，有海水影响的可能，海洋成分的不断加入对陆源元素含量起着"稀释"作用，反映了受海水影响的河湖沉积。

（7）39.16～19.16 m（0.78 Ma～0.03 Ma）

Si/Ti（106）和 Si/Al（5.03）开始以低值为特征，S，P，Mg，Ca 含量明显增高，在垂向图中呈密集峰展布。21 m 附近 S 高达 1 000～1 560 μg/g，几乎是中国东部上陆壳（160 μg/g）[31] 的 10 倍，呈现全剖面最高峰，与本段富含海相有孔虫相对应，印证了强烈的海洋（海水及海洋生物）来源特征，为 30 000a B.P. 的海侵（Q_3）；Ca/Mg 及 Ca/（Ca+Fe）均较下覆层段升高，反映了气候转暖，27～30 m 的低值区与侵蚀间断面可能代表浅海沉积相中短期的干旱事件。

(8) 19.16～14.72 m

Si/Ti(69)和Si/Al(4.01)持续呈低值的波谷,S,P,Mg和Ca虽仍有较高平均值,但自下向上开始逐渐降低,表明海源物质供应有所减少,Si的平均值虽然较低,却呈明显升高,Ca/(Ca＋Fe)(0.54)近于中国浅海沉积物(0.55)[32],反映海陆交互作用的潮下带沉积特征。

(9) 14.72～8.70 m

Ca(1.26%),Mg(0.81%)和P(404 μg/g)含量较下覆海相层有明显降低,陆源元素Al的平均值(7.76%)略高于总平均值(6.45%),近于中国东部上陆壳(7.81%)[31];Si(29.80%)也有所回升;Na(0.39%)低于海侵层,Si/Ti(64)比值位于长江和黄河沉积物之间[33],反映了"多源"混合特点的湖泊洼地相沉积。10.3～13.1 m的硬黏土层段,Mg和Ca含量均较上下层高,Al含量达全钻孔最高值(8.64%),Si/Ti与Si/Al均出现异常低值,分别变化在57～62和3.17～3.87,是干燥气候环境的反映。

(10) 8.7～0 m

总体上Ca(1.53%),Mg(0.85%),P(457 μg/g)和S(354 μg/g)均有增加,尤其在7.4 m和5.20～0.76 m处曲线波动较大,P,S,Mg等海洋元素增高,有孔虫分析显示轻度盐水环境,表明平原形成在全新世晚期时仍有海水组分加入;顶部(0.76 m以上)Si/Al降低而Ca/Mg升高,Ca/Mg比值(3.40%～3.53%,平均3.45%)接近风化基底,Mg/Al比值(0.185%～0.191%,平均0.188)达全剖面最高值,并位于黄河与长江沉积物之间[33],是典型的平原沉积特征。

五、结论

根据苏北盆地宝应钻孔沉积物的常量元素分析,得到如下几点认识:

(1) 宝应钻孔沉积物主元素质量分数分布模式为Si＞Al＞Fe＞Ca＞K＞Mg＞Na＞Ti＞Mn＞P＞S。与中国东部上地壳平均化学组成比较,其沉积物具有高Si,Ti,Mn,P,S,低Na,Mg,Al,K,Ca,Fe的特征。

(2) 钻孔中海相沉积物较陆相沉积物明显含有较高的Mg,Ca,P和S,而这些元素恰恰是海洋中的特征元素,与海洋生物及海水组成有着密切关系。海相沉积物常量元素组成与中国浅海沉积物相近,且高Mn,P和S,低Na。P含量介于半深海的冲绳海槽沉积物及属深海的西太平洋褐色黏土之间,显示了明显的海源(海洋生物及海水)特征。异常高的S含量则介于中国浅海沉积物与冲绳海槽沉积物之间,主要来自海盐及海洋生物。

(3) 陆相沉积物相对海相沉积物有较高的Si和Mn,其Al,Fe,Si含量则介于黄河、长江沉积物之间,意味着两河流域的碎屑物质对本区陆相沉积有着一定的影响。在岩石风化过程中,比较活泼的K,Na,Ca,Mg等碱金属和碱土金属元素很容易被淋滤出来,而Si和Al等稳定元素则在残余相(风化产物)中富集,被淋滤出来的K和Mg又很容易被风化产生的黏土矿物吸附或结合起来,从而相对富集于风化产物中,Na和Ca则比较容易流失而致使在风化产物中含量较低。

(4) 根据宝应钻孔沉积物常量元素含量的垂向变化,并结合沉积相、岩性、微古生物

分析和古地磁及 ^{14}C 测年，将苏北盆地 2.58 Ma 以来的海陆环境演化划分为 10 个阶段，每个阶段又可分出几个亚段。

参考文献

［1］凌申. 1994. 苏北全新世海进与古砂堤研究. 台湾海峡,13(4):338-345.

Ling Shen. 1994. Study on Holocene transgression and sandbars in North Jiangsu. *Journal of Oceanography in Taiwan Strait*, 13(4): 338-345.

［2］杨怀仁,谢志仁. 1984. 中国东部近 20 000 年来的气候波动与海面升降运动. 海洋与湖沼,15(1): 1-13.

Yang Huairen, Xie Zhiren. 1984. Sea-level changes along the east coast of China over the last 20 000 years. *Oceanologica et Limnologica Sinica*, 15(1): 1-13.

［3］朱大奎,傅命佐. 1986. 江苏岸外辐射沙洲的初步研究. 见:任美锷,朱季文,朱大奎等. 江苏省海岸带东沙滩综合调查报告. 北京:海洋出版社. 28-32.

Zhu Dakui, Fu Mingzuo. 1986. Morphological and surface-sedimentary characters in Dongsha Bank of Jiangsu coast. In: Ren Mei′e, Zhu Jiwen, Zhu Dakui *et al*. eds. Reports of the Comprehensive Survey on the Dongsha Beach of Jiangsu Coast Zone. Beijing:China Ocean Press. 28-32.

［4］凌申. 1990. 全新世以来苏北平原古地理演变. 黄渤海海洋,8(4):20-27.

Ling Shen. 1990. Changes of the palaeogeographic environment in North Jiangsu Plains since the Holocene. *Journal of Oceanography of Huanghai & Bohai Seas*, 8(4): 20-27.

［5］陈友飞,严钦尚,许世远. 1993. 苏北盆地沉积环境演变及其构造背景. 地质科学,28(2):151-160.

Chen Youfei, Yan Qinshang, Xu Shiyuan. 1993. Evolution of the sedimentary environments in North Jiangsu Basin and its tectonic setting. *Scientia Geologica Sinica*, 28(2): 151-160.

［6］朱诚,程鹏,卢春成等. 1996. 长江三角洲及苏北沿海地区 7000 年以来海岸线演变规律分析. 地理科学,16(3):207-214.

Zhu Cheng, Cheng Peng, Lu Chuncheng *et al*. 1996. Palaeocoastline evolution law of Yangtze River delta and coastal area of North Jiangsu Province since 7000a B. P. *Scientia Geographica Sinica*, 16(3): 207-214.

［7］杨达源,张建军,李徐生. 1999. 黄河南徙、海平面变化与江苏中部的海岸线变迁. 第四纪研究,(5):283.

Yang Dayuan, Zhang Jianjun, Li Xusheng. 1999. Yellow River southward diversion, sea level changes and coastline flux in the middle part of Jiangsu Province. *Quaternary Sciences*, (5): 283.

［8］王颖主编. 2002. 黄海陆架辐射沙脊群. 北京:中国环境科学出版社. 10-15.

Wang Ying ed. 2002. Radiative Sandy Ridge Field on Continental Shelf of the Yellow Sea. Beijing: China Environment Science Press. 10-15.

［9］凌申. 2003. 全新世以来硕项湖地区的海陆演变. 海洋通报,22(4):48-54.

Ling Shen. 2003. Changes of land and sea in the Shuoxiang Lake area since Holocene. *Marine Science Bulletin*, 22(4): 48-54.

［10］杨达源,陈可锋,舒肖明. 2004. 深海氧同位素第 3 阶段晚期长江三角洲古环境初步研究. 第四纪研究,24(5):525-530.

Yang Dayuan, Chen Kefeng, Shu Xiaoming. 2004. A preliminary study on the paleoenvironment during MIS 3 in the Changjiang Delta region. *Quaternary Sciences*, 24(5): 525 - 530.

[11] 王张华,陈杰. 2004. 全新世海侵对长江口沿海平原新石器遗址分布的影响. 第四纪研究,24(5): 537 - 545.

Wang Zhanghua, Chen Jie. 2004. Distribution of the Neolithic sites in the Changjiang coastal plains Holocene transgression impact. *Quaternary Sciences*, 24(5): 537 - 545.

[12] 陈志华,石学法,王湘芹等. 2003. 南黄海 B10 岩心的地球化学特征及其对古环境和古气候的反映. 海洋学报,25(1):69 - 77.

Chen Zhihua, Shi Xuefa, Wang Xiangqin *et al*. 2003. Geochemical changes in Core B10 in the southern Huanghai Sea and implications for variations in paleoenvironment and paleoclimate. *Acta Oceanologica Sinica*, 25(1): 69 - 77.

[13] Li X D, Shen Z G, Wai O W *et al*. 2001. Chemicalforms of Pb, Zn and Cu in the sediment profiles of the Pearl River Estuary. *Marine Pollution Bulletin*, 42(3): 215 - 223.

[14] Li X D, Wai OWH, Li Y S *et al*. 2000. Heavy metal distribution in sediment profiles of the Pearl River Estuary, South China. *Applied Geochemistry*, 15(5): 567 - 581.

[15] 杨慧辉,陈岚. 1998. 海坛岛海域表层沉积物中主量化学成分的地球化学. 海洋学报,20(3):47 - 55.

Yang Huihui, Chen Lan. 1998. Geochemistry of some major chemical composition in marine sediments of Haitan Island. *Acta Oceanologica Sinica*, 20(3): 47 - 55.

[16] 孟翔,刘苍字. 1996. 长江口区沉积地球化学特征的定量研究. 华东师范大学学报(自然科学版),36(1):73 - 84.

Meng Yi, Liu Cangzi. 1996. A quantitative study on the sedimentary geochemical characteristics of the Changjiang Estuarine region. *Journal of East China Normal University* (*Natural Science*), 36(1): 73 - 84.

[17] 陈安定. 2001. 苏北箕状断陷形成的动力学机制. 高校地质学报,7(4):408 - 418.

Chen Anding. 2001. Dynamic mechanism of formation of dustpan subsidence, Northern Jiangsu. *Geological Journal of China University*, 7(4): 408 - 418.

[18] 凌申. 2001. 全新世以来里下河地区古地理演变. 地理科学,21(5):474 - 479.

Ling Shen. 2001. Palaeogeographic changes of the Lixiahe district since the Holocene. *Scientia Geographica Sinica*, 21(5): 474 - 479.

[19] 杨怀仁,陈西庆. 1985. 中国东部第四纪海面升降、海侵海退与岸线变迁. 海洋地质与第四纪地质,5(4):59 - 80.

Yang Huairen, Chen Xiqing. 1985. Quaternary transgressions eustatic changes and shifting of shoreline in East China. *Marine Geology & Quaternary Geology*, 5(4): 59 - 80.

[20] 陈友飞. 1992. 苏北盆地与周围地区的新生代玄武岩及其形成的大地构造环境. 福建师范大学学报(自然科学版),8(1):94 - 103.

Chen Youfei. 1992. Cenozoic basaltic rocks in North Jiangsu Basin and surrounding area and their bearing on tectonic environment. *Journal of Fujian Normal University* (*Natural Science*), 8(1): 94 - 103.

[21] 吴向阳,牟荣,石胜群等. 1999. 苏北盆地火成岩发育与构造演化的关系. 中国石油勘探 4(1): 44 - 47.

Wu Xiangyang, Mu Rong, Shi Shengqun *et al*. 1999. The relation between volcanics development and the structure evolution in North Jiangsu Basin. *China Petroleum Exploration*, 4(1): 44 - 47.

[22] 王颖,张振克,朱大奎等. 2006. 河海交互作用与苏北平原成因. 第四纪研究,26(3):301-320.

Wang Ying, Zhang Zhenke, Zhu Dakui et al. 2006. River-sea interaction and the North Jiangsu Plain formation. *Quaternary Sciences*, 26(3):301-320.

[23] Wang Pinxian, Min Qiubao, Bian Yunhua et al. 1985. On micropaleontology and stratigraphy of Quaternary marine transgressions in East China. In: Wang Pinxian, Bian Yunhua, Cheng Xinrong et al. eds. Marine Micropaleontology of China. Berlin:Springer-Verlag. 265-284.

[24] Wang P X, Murray J W. 1983. The use of foraminifera as indicators of tidal effects in estuarine deposits. *Marine Geology*, 51(3-4):239-250.

[25] Wang Pinxian, Min Qiubao, Bian Yunhua et al. 1985. Characteristics of foraminiferal and ostracod thanatocoenoses from some Chinese estuaries and their geological significance. In: Wang Pinxian, Bian Yunhua, Cheng Xinrong et al. eds. Marine Micropaleontology of China. Berlin: Springer-Verlag. 229-242.

[26] 刘宝珺,张继庆,许敦松. 1986. 四川兴文四龙下二叠统碳酸盐风暴岩. 地质学报,60(1):55-67.

Liu Baojun, Zhang Jiqing, Xu Dunsong. 1986. Carbonate storm rock of Lower Permian in Silong, Wenxing area, Sichuan Province. *Acta Geologica Sinica*, 60(1):55-67.

[27] 吕炳全,韩昌甫,郑世培. 1987. 北部湾涠洲岛晚更新世风暴沉积. 科学通报,(5):362-365.

Lü Bingquan, Han Changfu, Zheng Shipei. 1987. Storm sediments of Late Pleistocene in Weizhou Island of the Northern Bay. *Chinese Science Bulletin*, (5):362-365.

[28] 赵希涛,王绍鸿. 1992. 江苏阜宁西园全新世风暴沉积与海岸沙丘的发现及其意义. 中国科学(B辑),00B(9):996-1001.

Zhao Xitao, Wang Shaohong. 1992. Holocene storm sediments and the significance of the coastal sand dune in FuningXiyuan, Jiangsu. *Science in China*(*Series B*), 00B(9):996-1001.

[29] 赵松龄. 1986. 近百年来中国东部沿海地区海平面变化研究状况. 见:国际地质对比计划第200号项目中国工作组编. 中国海平面变化. 北京:海洋出版社. 15-27.

Zhao Songling. 1986. Sea level changes in Eastern China littoral since 100a. In: International Geological Correlation Programme Project No. 200 China Working Group ed. Sea Level Changes in China. Beijing:China Ocean Press. 15-27.

[30] 杨怀仁,王健. 1990. 黄河三角洲地区第四纪海进与岸线变迁. 海洋地质与第四纪地质,10(3):1-14.

Yang Huairen, Wang Jian. 1990. Quaternary transgressions and coastline changes in Huanghe River (Yellow River) delta. *Marine Geology& Qauternary Geology*, 10(3):1-14.

[31] 鄢明才,迟清华. 1997. 中国东部地壳与岩石化学组成. 北京:科学出版社. 1-155.

Yan Mingcai, Chi Qinghua. 1997. The Chemical Compositions of Crust and Rocks in the Eastern Part of China. Beijing:Science Press. 1-155.

[32] 赵一阳,鄢明才著. 1994. 中国浅海沉积物地球化学. 北京:科学出版社. 179-198.

Zhao Yiyang, Yan Mingcai. 1994. Geochemistry of Sediments of the China Shelf Sea. Beijing: Science Press. 179-198.

[33] 赵一阳,鄢明才. 1992. 黄河、长江、中国浅海沉积物化学元素丰度比较. 科学通报,37(13):1202-1204.

Zhao Yiyang, Yan Mingcai. 1992. Comparison between of element abundances in sediments from Yellow River, Changjiang River and China shallow sea. *Chinese Science Bulletin*, 37(13):

1202 - 1204.

[34] 杨祝良,陶奎元,沈渭洲等.1998.苏北盆地隐伏早第三纪玄武岩地球化学及源区特征.岩石学报,
14(3):332 - 342.

Yang Zhuliang, Tao Kuiyuan, Shen Weizhou et al. 1998. Geochemistry and source characters of
the concealed Eogene basalts in North Jiangsu Basin. Acta Petrologica Sinica, 14(3): 332 - 342.

[35] Windom H L. 1976. Lithogenous Materials in Marine Sediments—Chemical Oceanography
(Volume 5, 2nd Edition). London: Academic Press. 103 - 135.

[36] Goldberg E D, Arrhenius G O S. 1958. Chemistry of Pacific pelagic sediments. Geochimica et
Cosmochimica Acta, 13(2 - 3): 153 - 198.

[37] 汪品先.2005.新生代亚洲形变与海陆相互作用.地球科学——中国地质大学学报,30(1):1 - 18.

Wang Pinxian. 2005. Cenozoic deformation and history of sea-land interactions in Asia. Earth
Science—Journal of China University of Geosciences, 30(1): 1 - 18.

[38] Gorbarenko S A. 1996. Stable isotope and lithologic evidence of Lateglacial and Holocene
oceanography of the northwestern Pacific and its marginal seas. Quaternary Research, 46(3):
230 - 250.

[39] 王菊英,刘广远,鲍永恩等.2002.黄海表层沉积物中总磷的地球化学特征.海洋环境科学,21(3):
53 - 56.

Wang Juying, Liu Guangyuan, BaoYong'en et al. 2002. Geochemical characteristic of TP in
sediments in Yellow Sea. Marine Enviornmetal Science, 21(3): 53 - 56.

[40] Nesbitt H W, Young G M. 1982. Earthly Proterozoic climate and plate motions inferred from
major element chemistry of lutites. Nature, 299: 715 - 717.

[41] Weaver C E, Postassium. 1967. Illite and the Ocean. Geochimica et Cosmochimica Acta, 31(11):
2181 - 2196.

[42] Nesbitt H W, Markovics G, Price R C. 1980. Chemical process affecting alkalis and alkaline earths
during continental weathering. Geochimica et Cosmochimica Acta, 44(11): 1659 - 1666.

[43] Peuraniemi V, Pulkkinen P. 1993. Preglacial weathering crust in Ostro-bothnia, Western Finland,
with special reference to the Raudaskylä occurrence. Chemical Geology, 107(3 - 4): 313 - 316.

[44] Darby D A, Naidu A S, Mowatt T C et al. 1989. Sediment composition and sedimentary processes
in the Arctic Ocean. In: Herman Y ed. The Arctic Seas: Climatology, Oceanography, Geology and
Biology. NewYork: Van Norstrand Reinhold Company. 657 - 720.

[45] 张明书,刘守全,林峰等.2000.粤北一个晚更新世复合洪积扇体的初步研究.海洋地质与第四纪地
质,20(4):30,36,62.

Zhang Mingshu, Liu Shouquan, Lin Feng et al. 2000. Study on the complex diluvial fans of Late
Pleistocene in Northern Guangdong Province. Marine Geology&Quaternary Geology, 20(4): 30,
36, 62.

河海交互作用沉积与平原地貌发育[*]

一、环境背景

中国位于亚洲东部与太平洋西缘,大陆与大洋相互作用、太平洋板块向亚洲大陆俯冲以及形成一系列岛弧环绕的边缘海是我国东部沿海基本构造格局,进一步发育的河海交互作用是东部沿海平原形成的主导因素。

河流是陆源物质(包括固体与溶解质)向海洋输送的主要动力,并且对海岸水体的沉积动力有巨大的影响。据估计,每年由河流输送入海的悬移物质达 20×10^9 t[1]。据不完全统计,源自亚洲大陆的泥沙约 4.6×10^9 t,而输入太平洋的约为 3.3×10^9 t[2],原因在于构造抬升造成的巨大地貌差异与湿润的季风气候——海陆相关作用、所形成的高侵蚀速率,尤其是大河流经大面积的未固结的黄土高原地区,以及长久历史的人类活动影响,这些因素的结合,使亚洲河流具有最高的陆源物质输送量。世界上有 8 条大河发源于世界屋脊青藏高原:黄河、长江、雅鲁藏布-布拉马普特河、恒河、怒江-萨尔温江、伊洛瓦底江、湄公河、印度河,其中 5 条河流是汇入亚洲-太平洋边缘海,泥沙输送量大(表1)[2-4]。

表 1　发源于青藏高原河流的入海泥沙量[*]

河流	黄河	长江	雅鲁藏布-布拉马普特拉河	恒河	怒江-萨尔温江	伊洛瓦底江	湄公河	印度河
年入海泥沙量/$\times 10^6$ t	1 115	461	540	520	100	260	160	59

[*] 本文系分析平原发育过程,有关河流水、沙资料采用未经水工建设前的。

区域特性导致河流输沙量的变异,例如:黄河的径流量约为密西西比河的 1/15,为亚马逊河的 1/183,为尼罗河的 1/2,但其输沙量是密西西比河的 3 倍,是亚马逊河的 2 倍,是尼罗河的 9 倍[2]。原因在于黄河流经未固结成岩的黄土高原,侵蚀产沙作用强,实际上,是黄河水系将黄土高原切割蚀低,继而将泥沙搬运至黄渤海堆积为华北平原与浅海。长江的径流量与降雨量呈线性相关,而长江的高输沙量形成于夏季与初秋的强径流时期。当代,长江上、中游兴建了一系列水库,中、下游又在实施分流,其结果必然改变长江的河性,上述的径流与输沙相关规律亦发生变化。长江口淡水与泥沙锐减,浪、潮、流作用效果相形增强,若干年后,可能在 21 世纪中后期,三角洲土地侵蚀、咸潮影响加剧等一系列前所未有的变化会逐渐显现。人们需了解自然河性,因势利导!

* 王颖,傅光翮,张永战:《第四纪研究》,2007 年第 27 卷第 5 期,第 674－689 页。
国家自然科学基金项目"河海交互作用与苏北平原成因研究"(批准号:40271004)外延成果。

中国海大陆架曾在晚更新世时是沿岸平原,河流是形成沿岸平原与输送陆源泥沙的主要动力[2]。在未兴建规模化的水利工程前,每年由中国河流(大河及上千条中小型河流)输入海中的泥沙约 2.01×10^9 t(表 2)[3],其中大约 1.21×10^9 t/a 堆积在渤海,14.67×10^6 t/a 堆积于黄海,0.63×10^9 t/a 堆积在东海,0.16×10^9 t/a 堆积在南海及台湾以东太平洋。因此,每年约相当于 $1~km^3$ 的泥沙是由中国河流输送到边缘海大陆架的。最高的沉积速率在渤海,年均沉积速率约为 8 mm[2],按此堆积速率,如不考虑海盆的下沉,平均水深约为 18 m 的渤海可能会在 2250 年的时间内被填满。但是,目前由于入海径流量减少,河口断流及中游水土保持效益日增,人为地改变自然过程,上述情况难以出现。可是,在中、新生代地貌发育的长期历史过程中,河流冲刷、搬运了西部山地与黄土高原的物质,堆积形成东部平原与中国海大陆架[5]。

表 2 中国主要河流径流量与输沙量参考值[3]

海区	河流	流域面积		平均径流量		平均输沙量		产沙量
		$/km^2$	/%	$/\times 10^8~m^3$	/%	$\times 10^4$ t	/%	$/t \cdot km^{-2} a^{-1}$
渤海	辽河	164 104	12.3	87	10.9	1 849	1.5	113
	滦河	44 945	3.4	49	6.1	2 267	1.9	505
	黄河	752 443	56.3	431	53.7	111 490	92.2	1 482
	上述三河总量	961 492	72.0	567	70.7	115 606	95.6	1 202
	所有入渤海河流总量	1 335 910	100.0	802	100.0	120 881	100.0	905
黄海	鸭绿江	63 788	19.1	251	44.8	195	13.3	31
	所有入黄海河流总量	334 132	100.0	561	100.0	1 467	100.0	44
东海	长江	1 807 199	88.4	9 323	79.7	46 144	73.1	255
	钱塘江	41 461	2.0	342	2.9	437	0.7	105
	闽江	60 992	3.0	616	5.3	768	1.2	126
	上述三河总量	1 909 652	93.4	10 281	87.9	47 349	75.0	248
	所有入东海河流总量	2 044 093	100.0	11 699	100.0	63 060	100.0	308
南海	韩江	30 112	5.1	259	5.4	719	7.5	239
	珠江	452 616	77.3	3 550	73.6	8 053	84.2	178
	上述二河总量	482 728	82.4	3 809	79.0	8 772	91.7	182
	所有入南海河流总量	585 637	100.0	4 822	100.0	9 592	100.0	164
	太平洋总计	11 760		268		6 375		467
	中国国境内河流汇入总量	4 311 532		18 152		201 375		

二、河海交互作用与沉积模式

我国河流源远流长，径流与泥沙特性各有不同，加之汇入海区的自然环境差异，河海交互作用所导致的沉积形式与结构各有特点。本文拟以 5 条不同类型的河流为代表，分析研究河海交互作用与沉积模式，将今论古，判断海陆过渡带环境的变化与稳定性（图 1 和表 3）。

表 3　中国 5 条代表河流基本情况[*][2]

河流	流域面积/km²	河长/km	年径流量/×10⁹ m³	年输沙量/×10⁶ t	平均泥沙浓度/kg·m⁻³	平均潮差/m
鸭绿江	63 788	859	27.8	4.75	0.33~0.42	4.48
滦河	44 945	870	3.89	24.08	3.94	1.50
黄河	752 443	5 464	48.5	1 190	37.7	0.80
长江	1 807 199	6 380	925	386	0.544	2.77
珠江	452 616	2 197	370	85~100	0.120~0.334	0.86~1.63

* 选用人为改变前的河流资料，以比较客观地分析平原发育的历史。

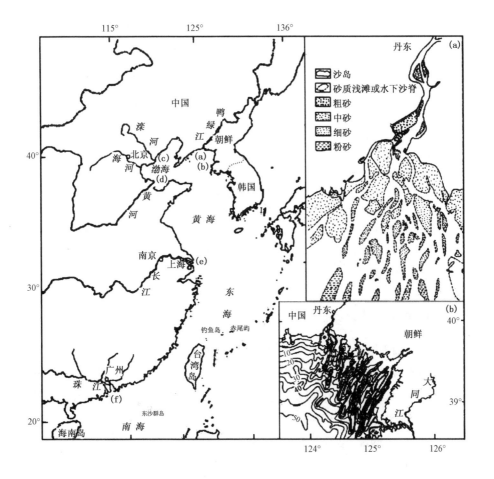

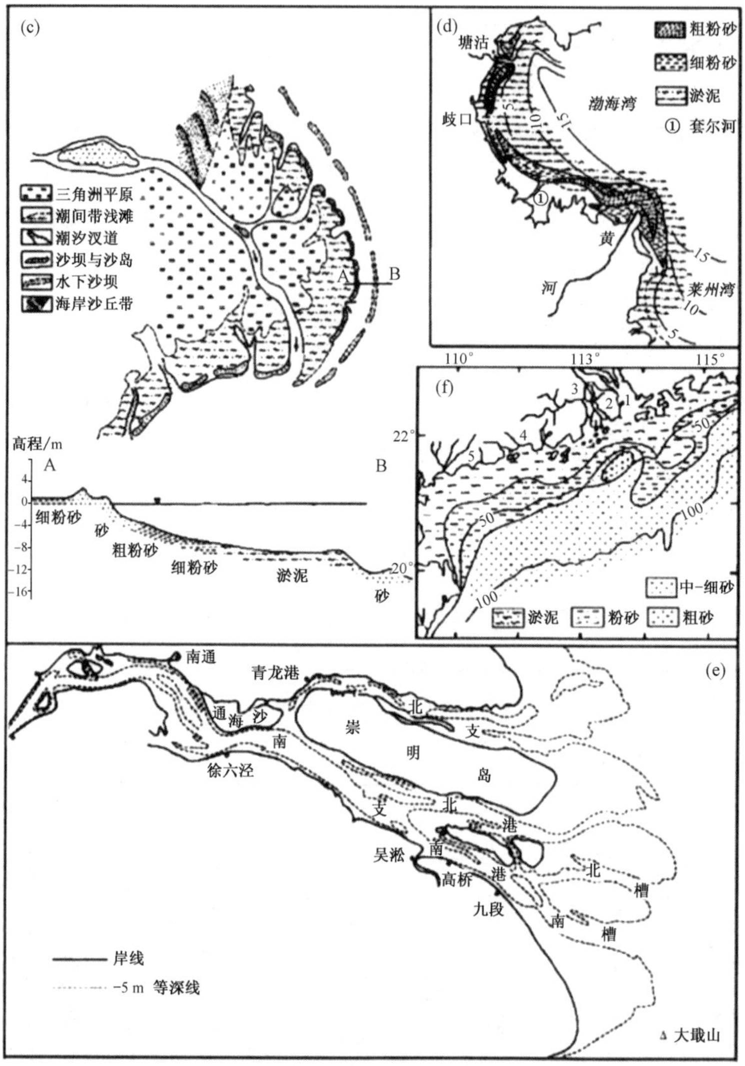

图1 中国5条代表性河流及其河口

（据文献[6,7]修改）

（a）强潮型的鸭绿江漏斗状河口；（b）黄海北端潮流脊；（c）滦河口—具沙坝环绕的双重海岸；

（d）黄河三角洲与渤海湾沉积分布；（e）长江口沙岛与河口分支；（f）珠江口外陆架沉积（1. 珠江；

2. 中山；3. 高要；4. 阳江；5. 电白）

（一）强潮型河口动力沉积

选择发源于长白山，汇入黄海北端的鸭绿江为例。河流的下游坡度陡，河道内仅有少量砂质沉积，占总径流量与输沙量的80%发生在夏季（6～9月）洪水期间。强潮作用使河口展宽成漏斗型（见图1a），大潮潮差可达6.9 m，强有力的潮流以平均为1.25～1.50 m/s的速度垂直向岸流动，涨潮时在河口处长，向上游传播时逐渐减小，而落潮时是上游长，而向河口减小。河口区的涨、落潮时差为15～18分，至上游丹东市的潮时差可达2小时28分钟。通常，河口区波高为1 m，但在河口外的大鹿岛，波高可达3.3 m。河口上游分布着砂砾和砾石;河口内沉积主要由砂质组成;口门堆积着由径流与潮流交互作用形成的沙脊浅滩,由中砂、细砂组成,其上部覆盖着薄层粉砂质淤泥;河口以外黄海北端的浅海内陆架延伸至20～30 km以外水深10 m处,堆积着细砂与中砂,强劲的潮流作用阻碍了堆积水下三角洲,泥沙受潮流携带,形成一系列沿潮流方向延展、彼此平行的沙脊和谷槽(见图1b),每一潮汛期间,涨潮时含沙量为0.42 kg/m³,而落潮时为0.33 kg/m³,每潮净沉积可达2 100 t。加之,潮水在喇叭口形河口湾内之辐聚与辐散,故而,每一潮汛均使潮流沙脊加积增高,形成黄海北端突出的沙脊群地貌。在内陆架浅海域,波浪扰动掀带起沙脊上的细砂,继之,又被从东向西的沿岸流携带搬运,结果,使沙脊末端呈现向西偏移之势态。

(二) 波浪动力主导的陆海交互作用沉积

以滦河为例。滦河口位于缓升的燕山与渤海沉降带之间的过渡地带,构造活动是制导滦河改道与三角洲发育的主要内动力因素。晚更新世以来,滦河自西向东发育了4个突出的三角洲瓣(图2),构成以滦县为顶点向东南展开、面积达4 500 km²的现代三角洲[8～10]。滦河是一条多沙性河流,它发源于河北省巴颜图古尔山麓,流经内蒙古高原、燕山山脉,至乐亭县的兜网铺入海,全长877 km,泥沙来源于山地变质岩与花岗岩的风化产物,含沙量曾高达3.9 kg/m³。滦河是季节性河流,6～9月夏秋季降雨期,径流量高达

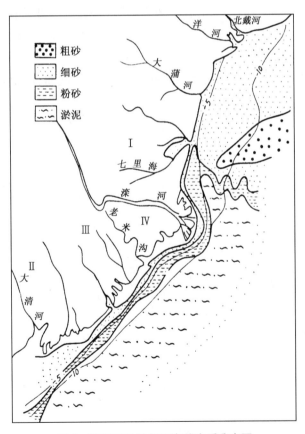

图2 滦河口及其邻近海域底质分布图

Ⅰ东三角洲瓣 Ⅱ西三角洲瓣 Ⅲ中三角洲瓣 Ⅳ新三角洲瓣

$34.3×10^8$ m³,占全年径流量的73%;夏季7~8月降水占总径流量的56%,沙随水行,季节性暴雨径流的平原河流尾闾多变迁改道。滦河沿岸潮差较小,平均为1.0~1.5 m,在南堡东西两侧形成两个小型的顺时针方向的旋转流,使南堡成为两侧泥沙会聚点,地貌呈三角"岬"形突向海。同时,南堡亦是东西两侧沙源分界点,东侧以滦河砂质为主,西侧为潮流粉砂黏土质沉积。滦河三角洲沿岸波浪作用强,常年盛行偏东风,ESE风与ENE风强盛,波浪最大,有效波高(1/3H)达3.8 m,有效周期为7.8s,波长平均30~31 m,最大达41 m,波浪作用可影响沿岸海底,破波水深可影响到5.5 m处。波浪掀沙力强,常年盛行的东向风浪阻止了滦河入海泥沙向海扩散,形成泥沙自海向岸横向搬运,发育了水下沙坝与海岸沙坝[8]。环绕河口的海岸沙坝,使海岸形成双重岸线,即内侧为原始的平原岸线,外侧为沙坝海岸,两者之间成为潟湖。当偏东向波浪与海岸斜交时,可携运泥沙沿海岸移动,形成纵向泥沙流,加之,海岸落潮流大于涨潮流,河口外部分泥沙,被落潮流主支向SW搬运,少量向NE方向搬运,据估计,泥沙向南输运量约为$384×10^4$ t/a,向北输移为$76.7×10^4$ t/a[10]。纵向输移的泥沙,继而又被波浪掀积成沙坝。所以,滦河平原沿岸均有沙坝环绕,这是外动力以波浪作用为主的河口三角洲沉积模式(见图1c和图2)。滦河口以南海岸带沉积组成为:一系列沙坝成为与岸平行的带状细砂沉积,分选佳(S_0=1.5~1.7);沙坝外围5 m深处的破浪带为粗粉砂沉积;向海至9 m以深处为深灰色的淤泥黏土质沉积,其中含有0.05%~2.00%的自生黄铁矿,反映该处为波浪作用影响小的还原环境,是现代海岸的下界。现代陆架沉积薄,不足1 m。滦河口以北,砂质沉积宽达20 km,从河口海滨至1.3 m深处为分选良好的细砂与中砂(1.3~1.8φ),中砂沉积延伸至水深13 m处,沙粒的磨圆度高,表面结构具有圆麻点,其粒级分配曲线(图3)表明该处宽广的

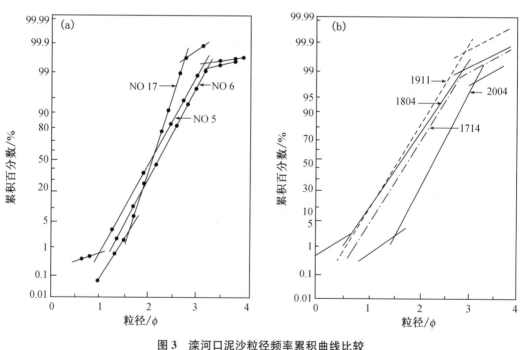

图3 滦河口泥沙粒径频率累积曲线比较

(a) 海滩沙丘砂(图中 NO17 等为样品编号)　(b) 内陆架砂(图中 1911 等为样品编号)

砂质沉积是低海面时的海滩沙丘带,是与滦河以北昌黎海岸的巨大沙丘相伴存在,是海岸沙丘的沙源地,因全新世海面上升而沉溺于海底[11]。由海岸沙坝围封的海岸逐渐成为浅水潟湖,河流泥沙堆积于沙坝内侧的浅水中形成粉砂与细砂的叠层。粉砂来自陆地,是河流泥沙在隐蔽的浅水环境的堆积,细砂系浪流越过海岸沙坝携带至潟湖中的沉积。原始的海岸由于沙坝围封而发育潮滩。这一系列沉积是由于波浪横向推移泥沙向岸,堆积环绕河口的沙坝而形成的。滦河泥沙影响的范围可达河口三角洲外坡水深 7 m 处,该处距河口约 2.5~5.0 km。在河口以外的海岸带,泥沙被沿岸水流搬运,自河口向 SW 方向运移,延长了沙坝长度,并使之具沙咀的型式,少量泥沙亦自河口向 NE 输运。

历史上滦河三角洲平原的发育,由于滦河迁徙,自晚更新世以来,在涧河与大蒲河之间发育了 4 个三角洲瓣[9](见图 2),它们分别是东、西、中、新三角洲瓣,均具有沙坝环绕的型式,反映出该海岸季风波浪作用稳定。三角洲体的规模大小与河流供沙量及下游河道、河口的稳定时间长短有关。但是,自 1979 年建设了潘家口、大黑汀水库,1983 年引滦河水供应天津以及 1984 年引滦入唐山后,滦河径流量减少为 18.4×10⁸ m³,入海径流比工程前减少 61%,其中 1980 年至 1984 年径流量为 3.55×10⁸ m³,减少量达 92%,滦河尾闾几乎干涸[10]。河水枯竭,入海泥沙减少,滦河三角洲由原来向海淤积延伸(最大达 81.8 m/a[10]),转变为受海浪冲蚀后退。最初几年,岸线平均后退约 3.2 m/a,最大 10 m/a,滦河口口门后退更为显著,曾达 300 m/a[10]。岸外沙坝宽度与长度均减小(20 m/a,400 m/a)[10],今后海岸蚀退的速度会减缓,但三角洲发育模式已有显著变异。

(三) 弱潮的径流河口沉积

以黄河为代表,黄河于渤海湾南部与莱州湾之间流入渤海,除强风暴期间外,河口外海域的波浪作用微弱。河口潮差小,仅 0.5~0.8 m,潮差向河口外两侧三角洲海岸逐渐增大至 2.0 m;潮流流速在河口最高可达 1.4 m/s,向两侧逐渐减低。与此相应,68% 的极细砂与 20%~30% 的粉砂堆积在河道内,粗粉砂(含量达 50%)及细粉砂(22%~30%)堆积在河道两侧的泛滥平原上,其中约 80% 的粗粉砂与细砂逐渐淤高成河道两岸的天然堤;流出河口的粗粉砂在河口外两侧淤积,逐渐形成向海伸出的指状沙咀。指状沙咀发育速度很快,尤其在新近改道的尾闾河口,曾在 1964 年记录到 10 km/a 的增长速度[6],沿河口向海伸展之指状沙咀是在弱潮及潮差较小的海域,具有以河流的外泄径流为主动力的沉积特点。在河口两侧被指状沙咀隐护的岸段则为细粉砂(40%~60%)和粉砂质黏土(30%)等悬移质淀积区,形似两个黏稠的淤泥湾,风浪平静。河口范围以外,黄河的泥沙被沿岸流搬运,形成伸向渤海湾南岸的淤泥舌(见图 1d)。历史上,黄河下游经多次迁移改道,范围北至天津,南徙夺淮入黄海,在渤海南部其尾闾迁徙尤为频繁,具有自淤高成的地面河向邻近低洼处改道之特点,其改道具有自 N 向 NE、向 E、向 SE、继而向 S,再折向 N 的顺时针方向的迁徙变化规律。据其变迁历史统计,大体上每 6~10 a 当行水河道淤高后,即发生迁徙[2],在新河口形成新的指状沙咀与烂泥湾。如:1996 年 8 月黄河自清八汊河口入海,沙咀与河道东迁,年均 1.9 km,年均造陆达 4.5 km²。而废弃河道的指状沙咀因无泥沙供应而受浪流冲刷后退,年均蚀退 450 m[12],未胶结的堆积体极易侵蚀,曾经测得冲蚀最大速度达 17.5 km/a[6]。冲蚀下来的粗粉砂则堆积于两侧的淤泥湾的细粒沉积上,而淤泥则堆积在被侵蚀的指状沙咀上,结果,河口的沉积剖面出现下部是泛滥平原粉砂与海相淤泥,而上层为海相淤泥与贝壳屑砂的交互沉积[6]。自1855年黄河北归夺大清河入渤海后至今,在100~150年的期间内决口改道50多次,新河口淤积迅速,废弃河口遭受侵蚀后退,如此过程,其结果形成一个扇形的总面积6 000 km²的三角洲,范围北起套尔河口,南至小清河,西起宁海(37°36′N,118°24′E)以东的1855年海岸线,东达水深15 m,海岸东伸约12~35 km(图4)[12]。晚更新世以来,黄河搬运了黄土高原的泥沙,堆积发育了下游大平原及河口大三角洲,沧海桑田,填海造陆作用巨大。但是,自20世纪70年代以来,开发三角洲油田及80年代市镇建设,已人为地控制了尾闾河道自由变迁,不允许黄河向北迁徙,避免淹及油田和城镇。在全球气候变暖、水源减少、沿河蓄拦引水情况下,入海径流量减少80%,小于携沙入海的低值径流量240×10⁸ m³[13],自70年代以来,黄河尾闾段断流频频,1997年断流达226天,断距达706 km[13],波及下游地段,水断沙绝,虽经人工调水入海,但杯水车薪,难改根本局势。在这种情况下,清水河口门指状沙咀不仅发育减缓,而且在NE向强劲的浪流作用下,沙咀北部受蚀,南侧淤积,呈现向南(下风向)偏移的型式。废弃的老河口持续受蚀后退,盐水入侵亦加剧,黄河三角洲体的发育途径已发生转折性变化。

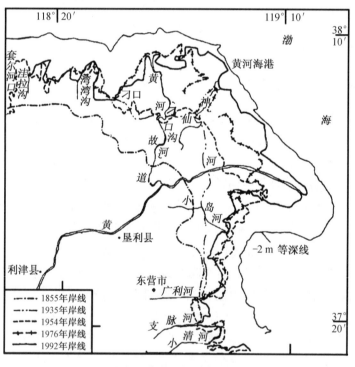

图4 黄河三角洲变迁图

(据文献[12]修改)

(四) 强径流与沿岸流结合的沉积模式

以长江口地区为例,长江是东海大陆架泥沙的主要来源。长江口地区为半日潮型,潮差0.17~4.60 m,属中等潮差区,平均涨潮流速0.95 m/s,最大涨潮流速2.08 m/s,平均落潮流速 1.11 m/s,最大落潮流速 2.5 m/s,落潮流大于涨潮流为该河口区特征。波浪受季风控制,夏、秋季盛行 SE 向风浪,冬、春季节为 NE 向风浪,台风期间多 NE 向大风。曾测到波高 2.3 m,波长 40 m,周期 5.6 s 的大浪。波浪影响主要在长江口外开阔海域。

长江年平均径流量 29 400 m³/s,年径流总量 9 250×10⁸ m³,年平均输沙量 486×10⁶ t,最大洪峰流量 92 600 m³/s,为最小流量 4260 m³/s 的 21.7 倍[2,14]。水沙的年内分配:洪季(5~10 月)水量占全年 71.7%,沙量占全年 87%[14]。长江河口段淡咸水交会,径流与潮流相互作用,在河口区形成一个高泥沙浓度的最大浑浊带。表层水的泥沙浓度变化于 0.1~0.7 kg/m³ 之间,底层含沙浓度高达 1~8 kg/m³,近底层形成饱含黏粒与粉砂、容重大于 1.04 g/cm³ 的浮泥层[14]。尤其在长江口拦门沙地区,在小潮汛与偏北向大风天后,浮泥层显著,反映出风浪在长江口浅滩区的掀沙影响。最大浑浊带沿长江口纵向延伸 25~46 km,在此范围内盐水楔活动频繁。高泥沙浓度带与洪季水沙量增多,与大潮汛时潮流对河床扰动增强密切相关。在这样的动力环境下,悬移质因絮凝作用而落淤加强,长江入海输沙量的 50% 在河口门堆积,发育拦门沙体系和水下三角洲[14]。长江径流长时期的水沙堆积作用,形成一系列边滩、沙洲与沙岛。随河道分岔、南岸边滩扩展、沙洲增大、北岸沙岛并陆、南槽加深、河道向南迁徙等过程,逐渐形成现代长江口沙岛、浅滩众多,出现三级分汊、四口分流的型式(见图 1e)[7]。长江口外,在 121°50′~122°20′E 间悬移质泥沙浓度大,表层泥沙浓度达 1 kg/m³,底层最大达 8 kg/m³。自 122°14′E 向东至水深 15 m 处堆积了现代水下三角洲,为黏土质粉砂沉积,顶部常现 2~9 cm 的浮泥层。²¹⁰Pb 定年该三角洲沉积速率大约为 5.3 m/ka[15],系全新世海面上升过程中近 7 000 年来形成的。调查发现,全新世水下三角洲叠置于晚更新世三角洲之上,后者范围约 40 000 km²,外边界可达水深 60 m 处,¹⁴C 测年表明其形成于 36 000 年前[5]。

长江口外大陆架有两股相向的海流活动:低温、高盐的沿岸流从北部的黄海经过 122°E 与 123°E 之间向南流;另一股是高温、高盐的黑潮主支在外陆架向东北方流过。两股海流之间有黑潮分支——台湾暖流经东海陆架向朝鲜海峡流去,其分支向西北流,影响本区(图 5)[7]。长江入海泥沙主要由沿岸流携运向南,成为沿岸带状沉积,并且是浙江港湾岸发育淤泥质潮滩的主要物质来源。同时,长江冲淡水亦携运了部分悬沙向 E 和 NE 方向输运,是黄海东部与朝鲜半岛西岸港湾岸的细颗粒泥沙源(图 6)[7]。河口内沙岛、浅滩堆积与分支河道、口外水下三角洲系

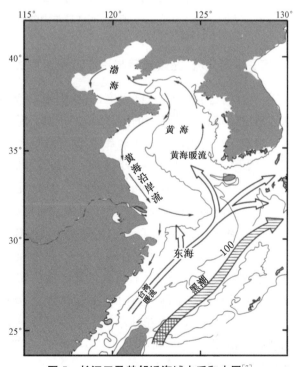

图 5　长江口及其邻近海域水系和水团[7]

与南部呈带状的泥沙堆积是强径流与沿岸流相互作用堆积的模式。

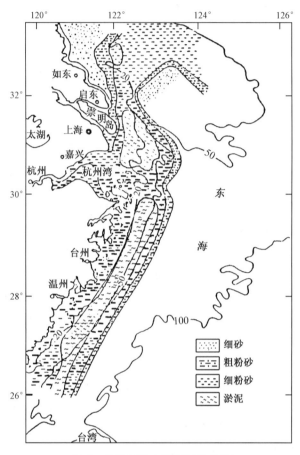

图6　长江口与内陆架泥沙分布

(五) 尾闾分流,充填港湾与沿岸流搬运模式

以华南珠江为例。区域降水多,植被繁茂,径流量大,含沙量较少(见表2),另一特点是珠江口位于基岩山地环绕的港湾内,该处为不规则半日潮,潮差0.86~1.63 m,属弱潮海区。发源于丘陵山地的珠江水系,汇集至隐蔽平坦的港湾中,形成诸多分支水流与多处会潮点,因而在河口区发育了指状沙咀、浅滩、沙洲、越流扇与潮流三角洲等,形成一种特殊的、介于河口湾与三角洲间的过渡型式。珠江泥沙充填着河口湾,堆积发育河口平原。来自珠江与外海的悬移质泥沙覆盖于上述砂质堆积上,形成泥帽[16,17]。与珠江相邻的陆架沉积可分为两个带(见图1f)[18],内陆架为河流输沙堆积带,宽度约35 km,外缘水深为20~30 m,从岸向海依次沉积着砂质粉砂、黏土质泥,与现代波浪分选堆积作用相适应。在沉积中含碳酸钙5%~10%,极少量重矿物(<0.55%)及大量有机质残体。外陆架沉积是分选好的细砂与中砂,含有2%的重矿和25%~50%贝壳屑以及大量有孔虫(在陆架坡折处的含量达90%),有机质小于1%,在细砂带中很少有有机物残留,外陆架系低海面时的砂质沉积,因未曾受到现代河流泥沙覆盖,呈残留砂出露于海底。内、外陆架沉积带之间有过渡沉积区[18]。

受海岸带季风沿岸流的影响,扩散到浅海的珠江悬移质泥沙被沿岸流携运终年自河口湾外向西南输移,成为邻近港湾的潮滩沉积物源(见图1f),夏季洪水期,悬浮的泥质可漂移到外陆架内侧[19]。

以上总观我国5条代表性河流,因区域基础地质、河流特性、河口及浅海环境与动力因素不同各有其独特的沉积模式。强潮动力区以发育潮流沙脊与潮流通道深槽为特征;波浪作用居优势的海岸带,则发育水下与沿岸沙坝,形成双重岸线,内侧发育潮滩,外侧多有越流扇沙体,而沙坝外侧出露低海面时的残留砂沉积;弱潮区河口以径流的入海射流堆积为特征,形成指状沙嘴与水下三角洲。但是,平原区河海交互作用亦有共同特征:下游与尾闾河道多易分流变迁,为平原发育提供泥沙沉积基础;径流与潮流在河口区双向作用,形成粗细粒径交叠沉积韵律层及口门淤泥质浅滩;口外海滨沉积均与沿岸流动力纵向输移泥沙作用有关;现代河流泥沙影响多限于水深20 m的内陆架区,外陆架普遍出露低海面残留砂;当代人类活动与水利工程改变着上述河流的自然发展规律,继而影响着河口海岸地区。

三、平原地貌发育探索

归纳分析现代5条河流河口河海交互作用过程与沉积效应,以探索三角洲平原发育的历史作用过程。黄河、长江贯通我国西东,是联系陆海的主动脉,黄河不仅冲刷搬运了黄土高原巨量泥沙入海,堆积了大平原,发育了淤泥质潮滩海岸与巨大三角洲,其影响遍及渤海与黄海,形成了与长江三角洲平原决然不同的景观:碱滩、泥沼、坦荡的黄土质原野,植被稀少。长江径流浩荡,淤填海湾,变沮洳为陆,併岛为山;扩展平原,发育了潮流迅猛的淤泥质粉砂海岸,形成富有特色的杭州湾涌潮、承袭古长江发育的洋口深水潮流通道、由沙脊组成面积达22 470 km² 的黄海辐射沙脊群,以及其相伴的动力、港口、土地与海洋资源,举世无双。限于篇幅,本文仅就河海相互作用与滦河平原的发育加以剖析。

滦河平原研究有基础,自20世纪80年代以来又深入研究,利用滦河三角洲最西侧曹妃甸沙坝与前方紧邻的25~30 m深槽,即自渤海湾北上的潮流通道(图7),建设20万~30万 t级码头泊位,填筑沙岛内侧向北伸展的潮滩浅海为港口堆场和陆域用地,实为利用海岸环境的天然优势,选建深水海港之范例;嗣后,又为港口发展,判断曹妃甸沙坝与深水航道稳定性以及对港区所在的滦河平原进行了深入研究①。

建港前的曹妃甸沙岛,系呈NE-SW向延伸,是与陆地海岸线相隔约19 km的岸外沙坝,亦名沙垒田岛。据1945年测量,沙岛东西向延伸,长1 500 m,南北宽750 m,高出海面约2 m,沙岛顶部有风成沙丘,高达4 m,其上筑有灯塔。据1959年测量,曹妃甸沙岛东西长1 500 m,南北宽度减小为450 m。沙岛组成物质为浅黄棕色石英、长石质细砂,顶部的沙丘为粉细砂,现代海滩出现中砂。细砂成分中含有普通角闪石、辉石、绿帘石、钛磁铁矿与锆石等重矿物以及贝壳屑,泥沙成分与滦河物质相同。曹妃甸的物质组成和排列方

---

① 王颖. 南京大学海岸与海岛开发教育部重点实验室"曹妃甸沙岛成因与深水航道稳定分析研究报告",2006。

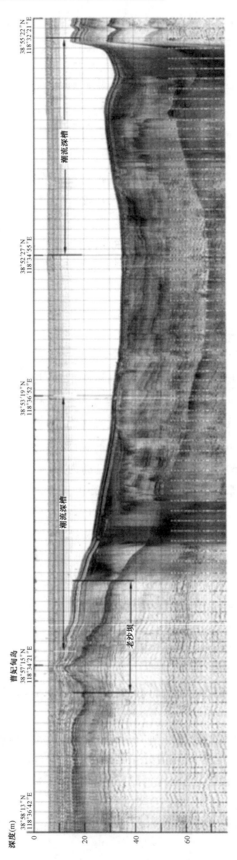

图 7　曹妃甸与相邻深槽地层剖面图

式与位于其东北方向的蛤坨、草木坨、腰坨等相同,是由古滦河自南堡与大庄河一带出口时的泥沙经波浪掀带以及浪流向岸推移而堆积成的海岸沙坝。地震剖面显示(见图7),组成曹妃甸岛的砂层厚度超过 40 m,为宽度 1.94 km 的大沙坝。全新世海面上升过程中,约 3 400a B. P. 时(表4),沙坝受蚀,如今仅余底部沙丘部分于海面上。曹妃甸沙坝南侧为渤海潮流主通道,其深槽宽度 6～7 km,水深 20 m,最深处超过 30 m。据1996年24号工程钻孔(38°55′N,118°30′E)揭示系由滦河沙源的河-海交互沉积,以海相潮滩相沉积为主,冲积层均含滦河泥沙(图8)。深水潮流通道紧临曹妃甸沙岛南侧分布,吸收了波能,减缓了海岸带波浪变形,一定程度上"保护"了沙岛少遭风浪侵袭。

表 4　曹妃甸-南堡-唐海区 ^{14}C 定年记录

样品编号		钻孔位置	层深/m	样品种类	^{14}C 测年 /aB. P.	校正年龄 /aB. P.	备注 (采样人、年代等)
孔号	测试号						
SK6孔		南堡盐场	14.7	海相层底部黏土	11 340±145		李元芳等[20],1982
海2孔		南堡咀东	22.5	海相层底部低盐泥	11 215±100		彭贵等[21],1980
SK1孔		南堡盐场	9.0	海相层下部黏土	9 945±150		李元芳等[20],1982
		南堡盐场	14.0	海相层底部黏土	8 785±150		李元芳等[20],1982
柏2孔		唐海柏各庄	24.4～24.8	有机质泥	9 220±250	10382 (10 732～9 920)	彭贵等[21],1980
柏3孔		唐海柏各庄	15.0	有机质泥 (黏砂土)	8 620±250	9637 (10 145～9 295)	林绍孟等[22],1979
柏3孔	CG335-2	唐海柏各庄	32.0	贝壳		16 275±1 000	林绍孟等[22],1979
CFD96-24		曹妃甸南侧潮流通道 (38°55′N, 118°30′E)	36.0	黏土	3 381±136		南京大学*,1997

＊南京大学海岸与海岛开发国家试点实验室,1997,曹妃甸港址码头工程预可行性研究报告——京唐港曹妃甸港区海洋动力地貌调查报告(未刊)。

在曹妃甸沙坝的东北方,沿滦河冲积平原各河口外围尚分布着一系列 NE-SW 向的海岸沙坝:大清河口的石臼坨、月坨,打网岗;湖林口与老米沟之间的灯笼铺、湖林口沙岗;稻子沟口的蛇岗;以及环绕现代滦河口的一系列沙坝。

据现代三角洲分析,滦河泥沙向外海扩散及沿岸运移过程中,15％的泥沙用于建造三角洲平原,65％用于陆上三角洲向海扩张,20％用于建设水下三角洲[10]。

2003—2006 年南京大学从事曹妃甸深水港建设项目研究认为:古滦河发育了两个大型三角洲体(图9):较新的三角洲体呈扇形,大体上以滦州以南的马城为顶点,西起西河口(溯河口),东至现代滦河三角洲,三角洲中心位置在大清河与王滩港之间,打网岗、石臼

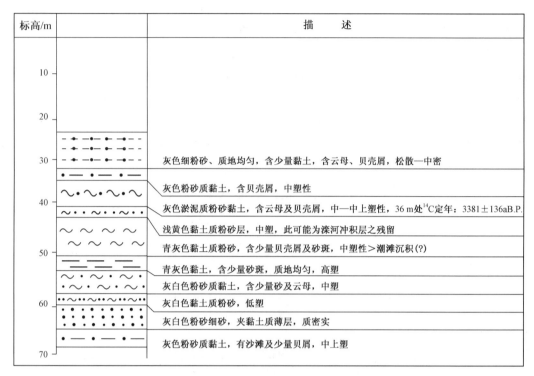

标高/m	描 述
10	
20	
30	灰色细粉砂、质地均匀，含少量黏土，含云母、贝壳屑，松散—中密
	灰色粉砂质黏土，含贝壳屑，中塑性
40	灰色淤泥质粉砂黏土，含云母及贝壳屑，中—中上塑性，36 m处^{14}C定年：3381±136aB.P.
	浅黄色黏土质粉砂层，中塑，此可能为滦河冲积层之残留
50	青灰色黏土质粉砂，含少量贝壳屑及砂斑，中塑性>潮滩沉积(?)
	青灰色黏土，含少量砂斑，质地均匀，高塑
	灰白色粉砂质黏土，含少量砂及云母，中塑
60	灰白色黏土质粉砂，低塑
	灰白色粉砂细砂，夹黏土质薄层，质密实
70	灰色粉砂质黏土，有沙滩及少量贝屑，中上塑

图 8 曹妃甸沙坝南侧潮流深槽钻孔剖面

坨、月坨为河口区的海岸沙坝；西部另一个老三角洲体界于洇河口与西河口之间，可能滦河曾于高尚堡、南堡、沙河口及洇河出口。曹妃甸、腰坨、草坨、蛤坨应形成于老三角洲发育时。南堡以北陆地平原上仍有一系列断流的干涸河道及小型河流(沙河、小戟门河、咀东河(双龙河)、小青龙河等的分流)入海。这些河流已非天然的河流水系，一些废弃河道被改造利用为输、排水渠道或平原水库。向三角洲平原北部的内陆方向追溯，至迁西与旧城之间观察到一些有意义的地质现象如下。

(1)滦河在唐山境内自北向南流，但在迁西与旧城之间，河流突呈直角拐弯向东流，至忍字口与尹庄间再向南流，经黑沿子河口，咀东河河口及西河(溯河)口入海；嗣后，又曾改道自尹庄向东至罗家屯再向南经大庄河、经大清河入海；最后又东移至湖林口、老米沟、浪窝口及现代滦河口入海。总之，自迁西突然转折向东，至罗家屯才南流入海。从卫星影像中判断，滦河流路的直角转向，是地震断层抬升阻挡而成。

(2)在滦河直角转弯向东流的白龙山村(海拔106 m；位于40°09.043′N，118°20.236′E与40°09.01′N，118°20.236′E之间)，仍保留着一段具有天然的河曲形式被废弃的老河道，现成为离地面高约12~14 m的阶地(海拔118~120 m)。此残留的废河道呈NE30° SW走向，代表着被袭夺前的滦河流向。阶地两侧背依黄褐色风化砂岩的基岩陡坎，陡坎两级，顶部一层的海拔高度140~142 m，比阶地面高出20~22 m。底下的一层陡坎高出阶地地面2.4 m，基岩陡坎延伸方向为NE20°，系原河道之基岩岸。

(3)阶地出露二元相的冲积层，从上至下为：顶部30 cm厚的极细砂与粗粉砂层，是河漫滩相沉积；2 m厚砂砾层，系河床相沉积；90 cm厚之砂砾层与黏土块的交叉沉积，为

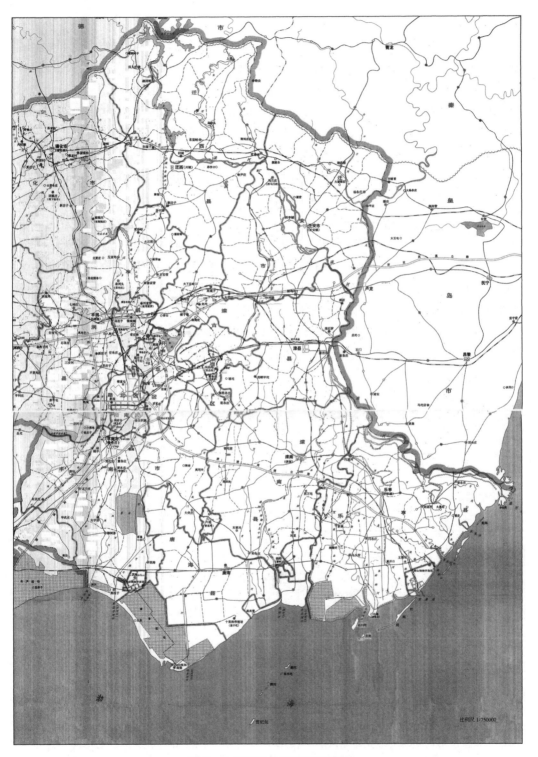

图9 古滦河三角洲体(附彩图)

侵蚀堆积层;最下部为砂层及砂砾层透镜体。此阶地剖面,证实目前已高出现代地面 12 m 的沉积层是改道前的古滦河沉积层,具有砂砾质底层与河漫滩相细砂黏土层之二元相沉积结构。

根据基岩陡坎古河岸的现代高程,可知,古滦河在白龙村处曾间歇地抬升 10 m 及 12 m,故而原基岩河岸高出现代地面 34 m,而河床成为 12 m 高阶地。判断是突发的断裂抬升事件,使古河道抬升成小丘,阻挡河流南下,而迫使滦河改道向东。两级基岩陡坎岸,反映出突发型的断裂抬升亦具有间歇之特点。而且在第一次抬升约 9 m 后,河流曾力图按原流向,即自 NE 向 SW 流,但第二次抬升(约 3 m)彻底结束了河流向南之流路,而直转向东。

判辨古滦河在迁西改道折向东流的时期,应是在曹妃甸沙坝发育后期。因为迁西以南约 119.6 km 宽的平原及以南堡为中心出口的三角洲体,应是改道前的古滦河输沙堆积形成,是滦河长期稳定所输运的泥沙堆积而成,堆积时代据曹妃甸沙岛及相关钻孔地层的 ^{14}C 定年(见图 8 和表 4)可能为更新世中期与晚期,即滦河向东流改道前。滦河不断地自西南向东北方迁移,估计为地震构造抬升的后续影响,地面因掀升而自西向东倾斜,河流顺势沿低地而行。迁西市及罗家屯以南的广袤平原,虽经开辟为农田,但是根据一些微地貌组合的残留遗迹与沉积组成的特性,仍可追溯平原的成因。

(1)断续的河道与干涸洼地,尚可辨识出自迁西向南,自忍字口与尹庄间向南,以及自罗家屯向南,古、今滦河三股主要的河道。在滦县以南,河道分流增多,呈现出两个三角洲体的水流分布势态。

(2)平原地势自北向南递减,组成物质的粒径亦发生变化。从北部的砂砾层与中砂、细砂层,向南逐渐变为细砂、粗粉砂以及极细砂,粉砂与粉尘沉积,至南堡已为黏土质淤泥沉积。但沉积物组成以石英、长石及深色矿物为主体,与滦河泥沙相同。

(3)迁安市以南,平原上残留着 9 列东西向排列的沙坨(表 5),判辨是古海岸沙坝(沙丘)的遗迹。

表 5　滦河平原沙坨遗迹

沙坨系列	名称	位置	排列方向	高度/m	物质组成	备注
I	野鸡坨	迁安南界(39°52.158′N,118°40.95′E)	NE-SW 近 EW	5～2 *	黄棕色粗粉砂质细砂,含淡水小螺、牡蛎碎片	低山丘陵环绕的平原中,可能为滦河流出山地时形成的海湾沙坝
	赤峰堡	野鸡坨以西	EW	3±	硬结砂层、黄棕色(泛红)、极细砂与粉砂	经湿热气候淋溶风化,半胶结,时代可能为中更新世(Q_2)
	龙坨、麻湾坨庄坨	崇家峪以南杨黄岭南	EW EW	3 2～3	极细砂、粉砂 极细砂、粉砂	古海湾沙坝 古海湾湾口沙坝

* 经过人工采砂挖掘后的高度。

（续表）

沙坨系列	名称	位置	排列方向	高度/m	物质组成	备注
II	黄坨→大庄坨→塔坨→西晒甲坨→东尖坨→晒家坨	古冶区滦县秦皇岛市	EW	2～3	极细砂、粉砂层	可能古海湾湾口沙坝
III	大茨榆坨、前龙坨、马坨店	滦县西南界秦皇岛市	EW	2～3	黄色细砂层、极细砂、粉砂、含贝壳屑	海岸沙坝*
IV	青坨营	滦南县西部,茨榆坨以南6.9 km处	EW	2～3	黄色细砂、粉砂层,含贝壳	海岸沙坝
V	辉坨、爽坨于家1坨	丰南市东南部,青坨营SW7 km处秦皇岛市与乐亭县之间	EW	2～3	黄色细砂、粉砂层,含贝壳	海岸沙坝
VI	孙家坨,西万坨,青坨,茨榆坨,胡家坨,翠坨,大黑坨	滦南县乐亭县	EW	2～3	黄色细砂、粉砂层,含贝壳	海岸沙坝(范围广)
VII	芝麻坨→东黄坨、西玉坨→坨里→东青坨→阁楼坨→溪家坨、前王坨子	滦南县西南部滦南县东南部乐亭县	EW	2～3	黄色极细砂,粉砂层,贝壳屑	大型海岸沙坝系列
VIII	高尚堡尖坨子	唐海县,南堡的NE方向南堡西北方	EW	3 3	贝壳沙堤,地表为粉砂淤泥层,底层有细砂	海侵型贝壳堤堆积
IX	南堡	咀东河口的西侧	EW	3	贝壳沙堤残留,地表为粉砂淤泥层,底层有细砂	海侵型贝壳堤堆积

表5中的第1列到第7列沙坨,是由海岸沙坝并陆后的残留地貌。其中第1列与第2列沙坨,是滦河流出山地至低缓丘陵区后,所发育的离岸坝与海湾湾口沙坝,当时泥沙充分,堆积地貌高大(野鸡坨与赤峰堡),其沉积层颜色较深,为黄棕色,估计经过湿热的气候影响,可能是更新世中期的堆积。第3列至第7列均为由滦河泥沙供给、经风浪与岸流作用形成的海岸沙坝。沙坝于外围环绕着原始海岸,拦阻了滦河泥沙,逐渐使浅海、潟湖填充而发育了冲积-海积平原,一列列的沙坝成为陆地上的沙坨(图10a)。平原上河渠多,粉砂质细砂与黏土沉积,经充足淡水灌溉多为绿色田野。至第7列沙坨——东黄坨、坨里、阁楼坨向南,平原景色截然不同,是缺乏淡水,缺乏林、草的粉砂和淤泥质盐土平原。第8和9列沙坨——高尚堡、南堡和尖坨子却是形成于盐土平原上的贝壳质海岸沙堤(图10b)。从贝壳堤中的砂质成分石英、长石、黑色矿物等分析对比,仍为来自滦河的粉砂与

---

* 曾被译为"离岸坎",原文为 offshore barrier。

极细砂。但是,第 8 和 9 列两处沙坨发育时,海岸坡度平缓,以潮流动力为主,发育了潮滩,嗣后,由于泥沙补给不足,潮滩遭受激浪冲刷,掏蚀潮滩沉积,于陆上岸线堆积形成贝壳堤[8]。据此分析,在第 7 列沙坝形成之后,滦河改道向东,第 8 和 9 列贝壳沙堤发育时,滦河泥沙的直接补给已中断,故而海岸发生冲刷形成贝壳堤。现代南堡、高尚堡等处外围为粉砂淤泥质潮滩海岸。

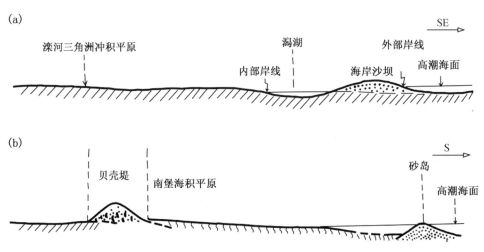

图 10 具沙坝环绕的滦河三角洲冲积平原海岸地貌综合图(a)和南堡海积平原海岸地貌图(b)[8]

总上述,① 滦河平原地貌始终以沙坝环绕的三角洲平原为特色,反映出:滦河的泥沙是构成平原的主要物源;海岸带动力以偏东向(NE,E 和 SE)的风浪与浪流为主,辅以潮流与沿岸流的输移泥沙作用;海岸泥沙运动以横向搬运与堆积为主,辅以次生的纵向泥沙流活动。沙坝的规模反映出古滦河的泥沙量远远大于现代滦河(1915 年改道后的)泥沙量,可能与晚更新世与全新世早期时较干寒的气候环境、岩层受风化剥蚀而产沙量大以及嗣后的海面上升过程中较强的海岸动力搬运与堆积作用有关。② 曹妃甸(原沙垒田岛)物质组成为棕黄色砂层,以细砂为主,海滩上分布着中砂、细砂及贝壳屑,而沙岛顶部风成沙丘处为极细砂与粉砂层,砂质沉积源于滦河入海之泥沙。据曹妃甸沙岛分布的位置,估计是滦河自迁西改道向东流之前与初次经忍字口与尹庄间向南流,经过迁安市以南的第 2 列塔坨、第 3 列大茨榆坨、第 4 列青坨营、第 7 列东黄坨等海岸淤进阶段,至高尚堡一带出口时形成。曹妃甸沙岛应是在岸外沙坝基础上由沙丘加积后发育的。其沉积质地与高尚堡贝壳堤不同,高尚堡、南堡贝壳堤时代晚于曹妃甸,是全新世海侵时增高的海面与海岸动力冲刷原砂质平原海岸及滩涂,由激浪与激浪流堆积潮滩中的贝壳与泥沙所形成的贝壳堤,而曹妃甸沙岛系形成于古滦河输沙量巨大、海平面较低时的砂质海岸带,由动力较强的波浪与激浪流所堆积的海岸沙坝。对比冀东海岸沙丘、沙滩广泛发育的时期,亦反映曹妃甸形成于晚更新世末低海面后期与全新世早期海侵时。而野鸡坨、龙坨、麻湾坨以及以南的冲积-海积平原主要是中更新世至晚更新世形成。南堡盐场和大清河盐场一带潮滩是全新世中期以来海侵改造基底平原海岸的产物。

四、结语

地质构造格局制导着中国主要河流从西部雪山、高原汇流入东部海域的基本流势,源远流长汇集成大河与大江。流域是经过长期发育形成的具有成因关联的有机体系,人为地改变任何一段江河环境的自然特性,必会形成后继的连续性变化以致遭灾。如:疏干沼泽湿地——减少水源涵养;梯级筑坝拦水——改变河道坡降与岸坡稳定均衡性,招致滑坡与回淤;多段分流引水——减少下游径流量与行水作用,招致断流,河口盐水入侵加剧与三角洲海岸蚀退……需要总结既往的教训,科学、系统地对待江河。

河流是贯穿东西联系海、陆的主动脉,河海交互作用过程与效应具有区域性差异,但是经历漫长的新生代地质历史阶段,是江、河冲刷输运的泥沙与海水动力的交互作用,“移山倒海”地堆积了东部平原,孕育了中华民族生息的大地。宜从长远利益审视江海的自然作用,研究其特性,因势利导地善待江、河、海洋,保护江、河、湖、海。

文章总结了5种河海交互作用与沉积结构型式,并拟将今论古,启迪思路研究区域平原的发育过程。虽然古今地质地貌过程不尽等同,本文总结的滦河平原的地貌结构发育:滦河的出口,泥沙量各时期有不同,并且是较大的变化。但是,海域的动力环境因素,尤其是季风波浪的因素恒定,形成了规模不同的沙坝环绕的三角洲岸线,为我们提供了一个海岸环境较类似的海岸发育范例。适用于发源于近海山地、多沙性的中、小型河流与优势的季风波浪动力、相互作用的海岸带,如:广东韩江三角洲有类似的地貌结构。不同类型的海陆交互作用地域有待进一步深入研究。

河海交互作用是中国——位居东亚大陆与太平洋边缘海之交,海陆交互作用带宽广,人类聚居与开发活动悠久——具有特色的重要研究课题,在21世纪的科学规划与工程建设中,宜将对陆地、对海洋及两者之间的海陆过渡带予以同样的重视!

参考文献

[1] Milliman J D, Syvitski J P M. 1992. Geomorphic/tectonic Control of Sediment discharge to the Ocean: The importance of small mountainous rivers. *Journal of Geology*, 100: 525 - 554.

[2] WangYing, Ren Mei'e, Syvitski J. 1998. Sediment transport and terrigenous fluxes. In: Brink K H, Robinson A Reds. The Sea, Vol. 10: The Global Coastal Ocean Processes and Methods. New York, Toronto: John Wiley & Sons. 253 - 292.

[3] 程天文,赵楚年. 1984. 我国沿岸入海河川径流量与输沙量的估算. 地理学报,39(4):418 - 427.
Cheng Tianwen, Zhao Chunian. 1984. The estimation of discharge and loads from the rivers flowing into littoral seas of China. *Acta Geographica Sinica*, 39(4): 418 - 427.

[4] Latrubesse E M, Stevaux J C, Sinha R. 2005. Tropical rivers. *Geomorphology*, 70: 187 - 206.

[5] 任美锷,曾成开. 1980. 论现实主义原则在海洋地质学中的应用——以中国海岸带及近海大陆架为例. 海洋学报,2(2):94 - 105.
Ren Mei'e, Zeng Chengkai. 1980. Late Quaternary continental shelf of East China. *Acta Oceanologica Sinica*, 2(2): 94 - 105.

[6] WangYing, Ren Mei'e, Zhu Dakui. 1986. Sediment supply to the continental shelf by the major rivers of China. *Journal of the Geological Society*, 143(3): 935 – 944.

[7] 恽才兴著.2004.长江河口近期演变基本规律.北京:海洋出版社.2,249.

Yun Caixing. 2004. Recent Evolution of the Changjiang River Estuary. Beijing: China Ocean Press. 2, 249.

[8] 王颖.1963.渤海湾北部海岸动力地貌.海洋文集,(3):54 - 56.

Wang Ying. 1963. The dynamic geomorphology of North Bohai Bay. *Proceedings of Marine Research*, (3): 54 - 56.

[9] 高善明,李元芳,安凤桐等.1980.滦河三角洲滨岸沙体的形成和海岸线变迁.海洋学报,2(4): 102 - 114.

Gao Shanming, Li Yuanfang, An Fengtong *et al*. 1980. The formation of sand bars on the Luanhe River delta and the change of the coastline. *Acta Oceanologica Sinica*, 2(4): 102 - 114.

[10] 钱春林.1994.引滦工程对滦河三角洲的影响.地理学报,49(2):158 - 166.

Qian Chunlin. 1994. Effects of the water conservancy projects in the Luanhe River Basinon Luanhe River Delta, Hebei Province. *Acta Geographica Sinica*, 49(2): 158 - 166.

[11] 王颖,朱大奎.1987.海岸沙丘成因的讨论.中国沙漠,7(3):29 - 40.

WangYing, Zhu Dakui. 1987. An approach on the formation causes of coastal sand dunes. *Journal of Desert Research*, 7(3): 29 - 40.

[12] 黄海军,季凡,庞家珍等.2005.黄河三角洲与渤黄海海陆相互作用研究.北京:科学出版社.29 - 81.

Huang Haijun, Ji Fan, Pang Jiazhen *et al*. 2005. Study on the Yellow Sea Delta and the River-sea Interaction in Bohai and Yellow Sea. Beijing: Science Press. 29 - 81.

[13] 王颖,张永战.1998.人类活动与黄河断流及海岸环境影响.南京大学学报(自然科学版),34(3): 257 - 271.

Wang Ying, Zhang Yongzhan. 1998. Human activities, break-off water discharge of the Yellow River and the impacts on coastal environment. *Journal of Nanjing University* (*Natural Sciences*), 34(3): 257 - 271.

[14] 沈焕庭,贺松林,潘定安等.1992.长江口最大浑浊带研究.地理学报,47(5):472 - 479.

Shen Huanting, He Songlin, PanDing'an *et al*. 1992. A study of turbidity maximum in the Changjiang Estuary. *Acta Geographica Sinica*, 47(5): 472 - 479.

[15] You Kunyuan, Sui Liangren, Qian Jiangchu. 1983. Modern sedimentation rate in the vicinity of Changjiang estuary and adjacent continental shelf. In: Proceedings of International Symposium on Sedimentation on the Continental Shelf, with Special Reference to the East China Sea. Beijing: China Ocean Press. 590 - 605.

[16] 任美锷.1964.珠江口与邻近地区地貌特点(俄文).南京大学学报(自然科学版),(1):80 - 90.

RenMei'e. 1964. The geomorphological characteristics of the Pearl River estuary and adjacent area (in Russian). *Journal of Nanjing University* (*Natural Sciences*), (1): 80 - 90.

[17] 李春初.1983.珠江口磨刀门的动力与沉积.热带地理,(1):27 - 34.

Li Chunchu. 1983. Dynamics and sedimentations of the Modaomen estuary. *Tropical Geography*, (1): 27 - 34.

[18] 中国科学院南海海洋研究所海洋地质研究室沉积组.1980.南海北部大陆架表层沉积物特征.南海海洋科学集刊,(1):35 - 50.

Sedimental Group of Marine Geology Department，South China Sea Institute of Oceanography，Academia Sinica. 1980. Characteristics of the surface sediments on the north shelf of the South China Sea. *Nanhai Studia Marina Sinica*，(1)：35 - 50.

[19] 赵焕庭. 1983. 珠江三角洲的水文特征. 热带海洋，2(2)：108 - 117.

Zhao Huanting. 1983. Hydrological characteristics of the Zhujiang(Pearl River) delta. *Tropic Oceanology*，2(2)：108 - 117.

[20] 李元芳，高善明，安凤桐. 1982. 滦河三角洲地区第四纪海相地层及其古地理意义的初步研究. 海洋与湖沼，13(5)：433 - 439.

Li Yuanfang, Gao Shanming, An Fengtong. 1982. A preliminary study of the Quaternary marine strata and its paleogeographic significance in the Luanhe delta region. *Oceanologia et LimnologiaSinica*，13(5)：433 - 439.

[21] 彭贵，张景文，焦文强等. 1980. 渤海湾沿岸晚第四纪地层 ^{14}C 年代学研究. 地震地质，2(2)：71 - 78.

Peng Gui, Zhang Jingwen, JiangWenqiang *et al*. 1980. Chronology of Late Quaternary deposits along the coast of the Bohai Sea. *Seismology and Geology*，2(2)：71 - 78.

[22] 林绍孟，祝一志. 1979. 唐山柏 3 孔的古气候及其时代初步探讨. 见：中国第四纪研究委员会第三届学术会议论文摘要汇编. 266.

Lin Shaomeng, Zhu Yizhi. 1979. Preliminary study of the paleo-climate and chronology of Bo3 core in Tangshan. In：Proceedings of the Third Chinese Quaternary Research Conference. 266.

宝应钻孔沉积物的微量元素地球化学特征及沉积环境探讨*

一、前言

沉积物微量元素特征对指示沉积盆地演化历史、沉积环境及沉积物的物质来源具有十分重要的示踪作用。研究表明,海岸带沉积盆地中沉积物的地球化学记录提供了过去环境变化的资料,沉积物的元素组成可以反映不同沉积环境的信息[1~5]。从元素地球化学方面探讨沉积物所记录的气候环境信息,已越来越多地用于全球气候变化研究[6~12]。

稀土元素(REE)作为特殊的元素在地球化学研究中占有相当重要的地位,稀土元素地球化学特征已经广泛应用于研究沉积物物质来源[12,13]、构造背景[14,15]、地质事件、界线剖面、地壳演化的关系[16]、沉积矿床的成矿作用及成因[17]、硅质岩成因[18]等方面。稀土元素含量及其特征参数随深度的变化不仅反映出各沉积相组成特征的变化,而且也常用来讨论沉积环境及古气候的演变过程[19~23],如在高寒半干旱地区湖相沉积物中,REE 的纵向变化可作为古气候波动的代用指标[24]。稀土元素总量和分布模式的差异可以确定地质体类型,它们之间的分异特征揭示地质体形成过程中元素的迁移、富集和环境变化。在海洋沉积物的研究中,稀土元素可以反映各类型沉积物的总体特征和差异,以及历史时期水体的地球化学特征和演化、大洋黏土的来源和火山活动的地球化学记录等。利用稀土元素来研究古海洋的氧化-还原性、海平面波动及沉积环境。近年来取得的研究成果表明,环境对沉积物中 REE 的影响也十分明显和重要[25~28]。

苏北平原的形成、发展与演化已引起人们越来越大的关注,有关苏北平原古地理环境演变的研究日益增多[2~9],取得了一系列有意义的认识。苏北平原宝应钻孔(BY1)2.58 Ma以来不同沉积相沉积物的主量元素地球化学特征分析已表明,本区是河-海交互作用下具有海相沉积环境与陆相沉积环境交互的元素分布特点[29]。本文试图通过对宝应钻孔沉积物微量元素和稀土元素地球化学的研究,来进一步印证该区物质来源的多源性,探讨苏北盆地的沉积环境变迁。

二、区域地质概况及宝应钻孔沉积相段划分

宝应位于苏北盆地中西部低洼处(33°14′21″N,119°22′41″E),东距现代海岸 120 km,

* 杨竞红,王颖,张振克,J. P. Guilbault,毛龙江,魏灵,郭伟,李书恒,徐军,季小梅:《第四纪研究》,2007 年第 27 卷第 5 期,第 735－749 页。

国家自然科学基金项目(批准号:40271004)资助。

处于华北陆块和扬子陆块拼贴的接壤部位。盆地中断裂构造十分发育,深大断裂控制了第三纪晚期以后的沉积作用。盆地中广泛分布着巨厚的中、新生代地层,晚白垩世以来受太平洋板块和印度板块的影响,形成了以箕状断陷为特点的复合型沉积盆地[30]。埋藏于盆地的玄武岩呈大面积广泛分布,在盆地中部、西部出露有第三纪钙碱性至碱性玄武岩。新生代玄武岩类常以互层形式赋存于下第三纪沉积地层中,以碱性系列为主。第四纪以来频繁的气候波动造成海平面变化和海岸线变迁,对苏北盆地的沉积环境有着深刻的影响,海陆交互作用十分明显[31~36],不仅发育了陆地泛滥平原的河湖相沉积,而且曾多次为海水所淹覆,发育了海相与滨海潮滩相砂和粉砂层,在地貌上经河-海交互堆积逐渐由盆地发展成平原。第四纪时该区主要有东台-海安和吕四-栟茶两个沉积区域,沉积厚度为100~340 m。

宝应(BY1)钻孔进深 144.98 m,实取沉积柱为 96.81 m,剖面揭示了约 2.58 Ma 的沉积历史[29,37]。岩芯古地磁年代测定[29,37]表明,95.22 m 深处为 Gauss 正极性期的顶界,B/M 界线位于 39.16 m 深度,45.63~53.83 m 处为 Jaramillo 正极性事件,76~82 m 处为 Olduvai 正极性事件,85~87 m 处为 Reunion 正极性事件。有孔虫鉴定结果证实钻孔中有多处层位出现一定量的不同种属的海相有孔虫组合[38~40],进一步印证了苏北平原曾经历多期海陆交互作用。

根据钻孔沉积物的颜色、组成和结构特点,将 BY1 钻孔沉积剖面自下而上划分为 10个沉积相段(表1)。2.58 Ma 以前(95.22~96.70 m)为平原基底,以泛红棕色的杂色硬质黏土层为主。基底上部 95.22 m 处为 M/G 界面,是 2.58 Ma 前的风化剥蚀面。岩相及岩性特征表明,2.58 Ma 以来本区沉积环境发生过较大变化,沉积物的颜色、组成、结构和微体古生物化石均表现出河湖相、潮滩相、淡水湖相与滨浅海相交替演变的环境。第四纪以来主要经历了 4 次较大海侵,即 1.95 Ma 左右(83.95~79.66 m),1 Ma 左右(65.4~58.0 m),0.78 Ma~3 MaB. P.(39.16~14.96 m)及全新世海侵(5.13~1.00 m),而以0.78 Ma~3 MaB. P. 间的海侵为最大,且晚期较早期规模大,与苏北平原上更新世(Q₃)发育巨厚海相层相一致[29]。杨怀仁等[41]对黄河三角洲地区的第四纪海陆变迁的研究也表明,早第四纪海进较弱,晚第四纪海进影响在范围和频率上有明显加强,即平原形成时曾遭受多次海水淹覆,海陆交互作用明显。地貌与沉积相综合研究证明[29,37],苏北平原内侧经历了浅海→潟湖→湖泊→潮滩→洼地,最后为河流冲积物覆盖成陆的过程,苏北平原整体上是在第四纪经河-海交互堆积形成。

表1　宝应钻孔沉积相段划分

沉积相段		颜色	岩性	深度/m
S10	平原湖泊低洼地	深黄棕,黄棕	黏土质粉砂,粉砂	0.76~0
	浅海与海陆过渡相	橄榄灰,中深灰,中黄棕	粉砂,黏土质粉砂,砂质粉砂	8.70~0.76
S9	陆相湖泊洼地	深黄棕,浅橄榄棕	黏土质粉砂,黏土,硬黏土	14.96~8.70
S8	潮滩浅海相	橄榄黑,橄榄灰,棕灰	砂质粉砂,粉砂,黏土质粉砂	19.16~14.96

（续表）

沉积相段		颜色	岩性	深度/m
S7	浅海相	橄榄黑，橄榄灰，橄榄棕	黏土质粉砂，粉砂，极细砂，粉砂质黏土	39.16～19.16
S6	陆相泛滥平原河湖相	中黄棕，灰黄棕，中棕	黏土，粉砂质黏土，含细砾、Ca质及铁锰结核	58.00～39.16
S5	浅海相	浅橄榄棕与浅橄榄灰互层	粗砂，砂质粉砂，黏土质粉砂	65.40～58.00
S4	陆相河湖相	中黄棕，绿灰，深黄桔	分选差的黏土-砂-粉砂，夹碎石	75.42～65.40
S3	泛滥平原河水沉积	暗棕，中黄棕，棕黑	砂质粉砂，粉砂质砂，含砾石	79.66～75.42
S2	海侵相	中黄棕，浅黄棕夹黄灰	粉砂质砂，黏土质砂质粉砂	83.95～79.66
S1	半干燥陆相河湖相	中棕色为主	黏土，砂质粉砂，黏土质粉砂	95.22～83.95
基底风化剥蚀面		暗棕色为主	含角砾岩的黏土质粉砂层	96.70～95.22

三、样品及分析方法

BY1钻孔各层位选取有代表性的样品共139个进行了微量元素、稀土元素含量分析；样品化学处理与测试均在南京大学成矿作用研究国家重点实验室完成。

样品自然风干后，用玛瑙研钵磨细至200目，在恒温箱中60℃下干燥24小时。称取50 mg粉末样品，在100级化学超净实验室利用HF-HNO$_3$混合酸和Teflon封闭反应罐进行溶样，用10 ppb的Rh作内标，在HR-ICP-MS（Finnigan Element Ⅱ）质谱仪上进行测定。分析中采用空白样及国家标准沉积岩标样SDO-1进行全程监控，各元素标准偏差（RSD）小于5％，误差小于±10％。

四、微量元素分析结果与讨论

（一）微量元素变化特征

宝应钻孔沉积物中微量元素变化范围和平均质量分数（μg/g）见表2。

用北美页岩NASC[42]进行标准化投影了17个微量元素（图1），包括大离子亲石元素LILE(Sr,K,Rb,Ba)、高场强元素HFS(Th,Ta,Nb,La,Ce,Zr,Hf,Sm,Ti,Sc,Yb)和过渡元素(Cr,Co)。

由图1a可以看出，宝应钻孔沉积物微量元素变化基本与NASC平均成分[42]接近，标准化值大都变化在0.5～1.5之间；不活泼元素Th有着极高的正异常，呈突出的峰型，但与平均上部地壳[43]Th的变化趋势一致；Nb表现出强烈的负异常，也是大陆地壳的特征[44]。相对于平均上部地壳变化曲线，研究区具有较低的Sr,K,Rb,Ta和Nb，而相对富

表2　宝应钻孔沉积物中微量元素变化范围和平均质量分数(μg/g)

沉积相	沉积相段	深度/m	样品个数	Li	Be	Sc	V	Cr	Co	Ni	Cu	Zn	Ca	Rb	Sr	Y	Zr	Nb	Mo	Cd	Sn	Cs	Ba	Hf	Ta	W	Pb	Bi	Th	U
全钻孔		96.7~0	Average(139)	25.5	1.62	10.1	75.8	79.9	12.8	35.1	21.1	68.8	14.9	98.7	187.8	26.2	241.3	10.9	2.23	0.19	2.55	5.70	727	6.20	1.02	6.76	22.04	0.26	10.34	2.81
			Max	48.2	6.29	19.8	440.5	416.9	73.5	230.6	80.5	827.9	21.9	144.2	353.4	222.8	718.1	17.0	37.37	1.41	4.75	10.91	15774	17.93	1.68	159.05	102.94	0.51	16.28	19.62
			Min	2.9	0.01	0.7	9.6	7.6	2.2	0.1	1.5	12.1	4.2	37.5	69.6	3.6	16.1	1.9	0.25	0.02	0.11	0.51	193	0.49	0.16	0.71	8.17	0.00	1.47	0.31
浅海相与海陆过渡相	S10	8.70~0	Average(28)	33.5	1.84	12.2	85.7	78.8	12.8	35.0	25.1	70.1	16.3	103.2	145.1	27.4	312.6	13.2	0.94	0.16	3.36	7.31	454	5.21	2.17	3.84	21.68	7.02	8.55	6.02
			Max	45.8	2.40	14.8	107.8	103.3	22.2	48.6	35.3	94.0	19.6	134.9	227.9	33.8	718.1	14.7	2.37	0.32	4.75	9.66	493	9.81	6.71	6.45	159.05	26.58	14.79	14.38
			Min	17.6	1.12	8.5	56.9	64.6	6.8	19.1	12.4	46.0	11.5	81.2	105.8	24.0	158.4	10.9	0.43	0.09	2.26	3.26	408	0.41	1.00	1.00	4.56	0.23	0.16	2.80
湖泊洼地沉积	S9	14.72~8.70	Average(10)	34.9	2.08	12.8	91.6	82.4	14.0	40.9	28.9	67.7	17.0	109.0	125.1	29.3	285.3	12.8	0.97	0.15	3.34	7.56	457	7.22	1.22	6.80	23.17	0.33	12.25	3.23
			Max	48.2	2.74	15.6	120.0	111.0	18.3	74.5	39.9	87.9	20.9	132.3	153.0	39.7	400.5	15.7	3.38	0.18	4.17	10.16	676	9.69	1.51	26.79	26.47	0.46	16.28	4.53
			Min	16.7	1.02	6.6	50.5	44.6	7.3	23.4	14.9	30.6	8.5	49.2	104.4	22.7	112.6	6.2	0.38	0.13	1.63	3.56	211	3.11	0.63	2.63	15.78	0.17	5.93	2.00
潮滩浅海相	S8	19.16~14.72	Average(10)	34.2	1.88	12.5	86.5	76.2	13.9	36.7	26.1	81.6	16.7	114.4	191.7	25.0	212.1	12.1	0.90	0.20	3.22	7.75	490	5.67	1.15	4.38	21.72	0.34	12.27	4.45
			Max	47.1	2.57	16.0	111.9	92.9	23.9	48.5	37.0	106.1	21.2	144.2	226.0	28.0	303.0	13.1	1.29	0.25	4.06	10.91	539	7.33	1.23	6.40	28.77	0.51	15.79	19.62
			Min	21.2	1.09	8.5	59.7	51.9	8.3	24.5	14.5	52.9	12.5	88.4	165.6	19.7	140.8	10.6	0.61	0.14	2.40	4.36	424	4.01	1.08	3.31	15.01	0.18	7.87	2.01
浅海相	S7	39.16~19.16	Average(34)	25.4	1.45	9.9	70.6	86.6	10.9	34.9	20.7	64.1	14.7	101.6	212.0	23.4	251.2	11.3	3.11	0.17	2.59	5.76	570	6.47	1.06	5.11	19.61	0.26	10.51	2.53
			Max	44.7	2.58	15.3	103.4	416.9	19.6	230.6	33.1	92.0	20.2	130.1	337.1	35.3	382.8	14.7	37.37	0.24	3.99	9.76	985	9.01	1.53	11.36	24.84	0.49	14.61	3.82
			Min	3.7	0.38	1.2	9.6	29.1	2.3	7.5	1.5	12.7	5.6	47.3	128.0	4.4	36.5	1.9	0.50	0.02	0.18	0.74	447	0.90	0.16	0.96	8.78	0.01	1.47	0.40
泛滥平原河湖相	S6	58.00~39.16	Average(20)	23.9	2.00	10.3	90.2	95.9	16.0	44.0	22.4	98.7	15.3	96.1	187.2	27.8	284.3	11.1	3.28	0.30	2.48	5.53	608	7.39	1.03	4.96	27.00	0.26	10.79	2.53
			Max	40.3	6.29	15.9	440.5	218.7	64.1	123.2	80.5	827.9	21.2	127.8	353.4	71.4	656.9	16.8	14.34	1.41	3.93	10.09	1140	16.08	1.54	8.48	102.94	0.45	15.99	4.41
			Min	5.8	0.10	1.4	12.3	50.1	4.9	13.2	3.2	25.2	9.4	60.2	76.3	9.7	74.1	2.6	0.72	0.04	0.31	0.89	345	2.03	0.26	0.71	8.17	0.03	3.22	0.54
浅海相	S5	65.4~58.0	Average(8)	17.5	1.06	6.8	51.8	113.3	10.3	42.9	15.3	53.4	12.9	93.7	246.4	18.3	203.0	8.2	5.20	0.12	1.72	3.52	823	5.44	0.77	5.21	19.91	0.17	7.96	2.86
			Max	37.9	2.45	13.9	93.6	155.0	20.1	70.9	25.3	94.2	19.6	111.7	327.6	26.7	331.7	11.7	10.67	0.20	3.26	7.08	1308	8.68	1.53	7.21	25.50	0.38	15.24	7.01
			Min	5.9	0.01	1.2	22.2	63.3	3.4	24.6	3.2	27.3	8.3	78.6	138.5	10.9	60.2	2.2	0.95	0.03	0.26	0.92	520	1.58	0.26	4.01	17.87	0.03	3.60	0.88

（续表）

沉积相	沉积相段	深度/m	样品个数	Li	Be	Sc	V	Cr	Co	Ni	Cu	Zn	Ca	Rb	Sr	Y	Zr	Nb	Mo	Cd	Sn	Cs	Ba	Hf	Ta	W	Pb	Bi	Th	U
滨海平原河湖相	S4	75.42~65.40	Average(8)	13.6	1.40	5.3	40.7	68.7	9.6	31.0	12.7	36.5	11.3	80.5	200.1	16.6	215.9	7.1	3.16	0.08	1.29	2.83	701	5.42	0.63	4.37	20.05	0.13	6.23	1.51
			Max	35.4	2.96	11.1	79.3	174.1	18.5	84.3	20.2	51.9	15.4	98.8	272.1	31.2	394.2	16.6	8.47	0.13	3.27	7.29	1111	10.07	1.49	8.80	40.43	0.35	15.23	4.77
			Min	2.9	0.07	0.7	9.7	15.4	2.2	4.0	1.7	12.1	4.2	41.2	69.6	5.5	104.5	2.0	1.28	0.02	0.11	0.51	193	2.35	0.20	1.28	9.17	0.00	1.93	0.37
泛滥平原河湖相	S3	79.66~75.42	Average(3)	9.6	1.45	4.7	92.8	41.5	23.2	26.9	11.8	58.1	12.8	77.6	213.2	12.7	94.7	5.0	2.58	0.33	0.99	2.55	889	2.50	0.48	14.52	18.04	0.13	5.59	1.67
			Max	18.2	4.33	10.6	346.6	120.6	73.5	77.6	27.1	115.7	21.9	100.1	291.3	30.4	343.0	9.7	6.23	1.18	2.42	5.74	1273	8.72	0.89	41.71	25.33	0.31	11.90	3.95
			Min	3.2	0.08	1.0	10.4	7.6	2.3	0.1	2.3	19.9	4.2	37.5	123.3	3.6	18.6	2.1	0.25	0.02	0.12	0.56	534	0.55	0.23	4.42	8.35	0.01	1.57	0.31
海侵相	S2	83.95~79.66	Average(3)	25.2	1.44	11.7	84.2	66.1	13.5	28.0	25.2	73.9	15.9	101.9	139.5	20.9	119.9	11.8	0.78	0.18	2.73	6.84	449	3.34	1.08	4.61	17.33	0.31	11.94	2.15
			Max	26.9	1.50	11.8	87.2	73.0	13.8	28.6	25.7	75.8	16.3	103.5	140.4	23.5	122.0	13.0	0.92	0.25	2.89	6.95	464	3.40	1.17	5.20	18.58	0.34	12.05	2.21
			Min	24.3	1.38	11.7	82.4	62.2	12.8	27.2	24.2	72.7	15.5	99.6	138.2	15.8	116.3	11.0	0.53	0.05	2.59	6.66	441	3.30	1.03	4.22	14.91	0.25	11.75	2.02
半干燥河湖相	S1	95.22~83.95	Average(19)	14.3	1.06	8.3	47.3	42.7	8.7	16.4	11.4	52.5	12.1	86.9	236.1	52.7	109.3	7.6	0.82	0.16	1.34	3.13	771	2.89	0.72	7.22	27.61	0.15	6.84	3.60
			Max	29.0	2.17	19.8	92.5	82.1	18.4	33.0	18.4	82.1	16.4	113.2	333.9	222.8	503.6	13.0	1.22	1.01	2.63	6.96	1248	12.63	1.18	42.08	67.17	0.30	12.07	14.37
			Min	5.8	0.27	1.6	14.9	12.4	3.3	2.7	2.9	28.1	7.8	49.8	97.8	6.8	16.1	2.3	0.28	0.02	0.21	0.78	314	0.49	0.24	1.60	10.50	0.02	2.51	0.47
基底风化剥蚀面	S0	96.70~95.22	Average(2)	22.6	2.38	9.4	67.0	59.3	13.3	27.6	25.6	70.5	16.2	91.9	149.8	18.5	104.3	10.5	0.63	0.30	2.27	5.16	467	3.02	0.90	3.97	26.38	0.28	10.81	1.95
			Max	24.1	3.06	9.8	67.5	63.9	14.4	31.8	26.3	73.3	16.4	96.0	167.6	19.0	116.2	12.2	0.82	0.42	2.56	5.87	531	3.44	1.06	4.02	28.16	0.32	12.25	2.17
			Min	21.1	1.70	9.0	66.4	54.8	12.3	23.5	24.9	67.6	15.9	87.8	132.1	18.0	92.3	8.8	0.43	0.18	1.98	4.45	403	2.59	0.75	3.92	24.59	0.24	9.38	1.73

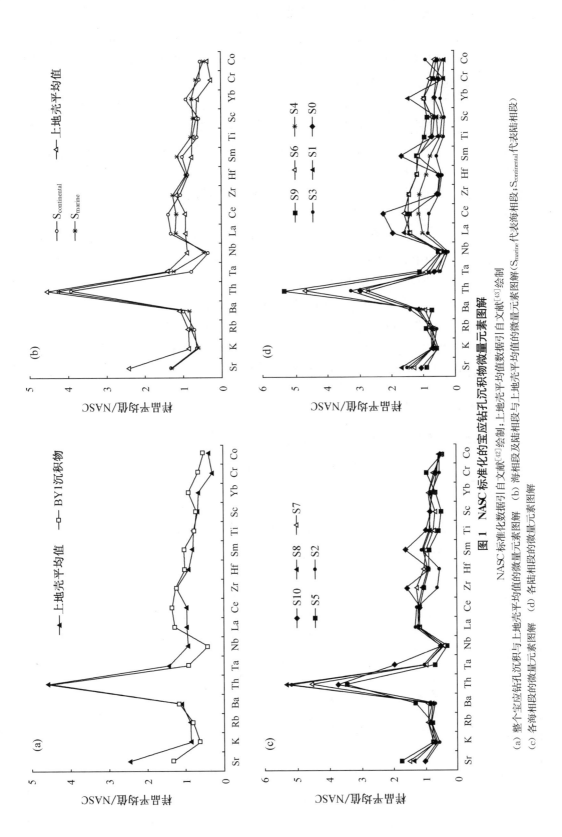

图 1　NASC 标准化的宝应钻孔沉积物微量元素图解

NASC 标准化数据引自文献[42]绘制；上地壳平均值数据引自文献[43]绘制

(a) 整个宝应钻孔沉积与上地壳平均值的微量元素图解　(b) 海相段及陆相段与上地壳平均值的微量元素图解（S_{marine} 代表海相段；$S_{continental}$ 代表陆相段）

(c) 各海相段的微量元素图解　(d) 各陆相段的微量元素图解

集 REE 以及 Cr 和 Co,显示了本区不同的沉积背景。

Ba,Th,Ta,Nb,REE,Cr 和 Co 在不同沉积相中有较大差异(见图 1b)。Th,Ta,Nb 和 Cr 在海相沉积中相对富集,而 Ba,REE 和 Co 则在陆相沉积中含量较高,反映了不同地质过程对微量元素成分的控制。

同一沉积环境下,不同时期的沉积中一些微量元素含量也有一定变化(见图 1c 和 1d)。活泼元素 Sr 在早更新世晚期的海侵层(S5)中最富集,在全新世海侵层(S10)中相对较低;且 S10 下伏陆相沉积段 S9 中 Sr 平均含量也为全钻孔最低;Sr 在海侵层与陆相层中均与 K 呈同步变化(见图 1c 和 1d),表现出与 K 的一致性(图 2a),反映了主量元素对微量元素的约束作用,而 Sr 比 K 有更宽的变化范围(见图 1c 和 1d),显示出微量元素对环境变化更为灵敏。

Sr 是碱土族元素中的分散元素,在暗色造岩矿物中含量低,主要分布在斜长石和钾长石中,而风化过程中 Sr 的活动性比 Ca 小,在黏土矿物中比较稳定。影响沉积岩中 Sr 含量的变化因素包括:产生沉积物的水中的 Sr/Ca 比、含 Sr 矿物(文石和方解石)的存在、有机体的生物作用效应、温度及溶液的含盐度等[45]。从研究区沉积物元素含量分析看,Sr 与 K 的相关性比 Sr 与 Ca 的要高(见图 2a 和 2b),在海侵相中也是如此(图 2c 和 2d)。区域中碱性玄武岩广泛分布,钾长石是其主要的造岩矿物,因此本区具有较低的 Ca 的背景(仅为平均上部地壳含量的一半)。钾长石风化后变为高岭石,黏土矿物对 Sr 有吸附作用。

(二)稀土元素变化特征

宝应钻孔沉积物的稀土元素特征参数见表 3。

表 3　宝应剖面各段沉积物稀土元素的特征参数

沉积段 (样品数)	ΣREE/μg·g^{-1}	LREE/%	MREE/%	HREE/%	LREE/HREE	δEu	δCe
BY1(139)	195.63	86.86	9.92	3.22	28.61	1.05	1.09
Max	658.43	90.45	12.82	6.75	59.84	4.24	1.95
Min	43.63	80.43	7.90	1.51	11.91	0.84	0.82
S$_{continental}$(65)	202.10	87.01	9.85	3.14	30.32	1.16	1.11
S9	218.85	86.50	10.14	3.37	26.64	0.92	1.09
S6	225.32	87.13	9.74	3.13	28.67	1.17	1.17
S4	157.13	87.66	9.45	2.90	31.57	1.21	1.12
S3	117.99	88.81	8.75	2.45	38.38	1.36	1.04
S1	222.43	87.16	9.66	3.18	33.20	1.24	1.03
S0	312.86	90.08	8.37	1.54	58.36	0.98	1.16
S$_{marine}$(74)	187.54	86.54	10.13	3.34	26.76	0.99	1.07
S10	204.00	86.14	10.30	3.55	24.30	0.92	1.17
S8	187.74	86.15	10.41	3.44	25.06	0.92	1.03
S7	181.60	86.54	10.11	3.36	26.44	1.01	1.05
S5	171.68	87.82	9.45	2.73	35.17	1.17	1.03
S2	192.12	86.96	10.05	2.99	29.33	0.97	1.00

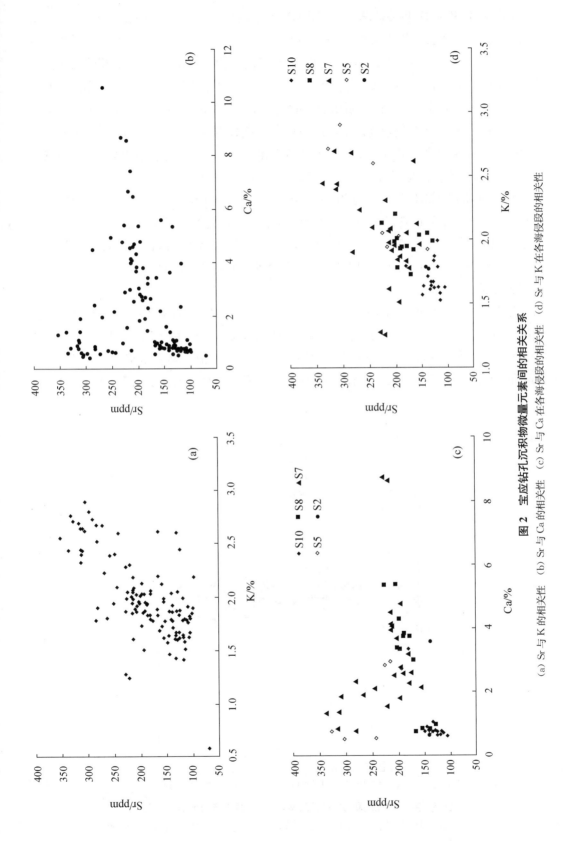

图 2　宝应钻孔沉积物微量元素间的相关关系

(a) Sr 与 K 的相关性　(b) Sr 与 Ca 的相关性　(c) Sr 与 Ca 在各海侵段的相关性　(d) Sr 与 K 在各海侵段的相关性

剖面中沉积物稀土元素总量（$\sum$REE）变化在 43.63～658.43 $\mu g/g$ 之间,平均为 195.63 $\mu g/g$,较北美页岩平均值(163.16 $\mu g/g$)[42]略高;高于上部地壳的 REE 平均含量(146.37 $\mu g/g$)[43]。与中国邻海表层沉积物 REE 的平均含量相比,本区沉积物稀土元素总量略低于渤海(229.29 $\mu g/g$),与南海(187.58 $\mu g/g$)相当,高于黄海(134.03 $\mu g/g$)和东海(140.03 $\mu g/g$)[46,47]。与中国内陆河流沉积物的稀土总量(长江为 167.11 $\mu g/g$,黄河为 137.74 $\mu g/g$)[48]相比含量较高。宝应沉积物中 LREE 占 86.86%,LREE/HREE 平均值为 28.61,即轻稀土相对富集。δEu 值变化范围在 0.84～4.24 之间,平均为 1.05,呈微小的正异常;δCe 值变化范围为 0.82～1.95,平均值为 1.09,略具正异常。

陆相沉积段 $\sum$REE 平均值为 202.10 $\mu g/g$,其中 LREE 占 87.01%,MREE 占 9.85%,HREE 占 3.14%,为轻稀土富集型。海侵沉积段的$\sum$REE 平均值为 187.54 $\mu g/g$,其中 LREE 为 86.54‰,MREE 为 10.13%,HREE 为 3.34%。海侵段的稀土含量比陆相沉积段略低。δEu 在陆相沉积中为 1.16,呈微小的正异常;在海侵沉积中为 0.99,基本不显示异常。

用北美页岩 NASC[42]对原子序数为 57 号 La 至 71 号 Lu 的镧系元素进行了标准化。由图 3 可以看出,宝应钻孔沉积物具有与 NASC 平均成分相近的 REE 型式,呈较平坦的近似直线形。La,Ce,Pr,Nd,Sm,Eu 和 Gd 较 NASC 平均成分略有富集,而 Dy,Ho,Er,Tm 和 Yb 较 NASC 平均成分稍有亏损,即宝应沉积物相对于重稀土元素(HREE),轻稀

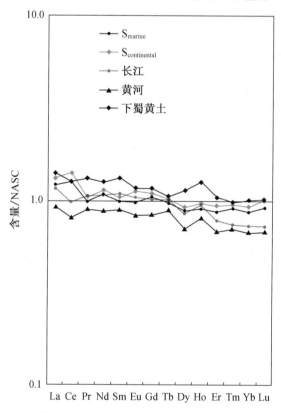

图 3 宝应钻孔沉积物与长江、黄河、下蜀黄土稀土元素分布型式

NASC 标准化数据引自文献[42];长江、黄河数据引自文献[48];下蜀黄土数据引自文献[49]

土元素(LREE)富集,并有 Ce 的正异常。分布曲线在轻稀土部位斜率较大,而在重稀土部位较为平坦,这种 LREE 富集型或平坦型反映的多是源区稀土元素地球化学特征[50]。

BY1 海侵相段(S_{marine})的 REE 型式呈轻微的右倾的近似平坦的直线形,具负斜率,即轻稀土元素相对富集;各海侵沉积段 REE 型式十分接近,呈狭窄的束状(图 4a),其中 LREE 的分布十分一致,且具平行变化,只是在 HREE 含量上稍有不同;而河湖相($S_{continental}$)整体亦表现出轻微的右倾的近似平坦的直线形,Eu 无明显异常显示。

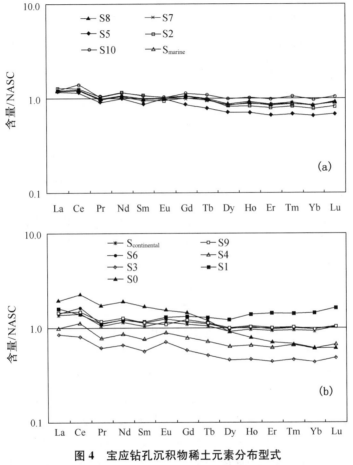

图 4　宝应钻孔沉积物稀土元素分布型式

(a) 各海侵沉积段稀土元素分布型式　(b) 各陆相沉积段稀土元素分布型式

各陆相沉积段 REE 型式相对海侵段较分散,呈较宽的带状(图 4b),表现了各段 ΣREE 丰度上的差异。除 S0 段外,各陆相段基本显示了平行变化的行为,只是在稀土总量上差别较海侵段大。S0 段的 REE 分布明显呈右倾型,即 LREE 比 HREE 具有明显的富集。

长江、黄河沉积物 REE 型式也同样具有较平坦的近似直线形的分布(见图 3),呈 LREE 相对稍富集的右倾状,并具弱的 Ce 和 Eu 负异常。长江流域广泛分布的中酸性岩浆岩及复杂的源岩决定了其沉积物中 REE 的含量较黄河沉积物中的高,LREE 比黄河富集,且 Ce 异常稍大于黄河沉积物[48]。

ΣREE,$(La/Yb)_N$,$(La/Sm)_N$,$(Gd/Yb)_N$,δCe 和 δEu 是表征稀土元素组成特点的

重要参数,它们的变化可以反映沉积物形成时的构造背景、物源和沉积环境的差别[51~56]。(La/Yb)$_N$反映了样品轻稀土与重稀土的分异状况,并可间接反映物质来源。(La/Sm)$_N$比值反映了轻稀土与中稀土之间的分馏程度,该值越大轻稀土越富集;(Gd/Yb)$_N$比值反映中稀土与重稀土之间的分馏程度,比值越小,重稀土富集程度越高。宝应沉积物ΣREE型式与长江、黄河相近,相比长江和黄河沉积物,宝应钻孔的(La/Yb)$_N$,(La/Sm)$_N$,(Gd/Yb)$_N$以及δEu和δCe与长江沉积物的值更接近,反映了物质来源上有着一定的继承性。

研究资料表明,来自我国西北的沙尘暴南界可到25°N左右,我国每年越过秦岭抵达长江中下游地区的浮尘日数在5天以上[57]。长江中下游地区分布广泛的下蜀黄土与宝应沉积物的REE配分模式也具有一定的相似性(见图3),具有与NASC[42]相近的REE组成,且略富集LREE:下蜀黄土ΣREE平均值为207.7 μg/g[49],十分接近宝应沉积物稀土总量,而明显高于黄土高原黄土,是与下蜀黄土中黏粒组分含量较高,对REE(特别是LREE)产生吸附作用有关[49]。

宝应陆相沉积物REE分布模式具有弱的正Eu异常。通常认为,斜长石是造成沉积物中Eu正异常的主要成分。从本区的区域地质背景看,苏北盆地中生代的火山岩呈分布广泛,多为致密熔岩,局部夹火山碎屑岩,大多数火山岩表现为喷溢相的平坦块状玄武质熔岩,其橄榄拉斑玄武岩的LREE适度富集,并具有弱的正铕异常的大陆玄武岩的特点[58];而含幔源橄榄岩包体的新生代钠质碱性玄武岩具有较强的富集LREE型式,第三纪碱性玄武岩LREE富化作用较显著[59],它们作为苏北平原形成的基底原岩,对宝应沉积物REE的分布特征必定产生一定的影响。高的δEu可判断出沉积物中原有的火山物质属中性或基性[60],宝应陆相沉积物δEu最高达4.24,平均值为1.16。X射线衍射分析发现,样品中均含有一定量的斜长石,沉积物中相对富集斜长石时会使Eu含量升高,推断本区沉积物很可能部分受到这些玄武岩火山物质的影响。

研究区海侵相沉积物δEu值稍有别于陆相段,最低值只有0.89,提示了它们沉积环境的差异。研究资料显示,海水的Eu异常是风成物或者热液的输入的影响[44],而研究区海侵的发生与暖湿气候环境有着密切联系,当古气候波动到温暖湿润时期时,会由季风带来大量风成物。对洛川黄土和兰州黄土的研究表明,其稀土分布模式明显的富集Ce组元素而相对亏损Y组元素[7];对比发现,南京黄土与新西兰Banks Peninsula、北美Kansas、Iowa及德国Kaiserstuhl黄土具有相近的REE分布模式,即以富轻稀土(LREE)和负的Eu异常为特点[7],与宝应沉积物的REE配分型式相一致。李徐生等的研究表明下蜀黄土的物质来源区可能是一个广泛而开放的空间范围[49],西北远源粉尘的直接搬运、黄土高原黄土的二次扬尘输送以及近源的粉尘物质均有可能参与其物源组成[61]。下蜀黄土粉尘颗粒在搬运过程中经过了高度的混合,其稀土元素特征主要继承了原始物质的特征。宝应沉积物REE分布模式与黄河、长江及黄土的稀土分布特征非常相近,因此可以推断,苏北平原在其形成及演化过程中,不仅受到黄河、长江两河流域陆源物质的影响,而且在第四纪强大季风系统影响下,很可能有黄土高原粉尘物质的参与,它们带有较高含量的稀土特征,从而使本区的稀土总量受其叠加的影响,不仅高于两河沉积物,而且Ce表现出小的正异常。

宝应钻孔沉积物海陆沉积相段均为 Ce 的弱正异常,是陆源物质的反映。研究资料也表明,海洋沉积物中海洋自生组分(如生物壳体、钙十字沸石、磷灰石、重晶石等)和海底火山物质蚀变成因的蒙脱石可使其具负铈异常[62,63],而海洋沉积物中的陆源黏土组分无铈异常[64]。

宝应海侵段沉积物ΣREE 略低于陆相段,反映了水动力分选作用等造成的粒级不同对 REE 浓度的影响。海侵段沉积物中含有的各粒级粉砂比例远大于其黏土含量[37],而在河湖相沉积段中黏土组分明显较高。Cullers 等研究发现含黏土的岩石比其他沉积物具有更高的 REE 含量[65],石英的存在对 REE 浓度起着稀释作用[15,66]。正是这个原因使得研究区沉积物在含砂较高的海侵段 REE 总量发生微小降低,而黏土比例较大的河湖段具有较高的 REE 总量。此外,重矿物,尤其是锆石、独居石和褐帘石的存在,也会对单一样品的 REE 型式起重要影响,使其表征源岩 REE 特征的意义下降[44,48]。

风化作用下稀土元素绝大部分呈矿物(独居石、锆石等)碎屑残留在沉积物中,因此,风化壳的 REE 含量较母岩高,且往往有着稀土含量的最大值。在风化壳中,碱性阶段占有明显优势,REE 的吸附量大大增加,大量的 LREE 发生富集而一些重稀土元素(如 Er,Lu)带出,沿风化剖面的淋失明显增加,并常有含量较高的 Ce[7]。所以风化剥蚀产物中以富含轻稀土元素为特征,宝应钻孔剖面底部 S0 因而有着较高的 LREE 和较低的 HREE。

尽管在海侵阶段存在海底自生作用和生物作用,但这些作用没有完全改变本区沉积物以陆架碎屑沉积为主体的物源特点,更不可能对黄海陆架区的陆源沉积产生明显的影响。因为长江和黄河这两条亚洲最大的河流,每年携带进入边缘海(包括渤海、黄海)的泥砂数量是非常巨大的,由此而引起的碎屑沉积作用与其他沉积作用相比,显然不可同日而语。然而从海侵相沉积物稀土元素分布模式所表现出的 Eu 的负异常来看,生物沉积作用对海侵相沉积物有着一定的影响,是海侵相段物质来源之一。

(三) 微量元素对比值和稀土元素特征参数的纵向变化

沉积物中的 Rb/Sr 比值变化曲线可反映物源及沉积环境的变化。Dasch[67]曾对各种母岩在风化作用条件下 Rb 和 Sr 的迁移规律进行了较详细的研究,指出 Rb/Sr 比值可指示母岩的风化作用强度;Gallet 等[68]对洛川黄土剖面中的 Rb/Sr 比值分布进行的研究发现,该比值可清晰地识别古土壤地层单元;陈骏等[69~71]的研究表明,Rb/Sr 值可指示黄土堆积物的风化成壤作用和物质的淋失量,Rb/Sr 曲线反映了 2.5 MaB. P. 以来的气候波动信息及其与区域性乃至全球性气候变化的耦合关系。已有的矿物学研究表明[72],粒径的粗细变化不会引起含钙和含钾矿物的显著分异。因此,原岩的物质组成成为制约 Rb 和Sr 元素在沉积地层中含量变化的主要因素,而物源与沉积环境密切相关,不同的沉积环境有可能带来不同的物源,物质来源的改变必将引起 Rb/Sr 比值的变化。与主元素 Ca具有相似地球化学性质的 Sr,在气候温热、海平面较高时,在海相沉积中的富集会随 Ca含量的升高而加强,有机体的生物作用也会使 Sr 得到不同程度的富集,此时的 Rb/Sr 比会以低值为特征;在气候湿冷而海平面较低时,陆相化学风化作用加强,Sr 作为微量组分主要赋存在碎屑矿物(长石、黏土矿物和白云石)中,因此河湖相沉积中 Rb/Sr 比值则相对较高。

宝应(BY1)钻孔沉积物稀土元素的特征参数（$\sum$REE，$(La/Yb)_N$，$(La/Sm)_N$，$(Gd/Yb)_N$以及δEu和δCe）在纵向剖面上随时间的变化存在着较明显的波动（图5）。总体可分为两段，以39.1 m(B/M界线)为界曲线的波动状况显示出明显不同；0.78 Ma以前各曲线的波动幅度较大，δEu多为大于1的正异常；而0.78 Ma以后波动幅度较低，仅有个别层段出现明显波动，大体上呈较平缓的变化趋势，δEu随沉积时代渐新而有所降低，多为小于1的负异常。

稀土总量$\sum$REE在纵向上的分布与δEu及δCe的变化在不同沉积相段中的表现不同。轻重稀土之间的分异程度$(La/Yb)_N$（北美页岩标准化后的La/Yb比值）与$\sum$REE变化在不同时段表现出不同的相关关系，而$(La/Sm)_N$和$(Gd/Yb)_N$的变化与$(La/Yb)_N$同步，反映了轻稀土元素含量的变化总体上决定了稀土总量的变化。

钻孔岩芯微量元素对比值及稀土元素特征值的纵向变化如图5所示，曲线波动较大的层位如下。

9 130~9 230 cm：$\sum$REE曲线出现全段最高峰（658.43 $\mu g/g$），δEu呈较尖的波谷（0.92），δCe亦为负异常的低谷（0.84），为一氧化环境，Rb/Sr曲线呈低值特征；$(La/Yb)_N$，$(La/Sm)_N$及$(Gd/Yb)_N$曲线均表现出不同程度的谷形，即轻稀土出现亏损，而与重稀土发生分异。副矿物独居石和褐帘石可引起LREE亏损[44]，可能受这些副矿物在该层位富集的影响。该段为含黏土的砂层，P和S元素含量较高，因而可能与暖湿气候波动有关，受到海水淹侵。

7 740~8 230 cm：与下段明显不同，$\sum$REE呈一较宽的波谷并具有全段最低值（43.63 $\mu g/g$），δEu出现较宽的波峰（1.74），δCe为变化较平缓的低谷（1.07），指示为一弱还原环境，Rb/Sr曲线呈波谷；与$\sum$REE变化相反的是$(La/Yb)_N$，$(La/Sm)_N$和$(Gd/Yb)_N$曲线均呈现较宽的波峰，表明LREE和MREE相对HREE富集，稀土总量的亏损与该段含粉砂质砂的稀释作用有关。该段为海侵相沉积，与气候转暖相对应。

5 582~6 268 cm：$\sum$REE曲线呈宽平的波谷（125.99 $\mu g/g$），而$(La/Yb)_N$，$(La/Sm)_N$及$(Gd/Yb)_N$均显示出波峰变化；δEu较宽的波峰（1.41）与δCe为较平缓的低谷（0.97）和Rb/Sr曲线的低缓波谷相对，反映了S5海侵段的沉积特征及当时的暖湿气候环境。

3 580~3 916 cm：$\sum$REE以较低值（50.39 $\mu g/g$）的波谷出现，Rb/Sr曲线也呈较尖的波谷；$(La/Yb)_N$，$(La/Sm)_N$和$(Gd/Yb)_N$均以同步的波峰变化相对应，轻稀土在稀土总量较低下发生一定富集；δEu波峰值为1.41，对应δCe的波谷值为1.06。该段紧邻B/M界线的底界，在泛滥平原潜育层中含有大量铁锰鲕粒，说明潮湿环境下沉积水体的pH-Eh经常变化，反映了相对暖湿的气候波动，对应海侵段S7的早期阶段。

2 716~3 090 cm：以$\sum$REE的小波谷（132.60 $\mu g/g$）和δCe的负异常（0.89）为特征，为一氧化环境；δEu（1.17）和$(La/Yb)_N$，$(La/Sm)_N$呈小的波峰，Rb/Sr曲线呈低谷，是宝应钻孔沉积相中最大的海侵段S7的中期阶段。

1 370~1 730 cm：Rb/Sr曲线、$\sum$REE曲线（131.3 $\mu g/g$）均出现小波谷，而$(La/Yb)_N$变化很小，即轻重稀土分异不明显；δEu上部出现小波峰（1.03），δCe曲线在1左右波动，对应海侵段S8，该层位发现了大量有孔虫，据^{14}C年代测定，是约3 Ma的海侵，且强度较大。

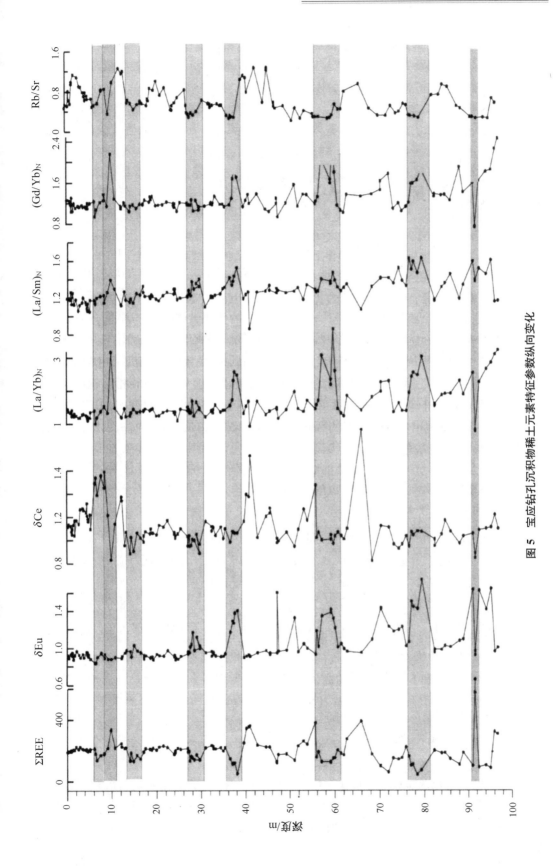

图 5 宝应钻孔沉积物稀土元素特征参数纵向变化

900～1 310 cm：$\sum$REE 呈小而尖的波峰（333.86 $\mu g/g$），而$(La/Yb)_N$，$(La/Sm)_N$ 和 $(Gd/Yb)_N$ 均以波峰变化相对应，表明稀土元素的富集以轻稀土和中稀土的富集为主。δEu 呈小的波谷（0.88），δCe 曲线（0.83）出现明显的尖状谷形，为一氧化的沉积环境，Rb/Sr 曲线在高背景下出现一个较尖的低谷，反映了陆相泛滥平原河湖相 S9 的元素变化特征。BY1 钻孔 10.3～13.2 m 处出现的含铁锰鲕粒的硬黏土层表明本区曾经历了潮湿与半干旱交替的气候变化。

630～870 cm：$\sum$REE 出现一个明显的小波谷（161.20 $\mu g/g$），$(La/Yb)_N$ 和 $(Gd/Yb)_N$ 均同步降低，即稀土元素的亏损主要表现在轻稀土的亏损。δEu 出现小波峰（0.94），δCe 则呈明显波峰（1.55），是一个较还原的沉积环境，Rb/Sr 曲线在高背景下出现一个小的低谷，对应海侵段 S10 早期。

综合稀土元素特征参数的纵向变化，发现当 $\sum$REE 和 δCe 出现波谷，而 δEu，$(La/Yb)_N$，$(La/Sm)_N$ 和 $(Gd/Yb)_N$ 呈波峰时，均为海侵段沉积，且与 Rr/Sr 的波谷相对应；当 $\sum$REE 呈波峰而 δEu 和 δCe 出现波谷时，对应河湖相沉积。Rb/Sr 的低值反映了母岩风化作用的强度[67]，且间接指示了气候变化。低的 Rr/Sr 比值与温热的气候有关，因此，这些稀土元素富集与亏损的现象反映了暖湿与冷干的气候环境。黏土的吸附作用造成了稀土的富集，冷湿的沉积环境引起介质条件改变（pH 升高），使元素 Ce 相对亏损，而硅质的砂粒组分会引起稀土含量的稀释效应。海侵相中石英砂的含量高，河湖相沉积中含有大量的黏土组分，REE 特征参数的纵向变化可作为古气候波动的代用指标，$\sum$REE 的高值段与 δEu 和 δCe 的低值段代表冷湿的气候环境，而 $\sum$REE 和 δCe 的低值与 δEu 的高值段代表暖湿润的气候环境信息。

五、结论

根据苏北盆地宝应钻孔沉积物的微量及稀土元素分析，得到如下几点认识：

（1）相对于平均上部地壳研究区具有较低的 Sr，K，Rb，Ta 和 Nb，而相对富集 REE 和 Cr 及 Co，显示了本区不同的沉积背景；Th，Ta，Nb 和 Cr 在海相沉积中相对富集，而 Ba，REE 和 Co 则在陆相沉积中含量较高，反映了不同地质过程对微量元素成分的控制。

（2）本区物质来源具有多源性，与黄河、长江具有类似的物源，并很可能受到基底碱性火山岩和黄土高原粉尘物质的影响。

（3）钻孔沉积物总体上属轻稀土适度富集、缓右倾斜型的稀土元素配分模式，具 Ce 的正异常；陆相段具有弱的正 Eu 异常，海侵段沉积物 $\sum$REE 略低于陆相段；稀土元素含量与沉积物粒度之间存在相关性，即随着粒度由粗变细，稀土元素含量增加；$\sum$REE 的高值段对应黏土含量较高，反之亦然。沉积物较高的稀土元素总丰度值反映了本区化学风化作用相对较强。

（4）海陆交互过渡带沉积物的 Rb/Sr 比值和 REE 的深度变化曲线与沉积相段的划分吻合，其变化与半干旱湿冷的气候及暖湿的环境波动有关，可作为古气候波动的代用指标。$\sum$REE 的高值段与 δEu 和 δCe 的低值段显示冷湿的气候环境；而 $\sum$REE 和 δCe 的低值与 δEu 的高值段具有温暖湿润的环境信息。

（5）苏北平原第四纪以来经历了多期海侵，也是气候环境多次冷湿—暖湿交替的演化过程，苏北平原的形成演化反映了古气候的波动。

参考文献

[1] 陈志华,石学法,王湘芹等. 2003. 南黄海 B10 岩心的地球化学特征及其对古环境和古气候的反映. 海洋学报,25(1):69-77.

Chen Zhihua, Shi Xuefa, Wang Xiangqin *et al*. 2003. Geochemical changes in Core B10 in the southern Huanghai Sea and implications for variations in paleoenvironment and paleoclimate. *Acta Oceanologica Sinica*, 25(1): 69-77.

[2] Li X D, Shen Z G, Wai O W H *et al*. 2001. Chemical forms of Pb, Zn and Cu in the sediment profiles of the Pearl River Estuary. *Marine Pollution Bulletin*, 42(3):215-223.

[3] Li X D, Wai O W H, Li Y S *et al*. 2000. Heavy metal distribution in sediment profiles of the Pearl River Estuary, South China. *Applied Geochemistry*, 15(5):567-581.

[4] 杨慧辉,陈岚. 1998. 海坛岛海域表层沉积物中主量化学成分的地球化学. 海洋学报,20(3):47-55.

Yang Huihui, Chen Lan. 1998. Geochemistry of some major chemical composition in marine sediments of Haitan Island. *Acta Oceanologica Sinica*, 20(3): 47-55.

[5] 孟翊,刘苍字. 1996. 长江口区沉积地球化学特征的定量研究. 华东师范大学学报(自然科学版),(1):73-84.

Meng Yi, Liu Cangzi. 1996. A quantitative study on the sedimentary geochemical characteristics of the Changjiang Estuarine region. *Journal of East China Normal University* (*Natural Science*), (1): 73-84.

[6] 黄汝昌. 1982. 陆相沉积中古气候演变及元素的迁移、聚集和演化. 见:中国科学院兰州地质研究所编. 中国科学院兰州地质研究所集刊. 北京:科学出版社. 12-18.

Huang Ruchang. 1982. Paleo-climate evolution and elements transfer,assemblage and evolution in terrestrial sediments. In: Lanzhou Geology Research Institute, Chinese Academy of Sciences ed. Collection of Lanzhou Geology Research Institute, Chinese Academy of Sciences. Beijing: Science Press. 12-18.

[7] 张虎才编著. 1997. 元素表生地球化学特征及理论基础. 兰州:兰州大学出版社. 15-18,160-165,395-434.

Zhang Hucai ed. 1997. Supergenical Elemental Geochemistry and Theoretical Principles. Lanzhou: Lanzhou University Press. 15-18, 160-165, 395-434.

[8] Heslop D, Langereis C G, Dekkers M J. 2000. A new astronomical timescale for the loess deposits of Northern China. Earth and Planetary Science Letters, 184(1): 125-139.

[9] Owen L A, Finkelb R C, Barnardc L P *et al*. 2005. Climatic and topographic controls on the style and timing of Late Quaternary glaciation throughout Tibet and the Himalaya defined by [10]Be cosmogenic radionuclide surface exposure dating. *Quaternary Science Reviews*, 24(12-13): 1391-1411.

[10] 韦桃源,陈中原,魏子新等. 2006. 长江河口区第四纪沉积物中的地球化学元素分布特征及其古环境意义. 第四纪研究,26(3):397-405.

Wei Taoyuan, Chen Zhongyuan, Wei Zixin *et al*. 2006. The distribution of geochemical trace elements in the Quaternary sediments of the Changjiang River mouth and the paleoenvironmental implications. *Quaternary Sciences*，26(3)：397 - 405.

[11] 戴纪翠，宋金明，李学刚等.2007.胶州湾沉积物中氮的地球化学特征及其环境意义.第四纪研究，27(3)：347 - 356.

Dai Jicui，Song Jinming，Li Xuegang *et al*. 2007. Geochemical characteristics of nitrogen and their environmental significance in Jiaozhou Bay sediments. *Quaternary Sciences*，27(3)：347 - 356.

[12] 杨守业，韦刚健，夏小平等.2007.长江口晚新生代沉积物的物源研究：REE 和 Nd 同位素制约.第四纪研究，27(3)：339 - 346.

Yang Shouye，Wei Gangjian，Xia Xiaoping *et al*. 2007. Provenance study of the Late Cenozoic sediments in the Changjiang delta：REE and Nd isotopic constraints. *Quaternary Sciences*，27(3)：339 - 346.

[13] 刘俊海，杨香华，于水等.2003.东海盆地丽水凹陷古新统沉积岩的稀土元素地球化学特征.现代地质，17(4)：421 - 427.

Liu Junhai，Yang Xianghua，Yu Shui *et al*. 2003. The REE geochemical characteristics of Paleocene-Eocene in the Lishui sag of the Donghai Basin. *Geoscience*，17(4)：421 - 427.

[14] Taylor S R，McLennan S M. 1985. The Continental Crust：Its Composition and Evolution：An Examination of the Geochemical Record Preserved in Sedimentary Rocks. Oxford London：Blackwell Scientific Publication. 1 - 301.

[15] Condie K C. 1991. Another look at REEs in shales. *Geochimica et Cosmochimica Acta*，55：2527 - 2531.

[16] 于炳松，裘愉卓著.1998.扬子地块西南部沉积地球化学演化与成矿作用.北京：地震出版社.26 - 52.

Yu Bingsong，Qiu Yuzhuo. 1998. Geochemistry Evolution and Mineralization of the Sediments in Southwest Yangtze Plate. Beijing：Seismological Press. 26 - 52.

[17] 朱如凯.1997.煤系高岭岩的地球化学判别标志.地质论评，43(2)：121 - 130.

Zhu Rukai. 1997. Geochemical discriminant criteria of the genesis of Kaolin rocks in coal measures. *Geological Review*，43(2)：121 - 130.

[18] 丁林，钟大赉.1995.滇西昌宁-孟连带古特提斯洋硅质岩的稀土元素和铈异常特征.中国科学(B辑)，25(1)：93 - 101.

Ding Lin，Zhong Dalai. 1995. REE and cerium anomalies in chert of the paleo-tethys in Changning-Menglian，Western Yunnan Province. *Science in China*（*Series B*），25(1)：93 - 101.

[19] XieYouyu，Guan Ping，Yang Shaojin *et al*. 1992. Geochemistry of sediments and environment in Xihu lake of Great Wall Station of China，Antarctica. *Science in China*（*Series B*），35（6）：758 - 768.

[20] 刘季花，张丽洁，梁宏峰.1994.太平洋东部 CC48 孔沉积物稀土元素地球化学研究.海洋与湖沼，25(1)：15 - 22.

Liu Jihua，Zhang Lijie，Liang Hongfeng. 1994. The REE geochemistry of sediments in core CC48 from the east Pacific Ocean. *Oceanologiaet Limnologia Sinica*，25(1)：15 - 22.

[21] 余素华，郑洪汉.1999.宁夏中卫长流水剖面沉积物中稀土元素及其环境意义.沉积学报，17(1)：149 - 155.

Yu Suhua，Zheng Honghan. 1999. REE of sediments of the Changliushui section at Zhongwei county of Ningxia Province and the environmental significance. *Acta Sedimentologica Sinica*，17 (1)：149－155.

[22] 谢周清,孙立广,刘晓东等. 2002. 近 2000 年来南极菲尔德斯半岛西湖沉积物中稀土元素 1/δEu 特征与气候演变. 沉积学报,20(2):303－306.

Xie Zhouqing，Sun Liguang，Liu Xiaodong *et al*. 2002. The characteristic of 1/δEu in the sediments of west lake with respect to climate change during the past 2000 years，Fildes Peninsula，Antarctica. *Acta Sedimentologica Sinica*，20(2)：303－306.

[23] 田正隆,戴英,龙爱民等. 2005. 南沙群岛海域沉积物稀土元素地球化学研究. 热带海洋学报,24 (1):8－14.

Tian Zhenglong，Dai Ying，Long Aimin *et al*. 2005. Geochemical characteristics of rare earth elements in sediments of Nansha Islands sea area，South China Sea. *Journal of Tropical Oceanography*，24(1)：8－14.

[24] 史基安,郭雪莲,王琪等. 2003. 青海湖 QH1 孔晚全新世沉积物稀土元素地球化学与气候环境关系探讨. 湖泊科学,15(1):28－34.

Shi Ji'an，Guo Xuelian，Wang Qi *et al*. 2003. Geochemistry of REE in QH1 sediments of Qinghai Lake since Late Holocene and its paleoclimatic significance. *Journal of Lake Sciences*，15(1)：28－34.

[25] Brenchley P J，Newell G. 1980. A facies analysis of upper Ordovician regressive sequences in the Oslo region—A record of glacioeustatic changes. *Palaeogeography*，*Palaeoclimatology*，*Palaeoecology*，31：1－38.

[26] Lijima A，Hein J R，Siever R eds. 1983. Developments in Sedimentology，36：Siliceous Deposits in the Pacific Region. Amsterdam：Elsevier Science Publishers. 193－210.

[27] Gibbs A K，Montgomery C W，O'Day P A *et al*. 1988. The Archean-Proterozoic transition：Evidence from the geochemistry of meta sedimentary rocks of Guyana and Montana. *Geochimica et Cosmochimica Acta*，52(3)：785－787.

[28] 陈衍景,邓健,胡桂兴. 1996. 环境对沉积物微量元素含量和分配型式的制约. 地质地球化学,223 (3):97－105.

Chen Yuanjing，Deng Jian，Hu Guixing. 1996. The constrain of environment to the compositions of trace elements and their distributions. *Geology-Geochemistry*，223(3)：97－105.

[29] 杨競红,王颖,张振克等. 2006. 苏北平原 2.58 Ma 以来的海陆环境演变历史——宝应钻孔沉积物的常量元素记录. 第四纪研究,26(3):340－352.

Yang Jinghong，Wang Ying，Zhang Zhenke *et al*. 2006. Major element records of land-sea interaction and evolution in the past 2.58 Ma from the Baoying borehole sediments，Northern Jiangsu Plain，China. *Quaternary Sciences*，26(3)：340－352.

[30] 陈安定. 2001. 苏北箕状断陷形成的动力学机制. 高校地质学报,7(4):408－418.

Chen Anding. 2001. Dynamic mechanism of formation of dustpan subsidence，Northern Jiangsu. *Geological Journal of China University*，7(4)：408－418.

[31] 杨怀仁,谢志仁. 1984. 中国东部近 20 000 年来的气候波动与海面升降运动. 海洋与湖沼,15(1)：1－13.

Yang Huairen，Xie Zhiren. 1984. Sea-level charges along the east coast of China over the last 200000 years. *Oceanologica et Limnologica Sinica*，15(1)：1－13.

［32］杨怀仁,陈西庆.1985.中国东部第四纪海面升降、海侵海退与岸线变迁.海洋地质与第四纪地质,5(4):59－80.

Yang Huairen, Chen Xiqing. 1985. Quaternary transgressions eustatic changes and shifting of shoreline in East China. *Marine Geology & Quaternary Geology*, 5(4): 59－80.

［33］朱大奎,傅命佐.1986.江苏岸外辐射沙洲的初步研究.见:任美锷,朱季文,朱大奎等.江苏省海岸带东沙滩综合调查报告.北京:海洋出版社.9－16.

Zhu Dakui, Fu Mingzuo. 1986. Morphological and surface-sedimentary characters in Dongsha Bank of Jiangsu coast. In: Ren Mei'e, Zhu Jiwen, ZhuDakui *et al*. Report of the Conprehensive Survey on the Dongsha Beach of the Jiangsu Coast Zone. Beijing: China Ocean Press. 9－16.

［34］陈友飞,严钦尚,许世远.1993.苏北盆地沉积环境演变及其构造背景.地质科学,28(2):151－160.

Chen Youfei, Yan Qinshang, Xu Shiyuan. 1993. Evolution of the sedimentary environments in North Jiangsu Basin and its tectonic setting. *Scientia Geologica Sinica*, 28(2): 151－160.

［35］凌申.2001.全新世以来里下河地区古地理演变.地理科学,21(5):474－479.

Ling Shen. 2001. Palaeogeographic changes of the Lixiahe district since the Holocene. *Scientia Geographica Sinica*, 21(5): 474－479.

［36］王颖主编.2002.黄海陆架辐射沙脊群.北京:中国环境科学出版社.10－15.

Wang Ying ed. 2002. Radiative Sandy Ridge Field on Continental Shelf of the Yellow Sea. Beijing: *China Environment Press*. 10－15.

［37］王颖,张振克,朱大奎等.2006.河海交互作用与苏北原成因.第四纪研究,26(3):301－320.

Wang Ying, Zhang Zhenke, Zhu Dakui *et al*. 2006. River-sea interaction and the north Jiangsu Plain formation. *Quaternary Sciences*, 26(3): 301－320.

［38］Wang P X, Murray J W. 1983. The use of foraminifera as indicators of tidal effects in estuarine deposits. *Marine Geology*, 51(3－4): 239－250.

［39］Wang Pinxian, Min Qiubao, Bian Yunhua *et al*. 1985. On micropaleontology and stratigraphy of Quaternary marine transgressions in East China. In: Wang Pinxian, Bian Yunhua, Cheng Xinrong *et al*. eds. Marine Micropaleontology of China. Beijing: China Ocean Press. 265－284.

［40］Wang Pinxian, Min Qiubao, Bian Yunhua *et al*. 1985. Characteristics of foraminiferal and ostracod thanatocoenoses from some Chinese estuaries and their geological significance. In: Wang Pinxian, Bian Yunhua, Cheng Xinrong *et al*. eds. Marine Micropaleontology of China. Beijing: *China Ocean Press*. 229－242.

［41］杨怀仁,王健.1990.黄河三角洲地区第四纪海进与岸线变迁.海洋地质与第四纪地质,10(3):1－14.

Yang Huairen, Wang Jian. 1990. Quaternary transgressions and coastline changes in Huanghe River (Yellow River) delta. *Marine Geology & Quaternary Geology*, 10(3): 1－14.

［42］Gromet L P, Dymek R F, Haskin L A *et al*. 1984. The "North American Shale Composite", its compilation, major and trace element characteristics. *Geochimica et Cosmochimica Acta*, 48(12): 2469－2482.

［43］Taylor S R, McLennan S M. 1981. The composition and evolution of the continental crust: Rare earth element evidence from sedimentary rocks. Philosophical Transactions of the Royal Society, A301: 381－399.

［44］Rollison H R 著.杨学明,杨晓勇,陈双喜等译.2000.岩石地球化学.合肥:中国科学技术大学出版

社. 54 - 68.

Rollison H R. Translated by Yang Xueming, Yang Xiaoyong, Chen Shuangxi *et al*. 2000. Rock Geochemistry. Hefei：Chinese University of Science and Technology Press. 54 - 68.

[45] 牟保磊编著. 1999. 元素地球化学. 北京：北京大学出版社. 169 - 177.

Mu Baolei ed. 1999. Elements Geochemistry. Beijing：Peking University Press. 169 - 177.

[46] 王金土. 1990. 黄海表层沉积物稀土元素地球化学. 地球化学，(1)：44 - 53.

Wang Jintu. 1990. REE geochemistry of surfical sediments from the Yellow Sea of China. *Geochimica*，(1)：44 - 53.

[47] 沈华悌. 1990. 深海沉积物中的稀土元素. 地球化学，(4)：340 - 348.

Shen Huati. 1990. Rare earth elements in deep-sea sediments. *Geochimica*，(4)：340 - 348.

[48] 杨守业，李从先. 1999. 长江与黄河沉积物 REE 地球化学及示踪作用. 地球化学，28(4)：374 - 380.

Yang Shouye, Li Congxian. 1999. REE geochemistry and tracing application in the Yangtze River and the Yellow River sediments. *Geochimica*，28(4)：374 - 380.

[49] 李徐生，韩志勇，杨达源等. 2006. 镇江下蜀黄土的稀土元素地球化学特征研究. 土壤学报，43(1)：1 - 7.

Li Xusheng, Han Zhiyong, Yang Dayuan *et al*. 2006. REE geochemistry of Xiashu loess in Zhenjiang, Jiangsu province. *Acta Pedologica Sinica*，43(1)：1 - 7.

[50] Johannesson K H, Hendry M J. 2000. Rare earth element geochemistry of groundwaters from a thick till and clay-rich aquitard sequence，Saskatchewan, Canada. *Geochimica et Cosmochimica Acta*，64(9)：1493 - 1509.

[51] Bhatia M R. 1985. Rare earth element geochemistry of Australian Paleozoic graywackes and mud rocks：*Provenance and tectonic control*. *Sedimentary Geology*，45：97 - 113.

[52] Bhatia M R, Crook KA W. 1986. Trace element characteristics of graywackes and tectonic setting discrimination of sedimentary basins. *Contributions Mineralogy Petrology*，92：181 - 193.

[53] 邵磊，李献华，韦刚健等. 2001. 南海陆坡高速堆积物的物质来源. 中国科学(辑)，31(10)：828 - 833.

Shao Lei, Li Xianhua, Wei Gangjian *et al*. 2001. Provenance of a prominent sediment drift on the northern slope of South China Sea. *Science in China* (Series D)，44(10)：919 - 925.

[54] 李双应，岳书仓，杨建等. 2003. 皖北新元古代刘老碑组页岩的地球化学特征及其地质意义. 地质科学，38(2)：241 - 253.

Li Shuangying, Yue Shucang, Yang Jian *et al*. 2003. Geochemical characteristics and implications of Neoproterozoic shales from the Liulaobei formation in North Anhui. *Chinese Journal of Geology*，38(2)：241 - 253.

[55] 刘锐娥，卫孝峰，王亚丽等. 2005. 泥质岩稀土元素地球化学特征在物源分析中的意义——以鄂尔多斯盆地上古生界为例. 天然气地球科学，16(6)：788 - 791.

Liu Rui'e, Wei Xiaofeng, Wang Yali *et al*. 2005. The geochemical characteristics of rare earth elements of the shale rock in the geologic signification of the analysis of the sedimentary provenance：An example in the upper Palaeozoic in the Ordos basin. *Nature Gas Geoscience*，16(6)：788 - 791.

[56] 张沛，郑建平，张瑞生等. 2005. 塔里木盆地塔北隆起奥陶系-侏罗系泥岩稀土元素地球化学特征. 沉积学报，23(4)：740 - 746.

Zhang Pei, Zheng Jianping, Zhang Ruisheng *et al*. 2005. Rare earth elemental characteristics of

Ordovician-Jurassic mudstone in Tabei uplift，Tarimbasin. *Acta Sedimentologica Sinica*，23(4)：740－746.

[57] Lu Huayu，Vandenberghe J F，An Zhisheng. 2001. Aeolian and palaeoclimatic implication of the 'Red Clay'(North China) as evidences by grain-size distribution. *Journal of Quaternary Science*，16(1)：226－232.

[58] 杨祝良，陶奎元，沈渭洲等.1998.苏北盆地隐伏早第三纪玄武岩地球化学及源区特征.岩石学报，14(3)：332－342.

Yang Zhuliang，Tao Kuiyuan，Shen Weizhou *et al*. 1998. Geochemistry and source characters of the concealed Eogene basalts in North Jiangsubasin. *Acta Petrologica Sinica*，14(3)：332－342.

[59] 陈道公，李彬贤，支霞臣等.1994.江苏六合橄榄岩包体的矿物化学、稀土元素组成及其意义.岩石学报，10(1)：68－80.

Chen Daogong，Li Binxian，Zhi Xiachen *et al*. 1994. Mineral chemistry and REE compositions of peridotite xenoliths from Liuhe，Jiangsu Province and its implications. *Acta Petrologica Sinica*，10(1)：68－80.

[60] 王中刚，于学元，赵振华等著.1989.稀土元素地球化学.北京：科学出版社. 45－75，88－93，133－359.

Wang Zhonggang，Yu Xueyuan，Zhao Zhenhua *et al*. 1989. Rare Earth Element Geochemistry. Beijing：Science Press. 45－75，88－93，133－359.

[61] Henderson P. 1984. General Geochemical Properties and Abundances of the Rare Earth Elements. NewYork：Henderson P. Elsevier. 1－30.

[62] Fleet A J. 1984. Aqueous and sedimentary geochemistry of the rare earth elements. In：Hendersion Ped. Rare Earth Element Geochemistry. Amsterdam：Elsevier Science Publishers. 343－421.

[63] Piper D Z. 1985. Rear earth elements in the sedimentary cycle：A summary. *Chemical Geology*，14：285－304.

[64] 刘季花，崔汝勇，卢效珍等.1999.中太平洋 CP25 岩心的矿物、稀土元素及 Sr、Nd 同位素组成——晚新生代海底火山活动的证据.海洋地质与第四纪地质，19(2)：55－64.

Liu Jihua，Cui Ruyong，Lu Xiaozhen. 1999. Compositions of minerals，REEs and Sr，Nd isotopes in core CP25 from the centeral Pacific Ocean——The evidences for volcanic process during Late Cenozoic. *Marine Geology & Quaternary Geology*，19(2)：55－64.

[65] Cullers R L，Barrett T，Carlson R *et al*. 1987. REE-earth element and minralogic changes in Holocene soil and stream sediment：A case study in the Wet Mountains，Colorado，USA. *Chemical Geology*，63：275－297.

[66] Cullers R L，Basu A，Suttner L J. 1988. Geochemical signature of provenance in sand-mixed material in soiland stream sediments near the Tobacco Root batholith，Montana，U1S. A. *Chemical Geology*，70：335－348.

[67] Dasch E J. 1969. Strontium isotopes in weathering profiles，deep-sea sediments and sedimentary rocks. *Geochimicaet Cosmochimica Acta*，33：1521－1552.

[68] Gallet S，Jahn B M，Torii M. 1996. Geochemical characterization of the Luochuan loess-paleosol sequence，China，and paleoclimatic implications. *Chemical Geology*，133：67－88.

[69] 陈骏，安芷生，汪永进等.1998.最近 800 ka 洛川黄土剖面中 Rb/Sr 分布和古季风变迁.中国科学(D辑)，28(6)：498－504.

Chen Jun，An Zhisheng，Wang Yongjin *et al*. 1998. Rb/Sr Distribution of Luochuan loess section and the changes of summer monsoon in recent 800 ka. Science in China（Series D），28（6）：498－504.

［70］陈骏，季峻峰，仇纲等. 1997.陕西洛川黄土化学风化程度的地球化学研究.中国科学(D辑)，27(6)：531－536.

Chen Jun，Ji Junfeng，Chou Gang *et al*. 1997. Geochemical studies of the intensities of chemical weathering in the Luochuan loess，Shanxi Province. Science in China（Series D），27（6）：531－536.

［71］陈骏，汪永进，季峻峰等. 1999.陕西洛川黄土剖面的 Rb/Sr 值及其气候地层学意义.第四纪研究，(4)：350－356.

Chen Jun，Wang Yongjin，Ji Junfeng *et al*. 1999. Rb/Sr varations and its climatic stratigraphical significance of a loess-paleosol profile from Luochuan，Shanxi Province. Quaternary Sciences，(4)：350－356.

［72］刘东生等著. 1985.黄土与环境.北京：科学出版社，1－48，208－219.

Liu Tungsheng *et al*. 1985. Loess and the Environment. Beijing：Science Press. 1－48，208－219.

沙漠·古海洋
——追溯塔克拉玛干沙漠砂源*

一、缘由——环境变化的感受

20世纪80年代赴欧洲开会,飞机自北京向西北横穿西部沙漠、高山,经沙迦飞巴黎再转德国。沿途景象引人注目:飞越北京不久,进入干旱的沙地区,黄河竭力地绕经河套再向南流去,遗留了一条条的废河曲沙带;经过塔里木沙漠,则是一片广垠、丘状起伏的沙海,沙迦迷漫于干热的黄尘中,地貌景象迥异;飞经欧洲,却见大片的森林。深感气候对环境影响巨大,是自然环境差异的直接因素。而导致气候变化的原因多样——纬度地带性?距海洋远近与通畅程度?构造运动导致喜马拉雅山地隆升,阻挡了海洋与水气之贯通,导致亚洲中部大陆度增强与我国西部沙漠的形成,等等,是区域的差异性。同时,存在着气候变化普遍性的证据:白垩纪的红色砂岩地层在中国、在北美、在东亚、在欧洲均有分布,应与当时干热的气候相一致;在更新世中期(Q_2)中国普遍存在湿润而炎热的气候,而晚更新世的冰期却具有世界可比性;现今全球变暖,虽具有纬度地带性特点,但变暖是不争的事实。不容置疑,气候是环境变化的主导因素。

目前,我们正经历着气候与气候带的变化过程。实际感受:位居40°N的北京气候与20世纪50年代不同,那时气候干燥,冬季十分寒冷,春季多风沙,夏季虽热,但不是炎热或酷热;现今北京冬季不如50年代冬季那么寒冷,但夏季却炎热度增高。北京市气象资料[1-3]表明夏季白天气温可达30℃,最高温度超过40℃,夏季平均温度为25℃(图1、图2和图3)。

相反地,人们感到,南京市(32°N)的气温变得较为温和——夏季不若50年代那么酷热,冬季不若50年代那么寒冷,积雪天数与厚度均减少。

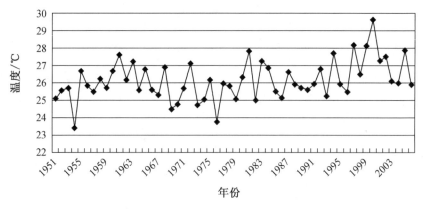

图1　北京夏季年平均气温年际变化

＊　王颖:《海洋地质与第四纪地质》,2011年第31卷第4期,第11-19页。

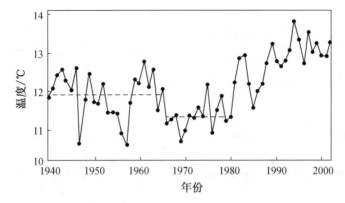

图2　北京站年平均地面气温(℃)变化特征(虚线为年代间的平均)[1]

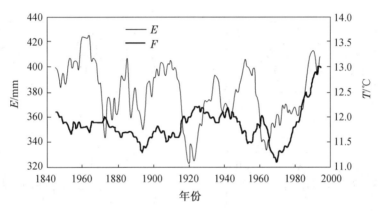

图3　北京近150年来蒸发和气温的变化[2](E:蒸发,T:温度)

　　南京夏季平均温度呈上升趋势,但是酷暑气温却有降低。据周曾奎[4]分析,"南京盛夏酷暑高温年在20世纪80年代到90年代正在减少",历史上南京酷暑年有8年,集中于50年代到70年代,而70年代至90年代仅有2年为酷暑;极端最高温度为43℃(1934年),依次为1959年的40.7℃,1978年的39.7℃。自80年代至今,极端高气温均在38.5℃或以下。年高温日数也在减少,而且以极端最高气温<36.8℃、≥35℃高温日≤10天定为冷夏,南京自80年代始每2～3年即出现冷夏;90年代以来几乎每隔一年即出现一次冷夏。气象资料与实际感受相同(图4)。上述表明:温度上升程度,增温的时段有地区差异,而且在一些地区气温上升的同时,也有一些地区气温有所下降。气候变化是区域与时间的双重概念,既具有同时异地的相似性,也具有同地异时的差异性。

二、联想——塔克拉玛干大沙漠沙层的起源

(一)概况

　　中国最大的沙漠塔克拉玛干界于37°～41°N之间,面积337 600 km²,为世界第二大流动沙漠(图5)。四周有山地环绕:西邻帕米尔高原,南部为昆仑山、阿尔金山,北部有天

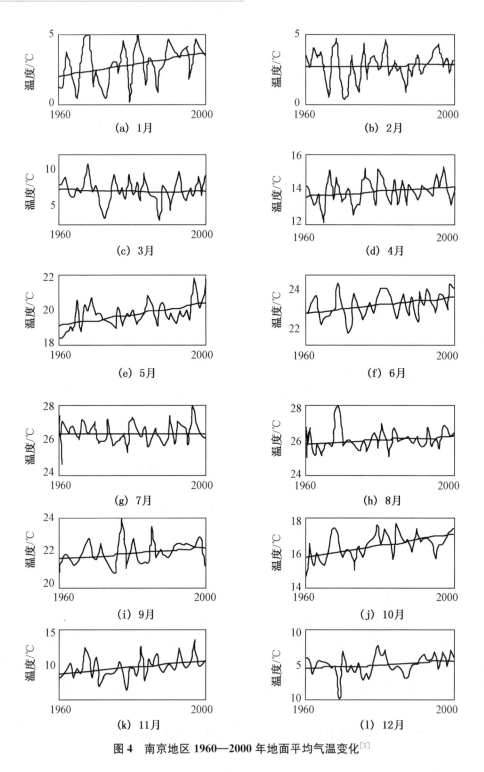

图 4　南京地区 1960—2000 年地面平均气温变化[5]

山横亘。从山地流下的河流——北部的阿克苏河、塔里木河及孔雀河,南部的叶尔羌河、和田河、克里亚河、车尔臣河,由于引水量大、蒸发量大及渗入冲-洪积扇下,沙漠区十分干燥。平均年降水小于 100 mm,沙漠中心约 10 mm,北部库尔勒 52 mm;西北部乌什 85 mm,阿克苏 57 mm;西部喀什 65 mm,南部和田 35 mm,若羌 17 mm。但蒸发量大,年平均蒸发量 250～3 400 mm。多风天,风向以 NW 与 NE 为主导,每年约 1/3 为风沙日,大风风速可达 300 m/s,风沙活动十分频繁。流动沙丘面积达 85%[5],沙丘向南移动,近千年来沙漠向南延伸约 100 km,昆仑山麓丝绸之路的一些城镇或村落已深埋在沙丘之下,甚至达数千米。

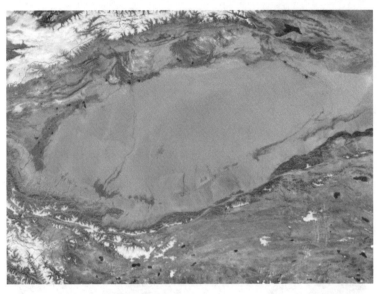

图5　塔克拉玛干沙漠卫星照片(据 www.izy.cn/travel-photo)(附彩图)

　　塔克拉玛干沙漠周缘与山地连接处多砾石戈壁,可能为山地洪积扇残留,其外缘地下水出露处有绿洲分布(图6)。塔克拉玛干沙漠沙层厚,多沙丘,一般高约 60～100 m,大沙丘高 200～300 m。沙丘形态具有三大类:沙丘脊与主风向大体平行的复合型沙垄,沙丘链(图7);沙丘脊与主风向垂直的新月型沙丘与沙丘链(图8);复合型及风力向上加积的金字塔型沙丘及穹形沙丘等(图9和图10)。

图6　塔克拉玛干沙漠边缘之山麓冲积扇(据 www.izy.cn)(附彩图)

图 7　塔克拉玛干沙漠沙丘链及波痕(据 www. xjb. cn/tklm)

图 8　塔克拉玛干沙漠密集的新月型沙丘链(据 www. xjb. cn/tklm)

图 9　塔克拉玛干沙漠陡然升起的复合型沙山(据 www. xjb. cn/tklm)

图 10　塔克拉玛干沙漠大面积鱼鳞状复合沙丘(据 www. xjb. cn/tklm)

　　沙丘纵横的大沙漠,沙浪滚滚,流动的砂是从哪来的? 砂源是关键,通过它追索沙漠成因。山地岩层的风化剥蚀是一来源,但分布不能如此广袤;河流堆积冲洪积扇,砾石留在当地,细颗粒泥沙被风吹扬堆积,却不可能有如此大量厚层之砂? 何况河流沙多沿河道成带状堆积,沿河尚有胡杨林与柽柳丛固沙(图 11)。如此广袤的盆地型大沙漠,厚层砂源只能是干涸的海盆残留! 当然海中沙来源于陆地——风化剥蚀的岩屑,被河流携运而来,被风吹入……但是,是海洋汇积了如此广袤深厚之沙层,海洋干枯了,而沙层裸露,倍受风力的吹扬、运移与再堆积。何况,塔克拉玛干沙漠中尚有海积残留之遗证岗——圣墓山(图 12),是两座红白分明的“沙丘”,实为风蚀红砂岩和白石膏相间的沉积层形成的风蚀残丘。砂岩与膏盐、膏泥间层是干涸海洋的沉积证据。

图 11　胡杨树(据 www. fengjingtupian. com)

图 12 圣墓山照片(据 www.xjb.cn/tklm)

(二) 古老海洋的证据

塔克拉玛干沙漠位于塔里木盆地(560 000 km²)之中,富含石油、天然气藏,是古老海洋的有力佐证。但古老海洋的沉积地层已成岩,现代沙漠却非其残留,砂岩层只构成现代沙漠砂之间接沙源。探索新生代古海洋遗留砂是本文之关键。据彭希龄、吴绍祖在新疆的系统研究成果阐明,在漫长的地质历史时期,塔里木曾多次历经海陆沧桑之变化[6-7]。塔里木盆地的基底距今 30 亿～25 亿年的古老变质岩,曾多次沉陷为海洋,至 8 亿年前(青白口纪末期)成为多个雏形陆块,位于 10°～30°S 范围内。至 4.3 亿年前才向北移与准噶尔陆块(30°N)逐渐合并为陆地。

1. 古生代期间主要为海洋环境

(1) 距今 5.7 亿～4.39 亿年的寒武、奥陶纪期间,为广阔的海域,沉积了厚 1 000～3 500 m 的石灰岩(后经变质成白云灰岩),含三叶虫、角石、笔石化石。西部巴楚县以北,在阿克苏至喀什公路一间房站的后山,出露着海绵礁块。寒武纪中晚期(距今 5.36 亿～5.14 亿年),塔里木海的中西部变成浅水潟湖,形成厚达数百米的膏盐层和红色膏泥层,成为下部泥质页岩、灰岩成油气烃源岩的封盖层,构成油储结构层。奥陶纪末,加里东运动,除南部仍为浅海,大部分地区上升为陆地。

(2) 志留-泥盆纪期间(4.39 亿～3.45 亿年前),为海陆交互相滨海平原与浅海洼地沉积环境。志留系为杂色砂质泥岩、粉砂岩夹砂岩、泥质灰岩,灰岩厚 500～1 000 m。含腕足、珊瑚、三叶虫、笔石、苔藓虫、腹足、瓣鳃类和植物化石。泥盆系在东部为陆相红色砂砾岩层,1 600～1 700 m 厚。在西部,下层为 200 m 厚滨海三角洲平原沉积之砂岩,含植物化石;中层为 500～600 m 厚潮滩相泥岩、粉砂岩、细砂岩,少量海相化石;上层为滨海平原河流相交错砂层夹薄层岩盐层,厚约 600～1 200 m。在盆地南部铁克里山地区,中泥盆统为浅海灰岩约 870 m 厚;上统为陆相。因此,泥盆纪时,塔里木盆地为陆地剥蚀平原山麓洪积与河流沉积,海域存在于南天山区与北昆仑区。

（3）石炭纪为古海洋。早期（约 3.45 亿年前），海水从西、北、南三面入侵，使晚泥盆纪之剥蚀平原淹没为海洋，范围与古生代早期相近。下石炭统沉积是红黑色相间的杂色砂岩、砾岩，夹石灰岩（500～1 200 m），含有很丰富的浅海生物化石：珊瑚、腕足、海百合、蜓科。中石炭统东北部沉积 300 m 厚灰岩。西南部为浅海及滨海沼泽，沉积灰岩、砂岩、粉砂岩、页岩，总厚 300～1 200 m，部分地区浅海煤系中含上述化石及植物化石。西部巴楚地区形成蒸发潟湖沉积，为 200～600 m 厚杂色膏泥及白色石膏层，顶部为 10 m 厚白云岩，均含海相化石。晚石炭纪为碳酸盐浅海沉积，灰岩 200～300 m 厚，含珊瑚、腕足类海相化石。石炭纪末（距今 2.95 亿年），海安运动中期，在北方的欧亚大陆与南方冈瓦纳大陆间形成古特提斯海（Tethys sea），塔里木为特提斯海东延部分。

（4）二叠世早期（距今约 2.9 亿年），海域主要分布在塔里木盆地的西部及南部，晚二叠世时（2.51 亿年前），海水全部退出。二叠纪海相地层，仅出现于早期及中期：西部巴楚与柯平地区出现 100～200 m 厚的薄层灰岩、泥灰岩与泥岩互层，含珊瑚、腕足、腹足、蜓类海相生物化石和瓣鳃类淡水化石；中层为杂色泥岩，砂岩及可采煤层；上部为黑色玄武岩与含群体珊瑚之灰岩堆积，总厚度达 300 m。为海底玄武岩喷发沉积；南部为厚 300～800 m 海相灰岩沉积。晚二叠世为陆地环境，堆积杂色泥岩、砂岩、粉砂岩。南部为山麓相砾岩。西部北部夹有数百米厚浅海喷溢的玄武岩。

总之，塔里木古生代海洋的沉积物已成岩，部分灰岩变质为白云岩。砂岩、粉砂岩、珊瑚礁及山麓洪积扇沉积，出露于地表经风化后，可提供沙源及戈壁滩砾石。但是，广袤的塔克拉玛干沙漠非为古生代海洋沉积的直接残留。

2. 中生代（2.5 亿～0.65 亿年前）主要为陆相沉积环境

曾在南部发现三叠纪海相管刺藻化石。在库车地区侏罗系中发现过海绿石，在喀什的侏罗系中发现海相叠层石。0.9 亿年前的晚白垩世初期，特提斯北支边缘海水沿东西向，近 400 km 长的阿莱山口（原为海峡，后抬升为山口）进入塔里木盆地，沉积了石膏及膏化泥岩，生物礁灰岩，泥质灰岩，富含海相化石，尤以浅海生物牡蛎为主。种种迹象表明，中生代时，塔里木区曾有短暂的海洋环境。

3. 新生代塔里木特提斯海

（1）古新世（0.65 亿～0.565 亿年前时）海水从塔里木西南进入，并经塔里木中部绕过巴楚隆起而到达北部。在海域边缘沉积不厚的浅水陆源泥沙层、钙质砾岩层，向海为沙坝生物礁块灰岩（厚度超过 100 m），海盆中沉积厚约 150 m 暗色泥岩夹薄层灰岩，间夹30～50 m 厚的杂色膏泥岩，潟湖沉积与生物礁块灰岩为沙坝体系沉积，其上又覆盖浅海沉积的暗色泥岩（30 m）。

（2）渐新世（0.565 亿～0.40 亿年前）为干热的滨岸泥滩与潟湖交互沉积。泥质红层、夹砂岩、石膏、砂砾岩和生物贝壳，可达 300～400 m 厚。塔西南在渐新世晚期仍有残留海，在渐新-中新统有海相介形类和有孔虫沉积。拜城地区山麓洪积扇前仍为膏盐湖，下部有盐层，上部为石膏与红泥层交互沉积。

（3）中新世（0.24 亿～0.12 亿年前）海进形成新特提斯海。入侵的海水使边缘湖盆盐度提高成为微咸或半咸水，介形类与有孔虫继续繁殖；塔北山前的库车—轮台一带形成膏盐潟湖，成为新近纪产盐洼地，岩盐、硬石膏与交互沉积的灰绿色及少量杂色泥质岩层，厚度达数百米至数千米。古新近纪盐湖，岩盐与石膏沉积层为海洋（新特提斯海）残留的遗证。大约在中新世晚期约 1 000 万～500 万年前结束了塔里木新特提斯海的历史。可能与青康藏高原持续隆升在 2 700 万年前已达 4 000 m 高度[8]并持续上升有关。

（4）上新世（0.05 亿～0.025 亿年前）塔里木成为一个闭塞的大型山间盆地，主要地貌是河流冲积平原及山麓洪积扇干三角洲堆积。

（5）第四纪（约 250 万～100 万年以来）初期，天山和昆仑山隆升加剧，山麓洪积扇砾石裾发育，为戈壁滩发育之基础。盆地中为沉积平原。早更新世末，喜马拉雅运动使青康藏高原隆升为今日的地貌，塔里木盆地进一步干燥，渐成为现代的沙漠环境。

所以，塔里木盆地是在中新世晚期结束了海洋环境，中新世海洋是塔克拉玛干大沙漠发育的基础。大沙漠是在第四纪发展成型的。其沙源有多种起源：风化剥蚀形成的砾砂，山地冰川剥蚀堆积的沙、石、粉尘，洪积扇砂砾层，风蚀，流水蚀积，湖泊沉积等，但主体是中新世新特提斯海残留砂，沙量最为可观。

古海相地层、石膏盐湖、巨厚沙层的流动沙漠、盆地的轮廓与抬升之盆地遗迹均反映出塔克拉玛干沙漠与海洋之渊源最深（图 13）。

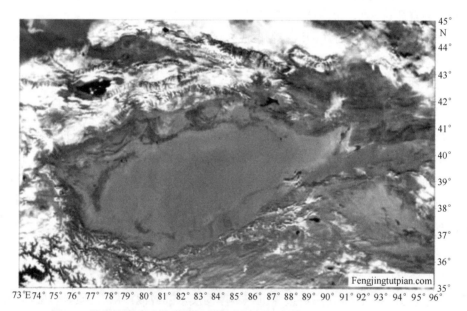

图 13　塔克拉玛干沙漠及塔里木盆地之组合（据 www. qinzd. com/npost）

三、沙漠砂表面结构的成因佐证

坚硬的石英砂表面铭刻了作用于它的动力过程烙印。它具有高处侵蚀、低处沉积之特点，也具有不同动力形成的特殊的形态标志，而标志的组合特征与相互关系，可以反映出颗粒所经历的不同环境。因此，石英砂表面标志，不仅仅反映着最后一次事件，而且包含着一系列历史过程[9]。石英砂表面结构分析的结果与沉积层结构、矿物与微体古生物鉴定结合起来，可作为判断沉积环境，解释沉积相的一个重要手段[10-11]。

采自塔克拉玛干沙漠的 10 个砂样①,代表着不同地点与不同部位(图 14)。均选用经过较长搬运作用过程的细砂粒级(2Φ),经清洁处理去掉有机质及氧化铁物质[10],每样任选 20 颗砂粒,置于金属托上,镀以 200A°厚度的金钯合金镁,然后置于扫描电镜下观察分析[9],其结果统计于表 1。

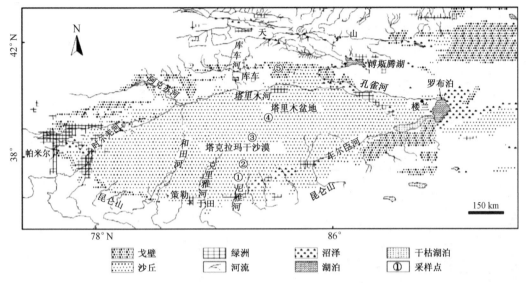

图 14 塔克拉玛干沙漠山河分布与采样点(据杨小平 2007)

(一) 共同特点

200 颗石英砂经过扫描电镜观测,了解到的表面结构共同特点显著(表 1)。

(1) 具有风化岩屑的特征。形态不规则,仍显示出明显的晶体控制形态——长单晶构成的方柱体,双晶或多晶构成的四方柱体或块体。54%颗粒上保存着贝状断口,39.7%颗粒上具碟形坑形态,均系从岩层中分解破裂之标志。

(2) 经过一定距离的搬运,棱角多经磨蚀,37.19%颗粒为尖棱→次棱,21%颗粒为次棱,17.6%为次棱→次圆的粒脊,少部分(4.5%)为次圆,但几乎无圆粒。同时,部分贝状断口已经磨蚀,发展为平行阶结构(26.6%)。

(3) 经过打磨作用,颗粒表面洁净。绝大部分颗粒具有大小不等的 V 痕(64.82%);约 14%的颗粒上有磨蚀撞击力强形成的大型 V 坑,表明经过波浪或波浪夹砂的冲击作用。V 痕在打磨圆滑的棱脊上保存完好,V 痕、V 坑叠加在贝状断口上,表明是在岩屑破碎后经浪流的作用过程。不少颗粒上,V 痕隐现,反映出残留标志之特点。

(4) 经过风力搬运作用,40%的石英砂具有砂粒相互撞击形成的圆麻点,而且细小的颗粒具有在风沙中悬浮搬运的方式——似螺旋状、梭状、水滴等流线形态,但不如黄土中石英砂粒经长途悬浮搬运所形成的扭曲流线形态。反映出,塔克拉玛干沙漠砂经风力搬运,但距离较近,时间较短。

---

① 砂样均由杨小平研究员提供。

表1　塔克拉玛干沙漠石英砂采样点及表面结构测试结果

样号		TK-1	TK-2	TK-3	TK-4	TK-5	TK-6	TK-7	TK-8	TK-9	TK-10	总计	百分数
采样点位置		塔克拉玛干沙漠南部,尼雅东河下游之沙丘砂	塔克拉玛干沙漠南部,采样点TK-1以北的沙丘砂	塔克拉玛干沙漠中部沙丘砂	塔克拉玛干沙漠中部TK-3以北之沙丘砂	塔克拉玛干沙漠北部戈壁沙漠砂	塔克拉玛干沙漠南部沙丘砂,位于TK-1北部	塔克拉玛干沙漠南部沙丘砂	塔克拉玛干沙漠南缘,克里亚河以东,金字塔型沙丘底顶部	塔克拉玛干沙漠南缘,克里亚河以东,金字塔型沙丘底部	塔克拉玛干沙漠南缘,克里亚河以东,金字塔型沙丘中部		
地理坐标	经度(E)	82°47′4″	82°55′15″	83°14′57″	83°51′1″	84°14′41″	82°49′13″	83°1′34″	81°54′42″	81°54′42″	81°54′42″		
	纬度(N)	37°8′46″	37°40′34″	38°29′45″	39°22′47″	41°4′12″	37°21′39″	37°49′32″	36°50′43″	36°50′43″	36°50′43″		
	高程/m	1390.0	1315.0	1164.0	1047.0	938.0	1363.0	1340.0	1522.0	1522.0	1522.0		
样品颗粒数		20	19	20	20	20	20	20	20	20	20	199	
石英砂表面结构 颗粒形态	不规则	11	15	9	9	6	8	10	16	9	15	108	54.27
	晶形控制	13	8	14	8	10	9	7	4	3	1	77	38.69
	光滑颗粒	0	2	6	6	0	0	0	0	0	0	14	7.04
	其他	近圆形(1)、水滴形(3)、流线形(1)	近圆形(3)、长水滴形(1)、方形(1)	长水滴形(1)、似纺锤形(2)	块体(2)、长柱体(1)、似水滴形(1)、似纺锤形(2)	长方形(4)、四方体(4)、方形(1)、水滴形(3)	长方柱体(4)、似纺锤形(2)、似梭形(2)、四方体(4)	柱体(9)、似梭形(1)、似水滴(1)	三角体(2)、长柱体(3)、似梭形(1)、似矛形(1)	四方柱体(4)、长方柱体(7)、三角体(2)、矛体(1)	挖掘平行坑(5)、块体(2)、长柱体(1)、矛形(1)	81	42.21
颗粒磨圆程度	尖棱	4	3	4	0	4	2	4	9	6	2	32	19.10
	尖棱—次棱	7	4	2	10	9	5	6	11	7	13	67	37.19
	次棱	4	5	7	5	1	4	5	0	6	5	37	21.11
	次棱—次圆	5	5	3	5	4	7	5	0	1	0	32	17.59
	次圆	0	2	3	0	2	2	0	0	0	0	8	4.52
	次圆—圆	0	0	1	0	0	0	0	0	0	0	1	0.50
	圆	0	0	0	0	0	0	0	0	0	0	0	0.00

（续表）

	样号	TK-1	TK-2	TK-3	TK-4	TK-5	TK-6	TK-7	TK-8	TK-9	TK-10	总计	百分数
表面标志	贝状断口	12	7	11	13	9	10	11	14	10	11	108	54.27
	平行阶	1	6	4	6	3	4	5	13	5	6	53	26.63
	擦痕	1	0	0	2	1	0	1	0	1	3	9	4.52
	碟形坑	7	13	4	5	8	5	14	8	5	10	79	39.70
	上翘面	0	2	0	0	4	6	4	6	4	4	30	15.08
	机械撞击的V坑	2	0	7	6	2	2	4	3	0	2	28	14.07
	V痕(机械撞击)	18	12	16	19	16	14	14	2	9	9	129	64.82
	撞击点	16	10	15	7	5	8	11	1	4	2	79	39.70
	弯曲沟	4	5	3	7	4	5	3	2	0	2	35	17.59
	刮痕线	0	0	0	0	0	0	0	0	0	0	0	0.00
	裂隙	1	1	1	6	8	16	14	11	12	13	83	41.71
	溶蚀孔	17	18	16	19	20	20	20	20	20	20	190	95.48
表面溶蚀形态	次生硅沉积	17	17	17	17	20	20	20	20	20	20	188	94.47
	硅质包装	0	0	0	0	0	0	0	0	0	0	0	0.00
	风化形态			表面清洁(3)、层状剥离(4)	层状剥离(5)、风化深(1)	层状剥离(1)	层状剥离(8)、光滑(1)	参次表面(1)		风化层面(2)	表面参次不齐(3)、风化甚久(1)	30	15.08

（5）几乎所有的石英砂粒表面皆具有溶蚀孔与次生硅堆积，表明砂粒已经过一定时期的堆积，有化学风化的效应。溶蚀孔多发育于 V 痕上，表明溶蚀发生在 V 痕形成之后。次生硅多堆积于低洼处。但是，未发现颗粒被硅包裹，是因在沙漠环境砂粒仍具移动性。

（二）独特结构

由于砂样部位不同，距砂源或搬运距离不同，除共同特点外，各砂样尚具有独特之结构：

（1）样品 TK‐1‐1（图 15），采自沙漠南缘，尼雅河下游的沙丘砂。晶体控制的长方体形态，少数砂粒略具水滴形。次棱→次圆棱脊。贝状断口残留，不明显的 V 痕、撞击点，具有溶蚀孔，次生硅沉积。

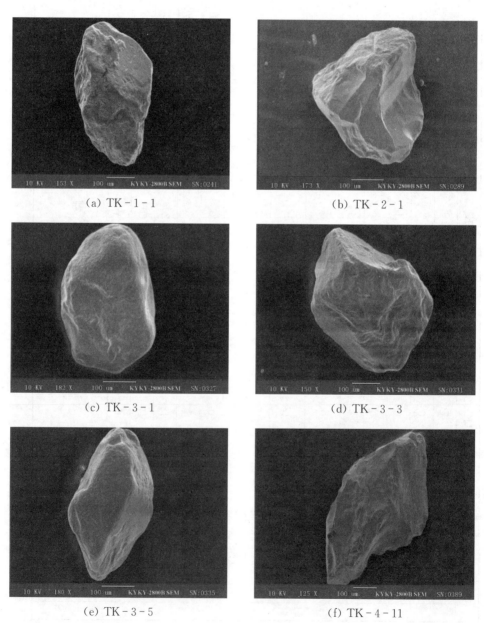

(a) TK‐1‐1　　　　　　　　　　(b) TK‐2‐1

(c) TK‐3‐1　　　　　　　　　　(d) TK‐3‐3

(e) TK‐3‐5　　　　　　　　　　(f) TK‐4‐11

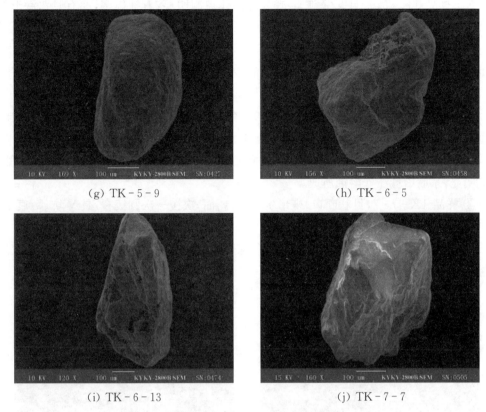

(g) TK - 5 - 9　　　　　　　　　　(h) TK - 6 - 5

(i) TK - 6 - 13　　　　　　　　　　(j) TK - 7 - 7

图 15　塔克拉玛干沙漠石英砂表面结构(附彩图)

（2）样品 TK-2-1,采自 TK1 以北之沙丘砂,晶体控制的不规则形态,次棱粒脊;平行阶,碟形坑,V 痕,撞击点,溶蚀孔及次生硅。

（3）样品 TK3,采自塔克拉玛干沙漠中部,其棱脊具有以次棱→次圆及次圆居多的特点(占 60%),部分颗粒具有似纺锤形态,表明经风力悬浮搬运。颗粒残留有贝状断口及平行阶;突出的特点是具有较多的 V 痕、V 坑及圆麻点,反映出明显的波浪作用过程叠加于风化剥离颗粒上,曾经风力搬运,具有溶蚀孔及次生硅沉积。TK3 为沙漠中部的沙丘砂(TK-3)反映出更为明显的主成因(海洋)标志:

① 样品 TK-3-1,晶体控制,圆滑颗粒;贝状断口,碟形坑,V 坑,V 痕,多撞击点,溶蚀孔,次生硅沉积。典型的海成砂及经风力作用的表面结构。

② 样品 TK-3-3,晶体控制,不规则形态,尖棱→次棱粒脊,贝状断口,碟形坑;V 痕,撞击点;溶蚀孔,次生硅沉积与层状剥离表面。

③ 样品 TK-3-5,晶体控制,呈纺锤形;次棱→次圆粒脊;残留的贝状断口;双 V 坑(表明波浪打击力强),砂棱脊上多 V 痕与撞击点;已发育了溶蚀孔与次生硅沉积;新鲜表面。典型的海成砂经风力搬运再沉积的表面结构。

（4）样品 TK-4,位于沙漠中部偏北,具有与 TK-3 类似特点。但流线形颗粒多(水滴形,似纺锤形),粒脊磨圆程度高,尖棱→次棱居 1/2,另有次棱,次棱→次圆居 1/2,较多的 V 坑、V 痕(99%);弯曲沟及圆麻点;化学风化的溶蚀孔与次生硅沉积普遍。TK-4-11,注意溶蚀孔沿 V 痕发育。

（5）样品 TK-5，采自戈壁滩上，颗粒受晶体控制明显，方柱体多，亦有具水滴形的。磨圆度混杂，具尖棱，尖棱→次棱（45%），但亦具有次圆颗粒；具贝状断口，平行阶及碟形坑，亦具上翘面；似有擦痕；多有 V 痕，亦具 V 坑；圆麻形撞击点较少；均有溶蚀孔与次生硅沉积，具层面剥蚀。采样点距山地较近，故砂粒具混杂作用特点。

（6）TK-6 与 TK-7 样品与 TK-2 与 TK-3 样品，具有类似特点：

① 样品 TK-6-5，仍具有晶体控制的颗粒形态，粒脊经磨圆，多 V 痕，V 坑，低洼处次生硅沉积具晶体形态，似为稳定环境下沉积。

② 样品 TK-6-13，V 痕叠加在上翘面上，反映波浪作用于剥蚀残积之后。V 型溶蚀坑明显，系后沿低洼处溶蚀所成。

③ 样品 TK-7-7，贝状断口、碟形坑、具上翘面，次棱→次圆棱脊，具 V 坑、V 痕与裂隙，有溶蚀洞及次生硅沉积。

（7）样品 TK-8、TK-9、TK-10 采自克里雅河以东沙漠南缘，金字塔沙丘的表层、底层与中层（图 16）。

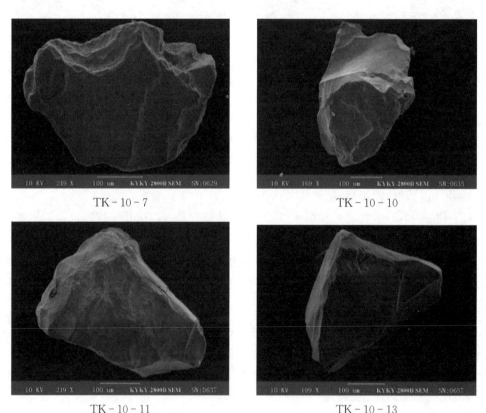

TK-10-7 　　　　　　　　　　　　　　TK-10-10

TK-10-11 　　　　　　　　　　　　　　TK-10-13

图 16　塔克拉玛干沙漠具有挖掘坑、无 V 痕或撞击点石英颗粒（附彩图）

其共同特点是：不规则形态，多柱体，少量为三角体及棱矛形，磨圆度较差，多半为尖棱或尖棱→次棱；具显著的贝状断口，部分样品显示出擦痕，皆具有溶蚀孔及次生硅沉积。TK-10 样品中出现平行的挖掘坑（TK-10-7），似经冰川作用标志，这是在 10 组样品中，唯一发现有冰蚀痕迹的。TK-10 样品中 V 痕稀少。

综上所述，对塔克拉玛干沙漠中线的不同部位，10 组砂样的石英砂表面结构分析后认为：砂源初始于基岩风化剥蚀，聚积于海盆，绝大部分砂曾经过波浪、浪流的搬运与撞击作用，普遍打上 V 痕烙印；后因海洋干涸，出露于地表，再经风力搬运，细粒砂部分被再吹扬与堆积，具流线型雏形，皆具撞击点；嗣后，经历过一定时期的沉积阶段，具溶蚀孔与次生硅沉积。

四、结论

（1）根据塔里木盆地的椭圆形构造轮廓，周边山地间谷地之"古海峡"通道抬升为山口的遗迹，沉积地层的结构、微古化石、沉积相与油气藏、塔克拉玛干沙漠的巨厚砂层以及石英砂表面结构标志的组合特征等，均反映出砂主要源于中新世晚期（约在 1 000 万～500 万年前）的新特提斯海，厚砂层是逐渐干枯的古海洋残留。由于青藏高原及塔里木盆地周边山地的持续隆升，隔断了新特提斯海之海水通道，并逐渐阻挡了海气影响，使海盆演变为内陆湖沼盆地以致干旱沙漠。

（2）广阔的古海洋积聚了自周边陆地风化剥蚀、侵蚀-冲积、磨蚀-海蚀以及生物遗骸堆积的巨厚活动砂层，当中新世晚期海洋干涸后，成为风动力肆虐的物质基础，形成沙丘起伏的沙漠。更新世时，沙漠周边局部有山地冰川砂汇入。沧海沙漠之巨变启迪着人们研究之兴趣，地球构造活动、气候变化影响着环境特征，受人类活动影响，环境的时代变迁亦很显著。

参考文献

[1] 何立富,武炳义,管成功. 2005. 印度夏季风的减弱及其与对流层温度的关系[J]. 气象学报,63(3)：365 - 373.

[2] 马柱国. 2005. 我国北方干湿演变规律及其与区域增暖的可能联系[J]. 地球物理学报,48(5)：1011 - 1018.

[3] 刘春蓁,刘志雨,谢正辉. 2004. 近 50 年海河流域径流的变化趋势研究[J]. 应用气象学报,15(4)：385 - 393.

[4] 周曾奎. 2000. 南京地区 50 年冬夏气温特征分析和演变趋势[J]. 气象科学,20(3)：309 - 316.

[5] 郑红莲,严军,张铭. 2002. 南京地区地面平均气温的变化[J]. 解放军理工大学学报(自然科学版),3(5)：87 - 91.

[6] 中国科学院《中国自然地理》编辑委员会. 1985. 中国自然地理总论[M]. 北京：科学出版社.

[7] 彭希龄,吴绍祖. 2001. 大漠古海——新疆曾经是海洋. 北京：海洋出版社.

[8] 吴珍汉. 青藏高原 2700 万年前已达到 4 000 米[R]. 人民日报(海外版),2007 年 7 月 23 日第 4 版.

[9] Vilks G and Wang Y. 1981. Surface texture of quartz grains and sedimentary processes on the Southeastern Labrador Shelf[R]. Current Research, Part B. Geological Survey of Canada. Paper 81 - 1B. 55 - 61.

[10] Wang Ying, Bhan Deonarine. 1985. Model Atlas of Surface Textures of Quartz Sand[M]. Beijing：Science Press.

王颖,B. 迪纳瑞尔. 1985. 石英砂表面结构模式图集. 北京：科学出版社.

[11] Krinsley, David Henry and Doornkamp, John Charles. 1973. Atlas of Quartz Sand Surface Textures[M]. Cambridge University Press.

当代海平面上升与海南海滩侵蚀 *

—

海平面因不同的作用过程，形成长、短周期不同的变化。

（一）长周期海平面变化

由于：① 全球气温的变暖变冷，改变了海水的总量或海盆的体积，或冰川形成，水量与水体密度变化，或由于构造运动、洋盆被沉积物充填等，这些均为水动型的海平面变化，或称冰川—水动型变化（glacier-eustatic）。② 地方性变化，是由于陆地的挠曲活动，或由于大陆冰盖后退，地壳反弹而形成区域上升或陆地沉降，此为构造-水动型（tectono-eustatic）。海平面继续在变化，根据来自验潮站水位记录分析，由于冰川的进一步融化、下滑入海，或海洋水体热膨胀，展现出世界范围的海平面上升。瑞典 Brest 站位自 1807年以来 200 多年的验潮记录反映出海平面上升与大气温度、海水表层温度变化趋势呈良好的相关，并且自 1930 年以后，上升速率增加。近 100 年来，全球海平面上升 19 cm，上升速率约 2 mm/a，而 2005 年以来海平面上升速率约 2.4 mm/a。国际气候变化组（IPCC）于 2014 年度评估报中，明确反映出海、气温度相关以及人类活动效应所造成的当代海平面变化的趋势（图 1）[1]。

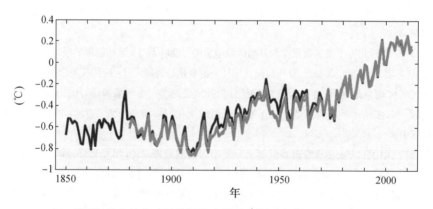

（a）观测与综合的全球平均陆海表层温度距平变化（1850—2012 年）

---

　　* 王颖，章华忠：2015 年 6 月 7 日在"2015 世界海洋日暨中国海洋宣传日"——建设生态文明海洋主题论坛上报告此文。

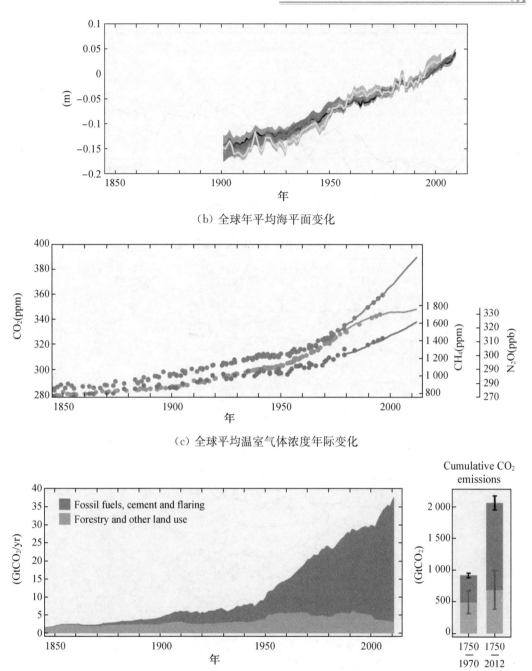

（b）全球年平均海平面变化

（c）全球平均温室气体浓度年际变化

（d）全球人为 CO_2 排放

图 1　海气温度相关——人类活动效应与海平面变化（1850—2012 年）[1]

构造活动与人类影响使陆地水准发生变化，使海平面上升速率形成地区性差异，如：纽约验潮站代表美国东部海岸，近百年海平面上升速率为 3 mm/a，是海面上升与陆地下沉之效果。南部得克萨斯站海平面平均上升值为 6 mm/a，是由于抽取地下水与原油引起地面沉降所形成之综合效应。

中国沿海海平面变化呈上升趋势（图 2 和图 3），据国家海洋局：1980 年至 2014 年，中国沿海上升速率为 3 mm/a，高于全球水平。

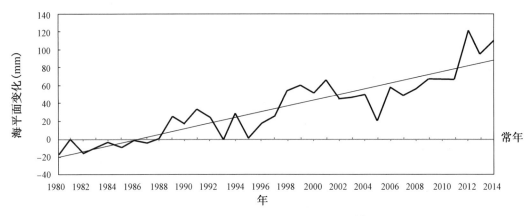

图 2　1980—2014 年中国海平面变化[2]

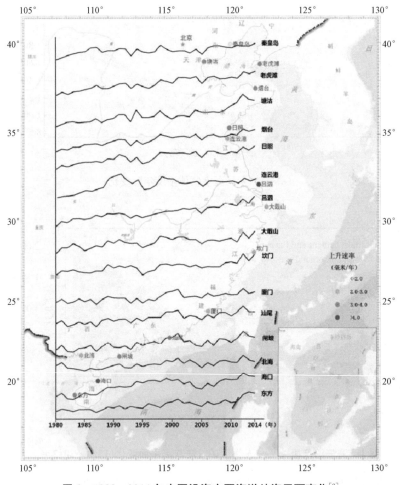

图 3　1980—2014 年中国沿海主要海洋站海平面变化[2]

(二) 短周期海平面变化

短周期海平面变化是由于大气与海洋作用过程的变化,如海水温度的地区性变化,海岸水流强度的改变,气压与风力方向改变等造成的日变化、季节变化或年度变化等。如:

太平洋的厄尔尼诺(El Nino)现象:在太平洋东岸赤道附近的岛屿验潮站重复记录到在一年内,海平面变化40~50 cm! 美国西海岸由于El Nino形成海平面高15~20 cm,俄勒冈州海岸因El Nino造成海平面在12个月内抬升60 cm。风暴潮造成海岸增水与减水,高海平面抬升了风暴增水的基础水位,风暴潮高潮位相应提高,波浪作用增加,排水受阻,致灾。在孟加拉湾形成100 cm的年海平面变化。世纪性的全球海平面变化促进风暴潮与El Nino现象频频发生。

人类活动影响:抽取地下水,建筑物重载使河口三角洲区大面积沉降,加大海平面上升,据国家测绘局1992年报道:天津新港码头自1966—1985年下沉0.5 m。

二

全球海平面上升,在海岸带的主要反应是:海滩侵蚀、海岸沙坝向岸陆位移。原因是:组成海滩与沙坝的沉积物主要是砂,是由波浪自水下岸坡海底掀动,并被浪、流向岸搬运至岸坡上部过程中,水流速度递减而逐次堆积的。泥沙来源,尤其是源于沿岸河流的入海泥沙,持续量大是最主要的;其次来源于区域海岸的侵蚀段落,或海底的古海岸堆积(1万年前气候寒冷低海面时的海岸堆积)。由于海平面持续上升,水深加大,使波浪对古海岸带扰动作用减小,而造成激浪自海底向岸的泥沙搬运量减少,却加强了激浪对上部海滩之冲刷;同时,海平面上升,减小了河流坡降而减少陆地河流向海运沙能力;加之,河流中上游筑坝拦沙,大大减少入海泥沙量;更由于,人工在河流与海岸取沙为建筑材料……所以,进入海岸带泥沙量大大减少,而风暴潮与El Nino加剧,因此,当代世界海岸普遍遭受侵蚀,海岸沙坝普遍向陆后退。近日在海口市西部宾馆与别墅区海岸观察到:因海水冲刷,海岸沙坝前坡形成高度超过1.5 m的陡崖,海滩底部出露基岩,海蚀陡崖距在沙坝顶上所建的舰标高塔不足50 m(照片1)。该处海岸侵蚀后退可能与近期在北面琼州海峡中建岛施工有关,海峡中水流湍急,人工建岛进一步束狭水流通道,束水流急,促成风暴潮时冲刷加剧,并形成持续性浪流对海岸侵蚀。应重视在海岸带施工效应的预研究。

海平面上升与海滩变化效应可应用Bruun定律计算获得。Bruun P. 阐明:随着海平面上升,海滩与外滨区的均衡剖面向上部与向陆地移动,海滨线的后退速率(R)与海平面增高(S)有关[3],即

$$R = \frac{L}{B+h} S \tag{1}$$

式中:h是近岸沉积物堆积的水深;L是海滩至水深h间的距离;B代表滩肩的高度。关系式(1)可表示为

$$R = \frac{1}{\tan\theta} S \tag{2}$$

$\tan\theta \approx (B+h)/L$是指沿着横向距离$L$的近滨平均坡度。

砂质与砂砾质海岸的坡度大部分界于1/100~1/200之间,即$\tan\theta \approx 0.01$至0.02,因此,据公式(2)可得到$R = 50 \sim 100\,S$,表明微小的海平面上升,可形成较大的海滨线后退[4](图4)。

照片 1　海口西海岸海滩遭受侵蚀（2015. 5. 17　蒋飞摄）（附彩图）

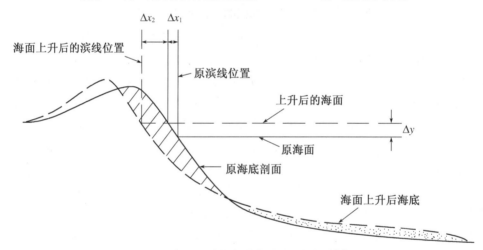

图 4　海面上升使海滩遭受淹没与侵蚀[4]

（Δy 为海面上升幅度，Δx_1、Δx_2 分别为海滨线因海滩遭受淹没和侵蚀

而产生的后退量；假定海面上升前后的海滩剖面均已达到平衡）

三

　　海平面持续上升，作者于 1995 年在国际海洋研究会（SCOR）第 89 工作组研究海平面上升与海滩侵蚀项目时，曾预测了当 21 世纪海平面上升 0. 5 m 时，对我国重要海滨的影响（表 1）。

　　表 1 反映在 20 世纪 90 年代时海南岛南岸港湾内的海滩侵蚀量最小，尤其在亚龙湾侵蚀后退量小。当时，湾内西侧小河与近岸珊瑚礁均供给一定数量的泥沙，沙坝位于湾内岛屿后的波影区，海滩处于相对的动态均衡。而同在南岸的大东海海滩，因湾口宽度大，湾深小，海浪冲刷虽较强。但大东海有附近老海岸沙坝供应，海滩砂粒粗，海滩砂对激浪抗蚀力亦强。

表 1　海平面上升 0.5 m 对我国重要海滨旅游区海滩的影响[4]

	海滨位置	现代海平面之海滩				海平面上升 0.5 m 之海滩响应预测									
						海滩淹没			海滩侵蚀			综合效应		
		长 (m)	平均宽 (m)	相对高 (m)	面积 (m)	损失面积 (m²)	损失率 (%)	滨线后退 (m)	损失面积 (m²)	损失率 (%)	滨线后退 (m)	损失面积 (m²)	损失率 (%)	滨线后退 (m)
大连	星海公园	2125	68.5	6.1	145 613	14 450	9.9	6.8	56 314	38.7	26.5	70 764	48.6	33.3
	东山宾馆	510	42.4	3.8	21 645	3 774	17.4	7.4	8 078	37.3	15.8	11 852	54.7	23.2
	大沙滩	756	56.3	3.8	42 560	5 972	14.0	7.9	18 674	43.9	24.7	24 646	57.9	32.6
	小　计	3 391	61.9		209 818	24 196	11.5	6.8~7.9	83 066	39.6	15.8~26.5	107 262	51.1	23.2~33.3
秦皇岛	北戴河	7 850	87.1	5.9	683 456	68 295	10.0	8.7	383 080	56.1	48.8	451 375	66.1	57.5
	西向河寨	3 124	223.6	6.4	698 466	20 930	3.0	6.7	129 650	18.6	41.5	150 580	21.6	48.2
	山东堡	756	88.2	3.5	66 672	5 670	8.5	7.5	19 202	28.8	25.4	24 872	37.3	32.9
	小　计	11 730	123.5		1 448 594	94 895	6.6	6.7~8.7	531 932	36.7	25.4~48.8	626 827	43.3	32.9~57.5
青岛	青岛湾	1 356	72.8	6.0	98 650	11 526	11.7	8.5	51 455	52.2	37.9	62 981	63.9	46.4
	汇泉湾	1 124	70.6	6.0	79 356	7 868	9.9	7.0	43 386	54.7	38.6	51 254	64.6	45.6
	浮山所口	1 625	193.1	5.4	313 857	14 462	4.6	8.9	42 932	13.7	26.4	57 394	18.3	35.3
	小　计	4 105	119.8		491 863	33 856	6.9	7.0~8.9	137 773	28.0	26.4~38.6	171 625	34.9	35.3~46.4
北海	外沙	2 530	60.8	6.2	153 750	17 254	11.2	5.8~9.5	70 587	45.9	27.9	87 841	57.1	33.7~37.4
	大冠沙至电白寨	5 516	258.4	5.0~9.2	1 425 588	41 926	2.9	5.4~9.8	265 335	18.6	48.1	307 261	21.5	53.5~57.9
	电白寨至白虎头	5 165	183.2	5.0~7.2	946 363	36 457	3.9	5.4~8.7	233 458	24.7	45.2	269 915	28.6	50.6~53.9
	小　计	13 211	191.2		2 525 701	95 637	3.8	5.4~9.8	569 380	22.5	27.9~48.1	665 017	26.3	33.7~57.9
三亚	大东海	2 650	81.5	5.9	215 905	20 935	9.7	7.9	32 330	15.0	12.2	53 265	24.7	20.1
	亚龙湾	8 880	166.1	5.4~13.4	1 475 184	74 592	5.1	6.8~9.8	112 776	7.6	12.7	187 368	12.7	19.5~22.5
	三亚湾	16 360	296.2	3.3~11.6	4 846 024	137 654	2.8	5.6~10.2	713 296	14.7	43.6	850 950	17.5	49.2~53.8
	小　计	27 890	234.4	3.3~13.4	6 537 113	233 181	3.6	5.6~10.2	858 402	13.1	12.2~43.6	1 091 583	16.7	20.1~53.8
	总　计	60 327	185.9	3.3~13.4	11 213 085	481 765	4.3	5.4~10.2	2 180 553	19.4	12.2~48.8	2 662 314	23.7	20.1~57.9

注：表中海滩面积指平均低潮海滨线以上包括沿岸沙岸沙坝现在内的海滨沙滩面积。

21 世纪以来,全球海平面持续性上升,是由于气候变暖导致海水增温膨胀、冰川与极地冰盖融化等因素造成。据联合国政府间气候变化专门委员会(IPCC)第 5 次评估报告:1951—2012 年全球表面气温上升速率为 0.12 ℃/10a;1971—2010 年,海洋上层 75 m 以浅的海水温度上升速率为 0.11 ℃/10a,全球海平面上升速率为 2.0 mm/a。同时,中国沿海气温与海温升高,气压降低,海平面升高。1980—2014 年沿海气温上升速率为 0.35 ℃/10a,海水温度上升速率为 0.19 ℃/10a,气压呈下降趋势,速率为 0.26hPa/10a。同期我国海平面上升,速率为 3.0 mm/a,较常年海平面高 111 mm(图 5)。

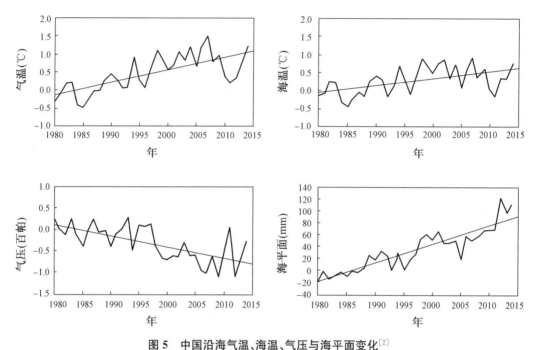

图 5 中国沿海气温、海温、气压与海平面变化[2]

当前,中国各海区海平面仍处于上升发展的趋势(图 6a),而且南海海平面上升居前列(图 6b)。

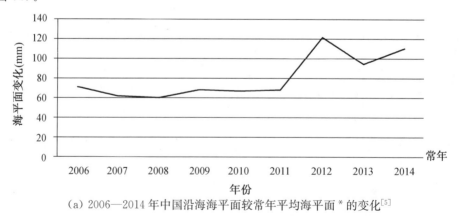

(a) 2006—2014 年中国沿海海平面较常年平均海平面 * 的变化[5]

---

* 依据全球海平面监测系统(GLOSS)的约定,将 1975—1993 年的平均海平面定为常年平均海平面(简称常年);该期间的月平均海平面定为常年月均海平面。

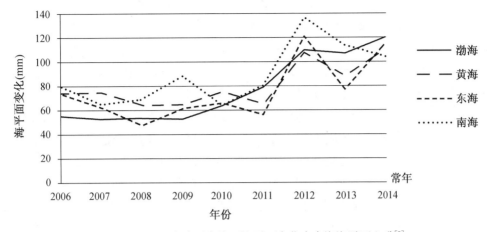

(b) 2006—2014 年中国各海区海平面变化中南海海平面上升[5]

图 6　2006—2014 年中国各海区海平面变化

国家海洋局统计沿海台站实测资料,表明 2012—2014 年,我国各省市海平面上升数值,海南省处于居高水平(图 7),我国各省市海域海平面在 2012 年均为高值年。

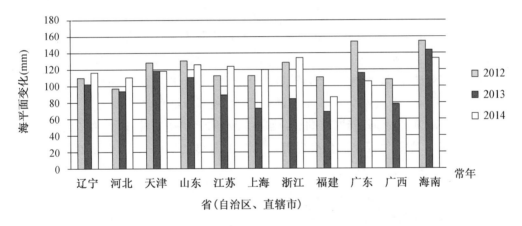

图 7　2012—2014 年中国沿海各省(自治区、直辖市)海平面变化[5]

相对而言,海南南岸港湾中海滩侵蚀还是比东岸或外岛海滩侵蚀量小,而且主要在风暴潮期间涨水高、侵蚀强,风暴过后,海滩再逐渐恢复。但是,近期与世纪性的趋势,海南岛海平面是持续性上升的,伴随着人工在海岸带的开发建设活动,海滩侵蚀与岸线后退具持续性。

以三亚湾与亚龙湾为例:

(一)三亚湾沙坝海岸[6]

现代三亚湾海岸是由一列复式大沙坝组成的。沙坝西部起于马岭的天涯海角——角岭海岬,向东直达三亚河河口,长 18 km,大体呈东西向。这列沙坝围封了原来的一系列港湾:角岭与洋岭间、洋岭与墓山岭间、墓山岭与光头岭间具有 6 条湾坝之海湾、光头岭与金鸡岭以及虎豹岭间的羊拦海湾、荔枝沟海湾,以及虎豹岭南的临春林海湾。诸多海湾由于堆积规模大小不等的海湾沙坝已改变了原始港湾的曲折岸线与陡峻岸坡的特征,并由三亚湾大沙坝构成统一的、砂质平坦海岸(图 8)[7]。现代海岸带激浪作用活跃,灰黄色的中细砂海滩目前普遍发育冲蚀的砂质陡坎。沙坝的西段从烧旗河口到海坡村,高度超过 10 m,宽度 200~350 m,两坡明显,沙坝沉积层紧密,为褐黄色中砂与粗砂,因海坡村位于此坝上,故定名为海坡沙坝。沿海坡沙坝向海一侧,海浪冲刷老沙坝,又重新分选堆积了中砂、细砂质现代海滩,所以,三亚湾沙坝西段是由海坡沙坝及现代背叠海滩所组成,是新老沉积并列的复式沙坝。三亚湾沙坝东段,大体从旧村向东长达 6 km 的一段沙坝宽坦低平,平均宽度为 600~700 m,最宽处达 1 km,坝顶高度约为 5 m 或略低,不具沿岸堤。三亚市主要建筑于这一段,三亚沙坝因三亚市建于其上而命名,是地质历史的最新时期——全新世形成的(图 8)。三亚湾内湾成陆,仍可分辨出几列沙坝分隔潟湖的原貌(照片 2)。三亚湾现代海岸滩坡平缓,湾口外有东瑁岛、西瑁岛之掩护,虽海平面上升,三亚湾海滩大体上连续完整,背叠于三亚大沙坝前坡,至 20 世纪 80 年代仍处于微冲,但基本上属于冲、淤动态平衡阶段。近二十年来,由于辟建滨海大道,在三亚沙坝上大量修建楼堂馆舍,消耗了海滨沙量,破坏自然滩坡,彻底改变了天然沙坝状态,坝高降低,失却了沙坝对海岸的自然防护作用。东段人工半岛建设阻挡沿岸流,形成局部回流对海滩的冲刷。当前三亚沙坝侵蚀显著:坝顶上在 40 年代建立的抗登陆碉堡已逐次坍沉于海(照片 3);市中金鸡岭路外侧海滩,经冲刷、沙质剥失而出露了海滩底部的海湾淤泥层(照片 4)。2014 年 8 月—2015 年 3 月,三亚市海洋渔业局采用南京水利科学研究院研定的局部段落"海滩喂养"措施,在三亚港北侧的市中心金鸡岭路口至光明路口全长 2.6 km 的外侧海滩部分人工补沙 22.3 万 m³,砂质取材为崖州宁远河河沙,是源于红色老沙坝的粗中砂,粒径为 0.3~0.4 mm,较原天然海滩砂(珊瑚-石英质细砂)质地略粗。补沙后,经过春、夏季 SE 与 S 向浪作用后观察,海滩形态仍有保存(照片 5),补沙的实践表明,在该处的人工海滩喂养措施是可行的。

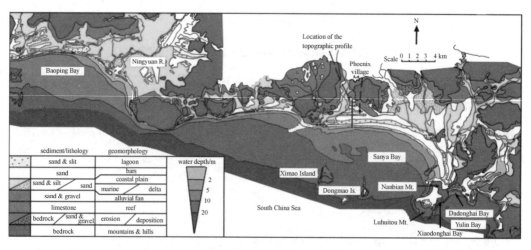

图 8　海南岛南部充填港湾的沙坝潟湖海岸(据 Y. Wang,P. Martini et al.,2001)[7](附彩图)

照片2　沿三亚湾发育的潟湖沙坝体系(附彩图)

照片3　海南三亚湾海滩侵蚀,碉堡相继入海
(2015.6.7　王颖摄)(附彩图)

照片4　三亚市金鸡岭路口外,在人工补沙施
工前的被侵蚀海滩(三亚海洋局　赵
壮卿摄)(附彩图)

照片5　三亚市金鸡岭路口至光明路口间,经海
滩喂养后,恢复的海滩现象(三亚海洋局
赵壮卿摄)(附彩图)

(二) 亚龙湾

位于海南岛南岸,三亚湾以东,是西起于虎头-六道岭,北达田独-田岸后大岭与东界珝瑯岭之间呈 NW-SE 走向的海湾,后内湾淤积成陆,目前的亚龙湾为原海湾的外湾部分,湾阔水深,沙滩与海水质地为南海最佳。基岩海湾,湾口界于白虎角(西侧)与珝琅角(东侧)之间,宽约 9.75 km,水深超过 30 m。海湾朝向 SE 方外海,湾深 4 km,湾内水深大部分超过 20 m,近岸水深 5~8 m。东侧湾口相邻分布着东洲与西洲两个小岛,后侧湾内有一基岩小岛——野猪岛,其西侧,相隔分布着一道与岸平行的岩礁,是古海湾岸蚀退过程中的残余(图9)。现代亚龙湾湾顶分布着一条呈 ENE-SWS 向的大沙坝,由厚层石英、长石与珊瑚质中、细砂组成,沙坝西高东低,适应海湾朝向 SE 方的动力强度分布。沙坝后侧陆上为潟湖洼地,已辟为农田与建筑用地,根据海湾背侧山地形式,仍可分辨出古海湾大势。2010 年以来,亚龙湾海滩冲刷明显,既基于海平面持续上升,也由于在沙坝顶上的砍伐木麻黄林,兴建一系列宾馆设施,使激浪失去可在沙坝顶端翻倾流入后侧洼地的消能作用,成片的人工建筑造成激浪进流(up wash)受阻,回流(back wash)加强,携沙向海

移运,使海滩侵蚀加剧(照片6、7)。人工改变沙坝自然安定的坡稳状况,失去沙坝对海岸的自然防护作用,却促进了海滩冲刷与海岸后退。

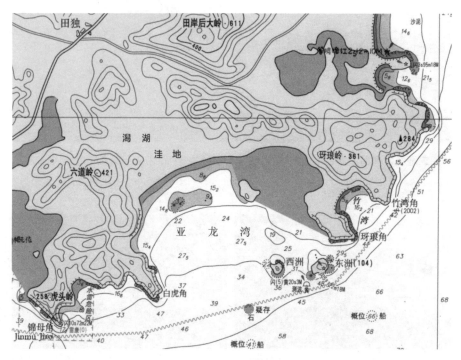

图9　亚龙湾概势图

照片6　沙坝顶端宾馆建筑阻流翻越　　　　照片7　在宾馆的沙坝前坡,回流冲刷移沙外流
　　(2014.夏　王颖摄)(附彩图)　　　　　　　(2014.夏　王颖摄)(附彩图)

更为引起警觉的是,自2012年以来,亚龙湾海滩侵蚀加剧,尤其是沙坝西段近潟湖出口处,高沙坝遭受侵蚀,自坝顶向海形成7~8 m高的海岸陡崖,木麻黄林坍塌坠海,海岸蚀退已危及沙坝上的宾馆楼舍建筑(照片8、9)。

照片 8 亚龙湾湾顶沙坝西段受蚀,沙坝顶上林木坍坠入海(2015.6.8 王颖摄)(附彩图)

照片 9 亚龙湾湾顶沙坝受蚀成 8 m 高陡崖,崖顶宾馆围拦绳网,禁止游人越过(2015.6.8 王颖摄)

探溯海蚀加强的原因:亚龙湾海岸沙坝西段正迎向东南方海湾出口,入射波浪直对西段海滩,2012 年台风在海南登陆,风暴浪与大潮叠加,是海岸侵蚀加剧的始因。但历经两年后至今,海滩未能获浪、沙动力平衡恢复,却加剧侵蚀,西段沙坝已蚀及坝顶危及宾馆安全,引起业主的紧急关注。而且,海滩侵蚀已从西段向湾中延伸。进一步探索海岸侵蚀发展的原因,认识到:人工建设改变了海湾的形态轮廓——海湾东部,自珩琅角向西建堤与东洲相联,再向西与西洲相联,继而向北与野猪岛间断相接,成 L 型合围成湾内港,主口门通道在野猪岛北侧与海岸之间。其结果改变了亚龙湾海湾轮廓型式,减少了约 1/3 的原海域面积(图 10),束狭了海湾口门水流通道,招致波浪传播变形,束水流急,侵蚀加剧,尤其是沙坝的西段,面对 SE 方湾口入射的波浪,侵蚀严峻。

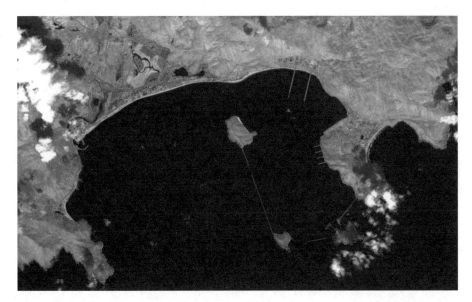

图 10 三亚亚龙湾地区遥感影像(据三亚中科遥感研究所,2013.11.19 高分 1 号融合)(附彩图)

亚龙湾进深小,水深大,岸坡陡,非人工补沙所能维持!而需海底工程补救,改变湾底坡陡与逐次消浪,预期可减轻海岸的侵蚀速度,但也改变了亚龙湾天然海湾沙滩海底结构。东侧新围港域若干年后会有泥沙回淤之虞。亚龙湾事例教育我们：必须加强海岸带行政管理与立法工作。今后，在海岸带大规模工程与围海措施，必须进行海岸海洋环境基础特性的先期调查研究，分析工程设施的预后效应，明确与论证无害后，才能批准施工，否则造成的不利影响难以弥补。

据国家海洋局中国海平面公报预测未来 30 年，南海海域海平面仍持续升高（图 11）。海南与南海海域近期的海平面变化尚具有区域性差异。

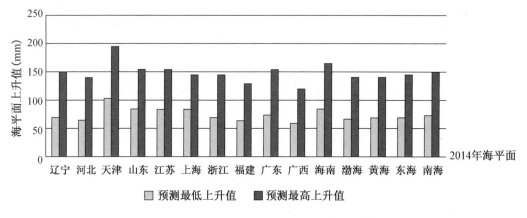

图 11 2014 年预测未来 30 年海平面变化[5]

（1）近期，海南岛周边海平面在 2014 年，比常年高 133 mm，比 2013 年低 10 mm。预计未来 30 年，海南沿海海平面将上升 85～165 mm[5]。

① 2014 年，海南东部沿海各月海平面均高于常年同期，其中，2 月、10 月和 12 月海平面分别高 156 mm、159 mm 和 216 mm；与 2013 年同期相比，1 月、5 月和 8 月海平面分别低 59 mm、55 mm 和 97 mm（图 12）。

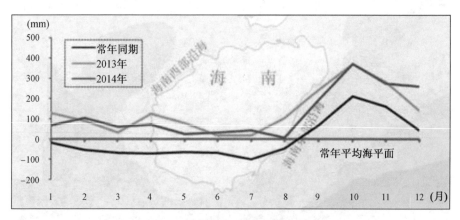

图 12 海南东部沿海海平面变化[2]

② 2014 年，海南西部沿海各月海平面均高于常年同期，4 月、10 月和 12 月海平面分别高 170 mm、171 mm 和 191 mm；与 2013 年同期相比，1 月、8 月和 9 月海平面分别低 66 mm、55 mm 和 53 mm（图 13）。

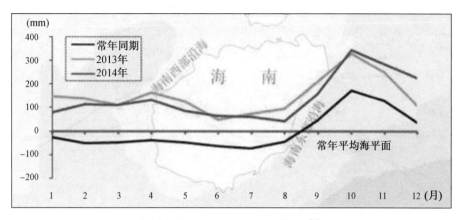

图 13　海南西部沿海海平面变化[2]

海南岛两岸秋冬季海面升高显著,反映出偏北向季风风成增水与季风风浪的影响效应。

(2)南海海域海平面变化状况与海南岛海平面变化有明显之差异,西沙以春夏季海平面高,南沙以夏秋季海平面高,季节差异是与海域不同季节盛行的季风风向有关,成因上仍属风成增水。

① 监测和分析结果表明三沙市海平面在 1990—2014 年期间,总体呈波动上升趋势。其中,西沙海域海平面上升速率较高,达 5.0 mm/a,远高于全球和中国沿海同期平均水平;南沙海域海平面上升速率相对较低,接近中国沿海同期平均水平。

2014 年,三沙市海域海平面季节变化区域特征明显。与多年(1997—2014 年)同期平均海平面相比,西沙海域 6 月海平面高 178 mm(反映出夏季风效应),4 月和 8 月海平面低 42 mm 和 45 mm(图 14)。

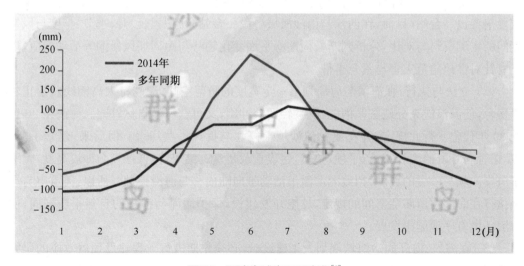

图 14　西沙海域海平面变化[2]

② 南沙海域 8 月海平面高 92 mm,9 月和 10 月海平面低 67 mm 和 45 mm(图 15)。

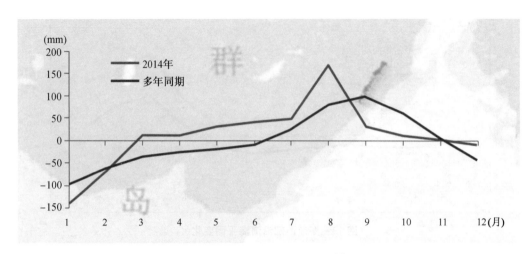

图 15 南沙海域海平面变化[2]

③ 海平面上升加剧了海南沿岸台风风暴潮和海岸侵蚀的灾害程度。

2014 年 9 月,海南沿海处于季节性高海平面期,台风"海鸥"于 16 日在文昌登陆,恰逢天文大潮,海口秀英站极值潮位达历史最高,海南沿海农林渔业和基础设施等遭受严重损失,直接经济损失超过 9 亿元[8]。

2009—2014 年,海口东海岸有 4.2 km 的岸段受到侵蚀,平均蚀退距离为 24.7 m,最大蚀退距离为 40 m,侵蚀总面积超过 10 万 m²;2013—2014 年平均侵蚀距离为 9.8 m,最大侵蚀距离为 18 m,侵蚀总面积约 4 万 m²。2007—2014 年,文昌铺前镇海南角东侧岸段有 6.24 km 的岸段蚀退,平均侵蚀距离 21.25 m,侵蚀总面积超过 13 万 m²[8]。

针对海南省海平面上升与海滩侵蚀现状,提出下列对策性建议:

(1) 建设统一规格与技术要求的海平面监测系统:宜在海南岛周边北、东、南、西岸开阔段落各设一台站(目前,在西沙与南沙海域各设一台站)。积累日、月、季节、年度与风暴期间的实测资料与对比,分析判断长周期海平面变化趋势与短周期风暴潮等活动效应,以期有针对性地防范灾害与紧急事件。

(2) 立法与执行:在海滨与海岛划出一定宽度的防范空间(超越特大高潮线),禁止建设房舍,使激浪能充分地翻倾消能,既减轻回流冲刷,又可获海沙自然补充。海南在 20 世纪 80 年代曾实行此规定,并在亚龙湾沙坝上停建与拆掉一些房舍。但后来,建立了"保卫"馆舍与外资的园林宾馆,平坝、垫基,建成重载楼堂。结果,减低沙坝与沙堤高度,宽缓了海岸沙坝与沙堤的坡度,招致海滨激浪活动带向陆内移与海滩侵蚀。所以,需重申与严格执行在海滨预留防范空间的规定,杜绝开发建设,保护海滩与沙坝的自然平衡剖面,使海滨环境有序地健康发展。

(3) 在海峡、海岛周边海域规划人工建设时,必须先期进行工程建设预后效应的研究与论证,防止改变波浪与水流的传播方向和速度所引起的蚀、淤与损害生态环境效应。严格执行规定,违建者需惩罚并承担海岸侵蚀损失与生态补偿之责。

(4) 对已侵蚀后退的海滩进行工程治理与人工海滩补沙喂养。各处海岸环境不同,海滩喂养与修复工程需进行专题调研并经批准后执行。

参考文献

［1］ IPCC. 2014. Climate Change 2014：Synthesis Report. Contribution of Working Groups I II and III to the Fifth Assessment Report of the Intergovernmental Panel on Climate Change［Core Writing Team，R. K. Pachauri and L. A. Meyer（eds. ）］. IPCC，Geneva，Switzerland，151 pp.

［2］ 国家海洋局. 2015. 2014 年中国海平面公报.
http：//www. soa. gov. cn/zwgk/hygb/zghpmgb/201503/t20150318_36408. html. 2015 - 05 - 25.

［3］ Bruun, P. 1988. The Bruun Rule of erosion by sea level rise：A discussion of large-scale two-and-three-dimensional usage. *Journal of Coastal Research*，4：627 - 648.

［4］ 王颖，吴小根. 1995. 海平面上升与海滩侵蚀. 地理学报，50(2)：118 - 127.

［5］ 国家海洋局. 2007—2015. 2006—2014 年中国海平面公报.
http：//www. soa. gov. cn/zwgk/hygb/zghpmgb/. 2015 - 05 - 25.

［6］ 王颖等. 1998. 海南潮汐汊道港湾海岸. 北京：中国环境科学出版社，91 - 115.

［7］ WANG Ying，Martini I. Peter，Zhu Dakui，Zhang Yongzhan，Tang Wenwu. 2001. Coastal plain evolution indicated by sandy barrier system and reef development in southern Hainan Island，China. *Chinese Science Bulletin*，46(supp.)：90 - 96.

［8］ 国家海洋局. 2015. 海平面上升——悄然发生的海洋灾害.
http：//www. coi. gov. cn/news/zhuanti/hpm/. 2015 - 05 - 26.

Study on the Quaternary Coastline in China[*]

Modern processes of coast erosion, deposition and sedimentary dynamics have been emphasized since late 50th in China, and using sedimentary evidence to identify ancient coastline. A series of subsidence coastlines with sedimentary facies of beach sands, sea shore shells, lagoon and delta deposits were found on the continental shelf of China Seas in the water depth of 8 m, 25～30 m, 50～60 m, 100 m, 120 m and 150 m. C‐14 data of these coastlines indicate that these were lower sea level coastlines formed during different stages of Holocene transgression.

By using a combination evidence of coast morphological features, such as sea notches, sea cliffs, wave cut platforms and sea stacks, it is able to identify the elevated coast terraces even though there are absence evidence of coast sediments. A series of terraces of old coastlines with heights of 5 m, 15～20 m, 40 m, 60～80 m, 120 m, 200 m, 320 m, 450 m and 600 m around the ancient island of Yuntai Mountain in Jiangsu Province can be distinguished and the features also found in the other areas along China Seas. The most of high terraces were marine erosional rocky platforms, but lower terraces were different types. The rocky benches of 5 m terraces are close to the location of storm benches of present time, but deposited 5 m terraces of sandy beaches are located in the adjacent coastal bay with unconsolidated well sorted sands and shell fragments, or beach rocks and coral reefs above modern high tidal level. C‐14 data of 5 m terraces are from 1 000 to 7 000 YBP, thus, these are old coastlines of Holocene higher sea level. The most of 10 m terraces were depositional formation, the sand layers are consolidated in the brownish colour, with C‐14 data in the rank from 7 000 or 8 000 YBP (Qinhuang Dao) to 28 000 YBP of the late Pleistocene (Hainan Island). However, in Hainan Island, the heights of same depositional terrace can be found from 0. 5 m to 15 m in the short distance along modern coastline, the changes were caused by local volcano eruption. The terraces of 20 m, 30 m and 40 m are mostly erosional type, but with a little of beach pebbles and ancient cultural remnants of early men. C‐14 data of several shell fossils from 30 m terrace in Hainan Island indicate a late Pleistocene coastline during 15 745±102 YBP.

It is true that there are old coastlines located in the different heights, formed during

＊ Ying Wang: International Union for Quaternary Research (INQUA), ⅩⅢ *International Congress Abstracts*, August 2‐9,1991, Beijing, China.

different time and with different genetic mechanism. Even though，same period of coastlines can be with different altitude and there is not any direct way using old coastline to indicate developing tendency of modern coast.

Several Aspects on Sea Level Rising and Its Effects on Coastline in China[*]

The State Survey Bureau of China has announced that (1) the rate of sea level rising in China Sea's is $2 \sim 3$ mm per year, and it is rising continually. (2) During recent hundred years period, sea level has risen 19 cm in the East China Sea and 20 cm in the South China Sea.

The effects of sea level rising on China's coastline vary in the areas:

1. Land subsidences apparently in the delta area, such as in Tianjin of the Haihe River delta, it was 80 mm subsidence in 1985, 64 mm subsidence in 1986, 43 mm in 1987, and the dock of Tianjin New Harbour at Tanggu subsidenced 0.5 m in 1985. As a result, the area suffer from storm surge often. Shanghai in the Changjiang River delta, the average rate of annual land subsidence was 4.0 mm $\cdot$ y^{-1} during 1977 to 1987 (Guo, Hsia-Chuang, 1989). In Laizhou Bay of Bohai Sea, because of over pumping ground water of 38×10^8 m³ during 1976 to 1989, ground water table went down for 15 m as the average, and forming a funnel-shaped water table in the subground area of 2 000 km², where has a 1 600 km² area lower than present sea level. As a result, salt water has intruded rapidly. The farm land becoming salinized soil, it has also happened in the Rizhao along the Yellow Sea.

2. Large rivers effects to Sea level changing predominately. In the Yellow River mouth, coastline prograding fast can reach $10 \sim 15$ km $\cdot$ y^{-1} as a first accumulated rate in the new river mouth of Bohai Sea, and the coastline at abandoned Yellow River mouth of the 1128－1855 delta suffers from erosion, because of lack of sediment supply. The regression is 17 km since 1855. Changjiang River runoff has influenced the monthly sea level rising for 6－11 cm and to form a seasonal difference to the variation (Wang, Baocan et al, 1989).

3. Coastal erosion as a result of sea level rising happens in the sandy coast plain, such as: $1.5 \sim 2$ m $\cdot$ y^{-1} retreat in the Shandong and Liaodong Peninsula; $2 \sim 5$ m $\cdot$ y^{-1} in Zhejiang and Fujian Province along the East China Sea. The reason of coast retreat is not only by the rising of sea level, but also by the artificial mining.

It may show the net result of coastline erosion by sea level rising along reddish sandy terrace coast, with the retreat rate at $0.7 \sim 1.5$ m $\cdot$ y^{-1} along Fujian and Guangxi

　　* Ying Wang: *3rd International Geomorphology Conference*, *Programme with Abstracts*, pp. 269. McMASTER UNIVERSITY, HAMILTON, ONTARIO, CANADA, 1993. 8. 23－28.

coasts.

It is limited erosion along bedrock coast as most bedrock coasts are granite and gneiss, which resist the coast erosion $0.07 \sim 0.1 \mathrm{~m} \cdot \mathrm{y}^{-1}$ retreating as the serious example along the East China Sea.

4. Relative stable sea level changing in Hainan Island influenced by tectonic uplifting.

Interruption of Flow in the Yellow River and Its Impacts on the Coastal Environment[*]

The Yellow River is characterized by three major features: One is the exceptionally high sediment load. The second is the marked seasonal change in flow and in regional distribution of sedimentary deposit. Third, the channels in the lower reaches of the river are frequently changing their course.

The total amount of water contained in the Yellow River drainage basin is 744×10^8 m³ and the annual water discharge is, including ground water(?) 580×10^8 m³. In years with abundant rainfall the river discharge is about 10. 3 kg/m³. The maximum annual sediment transport, recorded at the Sanmen canyon, is 3 910 Mt (million ton) while the minimum is only 1/8 of that mass, or 489 Mt.

Cessation of water flow in the lower reaches of the river occurred for the first time in 1960, and since 1991 flow has ceased completely during the dry season. Analysis by the present author shows that the main reason for this is excessive diversion of water upstream. Before the 1960's, even when the average precipitation over the drainage area was as low as 210 mm/year, the discharge volume into the Bohai Sea exceeded an average of 100×10^8 m³/year. The reason for the maintenance of uninterrupted river flow, even under these conditions, was that at the time the controlling factor, the artificial diversion of water upstream was still as low as 175×10^8 m³/year.

The cessation of flow phenomenon leaves an area of about 54 000 km² in the Yellow River delta without fresh water. The resulting loss of sediment deposition in the delta is leading to coastal erosion, loss of farmland and adverse ecological effects. Cumulative economic loss is estimated at 26. 8 billion (10^9) RMB (3. 4 billon U. S. ＄).

Over the last millennium, the sediment distribution in the middle-to lower reaches of the river has been determined by the occasional changes in its course, ranging from the present channel in the north to the Huaihe River channel in the south (ref. to map). Corresponding environment changes have affected the Bohai Bay and the Yellow Sea.

A strategy for sustainable management of the Yellow River and its drainage basin requires major efforts. An integrated plan has to be established for water management in

---

　　＊　Ying Wang: *International Coastal Symposium 2000*, *Conference Program and Abstracts*, pp. 28 - 29. Rotorua, New Zealand, 2000. 4. 24 - 28.

the river basin, limiting water diversion to on more than 200×10^8 m^3/year, and allowing discharge to reach the sea. Such a plan, to be effective, also requires a program to create public awareness of the environmental necessities and the value of the water resources.

River-Sea Interaction and the North Jiangsu Plain Formation[*]

　　The North Jiangsu Plain is located on the northern side of the Changjiang River and south of Lan Shan Tou Cape in the area $32°10'$ to $35°05'$ N and $118°40'$ to $120°30'$ E. The landform is mainly lowland declined from the Grand Canal in the west towards the Yellow Sea coast. The lowest part is the Xinhua-Sheyang lake area in the middle of the plain. A network of rivers, canals and lakes is the main feature of the plain with the abandoned Huanghe River-Huaihe River system in the north and the Changjiang River and numerous lakes in the south. Previous studies have discovered that the eastern part of the plain was formed during Holocene high sea level stand, evidenced by a series of shell beach ridges located in Longgang, Dagang, Dongtai and Hai'an over a length of 200 km north to south and about 60 km west of the present coastline. Carbon dating shows it was formed 6 500 to 5 600 B. P. The Fan Gong Dike was constructed on the shell beach ridges during 960 – 1127 AD to form a complete embankment. The aim of the study is to trace genetic formation of the plain inside this old dike.

　　Geomorphic features of the plain west of the Fan Gong Dike indicate: (1) The North Jiangsu Plain is circumscribed by a series of rocky hills as an arc-shaped"bay". (2) A series of lakes (Hongze, Gaoyu and Shaobo) located to the west of the Grand Canal are almost connected to each other by swamps. Several rivers enter these lakes with deltaic forms on the west shore. The Grand Canal was actually built partly using this system of interconnecting lakes. (3) A series of manmade islands were built based on natural bars or barriers of the lakes and swamps. These islands are named Duo, Dun or Wei, each meaning "manmade bar or crossway" in Chinese. They are aligned north to south and shell fragments support the notion that the area originated in the sea.

　　The BY1 core (145 m deep) from Wangzhigang Town, Baoyong County was collected between the lakes and old coastal ridges. The core penetrated almost all of the Quaternary sequence though owing to sand layers only 97 m of sediment was retrieved. A comprehensive study of sediment grain size, structure, facies, mineral and chemical element analysis, palaeomagnetic and palaeo-fossil analysis has indicated a river-sea interaction in the formation of the plain. Marine foraminifera found mainly from 39. 16

　　* Ying Wang, Zhenke Zhang, Dakui Zhu, Jinghong Yang, Longjiang Mao, Shuheng Li : *International Association of Geomorphologists—Regional Conference*, pp. 25 – 26, June 25 – 29, 2007, Kota Kinabalu, Sabah, Malaysia.

to 14. 70 m indicated a water depth of $20\sim50$ m, salinity of 3. 1 to 1. 5% and water temperature of about 150 ℃. The timing of this layer is thought to be Late Quaternary, earlier than the hard clay layer of the Changjiang delta area but later than 0. 78 Ma of the boundary at 39. 16 m. Carbon dating (AMS) gives dates of 26 470 ± 60 B. P. (16. 15 m); 32 910±170 B. P. (23. 90 m) and 39 385±170 B. P. (23. 90 m). Marine facies are found at 8. $7\sim1.$ 0 m, 65. $4\sim58.$ 0 m and 81. $00\sim79.$ 66 m. Although few forams are found from these layers it is clear that sea water inundated this area from time to time during the Late Pleistocene although it is now some 120 km from the coast. Terrigenous facies of alluvial plain or shallow lakes are found at $97\sim81$ m, 79. $66\sim$ 65. 40 m, 58. $00\sim39.$ 16 m, 14. $72\sim8.$ 70 m and also at the surface. Sedimentary facies reflect a change from coastal bay of a shallow sea with water depth less than 100 m to a wide plain during the Pleistocene. It became a fluvial plain in the Holocene. The current study traces the evolutionary history of the plain through geomorphic analysis combined with multidisciplinary analysis of sedimentary facies. The results can inform regional planning and development.

River-Sea Interaction and the Eastern China Plain Evolution[*]

River-sea interaction has impacted greatly on the formation of the eastern plains of North China. In particular, the Yellow River and its tributaries have transferred vast volumes of sediments from the Loess Plateau to the sea. From the Pleistocene to the present, great volumes of sedimentary materials have been deposited on the North China Plain. The North Jiangsu Plain located on the northern side of the Changjiang River is another example formed by river-sea interaction. The plain is 200 km wide as measured from a series of lakes in the west located at the foot of small hills and low mountains arranged in arc-shaped bay patterns towards the Yellow Sea. The Grand Canal was dredged through these lakes and depressions in the 7 - century some 1 400 years ago. On the east side of the plain, an artificial dike was constructed along a band of coastal chenier ridges during the Song Dynasty (1127 AD). But at present that dike system is 40 km inland to the west of the present shoreline. There is a sandy ridge field off the North Jiangsu Plain in the southern Yellow Sea. Geomorphological research indicates that this sand field was deposited around 30 000 a B. P. , at which time the Changjiang River had migrated north to enter the Yellow Sea in the Qianggang area of the present North Jiangsu Plain. The river deposits were reworked by strong tidal currents driven by > 9 m tidal ranges during the post-glacial transgression and a period of higher sea level. At present the sandy ridge field occupies an area some 200 km long from north to south, and 150 km wide from land to sea in a radiative pattern. The total area is ∼ 20 000 km^2 consisting of 70 ridges with deep channels between them. Water depth of the offshore sandy ridges is 10∼20 m, but with 1/10 of their area above sea level. With the protection offered by the offshore sandy ridge field, tidal flats developed along the coastline behind these ridges, even though the Changjiang River had shifted back to the East China Sea, and the Yellow River returned to the Bohai Sea. This formation mechanism may indicate the way that the North Jiangsu Plain formed in the past.

This present study was carried out on the inner plain to trace more evidence of an original west boundary of the Yellow Sea, and has been stimulated by the discovery of several salt mines, which were discovered buried in the inner part of the plain.

---

* Ying Wang, Dakui Zhu: *Joint AOGS 1ʳ Annual Meeting & 2ⁿᵈ APHW Conference*, *ABSTRACTS, Vol. I* , pp. 605, Natural Hazards 1, 57 - ONH - M176. Suntec Singapore International Convention & Exhibition Centre, Singapore, 5 - 9 July, 2004.

第二篇
中国典型区域海岸海洋研究

贝壳堤是平原海岸高潮岸线的标志,是浪流冲刷淤泥粉砂质潮间带浅滩,将贝壳与细砂物质堆积在高潮水边线附近,逐渐积累成海岸自然堤。岸堤的分布标志高潮岸线的位置,其物质组成(贝壳种类与泥沙粒度)反映当时的海岸环境(海岸动力、滩涂泥沙与相应的生物群体)。贝壳堤高、蔽浪,贝壳砂层与底部泥滩间聚积雨水为淡水源,为沿海渔民提供聚居环境,并反映着海岸线蚀、淤变化与稳定程度。贝壳堤是记录该段海岸演变的"天书"。

南黄海海底巨大的辐射沙脊群是海面上升过程中沉溺形成的巨大的古长江三角洲。沉积体蕴积着河海交互作用的蚀积过程历史,启示该环境演变的基本规律。

河海交互作用长江三角洲区域环境变化研究的实例。

海岛与港湾特点的研究实例。以南海,热带海洋环境为实例,探讨其地质与海洋环境特点、动力与沉积过程发展规律与适宜的开发利用。

渤海湾北部海岸动力地貌 [*]

1959 年夏季,当作者在北京大学地质地理系做研究生时,带领该系地貌专业部分同学参加了塘沽新港回淤研究的海岸动力地貌调查。是年冬季在导师王乃樑教授的指导下,将工作中的一些初步认识写成本文,供有关方面参考。

一、区域海岸发育因素

本文研究的范围是渤海湾北部,从大蒲河以南开始,经过滦河,沙垒田岛,南堡,到蓟运河口为止的海岸带。现就影响本区海岸地貌发育的基本因素分述如下。

(一) 地质基础

本区在大地构造上属于华北陆台渤海凹陷的一部分,从燕山运动以来一直下沉,堆积了巨厚的松散沉积层。据天津深井资料,在地下深达 850 m 处尚不见基岩,目前尚处下沉中。由于河流带来大量的物质的堆积,使海岸不断向海推进,所以下沉趋势不明显,下沉速度与堆积速度大致相当。

(二) 河流

1. 滦河

为本区的一条主要河流,发源于承德专区丰宁县西北的巴延图古尔山麓,流经燕山山地于乐亭县入渤海,全长约 877 km,流域面积为 4 900 km²。

滦河水流主要来自夏季降水,沿途汇注支流很多,其中常年有水的达 500 多条。滦河年平均流量为 136m³/s,历年瞬时最大流量曾达 3 000 m³/s,历年最枯流量为 20 m³/s。水文变率大,含沙量较高,年输沙可达 1 670 万 t,多集中于八月洪水期,此时输沙量约占全年总数的 63.5%,二月份最少,有时仅为全年的 0.1%。

滦河在罗家屯以上为山区,该段河床坡度约 1/500,上游河道沉积主要是卵石;罗家屯至滦县间为中游,所经为花岗岩流纹岩的低山丘陵,河谷展宽约 1 000 m,河床坡度约 1/1 000,为沙质河床;滦县以下是下游区,河道流经广大的冲积平原,河槽宽 2 000～3 000 m,河床坡度约 1/4 000,河床沉积为细砂物质,河流在此段冲淤甚快,改道变换很大。滦河的浑水入海后可远伸到距岸 20 km,刮西北风时滦河浑水向海中伸延最为显著。因此,滦河对沿岸带地貌影响是很大的。

---

[*] 王颖:国家海洋局海洋科技情报研究所,《海洋文集》,1965 年第 3 集,25 - 35 页。

原文"渤海湾北部海岸地貌"是王颖 1959 年 11 月在北京大学地质地理系攻读地貌专业副博士研究生期间的学年论文,导师是王乃樑教授。1965 年作少许修订后发表。

2. 蓟运河

发源于大水泉,到北塘以下入海,是一条常年有水的河流,河床比降为 3.5‰,蓟运河流量输沙量皆小,汛期流量约 1 300 m³/s,年平均输沙量约 60 万～100 万 t,主要是粉砂及淤泥质。由于河流泥沙较少,并受涌潮的影响,所以河口发育成三角港。

老米沟、大清河、大庄河、沂河等皆发源于冲积平原区,河流都很短小,水量不大,它们过去曾是滦河的支流,目前已日渐干涸,只是在河口段仍有些水流,河流本身对海岸的影响作用已很微弱。

(三) 风与波浪

本区冬季受蒙古高压控制,多北向风,夏季受夏威夷高压影响,多南向风。冬季盛行强大的西北风,显著地影响着岸边减水,当强烈西北风吹刮时,潮水往往后退很多。冬季尚有东北风,它是本区的一个吹经海面的强力风,也出现于春秋两季,对沿岸带波浪、水流影响很大,但它历经海面的吹程还不是很大的。夏季多东南风及西南风,东南风吹经较宽阔的海面,造成较大的风浪,并使沿岸发生增水现象,而南风是本区频率最高的常向风,因风力小且是陆风,所以,对海岸地貌的影响不及前两者大。必须提及此处沿海的海啸,它常常是由夏季强大的台风所引起的,当台风经过时,形成了强大的涨水及冲流,对海岸破坏很大。风的季节分布以春季(4、5月)最多,风速亦大,夏季时风少。现将昌黎、乐亭、塘沽等地风的资料总结成表 1。

表 1 风资料

季 节	常 风	平均风速	强 风	一般风速	最大风速
春 季	SW	4～5 m/s	NE	6～7.8 m/s	12～18 m/s
夏 季	SE	4 m/s	NE	4～4.5 m/s	10 m/s
秋 季	NW	3.5 m/s	NE	—	10～16 m/s
冬 季	NW	3～4.5 m/s	NE、NW	—	8～14 m/s

波浪与沿岸带的风有密切关系,在吹东北风、东东北风及东东南风时波浪最大。而以东南及东向的波浪最多。在夏季时东东南、东南向的波浪较多;秋季时以南南东及南向的波浪较多;冬季春季波浪多为东北及东东北方向的。东东北风所造成的波浪最大,其波高可达 3～4 m,波长平均为 30～31 m,最大者已测知的达 41 m,东南风所形成的波浪亦如此,这些波浪多半与海岸成斜交,其作用力皆可影响到沿岸带海底。所以,可掀起并携带泥沙沿岸移动。

(四) 潮汐与流

本区沿岸带均为不规则半日潮,一般是落潮延时大于涨潮延时。潮差由北往南逐渐加大,滦河口平均潮差为 1～1.5 m,大清河一带潮差不及 2 m,尖坨子湾岸处平均潮差约 1.3 m,而蓟运河一带平均潮差约 2.4 m,最大可达 4 m。

潮流的情况是,自大蒲河往南一直到滦河三角洲外围地带,涨潮流皆向西南,流速一般为 0.6～0.9 m/s,落潮流向东北。在南堡东西两侧湾岸中为两个小型的顺时针方向的转流。蓟运河到涧河一带,涨潮流方向为西北与西西北,流速 0.4～0.9 m/s,落潮流方向为东南及东东南。除滦河部分以外,各处都是涨潮流速大于落潮流速。由于流速不等,使涨潮所带入的泥沙,不能为落潮所带走,尤其是在水层的底部更显著,因此造成浅滩的淤积。

除潮流外,恒流也很重要,它与沿海的风情密切相关,由于东北风、东东北风、东风的势力强,所以从春季开始一直到入冬,此处沿海皆有一股强大稳定的恒流从东北往西南,其影响范围可自秦皇岛到达渤海湾北部湾顶,分布的深度从水面一直到底层,仅中层部分在五、六月时有些变化,在此恒流作用下,沿岸有大量泥沙不断自东北向西南运移。

其次一支恒流来自渤海湾南部,它是在南风、东南风及东东南风的作用下,从黄河口向西北伸,亦可到达渤海湾北部湾顶。此恒流的性质是底层比较稳定,从春季到夏季皆是,中层变化较大。这支恒流不及北部的强大,但底层稳定,而黄河带入海的泥沙主要是细粒的粉砂、淤泥,它们大多分布在底部水层,所以这恒流仍携运了大量的泥沙,成为沿岸带一个重要的动力因素。

由上分析可以看出,本区海岸泥沙运动与潮流恒流密切关联,这些水流皆向渤海湾北部湾顶处集中,而该处又极隐蔽,所以最利于淤泥浅滩的发育。

二、海岸类型

本区海岸长达 235 km,根据其成因、形态特征、动力特点、发育年龄及发展动态等指标可以划分为下列三个类型(图 1)。

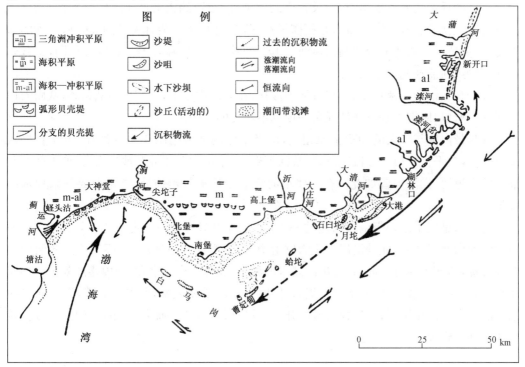

图 1　渤海湾北部海岸动力地貌图

(一) 大型的沙丘海岸

此类海岸北起大蒲河口[①],向西南延伸到滦河北岸,岸线全长约 25 km,沿岸皆分布着高大的风成沙丘,沙丘带宽约 2 km,沙丘多成垅岗形,靠海者多与岸线平行分布,内侧者大致与岸线呈斜交的方向,沙丘高大,一般在 20 m 以上,最高者可达 42 m,沙丘带的南段比北段高,在新开口以南沙丘多在 30 m 以上,往北到大蒲河一带沙丘高度即有所减低,分布亦零星。同时外侧靠海的沙丘比内侧的高,高大的沙丘多滨海分布,成为一列沙山(图 2)。

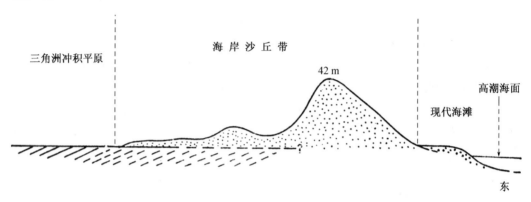

图 2 沙丘海岸地貌综合图示

沙丘的陡坡朝向西南,约 31°,缓坡为 25°,这反映着强劲的东北风对沙丘影响较大,但因西南风频率高,所以沙丘缓坡与陡坡坡度相差不大。

组成沙丘的物质主要是石英、长石质的中细砂,石英约占 60%,长石占 20%～30%,此外尚有磁铁矿、钛铁矿、绿帘石与柘榴子石等重矿物,亦夹杂有大量的贝壳屑。砂的磨圆度较好,这些物质的成分仍与沿岸河流上源的花岗岩、片麻岩等基岩成分相同,是基岩风化产物经河流冲刷携带入海的。

大沙丘上都没有植物生长,仍属活动沙丘,在海风作用下,它们向内陆移动,威胁农田与村落。所以,固定这些海滨沙丘是一项刻不容缓的工作。

沙丘带内侧与广大的冲积平原相接,流经平原的河流受阻于海滨沙丘,不能直接入海,而与沙丘平行流很长一段距离,当数河汇聚,力量增大后才冲开沙丘入海,如大蒲河及北面的人道河、洋河等皆如此。在一些小河流沿沙丘平行延伸处,常常形成一系列水泽洼地。

七里海是位于沙丘带南段内侧的一个潟湖,有很窄的通道与海相连。1813 年,滦河自大清河向北迁移后,此处沿岸冲积物供给增多,海滨沙丘迅速增长,逐渐使南北两岸的沙丘合拢,堵塞了通道。此后,七里海湖水即淡化,成为一个淡水湖,湖内盛产菱莲。1883年夏,由于滦河泛滥,支流北上到七里海,同时沿海因长期风暴,而发生海啸,海水向岸冲击,在河、海交汇作用下,大沙丘又被冲开,形成目前的新开口通道,七里海又与海相通,转变成潟湖直到现在。

---

① 沙丘海岸北达南大寺,大蒲河以北非本文研究范围,故不在此介绍。

据七里海潟湖平原上的浅井剖面,可见到海、河交互沉积的情况(图3)。

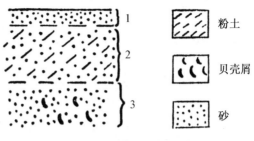

	粉土
	贝壳屑
	砂

图3　七里海沉积剖面图

1层:厚20 cm,潟湖相的灰蓝色砂层,主要为石英砂,内有贝壳砂及大量有机质,略具腐臭味。

2层:厚1 m,为黄色砂及粉砂层,中细砂,具水平层理,为河漫滩相沉积。

3层:黑灰色砂层,质粗,内夹有很多贝壳及螺化石,为海相沉积,厚度不详,出露厚度约1 m。

各层皆是逐渐过渡的。

七里海的通道——新开口,目前较狭窄,仅数百米,且遭受到海滩砂堵填。七里海中仅在高潮时水面开阔,已处于淤浅状态。

沙丘带外侧相毗连的是砂质海滩,其组成物质与沙丘物质相同。海滩的平均宽度约200 m,有些段落海滩很窄,沙丘即濒临水边。

大型沙丘海岸的岸线平直,沿岸无泥滩发育,水下斜坡较陡,平均坡度为1/200,波浪直接作用于岸边,为塑造本段海岸的主要动力。同时,风力的作用也很突出,海滨沙丘的形成就与此处自海吹向陆地的东北风密切相关。海风将海滩砂吹移,在有植物及其他障碍物阻挡处,砂子逐渐堆积形成海滨沙丘。此处各河流供给了丰富的冲积砂,沿岸带坡度适宜,有利于波浪的横向搬运形成砂质海滩,同时地势广平无阻,风力有用武之地,各种有利条件的结合使得此处形成了大型的海滨沙丘。因为沙丘是海风吹扬海滩砂而堆积的,所以靠海的沙丘系列要比内侧的高大。此处沙丘形成以后对沿岸带河流起了一定的阻碍作用,由河流带入海的冲积物的数量比沙丘形成的初期要少了,而风力由海向陆地吹移砂子的作用仍继续着,因此,渐渐形成了海岸带冲积物不足的现象,引起海岸的冲刷。目前,海滩是在后退之中的。

(二) 具沙坝环绕的滦河三角洲冲积平原海岸

此类海岸北部与沙丘海岸交叠衔接,最北到七里海,南部界限是大庄河一带。岸线长达100多km,呈东北—西南走向。其范围包括了滦河三角洲的全部。滦河是形成此段海岸的主要动力,其次才是波浪与海流。

由于三角洲的生长,海岸呈现向海突出的弧形,三角洲外缘环绕着一系列沿岸沙坝,因此形成了双重岸线的特点(图4)。根据三角洲各部分目前受滦河影响的大小及其发展状况,将本段海岸分为新、老三角洲两部分。

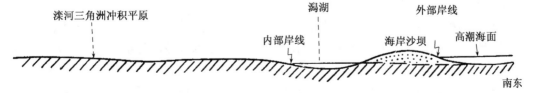

图4　具沙坝环绕的滦河三角洲冲积平原海岸地貌综合图示

1. 新三角洲

其范围起于此段海岸的北界,向南伸展到湖林口一带。湖林口之西南到大庄河是属于老三角洲的范围。

新三角洲部分,主要是滦河自大清河北迁后逐渐形成的,1813 年以前滦河在大清河一带入海,1813 年以后滦河北移流经老米沟一带入海,1915 年滦河才迁至今日的河道所在,但最初此段河道仅为滦河的一个分支,一直到 1938 年甜水沟主道废弃后,滦河主流才集中在今日的河道入海,所以新三角洲沿岸带主要是近百年来受滦河影响而发育的。

滦河下游流经松散物质组成的冲积平原,两岸地势低缓,河床宽浅,河水变率又大,河流含沙量很高,故淤积极盛,河床中发育有很多心滩、沙洲,河流多年经过的地方,河床即淤高很多。每当夏季汛期时,洪水猛然泄下,河床容纳不了就发生泛滥改道。因滦河改道频繁,当地居民称它为浪荡河。滦河长期的迂回泛滥,渐渐形成广大的三角洲冲积平原。平原上地势平坦,河渠繁多,并遗留很多废河道,沿废河道两旁相伴地分布着小型的滨河床沙丘(已固定),它们组成了平原上主要的微地貌形态。

滦河河口段为动态最活跃的部分,该处河道分汊多,出口改变极频繁。历史时期中,整个滦河下游河道有自西南往东北迁移的趋势。而河口部分,近数年来又渐渐往南迁,据访问知,1952 年河口位于现今位置以北 2.5 km 处,1953—1954 年南迁到现在河口北面的 1.5 km 处入海,1958 年迁移到现在的位置。河口伸展到那里,那里就淤积得多,岸线向前移就快。所以河口段附近岸线外形最为突出,是三角洲向海生长最快的部分,从 1915 年滦河发大水冲开了该处海岸带高 25 m、宽 200～300 m 的大沙丘,开辟了新河道入海后,三角洲即开始发育向海推进,尤其是 1938 年后三角洲生长更为迅速,20 多年来,三角洲已向外伸展 7～7.5 km,平均伸长速度约为 200 m/a。

向东突出的新三角洲是不对称的,北部宽短,南部窄长。滦河以北部分,由于岸线面迎东北风及涨潮流,潮水沿着一些河岔小凹地入侵,形成了规模不大而极为密集的潮水沟,它们分割岸线使外形呈锯齿状。而三角洲的南部,分布着几条较长大的河流——老米沟、臭泥沟、滦河岔河等,潮水沟不及北部密集,但因为是沿着长大的废河道发育成的,所以规模却较北部的大。近期由于海水的入侵,沿岸各河口及潮水沟口多呈现为开阔的喇叭形,潮水沿河上溯,在河流相的冲积层上覆盖了薄层的海相沉积。

沿三角洲的外缘,与岸平行分布着一系列沙坝,它们组成外部岸线。北部的沙坝数量少规模小,延伸不远。河口附近的沙坝因有上、下水流的影响,沙坝的规模也不大。河口以南的沙坝数量多,延伸很远,且自滦河岔河以下往西南规模逐渐加大,这些沿岸沙坝长约 2～6 km,宽度自数十米至 1 km,沙坝约高出海面 1～2 m,向海坡度一般为 6°左右,向陆坡度为 2°～3°,当内侧与泥滩相接时,其坡度更小,不到 1°。在沙坝中部,高潮水所不能到达的地方,发育了风成沙丘。

沙坝的组成物质主要是石英、长石质中细砂,磨圆度中等,同时沉积着较多重矿物;磁铁矿、绿帘石、柘榴石及金红石等,它们在沙坝外缘坡上部及坡脚处组成重矿富集带。沙坝物质的成分与滦河冲积物相同,自滦河口出来的物质在波浪及流的作用下,向南北移动,堆积成一系列海岸堆积体。因为此段海岸东北风浪及恒流作用强,且涨潮流向西南,所以自河口北上的物质不多,大约到王家铺北 3 km 处,物质的纵向运移已渐停止。而自滦河口向西南运行的沉积物数量多、规模大,使南部形成了一系列长大的沙坝,滦河物质供给丰富,目前这些沙都是在活动,在增长的,在沙坝西南端(滦河口以南的沙坝)及西北端(河口以北的)皆有新的水下堆积体延伸出去,在河口部分的沙坝是向海推进着,目前在河口外又生出一列新沙坝。所有沙坝都是不连续的,它们常为河流及潮水沟的出口所间断,这在三角洲的南部最明显,在缺口处,因潮水及波浪的横向作用,其尖端皆向内弯,虽然沿岸有一系列的河口水流作用,但沙坝仍不断增长,这反映着自滦河出口往西南的泥沙流是强大而稳定的。沙坝的生长对沿岸河口亦有影响,它们的延伸,迫使河口及潮水沟亦向西南方向偏移。

位于湖林口的大港沙咀是此处规模最大的海岸堆积体,长达 13.5 km,宽约 1 km。按其位置、形态、规模及物质成分来看,应是滦河在老米沟湖林口一带出口时,自河口向西南运行的一股沉积物流,由于海岸方向转变,容量降低而形成的堆积体,其年龄是和在新三角洲西南部,狼窝口和老米沟一带规模较大的沙坝相同,是古代东滦河在 1813 年滦河在老米沟一带出口时所形成的。但是在动态上是与狼窝口、老米沟一带的沙坝有区别。目前上述的沿岸沙坝皆是受滦河的沉积物补给在继续生长发育,而大港沙咀却是因沉积物补给缺少,遭受冲刷不断后退,在大港沙咀外缘坡脚已有泥滩出露。滦河附近的沙坝,共生成年代较近,是在滦河移到现在位置后生成的。

2. 老三角洲

湖林口以南到大庄河岸线向西转折,其外形向南南东方向略微突出,这是老的滦河三角洲部分(包括蛤坨到曹妃甸的沙岛在内),这一带的地貌形成是与滦河自 1453 年到 1813 年在大清河大庄河一带出口密切相关,沿岸陆上地貌与新三角洲的沿岸地貌相似,亦为细砂、粉砂质的三角洲冲积平原。其地势平坦,但河道不及北部密集,仅有的几条河,如大清河、大庄河及大苗庄河等大多已干枯,下游主要为潮水所占。同时,由于此处沿岸带堆积体遭受强烈破坏,平原掩护条件不及北部严密,所以不仅河口处海水入侵形成喇叭口的形态,沿岸带平原亦广泛受潮水的淹没,因而老三角洲冲积平原的边缘部分渐发展为冲积-海积平原。

老三角洲的外缘,自大清河口往西南亦分布着一列海岸堆积体,组成了外部海岸线,如石臼坨、月坨、蛤坨、曹妃甸及白马岗等。这些海岸堆积体成为长形小岛或圆形沙岛的形态,还有很多是水下沙堤和沙堆。小岛自东北向西南伸长并降低其高度,在曹妃甸以西主要是水下沙堤。在各沙堤之间多隔有较深的沟槽,这是潮水冲蚀而成,目前这些沟槽皆日渐加深。

组成沙岛的物质主要是石英、长石并含有磁铁矿、绿帘石、金红石等重矿物。但其颗粒较北面沙坝为小,是细砂并且磨圆度极好。从物质成分看,此砂仍应来自滦河,但搬运距离较远,并且沉积较早,日后又经受了波浪的再次磨蚀,所以颗粒细磨圆度好。它与北部沙坝区别的是含有大量当地海域的生物,因此有大量贝壳及贝壳屑沉积。贝壳的种类较多,主要有牡蛎——近江牡蛎(*Ostrea riuularis*)与猫爪牡蛎(*Ostrea [crassostrea] pestigris hanley*),青蛤(*Cyclina sinensis*),文蛤(*Meretkix linne*),毛蚶(*Arca subcrenata liscnke*),蜒螺(*Umbonium thomasi*),珠带锥螺(*Tympaotomus cingulatus*)与李氏金蛤(*Anomia lischkei dautzenbegos fisher*)等。

从沙岛的形态、位置及物质成分等各方面分析,这一系列堆积体,仍类似于今日滦河口外围的沙坝,也是滦河的冲积物受波浪与流的作用后,沿岸移动而堆积所形成。不过其堆积时代较早,大约相当于滦河从 15～19 世纪时在大清河大庄河一带出口时所形成,为老三角洲外围的堆积体。当时它的形态较完整,规模远较今日为大,滦河向东北方向迁移后,物质供给突然减少,使这里动力条件有所转化,由于沉积物流供给不足,冲刷作用大大加强了,近百年来这些老沙坝遭受着强烈的冲刷,处于日渐减小与后退中。例如曹妃甸,传说方圆曾达 20 km,但今日宽仅几十米。据我们访问,在 40 年前岛中部灯塔外缘有一曹妃庙,有僧人在庙旁种菜,后因沙岛受冲刷变小,海水渐至庙前,庙被迫迁往石臼坨,而今日曹妃庙墙基已全部在海水下面,庙后方的灯塔也已内迁,目前曹妃甸在大潮时已完全被海水淹没。其他的沙岛如蛤坨、石臼坨等也是受冲刷日渐变小。所以,从这一系列堆积体的动态看,也与北部的沙坝有所区别。但从残存沙坝排列方向来看,当时形成老沙坝的动力是以指向西南方向的波浪和海流占优势,由此可推论:在历史时期中,此处沿岸的水文气象条件并未发生多大变化。

在上述三角洲的内外岸线之间,为一浅水潟湖带,目前主要受潮水涨落的影响,由于外部沙坝的封锁作用,此处发生堆积,发育了粉砂淤泥质浅滩。浅滩的增长是自陆向海与沙坝内侧向陆地两方面同时进行的。据访问,浅滩堆积速度平均为 10 m/a。由于潮水的堆积作用及沙坝向内陆方向推掩,所以潟湖在日渐缩小中。

(三) 具有残留贝壳堤的冲积-海积平原海岸

从大庄河到西蓟运河岸线全长约 110 km。岸线方向从大庄河到南堡为东北—西南向,南堡到涧河为东南—西北向,涧河到蓟运河是由东到西后转为南北向。

这段海岸地势低平,除西部的蓟运河附近发育着冲积-海积平原外,其他地区都是河流作用微弱、陆源物质很少供给的海积平原。海水入侵平原甚广,土地皆盐渍化。平原边缘及南堡北面断续地分布着一些残留的贝壳质岸堤,岸堤是激浪形成的,后因淤泥浅滩的发展使岸坡变缓,岸边无激浪作用了。因此,目前该段海岸的基本动力因素是潮流。

该段海岸根据其东部西部在形态与年龄上的差异,可分为两部分。

(1) 新生的海积平原海岸:东部从沂河经南堡至尖坨子,岸线是往南突出成为三角形,南堡正在此三角形地带的南端。其北部与断续的贝壳堤相接,此贝壳堤自尖坨子向西与涧河上的贝壳堤相衔接,贝壳堤反映着老岸线的位置(图 5)。这三角形地带的坡度极缓,大潮时海水可淹没到岸线以上 20 km。除边缘较高部分有渔村以外,内部没有村落,目前大部分辟为盐田。组成这地带的物质,表层为粉砂淤泥,一米以下为细砂,其下又有淤泥及细砂的夹层,砂中都含有贝壳,所以主要是海相沉积。由此可推知,南堡的三角形地带是新生的海积平原,它是形成

图 5 南堡海积平原海岸地貌综合图示

在北部的贝壳堤以后,比尖坨子以西岸线发育的年代要年轻些。

海积平原的外缘发育着淤泥质浅滩,南堡以东到沂河口,岸滩平均宽约 2～3 km。南堡以西到尖坨子一带,岸滩平均宽约 3～4 km。南堡处最宽达 6 km。南堡以东是粉砂、淤泥粉砂及少量细砂,其重矿物主要是磁铁矿、绿帘石、金红石与柘榴子石等,这些物质仍属滦河系统。而南堡以西的浅滩物质颗粒较细,为粉砂淤泥,含有较多的碳酸盐,与滦河的物质有所区别,而类似黄土物质。岸滩沉积物由海向陆逐渐变细,近陆处淤泥沉积较多,这反映着潮水搬运分选的特征。

该处沿岸带外围有沙岛及水下沙坝的围绕,而突出地带的西侧又位居湾岸隐蔽区,有利于堆积作用,因而促使沿岸带泥滩迅速堆积。据当地渔民反映:该处岸滩平均每年以约数十米的速度向海推进,其中以北堡一带最为迅速。

(2) 西部自尖坨子起往西到蓟运河口,主要是海积平原,靠近蓟运河处为海积-冲积平原(图6)。平原外缘分布着贝壳质岸堤,它向东即与南堡北面的贝壳堤相连,在这里已受破坏很不完整。目前贝壳堤高约 1 m,宽度 1～30 m 不等。堤顶较平缓,向海坡约 10°左右,向陆坡 6°,当内侧有潮水沟平行分布时,朝陆坡受潮水冲刷变得较陡,有达 20°的,由于受到激浪作用,岸堤已被冲刷间断,在断口处是潮水的通道,高潮时涨潮流的蚀积改造作用使贝壳堤呈现为两端向陆弯曲的圆弧段(图7),这是贝壳堤的残留形态。组成贝壳堤的物质主要是贝壳及贝壳屑(贝壳种类同前述),另外尚有细粒的石英与长石质砂子和少量粉土。在近河口处海岸演化较快的地段,贝壳堤有分支的形态,如蓟运河口以北,贝壳堤分为四支,涧河口以东贝壳堤分为两支。本区贝壳堤分支的成因是:在河口处海岸向前伸展较快,岸堤也适应新的岸线不断地发展,而河口段以外的邻近地区,恰巧又遭受冲刷不断后退,贝壳堤亦不断遭受破坏,而河口段老贝壳堤却被保存下来,这样渐渐形成了分支形态(图8)。

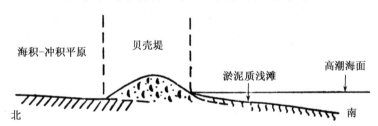

图6 具残留贝壳堤的海积-冲积平原海岸综合图示

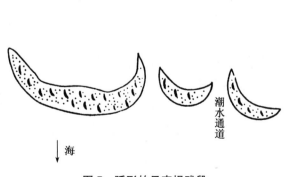

图7 弧形的贝壳堤残段

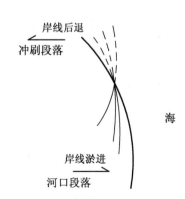

图8 分支贝壳堤形成图示

贝壳堤外缘与淤泥质浅滩相衔接,岸滩宽度在涧河以东平均为 3.5 km,而涧河以西宽约 1.5 km。其成分主要是含有大量碳酸盐的淤泥,并有窄带的粉砂沉积带。淤泥质岸滩西南延伸逐渐加宽,在岸滩的表层淤泥层下面是粉砂及粉砂淤泥的底部沉积。

淤泥质岸滩的坡度极缓(小于 1/1 000),目前此处只有潮水作用,波浪已达不到岸边了。可见,贝壳堤与淤泥质岸滩是不相适称的,岸滩非今日的产物。它是激浪流在海滩上部的堆积。形成岸滩时的岸滩坡度较今日为陡,那时激浪作用可充分发育。故形成贝壳堤的时代应早于淤泥滩,约相当于淤泥层下面的粉砂底质形成的时代,同时这些贝类在含有淤泥的粉砂沉积上最利于繁殖,因这种底质既透空气又有食料。所以在坡度较陡底质适宜的海滩段落发育了贝壳堤。这以后,由于一些携带大量淤泥物质的大河(在渤海湾主要是黄河),对渤海湾影响加强,沿岸发育了淤泥质岸滩,贝类繁殖受到窒息抑止,波浪作用亦因岸坡变小而达不到岸边,贝壳堤生长渐被中断,并遭受潮水的次生改造。

南堡海积平原与尖坨子以西虽为同一海岸类型,但这两部分海岸发育过程与发展阶段是不相同的。南堡突出地带岸外环绕着一列沙岛与水下沙堤,沙岛后形成波影区,使来自东面滦河方面与来自西面黄河方向的两股沉积物流,在岛后交汇发生了强烈的堆积,逐渐形成呈三角形突出的海积平原,其生成时代晚于西部岸线,当在北部贝壳堤形成之后。而尖坨子以西的岸线无沙岛环绕,水下沙堤也不发育,岸线微有冲刷,而岸滩仍在加积,加积的原因如下:

(1)从地形轮廓分析,此段海岸正位于渤海湾湾顶部分,环境隐蔽地势低平,波浪海流在此扩散,减弱了力量,有利于沉积。

(2)沿岸带南部有几条大河,如蓟运河、海河以至黄河,它们提供了大量泥沙,为岸滩堆积提供了丰富的物质。

(3)从水文气象资料分析看,潮流恒流的方向多自南北两面往此湾内集中,夏季风浪的方向亦向北。它们携带大量泥沙到此处沉积,尤其在南堡突出地带形成后,这里的淤积作用更加强了。

三、结　论

在讨论了本区海岸的类型、成因及发展状况后,我们有下列几点初步认识。

(1)历史时期中本区海岸经历数次巨大变化,其原因有二:

① 滦河改道的影响。滦河曾多次变迁改道,当其在大清河大庄河一带出口时(公元 1453—1813 年),即形成老的三角洲海岸,岸外还产生一系列沙坝,这些沙坝对南堡三角形海积平原的形成有密切关系,正是形成了老沙坝,才使得南堡三角形海积平原开始发育成长,它形成后使东西两侧形成湾岸,影响该处动力状况发生变化,促进了渤海湾湾顶淤积作用。

当滦河北迁后,又形成了新三角洲及沿岸一系列堆积体,目前使得此段海岸向前生长发展。

滦河的迁移对沙丘岸的发育亦有影响。

目前这些地带中,在接受滦河泥砂补给的海岸地段皆是向海推进的。较少得到供给或完全得不到供给的地段,则受到海水的冲刷逐渐后退。

由此可见,滦河发育变迁密切控制了北部海岸段落的生成和发展。

② 南面海岸受滦河及黄河双重影响。黄河是渤海湾最大的河流,年平均输沙量 9 亿 t,是渤海湾其他河流输沙量总和的 40 倍。因此,黄河对渤海湾的影响是巨大的,黄河长期作用形成了此处平原海岸的基础,当黄河改道由南部江苏省入海时,使此处形成适于贝壳堤发展的环境,发育贝壳堤,而当黄河又在渤海湾入海后,输入了大量淤泥粉砂物质,使沿岸广泛发育泥滩,海岸演变又进入一个阶段,目前南部岸滩发育仍与黄河的影响有关。

(2) 海岸各段落因动力条件的变化,目前尚未形成平衡剖面,各处的发展趋势是:

① 滦河三角洲部分今后仍将向前发展。

② 南堡海积平原仍要向海增长,而曹妃甸等沙岛继续受到冲刷破坏。随着这些沙岛变小后退,南堡海积平原的增长会逐渐减慢。

③ 在蓟运河东北一带,海岸遭到冲刷,但岸滩却在增长发展,这同南部不断地有物质供给有关,这里岸滩的发展在一段时期中可认为是稳步前进、变化不大的。

从海岸发展过程看,本区海岸的西段要较东段稳定些。

(3) 本区海岸升降状况:

因滦河带出的泥沙量大,三角洲不断在堆积发展,沉积很厚,而沉积较少的河口却都在发展成喇叭状河口及三角港(蓟运河),各潮沟口亦多呈喇叭型。河流沉积上多超覆着海相沉积。沿岸沙坝除邻近滦河口的以外,皆向后退,其外侧有泥滩出露。

由各种现象分析可知,本区海岸处于下沉中。下沉与下述三个原因有关:① 世界性海面上升;② 本区属渤海凹陷,地壳在下沉中;③ 本区沿岸巨厚的松散沉积在密实化沉陷过程中。

参考文献

[1] 塘沽新港回淤研究资料汇编. 1958 年.

[2] 渤海湾潮流系统分布图、余流矢量图. 见:科学院海洋研究所物理室报告,1959 年.

[3] 王曰翼,高培编. 昌黎县志(共 8 卷). 清康熙十四年(1675 年);陶宗奇等修,张鹏翱等纂. 昌黎县志(共 12 卷). 1993 年.

[4] 袁棻修. 滦县志. 共 18 卷,1937 年,1085 页.

[5] 史香崖,李润霖编纂. 乐亭县志. 1877 年.

渤海湾西部贝壳堤与古海岸线问题[*]

一、贝壳堤形成的基本原理

贝壳堤是由贝壳物质所堆积成的沿岸沙堤(或称沿岸堤、滩脊等),是发育在海滩上与海岸线平行排列的自然陇岗,形成在开阔的海岸段落,是激浪活动的产物。形成沿岸堤的海岸坡度界于 0.01～0.005,具有这样坡度的海岸主要是砂质海岸,间或是砾石或粗粉砂海岸。在这类海岸上波浪作用最完善,波浪在近岸浅水区破碎,形成激浪,激浪对海底具有强烈的冲刷扰动作用,因此往往掀起滩底泥沙,把它们向岸边携运,堆积在高潮线附近。这样长期地作用,岸边泥沙逐渐堆积加高而形成沿岸沙堤。

在坡度大于 0.01 的陡峭海岸上——主要是基岩港湾式海岸,波浪作用强烈而发生冲刷,不会形成堆积。因此,不可能形成沿岸堤。

在坡度小于 0.005 的海岸上,波浪向岸运行时要经过长距离的水下岸坡,沿途磨耗力量,在到达岸边时力量已消耗殆尽,所以也不会堆积成沿岸堤。

在腹背狭窄的海岸段落,由于激浪冲流不能完全向岸后倾流,并且常会在击撞海蚀崖后形成强力的向海运行的退流,对冲流所形成的堆积进行冲刷,因此不会形成双斜坡的堤状堆积。所以,沿岸堤主要是形成在开阔的以激浪作用为主的砂砾质海岸上。

贝壳堤是一种沿岸堤,它也要求有上述的形成条件。但是,它的组成物质不是砂或砾石,而主要是贝壳。贝类繁殖在近岸带滩底上,它要求透明度好的海水与细砂或粗粉砂底质,环境透光、通气,并含有一定的贝类的食物与营养盐,最利于大量贝类繁殖(然而,有一种毛蚶却是生长在淤泥中)。有大量淤泥与混浊淡水汇入的海岸带,不利于贝类生长。所以,贝壳堤是形成在缺少混浊淡水汇入的、沿岸海底繁殖有大量贝类的、以激浪作用为主的开阔的粗粉砂与细砂质海岸带。

二、渤海湾海岸地貌与贝壳堤

本文研究的范围是渤海湾西部沿岸地带。

本区主要是由粉砂、淤泥质粉砂组成的平原海岸。岸域开阔、地势平坦。岸线外围分布着宽阔粉砂与淤泥质的潮间带浅滩,在湾顶隐蔽处,河口外围以及在本区南部,淤泥的厚度大,而在开阔段落厚度小。岸滩的坡度很缓,平均为 0.001～0.000 3。因此,岸外海水较浅,波浪作用微弱,潮流作用活跃。近岸带的潮流速度低,只能携带细粒的淤泥,而不能营造形

* 王颖:《南京大学学报(自然科学版)》,1964 年第 8 卷第 3 期,第 424－440 页。本文由任美锷教授指导写成,其中插图由万瑛先生清绘,谨此致谢。

成沿岸沙堤。但是,在暴风天气时,强潮、大浪对海岸却有一定的破坏与改造作用。

目前,在沿岸平原上遗留着两列贝壳堤,它们的特点与分布情况阐述于后(图1)。

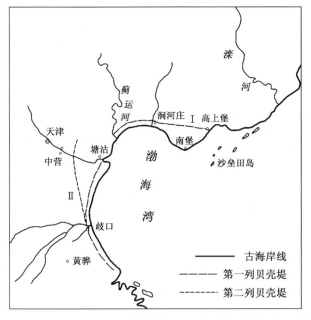

图 1　渤海湾西部海岸范围与贝壳堤分布略图

图 2　弧形贝壳堤

(一) 第一列贝壳堤的分布与结构特征

外侧的一条贝壳堤(以下简称Ⅰ贝堤)分布方向基本与现代海岸线方向一致,自高上堡至海河以北。在这一段贝壳堤被潮水冲刷与人工挖掘,已成为一段段的弧形小丘(图2),高 0.5～1 m,宽 20～30 m,向海坡度 6°,向陆坡度 5°～6°。这样的坡度是贝壳堤形成后为潮水冲刷改造所形成的,向陆坡较陡,局部地段潮水在堤内侧沿堤而行,常将堤顶冲刷成 20°～30°的陡坡(图3)。

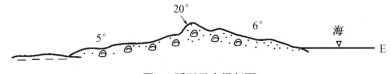

图 3　弧形贝壳堤剖面

Ⅰ贝堤在海河以北组成物质是贝壳(完整的壳、碎屑与贝壳砂)与灰黄色的长英质粗粉砂与细砂。贝壳种类多,以青蛤(*Cyclinna Sinensis (Gmelin)*)、白蛤、毛蚶(*Arca (Anadara) subcrenate Lischkei*)、蜗螺(*Umbonium thomosi (Crosse)*)为主,夹杂少量的李氏金蛤(*Anonia Lischkei Dautzenberg & Fisher*)、牡蛎(*Ostrea gigas Thunberg*)与纵带锥螺(*Batillaira fonlie (Brugnizrz)*)等,沉积层水平分选良好。蛏头沽附近贝壳堤沉积剖面,自上而下(剖面出露 2 m,未见底)为:

Ⅰ层:0.5 m 厚,贝壳碎屑层,贝壳主要是白蛤、毛蚶、蜗螺,具有良好的水平层次,层内夹有黄色细砂,细砂的主要矿物是石英、长石、磁铁矿。

Ⅱ层：0.16 m完整之白蛤层夹有砖瓦块，水平层次明显。

Ⅲ层：1.34 m厚，贝屑沉积。

Ⅰ贝堤到蓟运河口中断，至海河以南重新出现，一直到老马棚口。这一段Ⅰ贝堤受潮水冲刷成为一个个高约2 m、宽约50～60 m的弧形"小岛"，凸起在盐渍平原上。村落之间的一些贝壳堤残段，因无人工保护，已成为低平的贝壳海滩（贝壳沉积厚仅40 cm、自马棚口再向南，经歧口，一直到石碑河南岸的贾家堡，Ⅰ贝堤的分布就此终止，Ⅰ贝堤的这一段因受河流（歧口河、石碑河）泥沙在岸外淤积的保护，保存得最完整，成为高2 m、宽百米，绵延不断的陇岗，人们利用它修筑成一条挡潮堤。在堤的向海坡上目前尚受到浪流的作用，形成一些滩尖咀（照片1）。贾家堡以南由于海岸受蚀后退，Ⅰ贝堤已被破坏。海河南面Ⅰ贝堤的形成与北部不同，其贝壳种类少，主要是白蛤与毛蚶，夹有粉土，没有纯净的细砂（照片2）。驴驹河Ⅰ贝堤剖面自上而下为：

Ⅰ层：厚60 cm的贝壳层，在顶部、中部及底部有厚度为7 cm，5 cm，13 cm的完整贝壳密集层。壳顶多朝上，贝壳为白蛤及毛蚶。层间夹有贝壳碎屑与棕黄色粉土，层间有瓦砾、瓷片，亦为水平状沉积于内。

Ⅱ层：厚45 cm，棕黄色粉土，夹有少量贝壳层，具水平层次。

Ⅲ层：厚30 cm，贝壳屑密集层多为毛蚶屑，夹棕黄色粉土。

Ⅳ层：黄色淤泥粉砂，夹少量贝壳层，染有灰黄色细纹及锈斑。

根据分布位置与绵延方向一致的关系，可以确定：海河南北的贝堤是相连的一条堤。而南北贝堤物质的差异是由于各段海岸接受的物质来源不同的结果。北部海域接受了来自滦河的冲积物——燕山山地的花岗岩、变质岩破坏后所形成的长英质的中砂细砂，在这样的砂底上所繁殖的贝类，主要是蛏螺、扇贝、锥螺、文蛤等。北部贝堤中的贝壳种类就反映了这种情况。南部海域主要是黄河沉积的粉砂，极少中砂细砂。所以，海河以南Ⅰ贝堤的组成物质主要是毛蚶、白蛤及魁蛤。当然，黄河物质在北部海域亦有堆积，但滦河泥沙越过海河到达南部的却很稀少。

（二）第二列贝壳堤的分布与结构特征

在第一列贝壳堤的西面还分布着另一列贝壳堤（Ⅱ贝堤），它自海河以北开始，向南经过泥沽、老马棚口到歧口后与Ⅰ堤平行，向南到贾家堡（照片3）。在这长达70 km的距离内，Ⅱ贝堤始终呈现为连续绵延的堤岗，而自贾家堡向南，Ⅱ贝堤受到冲刷破坏，残留的贝壳堆积在海蚀崖的基部，成为断续的贝壳质的背叠海滩。大口河以南，贝壳堤又出现，并向北形成分支。

第二列贝壳堤规模大，沉积厚，高5 m，宽100～200 m。组成物质是种类丰富的贝壳与厚层的贝壳砂，夹有少量粉土[①]，贝壳的种类除白蛤、毛蚶外，尚有文蛤（*Meretrix Linne*）、蛏螺（*Umbonium thomosi*（*Crosse*））、珠带础螺（*Tympanotomus Cingulatus*（*Gmelin*））、强棘红螺（*Rapana rhomasiana Crosse* & *papna pechilionsis Graban et King*）、日本镜蛤（*Dosinia japonica Reene*）、太阳栉孔扇贝（*Chlamys solaris*（*Borm*））、扁

---

① 在海河以北此列贝堤内夹杂着黄色细砂（长英质）。

玉螺(*Neverita didyma*（*Balten*))等。沉积层具有极完好的水平层理,层中夹有经过海水磨蚀呈自然沉积状态的砖网坠(渔具)、瓷器碎片,沉积层坚实,已微胶结(图4歧口Ⅱ贝剖面,照片4)。

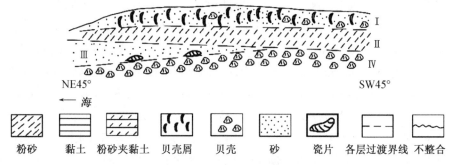

| 粉砂 | 黏土 | 粉砂夹黏土 | 贝壳屑 | 贝壳 | 砂 | 瓷片 | 各层过渡界线 | 不整合 |

图4 歧口Ⅱ贝堤剖面

Ⅰ层:厚80 cm,黄色粗贝壳砂屑层,有完整的毛蚶、白蛤、蜎螺及已磨蚀过之砖瓦片,此层夹杂沉积,沉积不甚明显。

Ⅱ层:厚56 cm,夹有灰黄色粉砂的贝壳屑砂层,多完整之毛蚶、蜎螺及破碎之文蛤,粉砂层有极完好之细理。已受挤压压得弯曲。

Ⅲ层:厚59 cm,黄灰色贝壳屑细砂层,水平层次好,贝壳蛤顶多朝上,有瓦砾与粗瓷片,亦成水平状态自然沉积。整个层从海(NE)向陆(SW)逐渐尖灭。已为钙质胶结。

Ⅳ层:厚50 cm,贝壳密集层,大小分选,已微胶结。此层下出现淡水,各层逐渐过渡。

在歧口村南贝壳堤内侧有一大水坑,露出高数米的贝壳堤沉积层,我们又在坑底打钻,钻孔深3 m,其中1 m为贝壳堤沉积,下部2 m为粉土平原之细粒沉积。图5即为水坑剖面及钻孔剖面连接而成。该剖面清楚地表明贝壳堤叠置在粉土平原上。

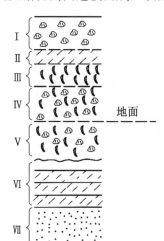

Ⅰ. 80 cm厚,贝壳层,种类丰富。

Ⅱ. 60 cm厚,棕黄色粉砂。

Ⅲ. 50 cm厚,贝壳屑,砂层,水平层次完好。

Ⅳ. 70 cm厚,灰黄色贝壳屑与贝壳层,层次分选好。

Ⅴ. 70 cm厚,灰黑色贝壳屑砂层,多小的毛蚶、白蛤。以上各层逐渐过渡。

Ⅵ. 75 cm厚,棕黄色含黏粒之粉土。上部贝壳层叠置在此层之上。

Ⅶ. 65 cm厚,灰黑色贝壳砂层,贝壳多深黑。

图5 歧口贝堤与内侧平原沉积剖面

根据Ⅱ贝堤的沉积特征,可以推断:在其形成时,海岸坡度大,海水清,激浪作用强;海底的粉砂层厚,贝壳繁殖旺盛。因此,经长时期激浪堆积作用,在高潮水边线附近形成了规模巨大的贝壳堤。同时,依据Ⅰ、Ⅱ贝壳堤的物质组成与今日海岸沉积物成分的对比,可以认为：历史时期中渤海湾西南部没有经历过砂质海岸的阶段。

贝壳堤形成后，由于它位置高，潮水不能经常地浸淹，而且受到降雨淡水的渗透淋洗，使得沉积层的盐分减轻，堤上常常生长着植物。Ⅱ贝堤上植物生长特别茂密，表层全为翠绿的灌丛所盖，而Ⅰ贝堤则植物稀少(有些耐盐的短草)，大多光秃。Ⅱ贝堤上的植物有灌木也有草本，主要是：刺枣(*Ziziphus Spinosus H. H. Hu*)、蔓荆(*Viter notundifolia r. f.*)、茵陈(*Artemisia Capillaris Thund*)、窝食(*lxcris repens A. Gray*)、艾蒿(*Artemisia Vulgaris L.*)、盐蓬(*Suaeda Ussuriensis Iljin*)、太阳花(*Erodium Stephanianun Willd*)、丝绵木(*Euonymus Bungeana Maxim*)、砂引草(*Tournefortia Sibirica L.*)、茜草(*Rubia Corifolia L.*)、米口袋(*Amblytropis Sp.*)、马兰(*Iris ensata Thunb*)等[1]，也有一些人工栽培的石榴、枣树、杜梨等果树(照片 5)。贝壳堤为砂层，其下为不透水的黏土粉砂层，使降水在砂层中聚起来，成为暂时的储水层，因此植物丰富。另外渔民掘堤为井，全村饮水取给于此。贝壳砂还可用来压碱、保摘，使堤后滨海平原可以种植蔬菜五谷。所以，本区历代渔民皆集贝壳堤而居。

Ⅱ贝堤的内侧是宽广的平原，平原沉积是微盐的粉砂黏土，地面上广布着一系列潟湖凹地，潟湖多已干涸，其长轴方向大多沿着海岸，当地称为"泊子"。凹地内盐质重，多半寸草不生，仅边缘生长盐蒿。有少数大潟湖仍有积水，1958 年后辟为蓄洪水库。潟湖凹地平原形成较早，久经雨水淋洗，盐分较轻，只要有足够的淡水即可开垦耕种，著名的"小站稻"就产在这种土地上。Ⅱ贝堤与潟湖凹地相伴存在，标识着一条老的海岸线。

在Ⅰ、Ⅱ贝堤之间是一个狭窄的盐土平原，原为海河冲积与浅海沉积所成的河口三角洲地带。平原自海河口向南北方向变狭，到歧口处平原尖灭。Ⅰ、Ⅱ贝堤的接触关系亦随之而有不同。在海河附近，Ⅱ贝堤不整合地叠置在潟湖凹地平原上，而盐土平原叠置在Ⅱ贝堤向海斜坡的基部，并向海延伸。Ⅰ贝堤呈不整合的关系叠置在盐土平原上，在Ⅰ贝堤外侧是现代潮间带浅滩(图6)。向南盐土平原逐渐变狭，Ⅰ、Ⅱ贝堤逐渐靠拢，到歧口后形成两级背叠的海积阶地，在Ⅰ堤的向海坡上叠置着现代岸滩，大潮时海水仍在堤足起作用(照片6、7)。Ⅱ贝堤的内侧是潟湖凹地平原(图7)。再到贾家堡的南部，Ⅰ贝堤已被破坏。Ⅱ堤只残留着一些堤基的沉积(照片8)，故Ⅰ、Ⅱ堤的接触关系已难查明(图8)。

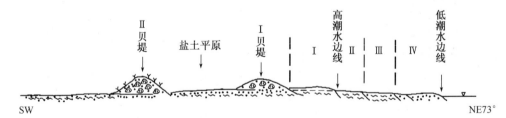

图 6　淤进的冲积海积平原海岸横剖面图示

从上述可看出：渤海湾西部沿岸有两列贝壳堤。它们与目前海岸动力、沉积条件是不相称的，各列贝壳堤的地貌结构亦互有差异(见表1)。可以确定，这两列贝壳堤是两条不同时代不同发育过程的古海岸线遗迹。

---

① 植物标本承南大生物系耿伯介先生鉴定。

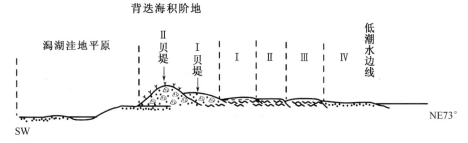

图7　稳定的贝壳堤潟湖凹地平原海岸横剖面图示

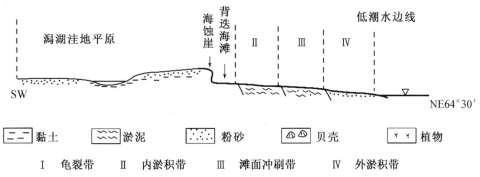

| ⊨⊨ 黏土 | ∿∿ 淤泥 | ∵∵ 粉砂 | ⌒⌒ 贝壳 | ⋎⋎ 植物 |

Ⅰ　龟裂带　　Ⅱ　内淤积带　　Ⅲ　滩面冲刷带　　Ⅳ　外淤积带

图8　有残留贝壳海滩的冲蚀潟湖凹地平原海岸横剖面图示

表1

堤名	位置	延伸范围（从北向南）	高度、宽度		形态	沉积组成特征		其他
			海河北	海河南		海河以北	海河以南	
Ⅰ贝堤	临近海滨分布，位于特大潮线附近。	高上堡→涧河庄→蛏头沽；海河以南→马棚口→歧口→贾家堡。	相对高度0.5～1.0 m	相对高度2 m。宽20～30 m	断断续续的弧形贝壳堤。在一些河口附近，贝壳堤有分支形态。在歧口以南为保存较完整的绵延陇岗。	①有较多种类的贝壳：白蛤、毛蚶、蜎螺、李氏金蛤、锥螺子…… ②沉积中夹有黄灰色的长英质细砂。 ③水平层次分选好，未胶结。	①贝壳种类少主要是白蛤、毛蚶。 ②夹杂有粉土层。 ③有瓦砾瓷片等片断。 ④具有良好的水平层次，疏松、未胶结。	长有少量的碱蓬等草本植物，大部分光秃。
Ⅱ贝堤	位于陆地上。①在海河一带距海约20 km。②上古林一带距海约20 km。③歧口以南，位于Ⅰ堤后侧，临近现代海岸。	泥沽→上古林→歧口→贾家堡；向南断断续续的直到大口河。		相对高度5 m。宽100～200 m。	基本上保持着完整的连绵的陇岗形态。贾家堡以南为断续的弧形海滩、贝堤冲蚀改造后的堆积。	同上，沉积层次较厚坚实。	①贝壳种类多除着白蛤、毛蚶外，尚有文蛤、日本镜蛤、蜎螺、太阳栉孔扇贝、锥螺等。主要是贝壳与贝砂沉积，粉土极少。 ②水平层次极好，夹有已经磨蚀过的砖、网坠、陶片等物。 ③沉积坚实，已微胶结。	生长着茂密的草类及灌木。堤内储有淡水。

（三）内陆古贝壳堤遗迹的调查与讨论

在Ⅰ、Ⅱ贝壳堤的西面，还有一些贝壳堤的遗迹，它们是更老的海岸线。这些遗迹有的已有报道[4]，经我们调查后有些不同的认识，兹简述如下。

（1）在北大港蓄洪水库中有很多淹没的小高地，它们多半是长条形的贝壳堤，向南延伸，经过南大港一直到石碑河两岸。自这条堤再往西，在黄骅县城附近的前苗村、后苗村一带，有更老的贝壳堤。这些在以前还缺乏报道。

（2）巨葛庄老贝壳堤的遗迹已有报道，但据我们观测，它是埋藏的，地形上已无表现，最近由于开挖河道而显露出来。在巨葛庄东面的洪泥河，两岸出露了贝壳堤的横剖面，贝壳堤沉积掩埋在半米厚的表土下面，剖面呈丘形（图9），出露的沉积在两端厚约 1 m，中部堤顶处厚 2 m，沉积物具有良好的水平层次，主要物质是贝壳，其种类有：白蛤、毛蚶、牡蛎、蝠螺、锥螺、织纹螺（*Aletrion Uarici fcnc*（*A. Adam*））、扁玉螺（*Neverita didrdyma*）等。贝壳皆经长期风化，表面失去光泽，成灰白色。

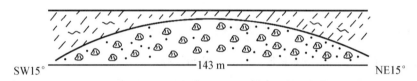

SW15°　　　　　　　　——143 m——　　　　　　　　NE15°

图 9　巨葛庄东，洪泥河所揭示之贝堤剖面

巨葛庄村西亦有新开的引河（此河通新河桥），切开了贝壳堤，在河的两侧出现了贝壳堤沉积层，宽 300 m，剖面出露厚 1.6 m（照片 9），各层的沉积特征见图 10。

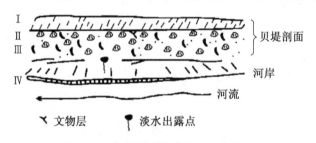

图 10　巨葛庄西，引河揭示之贝堤剖面

Ⅰ．0～40 cm，覆盖的壤土，下部与Ⅱ层交界处有大量的夹砂红陶的器皿残片，如豆、瓷棺、陶罐等残片，经李世瑜先生鉴定，是战国时的陶器。

Ⅱ．40～120 cm，是灰黄色贝壳与贝壳细层，粗细层相间，每层厚 3 cm，贝壳种类多，同前述。

Ⅲ．120 cm 以下为白蛤密集层，此处有淡水流出。

上述各层逐渐向西南倾斜。

据两剖面出露点，测得贝堤延伸方向为 NW - SE62°，与现代海岸横交。

巨葛庄以北：新河桥、荒草坨、范庄子一带直到东堤头，没有贝壳堤遗迹，因为：① 地表没有贝壳堤的地形表现；② 根据这些地点的天然露头，洼坑剖面以及我们用浅钻打钻结果，在地面下 5 m 以内没有发现贝壳堤的沉积层。

上述地点的地面上散布着一些细砂与少数贝壳碎屑，但它们是最近几年来（1958—1960 年）开河挖出来的，再从开挖剖面看来：地面下 2 m 深处细砂较厚，以石英、长石为主,石英磨圆度好,夹有较多的云母,在砂层中分散沉积着一些贝壳,主要是白蛤与蛏子(*Siliqua Pulchella*、*Solen grandis Dunker*)。有趣的是:蛏壳密闭,其内填满细砂,砂与壳外沉积一样皆具有锈斑、潜育条痕等。从贝壳散积于砂层中以及紧闭的蛏壳夹砂等情况:可以判明这不是激浪所形成的岸堤沉积,而是静水环境下的沉积。

据上述,可以认为:把巨葛的贝壳堤向北延伸,而定出又一条古海岸线,是缺乏科学根据的。

(3)有人据天津新县志记载:"育婴堂在城东北运河南,掘壕五尺见蛤壳"及华庄子一带的堤状地形,就确定了又一条古老的贝壳堤。据我们实地调查:在天津市南郊华庄子村西1 km处,地面上有一条高约1 m呈南北向的土堤,堤上长满灌木,有村屋残墟,在残墟中俯首即可拾到明瓷与战国夹砂红陶器碎片。当地老乡称此为"小堤子",说它是"老辈子的挡海堤,是海水晃荡出来的"。堤内有较多的石灰质砂薑与黏土结核,但经十余人分头挖掘寻找未见贝壳。看来,这可能是一条海堤的遗迹,而不是贝壳堤。也不能与育婴堂的1.7 m深处的"贝壳堤"联结为一条堤,因为同一时代的在同一地质构造与地理环境背景下所形成的沿岸堤,在距离不大的范围内,(育婴堂与华庄子相距约15 km)不可能在高度上有如此显著的差异。此外我们也在育婴堂旧址上(今志成道党校宿舍)进行了5 m的浅钻取样,同时又观察了该处的一个3 m深的土坑剖面,证实了该处没有贝壳堤状的沉积。天津县新志的资料是可信的,而后人据此所作的结论是错误的,因为天津一带的土地皆为河海交互沉积所成,沉积层中具有贝壳是不稀奇的。但是,具有贝壳的沉积层并不都是贝壳堤,贝壳堤沉积是一种由激浪流所形成的,具有双斜坡的上凸形的沉积体,贝壳密集,层次分选明显,堤中部沉积具水平层次,两翼分别向海向陆倾斜。而育婴堂的剖面不具有此特点,其钻孔剖面与中营(界于泥沽与张贵庄之间津塘公路之北)的钻孔近似。天津志成道育婴堂旧址钻孔剖面(由上而下):

Ⅰ层:100 cm厚,棕黄色粉土,并参夹细砂,中间有一层5 cm厚细砂层。

Ⅱ层:60 cm厚棕黄色粉砂黏土,具少量石灰粒(似贝壳风化物),有铁锰结核。

Ⅲ层:20 cm厚,棕黄色粉砂,水平层理极好,有铁锰结核。

Ⅳ层:棕黄色粉砂黏土有石灰质粗粒,80 cm厚。

Ⅴ层:黄棕色黏土,夹细砂与粉砂有完整之小白蛤,已风化脱粒。有芦苇根。80 cm厚。

Ⅵ层:40 cm厚,灰色夹锈黄色粉砂。

Ⅶ层:120 cm厚,青灰色(长英质与贝壳质)细砂,有贝壳及少量较细锈斑。

津塘公路中营村钻孔剖面(由上而下):

Ⅰ层:厚40 cm,灰褐色粉土,耕作层,有锈斑,40 cm处见潜水面。

Ⅱ层:厚100 cm,黄灰色粉土,黏土。有铁锰结核及潜育条根,夹少量贝壳。

Ⅲ层:厚20 cm,青灰色黏质粉土。

Ⅳ层:厚108 cm棕黄色,浅灰色粉砂黏土,有黄色锈斑及灰色潜育纹。

Ⅴ层:厚20 cm,黄色与灰色粉土,夹淤泥及少量极细砂,有贝壳。

Ⅵ层:厚80 cm青灰色淤泥粉砂。

Ⅶ层:厚100 cm(未至底)青灰色细砂,细砂为贝壳质,有贝壳屑及小蛤。

这两个钻孔是:1~3 m为棕黄色粉土与黏土,3~4 m是过渡层,4 m以下为青灰色粉砂与细砂层,此层中含有贝壳。如根据沉积中出现贝壳即定为贝壳堤,似乎中营沉积中贝壳更多些。而中营所在却被人①划为是没有贝壳堤分布的中间地带。实际上,这种含贝壳的粉砂层是一种均匀的近岸的浅海沉积。

总之,沿岸带两列贝壳堤可划定为两条岸线。内陆更老的贝壳堤资料较零散,还存在很多疑问,需进一步工作,掌握了确凿的地貌与沉积资料后,才能推断更老海岸线的分布位置。

三、贝壳堤所反映的地貌问题

（一）海岸类型及其动态

根据两列贝壳堤分布的位置、相互关系以及海岸地貌的其他特点,可将本区海岸划分为以下几个亚类型。从北向南,各类型海岸特征如下:

（1）新生的海积平原海岸:从大庄河到涧河庄,以高上堡到尖坨子的弧形贝壳堤为北界,向南到达以南堡为顶点的三角形地带。这是在Ⅰ贝堤形成后才淤长起来的,是最新的海积平原。目前当特大潮水时,海水仍可到达贝壳堤附近。

（2）淤进的冲积-海积平原海岸:从涧河庄开始向西,经过海河向南,一直到达歧口。Ⅰ贝堤滨海分布,Ⅱ贝堤在内陆,两堤间隔 1～20 km。这种情况表明了:Ⅰ贝堤所代表的新海岸线较Ⅱ贝堤的老岸线向海推进了一段距离,这带海岸自Ⅱ堤形成后是加积伸展的,淤积的速度在海河处最快,而向南北减缓,这显然与海河河口的泥沙堆积密切有关。盐土平原是冲积-海积形成的,是当年“海河”的三角洲,后逐渐淤高成为滨海的盐土平原。但是,在Ⅰ贝堤形成后,海岸淤积的速度减缓了,目前基本上是稳定的,所以直到现在,Ⅰ贝堤尚位于特大高潮线附近(图 6)。

（3）稳定的贝壳堤潟湖平原海岸:从歧口到贾家堡,两列贝壳堤并在一起,Ⅰ贝堤叠置在Ⅱ贝堤上,组成背叠的海积阶地。两列贝壳堤与堤后的潟湖平原都直接临海分布组成了现代海岸(图 7)。

Ⅰ贝堤的规模小,组成物质也不及Ⅱ贝堤纯,这反映着在Ⅰ贝堤形成时,海岸坡度与激浪作用都较以前减小,海底的组成也发生了变化,开始有淤泥汇入。但新老岸线位置最一致,没有位移变化,反映着这段海岸自Ⅱ贝堤形成时至今都是稳定的,冲淤变化不明显,目前海岸动态基本上仍与以前相同。若以此段海岸与其南北毗邻海岸段落的关系来看,它正是动态转折地区,也可说是冲淤平衡区。在转折区以北,岸线是淤进的;在转折区以南,海岸是冲刷后退的,而在转折区海岸是稳定的。

（4）有残余贝壳海滩冲蚀的潟湖凹地平原海岸:贾家堡以南一直到大口河,岸线略成西北—东南方向,这段海岸坡度陡,涨潮时海水即达岸边,浪流的冲蚀作用较强,Ⅰ贝堤已被蚀掉,Ⅱ贝堤也遭冲蚀,潟湖凹地平原前缘也被冲刷形成高 2～5 m 的海蚀崖,在崖麓残留着少量Ⅱ贝堤的蚀余堆积物,形成厚度不大的贝壳海滩(图 8)。目前残留的海滩及潟湖凹地平原仍遭受海水冲刷破坏,岸线不断向陆退却,但后退的速度不大,主要发生在风

---

① 见李世瑜的文章(“考古”1962 年 12 月),及其他根据李文所写之文章。

暴强潮相结合的时期。例如1939年阴历7月14日，沿海风暴海啸，此段海岸即被冲蚀后退，位于海滨的陈家堡被卷入海中，目前尚有小部分废墟残留在海滨。据访问，这一带岸线近50年来蚀退约1华里，刘家堡、范家堡、大辛庄村等，在近50年来均在向西迁移。

从上述可看出，贝壳堤可作为海岸演变的标志，对分析海岸发展过程与演变趋势，有很大的参考价值。

（二）贝壳堤形成时代与海岸发育史

分析这一段海岸演变历史必须从现代海岸着手，将今论古，来进行比较分析。

目前渤海沿岸海水浑黄，岸坡极缓，浅水面积大，粉砂淤泥质滩地宽广，活跃于海岸带的首要动力是潮流，波浪作用退居到海岸外围地区。显然，在这样的海岸上不会形成贝壳堤。其原因是渤海湾沿岸活跃着来自黄河巨量的淤泥物质。黄河年输沙量平均达9亿t，它汇入渤海湾后，浑化了沿岸海水，降低了波浪水流的冲刷力，形成了浅平的粉砂淤泥质海岸，并使岸线迅速地向海淤进，近40～50年来沿岸浅滩淤涨了0.5～1 km。可以想象当不受黄河的影响时，海岸会向相反趋势演变：泥沙来源骤减，波浪重新冲刷改造岸坡，坡度逐渐加大，激浪重新活跃在岸边地区，冲蚀岸坡，不断把泥沙推送到岸坡上部，集中堆积在高潮水边线附近，形成沿岸堤。同时，由于淤泥来源减少，原来堆积的淤泥被波浪作用掀带到外海深水地区，海岸带淤泥逐渐减少，在粉砂的底质上，贝类又重新繁殖，为激浪掀带到岸边堆积，形成贝壳堤。这就是贝壳堤形成的背景。

历史上，黄河曾有数次巨大的迁徙改道，离开渤海湾向南夺淮入黄海，对近期渤海湾海岸地貌的发展演变有巨大影响。我们从以下三点来讨论第二条贝壳堤的形成时代。

第一，根据考古资料。近年来，河北省文化局等先后在白沙岭、军粮城、泥沽、贾家堡东南与刘家堡西面的大双坨子等处贝壳堤的上面发现了一些坟墓、村舍遗址与文物，这些文物经验鉴定后，大部分是唐、宋的。因此，根据考古资料可以确定Ⅱ贝堤在唐时已经存在了，并且已具备了比较安定的生活条件，人类活动频繁，所以有较多的文物遗迹。但是，坟墓、村舍是在贝堤形成以后营建于堤上的，贝壳堤的形成时代还要早些。因此，考古的证据只帮助我们推断贝壳堤形成年代的下限，要确定其形成时代还要应用其他方法。

第二，根据Ⅱ贝堤的沉积组成与形态特点，可以分析它形成时的自然环境：贝壳种类丰富，主要是生活在粗粉砂与细砂中的；贝壳砂层厚，分选极好，质地纯，不夹杂黏土；这些反映着当时的海水清澈，不含有淤泥与黏土，海底组成主要是粉砂。同时，Ⅱ贝堤规模大，高度大，沉积厚，表明它形成时的岸坡坡度大，激浪流作用强烈；岸线稳定，岸堤继续堆积时间长。但是在唐朝以前，黄河一直是注入渤海的，未曾南迁，那么，为什么当时渤海湾沿岸没有大量的淤泥物质，岸坡较陡，与今日情况迥然不同呢？根据历史资料的分析，黄河的河性特点在历史时期是在改变，并非一直是饱含泥沙的。众所周知，黄河的巨量泥沙主要来自黄河中游山西、陕西黄土高原区。但是，在历史上，从春秋到秦，山陕黄土高原林木茂盛，原始植被未遭破坏，是良好的林牧场所，因此，水土流失极少，黄河的含沙量不大，水是清的。西汉时期，由于大量移民至边塞（即今山陕高原一带）从事农垦，变林牧为农耕，原始植被遭受到破坏，水土流失日益加强，黄河含沙量遂增，因而下游改道极为频繁，这种含沙量加大的影响亦波及河口及外围地区，但是由于变动时间不长（约100多年），尚不足以改变整个渤海湾的沉积组成。嗣后，东汉以至魏晋时，边疆少数民族大量入居塞内，在黄土高原一带还农为牧，植被又重新恢复，水土流失日趋减小，所以，在东汉以后五百多年期间，黄河一直是稳定的，入海的淤泥粉砂物质不多，海岸长期稳定，且沿岸海水清澈，激浪作用强，贝类宜于繁殖，这些条件均有利于形成高大的贝壳堤。

第三，从Ⅱ贝堤的分布特点来看：在歧口以北，它分布的方向近似于南北向，而在歧口以南，呈现为西北东南向，愈往南堤愈向东突出，可见，当时南部岸线是向海突出的，这显然是由于古黄河三角洲的影响。由上所述可以推断：Ⅱ贝堤形成在汉唐之际，当时黄河河口位于渤海湾南部。汉王莽始建国三年（公元 11 年），黄河第二次大改道，黄河自河南一带决口后的径流，使河从荥阳（今河南荥阳县东）到千乘入海。自此以至隋（相距 500 年多）唐（相距 800 多年）黄河长期稳定地注入渤海湾南部。根据以上三点，可以确定：第Ⅱ条贝壳堤开始形成在东汉明帝十三年王景治河成功以后，主要在东汉末年至隋时。唐朝时，贝壳堤已具有相当规模了。

关于Ⅰ、Ⅱ贝壳堤间盐土平原的形成时代。前已述及，盐土平原在中部海河一带东西宽度大（20 km），而向南北两侧逐渐收狭，最后尖灭。这反映着在Ⅱ贝堤形成以后，岸线的淤进是中部快，南北两侧慢。中部快与海河的堆积作用有关。整个平原的形态反映着这是一种口外三角洲的堆积，平原南北尖灭处，正是冲积物南北运移所及的终点。平原的沉积组成，在表层 18 m 以内为海积的黏土，基底是海陆过渡相的淤泥、粉砂等物质[1]，反映着这部分陆地是在水下三角洲的基础上发展起来的。一般地讲，河口外围的堆积作用是强盛的，但是，当时"海河"的泥沙量与堆积规模，远较今日巨大，这种泥沙骤增、堆积强盛的表现是与某些含砂量特大的河流（在渤海湾主要是黄河）的活动有关。所以，关于这带陆地的形成，我同意李世瑜先生的分析[4]：在宋庆历八年（1048 年）黄河三徙，其北支由天津一带入海后所形成的。

第一条贝壳堤开始形成于元至元年间（1336 年）黄河第五次大改道后，而盛于明弘治年间黄河全流入黄海时。这时，渤海湾沿岸的物质组成与动力因素又发生了改变，淤泥减少，岸坡变陡，波浪的作用又重新活跃在海岸带，有利于贝类繁殖。经历了相当时间后，海岸带又出现了新的贝壳质岸堤。这列贝堤分布的范围大，遍及整个渤海湾西岸。但是，唐代以后，由于黄河中游一带滥垦引起强烈水土流失，黄河含沙量增大，河性已发生改变，输入渤海湾淤泥增多，海底组成发生重大改变。虽然河口南迁，但沿岸的淤泥已难完全摒绝。故此时贝类较少——主要是能耐部分淤泥的白蛤与在淤泥中生长的毛蚶、魁蛤，淤泥沉积亦在贝堤中有反映。同时，由于海岸带仍有一定的淤泥物质，故岸坡较缓，使波浪对海岸的作用也较弱。因而Ⅰ贝堤具有高度低，宽度小；贝类少、夹有淤泥等特点。这也是Ⅰ贝堤虽形成较晚，但破坏程度却远较Ⅱ贝堤为大的原因。这种现象在渤海湾南部最为突出。海河以北，由于滦河的中细砂供应，Ⅰ贝壳堤组成与Ⅱ堤相近。

Ⅰ贝堤在明末时已经形成，清初已有渔民定居堤上。例如，驴驹河村的吴姓渔民是清初由海丰迁来，至此已定居了十辈，为时 200 多年（据吴姓家谱载：其第三辈是乾隆五年生于驴驹河的）。Ⅰ贝堤形成后，大部分岸线稳定，唯自 1885 年以来，黄河北归，流入渤海，

---

① 河北省地质局水文地质大队，天津新港钻孔剖面图。

岸滩又逐渐在淤长,这种趋向又以 1946 年以后为突出。但在本区北部,即自大庄河到涧河口一带,在Ⅰ贝堤形成后,岸线迅速向海淤进,这是因为:明景泰三年(1452 年)以后,滦河改道从大清河与大庄河一带出口,于岸外形成了石臼坨、曹妃甸等一系列海岸沙坝,它们围封了其后侧的海域,形成风浪平静的波影区,于是,滦河的泥沙与黄河的泥沙在此交汇,形成了大量堆积,在Ⅰ贝堤前方形成了南堡三角形滩地。1813 年,滦河东迁到老米沟入海,1915 年又向东北迁移至甜水沟入海,于是,大清河、大庄河外围地区泥沙供应断绝,海岸沙坝增长停止,并遭到风浪冲刷不断后退,对其后侧的保护作用作用减小,故南堡三角形地带的淤涨速度已渐减慢。

这就是关于渤海湾西部海岸形成演变过程的初步认识,在研究中国大面积的善冲、善淤,动态活跃的粉砂淤泥质平原海岸的形成演变时,必须着重分析大河因素的影响。

最后,关于Ⅱ贝堤以西更老的岸线形成时代,由于缺乏足够资料与证据,还不能确定。但是,可以提出一点:Ⅱ贝堤南段,南、北大港以西的陆地形成时代要早于周朝。兽骨化石与文物资料都可作为佐证。据北大港文物调查资料:在北大港西部沿岸,道口村北发现新石器晚期的陶制生活用具,属龙山文化遗物。这个珍贵的发现,足以说明,在新石器时代晚期,这一带已有人类活动。那么,成陆的时代应更早一些。在华庄子土堤、前后苗村贝堤中钙质结核较多,而东部较新的沉积中就没有,亦可作为对比。

四、结　论

(1) 贝壳堤是由贝壳物质堆积成的沿岸堤,形成在繁殖着大量贝类,以激浪作用为主的粉砂质与细砂质的海岸带。目前,渤海湾西部是潮流作用为主的粉砂淤泥质海岸带,不能形成贝壳堤。因此,贝壳堤的存在说明渤海湾西部海岸带的动力、沉积与地貌特征的变迁,其原因与黄河下游的变迁有关。贝壳堤分布位置与当时海岸线的位置一致,因此,根据老贝壳堤的位置可恢复古海岸线的位置;根据老贝壳堤与新贝壳堤之间的关系,可以分析海岸演变过程与目前动态。

(2) 根据地貌、沉积与历史地理的分析,初步确定了渤海湾沿岸两列贝堤的形成时代。Ⅰ贝堤开始发育在元朝至元年间(公元 1336 年)黄河第五次大迁徙以后,至明朝中叶已具有较大规模。Ⅱ贝堤开始发育于东汉明帝 13 年(公元 70 年)王景治河成功以后,形成在隋时。ⅠⅡ堤间的盐土平原主要形成在宋庆历八年(公元 1048 年)黄河三徙,由海河入海之后。Ⅱ堤南段南北大港以西的陆地形成在周以前。

(3) 根据Ⅰ、Ⅱ堤的分布关系,确定了各段岸线的近期动态。

① 大庄河到涧河口,是淤进最快的海岸。

② 涧河口到歧口,是缓缓前淤的海岸。

③ 歧口到贾家堡,是长期稳定的海岸。

④ 贾家堡到大口河岸线是微微蚀退的海岸。

(4) 从贝壳堤演变历史的研究,可见天津新港一带的海岸,自Ⅰ贝堤形成以后至今400 年来,海岸淤长不多(1 km 左右)。由于今后黄河不会再经海河入海,引起这里海岸迅速淤长的主要泥沙来源已经基本上消除。因此,我们有根据可以推论:天津新港的使用在近期不致因海岸的自然淤长,而受到不利的影响。

参考文献

[1] 邓子恢.1955.关于根治黄河水害和开发黄河水利的综合规划报告.人民日报,1955年7月31日第1版.

[2] 岑仲勉.1957.黄河变迁史.人民出版社.

[3] 谭其骧.1962.何以黄河在东汉以后会出现一个长期安流的局面——从历史上论证黄河中游的土地合理利用是消弭下游水害的决定性因素.学术月刊,(2):23-35.

[4] 李世瑜.1962.古代渤海湾西部海岸遗迹及地下文物的初步调查研究.考古,(12):652-657.

[5] 刘鹗.光绪十五年石印.(清)历代黄河变迁图考.

[6] 孙几伊.1935.河徙及其影响.金陵大学中国文化研究所丛刊甲种.

[7] 黄骅县文物资料.黄骅县文化馆(未刊).

[8] 沿海县志.

    ① 滦县志(明万历,清康熙,光绪,1937).

    ②天津府志(乾隆四年七月,光绪廿五年十一月).

    ③ 天津县志(乾隆四年七月).

    ④ 天津县志(1931年).

    ⑤ 沧县志(1932年).

    ⑥ 沧州志(乾隆八年).

    ⑦ 静海县志(同治十二年).

    ⑧ 盐山县志(同治六年三月).

照片1　歧口完整的Ⅰ贝堤与堤前坡的滩尖嘴

照片2　驴驹河村Ⅰ贝堤沉积剖面

照片3　泥沽村的Ⅱ贝堤

照片4　歧口Ⅱ贝堤剖面

照片 5　生长在 II 贝堤上的枣树

照片 6　歧口以南,相并排列的 I、II 贝壳堤

（注意,长草者为 II 贝堤,白色的是 I 贝堤）

I 贝堤外侧为泥滩（落潮时摄）

照片 7　I、II 贝堤

照片 8　贾家堡以南,冲蚀的海岸,贝堤已蚀尽

照片 9　巨葛庄埋藏的老贝壳堤

渤海湾西南部岸滩特征[*]

本文研究的地区是渤海湾西南部的河北省海岸,北起天津新港,向南经大沽、歧口等地,到河北、山东两省交界的大口河为止。全部岸线长 88 km。海岸调查的范围自水深－5 m 处开始,经过潮间带浅滩向陆延伸 5 km。本文着重讨论潮间带浅滩。文中所应用的资料是 1963 年 5 月—7 月在该地区浅滩上的地貌调查与一系列断面半定位观测所取得的。观测的项目包括:断面水准测量、地貌与微地貌观测、5 m 土钻钻探,用标志桩及标志层进行的冲淤观测、悬砂与底砂的沉积测定、应用指示砂观测滩上泥沙运动,以及贝壳种类与含量的测定等。这是一项集体研究工作,由任美锷教授领导,参加分析工作的还有何浩明和天津港务局回淤研究站蔡嘉熙。

一、区域概况

本区在大地构造上属于渤海拗陷中的黄骅临清拗陷,它在中生代末期,特别是喜马拉雅运动时期急剧下沉,因而第四纪沉积层深厚,据钻井记录,沧县、天津等处第四纪沉积厚度达 744～863 m。

地貌上,本区居华北平原的东部边缘,是新生代末期形成的冲积海积平原。北为海河三角洲,南为黄河三角洲,均向海凸出,故本区海岸呈现向西凹入的弧形平原海岸的特点。

根据塘沽气象站 1946—1960 年的记录,本区常风向为西南,五级以上大风的强风向为正东。平均风速以 4—5 月为最大,大风则以冬季最多,春季次之。本区冬季较冷,海水结冰期在塘沽(1946—1956 年平均)从 11 月 9 日开始,至翌年 3 月 22 日,长约 137 天。黄河口外由于海水盐度较小,冰期较长,3—4 月间,冰凌往往沿岸漂流北上,被推送至潮间浅滩,冰融后常在滩上留下一堆堆的浮泥。

风对沿海波浪与水流的影响较大,作用在本区海岸带的波浪主要是风波,各季节波向与控制该海面的风向是一致的(见表 1)。由于是风波,所以波浪周期小,波长短,形态不规则,易出现白浪,在正常天气下,一般波高约 30 cm,春季 3—4 月风浪大,平均波高达到 70 cm。本区东北和东风吹程大,在强劲的东北风(6 级)作用下,大口河一带波高可达 1.5 m。西北风因吹程小,虽然风力也大,但波高却是小的。波浪对本区海岸冲蚀和滩底扰动掀砂也起一定作用,例如:1963 年 5 月 22 日,连续 16 小时的 7 级东北风,结果在驴驹河波浪冲蚀掉大堤基底的滩面 4～7 cm。

---

　　* 王颖,朱大奎,顾锡和,崔承琦:《新港回淤研究》,天津新港回淤研究工作组,南京水利科学研究所,1965 年 10 月第 2 期,第 49 - 64 页。

表 1

渤海湾沿岸常出现的波浪频率(1952—1954 年统计)				
NNE 23%	ENE 20%	E 47%	NE 5%	ESE 5%

潮流是塑造本区岸滩的主要动力。据海河口大灯船 1953 年的观测资料,涨潮合成流向为 N37°W,最大流速 0.65 m/s,平均流速 0.54 m/s,落潮合成流向为 S38°E,最大流速 0.47 m/s,平均流速 0.41 m/s。这种流速足以冲蚀和搬运滩面淤泥。

二、海岸类型

本区海岸地势平坦,起伏微小,岸坡平缓,岸域辽阔。平原上地貌类型简单,只有河道,潮水沟,残余的潟湖凹地,贝壳堤与外围的潮间带浅滩(图 1),形成这些地貌的营力主要是:河流与潮流,间或也有风浪的改造作用。组成海岸的物质是淤泥、粉砂等松散沉积物(图 2)。它是典型的粉砂淤泥质平原海岸。根据地貌特征与结构,海岸形成的动力因素以及发育程度和发展动态的差异,可以进一步将本区海岸划分为三个类型。

(一) 淤进的冲积海积平原

如图 3 所示,这类海岸分布于本区的北部,从天津新港到头道沟子,岸线长 42 km,作北北东到南南西走向。这段海岸岸线平直,是一片坦荡的盐土平原,淡水与植物都很缺乏,目前大部分已开辟为盐田。平原的基底是海陆过渡相的三角洲沉积。主要沉积物是:黏土、黏质砂土、砂质黏土与粉砂。而组成地表以至 10 m 深的物质主要是海相的黏土[①],各种沉积层成楔形,交错叠置。这种构造与岩性特征表现着由冲积海积交互作用所形成的海河三角洲逐渐淤高成陆的过程。在平原的边缘残留着一列岛弧状的贝壳堤,它的相对高度是 2～4 m,组成物质是贝壳贝屑与棕黄色的粉土;贝壳的种类少,主要是白蛤与毛蚶,沉积层分选良好,层次分明。在上层有锈铁丝、砖瓦与瓷器残片等物,呈自然状态分布于水平层次内。

这列贝壳堤的宽度约 100 m,但它已被潮水冲断成为一段段的弧形小丘,断续地矗立在已成为盐水池的平原上,像一座座的小岛。沿岸的渔村如道沟子、驴驹河、高沙岭、白水头、唐家河等都建立在贝壳堤上。各村落之间有些贝壳堤段,因无人修缮,久受潮水冲刷,已成为高度不大的新月形贝壳质海滩。

在贝壳堤外侧,有一条人工修筑的挡海堤,堤高 2.7 m,高程为 5.29 m,堤外是广阔的潮间带浅滩。目前作用于海岸带的主要动力因素是潮流,而贝壳堤是以波浪为主的海岸带的堆积体,它是在过去岸坡较陡(0.01～0.005)的海岸段落上,由激浪作用堆积起来的。后来,海岸淤积变缓,波浪作用已达不到岸边,但贝壳堤却残留下来成为古海岸的遗迹。

---

① 参考河北省地质局水文地质队的钻孔剖面图。

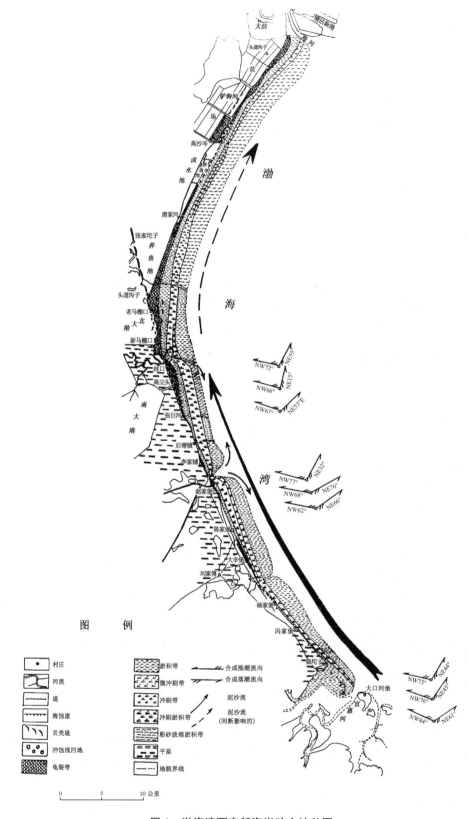

图1 渤海湾西南部海岸动力地貌图

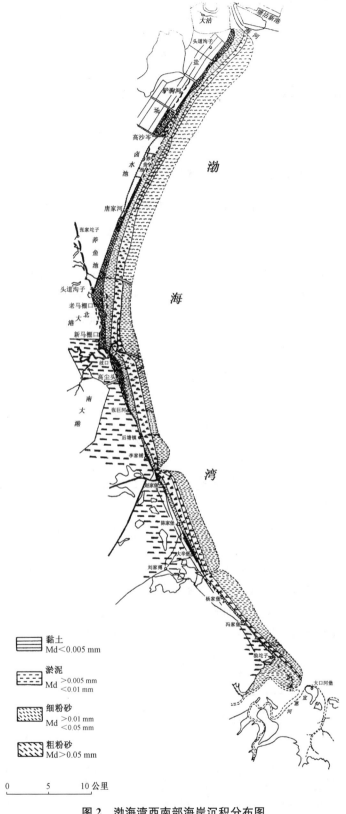

黏土
Md<0.005 mm

淤泥
Md >0.005 mm
<0.01 mm

细粉砂
Md >0.01 mm
<0.05 mm

粗粉砂
Md>0.05 mm

0   5   10公里

图 2 渤海湾西南部海岸沉积分布图

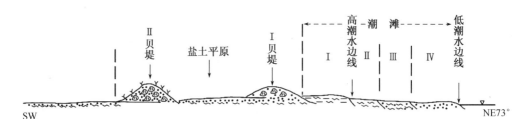

图3　淤进的冲积海积平原海岸

（二）稳定的双贝堤潟湖凹地平原海岸

如图4所示，在中段从头道沟子向南，经过老马棚口、歧口、张巨河、李家铺，一直到石碑河南岸的赵家堡、贾家堡，全长23 km。岸线方向在歧口以北是南北向，歧口以南，岸线转为西北—东南向，歧口正位于渤海湾向西凹入的中心部分。这段海岸的显著特点是分布着两列贝壳堤，外侧的一条贝壳堤（Ⅰ贝堤）是北段海岸的贝壳堤的向南延续，组成物质无变化，但由于此段海岸无海堤防护，它已被潮水冲刷蚀低，成为新月形的贝壳海滩，贝壳沉积厚度为20～40 cm。但在歧口河与石碑河以南，由于河流冲积物沿岸向南运移，保护了河口南面的岸滩，所以在这些地方，外侧的贝壳堤仍保存较好，堤高2～4 m，形态与剖面皆完整。

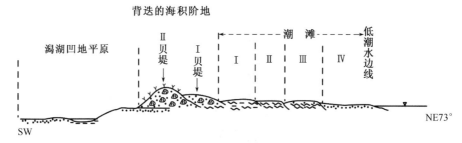

图4　稳定的贝壳堤潟湖凹地平原海岸

内侧是另一列较老的贝壳堤（Ⅱ贝堤），宽200 m，相对高度达5 m。组成物质纯净，主要是贝壳，有壳屑和大量的贝壳砂，夹杂着极少量的粉土、沉积层压实，并稍有胶结。贝壳种类丰富，沉积层次分选极好。例如歧口人工蓄水坑所开的剖面（图5）垂直海岸，是贝壳堤的向海一侧，出露宽度21.2 m，高3 m，从上到下可分为四层。

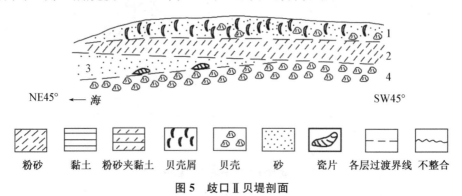

图5　歧口Ⅱ贝堤剖面

1层:厚 80 cm,黄色的粗贝壳砂屑层,有完整的毛蚶、碎白蛤、蝲螺及经海浪磨蚀过的砖网坠、陶器皿的残片,层次分选不很明显。

2层:厚 56 cm,夹杂有黄色粉土层的贝壳与屑砂层,贝壳中毛蚶和蝲螺完整,文蛤是碎片,还有一些小塔螺。粉土具有极好之微层理(1 mm),且已受挤压,变得微微弯曲。

3层:厚 59 cm,黄灰色细贝壳屑与砂层(碎屑长径 1~2 cm),水平层极好,中间夹着平铺成层壳顶向上的贝壳、瓦砾与粗瓷片。这一层向西南方向(向陆地)尖灭,沉积层很坚实,已为钙质胶结。

4层:出露 50 cm,下部为坡积土所掩。贝壳密集层,大小贝壳与粗细屑片相间成层,水平层理良好,稍胶结,贝壳种类是毛蚶、文蛤、蝲螺及少量础带锥螺。

这两列贝堤向北延伸可达上古林等地。

两列贝壳堤在头道沟子和马棚口相隔 1~2 km,向北去,间隔增大,最北在泥沽与道沟子之间,相隔 20 km。两堤之间是冲积海积盐土平原。自马棚口向南,二列贝壳堤逐渐靠拢,盐土平原渐渐变狭,到歧口,二堤合并。外侧低贝堤叠置在内侧的老贝堤上,组成二级背叠的海积阶地(照片 1)。多年来,人们不断地培修加固这条宽大的自然堤,成为一条高大的挡海堤。

照片 1 背叠的二级海积阶地

在Ⅱ堤后侧的海积平原上,绵延分布着一连串的潟湖凹地,如北大港、南大港、张家港、平家河、大泊以及赵家堡西南方的一些泊子,这些潟湖大部分已干涸成为浅平的凹地,凹地内是黑灰色的盐土,植物很少或完全没有,仅边缘生长了一些耐盐的草类,当地居民称为"泊"或"泊子",仍反映着潟湖的残余形态。但是,南、北大港由于辟为蓄洪水库,蓄存着大量淡水,所以其边缘地带或湖中高地生长着大量的芦苇和水草,潟湖凹地平原与Ⅱ贝堤相伴存在,是更老的一条岸线的标志。

贝壳堤带的外侧不整合地叠置着潮间带淤泥海滩,由于这段海岸坡度陡,岸滩较窄,除河口两侧部分外,平日涨潮时,潮水可达到堤脚,如遇大风天,潮流及风浪皆会对岸堤进行冲蚀,因而形成了一些海蚀崖,高约 2~5 m(照片 2)。但这种冲刷是局部的,主要发生在海岸线转折的地方。

此外,在这一段海岸上,人工开挖的大港排水河道较多,又有二条天然的河

照片 2 受冲蚀的贝壳堤

流——歧口河和石碑河,因此,河流对这段海岸带的影响比较显著。在经受淡水作用的河道两侧及浅滩上部,生长着碱蓬(*Suaeda glauca Bunge*)、芦苇(*Phragmites Oommunis*

Trin)及盐云草(*Limonium bicolor kuntie*)等植物[1],形成湿地型浅滩。同时,河流的冲积物也淤积在口外两侧的浅滩上,促进了岸滩的淤长。

(三) 具有残留贝壳海滩的冲蚀的潟湖凹地平原海岸

如图 6 所示,贾家堡以南,两列贝壳堤成交叉状集结在一起,海岸在该处受到冲刷,岸线自原来的西北—东南方向稍往北偏移,从此往南直至大口河,海岸的坡度都较陡,涨潮时,海水可达到堤下,波浪冲刷着海岸,贝壳堤受到冲蚀,Ⅰ贝堤已被冲去,岸线发生后退,

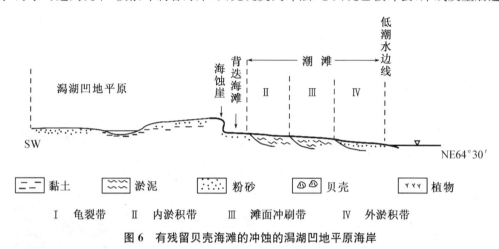

图 6 有残留贝壳海滩的冲蚀的潟湖凹地平原海岸

Ⅱ堤亦受到破坏,形成了小型的海蚀崖,高约 0.5~2.0 m(照片 3)。在崖脚的滩地上,蚀余的贝堤物质被波浪重新堆积起来,成为宽度不大的贝壳海滩。贝壳的种类多白蛤、文蛤、强棘红螺、蝈螺与锥螺,与内贝堤的物质相同,但是贝壳的表面多经磨损,并染有锈黄灰黑等杂色,表明了它不是新生的贝壳。此段岸滩坡度较陡,经常遭受波浪作用,滩上的沉积物主要是粉砂。

照片 3 南部海岸的小海崖

三、岸滩特征

岸线外围的潮间带浅滩(简称岸滩)是粉砂淤泥质平原海岸的一个重要的组成部分,这是一个特殊的地带,在高潮时,它被海水淹没;在低潮时出露成为滩地。在渤海湾西南部粉砂淤泥质平原海岸,潮间带浅滩特别发育,并且是动态最为活跃的部分。

本区岸滩平均宽 3 km,北部较宽,向南逐渐变窄,最宽的岸滩在新港南面,宽

---

[1] 植物标本承南大生物系耿伯介先生鉴定。

3 500 m,在大口河北面,岸滩最窄,宽只 1 500 m,大部分地区岸滩的宽度在 2 500 m 以上。岸滩的坡度介于 0.5‰～1.0‰ 之间,作用于岸滩上的主要动力是潮流,而波浪作用是比较微弱的。这里的岸滩具有明显的分带性,各带在动力、地貌和沉积上都存在着一定的差异,从陆向海,可以分为四个地带。

(一) 龟裂带

分布在特大高潮线和低高潮线之间,这个地带在特大潮水与高高潮时被海水淹没,平时出露,组成物质是黏土质淤泥,受海水浸泡时,黏土膨胀,干时,黏土又发生收缩,使滩面发生龟裂。龟裂的大小各地不同,在靠近特大高潮线附近,一般龟裂缝隙大,至低高潮线附近,则缝隙变小以至不明显。岸滩坡度平缓处,龟裂较宽,龟裂现象也较显著。龟裂带的宽度自百十米至数百米不等,最宽达 870 m(歧口),平均坡度 0.5‰～0.6‰。龟裂带分布以中段海岸最宽,北段海岸较窄,而在歧口以南。自张巨河向南,除在石碑河北部李家铺岸滩发现有龟裂带外,龟裂带基本上是不发育的,这主要是因为南段海岸的滩面窄、坡度陡,平日一上潮,海水就可达到岸边。

龟裂现象最发育的地方是北段海岸,在高沙岭村北 1 km 的岸滩上,龟裂缝隙宽达 1～2 cm,龟裂的地块宽度 15～50 cm 不等(照片 4)。中段海岸的龟裂带,在接近贝壳堤的地方多生长着小芦苇与碱蓬。

照片 4-1　渤海湾驴驹河贝壳堤与潮滩龟裂带　　　　照片 4-2　潮滩上部草滩(附彩图)

(二) 内淤积带

分布在高潮线以下,经常为潮水淹没,平均宽度 500 多 m,平均坡度 0.67‰,滩面沉积组成主要是淤泥,厚度约 1 m,松动如粥状,人行至此,经常下陷,在淤泥层的下面是厚约 2 m 的粗、细粉砂层。

此带是典型的淤泥浅滩,地形平缓单调,通行最为困难(照片 5)。内淤积带在本区普遍发育,形态要素变化不大。

照片 5　潮间带中部内淤积带:泥沼滩

据定位观测资料,潮流所带的淤泥与细粉砂主要沉积在这个地带。与其他各带相比较,它的特点是物质颗粒细,淤积厚度大。

(三) 滩面冲刷带

位于内淤积带的外侧,两带逐渐过渡。在本带的内缘,滩面上出现起伏不平的水坑,冲刷形态不甚明显。在中部逐渐出现各种滩面冲刷体,冲刷规模较大,形态显著。至外缘,冲刷体又隐伏下去。在各段海岸上,滩面冲刷带的形态略有差异。

在北段海岸上,滩面冲刷带的宽度不大,约 50m[①],坡度是 0.58‰,沉积组成是粗粉砂与淤泥的夹层。其形态表现为纵长的鳞形冲刷体(滩鳞)。冲刷体三面具陡坎,一面逐渐隐伏(照片 6、7),长 3~18 m 不等,宽约 0.5 m。长轴方向为 SE60°~70°,前端小陡坎朝向西北,即向岸。小陡坎的高度自 2 cm 到 12 cm 不等,向岸高度大于两侧高度。陡坎多发育在粉砂层中,因沉积组成是粗细夹层,因此小陡坎有一定的成层性。在各个鳞片之间,为具有薄层积水的小凹地。在凹地及鳞片的上面皆具有流水波痕。波痕的陡坡向岸,缓坡向海。根据鳞片明显的纵长方向性,可以推断它是水流作用形成的,是由岸向海的水流所冲刷成的,并且可以根据冲刷体的延伸方向,推测此股水流是朝向 S60°E 的。

照片 6-1　滩面冲刷带　　　　照片 6-2　淤泥层厚的潮滩:滩面冲刷带外淤积带

冲刷体形成在水层较厚、流势较大时,即潮流自龟裂带外缘退到岸滩中部,集中了较大的水量时,而冲刷体上的水流波痕是落潮后期薄层水流作用于表面的叠加形态。

照片 7　滩鳞　　　　　　　　照片 8　遭受 6 级东北风后的滩鳞

---

① 因有海堤阻挡。

在工作期间,我们遇到了连续16小时的7级东北风,结果冲刷体被稀泥所淹没,冲刷现象不明显了(照片8),大风后数天,又逐渐恢复。

应用标志层、指示砂等方法进行观测,发现这个地带是受冲刷的,物质是向两旁搬运的。

在中段海岸,冲刷带具有下列特点:

(1) 宽度大,超过1 km,坡度为0.402‰。

(2)上部组成物质主要是淤泥。

(3)冲刷形态变化较多,冲刷沟较明显。

在此带的内缘,冲刷现象不明显,是一些片状积水凹地,而松动的淤泥厚度是40 cm。

在此带的中部,出现明显的冲刷沟,宽0.5 m,曲折地以N60°E方向延伸,冲刷沟有单支的,也有丫形的,沟旁小陡坎高7~18 cm。沟内沉积有厚达45 cm的黄色淤泥。沟间地宽2~8 m,这里松动的淤泥厚度更大,达55 cm。

再向外,冲刷沟密度加大,成为平行排列的雏形潮水沟,沟宽约0.6~0.8 m,沟间地宽1 m,像平行的垄岗一样,与潮水沟方向一致,向东北延伸(照片9)。

照片9 滩面冲蚀潮水沟——中部海岸岸滩冲刷带

到冲刷带的外缘,冲刷现象又复减弱。

冲刷带的浮动淤泥厚度超过内淤积带,一般为50~60 cm,而下垫面崎岖不平,物质是夹有大量贝壳的粉砂。因此,在此带通行极为困难。从定位观测资料看,此带仍在进行着淤积作用,冲刷发生在滩地表面。

在南段海岸,冲刷带也具有宽度大,淤泥厚,下垫面不平的特点。而冲刷形态则不甚规则,有的为长条形垄岗,有的为向岸(在内侧)或向海(在外侧)的圆锥体,多向东或东北方向延伸。冲刷体陡坎高十数厘米,在冲刷体之间为形态不规则的潮水沟。冲刷现象仍然是在带的中部强,向两旁减弱。

滩面冲刷带是渤海湾西南部淤泥质岸滩上的特殊现象,在岸滩上普遍存在,没有间断。它以显著的地貌表现,给人以极为深刻的印象。但是,目前对它形成的原因还不够了解,只知它的存在与水流(可能是落潮流)的冲刷直接有关,从肉眼观察也可以看出冲刷带前缘坡(与内淤积带相接的)的坡度较大,但在水准测量中却缺乏明显的反映。因此揭示它的形成过程,还需要作进一步的研究。

(四) 外淤积带

分布于冲刷带外侧,一直到低潮线这个地带的宽度从500 m到2 000 m不等,通常在1 000 m左右,坡度为0.5‰,其组成物质主要是粉砂,经受潮流作用,在滩面上形成长3~7 cm,高1 cm、1.5 cm、2 cm,向NE方向的波痕(照片10、11)。在较宽的外淤积带上,各个部分波痕形态不同,在内侧,波痕向岸为陡坡,向海为缓坡,形态不对称;在中部,波痕是对称的;在外侧,则波痕向海陡,向岸缓。这反映着在外淤积带的各个部位上经受的涨落潮流力量对比的情况。但在接近低潮线的地方,在波痕上常常叠置着鱼鳞状微波痕。

照片 10　潮滩外淤积带-粉砂波痕带

照片 11　覆有薄层淤泥的外淤积带

这个地带仍以潮流作用为主,但也受到波浪与浪流的作用。由于经常为海水所浸泡,因此,沉积层表面多具有锈黄色。

外淤积带主要是进行着淤积作用,物质来自两个方面:

(1) 被潮流从外海中带来的泥沙;

(2) 从冲刷带带来的泥沙。

其中以从外海来的物质为主,这项物质主要是粉砂,由于潮流上滩后运行距离不远,力量磨损不大,所以沉积下来的主要是粗颗粒——在本区主要是粗粉砂。而落潮流自冲刷带带下来的主要是淤泥与细粉砂。

外淤积带在中段和南段靠近河口处,沉积有所变异,该处受河流影响,往往堆积着淤泥。但淤泥厚度小于内淤积带,仅为 10 cm 左右。在河口的外围地区,外淤积带的粉砂滩面上常盖有 2～5 cm 厚的流动淤泥。

这四个带是动力、地貌与沉积的综合反映,地貌的差异性表现得最为突出。岸滩分带性的特点在本区海岸普遍存在,各带的形态与沉积在不同的海岸段落上虽略有变化,但其分布规律则完全相同。

岸滩的分带现象是存在着季节变化的。在一年中,分带现象以春季最为明显,夏季时风浪大,冲刷带遭受破坏,变得模糊,淤泥厚度减薄,沉积物质地有粗化现象——淤泥减少,粉砂带扩大,但每次吹东风后,滩上流淤着薄层黄色的泥浆。秋季时风浪减小,冲刷带逐渐又恢复。秋末冬初,滩地淤涨,淤泥又堆积加厚,冬季结冰后,岸滩基本稳定,冰块对滩面有犁蚀作用[①]。冰中的浮土在冰融后亦大量的堆积在滩上。

---

① 关于岸滩季节变化详细情况,1964 年还要进行工作。

四、岸滩泥沙运动的初步分析

1963 年 5 月 20 日到 6 月 25 日,我们在驴驹河断面,歧口断面和李家堡断面进行了在不同潮性下岸滩冲淤与泥沙运动观测,其各项工作如下:

(1)利用标志桩、标志层,观测岸滩短期与长期的冲淤变化,其中标志层效果较好,而标志桩因桩径粗,对水层干扰大,观测效果不佳,拟暂不应用其资料。

(2)利用分向底砂捕砂器(具有十字隔板的沉降板,十字隔板分出Ⅰ、Ⅱ、Ⅲ、Ⅳ四个象限,分别表示 NE、NW、SW、SE 四个方向的沉砂状况)与沉降板观测岸滩底层泥沙运动与滩面落淤情况。

(3)利用分层的沉降斗(分成两组:10 cm、40 cm、90 cm 和 20 cm、50 cm、100 cm)与分向悬砂捕砂器(分北、东、南、西四个方向,每个方向分成高出滩面 40 cm 与 20 cm 两层)观测不同方向不同水层的含沙量与物质级配情况。

(4)利用红砖粉屑观测冲刷带物质运移情况。

下面就观测所得资料,分别阐述各断面滩面泥沙运动情况。

(一)驴驹河断面

断面位于海河口以南约 10.25 km 处。以驴驹河村为基点,向海延伸,与岸线垂直,断面方向是 SE58°,断面全长 3 496.7 m,此断面处于淤进的海积-冲积平原海岸上,作为此类型海岸的控制断面,也是新港港区南部岸滩的控制断面(图 7)。

岸滩各带沉积分布是:

外淤积带→滩面冲刷带→内淤积带→龟裂带

粗粉砂→粗粉砂与淤泥细粉砂之夹层→细粉砂、淤泥→淤泥与黏土淤泥

从 5 月下旬至 6 月下旬的观测(代表春季与夏季时该处岸滩的综合特点)资料分析,该处岸滩泥沙运动具有下列特点:

(1)从低潮水边线向岸堤、沉降板与分向底砂捕砂器上的底砂沉积主要是粉砂(Md 介于 0.012~0.07 mm),但底砂粒径有向岸变细的趋势,在外淤积带主要是粗粉砂,向岸递变为细粉砂与淤泥系细粉砂。淤泥颗粒的含量逐渐增加,如表 2。

底砂沉积的数量以外淤积带最大(如Ⅵ桩与Ⅷ桩的沉积结果),沉积数量向岸逐渐减少。次一沉积中心为内淤积带(Ⅲ桩所在),再向内,沉积量迅速减少。

(2)在涨潮时,外淤积带各水层的悬砂以底层(10~20 cm)为最大,落淤多。在 40 cm 高度以上,泥沙含量迅速减少,仅及底部之半,甚至不到其半。至冲刷带后,水层减薄,春季高潮时也不到 1 m,故泥沙至 1 m 高处已接近于零。将底部水层中各点含沙量作比较,亦是从低潮水边线向岸边逐渐减少。悬砂质的颗粒在近低潮水边线处较大,为细粉砂,至外淤积带内侧,颗粒变细,主要为淤泥。

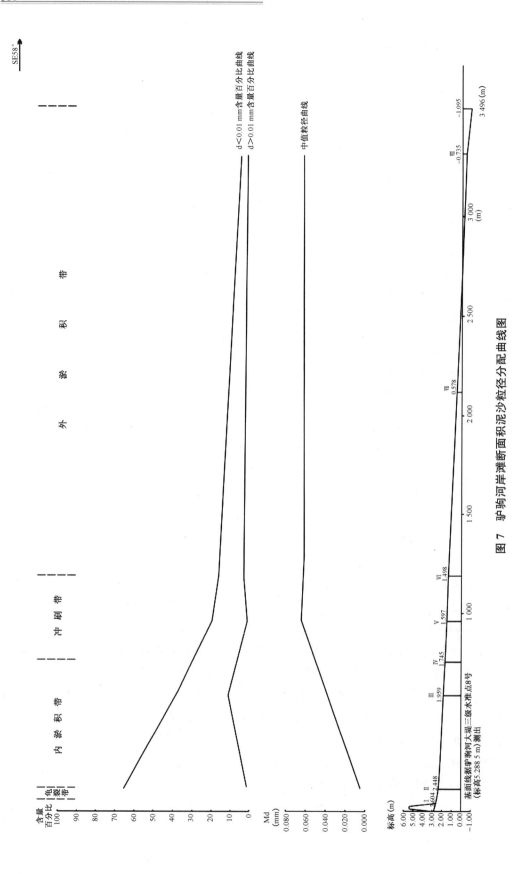

图 7 驴驹河岸滩断面积沙粒径分配曲线图

表 2　驴驹河岸滩沉降板沙样颗粒分析表

(a) 1963 年 5 月 26～27 日

桩号位置	$d>0.05$ 的百分数	0.05～0.01 的百分数	0.01～0.005 的百分数	0.005～0.001 的百分数	$d<0.001$ 的百分数	Md	定名
Ⅱ桩(龟裂带外缘)	28.64	52.68	6.69	7.32	4.67	0.029	细粉砂
Ⅳ桩(内淤积带外缘)	35.11	54.76	2.32	3.60	4.21	0.035	细粉砂
Ⅷ桩(外淤积带)	46.41	52.08	0.79	0.72	0	0.044	粗粉砂

(b) 1963 年 6 月 23 日

桩号位置	$d>0.05$ 的百分数	0.05～0.01 的百分数	0.01～0.005 的百分数	0.005～0.001 的百分数	$d<0.001$ 的百分数	Md	定名
Ⅲ桩(内淤积带)	28.81	55.76	0.19	9.12	6.12	0.028	细粉砂
Ⅵ桩(冲刷带外缘)	39.91	55.16	0.52	1.90	2.51	0.038	细粉砂
Ⅷ桩(外淤积带)	66.87	27.58	0.83	1.91	2.81	0.07	粗粉砂

(3) 将同一点上、同一潮汛、垂线上阻留下的悬砂与底砂相比较,底砂的颗粒比悬砂为粗。

(4) 应用红砖粉观测滩面冲刷带,其动态是:遭受水流冲蚀,泥沙向两侧移动,向岸方移动量大。1963 年 5 月 25 日于冲刷带中心(Ⅴ槽处)铺设红砖粉后,5 月份经数次观察,其最大移动方向是 NW 向,大量红砖粉散布于 NW28°～80°之间,主要集中在正西北,这种情况反映着涨潮流对此带滩面的冲刷与泥沙搬运起着较大的作用。

5 月 25 日在内淤积带(Ⅱ桩处)及外淤积带(Ⅵ桩、Ⅷ桩处)铺设了与滩面平齐的红砖屑层,5 月 28 日观测标桩Ⅱ处红砖层上已覆盖了厚约 0.2 cm 的淤泥、标志桩Ⅵ处覆盖了 0.5～1 cm 的淤泥,标桩Ⅷ红砖层已全被淤泥淹没,1964 年拟再继续观察。

(5) 断面上各点、各方向泥沙来量之比较:

① 分向悬砂:20 cm 层中以东向与南向来沙量较大,即泥沙主要为涨潮流自海中带入。其次是西向开口捕沙较多,部分地反映着落潮流的运砂情况。

40 cm 水层:在内淤积带(以Ⅲ桩为代表)主要来自西面(岸方的),在外淤积带以来自东北方向为多。

② 分向底砂沉积以西北向的沉积量最大,居绝对优势,其次是东南向沉积。

(6) 夏季强潮时,水势大,水层厚,泥沙含量较春季显著增大,滩面明显地反映出两个淤积中心——外淤积带与内淤积带。岸滩沉积物质一般较春季为粗。但是,在连续两日吹 3～4 级的东南风时,滩面即流覆着薄层的黄色淤泥。淤泥向岸流动,可达内淤积带。

(二) 歧口断面(Ⅲ断面)

断面位于黄骅县歧口河南岸,在歧口村与高尘头之间,以该处铁三脚架高标为基点,向海延伸至低潮水边线,断面全长 3 408.5 m,断面方向为 NE73°,与岸线垂直。本断面作为稳定的具双贝堤的潟湖凹地平原海岸的代表断面,同时控制渤海湾西南部海岸的中段地区(图 8)。

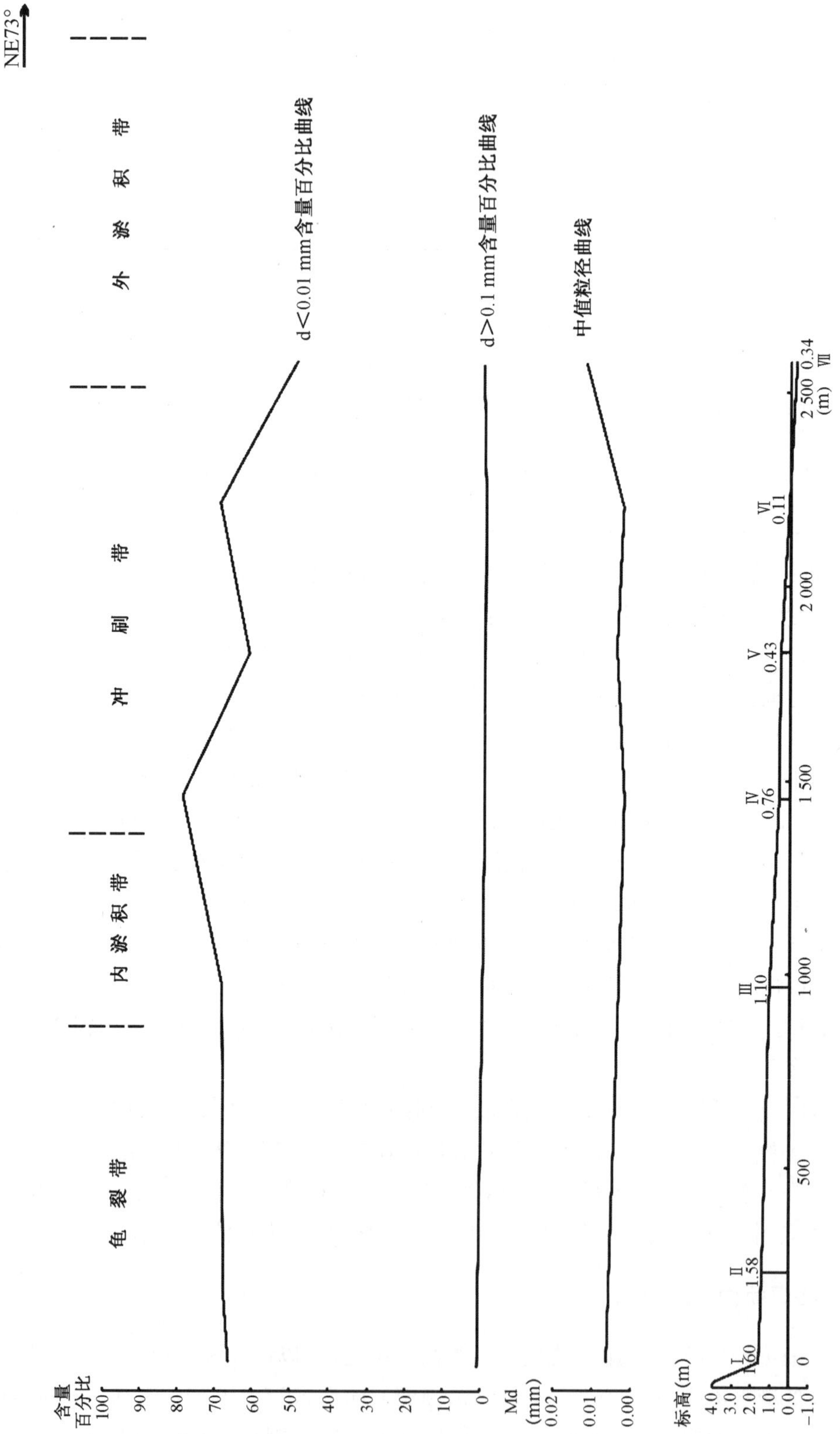

图 8 岐口岸滩断面泥沙粒径分配曲线图

歧口岸滩具有龟裂带宽、冲刷带宽与淤泥沉积范围广的特点。淤泥沉积自龟裂带开始遍及外淤积带，仅低潮水边线附近出现了粉砂带。此处淤泥广布是由于地居渤海湾湾顶，动力减弱，又位于河口近侧，接近局部泥沙供给地之故。

由于此处淤泥厚，搬运与通行极难，故仅作了标志桩、沉降斗与分向底砂观测器及指示砂的观测。岸滩泥沙运动观测是在6月上旬进行的，当时正逢小潮。

该处岸滩泥沙运移特点是：

（1）岸滩底砂沉积主要是细粉砂。沉积最大中心在滩面冲刷带内部（标志桩Ⅵ处），但是夏季时的歧口滩面经常覆盖着流动的淤泥，它掩埋了涨潮流所沉积下的粉砂，在吹东南风时，淤泥掩盖面积更大。可以认为：歧口滩面经常受着活跃于沿岸的淤泥流的影响。这股淤泥流不是来自歧口河（歧口河的冲积物主要是粉砂），而是来自南部的黄河。当地经常出海的渔民皆认为："黄泥汤子是刮东南风时由黄河口带来的。"经过实地勘测后，我们同意这种看法，源自黄河的淤泥流经常可达到歧口一带。

但是在吹西南风时，滩面沉积减薄，或无沉积，而反产生冲刷。

（2）此处的悬移质主要是淤泥，悬移质含量是近海测点大于近岸测点，但最大的含量是在冲刷带中部（Ⅵ桩处）。同时，冲刷带上的悬砂分布以50 cm水层中数量最大，其次为20 cm水层。1 m水层中含沙量大为减少。

（3）冲刷带中部铺红砖粉指示砂，从6月3日—6月10日，砖粉主要向WNW运移，速度与Ⅱ断面相近，但在刮东南风后，滩面上的红砖屑又为淤泥所掩埋。

总上三点看来，此处滩面冲刷带的淤积量最大，也是水流冲刷扰动强、水层浑浊度最大之处，其冲淤关系如何，还需进一步调查。

（4）此处龟裂带很宽，平时潮水不能全掩，故物质交换微弱，近贝堤处已生长了芦苇、碱蓬等短草。

（三）李家堡断面

断面位于黄骅县石碑河北面约1 km处。以李家堡为基点向海延伸，断面长2 896.2 m。断面线垂直岸线，方向为NE64°30′。本断面主要是控制地了解石碑河对其以北海岸的影响。1964年拟于赵家堡以南冲蚀岸段增设断面。对该断面的泥沙运动初步了解如下（图9）：

（1）从低潮水边线至岸底砂沉积：小潮时底砂沉积主要是细粉砂。底砂沉积具有两个中心：① 外淤积带沉积量最大；② 内淤积带沉积量次之。但在东南风作用下，滩面整个覆盖着流动性极大的浮泥。

（2）悬砂沉积物有两次观测：一次6月14日—6月16日，为E风，3～4级，小潮汛，分层分向悬砂捕砂器。在外淤积带与冲刷带上所捕获的主要是细粉砂，夹少量淤泥。另一次是6月16日—18日，SE风4级，亦为小潮，此时该两点分层分向悬砂器所捕获的主要是浮泥。后面这一情况是值得注意的，它又一次反映着：滩面有浮泥出现是与SE风有关系的，并且也可说明滩上浮泥层活动的厚度较大。又如附近的赵家堡，平时落潮后，贝堤坡脚的滩底是出露的，但在6月19日，SE－E风3级，滩上覆盖了淤泥浆（浮泥），厚约半米。

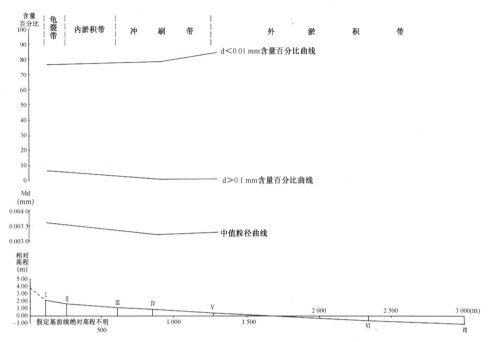

图 9　李家铺岸滩断面泥沙粒径分配曲线图

同一垂线上悬砂各层含量以近底层处最大,而各带水层悬砂含量以内淤积带最大。

分向悬砂沉积,西北向沉积量大于东、南向,反映着落潮流(合成落潮流向 NE)的悬砂含量是较大的。

(3) 其他情况与Ⅱ断面类似。

综上所述,可以看出:

(1) 岸滩泥沙运移的基本特征是与各带的动力特征相适应的。

外淤积带是主要的淤积场所,单位面积上淤积量大,且该带宽,故淤积总量亦大。淤积物颗粒较粗,主要是粗粉砂。自此带向岸,沉积数量与物质颗粒均减小。

冲刷带的泥沙是向两侧运移的,以落潮流带走的泥沙量较多;粗颗粒物质主要沉积在外淤积带,而细颗粒淤泥又被带入海中。

岸滩沉积的次高峰带是在内淤积带,这里沉积着涨潮流自外海所携带来的淤泥,形成了厚层淤泥积聚区。但也有部分细粉砂是来自冲刷带。

龟裂带的淤积量最小,并且是颗粒细小的黏土物质的沉积场所。

歧口以南,岸滩泥沙运动与沉积特性略有不同,主要是由于淤泥流经常活动,使滩面覆盖了一层淤泥,它掩藏了岸滩沉积分异的特点。

(2) 当滩上满潮时,近底层 20 cm 以内的水层的运移物质主要是粉砂。上部水层中是淤泥及少量细粉砂。而粉砂含量远远大于淤泥含量。据浅钻资料,粉砂是本区海岸的基本组成物质,它不断地自岸坡下部被运积到岸滩上,表明本区海岸剖面的塑造尚未达到均衡剖面阶段。

(3) 岸滩沉积物纵向分布具有差异性。北部海岸的粉砂沉积带宽广,向南至歧口逐渐束狭,而代之以宽大的淤泥沉积。这主要是由于南部靠近黄河口,接近淤泥补给地,而北部受黄河泥沙的直接影响已很微弱,即使淤泥能达到此地,因数量不大,很快即为浪流掀带至深海沉积。

淤泥的转运沉积主要依靠沿岸淤泥流。这股淤泥流自黄河向北,经常可达到歧口,因此,歧口岸滩在夏季吹东南风时,滩上经常覆盖泥浆。在夏季较强的东南风推动下,它亦能达到新港港区附近的岸滩,但已是强弩之末,其影响仅是局部的。

五、小　结

综合以上所述,我们对渤海湾西南部的海岸地貌和岸滩有下列几点初步认识:

(1) 本区是典型的粉砂淤泥质平原海岸,海岸动态很活跃。从动力地貌的研究,本区海岸可分三类:① 涧河口到歧口是缓缓淤长的海岸;② 歧口到贾家堡是长期稳定的海岸;③ 贾家堡到大口河是微微蚀退的海岸。新港位于①类海岸地段,该处在历史时期宋(庆历八年,公元 1048 年)黄河从海河入海时,岸线曾迅速向海淤进。当黄河南徙后,400多年来岸线基本稳定。因此,从海岸长期发展情况来看,新港所在地区岸线的自然淤长是比较缓慢的。

(2) 根据潮间带浅滩地貌、沉积与动态的研究,发现渤海湾西南部岸滩具有分带性。岸滩各带的水动力、地貌与泥沙运动状况各有不同,但目前整个岸滩均处在淤长过程中,其分布规律从海向陆是:

地貌:外淤积带→滩面冲刷带→内淤积带→龟裂带。

表层沉积:粉砂→淤泥与粉砂—淤泥→黏土质淤泥(组成目前岸滩的基底是粉砂,而淤泥是后来的叠置沉积)。

动力:潮流与微波→潮流(落流为主)→潮流(涨流为主)→潮流和日晒。

动态:大面积的强盛淤积→涨淤落冲→连续地淤积→微淤偶尔冲蚀。

这种分带现象是潮汐作用下粉砂淤泥质平原海岸的特点,它的形成显然与潮流活动有关,涨潮流沿滩上行时,受滩面摩擦,动能逐渐降低,因而在各个段落上卸下了不同粒径的泥沙。落潮时水流逐渐集中,当其顺岸滩坡面而下时,常在一定地带上造成冲刷。冲刷下来的泥沙,粗粒多沉积在外淤积带,细颗粒物质则被带入海中。

岸滩分带规律的研究有助于揭露岸滩剖面的自然塑造过程,从而了解新港所在浅滩的发展趋势。因此需重视分带性的研究,研究分带现象的成因、过程与季节变化。

Cheniers on the East Coastal Plain of China[*]

Introduction

Cheniers and chenier plains are geomorphological features widely distributed along low coastal areas of the world. Russell and Howe (1953) and Price (1955) are pioneers on this subject. They discussed the cheniers of southwestern Louisiana, their formation, and their environmental characteristics. Since then many scientists have investigated cheniers and chenier plains, in open as well as in sheltered coasts.

The areas under investigation included the Mississippi River plain, (west Louisiana-Texas coast), the coast of Guyana and Suriname in South America and the north coasts of Australia and New Zealand. These regions are mainly concentrated along the Atlantic coast and the Pacific coast in the southern hemisphere. Although some scientists have studied cheniers along the coast of China on the west coast of the Pacific Ocean margin as early as in the late 1950's, their work was completely independent from the investigations done in the western world (Wang, 1964) and hence unknown because reports were published mainly in Chinese within China.

There are many definitions and explanations of cheniers proposed in studies published so far. Gould and McFarlan (1959) associated the development of the chenier plain in southwestern Louisiana to pulses in sediment supply of the Mississippi River due to the lateral shifting of the river mouth position. Todd (1968) noted three necessary conditions required for the formation of cheniers: (1) a stable or lowering sealevel, (2) an effective longshore current and (3) available sediment supply from rivers. Greensmith and Tucker(1969) suggested two factors: (1) mass mortality of molluscs in the channel communities and (2) the reworking of pre-existing shell sheets interbedded with mud, silt and fine sand of the intertidal zone. Otvos and Price (1979) suggested that cheniers originate on shores with low to intermediate wave energies and small-to-large tidal ranges. Woodroffe et al. (1983) found individual cheniers in the Miranda coastal plain, New Zealand, developing initially from sand bars formed on the foreshore and migrating landwards through swash action. Chappell and Grindrod (1984) demonstrated that chenier systems along the Princess Charlotte Bay, Queensland, Australia, have an internal mechanism due to the geometry of the bay, which regulates

---

＊　Ying Wang, Xiankun Ke: *Marine Geology*, 1989, No. 90: pp. 321 – 335.

alternation between ridge and mudflat deposition. Sealevel fluctuation is not necessary to explain chenier ridge formation; however, shell availability in source areas is a primary factor. Augustinus (1980) described two types of cheniers: (1) medium to coarse sandy cheniers built up by longshore supplied sand, by beach-drift and washover processes and formed at or just above high tide level and (2) fine sandy cheniers beginning at approximately mean low tide level. The fine sand is winnowed out of the Amazon-borne mud by wave action and thus supplied by the shelf.

Summarizing, it appears that low to medium wave energy conditions and a tidal flat environment with abundant fine material, including a (fine) sand and shell component or in some cases a local sand supply, are the necessary conditions for the formation of cheniers.

This study on cheniers along the east coast of China shows that cheniers are coastal deposition forms. The coast is characterized by a low to medium wave energy, has optimum slopes of 0.9~3.0‰, and is composed of fine sediments such as silt or clayey silt or fine sand. The cheniers are a type of beach ridge, formed at high tide level, mainly consisting of marine shells and shell fragments partially mixed with coarse silt and fine sand. They therefore represent a distinctive coastal environment. The shells are reworked by the action of the breaking waves on the lower part of the tidal flat and accumulated by swash action to form cheniers on the upper part. However, this only occurs under certain wave energy conditions (e. g. , high tide).

The tidal flat itself, having very gentle coastal slopes, forms a favourable environment, with light and oxygen for benthic organisms and a supply of food and nourishing material for shellfish (Wang, 1983). Different species of molluscs are found on tidal flats composed of different sediments; e. g. , razor clams (*Solen gradis* Dunker and *S. gouldii* Conrad) only live on sandy flats.

Many types of molluscs (e. g. *Arca subcrenata*, *Meretrix meretrix*, *Umbonium townsi*, *Cyclina sinensis*, *Mactra veneriformis*, *Bullacta exerata*, etc.) live on wide silty flats with gentle slopes and can supply abundant shell material. Should a coast be retreating and such flats be eroded, the shells and coarser sediments, such as coarse silt or fine sand, will be transported landward by wave action, and cheniers then will be formed on the upper part of the tidal flats. This has actually occurred in the east coastal plain of China in the areas of the Jiangsu coast and the west coast of Bohai Bay (Fig. 1).

Cheniers are widely distributed along the coasts affected by fine-grained sediments supplied by large rivers, where well developed tidal flats are subjected to slight wave erosion, such as in several parts of the Yellow River delta and the coastal areas affected by Yellow River sediment. Whenever the river changes its course the sedimentation in the coastal area ceases. Erosion by waves follows and cheniers develop. For example, shell beach ridges of an altitude of 10 m have been recently formed along the abandoned modern Yellow River delta (Chenqi Chui, pers. commun., 1985). It is important to distinguish the shell beach ridges which are cheniers from the typical sandy beach ridges, even when the latter contain 20~30% of shell fragments. Typical sandy beach ridges represent a constructive and high-energy beach environment, whereas cheniers (shell beach ridges) indicate a low-energy environment of a slightly eroding tidal flat coast. Although several cheniers may appear on prograding coasts, such as Jiyunhe River mouth in the Bohai Sea (Fig. 1), this situation only indicates long-term evolution processes and does not contradict the principle that such features are the result of erosion of the former old tidal flat.

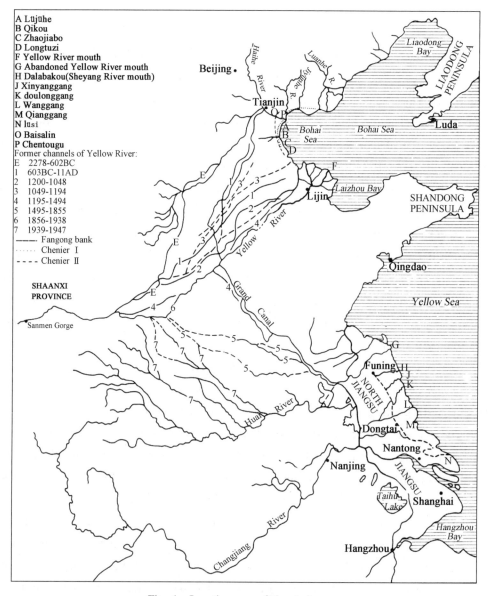

Fig. 1　Location map of the study area

Shell layers can also be found within the tidal flat sediments. They may represent either the former aggrading coastal environment or extreme events during the development of the tidal flat, rather than buried cheniers. In other cases, shell mounds built from shell remains by mollusc-eating peoples of prehistorical times in the ancient coastal areas of China, together with the oyster reefs growing on gravels or driftwood near river mouths, simply indicate a nearshore environment and should be distinguished from cheniers. Cheniers as a characteristic feature of tidal flat coasts represent erosional conditions of slightly retreating coastlines.

Methods of study

Field surveys and investigations of the dynamic geomorphology and sedimentology of the coastal plain have been carried out on three occasions: during the 1950's, the 1960's and the 1980's. These included drilling, sampling, hydrological measuring and monitoring. One-hundred samples of cores from the tidal flat were analyzed sedimentologically, including grain-size analysis, heavy mineral analysis, and the analysis of pollen, spores and foraminifera. Fifteen samples of shells, peat and wood were used for ^{14}C dating. A study was made of the historical records and archaeological data in order to gain a better understanding concerning the formation and evolution of the cheniers on the east coastal plain of China.

Areas of study

The studied areas discussed in this paper are located along the western border of the Pacific Ocean and adjacent to the two large marginal seas of China: the Bohai Sea and Yellow Sea (Fig. 1). The Bohai Bay of the Bohai Sea and the Jiangsu coast along the west Yellow Sea both have semidiurnal tides and an average tidal range of 3 m and 2.5~4 m each, and form part of the meso-macro tidal areas. The third largest river on the world, the Changjiang River (Yangtze River), flows into the sea in the southern part of the Jiangsu coast. The Yellow River affected the Jiangsu coast and the west coast of Bohai Bay during alternatively historical times. Most of the sediments from the two rivers are deposited on the inner shelf near the river mouth. The farthest distance that sediment has been transported into the deep water area of the continental shelf is 50 km in the east China Sea by the Changjiang River (Wang et al., 1986). The Yellow River sediments are deposited in 15 m of water, at the foreslope of the submarine delta. Thus the physical conditions of the large rivers have an important impact on the coastal environment, e. g., very gentle coastal slopes and a wide and shallow continental shelf.

The Changjiang River is the largest river in China with an annual water discharge of

9.25×10^{11} m³ and an annual sediment discharge of 4.86×10^8 tons (Table 1). The stratigraphic study showed that its river mouth constantly shifted from the north to its present position since the 6 000 yrs B. P. maximum transgression, and formed a huge fan-shaped delta (Ke, 1985). Deltaic tidal flats are therefore very well developed along the coast of the Changjiang River.

The Yellow River has the largest sediment discharge in the world: 11.9×10^8 tons/ yr (Table 1). Due to the internal instability of the river channel in its lower reaches, in the delta area it changes its course once every 10 years (Cheng et al., 1986).

This causes a change in the local sediment supply of the coastal area, hence controlling the erosion or progradation of the coast. According to historical records, the Yellow River has shifted its course eight times since 2278 B. C., and on the fourth, fifth and seventh occasion time the river migrated from the Bohai Bay to the Jiangsu coast (Wang, 1964; Zhu, 1982).

Table 1　Basic data on the Changjiang River and Yellow River (Wang, 1983)

River	Drainage area (km²)	River length (km)	Annual water discharge (m³)	Annual sediment discharge (ton)	Average sediment concentration (kg/m³)	Average tidal range (m)
Changjiang	1 807 199	6 380	9.25×10^{11}	4.86×10^8	0.544	2.77
Yellow River	752 443	5 464	48.5×10^9	11.9×10^8	37.7	0.80

Cheniers are very well developed in this area. There are mainly two series of cheniers along the modern west coast of Bohai Bay, even though a few remnants of older coastlines remain. Along the Jiangsu coast, at least four rows of cheniers including the ancient cheniers, have developed. These two areas form part of to the sheltered low tidal flat coast. The Bohai Sea is a marginal sea stretching inland, with very sheltered conditions being surrounded by land on three sides. The average maximum wave height at this location is only 1.9 m (Wang and Aubrey, 1987). Due to large quantities of sediment supplied by the Yellow River, wide tidal flats with four distinctive zones are developed along the west and north coast of the Bohai Bay (Table 2). The Jiangsu coast, facing towards the open Yellow Sea, appears at first glance to be an open coast, but in fact is also a sheltered coast with a weak wave regime because there is a zone of huge submarine sand ridges situated offshore which separate the coast from the open Yellow Sea. It appears that cheniers along the Jiangsu coast are formed in areas where wave action dominates over the tidal currents, whereas the tidal flats developed in areas with a dominant tidal influence (Table 3). The zones of these tidal flats are indicated in Table 2.

Table 2　The zonates of the tidal flats of the Jiangsu coast and the west coast of Bohai Bay

Area	Zone			
	HWSL			LWSL
Jiangsu coast	Grass zone	*Suaeda salsa* zone	Mud-silt mixing zone	Silt or sand zone
West coast Bohai Bay	Polygon zone silty-clay	Inner depositional zone, poor in mud	Erosional zone, mud-silt alternation	Outer depositional zone, silt ripples

Table 3　Environmental dynamic characteristics of tidal flats and cheniers on the in Jiangsu coast

Environment	Mean wave height (m)	Max. wave height (m)	Tidal current velocity (cm/s)	
			Max. flood	Max. ebb
Chenier coast	1. 5	4. 9	93	103
Tidal flats	1. 0	2. 9	97	165

Cheniers of the east coastal plain of China

The west coast of Bohai Bay

Two large cheniers (Ⅰ and Ⅱ) occur in this area; Chenier Ⅰ, is the closest to the coast (Fig. 1). The main depositional environments in this area are salt marshes and tidal flats (Table 4). The coastal sediments are predominantly clayey silt and sandy silt or silty sand, and the coastal slopes are in the range of 0. 1‰～0. 2‰ (Wang, 1964; Ren and Zheng, 1980). Chenier Ⅰ is situated along the modern coastline (Fig. 1). From Baisalin, Chenier Ⅱ continues directly towards the south as far as Qikou and Zhaojiabo, where it converges with Chenier Ⅰ. Further south, at Longtuzi, only one chenier (Chenier Ⅱ) continues along the shore.

Table 4　Cheniers in the coastal plain west of Bohai Bay

Name of chenier	Position	RH * (m)	Width(m)	Note
Chenier Ⅰ	Alongshore, above max. high tide level	0. 5～0. 1 (north of Haihe) 2. 0 (south of Haihe)	20～30	Thin vegetation, shell fragments mixed with mud. 300～600 yrs BP.
Chenier Ⅱ	40 km inland at Haihe River mouth, on the shore at Qikou and Zhaojiabo	5	100～200	Grass, bushes, freshwater, pure shell fragments. 1300～1900 yrs B. P.

* Relative height.

Chenier I is usually 0. 5～1 m high and 20～30 m wide. The seaward slope of the chenier is about 6° and the landward slope 5～6°. The composition of the chenier north and south of the Haihe River differs slightly. North of Haihe River, Chenier I is composed of shells (including whole shells, shell debris and shell sands), coarse silt and fine sand. The chenier profile near Chentougu indicates this clearly (location *P* in Fig. 1):

0～0. 50 m. Shell debris with good horizontal bedding laminated with fine sands. Shells are *Meretrix meretrix*, *Cyclina sinensis*, *Arca subcrenate* and *Umbonium thomosi*.

0. 50～0. 66 m. Layer composed of whole *Meretrix meretrix* and *Cyclina sinensis* containing pieces of bricks in good horizontal bedding.

0. 66～2. 0 m. Shell debris.

In the chenier deposit south of Haihe River the number of shell species, as compared to the northern part is smaller and fine sandy laminae are absent. This is indicated in the profile of Chenier I near Lujihe village:

0～0. 6 m. Shell layer of *Meretrix meretrix*, *Cyclina sinensis* and *Arca subcrenate*. There are three layers of whole shells with thicknesses of 5, 7 and 13 cm in the top, middle and bottom part of the layer. Between the shell layers, there is shell debris and silt.

0. 6～1. 05 m. Silt containing a few shell debris in horizontal bedding.

1. 05～1. 35 m. Compacted *Arca subcrenate* shell debris containing silt.

＞1. 35 m. Muddy silt containing few shells.

The orientation, location and composition of the chenier gives the impression that the cheniers on both sides of the Haihe River were in fact one set. The difference in composition between the two parts of Chenier I is caused by the difference in sediment supply to the coast. North of Haihe River, sand is supplied by the Luanhe River (Fig. 1), which drains a mountainous area. In the related sandy environment, shellfish such as *Umbonium thomosi*, *Chlamys solaris* (Borm) and *Batillaira fonlie* (Brugnizrz), as well as *Meretrix meretrix*, thrive very well. In the southern part, however, the coast is mainly composed of silt. Occasionally sand occurs. In this environment the shellfish population is restricted to a few species only, mainly *Arca subcrenate* and *Cyclina sinensis*. This indicates that the sediment supply determines the coastal sedimentary pattern and this in turn is related to the (number of) shellfish species which will thrive in such an environment. The combination of these factors determines the composition of the chenier sediments, their grain size and the prevailing shellfish species.

Chenier II is larger than Chenier I. Its sediment composition shows an abundance of shells, thick layers of shell sand and some silt (Fig. 2). Besides shells such as *Cyclina sinensis* and *Arca subcrenate* other species occur such as *Meretrix linne*, *Umbonium thomosi*, *Tympanotomus cingulatus*, *Chlamys solaris*, etc. The chenier deposit shows a well-developed horizontal bedding, contains artificial materials, and is slightly lithified. Cores have demonstrated that near Qikou 3. 2 m of chenier sediments are underlain by silty salt marsh deposits (Fig. 2).

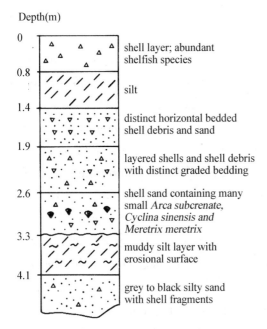

Fig. 2　Drilling profile of Chenier Ⅱ near Qikou

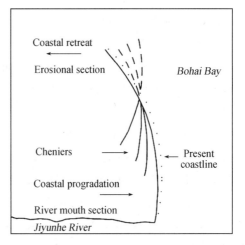

Fig. 3　Sketch map of the formation of the diverging cheniers in the Jiyunhe River mouth

Near the Jiyunhe River mouth (Fig. 1) is another type of small scale branched chenier which is in fact a type of shell beach deposit. All the branches diverge from one point. Closer to the river mouth, more cheniers appear. The reason for this is that the coast near the river mouth prograges more rapidly, as does the chenier formation. However, on both sides of the river mouth in the area, with no sediment available, erosion and retreat of the coast occurs constantly; thus the chenier gradually erodes and destruction follows (Fig. 3).

Examining the pattern and coastal profiles of the west coast of Bohai Bay (Fig. 4), it is possible to trace the development of the coast using the chenier as an indicator of the coastline. For example, in the coastal area around the Haihe River, due to the progradation of the Haihe River deltaic plain, the two cheniers are separated by a 20 km strip of low land, which had been accreted rapidly in an earlier stage. In the coastal area of the middle section of Bohai Bay between Qikou and Zhaojiabo, the coast is fairly stable, and thus the two cheniers are closely spaced, which implies little coastal progradation since the formation of Chenier Ⅱ. To the south of Zhaojiabo, the coast is erosional where Chenier Ⅰ has disappeared, and only fragments of Chenier Ⅱ remain; thus the coast is constantly eroding and retreating.

Historical and archaeological studies have identified Chenier Ⅱ as being formed in 70 AD (i. e. ~1 910 yrs B. P. and Chenier Ⅰ as being formed at 600~300 yrs B. P. (Wang, 1964). On Chenier Ⅱ, many graves and village relics of the Tang Dynasty (618

～907 AD) have been found (Li，1962)，indicating that the chenier was formed before the Tang Dynasty. This also demonstrates the importance of cheniers to human habitation on the salt marsh plain，only the chenier providing dry land for occupation space and access to drinking water. According to historical records，during a period of 500 years from 70 A. D. to 600 A. D.，the Yellow River channel was quite stable and had a lower sediment concentration owing to lower farming activity on the loess plateau of its drainage basin. Therefore，the coastal environment was different at this time: clear seawater with a small sediment supply from the river and under breaking-wave processes in the coastal zone，these conditions just meeting the requirements for chenier formation. According to detailed historical records，Chenier Ⅰ formed from 1336 A. D. (Yuan Dynasty)，after the fifth change of the Yellow River，until 1506 A. D. (Ming Dynasty). A few fishermen lived on Chenier Ⅰ at the beginning of the Qing Dynasty (1644 A. D.). The recent ^{14}C dating has confirmed our earlier dates of the chenier as found in the literature and from archaeology. The ^{14}C date of the shells of Chenier Ⅱ near Qikou is 2 020±100 yrs B. P. (bottom) and 1 080±90 yrs B. P. (top)，and 1 460±95 yrs B. P. in Baisalin，north of the Haihe River (Zhao et al.，1979；Zhu，1982).

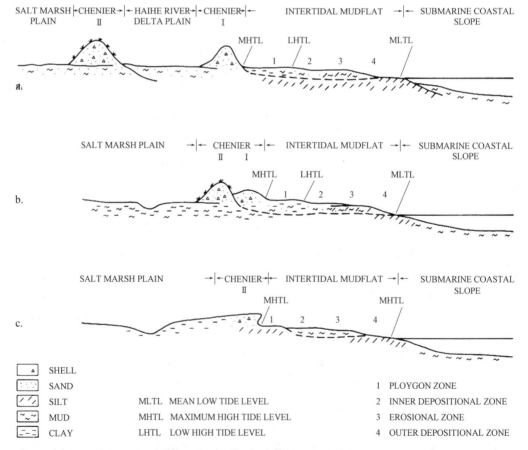

Fig. 4 Coastal sections through the west coast of Bohai Bay. a. Near the mouth of the Haihe River, strongly prograding part. b. Near Qikou, stable part. c. South of Qikou, slightly eroding part.

Jiangsu coast

On the Jiangsu coastal plain, the area between the modern Changjiang River and the old Yellow River, are three clusters of abandoned cheniers and one modern chenier (Table 5). The three clusters of old cheniers are mainly composed of fine sands and shells (oysters and clams) together with calcic concretions and are located 50~60 km westward of the modern coastline. They run almost parallel (NW – SE) to the modern coastline (Fig. 5). They are different from the cheniers of the Bohai Bay. Sand is the main component of the cheniers here, while the shells and concretions are often interbedded with the sand. The sand is fairly clean. The oyster shells can be more than 20 cm long, and the rounded concretions are several centimetres to more than 10 cm in diameter. The many shells and concretions in the cheniers are quite noticeable. The local population use the sand as building material and the shells to make lime, but leave the concretions, forming high mounds. The species of shells in the cheniers of the Jiangsu coast are mainly *Meretrix meretrix*, *Ostrea denselamellosa*, *Ostrea rivularis*, *Mactra veneriformis* and *Cyclina sinensis*. The ^{14}C dating of the carbonate materials in these cheniers indicated that the three old cheniers of Xigang, Zhonggang and Donggang were formed at 6 000~5 000 yrs B. P., 4 600 and 3 500 yrs B. P. (Table 6).

Table 5　Cheniers of the Jiangsu coastal plain

Name of chenier	Position	RH* (m)	Width (m)	Sediment type	Age (yrs B. P.)
Xigang	Xiyuan-Longgang-Dagang (inland)	5	600	Fine or medium sand	6 000~5 000
Zhonggang	Funing-Sawang-Longgang (inland)	2. 5	400	Medium or fine/less coarse sand	4 600
Donggang	Funing-Shanggang-Yancheng (along Fangong Bank)	4. 5~5. 5	500	Fine sand	3 500
Modern chenier	Near HWL	0. 2~0. 3	60	Shell and fine sand	

Table 6　^{14}C dating of cheniers of the Jiangsu coast

Position	Boring depth (m)	Material (yrs B. P.)	Date	Source
Xigang				
Funing Xiyuan	2. 8	Peat	4 480±80	(1)
	3. 2	Oyster	6 540±70	(2)
	1. 0	Wood	5 533±70	(2)
	2. 5	Wood	5 600±80	(1)

（续表）

Position	Boring depth（m）	Material（yrs B. P.）	Date	Source
Yancheng Longgang		Shell	7 654±75	(3)
Yancheng Dagang	1. 6	Shell	5 677±75	(2)
		Shell	6 832±106	(2)
	1. 6	Shell	5 600±80	(2)
	1. 5	Oyster	7 020±200	(1)
Zhonggang				
Sawang	2	Oyster	4 650±100	(1)
		Wood	4 610±100	(1)
Donggang				
Shanggang	2. 5	Shell	3 882±69	(3)
South of Shanggang	1. 5	Shell	3 310±80	(1)
Xingxin		Shell	2 483±80	(2)
		Wood	4 610±100	(2)
Erdun		Shell	3 472±121	(2)

1—Guo (1984); 2—[14]C laboratory, Nanjing University; 3—Guo (1982).

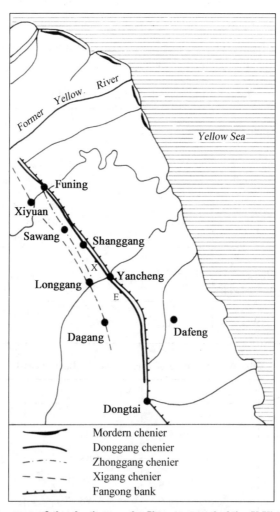

Fig. 5　Location map of the cheniers on the Jiangsu coastal plain. *X*-Xingxin; *E*-Erdun.

The modern chenier can only be found on the two sides of the old Yellow River mouth on the Jiangsu coast, forming only on a small scale along the high tide level overlying the erosional basement of the salt marshes (Fig. 6 and Table 7). There are no cheniers on the coast near the former Yellow River mouth where serious coastal erosion is taking place and the coastline is retreating at 130 m/yr (Zhu and Xu, 1982). This indicates that cheniers can only develop in a stable or slightly erosional coast. There are therefore restrictions as far as the slopes and stability of the coastal profiles are concerned. Comparing the different coastal profiles of the abandoned Yellow River from the north towards the south on the Jiangsu coast (Table 8), it may be concluded that the optimum coastal slope for the development of a modern chenier is between 0. 9 and 3. 0‰. On the Jiangsu coast, miniature cheniers (deposits of shells) also form during typhoons on the higher part of the mudflat near Yangkou where under normal weather conditions the tidal flat aggrades (Ren et al. , 1983). These miniature cheniers usually cannot develop into a large chenier because they remain mostly in an accumulative environment. Even though such miniature cheniers (better termed shell debris) may be preserved in the strata of a typical tidal flat sediment, they cannot really be called shell beach ridges (cheniers) because they only indicate an extreme event or catastrophic occurrence in the aggradation processes of the tidal flat.

Table 7　Morphological characteristics of modern cheniers on the Jiangsu coast

Altitude (m)	Width (m)	Thickness (m)	Slope	
			Seaward	Landward
+3 to +4	50~60	0. 3~1. 6	0~0. 12	0~0. 04

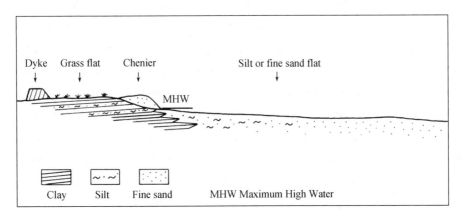

Fig. 6　Typical profile of the modern chenier coast at Dalabakou tidal flat, Jiangsu coast

Table 8   Coastal slopes of the abandoned Yellow River delta on the Jiangsu coast

Position	Slope (‰)	Coastal retreat rate (m/yr)	Coastal environment
Tuangang	2. 85*	20～30	Chenier formed
Xinhuai saltern	3. 11*		
Old Yellow River mouth	3. 5～5. 01	130	Serious coastal erosion, no cheniers
Damatou	0. 97*		Chenier formed
Dalabakou	1. 23	15	

* Gao and Zhu (1988).

In order to elucidate the relationship and contacts between cheniers and the underlying strata, a core was drilled directly through the top of the old Xigang Chenier at Longgang, Yancheng City and was examined sedimentologically in the laboratory (Fig. 7). The results proved that the strata 10 m below the surface are a set of typical tidal flat sediments with sedimentary and tidal flat lithological features. On the top of the tidal flat sediment, are 5 m of swamp or salt marsh silty clay deposits, and the sandy chenier overlies the grey-black clay, forming a distinctive unconformity. The grain-size analysis indicates that the sediments of the cheniers can only be derived from the reworked materials of the silt or fine sand flat near the low water level and on the submarine slope (Fig. 7 and Table 9). This slope of the lower part of the tidal flats mainly received sediments from the Changjiang River mouth during the Quaternary when this river mouth was towards the north of its present location and a huge submarine delta was formed. Only a small fraction is provided by coarser particles of the Yellow River sediments. When there are no sediments available on the coast, coastal erosion will occur: the fine materials in the upper part of the tidal flats will be eroded and carried away by tidal currents, and the coarse sediment such as shells, concretions and fine sands in the lower part will be washed out and then transported to the high water level by wave action. Although the important role of the Yellow River is to be stressed, especially the shifting of its course, it should be emphasized that cheniers on the Jiangsu coastal plain were not formed by Yellow River sediments directly. The shifting of the Yellow River can cause a sudden coastal environmental change which may control the erosion or progradation of the coast, thus influencing the chenier formation.

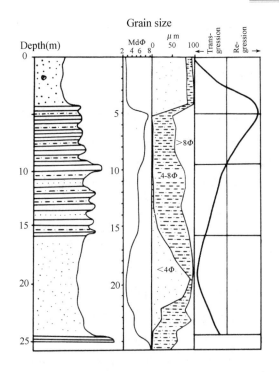

Grain size

Fine to very fine sand,
occasional wavy bedding,
much shell debris near the bottom
——————— 4 750±200 Yr BP* ———————
Grey-balck silty clay with
shell fragments. Compacted
with many small rounded
concretions. Wormtubes

Alternating silty clay and sand in
horizontal, cross and wavy bedding.
Thickness of clay laminae increasing
upwards

Fine sand/silt, with shell and plant
remians. Grain size grading;
horizontal, cross and flaser bedding

Silty clay with calcareous concretions
(size: 2×1.5×1 cm).
23 530±1900 Yr BP (Black clay)* and
18 960±820 Yr BP (calcareous
concretion)*

Fig. 7 Schematic profile of the Longgang core on the Jiangsu coastal plain

* ¹⁴C data from the ¹⁴C laboratory, Nanjing University.

Table 9 Grain-size distribution of cheniers and related tidal flat environment

Position/environment	Weight percentage					
	0~1φ	1~2φ	2~3φ	3~4φ	4~8φ	>8φ
Cheniers						
Xigang:						
Funing Xiyuan	0.8	50.8	45.5	1.7	← 1.2 →	
Yancheng Dagang	0.4	0.7	75.1	17.3	← 6.5 →	
Yancheng Longgang	3.0	15.0	78.0	4.0		
Longgang Guoyan	0.2	0.6	27.89	83.21	10.4	6.4
Donggang:						
Xingxin	0.3	25.4	66.2	3.5	4.6	
Tidal fiat south of Dalabakou						
Grass flat			← 10 →		80	10
Suaeda Salsa flat			← 10 →		>85	10
Mud-silt flat			← 10 →		60	30
Silt-fine sand flat			← 30 →		60	10
Changjiang River mouth*			← 52 →		41	7
Yellow River mouth**			← 12 →		79	10

* After Li et al. (1983).

** After Ye (1982).

The Jiangsu coastal plain has been in constant a state of subsidence since the beginning of the Cenozoic, and has received large quantities of sediment supplied by rivers. After the last glacial, the climate changed, there was a rise in sealevel and an extensive transgression on the coastal plain took place, which reached its maximum (to the west of Xigang chenier) at about 7 200 yrs B. P. (Yang and Xie, 1984). Since that period the whole area was in a state of regression, during which the sediments supplied by the Yellow River to the Jiangsu coast had their greatest influence on the regression process. Hence, the coast prograded to the sea with a certain periodicity, and three cheniers were formed in this period. The modern chenier developed on the coastal plain is very similar to the old cheniers. When the Yellow River shifted its course to the Bohai Bay in 1855, a modern chenier developed after a short interval. This chenier could not develop in the rapidly eroding river mouth, instead developing on the only slightly erosional coast. This implies that the formation of the cheniers occurs under special dynamic and environmental conditions. Shell ridges may be formed at the prograding coast by breaking waves during storm events. However, they cannot form large-scale ridges if the coast continues to prograde and in this case they will be buried in the tidal flat sedimentary sequence.

The fact that the history of chenier evolution on the eastern coastal plain of China is associated with sediment supply by the Yellow River and with the shifting of the river mouth is interesting. During historical times there have been eight major shifts in the course of the Yellow River (Wang, 1964; Pang and Si, 1979; Shen, 1979), which at times has discharged into Bohai Bay and at other times on to the north Jiangsu coast of the Yellow Sea. Coastal progradation occurs in the vicinity of the active delta of the Yellow River. On the contrary, along the abandoned deltaic coasts, on the other hand, erosion occurs and cheniers form (Fig. 8). Coarse shelly chenier

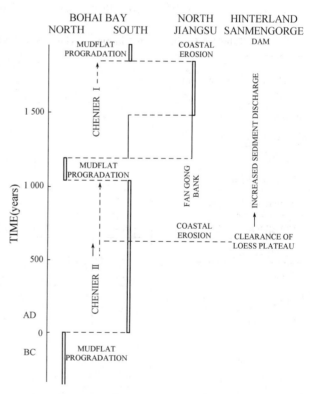

Fig. 8　Summary of the changes in the course of the lower reaches of the Yellow River in the last 2000 yrs and their effects on the coastal development (Wang, 1983).

sediments occur on the coastal plain of both Bohai Bay and the north Jiangsu coast, and must have been deposited under erosional and wave-dominated processes, rather than under a depositional tidal current regime.

In early historical times (prior to 11A. D.) the Yellow River entered northwestern Bohai Bay near the mouth of the present Haihe River. In 12 AD the river shifted to near its present course, entering southwestern Bohai Bay. Erosional retreat of the former delta led to the formation of Chenier Ⅱ on the northwest coast early in 70 A. D. (the time of the Eastern Han Dynasty), after Wang Ching (a famous hydrographical officer of the period) had successfully controlled the Yellow River. In 1049 the Yellow River again shifted to the northwest part of Bohai Bay, thereby terminating the accumulation of Chenier Ⅱ and causing renewed progradation.

In the 9th~11th centuries A. D. , the 580 km long Fangong Bank (Dyke) was built on the coast of Chenier Ⅲ (Donggang) to protect the Jiangsu coast from tidal flooding and erosion (during this time Yellow River sediment was being diverted to the Bohai Sea). After 1194 the Yellow River changed its course from debouching into the Bohai Sea to flowing into the Yellow Sea and rapid progradation of the Jiangsu coast occurred as the sediment supply increased. By 1453 AD, a strip of land 15 km wide had been formed seaward of the Fangong Bank. Subsequently, during the next seven centuries, the Yellow River prograded 90 km into the Yellow Sea, forming a deltaic plain area of 15 700 km². Since then the Fangong Bank has acted as a road, naturally a long distance from the shore.

Chenier Ⅰ began to form in Bohai Bay when the Yellow River changed its course to the Yellow Sea (Wang, 1964; Zhao et al. , 1979). During this time the river did not discharge sediment in the coastal area of the Bohai Sea and thus the tidal flats were subjected to erosion, and shell ridges were formed along the coast.

The Yellow River returned to the Bohai Sea again in 1855, forming mudflats on the coastline adjacent to Chenier Ⅰ. On the abandoned Yellow River deltaic tidal flat of the Jiangsu coast, nearly 1 400 km² of land along 150 km of coast has been eroded since that time, and the old Yellow River deltaic coastline of 1959 A. D. is now located on the submarine slope in 1 m of water. The coast in this area has essentially reached equilibrium, with slight erosion in some locations so that modern cheniers are forming.

Conclusion

(1) Cheniers are widely distributed along the low tidal flat coast of eastern China where enormous quantities of sediment are supplied by large rivers. They are of an accumulative morphology from eroded tidal flats, being formed when the coasts are retreating while no modern river sediment is available.

（2）Cheniers are usually formed near the high tide level and their formation requires a suitable coastal slope, and abundant sediment of shells and silt. In the area of very gentle coastal slopes, wave action is well offshore and its energy may gradually diminish when it passes over the wide and gentle tidal flats; thus it cannot transport coarser silt and shells from the lower part of the tidal flat or sub marine slope to the upper part of the coast. On the other hand, if the slope is too steep and wave action too strong, the coast will rapidly retreat and the silty and shell material will not be accumulated on the coastline to form beach ridges. In China the optimum coastal slope for chenier development is between 0.9‰ and 3.0‰.

（3）In the coastal plain area near large rivers, river sediment supply has played a major role in the formation of cheniers. In the period 5 000 yrs B. P. , the influence of sealevel changes probably took second place.

（4）Owing to the morphological and sedimentary characteristics of cheniers, we are able to explain the evolutionary history of the coast, determine the coastal dynamic processes, and to identify the developing trends of the coastal area.

Acknowledgement

We would like to thank Drs. Leonie van der Maesen for editing this manuscript.

References

[1] Augustinus, P. G. E. F. 1980. Actual development of the chenier coast of Suriname (South America). *Sediment. Geol.* , 26: 91 - 113.

[2] Chappell, J. and Grindrod, J. 1984. Chenier plain formation in northern Australia. In: B. G. Thom (Editor), Coastal Geomorphology in Australia. Academic. Press, Sydney. 197 - 231.

[3] Cheng Guodong, Ren Yucan and Li Shaoquen. 1986. Channel evolution and sedimentary sequence of the modern Huanghe River delta. *Mar. Geol. Quat. Geol.* , 6: 1 - 15 (in Chinese with English Abstr).

[4] Gao Shu and Zhu Dakui. 1988. The profile of the mud coast of Jiangsu. *J. Nanjing Univ.* , 24: 75 - 84.

[5] Gould, H. R. and McFarlan, E. 1959. Geologic history of the chenier plain, Southwestern Louisiana. *Trans. Gulf Coast Assoc. Geol. Soc.* , 9: 261 - 270.

[6] Greensmith, J. T. and Tucker, E. V. 1969. The origin of Holocene shell dosits in the chenier plain facies of Essex (Great Britain). *Mar. Geol.* , 7: 403 - 425.

[7] Guo Ruixang. 1982. Jiangsu coast evolution during the historical period, and modern coastal morphology. *Jiangsu Hydrogr.* , 2: 25 - 39 (in Chinese).

[8] Guo Ruixang. 1984. Sandy beach ridges distribution and coastal evolution of North Jiangsu coastal plain. In: Proc. Coastal and Tidal Flat Resources. Jiangsu Press, Nanjing. 4, 275 - 286.

[9] Ke Xiankun. 1985. Holocene sedimentary environmental changes of Jiangsu coastal plain. MA Thesis, Nanjing Univ. 98－108 (in Chinese with English Abstr).

[10] Li Congxian, Li Ping and Cheng Xinrong. 1983. The influence of marine factors on sedimentary characteristics of Yangtze River channel below Zhengjiang. *Acta Geogr. Sin.*, 38: 128－140 (in Chinese, with English Abstr).

[11] Li Shiyu. 1962. Preliminary study on relics of coast and culture in the west coast of Bohai Bay. *Acta Archaeol. Sin.*, 6: 652－657 (in Chinese).

[12] Otvos E. G., Jr. and Price, W. A. 1979. Problems of chenier genesis and terminology—an overview. *Mar. Geol.*, 31: 251－263.

[13] Pang, J. S. and Si Sheheng. 1979. The estuary changes of Huanghe River, I. Changes in modern time. *Oceanol. Limnol. Sin.*, 10: 136－141 (in Chinese with English Abstr).

[14] Price, W. A. 1955. Environment and formation of the chenier plain. *Quaternaria*, 2: 75－86.

[15] Ren, M. and Zheng, C. 1980. Late Quaternary continental shelf of East China Sea. *Acta Oceanogr. Sin.*, 2: 106－111 (in Chinese with English Abstr).

[16] Ren Mei-e, Zhang Renshun and Yang Juai. 1983. Sedimentation on tidal flat of Wanggang area, Jiangsu Province, China. In: Collected Oceanic Works. Oceanic Press, Beijing. 6, 84－108.

[17] Russell, R. H. and Howe, H. V. 1953. Cheniers of southwestern Louisiana. *Geogr. Rev.*, 25: 449－461.

[18] Shen, H. W. 1979. Some notes on the Yellow River. *Trans. Am. Geophys. Union*, 60: 545－546.

[19] Todd, T. W. 1968. Dynamic diversion: influence of longshore current tidal flow interaction on cheniers and barrier islands. *J. Sediment. Petrol.*, 3: 734－746.

[20] Wang Ying. 1964. The shell coast and ridges and the old coastlines of the west coast of the Bohai Bay. *Acta Sci. Nat. Univ. Nankinesis*, 8: 424－442 (in Chinese with English Abstr).

[21] Wang Ying. 1983. The mudflat of China. *Can. J. Fish. Aquat. Sci.*, 40: 160－171 (Suppl. 1).

[22] Wang Ying, Ren Mei-e and Zhu Dakuei. 1986. Sediment supply to the continental shelf by the major rivers of China. *J. Geol. Soc. London*, 143: 935－944.

[23] Wang Ying and Aubrey, D. 1987. The characteristics of the China coastline. *Cont. Shelf Res.*, 7: 329－349.

[24] Woodroffe, C. D., Curtis, R. J. and McLean, R. F. 1983. Development of a chenier plain, Firth of Thames, New Zealand. *Mar. Geol.*, 53: 1－22.

[25] Yang Huairen and Xie Zhiren. 1984. Sealevel changes along the east coast of China over the last 20,000 years. *Oceanol. Limnol. Sin.*, 15: 2－13.

[26] Ye Qingchao. 1982. The geomorphological structure of the Yellow River Delta and its evolution model. *Acta Geogr. Sin.*, 370: 349－363 (in Chinese with English Abstr).

[27] Zhao, X., Guang, S. and Zhang, J. 1979. The sealevel fluctuation during the last 20,000 years in East China. *Acta Oceanol. Sin.*, 1: 269－280 (in Chinese with English Abstr).

[28] Zhu Dakui and Xu T. 1982. The coastal development and exploitation of middle Jiangsu. *Acta Sci. Nat. Univ. Nankinesis*, 3: 799－818 (in Chinese with English Abstr).

[29] Zhu Kezhen. 1982. Physical Geography of China. Historical and Geographical Science Press, Beijing. 230－232. (in Chinese).

中国粉砂淤泥质平原海岸的发育因素及贝壳堤形成条件[*]

一、中国粉砂淤泥质平原海岸的特点

中国粉砂淤泥质平原的海岸线较长(1 600～1 800 km),主要分布在钱塘江以北各大河口附近,如:辽河口的松辽平原海岸、黄河下游的华北平原海岸及淮河、长江下游的苏北平原海岸等。海岸外围无沙坝和沙嘴的环绕,直接面向开敞的外海,反映其形成的特殊性。均发育在海积-河积平原上,组成物质主要是粉砂(0.05～0.005 mm)和淤泥(小于0.005 mm)。

平原海岸在平原上有河流遗留的废河道和滨岸沙丘等,其外缘是潮水与波浪作用所形成的浅滩,浅滩坡度平缓(小于1/1 000),堆积着厚层的淤泥,分布着潮水沟,部分浅滩内侧发育着贝壳质岸堤。根据发育特性又可将海岸分为两种类型:

1. 泥滩岸

为新生的粉砂淤泥质海滩平原,地势极平缓,没有草类生长,属淤积型,有大量的细粒泥沙供给,岸线前进速度快,草类来不及繁生,仍保持原始岸滩形态。例如,苏北平原南部及黄河三角洲一带,该处海岸前伸速度平均250～500 m/a。

2. 草滩岸

分布在潮间带内,生长着茂密的植物,其中特多耐盐的短草,如碱蓬、狗芽根、白茅等,也有芦苇。草滩岸有属淤积型的,也有属冲刷型的。在我国草滩岸多是受到海侵冲淹的海岸平原,是属冲刷型的,如苏北平原的北段海岸岸线冲刷后退速度平均100～250 m/a。

此类海岸在实际利用上存在着以下一些问题:

(1)岸坡平缓,沿岸水浅。利用河口建港,由于潮水上溯,淤积严重;同时在挡潮闸下游容易淤积,影响平原河流下游排水不畅,发生涝灾。

(2)岸线冲淤变化迅速,或部分农田冲毁,或部分盐田受淤。

二、粉砂淤泥质平原海岸的发育因素

1. 发育因素

在沉降凹陷区(松辽凹陷、渤海凹陷和苏北凹陷),从第三纪以来即处于缓慢稳定的下

　　* 王颖:中国地理学会地貌专业委员会编辑,《中国地理学会1961年地貌学术讨论会论文摘要》,112-114页,科学出版社,1962年11月。注:当时的学术年会论文均以摘要形式正式出版。

沉中,堆积着巨厚的第三系、第四系沉积物,它们组成了平原海岸的基底。

自沉降凹陷区出口的河流(如黄河、辽河等),源远流长,含沙量很大,泥沙颗粒细,堆积作用强烈,影响岸线增长迅速,而未胶结的泥沙又不耐冲蚀,一旦泥沙补给中断时,在海水冲刷下,海岸崩塌后退很快。

2. 动力因素

(1) 河流作用　含沙量大的河流,其对海岸塑造的作用是:供给海岸带大量泥沙,使近岸地区即分布着细粒沉积;河口浑水流直接输送泥沙,参与海岸泥沙流运动。如黄河平均每年汇入渤海的泥沙约 9 亿 t(相当整个渤海湾河流输沙量的 40 倍),它直接促成了渤海岸广大泥滩形成;黄河的浑水流又在风力推动下,可把泥沙送到渤海湾的北部。

(2) 潮水作用　潮水作用在海岸发育中占有特别重要的地位,因为潮汐涨落的水位变化平均相差 3 m,由于海岸地势平缓,由涨落潮作用所形成的潮间带浅滩宽度可达数千米;潮流不仅冲刷岸基,而且携带泥沙,岸滩泥沙自海向陆逐渐变细,反映潮流对泥沙运送过程中的分选性。

(3) 波浪和风成流　两者均受季风的影响,发生恒定持久的东南向波浪与风成流以及强劲的东北向波浪与风成流。其作用是:波浪掀起泥沙,供给潮流与风成流的携运物质。

三、粉砂淤泥质平原海岸上贝壳堤形成条件

从苏北射阳河以北到灌河以南,从黄河口以北到大清河以南,数百千米岸线内分布着由贝壳、壳屑与细砂组成的海岸自然堤。贝壳繁殖的环境须要有透明度较大的海水,阳光充分,并含较多的营养盐,才有利贝壳生长。淤泥质粉砂所组成的岸滩海底食料多,孔隙度利于呼吸。贝壳受激浪与浪流的冲刷,搬运在开敞的海滩堆积成堤。适宜于贝壳堆积的岸滩要坡度不小于 1/1 000,使波浪作用达到岸边;同时岸线变化不宜太快,须具有一定的稳定性,才利于岸堤堆积增高。分支状贝壳堤多发育在前进海岸(淤)与后退海岸(冲)相互转换的地区。

渤海湾西部贝壳堤与古海岸线问题[*]

（摘要）

贝壳堤是激浪与浪流冲刷沿岸滩底的贝壳与泥沙带到海滩上部所堆积成的滨岸堤[①]。与其他的滨岸堤（沙堤或砾石堤）的成因是一样的。贝壳堤形成在岸坡较陡（大于千分之一）以波浪作用为主的海岸段落。但贝壳堤不仅形成在沙质海岸上，且更广泛地发育在缺少淡水汇入以粉砂与淤泥粉砂沉积为主的海岸段落。

目前在渤海湾沿岸分布着二列贝壳堤。外侧的一条规模小（高 1～2 m，宽 100 m），其分布方向与现代海岸线方向近于一致。它北起大庄河附近的高上堡，向西经涧河口、大神堂到蓟运河口，断断续续地分布着此列贝堤在海河口处中断。向南从道沟子开始经驴驹河、高沙岭、白水头、唐家河、老马棚口、歧口，一直到石碑河南岸的贾家堡。沿海渔村皆在此列贝堤上。此列贝堤在海河以北高 1～2 m，组成物质是白蛤、青蛤、毛蚶、文蛤、蝺螺等，以及石英、长石、磁铁矿等细砂。细砂的来源与滦河水系的冲积物有关。在海河以南贝壳堤高 2 m，沉积主要是贝壳碎片及贝壳砂，以白蛤、麻蛤为主，纯粹是沿海滩底上的沉积。这列贝堤已为潮流破坏，改造成为一段段的弧形小岛。贝堤外侧的粉砂滩地上已被淤泥所覆盖，波浪作用已达不到此，它已演化成为现代海岸的内缘了。

在上述第一列贝壳堤西面，还分布着一列较老的第二条贝壳堤，老贝堤方向为北西—东南向，北起白沙岭，向南经过泥沽、小站、上古林、老马棚口到歧口与第一列贝壳堤相并接，向南经赵家堡到贾家堡二列堤相交了，再向南断断续续分布第二列贝堤被破坏下的残余物。内侧这条堤规模大、高 5 m、宽约 150～200 m，沉积层次厚，为分选极好的贝壳与贝壳砂。贝壳种类较丰富，除白蛤、毛蚶外，尚有文蛤、日本镜蛤、强棘红螺、锥螺、太阳栉孔扇贝等。沉积层坚实，已初步胶结。贝壳堤上面生长了茂密的植物。

此两列堤在歧口以北是相隔有一定距离的，从歧口到赵家堡二堤相并接，第一堤叠置在第二条堤上，贾家堡处二堤相交，向南只断续分布第二条堤。这表明了各段岸线动态不同。歧口以北岸线是淤积前进的。故两堤分离，各适应其岸线所在。歧口至贾家堡段，岸线基本上是稳定的，变化不大，故新贝壳堤叠置发育在老贝堤上。而贾家堡以南，海岸受冲刷后退，新贝壳堤已被破坏，仅留老堤残余了。

在老贝壳堤以西一些地点还见到一些贝壳堆积层，但它们有的埋藏，有的出露，发现地点比较零散，还需要进行深入的工作才能确定更老的岸线。

---

　　* 王颖：中国海洋湖沼学会编辑，《中国海洋湖沼学会 1963 年学术年会论文摘要汇编》，57 页，科学出版社，1964 年 9 月。注：当时的学术年会论文均以摘要形式正式出版。

　　① 滨岸堤在形态上是呈与当时岸线方向一致的陇岗，在沉积上具有粗细分选，层理清楚的特点。

渤海湾西南部岸滩特征[*]

（摘要）

一、区域概况

　　渤海湾西南部在构造单元上属渤海拗陷西部的南部凹陷,地貌上居华北平原的东部边缘,是新生代末期的海积平原,北界海河口,南部为庞大的向海突出的扇形三角洲。海岸受古老基底构造线、大地貌单元,特别是黄河三角洲发展的影响,所以具有向西凹入的弧形平原海岸的特点。

　　风对沿岸波浪、水流影响较大,本区常风向为东向及西南向。波浪主要为风波,各季节波向与主要风向一致。当风平行海岸时岸流流速大,含沙量亦大,而风垂直海岸时,流速与含沙量均小。东向风时岸滩涨潮快、落潮慢,吹西风时反之。偏南风时表层流向西北,将黄河口的泥沙成淤泥流北上,在春末夏初经本区岸滩,对岸滩地貌有重要意义。潮流是搬运泥沙以及形成岸滩冲蚀地貌的最主要的因素。本区沿岸河流中以黄河与海河的搬运泥沙作用最大,而对岸滩的塑造仅次于潮流。

二、海岸类型

　　本区是典型的粉砂淤泥质平原海岸,根据地貌、动力与动态特征,可将本区海岸划分为三个亚类。

　　1. 淤进的海积平原岸

　　位于北段,岸线平直呈北北东走向;沿岸是一片坦荡的盐土平原,目前大部分面积为盐田。该处岸线在晚近时期中不断东进,海积平原为晚近时期形成的。

　　2. 稳定的贝壳堤潟湖岸

　　位于中段,岸外侧有新老二列贝壳堤,内侧为海积潟湖平原,潟湖已干涸成浅凹地,它与老贝堤相结合代表了一条古岸线,故该段海岸稳定。

　　3. 冲蚀的潟湖凹地平原岸

　　位于南段,呈西北走向,海岸坡度较大,涨潮时岸边受冲刷,新贝堤已蚀,老贝堤亦在破坏中,岸线后退。

---

　　* 王颖,朱大奎,顾锡和:中国海洋湖沼学会编辑,《中国海洋湖沼学会 1963 年学术年会论文摘要汇编》,55－56 页,科学出版社,1964 年 9 月。注:当时的学术年会论文均以摘要形式正式出版。

三、岸滩特征

岸线外围的潮间带浅滩(简称岸滩)是粉砂淤泥质平原海岸的重要组成部分。它在高潮时被海水淹没,低潮时出露成滩地。本区岸滩特别发育、动态活跃,北段较宽,沿岸向南变狭。岸滩上主要动力是潮流,而波浪作用是较微弱的。我们于 1963 年 5—7 月在若干断面上进行了微地貌、沉积特性、泥沙运动等项目的定位观测,根据观测所得岸滩上地貌、动力与动态的差异性,而将岸滩从岸向海划分出下列四个带。

1. 龟裂带

分布在特大高潮线与低高潮线之间,它在特大潮水与高高潮时被海水淹没,平时出露疏干,沉积物为黏土或淤泥质黏土,海水浸泡时淤泥质膨胀,而干露日晒时黏土又收缩,使滩面发生龟裂。在靠近特大高潮线附近,龟裂隙较大,而向低高潮线逐渐减弱而不明显;在岸滩坡度平缓处,龟裂较宽,龟裂现象也较明显。龟裂带以中段海岸处最宽,北段较窄,而南部发育最差,这是南段岸滩坡大之故。

2. 内淤积带

分布在高潮线以下经常为海水淹没。沉积组成主要是淤泥或黏土质淤泥,表层是厚 20～40 cm 的粥状淤泥。泥沼状滩面平坦单调。据定位观测资料,潮流沿滩带进的泥沙主要沉积于此,其特点是颗粒细、沉积量大。

3. 滩面冲刷带

位于内淤积带的外侧。在该带的内缘滩面上出现起伏不平的水坑,冲刷形态不甚明显,在中部逐渐出现各种冲刷体——鳞状长条的冲刷体、积水凹地、冲刷沟壑或垄岗状冲刷体,冲刷规模较大,形态显著。而向外侧冲刷体又隐伏下去了。据其微地貌特征结合动力分析,可确定它们是落潮流作用形成的,其走向与落潮流方向一致。用标志层、指示砂等方法观测,发现这个地带是受冲刷的,物质向两旁搬运。

4. 外淤积带

最外侧一直到低潮水边线,是粉砂、细砂质,在潮流作用下滩面满布流水波痕,长 3～7 cm,波痕在各部位形态各异,反映了落潮流及波浪作用特性。外淤积带物质来自两方面:① 涨潮流从外海带来;② 落潮流从冲刷带带来。据实测,前者主要是粉砂、细砂,后者是淤泥。

由上述可知,渤海湾西南部潮间带浅滩上有明显的分带性,其分布规律从海向陆是:

地貌:Ⅳ.外淤积带→Ⅲ.滩面冲刷带→Ⅱ.内淤积带→Ⅰ.龟裂带。

沉积:粉砂、细砂→淤泥粉砂→淤泥→黏土或黏土质淤泥。

动力:潮流和微波→潮流(落流为主)→潮流(涨流为主)→潮流和日晒。

动态:大面积薄层淤积为主→涨淤落冲→强盛淤积→稳定微淤偶尔冲刷。

渤海曹妃甸深水港动力地貌研究[*]

　　曹妃甸沙岛是滦河三角洲的岸外沙坝,向海是水深36m的曹妃甸岸外深槽,向陆是宽23 km的潮带浅滩。经多年勘测研究,证实深槽沙坝是稳定的,可建深水港,现在潮间带浅滩填海造地,沙坝前缘建25万～30万t级深水码头,整个海岸区域开发为河北省唐山市曹妃甸工业区。有钢铁、石油、化工、煤炭、电力等大型企业进入,成为我国"十一五期间"最大的工业群项目。国家领导人胡锦涛等均到过工地现场,视察指示。2007年5月1日,温家宝总理在工地现场听取曹妃甸建设汇报时说:"曹妃甸是北方的天然深水良港,建设好这个大港,将促进河北、环渤海地区及北方地区经济社会的发展,在短短几年间,这个荒凉的小岛发生了翻天覆地的变化,看了令人鼓舞。"他特别提出"曹妃甸的建设是从一张白纸开始,好画最新最美的图画,因此要科学规划,有长远眼光,坚持高标准,高质量,高水平,把曹妃甸建设成环渤海地区耀眼明珠"。

　　曹妃甸工业区的建设规模:初期(～2010年)填海造地105 km²,工业区投资规模1 500亿,基本建成钢铁、电力、物流三功能区。中期(2011～2020年)再填海150 km²投资规模5 000亿。主要有:2 500万t精品钢,逐步形成4 000万t钢铁产业群。石油化工,1 000万t炼油,100万t乙烯,及其下游化工企业。4座100 kW超临界燃煤发电机组,2座30万kW燃气发电机组。利用大秦铁路输出煤炭5亿t/a。远期(2021～2030年)完成310 km²填海造地,建设中国北方地区最大的深水港区和铁矿石、煤炭、原油、液化天然气储运基地,形成世界级规模和水平的重化工基地,循环经济示范区,达到国际生态建设标准的新型工业城市。

　　深水港是建设曹妃甸工业区的前提条件,能通航25万～30万t级货轮。在选址勘测研究与规划建设中,要解决主要关键问题是:① 岸外深槽的稳定性;② 曹妃甸海岸演变动态及稳定性;③ 潮间带浅滩的形态演变及围海造地合理规划利用。本文将综合南京大学长期的勘测研究,着重介绍论证:① 曹妃甸沙坝的成因;② 曹妃甸槽的稳定性和沉积速率;③ 曹妃甸潮间带浅滩的冲淤变化;④ 曹妃甸潮水沟演变与动态;⑤ 深水港建港条件分析。

　　*　朱大奎,王颖:《中国地理学会2007年学术年会(摘要集)》,7专题14页,南京,2007.11。

Submarine Sand Ridges: Unique Marine Environment and Natural Resources[*]

Submarine sand ridges are large scale depositional sandy bodies on the inner shallow water continental shelves, which are mostly distributed in the macro tide area where the mobile sandy bodies are controlled by air-sea interaction dynamics. Study on its environment conditions, moving trend or stability is of important effect on the safety of marine engineering projects (e. g. prevention of fall out of petroleum platform and break of submarine pipelines) and the regulation of sedimentary trap of oil and natural gas in the sandy bodies.

Submarine sand ridges are widely distributed in China: Liaodong Shoal in the west side of the Bohai Strait, three ranks of sand ridges in the west side of the Liaodong Bay of the Bohai Sea (Fig. 1), and finger-shaped sand ridges in the west side of Qiongzhou Strait of the South China Sea (Fig. 2).

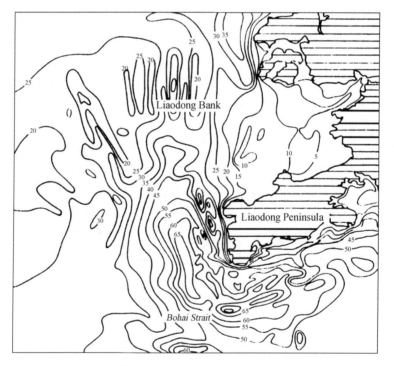

Fig. 1　Liaodong Bank of the Bohai Sea

---

　* Ying Wang, Dakui Zhu: Orville T. Magoon, Hugh Converse, Virginia Tippie, L. Thomas Tobin, and Delores Clark eds, *COASTAL ZONE* '91, 1991, Vol. 4, pp. 3377 - 3380.

　1995 年载于: *XVIII Pacific Science Congress, Collection of Abstracts*. pp. 116, Beijing, China, 1995. 6.

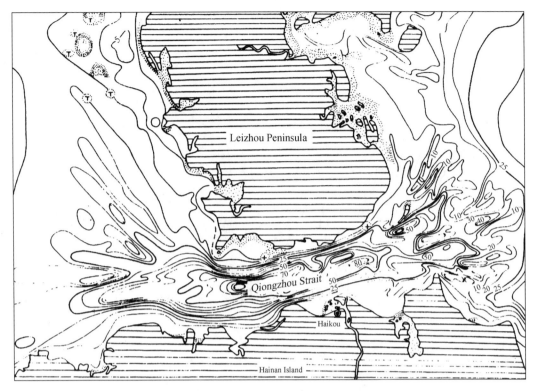

Fig. 2 Finger-shaped sand ridges in the west side of Qiongzhou Strait of the South China Sea

The largest one of sand ridges in China is the radiate submarine sand ridges in the offshore of Jiangsu coast, west part of the Yellow Sea (Fig. 3). It is a sand ridge field[1], and composed of more than 70 sand ridges and tidal channels between the ridges. The total area of the sand ridge field is 2 125.5 km², and located in the area from 31°55' to 34°20' North latitude, and 120°49' to 122°32' East longitude. The water depth in the ridge field is 0~25 m, but reaches to 50 m in the channels. Most are submarine ridges and only 10% of ridges is above the local sea level. From Qiang Gang as a centre, the ridges are distributed in the radiate pattern, which is the result of responsing to the rotary tide of the area. Two types of tidal waves; the progressive tidal waves of Pacific which is prograded from southeast, and a local reflected tidal waves formed by the obstraction of Shandong Peninsula in the North. The two tidal waves have converged along Jiangsu coast in the west part of Yellow Sea, and have formed the clockwise rotary tides with a centre in the non-tidal point just off the Qiang Gang. The semidiurnal rotary tidal currents are very strong with an average velocity of spring tide is 4 knots, and the maximum velocity is more than 5 knots. The tidal ranges are normally 4~6 metre, but up to 9.3 metre in the deepest channels. Prevailing waves are controlled by monsoon winds. North and Northeasterly waves prevail in the winter, and Southeasterly waves prevail in the summer. Wave energy has been decreased while transiting through the submarine sand ridge field, such as, the average wave height is

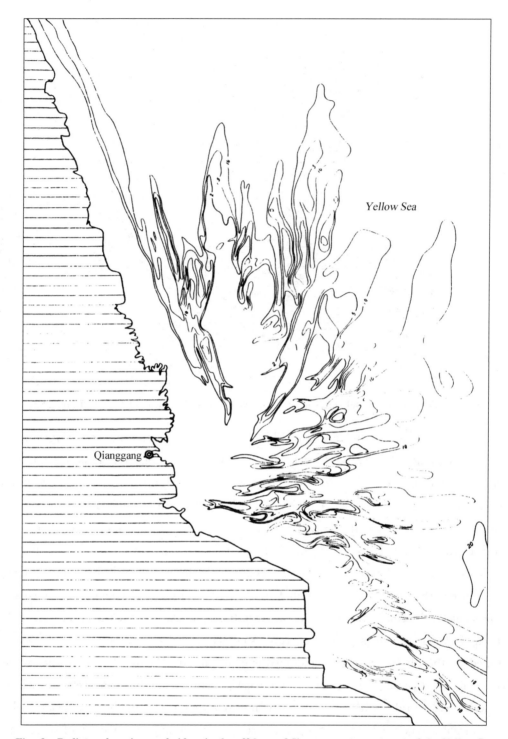

Fig. 3　Radiate submarine sand ridges in the offshore of Jiangsu coast, west part of the Yellow Sea

0. 4 m and with only 5% appearance of 2 m wave height in the field area，but up to 7～ 9 m wave height in the vicinity of open sea. Under the circumstances，the outer part of sand ridge field suffers from wave erosion. As a result，there are 2.02×10^9 tonne[2] of sediment which have been carried in the field，and it is the major sedimentary source to

form the accumulated tidal flats and sand ridges landward along shoreline. The sediments of sand ridges consist of fine sand, silty sand and coarse silt. The mineral complex is mainly quartz, feldspar, mica and also contains $2\% \sim 5.5\%$ of heavy minerals such as hornblende, magnetite, apatite and zircon. The clay minerals are illite, chlorite, kaolinite and montmorillonite. Besides, there are apparently cold and dry climate pollen and spore assemblages in the sediments. Geomorphological and sedimentary evidences have shown that radiate sand ridges were a submarine delta of the Changjiang River when it entered the Yellow Sea from Qiang Gang 10 000 Y. B. P. [2] The delta have been submerged and reformed by tidal currents during Holocene transgression. Thus the radiate sand ridges are a formation of river-sea interaction.

The larger area of submarine sand ridge field is a dangerous maze for navigation, but a great potential land resource for farming, agriculture, and offshore oil industry land uses. Even though, the 10% of subaerial sand ridges, i. e. 2 000 km^2 area of land and the accumulated tidal flats behind the sand ridge field, which is also a 2 600 km^2 area of land, have great economic value for the dense populated coast regions in China. Recent studies indicate that three major deep water channels in the sand ridge field are stable in position since early time of the transgression, and the water depth there can be used for setting up 50 000 to 100 000 Tonnes sea harbour. The result has brought good future for people who live in the 1 000 km long tidal flat coast of eastern plain area of China and would change the traditional idea of no favourable conditions for setting up deepwater sea harbour in such shallow water low coast. The study on environment, dynamics and sedimentation model of the submarine sand ridges will promote the foundation of a new frontier subject "Coastal Dynamic Sedimentology". We are looking forward to carrying out international cooperation on this topic.

References

[1] Ren, Mei'e and Zeng Chenkai. 1980. Late Quaternary Continental Shelf of East China. *Acta Oceanologica Sinica*, 2(2): 109 - 114.

[2] Zhu, Dakui and Fu, Ming Zuo. 1986. Primary Study on Offshore Radiate Sand Ridges of Jiangsu. In: Proceedings of Jiangsu Coast Studies. Ocean Press. 28 - 32.

南黄海辐射沙脊群沉积特点及其演变[*]

　　潮流沙脊群是大陆架浅海大型的海底堆积体,分布于有丰富砂质沉积与强潮流作用的大陆架浅海域[1,2]。南黄海处于半封闭浅海,太平洋前进潮波与黄海驻潮波辐聚、辐散,海区内潮差变化大,潮流作用强,在古河口、古河道砂质富聚区,具备有形成潮流沙脊群的良好条件[3~6]。

　　南黄海辐射沙脊群分布于江苏岸外,南北长 199.6 km,东西宽 140.0 km,呈褶扇状向海,由 70 多条沙脊与潮流通道组成,脊槽相间,水深界于 0~25 m。据最近测量,其面积为 22 470 km²。其中 0 m 线以上面积为 3 782 km²(图 1)。

　　南黄海海域为半日潮型,涨潮时,潮流自北、东北、东和东南方向涌向弶港海岸;落潮时,潮流以弶港为中心,呈 150° 的扇面向外逸散,形成以弶港为中心的放射状潮流场,沙脊群的地形与潮流场完全吻合。

一、沉积物组成与物质来源

(一) 粒度组成

　　辐射沙脊群沉积物主要成分为分选良好的细砂,含量在 90% 以上,表面洼地含有粉砂,使沙脊沙体中有局部粉砂薄层。

　　潮流通道内段、中段为细砂,粉砂含量仅占 2%～6%;外段、口门及一些大潮流通道(黄沙洋、烂沙洋)中段,水深大于 15 m,粉砂含量增多,出现粉砂质砂或砂质粉砂。

　　若干大通道口门段,如烂沙洋主要为砂砾质,"砾石"(直径＞4 mm)是钙质胶结砂粒而成的结核体,含量达 29%,是强潮流作用下的堆积物。

　　北部潮流通道(西洋)粉砂含量较南部增多,从通道顶端向口门(自南向北)依次出现粉砂质砂、砂质粉砂、砂、粉砂、黏土,黏土含量高达 30%。反映出来自北部海域废黄河三角洲的冲刷产物。

　　黄河口的床沙为砂质粉砂,悬沙为黏土质粉砂;长江口的床沙为粉砂质黏土,悬沙为黏土质粉砂[7]。黄河与长江泥沙均有自中下游向海沿程细化的规律,床沙比悬沙更明显;黄河泥沙沿程缓慢变细,长江河口段泥沙突然异常细化,与咸淡水混合产生絮凝作用,使细颗粒物质大量落淤有关,是双向动力环境的堆积特征。黄河汇入黄海、渤海的泥沙以含黏土的粉砂为主,而长江则为细砂及粉砂质黏土。

　　[*] 王颖,朱大奎,周旅复,王雪瑜,蒋松柳,李海宇,施丙文,张永战:《中国科学(D辑)》,1998 年第 28 卷第 5 期,第 385－393 页。

　　国家自然科学基金资助项目(批准号:49236120)。

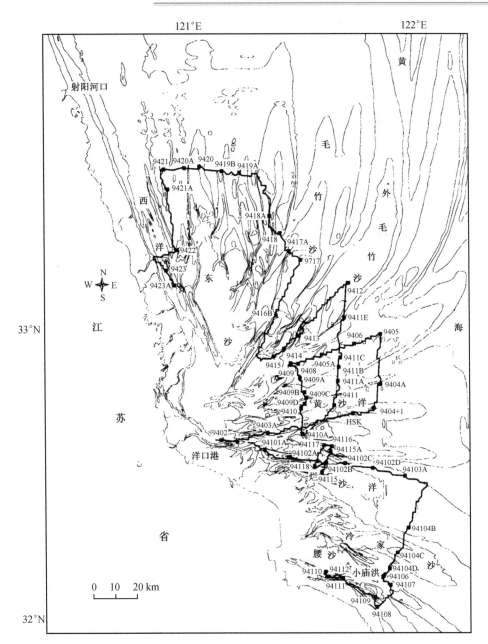

图1 辐射沙脊群海域图

（二）矿物组成

辐射沙脊群沉积物的轻矿物主要是石英与长石，石英含量达 30%，颗粒形态多种，泛红色；长石约 20%，含有一定数量的方解石。

辐射沙脊群重矿物含量约 2%，北部重矿物含量低于南部，碎屑矿物多为凝灰岩、磁铁矿、石英衍生之碎屑，经风化、含水变色或裂隙中具充填物，反映沉积年代久，非现代沉积，按重矿物组合特点可分北部、东部、东南部及南部 4 个区域。

黄河泥沙的矿物组合为角闪石、黑云母、绿帘石、赤褐铁矿、柘榴石、榍石、磁铁矿与锆石,其中以黑云母为特征矿物;长江泥沙的主要重矿物为角闪石、绿帘石、赤褐铁矿、柘榴石、辉石、绿泥石、锆石与磷灰石,其特征矿物为柘榴石、锆石、磷灰石(含量分别为 4.0%、3.4%、2.5%)以及辉石与绿泥石,这类重矿物在黄河泥沙中极少,此外,长江泥沙中稳定矿物含量高,绿帘石含量高达 31.8%,是黄河沉积物的 6 倍。

根据沙脊区及黄河、长江黏土矿物分析,沙脊群区域伊利石含量 70%,蒙脱石含量 16%,高岭石 7%,绿泥石 7%~8%,近似黄河的黏土物质组合。

大量矿物分析表明,辐射沙脊主体是长江系统的细砂物质,而细粒的黏土、粉砂物质明显地受到废黄河(北部)及现代长江(南部)的补给。

二、沙脊群沉积结构

应用多频道地脉冲剖面仪(Geopulse)探测辐射沙脊群,获得 600 km 的地震剖面(图1),反映出海底以下 50 m 厚度的沉积层结构,同时,在邻近潮滩上作钻探、分析、定年,解译地震剖面图像。

(一) 钻孔分析

钻孔在辐射沙脊区的轴心位置,如东县三明村岸外潮滩,孔深 60.25 m,分三段(图2)。

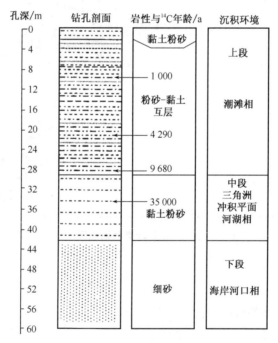

图 2 三明孔柱状图

1. 上段

厚 29 m,潮滩相,保存着有序的潮滩分带结构[8,9];上部泥滩堆积层具有良好的水平纹层(浅色细粉砂层 1~2 mm,深色黏土层 0.5~1 mm),含有孔虫、龟裂楔劈、泥砾、虫粪、反卷层和沙涡旋等微结构；中部为砂泥交互带状沉积层（黏土与粉砂、粉砂与细砂），泥带层厚 2～50 mm 或 200 mm 不等，粉砂带层厚 2～200 mm，砂带层厚 400～700 mm 不等，夹有极细砂、砂透镜体、交错层理与斜层理等镶嵌的潮水沟沉积，亦有泥饼；下部为粉砂—细砂堆积层，含极细砂沉积，多交错层理与砂质透镜体，有侵蚀间断面与泥砾沉积，各层中皆具有风暴沉积的砂质丘状流，或沙涡旋微结构，并含有贝壳屑，或受冲刷而间断保留部分钻孔剖面分带结构，如：水平纹层、砂泥交互层或粉砂细砂层，如此反复三次，反映海平面的振荡性变化，但总厚度约 30 m，相当于一个潮滩环境旋回[10,11]。上段含有孔虫 *Ammonia beccarii* var.，*Ammonia convexidorsa*，*Elphidium magellanicum*，*Elphidium advenum*，*Cribrononion subincertum* 及介形虫 *Tanella opima*。上段潮滩相沉积中含小河堆积物，4～5 m 厚细砂层，夹薄层粉砂，具交错斜层理（鱼脊状），层顶夹有泥片、泥块或泥豆堆积，并含卤淡水交互相的小贝壳（白蛤）及毕克卷转虫等[①]，据[14]C 定年，表层 10 m 以内年代小于 1 000 a（BA94013，BA94014），该年代可能因样品量过少、集炭困难而有误差，在辐射沙洲北部王港潮滩钻孔的相同潮滩层 21 m 深处贝壳定年为（4 290±150）a[②]。如东组下段（Q$_4^l$）约 30 m 深处[14]C 定年为（9 680±520）a[12]，三明孔约 30 m 厚的潮滩旋回沉积相当于全新世海侵以来的堆积。

2. 中段

深度界于 29.0～41.4 m。上部厚 4 m，为青灰色、黄绿色的杂色黏土屋，夹细粉砂层，中部为 3 m 厚的灰色黏土层、黏土粉砂层及灰褐色粉砂细砂层，黏土粉砂层具水平层理及斜层理，粉砂细砂层呈透镜体（层厚 0.5 m）或侵蚀谷嵌入型沉积，内含密集的贝壳及牡蛎碎片，有淡水腹足纲的前鳃亚纲中腹足目的田螺科，肺螺亚纲的扁卷螺科以及瓣腮纲真瓣腮目蚌壳科的珠蚌亚科，在 33.3 m 深处取扁卷螺等，据[14]C 定年（BA94106）早于 35 ka。下部厚 7 m，浅灰色黏土质粉砂层（2 m 厚）有不明显的水平层理，黑灰色粉砂与黏土互层，具斜层理的细粉砂层（1 m 厚）以及 3.1 m 厚的黑灰色—灰色粗粉砂与极细砂层，具扁球状同心圆结构、沙团、沙涡旋及瓣状斜层理，黏土含量少，且分布不均匀，似为含异重流的湍急水流堆积，中段有孔虫含量少，个体小（90～170 μm），破损严重，有 *Lagena* spp.，*Epistominella naraensis* 及少量个体细小的浮游有孔虫，结合钻孔中段物质从粗（下部）渐细（上部），具不明显的二元结构，认为该段系三角洲冲积平原河湖及漫滩相堆积，时代约为晚更新世末期，它与上段潮滩沉积以侵蚀面接触，与下段为间断接触面。

3. 下段

深度 41.4～60.25 m，未穿透，为青灰色细砂，质地均一，无明显层理，各层中皆有云母、贝壳，薄壳小蛤及稀少的有孔虫，如：粒粗较重的 *Ammonia compressiuscula*、*Florilus decorus* 及冷水种 *Rosalina brady* 等，为滨海河口湾砂层，与三明孔同一区位的北渔乡北坎孔，全新世黏土与粉砂层厚度为 36 m，晚更新世冲积-海积相砂层厚 70 m，其底有厚度

超过 20 m 的砂层,该砂层是形成辐射沙脊群的物质基础。

江苏沿海苏北平原晚更新世与全新世沉积层以海陆过渡相为主[12,13],夹有海相沉积层,其中上更新统厚度 70~90 m,为海滨冲积平原河湖相及浅海相;全新世沉积厚度 25~30 m,潮滩相,底部层[14]C 定年为(9 680±520)a。三明孔地层与该区域地层一致,其层位与年代数据可作为分析辐射沙脊群发育与古环境的依据。

(二) 地震剖面

按地层结构,可分为四个区域。

1. 东南区(沙脊群枢纽区域:黄沙洋与烂沙洋)

潮流通道沉积剖面中普遍出现谷中谷现象,烂沙洋—黄沙洋汇合处 V 型谷地中为全新世沉积,具斜层理砂层,厚约 20 m;晚更新世沉积层中,有 V 型谷地,宽 2.2 km,埋深 37 m。烂沙洋中部沿晚更新世老谷地发育,谷地埋深 40 m,谷地宽达 10.42 km,北岸高差 20 m,南岸 33 m,谷地内有河床、河漫滩及沙波等(图 3)。

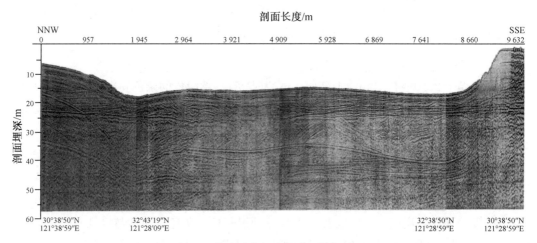

图 3 黄沙洋中段地震地层横剖面

烂沙洋口门段埋深 30 m 的晚更新世谷地宽 3 km,出现于 30 m 厚的沉积层下,上叠全新世谷地宽 6.5 km,埋深 18 m,谷地内具水平层理之砂层与粉砂层,为一大型的古河道遗址。

上述表明,晚更新世末与全新世初,该区具有河谷系统,古谷地中有河槽、河漫滩、阶地、潮流深槽、沙波与沙脊等,反映出潮汐河道特征,古长江在晚更新世末与全新世初在苏北入海①,黄沙洋曾是古河道。

2. 南部区域(冷家沙、横沙及小庙洪潮流通道)

冷家沙、横沙是沙脊区南部大型沙脊,长度 40~60 km,宽度 13 km,沙体厚 60 m。

冷家沙沙体主干部分为老沙脊,地层剖面为:

0~1 m:现代海底沉积,为粉砂质细砂、砂质黏土;

1~16 m:含淤泥的粉砂细砂层,具水平层理,为全新世中—晚期沉积;

---

① 贾建军,朱大奎. 长江入海流路的演变及其机制研究. 南京大学硕士研究生论文,1996.

16～30 m：砂层，有潮流冲刷槽，是全新世早期沉积；

30～78 m：均质粗、中砂，为沙脊主体，顶宽 3.7 km，底宽 5.9km，顶部有侵蚀沟，沟宽 150～450 m，深 4～8 m。冷家沙主体为晚更新世形成的沙脊，当时水深较大，沙脊的基底为起伏 7～12 m 的波形砂层，但尚未形成明显的沙脊与深槽地形。

横沙是发育于全新世冲积平原上的雏形沙脊，横沙南的潮流深槽发育于全新世晚期砂层上，沙脊与深槽均十分年轻。

小庙洪潮流通道发育于全新世晚期地面上，砂、粉砂、淤泥交互堆积具明显水平层，有埋藏河谷，下部为水平层与交错层，覆盖了底部的起伏地形，小庙洪内段的剖面出现全新世早期与中期两层大型河谷，分别埋深于 23 m 和 36 m（图 4）。

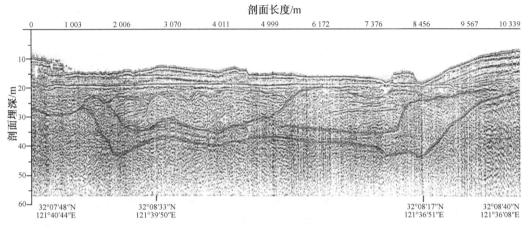

图 4　小庙洪埋藏谷地地震地层剖面

3. 东北区（蒋家沙、苦水洋、毛竹沙、草米树洋、陈家坞槽）

蒋家沙是辐射沙脊群中的主单向倾斜形结构的两种沙脊，目前位于 35～45 m 厚的沉积层之下，全新世早、中期沙脊层厚度小于 20 m，位于海底 18 m 厚的沉积层之下，发育为叠覆的穹形沙脊或为冲刷改造老沙脊形体沙脊，是叠覆式的大型沙脊，晚更新世沙脊宽 3 km，相对高度可达 20 m，具圆穹形结构与单向倾斜形结构的沙脊，全新世晚期砂层沉积叠覆于上，有明显的斜层理与双向斜层理，目前沙脊堆积过程尚在继续（图 5）。

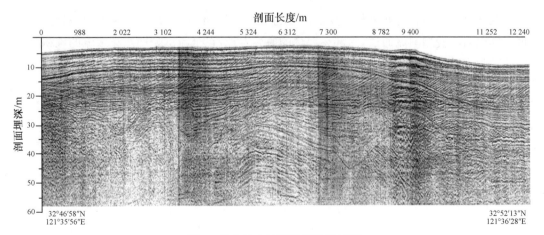

图 5　蒋家沙大沙脊地震地层剖面

苦水洋是介于蒋家沙与外毛竹沙之间的潮流通道,通道具有一致性的沉积结构,系全新世中期与晚期冲刷槽,晚更新世砂层为埋藏于25～36 m深处、厚度15～20 m的颗粒较粗的砂层,地面起伏小,有一些洼地;全新世早期沉积厚4～8 m,为略具水平层理的砂质层,上下两层间为明显的侵蚀界面;全新世中期沉积厚12～14 m,为急湍水流沉积,砂粒较粗,剖面中有填充的侵蚀谷、沙脊与沟间地等,具交错层理与水平层理,并常为沟谷剖面所中断,层面上部受侵蚀,局部保存着沟壑剖面,全新世晚期沉积厚6～10 m,多斜层理、鱼脊状交错层理等。

4. 北部区(大北槽、平涂洋、东沙、西洋)

大北槽是一狭窄的潮流通道,海底沉积层中埋藏着四个时期的谷中谷:晚更新世末与全新世早期的谷地埋藏于海底以下40～44 m,为不对称的W型谷,宽500 m余,相对深度5 m,谷内堆积着水平沉积层;全新世早、中期谷地位于海底30～33 m,不对称谷,宽1 000 m,深5～8 m,谷内沉积层厚度20 m,已受侵蚀成为脊、槽;全新世晚期谷地叠置在老谷的10 m厚沉积物之上,相对深度5 m,谷内为具水平层理的沉积层;全新世晚期谷地范围较大,似为一河谷系统,而后堆积了具良好水平层理的沉积,连续过渡至今海底,该层应为大北槽潮流通道的堆积,大北槽形成于全新世晚期(图6)。

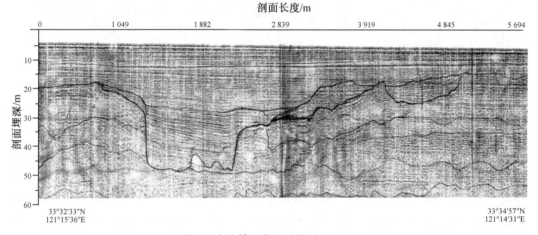

图6　大北槽埋藏谷地震地层剖面

西洋是辐射沙脊群北部最主要的潮流通道,呈南北走向,地层剖面反映出该处为全新世中期以后海侵型潮流通道,深部无埋藏的原始谷地,在全新世晚期西洋水流的堆积与冲刷活动强烈,海底多冲刷穴,掏蚀为涡穴状,亦有厚度近10 m的沉积层。

三、黄海辐射沙脊群形成与演变

(一) 形成条件

主要是辐合潮波与巨量泥沙供应,太平洋前进潮波与黄海受海岸地形影响形成反射潮波在江苏弶港相会,构成辐合潮流场,辐合带发生潮涌,潮差最大①,弶港平均潮差4.18 m,沿海岸向南、向北逐渐减少,辐合带核心区黄沙洋实测最大潮差9.28 m,成为中国沿海最大的潮差记录[3]。辐射沙脊群地区为正规半日潮,沙脊间潮流通道中为往复型,而整个区域涨潮时辐聚,平均流速1.2～1.3 m/s;落潮时辐散,平均流速1.4～1.8 m/s。持续上升的海平面使潮流动力与潮流系统加强,形成为全球有代表性的辐合潮流系统海域,河流供沙是沙脊体得以发育的重要物质来源,晚更新世末古长江经苏北弶港、新川港、遥望港一带入海,堆积了厚层的沙体,是形成潮流脊的物质基础,沙脊、深槽地形的塑造具三个阶段,即:晚更新世末与全新世初,全新世中期,全新世晚期与近代,反映出海平面上升过程造成的动力加强的时期;沙脊体发育的适宜水深是10 m左右,属于波场作用的范围,最深不超过20 m,即使晚更新世古沙脊分布的最深范围亦位于现代海底20 m的水深范围内,这表明辐射沙脊既是以潮流动力为主的堆积体,同时有波浪作用参与,使沙体宽大,与单纯的线性潮流脊的成因[14~16]有明显区别。

(二) 形成过程

辐射沙脊群是大型海底地貌组合体系,由辐射潮流沙脊(东北部)、改造的古河道堆积沙体(枢纽区与南部区)、侵蚀-堆积成因的沙体所组成。

辐射沙脊群物质的主体是细砂与粉砂,主要来源于古长江,全新世晚期沉积层具水平层理,黏土质成分增多,反映出受黄河的影响,现代粉砂与黏土粒级物质受到长江与废黄河泥沙补给的双重影响。

潮流通道,即沙脊间的潮流深槽是在海面上升过程中形成的潮流动力载体,亦是泥沙运动、沙脊—深槽堆积—侵蚀演变的主动脉,它们亦有三种类型。

(1) 沿古河谷发育的潮流通道,如黄沙洋与烂沙洋,从通道的内端至口门仍保留着晚更新世末期以来的各期老河谷,其规模较现代长江口谷地小,可能是数个出口河道之一。

(2) 承袭谷地,晚更新世末或全新世中期已有谷地或低洼地,以后的潮流通道大体承袭埋藏的老谷地,表明该处始终为低地,如大北槽下38 m深处有晚更新世末一支NE-SW向的老河谷被埋藏于东沙之下,可能是东沙的供沙源渠道;陈家坞槽在15 m厚的砂层下有深度为25 m的古河谷,大型沙体之间必有埋藏的古谷地,它们是大型沙脊的泥沙供应通道,小庙洪深槽则是复合型的:潮流通道内段是全新世冲刷槽,中段谷底亦有埋藏谷,方向与现代潮流通道斜交,可能是古长江在南移过程中,在遥望港出口时的遗留谷地,它是冷家沙的泥沙供应源。

(3) 全新世潮流冲刷槽,谷底多冲刷穴、深潭与冲刷槽,如西洋与苦水洋,两通道均是全新世中后期海面上升中形成的。

现代大型沙脊体的主体以已埋藏于海底的晚更新世末的沙脊为主干,上面又叠加着全新世的堆积,例如蒋家沙是晚更新世与全新世的复式大沙脊,其中晚更新世末沙脊位于海底以下35～45 m处,规模大,宽度达3 km,相对高度可达20 m,表明泥沙供应丰富,而

---

　　① 张东生、张君伦、张长宽,等.潮流塑造—风暴破场—潮流恢复——试释黄海海底辐射沙脊群形成演变动力机制(研究报告).1997.

全新世早、中期叠加其上的沙脊规模逐渐减小。

辐射沙脊区的枢纽区开始发育的时代较老,砂质堆积可能始于 30 ka B. P. 的副间冰期,海岸带位于现代海岸以东,长江流经海安—李堡一带入海[①][17],晚更新世沙脊埋藏的深度(30～40 m)与沙脊发育的适宜水深为(10～20 m)均适应;东北区与南部区的时代较枢纽区稍晚,与全新世李堡、遥望港一带供沙有关,脊槽地形主要是全新世高海面时形成;北部最新,为全新世后期与现代产物,地层剖面中的侵蚀时期,可与江苏海岸晚更新世末以来相对海面变化曲线[②]相对比,全新世初期、中期、晚期的侵蚀界面分别为 10 kaB. P. 与 8 kaB. P. ,6 ka～5 kaB. P. 和 1 kaB. P. 海面上升的结果。

(三) 演变规律

辐射沙脊地层剖面中出现数次沙脊与深槽相叠的记录,反映出发育过程的阶段性:泥沙堆积与侵蚀造型,即:河流堆积沙体、砂层;潮流侵蚀沙体与砂层使之成沙脊与深槽;雏形沙脊受到波浪的扰动、冲刷与泥沙的横向再分布,改造线形潮流脊成为宽体的带状沙脊。

"海侵—动力—泥沙"这是沙脊群得以形成发育的基本成因因素,冰期与寒冷期海平面降低,河流下切携运着丰富的"粗"粒泥沙入海,低海面时为"泥沙堆积期",冰后期或温暖期,海平面上升至－20 m 时,潮波辐聚形成辐射流场,提供了潮流沙脊发育的动力条件,持续上升的海面,使潮流动力与波浪作用增强,侵蚀改造波场的厚层泥沙堆积,并进一步塑造成沿辐射流场分布的宽大沙脊与潮流深槽的地貌组合。

因此,低海面堆积砂层,高海面时侵蚀砂层塑造脊、槽,这就是在南黄海这一特定海域环境、地质历史过程条件下,大型的辐射沙脊

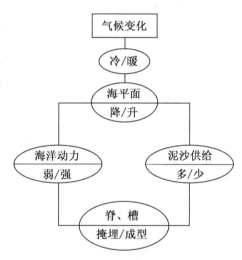

图 7　辐射沙脊形成机制过程

群发育的基本成因基础,可概括为一图式(图 7)。海平面持续上升导致沙脊体向海侧受侵蚀,粗颗粒向岸运移,堆积在靠近陆地的沙脊上,使之逐年扩大增高为沙洲,细颗粒向岸使潮滩淤长。

参考文献

[1] Collins M B, Shimwell S J, Gao S, et al. 1995. Water and sediment movement in the vicinity of inner sand banks: the Norfolk Banks, Southern North Sea. *Marine Geology*, 123: 125 - 142.
[2] Off T. 1963. Rhythmic liner sand bodies caused by tidal currents. *Bull AAPG*, 47(2): 324 - 341.

---

① 贾建军,朱大奎. 长江入海流路的演变及其机制研究. 南京大学硕士研究生论文,1996.
② 张永战. 江苏淤泥质平原海岸环境变迁研究. 南京大学硕士研究生论文,1996.

［3］任美锷主编.1986.江苏省海岸带和海涂资源综合调查报告.北京:海洋出版社.

［4］朱大奎,傅命佐.1986.江苏岸外辐射状沙洲的初步研究.见:江苏海岸带东沙滩综合调查文集.北京:海洋出版社,28－32.

［5］Wang Ying, Ren Mei'e, Zhu Dakui. 1986. Sediment supply to the continental shelf by the major rivers of China. *Journal of Geological Society*, *London*, 143:935－944.

［6］Milliman J D and Meady R H. 1983. World-wide delivery of river sediments to the oceans. *Journal of Geology*, 91:1－21.

［7］王腊春,陈晓玲,储同庆.1997.长江黄河泥沙特性对比分析.地理研究,16(4):71－79.

［8］Wang Ying, Collins M B, Zhu Dakui. 1990. A comparative study of open coast tidal flats: the Wash (U. K.), Bohai Bay and West Huanghai Sea (Chinese mainland). In: Proceedings of ISCZC 1988. Beijing: China Ocean Press. 120－134.

［9］Robert W D. 1992. Tidal depositional systems. In: Roger G W, Noel P J eds. Facies Models-Response to Sea Level Change. Geological Association of Canada, Love Printing Service Ltd. 195－218.

［10］朱大奎,许廷官.1982.江苏中部海岸发育和利用问题.南京大学学报(自然科学版),(3):799－818.

［11］Zhu Dakui, Martini I P, Brookfield M E. 1988. Morphology and land-use of coastal zone of North Jiangsu plain. *China Journal of Coastal Research*, 14(2):591－599.

［12］张宗祜,周慕时,邵时雄等.1990.1:25 000《中华人民共和国及其毗邻海区第四纪地质图》及说明书.见:中国地图出版社中国科学院遥感应用研究所.1991.中国卫星影像图(1:6 000 000).北京:科学出版社.

［13］陈报章,李从先,业治铮.1995.冰后期长江三角洲北翼沉积及环境演变.海洋学报,17(1):64－75.

［14］Swift D J P. 1975. Tidal sand ridges and shoal retreat massifs. *Marine Geology*, 18:105－134.

［15］Swift D J P., Gerardo P, Nestor W L, et al. 1978. Shoreface-connected sand ridges on American and European shelves: a comparison. *Estuarine and Coastal Marine Science*, 7(3):257－273.

［16］Swift D J P and Field M E. 1981. Evolution of a classic sand ridge field: Maryland sector, North American inner shelf. *Sedimentology*, 28:461－482.

［17］杨怀仁.1996.长江下游晚更新世以来河道变迁的类型与机制.见:环境变迁研究.南京:河海大学出版社.195－211.

南黄海辐射沙洲成因的浅层地震与有孔虫证据[*]

一、问题的提出

南黄海辐射沙洲群①分布于江苏岸外,南北长 199.6 km,东西宽 140.0 km,呈褶扇状向海,由 70 多条沙脊与潮流通道组成,脊槽相间,水深界于 0~25 m。据最近测量,其面积为 22 470 km²,其中 0 m 线以上面积 3 782 km²。

辐射沙洲以其巨大的规模、独特的形态特征以及作为潜力巨大的土地、渔业和港口航运资源的开发利用对象,具有重要的理论和实际意义,引起了众多学者的重视和研究。辐射沙洲的成因是一个研究最多的问题,目前主要有两种成因解释。一种认为它是由长江北上的近岸流及其携带的泥沙与废黄河口南下的近岸流和泥沙相遇并向海辐散而形成。另一种认为它是以古长江河口沙坝为初始形态发育而成的,即曾有古长江在辐射沙洲中心(弶港附近)入海。这两种解释至今均因各自证据的局限性和多解性而不能完全否定对方、肯定自己。因而,是否存在过古长江口实质上就成了辐射沙洲成因解释的关键问题。诚然,由于现代长江口和辐射沙洲在地貌、沉积、水动力条件等方面的相似性,两者在沉积物的成分、结构、重矿物组成和有孔虫的常规特征等方面比较相似,不易区别。

鉴于辐射沙洲和现代长江口在沉积动力环境的相似性和区分的困难性,我们试图通过密集的浅地层地震剖面结构调查和有孔虫沉积动力学方法,寻求两者的辨别途径,从而解释辐射沙洲的成因。

二、材料与方法

应用多频道地脉冲剖面仪(Geopulse)探测辐射沙洲,获得 600 km 的地震剖面,揭示了海底以下 50 m 厚度的沉积层结构。同时,在邻近潮滩上作钻探、分析、定年,解译地震剖面图像。钻孔位置在辐射沙洲区的轴心部位,即如东县三明村岸外潮滩上(图 1),简称 SMK 孔。

---

* 王颖,朱晓东,邹欣庆,朱大奎:(陈颙等主编)《寸丹集——庆祝刘光鼎院士工作 50 周年学术论文集论文集》,113 - 120 页,北京:科学出版社,1998 年。

国家自然科学基金"八五"重点资助项目(49236120 号)。

① 王颖再校注:在 20 世纪 90 年代初期研究该处时,仍沿用"辐射沙洲"术语。随着研究的深入,认识到包括沙洲(出露水面的堆积体)、水下沙脊及潮流通道是一个体系,故定名为"南黄海辐射沙脊群"。此文研究内容是沙脊群。2016.12.13 注。

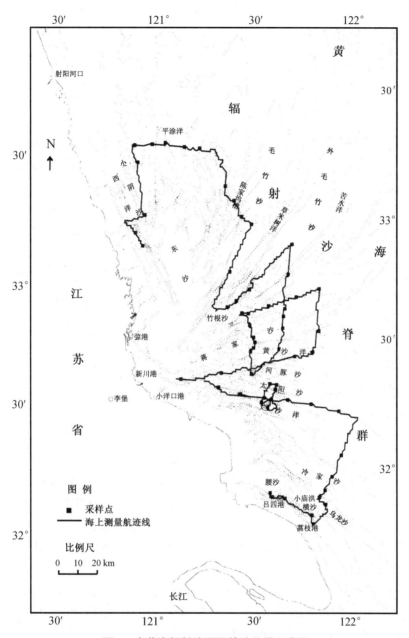

图 1　南黄海辐射沙洲及钻孔位置示意图

　　现代长江口内有孔虫埋葬群是由潮流搬运作用形成的异地埋葬群。辐射沙洲区的有孔虫埋葬群由原地生活群和由潮流从外海向沙洲区搬运来的两类混合而成,属混合埋葬群。根据有孔虫壳体沉降试验分析,这两类埋葬群在壳体当量直径及其与沉积物粒径的关系是有显著区别的,这是本文有孔虫分析的方法依据(朱晓东,1990;朱晓东等,1997)。

三、结果与讨论

（一）地震剖面

1. 东南区

该区位于沙洲枢纽,主要是黄沙洋与烂沙洋。潮流通道沉积剖面中普遍出现谷中谷现象。烂沙洋—黄沙洋汇合处埋深 37 m 谷地中为全新世沉积,具斜层理砂层,厚约 20 m;晚更新世沉积层中,有 V 型谷地宽 2.2 km。烂沙洋中部是沿晚更新世老谷地发育,谷地埋深 40 m,谷宽达 10.42 km,北岸高差 20 m,南岸 33 m,谷地内有河床、河漫滩及沙波等。烂沙洋口门段埋深 30 m 的晚更新世谷宽 3 km,出现于 30 m 厚沉积层下,上叠全新世谷地宽 6.5 km,埋深于 18 m,谷地内系具水平层理之砂层与粉砂层,为一大型的古河道遗址。这表明晚更新世末与全新世初,该区具有河谷系统,古谷地中有河槽、河漫滩、阶地、潮流深槽、沙波与沙洲等,反映出潮汐河道特征。古长江在晚更新世末与全新世初在苏北入海(贾建军,1996),黄沙洋曾是古河道。

2. 南部区域(冷家沙、横沙及小庙洪潮流通道)

冷家沙、横沙是沙洲区南部大型沙脊,长度 40～60 km,宽度 13 km,沙体厚 60 m。

冷家沙沙体主干部分为老沙脊。地层剖面为:

0～1 m,现代海底沉积,为粉砂质细砂、砂质黏土;

1～16 m,含淤泥的粉砂细砂层,具水平层,为全新世中-晚期沉积;

16～30 m,砂层,有潮流冲刷槽,是全新世早期沉积;

30～78 m,均质粗、中砂,为沙脊主体,顶宽 3.7 km,底宽 5.9 km,顶部有侵蚀沟,沟宽 150～450 m,深 4～8 m。冷家沙主体为晚更新世形成的沙脊,当时水深较大。沙脊的基底为起伏 7～12 m 的波形砂层,但尚未形成明显的沙脊与深槽地形。

横沙是发育于全新世冲积平原上的雏形沙脊,横沙南的潮流深槽发育于全新世晚期砂层上,沙脊与深槽均十分年轻。

小庙洪潮流通道发育于全新世晚期地面上,砂、粉砂、淤泥交互堆积具明显水平层,有埋藏河谷,下部为水平层与交错层,覆盖了底部的起伏地形。小庙洪内段的剖面出现全新世早期与中期两层大型河谷,分别埋深于 23 m 和 36 m。

3. 东北区(蒋家沙、苦水洋、毛竹沙、草米树洋、陈家坞槽)

蒋家沙是辐射沙洲中的主体沙脊,是叠覆式的大型沙脊。晚更新世沙脊宽 3 km,相对高度可达 20 m,具圆穹形结构与单向倾斜形结构的两种沙脊,目前位于 35～45 m 厚的沉积层之下。全新世早、中期沙脊层厚度小于 20 m,位于海底 18 m 厚的沉积层之下,发育为叠履的穹形沙脊或为冲刷改造老沙脊形成的单斜式沙脊。全新世晚期砂层沉积叠覆于上,有明显的斜层理与双向斜层理,目前沙脊堆积过程尚在继续。

苦水洋是介于蒋家沙与外毛竹沙之间的潮流通道,通道具有一致性的沉积结构,系全新世中期与晚期冲刷槽。晚更新世砂层为埋藏于 25～36 m 深处、厚度 15～20 m 的颗粒较粗的砂层,地面起伏小,有一些洼地;全新世早期沉积 4～8 m 厚,为略具水平层理的砂质层,上下两层间为明显的侵蚀界面;全新世中期沉积厚 12～14 m,为急湍水流沉积,砂粒较粗,剖面中有填充的侵蚀谷、沙脊与沟间地等,具交错层理与水平层理,并常为沟谷剖面所中断。层面上部受侵蚀,局部保存着沟壑剖面。全新世晚期沉积厚 6～10 m,多斜层理、鱼脊状交错层理等。

4. 北部区(大北槽、平涂洋、东沙、西洋)

大北槽是一狭窄的潮流通道,海底沉积层中埋藏着四个时期的谷中谷;晚更新世末与全新世早期的谷地埋藏于海底以下 40～44 m,为不对称的 W 型谷,宽 500 多 m,相对深度 5 m,谷内堆积着水平沉积层;全新世早、中期谷地位于海底 30～33 m,不对称谷,宽 1 000 m,深 5～8 m,谷内沉积层厚度 20 m,已受侵蚀成为脊、槽。全新世晚期谷地叠置在老谷的 10 m 厚沉积物之上,相对深度 5 m,谷内沉积着具水平层理的沉积层。全新世晚期谷地范围较大,似为一河谷系统,而后,堆积了具良好的水平层理沉积,连续过渡至今海底。该层应为大北槽潮流通道的堆积,大北槽形成于全新世晚期。

西洋是辐射沙洲北部最主要的潮流通道,呈南北走向。地层剖面反映出该处为全新世中期以后海侵型潮流通道,深部无埋藏的原始谷地。西洋在全新世晚期水流的堆积与冲刷活动强烈,海底多冲刷穴,掏蚀为涡穴状,亦有厚度近 10 m 的沉积层。

(二)钻孔地层的沉积学与有孔虫埋葬群特征

根据 SMK 孔岩芯特征(颜色、粒度、结构和构造等)及其变化规律和有孔虫埋葬群特征,所钻探地层可分为上、中、下三段。

1. 上段

厚 30 m。潮滩相,保存着有序的潮滩分带结构(Wang et al.,1990;Dalrymple,1992):上部泥滩堆积层具有良好的水平纹层(浅色细粉砂层 1～2 mm,深色黏土层 0.5～1 mm)、有孔虫、龟裂楔劈、泥砾、虫粪、反卷层、沙涡旋等微结构;中部的砂泥交互带状沉积层(黏土与粉砂、粉砂与细砂),泥带层厚 2～50 mm 或 200 mm 不等,粉砂带层厚 2～200 mm,砂带层厚 400～700 mm 不等,夹有极细砂、砂透镜体、交错层理与斜层理等镶嵌的潮水沟沉积,亦有泥饼;下部的粉砂-细砂堆积层,含极细砂沉积,多交错层理与砂质透镜体,有侵蚀间断面与泥砾沉积。各层中皆具有风暴沉积的沙丘状或沙涡旋微结构,并含有贝壳屑。或受冲刷而间断保留部分分带结构,如:水平纹层、砂泥交互层或粉砂细砂层,如此反复三次,反映海平面的振荡性变化,但总厚度约 30 m,相当于一个潮滩环境旋回(朱大奎,许廷官,1982;Zhu and Martini,1998)。上段潮滩相沉积中含小河堆积物,4～5 m 厚细砂层,夹薄层粉砂,具交错斜层理(鱼脊状),层顶夹有泥片、泥块或泥豆堆积,并含卤淡水交互相的小贝壳(白蛤)及毕克卷转虫等。[14]C 定年,表层 10 m 以内年代小于 1 000 年(北京大学BA94013、BA94014),该年代可能因样品量过少,集碳困难而有误差。根据在辐射沙洲北部王港潮滩钻孔的相同潮滩层的 21 m 深处贝壳定年为 4 290+150a[①],如东组下段(Q_4^1)约

────────────────
① 南京大学海岸与海岛开发国家试点实验室,1993:江苏省大丰县岸外西洋潮流通道稳定性及王港建港可行性研究报告,67 页。

30 m深处^{14}C定年为9 680±520a(张宗祜等,1990),可将三明孔约30 m厚的潮滩旋回沉积相当于全新世海侵以来的堆积。

本段有孔虫埋葬群也同样反映出沉积环境的特点,与现代辐射沙洲潮滩埋葬群相似,即:数量丰富,平均每50 mL沉积物中含1 600枚以上,分异度也较大,平均每个样品含19种以上,主要的属种有:*Ammonia beccarii* var.,*A. maruhasii.*,*A. Convexidorsa*,*Elphidium advenum*,*E. magellanicum*,*E. simplex*,*Epistominella naraensis*,*Cribrononion poeyanum*,*C. porisutualis*,*C. Subincertum*,*Florilus decorus*,*Spiroloculina laevigata*和*Quinqueloculina seminula*及壳体细小的浮游有孔虫等,下部泥质沉积中的有孔虫壳体较小,平均176 μm,而中部则稍大,平均达211 μm,而到上部(孔深4.5 m以上)又表现为个体细小(平均壳径166 μm),数量丰富(平均每50 mL沉积物中含4 137枚),分异度较高(平均每个样品中有19种以上),浮游类等外海分子经常出现,这是典型的潮汐作用较强的潮滩环境的有孔虫埋葬群特征。介形虫主要是近岸种丰满陈氏介*Tanella opima Chen*。上述特点与沉积结构特征一致,均反映出潮滩旋回沉积。

2. 中段

深度界于30~41.4 m。上部厚4 m,为青灰色、黄绿色的杂色黏土层,夹细粉砂层。中部为3 m厚灰色黏土层、黏土粉砂层及灰褐色粉砂细砂层。黏土粉砂层具水平层理及斜层理,粉砂细砂层呈透镜体(层厚0.5 m)或侵蚀谷嵌入型沉积。内含密集的贝壳及牡蛎碎片,属淡水生态环境的腹足纲(Gastropoda)的前鳃亚纲(Prosobran Chia)、中腹足目(Mesogastropoda)的田螺科(Viviparidal);肺螺亚纲(Pulmonata)的扁卷螺科(Planorbidae);以及瓣腮纲(Lamellibranchia)真瓣腮目(Eulamellibranchia)蚌壳(Vrionidal)的珠蚌亚科(Vnioniae)。在33.3 m深处取扁卷螺等,经^{14}C定年(BA94106)系>3.5万年。下部7 m厚,浅灰色黏土质粉砂层(2 m厚)有不明显的水平层理,黑灰色粉砂与黏土互层及具斜层理的细粉砂层(1 m厚),以及3.1 m厚黑灰色-灰色粗粉砂与极细砂层,呈扁球状同心圆结构、沙团、沙涡旋及瓣状斜层理,黏土含量少,且分布不均匀,似为含异重流的湍急水流堆积。这些沉积当属三角洲冲积平原河湖有漫滩相堆积,时代约为晚更新世末期沉积。它与上段潮滩沉积以侵蚀面接触,与下段为间断接触面。

本段地层有孔虫埋葬群的证据表明属非海相环境的沉积记录,仅出现少量(个数和种数均很少)有孔虫,平均壳径165μm,壳体破损较强,主要见*Ammonia beccarii* var.(幼小个体);*Ammonia convexidorsa*,*Bolivina chchei*,*Bolivina* sp.,*Brizalina striatula*,*Buccella frigida*,*Bulimina marginata*,*Cassidulina carinata*,*Cribrononion* spp.;*Elphidium magellanicum*,*Epistominella naraensis*,*Florilus decorus*;*Guembelitria vivans*,*Lagena substriata*,*Protelphidium*,*tuberculatum*,*Quinqueloculina* sp.小个体浮游类5枚。上述有孔虫的特点是丰度和分异度都较小,而其代表的生活环境却应属浅海;壳体细小(平均范围为100~170 μm),磨损、破碎较多,有较强的分选性。这些特征表明其当属异地埋葬群,而非原地或混合埋葬群,与现代长江口崇明岛以上河道中发现的有孔虫埋葬群极为相似(汪品先等,1986;朱晓东,1990),沉积物亦一致,皆属细砂,而与现代辐射沙洲区的有孔虫不同。因此可以认为SMK孔下部,是离河口有一定距离的河床沉积环境,大致相当于现代长江口崇明岛以上至江阴附近之间,而不含有孔虫的样品可能指示离口门更远。现代长江在潮流界以上河道不再含有孔虫(汪品先等,1986)。

3. 下段

深度41.4~60.25 m。为青灰色细砂,质地均一,无明显层理,各层中皆有云母、贝壳,薄壳小蛤及稀少的有孔虫,如:粒粗较重的压扁卷转虫(*Ammonia compressiuscula*)、优美花朵虫(*Florilus decorus*)及冷水种的布氏玫瑰虫(*Rosalinabrady*)等,为滨海河口湾砂层,此为辐射沙洲形成的物质基础。

纵观全孔揭示的地层,可以发现沉积物质、结构具明显的二元结构特征,即从粗(下部)渐细(上部)的旋回变化,其成因机理是在感潮河口区,由于既受较强的潮汐作用,又仍以河流作用为主导,并随着河流主流线的横向摆动,从而在垂向剖面上形成这种典型河口二元沉积结构特征。

(三) 辐射沙洲成因讨论

根据上述钻孔分析结果可对该地区的古沉积环境演变作如下讨论:晚更新世晚期至全新世前(相当于SMK孔60.25~30 m层段),该地区没有出现真正海洋环境,仅在钻孔下部出现少量有孔虫,滨海河口湾砂层,属河口段河床沉积环境。之后,为一持续的相对海退过程,这在SMK孔被记录在上至埋葬深度为30 m的"硬黏土层"为止,该段地层内很少含海相生物。所记录的沉积环境可能是潮流界以上的古河流环境(包括河道或河床、河流边滩或河漫滩,以及泛滥平原乃至泛滥湖沼等陆相环境),即晚更新世(3.5万年以前)的低海面时期。进入全新世,即相当于SMK孔"硬黏土层"以上地层,埋深为30 m以上直至钻孔顶面,该段地层均含有孔虫等海相生物,沉积环境先后依次为:海滩环境、河口环境(包括河口河床或汊道、河口边滩)、浅海潮流沙洲、潮滩环境。由此可见,在辐射沙洲沿岸的东台—海安—如东海岸地区,曾经有古河流出现,其时代分别在晚更新世和全新世早期。沉积物的矿物成分和卫星遥感图像表明该河流正是古长江(中国科学院遥感应用研究所,1991)。

在上述沉积环境背景条件下,辐射沙洲形成的条件主要是辐合潮波与巨量泥沙供应。南黄海海域为半日度潮型,涨潮时,潮流自北、东北、东和东南方向通向弶港海岸;落潮时,潮流以弶港为中心,呈150°的扇面向外逸散,形成以弶港为中心的放射状潮流场。沙洲的地形与潮流场完全吻合。太平洋前进潮波与黄海受海岸地形影响形成反射潮波在江苏弶港相会,构成辐合潮流场,辐合带发生潮涌,潮差最大,弶港平均潮差4.18 m,沿海岸向南向北逐渐减小。辐合带核心区黄沙洋实测最大潮差9.28 m,成为中国沿海最大的潮差记录(任美锷,1986)。辐射沙洲地区为正规半日潮,沙脊间潮流通道中为往复型,而整个区域涨潮时辐聚,平均流速1.2~1.3 m/s,落潮时辐散,平均流速1.4~1.8 m/s。持续上升的海平面使潮流动力与潮流系统加强,形成为全球有代表性的辐合潮流系统海域。河流供沙是沙脊体得以发育的重要物质来源。晚更新世末古长江经苏北弶港、新川港、遥望港一带入海,堆积了厚层的沙体,是形成潮流脊的物质基础。沙脊、深槽地形的塑造具三个阶段,即:晚更新世末与全新世初、全新世中期、全新世晚期与近代,反映出海平面上升过程造成的动力加强的时期;沙脊体发育的适宜水深是10 m左右,属于波场作用的范围,最深不超过20 m,即使晚更新世古沙脊分布的最深的范围,亦位于现代海底的－20 m的水深范围内,这表明辐射沙脊既是以潮流动力为主的堆积体,同时有波浪作用参与,使沙

体宽大,与单纯的线性潮流脊的成因有明显区别(Swift,1975;Swift and Field,1981;
Swift et al.,1978)。

参考文献

[1] 贾建军.1996.长江入海流路的演变及其机制研究.南京大学硕士研究生论文.

[2] 任美锷(主编).1986.江苏省海岸带和海涂资源综合调查报告.北京:海洋出版社.

[3] 汪品先,闵秋宝,卞云华,成鑫荣,朱晓东.1986.河口有孔虫的搬运作用及其古环境意义.海洋地质
与第四纪地质,6(2):53-66;6(3):83-92.

[4] 杨怀仁.1996.长江下游晚更新世以来河道变迁的类型与机制.见:环境变迁研究.南京:河海大学出
版社.195-211.

[5] 张宗枯,周慕时,邵时雄等.1990.1:25 000《中华人民共和国及其毗邻海区第四纪地质图》及说明
书.见:中国地图出版社中国科学院遥感应用研究所.1991.中国卫星影像图(1:6 000 000).北京:科
学出版社.

[6] 朱大奎,傅命佐.1986.江苏岸外辐射状沙洲的初步研究.见:江苏海岸带东沙滩综合调查文集.北
京:海洋出版社.28-32.

[7] 朱大奎,许廷官.1982.江苏中部海岸发育和利用问题.南京大学学报(自然科学版),(3):799-818.

[8] 朱晓东.1990.长江河口三角洲区有孔虫沉降速度试验.海洋地质与第四纪地质,10(3):47-58.

[9] 朱晓东,任美锷,王颖.1997.江苏海岸带沉积环境中的有孔虫埋葬群特征.海洋科学,2:52-56.

[10] Collins, S. J. Shimwell, S. Geo, H. Powell et al. 1995. Water and sediment movement in the vicinity of
liner sand banks: the Norfolk Banks, Southern North Sea. *Marine Geology*, 123: 123-142.

[11] Milliman, J. D. and R. H. Meady. 1983. Word-wide delivery of river sediments to the oceans.
Journal of Geology, 91: 1-21.

[12] Off. 1963. Rhythmic liner sand bodies caused by tidal currents. *Bull. AAPG*, 47(2): 324-341.

[13] Robert W. Dalrymple. 1992. Tidal depositional systems. In: Roger G. Walker and Noel P. James
(eds), Facies Models-Response to Sea Level Change. Geological Association of Canada, Love
Printing Service Ltd. 195-218.

[14] Swift, D. J. P. and M. E. Field. 1981. Evolution of a classic sand ridge field: Maryland sector
North American inner shelf. *Sedimentology*, 28: 461-482.

[15] Swift, D. J. P. 1975. Tidal sand ridges and shoal retreat massifs. *Marine Geology*, 18: 105-134.

[16] Swift, D. J. P., Gerardo Parker, Nestor W. Lanfredi et al. 1978. Shoreface-connected sand
ridges on American and European shelves: a comparison. *Estuarine and Coastal Marine Science*,
7: 257-273.

[17] Wang Ying, Collins, M. B., Zhu Dakui. 1990. A comparative study of open coast tidal flats: the
Wash (U. K.), Bohai Bay and West Huanghai Sea (mainland China). In: Proceedings of ISCZC
1988. China Ocean Press, Beijing. 120-134.

[18] Ying Wang, Mei-e Ren and Dakui Zhu. 1986. Sediment supply to the continental shelf by the major
rivers of China. *Journal of Geological Society*, London, 143: 935-944.

[19] Zhu Dakui and I. P. Martini. 1988. Morphology and land-use of coastal zone of North Jiangsu plain,
China. *Journal of Coastal Research*, 14(2): 591-599.

苏北辐射沙洲环境与资源特征
及其可持续发展初步研究[*]

一、引　言

在人类即将跨入 21 世纪这一海洋新世纪之际，开发海洋的热潮正在全球、全国兴起[1]。江苏省 1997 年末常住总人口达 7 147.86 万人，人口密度 698 人/km²，高居各省区之首，土地供需矛盾尤为突出。位于苏北海岸与南黄海内陆架的辐射沙洲具有极其丰富的土地资源，同时还拥有丰富的生物、航运、旅游等其他海洋资源类型[2]。然而，苏北辐射沙洲这一特殊巨型沉积地貌体系，在环境方面，存在波浪与潮流作用活跃，泥沙搬运，岸线冲淤不稳，生态脆弱等诸多不利因素和复杂性；在资源的开发利用方面业已存在许多问题。因此，辐射沙洲及其沿岸资源的开发唯有依据其环境特征及其演变规律进行，才能实现环境资源和经济的可持续发展。为此，本文在认识辐射沙洲形成演变的自然规律基础上[3~6]，分析辐射沙洲环境、资源特征，提出可持续发展途径，从而为政府及有关产业部门提供决策依据，为加快"海上苏东"建设服务。

二、辐射沙洲地质地貌特征及其成因、演变

辐射沙洲位于苏北海岸带中部岸外的南黄海西南隅，面积 2 万多 km²，由 70 多条沙脊（水深 0～25 m）与潮流通道（水深 10～50 m）组成，其形成演变是受区域地质构造背景控制的，是第四纪以来海面变化与受大河影响的海岸带沉积作用的结果。地质构造运动和海面变化是控制辐射沙洲发育的两个重要地质因素。东亚大陆板块与西北太平洋板块自中生代以来相向运动，造成了我国和东亚地区独特的地质构造体系——新华夏构造体系，表现为一系列呈雁行排列的北北东向的隆起带和沉降带。辐射沙洲就位于其中的南黄海-苏北沉降带上，这一大地构造位置决定了辐射沙洲地区的新构造运动以下沉为主，加之长江、黄河的巨量泥沙来源，使得苏北沿海发育了 400 多 m 的第四纪地层，发育了粉砂质淤泥海岸带与辐射沙洲。组成辐射沙洲的沉积物在沙脊和潮流通道因沉积动力条件差异而不同，在沙脊主要为分选良好的细砂，含量在 90% 以上。沙脊表面低洼处亦可出现局部有粉砂与淤泥质沉积。而潮流通道主要为粉砂质砂或砂质粉砂。辐射沙洲沉积物的矿物组成以轻矿物为主，主要是石英和长石，含量达 30%，颗粒形态多种；长石约 20%。此外含有一定数量的方解石和云母。而重矿物含量在 2% 左右，且呈北部重矿物含量低

---

　　* 朱晓东，朱大奎，王颖：中国地理学会地貌与第四纪专业委员会编，《地貌·环境·发展：1999 年嶂石岩会议文集》，310 - 314 页，北京：中国环境科学出版社，1999.9。

于南部的分布态势[7]。

王颖等[7]应用多频道地脉冲剖面仪(Geopulse)探测辐射沙洲,获得 600 km 的地震剖面,反映出海底以下 50 m 厚度的沉积层结构。按地层结构,可分为东南区(沙脊群枢纽区域,包括黄沙洋与烂沙洋)、南部区域(冷家沙、横沙及小庙洪潮流通道)、东北区(蒋家沙、苦水洋、毛竹沙、草米树洋、陈家坞槽)和北部区(大北槽、平涂洋、东沙、西洋)四个区域,反映出由潮流沙脊、古河道砂体、后期侵蚀-堆积的砂体三类沉积体,揭示出低海平面古长江在苏北入海时的堆积体受辐合潮波改造而成。而当前的沙脊与深槽主要形成于全新世高海面时,海平面持续上升使沙脊体向海侧受侵蚀,粗颗粒堆积在靠近陆地的沙脊上,使沙洲逐年扩大增高,细颗粒向岸使潮滩淤长。

最近我们对南黄海辐射沙洲中心沿岸地区两个钻孔进行了有孔虫和沉积学分析,恢复了晚更新世以来的沉积环境演变历史:晚更新世为非海相沉积环境,有孔虫仅出现在钻孔底部,指示古长江河口环境,继之为不含有孔虫的远离河口(潮流界以上)的河道沉积环境,可能有河床、河漫滩、泛滥平原和湖沼等亚环境。其中暗绿色硬黏土即泛滥平原和湖沼的产物。全新世沉积记录的是与海相关的环境,从全新世开始的潮滩环境演变到河口环境,至近海潮流沙洲和潮滩,最终成为陆地。

三、辐射沙洲沿岸的水文、气候和土壤特征

辐射沙洲的不断并岸成陆,成为大片开发潜力巨大的新生土地资源。土地资源的开发利用首先在很大程度上依赖于其水文、气候和土壤特征。

潮汐作用是辐射沙洲的主要动力因素,它们对于辐射沙洲及其沿岸的冲淤变化、沉积物的运动、土壤乃至生态环境有重要影响。江苏沿海的潮汐在远离岸线的外海为正规半日潮,而近岸和河口因受局部地形、陆地径流的影响,多表现为不正规半日潮,落潮历时比涨潮历时长——这是江苏海岸带潮汐的一重要特点,很大程度上决定了辐射沙洲沉积物质的运动基本态势。至于潮差,因地而异,地区差别十分显著。由太平洋传来的潮波,经东海以前进波形式自南往北推进,致使江苏海区北部广大海域,受黄海、渤海两个半封闭的大型海湾湾顶的反射,形成黄海潮波系统,这一潮波系统同时在科里奥利力的影响下,由太平洋进入黄海的潮波便形成了一系列逆时针旋转的驻立波,在 35°N,122°E 处海域形成无潮点。废黄河口及苏北灌溉总渠的入海口扁担港一带,距离此无潮点最近,再加之这里的海岸向海突出,因而这里平均潮差最小,一般为 1.5~1.7 m。由此向南北两向,潮差普遍增大。潮差的大小往往决定着潮流流速大小、水动力强弱,因而直接控制着辐射沙洲及其沿岸沉积物的颗粒大小、坡度的陡缓。

辐射沙洲及其沿岸海区的海流属于受黄海环流控制的沿岸流。黄海环流是由黑潮分支对马暖流,受阻于朝鲜半岛,形成一个沿朝鲜半岛西岸北上的分支,过辽东半岛进入渤海,再沿河北、山东海岸南下,进入江苏沿海,最后经辐射沙洲群外侧向东海北部流去。江苏沿岸流具有低温低盐的特征,这是因为一方面它来自北方高纬度海区,另一方面受大陆径流的不断冲淡。江苏沿岸流的这一特征对该海区的海洋生物的生态分布起重要控制作用。

辐射沙洲及其沿岸地势低平,雨量充沛,潮汐作用强,地下水位较高,夏季为 0.2～0.5 m,春秋季为 1 m 左右,而冬季为 1.5～2.5 m[3]。该地区的气候具有显著的季风气候特征:夏季盛行偏南风,冬季盛行偏北风。

辐射沙洲及其沿岸的土壤发育于滩涂沉积物,这是海洋潮汐、波浪搬运作用的产物,虽然其最初来源是江河入海的泥沙,但经过海岸动力的分选、粗化作用,以粉砂粒级(0.05～0.001 mm)为主,使得土壤蓄水保肥性降低,易于水土流失,故该地区土壤类型属高盐低肥的草甸海滨盐土和潮滩盐土。

四、生态系统类型和特征

辐射沙洲及其沿岸的生态系统可分为盐土生态系统、沼泽生态系统、沙丘生态系统和水体生态系统四类[8]。

盐土生态系统多发育于潮上带,主要有盐角草(*Saliconia herbacea*)、盐蒿(*Suaeda salsa*)、碱蒿(*Suaeda glauca*)、獐毛草(*Aeluropus littoralis* var. *sinensis*)、大穗结缕草(*Zoysia macrostachys*)、茵陈蒿(*Artemisia capillaris*)、白茅(*Imperata cylindrica*)等系统类型。该系统作为先锋植物在开发荒滩裸地中有重要作用。盐蒿生态系统是定居滩涂的先锋,常见于重盐土,土壤含盐达 0.79%,伴生有盐角草、碱蒿等。碱蒿生态系统分布在次生盐渍地,土壤含盐量 0.76%。獐毛草、大穗结缕草、茵陈蒿生态系统的土壤含盐在0.33%～0.37%之间。白茅生态系统含盐约 0.27%,属轻盐土,发育中期有机质达0.96%,发育后期土壤基本脱盐,有机质含量高,可开垦利用。

沼泽生态系统分布于潮间带和湖上带的低洼织水地,有芦苇(*Phtagmitas australis*)、米草(*Spartina* spp.)、糙叶苔(*Carex scabrifolia*)、扁杆镳草(*Scirpus planiculnis*)等类型。芦苇生态系统属淡咸水类型,含盐 0.25%,伴生植物少,可开垦利用。米草生态系统分大米草(*S. anglica*)和互花米草(*S. alterniflora*)两种,均为人工生态系统,种植于潮间带,是我国目前耐盐最强的类型,土壤含盐达 0.93%～1.96%。糙叶苔位于常年积水的低地,土壤含盐为 1.2%,少伴生种。沼泽生态系统的演替系列为:糙叶苔生态系统→扁杆镳草生态系统→芦苇生态系统或米草生态系统→芦苇生态系统。

沙丘生态系统分布于砂质滩地,面织小,分布区局限,主要由筛草(*Carex leobomugii*)、窝食(*Ixeris repens*)、蔓荆(*Vitex rotunlifolia*)等类型。

水体生态系统分布于常年积有盐水的塘、沟、渠内,类型和组成简单,主要有川蔓藻(*Ruppia rostellata*)和狐尾藻(*Myriophyllum spicatum*)类型。

五、辐射沙洲的自然资源特征

辐射沙洲及其沿岸的海洋空间资源主要是土地资源和深水港口与航道资源。利用辐射沙洲西洋水道建设的大丰港,已被列为国家"九五"港口发展计划和江苏省"九五"重点基础设施建设项目。大丰港引堤工程和大丰港一期工程 2 个 5 000 t 级码头将于 1999 年底建成。南京大学王颖主持的国家自然科学基金重点项目通过大量现场调查和室内分析、模拟研究证明辐射沙洲区拥有很好的深水码头与航道等建港条件,例如处于该区域的黄沙洋可开辟为 20 万 t 级深水航道。1995 年,美国商业部与我国签订了利用辐射沙洲黄沙洋深水道建立天然气发电厂协议,同时需建设一深水港,其第一期投资达 24 亿美元。

辐射沙洲及其沿岸的土地资源丰富,全省海涂年淤长面积近 2 万亩,就主要集中在大丰、东台、如东等市县的辐射沙洲及其沿岸。该地区土地资源具有资源数量大、自然属性优越、自然条件组合特征有利于开发等特点,其中尤其以土地资源数量丰富且呈淤长性、利用潜力大为特色,且具有多宜性特点,农业、渔业、林业、畜牧业、盐业以及旅游业等均有广阔的发展前景。

辐射沙洲及其沿岸气候资源优越,气候宜人,光能源和热能资源都比较丰富,具有海洋性与大陆性气候资源的特点。

目前,辐射沙洲及其沿岸最主要的海洋物质资源是其丰富的动物资源和植物资源以及滩涂农业生物资源。就植物资源而言,整个滩涂地区的木本植物种类有 50 多科近 100 属 230 多种。沿海滩涂地区的动物资源也很丰富,陆生脊椎动物有 146 种,其中两栖类 9 种,爬行类 18 种,鸟类 104 种,兽类 15 种。各种类型的草滩、芦苇滩以及海岸林地,均为一些陆生脊椎动物提供栖息地。沿海珍禽鸟类较多,仅珍稀鹤就有丹顶鹤、白鹤、白头鹤和灰鹤等 5 种。该地区主要经济生物多系以含砂滩涂为栖息生活条件的种类,如文蛤、四角蛤蜊、青蛤、西施舌、大竹蛏、缢蛏、泥螺等。文蛤为本区的绝对优势种,分布广,产量高,居全国之首。这将是滩涂贝类增殖的重要区域。该地区的泥螺是又一种特色宝贵资源,高蛋白、低脂肪,微量元素丰富。近年来该地区泥螺产量占全国的 90% 以上,已成为地区经济发展的一个支柱产业之一。滩涂区紫菜养殖也已获得成功。随着以碱蓬为代表的盐生作物筛选和种植的成功,辐射沙洲及其沿岸滩涂种植业将是 21 世纪大有前途的开发产业。

辐射沙洲及其沿岸拥有比较丰富的独特的能源资源,其特点是可再生性和无污染,主要有海洋潮汐能、波浪能、海水温差能和盐差能等,同时风能和太阳能亦十分丰富。但目前这些资源尚未得到开发利用。

六、辐射沙洲的开发利用现状和存在问题

江苏海岸带开发利用历史悠久,至迟自范公堤修筑以来,苏北沿海的发展历史就是海岸带开发的历史。但辐射沙洲的大规模开发利用主要是在最近 20 年来的改革开放时代。目前农林渔盐畜牧各业都有一定发展,已初步建成粮棉、林果、畜牧、对虾、鳗鱼、淡水鱼、河蟹、甲鱼、文蛤、紫菜、盐业、芦苇等有海岸湿地特色的、多产业的商品生产和出口创汇基地。

随着该地区开发活动的日益推进,目前资源与环境面临的问题日益突出,主要有:

(1) 辐射沙洲地区自然演变频繁,局部冲淤变化强烈,不查明各地滩涂演变规律,将使有关开发工程耗资巨大且留下隐患,难以正常生产与持续发展。

(2) 因行业条块分割,各产业部门从各自短期利益出发,对资源环境呈粗放、低效甚至破坏性开发,缺乏一体化综合利用资源环境的开发科学理论依据、规划和开发模式。

（3）小规模、零散开发，按照旧模式经济单一发展，局限于当前效益，未形成一体化的规模开发，不能发挥持续发展的高效益。

（4）渔业资源状况及可持续发展的后劲不容乐观。多年来，由于捕捞强度过大，主要经济鱼类资源的恢复十分艰难。

（5）养殖种类或类群单一，增养殖技术落后，过度密集养殖区病害肆虐，使海岸带种植业和养殖业风险性增加，同时加剧对生态环境的压力，养殖环境日益恶化，生态系统失衡。

（6）近几年来，随着出海渔船数量激增，捕捞能力迅速膨胀，年年狂捕滥捞，渔业资源严重衰退，去年春秋两汛出现了渔业增产，渔民却并不增收的尴尬局面。由于酷捕滥采，不合理开发导致贝类等栖息生存生态环境的破坏等原因，滩涂贝类（特别是重要的经济种类，如文蛤）的自然资源都遭到严重的破坏，近年来滩涂贝类大批死亡时有发生。

（7）土地盐碱度高、肥力低，是限制农业发展的一个重要制约因素，急需因地制宜地探索快速改良土地的生态工程方式。

七、可持续发展对策

从上述讨论可见，辐射沙洲环境资源开发潜力与压力并存，加强科学研究和管理，合理开发与保护，走可持续发展之路是当务之急。根据我国国情和江苏省省情，辐射沙洲及其沿岸是一种必须开发的巨大资源，而不可能像欧美国家那样仅用作自然保护区。在我国"发展是硬道理"。目前多数开发项目还仅属初步的低层次开发，在广度和深度上仍有很大潜力。主要在土地资源和综合农业和海洋生态工程两方面。在土地资源方面，已围垦的潮上带中还有 20 多万亩尚未开发，潮上带高滩还有 90 多万亩未围垦，而且每年还以 2 万亩的速度在淤涨。590 万亩潮间带和辐射沙洲中，尚有 200 多万亩可用于贝、藻类养殖。至于在辐射沙洲及其沿岸的综合生态农业开发利用方面，潜力更为巨大，而且对环境资源实现可持续发展利用，即基于湿地的自然生态规律，在不破坏或损害其环境资源利益的前提下，充分利用辐射沙洲及其沿岸的土地、生物、水源、太阳光热能进行综合农业开发，营造一种自然辐射沙洲及其沿岸的系统和人工湿地农业生态系统协调、和谐的复合生态系统。

针对辐射沙洲的海洋沉积动力特征和演变规律、生态特点，应当因地制宜地建立生态农业和自然保护区。特别应当注意的是在基于辐射沙洲自然演变规律，运用一体化综合利用资源环境的开发科学理论，进行科学合理规划，实施可持续发展开发。通过地貌与沉积动力学研究和 GIS 技术方法，确定合理围垦时空范围。对海洋空间资源在港口航运、农业开发与城镇建设用地进行统筹规划，寻求科学依据、优选开发方案，最终实现土地空间资源的可持续发展开发。

坚持科技兴海，走集约化经营道路。把沿海滩涂开发定位在科技含量高，现代化水平较高的层次上，加快实施沿海滩涂养殖规模化。首先从保护渔业资源入手，严格控制近海捕捞强度。严格捕捞渔船的监督管理和检验，淘汰一批安全性能差、破坏资源的小型渔船，对不符合国家规定的出海渔船不发出海捕捞许可证，同时，通过制定和实施优惠政策，加快调整捕捞结构,引导一部分近海捕捞渔船转向养殖业或拓展新门路,鼓励扶持大马力渔船发展外海远洋生产。引进先进技术,建设一批水产品系列加工基地,开展水产品深度加工和综合利用,兴办一批水产品加工龙头企业,走贸工渔一体化之路,努力提高水产品加工质量,提高水产品的附加值,增进效益,弥补海洋捕捞实现"零增长"带来的损失。发展观赏鱼养殖业,使休闲渔业与旅游业有机结合,促进海洋经济发展。

加强现有自然保护区的建设,特别是大丰和射阳的麋鹿、丹顶鹤自然保护区。发挥湿地净化功能,减轻和防止环境污染。同时,随着全球性气候变暖和海平面上升,辐射沙洲亦将面临严峻的问题,对此应有所准备,做到未雨绸缪。

参考文献

[1] 王志雄,朱晓东.1996.浅论海洋作为人类 21 世纪生存与发展的自然资源.见:江苏省科学技术协会编,建设海上苏东的科学之路.北京:中国科学技术出版社,36 - 40.

[2] 朱晓东,施丙文.1998.21 世纪的海洋资源及其分类新论.自然杂志,20(1):21 - 23.

[3] 任美锷(主编).1986.江苏省海岸带和海涂资源综合调查报告.北京:海洋出版社.

[4] 朱大奎,傅命佐.1986.江苏岸外东沙滩地貌和表层沉积特征.见:江苏海岸带调查论文集.北京:海洋出版社,15 - 27.

[5] 朱大奎,傅命佐.1986.江苏岸外辐射沙洲的初步研究.见:江苏海岸带调查论文集.北京:海洋出版社,28 - 33.

[6] 朱大奎,安芷生.1993.江苏岸外辐射状沙洲的形成演变.见:任美锷教授八十华诞地理论文集.南京:南京大学出版社,142 - 145.

[7] Wang, Y., Zhu, D., You, K., et al. 1999. Evolution of radiative sand ridge field to the South Yellow Sea and its sedimentary characteristics. *Science in China* (*Series D*), 42(1): 97 - 112.

[8] 朱晓东,安树青,朱大奎等.1998.江苏海岸湿地环境资源特征及其可持续发展策略.见:陆健健等主编,中国湿地研究和保护.上海:华东师范大学出版社,317 - 323.

Evolution of Radiative Sand Ridge Field of the South Yellow Sea and Its Sedimentary Characteristics[*]

A sand ridge field is a large-scale group of submarine sandy bodies of the continental shelf shallow sea formed by abundant sandy deposits and strong tidal currents[1, 2]. In the South Yellow Sea, progressing tidal waves of the Pacific Ocean have interacted with the reflected tidal waves from the Shandong Peninsula by converging and diverging to form the large-scale tidal ranges and strong tidal currents. The strong, ever changing tidal currents combined with the rich sand deposits at the mouths and channels of the ancient river provide good conditions for the formation of the sand ridge field[3~6]. The South Yellow Sea sand ridge field is located offshore the northern Jiangsu coasts in a radiative fan pattern. The length of the sand ridge field is 199.6 km from north to south and the width 140.0 km from east to west. The sand ridge field fans out to the sea consisting of more than 70 sand ridges and tidal channels. The water depth of the area ranges from 0 to 25 m. According to the latest measurements, the total area encompasses 22 470 km², of which 3 782 km² is above 0 m level (Fig. 1).

During the semidiurnal tides of the South Yellow Sea as well as its flooding period, tidal currents flow from north, northeast, east and southeast to the west coast of the Yellow Sea, eventually converging at the shore of Jianggang. During the ebb tide period, the tide flows outward from Jianggang as its center, forming a radiative tidal current field in the shape of a fan with an angle of 150 degrees. The landscape of the sand ridge field has come to fit with tidal current field.

1　Sediment composition and its origin of the sand ridge field

1.1　Size composition

The sediments of the radiative sand ridge field mainly consist of well-sorted fine sand at the percentage more than 90.

＊　Ying Wang, Dakui Zhu, Kunyuan You, Shaoming Pan, Xiaodong Zhu, Xinqing Zou, Yongzhan Zhang: *Science in China* (*Series D*), 1999, Vol. 42, No. 1, pp. 97 – 112.

Project supported by the National Natural Science Foundation of China (Grant No. 49236120). Project coding: SCIEL21198103.

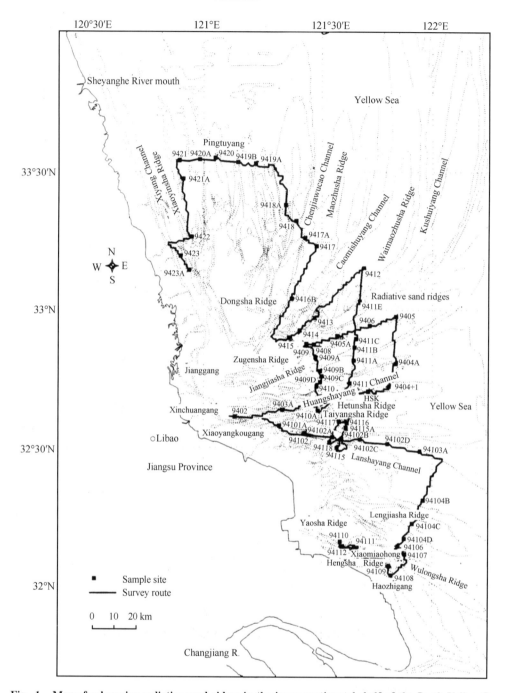

Fig. 1 Map of submarine radiative sand ridges in the inner continental shelf of the South Yellow Sea

1) Sand ridge is mainly composed of fine sands. Silts can be found in some lower depressions which form multiple thin layers in the sediment strata.

2) The sediments of the inner and middle parts of the tidal channels are mainly fine sands containing silts ($2\% \sim 6\%$). In the outer part, entrance or the middle part of some large channels (such as Huangshayang, Lanshayang), where water depth exceeds 15 m, the silt content increases, and sandy silt as well as silty sand appears. The

sediment composition in the tidal channel changes as the water depth changes. Coarser composition can be found near shore, which reflects the characteristics of the wave process. This observation indicates that tidal current is a major dynamic of the sand ridge field but it also has significant influences by wave actions in the shallow water area. Under normal weather conditions, the decay limits of wave winnowing processes occur in water depth of 15 to 20 m.

3) The sediments at the entrance part of several large channels, such as the main stream of Lanshayang channel, are sands and gravels. The content of "gravel" (calcic cemented sand grain nodules) is up to 29%, its diameter is >4 mm. The water depth of Pingtuyang is > 15 m, and the fine sand sediment contains 5% "gravel". These sediments are deposited in the strong tidal currents environment.

4) In the northern part of the sand ridge field, the sediments in the tidal channel (Xiyang) consist mainly of fine sands but contain more silts. Silt increases at the water depth of 10 m. Under the water depth of 20 m, clay deposits can be found. From the channel's end towards its entrance (from south to north), the sediments appear in the order of silty sand, sandy silt, and sand-silt-clay. In the north of the sand ridge field, the clay content can reach up to 30%, indicating that the provenance of the clay sediments is the abandoned Yellow River Delta.

The sandy silt is the major type of the bottom sediments of the Yellow River mouth, the suspended particle material (SPM) breaks into clayey silt, and the bottom Sediments of the Changjiang estuary consist of silty clay with SPM as clayey silt[7]. The sediment distribution of both the Yellow River and the Changjiang River shows clearly that sediment size gradually decreases from the lower reaches toward the sea, but the same phenomenon at the Changjiang estuary occurs much faster because of the flocculation process, as effected by tidal currents, causing fine grain of clay material to settle quickly. It is a sedimentary facies of bio-dynamic environment, and does not show that there is no fine sand being transported into the sea. The Changjiang River has transported large quantities of fine sands in the sea especially during the flood season. The sediments of the Yellow River discharged to the Yellow Sea and the Bohai Sea are mainly clayey silt, but those of the Changjiang River discharged to the Yellow Sea and the East China Sea are fine sands and silty clay.

1.2 Mineral particles

The light minerals in the sand ridge sediment are mainly quartz and feldspar. The content of the quartz reaches 30% in numerous grain shapes, in reddish color. The feldspar content is about 20%, containing a certain amount of calcite.

The total amount of heavy minerals in the sand ridge sediment is less than 2%, being unevenly distributed with more heavy minerals in the south. Most of the rock fragments are tuff, magnetite, and derivatives of quartz. The old age of the heavy mineral deposits is evidenced by many weathering features and the large amount of fissure fillings.

The heavy materials are distributed in the following four regions with different characteristics.

1) The North Region (Channels of Xiyang, Pingtuyang, Dabeicao). The groups of heavy minerals can be ordered according to the quantity of minerals in their composition: hornblade, epidote, limnite, ilmenite. Among them, the quantity of hornblade reaches up to 60%~67%, that of epidote, liminite, ilmenite is less than 20%; that of liminite and ilmenite is more than 5%, and that of garnite is less than 2%. The heavy minerals in Pingtuyang sediments contain more zircon (6.49%) and a small amount of rutile.

2) The East Region (Sand ridges of Waimaozhusha, Jiangjiasha). The heavy minerals are not high in content, but rich in varieties: hornblade, epidote, granite, limnite, ilmenite, zircon, leucoxene, apatite, and fresh mica, chlorite, and calcite. The quantity of hornblade is 50% (77.63% in Jiangjiasha), that of epidote is more than 20% (11% in Jiangjiasha), and the quantities of granite and limnite are between 2% to 5%.

3) The Southeast Region (Channels of Huangshayang and Lanshayang). The quantities of hornblade and epidote reach up to 80% and 60%, respectively. The second mineral group contains ilmenite, zircon, epidote, magnetite, apatite, rutile, tourmaline, fresh mica, chlorite, and calcite.

4) The South Region (Lengjiasha Ridge and Xiaomiaohong Channel). The quantities of hornblade and epidote reach up to 72% and 86%, respectively. The high quantity of hornblade is a predominant characteristic of the radiative sand ridge field. The quantity of hornblade in the heavy mineral decreases from north to south, but the epidote content increases from north to south. This gives evidence to different quantities of major heavy mineral in different provenance areas. The heavy minerals of the Lengjiasha Ridge contain a large portion of ilmenite, magnetite, as well as limnite, in a total quantity of more than 2%. This is apparently different from the other regions. The quantity of ilmenite in the Lengjiasha Ridge (10.12%) is larger than that in the Xiaomiaohong Channel (3.4%). The quantity of limnite in the Xiaomiaohong Channel is 8.11%, which is larger than that of the Lengjiasha Ridge (4.21%). The higher proportion of the ilmenite in the heavy mineral is an important feature in the southern area of the radiative sand ridge field.

The heavy mineral composition of the Yellow River sediment is: hornblade, mica, epidote, reddish limnite, granite, magnetite, among which mica is the key mineral. The heavy mineral composition of the Changjiang River sediment is: hornblade, epidote, reddish limnite, granite, pyroxene, chlorite, zircon, and apatite. The majority of the heavy minerals are granite (4.0%), zircon (3.4%), apatite(2.5%), pyroxene, and chlorite. The Yellow River sediments have only a small quantity of heavy minerals. The quantity of epidote is especially high，reaching up to 31.8% in the Changjiang River sediment，which is 6 times larger than the epidote content in the Yellow River sediment. The heavy mineral composition of the Changjiang River contains large portions of stable minerals which reflects that fine sands make up a large part of the Changjiang River sediment discharges.

From the heavy minerals distribution，it can be seen that the fine sand material of the Changjiang River is a major portion of the radiative sand ridge field. The core parts of the sand ridge field such as Maozhusha Ridge，Waimaozhusha Ridge，Huangshayang Channel，Lanshayang Channel have received sediment supplies both from the Changjiang River and the Yellow River. The northern part of the sand ridge field has more sediment supply from the abandoned Yellow River，while the southern part of the sand ridge field is dominated by the Changjiang River sand supply.

1.3 Clay mineral

A study comparing the clay mineral content and regional distribution shows that the illite content reaches up to 70%，which is similar to the clay minerals content of the East China Sea. The content of montmorillonite is 16%，which is higher than that in the Changjiang River region，but closer in quantity to the Yellow River sediment；kaolinite (7%)，and chlorite (7%~8%) contents are low，similar to clay mineral of the Yellow River sediment. The fine grain portion of the sand ridge field sediments not only comes from the Changjiang River，but also has sediments supplied from the Yellow River. The majority of the sediments of the radiative sand ridge field consist of the fine sands from the ancient Changjiang River，but the fine sediments of the clay material apparently receive sediment supply both from the modern Changjiang River and the abandoned Yellow River.

2　Sediment strata of the sand ridge field

A 600-km seismic profile has been recorded by using a multi-channel Geopulse seismic profiler，which reflects 50-m thick sub-bottom structure of the sediment strata. At the same time，sediment core has been drilled up to 60 m in the vicinity of the tidal flat. The core has been analyzed and carbon dated to interpret the seismic records.

2.1　Core analysis

The core is located on the tidal flat outside the coastal bank of Sanming Village in Rudong County of Jiangsu Province. The total entry is 60.25 m deep，and can be divided into three sections (Fig. 2).

1) Upper section, 29 m thick. The tidal flat facies of deposition can be divided into three zonations: inner, middle and outer tidal flat strata. There are 3-times repeatings of the tidal flat strata appearing in this section which shows the changes of sea level, but all in one cycle of tidal flat environment as the total thickness of the section is about 30 m[8~11]. A few fossils of foraminifera are *Ammonia beccatii var*, *Ammonia convexidorsa*, *Elphidum magellanicum*, *Elphidum advenum*, *Cribrononion subincertum* and *Ianellaopima* of Ostracoda. [14]C datas of 21 m and bottom layer are (4 290±150) a B. P. and (9 680 ±520) a B. P. [12] from tidal flat cores in Wanggang and Rudong of the sand ridge field. Thus the section is the deposition of Holocene transgression.

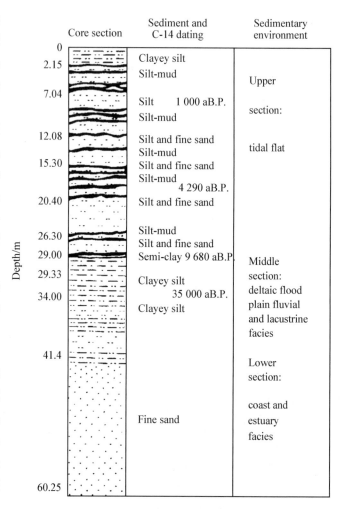

Fig. 2 Vertical profile of Sanming core.

2) Middle section, 12-m thick (at 29.0 to 41.4 m depth) yellowish grained hard clay with thinner layers of silty and lenticular fine sand. The sandy layer contains fresh water fossils of *Viviparidal spp.* of Gastropoda, *Planorbidae spp.* of Pulmonata and *Vnioniall spp.* of Eulamellibran chiu, and a little amount of foraminifera fossils of *Ammonia beccarii var*, *Lagena spp*, *Epistominella naraensis* and a few of smaller planktonie forams which have been found in the middle section. All of the features reflect the facies of deltatic alluvial plain and deltatic lake sedimentation. By carbon 14 dating of the fossils at a layer of 33.3 m deep, the age is greater than 35 000 a B. P. There is an erosional surface between the late Pleistocene stratum and the upper section as well as an unconformity with the bottom section.

3) Lower section, greenish grey color homogeneous fine sand strata without any apparent stratification at an unpenetralble depth of −41.4 to −60.25 m. The total thickness of the unpenetrable sand strata is 70 m according to the core of Beikan from

nearby tidal flat. There are estuary-seashore facied sand strata with abundant shell fragments and several foram species of *Ammonia compressiuscula*, *Florilus decorus*, and cold water species as *Rosalina brady* in the sediments.

The proof of the Sanming core shows that the late Quaternary strata of radiative sand ridge field can be compared with the one of the strata of the North Jiangsu plain and Changjiang River delta[12,13]. According to a number of the sediment cores along the North Jiangsu coast, it can be known that the strata of the late Pleistocene are 70~90 m thick seashore-lagoon-estuary facies and shallow sea facies. The Holocene stratum is an about 30 m thick estuary-sea coastal facies. ^{14}C data of its bottom is (9 680±520) a B. P. Thus the Sanming core is well in accordance with the regional sediment strata. The data (depth, dating, etc.) of Sanming core can be used to explain the evolution and the environment of the sand ridge field.

2.2　Seismic profile

The 600 km long seismic profile indicates that the radiative sand ridge field can be divided into four regions according to the structure of the preserved sediment strata.

2.2.1　The southeastern region (the central part of the sand ridges, the channels of Huangshayang and Lanshayang).

The phenomenon of the valley-in-valley is prevalent in the sand ridge field. At the convergence section of the Lanshayang and Huangshayang channels, the seismic profile shows V-shaped valleies in the inner part of the sand ridge field: one is a 2.2 km wide valley with 20 m thick early Holocene sediments in the slanted sandy bedding forms. A 2.2 km wide V-shaped valley of the late Pleistocene is buried underneath 37 m thick sediments. The old valley of the late Pleistocene at the middle part of tide channel is buried 40 m below the present sea bottom. The widest part of the old valley reaches 10.42 km, whose northern bank is 20 m above the old valley bottom while the southern bank is 33 m above the old valley. In the valley, there are still preserved narrow river bed, small-sized flood plain and sandy ripples (Fig. 3). In the outer part of the southeastern region (121.817°E, 32.534°N towards east), where the sand ridges gradually disappear and the thickness of sandy strata decreases, there is a Holocene valley buried in the 20-m thick sediment which is 6 km wide, with slanted and laminae sandy beddings. Another large, 8.1 km wide valley is buried underneath 38-m thick sediment below sea bottom, with a sandy ridge (10 m high and 1.8 km wide) preserved within.

A valley-in-valley appears in the seismic profile taken at the entrance section of Lanshayang (between 121.938°E, 32.384°N, and 121.931°E, 32.363°N), in which a 3 km late Pleistocene valley is located underneath a 30 m thick sediment and superimposed on top of the Pleistocene valley is a 6.5 km wide Holocene valley now buried underneath

18 m thick sediments. Inside of all the valleys, there are laminae bedding of sand and laminae bedding of sand and silt preserved. This phenomenon shows that they are the remnants of an old large river channel.

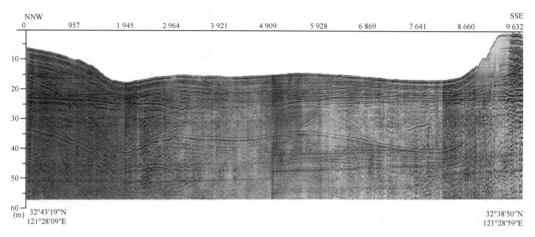

Fig. 3 A cross-section seismic profile of the middle Huangshayang channel

(Abscissa: relative length of the profile; unit: m. Ordinate: relative depth of the profile; unit: m)

The above results indicate that during the end of the late Pleistocene and early stage of Holocene, there are river systems in this area. The narrow river bed, the flood plain, sandy ripples, and deep tidal channels in the old valley reflect the characteristics of a tidal river. Together with scientific results obtained on land[①], it can be determined that the old Changjiang River entered the sea through the North Jiangsu plain during the end of the late Pleistocene and early stage of the Holocene. Therefore, the Huangshayang and Lanshayang were among the valleys of the old Changjiang River.

2.2.2 The southern region (Lengjiasha ridge, Hengsha ridge, and Xiaomiaohong tidal channel). Lengjiasha, Hengsha ridges are the large-sized sand ridges in the southern part of the sand ridge field. The ridge is 40~60 km long, 13 km wide, consisting of 60 m thick sandy material.

The Lengjiasha body is composed of old and new ridges superimposed upon each other with the main body being the old ridge. Seismic profile of the sandy ridges composition shows the following profile in the order from the present to the old. The present sea bottom sediment is 1 m thick greenish black silty fine sands with abundant shell fragments. To the north of sea bottom, sediments are sandy clay, and to the south, they are yellowish grey silty mud. The difference indicates that there are different material resources. The second layer is 10~16 m thick silty fine sands with laminae bedding deposited during the middle and late Holocene period. The third layer

---

① Jia Jianjun, Zhu Dakui, Evolutionary mechanism of the Changjiang River channels into the sea, 1996.

is 8~15 m thick sands superimposed on the main part of the sandy ridges. There are tidal current gullies located on both sides of the ridges. The northern gully is located 26 m while the southern gully is located 22 m underneath the present sea bottom. This layer is presumably that of early Holocene transgression and the deposition. The fourth layer is the main body of the large sand ridges, 48 m thick homogenous mixture of coarse- and medium-grained sands. The sand ridges are 3.7 km wide on top, and 5.9 km wide on the bottom with gullies (4~8 m deep, 150~450 m wide) formed on the top, and the matured deep channel developed on both sides of the ridge, and buried 38 m (north)~41 m (south) below the sea bottom. The base ground of the sand ridge is sandy strata in 7~12 m amplitude wave forms, located 40~61 m deep. Thus, while the Lengjiasha ridge formed during the late Pleistocene, at that time, the southern part of the valley was deep and large.

Hengsha ridge is the young stage sand ridge of the late Holocene, originally developed on the alluvial plain of the early and middle Holocene times. The tidal channel on the southern side of the Hengsha ridge has developed on the surface of the late Holocene sandy sediments. And the sandy sediment has been washed away by erosion and then accumulated in the sea bottom depressions.

The Xiaomiaohong tidal channel developed on the surface of late Holocene ground. The sediment profile shows clearly laminae bedding strata, which consist of sand, silt, muddy intervals with distinct valley forms. The sediment structure of the middle Holocene strata consists of the level bedding and a cross section covering the undulant land form; the bottom part of the early Holocene sandy layers has been eroded heavily into ridges and valleys with a height of 10~20 m.

In the profile of the inner part of the Xiaomiaohong channel appears a large sized valley of the early and middle Holocene period. The depths of the two valleys buried underneath are 23 and 36 m respectively. Early Holocene sandy layers through the process of erosion have now shown clearly valleys and ridges. The valleys have alluvial deposits (Fig. 4). The late Pleistocene strata had been eroded to form a 9.5 km wide river valley. Inside the valley there are sand ridges, sand ripples, and a network of channels, with the thickness of the sand layers in the sand ridges being greater than 20 m. This buried valley intersects the Xiaomiaohong channel at either a perpendicular (SN direction) angle or a cross (NW – SE direction) angle. The size of the buried erosion valley indicates that there was a large river passing through the area of the Xiaomiaohong channel perhaps during the time from the end of the Pleistocene to the middle Holocene. It may be assumed that Yaowanggang may have been another exit of the old Changjiang River after Lanshayang during the Holocene time. The sandy material of Lengjiasha ridge and Yaosha ridge may have been supplied by the old Changjiang River from Yaowanggang. But the Xiaomiaohong channel is a tidal erosional valley formed during the late Holocene time.

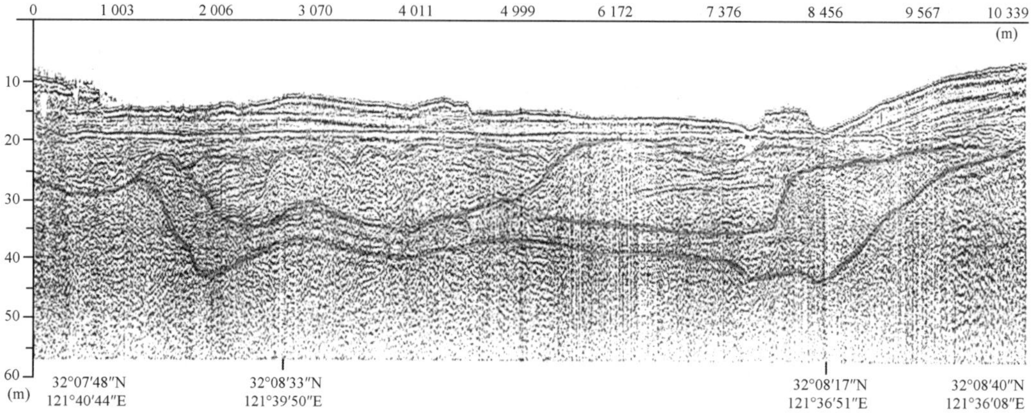

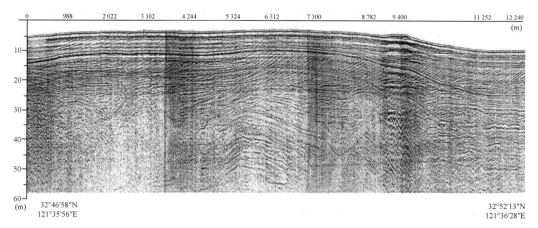

Fig. 4　The seismic profile of the buried valley of Xiaomiaohong channel

（Abscissa：relative length of the profile；unit：m. Ordinate：relative depth of the profile；unit：m）

2.2.3　The northeast region（Jiangjiasha ridge，Kushuiyang，Maozhusha ridge，Chengjiawucao channel）.

Jiangjiasha ridge is one of the main sand ridges of the radiative sand ridge field. It was deposited during the late Pleistocene and superimposed by the Holocene period sandy deposits. The late Pleistocene ridge is 3 km wide and approximately 20 m high. The structure of the ridge is a dual structure of dome-shaped ridge and monoclinic ridge. At present，the Pleistocene ridges are located below $35 \sim 45$ m sand layers. The thickness of the sand layers deposited by the early Holocene and middle Holocene is less than 20 m，and located 18m under the sea bottom. The dome-shaped ridges are eroded and reformed to the monoclinic structure. Late Holocene sand sediments are superimposed on the top of dual directional slanted bedding. The sedimentary process of sand accumulation still continues，but on a small scale（Fig. 5）.

Fig. 5　The seismic profile of the Jiangjiasha sand ridge

（Abscissa：relative length of the profile；unit：m. Ordinate：relative depth of the profile；unit：m）

The Kushuiyang channel is a tidal channel between the Jiangjiasha and the Waimaozhusha ridges. Seismic profile shows the channel with the same sedimentary structures as an erosional channel formed during the middle and late Holocene times.

The late Pleistocene sandy layer of the Kushuiyang composed of $15 \sim 20$ m thick coarse sands is buried $25 \sim 30$ m below the present sea bottom, the layers are relatively flat with only a few valleys. The early Holocene deposits are $4 \sim 8$ m thick with continuous sandy level beddings. The middle Holocene sediment is a $12 \sim 14$ m thick rapid current deposit of coarse sands with the structures of filled gullies, sand ridges and the flat ground consisting of cross bedding and level beddings. The top of early Holocene sediments was eroded by gullies. The late Holocene sediments are $6 \sim 10$ mthick and preserved with slanted and across beddings.

2.2.4　The northern region (the channels of Dabei trough, Pingtuyang, Xiyang and Dongsha ridge).

The Dabei trough is a narrow tidal channel. The seismic profile reveals that there are valleys-in-valley of 4 periods buried in the sea bottom sediments. (i) Valley of the end of the late Pleistocene to early Holocene, are preserved under the sea bottom at a depth of $40 \sim 44$ m, and asymmetrical W shaped, is more than 0.5 km wide, and relatively 5 m deep with the stratified sediments in the valley. (ii) The early to middle Holocene valley, asymmetrical and located $30 \sim 33$ m under the sea bottom, is 1 km wide and $5 \sim 8$ m deep. The sediments accumulated in the valleys are more than 20 m thick, and have been eroded in the forms of ridges and troughs. (iii) There are middle Holocene troughs in the accumulated sediments. Inside the troughs, the sediments under constant pressure show a convex form. But sediments in the ridges were disturbed to become discontinuous series of layers. (iv) There are late Holocene valleys superimposed on the old valleys and separated by 10 m thick sediments. The relative depth of this valley is 5 m and it contains stratified layers. It seems that the late Holocene valleys were river tributaries in a large area. The tributary remnants were then covered by 8 m thick sediments of the Dabei trough deposits, which are characterized by well preserved level and cross beddings to link with the present sea bottom. Therefore, the Dabei trough was formed during the late Holocene (Fig. 6).

Xiyang channel is the main tidal channel in the north of the sand ridge field. It flows out from south to north and eventually divides into two branches: the eastern and western Xiyang channels. The eastern Xiyang channel is located between the ridges of Liangyuesha and Xiaoyinsha. The profile shows that there is only one transgressional tidal channel after the middle Holocene without any original valley underneath. There are scour pots on the bottom of the present eastern Xiyang channel, indicating that the erosional process started during the late Holocene.

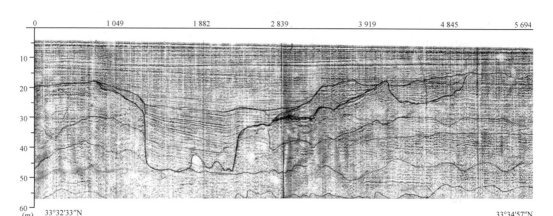

33°32'33"N
121°15'36"E

33°34'57"N
121°14'31"E

Fig. 6　The seismic profile of the buried troughs of Dabei trough

（Abscissa：relative length of the profile；unit：m．Ordinate：relative depth of the profile；unit：m）

The western Xiyang channel is a young stage erosional type tidal channel formed by continuous erosion on the original land-and-sea bottom. The present channel bottom has many erosion features, and the erosional troughs can be several hundred meters wide and 1～2 m deep.

3　Formation and evolution of radiative sand ridge field of the Yellow Sea

3.1　Formation conditions

The radiative sand ridge field is developed along the western coast of the southern Yellow Sea as a result of ideal environmental combination. The dynamic factors and sediment materials are the related process patterns of "transgression-dynamic-sediment". The Yellow Sea is surrounded by Korean Peninsula, Liaodong Peninsula, Shandong Peninsula and northern Jiangsu coast from east to west to form an open bay environment. The progressive pacific tidal waves have changed in such environments, and the sea level rising intensified the tidal dynamics and systematic current process. It forms a dominant tidal dynamics sea area in the world. The progressive tidal waves of the Pacific Ocean run through the East China Sea toward the Yellow Sea, then they are blocked by the Shandong Peninsula to form reflected tidal waves from north to south. The following progressive tidal waves of the Pacific converge with the continuing reflected tidal waves at Jianggang to form convergent tidal current fields with large tidal bore and tidal ranges[①]. The average tidal range is 4. 18 m at Jianggang, and gradually decreases on both sides toward south and north. In the Huangshayang area, the center of the convergence tidal fields, the maximum tidal range recorded has been 9. 28 m along the China sea [4]. There are regular semi-diurnal tides in the radiative sand ridge field and linear tidal currents in the tidal channels. During the flood tide period, tide converges in the whole area. The average current velocity is 1. 2~1. 3 m/s, the average current velocity of the diverging tide during ebb tide period is 1. 4~1. 8 m/s. The strong converging tidal fields are the dynamic factor to form radiative sand ridges. The large sand ridges such as Jiangjiasha ridge, Dongsha ridge, Lengjiasha ridge and Maozhusha ridge exist with the late Pleistocene river valleys distributed in the nearby sea bottom, the river sand is important material resources for sandy ridges development. The old Changjiang River ran through northern Jiangsu and entered the sea through Jianggang, Xinchuangang and Yaowanggang during the end of the late Pleistocene. The river accumulated thick sand layers there, which gave materials to develop tidal current ridges. The formation of the ridge and trough landscape passed several stages: the end of the late Pleistocene through early Holocene, middle Holocene, late Holocene to the present. The stages indicate that the tidal dynamic was strengthened during the sea level transgressions; optimum water depth of the sandy ridge development is around 10 m, which is under the wave dynamic fields. The maximum water depth does not exceed 20 m, even though the far sides of the late Pleistocene sand ridges are still located in the area of −20 m. This situation indicates that the tidal current is a dominant dynamic agent to form channels and sand ridges. However, the wave actions modify the form of the sand ridges by winnowing and redistributing the sands. The seismic records also show that the early-stage sandy bodies are large and are distributed in a limited area. That is why the sandy ridges are broad and flat which is different from the linear tidal current ridges [14~16]. Thus, radiative sand ridges are large inner continental shelf sandy bodies formed by the dynamics of the tidal current combined with wave actions. As a whole, the sand ridge field is completed during the high sea level of the Holocene.

3. 2　Formation processes

The radiative sand ridge field is a large bottom geomorphological system. It consists essentially of three parts of radiative tidal current ridges in the northeast, reformed old river depositional bodies in the central and southern parts, and eroded-depositional sandy bodies such as large ridges in the northern part, outer part, and of

---

① Zhang Dongsheng, Zhang Junlun, Zhang Changkuan et al. , Formation of tidal currents-destruction by storm surge-recovery of tidal current-Dynamic mechanism of the evolution of the subterranean radiative sand ridge fields in the Yellow Sea. 1997.

small ridges out of the large one. The differences of the three parts are caused by the genetic process, sediment resources(quantity of sediment supply or the distance to the sediment provenance), and the development stages.

3.2.1 The difference of the genetic processes.

The tidal channels or troughs between the ridges are the major geomorphological component of the radiative sand ridge field. It has three different types.

1) The tidal channel developed along the old river valley type, such as the Huangshayang channel and Lanshayang channel.

From the inner part to the end the channel still preserves different stages of the old river valley of the end of the late Pleistocene to the present. The old Changjiang River valley during the late Pleistocene was smaller than at present. The channels of old river valley type may indicate that there were three or four tributaries during that time, among which the Huangshayang ridge may be the major one.

2) Patrimonial valley type.

There were mature valleys or depressions during the end of the Pleistocene or the early Holocene. Even though the old valley was buried by the sediment, the locations of later stage valleys are still similar to the buried old valley, which means that there are always the low land areas, even though the locations of the later valleys are slightly different from the buried valley, or across to each other. Basically, they are buried valleys underneath the present tidal channels, such as the late Pleistocene valley located 38 m underneath the Dabei trough, and the old valley stretching out in an angle direction with the Dabei trough. In the Chenjiawucao trough, there is an old valley buried under 15 m thick sediment. The depth of the old valley is greater than 25 m. In the Caomishuyang area, there is a V-shaped valley buried under 25 m thick sand sediment. The depth of the valley bottom reaches 30 m. Even this old valley was buried somewhere under the large sand ridges such as Maozhusha and Waimaozhusha, because the valleys were the provenance of the sandy material of the sandy ridges. Xiaomiaohong channel is a complex type: its present, inner part is formed in the erosional process during the Holocene period. But the middle part under the present channel bottom is buried old river valleys, in the direction obliquely crossing the present channel. It is estimated that the old Changjiang River had migrated from north to south, passing through Yaowanggang channel exit to the sea. These buried valleys underneath the present Xiaomiaohong channel were the remnants of the old Changjiang River. These valleys were also the provenance of the sandy material to support the development of the large sandy body of Lengjiasha ridge.

3) The type of tidal current erosional trough of the Holocene period.

The feature is indicated by the erosional pot holes, deep ponds, and erosional depressions distributed on the bottom of the present tidal channel, such as Xiyang channel and Kushuyang channel. These two channels were formed during the later part of the middle Holocene sea-level transgression.

3.2.2　The difference between sand ridges genetic processes.

The common features of genetic processes of the sand ridge are that the large sand ridges exist along with the buried valleys, the core part of the present large sand ridges consist of the buried late Pleistocene sand bodies, and the upper part is superimposed by the Holocene deposits. The differences between the sand ridges are as follows.

1) The Jiangjiasha ridge is a large sand ridge, which consists of late Pleistocene sand ridges superimposed by the Holocene ridge. The sand ridges of the end Pleistocene presently located 35~45 m under the sea bottom are large in size (3 km wide, relative height reaches 20 m), indicating that there was abundant sand supply during that time in that area. The old sand ridges gradually decreased its size during the early Holocene and the middle Holocene.

2) The Lengjiasha ridge is composed of superimposed sand ridges. The late Pleistocene ridge is large in size and is buried 30 m under the present sea bottom. However, sand ridges stayed in the same conditions during the middle Holocene (16 m).

3) The Waimaozhusha ridge is a large sand ridge which was formed during the end of the late Pleistocene but underwent erosion during the early and middle Holocene, only superimposed by the clay and sandy material during the late Holocene period. At present, the Waimaozhusha ridge has retreated back towards land by erosional process. The main part of the Maozhusha ridge, consisting of late Pleistocene ridge, then suffered from erosion. Only the inner parts of the Maozhusha, such as the areas of Tiaozini, Shichuanyan, and Zhugensha ridges, have received accumulation during the Holocene time. The thickness of the sand layers in the ridge reaches 30 m.

4) Liangyuesha ridge is a mid-Holocene accumulated sand body; Taipingsha ridge is the late-Holocene accumulated sandy body. The thickness of the Holocene sediments is more than 40 m, implicating that there was a new sediment resource in the north during the Holocene time.

5) Dongsha ridge is located on the land side of Liangyuesha ridge and Taipingsha ridge. The large sand bodies consist of late Pleistocene ridges superimposed by the Holocene ridges. It received the offshore erosion sediments and gradually merged with the coastline.

6) Xiaoyinsha Ridge is small in size, with its bottom as the remnants of the late Pleistocene period, and is superimposed by thicker sandy strata of the middle and late Holocene period.

3.2.3　The geomorphological evolution of the ridges and channels of the radiative sand ridge field.

1) The southern and axial parts of the radiative field are under the control of old rivers, and all the ridges and troughs distributed along the old river valleys. But the northern part and north-eastern part are under the control of the radiative tidal currents, the ridges and troughs distributed in the directions of the ebb tide currents. From the distribution pattern we may understand the genetic formation process.

2) The sediment of the radiative sand ridge field is mainly fine sands and silts, originally from the old Changjiang River deposits. The late Holocene sand straum is well preserved laminae beddings with more clay particles reflecting the influences from the Yellow River. The present silt and clayer materials are the sediment supply both from the Changjiang River and abandoned Yellow River delta. The fine sand, especially during the late Pleistocene period, was a major component in the formation of the sand ridges mainly from the old Changjiang River sediments. In the northern part of the radiative sand ridge field, the sandy material increased during the later part of the middle Holocene time, but not in the southern part. This indicates that the northern part had a new sand source during that time. Whether the sand came from the Yellow River or from the old Changjiang River needs further studies.

3) Old river valley type and patrimonial valley type channels and old sand ridges are relatively ancient. They might be formed during the late Pleistocene high sea-level period (around 30 ka B. P., at that time the sea level was at -20 m) or during the end of the late Pleistocene and early Holocene transgression. Erosional tidal channels and smaller sand ridges inside the big sand ridges with younger ages might be formed during the later part of the middle Holocene. However, determination of the time periods of the developing sand ridge field still need further studies on the core samples from that area. Presently, the dating of the time is mainly by relying on the evidence of the structure and layer depth of the sediment strata, compared with the boundary features of the seismic profile and study on the regional sediment strata. (i) The development of axial part of the radiative sand ridges started the earliest. The sand deposits may have started at 30 ka B. P. para-interglacial high sea level. The sea level at that time was located 20 m less than the present, and the Changjiang River entered the sea from the area of Haian and Libao[①][17]. The coast zone at that time was to the east of the present coast. The buried depth of the late Pleistocene sand ridges is -30 to -40 m. The location matches well with the fine sand structures of the coastal cores, ancient sea level, and optimum water delta of $10\sim20$ m for sand ridges development. (ii) Time periods of the northeast region and the south region development are later than the axial region, they are related to the old Changjiang River sand supply from Libao,

---

① Jia Jianjun, Zhu Dakui, Evolutionary mechanism of the Changjiang River channels into the sea, 1996.

Yaowanggang area during the Holocene time. (ⅲ) Ridge and trough landforms of the northeast and south regions were mainly formed during the Holocene high sea level. (ⅳ) The northern part of the radiative sand ridge field is formed last, during the late Holocene period and the present time. All the erosional periods of the sediment strata fit well with the changing sea-level curve of the Jiangsu muddy plain coast since the late Pleistocene[①]. The initial Holocene erosional boundary may be the result of the 10 ka B. P. transgression, and the early Holocene erosional surface may be correspondent to the 8 ka B. P. High sea-level. The middle Holocene erosional surface may be correspondent to the process of 6 ka~5 kaB. P. high sea leve. The late Holocene erosional surface may be correspondent to the period of 1 ka B. P.

4　Conclusion: the evolution regularity of the radiative sand ridge field

The ridge-trough imposed records have been seen several times in the seismic profiles. This phenomenon indicates the developing process of the sand ridges has a periodic nature: the period of sediment accumulation and the period of erosional formation; that is, the river deposited the sandy bodies and sandy sediment strata as the first period. Then, tidal current eroded sandy bodies and sandy layers to form ridges and separate channels, and also transported the sediments to produce radiative pattern of tidal current regions as the second period. The initial sand ridges reacted with wave action of winnowing, erosion and redistribution of the sediments horizontally to change the linear tidal current ridges to be broader and belt-shaped sand ridges.

The river-sea interactive process of the Yellow Sea area is closely related with the climate and sea-level changes. The climate change is the initial cause of sea-level changes. The rising and falling of the sea level is a key agent to start the coast zone land-sea interactive dynamic process. Thus, the fundamental factors of sand ridge fields formation are the "transgressions-dynamics-sediments".

During the glacial and cold epoch, the sea level fell, the stream was down-cutting, bringing abundant coarse sands into the sea. During the low sea-level time, most parts of the present Yellow Sea were dry. There were no water dynamic conditions for forming of radiative tidal current fields. As a result, the low sea-level epoch was a period of sediment accumulation.

During the para-interglacial epoch and post-glacial epoch or the warm epoch, the sea level approached 20 m below the present level. The Pacific tidal waves could have

① Zhang Yongzhan, Study on environment changes of tidal flat coastal zone of Jiangsu Province. 1996.

been transported to the Yellow Sea; the bay-shaped coastline, especially the Shandong Peninsula, has blocked the progressive tidal waves from the Pacific to form reflected tidal waves from the north. The reflected tidal waves converge with the later advancing tidal waves to form radiative current fields. As a result, this offers the dynamic conditions for developing tidal current ridges. The continuing rise of the sea level strengthened the tidal current and wave actions. The currents eroded the thick layer of sandy deposits and furthermore reformed the sandy bodies into the geomorphological complex of large sand ridges and current channels or troughs distributed along the radiative current fields. If there was no tidal channel, then there would not exist the tidal current sand ridges. Therefore, the tidal currents eroded the old sandy bodies, transported sediments and deposited it to from radiative sand ridges. It is "the moulding period of sand ridges and tidal channels". It must be pointed out that without the wave actions of winnowing and redistributing the sediment, i. e. "flatting" and "broadening" process, it is impossible to form large, wide sand bodies. It is also the genetic reason for developing the tidal current sand ridges in the coast wave active area. As a result, the belt-shaped sand ridges are different from the linear tidal sand ridges.

Therefore, sands were deposited during the low sea-level period, and the ridges and troughs were carved out during the high sea-level period while the sandy sediments were being eroded by currents and wave actions. In such a special environment of the South Yellow Sea and historical geological conditions, the large-scale radiative sand ridge field has been completed.

The continued rise of the sea level would cause the erosion of the sandy ridges, thereby decreasing the size and height of ridge fields. The outer part of the sand ridge fields will gradually diminish through the process of erosion. The sediments washed away will be carried by the currents to accumulate on shore. The long shore tidal flats would be extended by this process and eventually the ridge field will connect with the coast land. This evolutionary process will be long and slow because during the last hundred years the average rate of sea level rise has been only 1 mm/a. The formation process described above can be summarized in Fig. 7.

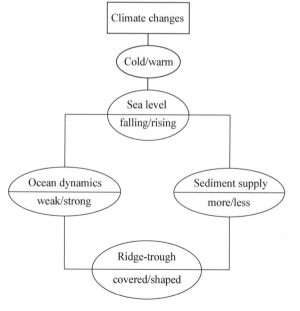

Fig. 7 The formation process of the radiative sand ridge field

References

[1] Collins, M. B., Shinwell, S. J., Gao, S. et al. 1995. Water and sediment movement in the vicinity of liner sand banks: the Norfolk banks, southern North Sea. *Marine Geology*, 123(3~4): 125 – 142.

[2] Off, T. 1963. Rhythmic linear sand bodies coused by tidal currents. *Bull. AAPG*, 47(2): 324 – 341.

[3] Milliman, J. D., Meade, R. H. 1983. World-wide delivery of river sediments to the oceans. *Journal of Geology*, 91(1): 1 – 21.

[4] Ren Mei'e ed. 1986. The Comprehensive Investigative Report on the Resources of the Coastal Zone and Tidal Flats of Jiangsu Province. Beijing: China Ocean Press.

[5] Wang Ying, Ren Mei'e, Zhu Dakui. 1986. Sediment supply to the continental shelf by the major rivers of China. *Journal of Geological Society*, *London*, 143(6): 935 – 944.

[6] Zhu Dakui, Fu Minzuo. 1986. The preliminary study on the offshore radiative sand ridges of Jiangsu Province. In: *Collection of Comprehensive Survey of Dongsha Ridges in Jiangsu Coastal Zone*. Beijing: China Ocean Press. 28 – 32.

[7] Wang Lachun, Chen Xiaoling, Chu Tongqing. 1997. A contrast analysis on the loads character of the Changjiang River and the Yellow River. *Geographical Research*, 16(4): 71 – 79.

[8] Zhu Dakui, Xu Tingguan. 1982. The problems of coast development and utilization in the middle pan of Jiangsu Province. *The Journal of Nanjing University* (*Natural Science Edition*), (3): 799 – 818.

[9] Wang, Y., Collins, M. B., Zhu, D. K. 1990. A comparative study of open coast tidal flats: the Wash(U. K.), Bohai Bay and West Huanghai Sea (mainland China). In: *Proceedings of ISCZC 1988*. Peking: China Ocean Press. 120 – 134.

[10] Dalrymple, R. W. 1992. Tidal depositional systems. In: *Facies Models Response to Sea Level Change* (eds. Walker, R. G., James, N. P.). Geological Association of Canada, Love Printing Service Ltd. 201 – 231.

[11] Zhu, D., Martini, I. P., Brookfield, M. E. 1998. Morphology and land-use of coastal zone of the North Jiangsu Plain, Jiangsu Province, eastern China. *Journal of Coastal Research*, 14(2): 591 – 599.

[12] Zhang Zhonggu, Zhou Mushi, Shao Shixiong et al. 1990. The Quaternary Geological Map of People's Republic of China and Nearby Sea Area (1 : 25 000) and its Direction. Peking: The Cartographic Map Press of China.

[13] Chen Baozhang, Li Congxian, Ye Zhizheng. 1995. Deposition and its environmental evolution in north part of the Changjiang River delta during post-glacial period. *Acta Oceanologica Sinica*, 17(1): 64 – 75.

[14] Swift, D. J. P. 1975. Tidal sand ridges and shoal retreat massifs. *Marine Geology*, 18(3): 105 – 133.

[15] Swift, D. J. P., Parker, G., Landfredi, N. W. et al. 1978. Shoreface-connected sand ridges on American and European shelves: a comparison. *Estuarine and Coastal Marine Science*, 7(3): 257 –

273.

[16] Swift，D. J. P. ，Field，M. E. 1981. Evolution of a classic sand ridge field: Maryland sector，North American inner shelf. *Sedimentology*，28(4): 461 – 482.

[17] Yang Huairen. 1996. Type and mechanism of lower Changjiang River migration since the Late Pleistocene. In: *The Study on Environment Change*. Nanjing: Hohai University Press. 195 – 211.

河海交互作用与黄东海域 古扬子大三角洲体系研究[*]

一、序　言

河海交互作用与堆积型大陆架的发育是中国海的特色。发源于青藏-云贵高原的大河之中,有 5 条(长江、黄河、澜沧江-湄公河、红河、西江-珠江)流入中国海(China Seas)。现代中国边缘海大陆架在晚更新世时期曾是海岸平原,海平面的大幅度变化,导致岸线的大范围退进,河流的长距离伸缩和大尺度迁移,河-海交互作用是形成平原海岸与浅海输积泥沙的主要作用过程[1,2]。源自隆升高原与山地的河流不断下切并自西向东,或折向东南,从亚洲内陆向东南海域输送了巨量泥沙,形成堆积大平原,填充着边缘海,发育着堆积型大陆架[1,3]。亚洲-太平洋北部的边缘海(鄂霍次克海、日本海),因无大河泥沙汇入而未能形成堆积大陆架;南亚虽有大河汇入,但由于孟加拉湾沉陷,恒河入海泥沙滑落而发育了深海扇[4]。因此,只有我国黄、东海因大河泥沙汇入陆缘浅海,堆积为广袤的大陆架而独具特色。最近研究表明[5]:东海大陆架与南黄海大陆架实由"古扬子大三角洲沉积体系"(Paleo-Yangtze Grand Delta System)所组成。它是由源于中国内陆的长江与黄河历经若干万年长期侵蚀、搬运、输入边缘海的泥沙,经季风波浪及潮流动力相互作用形成的巨大堆积地貌体系。它包括:基底的古扬子大三角洲、上叠的古江河三角洲、南黄海辐射沙脊群、全新世-现代长江三角洲以及历史时期的废黄河三角洲 5 个部分[5]。这一巨大的三角洲复合体系,是研究晚更新世以来,东亚-太平洋边缘海之河-海交互作用最佳的地质载体。陆海交互作用及环境效应与人类生存发展密切相关,是当前地学领域的前沿课题,备受国内外学术组织与研究计划的重视。

二、前人研究

前人在相关论著中曾提到"扬子浅滩"或"扬子大沙滩"[6,7]、"长江口大浅滩"或"长江大沙滩"[8,9],但尚未谈及大三角洲体系。

陈吉余等[10,11]最早在对长江口外水下地形分析中肯定地提出"长江口外具有显著的

＊　王颖,邹欣庆,殷勇,张永战,刘绍文:《第四纪研究》,2012 年第 32 卷第 6 期,第 1055 - 1064 页。

2012 年以中文摘要载于:《中国海岸与海岛资源环境学术研讨会会议日程与论文(摘要)汇编(2012 年联合年会)》,1 页,厦门,2012. 10. 13 - 21。

国土资源部公益性行业科研专项项目(批准号:201011019)与海洋公益性行业科研专项项目(批准号:201005006)共同资助。

呈扇形分布的古代水下三角洲,面积达 7 000 km², 前缘水深在海面下 50 m, 它的平面中心在 32°18′N, 与现代长江三角洲主泓——南泓道的出口显然不符合, 说明这个水下三角洲的长江主泓在现代主泓之北, 并具有一定的稳定性。从等深线分布看来, 这个古代三角洲之上, 还叠复着一个近代水下三角洲, 其前缘约与-10 m 的等深线相符, 却与古代三角洲不符, 说明长江的水下三角洲在发展过程中是向南移动的"。任美锷[12]在《江苏省海岸带和海涂资源综合调查报告》中明确提出"辐射状沙脊群叠置发育在古黄河和古长江三角洲上", "古长江三角洲延伸于东海和黄海之间, 其外缘有明显的地形坡折, 坡折线水深为40 m 左右。古三角洲表层大部分被残留细砂所覆盖, 厚度可达 0.5~1.0 m 以上, 其下为较老的河湖相黏土等沉积。残留砂研究表明, 古长江三角洲在玉木冰期末—冰后期初形成和后退中, 遭受了古滨岸海滩化动力改造", 此分析深入。国家海洋局第一海洋研究所和中国科学院地理科学与资源研究所 1984 年出版的《渤海黄海地势图》[13], 展示出古长江中期三角洲和古黄河、古长江早期三角洲。古长江中期三角洲保存尚属完整, 其外界范围在北部为-20 m, 在中部辐射沙脊群外围为-45 m, 在东南部古三角洲远端的外围水深为 50 m。秦蕴珊[14]在《黄海地质》研究中, 明确指出"旧黄河—古长江复合三角洲地貌位于长江口以北, 射阳至弶港以东海域, 其外缘界限北面至-30 m 等深线, 在东及东南面为-60~-65 m 等深线"。Liu 等[15]通过浅地层剖面和 4 个钻孔推测了南黄海有一个大型古三角洲的存在, 时代为 MIS 3, 向海分布可达-50 m 等深线。李全兴[16]认为位于长江口外东侧的扬子浅滩与北部的废黄河三角洲这片不规则的扇形隆起(-50 m 等深线)是古长江—古黄河复合三角洲。金翔龙[17]在《东海海洋地质》中明确界定了长江口从北向南迁移过程与现代三角洲之发育。刘振夏[6]指出扬子浅滩水下外界是水深 25~55 m。上述表明, 从 20 世纪 50 年代以来, 前人研究多次提出在黄海、东海之间存在着古长江与古黄河的巨型复合沉积体, 有些学者已提出水下复合三角洲概念, 但所划定的范围与外缘水深不同, 尚未涉及到水下复式三角洲的整体结构与范围。此外, 对于古三角洲沉积体系的成因也存争议。一般认为是陆架残留的前期沉积, 属晚更新世末期海退沉积, 虽经海侵作用影响, 但由于物质来源匮乏, 未被覆盖, 尚保留原来的沉积特征[14,17]; 也有学者[6,18]认为扬子浅滩不是古长江三角洲, 也不是陆架残留沉积, 而是典型的现代潮流沙席。

国际上, 对大河的研究, 反映出不同的河海相互作用效果: 密西西比河汇入沉降中的墨西哥湾, 泥沙滑落入深海, 只在沿河流口门两侧堆积了鸟足状三角洲[19~22](图 1), 且由于河流输沙量的减少及海平面上升, 该三角洲进一步沉溺[23]。亚马逊河流量大, 但向海输送的泥沙量未及长江、黄河丰富, 其泥沙主要输入大西洋, 并沿海底堆积为深海平原[24~27]。只有中国既有发育于岛弧-海沟系后侧的边缘海, 又有来自亚洲内陆大河输入的泥沙, 才能形成由大三角洲体系组成的宽阔的堆积型陆架以及孕育于沉积层中的油气矿藏等。

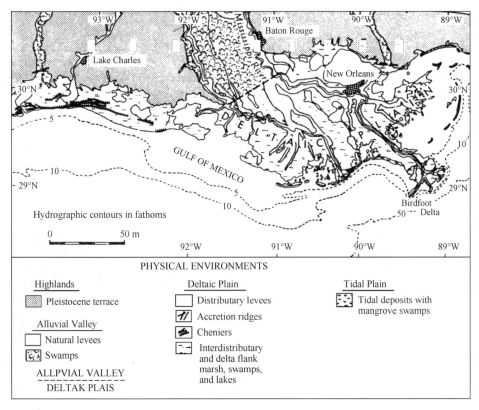

图 1　密西西比河现代鸟足状三角洲[22]

三、科学问题的提出

南京大学海岸地貌与沉积研究室成员在"江苏海岸带调查"时(1980—1984 年)认识到:"江苏的潮滩主要在淤涨,是新生的土地资源,淤涨显著的岸段,位于南黄海辐射沙脊群波影区——射阳河口向南至吕四段,与沙脊群的蔽浪作用有关。"1993—1996 年,本文作者在从事"八五"国家自然科学基金重点项目(No. 49236120)"南黄海辐射沙脊群调查"时,发现:"南黄海辐射沙脊群是全新世海侵由潮流改造的古长江三角洲堆积体","沙脊群北部的废黄河水下三角洲受蚀,细颗粒物质沿岸向南输运;沙脊群中部枢纽区主潮流通道是承袭古长江河谷发育的;伴随着现代海平面上升,在波浪作用下,沙脊展宽的同时,沙脊群形成沿岸潮滩淤积的重要泥沙来源。"[28]但是,在海平面持续上升的背景下,至 2006—2010 年"十一·五"期间"中国近海海洋综合调查与评价"(908 专项)调查发现,辐射沙脊群有冲有淤,或时冲时淤,总体上却趋于蚀淤动态平衡,并未出现整体规模的显著缩小或向岸迁移。显然,近海海域存在巨大的沙源供给。先期研究启迪我们向外海探索,找寻新沙源。在从事"中国近海海洋综合调查与评价"专项研究中,对辐射沙脊群进行了多孔钻探,进一步了解到古长江在江苏海岸的出口较已知的黄沙洋—烂沙洋(辐射沙脊群中部枢纽区)更北;而黄河不仅在历史时期曾直流入黄海,在辐射沙脊群北部钻孔揭示的晚更新世沉积中,多次出现较厚的黄色粉砂与黏土沉积。结合遥感与测量的海底制图中所提供的海底地形地貌特征信息,以及在《江苏省地图集》[29]的遥感影像图中清晰展现出——在辐射沙脊群外侧的南黄海内陆架上还有一个更大的三角洲,分布范围覆盖整个苏北海域,北部达到海州湾,宽度小,外围水深约 30 m;中部宽大,位于辐射沙脊群外围,水深 30 m 及 50 m,陡坡界限明显,向南宽度变小,延伸至长江口外,呈现受蚀残缺形态;该三角洲外缘-30 m 与-50 m 等深线形成地形转折,界限明显,显然为古长江与古黄河加积而成。

南黄海海域开阔,西与苏北平原西侧地质时期的新生代古海湾相衔接[30]。中国的两条大河——黄河、长江也是世界级的大河,曾先后在南黄海海域入海,堆填了古海湾为苏北平原,在现代海岸线以外的陆架海域 30~50 m 水深处遗留着曾由长江、黄河相互堆积的大型三角洲体,可定为古江河大三角洲,其上依次叠置着辐射沙脊群、全新世—现代长江三角洲及废黄河三角洲(图 2)。

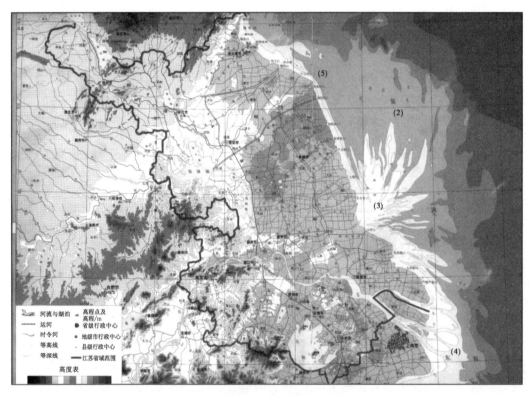

图 2 南黄海古江河大三角洲(2)[29]

图中:(3) 辐射沙脊群;(4) 全新世—现代长江三角洲;(5) 废黄河三角洲

古江河三角洲体,大体上以弶港(32°39′50″N,120°54′25″E)为中心,向东延伸至-30 m 和-50 m 等深线处;东北侧为三角洲的突出部分,地理位置约在 34°17′43″N,122°14′14″E;北端可达连云港区东西连岛以东的外海(34°45′58″N,119°41′25″E)及 30 m 水深处(34°50′8″N,119°56′58″E);达山岛、车牛山岛当时为海岸带的岛屿,而如今距现代海岸线为 66.99 km 和 55.37 km,该三角洲的东南侧地理位置在 32°0′18″N,122°43′59″E,以东可达 50 m 水深;而南端定位在长江口南岬(30°52′26″N,121°52′55″E)。大体上以-30 m 等深线为外界,将上述各外缘点链接,呈现出一幅"完整"的褶扇扇面(图 2)。其面积量计的结果是:-30 m 等深线所圈定的范围约为 65 330 km²,-20 m 等深线以内面积约为51 330 km²，−20 m和−30 m等深线与古江河三角洲体的北部和东北部外界基本吻合，但在中心枢纽部位，因43 000～35 000年前古长江在黄沙洋—烂沙洋地区出口侵蚀缺失范围较大[28]；而−5 m等深线，即：岸滩与沙脊外界明显处的范围约为11 640 km²。这个三角洲体明显地表现在海底地形分层设色图上（图2），其范围远比现代的长江三角洲、废黄河三角洲大，比南黄海辐射沙脊群的面积大出3倍。该海域底质主要是细砂、粉砂、砂质粉砂、粉砂质砂。北部多黏土质粉砂，富含结核（粒径长约数毫米至数厘米，钙质结核和软体动物遗壳），表明系黄河携运之泥沙。细砂之重矿组成表明沉积物源主要来自长江[28]。

在三角洲体北部叠置着废黄河三角洲，中枢地区叠置着南黄海辐射沙脊群，南部叠置着全新世—现代长江三角洲。本文拟将此复合三角洲定名为南黄海古江河三角洲体。三角洲体形成的年代尚未获得系统的年代测定，但是，其形成必然在上叠的三角洲之前。据辐射沙脊群各钻孔定年资料[28]，目前已知的年代是>43 000年。因此，古三角洲体必然比43 000年更老，应当是在晚更新世低海面时期，当古长江在苏北入海时，因寒冻风化强而形成的泥沙（细砂）巨大。海区钻孔的沉积中多次出现黄色粉砂-黏土沉积层，据此初步认为：自晚更新世以来，黄河曾多次从苏北入海，遗留下黄土质沉积层。而1128—1855年黄河夺淮入黄海仅是历史时期最近一次夺淮入黄海，而非唯一的一次。

最近研究发现，古江河三角洲体是叠置在外围的一个更大的三角洲上。这也是辐射沙脊群现今在陆地江、河泥沙补给已急剧减少，而海平面日益上升的背景上，仍能保持蚀、积动态平衡之重要原因，即海域外围有大三角洲的砂质补给。从遥感影像与海图分析，该大三角洲范围北起灌河口（34°30′N），南达马祖列岛南部（26°N），东部海域界限相当于−100～−150 m等深线，接近东海陆架边缘。大体上，以现代长江口与杭州湾为轴心，呈230°弧形向海扩展，覆盖黄东海大陆架，面积约38×10⁴ km²，基本组成物质是细砂与粉砂，是全球罕有，尚保留于大陆架的浅海地貌，可定名为"古扬子大三角洲体系"（图3）[31]，此一完整的大三角洲地貌在南黄海—东海海底地形

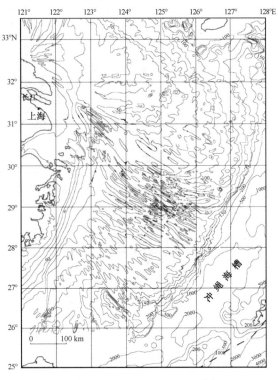

图3　古扬子大三角洲体系[31]

图中得到印证（图4）[32]。以它为基底，上面叠置着古江河三角洲体，形成于3万～4万年前的南黄海辐射沙脊群，全新世—现代长江三角洲，以及历史时期的废黄河三角洲，泥沙与水动力互有关联，五部分组成巨大的三角洲复合体系，至今尚未进行过系统研究。古扬

子大三角洲体系是东亚—太平洋边缘海河海相互作用的最大地质载体,保留着中、晚第四纪以来河海交互作用、海陆变迁之环境变化的重大信息,探索其形成发展,将是对地球科学理论的重大贡献。大三角洲体系的组成物质,充分反映了来自我国内陆的泥沙,在地质历史长河中的积累,将是大陆架划界归属的有力证据。故以此文阐明开展一体化研究的意义与首次全方位工作的建议。

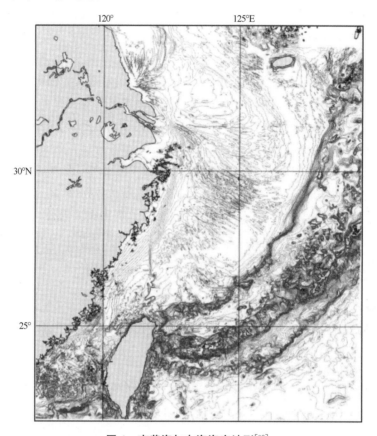

图 4　南黄海与东海海底地形[32]

四、研究设想与建议

(一) 大河入海泥沙的沉积形式和实例

纵观世界大河入海泥沙的沉积形式,各具特点,与海区动力活动强弱、海底原始地形结构、陆源输沙量及泥沙粒径性质有关。

1. 黄河

黄河携运来自黄土高原与下游冲积平原的细粉砂与黏土,以悬移质量大。随径流入渤海湾与莱州湾浅海,该处潮差小,波浪力弱,黄河入海泥沙中的细粉砂与少量极细砂首先淤积于河口两侧以突出沙嘴状(亦被称为鸟嘴状)直伸入海,堆积迅速(图5);而悬移质黏土除向深水区扩散外,很大部分被沿岸堆积于河口两侧因沙嘴外突而形成的凹湾中,成为典型的烂泥湾[3]。当入海河道变动,新沙嘴与另一对烂泥湾发育时,则旧河口受蚀,岸线后退,烂泥湾被粉砂所掩盖。随河道的多次变迁,而发育了扇形三角洲(图6)。由于人工限制河道北迁以保护油田与东营市,促使河口向南偏转。又经人工引导河流向北湾出口,并由于沿河流筑坝蓄水及引水灌溉,入海径流量与泥沙量均减少,河口沙嘴伸展变缓,三角洲体遭受侵蚀。

图5　现代黄河三角洲遥感影像①(附彩图)
(新改道的黄河口鸟嘴状三角洲及旁侧烂泥湾)

图6　黄河口偏移与扇形三角洲(附彩图)
(据黄河口遥感影像)

2. 密西西比河

密西西比河为富含细砂与黏土质粉砂河流。汇入水深较大的墨西哥湾,入海泥沙除向深海扩散外,主要沿河道两侧堆积成鸟足状三角洲瓣。随着下游河道的侧向摆动与向深水延伸,逐渐发育了串珠形分布的簇状三角洲瓣体(图7)。

---

① 据东营市海洋与渔业局提供的遥感图像,2010。

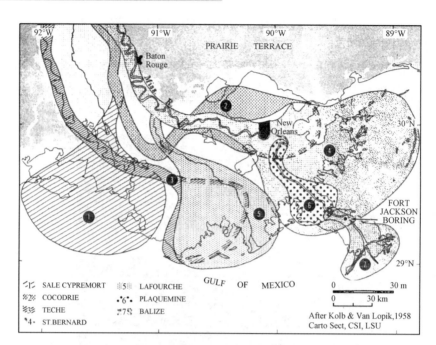

图7 密西西比河三角洲瓣体[22]

3. 恒河与印度河

恒河与印度河分别汇入孟加拉湾与阿拉伯海,处于喜马拉雅隆起的山前拗陷带,水深大、坡度陡,泥沙入海滑落于深水区形成深海扇,而陆架堆积带不宽(图8)。

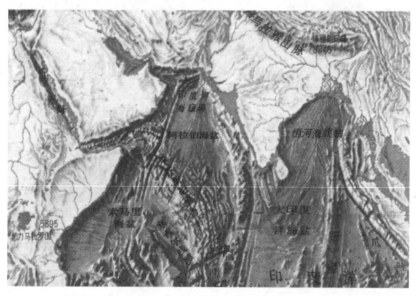

图8 分布于印度洋海盆的恒河海底扇与阿拉伯海盆的印度河海底扇[33](附彩图)

4. 亚马逊河口

亚马逊河口河流径流量丰盛(年均 $69\,300\times10^{8}\ \text{m}^3$),因途经茂盛的热带林区,泥沙含量相对较少,加之,外海分布着与海岸大体平行的断陷谷地,泥沙滑落处,沿河流入海外缘发育了不宽的堆积陆架(图9)。

图9 亚马逊河口外侧陆架(图片来源于 Google Earth)(附彩图)

上述大河入海堆积体各具特色,与各河流水、沙特点及海域动力及海底结构密切相关,相比较之,则古扬子大三角洲体系独具特色:位居长江口外南黄海—东海边缘海中,外侧有日本—冲绳—中国台湾岛弧围栏;岛弧内侧边缘海为构造沉陷带;长江源远流长,贯通中下游汇入东海,可能在中晚更新世,历经超过10余万年的泥沙搬运与沉积过程,发育了巨大的三角洲体,构成东海陆架的主体沉积层;其上叠置着中、晚更新世以来各时期海平面变化的三角洲遗证,以及全新世海侵所塑造的潮流沙脊群地貌。这一丰富的海底大三角洲体系,实为地质奇迹。研究其范围、组成结构、成因机制、现代动力过程与发展变化趋势,具有重大的科学意义与应用价值。

(二) 研究步骤与专题研究内容建议

本文研究的对象是遍及南黄海和东海海域的巨型海底地貌体系,蕴藏着中、晚更新世以来气候与海平面变化的遗证以及其生态环境效应。如何进行研究? 关键是:组织多学科结合的系统一体化调查,掌握全面细致的客观观测资料;分别进行深入细致的科学实验分析;总结实证结果,做出一步又一步的深入的成因演变分析与科学理论总结;并进一步应用于经济与国防实践。本文提出研究工作内容与步骤不仅是此专项研究的建议,而且是希望可与同类海岸海洋调研工作相互衔接与比较。

首次调查研究古扬子大三角洲体系,需阐明其分布范围与体系的各个组成部分。以大范围的海底测量(测深与底质)工作作为基础,关键是统一基准面,统一比例尺,统一工作方法与严密的互应校正。目前研究可采取:汇集已有的区域海图资料,在统一基准面、叠加分析的基础上,布置若干垂直于三角洲体的控制断面;进行连续的地震剖面测量与平面控制的测深点底质采样。由于长三角、辐射沙脊群及废黄河三角洲已有前期工作基础，本次调研，宜先着重东海与外海，然后内延。

其次，平面展布测深点与底质资料，并插入控制断面的地震剖面，加以研究分析后，再确定控制性钻孔的位置与进尺深度，多船次、同步海洋水文与水动力断面观测，与海洋水文资料汇集后的补充性走航观测等。

海上勘察工作的同期与后继进行实验室各项分析，主要涉及：钻孔沉积柱（或岩芯）的沉积相分析；沉积物粒度、重矿组合、石英砂表面结构、包裹体及地球化学元素等分析与鉴定；生物化石与微体古生物鉴定与测年；水位与水深订正，浪、潮、流等观测数据计算、定性与建模等。在海量的数据资料基础上，才能进行客观的总结，分析南黄海-东海陆架形成发育的最后阶段历史与发展规律。

通过上述调查研究可望达到：

（1）阐明古扬子大三角洲体系地貌与沉积结构特征，古气候与海平面变化的地质记录与大三角洲体系发育历史过程及趋势性分析。

（2）江、河水沙输入与黄—东海海洋动力相互作用与物质交换特点，沉积动力过程、机制与海岸变迁及陆架堆积效应。

（3）初步建立河海交互作用过程与边缘海沉积模式规律。

（4）古扬子大三角洲海岸海洋开发利用与生态、环境保护的相宜途径探讨。

（5）东海陆架沉积泥沙来源、海岸变迁与海洋疆界划分的依据。

理论与应用并重为研究目的，全方位调查与多学科结合的研究途径是达到预期目标的重要手段。开展本项研究可在"十二·五"科学规划期间为我国国力增强做出重要贡献。

参考文献

[1] Wang Ying, Ren Mei-e, Syvitski James. 1998. Sediment transport and terrigenous fluxes. In: Brink Kenneth, Robinson Allan eds. The Sea (Vol. 10): The Global Coastal Ocean Processes and Methods. Chichester: John Wiley & Sons, Inc., 253 - 292.

[2] 王颖，傅光翮，张永战. 2007. 河海交互作用沉积与平原地貌发育. 第四纪研究，27(5)：674 - 689.

[3] Wang Ying, Ren Mei-e, Zhu Dakui. 1986. Sediment supply to the continental shelf by the major rivers of China. *Journal of the Geological Society*, London, 143(6)：935 - 944.

[4] Curray J R, Moore D G. 1971. Growth of the Bengal deep-sea fan and denudation in the Himalayas. *Geological Society of America Bulletin*, 82(3)：563 - 572.

[5] 王颖主编. 2012. 中国区域海洋学——海洋地貌学. 北京：海洋出版社，169.

[6] 刘振夏. 1996. 对东海扬子浅滩成因的再认识. 海洋学报（中文版），18(2)：85 - 92

[7] 叶银灿，庄振业，来向华，刘奎，陈小玲. 2004. 东海扬子浅滩砂质底形研究. 中国海洋大学学报（自然科学版），34(6)：1057 - 1062.

[8] Chen Jiyu, Zhu Huifang, Dong Yongfa, Sun Jieming. 1985. Development of the Changjiang estuary and its submerged delta. *Continental Shelf Research*, 4(1～2)：47 - 56.

[9] 陈吉余. 2007. 中国河口海岸研究与实践. 北京：高等教育出版社，156 - 199.

［10］陈吉余.长江三角洲江口段的地形发育.地理学报,1957,23(3):241-253.

［11］陈吉余,虞志英,恽才兴.1959.长江三角洲的地貌发育.地理学报,25(3):201-220.

［12］任美锷主编.1986.江苏省海岸带与海涂资源综合调查报告.北京:海洋出版社,120-133.

［13］国家海洋局第一海洋研究所,中国科学院地理研究所.1984.渤海黄海地势图.北京:地图出版社.

［14］秦蕴珊主编.1989.黄海地质.北京:海洋出版社,24-30.

［15］Liu Jian, Saito Yoshiki, Kong Xianghuai *et al*. 2010. Delta development and channel incision during marine isotope stages 3 and 2 in the western South Yellow Sea. *Marine Geology*, 278(1～4): 54-76.

［16］李全兴主编.1990.渤海黄海东海海洋图集(地质地球物理).北京:海洋出版社,2.

［17］金翔龙主编.1992.东海海洋地质.北京:海洋出版社,525.

［18］Liu Zhenxia. 1997. Yangtze shoal—A modern tidal sand sheet in the northwestern part of the East China Sea. *Marine Geology*, 137(3～4): 321-330.

［19］Fisk H N, McFarlan E, Kolb C R *et al*. 1954. Sedimentary framework of the modern Mississippi Delta. *Journal of Sedimentary Petrology*, 24(2): 76-99.

［20］Roberts Harry H. 1997. Dynamic changes of the Holocene Mississippi River Delta Plain: The delta cycle. *Journal of Coastal Research*, 13(3): 605-627.

［21］Coleman J M, Roberts Harry H, Stone G W. 1998. Mississippi River Delta: An overview. *Journal of Coastal Research*, 14(3): 698-716.

［22］Shepard F P. 1973. Submarine Geology(Third Edition). New York, Evanston, San Francisco, London: Harper & Row, Publishers, 164-165.

［23］Blum M D, Roberts H H. 2009. Drowning of the Mississippi Delta due to insufficient sediment supply and global sea-level rise. *Nature Geoscience*, 2(7): 488-491.

［24］Milliman John D. 1979. Morphology and structure of Amazon upper continental margin. *American Association of Petroleum Geologists Bulletin*, 63(6): 934-950.

［25］Damuth J E, Flood R D. 1984. Morphology, sedimentation processes and growth pattern of the Amazon deep-sea fan. *Geo-Marine Letters*, 3(2～4): 109-117.

［26］Damuth J E, Flood R D, Kowsmann R O, *et al*. 1988. Anatomy and growth pattern of Amazon deep-sea fan as revealed by long-range side-scan sona（GLORIA）and high-resolution seismic studies. *American Association of Petroleum Geologists Bulletin*, 72(8): 885-911.

［27］Figueiredo J, Hoorn M C, van der Ven P, Soares E. 2009. Late Miocene onset of the Amazon River and the Amazon deep-sea fan: Evidence from the Foz do Amazonas Basin. *Geology*, 37(7): 619-622.

［28］王颖主编.2002.黄海陆架辐射沙脊群.北京:中国环境科学出版社,230-260.

［29］史照良主编.2004.江苏省地图集.北京:中国地图出版社,8-9.

［30］王颖,张振克,朱大奎,杨竞红,毛龙江,李书恒.2006.河海交互作用与苏北平原成因.第四纪研究,26(3):301-320.

［31］Berne S, Vagner P, Guichard F, *et al*. 2002. Pleistocene forced regressions and tidal sand ridges in the East China Sea. *Marine Geology*, 188(3～4):293-315.

［32］李家彪主编.2008.东海区域地质.北京:海洋出版社,81.

［33］中国地图出版社翻译出版(埃塞尔特地图公司编绘制版).1991.埃塞尔特世界地图集.北京:中国地图出版社,206.

The Sand Ridge Field of the South Yellow Sea: Origin by River-Sea Interaction[*]

1　Introduction

Sand ridge fields consist of large, elongate sand bodies and channels formed by tidal dynamics in shallow seas with abundant sediment. The ridges are linear-shaped deposits, aligned parallel to the direction of the tidal currents. They reach heights of several meters to several tens of meters, widths of several hundred meters to several kilometers, and lengths of several kilometers to tens of kilometers. Normally, the ridges are distributed in groups or fields in shallow seas. Such sand ridge fields occur widely in Chinese shelf seas (Fig. 1); for example, the radiating sand ridge field in the South Yellow Sea; the finger-shaped sandy ridges of Liaodong bank in the Bohai Strait (Fig. 1D); the parallel sandy ridges lying offshore of the Yalu River Estuary and the West Korean Bay (Fig. 1B); and the finger-shaped sandy ridges in the Qiongzhou Strait of the South China Sea (Fig. 1G). They have also been found on the continental shelf of the East China Sea (Fig. 1E) (Liu et al., 2000). The largest ridge complex along the Chinese coast is the field of radiating sand ridges off the north Jiangsu coast in the South Yellow Sea (Fig. 2) where the two largest Chinese rivers, the Changjiang River and the Yellow River, have supplied huge quantities of sediment since early Cenozoic times. These sand ridges fan out for $30 \sim 110$ km, covering an area of 22 470 km², the water depth generally ranging from 0 to 25 m, but maximum depths can reach 50 m in certain parts of the tidal channels.

Sand ridges are in principle a result of tidal current action, but are also influenced by pre-existing morphology, river discharge (e. g., Giosan et al., 2005), and storm wave processes (Li and King, 2007). The deposits may preserve records of sea level and other environmental changes on the continental shelf and the flux of sediment from land to sea. As in other highly populated coastal areas of the world, land resources along the Jiangsu coast are very scarce. The radiating sand ridges are thus considered to be a potential land resource through reclamation. The deep tidal channels between the sand ridges are, furthermore, regarded as ideal locations for deep-water harbors in response

---

　　* Ying Wang, Yongzhan Zhang, Xinqing Zou, Dakui Zhu, David Piper: *Marine Geology*, 2012, Vol. 291 - 294, No. 1: pp. 132 - 146.

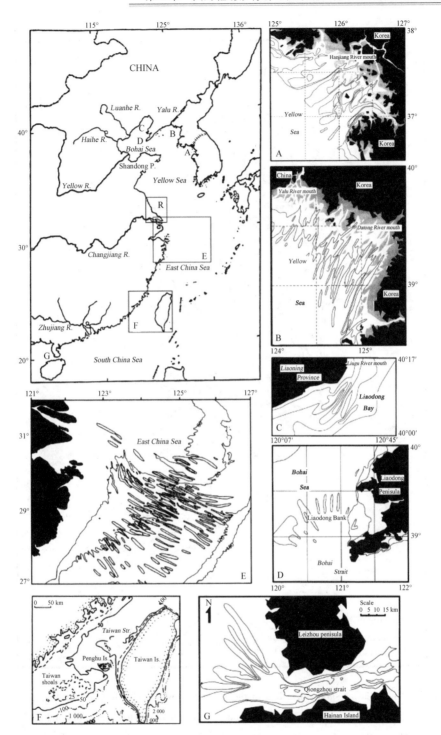

Fig. 1 Map of China and the China seas showing the major rivers and representative examples of sand ridge fields. P: Peninsula. R: Location of the study area (Fig. 2). A: Sandy ridges off the Hanjiang River mouth. B: Sandy ridges off the Yalu River Estuary and the West Korean Bay (Datong River mouth). C: Sandy ridges off the Liugu River mouth in the Liaodong Bay. D: Sandy ridges of Liaodong bank in the Bohai Strait. E: Large-scale sandy ridges on the continental shelf of the East China Sea. F: Sandy ridges of the Taiwan shoals in Taiwan Strait. G: The finger-shaped sandy ridges in the Qiongzhou Strait. R: Location of the study area (Fig. 2).

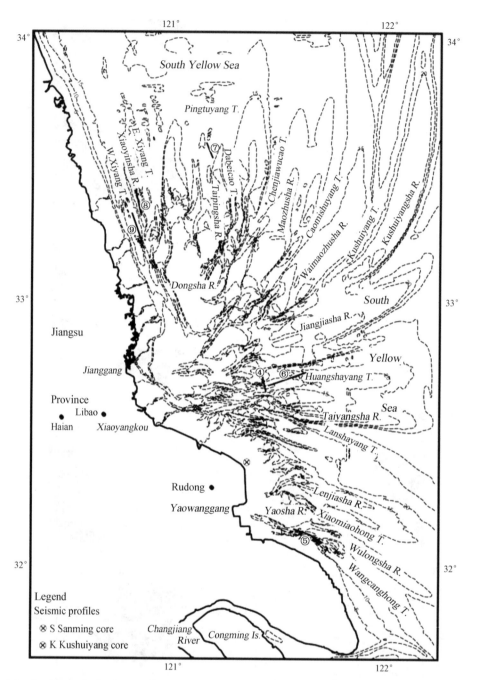

Fig. 2 **The radiating sand ridge field in the South Yellow Sea (for location see Fig. 1). Also shown are the position of the Sanming and Kushuiyang boreholes (⊗ S and ⊗ K) and the locations of seismic sections numbered in correspondence to the figure numbers illustrating them. T: trough; and R: ridge.**

to the demands of local economic development. The looming conflict between "economic development" and "conservation" thus calls for in-depth scientific studies in this area.

Sand ridges have been studied elsewhere since the 1960s. Good examples are those in the English Channel and the Celtic Sea between Ireland and France (Houbolt, 1968), in the south-eastern North Sea (Collins et al. , 1995), and along the Atlantic coast of North American (Off, 1963). These regions have been continuously investigated over the last 40 years (e. g. , Caston, 1972; Dyer and Hunltey, 1999; Kenyon, 1970; Swift, 1975). As far as is known, the largest sand ridge field is located along the shelf edge of the Celtic Sea southwest of England (Houbolt, 1968; Tessier et al. , 2000). It is situated in water depths of $100\sim170$ m, the parallel ridges being spaced 16 km apart and reaching 50 m in height, $5\sim7$ km in width, and $40\sim120$ km in length. These ridges formed duringthe last glacial maximum at 18 ka BP when sea level was about 120 m lower than at present. The relatively shallow water at that time and the large sediment supply, coupled with strong tidal currents, favored the formation of a large sand ridge complex reaching up to 50 m in thickness. The field has passed through several stages of development following changes in sea level, preserving coastal changes in the process. The fundamental factors controlling sand ridge development are thus the availability of large quantities of sediment that can be moved by strong tidal currents flowing at velocities of $0.25\sim2.5$ ms^{-1} (Off, 1963). The currents shape the sediments into sand bodies aligned in the direction of the tidal current to form erosive valleys and linear sand bodies or, in the presence of older valleys or depressions, these are further excavated, the eroded sediment being then deposited on adjacent sand ridges. The formation processes of ridges and troughs are related to a combination of horizontal and transverse currents developed in the troughs, which move sand from the troughs to the ridges. Due to a positive feedback, the development of the ridge and trough system is enhanced as it grows (Houbolt, 1968). Because the velocity of the tidal current is larger in the deep troughs than on the top of the ridges, the difference produces transverse water circulation, the bottom current in the trough diverging outwards, while converging inwards on top of the ridges. This helicoidal water circulation moves sediment eroded in the troughs to the crests of the ridges, thereby increasing their height. Caston (1972) and Stride (1974) showed that, in the southern North Sea, flood and ebb tidal currents were dominant on opposite sides of the ridges, respectively, the different velocities causing the ridges to slowly migrate laterally.

The radiating sand ridge field of the Yellow Sea was named as Five Sands on British Admirality Charts of the 1930s. Hydrodynamical and geomorphological surveys have been undertaken in this region since the early 1960s by the Institute of Oceanology of the Chinese Academy of Sciences (Li and Li, 1981), and the Bureau of Marine Geological Surveys (Yang, 1985; Zhou and Sun, 1981). Major studies in marine hydrology, meteorology, geomorphology, geology, and biology were carried out along the Jiangsu coast in the period from 1980 to 1985. Offshore, the scientific focus was initially on the sand ridge field, but later included several tidal channels for the purpose of a harbor site selection. A systematic study of the sand ridge field was carried out from 1993 to 1996 by Nanjing University, jointly with Hohai University, Tongji University and the Institute of Oceanology of the Chinese Academy of Sciences (as a Key Study Project supported by the National Natural Science Foundation of China). The study involved the generation of a detailed bathymetric chart of the whole area, the acquisition of more than 600 km of seismic profile data, the retrieval of several hundred seabed samples, and the collection of hydrological data on tides, waves, currents and suspended sediment to improve the fundamental understanding of the sand ridge field (Wang et al., 1999). The results have been published in Chinese in a scientific volume under the title "Radiating sand ridge field of the South Yellow Sea" (Wang et al., 2002). The present study is primarily based on those results, but has been supplemented by new data collected during July—November of 2007 when 1050 km of additional seismic profile data and 9 long sediment cores were obtained in the field.

The purpose of this study is to comprehensively assess the sedimentary processes and the coastal geomorphological evolution in some key areas, especially the axial area and the northeastern part of the ridge field.

2　Geological and oceanographic setting

The radiating sand ridge field is located seaward of the large deltaic alluvial plain built by the Changjiang and Yellow rivers into the central Yellow Sea (Figs. 1 and 2). Its apex is located near Jianggang and Xiaoyangkou on the north Jiangsu coast where the trend of the modern coastline changes from NNW in the north to WNW in the south, being situated about 80 km north of the late Holocene path of the Changjiang River. Satellite imagery suggests that two former distributary channels of the Changjiang discharged near the apex of the radiating sand ridge field. The modern coastline is fringed by tidal flats typically $5 \sim 10$ km wide. The coastline has prograded ~ 40 km over the last thousand years, most rapidly between 1128 and 1855 when the Yellow River mouth migrated from the Bohai Sea southwards to the north Jiangsu coast.

The semi-enclosed Yellow Sea is influenced by two types of tidal waves, a progressive tidal wave from the Pacific, which propagates from the southeast towards the North Yellow Sea, and a local reflected tidal wave formed by the obstruction of the Shandong Peninsula in the northwest. The two tidal waves converge along the Jiangsu coast in the western Yellow Sea, forming a "standing wave" (Zhang, 1998) that rotates clockwise about an amphidromic center 200 km north of Jianggang (Fig. 3A). In the radiating sand ridge field the semidiurnal rotary tidal currents are very strong, average spring tidal current velocities being ~ 2 ms^{-1} and maximum tidal current velocities exceeding 2.5 ms^{-1}. Tidal ranges are normally $4 \sim 6$ m, but can reach up to 9.28 m in the deepest channel of Huangshayang, which is the maximum tidal range recorded in the

China Sea (Ren et al., 1986). Uehara et al. (2002) have modeled the tides of the middle and early Holocene at 6 ka and 10 ka BP (Fig. 3B and C). In the middle Holocene, tidal conditions were similar to the present, but when sea level was 45 m lower in the early Holocene, no amphidrome was developed.

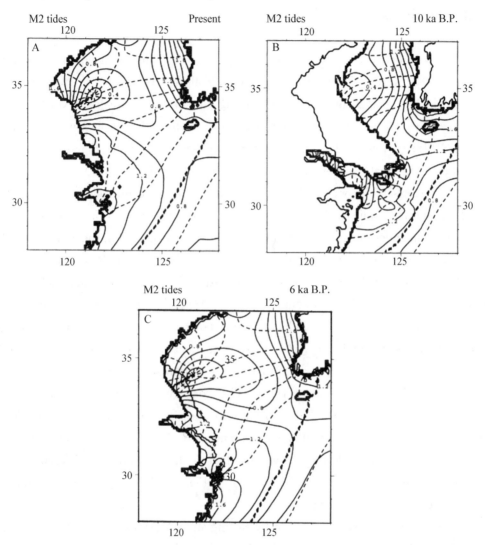

Fig. 3　(A) Calculated isolines of modern constituent M2 tides in the Yellow Sea (Professor Dongsheng Zhang, Hohai University, Nanjing, China, pers. comm. 2009). (B) and (C) Modeled isolines of constituent M2 tides at 10 ka BP and 6 ka BP, respectively (modified after Uehara et al., 2002).

Wave climate is controlled primarily by the Monsoon winds. Waves from a northerly and northeasterly direction prevail in winter, but waves from the southeastern quadrant prevail in summer. Wave energy progressively decreases while crossing the sand ridge field, to the extent that wave heights of 7～9 m in the open sea are reduced to 0.4 m, 2 m high waves occurring with a probability of only 5%. As a result, the outer parts of the sand ridge field suffer from wave-induced bottom erosion, to the extent that about 2.02×10^9 t of sediment has been transported into the sand ridge field from the offshore seabed (Zhu and Fu, 1986).

3 Methods

Sea floor bathymetry is based on published Chinese navigational charts. Altogether, 324 surface sediment samples were collected by means of short cores and single or double clam weight samplers during the first large-scale survey along the Jiangsu coast (1980—1984) carried out in the period when a key project was financed by the NSFC (1993—1996). This data set was complemented in the course of a second, more recent large-scale survey along the Jiangsu coast (2006—2010) when 125 more surface sediment samples were collected by mean of double clam weight samplers, and some 1650 km of seismic profiles were collected with penetrations up to 50 m into the sub-bottom sandy deposits. Of the total seismic profiles collected in the area, more than 600 km were acquired with a Geopulse seismic profiler (Ocean Research Equipment Inc.), which covered the whole radiating sand ridge field during the summer of 1994, and about 1 050 km with a Boomer system (Applied Acoustic Engineering, UK) covering different sections along individual sand bodies or tidal channels in the summer and autumn of 2007. The boomer surveys were conducted at energy source levels of 300 or 400 J. To assist the interpretation of the sub-bottom seismic data, 2 long sediment cores were drilled in September 1992. The core that provided the most useful stratigraphic control penetrated 60.25 m into the supra-tidal flat just outside the protective coastal dike near the village of Sanming in Rudong County of Jiangsu Province (Fig. 2). Nine additional deep cores were drilled in shallow water in different parts of the radiating sand ridge field during 2007—2008 (Fig. 2). These penetrated to the sub-bottom depths of 30.8~70.9 m.

Grain-size analysis was carried out by sieving and a Mastersizer 2000 laser analyzer. Sediment classification is based on the triangular scheme of Shepard (1954). The 0.063~0.125 mm fraction was separated for mineral analysis. Heavy minerals were separated using tribromomethane (SG = 2.89) and an electronic magnetic separator, and were subsequently counted under a petrographic microscope. Radiocarbon ages were calibrated using the Fairbanks 0107 calibration (Fairbanks et al., 2005) with a -450 years marine reservoir correction. The half life of ^{14}C was set at 5 568 ka BP, counting backward from the year 1950.

4　Results and discussion

4.1　Seabed character of the sand ridge field

4.1.1　Geomorphology

The sand ridges radiate from near the coastal city of Jianggang town (Fig. 2), the fan pattern extending through an angle of 150°. The field is composed of more than 70 sand ridges separated by tidal channels that jointly cover an area of ~22 470 km² from nearshore to inner continental shelf. Recent surveys indicate that ~3 782 km² lies above the chart datum, ~2 611 km² is shallower than 5 m, ~4 004 km² is between 5 and 10 m deep, ~6 825 km² lies between 10 and 15 m, and some 5 045 km² is deeper than 15 to 20 m, i. e. reaches water depths of up to 30 m off the northeastern part. The most pronounced geomorphic expression of the sand ridges thus occurs in the depth interval from 5 to 20 m.

The morphology of the sand ridges is quite unique in comparison with their linear counterparts. In the South Yellow Sea they consist of fine sand mainly supplied by the paleo-Changjiang River. They are aligned with the direction of the radiating tidal current, especially in the northeastern part of the sand ridge field, but are also affected by wave winnowing processes, especially during the winter Monsoon which is associated with high NE waves. As a consequence, the ridges are flat-topped. In response to the present sea-level rise, the outer parts of the sand ridge field are being eroded by wave action, the sediment being transported landward by the tidal currents. As a result, the ridges in the inner part amalgamate to form sandy island and the tidal flats along the coast accrete and prograde rapidly seaward. Most of the tidal channels between the ridges follow the inherited morphology of old, underlying river valleys. Especially in the axial part, the main tidal channels follow the paleo-Changjiang River distributary valleys, although some channels have also been formed by current erosion, e. g. the Xiyang channel in the north. This demonstrates that, from the later Pleistocene to the present, the geomorphology of the radiating sand ridge field has essentially been controlled by river-sea interaction.

4.1.2　Sediment texture and provenance

The sediment mainly consists of well-sorted fine sand (>90%). Silt can be found in several of the lower depressions in channels and stratigraphically as multiple laminae intercalated in the sandy sediment. The textural composition in the tidal channels varies with the water depth. From the seaward mouths towards the middle parts of large channels where water depth exceeds 15 m, e. g. in the Huangshayang and Lanshayang channels, the silt content increases, the seabed sediment consisting of sandy silt or silty sand. The sediment between the inner and middle parts of the tidal channels is mainly composed of fine sand containing $2\%\sim6\%$ silt. Coarser textures can be found nearer to the shore, reflecting the significant influence of wind-generated waves. Under normal weather conditions, wave-induced winnowing processes do not occur in water depths greater than 20 to 25 m.

The gravel fraction comprises calcite-cemented sand intraclasts that constitute up to 29% of the sediment at the entrance of several large channels such as the main Lanshayang channel. In the Pingtuyang channel, by contrast, where the water depth exceeds 15 m, the sediment consists of fine sand containing up to 5% of gravel-sized intraclasts.

In the northern part of the sand ridge field, the sediment is mainly composed of fine sand or muddy fine sand with a clay content of up to 30%. At water depths of 10 m, silt contents begin to increase, while clay appears at depths of 20 m. In the northeast of the sand ridge field, sediment composition evolves landward from sand-silt-clay to sandy silt and silty sand.

4.1.3 Sediment mineralogy

Light minerals in the sediments of the northern radiating sand ridge field comprise $30\%\sim40\%$ of reddish colored quartz of various grain shapes, $\sim20\%$ feldspar, frequent rock fragments and minor amounts of calcite ($<2\%$). Heavy minerals, by contrast, contribute less than 2% to the sediment, these being unevenly distributed with higher concentrations in the south.

Based on heavy mineral distributions (Table 1), the sand ridge field can be divided into four regions:

Region I　In the northern channels of Xiyang, Pingtuyang and Dabeicao, hornblende accounts for $60\%\sim67\%$ of the heavy mineral fraction, epidote, limonite, and ilmenite contributing $<20\%$. Limonite and ilmenite typically make up $>5\%$, but garnet $<2\%$. The Pingtuyang sediments contain more zircon (6%).

Region II　For the northeastern sand ridges of Waimaozhusha and Jiangjiasha, heavy minerals are not abundant but include a variety of species including hornblende, epidote, garnet, limonite, ilmenite, zircon, leucoxene, apatite, and fresh mica, chlorite, and calcite. Of the heavy mineral suite here, hornblende comprises about 50% (78% in Jiangjiasha), epidote $>20\%$ (11% in Jiangjiasha), and garnet and limonite account for $2\%\sim5\%$.

Region III　In the axial part, eastward channel of Huangshayang, southeastward channel of Lanshayang and Taiyangsha ridge, hornblende and epidote contribute 80% and 60%, respectively, to the heavy mineral suite. Secondary species include ilmenite, zircon, magnetite, apatite, rutile, tourmaline, fresh mica, chlorite and calcite.

Table 1　Heavy mineral distribution in submarine sandy ridge field

Region	I	II	III	IV
Heavy mineral constant(%)	Northern Channel of Xiyang, Pingtuyang and Dabeicao	Northeastern sandy ridges of Waimaozhusha and Jiangjiasha	Axial part channels of Huangshayang, Lanshayang and Taiyangsha ridge	Southern sector of Lengjiasha ridge and Xiaomiaohong channel
Hornblende	60.29~67.48	56.04~77.63	32.06~60.19	20.52~59.93
Epidote	8.72~13.66	10.75~21.96	20.14~30.46	25.85~50.51
Limonite	2.71~16.66	2.08~4.79	3.60~6.39	4.21~8.11
Ilmenite	5.67~13.10	1.60~2.08	6.39~12.94	3.45~10.12
Garnet	1.07~1.96	1.36~2.65	2.73~3.60	0.80~1.69
Zircon	0.47~6.49	1.04~2.39	2.40~8.21	0.31~3.19
Magnetite	3.53~11.48		3.44	
Apatite	0.31~1.15	1.28	0.80~5.13	0.06~0.70
Rutile	0.07~0.35	0.16	0.10~0.17	0.01~0.04
Tourmaline	0.90~0.12	Few	0.97	0.12~0.80
Mica	—	√	√	—
Chlorite	—	√	√	—
Calcite	—	√	√	—

Region IV　In the southern sector, comprising the Lengjiasha ridge and Xiaomiaohong channel, hornblende and epidote account for 72% and 86%, respectively, of the heavy mineral content. Hornblende abundance in the southern region decreases from north to south, whereas epidote content increases. The heavy minerals of the Lengjiasha Ridge contain a comparatively large proportion of ilmenite, magnetite and limonite, exceeding 2% of the total sediment. This differs from the other regions, as the amount of ilmenite on the Lengjiasha Ridge (10% of heavy minerals) is significantly larger than that in the Xiaomiaohong Channel (3%). The reverse holds for limonite, which contributes 8% to the heavy mineral fraction in the Xiaomiaohong Channel, but only 4% on the Lengjiasha Ridge. The higher ilmenite content is a distinctive feature of the southern region.

5 Stratigraphy of the radiating sand ridge field

5.1 Seismic stratigraphy

Sequence stratigraphic concepts were used to describe the stratigraphy in the seismic reflection profiles. All seismic reflection profiles show stratified near-surface sediments overlying a prominent, generally planar erosion surface at sub-bottom depths of 5~20 m that can be correlated throughout the seismic grid. This erosion surface has the general form of a ravinement surface (labeled RS1). Some seismic profiles show an underlying, highly irregular erosion surface with a relief of up to 30 m that is interpreted as a lowstand sequence boundary (labeled SB1). In some seismic profiles, irregular erosion surfaces are also visible at greater depths. Two seismic profiles (Figs. 4 and 5) cross channels and adjacent ridges north and south of the Sanming borehole, respectively. The northern profile, which crosses the middle Huangshayang channel, shows surface RS1 at 22~27 m depth, assuming a sound velocity of $v = 1\ 500$ m/s (Fig. 4). Above RS1, the floor of the channel shows complex cut-and-fill structures. On the adjacent ridge to the north, the part of the section above the pronounced erosion

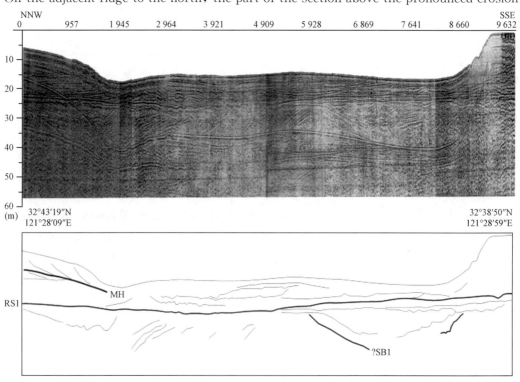

Fig. 4 Seismic profile across the upper-middle Huangshayang channel. SB1, RS1 and MH are prominent erosion surfaces discussed in text

surface（MH）shows southward dipping reflectors in the upper part of the ridge that prograde into the channel, whereas the lower part of the ridge is composed of horizontal reflectors. Beneath RS1, complex cut-and-fill patterns are visible, but the acoustic signal dissipates at 40 m depth, suggesting a predominantly sandy section. Individual channel-like features are 1～2 km wide.

The southern profile (Fig. 5) displays a planar RS1 erosion surface at 18～20 m depth. It is also underlain by cut-and-fill structures in the form of relatively narrow channel-like features. In this case, the acoustic signal dissipates at a depth of 45 m. As in Fig. 4, dipping clinoforms suggest that ridge sediments prograde into the channel.

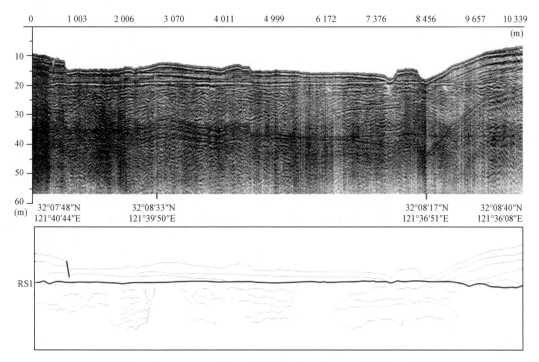

Fig. 5　Seismic profile across the Xiaomiaohong trough

Farther seaward, beneath the Huangshayang channel (Fig. 6), a buried channel some 10 km wide underlies the planar RS1 erosion surface, which is located at a depth of 28～30 m. At the SW end of the profile, the stratal geometry immediately above the RS1 surface is characterized by complex cut-and-fill structures. This unit is truncated by an erosion surface that appears to correlate with the MH erosion surface in Fig. 4. This erosion surface is located a few meters above the RS1 surface in the center of the channel. The MH surface is overlain by a unit of obliquely stratified sediments that are truncated at the top by a younger erosion surface overlain by a unit comprising continuous parallel reflections. The oblique strata above MH at the SW end of the profile are truncated by the modern seabed, which has also eroded several meters into the RS1 surface at the NE end of the profile.

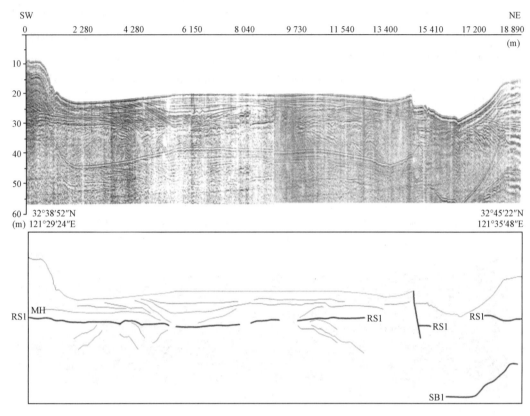

Fig. 6 Seismic profile across the lower-middle Huangshayang channel

Farther north in the Dabeicao trough (Fig. 7), the RS1 surface can be identified at a depth of 18～27 m. It is overlain by horizontally bedded sediments. Below RS1, a prominent erosion surface, labeled SB1, outlines a steep-sided 25 m deep and 1 km wide channel. This channel is filled with sub-parallel reflections that deepen northward. Laterally, the SB1 reflector merges with the RS1 surface, while one or more irregular erosion surfaces can be seen below the former two.

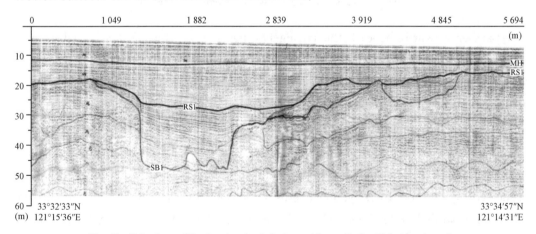

Fig. 7 Seismic profile showing buried channel beneath the Dabeicao trough

Beneath one of the ridges in the northern part of the sand ridge field (Fig. 8), the RS1 surface is located at a depth of 28～32 m. It is overlain by horizontal or gently oblique, 10～15 m thick strata that are cut by an irregular channeled surface, the channels typically being 100 m wide and 5 m deep. This channeled surface, which is overlain by horizontally stratified ridge sediments, probably correlates with the MH erosion surface farther south.

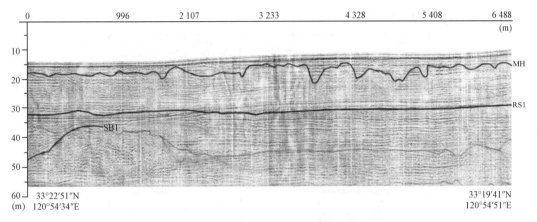

Fig. 8　Seismic profile of a ridge near the Xiyang Channel

The channels in the northwestern part of the sand ridge field appear to be very young features (Fig. 9). The widespread surface RS1 is overlain by horizontal or gently inclined strata all the way up to the modern ridge crest (northern end of Fig. 9). This succession is abruptly cut by channels that are locally incised to more than 13 m below the RS1 surface.

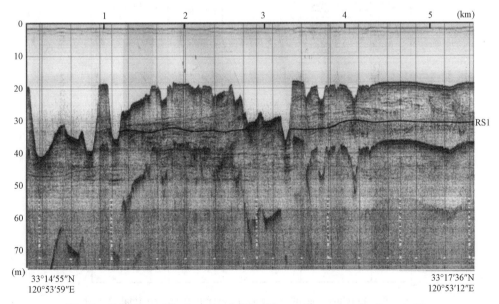

Fig. 9　Seismic profile along part of the Xiyang Channel and adjacent ridges

5.2　Core control on the seismic section

The Sanming core (32°27.8′N,121°18.6′E), which is located on the upper tidal flat in the axial part of the radiating sand ridge field, and which represents the land-sea transition between the shallow sea and the coastal lowland of this region, can be divided into three units (Fig. 10) on the basis of the sedimentology, fauna and radiocarbon dating (Wang et al., 1999).

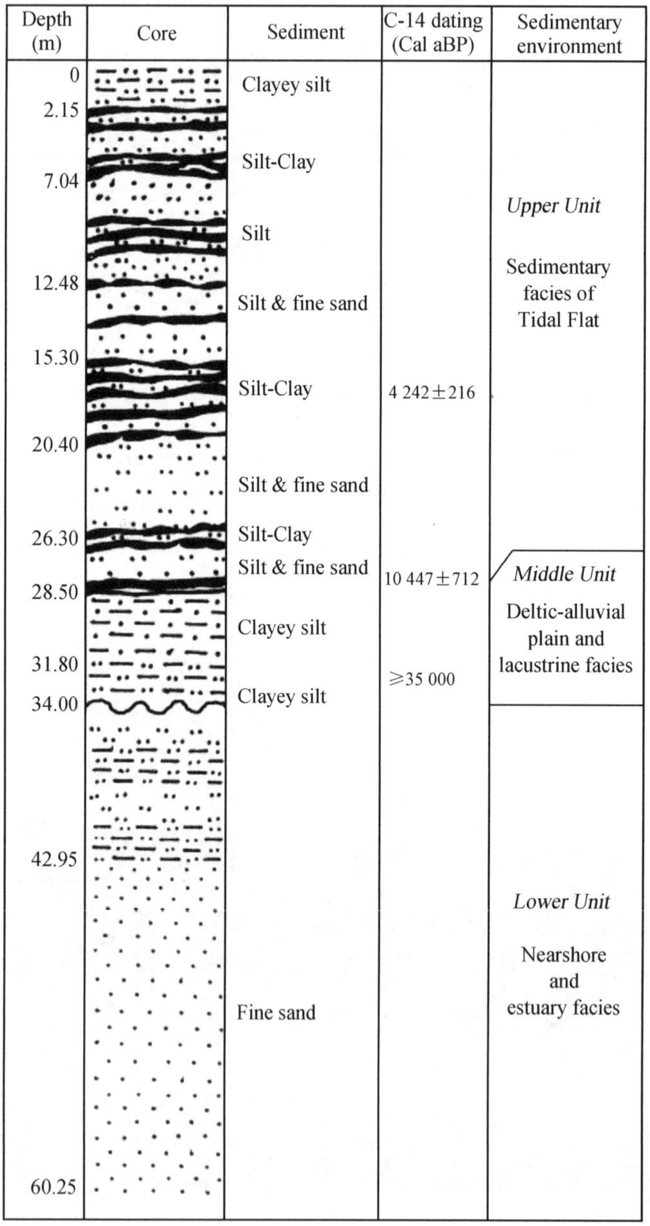

Depth (m)	Core	Sediment	C-14 dating (Cal aBP)	Sedimentary environment
0		Clayey silt		
2.15				
7.04		Silt-Clay		Upper Unit
		Silt		Sedimentary facies of Tidal Flat
12.48		Silt & fine sand		
15.30		Silt-Clay	4 242±216	
20.40		Silt & fine sand		
26.30		Silt-Clay		
28.50		Silt & fine sand	10 447±712	Middle Unit
		Clayey silt		Deltic-alluvial plain and lacustrine facies
31.80			≥35 000	
34.00		Clayey silt		
42.95				Lower Unit
		Fine sand		Nearshore and estuary facies
60.25				

Fig. 10　Borehole log of the core recovered from the tidal flat at Sanming. The location of the borehole is labeled ⊗ S in Fig. 2.

The *Upper Unit* of the core (0～29 m) comprises three repetitive sequences of clayey fine silt overlying fine sand that reflect a threefold change from a muddy supratidal flat to an intertidal flat comprising silt and clay or silt and fine sand alternating with silty and sandy subtidal flat sediment, the units even incorporating partly eroded sections. The succession contains a few foraminifera, including *Ammonia beccarii var*, *Ammonia convexidorsa*, *Elphidium magellanicum*, *Elphidium advenum*, *Cribrononion subincertum*, and the Ostracod *Ianellaopima*. Calcareous nannofossils are also present (Zou et al., 1999). Two ^{14}C dates were obtained from the same facies in the other two cores from this area. Shell fragments from a depth of about 21 m in a core located on the upper tidal flat to the north of the Sanming borehole, yielded a radiocarbon age of 4 290±150 ka BP (4 242±216 cal ka BP) (Wang et al., 2002). The second date was obtained on shell fragments from the bottom of this unit at 29 m, giving a radiocarbon age of 9 680±520 ka BP (10 447±712 cal ka BP) (Zhang et al., 1990). These dates clearly document that the upper unit evolved in the course of the Holocene.

The *Middle Unit* of the core, which comprises a 5.5 m long interval from 28.5 to 34.0 m depth, consists of yellowish, grainy hard clay in which thinner layers of silty and lenticular fine sand are intercalated. The sandy layers contain freshwater fossils such as *Viviparidae* (Gastropoda), *Planorbidae* (Pulmonata) and *Unionidae* (Bivalvia), but also a few foraminifers such as *A. beccarii var*, *Lagena* spp, *Epistominella maraensis* and small planktonic ones. The lithofacies and fauna are indicative of a deltaic alluvial plain with some lakes. A ^{14}C-date on freshwater shells from a depth of 33.3 m gave an age exceeding 35 000 years BP. There thus appears to be a hiatus, either comprising an erosional unconformity or a period of non-deposition, between this late Pleistocene unit and the upper Holocene unit.

The *Lower Unit* of the core (34.0 to 60.25 m) consists of clayey silt in its upper part and homogeneous fine sand lacking any apparent stratification in its lower part, except for some very fine yellowish laminae intercalated at larger intervals. The upper subunit consists mainly of horizontally bedded, greenish-gray clayey silt. The clay content increases upward in this subunit, whereas the silt content decreases. The lower subunit comprises gray homogeneous fine sand without any bedding. It contains abundant shell fragments and several foraminifer species, e. g. *Ammonia compressiuscula*, *Florilus decorus*, and some cold water species such as *Rosalina brady*. The lithofacies and microfauna indicate an estuarine-shoreface depositional environment.

The Kushuiyang core, extracted from the channel bed at a water depth of 12.0 m, is located in the northeastern part of the radiating sand ridge field (Fig. 2). The core (Fig. 11) is centered at 32°27.178′N, 122°5.617′E, and penetrated to a depth of 30.8m, but yield only 24.8 m of preserved sediment strata. As in the previous case, three sedimentary facies units were distinguished.

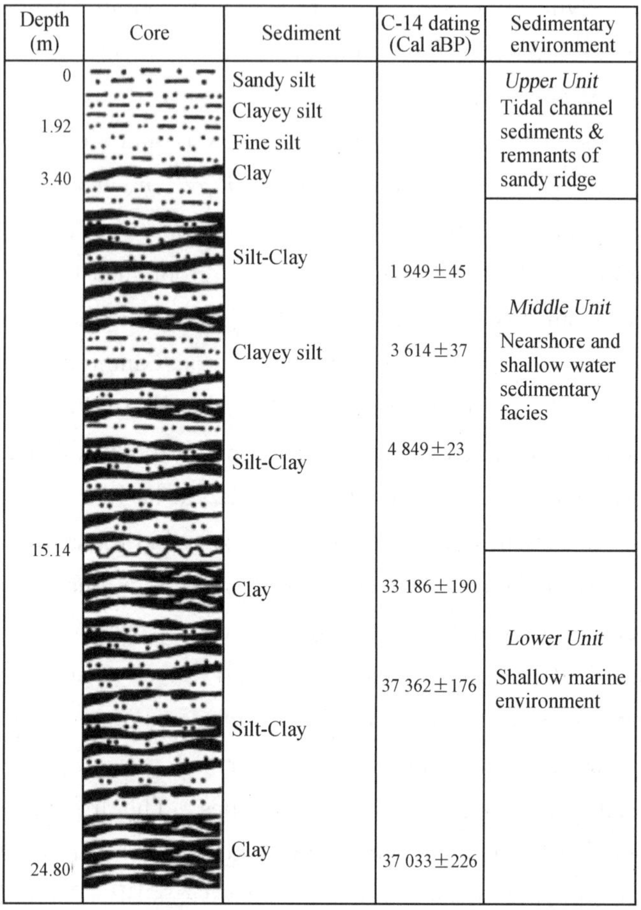

Depth (m)	Core	Sediment	C-14 dating (Cal aBP)	Sedimentary environment
0 1.92 3.40		Sandy silt Clayey silt Fine silt Clay		*Upper Unit* Tidal channel sediments & remnants of sandy ridge
		Silt-Clay	1 949±45	*Middle Unit* Nearshore and shallow water sedimentary facies
		Clayey silt	3 614±37	
		Silt-Clay	4 849±23	
15.14		Clay	33 186±190	*Lower Unit* Shallow marine environment
		Silt-Clay	37 362±176	
24.80		Clay	37 033±226	

Fig. 11 Borehole log of the Kushuiyang core recovered from a tidal trough in the northeastern part of the radiating sand ridge field. The location of the borehole is labeled ⊗ K in Fig. 2.

The *Upper Unit* extends from the seabed to a depth of 3.4 m. It consists of modern tidal channel sediments and remnants of eroded sandy ridge material. The uppermost 0.5 m consist of dark yellowish-brown, well-sorted homogenous fine sandy silt containing abundant shell fragments. It represents typical modern tidal current deposits. From 0.5 to 1.9 m, the core comprises tidal channel deposits of alternating olive-gray to light olive-gray fine silt, fine sandy silt and clayey silt laminations displaying ripple cross-bedding typical of tidal channel deposits. From 1.92 to 3.40 m, the sediment consists of light olive-gray fine silt and dark yellowish-brown clayey silt, or olive-gray to olive-black clay displaying tidal bedding of alternating silt and clay laminae covering monthly lunar cycles. The clay layers display insect and worm holes that were filled with silt. This unit is interpreted to represent the remnants of ridge basement deposits that were subaerially exposed some time in the recent past. It is separated from the middle unit below by an erosion surface. The *Middle Unit* occupies the depth interval between 3. 40 and 15. 14 m. It represents a nearshore, shallow water sedimentary facies. Olive-black or light olive-gray clay is the main component, followed by dark yellowish-brown silty clay and silt, or clayey silt. The alternating lamellae display tidal bedding with micro-wave marks. The unit is characterized by bivalve shells and foraminifers, including *A. beccarii*, *Rotaliidas*, *Ammonia annectens*, *Nonion schwageri*, *Globigerina bulloides w*, *Cribronion frigidum* and *Nonion grateloupi*, which are mostly nearshore-shallow water species. The bivalve *Dosinia* (*phacosoma*) *gibba* Adams, recovered from a clayey silt layer at a depth of 12. 1~12. 3 m, typically lives in water depths ranging from the nearshore down to 60 m, whereas *Corbicula leana* and *Area* (*Arcp*) *sp.*, recovered from a clay layer between 16. 10 and 16. 5 m, are prime examples of semisaline to fresh water species.

Several AMS ^{14}C-dates were obtained from this unit. Shell fragments from 6. 6 to ~6. 7 m were dated at 2 005±40 ka BP (1 949±45 cal ka BP) (Peking University, lab no. BAO 90484), shells from a depth of 9. 4 m at 3 375±30 ka BP (3 614±37 cal ka BP) (lab no. BAO 90485), and a shell of D. (*phacosoma*) *gibba* Adams from a depth of 12. 1~12. 3 m at 4 290±40 ka BP (4 849±23 cal ka BP) (lab no. BAO 90486). The transition between the middle and lower units is marked by an erosion surface.

The *Lower Unit*, extending from 15. 14 to 24. 80 m, represents a shallow-marine environment that received an abundant terrigenous sediment supply. It mainly consists of dark yellowish-brown clay with dusky-brown silty clay or light olive-gray clay containing some silt. The unit contains shell fragments, and displays wave marks and other small sedimentary structures. Shells of *C. leana*, a coastal ocean species recovered at a depth of 16. 1 m, was ^{14}C-dated at 27 830±100 ka BP (33 186±190 cal ka BP) (lab no. BAO 90487).

A gray to brownish-gray volcanic ash layer was identified at a depth of 21. 4~23. 7 m. Similar deposits have been found in cores from the South Yellow Sea. ^{14}C-dates obtained from organic clay recovered at depths of 19. 3 and 24. 4 m yielded ages of 31 975 ±100 ka BP (37 362±176 cal ka BP) (lab no. BAO 90488), and 31 655±170 ka BP (37 033±226 cal ka BP) (lab no. BAO 90489) respectively.

6　Evolution of the radiating sand ridge field of the South Yellow Sea

6. 1　Provenance of surficial sediments in the radiating sand ridge field

Comparing the sediments of the sand ridges off the north Jiangsu coast with those of the larger rivers entering the Yellow Sea area, sandy silt is seen to dominate both the sediments of the Yellow River mouth in the Bohai Sea (Wang et al., 1997) and the abandoned Yellow River mouth on the north Jiangsu coast of the South Yellow Sea. The suspended particulate matter (SPM) of the Changjiang River, by contrast, disaggregates into clayey silt, and the sediment off this river mouth thus consists of silty clay and clayey silt (Wang et al., 1997). The textural trends off the Yellow and Changjiang rivers clearly show that mean grain size decreases from the lower reaches toward the sea. The sediment of the Yellow River discharging into the Bohai Sea comprises clayey silt, whereas those of the Changjiang River discharging into the Yellow and East China Seas consist of fine sand and clayey silt.

The major heavy minerals supplied by the Yellow River are hornblende (37%), biotite (54%), epidote (5%), limonite (2.5%), garnet (0.6%), ilmenite and magnetite (0.2%), and zircon (0.2%). The abundance of biotite is particularly distinctive in comparison with the Changjiang River (Wang et al., 1997, 2002). The major heavy minerals of the Changjiang River are hornblende (40%), epidote (30%), limonite (14%), garnet (4%), pyroxene (1.2%), chlorite (1.2%), zircon (3.4%), and apatite (2.5%) (Wang et al., 1997). The epidote content in the Changjiang River sediment is six times higher than in that of the Yellow River. The stable minerals garnet, zircon, and apatite, together with pyroxene and chlorite, are also more abundant in the Changjiang River sediment than in that of the Yellow River. The abundance of diagnostic heavy minerals in the southern part of the sand ridge field is generally similar to that of the Changjiang River, i.e. a high stable mineral content and abundant epidote. These minerals are less abundant in the northern part of the sand ridge field, suggesting a larger input of sand from the Yellow River.

A previous study of the clay mineral content of the radiating sand ridge field (Wang et al., 2002) has shown that the illite content reaches up to 70%, which is similar to the clay mineral content of the East China Sea. The montmorillonite content is 16%, which is higher than that in the Changjiang River region, but closer to that in the Yellow River sediment. Kaolinite (7%) and chlorite (7%~8%) contents are low, showing similarity to the clay mineralogy of the Yellow River sediment. Thus, clay mineral contents suggest a predominant supply from the Yellow River with a lesser contribution from the Changjiang River. The greater abundance of clay in the northern part of the sand ridge field also suggests that the provenance of the clay is principally from the Yellow River delta located to the north.

Sediment analyses thus reveal that the bulk of the sandy sediment of the radiating sand ridge field consists of fine sand derived from the Changjiang River, with a small contribution from the Yellow River in the northern part of the field. The clay material, by contrast, was supplied from both rivers, the contribution from the Changjiang River being overall smaller than that of the Yellow River. The analyses on geochemical compositions in the fine-grained fraction ($<63~\mu m$) also showed the same influences of the Yellow River (Yang et al., 2002).

6.2　Seismic stratigraphy and chronology of the radiating sand ridge field

The stratigraphic section in the Sanming core is similar to the late Quaternary stratigraphy of a number of sediment cores recovered along the north Jiangsu coast and the Changjiang River delta (Chen et al., 1995; Zhang et al., 1990). These demonstrate that late Pleistocene sediments are represented by at least 70~90 m thick deposits comprising coastal-lagoon-estuarine and shallow marine facies. The Holocene sediment comprises about 30 m of estuarine, tidal flat and shoreface deposits. The radiocarbon date of 10 447 cal ka BP at the base of the upper unit in the Sanming core is consistent with a eustatic sea level of −30 to −35 m according to published sea-level curves (e.g., Peltier and Fairbanks, 2006).

The three sand to clayey silt sequences in the upper unit of the Sanming core (Fig. 10) correspond to typical tidal flat deposits characterizing the Chinese coast and which can be divided into three zones, namely an inner zone comprising clayey silt or silty clay, a middle zone consisting of silt-dominated mud, and an outer zone consisting of silt and fine sand (e.g. Dalrymple, 1992; Wang et al., 1990). The stratigraphic sequences in the upper unit thus represent a balance between tidal flat progradation and rising sea level, the sediment sequence in this area being about 30 m thick (Zhu and Xu, 1982; Zhu et al., 1998). The possibility was considered that the 10 447 cal ka BP age at the bottom of the upper unit could be too young and that the Holocene transgression actually took place during the Younger Dryas slowing-down of sea-level rise. In this case, the lowest tidal flat sequence would represent a Younger Dryas progradation (Fig. 12). The middle sequence, constrained by the 4 242 cal ka BP date, represents renewed flooding by the early Holocene transgression followed by middle Holocene progradation. The upper sequence, in turn, could represent local deepening due to the migration of a now buried tidal channel, followed by renewed late Holocene progradation. An analogous shift of a tidal channel is shown at the northern end of Fig. 4 where horizontally bedded strata in the middle of the Holocene succession are truncated by a local erosion surface over which a tidal ridge has prograded.

The RS1 erosion surface, which can be traced through all the seismic reflection profiles, is generally planar, truncating complex underlying deposits. In nearshore seismic profiles (Figs. 4 and 5), it overlies an apparently sandy facies with abundant small channels, probably representing a subaerial delta plain. Only locally does the erosion surface become shallower where it encounters more resistant underlying sediments (e.g., Fig. 7). It is slightly deeper in paleo-channels than on surrounding highs (Fig. 7). Such features are characteristic of transgressive or ravinement surfaces (e.g. Catuneanu et al., 2009), and it can be correlated with the ravinement surface

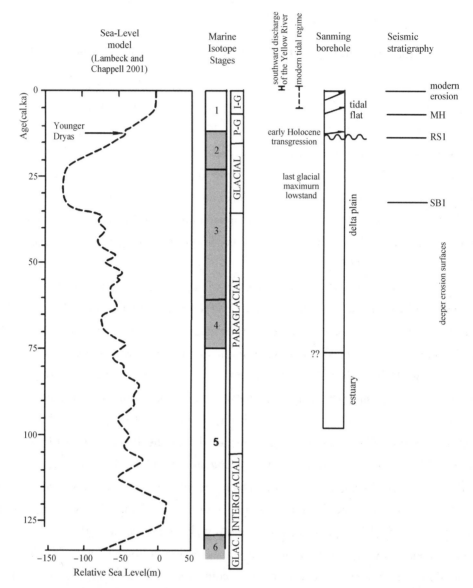

Fig. 12 Chronological model illustrating sea-level changes and age interpretations of the Sanming core and the seismic reflection profiles

P‐G: Paraglacial. I‐G: Interglacial.

between the upper and middle units in the Sanming core. Although the erosion surface is a little deeper in the Sanming borehole (29 m) compared with the seismic profiles (18~27 m), the inferred facies change in both the core and the more inshore seismic profiles is quite similar.

The prominent channels illustrated in Figs. 6 and 7 are deeply incised into sediments that show a variety of acoustic styles. These erosion surfaces are identified as SB1 in the seismic profiles. The up to 25 m high steep slopes and the flat floor are typical of incised lowstand valleys (e. g. , Catuneanu et al. , 2009). However, the depth of their incision is greater than that of other buried channel-like features, and there is little sediment accumulation between the tops of their sides and the RS1 ravinement surface (e. g. , Fig. 7). For this reason they are correlated with the 20~25 cal ka BP lowstand of the sea. The size of these channels is comparable to that of the modern Changjiang River above its estuarine reach. The large buried channels beneath the modern Huangshayang and Lanshayang channels outlined by the SB1 erosion surface (Fig. 6) are therefore interpreted as distributaries of the paleo-Changjiang River during the late Pleistocene. The lower river probably had three to four distributaries at that time, of which the Huangshayang channel may have been the major one. At the time of this major lowstand, there would have been little sediment accumulation on the adjacent deltaic plain because of very rare overbank flooding of the deeply incised river channels.

The middle unit in the Sanming core consists of delta plain sediment and probably correlates with the sandy facies seen in seismic profiles from the inner part of the radiating sand ridge field (Figs. 4 and 5). Its age is greater than 38 cal ka and it probably corresponds to sedimentation during the early part of marine isotope stage (MIS) 3 and MIS 4 when sea level would have been around −50 to −70 m (Lambeck and Chappell, 2001), i. e. only 15~25 m below the level of the unit in the Sanming core. This allowed some delta plain sediments to accumulate.

The lower unit in the Sanming core, which consists of coastal and estuarine sands below −41 m, implies a sea level above ~−50 m, a feature characteristic of MIS 5 (Fig. 6). The abundance of sand in this section suggests direct supply from the paleo-Changjiang river at that time.

7 Types of tidal channels and ridges in the radiating sand ridge field

The tidal channels or troughs between ridges are the major morphological component of the radiating sand ridge field. Three types have been identified:

(i) Tidal channels developed within ancient river valleys (Fig. 6) such as the Huangshayang and Lanshayang channels, and which still preserve remnants of the incised river valleys dating from the late Pleistocene MIS 2 lowstand of sea level.

(ii) Tidal channels that may have inherited their position from older channels. Even though a pre-existing valley was infilled with sediment, the locations of later valleys are within the original buried valley, as can be seen in the Dabeicao (Fig. 7) and the Caomishuyang troughs. In the case of the Xiaomiaohong channel, its present inner reach was formed by erosion during the Holocene, but its middle reach obliquely overlies older buried paleo-river valleys.

(iii) Tidal channels that are troughs eroded by tidal currents. These features are characterized by erosive pot holes, deep hollows and erosive depressions distributed over the bottom of the present tidal channel as, for example, the Xiyang and Kushuiyang channels. The seismic profiles (Fig. 8) show a prominent erosion surface at a stratigraphic level estimated to be of middle Holocene age. The erosion surface in Fig. 8 probably represents a widening or shifting in the position of the channels; the scale of the erosion features is similar to that imaged by a seismic profile along the axis of the modern Xiyang channel. This implies that the formation of the ridges above the planar RS1 surface, represented by obliquely dipping strata (Fig. 8), took place during the early Holocene transgression.

The southern and axial parts of the radiating sand ridge field evolved in an area of paleo-Changjiang River distributaries, the ridge sand troughs being generally aligned along the paleo-river valleys and sediment accumulation in the troughs above RS1 being rather thin (Figs. 4~6). The northern and northeastern sectors, by contrast, truncate any buried paleo-river valleys and there are generally thick Holocene sediments above RS1 (Fig. 7). Therefore, the pattern of tidal channels and ridges in this area evolved under the control of radiating tidal currents, the ridges and troughs being aligned along the directions of the ebb tidal currents.

Three genetic types of sand ridges can be distinguished:

(1) In the central and southern parts of the radiating sand ridge field, the ridges are on the site of late Pleistocene interdistributary erosional highs or depositional bodies capped with Holocene sediments.

(2) The radiating current ridges in the northeast located above the RS1 erosion surface formed during early and middle Holocene times by the action of tidal currents.

(3) Erosional-depositional sandy bodies in the northern and outer parts of the radiating sand ridge field, and small ridges extending seaward from the central complex, were formed mainly during the early Holocene transgression.

7.1 Historical evolution of the radiating sand ridge field

The development of the radiating sand ridges began in the axial sector. In this area, sand was deposited before 35 ka BP during the para-interglacial high sea-level period from marine isotope stage 3 to early stage 5 (Fig. 12). The sea level at that time was located 20~50 m lower than at present, and the Changjiang River entered the sea in the area of Haian near Libao town, north of the sand ridge field (Yang, 1996). The shoreline at that time was located east of the present coast. Buried ridges imaged in seismic profiles at depths of -30 to -40 m (Figs. 4~7) probably have multiple origins. Many appear sandy from their acoustically incoherent seismic character. Some appear to be erosion remnants between distributary channels (e. g. Fig. 4), but deeper sands may correspond to the thick estuarine sands sampled in the lower unit of the

Sanming core (Fig. 10). Throughout this area, widespread erosion during the formation of the early Holocene ravinement surface provided abundant sand. Excavation by modern tidal channels below the ravinement surface (Figs. 6 and 9) provided further sand. Tidal channels re-occupied the buried valleys created by the Changjiang River distributaries at the 20 ka～25 ka lowstand of the sea (SB1) and tidal sand ridges therefore formed above the interdistributary highs.

In the northeastern and southern regions, the tidal ridge and channel landforms do not conform to the underlying topography (Pleistocene distributary channels). The early Holocene ravinement surface is overlain by horizontally bedded sediments (Figs. 7 and 8), the formation of ridges and channels having taken place in the middle Holocene. Even in the axial region, there was an intensification of tidal erosion at that time (Fig. 4). This probably corresponds to the onset of the modern tidal regime (Uehara et al., 2002) as sea level reached its maximum elevation.

The northwestern part of the radiating sand ridge field evolved last, apparently in the late Holocene (Fig. 9), the intense tidal erosion continuing to the present. This intense tidal reworking is further indicated by poorly preserved nannofossils, which also occur in low abundance (Zou et al., 1999). This erosion is possibly related to the rapid progradation of the northern Jiangsu coastline over the past thousand years as a result of the southward displacement of the Yellow River mouth. This is also consistent with the results of two numerical models. A present-day tidal model focused on a tidal-current field in the ridge area indicates that the magnitude of tidal currents during spring tides is strongest (exceeding 1.0 m/s) in the northwestern part (Wang et al., 2002). In addition, Uehara and Saito (2003) suggested that the area of intense tidal current strength emerged at its modern location after the northern Jiangsu coastline had prograded to its present position. Liu et al. (2010) also suggest that there was a significant contribution of the Yellow River sediments to the Changjiang subaqueous delta during the last ～600 years.

7.2　Climate variability and evolution of the radiating sand ridge field

The river-sea interactive processes in the South Yellow Sea are closely related to climate and sea-level changes (Wang et al., 1986). Global climate change is the ultimate cause of sea-level change, which in turn drives transgressive and regressive cycles. During the glacial stages, e.g. marine isotope stages 2, 4 and 6, sea level fell, exposing most of the Yellow Sea shelf. Rivers were incised and brought abundant coarse sand to the paleo-shore. During these periods, sediment accumulated in terrestrial environments of the study area and, as a consequence, was not reworked by tidal currents.

During para-interglacial stages such as marine isotope stage 3 and parts of stage 5, together with the early Holocene, the sea level fluctuated between 50 and 20 m below the present level (Fig. 12). Although the Pacific tides reached the Yellow Sea, the bay-shaped coastline, especially Shandong peninsula, would have blocked the progressive tidal waves from the Pacific to form a reflected wave propagating from the north. This reflected wave converged on the advancing Pacific tidal wave and formed a rotational tide that created the dynamic conditions for developing the tidal current ridges (Fig. 3). As a result, it was only as sea level approached the present level in mid-Holocene times that tidal conditions developed that were able to generate tidal ridges (Uehara et al. ,2002), as shown by the mid-Holocene erosional surface (Figs. 4 and 8). The currents eroded older sandy deposits from the paraglacial and glacial stages and reworked the sands into the large, radiating sand ridge and channel pattern. Without the tidal channels, the tidal current sand ridges would not have formed. Waves act to winnow and redistribute sand on the ridges, resulting in a flattening and broadening of the shallow-water parts of the ridges (Fig. 2).

Fig. 13 summarizes the key differences between glacial and interglacial stages. Sand was deposited in the region during late Pleistocene sea-level lowstand periods, whereas the tidal sand ridge sand troughs were carved out by tidal currents and wave action when the sea level reached its modern highstand in the mid-Holocene. The changing tidal regime of the Yellow Sea in the course of late Pleistocene and Holocene sea-level changes has created the special conditions in which the sand supplied by the paleo-Changjiang River could be reworked into the large-scale radiating sand ridge field when the sea level reached its modern highstand.

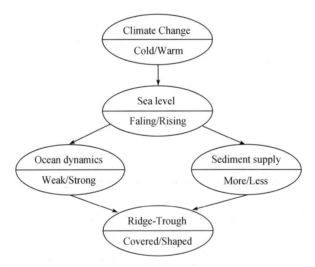

Fig. 13 Summary of the role of climate in influencing sea level, ocean dynamics,sediment supply and the evolution of sand ridges

8 Conclusions

(1) The radiating sand ridge field off the north Jiangsu coast comprises sandy

sediment largely derived from the Changjiang River, although some silt and most of the clay were derived from the Yellow River.

（2）The abundance of sand in the radiating sand ridge field is related to late Pleistocene deposition of fluvial and estuarine sands from Changjiang river distributaries at times of lowered sea level.

（3）A widespread early Holocene ravinement surface can be traced in seismic reflection profiles throughout the sand ridge field at sub-bottom depths of 15~30 m. It is generally overlain by horizontally bedded strata interpreted as tidal flat sediments.

（4）Over much of the sand ridge field, a mid-Holocene erosion surface has been identified that represents the onset of the modern tidal regime and the development of tidal channels and adjacent ridges in the northern and southern parts of the field.

（5）In the central part of the sand ridge field, tidal channels generally follow incised fluvial channels that date from marine isotope stage 2.

（6）In the inner part of the sand ridge field, wave action has flattened and broadened the tidal ridges.

Acknowledgements

The study was funded by grants from National Natural Science Foundation of China (Grant no. 49236120 during 1990s and Grant no. 40776023 during 2008~2010) and State Oceanic Administration of China during 2006~2010 (Grant no. JS-908-01-05). We are particularly grateful to Professor Terry Healy (University of Waikato, Hamilton, New Zealand) for commenting on an early draft of this paper, Professor Michael Collins (University of Southampton, UK) for commenting on the preface of the introduction part, Professor Burg Flemming (Senckenberg Institute, Germany) and two anonymous reviewers for providing constructive comments on this paper.

References

[1] Caston, V. N. D. 1972. Linear sand banks in the southern North Sea. *Sedimentology*, 18:63-78.

[2] Catuneanu, O. , Abreu, V. , Bhattacharya, J. P. , Blum, M. D. , Dalrymple, R. W. , Eriksson, P. G. , Fielding, C. R. , Fisher, W. L. , Galloway, W. E. , Gibling, M. R. , Giles, K. A. , Holbrook, J. M. , Jordan, R. , Kendall, C. G. , St, C. , Macurda, B. , Martinsen, O. J. , Miall, A. D. , Neal, J. E. , Nummedal, D. , Pomar, L. , Posamentier, H. W. , Pratt, B. R. , Sarg, J. F. , Shanley, K. W. , Steel, R. J. , Strasser, A. , Tucker, M. E. , Winker, C. 2009. Towards the standardization of sequence stratigraphy. *Earth-Science Reviews* 92(1): 1-33.

[3] Chen, B. Z. , Li, C. X. , Ye, Z. Z. 1995. Deposition and its environmental evolution in north part of Changjiang River delta during post-glacial period. *Acta Oceanologica Sinica*, 17 (1), 64-75 (In Chinese with English abstract).

[4] Collins, M. B. , Shinwell, S. J. , Gao, S. , Powell, H. , Hewitson, C. , Taylot, J. A. 1995. Water

and sediment movement in the vicinity of liner sandbanks: the Norfolk Banks, southern North Sea. *Marine Geology*, 123(3~4):125 – 142.

[5] Dalrymple, R. W., 1992. Tidal depositional systems. In: Walker, R. G., James, N. P. (Eds.), Facies Models Response to Sea Level Change. Geological Association of Canada, Love Printing Service Ltd, p. 195.

[6] Dyer, K. R., Hunltey, D. A. 1999. The origin, classification and modelling of sand banks and ridges. *Continental Shelf Research*, 19(10): 1285 – 1330.

[7] Fairbanks, R. G., Mortlock, R. A., Chiu, T. C., Caoa, L., Kaplan, A., Guilderson, T. P., Fairbanks,T. W., Bloom, A. L., Grootes, P. M., Nadeau, M. J. 2005. Radiocarbon calibration curve spanning 0 to 50,000 years BP based on paired ^{230}Th/ ^{234}U/ ^{238}U and ^{14}C dates on pristine corals. *Quaternary Science Reviews*, 24(16~17):1781 – 1796.

[8] Giosan, L., Donnelly, J. P., Vespremeanu, E., Bhattacharya, J. P., Olariu, C., Buonaiuto, F. S. 2005. River Delta Morphodynamics: Examples from the Danube Delta: SEPM. Special Publication No. 83.

[9] Houbolt, J. J. H. C. 1968. Recent sediments in the southern bight of the North Sea. *Geologie en Mijnbouw*, 47(4): 245 – 273.

[10] Kenyon, N. H., 1970. Sand ribbons of European tidal seas. *Marine Geology*, 9: 25 – 39.

[11] Lambeck, K., Chappell, J. 2001. Sea level change through the last glacial cycle. *Science*, 292 (5517): 679 – 686.

[12] Li, M. Z., King, E. L., 2007. Multibeam bathymetric investigations of the morphology of sand ridges and associated bedforms and their relation to storm processes, Sable Island Bank, Scotian Shelf. *Marine Geology*, 243(1~4): 200 – 228.

[13] Li, C. Z., Li, B. C. 1981. Studies on the formation of sandbanks along north Jiangsu coast. *Oceanologia et Limnologia Sinica*, 12 (4): 321 – 331 (In Chinese with English abstract).

[14] Liu, Z. X., Berné, S., Saito, Y., Lericolais, G., Marsset, T., 2000. Quaternary seismic stratigraphy and paleoenvironments on the continental shelf of the East China Sea. *Journal of Asian Earth Sciences*, 18(4): 441 – 452.

[15] Liu, J., Saito, Y., Kong, X. H., Wang, H., Xiang, L. H., Wen, C., Nakashima, R., 2010. Sedimentary record of environmental evolution off the Yangtze River estuary, East China Sea, during the last ~13,000 years, with special reference to the influence of the Yellow River on the Yangtze River delta during the last 600 years. *Quaternary Science Reviews*, 29(17~18): 2424 – 2438.

[16] Off, T. 1963. Rhythmic linear sand bodies caused by tidal currents. *Bulletin of American Association of Petroleum Geologists*, 47 (2): 324 – 341.

[17] Peltier, W. R., Fairbanks, R. G., 2006. Global glacial ice volume and Last Glacial Maximum duration from an extended Barbados sea level record. *Quaternary Science Reviews*, 25(23~24): 3322 – 3337.

[18] Ren, M. E., Xu, T. G., Zhu, J. W., Chen, B. B., et al. (Eds.). 1986. The Comprehensive Investigative Report on the Resources of the Coastal Zone and Tidal Flats of Jiangsu Province. China Ocean Press, Beijing, pp. 25 – 37 (InChinese).

[19] Shepard, F. P. 1954. Nomenclature based on sand-silt-clay ratios. *Journal of Sedimentary*

Petrology，24(3)：151－158.

[20] Stride，A. H.，1974. Indications of long term，tidal control of net sand loss or gain by European coasts. *Estuarine and Coastal Marine Science*，2(1)：27－36.

[21] Swift，D. J. P.，1975. Tidal sand ridges and shoal retreat massifs. *Marine Geology*，18(3)：105－133.

[22] Tessier，B.，Reynaud，J.，Marsset，T.，Dalrymple，R. W.，Proust，J.，2000. Transitory tidal resonance during the last transgression as recorded in the deep shelf sand banks of the Celtic Sea. In：Park，Y. A.，Chun，S. S.，Choi，K. S. （Eds.），TIDALITE 2000：Dynamics，Ecology and Evolution of the Tidal Flats，Abstracts of 5th International Conference on Tidal Environments. Hoam Convention Center，Seoul National University，Seoul，Korea，pp. 149－153. June 12～14.

[23] Uehara，K.，Saito，Y. 2003. Late Quaternary evolution of the Yellow/East China Sea tidal regime and its impacts on sediments dispersal and seafloor morphology. *Sedimentary Geology*，162 (1～2)：25－38.

[24] Uehara，K.，Saito，Y.，Hori，K. 2002. Paleotidal regime in the Changjiang (Yangtze) Estuary，the East China Sea，and the Yellow Sea at 6 ka and 10 ka estimated from a numerical model. *Marine Geology*，183(1～4)：179－192.

[25] Wang，Y.，Ren，M. E.，Zhu，D. K. 1986. Sediment Supply to the continental shelf by the major rivers of China. *Journal of Geological Society*，London，143(6)：935－944.

[26] Wang，L. C.，Chen，X. L.，Chu，T. Q. 1997. A contrast analysis on the loads character of Changjiang River and the Yellow River. *Geographical Research*，16 (4)：71－79 (In Chinese with English abstract).

[27] Wang，Y.，Zhu，D. K.，You，K. Y.，Pan，S. M.，Zhu，X. D.，Zou，X. Q.，Zhang，Y. Z. 1999. Evolution of radiating sand ridge field of the south Yellow Sea and its sedimentary characteristic. *Science in China* (*Series D*)，42 (1)：97－112.

[28] Wang，Y.，Xue，H. C.，Zhu，D. K.，Yan，Y. X.，Zhang，D. S. （Eds.）2002. Radiating Sand ridge fields on Continental Shelf of the Yellow Sea. China Environmental Science Press，Beijing. pp. 1－15，29－90，170－374 (In Chinese).

[29] Yang，C. S. 1985. Studies on the formation of radiating sand ridges at Jianggang. *Marine Geology & Quaternary Geology*，5 (3)：35－43 (In Chinese with English abstract).

[30] Yang，H. R. 1996. Type and mechanism of lower Changjiang River migration since the Late Pleistocene. The Study on Environment Chang e. Hehai University Press，Nanjing，pp. 195－211 (In Chinese).

[31] Yang，S. Y.，Li，C. X.，Jung，H. S.，Lee，H. J. 2002. Discrimination of geochemical compositions between the Changjiang and the Huanghe sediments and its application for the identification of sediment source in the Jiangsu coastal plain，China. *Marine Geology*，186(3～4)：229－241.

[32] Zhang，D. S. 1998. Tidal current formation-storm history-tidal current reform：explanation the dynamic mechanism of radiating sand ridges of the Yellow Sea. *Science in China* (*Series D*)，28 (5)：394－402 (In Chinese with English abstract).

[33] Zhang，Z. G.，Zhou，M. S.，Shao，S. X.，et al. （Eds.）. 1990. The Quaternary Geological Map of People's Republic of China and Nearby Sea Area (1：25 000) and its Direction. The Cartographic

Map Press of China，Beijing（In Chinese）.

[34] Zhou，C. Z. ，Sun，J. S. ，1981. Discussion on the formation offshore sand banks along north Jiangsu coast. *Research on Marine Geology*，1（1）：83 - 92（In Chinese with English abstract）.

[35] Zhu，D. K. ，Fu，M. Z. 1986. The Preliminary Study on the Offshore Radiating Sand ridges of Jiangsu Province. The Collected Works for the Comprehensive Investigation on Dongsha Sandbank at the Coastal Zone of Jiangsu Province（Office for Tidal Flat in Science Commission in Jiangsu Province ed. ）. Ocean Press，Beijing，pp. 28 - 32（In Chinese）

[36] Zhu，D. K. ，Xu，T. G. 1982. The problems of coast development and utilization in the middle part of Jiangsu Province. *The Journal of Nanjing University（Nature Science）*，3：799 - 818（In Chinese with English abstract）.

[37] Zhu，D. K. ，Martini，I. P. ，Brookfield，M. E. 1998. Morphology and land-use of coastal zone of the north Jiangsu Plain Jiangsu Province，eastern China. *Journal of Coastal Research*，14（2）：591 - 599.

[38] Zou，X. Q. ，Shi，B. W. ，Xu，Y. H. ，Li，H. Y. 1999. The distribution of coccoliths in submarine ridges of Yellow Sea and the research on the submarine sedimentation type. *Acta Palaeontologica Sinica*，38（2）：255 - 259（In Chinese with English abstract）.

黄海海底辐射沙洲形成演变研究[*]

一、海底沙脊群的形成

海底沙脊群是浅海内陆架的大型堆积地貌,主要分布于强潮水流与风浪交织的水下古三角洲,或海峡两端的海域。研究海底的沙体受海气动力作用而移动变化、海底沙脊群的形成演变规律、结构特点、沙脊及沙脊间潮流通道的稳定性等,是海洋地质学、大陆架地质过程、海洋工程及油气资源开发等方面的研究热点。如英国与荷兰对北海海底沙脊群成因分布与移动规律的研究,提出外海潮波至近岸,反复往返的潮流是形成海底沙脊的基本条件。美国、荷兰与韩国对西朝鲜湾潮流沙脊的研究,提出横向环流作用机制,美国在大西洋海底研究中提出大陆架风暴作用对海底巨型沙波的沉积动力学分析,及其移动规律的研究,在理论上及应用上均取得令人注目的进展。

海底沙脊群是一种独特的大陆架沙体。大陆架沙和沙体的形成和改造是冰后期海侵和浅海动力体综合作用的产物,它系统地记录了大陆架环境演变的信息。因此,海底沙脊群(沙体)的形成发育是全球变化研究的重要组成部分。浅海沙体具有良好的生储油组合,常有丰富的油气蕴藏,是勘探海底油气的重要目标,是以现代大陆架沙体与古沙体对比的重要识别标志。海底沙体的稳定性与移动规律又为各种海底工程项目所关注。

二、海底沙脊群的分布

中国近海海底沙脊群分布广泛:渤海海峡老铁山水道西侧的指状沙脊群——辽东浅滩,辽宁六股河口外的海底三道岗,鸭绿江口外与岸斜交的海底沙脊群,西朝鲜湾大范围的潮流脊,琼州海峡沿潮流扩散方向延伸的指状沙脊等。其中以黄海海底沙脊群的形成发展最为科学界注目,南黄海的辐射沙脊群,其规模最大、型式独特、脊槽分布变化复杂,是国际海洋学界关注的"海底迷宫"。

南黄海辐射沙脊群分布于江苏海岸外,范围界于 $31°55'N \sim 34°24'N$、$120°49'E \sim 120°32'E$ 之间,总面积达 2 万 km^2。由于太平洋前进潮波与黄海当地反射的潮波,在江苏岸外聚合,古三角洲堆积体被改造成辐射状的沙脊群,共有沙脊 70 多条,其中 0 m 以上(低潮出露水面)的沙洲有 8 条,0 m 以上的面积为 2 125.5 km^2,其余为水下沙脊与潮流槽,水深 15～30 m,最深 48 m。这是世界上罕见的海洋环境,是中国主要的渔场、潜在的

---

　　* 王颖,朱大奎:江苏省科学技术协会编,《建设海上苏东的科学之路》,70-72 页,中国科学技术出版社,1996。

巨大土地资源与南黄海油气田,以及平原海岸急需而珍稀的深水航道资源。

黄海辐射沙脊群的结构、成因、分布、变化规律的研究,将提供研究河海交互作用的堆积型大陆架形成发展的关键性实例。揭示了在海底构造控制的基本框架的基础上,沉积动力作用塑造大陆架的机制,有助于恢复古海洋环境,推动浅海海洋地质学的发展。大陆架演化过程的学说,经历了从 Johnson(1919 年)的动力响应模式,到 Shepard-Emery (1963—1968 年)的海侵模式,至 Swifi(1972 年,1974 年,1968 年)的海侵-动力模式,在这一学说的深化过程中,大陆架沙与沙体的研究起了关键性的作用。大陆架沙和沙体的形成和改造是冰后期海侵和浅海动力体系综合作用的产物,其中系统地记录了大陆架环境演变的信息。因此,研究海底沙脊群(沙体)的形成发育是全球环境变化研究的一个重要组成部分。浅海沙体具有良好的生储油组合,蕴藏着丰富的油气资源,已成为重要的勘探目标,石油天然气的勘探需要从现代大陆架沙和沙体寻求古代同类沉积体的识别标志和对比模式,进而推动了现代大陆架沙和沙体的研究。黄海辐射沙脊群是世界罕见的巨大的大陆架沙体,对它的研究将推动大陆架沉积动力学的研究,使中国在这一领域跻身于国际的先进行列。

三、关于辐射沙脊群的研究成果

1980—1985 年,南京大学等单位在江苏省海岸带综合调查研究中,获得了有关辐射沙脊群分布、组成与海洋环境基本特点的成果,并根据大量地形、沉积物成分等的对比与数学模型分析等数据资料,提出辐射沙洲是古长江三角洲被辐合潮波改造形成的假说。80 年代中后期,南京大学与河海大学等单位合作研究,进一步提出古长江三角洲与沙脊群中最大的潮流通道——黄沙洋系古河道遗留的证据。由此一系列的勘测研究论证了这主要潮流通道的稳定性,从而提出利用这深水通道建设为通海深水航道的可能性,经过近 10 年的持续研究,现已利用黄沙洋—烂沙洋潮流通道,开辟了 10 万 t级深水航道,建设了深水码头及燃气轮电厂,并通过了国家级鉴定,第一期工程投资 24亿美元,各项建设工作正在积极进行中。同样,北部主要的潮流通道——西洋,已研究论证建设开发深水航道及中型港口,其中大丰港前期开发已批准立项,正在积极进行中。浅平的淤泥质海岸,历年来难以建设海港,致使江苏将近 800 km 的海岸线无现代化海港,对海外交通的闭塞使江苏沿海成为经济落后的区域。辐射沙洲及其潮流通道形成演变的研究,发现并论证其深槽的稳定性,在稳定的深槽中开辟深水航道,建设深水港,无疑是我国建港史上的一项重大突破,这对促进江苏沿海经济发展具有重要意义。

江苏岸外辐射沙洲的研究在理论上、应用上均具有重要意义,随着研究的深化其效应更为扩大。自"八五"以来一直为国家海洋科学研究的重点之一。其研究的途径是从浅海环境场入手,进一步研究辐合潮波形成、分布与变化的特点;从动力机制与模型研究深入联系到沉积与地貌反应;从沙脊与潮流槽的动态平衡规律探索古长江深槽移迁规律,以及黄海东海古海洋环境恢复;由此进一步阐明河海体系与堆积型大陆架的发展理论。研究中组织了浅海地质、海岸动力、地球化学、黄河口与长江口演变、测量与海图、数学模型与地理信息系统等多种学科,开展综合调查与研究。研究中采用国际先进的技术手段,采用全球定位系统与海岸微波定位系统的结合,应用双频道测探仪、旁侧声纳、地脉冲剖面仪以及钻探等方法进行沙脊群结构物质组成的详查,以现场工作系统资料与样品作为多项实验分析的基本依据,对比历史资料,并以数学模型方法模拟潮流场、波浪场、风暴潮,以地理信息系统方法分析总结多项成果,建立综合数据库,综合研究分析以取得辐射沙洲区域理论及应用的系统成果。

河海交互作用与海岸海洋沙脊群发育
——以南黄海辐射沙脊群为例[*]

　　沙脊群是海岸海洋的大型堆积体,沿中国海海岸与大陆架浅海区,分布着一系列沙脊群:渤海西部六股河口三道沙岗、渤海海峡西侧指状沙脊群、黄海北部鸭绿江口与西朝鲜湾沙脊群、黄海东部汉江口沙脊群、南黄海西部辐射沙脊群、东海外陆架沙脊群、南海琼州海峡西侧指状沙脊群。其中以黄渤海域沙脊群发育普遍,与海域环境特性有关。黄海是位于中纬度地区的半封闭浅海,太平洋前进潮波与受湾岸陆地地形影响所形成的反射潮波辐聚与辐散作用显著,形成移动型驻波(张东生 1998),海域内潮差变化大,潮流作用强,在河口或古河道砂质富积区,发育形成多处潮流沙脊群。中国海潮流沙脊群具有不同的动力作用类型:

　　(1) 海湾湾顶潮波辐聚,泥沙沉降,受往复性潮流作用,潮流脊发育受潮流控制形成与岸线直交或斜交的线型脊,如黄海北端的沙脊群。

　　(2) 前进潮波与反射潮波辐聚的旋转潮流作用区,如南黄海辐射沙脊群。

　　(3) 海峡束水流急,"狭管效应",于海峡末端所形成的指状沙脊群,如渤海海峡老铁山水道西侧的辽东湾浅滩、琼州海峡西侧指状沙脊群。

　　(4) 潮流与沿岸流综合作用,改造古河口堆积所形成的弧形沙脊群——三道沙岗,或线形沙坝体等。

　　总之,沙脊群形成于以潮流动力为控制作用的、有丰富砂质或粉砂堆积的内陆架浅海区(Off, 1963, Collins, 1995),是发育于冰后期海侵过程中,河海交互作用的产物(Wang, et al, 1998)。

　　辐射沙脊群以新川港为主轴,自西向东呈褶扇状辐射向海,由 70 多条沙脊与潮流通道所组成,分布于南黄海西部水深 25 m 以浅,南北范围界于 $32°00'N\sim33°48'N$,长达 199.6 km,东西范围界于 $124°40'E\sim122°10'E$ 之间,宽达 140 km,所占海域面积为 22 470 km²。主干沙脊的形态为对称的坦峰,沿潮流通道侧的陡度加陡。据其成因可分为三种类型:

　　(1) 被改造的老河道沉积沙体,具有承袭形态的沙脊,由全新世沙层叠置的晚更新世的沙体所组成,分布于辐射沙脊群的枢纽部及南部。

　　(2) 由潮流辐射所成型的、分布于东北部呈辐射状的沙脊,形成于早、中全新世。

　　(3) 冲蚀-堆积成因的,分布于北部、外侧及一些残留的小沙脊,主要形成于全新世。

　　沙脊间的潮流通道亦有三种类型:① 沿古河谷发育的潮流通道,如黄沙洋、烂沙洋,其发育自晚更新世末至今;② 承袭型,沿晚更新世末或早全新世的洼地或谷地所形成;

---

　　* 王颖,朱大奎:中国地理学会地貌与第四纪专业委员会编,《地貌·环境·发展:1999 年嶂石岩会议文集》,29－30 页,北京:中国环境科学出版社,1999。

③ 全新世潮流冲刷槽。沙脊群的地貌成因类型反映着晚更新世末至全新世期间海平面变化及相应的河海交互作用过程：大约始自 30 kaB. P. ，当时海面位于－20 m 处，长江于海安—李堡处入海；当低海面时，河流堆积沙体与沙层；海侵过程中潮流侵蚀沙体与沙层使成脊、槽相间；潮流进一步搬运，分布泥沙渐塑造形成目前形态。沙脊组成中的几个侵蚀界面：初期的侵蚀界面可能相当于 10 kaB. P. 与 8 kaB. P. 海侵的结果；中期侵蚀面约相当于 6 ka～5 ka B. P. 时期的高海面作用；晚期完善界面约相当于 1 kaB. P. 。该区河流-海洋相互作用过程与气候变化密切相关。海平面升降是海岸带控制陆地-海洋相互作用动力过程的关键因素。其"海侵（退）—动力—沉积"作用过程模型可概括如下：

$$\text{气候变化(冷/暖)—海平面降/升—} \begin{bmatrix} \text{海洋动力作用} \\ \text{沉积物来源} \end{bmatrix} \text{—沙脊形态}$$

River-Sea Interaction and Coastal Ocean Sandy Ridges Evolution: Examples From South Yellow Sea[*]

Sand Ridges are distributed widely along the coastal ocean of China seas especially in the Yellow Sea area such as finger-shaped ridges off west side of Bohai strait, three ranks of sand ridges off Liu Gu River mouth in the Bohai Sea; sand ridge field off Yalu River mouth and the north end of Yellow Sea; sand ridges field off Hanjiang River mouth of Korea in the east side of Yellow Sea; radiative pattern sand ridges field off west coast of the south Yellow Sea, residual Sandy ridges of the outer continental shelf of East China Sea; and finger shaped sand ridges off west side off west side of Qiongzhou Strait in the South China Sea. There are Larger Sandy bodies developed in the area with abounded sandy deposits and predominant tidal dynamics in the inner continental shelf environment (Off, 1963, Collins, 1995). The semi-closed Yellow Sea is located in the mid latitude zone, there, the progressive tidal waves of the Pacific are transferring towards the north Yellow Sea and reflected by Shandong Peninsular, thus, the reflected waves meet and converge with the continuing progressive tidal waves to form movable "standing wave" (Zhang Dongsheng, 1998). As a result, the largest tidal ranges up to 9. 28 m and the strong tidal currents developed in the area. The transgressive tidal dynamics of the Holocene time perform on the coastal sandy deposits, accumulated by the ancient Changjiang River and others during lower sea level of late Pleistocene time, to evolve the larger Sandy Ridge fields. Tidal dynamics in China seas are strengthened by the convergence processes either in the embayed of north Yellow Sea or by reflected waves, and also through the narrow sea strait effect.

---

*　Ying Wang, Dakui Zhu: *MONSOON CLIMATE, GEOMORPHOLOGIC PROCESSES AND HUMAN ACTIVITIES-IAG 2000 Thematic Conference, Schedule and Abstracts*, pp. 107 – 108, Nanjing, 2000. 8. 25 – 29.

1999 年以 "EVOLUTION OF COASTAL OCEAN SANDY RIDGES BY THE PROCESSES OF RIVER-SEA INTERACTION-EXAMPLES FROM SOUTH YELLOW SEA" 为题载于: *Fourth International Conference on Asian Marine Geology* (第四届亚洲海洋地质学大会)— *Geology of the Asian Oceans in the 21st Century, ABSTRACTS, PROGRAMS & DIRECTORY*, pp. 154, 1999. 10. 14 – 18, Institute of Oceanology, Chinese Academy of Sciences, Qingdao, China. 本文稍作修订。

2008 年以 "RIVER-SEA INTERACTION AND TIDAL SANDY RIDGE FIELD EVOLUTION-EXAMPLE FROM THE SOUTH YELLOW SEA, CHINA" 为题载于: *Proceedings of 7th International Conference on Tidal Environments*, pp. 2, 2008. 9. 25 – 27, Qingdao, China, 有缩减。

Tidal sandy ridges are the larger sediment bodies in the continental shelf environment, the bodies preserve the geological history records of Sea level changes, land-sea environment changes and the shelf evolution. It is a kind of new land and navigation channel resources and biological habitat. Thus, it has great scientifical and utilizable value for study. Following is the example from South Yellow Sea.

The South Yellow Sea sand ridge field is located offshore of the north Jiangsu Coast of China in a radiative fan pattern. The length of the sand ridge field is 199. 6 km from north to south, and 140. 0km in width from east to west. It consists of more than 70 sand ridges and tidal channels in the water depth from 0 to 25 m. The total area encompasses 22 470 km², of which 3 782 km² is above 0 meter.

Fine sands and silts, originated from old Changjiang River sediment during late Pleistocene period, are major sediment components of the sand ridge field. Late Holocene sand strata have well-preserved liminal bedding with more clay particles reflecting the influence from the Yellow River. Present, silt and clayey materials are supplied both by Changjiang River sediments and the abandoned Yellow River delta.

The radiative sand ridge field consists of essentially three genetic types according to seismic profiles and sediment core data. (1) Reformed old river depositional bodies in the center and southern parts, consisted of late Pleistocene sand ridges and superimposed by Holocene sediments. (2) Radiative current ridges in the northeast formed during early and middle Holocene. (3) Eroded-depositional sandy bodies in the northern part, outer part and small ridges out of the large one formed mainly during Holocene time.

The tidal channels have also three types. (1) The tidal channel developed along the old river valley (Changjiang), which were formed during the end of late Pleistocene to the present. (2) Patrimonial valley types were mature valleys or depressions during the end of the Pleistocene or early Holocene. (3) The tidal current erosional trough, formed during the late part of middle Holocene transgression.

Several times of the ridge-trough imposed records have been seen in the seismic profiles, all indicated the developing process of

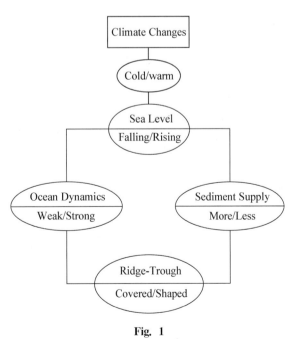

Fig. 1

the sand ridges has a periodic nature. That is, the period of sediment accumulation by rivers started during 30 ka B. P. of cold epoch while low sea level of $-20\,\mathrm{m}$, at that time the accent Changjiang River was entered Yellow Sea from Haian-Libao in the north Jiangsu provenance. Then the erosional formation period by tidal currents during warm epoch of Holocene transgression, such as, several erosional boundaries preserved in the strata of sandy ridges may indicate the transgression of sea level rising during 10 ka B. P. and 8. 0 ka B. P. of the initial and early Holocene time; 6 ka~5 ka B. P. of mid-Holocene high sea level, and the latest erosional period of 1 ka B. P. sea level. Thus, the river-sea interactive process in this area is closely related with the climate change, the rising and falling of the sea level is the key agent to free the coast zone land-sea dynamic interactive processes. It can be summarized as "transgression-dynamic-sedimentation" process pattern.

River-Sea Interaction and Paleo-Yangtze
Giant Delta System Studies[*]

Recent researches indicated that the Paleo-Yangtze giant delta system distributed in South Yellow Sea and East China Sea was formed during mid-late Pleistocene, originating from the huge river sediments, discharged from ancient Changjiang and Yellow River, and being shaped by monsoon waves and tidal currents. Upon the huge delta base, there are four delta unions superimposed on the top, and developed in different geological time: Paleo Changjiang-Yellow river delta, radiative sand ridge field, Holocene-modern Changjiang river delta and abandoned Yellow River delta. The Paleo-Yangtze giant delta system geographically covers most areas of East China Sea and South Yellow Sea, which also presents the characteristics of local continental shelf sediments. A pioneering multi-dimensional survey to this region is proposed in the paper, which includes RS analysis, water depth and hydro-dynamic survey, seismic profile survey on bottom stratigraphic structure, sea-bottom surface and sedimentary core sampling, followed by multi-disciplinary scientific studies of marine geology, marine sedimentology, marine dynamics, land-sea interaction, sea-level change and marine GIS etc. Through the comprehensive study, it is expected that the geographical range of paleo-Yangtze giant delta system be identified, the sedimentary structure and geomorphologic features of the delta system be revealed. Furthermore, the study will re-build the land-sea interaction process and dynamic mechanisms during mid-late Pleistocene in the region, and supply case study on the development of accumulative continental shelf, which will also provide theoretical input to marine geology. Further study of this giant delta system will help to understand the developing trend of eroding-depositing dynamics of modern tidal flat, to legitimately plan potential land resources in the region, and to contribute important scientific evidences to guard the National territorial sea-rights.

---

* WANG Y., ZOU X., YIN Y., ZHANG Y., LIU S.: *8th International Conference (IAG) on Geomorphology*, "*Geomorphology and sustainability*", *ABSTRACTS VOLUME* pp. 928, S22 – Submarine geomorphology, 27 – 31 August 2013, Paris.

长江三角洲水资源水环境承载力、发展变化规律与永续利用之对策研究[*]

长江三角洲经济区的地理范围介于 28°45′~33°25′N,118°20′~123°25′E,行政区域包括:上海市;江苏省的南京、镇江、扬州、泰州、南通、苏州、无锡、常州 8 市;浙江省的杭州、嘉兴、湖州、宁波、绍兴及舟山 6 市,共计 15 个市。地势从北向南,从东向西增高。三角洲平原的西南侧为低山丘陵所环绕。

长江三角洲土地总面积为 99 610 km²,占全国土地面积的 1‰;全区总人口约 7 404.71 万,占全国人口的 6.05%。2001 年,全区国民生产总值(GDP)约 16 981 亿元,占全国的 17.7%;完成财政收入 3 350 亿元,占全国的 20.4%;年进出口总额占全国的近 1/3,人均 GDP 达 22 537 元,相当于全国人均的 3 倍,发展的势头坚挺[1,2]。

长江三角洲城市群体、工农业与经贸文化水平在全国具有举足轻重的地位。但是,区域的自然环境承载力与经济发展存在着严重的不平衡。人均土地面积为全国平均水平的 1/6 略多,人均耕地为全国平均水平的 1/2[3]。随着城市化发展,耕地面积逐年减少;"三废"排放使大气污染,酸雨增多,水体污染;河流淡水富营养化,赤潮频发,水质恶化,地表淡水呈现水质型缺水;地下水水位大幅下降,土地沉陷,形成互为因果的系列性反应;海平面上升,海侵与风暴潮频频,河道排洪受阻,内涝积水;海岸蚀退,土地减少,沿江沿海土壤质地恶化等,区域生态环境脆弱,环境质量恶化,威胁到生存安全与经济发展,损害着我国投资环境与国防地位[4,5]。长江三角洲面临着严重的资源与环境压力的挑战!

1　长江三角洲地理环境

长江三角洲地处沿海,属于亚热带季风气候区,四季干湿冷暖分明,光照充足(1 800~2 250 h/a),热量充沛,平均太阳总辐射量达 4 187~5 024 MJ/m²,年平均温度 14.6 ℃~17.0 ℃,无霜期长(212~268 d),气候条件较优越。

1.1　河湖水系分布

(1) 长江三角洲水系主要分属长江、钱塘江与淮河 3 条水系,东南部沿海与舟山群岛为独立入海的甬江水系。平原水系虽不及南部山区水系繁茂,但具有人工河道网络的特色(图 1)。

＊　王颖,王腊春,王栋,陈文瑞:《水资源保护》,2003 年第 6 期,第 34 - 40 页。

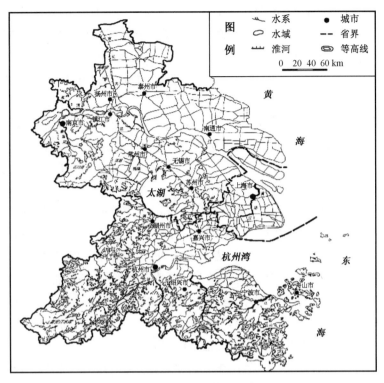

图1　长江三角洲范围及水系

属淮河水系的位于扬州至海安的通扬运河以北,水量丰富,河湖水网密集,从东向西为次一级的高宝湖水系,里运河水系和里下河低洼地水系。

南京至常熟徐六泾段为长江下游部分,其东为河口段。江北有滁河、通扬运河,以及京杭大运河贯穿南北,江南有秦淮河、锡澄运河、张家港河、望虞河、浏河与黄浦江。

钱塘江以桐庐分为中、下游,中游山区河流建有新安江与富春江水库,下游为感潮河道,有涌阳江与曹娥江。

(2) 湖泊主要集中于以太湖为中心的水网系统。太湖以西为上游地区,有苕溪、洮滆等水系及洮湖、滆湖及宜兴三九湖;太湖以东为下游区,有黄浦江水系,及阳澄湖、澄湖、淀山湖和昆承湖等,可能是太湖退缩过程的残留。三角洲湖泊还有南京市的石臼湖、固城湖,扬州市高邮湖、白马湖、邵伯湖,宁波市的钱湖及绍兴市的小湖群。

本区河湖水系密布,反映出长江三角洲低地区降水、径流与入海流量间的平衡方式。

1.2　地下水

长江三角洲地下水以第四系松散沉积物孔隙水为主,水质好,含有数层承压水层。向东部滨海区,水质较差。西部宁镇扬丘陵区、杭湖地区地下水属孔隙水、裂隙水及喀斯特类型。甬绍舟丘陵、盆地与岛屿多为裂隙水、孔隙水,来自白垩系碎屑岩中的裂隙水,矿化度较高,不宜饮用。

1.3　降水蒸发与气候灾害

长江三角洲区降水量丰富,尤其是每年6月中旬至7月中旬的梅雨型降水,持续时间长,笼罩范围广,降雨总量大,占全年降水量 20%～30%。7月下旬至 9月中旬为台风暴雨型降水。此外,尚有春雨及夏季雷雨降水。年大气降水仍属均匀情况,区域内平均降水量 1 000～1 465 mm/a,其中,江苏与上海一带降水量为 1 000～1 200 mm/a,浙东北为 1 100～1 400 mm/a,淳安、昌化达 1 430～1 465 mm/a,而启东与崇明降水量偏少,约 1 000 mm/a。

受季风活动影响,降水的年际与月际变化大。如上海年最大降水量为 1 568.7 mm (1977年),最小为 772.3 mm(1978年);杭州年最大降水量为 2 356.1 mm(1954年),最小为 954.6 mm(1967年);吴县年最大降水量为 1 467.0 mm(1960年),最小为 604.2 mm (1978年)。最大最小之差别为 1～2.5 倍[6]。

长江三角洲河湖水体的多年平均蒸发量为 950～1 100 mm,夏季蒸发量大,占全年总量的 40%;冬季蒸发量小,占全年总量的 12%;春秋两季蒸发量各占 24% 左右。本区陆面蒸发量 600～800 mm,相当于水面蒸发量的 70%～80%,区内蒸发量大致是南部大于北部,平原大于山区[7]。因此,河湖水体降水量多消耗于蒸发,70% 陆地降水消耗于蒸发。但是,水气资源总量是相对平衡的,因此,利用蒸发水汽是补充水源的一条重要途径。

夏秋季洪涝灾害:经历过 1954年,1962年,1991年与 1998年份大水,尤以 1991年灾害严重,太湖平原与里下河平原洪灾损失达 200 亿元。

干旱灾害:干旱灾害发生于伏秋两季。伏旱影响水稻、棉花与果木生长,秋旱影响三麦发育。

寒潮与霜冻:每年约受 3～4 次寒潮,于 48 小时内降温 10 ℃ 以上,伴以大风。太湖平原于 1955年和 1997年受寒潮影响,低温为 −10 ℃～−15 ℃,湖水结冻,油茶、三麦、柑桔大面积冻坏。

阴湿害:一年四季均可发生,由于日照少,气温低,湿度大,地下水水位高等原因,造成作物霉烂与病害。

以上综述反映出长江三角洲区域除河水以外的水环境资源的自然特点。

2　长江三角洲水资源现状与问题

2.1　水资源量

(1) 当地水资源量。长江三角洲多年平均当地径流量为 508.4 亿 m³,其中浙东北约占 70%,江苏部分占 26%,上海市占 4% 左右。

地下水也是本区的重要水源。区内多年平均地下水水量 177.71 亿 m³,其中浙东北为 87.72 亿 m³,江苏部分为 70.66 亿 m³,而上海市为 19.33 亿 m³。

长江三角洲多年平均水资源量达 573.79 亿 m³。其中上海市 32.07 亿 m³,江苏 76.57 亿 m³,浙东北 365.15 亿 m³(表 1)。本区单位面积水资源量 57.6 万 m³/km²,比较丰沛。但人均水资源量只有 774.9 m³,仅相当于全国平均水平(2 420 m³)的 1/3。这与年降水分配不均,雨季洪水排泄,蒸发及污染水质性缺水等综合原因有关。

在进行水资源量计算时,将地表水与地下水资源量之和减去河川基流量(计算中的重复水量),即为当地水资源量。

表1 长江三角洲多年平均当地水资源量[7]

地区	年降水量 /mm	年径流深 /mm	当地径流量/亿 m^3	地下水量 /亿 m^3	河川基流量/亿 m^3	当地水资源量/亿 m^3	水资源量/(万 m^3·km^{-2})	人均当地水资源量/m^3
上海市	1 141	301	1 866.00	19.33	5.92	32.07	50.58	245.85
苏中南	1 040	254	134.43	70.66	28.52	176.57	36.55	457.55
浙东北	1 465	737	355.31	87.72	77.88	365.15	81.21	1 629.25
全 区	1 266	501	508.40	177.71	112.32	573.79	57.60	774.90

年径流深反映出一个地区水资源量的丰寡(图2)。

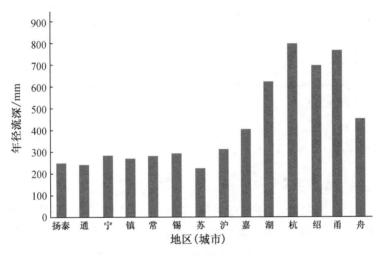

图2 长江三角洲地区多年平均径流深

(2) 外来水。外来水包括过境水和引江水。本区处于几条大河下游,外来水资源丰富。仅长江干流多年平均过境水量就有9 730亿 m^3。1954年年过境水量达13 590亿 m^3,为均值的1.4倍,最枯年(1928年)也有6 320亿 m^3。长江具有较好的供水条件,即使在枯水年份,也能满足供水要求,如大旱之年1978年,苏南引江水112亿 m^3。另外,上海黄浦江年均进潮量有409亿 m^3。上海市的潮水量是当地水量的22倍。长江三角洲的外来水丰富,是区内开发利用水资源十分有利的客观条件。

2.2 水资源的时空分布

长江三角洲水资源量的时空分布与降水一致,大致呈南多北少,山区多平原少的特点。但是,平原区外来水较为丰富,水资源开发具有较大潜力。

由于本区为典型的季风气候,降水季节变化明显,导致了水资源的年内分配不均。每年降水主要集中在春夏之间的梅雨和夏秋之间的台风雨。相应地,5—9月份的径流量占全年径流量的60%～70%。同时,本区降水与径流的年际变化可达2～5倍(表2、表3)。丰水年雨水过多,造成洪涝灾害;而少水年雨水过少,以致干旱缺水。

表 2 长江三角洲部分水文站降水量年际变化[8]

水文站	降水量最大值/mm	降水量最小值/mm	平均值/mm	最大值与最小值之比
嘉兴	1 719.4(1954)	723.1(1978)	1 189.9	2.4
绍兴	2 182.3(1954)	922.5(1967)	1 420.0	2.4
诸暨	2 171.3(1937)	903.8(1978)	1 449.7	2.4
岱山	1 589.9(1977)	701.0(1967)	1 133.9	2.3

注:括号中数字为发生年。

表 3 钱塘江等河流入海水量年际变化[8]

河流	集水面积 /km²	多年平均入 海水量/亿 m³	水量最大值 /亿 m³	水量最小值 /亿 m³	最大值与最小值之比
钱塘江	41 700	373.0	692.0(1954)	179.00(1979)	3.87
曹娥江	6 046	42.8	72.2(1954)	19.60(1967)	3.68
甬江	4 294	28.6	41.9(1952)	9.96(1967)	4.82

注:括号中数字为发生年。

2.3 水体质量

2.3.1 长江沿岸水体质量

长江下游总体水质尚可,一般能达到 GHZB1—1999《地表水环境质量标准》Ⅱ~Ⅲ类水质。2000 年,长江江苏段长江干流共布设 12 个水质监测断面,监测结果表明:12 个监测断面中有 4 个断面受石油类影响,水质处于Ⅳ类,其余断面均能符合Ⅱ类标准。石油类是影响江苏长江干流水质类别的主要污染指标,石油类超标现象较为普遍。镇江段的 DO、镇江内江段的 COD$_{Mn}$ 和非离子氨等超标现象也时有发生(表 4)[9]。

表 4 2000 年长江江苏段水质评价[9]

河 段	综合污染指数	水质类别	主要污染指标
南京段(江宁河口)	2.52	Ⅳ	石油类、非离子氨
镇江外江(龙门口)	2.00	Ⅱ	石油类、DO
镇江内江(三号码头)	4.74	Ⅳ	石油类、非离子氨、DO、COD$_{Mn}$
南通段(姚港)	2.82	Ⅱ	COD$_{Mn}$、非离子氨、石油类

长江近岸水域污染程度比主流深泓水域高。南京段近江边局部存在宽窄不等的污染带,最大面积达 1.26 km²,有些还影响到自来水水源地的水质。环境规划划定的 9 个饮用水源保护区中有 1/3 水质达不到国际标准。

沿江各市其他河道污染严重,除少数几条紧靠长江的河流之外,几乎全为Ⅳ~Ⅴ类水质。部分河段已达不到农业灌溉要求。如南京市内河水域(内外秦淮河、金川河)水质均劣于Ⅴ类,终年发黑发臭,属有机污染,主要污染源为生活污水。

因水源污染,南京郊县在枯水季节多次出现水荒。又如南通市内河水源主要来自长江,1995 年水质综合污染指数 7.23,比 1990 年上升了 28.2%。按 1995 年监测结果,只有沿海运河各项指标能达Ⅱ类水质标准,新通扬运河因石油类和 DO 超标,仅符合Ⅴ类水质标准,其余河流均为Ⅳ类水质。该市内河主要污染指标为非离子氨,BOD_5,DO,COD_{Mn}。

2.3.2 太湖流域水体质量

太湖是上海市、无锡市和苏州市的主要饮用水源。近几年来,其水质一直在下降。按照 GHZB1—1999 标准评价,2000 年太湖总体水质为Ⅳ类,其中湖岸苏州片区、湖州长兴片区及湖心区水质相对较好;湖岸无锡片区和宜兴武进片区水质较差,五里湖和梅梁湖分别受非离子氨和 TP 影响,水质劣于Ⅴ类标准,污染明显重于其他湖区。

由于太湖流域地面水体流向、流速受闸坝控制程度较大,太湖湖体枯水期水质好于平水期和丰水期。丰水期由于大量污染物的汇入而导致太湖湖体水质明显下降。

目前太湖富营养化仍然较为严重,水体中 TP、TN 含量偏高。苏州片区、湖州长兴片区及湖心区处于中富营养水平,其余湖区均处于富营养状态。太湖枯水期水体为中富营养,平水期和丰水期为富营养。

根据 2000 年监测结果,阳澄湖和滆湖总体水质为Ⅳ类,主要超标(Ⅲ类)项目有 TP,非离子氨,COD_{Mn},BOD_5(表 5)。

表 5 2000 年太湖等湖泊水质监测结果[9]

湖泊	DO/(mg·L^{-1})	超标率/%	COD_{Mn}/(mg·L^{-1})	超标率/%	BOD_5/(mg·L^{-1})	超标率/%	TP/(mg·L^{-1})	超标率/%	TN/(mg·L^{-1})
太湖	9.4	0	5.9	32.3	2.9	19.8	0.107	73.8	1.91
阳澄湖	8.8	7.1	5.6	23.8	3.4	31.0	0.127	97.6	2.71
滆湖	7.0	12.5	5.2	8.3	2.2	0	0.094	58.3	1.38

注:表中超标率按地面水Ⅲ类标准统计。

在上海,除长江口大多数指标能达到Ⅱ类水质标准外,地面水已无Ⅰ、Ⅱ类水体。Ⅲ类水体也仅限于淀山湖及太浦河,上海市已经成为长江三角洲水质型缺水最严重的地区。

另外,属于太湖流域的东西苕溪超标项目主要为 COD_{Mn} 和非离子氨。按 1997 年监测结果,属于Ⅰ~Ⅱ类水体的占 17%,Ⅲ类占 60%,Ⅳ~Ⅴ类占 17%。劣于Ⅴ类的有 6%。东苕溪水质优于西苕溪,受污染河段主要分布在西苕溪的塘浦河段、东苕溪的青口和青山水库东河段,以及长兴平原水系的下菶桥河段[10]。

2.3.3 钱塘江、曹娥江及甬江水质

钱塘江、曹娥江和甬江水质以Ⅰ~Ⅱ类为主。钱塘江水质较差河段主要分布于东阳南江的城头河段、东阳江的后金渡河段、江山港的双港口河段、浦阳江的安华和浦阳江出口河段,主要超标项目为 COD_{Mn}、BOD_5、挥发酚、非离子氨和石油类。

曹娥江污染河段主要分布于下游桑盆殿和东江闸一带以及上游新昌江的水文站河段。水质主要超标项目为 BOD_5、非离子氨、挥发酚和石油类。

甬江Ⅳ～Ⅴ类水体占14％,主要分布在姚江的下陈渡河奉化江的翻石渡河段,主要超标项目有COD$_{Mn}$、DO(图3)。

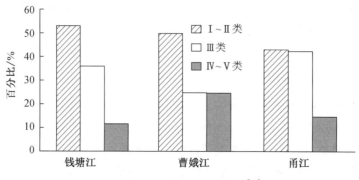

图3　钱塘江等水质监测结果[10]

2.3.4　京杭大运河

京杭大运河在长江三角洲范围内水质大多为Ⅴ类,尤以镇江段和无锡段污染最为严重(表6),为劣Ⅴ类标准,属重污染,主要污染指标是BOD$_5$、COD$_{Mn}$、DO、挥发酚。

表6　京杭大运河水质监测结果[9,10]

河段	综合污染指数	水质类别	主要污染指标
扬州	3.07	Ⅳ	石油类、BOD$_5$、COD$_{Mn}$
镇江	5.58	劣于Ⅴ	BOD$_5$、COD$_{Mn}$
常州	4.96	Ⅴ	石油类、挥发酚
无锡	9.17	劣于Ⅴ	DO、非离子氨
苏州	5.46	Ⅴ	DO、BOD$_5$、挥发酚
嘉兴、杭州		Ⅳ～Ⅴ	COD$_{Mn}$、BOD$_5$、石油类、非离子氨

2.3.5　近海水质

近几年来,长江口外、杭州湾和舟山渔场海水综合水质大多劣于Ⅲ类海水标准(图4、图5)。长江三角洲近海水体中无机氮、无机磷和石油类超标尤为严重。其污染源主要是沿海工业废水、生活污水和船舶漏油。另外,近海养殖所产生的养殖废水和沿海滩涂农业开发排放的化肥、农药等污染物也是影响近海水质的因素之一。

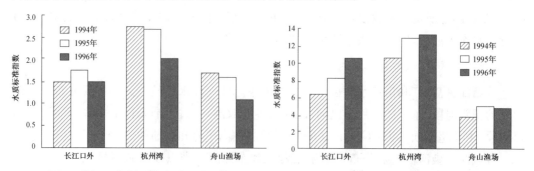

图4　长江三角洲近海水质(无机磷)　　　　**图5　长江三角洲近海水质(无机氮)**

2000 年,按 GB3097—82《海水水质标准》进行水质评价,苏北沿岸黄海水质,除个别河口区污染较重外,大部分仍属Ⅱ类海水。如南通大洋港海区受活性磷酸盐影响水质属四类,小洋口海区水质属Ⅱ类。

1996—2000 年,南通两个海区总体水质略呈好转趋势(图 6)。与"八五"期末相比,石油类、活性磷酸盐、硝酸盐氮和铅等项目污染有不同程度的减轻。

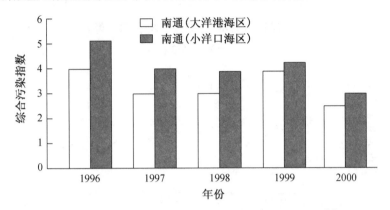

图 6　1996—2000 年江苏南通近岸海域水质变化趋势[9]

2.4　水问题的主要症结

2.4.1　水资源紧缺矛盾比较突出

长江三角洲名曰水乡,但由于水资源分布不均,时空变化大,组合很不平衡。全区人均当地水资源量 774.90 m³,仅为全国平均水平的 1/3。其中江苏地区和上海市,人均只有 245.85 m³ 和 457.55 m³,属水资源紧缺地区。过境水为区内具有很大开发潜力的水源,由于开发资金和水质问题,利用量较少。

在沿海岛屿和沿海垦区,水资源紧缺矛盾更为突出。舟山群岛人均水资源量582.9 m³,相当于全国平均水平的 1/5。目前,有 30 万人的饮水存在不同程度的困难。其中特别困难的有 20 万人,占人口总数的 20.3%[11]。余姚、慈溪两市地处浙东北沿海,若遇特枯年份,其年径流量仅为平水年的 55.9%,灌溉用水将欠缺 6 亿 m³[7]。

2.4.2　水质日趋恶化

(1)水质恶化与湖泊富营养化。由于区内湖泊所接纳的废水、污水量逐年增加,营养盐含量逐渐上升,浮游植物的生物量有了较大幅度的提高,许多湖泊已达富或中富营养化状态。藻类过度繁殖而使水体着色或在水面聚集形成条带或片状的现象。其发生时间主要集中在 4—10 月份,局部水域 2—3 月份就出现,夏季几乎全湖都能看到。春夏季以蓝藻、硅藻占优势,秋季则以蓝藻占绝对优势。太湖藻类水华在空间分布上也极不均匀,由重至轻依次为五里湖、梅梁湖和竺山湖、大太湖、东太湖[12]。

长江三角洲水质的变化趋势可以太湖为例(图 7、图 8)。

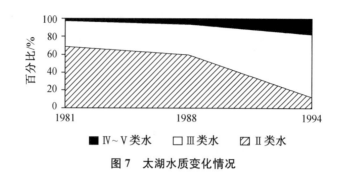

图 7　太湖水质变化情况

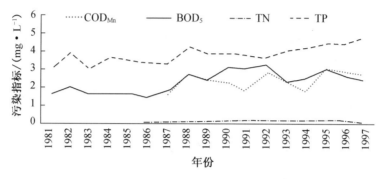

图 8　太湖水质主要指标逐年变化趋势[13]

20 世纪 80 年代初,太湖以Ⅱ类水为主(约占 69%),Ⅲ类水体占 30%,仅有 1%为Ⅳ类水。Ⅲ、Ⅳ类水主要分布在入湖水道及东太湖沿岸水域。1981 年富营养化水体所占面积仅 16.8%。80 年代后期,太湖水体污染加重,1988 年Ⅲ类水增加为 36.6%,Ⅳ类水占 3.2%,且已出现重污染水体(Ⅴ类),主要分布在梅梁湖区间江口附近及五里湖区部分水域,面积约占 0.8%。富营养化水体所占面积已达到 40%。至 90 年代中期,太湖水质以Ⅲ类为主(占 70%),Ⅱ类水仅占 15%,Ⅳ类水也扩大为 14%,主要分布在梅梁湖、太湖西岸和南岸。同时,该期Ⅴ类水体所占面积也增加为 1%。自 1995 年以来,太湖大部分水域属Ⅲ类水体,Ⅱ类水体所剩无几。全太湖大部分为富营养化状态,其中五里湖和梅梁湖已达重富营养化水平。1996—1999 年,太湖总体水质趋于好转,但 2000 年水质有所下降,其中 COD_{Mn} 污染明显加重。这与该年汛期太湖流域水量较往年偏枯,加之前两年直湖港和武进港断流建闸,而该年正常启闭,污染物入湖总量加大有关(图 9)。

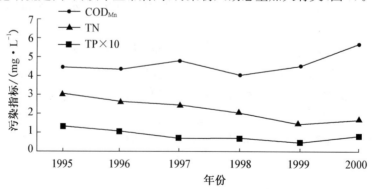

图 9　太湖水质三项主要指标逐年变化示意图[9]

1994—1999 年,太湖富营养化程度一直处于富营养水平,1998 年以后富营养指数下降,水体富营养化严重的局面有所缓解,但是 2000 年水体富营养化程度又由 1999 年的中富营养水平上升至富营养水平(图 10)。

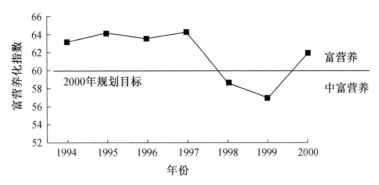

图 10　太湖湖体富营养化指数变化趋势[9]

(2) 近海赤潮。大量工业废水和生活污水,以及含有化肥等的农业灌溉排水,通过各种途径注入海洋,使得浅海营养盐不断增加,造成某些海域富营养化,赤潮频频发生。

根据国家海洋局第二海洋研究所遥感室的研究,本区赤潮主要发生在长江口外海域、南通市东面海域、舟山群岛附近海域以及三门湾附近海域(图 11),其中长江口外海域赤潮发生最为频繁。1993 年,在该海域共发现 4 次赤潮,分布面积最大达 50 km²。1995 年 5 月 27 日,该海域发现棕红色、呈条带状分布的赤潮带,面积达数百平方千米[14]。

图 11　长江三角洲赤潮分布示意图[13]

(3) 洪涝灾害。梅雨和台风雨是导致本区洪涝灾害形成的重要气候因素。洪水类型相应可分为两种:一是梅雨型,一般发生在 6—7 月间,总量大,历时长,范围广;另一种是台风雨型,一般发生在 7 月下旬至 9 月中旬,强度大,历时短,范围小。太湖平原地形极为低洼,地势周高中低呈洼形,总地势自西向东缓降,海拔 2～10 m,大多为 3～8 m。西部与小丘、阶地相接,南北两侧分别由钱塘江、长江的堤岸构成,地势高爽,东临滨海滩地,全区水网稠密,中部湖荡众多,水域开阔。上游来水易于急速汇集,而下游河道束窄,地形平坦,泄洪滞缓,易于积水成涝。

里下河地区也是周边高、中间低的碟形洼地。该区西部为运西平原,地面海拔为 6～10 m,以东为海积平原,地面海拔多为 2.5～4 m,以南为三角洲平原,海拔 6～9 m,以北为黄泛平原,海拔 6～7 m 以上,唯中部里下河低地海拔基本在 2 m 以下,中部最洼处仅1.1 m。这种地形条件容易积涝成灾,一遇大水则整个里下河便成泽国。

2.4.3　人类活动不当导致三角洲地区灾害加剧

太湖流域湖荡围垦是在 1985 年之前进行的。36 年间,太湖流域为围垦共建圩 498座,面积 538.55 km²(表 7),占 20 世纪 50 年代初期原有湖泊面积的 13.6%。平均每年因围垦而减少的湖泊面积为 14.69 km²。因围垦而消失的湖泊计 165 个,面积161.3 km²[15]。

表 7　太湖流域湖荡围垦动态变化分片统计[15]

湖荡片	50 年代		60 年代		70 年代		80 年代		合计	
	圩数/座	面积/km²	圩数/座	面积/km²	圩数/座	面积/km²	圩数/座	面积/km²	圩数/座	面积/km²
洮滆片	3	2.01	28	28.63	91	147.32	4	4.18	126	182.14
太湖片	7	9.23	39	67.73	68	82.16	2	1.05	116	106.17
阳澄片	2	0.94	8	28.08	24	19.05	2	0.30	36	48.37
淀柳片	1	0.45	26	17.36	84	58.63	5	2.05	116	78.49
嘉平片	1	0.35	10	4.41	13	8.91	4	6.88	28	20.55
杭嘉湖片	2	0.37	14	10.76	54	24.94	6	2.76	76	38.83
合计	16	13.35	125	156.97	334	341.01	23	17.22	498	528.55

湖荡的围垦导致太湖水位抬高 9～14 cm,湖西洮滆片抬高 15～20 cm,湖东运河片抬高 10 cm 左右,淀柳片抬高 7～10 cm,澄锡虞、阳澄片及杭嘉湖片抬高 5 cm 左右[15]。围湖之后,不仅削弱了湖泊的调蓄能力,同时也切断了原与湖泊相通连的河道,而且有的圩区位于湖泊下游排水尾闾地段,使过水断面束窄,阻洪碍洪,太湖 1991 年的排水量仅相当于 1954 年出水流量的 1/3[16]。

其次是过量开采地下水,引起地面沉降,形成了新的低洼地,大面积低洼地的出现改变了河流的径流条件和流向;同时,地面沉降降低了防洪工程标准,加重了洪涝灾害的险情,加剧了水淹损失。1991 年水灾中,苏锡常地区因地面沉降扩大受淹面积达 1 300 km²,使灾情明显加重,且受淹区多在工业区和新建住宅生活小区,经济损失甚大。

3 探索长江三角洲水环境水资源优化与良性发展的途径

清洁的淡水、新鲜的空气及营养的食物是人类健康的基本要素。水量供需平衡、减轻水体污染和洪涝灾对水质水环境的损害,是当前亟待解决的三个重要问题。需在现有的基础上,对长江三角洲全流域以及相关的毗邻区进行整体的、系统的深入研究,为长江三角洲优化水质、增补水量与导致水环境的良性发展做出宏观控制规划、阶段性的具体措施以及监督实施与管理的制度,即:一体化的规划、一体化的实施与一体化的管理(图12)。

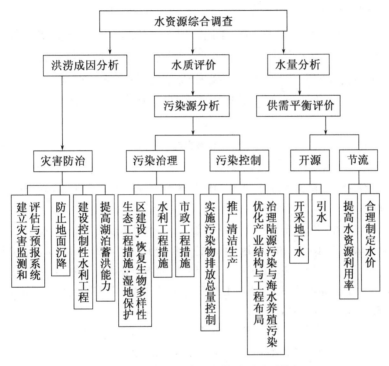

图 12 长江三角洲水资源可持续利用对策框图

科学的规划与措施,必须包含关键问题的解决途径、成功的标志、发展的导向与前景。关键问题包含以下几方面:

(1)首先需阐明三角洲区当代长江干流的水文与河道特性,变化状况、原因与活动规律。据此,所做出的对长江兴利避害、发挥其优势作用的宏观指导思想。

(2)对长江三角洲水循环与水环境的承载力、其特点与发展变化的趋势性等做出量值分析。在当代自然条件与全球变化的前景下,应该分别做出长江三角洲水环境水资源所能支持民众健康生存,农、工、城、交、旅、贸社会经济发展的特点(类型组合)与承载能力,及保持生态多样性的状况下的纳污能力的最佳、适宜与极限值 3 个档次的分析。应对以下两方面做出评价:① 今后 20 年中,当工程技术有相当程度应用于污染治理、环境修复,增加水的重复利用比率,以及水资源环境得以宏观调控的情况下,长江三角洲人口、产业(类型、规模)、经济发展的承载力分析。② 2050 年,在新技术广泛应用,人民群众的环境意识与公共道德水平普遍提高的情况下,人口与发展、产业类型与城乡规模、生态环境与生存方式等的变化及承载力水平分析。

（3）现有水环境的治理（控制减少工、农业污染源，就地解决城镇污染）、污染防治、清洁水源的有效获取、有效利用与有偿使用及清洁排放水流的措施与管理系统。

（4）三角洲流域的整体规划，疏导江、河、湖、塘与入海体系，控制地区间的人工建筑与统一水利工程设施，使水体流畅与良性循环，有计划地、分阶段、有效地防治洪水、内涝与潮侵灾害。

（5）全区的生态环境建设与发展规划，以流域为整体，利益共享，风险共担；按地区条件、特点不同，各地区互有分工，互相补充，发展形成长江三角洲低地环境特有的良性生态系统、生态景观与生态经济发展区。

（6）实施区域水环境、水资源的一体化管理方案、实施办法与规章制度原则。

（7）公众的教育。使广大群众充分认识到淡水资源有限，要珍惜淡水，形成节约用水、人人有责的观念、习俗与公共规范。发展形成社会性多途径地利用自然水循环增水与补水，发扬良好的用水传统，重复有效地用水，保证清洁水流入海洋。

参考文献

［1］国家统计局城市社会经济调查司.2002.中国城市统计年鉴(2001)[M].北京:中国统计出版社,118-119,480-539.

［2］国家统计局.2002.中国统计年鉴(2001)[M].北京:中国统计出版社,52-53,251-258.

［3］陈文瑞,朱大奎.1998.长江三角洲土地资源可持续利用[J].自然资源学报,13(3):261-266.

［4］任美锷.1996.长江三角洲可持续发展的若干问题[J].世界科技研究与发展,18(3-4):97-100.

［5］朱大奎.1999.长江三角洲的特征与当前主要任务[A].严东生,任美锷.论长江三角洲可持续发展战略[C].合肥:安徽教育出版社,139-151.

［6］严济道.1986.长江三角洲自然季节降雨趋势[J].地理研究,5(1):51-57.

［7］佘之祥.1997.长江三角洲水土资源与区域发展[M].合肥:中国科技大学出版社.

［8］中国农业丛书编辑部.1997.中国农业全书(浙江卷)[M].北京:中国农业出版社.

［9］江苏省环境监测中心.2001.江苏省环境质量报告书(2000年)[R].

［10］浙江省环保局.1998.浙江省环境状况公报(1997年)[R].

［11］李植斌.1997.浙江省海岛区资源特征与开发研究[J].自然资源学报,12(2):139-145.

［12］吴瑞金.2001.我国湖泊资源环境现状与对策[J].中国科学院院刊,(3):176-181.

［13］陈文瑞.1999.长江三角洲水土资源可持续利用研究[D].南京:南京大学博士学位论文.

［14］国家海洋局.1996.一九九五年中国海洋环境年报[R].

［15］梁瑞驹,李鸿业.1993.1991年太湖流域洪涝灾害[M].南京:河海大学出版社,97-103.

［16］王腊春.1996.太湖流域洪涝灾害模拟及预测[D].南京:南京大学博士学位论文.

对南京市港口与沿江风光带发展建设意见[*]

第一部分 概 况

2003 年 9—10 月间,在南京市委组织部的精心组织和安排下,本"沿江景观风貌和港口建设专题组"的考察工作分两个阶段进行。

第一阶段是专题组随院士专家沿江考察全体成员一同考察。

9 月 5—6 日两天的考察在沿江北岸进行,路线自佛手湖→珍珠泉→大吉度假村→老山国家森林公园→经珠江镇→浦口老城区;再从六合桂子山石柱林→金牛湖风景区→南京化学化工园(图 1)。

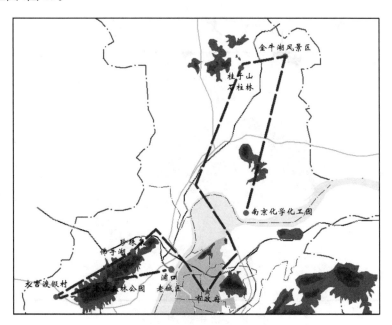

图 1 沿江北岸考察路线示意图

9 月 29—30 日两天集中在沿江南岸考察,路线自栖霞区仙林大学城→金陵石化烷基苯厂→八卦洲→新生圩深水码头→南京经济技术开发区→幕府山植被修复工程→燕子矶三台洞→下关区政府;后从鼓楼区鬼脸城及外秦淮河→龙江市内商业街规划→江东宝船厂遗址→凤凰西街红二楼居民区→建邺区奥体中心→江东管理区河西指挥部→江心洲污水处理厂(图 2)。

---

* 景观风貌与港口建设调研组(王德滋,王颖,朱大奎):《南京市沿江开发若干问题咨询调研报告》,112 - 130 页,2004 年 2 月。

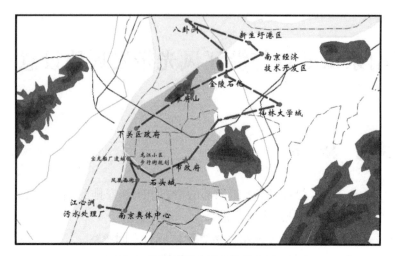

图 2 沿江南岸考察路线示意图

第二阶段为专题组考察。

于 10 月 25、26、30 日进行。再次对浦口区老山国家森林公园及其附近景观和沿江港口码头作了重点调研。其路线见图 3。

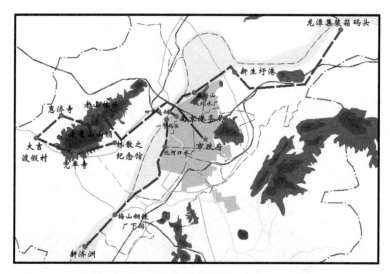

图 3 专题考察路线示意图

专题组成员在老山国家森林公园,详细听取了老山林场开发规划的情况介绍,并对老山顶峰的老鹰山顶瞭望塔及附近的兜率寺、惠济寺等作了调研。在下关南京港务局,听取了负责同志对整个南京段码头港口的全面介绍后,自下关码头乘船进行了沿江考察。沿江下行至新生圩港后驱车到达龙潭深水港工地;沿江上行经板桥至梅山钢铁厂。在沿江观察两岸景观现况的同时采集了江水样品。

考察合计历时 7 天,涉及沿江 6 个区,在所到地区有关方面的全力支持下,对南京沿江开发的总体情况,包括基础设施、环境、经济、旅游、资源等很多方面的开发建设思路、规划和实施情况有了较全面的初步了解和印象。

第二部分 专题调查研究初步意见

1. 关于南京市的城市定位设想

南京市位居长江三角洲的顶点,居内陆与江海联运的枢纽。受潮水顶托、江潮相互作用,长江在南京地区分叉形成新生洲、新济洲、江心洲、八卦洲及潜洲等一系列沙岛与浅滩。沙岛、浅滩使江水分为夹江与主河道。陆地上河、渠、湖、塘呈星网状分布,反映出三角洲地貌特点。沿江两岸除南部有幕府、栖霞低山丘陵与北部老山、龙王、灵岩山地遥相夹峙外,均为坦荡的平原河段(参见图4)。长江流贯东西,孕育了两岸肥田沃野,赋予南京市生存活力。六朝古都历史悠久,华东重镇人文荟萃,江南大都市生机勃勃,南京的发展必须突出地体现其地理环境特色,建设成长三角山水生态园林历史文化名城。南京自然风光很美,有悠久的人文历史,有山、有水、有泉。

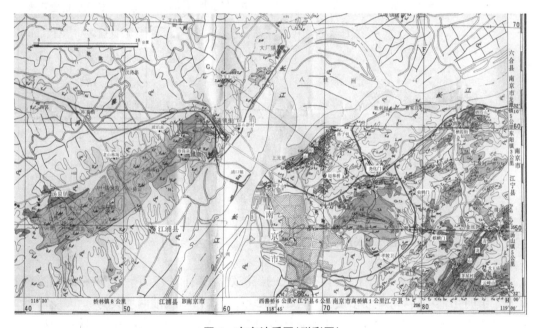

图4 南京地质图(附彩图)

2. 沿江发展港口是龙头

南京以下长江长达366 km(参见图5、表1)。南京段长江江宽约1.5 km,最宽处3.5 km(新济洲尾东北方),最窄处为长江大桥一带(1.4～1.5 km),江心洲夹江河道仅250 m宽。水深一般15～30 m,最深约70 m,是一条黄金水道,天然水深可通行5万t级海轮。但是,以下两个因素影响着深水港的发展。

(1) 长江口拦门沙淤浅,天然水深－6.5 m,经一期工程浚深至－8.5 m,趁潮(平均潮差2.6 m)可通行5万t级海轮。长江口二期整治预期主航道水深为－12.5 m,趁潮可通行7万～10万t级海轮。南京以下航道维护水深10.5 m,据南京港务局长江航道整治报告(2002.9.4):南京以下长江航道的几处浅点已成为长江下游航道的"瓶颈",制约着大型海轮进入南京港。下游航道中,影响最严重的是福姜沙水道和焦山丹徒水道。福姜沙南水道为海轮航道,航道弯曲狭窄,上口最窄处仅有 120 m,福姜沙北水道航道顺直,但河床变化较大,江心多暗滩,水深较浅,最浅处水深仅 4.5 m。焦山丹徒水道主航道在右汊,航道弯窄,洲头处航路曲折,10 m 深槽窄处仅 300 m 宽,左汊经冲刷深泓变深,水深在 15 m以上,但未辟作航道。如结合长江整治的总体方案统筹考虑对南京以下长江航道的综合治理,将福姜沙、焦山丹徒水道改造成航宽 350 m,维护水深在 15 m 的航道,可满足 7 万 t海轮满载直抵南京港,以充分发挥长江"黄金水道"的社会效益和经济效益。该报告建议十分重要,必须整治浚深镇江附近的焦山丹徒水道浅滩与江阴以下的福姜沙水道浅滩,疏通长江深水航道。当长江口疏深为－12.5 m 水深的深水航道工程完成后,则 7 万 t 级满载海轮可直达南京,10 万 t 级可乘潮满载到南京。

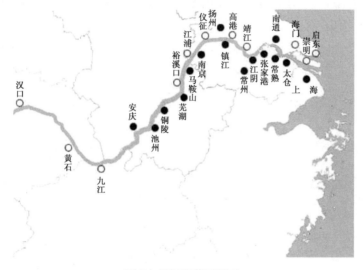

表 1　长江航道里程表

起讫航段	里程(km)
吴淞口—太仓	74
吴淞口—南通	120
吴淞口—镇江	279
吴淞口—南京	366
吴淞口—芜湖	460
吴淞口—安庆	690

图 5　长江航道历程图

疏通南京长江深水航道工程,具有很大的经济效益。目前 15 万～20 万 t 级海轮铁砂矿在宁波舟山港中转,每吨要多支出 30 元,以 2002 年矿砂与石油两项共为 2 400 万 t 计,将多开支 7 亿元,随着南京及长江中游经济快速发展,大宗货物运量会快速增加,其经济效益将更大。此外,南京以下长江深水航道建成后,南京港可通行 6 000 个标箱集装箱船(乘潮吃水 15 m),除对直接货运的经济效益以外,则对南京地区经济发展将起巨大的推动作用,其社会经济效益不可估量。

(2) 跨越下关至浦口之间的南京长江大桥净空 24 m,阻挡了 5 万 t 级海轮向上游通行。因此,目前仅在大桥以东才能建设万吨级以上的江海联运深水港;只有低桅杆、吃水浅的 5 000 t 级宽体船可上行达武汉。因此,长江大桥以东再建设桥梁必须重视桥下净空不再碍航的问题。

南京港是长江直接通海的江海港的顶端,其岸线资源十分宝贵。南京港万吨级轮船码头主要分布于弯道、江面束狭的深水岸段。如新生圩石油、机械集装箱码头,龙潭集装箱深水码头及仪征石油化工码头三处是适应深水深用原则的合理安排,码头前沿水深为

−11～−13 m。沿江还分布着许多小码头,尤其是采沙装沙码头(后附照片10～12),需加以整治。建议作为交通运输的长江南京段各港口码头,统一规划整治,统一归属市港务机构管理,深水深用,浅水浅用,以便合理利用长江干线资源、长江航道资源,以结束目前一些岸线码头经营中存在的多头管理的状况。

3. 关于建设沿江风光带的意见

江阔水深、浩浩荡荡、气势磅礴的长江带给南京市无限生机,可陶冶胸怀、焕发活力、令人鼓舞竞进。但是,由于战乱、血吸虫病害与过去无序的开发,留下了荒芜、杂乱与破旧的景象(后附照片)。南京大都市的建设必须重视滨江两岸的规划,改变城不见江,江不见城的现状,建设自然生态环境与历史人文要素和谐发展的沿江城市风光带。

对比分析国内外滨水城市地区的规划与建设特点(见照片),水体在城市发展中具有:景观功能,水源功能,交通运输功能,游乐运动功能,生态环境功能以及继文化历史渊源、发展文化与最宜人居的功能等。

滨水地段是城市的珍贵资源,需以人为本,很好地从"人地和谐相关"与"生存环境持续发展"的观点指导进行城市规划与建设。

国内外滨水城市建设典范照片

世界名城——悉尼

东亚模范——厦门

动感之都——香港

东方明珠——上海

今日珠江——广州

设想中的南京沿江风光带宜由六个板块及连通的绿树廊道组成串珠状沿江分布,见图6。

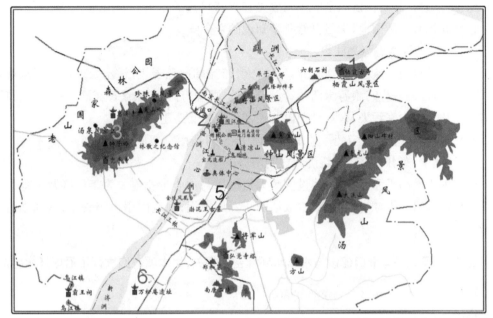

南京大学海岸与海岛开发教育部重点实验室绘制

图6　沿江六大板块分布图

风光带大体上按建设先后之序:

(1) 基岩岸,主要由古老的石灰岩组成,悬崖临水、岩洞成群。燕子矶位于南岸直渎山,海拔36 m,是长江中下游三大矶头之一(燕子矶、采石矶、城陵矶),石峰突兀江上,似基岩深水岸段——燕子矶、幕府山、栖霞山风光带。

以长江二桥为中心点,此带由两个中心组成:燕幕为一,另一为九乡河—栖霞山。

① 西侧为燕子矶与幕府山乳燕展翅,矶头远眺,江亭御碑、文人辞赋,浑然一气。

据我们实测,三台洞记录着四次长江洪水位的珍贵遗迹,标高8.5 m,9.4 m,10.5 m,12.9 m,断续相延保留于头台洞至燕子矶之间,高出于现代长江江面之上(长江标高3.498 m)。

幕府山植被修复将形成沿江绿色丘陵。

② 二桥东侧,栖霞山麓江水蜿蜒,丹枫秋色与古寺、石刻交相辉映,为古城及龙潭新港的休闲山野。

(2) 下关—浦口老轮渡码头建设南京外滩中心风光带

仿效广州海珠桥两岸滨江带与天河码头的建设格局,发展两岸轮渡与江上夜游,繁荣沿江中心带的经济。

下关地区人文历史资源丰富:狮子山、阅江楼、明城墙、静海寺、江南水师学堂、英国使馆、渡江战役纪念碑以及小桃园、绣球公园景区。

(3) 江北"一山三泉"—森林地质生态景观带

将长江北岸老山森林、珍珠—琥珀—汤泉自然风光与沿江民居、古文化遗迹融为一体,自然景观列入新市区建设的重点之一。老山既是森林公园,又是地质公园,其岩层是南京最古老的岩层,是6亿年前震旦纪的浅海沉积形成的。对兜率寺和惠济寺的建设,应着手保护性整修,尽量保留周围的田园风光和自然村落的本色。老山与"三泉"可为中小学生建造知识和生活园地,学习自然科学知识的夏令营。

（4）江滩沙洲生态绿岛建设

潜洲绿林草地自然景观:江心洲景色宜人、芦花翻白燕子飞,污水处理厂加强环境管理;建设八卦洲四季生态旅游绿洲:"春游芳草地,夏绕碧荷池,秋饮黄花酒,冬吟白雪诗",阻滞北岸化工园大气污染,在八卦洲以北夹江沿岸的低丘建树林带。

（5）河西新民居区—宝船厂遗址—奥体中心,打造南京新形象。

（6）板桥—梅山工业区与新济洲之间两岸。

保持着天然的平原江岸,可建设柳荫大道绿廊与中心带相连。为工厂区及南郊山村居民提供江滨风光带,严格制止江中取沙防止江岸坍塌;规范民用小码头建设,加强沿江地带管理。

4. 长江风光带与港口建设与水质水环境密切相关,为本项咨询进行了沿岸江水检测

采样时间:2003/10/26,2003/10/30

检测站位:见图7和表2

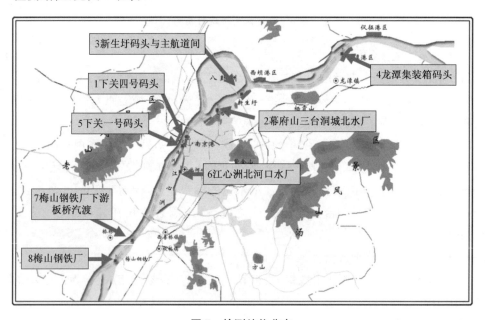

图7　检测站位分布

表 2　长江江水采样点位置

样品编号	采样地点	时间
1	下关 4 号码头前沿	2003 - 10 - 26　9：20
2	幕府山三台洞城北水厂	2003 - 10 - 26　10：50
3	新生圩码头与主航道间	2003 - 10 - 26　11：15
4	龙潭集装箱码头	2003 - 10 - 26　12：30
5	下关 1 号码头前沿	2003 - 10 - 30　13：30
6	江心洲北河口水厂	2003 - 10 - 30　13：56
7	梅山钢铁厂下游板桥汽渡	2003 - 10 - 30　15：01
8	梅山钢铁厂	2003 - 10 - 30　15：20

检测指标：水温、pH、溶解氧、COD（化学耗氧量）、石油类、总磷、氨氮、铜、锌、铅、镉、六价铬共 12 项（见表 3、表 4）。

表 3　检测参数（单位：mg/L）

指标	铜	锌	铅	镉	铬	氨氮	总磷	COD	石油类	溶解氧	pH
梅山钢铁厂	0.035	<0.04	0.021	0.000 2	<0.004	0.124	0.079	5.6	0.12	6.89	7.65
板桥汽渡	0.018	<0.04	0.005	0.000 1	0.012	0.368	0.129	3.0	0.11	7.23	7.74
北河口水厂	0.003	<0.04	0.004	0.000 7	<0.004	0.222	0.173	<2	0.10	7.04	7.82
下关 1 号码头	0.002	<0.04	0.003	0.000 2	<0.004	0.214	0.118	<2	0.14	7.60	7.75
下关 4 号码头	<0.002	<0.04	0.003	0.000 3	<0.004	0.244	0.115	2.0	0.20	7.58	7.62
三台洞	<0.002	<0.04	0.004	0.000 3	<0.004	0.224	0.067	<2	0.05	7.60	7.81
新生圩	<0.002	<0.04	0.001	0.000 2	<0.004	0.502	0.079	<2	0.20	6.96	7.73
龙潭码头	<0.002	<0.04	0.006	0.000 3	<0.004	0.388	0.121	<2	0.11	7.69	7.52

表 4　单因子指数评价表（以 GB 3838—2002 Ⅱ 类水为标准）

指标	铜	锌	铅	镉	铬	氨氮	总磷	COD	石油类	溶解氧	pH	水质类别
梅山钢铁厂	0.04	<0.04	1.20	0.05	<0.08	0.25	0.79	1.40	2.40	0.73	0.33	Ⅲ
板桥汽渡	0.02	<0.04	0.50	0.02	0.24	0.74	1.29	0.75	2.20	0.62	0.37	Ⅲ
北河口水厂	0.00	<0.04	0.40	0.13	<0.08	0.44	1.73	<0.5	2.00	0.68	0.41	Ⅲ
下关 1 号码头	0.00	<0.04	0.30	0.05	<0.08	0.43	1.18	<0.5	2.80	0.51	0.38	Ⅲ
下关码头	<0.00	<0.04	0.30	0.05	<0.08	0.49	1.15	0.50	4.00	0.51	0.31	Ⅲ
三台洞	<0.00	<0.04	0.40	0.05	<0.08	0.45	0.67	<0.5	1.00	0.51	0.41	Ⅱ
新生圩	<0.00	<0.04	0.10	0.05	<0.08	1.00	0.79	<0.5	4.00	0.70	0.37	Ⅲ
龙潭码头	<0.00	<0.04	0.60	0.05	<0.08	0.78	1.21	<0.5	2.20	0.48	0.26	Ⅲ

评价标准:GB 3838—2002地表水环境质量标准。

根据长江南京梅山—龙潭码头段(以下简称"长江南京段")水体功能,应采用 GB 3838—2002地表水Ⅱ类水标准对水质进行单因子评价,计算所得水质单因子评价指数及评价水质类别见表4。

(1)重金属污染

铅超标,出现于梅钢段水域,南京北河口取水口位于其下游。

铜:实测值范围为0.002~0.035 mg/L,断面超标率为0,最高值出现在梅钢段水域。

锌:实测值全部低于检测限,即小于0.04 mg/L,断面超标率为0。

铅:实测值范围为0.001~0.021 mg/L,断面超标率为12.5%,最高值出现在梅钢段水域,超标1.2倍。

镉:实测值范围为0.000 1~0.000 7 mg/L,断面超标率为0,最高值出现在北河口水厂段。

六价铬:实测值范围为0.004~0.012 mg/L,断面超标率为0,最高值出现在板桥汽渡段。

(2)营养盐污染

氨氮:实测值范围为0.124~0.502 mg/L,断面超标率为0,最高值出现在新生圩段,单因子指数为1.00,处在超标的边缘。

总磷:实测值范围为0.067~0.173 mg/L,板桥汽渡段、北河口水厂段、下关一号码头段、下关码头段、龙潭码头段共5个断面超标,断面超标率达到62.5%。最高值出现在板桥汽渡段,五个断面平均超标倍数为1.31倍。各断面浓度比较见图8。

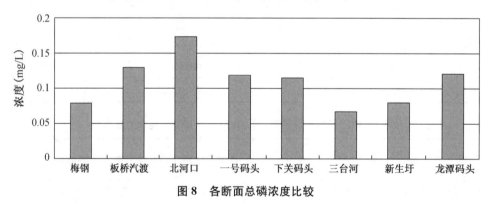

图8　各断面总磷浓度比较

(3)有机物污染

COD:实测值范围为2~5.6 mg/L,断面超标率为12.5%,最高值出现在梅钢段,超标1.4倍。

石油类:实测值范围为0.05~0.20 mg/L,除三台洞段外其余7个断面都超标,断面超标率达到87.5%。最高值出现在下关码头段,7个断面平均超标倍数为2.8倍。各断面浓度比较见图9。反映出南京长江段大部分遭受石油污染。

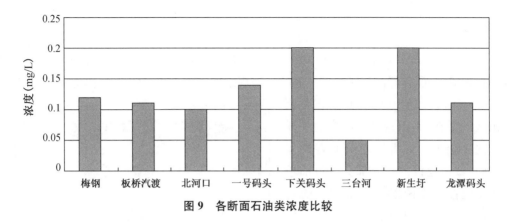

图9 各断面石油类浓度比较

（4）其他

pH、溶解氧：未超标。

水温：监测温度介于19℃±0.3℃之间，经调查，基本不存在人为造成水温的变化，该指标全部达标。

南京长江段水质检测及评价结论：

（1）所测试的12个指标中，八个达到GB 3838—2002地表水Ⅱ类水标准，分别是水温、pH、溶解氧、氨氮、铜、锌、镉、六价铬。出现超标的因子有：石油类、总磷、COD、铅（见图10）。

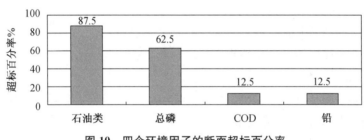

图10 四个环境因子的断面超标百分率

（2）石油类、总磷、化学耗氧量、铅超标断面百分率分别是：87.5％、62.5％、12.5％、12.5％。

（3）石油类、总磷、化学耗氧量、铅平均超标倍数分别是：2.8倍、1.3倍、1.4倍、1.2倍。

（4）除三台洞段为Ⅱ类水外，其余七个断面水质均为Ⅲ类水。

（5）检测表明南京长江段江水大部分遭受石油污染，可能与船舶泄漏及洗舱有关；铜、铅、磷与有机物污染，主要源于梅山钢铁厂水域。

（6）加强长江南京段水体保护，已到刻不容缓的关键时刻。南京市在城市建设发展与沿江开发中，从关注南京段长江的水质出发，应加强对已建的化工厂所排废水、废气之处处理与监测，即使达标排放，亦会有长年积累的危害；尽可能地限制在沿江兴建化工、造纸等大量排放废弃水、气的工厂与产业。切实保护南京市民的饮水源是当务之急，也是生存环境可持续发展之关键。

5. 结语

（1）从南京市市域环境与历史发展特色，统筹考虑沿江风光带建设、经济发展与生态环境保护等方面，作好沿江风光带发展的详细规划融入南京市总体规划中，按规划实施。

（2）浦口区老山林区，其面积是紫金山区的2倍，是南京主城区内的主要绿肺之一（图11），需将老山的森林保护，列入城市建设的一个部分。目前单纯由企业管理已不能适应。沿江生态环境建设中，将老山山林区列入重点工作之一，将老山山区与紫金山区两个城市的绿肺共同重视，开发建设为生态休闲旅游度假与地质地理生态环境教育基地，为400万南京市民多一块可作假日走向自然的去处。

（3）切实加强南京长江段水质保护，确保沿江带水环境安全。

（4）南京市地处长江三角洲陆地端点，其地理环境与所面临的发展挑战等，与广州市相似。自2000年广州市实施多学科交叉综合所完善的城市总体规划以来，发挥珠江活力，疏导老市区，建设珠江风光带与新市区开发中，成果显著。建议南京市与广州市结为姊妹城市，相互学习，共同前进。

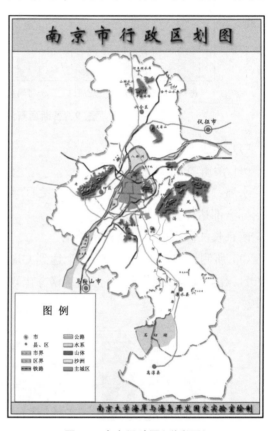

图11　南京绿肺图（附彩图）

照片1　港口与沿江风光带考查组全体成员
（2003.10.25）（附彩图）

照片2　江上考察指挥（2003.10.30）（附彩图）

照片 3　沿江采取水样(2003.10.30)　　　照片 4　长江大桥桥下净空限制了万吨级海轮上行
　　　　　(附彩图)　　　　　　　　　　　　　　　　(2003.10.26)

照片 5　新生圩深水港区(2003.10.26)　　　照片 6　建设中的龙潭深水港集装箱码头
　　　　　(附彩图)　　　　　　　　　　　　　　(2003.10.26)(附彩图)

照片 7　冲蚀的江岸与堆放的石材　　　　　照片 8　江岸的民居(2003.10.30)
　　　　(2003.10.30)(附彩图)

照片 9 夹江北河口取水（2003.10.30）（附彩图）

照片 10 沿江采沙堆场（2003.10.26）（附彩图）

照片 11 平坦的江岸无序的建筑（2003.10.26）

照片 12 又一处采沙码头（2003.10.26）

照片 13 烟尘笼罩的梅山段沿江（2003.10.30）

照片 14 烟尘笼罩的梅山段沿江
（2003.10.30）（附彩图）

照片 15 江心洲洲头（2003.10.30）（附彩图）

照片 16 新济洲自然风光江岸（2003.10.30）

长江南京段历史洪水位追溯*

　　在长江幕府山、燕子矶岸段的仑山灰岩岩壁上保存着数道因江水长期浸淹而形成的水位痕迹。该痕迹呈深灰色或灰色水平直线状沿长江南岸延伸分布,沿痕迹线有密集的溶蚀小孔发育,有的已发育成穹状或直立状雏形洞穴,其中以南京幕府山三台洞处水位痕迹保存得最为清晰完整。经进一步深入调查研究,在幕府山以西,安徽长江段南岸的采石矶岩壁上也存在着同样的水位痕迹(表1),两处水位痕迹高程大致相等且目前都已脱离长江江面4～12 m,反映出沿江南岸崖壁上的水平痕迹是长江下游区域性洪水位遗迹。

表1　长江下游南岸悬崖崖壁及岩洞洪水位遗迹表

水位遗迹坐标	现代洞底高程/江面标高	水位线编号	水位高程(m)	水位遗迹特征	相对于现代江面高度(m)
三台洞 N32°08.418′ E118°47.955′	7.264/3.498	FL_1	8.541	沿岩洞底平行发育,深灰色联线,有小洞穴,沿线成水平状分布	5.016
		FL_2	9.425	沿洞顶发育,黑灰色,水平状相连,有小洞穴发育,清晰明显	5.927
		FL_3	10.474	深灰色,有小洞穴发育,水平状相连	6.976
		FL_{4-3}	11.659	水平印痕不明显,但存在着似水平状的小孔	8.161
		FL_4	12.844	灰白色,断续相接,磨蚀,不太明显	9.346
头台洞 N32°08.502′ E118°47.880′	8.711/3.498	FL_{4-1}	12.341	与洞顶平行,深水白色,水平状相连,较为明显	8.843
		FL_4	12.677	灰白色,断续相连,不甚明显	9.179
燕子矶 N32°08.502′ E118°47.880′	3.498(江面)	FL_{1-1}	约5.5	深灰白色,水平状相连,清晰明显	约4.0
		FL_1	约7.5		约5.0
		FL_3	约10.5	灰白色,断续相连,不甚明显	约7.0
		$FL_4^{(1)}$	约12.5		约9.0

　　*　何华春,王颖,李书恒:《地理学报》,2004年第59卷第6期,第938－947页。

　　基金项目:国家自然科学基金项目(40271004)〔Foundation:National Natural Science Foundation of China,No.40271004〕

（续表）

水位遗迹坐标	现代洞底高程/江面标高	水位线编号	水位高程（m）	水位遗迹特征	相对于现代江面高度（m）
采石矶 31°38.958′N 118°27.052′E	4.057（江面）	FL₁₋₁	5.6	深灰白色，水平状相连，水位线密集且清晰明显，	约1.5
		FL₁	6.5	发育有小的洞穴	约2.5
		FL₅	14.145	灰白色，水平状连续分布	10.088
		FL₆	16.568	沿水位线有小型植物生长	12.511
饮用水源一级保护区取水站 31°38.508′N 118°27.075′E	4.057（江面）	FL₁₋₁	5.667	黑黄色，水位痕迹明显清晰	1.61
		FL₁₋₂	6.597	沿水位线有黏土沉积	2.54
		FL₁	8.387		4.33
		FL₂	9.507		5.45

注：所用高程依据为吴淞基面；（1）：据相邻陆地高程估计，洪水遗迹位于江面陡壁上，无法准确测量

南京位居长江三角洲的顶点。长江下游南京河段自西南向东北流贯南京市境，河段南岸全长98 km，北岸全长88 km。江面宽1.1～2.5 km，平均水深20～30 m，深槽最深处—72 m，多年平均流量28 800 m³/s[1]。受潮水顶托、江潮相互作用，长江在南京地区分叉形成新生洲、新济洲、江心洲、八卦洲及潜洲等一系列沙岛与浅滩。沙岛、浅滩使江水分为夹江与主河道。陆地上河、渠、湖、塘呈星网状分布，沿江两岸除南部有幕府、栖霞低山丘陵与北部老山、龙王、灵岩山地遥相夹峙外，均为坦荡的平原河段（图1）。北岸：老山由6亿年前震旦纪的浅海沉积岩组成，山体为森林所覆盖；远郊灵岩山为由数层玄武岩层覆盖

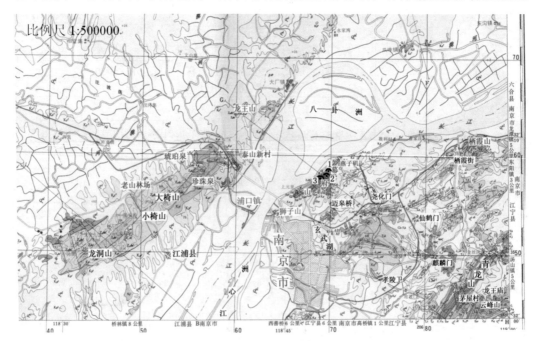

图1　南京市及幕府山地理位置示意图（1—头台洞 2—二台洞 3—三台洞）（附彩图）

的更新世砂砾层组成,目前人工开采砂砾层,使山体变化大。南岸:幕府山主体由古老的下奥陶统仑山灰岩组成,沿长江南岸由东北向西南延伸分布,长 6 km,宽 1 km,海拔205 m。沿幕府山悬崖峭壁的深水岸分布着一系列的石灰岩岩洞,其中三台洞临江一侧的崖壁上有明显的水位痕迹,经多次实地研究考证,确定为洪水遗迹。

1 洪水位遗迹特征及意义

1.1 洪水位遗迹特征

燕子矶位于幕府山东北端,海拔 36 m。沿幕府山至燕子矶之间分布着头台洞、三台洞等一系列的石灰岩溶洞,溶洞发育在江边的震旦系及寒武系白云岩及白云质灰岩之中,大致可分三层[2]。三台洞洞底高程 7.3 m。头台洞位于三台洞东侧 470 m,洞底高程8.73 m。二台洞位于三台洞与头台洞之间,洞底高程 28.3 m。三台洞正北 200 m 即为长江,目前由沿江路堤与江水相隔,江面标高 3.498 m(2003 年 11 月 29 日)。

洪水位遗迹分布于临江的石灰岩崖壁上,呈深色水平线沿崖壁及洞穴壁内延伸分布,沿水位线发育了一系列溶蚀孔,有的已发育成雏形溶穴,反映着曾经久历江水浸淹。保存较好的水位痕迹有四道,以三台洞处保存完好(图 2)。燕子矶和头台洞临江的崖壁上也有洪水位遗迹存在。燕子矶直临江上,遭受长江冲刷侵蚀,洪水位遗迹呈灰白色水平状延伸;头台洞由于人类活动干扰严重,洪水位遗迹不甚清晰明显,呈浅灰色水平状断续分布;二台洞位于三台洞之西的山坡上,因民房建筑,洞穴已破坏荒弃。三台洞为南京市幕燕风光带的自然岩洞景点,未遭近期城市改造的破坏,崖壁及洞壁中清晰地保存着四道平行的洪水位遗迹。

图 2　三台洞崖壁及洞穴内洪水位遗迹图(附彩图)

(1) 第一道洪水位遗迹(Flood Level 1,FL1)高程约 8.5 m,高出现代江面 5 m,呈深灰色水平直线状侵蚀凹痕,紧贴水位凹痕有密集的小孔穴发育,孔穴呈水平状沿凹痕分布,具小孔的岩壁上因水渗有苔藓生长,水位遗迹色深。目前三台洞洞底与洞内水潭位于此高度上。记录表明,现代洪水(如 1937 年或 1954 年的洪水)仍可到达此高度。分析认为该水位遗迹的形成时代要早,因为石灰岩岩壁已侵蚀出明显的水平凹痕。

（2）第二道洪水位遗迹(FL2)高程约 9.4 m,呈黑灰色水平直线状延伸,于崖壁上保存清楚。特点是水平小孔成线分布。有洞穴因崖壁塌陷,遗迹不明显。

（3）第三道洪水位遗迹(FL3)高程 10.5 m,这道深灰色水位遗迹最完整,清晰地显现于岩壁与洞穴内,特点是已形成水平的侵蚀凹痕,沿水位线发育了一系列的小孔与雏形洞穴,洞穴呈穿状或直立状,反映出在该高程的水位为时已久,可能是当时当地的潜水基面,故而沿地下水渗流通道形成垂直状洞穴。

（4）第四道洪水位遗迹(FL4)高程约 12.8 m,为断续相连的灰色水平印痕与孔穴,特点是沿水平线上方有小型穿状洞发育,并呈现高低水位之变动。在三、四两道洪水位遗迹间,相当于标高 11.7 m 处,还有一道水平印痕不明显的水位遗迹(FL4-3),但沿线存在着似水平状的小孔,并为小型穿形洞之洞底。但是,它已被日后的垂直洞穴所凿穿,垂直洞穴(＊)系潜水注入 FL3 之通路。上述现象表明,FL3 发育于 FL4 与 FL4-3 之后,而 FL4 不及 FL3 水位持续稳定。

明确三台洞 4~5 道洪水位遗迹后,以此为依据,沿长江向东(下游方向)向西(上游方向)两侧进行追寻。结果发现:

（1）在三台洞以东的头台洞也发现有同样的水位遗迹存在:标高 12.677 m,深灰色水平印痕,沿水位线有小型穿形洞及小孔穴分布。与洞顶相当高处,还有另一条明显的呈水平状向东西延展的水位线(图 3)。事实说明相当于 FL4 的水位曾经历过两次水位变动。由此证实三台洞 FL3 与 FL4 之间不明显水位遗迹的存在,其编号暂定为 FL4-3。

（2）在三台洞以东,沿长江的下游方向,燕子矶崖壁上,其水位遗迹分为上下两层,距江面 2~5 m 经常遭受江水冲刷的崖壁上其水位遗迹清晰可辨;而高出江面 10~12 m 处的水位遗迹则呈灰白色水平断续状延伸分布,其高程分别为 5.5 m、7.5 m、10.5 m 和12.5 m(图 4)。

图 3 头台洞洪水位遗迹图(附彩图)

图 4 燕子矶崖壁洪水位遗迹图(附彩图)

（3）沿长江南京段往西,直至安徽境内,于长江南岸翠螺山西麓采石矶的悬崖壁及相邻岸段多处存在着清晰可见的洪水位遗迹(图 5)。该处山体海拔 50 m,主要由侏罗纪含钙质石英砂岩及粉砂质灰岩组成。江面标高 4.057 m(2004 年 3 月 26 日)。高出江面1~3 m 的崖壁上分布着多条清晰的水位遗迹,距江面 4~12 m 高度处也同样有水位遗迹存在,其中清晰的有 4 道,其高程分别为 8.6 m,9.5 m,14.1 m 和 16.6 m。此外于采石矶饮用水源一级保护区取水站 10 m 高水泥柱上(1997 年建造)仍清楚地保留着 1998 年长江洪水位变动之泥水印痕(FL2 主要成分为黄色黏土及粉砂)(图6)。

图5 采石矶崖壁水位遗迹图(附彩图)

同时,发现在三台洞洞穴中(高程 10.38 m,32°08.255′N,118°47.590′E)、采石矶崖下河漫滩上(高程 4.8 m,31°38.961′N,118°27.044′E)及在相邻岸壁的河蚀凹穴中(高程 8.2 m,31°38.958′N,118°27.052′E),有洪水泛滥之淤泥粉砂质沉积。粒度分析结果表明,长江河漫滩处物质成分组成与洞穴内物质成分组成基本一致,主要以粉砂为主,其含量达到80%以上,平均粒径 6.02φ,频率曲线为对称的单峰型曲线。由此进一步证明 FL3 及 FL1 是确实存在的长江洪水位遗迹。

图6 采石矶饮用水源一级保护区取水站水位遗迹(附彩图)

上述研究表明,在南京长江南岸的石灰岩崖壁上,保存着长江洪水期的水位遗迹,根据其水平方向的连续分布与高程一致性,目前可以明确地定出长江南京段 4 道古洪水位遗迹(图7,表1)。

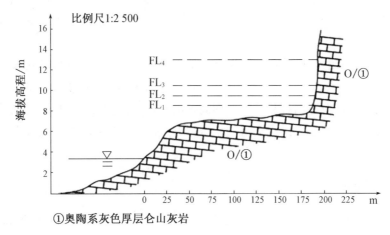

图7 三台洞水位遗迹分布剖面图

1.2　洪水位遗迹意义

（1）于三台洞、燕子矶、头台洞及采石矶沿江崖壁上均发现有洪水位遗迹存在且沿水位线有小的溶蚀孔洞分布。其中长江南京段沿岸保存尚为完整的高洪水水位遗迹约为12.5 m（FL4），最明显的是10.5 m高的洪水位线，遗迹清晰且发育了小型穿状与垂直洞穴，反映该水位稳定。曾在相当长的一段时期内为当地潜水基面，地面水渗流至此高度转为水平，这两级水位分布普遍，反映出幕府山沿长江一带曾经历过高的水位且保持了相当长一段时间，以至在沿江的石灰岩岩壁上形成了目前水平状连续分布的水位遗迹，同时，在FL4与FL3之间（FL4-3），曾经历过0.3～1 m左右的水位波动期。

（2）对比4条洪水位遗迹的高程，FL2与FL1之水位差为0.916 m，FL3与FL2水位差为0.949 m，FL4与FL3之水位差为2.370 m（三台洞），2.0 m（燕子矶），FL4与FL4-3之水位差为1.184 m（三台洞），0.33 m（头台洞），因此可以认为，上述多时期水位变动多半界于1～2 m之间，已知最小变动值为0.33 m，最大变动值为2.37 m。

（3）洪水位遗迹保留在临江的由古老的下奥陶统仑山组石灰岩组成的稳定基岩岸段，记录长江当时的水位，南京市防洪标准为100年一遇，堤防按下关设计水位10.6 m，江堤超高2 m，即12.8 m[3]，相当于洪水位遗迹FL4水位高程，古洪水位遗迹的研究为现代长江南京段防洪决策提供新的科学依据。

（4）古洪水位遗迹的研究及水位年代的确定将为长江南京段无文字记录水文资料（南京下关站自1912年以来有洪水位文字记录）提供新的洪水位证据。

2　长江洪水位遗迹的年代分析

南京市地处三角洲平原的顶点，市域内江、河、湖、塘水系众多，长江自西向东穿越，浦口与主城区隔江相望；秦淮河自东向西蜿蜒于城南后转向北汇入长江，另外尚有滁河入江水道马汊河和朱家山河。沿江河地带地面高程6～9 m。同时，长江河口受潮汐影响，范围向西可达安徽省的大通。南京地处感潮河段，沿江水位受洪水及自长江口传来的潮波双重影响，加之，淮河入江水量对江水下泄有一定的顶托影响。因此，大洪水年份，易在长江南京河段形成高水位，其中汉口—大通区间的洪水对长江南京段大洪水的形成起着重要作用[4]。

南京段长江实测记录的最高水位为10.22 m（1954年8月17日），最低水位为1.54 m（1956年1月9日）。多年平均水位（1950—1982年）高潮5.48 m，低潮4.97 m[5]。

根据量测的水位遗迹高程与长江水面高差的对比，通过查阅古代文献、历史档案及地方方志中有关长江下游岸段洪水灾害的记载，初步将长江南京段高水位及洪水发生归纳为三个时段。

2.1　全新世时期

（1）6 000 aB. P. ～5 000 aB. P.，由于气候变暖，海平面上升，河流基面抬高，长江下游平原曾发生较大洪水。当洪水时期，河流流量增大，河型发生变化，由寒冷低海面时代的下切转而加强旁侧侵蚀，河谷加宽，岸上茂密的亚热带林木不时地坍入摆动的河床中。长江沿岸自宜昌、安庆至南京附近，在埋藏的河床沉积物中，均发现埋藏的亚热带阔叶林树干。据[14]C 测定，其年代为 6 000～5 000 aB. P. 大暖期前后[6]。

（2）4 400 aB. P. ～4 000 aB. P.，中国全新世第二次大洪水。因海面较高，降水量大，而发生于长江下游低下平原地区。据古籍记载，河水漫溢，淹没大量平地，损坏了原始农垦，造成严重灾害[6]。江北浦口区砂砾层的古木[14]C 定年为 4 085±95 aB. P. ～4 090±100 aB. P. 与 3 730±90 aB. P.[7]，反映着这段时期的洪水变化。

2.2　历史时期

地质时期与历史文献反映出长江洪水泛滥多次发生，为时已久，但没有长江沿岸洪水水位确切记录（表 2）。

表 2　南京及江苏历史洪水记录

时间	洪水特征及洪灾描述
公元 215（太元元年七月）	长江洪水漫溢南京城，"水深八尺"[8]
公元 351（永和七年七月）	长江洪水突袭石头城，"溺死数百人"[8]
公元 404（东晋元兴三年二月）	涛水入石头，大航流败，商旅方舟万计，骸胔相望[9,17]（"涛水入石头"估计其水位当在±8m 按：南京地面高程较低，一般在 7～9 m，公元 351—404 两次时间间隔为 53 年，可能反映着 FL₁、FL₂ 间变动之重现期）
公元 1028（宁天圣六年七月）	江宁府、扬、真、润三州水溢，坏官私庐舍[10]
公元 1170（宋乾道六年五月）	建康大水，江东城市有深丈余者，漂民庐，溃圩堤徙[10]
公元 1298（元大德二年）	江水大溢，高四、五丈，江宁府水灾，漂没庐舍[10]
公元 1502（明弘治十五年七月）	"南京江水泛溢入城五尺余深"，军民房宇倒塌一千多间，孝陵内一些墙垣、桥梁也未能幸免[10]
公元 1560（明嘉靖三十九年）	江水涨至三山门，秦淮民居水深数尺[10]
公元 1589（明万历十七年）	洪水泛滥，应天、太平等府平地水深丈余，田庐没为巨浸[10]
公元 1608（万历三十六年六月）	南京大雨连下近二十日，"水灾异常，百姓浸没……盖二百年来未有之灾"[9]
公元 1755（清乾隆二十八年夏）	江苏大雨，江水涨 40 多天始退[10]
公元 1848（道光二十八年夏）	长江中下游沿江各省普遍多雨，"江湖并涨，堤圩冲决甚多，江苏江海暴溢，长江两岸水深数尺"[11]

2.3　20 世纪洪水位记录

确切的洪水位记录始自 1912 年进行水文观测以来（表 3），记录表明长江南京河段出现水位超过 9.0 m（南京下关百年一遇警戒水位为 8.50 m）的大洪水年共 12 次。水位高程表明与 FL2 相当，尚未达到 FL3。

表3　20世纪南京下关站洪水记录（按水位高低排序）

20世纪中的排序	年份	年最高水位(m)	出现日期（月·日）	≥9.0 m的天数(d)	≥9.5 m的天数(d)
1	1954	10.22	8·17	87	62
2	1998	10.14	7·29	42	17
3	1983	9.99	7·13	27	11
4	1991	9.69	7·13	17	6
5	1977	9.30	7·2	6	—
6	1931	9.29	9·15	45	—
7	1969	9.20	7·8	9	—
8	1980	9.20	8·28	9	—
9	1973	9.19	7·20	7	—
10	1949	9.17	7·25	1	—
11	1989	9.08	7·21	3	—
12	1992	9.05	7·16	4	—

1954年江苏省遭遇百年未遇的特大洪水，江淮沂沭泗并涨，加之长江、淮河上游洪水下泄，又受海潮顶托，致使河湖水位猛涨。长江大通站出现历史最大流量92 600 m³/s，南京高潮位达10.22 m，是有记录以来的最高值，并且历时最长。当代洪水记录是南京江防高程的科学依据。

3　洪水位遗迹形成原因及时间分析

3.1　长江南京段高水位原因

（1）长江下游为广坦的冲积平原，支流众多，江淮准静止锋于春末夏初停滞于长江流域下游，形成梅雨连绵；继之，夏秋雨季降水丰沛加之台风暴雨与大潮叠加，长江洪水难以下泄，易在感潮段，特别是在南京与浦口两岸间山地夹峙河段形成高水位。据记录，自上游向下游传播的洪水流量是形成南京最高洪水位的主要原因，南京最高水位与大通年最高水位的相关系数达0.97，与吴淞年最高潮位的相关系数仅多0.2[4]。

（2）蓄洪能力下降导致长江持续性高洪水位，幕府山以东至镇江河段是长江中下游淤积量最严重的河段，长期而大量的泥沙淤积，河道淤浅，滩洲淤积，导致蓄、泄洪水能力下降，甚至滞水壅水，以致造成长时间高水位[12,20]。

（3）大通站大洪水与大通—南京区间特大暴雨遭遇以及淮河下游特大洪水的顶托作用，将对长江南京段高水位产生影响。据推算，当淮河入江水道泄洪流量达12 000 m³/s的设计流量时，可使南京下关站洪水位抬高约0.15 m[13]。

（4）湖泊淤浅围垦增大，湖泊面积减少，池塘填没，使洪水宣泄受阻，调蓄能力下降及下垫面性质改变,蓄泄能力减少是产生高水位的另一直接原因。

3.2 洪水位遗迹形成时间推算

对长江南京站逐年各月潮水位资料,按 10 年和 50 年分别进行统计,并绘制了南京站年最高(低)潮位曲线(图 8)。该潮位曲线表明了长江南京站有确切水位记录以来的水位变化趋势。

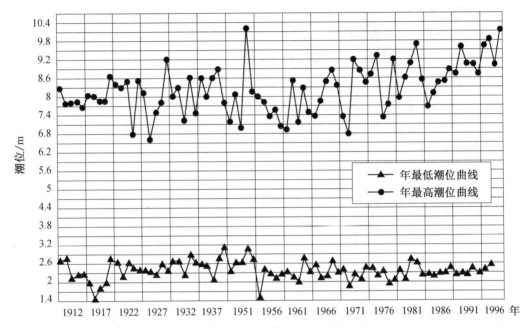

图 8　南京站年最高(低)潮位曲线[14]

另外,对年最高水位序列采用 15 年滑动平均,根据计算结果绘出,$\overline{H}_{15} \sim t$ 过程(图9),从图 4 和图 5 分析可知,长江水位正经历着一个周期变化,从 20 世纪初期到 60 年代中期,年最高潮位逐年下降,然后维持在一定的水位上,前期虽有波动却基本平稳,大约从 70 年代开始水位持续上升,在 1987 年以后,长江水位稳步上升。水位降低变化与 20 世纪 60 年代"寒冷期"气象时段的出现相呼应。

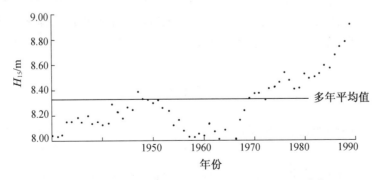

图 9　南京下关站年最高水位 15 年滑动平均过程[15]

幕府山沿长江岸最高水位遗迹高程 12.844 m,超出长江南京段有洪水文字记录以来最高值 2.646 m,三台洞最高水位与头台洞、燕子矶水位遗迹高程相当,反映出 FL4 是发生于有确切水文记录以前的长江南京段的最高水位,其余三条洪水遗迹高程在 8.5～10.5 m 之间,且水平的洪水位遗迹线颜色清晰,同有明显倾角的灰岩层理相交切。沿水位痕迹发育了细微密集的溶蚀凹痕或孔穴。于古洪水位线孔穴内及现今长江河漫滩上采样并作沉积物粒度分析,表明它们属于洪水沉积的同一粒级的粉砂。此外,各处崖壁上古水位线的延伸高度大体一致,且在西南 40 km,上游的安徽采石矶崖壁上亦可追溯到相同的水位遗迹,由此可推断古水位发生的大致年代。

(1) FL4 海拔高度 12.844 m,为四道洪水位遗迹之最高,发育时间早,部分水位线已模糊,推论该水位相当于全新世中期 6 000 aB. P. ～5 000 aB. P. 时高海面的遗迹。

长江下游在全新世中期曾普遍发生较大洪水,且在南京市北岸海拔约 10 m 的二级阶地区,存在全新世高温期冲积急流相的沉积地层,大量炭化木、具冲积特征的埋藏古树和砂砾层表明这段时间经历了高温多雨的洪水环境,从沉积物粒度来看,这一时期沉积物中砂、粉砂含量相对较高而黏土含量较低[16]。全新世中期 6 000 aB. P. ～5 000 aB. P. ,长江在镇江附近入海,镇江水位为 -2～-3 m[17]。按现时南京 6～8 月平均高潮位 7.15 m,南京至长江口平均坡降 $0.5×10^{-5}$～$1.0×10^{-5}$[18](南京至长江口距离约为 480 km),可推算出全新世中期南京附近高水位在 6～8 m 左右。大洪水年份在上游洪水与下游潮水顶托的共同作用下,南京地区长江古洪水位应能达到 FL4 之海拔高度。

(2) 若文献记载 1298 年"江水大溢,高达 4～5 丈"属实,推算当时水位亦超过 12 m,可能是 FL3 水位在历史时期的记录。历史记录中,关于 1170 年、1502 年与 1589 年,"江宁大水深达丈余"、"入城 5 尺余"、"水深丈余"及"宁、苏、松、常镇诸府皆被淹"等,可能是 100～200 年周期高水位的印证。

(3) FL2 与 FL1 是与现代长江洪水位相当的古洪水位遗迹,目前已脱离长江江面,在沿长江基岩岸已侵蚀出深灰色水平直线凹痕,但现代长江大洪水仍可到达此高度。由图 4 可知,长江水位在 1954 年达到最高,从 20 世纪 70 年代开始水位持续上升,年最高潮位均在 9.3 m 左右波动,重现期为 10～50 年。

此次古洪水位遗迹研究中,采用历史文献史料、长江南京段逐年水位记录与水位遗迹实测相结合的方法进行[19-26]。进一步确定洪水位遗迹发生的年代,尚需探采研究洪水期相关沉积[27-28],进行石英砂表面模式结构对比分析[7,29]以及沉积层定年分析[30-36]等工作。

参考文献

[1] 潘凤英. 1990. 中全新世以来长江南京河段的河床变迁. 南京师范大学学报(自然科学版),13(4):81－88.

[2] 夏邦栋. 1998. 下扬子前中生代构造演化. 成都理工学院学报,25(2):145－152.

[3] Committee of Yangtze River Water Resources of the Ministry of Water Resources (ed.). 2001. Atlas of Flood Control of Yangtze River. Beijing:Science Press. 130－131.

[4] 芮孝芳. 1994. 长江南京段洪水成因及趋势的水文学分析. 水利水电技术,(10):2－7.

〔5〕 Institute of Geography，Chinese Academy of Sciences. 1985. Formation and characters of river channel along middle-lower reaches of the Yangtze River. Beijing：Science Press. 6－46.

〔6〕 Yang Huairen. 1996. Research on Environment Changes. Nanjing：Hohai University Press. 366－373.

〔7〕 朱诚. 1997. 南京江北地区全新世沉积与古洪水研究. 地理研究，16(4)：23－30.

〔8〕 Water Resources Bureau，Jiangsu Province. 1976. Disaster Tables of Flood & Drought in the Jiangsu Province over the Last 2000 Years. Nanjing：Water Resources Bureau of Revolutionary Committee，Jiangsu Province. 26－47.

〔9〕 Hydrological Department of the Ministry of Water Resources. 1997. Annals of Chinese Water Resources. Beijing：China Water Resources & Hydroelectricity Press. 463－489.

〔10〕 Song Zhenhai (ed.). 1992. Atlas of Anomaly Chronology & Heavy Natural Disaster in Ancient China. Guangzhou：Guangdong Education Press. 305－317.

〔11〕 Committee of Yangtze River Water Resources of the Ministry of Water Resources (ed.). 2002. Flood & Drought in the Yangtze River Basin. Beijing：China Water Resources & Hydroelectricity Press. 50－58.

〔12〕 许厚泽. 1999. 长江流域洪涝灾害与科技对策. 北京：科学出版社，270－274.

〔13〕 芮孝芳. 1996. 长江下游感潮河段大洪水和特大洪水的形成及趋势. 水科学进展，7(3)：221－225.

〔14〕 朱红耕，黄红虎. 2002. 长江下游水情变化趋势分析. 黑龙江水专学报，29(4)：18－20.

〔15〕 李国芳，黄振平等. 1999. 长江防洪堤南京段设计洪水位风险分析. 河海大学学报，27(2)：22－27.

〔16〕 张强. 2001. 南京江北地区晚更新世以来环境演变研究. 地理科学，21(6)：498－504.

〔17〕 李从先，闵秋宝. 1981. 全新世长江三角洲顶部的海进时间和海面位置. 同济大学学报，(3)：104－108.

〔18〕 Chen Zhongyuan，Li Jiufa，Shen Huanting et al. 2001. Yangtze River of China：historical analysis of discharge variability and sediment flux. *Geomorphology*，41：77－91.

〔19〕 陈雪英. 1999. 长江流域重大自然灾害及防治对策. 武汉：湖北人民出版社，30－57.

〔20〕 Hu Mingsi，Luo Chengzheng (eds.). 1992. Chinese Historical Large Floods (V. II). Beijing：China Book Store. 275－362.

〔21〕 黄兰心. 1999. 近40年来长江下游干流洪水位变化及原因初探. 湖泊科学，11(4)：99－104.

〔22〕 江苏省水文总站. 1984. 江苏省水文统计. 南京：江苏省水文总站. 746－770.

〔23〕 Hydrology Department of Yangtze River Planning Office. 1958. Characteristic Handbooks of Hydrological Information in Yangtze River Basin. Shanghai：Hydrology Department of YangtzeRiver Planning Office. 26－27.

〔24〕 Hydrology Station of Water Resources Bureau of Revolutionary Committee，Jiangsu Province. 1976. Hydrological Handbooks of Jiangsu Province. Nanjing：Hydrology Station of Water Resources Bureau，Jiangsu Province. 166－167.

〔25〕 梅金焕. 2001. 长江中下游干流堤防设计洪水位分析. 人民长江，32(12)：12－14.

〔26〕 黎明. 1999. 近百年来长江洪水变化的初步分析. 西南师范大学学报(自然科学版)，24(1)：97－102.

〔27〕 张建新，丁贤荣. 1995. 古洪水调查初探. 江苏地质，19(4)：209－212.

〔28〕 杨达源，谢悦波. 1997. 古洪水平流沉积. 沉积学报，15(3)：29－32.

〔29〕 王颖，B. 迪纳瑞尔. 1985. 石英砂表面结构模式图集. 北京：科学出版社，6－30.

〔30〕 谢悦波，费宇红. 2001. 古洪水平流沉积与水位. 地球学报，22(4)：320－323.

〔31〕 谢悦波，姜洪涛. 2001. 古洪水研究. 南京大学学报(自然科学)，37(3)：390－394.

［32］詹道江,谢悦波. 1997. 洪水计算的新进展. 水文,(1):1-6.

［33］Zhan Daojiang,Qian Tie. 1989. Extreme calculation frequency analysis with paleoflood. In:B C Yen,Proceedings of the International Conference on Channel Flow and Catchment Runoff. 1004-1012.

［34］Zhan Daojiang. 1995. Paleoflood study in China. Annual Report of Science and Technology. *Journal of Hohai University*,(suppl.):7-10.

［35］Baker V R,Kochal R C. 1998. Flood Geomorphology. The University of Arizona,Tucson, Arisona U S A. 124-129.

［36］谢悦波,王文辉等. 2000. 古洪水平流沉积粒度特征. 水文,20(4):18-20.

Water Resource Capacity,
Regulation, and Sustainable Utilization
of the Changjiang River Delta [*]

INTRODUCTION

Exiting from the Nanjing-Zhenjiang hilly terrain, the Changjiang River flows 300 km eastward passing through lowlands of the deltaic plain, to enter the East China Sea. Landforms of the delta gradually decrease in elevation from West to East, and increase from North to South, to link to the mountainous-to-hilly terrain to the southwest. The delta river systems have evolved from the Changjiang River in the central plain area, the Qiantangjiang River in the south, and the Huaihe River in the north, but with an independent river system of the Yongjiang River along the southeast coastal region. Several small rivers occur in the Zhoushan Islands. Mountain rivers are characterized by plentiful discharge, while the plain rivers are often confined to an artificial network of canals and dike structures (Figure 1).

Climate of the delta region is subtropical monsoon with four seasons changing from dry, wet, and cold, to warm (Liu and Cui, 1991). Average annual precipitation is 1 000~ 1 456 mm. Evaporation loss is 950~1 100 mm p. a. on the water surface, and 600~ 800 mm p. a. on land. Annual sunshine is about 1 800~2 250 hours, with sun radiant heat attaining $4\ 187\times10^6\sim5\ 024\times10^6\ J/m^2$. Annual average temperature is 14. 6 ℃~ 17. 0 ℃, with a frost free season of 212 to 268 days. As a whole, the Changjiang River delta possesses favorable climatic conditions, especially for mixed farming (Yan et al., 1986; Yang and Wang, 1996).

Located between 28°~33°N and 118°~123°E, the Changjiang River delta includes the urban centers of Shanghai City, Nanjing, Suzhou, Wuxi, Changzhou, Zhenjiang, Yangzhou, Taizhou, and Nantong (Tongzhou) cities of Jiangsu Province, and Hangzhou, Jiaxing, Huzhou, Ningbo, Shaoxing and Zhoushan cities of Zhejiang Province. Total land area of the delta is 99 610 km², which occupies 1% of the nation's total. The population is about 70million, accounting for 6. 05% of China's total. The

[*] Ying Wang, Terry Healy, Wenrui Chen, Dong Wang, Aimee Bishop: *Journal of Coastal Research*, 2004, Special Issue No. 43, pp. 75 – 88.

local GDP from 2001 data was 16 860 Billion Yuan, accounting for 17. 6% of the national total, and representing 30% of the import and export value. Development continues strongly (China Statistical Yearbook—2000, 2000; Urban Statistical Yearbook of China—2000, 2001).

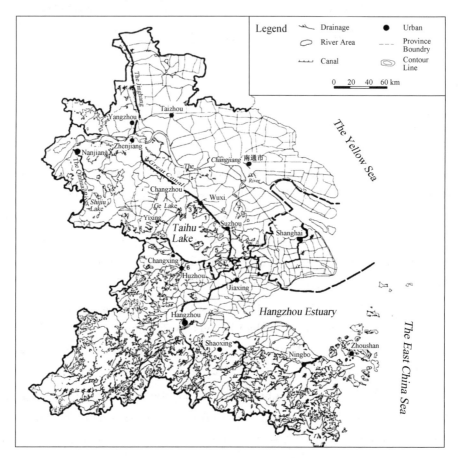

Figure 1 River systems and provincial boundaries of the Changjiang Delta region. Index for lakes mentioned in text: 1. Lake Gonghu 2. Lake Wuli 3. Lake Meiliang 4. Lake Zhushan 5. East Lake Taihu 6. Shaoxi River 7. Lake Yangcheng.

With the development of conurbations, industrial development, intensive agriculture, busy trade and commence, and cultural historical importance, the delta occupies a leading position in China. However, there is a serious imbalance between the capacity of the natural environment and the regional economic development. For example, in the delta, the average land available per person is only 1/6 of the national average, and the farm land per person is 1/2 of the national average (Chen and Zhu, 1998), yet the land area under farming is decreasing consequent upon urban industrial expansion. Waste effluent discharges of various sorts cause air pollution, increase acid rain, pollute waterways and induce eutrophication of rivers and freshwater lakes. In addition, artificial channel networks constructed in the deltaic lowlands have slowed distributary flow rates, even to the point of stagnation. Frequent algal blooms inducing red tides, and deteriorated water quality all reduce the available quality freshwater in the deltaic area, except, of course, for the main channel of the Changjiang River. Consequently, following lowering of the groundwater table and deltaic land subsidence, there have been a series of serious environmental effects, exacerbated by accelerated relative sea level rise enhancing coastal erosion, and frequent storm surges causing coastal flooding. As a result of river and estuarine re-charge of adjacent ground water tables causing serious waterlogging, the regional ecological environment and general environmental quality has deteriorated to the extent of threatening economic development by discouraging the regional supply of capital (Ren, 1996; Zhu, 1999). Thus, the Changjiang River delta faces serious challenges of resource management and environment pressure!

In this paper we review the water budget of the Changjiang River delta, review the water quality, outline the variety of water related problems, and present the argument for an integrated sustainable resource management regime.

WATER RESOURCES OF THE CHANGJIANG RIVER DELTA

There are systematic hydrological monitoring stations located along the Changjiang River main channel, surrounding lakes, and distributaries. Monitoring data have been collected since the 1950s, and the data here is only available for public dissemination in the form of annual reports. The following are the prime data sources for this paper: National Marine Environment Monitoring Networks Office of China (1995, 1996, 1997), The Center for Environment Monitoring of Nanjing City (1996), Water Resource Protection Bureau of the Taihu Basin and the Center for Environment Monitoring of Shanghai City (1997), Environment Protection Bureau of Shanghai City (1998), Environment Protection Bureau of Jiangsu Province (1998), Environment Protection Bureau of Suzhou City (1998), The Center for Environment Monitoring of Jiangsu Province (1997, 1998, 2001).

Available Local Water Quantity

Perennial available fresh water quantity for the Changjiang River delta is $508.4 \times 10^8 \ m^3$, of which 70% is in the northeast Zhejiang Province, 26% in Jiangsu Province, and 4% in the Shanghai area. Additional ground water resources total about $177.71 \times 10^8 \ m^3$, of which $87.72 \times 10^8 \ m^3$ are in Zhejiang Province, $70.66 \times 10^8 \ m^3$ in Jiangsu Province and $19.33 \times 10^8 \ m^3$ around the city of Shanghai.

Local available water resource is $573.79 \times 10^8 \ m^3$ on average, with $32.07 \times 10^8 \ m^3$ in Shanghai, $76.57 \times 10^8 \ m^3$ in Jiangsu, and $365.15 \times 10^8 \ m^3$ in Zhejiang Province (Table

1). The delta abounds with water resources of 57.6×10^4 m^3/km^2, but average water quantity per person is 774.9 m^3 which is only one third of the Chinese national average of 2 420 m^3 per person.

Table 1　Local water resources of the Changjiang River delta.

Area	Annual Precipitation (mm)	Annual Runoff Depth (mm)	Local Runoff Quantity (10^8 m^3)	Underground Water Quantity (10^8 m^3)	River Water Discharge (10^8 m^3)	Local Water Resources Quantity (10^8 m^3)	Per Area Water Resources (10^4 m^3/km^2)	Local Per Capita Water Resources (m^3)
Shanghai City	1 141	301	18.66	19.33	5.92	32.07	50.58	245.85
Central and southern Jiangsu Province	1 040	254	134.43	70.66	28.52	176.57	36.55	457.55
Northeast Zhejiang Province	1 465	737	355.31	87.72	77.88	365.15	81.21	1 629.25
Total Area	1 266	501	508.40	177.71	112.32	573.79	57.60	774.90

Source：SHE (1997).

Water Input from External Sources

Input of water from delta-marginal catchments is abundant, including water transported and diverted, as occurs in the lower reaches of several large rivers. Water arriving from the middle reaches of the Changjiang major channel is about $9\,730 \times 10^8$ m^3 per year on average. The river provides sufficient water supply even in drought years (1928). During the drought year of 1978, for example, South Jiangsu diverted 112×10^8 m^3 of water from the Changjiang River. On the other hand, the Huangpujiang River from Shanghai has diverted 400×10^8 m^3 of tidal prism annually and the volume of tidal prism is about 22 times that of the local water quantity. The enormous water quantity available from the outer reaches provides a great advantage for local water supply.

Water Quality along the Lower Reaches of the Changjiang River

In China, surface water quality is classified into 5 Classes. Water quality of Class I is applicable to headwaters and national natural protection areas. Class II water is applicable to the 1st rank of protected drinking water resources, and the habitats for rare aquatic animals and water plants, for spawning, breeding, and feeding of fish, shrimp, and aquaculture. Class III water quality is applicable to 2nd ranking protected areas for surface drinking water, but also to fish migration channels and aquaculture, fishing, and swimming grounds. Class IV water applies to industrial water use and is not suitable for human use. Class V water is applicable to agricultural use and general scenery purpose (GB3838 – 2002, National Standard of Surficial Water Quality, China, see Table 2).

Table 2　Limiting value of some State Fresh Surface Water Quality Standards (units: mg/L)

		Class I	Class II	Class III	Class IV	Class V
DO	$\geqslant$	7.5	6	5	3	2
KMnO$_4$	$\leqslant$	2	4	6	10	15
COD	$\leqslant$	15	15	20	30	40
BOD$_5$	$\leqslant$	3	3	4	6	10
TP	$\leqslant$	0.02	0.1	0.2	0.3	0.4
TN	$\leqslant$	0.2	0.5	1.0	1.5	2.0
volatile phenol	$\leqslant$	0.002	0.002	0.005	0.01	0.1
petroleum products	$\leqslant$	0.05	0.05	0.05	0.5	1.0

Source: GB3838 – 2002, National Standard of Surficial Water Quality, China.

Water quality of Class II or III occurs along most of the Changjiang River in the deltaic reaches. According to the monitoring along the Changjiang channel in that section of Jiangsu province, it was discovered that 4 places were affected by petroleum contamination. For these areas water quality was ranked Class IV, but for the other parts of the Jiangsu Province sections, water quality was still ranked Class II (Table 3).

Table 3　Water quality Classification of the lower reach of the Changjiang River in 2000.

Section	Comprehensive Pollution Indicator	Classification of Water Quality	Main Pollutant
Nanjing section	2.52	IV	Petroleum products, ammonia
Outer Zhenjiang branch	2.00	II	Petroleum products, DO
Inner Zhengjiang branch	4.74	IV	Petroleum products, ammonia, DO, KMnO$_4$
Nantong section	2.82	II	KMnO$_4$, ammonia, petroleum products

Source: THE CENTER FOR ENVIRONMENT MONITORING OF JIANGSU PROVINCE (2001).

Other distributaries of the Changjiang River have also become seriously polluted, almost to Class IV-V level except for a few distributaries near the major channel of the Changjiang. Ammonia is a major pollutant, and Biochemical Oxygen Demand (BOD$_5$), KMnO$_4$ and dissolved oxygen(DO) levels indicate that these waters are in the seriously polluted range—to the degree that they are too polluted to be used for agricultural irrigation.

Water Quality of the Lake Taihu Area

Lake Taihu is the major drinking water resource for Shanghai, Wuxi and Suzhou cities. However, water quality of this lake has continuously decreased (She, 1997; Cai, 1998; Wu, 2001). Water quality in Lake Taihu was ranked class III – II in 1981, Class IV overall in 2000 and ranked class IV – V in 2001. Only in the central portions and the lakeshore near Suzhou, Huzhou and Changxing is water quality relatively better, but the lake shore near Wuxi, Wujin and Yixing exhibits poor quality water. The sub-lakes of Wuli and Meiliang possess worse water quality even than Class V due to the influence of ammonia and total phosphorus (TP) originating from agricultural runoff (Figure 2).

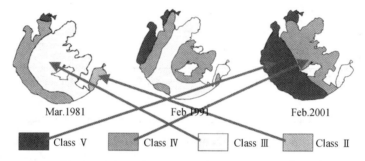

Figure 2　Changes in surface patterns of water quality since 1981 for Lake Taihu
Source: after Huang *et al*. (2001)

At present Lake Taihu is still seriously eutrophic, containing high levels of total phosphorus (TP) and total nitrogen (TN). Along the lake shore at Suzhou, Huzhou, Changxing and the central lake area a medium level of eutrophication is evident. During the dry season, the water quality of Lake Taihu is characterized by a medium level of eutrophication, but higher eutrophication levels are evident during both average and higher water level periods because of re-mobilized total phosphorus (TP) and total nitrogen (TN) from farm runoff.

According to the monitoring results for the year 2000, the water bodies of Lake Yangcheng and Lake Ge exhibited water quality of Class IV. Major pollutants and pollutant indicators were TP, ammonia, $KMnO_4$ and BOD_5 (Table 4).

Around Shanghai water quality become the most seriously deteriorated. Here no Class I or II water occurs except for the main channel of the Changjiang River mouth. Even Class III water is limited only to Lake Dianshan and the Taipu River.

According to 1997 monitoring data, water bodies of the east and the west Shaoxi River of the Lake Taihu region had 17% of Class I – II water quality, 60% of Class III, 17% of Class IV to V, and 6% of the water bodies had water quality lower than lass V. The east Shaoxi River was relatively better than the west Shaoxi River (Environment Protection Bureau of Zhejiang Province, 1998).

Table 4   Water quality monitoring results for Lake Taihu, 2000.

Area	Value / Elements	DO	KMnO₄	BOD₅	TP	TN
Lake Taihu	average (mg/L)	9. 4	5. 9	2. 9	0. 107	1. 91
	over standard value (%)	0	32. 3	19. 8	73. 8	
Lake Yangchenghu	average (mg/L)	8. 8	5. 6	3. 4	0. 127	2. 71
	over standard value (%)	7. 1	23. 8	31. 0	97. 6	
Lake Gehu	average (mg/L)	7. 0	5. 2	2. 2	0. 094	1. 38
	over standard value (%)	12. 5	8. 3	0	58. 3	

Source: THE CENTER FOR ENVIRONMENT MONITORING OF JIANGSU PROVINCE (2001). The "over standard" relates to values poorer than Class.

Water Quality of the Qiantangjiang, Caoejiang, and Yongjiang River

Water quality of these three rivers is generally of Class I and II. The only relatively poor reaches of the Qiantangjiang River are the Chengtou section of the Dongyang-Nanjiang River, the Houjindu section of the Dongyangjiang River, the Jiangshangong section of the Shuanggangkou River, and the Anhua and river mouth of the Puyangjiang River. The pollutants and indicators are KMnO₄, BOD₅, volatile phenol, ammonia and petroleum (Figure 3).

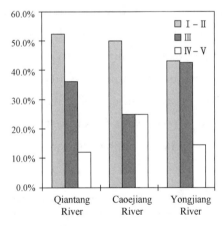

Figure 3   Proportions of water quality classes of the Qiantangjiang; Caoejiang and Yongjiang rivers, 1997.

Source: Environment Protection Bureau of Zhejiang Province(1998).

Water Quality of the South Section of the Grand Canal

The average water quality for this area is categorized as Class V. The poorest quality waters are contained in the Zhenjiang and Wuxi sections of the canal. This area is heavily polluted. Analyzed water samples revealed several pollutants and pollutant indicators including BOD_5, $KMnO_4$, DO, and volatile phenol (Table 5).

Table 5   Water quality of the Grand Canal, 2000.

River Section	Comprehensive Pollution Index	Class Water Quality	Main Pollutant
Yangzhou	3.07	IV	Petroleum products, BOD_5, $KMnO_4$
Zhenjiang	5.58	Worse than V	BOD_5, $KMnO_4$
Changzhou	4.96	V	Petroleum products, volatile phenol,
Wuxi	9.17	Worse than V	DO, ammonia
Suzhou	5.46	V	DO, BOD_5, volatile phenol
Jiaxing & Hangzhou, 1997 *	~	IV~V	$KMnO_4$, BOD_5, Petroleum products, ammonia

Source: THE CENTER FOR ENVIRONMENT MONITORING OF JIANGSU PROVINCE (2001).

* Source: ENVIRONMENT PROTECTION BUREAU OF ZHEJIANG PROVINCE (1998).

Water Quality of the Nearshore Changjiang River Delta

The nearshore seawater standard is different to the fresh ground and surface water standard. Depending upon the utilization function and protection purpose of the marine waters, the nearshore seawater standard can be subdivided into four Classes (GB3097 - 1997, National Standard of Sea Water Quality, China). Class I seawater is applicable to areas for marine fishing, marine natural protection areas, and protection areas for rare and endangered species. Class II seawater applies to marine aquaculture, and swimming and recreational areas. Class III seawater is applicable to industrial water use and seashore scenic tourism. Class IV is applicable for harbor and ocean development construction (Table 6).

Table 6   Limiting values for state sea water quality standard (units: mg/l).

		Class	Class	Class	Class
DO	>	6	5	4	3
COD	≤	2	3	4	5
BOD_5	≤	1	3	4	5
TP	≤	0.02	0.1	0.2	0.3
TN	≤	0.2	0.5	1.0	1.5
volatile phenol	≤	0.005		0.010	0.050
petroleum products	≤	0.05		0.30	0.50

Source: GB3097 - 1997, National Standard of Sea Water Quality, China.

Water quality (as indicated by Inorganic Phosphorus level) outside of the Changjiang River mouth, Hangzhou Estuary and Zhoushan fishing grounds has deteriorated and the water quality index is less than 2 (Figures 4 and 5).

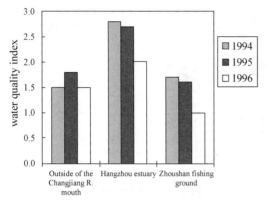

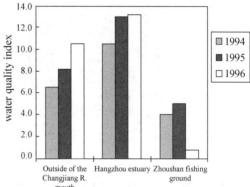

Figure 4   Change in water quality index as indicated by inorganic phosphorus in the nearshore off the Changjiang River mouth, 1994—1996.
Source: WANG (1996).

Figure 5   Change in water quality index as indicated by inorganic nitrogen in the nearshore off the Changjiang River mouth from 1994—1996.
Source: WANG (1996).

Major pollutants are inorganic nitrogen and phosphorus, and petroleum products. The main source of pollutants are the liquid waste from coastal industry, wastewater from cities, and oil spilled from ships. In addition, there are effluents from chemical fertilizers, and pesticides resulting from tidal flat and coastal aquaculture developments.

During 2000, the Yellow Sea water quality along the Nantong coast of North Jiangsu was categorized predominantly as Class II, except for the Dayangkou coast (Class IV). The Dayangkou area is influenced by active phosphate. However, water quality in the Xiaoyangkou area remains as Class II. Part of the seawater quality standard is listed in Table 6 (GB3097 – 1997, National Standard of Sea Water Quality, China).

The seawater quality, as indicated by the comprehensive pollution index, appeared to improve at the two nearshore areas of Dayangkou and Xiaoyangkou between 1996 to 2000, compared to data collated between 1990 to 1995 (Figure 6). The improved water quality resulted mainly from a decrease in contaminants such as petroleum products, active phosphate, nitrogen and lead contaminants.

PRESSING WATER RESOURCE PROBLEMS OF THE CHANGJIANG RIVER DELTA

Shortage of Available Fresh Water Supply

Virtually all developing countries, even those with adequate water in the aggregate, suffer from debilitating regional and seasonal shortages of water (Mou, 1995). Likewise

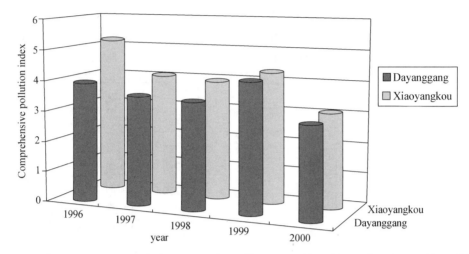

Figure 6   Seawater quality trends in the nearshore off Nantong City, Jiangsu Province, 1996—2000.
Source: THE CENTER FOR ENVIRONMENT MONITORING OF JIANGSU PROVINCE
(2001).

there is an uneven distribution of water resources in the Changjiang River Delta. The average available local water resource is 775 m^3/person, which is about 1/3 of the national average. The water availability is considerably lower for the Jiangsu and Shanghai area, where on average there are 255 m^3/person and 458 m^3/person respectively. Such low values suggest that these areas suffer from a water shortage.

River discharges are an important potential water resource, which have not been utilized due to the poor water quality and the lack of funds to support appropriate water treatment to make it suitable for human consumption.

Coastal areas and nearshore islands are especially short of fresh water resources. For example, personal average annual water availability in the Zhoushan Islands is only 570 m^3, which is 1/5 of the national average. At present, there are 0.3 million people who lack suitable drinking water, and about 0.2 million, who face severe difficulties in obtaining freshwater (Li, 1997).

Water Deterioration and Lake Eutrophication

As a result of continuing and increasing waste and polluted water discharges into lakes, there has been accumulation of nutrients. Consequently, many lakes are strongly eutrophic, as indicated by phytoplankton blooms (Goldman and Horne, 1994). Phytoplankton blooms occur mainly during the summer months of April to October and to a lesser extent during February to March. Blue algae and diatoms predominate during spring and summer, while blue algae predominate during autumn. There are strong algae blooms in Lake Wuli, but the strength of the blooms decreases from Lake Meiliang, to Lake Zhushan, to grand Lake Taihu, through to the East Lake Taihu

(Wu, 2001).

In the early 1980's, some 60% of the water in Lake Taihu was categorized as Class II quality, 30% as Class III, and only 1% as Class IV. Class III and IV water quality were distributed near the entrance and along the lake shore of East Lake Taihu, while eutrophic water occupied about 17% of the total lake area.

During the late 1980's, Lake Taihu became more heavily polluted and this impacted on the water quality. Class III water area coverage increased by 7%, Class IV affected water tripled in size. In addition, Class V water occupied approximately 1% of the Lake Meiliang and the Lujiang river mouth, and part of Lake Wuli. The total eutrophic area of the water body reached a staggering 41%.

By the mid 1990s, the lake water quality had deteriorated as a consequence of eutrophication. Class II water had decreased to only 15% of the total water area, while 70% of the water body had been categorized ed as Class III. In Lake Meiliang and along the west and south lake shore of Lake Taihu, Class IV water had increased to occupy 14% of the total lake area.

Between 1996 to 1999, Lake Taihu water quality generally improved, but declined again in 2000. The potassium permanganate index (PPI) increased significantly during that year because of the lower incidence of heavy rainfall and flooding. And in addition two sluicegates of distributaries in the surrounding area of Lake Taihu near Changzhou were opened, thereby increasing the volume of pollutants that entered and dispersed through Lake Taihu (Figure 7).

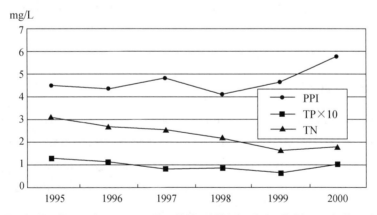

Figure 7　Change in water quality 1995—2000 for Lake Taihu, as indicated by the potassium permanganate index (PPI), total phosphorus (TP), and total nitrogen (TN).

Source: THE CENTER FOR ENVIRONMENT MONITORING OF JIANGSU PROVINCE (2001).

Between 1994 to 1999, Lake Taihu was in a stable eutrophic condition. The eutrophic index decreased after 1998, before steadily increasing again from 1999 to 2000

(Figure 8).

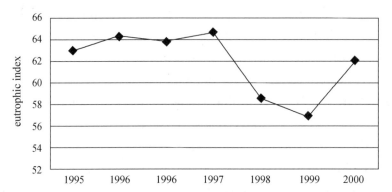

Figure 8　Eutrophication trend for Lake Taihu, 1994～2000. The eutrophic index indicates the degree of eutrophication. A value > 60 represents a high eutrophic level, and < 60 represents a medium eutrophic level.

Source: The Center for Environment Monitoring of Jiangsu Province (2001).

Nearshore Red Tides

Nearshore red tides develop as a result of nutrient enrichment. Nutrient enriched water is discharged into the shallow coastal ocean off the Changjiang River Delta. Red tides appeared initially in the coastal water located near the Changjiang River mouth, then spread to include the coastal waters to the east of Nantong City and the Zhoushan Islands. Remote sensing data identified that red tides occurred on four separate occasions in this area. In 1993, the largest red tide covered 50 km² of coastal water and in 1995 the reddish brown colored seawater appeared as a belt that extended over several hundred km²(State Oceanic Administration of China, 1996).

Flood Impacts and Waterlogging

'Plum rains' and typhoons are the major climatic conditions which induce flooding and waterlogging of low-lying areas in the river delta. 'Plum rains' occur during late May and early June, while typhoons occur during summer and autumn.

Lake Taihu is located in a depression on the southern side of the main Changjiang River channel, in the delta area which from west to east is low-lying and is elevated between 2～10 m above msl with many low hills and terraces. The plain is, connected to the Changjiang River in the north, and the Qiantangjiang River in the south. There are low hills by the headwaters, while canal and lake networks characterize the central parts, and tidal flats are distributed along the east coast. When summer and autumn floods overflow from the river networks and onto the lowlands, they may induce waterlogging (Yang and Chen, 1995; Yang and Yao, 1997).

On the north side of the Changjiang River the deltaic plain also contains a large depression elevated <2 m above MSL datum. The deltaic plain in the north merges with the Grand Canal plain in the west. These lowland deltaic plains are comprised of an ancient marine depositional plain, which to the east is elevated $2.5 \sim 4$ m above MSL. The depressional plain connects to the delta plain on the southern side, where the elevation is $6 \sim 9$ m above MSL. To the north side of the depressional plain is the Yellow River floodplain (elevated $6 \sim 9$ m above MSL datum). The elevation at the center of the depressional plain is only 1.1 m above datum, and thus the area suffers frequently from flood inundation and waterlogging (Chen et al, 1996).

Human activity has exacerbated the flooding problems (Liang and Li, 1993; Kundzewicz, 2003). For example, reclamation of part of Lake Taihu commenced in 1985. During the last 36 years, 539 km² have been reclaimed which equates to 13.6% of the original lake area, with an average loss of 14.7 km² per year. The reclamation resulted in the disappearance of 165 small lakes with a combined water surface of 161.3 km². The ability to store flood water and regulate the river-lake discharge system has been weakened by the lake reclamation. To compound the problem, the artificial dam and protective embankments that surround the lakesides and low lying land, have reduced channel discharge and groundwater recharge, and consequently the river channel has narrowed. In addition the over-pumping of groundwater has induced land subsidence, in some areas by as much as 5 mm/a. Land subsidence has resulted in the formation of several new depressions (Shi et al, 2000). Contemporary relative sea level rise (average rate of 2 mm/a) has increased the risk of storm surge erosion, flooding, and waterlogging of soils. Some 1 300 km² of land is subject to land subsidence, which has increased the risk of economic loss in flood disasters (Li, 1998).

THE NEED FOR INTEGRATED MANAGEMENT OF DELTA WATER RESOURCES

Fresh air, clean water and healthy food are three fundamental elements required for human existence. Here in the Changjiang River delta the quality and quantity of the water environment and water resources produce a major challenge. The quality and quantity of freshwater supply needs to be improved to meet the safety requirements and national demand. Essentially water body pollution, flooding and waterlogging issues need to be addressed. The situation of the Changjiang River delta is quite unique with respect to rivers both in China and Asia, for example, the Yellow River (the second largest river in China) has an average annual runoff volume of 49.3×10^9 m³, which discharges to the Bohai Sea. The river flow ceases intermittently in parts of the lower reaches; this has happened frequently since 1972 (Wang et al, 2001). The longest period without any water flow occurred in 1997, and lasted for 226 days and extended

700 km upstream from the river mouth.

The Mekong River in southeast Asia is the eighth largest river in the world on account of its average annual water discharges of 470×10^9 m³ (Milliman and Meab, 1983). As an international river, it passes through six countries including China, Burma, Kampuchea, Laos, Thailand and Vietnam. There is no integrated resource management plan for the Mekong River, and consequently that river environment has been adversely affected by misguided, short-term local management decisions. For example, the upper stream was dammed. Dams can decrease the threat of floods to the lower reaches and aids in protecting wetland and deltaic fish spawning/nursing grounds. Conversely dams can result in reduced fish diversity and production, because the dam blocks the route for fish migration. This adversely affects communities dependent upon the river delta. For example, more than 30% of poor families in the Mekong river delta are dependent on fishing and farming for their livelihood. In addition, salt water intrusion can occur upstream as far as 70 km inland during the dry season, and as a result rice production has decreased between 30%~50% over a 30 year period (UNDP, UNEP, World Bank and World Resource Institute, 2002).

The Changjiang River delta is characterized by high precipitation values, a large area off farmland and aquaculture ponds, high population density, high chemical fertilizer application to the soils, rapid development of industry in rural areas, and urban expansion—all causes of water environmental deterioration and water resource reduction by pollution.

To address this multitude of wide ranging disparate issues, a regional study on the Changjiang River delta and related area needs to be carried out. The regional study should aim to seek solutions to improve water quality, increase water quantity supplement, and to generally improve the efficiency of water utilization within the framework of integrated planning, integrated implementation, and integrated management.

At this point, several case studies can be adopted as references, for example, integrated catchment management in Australia (Cunningham, 1986; Mitchell and Hollick, 1993; Mitchell, 1990, 1997). These examples highlight several principal issues of concern for integrated management of fresh water environments, including turbidity, salinity and nutrient enrichment, and the role of storm water management in Sydney's urban rivers (Brizga and Finlayson, 2000).

"Integrated water resources management" has introduced a growing complexity. The concept now involves numerous aspects (environmental management, economic development, public health and social well-being), multiple goals, the mitigation of side effects and unforeseen impacts, and multiple users (Ubbels and Verhallen, 2001). Based on the above studies, and to rationally address water resource management issues and problems in the Changjiang River delta, the following components of an integrated water resources management plan are suggested:

(1) To integrate the widely disparate available knowledge, and where necessary to undertake additional research on the hydrologic and geomorphic characteristics of the Changjiang River major channel and basin characteristics, including the groundwater budget, and to develop this data into a sub-basin hydrological operational model. With this modern state-of-the-technology tool, a regional plan can be produced for water resource utilization and management of the water environment of the delta in a similar way to that advocated by Acreman (2001) and Marino (2001).

(2) To construct a numerical model of water capacity of water circulation and water environment of the delta region, for present and future scenarios of water demand from a variety of water consumers, including irrigation, urban use, industrial use, forest use, conservation, including influences such as under present natural, and global, change trends (Fashchevsky and Fashchevskaya, 2003). The model scenarios need to include the load capacity of water resources and the water environment for human consumption, agriculture, industrial, urban, transportation, tourism, and economic trade development uses, as well as retaining the status of a bio-diverse ecological system.

The modelling should address the scenarios of the *best practical option*, the *ultimate potential*, and the *limiting case*.

To research a scenario for the year 2020, the modelling needs to incorporate engineering projects to control pollution and environmental restoration, increase the proportion of water re-used, and govern control on water resource availability, in order to ascertain the water capacity load for the deltaic population and economic development (Datta et al, 2001). This would include the situation for wide application of new technology, and assume that environmental awareness and public morality in terms of acceptable resource utilization practices have been improved.

(3) To establish procedures and management systems for control of existing water resources and environment, preventing pollution, conserving clean water resources, and enhancing effective and beneficial usage, as advocated in a similar situation by Sharma (2003).

(4) To establish a regional master plan for integrated drainage management and control of rivers, lakes, canals and ponds discharging to the East China Sea. This would involve control over artificial constructions and water conservancy projects, leading to controlled mitigation of flooding, waterlogging and tidal intrusion episodes.

(5) Planning for environmentally friendly construction and development for the delta region as a whole, while recognizing different conditions and characteristics of each area, and restoring the lowland delta environment to high quality ecosystem, eco-

landscape, and an economic development region sensitive to maintaining ecological values.

(6) To raise public awareness on limitations of water resources, and to encourage responsibility for cherishing and saving fresh water, while gradually producing new regulations, and even developing a whole new social tradition for use of natural water systems.

CONCLUSION

The quality and quantity of fresh water resources of the Changjiang River delta poses a major problem that needs urgent consideration by state and local planners, and water managers. A multifaceted strategy for attaining an appropriate water management regime for sustainable resource utilization of the Changjiang River delta has been presented (Figure 9). Such a study may offer ideas of potential application to other similar deltaic plains.

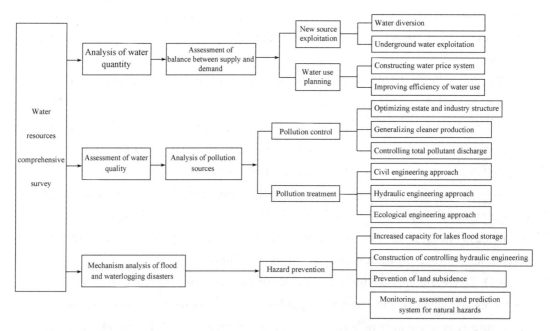

Figure 9　A strategy for water resource sustainable management for the Changjiang River delta

LITERATURE CITED

Acreman, M. 2001. Towards the sustainable management of groundwater-fed catchments in Europe. In: Acreman, M. C., Davis, R., Marino, M. A., Rosbjerg, D., Schumann, A. H., and Xia, J. (eds.), *Regional Management of Water Resources*. IAHS Publication 268. 123 – 130.

Brizga, S. and Finlayson, B. 2000. *River Management*. New York: John Wiley and Sons. 16 – 170.

Cai, Q. 1998. *Research on Environment and Ecology of Lake Taihu*. Beijing: Meteorology Press, 228 p (In Chinese).

Chen, W. and Zhu, D. 1998. Sustainable utilization of land resources in the Yangtze River delta. *Journal of Natural Resources*, 13(3): 261 – 266 (In Chinese).

Chen, X., Wang, L., and Zhu, D. 1996. System of northern Jiangsu lowlands and its complex responses to the sea level rise. *Acta Geographica Sinica*, 51(4): 340 – 349 (In Chinese).

China Statistical Yearbook—2000, 2000. Beijing: China Statistical Press. 52 – 53 (In Chinese).

Cunnigham, G. M. 1986. Total catchment management—resource management for the future. *Journal of Soil Conservation, New South Wales*., 42(1): 5.

Datta, P. S., Rohilla, S. K., and Tyagi, S. K. 2001. Integrated approach for water resources management in the Delhiregion: problems and perspective. In: Acreman, M. C., Davis, R., Marino, M. A., Rosbjerg, D., Schumann, A. H., and Xia, J. (eds.), *Regional Management of Water Resources*. IAHS Publication 268. 65 – 72.

Environment Protection Bureau of Jiangsu Province. 1998. *Annual Report on Provincial Environment Status of 1997*. 2 – 15 (In Chinese).

Environment Protection Bureau of Shanghai City. 1998. *Annual Report on Shanghai City Environment Status of 1997*. 2 – 13 (In Chinese).

Environment Protection Bureau of Suzhou City. 1998. *Annual Report on Suzhou City Environment Status of 1997*. 3 – 14 (In Chinese).

Environment Protection Bureau of Zhejiang Province. 1998. *Annual Report on Provincial Environment Status of 1997*. 2 – 15 (In Chinese).

Fashchevsky, B. and Fashchevskaya, T. 2003. Water management budget as a basis for assessing water priorities in a catchment. In: Bloschl, G., Franks, S., Kumagai, M., Musiake, K., and Rosbjerg, D. (eds.), *Water Resources Systems—Hydrological Risk, Management and Development*. IAHS Publication 281. 322 – 326.

GB3097 – 1997. National Standard of Sea Water Quality, China. State Environmental Protection Administration of China and Quality Supervision, Test and Quantitative Administration of China. In: *Legislation on Environment Impact Assessment*. Beijing: China Environment Science Press. 1729 – 1737 (In Chinese).

GB3838 – 2002. National Standard of Surficial Water Quality, China. State Environmental Protection Administration of China and Quality Supervision, Test and Quantitative Administration of China. In: *Legislation on Environment Impact Assessment*. Beijing: China Environment Science Press. 1167 – 1170 (In Chinese).

Goldman, C. and Horne, A. 1994. *Limnology 2nd Edition*. New York: McGraw-Hill. 457 – 475.

Huang, Q., Fan, C, Pu, P., Jiang, J., and Dai, Q. 2001. *Water Environment and its pollution control of Taihu Lake*. Beijing: Science Press. 94 – 142 (In Chinese).

Kundzewicz, Z. W. 2003. Extreme precipitation and floods in the changing world. *In*: Bloschl, G., Franks, S., Kumagai, M., Musiake, K, and Rosbjerg, D. (eds.), *Water Resources Systems—Hydrological Risk, Management and Development*. IAHS Publication 281. 32 – 39.

Li, D. 1998. The fundamental way out to prevent the land subsidence in Lake Taihu Plain. *Hydrogeology and Engineering Geology*, (4): 16 – 18 (In Chinese).

Li, Z. 1997. Research on the characteristics and exploitation of the island resources of Zhejiang province. *Journal of Natural Resources*, 12(2): 139 – 145 (In Chinese).

Liang, R. and Li, H. 1993. '91 *Flood and waterlog of the Taihu drainage basin*. Nanjing: Hohai University Press. 97 – 103 (In Chinese).

Liu, G. and Cui, Y. 1991. The water balance of China and its large river basins. *In*: Gutknecht, D.; Loucks, D. P.; Salewicz, K. A., and Van De Ven, F. H. M. (eds.), *Hydrology for the Water Management of Large River Basins*. IAHS Publication 201, pp. 153 – 162.

Marino, M. A., 2001. Conjunctive management of surface water and groundwater. *In*: Acreman, M. C., Davis, R., Marino, M. A., Rosbjerg, D., Schumann, A. H., and Xia, J. (eds.), *Regional Management of Water Resources*. IAHS Publication 268. 165 – 173.

Milliman, J. and Meab, R. H. 1983. World-wide delivery of river sediment to the ocean. *Journal of Geology*, 91: 1 – 20.

Mitchell, B. 1990. Integrated Water Management. *In*: Mitchell, B. (ed.), *Integrated Water Management: International Experience and Perspectives*. London: Belhaven. 1 – 21.

Mitchell, B. 1997. *Resource and Environmental Management*. England: Addison Wesley Longman Limited. 240 – 259.

Mitchell, B., and Hollick, M. 1993. Integrated catchment management in Western Australia: transition from concept to implementation. *Environmental Management*, 17(6): 735 – 743.

Mou, H. 1995. Zero increase mode for water resources sustainable development in China. *Geographical Research*, 14(1): 80 – 84 (In Chinese).

National Marine Environment Monitoring Networks Office of China. 1995. *Annual Report on Nearshore Environment Quality of China of 1994*. 2 – 12 (In Chinese).

National Marine Environment Monitoring Networks Office of China. 1996. *Annual Report on Nearshore Environment Quality of China of 1995*. 3 – 10 (In Chinese).

National Marine Environment Monitoring Networks Office of China. 1997. *Annual Report on Nearshore Environment Quality of China of 1996*. 2 – 11 (In Chinese).

Ren, M. 1996. Several issues concerning sustainable development of the Changjiang River delta. *World Science and Technology Research and Development*, 18(34): 97 – 101 (In Chinese).

Sharma, U. C. 2003. Impact of population growth and climate change on the quantity and quality of water resources in the northeast of India. *In*: Bloschl, G., Franks, S., Kumagai, M., Musiake, K., and Rosbjerg, D. (eds.), *Water Resources Systems—Hydrological Risk, Management and Development*. IAHS Publication 281. 349 – 357.

She, Z. 1997. *The Water-Land Resources and the Regional Development In Yangtze River Delta*. Hefei: University of Science and Technology of China Press. 2 – 128 (In Chinese).

Shi, Y., Zhu, J., Xie Z., Ji, Z., and Yang, G. 2000. Predicting and preventing countermeasures of sea level rise effect on the Changjiang River delta and the adjacent area. *Science in China* (*series D*), 30(3): 226 (In Chinese).

State Oceanic Administration of China. 1996. *Annual Report on Sea Environment of China—1995*. 3 – 15 (In Chinese).

The Center for Environment Monitoring of Jiangsu Province. 1997. *Annual Report on Provincial Environment Status of 1996*. 2 – 12 (In Chinese).

The Center for Environment Monitoring of Jiangsu Province. 1998. *Annual Report on Provincial Environment Status of 1997*. 4 - 11 (In Chinese).

The Center for Environment Monitoring of Jiangsu Province. 2001. *Annual Report on Provincial Environment Status of 2000*. 3 - 14 (In Chinese).

The Center for Environment Monitoring of Nanjing City. 1996. *Annual Report on Nanjing City Environment Status of 1991~1995*. 2 - 13 (In Chinese).

Ubbels, A. and Verhallen, A. J. M. 2001. Collaborative planning in integrated water resources management: the use of decision support tools. *In*: Marino, M. A. and Simonovic, S. P. (eds.), *Integrated Water Resources Management*. IAHS Publication 272. 37 - 43.

UNDP, UNEP, World Bank and World Resource Institute. 2002. *World Resource Report 2000~2001*. Beijing: China Environment Science Press. 206 - 209 (In Chinese).

Urban Statistical Yearbook of China—2000, 2001. Beijing: China Statistical Press. 118 - 119 (In Chinese).

Wang, L. 1996. Simulation and Prediction of Flood and Waterlogging Disasters of Lake Taihu Basin, P. R. China, Nanjing: Nanjing University, Ph. D. thesis. 298 (In Chinese).

Wang, Y., Arrhenius, G., and Zhang, Y. 2001. Drought in the Yellow River—an environmental threat in the coastal zone. *In*: Healy, T. R. (ed.), *Challenges for the 21st Century in Coastal Science, Engineering and Environment*. Journal of Coastal Research, Special Issue 34: 503 - 515.

Water Resource Protection Bureau of the Taihu Basin and the Center for Environment Monitoring of Shanghai City. 1997. Report on Survey and Analysis of the Pollution Sources of the Upper Stream of the Huangpu River and the Boundary Area of Jiangsu Province, Zhejiang Province and Shanghai City. 2 - 19 (In Chinese).

Wu, R. 2001. Lake Water Resources and Environments in China and Countermeasures. *Bulletin of the Chinese Academy of Sciences*, (3): 176 - 181 (In Chinese).

Yan, J., Xu, W., Zhu, J., and Shu, J. 1986. An investigation of the inter annual rainfall variations for different natural seasons in the Yangtze delta. *Geographical Research*, 5(1): 51 - 57 (In Chinese).

Yang, G. and Wang, B. 1996. A study on the developing state and regional divergence of various development zones in the Yangtze River deltaic area. *Resources and Environment in the Yangtze Valley*, 5(3): 193 - 198 (In Chinese).

Yang, S. and Chen, J. 1995. Formation and evolution of flood and waterlog disasters of the Taihu Lake area. *Scientia Geograpfica Sinica*, 15(4): 307 - 314 (In Chinese).

Yang, S. and Yao, Y. 1997. Causes of flood and strategy of flood loss reduction in Taihu Lake reaches. *Journal of Catastrophology*, 12(3): 34 - 37 (In Chinese).

Zhu, D. 1999. The characteristics and current tasks of sustainable development of the Changjiang River delta. *In*: Yan, D., and Ren, M. (eds.), *Strategy on Sustainable Development of the Changjiang River Delta*. Hefei: Anhui Education Press. 139 - 151 (In Chinese).

长江三角洲水资源现状与环境问题[*]

　　三角洲是河流入海口的堆积体,长江三角洲是海拔小于 6 m 的冲积平原,面积 22 800 km²,包括江苏的苏、锡、常,上海,浙江杭、嘉、湖平原。本文所指长江三角洲是区域经济上的称谓,行政上包含上海市;江苏省的宁、镇、扬、泰、通、苏、锡、常八市;浙江省的杭、嘉、湖、甬、绍、台及舟山七市。地形有平原、丘陵、山地及海岛。

　　长江三角洲(图 1)地处沿海,属于亚热带季风气候区,四季干湿冷暖分明,多年平均降水量 1 000～1 465 mm,受季风影响,降雨的年际与年内变化大。多年平均水面蒸发量为 950～1 100 mm,陆面蒸发量 600～800 mm,夏季蒸发量占全年总量的 40%,冬季占 12%,春秋季各占 24%,大体上蒸发量是南部大于北部,平原大于山区。区内光照充足(1 800～2 250 h/a),热量充沛,平均太阳总辐射量达 4 187×10⁶～5 024×10⁶ J/m²,年平均温度 14.6 ℃～17.0 ℃,无霜期长(212～268 d),气候条件较优越[1]。

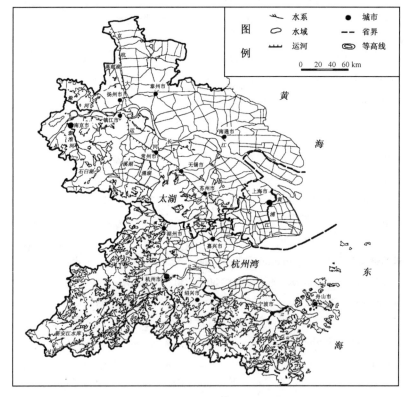

图 1　长江三角洲范围及水系图

---

　　* 王颖,王腊春,朱大奎:《科技通报》,2010 年第 26 卷第 2 期,第 171－179 页,188 页。
本文为 2009 年 5 月杭州市"杭州湾水库与长江三角洲水资源研讨会"会上报告。

长江三角洲河流水系主要分属长江、钱塘江与淮河三条水系。东南部沿海与舟山群岛为独立入海的甬江水系,总体上具有平原水系与人工河道网络之特征。属淮河水系的位于扬州至海安的通扬运河以北,水量丰富,河湖水网密集。从东向西为次一级的高宝湖水系、里下河水系和里下河洼地水系;南京至常熟徐六泾段为长江下游部分,其东为河口段,江北有滁河、通扬运河,以及京杭大运河贯穿南北,江南有秦淮河、锡澄运河、张家港河、望虞河、浏河与黄浦江;钱塘江以桐庐分为中、下游,中游山区河流建有新安江与富春江水库,下游为感潮河道,有涌阳江与曹娥江。

长江三角洲湖泊主要集中于以太湖为中心的水网系统。太湖以西为上游地区,有苕溪、洮滆等水系及洮湖、滆湖及宜兴三九湖;太湖以东为下游区,有黄浦江水系,及阳澄湖、澄湖、淀山湖和昆承湖等,可能是太湖退缩过程的残留;此外尚有南京市的石臼湖、固城湖,扬州市高邮湖、白马湖、邵伯湖,宁波市的钱湖及绍兴市的小湖群。本区河湖水系密布,反映出长江三角洲低地在降水、径流与入海流量间的自然平衡状态。

长江三角洲土地总面积为 109 600 km²。占全国土地面积的 1.1%;全区总人口约 8 536.6 万人,占全国人口 6.3%。2007 年,长三角地区 16 城市共实现地区生产总值(GDP)46 672.07 亿元,占全国经济总量的 18.9%。

长江三角洲城市群体、工农业与经贸文化水平在全国具有举足轻重的地位。但是,区域的自然环境承载力与经济发展存在着严重的不平衡。人均土地面积为全国平均水平的 1/6 略多,人均耕地为全国平均的 1/2[2]。随着城市化发展,耕地面积逐年减少。"三废"排放使大气污染,酸雨增多,水体污染,河流淡水富营养化,赤潮频发,水质恶化,地表淡水呈现水质型缺水,地下水位大幅下降,土地沉陷,形成互为因果的系列性反应。加之,海平面上升,海侵与风暴潮频频,河道排洪受阻,内涝积水,海岸蚀退,土地减少,沿江沿海土壤质地恶化等,区域生态环境脆弱,环境质量恶化,威胁到生存安全与经济发展,损害着我国投资环境与国防地位[3,4]。长江三角洲面临着严重的资源与环境压力的挑战!

1　长江三角洲水资源现状

1.1　水资源量

1.1.1　当地水资源量

长江三角洲多年平均的当地径流量为 508.4×10⁸ m³,其中浙东北约占 70%,江苏部分占 26%,上海市占 4% 左右。地下水也是本区的重要水源,以三角洲平原沉积物孔隙水为主,水质好,含有数层承压水层。东部滨海区,水质较差。西部宁镇扬丘陵区、杭湖地区地下水属孔隙水、裂隙水及喀斯特类型。甬绍舟丘陵、盆地与岛屿多为裂隙水、孔隙水,来自白垩系碎屑岩中的裂隙水,矿化度较高,不宜饮用。区内多年平均地下水量 177.71×10⁸ m³,其中浙东北为 87.72×10⁸ m³,江苏部分为 70.66×10⁸ m³。而上海市为 19.33×10⁸ m³。

长江三角洲多年平均当地水资源量达 573.79×10⁸ m³。其中上海市 32.07×10⁸ m³,江苏 76.57×10⁸ m³,浙东北 365.15×10⁸ m³(表 1)。单位面积水资源量丰沛,为 57.6×

$10^4\ m^3/km^2$。但人均水资源量只有 $774.9\ m^3$,仅相当于全国平均水平($2\ 220\ m^3$)的三分之一。

表 1　长江三角洲多年平均当地水资源量[3]

地区	年降水量/mm	年径流量/mm	当地径流量/$10^8\ m^3$	地下水量/$10^8\ m^3$	河川基流量/$10^8\ m^3$	当地水资源量/$10^8\ m^3$	单位面积水资源量/$10^4\ m^3/km^2$	人均当地水资源量/m^3
上海市	1 141	301	1 866	19.33	5.92	32.07	50.58	245.85
苏中南	1 040	254	134.43	70.66	28.52	176.57	36.55	457.55
浙东北	1 465	737	355.31	87.72	77.88	365.15	81.21	1 629.25
全区	1 266	501	508.40	177.71	112.32	573.79	57.60	774.90

1.1.2　外来水

外来水包括过境水和引江水。本区处于几条大河下游,外来水资源丰富。仅长江干流多年平均过境水量就有 $9\ 730\times10^8\ m^3$,最枯年(1928 年)仍有 $6\ 320\times10^8\ m^3$ 。长江具有较好的供水条件,即使在枯水年份,也能满足供水要求,如大旱的 1978 年苏南引江水 $112\times10^8\ m^3$ 。另外,上海黄浦江年均进潮量有 $409\times10^8\ m^3$ 。上海市的潮水量是当地水量的 22 倍。2007 年长江入海水量 $7\ 913\times10^8\ m^3$,钱塘江径流量 $321\times10^8\ m^3$ 。长江三角洲的外来水丰富,是区内开发利用水资源十分有利的客观条件。

1.2　水资源时空分布

长江三角洲水资源量的时空分布与降水一致,大致呈南多北少、山区多平原少,但平原区外来水较为丰富。

由于本区为典型的季风气候,降水的季节变化明显,导致了水资源的年内分配不均。每年降水主要集中在春夏之间的梅雨和夏秋之间的台风雨。相应地,5~9 月份的径流量占全年径流量的 $60\%\sim70\%$ 。同时,本区降水与径流的年际变化可达 $2\sim5$ 倍(表 2,表 3)。丰水年雨水过多,造成洪涝灾害,而少水年雨水过少,以致干旱缺水。

表 2　长江三角洲部分水文站降水量年际变化

水文站	历年最大降水量/mm	历年最小降水量/mm	多年平均值/mm	最大值最小值之比
南京	1 612.3	567.6	1 001.8	2.8
南通	1 394.3	641.3	1 066.8	2.2
扬州	1 930.6	424.1	1 030.0	4.5
杭州	2 530.0	797.3	1 411.5	3.2
嘉兴	1 719.4	723.1	1 189.9	2.4
绍兴	2 182.3	922.5	1 420.0	2.4
诸暨	2 171.3	903.8	1 449.7	2.4
岱山	1 589.9	701.0	1 133.9	2.3

表3　钱塘江等河流入海年际变化*

河流	集水面积/km²	多年均入海水量/10⁸ m³	水量最大值/10⁸ m³	水量最小值/10⁸ m³	最大值与最小值之比
钱塘江	41 700	373.0	692.0(1954)	179.00(1979)	3.87
曹娥江	6 046	42.8	72.2(1954)	19.60(1967)	3.68
甬江	4 294	28.6	41.9(1952)	9.96(1967)	4.82

* 据中国水资源公报,1997。

1.3　与水汽相关的灾害

(1) 台风暴雨与风暴潮:长江三角洲地区夏秋季节洪涝灾害严重,经历过 1954 年、1962 年、1991 年、1998 年、1999 年大水,尤以 1991 年灾害严重,太湖平原与里下河平原洪灾损失达 200 亿元。

(2) 干旱灾害:长江三角洲地区的干旱灾害发生于伏秋两季。伏旱影响水稻、棉花与果木生长,秋旱影响三麦发育。

(3) 寒潮与霜冻:长江三角洲地区每年约受 3～4 次寒潮,于 48 h 内降温 10 ℃以上,并伴以大风。太湖平原于 1955 年和 1997 年受寒潮影响,低温为 −10 ℃～−15 ℃,湖水冻,油茶、三麦、柑桔大面积冻坏。

(4) 阴湿害:阴湿害在长江三角洲地区一年四季均可发生,由于日照少,气温低,湿度大,地下水位高等原因,造成了作物的霉烂与病害。

2　长江三角洲水问题的主要症结

2.1　水资源紧缺矛盾比较突出

长江三角洲由于水资源分布不均,时空变化大,组合很不平衡。全区人均当地水资源量为 774.90 m³,仅为全国平均水平的三分之一,其中江苏地区和上海市,人均只有 245.85 m³ 和 457.55 m³,属于水资源紧缺地区。过境水是区内具有很大开发潜力的水源,但是由于开发资金和水质问题,利用量较少。

在沿海垦区和岛屿,水资源紧缺矛盾突出。舟山群岛人均水资源量为 582.9 m³,相当于全国平均水平的五分之一。海岛区约有 30 万人的饮水存在不同程度的困难,其中特别困难的有 20 万人,占人口总数的 20.3%[4,5]。

2.2　水质日趋恶化

2.2.1　河道水质污染严重

2007 年长江水质总体较好,基本能满足Ⅱ、Ⅲ类水。甬江上游溪口为Ⅱ类水,鄞江Ⅳ类水,东江下游、县江中下游及奉化江为Ⅴ类水,甬江干流为劣Ⅴ类水,主要超标项目为溶解氧、氨氮、总磷等。钱塘江流域新安江水库以上为Ⅰ类水,富春江为Ⅲ类水,东阳江和金华江为劣Ⅴ类水,曹娥江为Ⅳ、Ⅴ类水。

2.2.2　湖泊水质恶化与湖泊富营养化

由于区内湖泊所接纳的废水、污水量逐年增加,营养盐含量逐渐上升,浮游植物的生物量有了较大幅度的提高,许多湖泊已达富-中富营养化状态。其发生时间主要集中在4—10月份,局部水域2、3月份就出现。春夏季以蓝藻、硅藻占优势,秋季则以蓝藻占绝对优势。太湖藻类水华在空间分布上也极不均匀,由重至轻依次为五里湖、梅梁湖和竺山湖、大太湖、东太湖[6]。

20世纪80年代初,太湖以Ⅱ类水为主(约占69%),Ⅲ类水体占30%,仅有1%为Ⅳ类水。Ⅲ、Ⅳ类水主要分布在入湖水道及东太湖沿岸水域。1981年富营养化水体所占面积仅为16.8%。

80年代后期,太湖水体污染加重,1988年Ⅰ、Ⅱ类水增加为36.6%,Ⅳ类水占3.2%,且已出现重污染水体(Ⅴ类),主要分布在梅梁湖区闾江口附近及五里湖区部分水域,面积约占0.8%。富营养化水体所占面积已达到40%。

至90年代中期,太湖水质以Ⅲ类为主(占70%),Ⅱ类水仅占15%,Ⅳ类水也扩大为14%,主要分布在梅梁湖、太湖西岸和南岸。同时,该期Ⅴ类水体所占面积也增加为1%。

1997—2005年9年间太湖水质基本保持稳定,在总磷和总氮不参加评价的情况下,全湖综合评价为Ⅲ类,分湖区达标率在90%以上[7]。若考虑总氮和总磷,1997—2006年全太湖水质均不达标,1999—2001年全湖综合评价为Ⅴ类,其余年份均为劣Ⅴ类,TN超标严重。

2006年太湖水质有所下降,整体评价为Ⅳ类[8],9个湖区中超标湖区水体面积占32.5%,其中五里湖、竺山湖和西部沿岸区水质为劣于Ⅴ类,共占11.7%,约274 km²;梅梁湖水质为Ⅴ类,占5.3%,约124 km²;南部沿岸区水质为Ⅳ类,占15.5%,约363 km²;其余为Ⅲ类,共占67.5%,约1 577 km²[9]。

从水质单项指标变化趋势来看。有机污染指标中,NH_3- N浓度在2000年达到最低,之后明显增长,至2003年之后趋于稳定,COD_{Mn}变化趋势平缓,从2001年开始呈缓慢增长,TP浓度在2000年后有所下降。至2002年出现最低值,2002年后又逐年上升。总体上看,COD_{Mn}、NH_3- N、TN和TP等指标在1997—2006年间均出现先抑后扬的趋势,总体呈逐年升高趋势。叶绿素a近年来有加速增长的趋势。DO和BOD指标基本持平(图2)。

根据1997—2006年太湖富营养化监测结果,以年平均值进行评价,太湖整体已由轻度富营养化升至中度富营养化。中度富营养化所占比例不断上升,轻度富营养化水域面积由1997年的1 995 km²降至2006年的157.5 km²[8](见图3),可见近年来太湖富营养化的趋势已经较为明显,正在由轻度富营养水平发展为中度富营养水平,而且富营养化程度在逐年上升。叶绿素a的含量的逐年增加。表明湖区藻类发生量在逐年增加,致使太湖蓝藻频繁爆发,而且其变化趋势已经脱离了营养盐的走势[9]。

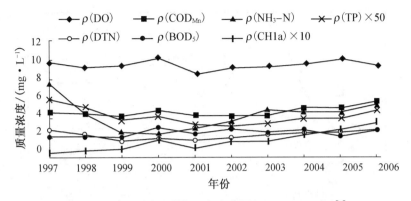

图2　太湖水质主要指标逐年变化图(1997—2006年) [9]

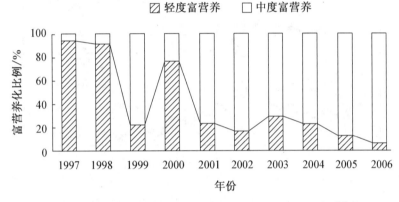

图3　太湖湖体富营养化比例年际变化(1997—2006年) [9]

其他湖泊 2007 年污染均为严重,淀山湖、西湖、鉴湖全年均为劣Ⅴ类,东钱湖为Ⅳ类。淀山湖、鉴湖为中度富营养化,西湖、东钱湖为轻度富营养化。

2.2.3　近海赤潮

大量的工业废水和生活污水,以及含有化肥等的农业灌溉排水,通过各种途径注入海洋,使得浅海营养盐不断增加,造成一些海域富营养化,赤潮频频发生。

根据国家海洋局第二海洋研究所遥感室的研究,本区赤潮主要发生在长江口外海域、南通市东面海域、舟山群岛附近海域以及三门湾附近海域。其中长江口外海域赤潮发生最为频繁(图4)。1993 年,在该海域共发现 4 次赤潮,分布面积最大达 50 km²。1995 年 5 月 27 日,该海域发现棕红色、呈条带状分布的赤潮带,面积达数百平方千米。2004 年 4 月开始,东海区发生多起赤潮,从福建沿海向北发展到浙江海域,其中中街山海域赤潮面积达 2 000 km²,渔山列岛海域的赤潮面积达 1 000 km²。赤潮爆发呈逐年增加之趋势。

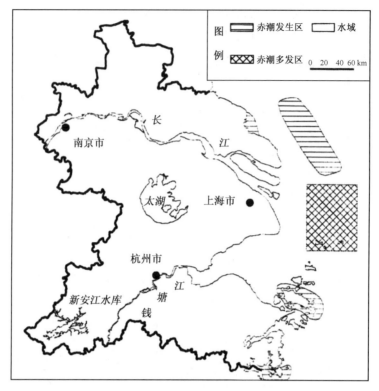

图4　长江三角洲赤潮分布示意图

2.3　洪涝灾害

梅雨和台风雨是导致本区洪涝灾害形成的重要气候因素。太湖平原地形极为低洼，地势周高中低呈洼形，总地势自西向东缓降。海拔2～10 m，大多为3～8 m。西部与小丘、阶地相接，南北两侧分别由钱塘江、长江的堤岸构成，地势高爽，东临滨海滩地，全区水网稠密，中部湖荡众多，水域开阔。上游来水易于急速汇集，而下游河道束窄，地形平坦，泄洪滞缓，易于积水成涝[10]。

里下河地区也是周边高、中间低的碟形洼地。该区西部为运西平原，地面海拔为6～10 m，以东为海积平原。地面海拔多为2.5～4 m，以南为三角洲平原，海拔6～9 m，以北为黄泛平原，海拔6～7 m以上。唯中部里下河低地海拔基本在2 m以下，中部最洼处仅1.1 m。这种地形条件容易积涝成灾，一遇大水则整个里下河便成泽国[11]。

因此，研究长江三角洲区水土环境特征、环境容量、资源承载力与发展变化规律，求得可容许改变的指标体系(LAC)，提出相应的解决对策，是迫切需要进行研究的重大课题。

3　长江三角洲核心区——太湖流域剖析

太湖流域地处长江三角洲核心，人口稠密，工农业发达。对太湖流域水资源与水环境问题的剖析，对了解、分析和解决长江三角洲的水资源与水环境问题具有典型意义。

3.1 水资源现状

太湖流域面积 36 985 km²,2007 年人口约 4 917 万人,流域国内生产总值为 28 648 亿元,约占全国的 12%,人均 5.8 万元,为全国的 3.1 倍。

太湖流域水系发达,是以太湖为中心的湖泊河网系统,属长江最下游的一个支流水系。流域平均水资源量为 176×10^8 m³,流域人均水资源量为 358 m³。2007 年流域年降雨量为 1151 mm,水资源总量为 172.7×10^8 m³(表 4)。此外,2007 年沿长江口门和钱塘江口门引水 101.9×10^8 m³。2007 年流域总用水量为 372.7×10^8 m³,其中工业用水占 68.5%,农业用水占 27.4%,生活用水占 4.1%。

表 4　2007 年太湖流域水资源总量　　　　　　　　　　　　　　　　单位:10^8 m³

分区	年降水量	地表水资源量	地下水资源量	重复计算量	水资源总量	产水系数
江苏省	210.4	68.4	17.0	5.4	80.0	0.38
浙江省	151.2	62.7	18.8	13.9	67.6	0.45
上海市	60.7	23.4	7.9	7.1	24.2	0.40
安徽省	2.4	0.9	0.1	0.1	0.9	0.38
太湖流域	424.7	155.4	43.8	26.5	172.7	0.41

3.2 水环境现状

2007 年流域内污水排放总量为 63.0×10^8 t,其中城镇居民生活污水排放量为 16.5×10^8 t,第二产业污水排放量为 35.7×10^8 t,第三产业污水排放量为 10.8×10^8 t。2007 年仅东太湖为 Ⅳ 类水,占湖面 7.4%,东部沿岸为 Ⅴ 类水,占 11.5%,其他劣 Ⅴ 类水占 81.1%,主要超标项目为 TN、TP。淀山湖、西湖全年均为劣 Ⅴ 类水,主要超标项目为 TN、TP。2007 年太湖蓝藻水华在梅梁湖、贡湖、竺山湖和太湖西部发生较频繁。

3.3 河流水质现状

2007 年评价河长为 2 510 km。全年期 85.7% 的评价河长劣于 Ⅲ 类(图 5 和图 6),主要超标项目包括氨氮、高锰酸钾指数、溶解氧、五日生化需氧量、石油类、总磷和化学需氧量等。

3.4 水资源主要问题

3.4.1 防洪问题

太湖流域现有的骨干工程主要为 50 年一遇防洪标准。随着流域经济社会的快速发展,城镇建设面积不断扩大,公路等基础设施逐步完善,行政区域内城市、圩区防洪除涝标准也随之提高,流域下垫面发生了很大的变化。但是流域目前的防洪工程体系尚且无法适应经济社会发展的要求,一旦受淹仍然会造成很大的损失。

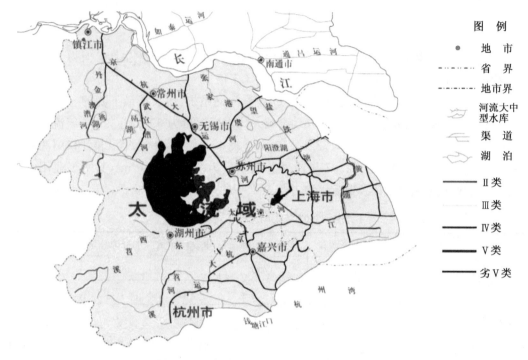

图5 2007年太湖流域水环境状况(附彩图)

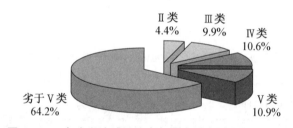

图6 2007年太湖流域河流全年期水质类别比例(附彩图)

(1)海平面上升、地面下沉

目前采用的防洪标准多以50年或100年一遇的洪水位设防为主。但是由于建设面积的增加,地面负重加大,再加上不合理的地下水开采,导致地面沉降。而因地面沉降,导致实际发生洪水时,与地面的水位差远大于设计时预计的水位差,造成设计标准越来越高,而洪水威胁却越来越大。同样,因为海平面上升,而设计洪水位基准面没有相应改变。使得设计防洪水位小于实际相应洪水位,更增加了洪涝灾害的威胁。

(2)围垦

从20世纪50年代起至80年代,太湖流域围垦建圩共498座[12],围垦面积达到528.55 km²,因为围垦而消失的湖泊共计165个,面积为161.3 km²。围垦导致太湖洪水水位抬高了9~14 cm。围湖不仅削弱了湖泊的调蓄能力,同时也切断了原与湖泊相通连的河道,使过水断面束窄,阻洪碍洪[13]。此外,城镇道路建设填占河道,也造成了行洪不畅,雍高洪水位。

3.4.2 水污染和水质型缺水问题

由于污水处理能力远远滞后于流域的经济发展速度,致使流域水体污染日趋严重,流域 85% 的河道水体已被污染。太湖作为流域最大的湖泊和水源地,其水域面积有 81% 已经达到了中度富营养化的水平。水污染引起了水环境的恶化,加上全区人均当地水资源量仅为全国平均水平的三分之一,因此导致区域内水质型缺水严重,更加剧了流域内水资源的供需矛盾。

（1）污水排放量大

太湖流域正处在经济高速发展的初期,经济的粗放发展导致用水量大大增加,同时由于水资源重复利用程度较差,使得污水排放量也随之大量增加。从 1980 年到 2007 年的 28 年间,太湖流域的用水量增加了 139×10^8 t。从 2003 年到 2007 年污水排放总量和 GDP 的相关关系中可以看出,流域内的污水排放总量随着 GDP 的增长明显增加(表 5,图 7)。可见,提高用水效率,减少单位 GDP 的用水量已迫在眉睫。

表 5　太湖流域用水量表(1980—2007)　　　　　　　　　　　　单位:10^8 t

年份	1980	1985	1990	1995	2000	2005	2007
用水量	234	250	272	292	316	354	373

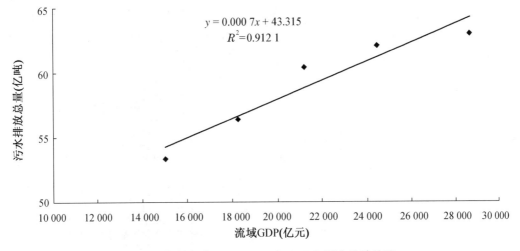

图 7　太湖流域 2003—2007 年 GDP 与污水排放关系

（2）水环境容量小

太湖流域多年平均水资源量为 176×10^8 m³,但由于用水量大,而可用于稀释污染物和水体自净的水量很少,加上山丘区兴建了大量水库蓄水,可用于水环境稀释和自净的水量更少。例如仅流域内的天目湖、大溪、横山、赋石、老石坎、青山、对河口七大水库拦水面积就达 1 745 km²,蓄水量大于 10×10^8 m³。

（3）太湖水体流动性差

由于流域用水量大,加上山丘区的蓄水,大大减少了入出湖水量,进而降低了水体的流动性,容易造成蓝藻暴发。引江济太工程把江水引入望亭入湖,可增加东太湖水体的流动性,但是对西太湖作用不大,因此应增加从西太湖入湖的引水通道。

（4）水质性缺水

太湖流域本地水资源少,而污水排放量大。如果全部以国家城镇二级污水处理厂

COD_{Cr} 的一级排放表准测算,仅考虑稀释,要将 2007 年排放的 $63×10^8$ t 污水稀释到Ⅲ类水,需要新鲜的纯净水为 $126×10^8$ m^3,然而 2007 年流域地表水资源总量仅为 $155×10^8$ m^3,由此可见,太湖流域水质性缺水的形势十分严峻。

3.5 研究重点

长江源远流长,水量丰富,主干道贯穿本区,联通中西部与海外。因此要想防治太湖水污染,改善太湖水环境,解决太湖流域的缺水问题,需要对大区域,即长江三角洲整个地区的长江干流主导作用与历史时期变化的原因及水资源水环境效应进行分析总结,将兴利避害作为重点研究。

4 探索长江三角洲水环境水资源优化与良性发展的途径

清洁的淡水、新鲜的空气及营养的食物是人类健康的基本要素。水量供需平衡、水体污染和洪涝灾害,是当前迫待解决的三个重要问题。需要在现有的基础上,对长江三角洲全流域以及相关的毗邻区进行整体的、系统的深入研究,为长江三角洲优化水质、增补水量与导致水环境的良性发展做出宏观控制规划、阶段性的具体措施以及监督实施与管理的制度,即一体化的规划、一体化的实施与一体化的管理(图 8)。

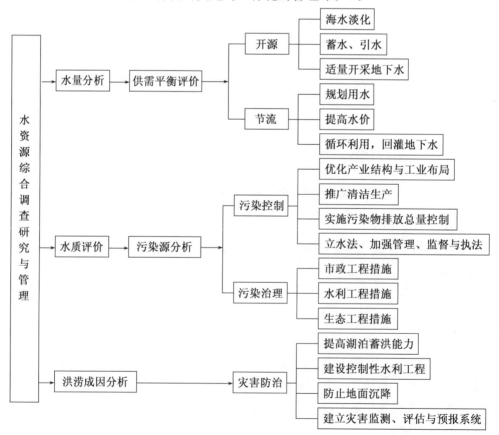

图 8 长江三角洲水资源可持续利用对策框图

长江三角洲水环境水资源优化与良性发展的途径必须有科学的规划与治理措施,需要解决以下几个关键的问题[14,15]:

(1) 要阐明三角洲区域当代长江干流的水文与河道特性、变化状况、原因以及活动规律。据此,做出对长江兴利避害、发挥其优势作用的宏观指导思想。

(2) 对长江三角洲水循环与水环境的承载力、可容许变化的限值与发展趋势做出分析,主要包括以下几个方面:

① 在当代自然条件与全球变化的前景下。长江三角洲水环境水资源所能支持民众健康生存,农、工、城、交、旅、贸社会经济发展的特点(类型组合)与承载能力,以及保持生态系统多样性的状况下的纳污能力。应该分别做出最佳、适宜与极限制值三个档次的分析。

② 分析在今后的20年中,当工程技术相当应用于污染治理、环境修复,增加水的重复利用比率,以及水资源环境得以宏观调控的情况下,长江三角洲区域的人口、产业类型、产业规模,以及经济发展的承载力。

③ 分析当2050年的时候,在新技术广泛应用,人民群众的环境意识与公共道德水平普遍提高的情况下,人口与发展、产业类型与城乡规模、生态环境与生存方式等的变化以及承载力。

(3) 探索现有水环境的治理(例如控制减少工、农业污染源,就地解决城镇污染)、污染防治、清洁水源的有效获取、有效利用与有偿使用,以及清洁排放水流的措施与管理系统。

(4) 编制长江三角洲流域的整体规划,疏导江、河、湖、塘与入海体系,控制地区间的人工建筑与统一水利工程设施,使水体流畅,并形成良性循环,有计划地、分阶段、有效地防治洪水、内涝与潮侵灾害。

(5) 全区的生态环境建设与发展规划,应以流域为整体,利益共享,风险共担,按照地区条件、特点的不同,各地区互有分工,互相补充,协同发展形成长江三角洲低地环境特有的良性生态系统、生态景观与生态经济发展区。

(6) 实施区域水环境、水资源的一体化管理方案、实施办法与规章制度原则。

(7) 公众的教育。应利用报纸、电视、广播等多种新闻媒体手段对群众进行宣传教育。使广大群众充分认识到淡水资源有限,要珍惜淡水资源,形成节约用水、人人有责的观念、习俗与公共规范。发展形成社会性的多途径的利用自然水循环增水与补水,发扬良好的用水传统。重复有效地用水,提高用水效率,保证清洁水流入海。

总之,组织长江三角洲地区有关水环境水资源的高校与研究机构的地学、工程与技术力量,与行政管理机构结合,联合申报国家级大课题,开展深入研究,促进水质与水环境优化。必然会对长江三角洲地区经济建设的腾飞发展,以及推动长江流域的发展进步做出贡献,并且能够有力地支持富民强国宏伟目标的实现!

参考文献

[1] 严济道.1986.长江三角洲自然季节降雨趋势[J].地理研究,5(1):51-57.

[2] 佘之祥.1997.长江三角洲水土资源与区域发展[M].合肥:中国科学技术大学出版社.

[3]《中国农业全书》总编辑委员会.1997.中国农业全书(浙江卷)[M].北京:中国农业出版社.

[4] 李植斌.1997.浙江省海岛区资源特征与开发研究——以舟山群岛为例[J].自然资源学报,12(2):139-145.

[5] 朱大奎,王颖,王栋,王腊春.2004.长江三角洲水环境水资源研究.第四纪研究,24(5):486-494.

[6] 王颖,王腊春,王栋,陈文瑞.2003.长江三角洲水资源水环境承载力、发展变化规律与永续利用之对策研究.水资源保护,(6):34-40,49.

[7] 陆铭峰,徐彬,杨旭昌.2008.太湖水质评价计算方法及近年来水质变化分析[J].水资源保护,24(5):30-33.

[8] 吴瑞金.2001.我国湖泊资源环境现状与对策[J].中国科学院院刊,(3):176-181.

[9] 毛新伟,徐枫,徐彬,高怡.2009.太湖水质及富营养化变化趋势分析[J].水资源保护,25(1):48-51.

[10] 徐枫,徐彬.2006.太湖流域省界水体十年水质变化分析报告[R].无锡:太湖局水文水资源监测局,39-44.

[11] 杨世伦,陈吉余.1995.太湖流域洪涝灾害的形成和演变[J].地理科学,15(4):307-314.

[11] 陈晓玲,王腊春,朱大奎.1996.苏北低地系统及其对海平面上升的复杂响应[J].地理学报,51(4):340-349.

[12] 杨世伦,陈吉余.1995.太湖流域洪涝灾害的形成和演变[J].地理科学,15(4):307-314.

[13] 梁瑞驹,李鸿业,王洪道等.1993.1991年太湖洪涝灾害[M].南京:河海大学出版社.

[14] 朱大奎.长江三角洲特征与当前主要的任务[C]见:严东生,任美锷主编.论长江三角洲可持续发展战略.1999.合肥:安徽教育出版社,139-151.

[15] 陈文瑞,朱大奎.1998.长江三角洲土地资源可持续利用[J].自然资源学报,13(3):261-266.

南京三台溶洞地貌形成与长江古水面关系初探[*]

　　长江南京段南岸的幕府山紧邻著名的燕子矶,其构造为一复式背斜,被多条正断层顺山体延伸方向所切过,岩性复杂。临江一侧主要分布下奥陶系仑山灰岩(白云质灰岩,李希霍芬 1877 年命名,参见中国国土资源网 http://www.clr.cn/front/read/read.asp? ID=143599),喀斯特地貌比较发育,有多层溶洞分布。其中,三台溶洞(在本文中指头台洞、二台洞和三台洞)作为风景名胜区已有数百年历史,洞穴内烟火熏染,深受人类活动影响,但洞壁尚存地下水侵蚀、淀积的遗迹。至今,尚无人对这些洞穴的成因、年代进行过系统研究。

　　2005 年 5 月,我们在为三台溶洞区长江洪痕[1]做年代取证时,意外发现了洞穴"悬钙板"。所谓"钙板",指洞穴流水在二氧化碳逸出后过饱和而析出的成层状碳酸钙沉积;如果水流再次活动,掏蚀钙板下面的充填物形成新洞穴,而前期沉积的钙板则悬于新生洞穴顶部,此便称为"悬钙板"[2]。从悬钙板产生的过程可知,如果测定其年代,并结合地貌部位及与围岩的关系,便可为据探讨地貌形成或水位变化的年代界限。

　　然而幕府山临江山壁有太多的人工开凿痕迹(如残留的炮眼),已难于探讨该处山壁地貌的自然形成过程。2009 年实施"南京市三洞一阁景区改造工程"时,工程队已将所有知名洞穴的洞壁喷上油漆,几乎完全掩盖了原始洞壁的基质,分辨不出哪些是基岩、哪些是石钟乳或钙板,这些人工活动损害了灰岩崖壁与溶洞的天然面貌,加之本区有不少人工开挖洞穴,给我们研究自然变化带来极大困难。本文就野外踏勘所获的可确认证据,结合实验室测试数据,探讨三台溶洞地貌的形成过程、动力条件及年代界限,初步提出研究长江地貌发育的"沉积—再侵蚀交叉论证"年代学方法。

一、洞穴参数和特征

　　三台溶洞群的位置及与长江之关系见图 1。区内溶洞分布成层性明显(表 1):如三台洞高程 7.56 m,其上玉皇阁洞 20.6 m,望江亭洞 25.1 m(图 2)。洞分 3 层,说明该处山体曾有过缓慢而断续的上升过程,长江水面下落,故使玉皇阁洞以垂直通道与三台洞相通。而三台洞发育时,长江基面已基本稳定,发育的洞穴规模宽大。没有长江基准面在当地的变化,不可能形成 3 层洞穴。

---

　　[*] 谭明,王颖,何华春,程海:《第四纪研究》,2010 年第 30 卷第 5 期,第 877-882 页。2011 年载于:《中国科学院地质与地球物理研究所第十届学术年会论文集(下)》2011 年 1 月,北京国家自然科学基金项目(批准号:40271004)资助。

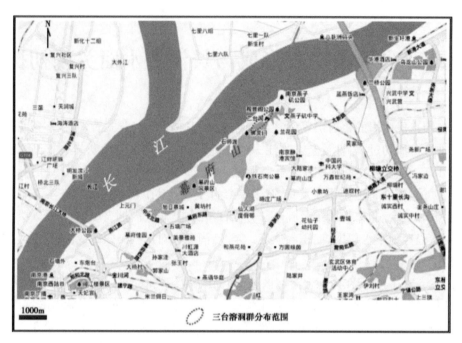

图1 三台溶洞群位置及与长江关系

表1 各洞穴量测数据(按照高程从上往下列出)

洞穴名称	洞口朝向	高程/m	洞宽/m	洞高/m	洞纵深/m
望江亭洞	315°(北西)	25.10(洞口)	3.5	2.1	2.97
观音洞	28°(北东)	22.10(洞口)	2.2	2.1	8.0
玉皇阁洞	255°(南西)	20.60(洞口)	10.5	4.2	6.1
二台外洞	290°(北西)	16.95(洞底)	10.0	9.0	7.0
二台内洞	295°(北西)	16.90(洞底)	1.7	3.5	3.0
头台洞	330°(北西)	8.63(洞口)	15.1	4.0	13.4
三台外洞	295°(北西)	8.30(洞口) 7.56(洞底)	8.0	8.5	13.0
3号洞	265°(西)	8.30(洞口水面)	5.0	3.0	3.2
三台内洞	195°(南西)	7.31(洞底)	1.7	3.6	6.0

溶洞的宽度多大于纵深,或"宽×高"大于"纵深",洞口向内收缩封闭,均为袋状洞穴(图3),属于喀斯特洞穴分类中的"脚洞"。所谓"脚洞",指沿地下水面或河水面发育形成的水平洞穴,多由泛滥洪水侵蚀、溶蚀而形成于石峰脚下[3]。三台洞洞口内西侧与上部玉皇阁洞相通的甬道石阶处,洞顶可见许多悬垂如石钟乳,但为基岩质的岩吊。洞顶的岩吊既有水流下渗的作用,亦会因在洞穴发育的某个阶段,水流充满洞腔并沿裂隙向上溶蚀洞穴顶板而形成,属于蚀余形态。此外,一些洞穴洞口顶端的崩塌作用也很明显。

图 2　三台溶洞群的洞穴分层（三层溶洞的洞　　图 3　典型的袋状脚洞"望江亭"洞（位置见图 2）
口均位于建筑物后）

二、定年材料和数据

自三台洞洞口中心向东约 77 m 处有一小洞（表 1 中的 3 号洞），洞顶保存有悬钙板（位置：$32°08'N,118°47'E$；高程约为 10.3 m，见图 4）。继之，在 3 号洞附近、三台洞内以及二台洞东侧山脚陆续发现填充在基岩裂隙中的钙板。当发现洞顶悬钙板时，自然联想到两个问题：钙板在何时堆积形成？钙板与其所在洞穴以及长江的关系？于上述几处钙板处取样，在美国明尼苏达大学同位素实验室进行质谱铀系定年。由于 3 号洞顶的钙板结晶较好，^{232}Th 含量较低，无重结晶，终获得有意义的年代数据（表 2）：图 4 中，下钙板 a 的年龄为（560±92）ka（表 2 中编号 05-5-14-No.3-1 样品），上钙板 b 的年龄为（470±40）ka（表 2 中编号 05-5-14-No.3-2 样品）。从钙板下老上新的关系看，两个年代没有倒序。但两钙板上下紧相邻（垂直距离约 12.7 cm），故在误差范围内，两个年龄数据是交叉的。由于上钙板 b 的年代数据（精度为 8.5%）明显好于下钙板 a 的年代数据（精度为 16%）。所以，笔者采信于 b，即认为钙板沉积于距今 430 ka～510 ka，并以此作为这个位置的钙板年龄范围。

表 2　铀系定年结果，偏差为 2σ

Sample Number	^{238}U/ppb	^{232}Th/ppt	^{230}Th/^{232}Th ×10^{-6} atomic	^{234}U* (measured)	^{230}Th/^{238}U (activity)	^{230}Th Age/a (uncorrected)	^{230}Th Age/a (corrected)	^{234}U$_{Initial}$** (corrected)
05-5-14-No.3-1	176.0±0.3	2486±9	1170±6	5.1±1.5	1.0012±0.0039	560 000±92 000	560 000±92 000	25±10
05-5-14-No.3-2	146.5±0.2	4 260±20	568±3	9.1±1.8	0.9993±0.0042	473 000±40 000	473 000±40 000	34±8

$\lambda_{230}=9.1577\times10^{-6}a^{-1}$，$\lambda_{234}=2.8263\times10^{-6}a^{-1}$，$\lambda_{238}=1.55125\times10^{-10}a^{-1}$

* $\delta^{234}U=([^{234}U/^{238}U]_{activity}-1)\times1000$

** $\delta^{234}U_{initial}$ was calculated based on ^{230}Th age (T)，i.e.，$\delta^{234}U_{initial}=\delta^{234}U_{measured}\times e^{\lambda234\times T}$

Corrected[230]Th ages assume the initial ^{230}Th/^{232}Th atomic ratio of $4.4 \times 10^{-6} \pm 2.2 \times 10^{-6}$. Those are the values for a material at secular equilibrium, with the bulk earth ^{232}Th/^{238}U value of 3.8. The errors are arbitrarily assumed to be 50%.

图4　3号洞、洞顶悬钙板及定年样品取样位置(附彩图)

a—表2中05-5-14-No3-1　b—表2中05-5-14-No3-2的取样位置

三、洞穴发育条件与演化分析

分析各个洞穴的洞口朝向(见表1),除个别洞穴(观音洞)外,不同高度的洞穴洞口多数朝西,即迎向长江水流上游方向,迎水迎浪受冲击力大,反映洞穴多为江水冲蚀而成。观音洞很可能由于受江流下泻时侧向横流所蚀,故朝向不同于多数洞穴。继之,以江水侧蚀作用而使洞穴扩展。但江水流入山体后能量滞减,以致洞穴发育向内变小收缩,形成大洞内套小洞之"洞中洞",故三台洞、二台洞、头台洞和观音洞均有此现象。

3号洞及其附近山壁岩相为钙板充填空隙的角砾堆积(图5)。由于没有发现岩粉、断层泥等胶结物,也没有发现构造透镜体、挤压片理、断层褶皱等现象,故可排除其为断层角砾岩,而将这套混杂岩定为溶塌角砾岩。由于混杂岩易被水流差异侵蚀与溶蚀,故而形成了多孔岩壁(图6)。

图5　3号洞附近岩壁的岩相(附彩图)

(a为钙板,b为角砾)

图6　3号洞(左下)附近的多孔石壁，
明显区别于附近纯基岩构成的石壁面貌

图7　三台洞西侧的天窗(仰拍)

古暗河或溶洞周边应力松弛导致围岩的松动崩塌，形成溶蚀塌陷堆积之溶塌体。而后空隙被钙板填充，形成特殊的溶塌角砾岩，亦代表老一代洞穴的彻底消亡，图5显示的就是这一幅图景。继而，当钙板沉积之后，长江在溶塌角砾岩中侧蚀出3号洞，部分钙板就成为新一代洞穴的围岩部分而悬于洞顶，结果正如图4显示的图景，同时表明了新老两代洞穴的关系。根据洞穴悬钙板的地貌学意义和定年数据，可以肯定，距今(470 ± 40)ka以来，长江主泓靠近幕府山对崖岸冲蚀加强并进一步由涡流掏蚀出3号洞。该洞底面高程与三台洞底面相当，推断与三台洞约同期形成。但三台洞更为宽大(表1)，其洞顶有天窗发育(图7)，反映江面在该时段稳定，历时长，并且有垂向重力崩塌作用使洞穴扩大(头台洞也有斜井将其与上面的观音洞联系)。

四、长江古水面与洞穴演化之关系的讨论

根据测年数据，目前幕府山位置最低的一系列洞穴(洞口高程为8m左右洞穴)形成时间不超过(470 ± 40)ka，其年代上限的意义在于解决了一个问题，即"长江在什么年代前侵蚀出高位洞穴?"

首先需要明确一点，脚洞被江水侧向侵蚀出来后，由于洞道较短，而且洞穴与顶板水文连通性不好(有天窗者除外)，因此可推测很难有过饱和地下水从中流过而形成钙板。若有钙板沉积，所充填的应该不是脚洞，而是具有地下排水功能的洞隙。在这些洞隙中地下水流程较长，有足够的水岩相互作用时间，才可能溶解较多的基岩碳酸钙，也才可能有钙板沉积。所以认为，钙板所填充的洞隙应该是过去的排水道，而排泄基准很可能就是近侧之长江。那么，在40万～50万年前当钙板沉积时，长江水面高度如何? 如果洞中水流以长江为排泄基准，意味着排水管道高于江面，即钙板沉积时(冰期)江面或地区潜水面比现在要低。而基于此假设的一个合乎逻辑的推断是，位置较高的二台洞、观音洞、玉皇阁洞的形成时间远老于40万～50万年。

另一种可能是位于3号洞的已消亡的古洞穴为过去的深部岩溶[4~6]，发育在区域排

泄基准面以下,那么当钙板沉积时,长江高于 3 号洞位置。或者说,当长江在较高位置侧蚀出二台洞、观音洞等洞穴时,在长江之下也有洞穴发育,而后崩塌填充消亡,这种情况下钙板在潜水面以下的空洞中形成堆积。如果这一推测成立,高位洞穴的形成年代也远老于 40 万～50 万年。

因为高程为 10.5 m 的洪水位遗迹[1]高于 3 号洞(图 4),而这道痕迹形成于至多不过数百年前[1]。可见 3 号洞的形成,包括三台洞下部的形成应该是自远古持续到近千百年历史时期的事情,因为它们还未能完全脱离数百年前江水的侵蚀改造作用。最后,如果考虑到长江南京段已经处于下游,水力坡度非常小(约为 1/275 000)[7],因而下蚀力极其微弱,在这种条件下,并结合前面所分析的两种可能性,由此推测长江水面在三台洞或 3 号洞所在高程附近上下波动时间可能以 10 万年计。

五、初步结论

从研究区域各洞穴的规模、大小比较,高层不如中层,中层不如下层(见表 1),因此,如果岩性、构造基本一致,确认无崩坍毁损,这种差别似乎表明气候-水文动力在第四纪,至少在中晚更新世不断增强。各层溶洞按照高程由下至上大致可以分为 7～8 m、17 m 左右、20～22 m 和 25 m 等 4 层,但由于溶洞附近长江几级阶地已经被人类活动深刻改造,目前已难与其一一对比。

长江南京段南岸幕府山临江一侧溶洞群类型主要为脚洞,成层分布,最底层溶洞仍然被距今不远的历史时期水流作用持续侵蚀改造,只是因为人为修筑堤防才使得这些洞穴与长江隔开。三台溶洞群普遍存在“洞中洞”现象,是为长江向内水动力减弱的证据,也是将此处洞穴定为脚洞的依据之一。

高位洞穴的形成年代至少老于 40 万～50 万年。如果这些洞穴的形成与长江水动力有关,则长江在南京附近流动已超过 50 万年。但由于存在“深部岩溶”的可能性,所以尚不能确认定年钙板沉积时长江水面与之关系。

根据钙板年代数据、洞穴围岩岩相特征、洞穴发育形态以及新老洞穴关系等综合证据,初步推断三台溶洞区底层溶洞形成后,长江三角洲已发育,水位稳定、水力坡度小,下蚀力极其微弱而侧蚀力较强,江面有可能在幕府山底层洞穴所在之高程附近上下波动超过 10 万年。

幕府山附近的燕子矶长江段因崖岸夹峙,河床受约束,难以摆荡,已有 40～50 万年的历史,表明是古老稳定的河段。现代江水涨落在崖岸上留下明显的洪痕,洪水水位的变动可达 3～10 m,与水动型的海平面上升及下游江水顶托有关。

观音阁所在山岩由白云岩组成,也形成了多孔状岩壁,是强烈的岩溶作用造成糖粒状风化所致,与三台溶洞区的多孔状岩壁成因不同。

本文的若干推论仍基于假设,存在很多不确定性,而本文的主旨在于开启新的研究方法或思路,因为在万里长江沿岸有众多喀斯特山地,类似于本文研究的可准确定年并能揭示地貌发育的喀斯特沉积一定不少,如果将来进一步做系统的分析,也许能推动解决一些长期悬而未决的长江发育难题。

致谢 感谢段武辉协助野外洞穴测量工作;感谢陈中原教授提出重要的修改意见以及杨达源教授富有启发的讨论;感谢审稿人建设性的修改意见。

参考文献

[1] 何华春,王颖,李书恒.2004.长江南京段历史洪水位追溯.地理学报,59(6):938-947.

He Huachun, Wang Ying, Li Shuheng. 2004. Tracing flood water level along Nanjing cliff bank of the Yangtze River. *Acta Geographica Sinica*, 59(6): 938-947.

[2] 谭明.1993.喀斯特水文地貌学——理论、方法及应用研究.贵阳:贵州人民出版社,109-111.

Tan Ming. Hydrogeomorphology of Karst—The Theory, Method and Their Application. 1993. Guiyang: Guizhou People's Press, 109-111.

[3] 中国地质科学院岩溶地质研究所.1988.《桂林岩溶地质》之五:桂林岩溶地貌与洞穴研究.北京:地质出版社,1-249.

Institute of Karst Geology, Chinese Academy of Geological Sciences. Chapter Five-Karst Geology of Guilin—Study on Guilin Karst Landforms and Caves. 1988. Beijing: Geological Publishing House, 1-249.

[4] 任美锷,刘振中,王飞燕等主编.1983.岩溶学概论.北京:商务印书馆,1-331.

Ren Mei'e, Liu Zhenzhong, Wang Feiyan *et al*. eds. 1983. Conspectus of Karst. Beijing: The Commercial Press, 1-331.

[5] 何宇彬,邹成杰.1997.关于喀斯特洞穴发育深度问题.中国岩溶,16(2):167-175.

He Yubin, Zou Chengjie. On the depth of karst cave development. 1997. *Carsologica Sinica*, 16(2): 167-175.

[6] 张英骏等编著.1985.应用岩溶学及洞穴学.贵阳:贵州人民出版社,62.

Zhang Yingjun *et al*. eds. Applied Karstology and Speleology. 1985. Guiyang: Guizhou People's Press, 62.

[7] Chen Zhongyuan, Li Jiufa, Shen Huanting, Wang Zhanghua. 2001, Yangtze River of China: Historical analysis of discharge variability and sediment flux. *Geomorphology*, 4: 77-91.

三亚湾海岸地貌的几个问题*

一、概 况

三亚湾位于海南岛南部,海岸走向略呈弧形,其范围西起马岭,东至三亚,再向南至鹿回头。岸外有东瑁洲、西瑁洲两岛及近岸的鹿回头岭为屏障,是一半开阔的海湾。

该区属热带海洋性气候,雨量充沛。一年无四季之分,而干、湿季明显。湿季(5月至10月)炎热多雨,干季(11月至次年4月)凉爽干燥。作用于海岸带的风力以西风与西南风为主,风速较大,一般为10~20 m/s。其盛行之时为6~9月,恰当台风频繁之际,且两者风向相同,台风风力高达8~12级,对三亚湾影响甚大。影响三亚湾沿岸的风浪亦以西向与西南向为主,其次为偏南向的风浪与涌浪。东南向波浪因有鹿回头半岛之屏障,故而对本湾影响较小。本区的潮汐是以日潮为主的混合潮,潮差2.13 m。半日潮天数平均为11天。日潮时一天中涨潮为16~17 h,落潮时极短,仅7~8 h。大潮汛对应着日潮,小潮汛对应着半日潮。沿岸带潮流系往复流,涨潮流为东南向,落潮流向西北,自表层至底层均为落潮流流速大于涨潮流流速。

三亚湾现代海岸绝大部分位于宽平的海岸大沙坝上,是激浪活跃的砂质堆积海岸;湾的两端是海蚀的基岩岸,且不同程度地分布着珊瑚礁。在整体上,三亚湾为一大型的较开阔的港湾海岸,波浪作用为主,但在某些岸段潮流作用与生物活动亦为一极活跃的因素。所以,三亚湾海岸地貌类型丰富。本文拟对三亚湾海岸作一剖析,对若干海岸地貌问题进行初步讨论(图1)。

二、陆连岛与岸礁

鹿回头半岛位于三亚湾的东南部,呈东北—西南向伸入海中约4 km,构成了三亚湾的东南"岬角"。在地貌上,它是一个中型的陆连岛。

鹿回头岭高270多 m,是由花岗岩与花岗闪长岩组成的陡峻山地,它突立于外海,水深浪大,沿岸海崖陡立(图2)。向陆侧山坡(东北坡)保留着明显的断层三角面。此情况适与对面的南边岭山坡相对应,南边岭西南坡亦存有显著的断层崖壁,表明两山间为一NW向的断裂地堑带,断裂使得鹿回头岭曾与陆地分离,突立海中为岛。小洲、白排为此断裂陷落部分在海中的残留岩体。后由于偏南向的波浪作用,而且主要是西南方向的波

---

* 王颖,陈万里:《海洋通报》,1982年第1卷第3期,第37-45页。

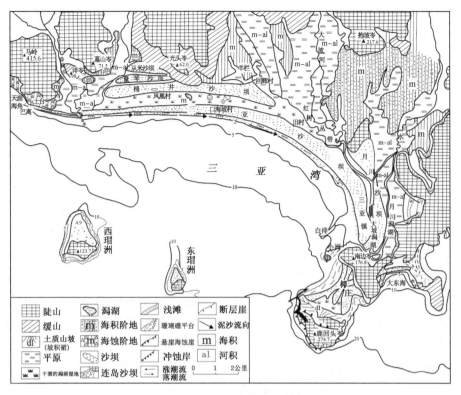

图1　三亚湾海岸动力地貌图

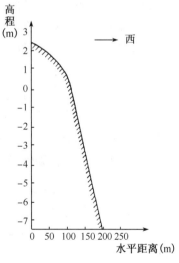

图2　鹿回头海蚀岸水下岸坡剖面图

浪作用[①]，它不断冲蚀岛屿沿岸并将泥沙于岛后波影区堆积，逐渐形成三条小型连岛沙坝，沙坝发育后不断接受来自两侧海底与陆上的泥沙，渐次合并成为一条长约2.2 m、宽1.2 km的沙坝（按其地名而称为"椰庄沙坝"），并将鹿回头与南边岭陆地相连，成为今日的陆连岛。连岛沙坝组成物质为浅色的岩屑、石英质砂砾以及珊瑚礁块屑，其地势平坦，高程为2～4 m，东北部略高，西南部低矮处为残留的潟湖洼地。看来，连岛沙坝形成为时不久。

华南港湾海岸虽在成因上与北方港湾岸基本一致，是在冰后期海面上升的背景上，受岩性、构造控制，以波浪作用为主。但是，华南沿海高温多雨，化学风化作用强烈，岩石遭受风化分解的速度快，故而海岸带泥沙供应较北方海蚀岸段数量大，海岸堆积作用发展迅速，堆积地貌发育普遍，规模亦较大，其中尤以海岸沙坝与连岛坝最为突出。原因在于华南海岸面临开阔的外海，浪大水深；海岸带不仅有较多的泥沙供应，而且海积作用是在海湾内或在岛礁等蔽障物后侧的波影区，

---

①　据实测，东南方向波高为2 m的外海波浪，经鹿回头岭发生折射与绕射后，波高降至6 m以下，波向转为西南，波长加大，故波浪作用力显著减小。

这样才得到了最充分的发展条件。

在鹿回头的背侧，沿着椰庄沙坝的两侧岸边，并延伸到南边岭岬角，发育着裙状的珊瑚岸礁平台。该区珊瑚礁已有专文研究[1][2]，这里据我们实测，再作些报道。该区礁平台分布于高潮线以下至水深4m的范围内，其宽度自100m至400m不等，大体是湾内的较岬角处略宽些，一般宽度为300m。礁平台在落大潮时可露出，台面崎岖不平，大部分为死礁体，少量的为活珊瑚，活珊瑚体质软、色褐，高约

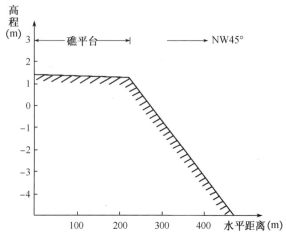

图3　椰庄湾水下岸坡剖面

20～30cm，呈帽状或蘑菇状。礁平台上生长着较多质硬透明的褐色海藻，并且已沉积了薄层的珊瑚、贝壳碎屑与砂砾层，此沉积色黑微具臭味，表明该处浪流活动已较微弱。礁平台外缘以2%坡度下降至4m水深（图3）。礁平台外缘因受波浪冲蚀，形成一些珊瑚质砂砾，后者又被激流叠复在岸礁平台的上部，形成珊瑚砂砾海滩。海滩宽度不大，约20～30m，但洁白松软，蔚为壮观。

三亚湾的珊瑚礁主要是沿基岩岸段发育的岸礁，在水深浪大面临外海的岬角处岸礁难以扩展，但在岛侧邻接岬角的基岩港湾岸段，岸礁发育较宽。其次，在海中的岛、礁背侧的波影区，珊瑚礁亦得到发育。如：西瑁洲、东瑁洲两岛的岛后发育着由岸礁与珊瑚形成的波影区沙坝；三亚港外白排礁后侧亦拖着一条长达1.2km的以珊瑚礁为主体的沙尾。但从整体看，珊瑚礁在三亚湾仅为局部发育的岸礁，目前发育并不旺盛。

三亚湾西端马岭、天涯海角一段，有大片岩滩，高潮线以下的海滩物质——珊瑚碎屑、贝壳屑、花岗岩风化的砂砾——已被钙质胶结为"海滩岩"。土台附近"海滩岩"（图4）每层厚20cm，充填于岩滩的礁石间，需用铁锤敲打方能击破；天涯海角的海滩岩是岩脉岩礁中间充填的珊瑚碎屑，砂砾胶结而成，一直分布到高潮线。

▐▐▐ 花岗岩　▨▨▨ 现代海滩砂岩

图4　巴离以东土台之下海滩砂岩

据格里舍（A. Guilcher）与金（C. A. M. King）等人研究[3]，海滩岩主要发育在珊瑚繁殖区，在高温气候下，海滩物质由文石、方解石胶结。海滩岩只形成在潮间带，其厚度取

①　蔡爱智等（1964）.海南岛南岸珊瑚礁的若干特点，海洋与湖沼，6卷2期。
②　黄金森，关于南海珊瑚礁研究的专著。
③　C. A. M. King (1972), Beaches and Coasts, 391-393.

决于潜水面的波动,在潮差小的海岸上分布在平均海平面附近,在潮差大的海岸上则分布在高潮线附近。海滩岩以中等能量的、后退海岸的海滩上发育得最好。

三亚湾海滩岩主要分布在西端马岭天涯海角的海蚀岸段。此段是海蚀后退海岸,有大片岩滩,这些海滩岩的形成条件与该地现代环境一致,是"现代的"海滩岩,正在发展中。海滩岩形成后,抗风浪侵蚀力强,对后侧海岸具有一定保护作用。因此,三亚湾西端海滩的分布,说明西端泥沙供应已减少,泥沙沿岸运动基本停止。

三、海岸沙坝与海岸发育动态

(1) 在上述东西岬角之间的三亚湾内,沿着海岸是一条长达 19 km 的海岸大沙坝。这条海岸沙坝形成于现代海岸发育的初期,由原来的古海岸带受冲蚀破坏而形成,大量堆积于近岸带海底的泥沙,受波浪长期作用,重新向岸推移,逐渐堆积成一条与海岸平行的大型砂质堆积体——海岸沙坝。据其地理位置,称之为三亚沙坝。

三亚沙坝的高度与宽度,从西向东由窄变宽,由高到低(见表1,图5～图9)。

表 1　三亚沙坝的形成及物质组成

地　名	宽　度	高　度	组成物质
1. 西段:烧旗港 　　——凤凰村	250 m	10 m 以上, 最高为 13 m	石英贝壳粗砂(凤凰村南) 花岗岩石砾粗砂(烧旗河口)
2. 中段:海坡村	450 m	10 m	石英质细砂为主,含粗中砂贝壳。
3. 东段:三亚一带	750～800 m	5 m±	表层为石英、长石质细砂,含粉砂颗粒,砂砾稍经磨蚀,仍具棱角。下层为石英质中粗砂,交互成层。有贝壳。

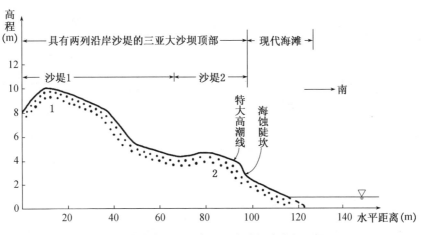

图 5　烧旗水河口以东三亚大沙坝海岸剖面图

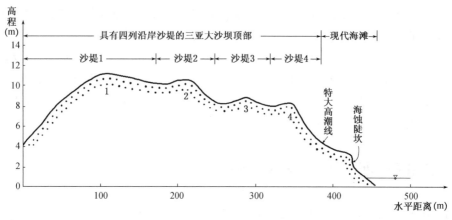

图6　凤凰村东三亚大沙坝海岸剖面图

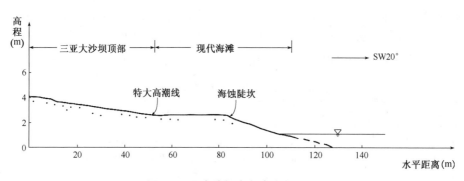

图7　三亚大沙坝中段海岸剖面图

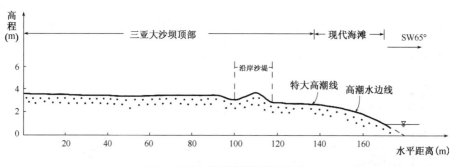

图8　三亚大沙坝东段海岸剖面图

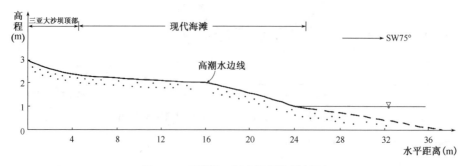

图9　三亚附近三亚大沙坝海岸剖面

沙坝的高、宽形态标志与海滩的形态标志一样,既反映着当地供沙量多少与泥沙粒度,同时,亦反映着因海岸朝向不同所经受的波浪力的差异。三亚沙坝西段,面临开敞大海,风浪大,激浪作用活跃,并且靠近花岗岩山地,泥沙来源丰富,因此沙坝的高度大,坡度陡,泥沙颗粒粗(粗砂);而东部的三亚,有东瑁洲、西瑁洲与鹿回头半岛为屏障,外海传来的波浪经过岛屿折射后,至岸边动力减弱,故东段沙坝宽缓,泥沙颗粒小,主要为细砂。

在三亚沙坝西段的顶部,叠加着四条与岸线平行的沿岸沙堤,其高度自海向陆略有增高。这种现象反映着西段海岸接近供沙源地,泥沙供应充分,岸线淤进较快。波浪横向堆沙作用强盛,且稳定持久,它不断地将泥沙推向海滩顶部,因而堆积成沿岸沙堤。至三亚大沙坝中段,沿岸沙坝数目减少,高度降低,沙堤带宽度亦变窄。而沙坝东段沿岸沙堤则不发育,这是由于距沙源地的距离加大,泥沙供应逐渐减少的缘故。

大沙坝的物质组成:表层主要为石英长石质细砂,含有贝壳及珊瑚碎屑。但自海坡村向西,颗粒逐渐变粗成中砂,至烧旗水河口一带为粗砂。各段海岸泥沙颗粒均自高潮线向下依次变细,其重矿组合亦反映与花岗石类有关,主要为钛铁矿,其次为绿帘石、角闪石、锆石及电气石(表2)。

沙坝的底层物质,据三亚港地区钻孔自上而下综合如下:

① 表层为 5 m 厚的石英、长石质粗细砂。

② 约 1 m 厚的珊瑚枝块。

③ 12 m 厚的青灰色淤泥质亚黏土。

④ 厚 6 m 的粗砂层。

⑤ 底层为黏土层。

因此,①、②层为海岸相堆积,即沙坝主体堆积物。③层为浅海相沉积,即三亚沙坝的基底层。③层以下为更老的海岸堆积。

此沙坝规模大,堆积厚,表明在形成沙坝时,沿岸带泥沙来源丰富。同时,由于西南向的风浪作用持久,因而形成一条规模巨大与当地基本岸线平行的海岸沙坝。

大沙坝的后侧为一宽约 500 m 的低湿洼地,洼地分布方向大致与沙坝平行。洼地大部分干涸,部分已辟为稻田,局部为低湿沼泽地。据浅钻取样,洼地沉积自上而下组成如下:

① 表层 0.8 m 厚,为棕黄色夹白色的细砂、粉砂,砂为石英质,质地均一,皆有锈斑。

② 1 m 厚的杂色细粉砂,有水平层理,底部夹有少量淤泥。

③ 1.8 m 以下至 3 m,皆为青灰色细砂,以石英、长石为主,有云母、贝壳以及棕黄色铁质结核。

根据地貌、沉积与分布特点综合分析,这一系列洼地为干涸的潟湖。由于大沙坝的形成,使其内侧海湾与大海隔离,成为潟湖。沙坝发展加宽,逐渐阻隔了潟湖与大海之间的海水交换,结果,潟湖逐渐干涸成为沼泽洼地。在大沙坝东端的三亚一带,潟湖洼地中仍有潮流往返作用,故潟湖形态保存至今。

(2)穿过潟湖洼地,再向北,又为一列沙坝,公路即沿沙坝顶部而建,此沙坝从羊栏到桶井呈东西向延伸约 9 km,可称桶井沙坝,此沙坝规模大,宽 700 m,高 15 m 以上。沙坝组成物质为石英、长石质细砂及长径为 0.6～10 cm 的石英质砾石,砂砾具一定程度的磨

表 2　三亚湾泥沙的重矿物组成

百分含量（矿物名称）\采样地点（序号）	1 天涯海角海滩上	2 巴离岬角	3 烧旗港口外沙坝	4 凤凰村东海岸	5 栈桥旧飞机场外海岸	6 三亚镇水下碉堡西北	7 三亚港航道之北	8 三亚港航道之南	9 三亚河河口	10 神州浅滩之东航道	11 神州浅滩之西航道	12 航道口门与神州之间	13 三亚港航道丁门口	14 大坡河上游	15 月川河上游	16 三亚崖县县委沙坝上	17 三亚港务处宿舍（三亚沙坝）
磁铁矿	10.95	10.02	16.54	6.62	5.09	5.73	10.84	7.75	15.54	10.03	16.27	3.98	11.57	18.31	8.98	7.62	7.82
钛磁铁矿	8.69	14.72	9.52		4.20	7.16	7.34	4.07		2.53	7.19	2.86	1.73	0.19	14.55	6.34	7.41
钛铁矿	63.59	29.00	27.44	20.57	28.39	29.06	31.50	35.00	32.05	38.58	40.06	36.26	34.78	15.06	44.08	19.59	46.89
赤铁矿	0.02	1.17	1.00	3.31	2.86	3.95	3.26	0.47	2.48	3.37	1.93	2.88	1.09	5.33		0.58	5.70
褐铁矿		2.36	1.55	3.35	2.21	1.17	1.34		0.11	3.27	3.90	2.22	2.58	9.77	2.92		4.75
黄铁矿							0.58		1.79	0.54			1.54	0.75			0.47
石榴石	8.63	0.39		0.41	0.22				0.18	0.82	0.68	1.33	1.37		3.17		
锆石		9.41	3.25	7.27	10.78	5.42	17.85	18.48	26.30	2.73	2.76	12.66	5.32	1.24	6.58	1.70	4.04
磷灰石		1.37		2.42	1.51	2.59	0.77		0.18		1.61	1.99	1.03	0.99	1.70	1.42	1.90
电气石	1.24	6.67	1.29	4.63	3.14	14.10	6.53	11.40	2.88	7.93	5.02	6.88	8.24	0.49	0.25	30.40	10.20
金红石	0.41	0.59	0.52	1.42	1.29	0.70	0.96	1.17	0.54	0.82		0.66	0.17	0.75		1.42	0.24
锐钛矿		0.39												0.49			
黄玉	0.41																
绢石	1.64	1.96	1.55	1.42	0.87	0.47	2.67	0.47	1.08	1.00	3.68	1.10	0.51	0.49	1.95	1.99	0.48
绿帘石	0.82	11.57	14.97	14.37	18.30	0.94	7.67	1.17	10.08	13.38	7.95	17.55	0.86	11.61	8.77	12.20	0.47
角闪石		5.67	13.42	18.13	7.90	14.80	1.34	10.95	1.62	4.10	5.27	5.34	3.09	9.38	3.41	3.69	4.51
阳起石			0.25		0.65	5.17		2.34					0.51	0.49			1.90
透闪石					1.08		0.19		0.18					0.49		0.28	
十字石	0.62			0.22	0.22					4.10							
矽线石	0.41	1.65		0.66			0.58		0.18								
矽灰石	0.62	1.57	0.25			0.28	1.34	0.94	0.18			0.45	0.35		0.73	0.58	0.95
紫苏辉石		1.17		3.57	1.08	0.47		0.23		0.28		0.89	0.17	2.71	2.19	0.28	1.19
红柱石				0.89	1.29	2.40	0.96							3.21			
黑云母				0.44					0.36							1.99	
绿泥石		0.59		0.89	1.29	2.40	0.96	2.34		2.46			1.72		0.73	7.09	
其他	2.09		7.98	9.05	1.75	3.36	4.27	3.51	2.41	3.28	3.66	3.33	23.33*	18.77			1.19

备注　（1）本表分析成果只是根据 0.25～0.10 mm 细砂粒级分析而成。（2）其他矿包括：混杂在重矿内的石英、长石（深风化）及贝屑等非矿物质。（3）该重矿物样品中见到较多量呈圆球状的物质，在镜下呈星隐晶质，暂定为泥质钙质结核（中间有中心核）。

* 该重矿分析成果只是根据……

圆,沉积物中还保存着少量贝壳碎片。由于经过长时间的风化淋溶作用,贝壳、钙质物多已分解,但手摸砂砾面仍具滑腻感。沙坝内局部积水为乳白色,表明沙坝内含钙、碱物质较多。物质组成亦表明其为海岸相的堆积体。组成此道沙坝的沉积层已胶结坚定,颜色呈肉红色。据此看来,桶井沙坝形成时代较三亚沙坝早。

桶井沙坝以北,在回辉村一带是与一古海湾淤积的海积平原相连,此平原现已辟为稻田。在桶井村北,羊栏沙坝邻接的为一宽约百米的小型干涸潟湖洼地。

(3) 越过洼地又为一列海岸沙坝名为量琴沙坝。此沙坝高度在 20 m 以上,宽 500 m,两侧具有明显的陡坡,但向海坡略缓于向陆坡,坝顶平坦。此列沙坝为砖红色砂组成,其质地单一,分选好,砂层内亦夹有 1～2 mm 粒级的小石英粒,石粒磨圆好,表面亦染成红色。整个砂层均胶结硬实。此列沙坝系较羊栏沙坝更古老的沙坝。

量琴沙坝后侧又为一小型的干涸潟湖洼地。洼地后侧还有一条沙坝,高约 15 m,为砖红色砂层组成,称之为从米沙坝。

这一系列沙坝与潟湖洼地相伴分布的状况,表明本区海岸带波浪作用方向自形成从米、量琴沙坝起至今,变化不大。本区在较长时期内,以波浪横向搬运泥沙向岸堆积沙坝的过程为特征。

其次,沙坝的长短与供沙的范围大小有关,沙坝高度反映沙坝形成后的地体高度变化,但主要反映形成沙坝时的海岸原始坡度的陡缓。

(4) 在上述一系列沙坝与潟湖洼地组合的背侧,毗连着光头岭、墓山岭、洋岭、角岭等花岗岩山地,这些山地向海侧,皆为陡崖(坡度达 40°～45°),崖麓保留着凹穴及磨蚀平台,有的崖壁底部覆盖了坡积物,在崖壁的前方或突出山咀(多朝南面向大海)的前方,还保留着一系列圆浑的巨块孤石(石柱),孤石之间已为坡积物所填充。这一系列地貌皆为古海岸的遗迹。突出向南的山咀,即当时伸入海中的岬角;孤石是岬角被波浪冲蚀后退过程中的残留物,即海蚀柱;崖脚具有凹穴的陡壁是当时临海的海蚀崖。这一系列地貌反映着角岭、洋岭、墓山岭、光头岭一带是目前尚保留的本区最早的海岸线。当时,南边岭、虎豹岭、金鸡岭、光头岭、墓山岭、洋岭、角岭、马岭、发财岭、红塘岭以至最西边的南山,都是临海的山地,其山脚伸入海中部分皆成为陡立的岬角,而鹿回头和东瑁洲、西瑁洲一样是突立于海中的小岛。而现代的河流冲积平原,如月川水、大陂河、羊栏河、冲会河、烧旗河和担油港等原来皆为海水淹没的港湾。当时河流皆没有今天这么长,而是短小流急、冲刷力强,河流将山地风化剥落的岩屑不断地冲刷携带到各海湾。波浪冲刷海岸亦形成了大量岩屑与砂,加之波浪的折射作用,泥沙亦由两侧岬角向海湾内堆积。这样在风平浪静的海湾形成了最好的堆积场所,经长年累月不断淤积,使得海湾淤填为平原。海湾两侧岬角逐渐蚀退。原来位于岸外的岛屿(如鹿回头),因靠近陆地,泥沙来源丰富,在岛后波影区亦形成大量泥沙堆积场所。泥沙堆积渐使岛屿与陆地相连。由于岬角蚀退,海湾填平,岛屿与陆地相连,这样长期的自然演变结果,使得本区海岸由原始的岬湾曲折的海蚀基岩港湾岸逐渐演变为平坦的砂质堆积岸。长期以来,本区岸线是逐渐淤积前进的。但是,由于海岸淤进,加长了河流延长,减小了河床坡度,因而降低了河流负载携运泥沙的能力,加之,由于河口与原海湾口堆积的泥沙被海浪向岸推移成为海岸沙坝,沙坝的形成减缓海岸坡度,同时又阻挡河流泥沙向海推送,而多栏积于沙坝内侧的海湾或洼地中,这一系列因素交互影响的结果,使得本区入海泥沙较过去大为减少。泥沙来源减少,反过来又影响了岸线向海的淤进过程。所以,本区海岸今后不会再像去那样大规模地淤进。并且,由于入海泥沙量减少,而波浪却不断对岸坡进行冲蚀改造,这样必然形成补给少于支出,岸坡将被冲刷变陡,而陡的岸坡又将促使波浪对海岸作用的进一步加强,从而导致海岸受冲刷后退。由于本区尚未完全断绝泥沙供给,所以冲蚀后退的过程很缓慢。这就是本区海岸自然发展的总趋势。

(5) 现代海岸线是沿三亚大沙坝向海坡发育而成。在特大高潮线以上海水已作用不到的部分,多已植树造林,固定风沙。但高潮线以下的现代海滩,目前的动态是普遍遭受轻微的冲刷。其冲蚀变化情况是:

① 海滩有季节性的冲淤变化。大约相当于夏秋季节,在西南向风浪作用时,海滩受冲蚀形成陡坎,冲刷强度是自东向西,即自隐蔽岸段向开敞岸段增大,冲蚀陡坎的高度由15~20 cm 增至 30~40 cm。特别是在台风季节,岸滩受蚀后退强烈。在冬季,风浪小,且为偏北向离岸风,此时岸滩又逐渐淤积而恢复平衡。

② 局部地区海岸略有蚀退。如三亚港以北海滩上有一碉堡,系 1940 年日寇侵华时修建的,当时碉堡位于高潮线以上的海滩顶部。至 1966 年已位于低潮水边线附近,落大潮时碉堡才干出水面,碉堡距高潮水边线约 17 m,故该处海滩确有蚀退,后退速率为65 cm/a。此外,碉堡以南,如栈桥旧飞机场一带,海岸也略有蚀退。

但是,在三亚大沙坝末端,因靠近三亚河口,岸滩略有淤进。

四、三亚潮汐汊道

潮汐汊道是海洋伸向陆地的支汊。地貌上表现为狭长的海湾或开阔的潟湖,在口门处具有一狭窄水道与海洋沟通,潟湖外侧常具有沿岸伸展的海岸沙坝作为掩体,汊道内侧没有河流汇入,或者河流甚小。汊道主要是潮流作用为主的海洋环境。维持、稳定汊道的水深与位置,有效地利用潮汐汊道,在开发海运事业上具有重要意义。

三亚湾潮汐汊道位于三亚大沙坝南端,是一个以沙坝为外掩体的潟湖型潮汐汊道。汊道口门在三亚港与南边岭之间,为一 80 m 宽的狭窄通道,通道内侧为一宽度超过 1 km 的潟湖洼地,潟湖内侧分汊并与大陂水与月川水(东支)相接。但是,上游河流除 7—9 月由于热雷雨及台风引起洪水下泄外,其他季节流量极小。河流流量在三亚镇一带约 5~6 m³/s,洪峰流量为 2 375 m³/s(50 年一遇)。洪水一般在暴雨后 2~3 h 到来,但数日即落。而枯水位流量仅 0.3 m³/s。大陂水上游花岗岩山地,林木繁茂,水土流失少。上游山区河谷宽 40~50 m,河岸高 3 m。河床沉积是长径为 10~20 cm 的砾石与粗砂,砂砾沉积厚 1~1.5 m,平时河床干枯,卵石心滩出露,河谷中已生长植物。大雨后洪水下泄,但河水下沙不多,尤其在大陂水下游接近三亚镇的河谷中,长满了红树灌木丛,河水流经茂密的红树丛,红树丛像梳篦一样将大陂水挟带的泥沙拦截滤掉,因而,大大减少了大陂水向口门地区的输沙。月川水下游主要受潮流冲刷加宽了河道,涨潮流沿月川水上溯,可达水口坡一带,沿月川水两岸开辟了大片盐田。水口坡以上,月川水蜿蜒曲流于平原稻田中,由于上游打坝拦水及沿途稻田用水,所以月川水河谷中河水枯竭,平原上已无完整的河道。因此,月川水实际无淡水向河口下泄。

所以,在大陂水与月川水汇合以下的"三亚河口地区",实为一潮汐汊道,潮流起主要作用。本区涨潮时间长(平均 16～17 h),进潮量大,潮水沿河上溯可达 7 km。而落潮时间短(7～8 km),流速急,大于涨潮流速约 1～2 倍。尤其在口门段落,由于有三亚大沙坝外掩阻拦,口门狭窄,东水流急,致使该处落潮流最大表层流速为 0.9 m/s,最大底流速为 0.5 m/s,加之,水位近口门下降,因此在口门狭窄段造成底部冲刷,形成东西向长达 400 m、南北宽约 50 m 的潮流冲刷槽,槽底水深皆大于 3 m,最深处为 -9.4 m(据当地零点计)。汊道口门段的潮流冲刷槽在地形与沉积上与两侧海底有明显差异,三亚港老码头即利用此天然深槽的有利条件,保证了使用水深。

由于三亚河上游下沙不多,三亚大沙坝沿岸也无大量来沙,大沙坝末端淤进变动微小,故该潮汐汊道是稳定的,水流通畅,在冲刷槽的末端海底也未形成砂质的潮流三角洲堆积。根据潮汐汊道这些特性,顺落潮流方向在口门外海底开挖航道,将能保持着足够的使用水深。

在华南,潮汐汊道广泛发育,研究潮汐汊道的特点与形成演变规律,是研究与开发我国海岸的一个重要课题。

Characteristic of Tidal Inlets Designated for Deep Water Harbour Development, Hainan Island, China[*]

1.

The Hainan Island is the second largest island in China with the total area of 33 920 km², 258 km from the west to the east, and 180 km from the south to the north. It is a tropical island located between $18°10'$ N to $20°10'$ N in the South China Sea (Fig. 1). The total length of coastline is about 1 500 km with 60 embayments. The major transportation of the island to contact the outside is by sea, and with a few highway system and a west coast railway for inside transportation.

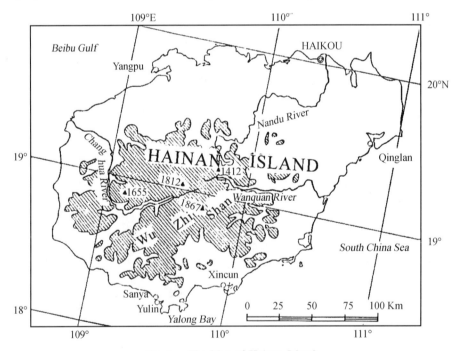

Fig. 1   The Map of Hainan Island

* Ying Wang, Charles Schafer, John N. Smith: Nanjing Hydraulic Research Institute, *1987 Proceedings of Coastal and Port Engineering in Developing Countries*, Vol. I, pp. 363 – 369.

The pre-Tertiary coastline of Hainan Island consisted of bedrock embayments that were surrounded by the steeper coastal slopes, deeper water, more irregular coastlines and eroded by higher wave energies than existents today. Tectonic movements in the coastal zone have become more frequent since late Tertiary time. In the northern part of the island, volcanic activity has produced both basalt lava coasts and shorelines controlled by volcanic cone morphologies. Along the coasts of the southern part of the island, sand barrier-lagoon systems have enclosed the previous embayment coast to form a flat sandy shoreline. In this part of the island, there is a series of four major sand bar systems that lie beyond the headlands of embayments with several exceptions having seven bars including some bars located inside of the old bays. The sediments composing of these bars are the result of the erosion of coastal bedrock outcrops, or can be correlated to the sediment transported by river draining of hinterland mountain valleys.

Both the volcanic shoreline and the sand barrier shoreline have been lifted above sea level to form raised terraces at elevations of 40 m, 30 m, 10~15 m and 5 m. Since postglacial time, the climate has warmed and coral reefs and beach rocks (cemented beach sediments) have developed along the coast. The Holocene sea level transgression submerged the sand barrier/lagoon systems, drowned depressions formed between volcanic rock and old sandy coastal deposits (i. e. terraces), and flooded mountain valley or old river channel systems.

Thus, there are three types tidal inlet embayments according to genetic pattern:

(1) The tidal inlet developed along structural fault zone, where is also boundary weak zone of volcanic rocks and sandy deposits of old terraces, such as Yangpu inlet in the northwest part of the island.

(2) Tidal inlet formed by sand barrier enclosed embayment, such as Sanya, Xincun and Qinglan inlets. It is a major type of the tidal inlet developed along southern and eastern coast of Hainan Island.

(3) The Holocene transgression flooded mountain or river valleys, Yulin and Yalong bays belong to this type.

All three genetic type of tidal inlet are suitable for setting sea harbour and with several common natures:

(1) Narrow embayment as a branch of open sea is always with deep water depth and larger area of sea water. River input is minimal, and the inlet environment is mainly maintained by tidal prism, i. e. tidal currents play major parts of dynamic processes, and fine grain materials of silt can be deposited in the bay and mixed with sandy sediments.

(2) There are tidal deltas developed at the both ends of tidal channel which formed by powerful currents erosion through the narrow pass of tidal inlet. We call the delta as "blocked-gate shoal" in China. The inner shoal or flood delta consists of fine sediment

normally with flat forms and it is still navigable waters. The outside shoal consists of coarser sediments and deposited by ebb tidal currents as the current velocity decreased suddenly while it passed over the narrow channel and entered into the open sea, even very powerful ebb current, there is still possibility to develop the ebb delta if the sediment supply is abundant. Some of the outer shoals were formed by longshore drifting passing the inlet mouth. However, the outer blocked-gate shoal may cause the problems to the navigation.

It is two factors to maintain the inlet bay and the water depth of navigation route:

(1) The factor of tidal prism, which is decided by tidal range and the duration ratio of flood and ebb tides. Large tidal range or a longer period of flood tide brings huge quantity of sea water which produces powerful current during short duration or ebb tide. The current is natural dredger washing out sandy sediments to benefit the navigation channel. It is very important to keep or extend the larger area of inlet bay for retaining larger quantity of flood sea water.

(2) To determine the sediment sources along coastline and tidal inlet, and to decrease the coast sediment supply are important step to produce natural flushing out the sediment, or sediment bypassing of long shore drifting.

2.

The coastal tide of Hainan Island is irregular diurnal tide with small range less than 1 m tidal range in the most area, except in the Yangpu Bay where is about 2 m. The average duration of flood tide is 17 hours, as a result, large quantity of sea water can be stored in the bay, and consequently, ebb tidal speeds are high as the short period of $7 \sim 8$ hours ebb tide. The natural conditions limit the growth of ebb tidal deltas but is suitable for retaining water depth of navigation.

Wave condition is one of the important factors to influence the safety of berth and the sediment movement of the coastal zone. Both wind wave and swell are active along Hainan Island. North and northeasterly wind wave prevail in winter and spring, and the heavy seas are always in company with northeasterly wind waves. Southwesterly and southeasterly waves are dominant during summer. The average wave height is 0.2 m in the north coast of Hainan Island, and the maximum wave height there according to records is 1.8 m; the average wave is 0.6 m in the southeast coast; 0.8 m in the west coast area; and 0.7 m wave height as average in the south but 4m wave height has been recorded during storm period of typhoon. Normally there are $4 \sim 6$ seconds of short wave period, but with several of 25 seconds long period and small steepness swell. Typhoon has a great influences to the east and south coasts of the island. There are nearly two or three times each year of typhoon passing on land in the east, and five times of typhoon storm influences in the south. There is no typhoon on land in the west and northwest parts of the island. The maximum wind speed of typhoon is 50 m/s (1960, Qinglan) and 100 m/s (1973, Qionghai) recorded in the east coast. The maximum wind speed of typhoon recorded in the south is 40 m/s, while typhoon passing on land is always with northerly winds, waves and storm heavy seas. However, with the advantageous characteristics of long, narrow bay inland and shelter area for berth, the tidal inlets can be avoided the disaster of storm waves. Thus, the tidal inlet embayments of Hainan Island are the ideal sites for setting up deep water new harbours. However, there were defeated examples shown that a detail study of coastal dynamics, sediment movement, coast geomorphology and evolution is very important.

Let us take Xincun harbour as an example. The harbour is located in a bedrock embayment of the southeast end of Hainan Island. The bay is shallow, and the total area is about 20 km² with only a tidal channel to connect the sea. The tidal prism is about 15 000 000 m³. It is a small fishing harbour and also a suitable site for aquaculture. The bay mouth channel faces southwestward open sea with an ebb tidal delta acrossing channel and the water depth there is only 2 m, thus apparent breakwave zone has always appeared there. In 1960's, local people dredged a navigation route with the water depth of 5 m on the sand shoal. But the new navigation route had been silted up immediately only after a spring storm, because there is a strong longshore drift from the east to the west acrossing the inlet, and also, the active southwest waves move nearshore sands transversely back to the seashore break zone. It was a practical lesson to tell people that how important to study the natural environment before constructing anything to reform it. Nevertheless, a successful example of using tidal inlet to build up medium size (berth for 5 000 tonne ships) new-harbour of Sanya where is 50 km away in the west to the Xincun.

Because of that Marine Geomorphology and Sedimentology Laboratory (MGSL) has done detail study of coastal dynamic geomorphology beforehand, we can predict coastal erosion, deposition and equilibrium developing tendency, sediment source, budget and sedimentary rate. Thus, the water depth of new berth and navigation route is nearly no change, for instance, an outer navigation route had been dredged passing through a sandy shoal from 2 m water depth to 4 m in 1966, and 20 years later, the total siltation is only 0. 4 m. It was super-result to satisfy the original requirement of four years using the route. In 1978 the navigation route had been extended and deepened to 6 m water depth, since then, there is no siltation. The reason is that there is no large quantity of coastal sediment supply, and the outer shoal is a residual beach with a veneer of sandy deposits. Thus, the dredged route can be maintained by powerful ebb current to flush out silted sediment.

Lately, MGSL has studied coastal environment and sedimentary dynamic processes of the tidal inlet of Yangpu Bay, and gives sufficient proof of the exploitation of the inlet to set up a deep water harbour. As a result, the study has been accepted by the department of transportation, and will start the new harbour project there in the near future.

Based on above research, the paper will give a review of the characteristics of three types tidal inlet and their conditions for harbour construction.

3.

(1) Sanya harbour

The harbour is located between $18°13'N$ to $18°14'25''N$ and $109°29'E$ to $109°30'E$, and it is the nearest one to the international route of China's harbours in the South China Sea. The harbour has been set up on the southern end of a flat bay mouth bar, and it holds the mouth of inlet. The harbour is protected well to avoid wind wave attack because there is a mountain range located at hinterland in the north, which prevents northeastly gale of cold wave or typhoon. Two islands are in the west and with nearshore reefs of Baipai & Dazhou, they decrease the wave energy from the west and southwest. And also, there is the Luhuitou mountain range in the south. Thus, the wave energy is small in the harbour area. The average wave height is less than 0.5 m and period is shorter than 4 seconds. An instaneous maximum wave height caused by typhoon reached 3 m. However, there is not serious wave erosion or sediment transportation around the harbour area by wave action. Sediment supply from southern part of bedrock coast is minimal because of little coast erosion and the fringing reefs has blocked the longshore sediment movement from the south. Sandy coast in the north of the harbour slightly suffers erosion by the southwesterly wave from open sea. But the longshore sediment supply has been stopped by a submerged bank which built up in 1970's to connect the Baipai reef with the shoreline and the top height of the submerged bank is ± 0 m above sea level. Thus, the outer part of Sanya harbour has little sediment supply and without serious wave turbulent to stir the bottom sands.

The landside of Sanya harbour consists of a long and narrow lagoon depressions, actually it is a branch of outer sea, i. e. a tidal inlet. River discharges from inland mountain range are minimal, it is about 170 000 000 m^3 runoff annually and the annual sediment discharge about 2.4×10^3 tonne which mainly deposited in the inner bayhead. Fresh water actually is only run out during flood season and Sanya harbour is maintained only by sea water.

The tides in Sanya harbour is the pattern of mixed diurnal with 11 days of semidiurnal tides per month as an average. The duration of flood is 8~17 hours and the ebb tide only lasts $4 \sim 8$ hours. The average tidal range is less than 1 m and the maximum tidal range is about 2 m. The longer period of flood tides combined with the factor of a definite tidal ranges causes the tidal prism reaching 1 500 000 to 4 000 000 m³/ per tide. During the short period of ebb tides, the current speed is 100 cm/s which is twice of flood current speed. While the ebb currents pass through the narrow channel at the southern end of the sand bar, the convergence of water causes a downward erosion by the powerful current. As a result, a 700 m long and 6 to 9 m deep channel has been formed at the southern end of the bar, the channel follows the direction of the ebb current (NW). The navigation route has been dreged following the same direction and extended the channel seaward to use the natural flushing out to keep the water depth. It is clear that harbours must be designed with a view to maintaining or enhancing the natural flushing process, i. e., it is important to keep the larger area of inner bay (or lagoon depressions). Developments such as land reclamation for new buildings, the construction of evaporating ponds for sea salt recovery both of which decrease the volume of the tidal prism and destruction of inner harbour mangrove communities must be avoided to prevent relatively rapid siltation of outer harbour areas.

The tidal inlet of Sanya can serve as multipurpose. The inner bay or lagoon is suitable for small fishing craft and pleasure boats. In its natural state, the outer bay is most appropriate for medium size (up to 5 000 tonne) commercial ships, additionally it can be also opened as base-harbour for operating the offshore oil and natural gas fields.

The outer harbour can be used for ships of up to 10 000 tonnes if the nearshore reef of Baipai or islands can be joined to the mainland coast by an artificial pier, then the railway can be built on the top of the pier and leeward sides of these structures can accommodate ships of up to 30 000 or 50 000 tonnes.

(2) Qinglan harbour

The harbour is located in the east coast of Hainan Island, and it is also a tidal inlet semienclosed by sand barrier systems. The area of the lagoon-inlet system is 40 km² and the tidal prism is 2×10^7 m³. It has a tidal channel with the water depth more than 10 m. The inlet can be reformed as medium and small harbours served for local transportation. The problem of Qinglan harbour is the ebb delta where the water depth is only 3 m, as a result, it is a barrier to block the navigation of medium size ships. Also, the destruction of natural coral barriers on the perimeter of the outer harbour enhanced siltation and longshore sediment drift at the seaward end of the navigation channel.

(3) Yangpu harbour

Yangpu harbour is located in a larger tidal inlet in the northwest part of Hainan Island. The tidal inlet was formed under the control of NE trending faults and associated volcanic activity. An old river has subsided 25 m since 8 500 Y. B. P. , forming an inner estuary and outer bay. The subsided deep channel connects the estuary, bay and the shallow Beibu Gulf. The river flowing into the embayment of tidal inlet has an annual sediment discharge of 1. 6×10⁵ tonne, 40% of which is fine sands and shingles deposited in the estuary. Silt and clay leave the bay by tidal currents. Sediment supply from the eroded northern basalt coast is 2 400 m³/yr. , from the south coast, having an elevated 10 m high old beach, the supply is 22 000 m³/yr. , and from eroding barrier coral reefs outside of the bay, it is 4 000 m³/yr.. Combining river sediment discharge with coastal erosion the total sediment supply is 90 800 m³/yr.. Dominant southwest waves stir up sediments, which are deposited in the estuary by the flood tidal currents. Diurnal tides with a range of 1. 82 m creates a tidal prism in the estuary of 10⁸ m³ and when this is combined with the river's runoff about 10⁶ m³, ebb current speed is normally stronger than the flood current, even though the ebb period is longer than that of the flood. Also, the current speed at the surface and median layers is larger than the bottom layer. The deep channel maintains its depth through the tidal action.

The tidal inlet of Yangpu is an excellent deep water bay with large area of berth which is protected from wave attack by the hilly hinterland. Many locations of the bedrock coastline are distributed along the deep channel, which are ideal sites for building up new harbour. The problem is a sandy shoal outside of the channel. The shoal is 400 m long, 85~100 m in width and the water depth is 5 m.

It is an ebb delta formed since 8 500 Y. B. P.. The sedimentation rate of the sand shoal averaged 0. 35 cm/yr. during the period from 8 500 to 6 350 Y. B. P. according to 14-C data, 0. 1 cm/yr. during 6 350 to 3 250 Y. B. P. and it is 0. 08 cm/yr. from 3 200 Y. B. P. to the present. Monitering data indicates that the sediment concentration of sea water is nearly same (0. 05 kg/m³) both at the deep channel and the sand shoal, and current speeds have shown a sequence of decrease from the channel to the sand shoal. Even though, the maximum current speed of 81 cm/s is strong enough to transport the silty sediment in suspension passing over the sand shoal.

However, most parts of the tidal inlet are in a state of dynamic balance except in the inner estuary which is generally accumulating the sandy materials. It might be possible to use the natural deep channel and dredge the sand shoal for navigation to establish a deep water harbour in Yangpu Bay.

(4) Yulin and Yalong bays

The two bays are the type of tidal inlet developed in the submerged mountain valleys, where are flooded during Holocene transgression. These are deep water bays and stretch for long distance inland without serious siltation problem. Thus, it is the best type of the tidal inlets for harbour construction.

Presently, studies of the sediment dynamic processes and modern sedimentation rates of all harbour areas of tidal inlet systems are in progress to optimize the planning and development of inner harbour areas. These studies will provide guidelines to engineers to insure the environmentally compatible development of these facilities for the increase in commercial marine traffic which is anticipated for this part of China over the next seral decades.

海南岛西北部火山海岸的研究[*]

　　火山海岸是指沿火山锥、火山口和喷溢带所发育的海岸。海南岛火山海岸主要发育于西北部儋县一带(图1)。关于海南岛的火山活动最早为1929年李承三所报道[1],1958年边兆祥提出了第四纪火山活动[2],1964年丁国瑜等论述了第四纪火山及玄武岩的构造地貌[3],同年黄金森①、钟晋樑②等讨论了火山地貌等问题。1984—1985南京大学在进行洋浦港选址研究中,对该区火山海岸做了考察,本文着重讨论火山海岸的特征以及火山活动对海岸发育的影响。

图1　海南岛西北部海岸地貌图

　　[*] 王颖,周旅复:《地理学报》,1990年9月第45卷第3期,第321-330页。

　　1993年以中文摘要载于《中国第四纪南北对比与全球变化——第六届全国第四纪学术讨论会论文摘要汇编》,151-152页,广东高等教育出版社。

　　① 黄金森、汪国栋,海南岛海岸地貌与第四纪地质的若干资料,南海海洋研究所地貌学论文集,第二集,1964,1-28。

　　② 钟晋樑,海南岛火山地貌,南海海洋研究所地貌学论文集,第二集,1964,201-215。

一、火山活动

海南岛西北部火山区处于雷琼拗陷南部的邻昌凹陷内[4]。有两条主要的断裂成X相交:北西向的兵马角—木棠—儋县断裂,北东向的松林—木棠—干冲断裂。沿此X相交的断裂有多次火山活动(图2)。

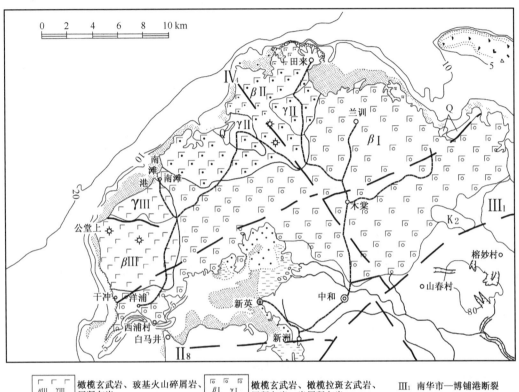

βIII γIII 橄榄玄武岩、玻基火山碎屑岩、层凝灰岩	βI γI 橄榄玄武岩、橄榄拉斑玄武岩、层凝灰岩、岩屑凝灰岩
βII γII 橄榄玄武岩、岩屑凝灰岩、层凝灰岩	✿ 火山口

III₁ 南华市—博铺港断裂
IV₄ 东澳港—峨蔓断裂
II₈ 王五—文教断裂

图2　海南岛西北部地质概况图

根据地貌演化与沉积地层学原则,结合岩性、火山活动性质与测年资料,可将该区火山活动划分出晚更新世与全新世两期,共5次喷发(表1～表3)。

表1　火山活动与地质地貌特征

喷发期次		时　代	岩石特征	分布与地貌特征
第一期	第一次	晚更新世玄武岩烘烤层热释光定年为52 000±400 a B. P.	气孔状玄武岩,岩流层厚5～10 m,经鉴定为橄榄玄武岩及橄榄拉斑玄武岩	洋浦,沿NEE向断裂多中心溢流喷发,形成整个火山区的玄武岩台地和丘陵,向海盖在海相沉积层上,火山弹、火山碎屑堆积易侵蚀成低地,岩流形成8 m高海岸阶地。有多个喷发孔,使地面起伏为火山丘陵

（续表）

喷发期次		时代	岩石特征	分布与地貌特征
第一期		第二次	气孔状橄榄玄武岩 橄榄玻基玄武岩 岩屑玄武岩 层凝灰岩	峨蔓为中心,沿NW向断裂成多火口喷发,喷发剧烈,形成20～40 m崎岖不平熔岩低丘,沿断裂带成一系列火山锥,笔架山三峰高度为208 m,121 m,171 m,春历岭100～167 m,火山锥兀立于第一期玄武岩台地上
		第三次	橄榄玄武岩	干冲断裂北侧及德义岭,成盾形火山及熔岩台地,穿插、覆盖在第二玄武岩上,德义岭火山高97.4 m,火山口完整、直径200 m
		第四次	火山岩盖在海相潮滩沉积层上,其中的扇贝[14]C定年为26 100±960 aB. P.,喷发时代在更新世末	多孔的熔岩 拉斑玄武岩 玻屑凝灰岩 玻璃火山碎屑岩及海水 具微层理与波痕结构的凝灰质砂砾岩	莲花山及德义岭以北,龙门以南,喷发强烈,黏滞性大。火山锥兀立,坡度大,穿插于熔岩台地(第三次)间,莲花山有一中心火山喷发口,高65 m。四周有几个火山喷发孔
第二期	第五次	经热释光定年为4 000±300 a B. P.	炉渣状熔岩 气孔状玄武岩 拉斑玄武岩	母鸡神、龙门、沿海岸断裂带,成串珠状小火山喷发,孔裂隙式喷发,全新世沉积受火山活动而抬升构成现代火山海岸

表2 各期玄武岩标准矿物含量

喷发顺序		岩石名称	ap	IL	Or	Ab	en	mt	di Wo	di en	di fs	Hy en	Hy fs	Ol fo	Ol Fa	q	Cc	产地
期	次																	
I	1	橄榄拉斑玄武岩	0.63	2.85	6.92	24.32	22.47	6.48	5.73	3.98	1.27	15.86	5.08			2.39		洋浦
	2	橄榄玄武岩	1.01	3.04	13.91	35.13	15.58	2.78	5.0	2.90	1.88	7.21	4.65	7.38	5.22			峨蔓
	3	橄榄玄武岩	1.11	3.61	7.68	30.83	19.53	2.50	7.20	4.08	2.82	8.98	6.25	2.0	1.05			德义岭
	4	拉斑玄武岩	0.34	3.04	3.90	24.64	23.64	3.01	6.0	3.70	2.61	11.86	9.66			5.79		莲花山
II	5	拉斑玄武岩	0.38	2.94	4.23	25.69	21.33	1.62	7.90	3.94	3.80	12.29	11.82			2.63		龙门
		拉斑玄武岩	0.37	2.91	5.57	23.27	21.75	1.92	7.77	4.14	3.39	10.52	8.62			5.21	2.48	母鸡神

符号名称:ap磷灰石、IL钛铁矿、Or正长石、Ab钠长石、mt磁铁矿、Wo硅灰石、en顽火辉石、fs斜铁辉石、di透辉石、Hy紫苏辉石、Ol橄榄石、fo镁橄榄石、Fa铁橄榄石、q石英、Cc方解石。

表3　各期玄武岩化学成分

喷发顺序		岩石名称	SiO₂	TiO₂	Al₂O₃	Fe₂O₃	FeO	MnO	MgO	CaO	Na₂O	K₂O	P₂O₅	H₂O	判别系数 $\frac{Na_2O+K_2O}{SiO_2-39}$	产地
期	次															
I	1	橄榄拉斑玄武岩	51.46	1.49	14.28	4.49	6.64	0.18	7.98	7.63	2.87	1.16	0.29		0.32	洋浦
	2	橄榄玄武岩	53.96	1.60	15.11	1.93	7.18	0.112	5.68	6.09	4.14	2.40	0.446	0.43	0.44	峨蔓
	3	橄榄玄武岩	51.66	1.88	14.58	1.88	8.40	0.146	6.38	8.01	3.62	1.28	0.427	0.60	0.39	德义岭
	4	拉斑玄武岩	53.16	1.64	14.17	2.07	8.88	0.161	6.06	7.87	2.92	0.69	0.210	0.93	0.25	莲花山
II	5	拉斑玄武岩	53.14	1.54	14.48	1.12	9.28	0.153	6.52	8.29	3.06	0.73	0.210	0.70	0.27	龙门
		拉斑玄武岩	52.36	1.52	13.56	1.33	8.38	0.147	5.90	9.73	2.74	0.90	0.208	0.95	0.27	母鸡神

注:判别系数>0.37为碱性玄武岩,判别系数<0.37为拉斑玄武岩。

二、现代火山海岸

现代火山海岸是沿着全新世火山活动带发育的,火山大多为裂隙多孔喷发,以母鸡神和龙门两个岸段为典型。

(一) 母鸡神火山岸段

该段长1.8 km,兼有第4、5两次火山喷发活动,由火山颈、火山口、喷发岩构成的海蚀柱、海蚀岩墙、海蚀崖以及受火山构造活动影响抬升的古海岸沙坝、古海滩等组成。

"母鸡神"是高10 m玄武岩构成的海蚀柱,顶部系火山弹、火山渣堆积。下部为玄武岩,玄武岩各层不连续,是经3~4次喷发升起的。海蚀柱上有三层海蚀穴,朝向西北其高程为1.7 m、4.0 m、6.0 m。向陆一侧,有残留的火山喷发口,海蚀后成环状的岩墙及布满角砾的磨蚀台地,母鸡神附近多这类火山岩墙与磨蚀台地(图3)。

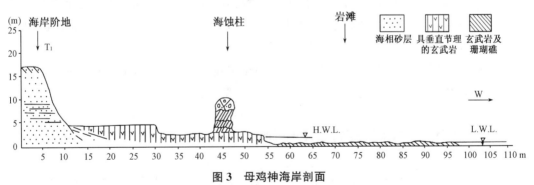

图3　母鸡神海岸剖面

沿着母鸡神附近海岸线为一断裂带,有15个火山口呈串珠状。喷发口大多为椭圆形,直径7~14 m,内壁为致密玄武岩,外围是气孔玄武岩,具垂直节理,内壁表层是烘烤层,口内堆积着熔渣状玄武岩与集块岩。这些小火山孔喷发物不整合覆盖在莲花山火山堆积物(第4次喷发)之上。向海的小火山口喷发物与海相沉积物成互层,层理清晰。这些火山孔在海岸成岬角,海蚀后成崎岖的岩滩及海蚀柱(图4,照片1)。

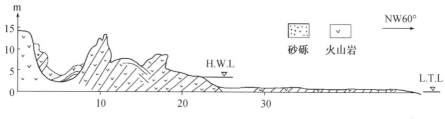

图4　母鸡神北侧高炉状火山口海岸剖面

母鸡神以南为一小港湾,湾顶是棕黄色石英长石质海滩,向陆是10 m高的海积阶地。海滩四周均系玄武岩海蚀崖,故海滩物质并非来源于沿岸海蚀,而是火山活动以前的河流供应的。该港湾原为一河口湾,德义岭与莲花山火山喷发使地形改变阻挡了河流入海,河口湾成小海湾,湾外沿岸仍保留着当时河流供应物形成的大型海岸沙坝。第5次喷发(母鸡神)使小港湾抬升成10 m的海积阶地,海岸沙坝抬升成高大的海岸带沙山,同时沿岸的海相沉积层被抬高到20 m高。因此,火山口形成的海蚀柱、海蚀崖与岩滩相伴生的最新海相沉积层,是这段火山岸的特征。

(二) 龙门火山岸段

该段长3 km是北西向排列的笔架山火山喷发物所组成,现已抬升为高20 m海岸阶地,阶地表层有残留的海相砂砾层,岬角之间的老港湾也被抬升,湾底成为高5 m堆积阶地。龙门兵马角火山是一盾形火山,叠加在20 m海岸阶地上。火山向海一侧成海蚀崖,其剖面为黑色玄武岩与黄灰色凝灰岩互层。该段海岸由一系列火山口与数十个小港湾组成。火山喷发口直径15～20 m,口内底部保留着熔融岩浆沸腾时的形态,熔融岩浆溅起高出熔岩流表面40 cm,犹如沸腾的岩流骤然冷却凝固的形态。此火山熔岩经热释光测定年龄4 000 aB. P. ,一些火山喷发口受海蚀成圆盆形、弧形的斜坡,弧形悬岩及箱形"巷道",将岩滩分割,或成柱状兀立岩滩上(照片2、3、4)。另外,沿玄武岩节理(走向N10°E,N30°W)发育海穹、海蚀穴,"龙门"即沿节理产生的高8 m、宽10 m的海穹,目前海穹底部已高出现代岩滩4m(照片5)不受海水作用。龙门岸段一系列海蚀穴,海穹底部高出现代岩滩4～5 m,反映这4 000 a以来的抬升量。

因此,龙门火山岸段是密集的火山口群,由熔岩流与岩块堆积构成,年代为全新世中期,火山活动使海岸普遍抬升,海蚀强烈,成为很典型的海蚀火山港湾岸。

三、火山活动对海岸发育的影响

构成海岸带的火山锥、喷发口、熔岩流等海蚀堆积地貌,同时还影响到海岸的发育,主要有活动与效应。

(一) 火山活动使海岸抬升

形成海岸阶地、海蚀地貌,改变海岸坡度、海岸带动力状况,从而影响到海岸的沉积类型与演化特点。

　　母鸡神附近海蚀崖剖面是很典型的(图5,照片5)。该剖面长150 m,高19 m,出露地层共9层,由上而下:

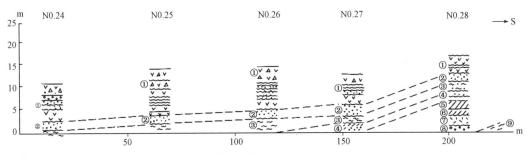

图5　母鸡神高炉火山口海岸沉积剖面

　　(1) 火山堆积物,厚4~10 m,分三层,上层玄武岩;中层凝灰质砂砾岩,具良好的水平纹层及数厘米厚的粪丸砂质沉积,是火山砂砾在潮滩环境下的堆积;底层为火山角砾岩。

　　(2) 灰黄色细砂层,厚2 m,含贝壳,表层为水平纹层粉砂层,受火山岩烘烤成红色。

———————不整合———————

　　(3) 灰绿色黏土层与锈黄色粉砂层,厚2.4 m,具良好的水平纹层及虫穴构造,另有竖立的钙质结核,似为沿虫穴的次生的淀积物,而锈黄色粉砂是沿龟裂带氧化的,此层为潮滩沉积。

　　(4) 锈黄色极细砂与粉砂层,厚1.6 m,夹两层钙质砂礓,沙礓块体5 cm。

　　(5) 粗砂砾石层,厚0.4~0.7 m,为海滩岩,其中含有化石,栉孔扇贝类(*Chlamys* sp.)曼氏孔楯海胆(*Astriclypeus Manni Varrill* Mio-Rec)、笠贝(*Aemaea*)等,含化石砂层具斜层理。

———————不整合———————

　　(6) 灰褐色夹黄色锈斑的粉砂黏土层,厚2.8~4.0 cm,是水平纹层的潮滩沉积。

———————不整合———————

　　(7) 灰褐色泥质粉砂与钙质沙礓层,厚2.5 m,砂礓有二层,水平状,含有钙质或铁质的管状沉积,当时接近潜水面成水平层,属潮滩沉积。

　　(8) 黄褐色细砂、中砂与砾石层,厚2 m,含贝壳层,主要有偏胀蚶(*Arca Ventricosa Lamarck*)、船咀须蚶(*Barbatia Rostrata Lan*)、笠贝、猫爪牡蛎(*Ostrea pestifris* Hanley)、拙脊卵蛤(Pitar)、李氏不等蛤(*Anomia Lischkei Dautzenberg et Fischer*),此层中有较多的铁盘状积淀物。沿虫穴皆有铁质沉积,底层铁质胶结为10 cm厚的铁质砂岩,层2、3似经历较长期的水下环境。

　　(9) 夹砂砾层的粉砂黏土层,厚2 m,上部纹层状黏土层,厚30 cm;中部砂砾层夹岩块,厚40 cm;下部泥质粉砂层,厚130 cm,水平纹层理,有虫穴构造,此层系潮滩沉积,所夹薄砂层系风暴沉积物。

　　该剖面反映了本段海岸经历了三次潮滩环境与二次海滩环境,是火山活动改变了海岸坡度引起的。火山活动使地形变化,其影响范围比构造断裂影响的范围要小得多,甚至仅在火山喷发口几十米的范围内。坡度增大时波浪作用为主,堆积砂砾物质,发育为海滩。当坡度减小,为平缓宽阔的潮间带时,潮流作用为主,形成潮滩沉积。该剖面16 m厚的海相沉积,也说明海岸经历过沉降,其间有轻微的回升,至第 5 次火山活动,抬高为高出海面的阶地,目前为海蚀崖。

公堂上村剖面(图 6,照片 6)从上向下有 6 层:

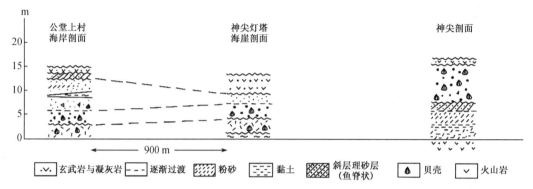

图 6　母鸡神南部海岸沉积剖面

(1) 凝灰岩与玄武岩,厚 4 m。

~~~~~~~~不整合~~~~~~~~

(2) 灰黄色粗砂层,厚 2 m,是含贝壳的海滩岩

~~~~~~~~不整合~~~~~~~~

(3) 黄色砂层,厚 3.3 m,含栉孔扇贝、海胆(Echinoidea)及绿色条纹的泥球泥饼夹层。

(4) 灰绿色黏土,厚 0.2 m,向北尖灭,系岬角段潮滩沉积残留部分。

(5) 黄色砂层,含凝灰质砂粒及丰富的贝壳,如牡蛎、海菊蛤(Spondylus)、栉孔扇贝及皱红螺(Rapana bezoar)等,底部粒度变化呈花边状,系滩角构造(Beach cusps),这层是海滩沉积。

(6) 黄灰褐色黏土质粉砂层,含扇贝、海胆及大量有孔虫(分布于热带浅水底栖种)。扇贝经 ^{14}C 测年为 26 000±960 aB. P.(样号南大 84062),这层为浅水海湾相,相当于上述的第二潮滩层(层 2),形成于晚更新世末。

神尖灯塔基底为高 12 m 海蚀崖,由海岸沉积与火山岩构成,被一系列断层分隔错动,断层走向 N80°E,倾角 10°,与公堂上村剖面相比较,此处潮滩沉积已被侵蚀殆尽,玄武岩直接盖在具滩角构造的海滩沉积上(照片 7),自上而下有 4 层:

(1) 玄武岩与凝灰岩,厚约 4 m。

(2) 黄色细砂层,厚 2 m。

(3) 砂砾层,厚 3 m,层内有完整的贝类化石,主要是栉孔扇贝、曼氏孔楯海胆、笠贝及舌骨牡蛎(Ostreahyotis Linnaeus)等。牡蛎化石 ^{14}C 测年为 25 320±960 aB. P.(南大 84061)。

(4) 棕黄色粉砂层,厚 3 m,均匀无层理,多有孔虫,并含大量绿色泥球,此层为浅海湾或潮下带沉积,此层下出露灰绿色黏土层。

神尖是海蚀残留的小岛,由海相沉积组成,顶部盖有凝灰岩(图 6,照片 8):

(1) 表层为凝灰岩及凝灰质砂岩,已侵蚀残缺。

（2）浅棕黄色砂岩，中砂夹贝壳，厚 9 m，为海岸沙坝相。上部已胶结成海滩砂岩，其中扇贝[14]C 测年为 23 830±1 440 aB. P.（样号南大 84059）。

（3）棕黄色石英砂及贝壳层，具斜层理，已胶结。

（4）粉砂黏土层，厚 5.5 m。上部 3 cm 是黄色粉砂黏土，夹有密集的有孔虫（盖虫）透镜层及灰绿色黏土块；中部厚 0.5 m，灰色黏土，亦含大量盖虫；下部为红褐色黏土质粉砂，无明显层理。贝壳化石丰富，主要是栉孔扇贝、曼氏孔楯海胆及众多的蛏壳印模、盖虫化石及印模。扇贝经[14]C 测年为 27 300±1 200 aB. P.（样号 ZK - 1056）。这是晚更新世浅海湾及潮下带沉积，以后海水变浅，波浪作用活跃，渐渐发育了上部海滩沉积。

这三个剖面所在区域与母鸡神海岸相同。火山口沿北东向断裂呈带状分布，晚更新世至全新世初，海岸带动力因火山活动而逐渐增强，经全新世火山活动而抬升为海岸阶地。向海一侧残留着一些火山海蚀的形态。

（二）火山活动改变了沿岸陆地形态，改变了河流流向，影响到海岸沉积物的来源与性质

沿岸有许多体积巨大的砂质堆积体，其物质成分与周围的火山岩不同，是火山喷发以前由河流堆积的。如干冲以北沿岸宽 1 000 m 高出高潮位 10 m 的沙坝群，其石英长石质砂来自三都的河流，目前河道已被德义岭火山喷发物阻塞，河流已消失。但在三都至长地还残留着长 9 km、宽 200 m 的古河谷，在航空相片上也清晰可辨。北部沿岸，德义岭火山喷发前，有河流自东南向西北方向入海，河流带来大量的由花岗岩山地侵蚀而来的长石石英质砂，形成海滩。德义岭、莲花山火山锥形成后，改变了地形，但尚有涓涓流水供应泥沙，第 5 次火山喷发后，河道完全消失。神尖灯塔南有一群白色的海岸沙丘，宽 200～600 m，高 10 m，砂粒为石英长石细砂与贝壳珊瑚砂砾，这些粗粒物质不是风力堆积，而是激浪作用形成的海岸沙坝，按目前风浪强度及临近岸段沙坝比较，沙坝高度原来约 3 m，是第 5 次火山喷发使沙坝抬高到 10～11 m。沙坝内侧原为河流，供应泥沙，火山活动使海岸带抬高，河流断流，河口形成潟湖，河口段在抬升过程中使河流下切，形成深切曲流。

四、结论与讨论

（1）海南岛西北部沿岸有更新世晚期及全新世中期两期火山活动，共 5 次喷发。现代火山海岸是沿着北东向断裂带的全新世火山活动带发育的，这一期是密集成串的小火山口发生裂隙式喷发，火山喷发物未形成显著的火山锥，而是呈岩墙或依附于早期火山的岩壁上。堆积过程中经历构造变动，多断层，再经海蚀成崎岖的海蚀型火山港湾岸。因火山喷发物抗蚀程度较低，故岩滩广泛发育，并有众多的海蚀形态——海蚀柱、海穹等。岩滩的外缘常成为珊瑚礁发育的基础，珊瑚堡礁——大铲岛、小铲岛就是在北东向玄武岩暗礁基础上发展起来的。神尖至龙门整个火山海岸的外缘，发育着珊瑚礁坪，火山海岸均表现出明显的构造抬升，在火山口密集处抬升幅度大，全新世中期以来，海岸抬升高度为10～20 m。

（2）该区火山活动具继承性、多发性。从第 1 次喷发至第 5 次喷发，有从陆地向海，即向西转移的特点，规模也渐减小。火山活动在短期内改变了区域环境特征。整个海岸带的地势、动力条件、沉积物特征及海岸演变均受到火山构造活动的控制。

（3）晚更新世火山喷发为橄榄玄武岩，近岸火山喷发物为拉斑玄武岩。全新世喷发物为拉斑玄武岩，与地幔物质有关[5,6]。根据 $MnO\text{-}TiO_2\text{-}P_2O_5$ 的判别图式，洋浦玄武岩界于洋岛玄武岩和大洋中脊玄武岩区间界线上；峨蔓、德义岭玄武岩在洋岛玄武岩区内；而莲花山、龙门、母鸡神玄武岩位于洋中脊玄武岩区内。这是很有意义的现象。它与火山活动自陆向海转移的趋势相一致，即向西时代更年轻。结合南海构造的分析，这些是否反映了北部湾具有海底扩张的趋势。在地壳减薄开裂处，地幔物质沿断裂喷发溢出。北部湾的涠洲岛、斜阳岛则是扩张带上的火山岛。

（4）珊瑚礁在海南岛西北部发育很好，分布到 $19°40'N$。涠洲岛珊瑚生长亦较好，该处的海洋环境对珊瑚生长是偏暖凉的，那里珊瑚发育良好，除了有适宜的岩石基底外，可能与近代火山活动区有较高的海底地热流有关。这个区域的珊瑚礁与火山岸相伴生，为研究该区域地壳构造、边缘海的演化提供了线索。

参考文献

[1] 李承三. 1929. 广东海南岛北部地质矿产（上册）. 两广地质调查所年报，2.

[2] 边兆祥. 1958. 海南岛第四纪火山. 中国第四纪研究，1(1)：250-251.

[3] 丁国瑜等. 1964. 海南岛第四纪地质的几个问题. 见：中国第四纪地质问题. 北京：科学出版社. 207-233.

[4] 王颖. 1977. 南海海底. 见：海洋地理. 北京：科学出版社. 33-35.

[5] Nicholls, I. A. and Ringwood A. E. 1973. Effect of water on olivine stability in tholeiites and the production of silicasaturated magmas in the island-arc environment. *Jour. Geology*，81(3)：285-330.

[6] 周新民等. 1981. 我国东南沿海新生代玄武岩的成分和演化特征. 海洋文集，4(1)：100-109.

照片1　高炉火山口海蚀柱与岩滩

照片2　海蚀形成的火山峰巷道

照片 3　小喷火口

照片 4　火山岩岩滩

照片 5　火山喷发抬升的海岸沉积剖面

照片 6　公堂上村沉积剖面

照片 7　神尖灯塔海岸沉积剖面

照片 8　神尖海洋沉积剖面

洋浦港海岸地貌与海岸工程问题[*]

0　前　言

洋浦港位于海南岛西北部儋县,由新英湾及洋浦湾组成,是海南岛急待开发的天然优良港湾。1983年冬南京大学应交通部水运规划设计院、海南港务局邀请,作洋浦港选址初步可行性研究,朱大奎、王飞燕及研究生傅命佐、柯贤坤、高抒、张红霞等做现场工作,完成研究报告。1984年夏至1985年冬,为洋浦港一期工程,王颖带领教师10人,研究生十几人,两次赴现场作了近40船次全潮水文泥沙测验,大面积的海岸地貌地质调查,分析海底表层及钻孔样近千个,完成了"洋浦港泥沙来源、数量、回淤条件及航道稳定性研究"报告。1988年10—12月,以王颖与C. T. Schafer博士为首的中国—加拿大海洋科技合作计划,对洋浦港又做了地震剖面、柱状取样、海流及温盐密等测验。工作期间朱大奎应邀作了"洋浦港动力地貌与港口建设"报告。本文材料取自上述几次的工作。目的是着重介绍在洋浦建港的一些优越条件与注意事项,同时想以此为例说明地貌学在海岸工程建设中的作用。

1　海岸地貌

(1) 洋浦港水域是由新英湾及洋浦湾组成(图1),该处是三列构造体系之交,构造活动频繁,NE向断裂与隆起是南海构造的主干,自晚第三纪以来,NE向断裂多次发生[1][2]。其次是北西向构造。洋浦港在雷琼断陷南部,上新世至第四纪沉积了厚度超过3 000 m的滨海-浅海相碎屑岩建造(湛江组),洋浦港受到第三列构造——EW向断裂的控制,港湾沿此断裂产生,也即沿着地层交换的薄弱带,在港湾北部底部是湛江组,上部是大面积的玄武岩流构成的台地,而南部是湛江组沉积的台地,玄武岩流的年代经测定为52 000年,湛江组的上部是未胶结的海岸相砂砾层。

(2) 新英湾,是纳潮水域,面积50 km²,出口处(白马井峡)宽度仅550 m,新英湾主要是堆积岸,两条主要的河流,大水江、春江在湾顶注入,其流域面积合计为1 419.44 km²。河流泥沙在河口堆积,岸线随三角洲发展而推进,据历史记载,中和镇和新州镇在唐朝时均为海港,后逐淤积成陆。至清朝海港移至新英镇,而今海船(渔轮)在白马井(湾口),十几吨的木船要高潮才能航行至新英镇,目前,大水江、春江三角洲的面积占新英湾面积的50%,潮流作用将湾底侵蚀成一条条水道,同时将河口堆积中<0.01 mm的细粒泥沙带走,使河口堆积物中>0.01 mm的沙粒占90%以上,当河口浅滩逐渐增高,外海潮流带来

＊　王颖、朱大奎:《南京大学学报(地理学专辑)》,1990年总第11期,第1-13页。

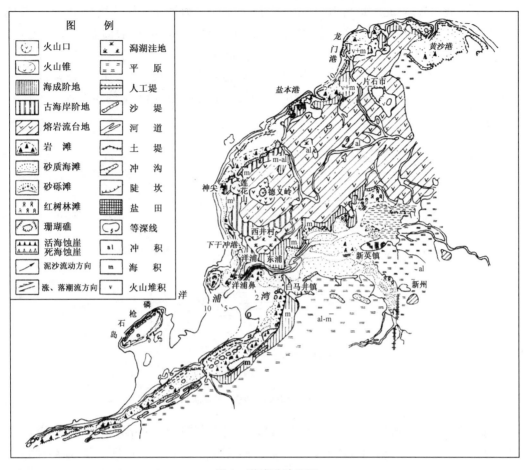

图 1　洋浦港地貌图

的悬移质泥沙沉积其上,浅滩上有成片密集的红树林,红树捕捞悬浮体促进了浅滩发育,形成了下部冲积相砂砾层,顶部薄层海相淤泥层的二元相河口潮滩沉积。湾内北岸是玄武岩港湾岸,岬角处有宽平的岩滩及海蚀崖,港湾内有薄层堆积,红树林丛生,海蚀崖已死亡,反映了北岸处于海蚀转为堆积的过程(图 2)。

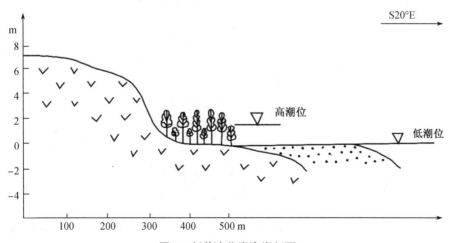

图 2　新英湾北岸海岸剖面

（3）洋浦湾海岸，北岸是高 15 m 玄武岩台地（图 3），海蚀崖大体连续分布，玄武岩平台上覆盖着一层海相沙层，黄色中细砂，含较多的生物碎屑，石英砂表面多 V 形刻痕，与现代沙坝石英砂形态一致，但有较多的溶蚀形态，该沙层经¹⁴C 定年为 4 900±200 aB. P.，是老沙坝沉积，老沙坝之间有潟湖沉积，故北岸是海成阶地。北岸玄武岸是稳定的。洋浦村西崖前的海滩岩，其年代为 3 130±195 aB. P.，而崖前的深槽年龄为 8 500±230 aB. P.。

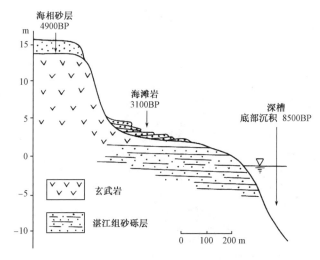

图 3　洋浦湾北岸海岸剖面（玄武岩台地）

南岸是湛江组砂砾层构成的台地（图 4），按成因沿岸陆地可分为两部分，沿岸宽约 1 km 是 12 m 高海岸阶地，组成物有两类：棕黄-橘黄色粗中砂，系古沙坝沉积，地势高起成垄岗状；灰白色细砂中砂为潟湖沉积，夹在沙坝之间，构成数列沙坝-潟湖地貌沉积体系，也即湛江组的顶层沉积。台地内侧为海积-冲积物亦为沙坝潟湖沉积，但受近代河流作用改造。整个台地由松散碎屑物质组成，是向海域供沙的主要源地，海蚀岸线长 2.5 km（寨基—小松鸣），海蚀崖高 6 m，崖前海滩宽仅 20 m，1984 年 9 月上旬至中旬暴雨期间，2.5 km 岸线海崖崩落 11 处，海岸崩落后退 0.8～1 m，崩塌堆积布满高海滩，至 10 月初，崩塌堆积全被海蚀带走。据 1976—1984 年测记，海蚀崖平均后退 0.5 m/a。

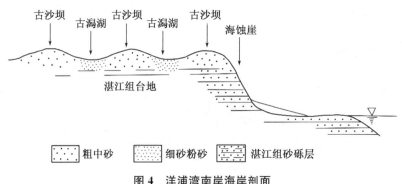

图 4　洋浦湾南岸海岸剖面

（4）珊瑚礁平台。排浦—超头市岸线长 7.8 km 是湛江组砂层台地，崖前是珊瑚礁平台，宽度 1.8 km，自高潮线向海有 6 个带：① 高潮海滩，宽 50～100 m，粗砂-中砂富含生

物碎屑,海滩基部向海侧已出露潟湖沉积,表明高潮线受侵蚀。② 礁块砂砾带,宽600 m,
底为湛江组,滩面为砂砾薄层,厚 20 cm,以及直径几十厘米的礁块,砾石成分为板岩、石
英岩,滩面多波痕。③ 藻垫层,宽490 m,滩面为砂砾,礁块及圆盘状的藻垫,该带中央藻
垫密集,约占整个滩面积的 1/5。④ 活珊瑚带,宽 400 m,有鹿角珊瑚(Acropora)、圆柱藻
叶珊瑚(Scapopyllia cylindrica)、菊花圆柱珊瑚(Goniastrea)等。随水深增加,活珊瑚数
量及种属增多,滩面沉积主要为珊瑚碎屑。⑤低潮位碎屑沙坝带,宽 200 m,沙坝起伏高
度 0.4~1 m,⑥ 水下岸坡活珊瑚带,宽度大于300 m,分布至水深−4 m(图5),珊瑚礁平
台是发育在湛江组海蚀平台上,潮间带中下部及水下岸坡均为活珊瑚生长带,珊瑚是生长
在清澈的暖水中的生物,若海水浑浊,珊瑚虫将被窒息致死,因此礁平台上活珊瑚的发育,
表明沿岸无泥沙流活动。

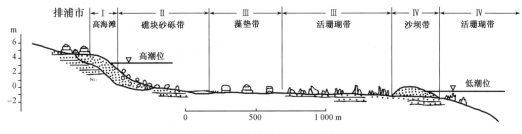

图 5　排浦珊瑚礁平台剖面

（5）洋浦大浅滩,是湛江组地层构成的侵蚀-堆积阶地,基底原始坡度 18%,上覆
0~2.2 m 厚海相砂砾、粉砂黏土沉积。砂砾的矿物组成与湛江组相同,为钛铁矿、锆石、辉
石、褐铁矿、电气石,其中锆石含量高,铣石、重晶石、电气石作为湛江组的指示矿物,在大浅
滩均有较高的含量。大浅滩表层为砂砾,向深槽边缘松散砂砾在潮流波浪作用下形成沙坝
状堆积。浅滩高出海图 0 m 的面积为 2.5 km²,水深<−2 m 的面积占洋浦湾40%(23km²)。
据地图对比,1955—1985 年,0 m 线以浅的面积在增大,增加了 0.7 km²,而水深 0~2 m 的面
积在缩小,水深 5 m 的面积减少了 1 km²,即洋浦湾面积未变,水深较大的面积在缩小,而浅
滩在增加。但这大浅滩并非堆积的,是湛江组被侵蚀的平台,沿岸砂砾层被侵蚀,海蚀崖后
退,崩落物质被波浪潮流搬运堆积,而细粒物质被带走,使浅滩顶部物质粗化。

（6）洋浦深槽与拦门沙浅滩,深槽从新英湾口(白马井)以内至拦门沙浅滩,长10 km,
宽 400~500 m,水深 5~25 m,呈明显的河谷形态,深槽沉积层有三,底部为湛江系亚黏
土,中层为河流砂砾沉积层,上层为海相砂质淤泥或淤泥。若把湛江组地层的顶面作为深
槽的原始谷底,那么,白马井角−24.34 m 至洋浦灯标−28.03 m,拦门沙为−30.13 m,其
平均坡降为 0.58‰。深槽切在湛江组层中,深槽走向与该处断裂构造走向(NE,NW)一
致,故是沿构造断裂谷发育的河谷。全新世海面上升过程中,河谷上游段(新英湾内)及下
游段(拦门沙及向西的海域)逐渐被海相沉积所充填,目前的深槽段是由潮流作用,特别是
新英湾内蓄纳的潮量的落潮时冲刷所维持的。

拦门沙浅滩是深槽通向海域的浅水段,水深 5 m 左右,长 400 m,宽 85~150 m,向海坡
3%,与洋浦向海的自然坡度一致。拦门沙的沉积层与深槽相同(图6),湛江系砂质亚黏土
为基底,上覆河流相砂砾层及海相淤泥层,拦门沙基底的原始地形亦具河谷形态,是深槽古

河谷向外淤积而成。其地貌发育经历三个阶段：① 古河谷阶段，河谷切割在湛江组地层中，河谷中堆积了厚 7 m 的砂砾层，其年代在 8 000 年前为全新世早期沉积。② 海浸，古河谷淤积阶段，全新世海面上升过程中洋浦湾新英湾平原成为海湾，古河谷转为潮汐通道，水道中沉积了新英湾带出的细粒砂质淤泥，其年代在 6 000～8 000 aB. P.。③ 拦门沙形成阶段，6 000 年以来拦门沙逐渐淤高，这阶段以出现一粗砂层为标志（粗砂含细砾），在拦门沙南部各钻孔中均有此细砾粗砂层，而北部相应层位是中砂。可见拦门沙形成时期，南部湛江组地层有过强烈侵蚀（当海面上升到目前位置或略高时，湛江组地层的海蚀崖受强烈侵蚀，这时期也即大浅滩形成阶段），粗粒物质进入深槽入拦门沙。而北岸，古沙丘侵蚀的中砂一般只分布在洋浦深槽以上，从拦门沙钻孔岩芯可知，随拦门沙的发展，进入该区的物质逐渐变细，即由粗砂为主的黏土质砂变为细砂为主的黏土质砂。即初期有大浅滩形成时期的粗粒物质进入，逐渐为主要是深槽输送的细粒物质。由地层剖面及年代测定可知，拦门沙发展到目前这种突起碍航形态，是近 3 000 年来缓慢形成的。拦门沙的沉积速率在 0.1 mm/a 左右，最大值为 4.3 mm/a，自然沉积速率很小，或可说该堆积体是稳定的。

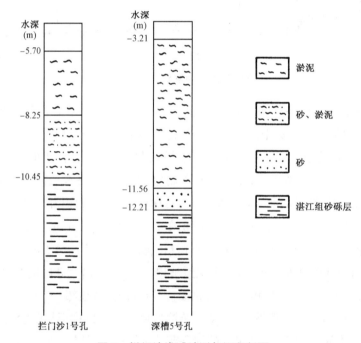

图 6　拦门沙浅滩地形与沉积剖面

2　沉积物来源、运动和沉积过程

2.1　沉积物来源与数量的分析

洋浦湾新英湾沉积物来源，主要是河流的、海岸侵蚀的及珊瑚礁生物的三种来源。

河流：沿岸河流主要有大水江（流域面积 648.34 km²）、春江（面积 577.76 km²）及沿岸小河（81.25 km²）。该地区无实测水文资料，1984 年据航空相片与实地调查相结合，采用自然条件相似的,南渡江龙塘水文站以上的流域侵蚀模数,来推求该两河流的输沙量。1985年吕明强、陈惠芬等[①]以降水量指标 K 值来推求年输沙量。

求得:大水江年输沙量,7.39×10^4 t/a;春江年输沙量,6.72×10^4 t/a;沿岸小河输沙量,1.5×10^4 t/a。合计 15.6×10^4 t/a(6.3×10^4 m³/a)。

大水江、春江所携带的沉积物除少量被潮流搬运或洪水时河流搬运直接向新英湾外运动以外,大部分在河口堆积,发育了河流三角洲,在河口外的潮间带和潮下带,细粒物质堆积成大片红树林、草滩和泥滩。一部分砂砾物质堆积在潮汐水道中,在水流作用下重新搬运,成为其他堆积体的物质来源。

海岸侵蚀:海岸侵蚀是沿岸输沙的重要来源,主要在洋浦湾南湛江组地层的侵蚀产物及北岸玄武岩岸的侵蚀。

南岸的侵蚀量,据沿岸地貌调查,有 2.5 km 侵蚀岸,岸线平均后退 0.5 m/a,岸上部分年侵蚀量约 7 500 m³。应用 1977—1979 年三年波浪观测资料,按交通部《海洋水文》规范求得该段海岸沿岸输沙量为 2.2×10^4 m³/a,自南向北沿岸输送。

北岸玄武岩海岸的侵蚀量,是以海蚀崖后退速率估算的。洋浦深槽底部沉积物年龄为 6 000 年,若以深槽谷边缘 -5 m 等深线作为 6 000 年前古河谷的谷坡位置,则 6 000 年来岸线后退到今日位置,在 12 km 岸线上共侵蚀掉 $1 444 \times 10^4$ m³,平均 2 400 m³/a。沿岸地貌观测表明,北岸的侵蚀量是轻微的,目前小于此数字。

神尖—干冲沿岸有巨大的砂质堆积体分布,是否有海蚀泥沙从神尖—干冲沿岸南下,通过小铲—洋浦鼻之间水道进入洋浦湾、深槽、拦门沙浅滩? 这是建港中关心的问题。

神尖附近为火山带,地层剖面从表层向下为:① 黑色玄武岩、褐色凝灰岩厚 4 m;② 黄色砂层含贝壳砾石及珊瑚屑,厚 3 m,是已胶结的海滩岩;③ 褐黄色石英砂及贝壳层。已胶结成砂岩,灰色黏土贝壳层,贝壳层年龄为 23 830±144 aB. P. 。神尖南有古沙坝堆积,高 10.3 m,古沙坝相当于神尖层②③层,受火山活动抬升到今高度。这古沙坝的砂砾并非来自海岸侵蚀,而是河流入海处的堆积,火山活动河流改道,而断绝沉积物供给,所以,这高 10 m 巨大的砂质堆积体是晚第四纪残留的沉积物。

干冲镇附近(下兰村)10 m 海岸阶地前缘有宽 500 m 沙坝堆积,沙层厚 6 m。该沙体亦系古河流堆积、德义岭等火山活动(26 100 aB. P.),三都附近地面隆起,三都下兰古河道断流,下兰河口堆积改造为沙坝。

神尖—干冲—洋浦鼻沿岸是岬弯相间,海岸阶地前方有 200～400 m 宽岩滩,玄武岸侵蚀形成表面崎岖,海蚀柱岩渣残留其上,有巨大玄武岩块堆积,而仅在岬湾顶部有少量砂质堆积。所以这段海岸以海蚀产沙少,仅做横向运动,巨大砂质堆积系古河谷残留沉积,目前无泥沙沿岸南下。

小铲—洋浦鼻间水道低潮时可涉水而过(1985 年 2 月 3 日 13 点 45 分,水深45 cm),水底均为活珊瑚,珊瑚群体间为珊瑚碎屑珊瑚砂,从水道中央向岸其横剖面如下:Ⅰ. 活珊瑚带,宽 100 m,鹿角珊瑚为主,低潮时水深仅 40 cm,水底多巨岩块(玄武岩块、珊瑚礁块)有珊瑚肢体固定于岩块上水底为珊瑚砂。Ⅱ. 藻垫珊瑚带,宽 500 m,滩面由藻垫珊瑚

---

① 陈惠芬.洋浦港河流输沙量计算.南京大学水文专业毕业论文,1985.

岩块及珊瑚砂构成,无陆源碎屑堆积。Ⅲ. 珊瑚块体沙坝带,宽 200 m,由>10 cm 的珊瑚碎块碎枝与珊瑚砾石堆砌成沙坝,高出高潮面 2.5 m,无陆源碎屑。Ⅲ. 礁湖带,小铲岛中央向南开口水底为珊瑚碎屑、砂和泥,低潮时水深 20 cm,有部分活珊瑚,这些表明,基本不会有泥沙通过小铲水道。

因此,洋浦鼻以北海岸无海蚀泥沙南下供应航道及拦门沙。

生物沉积物:洋浦港水域,年平均水温 25.1 ℃,适于珊瑚生长,岸外有小铲(2 km²)、大铲(4 km²)两个珊瑚岛,岛礁四周均为珊瑚沉积,为确定珊瑚物质在洋浦沉积物中含量比例。我们对不同海域样品作 $CaCO_3$ 含量测定,拦门沙浅滩中 $CaCO_3$ 含量 33.1%,洋浦湾南寨基村外水深-2 m 外 $CaCO_3$ 含量 21.6%,而新英湾内大水江河口 $CaCO_3$ 含量仅0.6%。这表明珊瑚礁生物沉积物是礁体附近沉积物中一项重要的物质来源,而对拦门沙有影响的是小铲。据 D. Stodart[①] 等研究,太平洋热带珊瑚礁生长速度 2~7 m/ka,小铲面积 2 km²,则小铲每年产碳酸钙沉积物为 1 400~4 000 m³。事实上小铲活珊瑚带面积小于 2 km²,珊瑚生长速度远小于 7 m/ka,所以,珊瑚形成的生物碎屑远小于 4 000 m³/a,此处仍采用极大值 4 000 m³/a。

这样,洋浦港海域泥沙来源为,河流供沙 15.6×10⁴ t/a(6.24×10⁴ m³/a),海蚀来源为 2.44×10⁴ m³/a(2.2+0.24=2.44),生物来沙 0.4×10⁴ m³/a,共计 9.08×10⁴ m³/a。

2.2 沉积动力作用

1984 年 1 月至 1985 年 1 月,分三次对洋浦港作 14 个断面 37 个船次的全潮水文泥沙测验,取得洋浦港潮流含沙的基本特征。

(1) 新英湾面积 50 km²,口门(白马井角)断面 5 000 m²,平均潮差 1.81 m,实测最大潮差 3.8 m,纳潮量 1 亿~2 亿 m³,口门流速表层>1 m/s,底层>0.25 cm/s。这巨大水量与流速是维持洋浦港水道的主要动力,提供了良好条件。

(2) 洋浦港涨潮流速<落潮流速,涨潮历时(13 小时)>落潮历时(12 小时),从新英湾口—洋浦深槽—拦门沙外海,潮流流速稍有减少,但总的看,表层及底层流速均较大,能搬运该海区各粒级的泥沙。

表 1 洋浦湾的潮流

地点	潮沙	含沙量(g/L)	表层流速(cm/s)	中层流速	底　层
白马井	涨潮	0.088	51	88	84
	落潮	0.103	84	86	97
深槽(洋浦村)	涨潮	0.102 4	42	67	56
	落潮	0.113 5	68	53	45
拦门沙	涨潮	0.115 6	57	49	22
	落潮	0.119 0	54	75	56

---

① D. Stoddart. 珊瑚礁的发育与海面变化年代学(在南京大学的讲演稿). 地理科技资料,1981.

（3）涨潮流与落潮流有不同流，涨潮历时大于落潮历时，特别是表层流，涨潮历时大于落潮历时为 2～4 h，所以涨潮水流从洋浦大浅滩、深槽向新英湾均匀进水，进入新英湾后分两支，南支深槽是主要的进潮水道，而北支以落潮为主。落潮流出新英湾后，水流主要沿洋浦深槽外泄，其中经深槽的水量约占 3/4，而 1/4 水量经浅滩呈 NE-SW 向向外海排送，其流速也以深槽为主（图 7）。

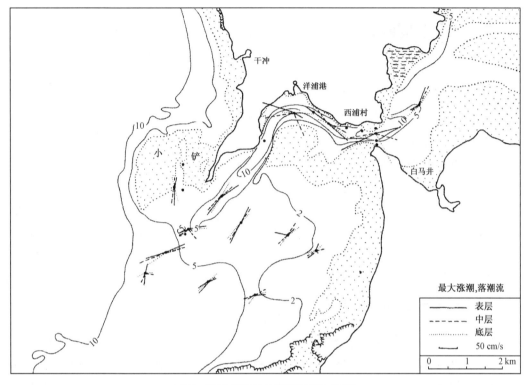

图 7　洋浦港潮流图（涨落潮流路图）

（4）洋浦湾内悬浮体含量＜0.1 kg/m³，其成分为石英，呈片状，易于悬浮搬运，占 67%，其次为微体生物、生物碎屑和絮状碳酸盐占 15.2%，其余为褐铁矿、玄武岩、辉石、锆石等碎屑。絮状碳酸钙粒径小于 0.008 mm 的微粒集合体。

2.3　沉积物的分布与运移

根据粒度分析制作的底质图，粗砂砾石分布在河口区及洋浦大浅滩的水边缘线处。中砂细砂沉积物主要分布新英湾浅滩及水道、洋浦大浅滩及洋浦湾沿岸。黏土分布于新英湾顶红树林滩地，深槽深泓线。从深槽至拦门沙的粒度是砂质黏土—黏土质细砂—细砂质黏土。

这些分布表明：

（1）河口堆积体发育，由河口至海湾外，物质从粗逐渐变细，这说明河流来沙是新英湾沉积物的主要来源。

（2）新英湾水道中砂占 80%，水道间浅滩砂占 50%～70%，即新英湾向湾外输送的主要是粉砂、黏土等细粒物质。

（3）深槽、拦门沙主要是黏土粉砂。大浅滩是砂质，其粒度、矿物成分同湛江组地层一致，是湛江组受侵蚀供应的。大浅滩与拦门沙之间有一片细粒沉积分布区（黏土），将浅滩与拦门沙分隔开。拦门沙黏土含量 40%～50%，与深槽黏土含量相似，呈递变的，是通过深槽向拦门沙输入。拦门沙的砂质含量为 20%～30%，与大浅滩不连续也稍高于深槽的砂质含量，而拦门沙沉积物中 33% 为 $CaCO_3$，与小铲的有关，故拦门沙的砂质主要来自小铲。

3　海岸工程问题

3.1　洋浦建港的自然条件分析

洋浦湾是海南岛天然的避风良港。洋浦深槽深度 10～23 m，宽 500～800 m。由于有巨大潮流量保持深槽不淤，是洋浦港天然的通海航道，洋浦北、东及南面是 10 m 高的海岸阶地，南面有大浅滩，阻挡了 N，NE，E 及 S 向风浪，使洋浦湾水域有良好的掩护条件。大风出现频率小，平均 ≥6 级以上的风频率仅 0.42%，波高 H1/10≤0.5 m 频率为 99.1%，H1/10≤0.8 m 为 99.88%，波高 H1/10≥1 m 的仅占 0.06%，方向为 SSE，这是湾内的强浪向，波高 ≥1.2 m 的频率为 0，所以，洋浦是一天然的避风港。同时港区为玄武岩台地及结构良好的湛江组地层，沿岸石料丰富，陆域开敞宽阔，建港的自然条件优良。存在的主要问题是湾口外有一片水深 5 m 长约 400 m 的拦门沙浅滩。

3.2　深槽及拦门沙航道的沉积速度

洋浦港海域为一潮汐汊道，新英湾是汊道体系中的纳潮水域，深槽是潮流通道，拦门沙是落潮流堆积体。据钻探的地层及区域地貌分析：8 500 年前，深槽及拦门沙原为一古河道，河谷的谷底标高在 −25 m～−30 m 处，河流沉积层的顶面为 −20 m，古河谷纵比降 0.7‰，大水江、春江在今新英湾内汇入古河谷，通过深槽、拦门沙段，至水深 −20 m 处注入北部湾。随着冰后期海平面上升，沿岸平原淹没为洋浦湾新英湾，古河谷被海相沉积所充填，而新英湾、洋浦湾的纳潮量，涨落潮流的冲刷，使古河谷保留了深槽段的水深，而新英湾内及拦门沙段的古河谷被淤积而消失。

据拦门沙沉积的年代测定，8 500 年来的平均沉积速率为 0.16 cm/a，其中近 3 000 年以来约为 0.1 cm/a，这与世界海面变化过程相适应。因此，拦门沙的自然形成过程是缓慢的。

近几十年来的变化，主要是用 1947—1983 年 6 幅海图及水深地形图的量测比较，将各图的基面统一换算成理论深度基准面，按拦门沙、深槽、大浅滩等类型，用地形断面法量计各阶段的体积求出冲淤变化量。得出：拦门沙厚度变化为，1947—1974 年冲刷 2～10 cm/a，1974—1983 年淤积 2.2～4.5 cm/a，深槽中 1947—1974 年冲刷 1.2～5 cm/a，1974—1983 年淤积 5 cm/a，而总的看近 50 年来，拦门沙及深槽均在微弱受冲刷，拦门沙冲刷总量 $100×10^4$ m³，深槽冲刷总量为 $66×10^4$ m³。

因此，从 8 000 年以来的历史看，拦门沙及深槽的沉积速率很低，从近 50 年来的地形变化，表明拦门沙及深槽基本是稳定的，时有淤积冲刷，变化幅度均小，对工程并不构成危害。

3.3　拦门沙开挖航道的可行性分析

拦门沙浅滩是洋浦港阻碍航运的关键,若洋浦港按二万吨级船型建造,航道水深需保证 $-9.2\,\mathrm{m}$(理论深度基准面水深),乘潮水位 $+2.1\,\mathrm{m}$。拦门沙浅滩天然水深不足 $9.2\,\mathrm{m}$ 的航道长度 $4\,\mathrm{km}$,其中最小水深 $5\,\mathrm{m}$,长约 $400\,\mathrm{m}$。据多次水文测验,拦门沙段涨潮流方向 $45^\circ\sim50^\circ$,流速 $0.3\sim0.8\,\mathrm{m/s}$,落潮流方向 $210^\circ\sim230^\circ$,流速 $0.8\sim1.2\,\mathrm{m/s}$,为使航道尽量利用天然深槽,顺直的最短距离通过拦门沙;利用涨、落潮流减少航道回淤,使航道走向与强风向、强浪向及流向的夹角较小(小于 70°)。为此,洋浦港通海航道走向在 $45^\circ\sim225^\circ$ 附近为宜。

拦门沙浅滩开挖航道后的回淤量,取决于以下情况:① 新英湾下泄泥沙数量,是否在拦门沙堆积;② 洋浦鼻以北沿岸海蚀泥沙有否通过小铲水道进入拦门沙;③ 洋浦大浅滩泥沙是否在风暴天气被波浪掀起,带入拦门沙;④ 开挖航道后边坡稳定性及回淤量。

新英湾沙主要来自河流,年输沙 $6.3\times10^4\,\mathrm{m}^3$,其中粗颗粒($\geqslant0.125\,\mathrm{m}$)约占 40% 以上。根据底质图,新英湾内沉积物中粗颗粒占 $60\%\sim100\%$,即河流来沙中粗颗粒主要堆积在新英湾内。据历次水文测验,洋浦港水域落潮流速(平均值、最大值)均大于涨潮流速,泥沙向外海输送;各站含沙量大体一致,落潮流含沙量大于涨潮流含沙量;流速沿程变化均匀,在拦门沙最大落潮底流速达 $81\,\mathrm{cm/s}$(流向 255°),在拦门沙没有发生流速骤变小的现象,其余流及底质粒度均与深槽连续递变的,这些说明:拦门沙是湾内悬移质泥沙通过区,没有发生大量沉积的条件。

洋浦鼻以北沿岸无泥沙通过小铲通道进入拦门沙已有前述,而来自小铲的生物沉积物每年的总量为 $0.4\times10^4\,\mathrm{m}^3$。

洋浦大浅滩的泥沙,是否能在风暴天气被波浪掀起,带入拦门沙,引起航道的骤淤。对大浅滩掀沙作用主要是 SW-W 向风浪,最大平均波高 $0.83\,\mathrm{m}$,周期 $5.4\,\mathrm{s}$,最大波高 $3.4\,\mathrm{m}$,周期 $4.2\,\mathrm{s}$,其波浪破碎带在 $-0.8\sim-1\,\mathrm{m}$ 处,该处底质为中砂-细砂,粒径 $\geqslant0.125\,\mathrm{mm}$ 占 $92\%\sim100\%$,显然这些颗粒泥沙主要是向浅滩顶部堆积,而不是被水流向外海搬运。据底质图,大浅滩为沙,拦门沙为粉砂黏土,两者间有大片黏土区分布,这亦表明浅滩泥沙未向拦门沙方向输送。据水文测验,潮流余流均指向西南,洋浦海域及大浅滩的细粒泥沙是向西南,向外海输送的。

考虑到拦门沙航道是涨落潮流的主要通道,流速足以搬运粉砂细砂($0\sim81\,\mathrm{cm/s}$),泥沙来源少(整个洋浦港水域为 $10\times10^4\,\mathrm{m}^3/\mathrm{a}$);海水含沙量低(平均值 $0.09\,\mathrm{kg/m}^3$,最大值 $0.15\,\mathrm{kg/m}^3$);天然沉积速率小($0.4\,\mathrm{cm/a}$)。近 50 年来水下地形变化幅度在 $1\mathrm{m}$ 以内(1955—1963 年为 $0.8\,\mathrm{m}$)。按目前资料,航道开挖后,没有发生骤淤现象。航道回淤量主要将是人工开挖后边坡的自然平衡调节过程及航道附近底移质泥沙的回淤。根据洋浦港航道的各种回淤条件、状况,估算洋浦拦门沙航道开挖后的回淤量为 $0.5\,\mathrm{m/a}$。

参考文献

[1] 丁国瑜等. 1964. 海南岛第四纪地质的几个问题. 见:中国第四纪地质问题. 北京:科学出版社,207-233.

[2] 王颖. 1977. 南海海底. 见:海洋地理. 北京:科学出版社,33-35.

Marine Geology and Environment
of Sanya Bay, Hainan Island, China [*]

INTRODUCTION

The coastal zone is the interface between terrestrial and marine environments. It is perhaps the most important of the major environments of southeast Asian islands because of the role it plays in economic and transportation activities. Because of their dependence on the local marine resources, and on their use of marine transportation for the import and export of local products, a large proportion of indigenous island peoples live in the coastal zone. However, unlike many inland situations, the open coast environment has a relatively greater exposure to climatic processes. This factor dictates that ports must be designed with a view to both commercial shipping capacity, and as a potential shelter for a variety of commercial and recreational vessels. Port maintenance is another important design consideration because ambient siltation and sedimentation during storms can generate a need for frequent dredging of anchorages and navigation channels.

In this paper, we describe the geological setting, and several important sedimentation processes, in the vicinity of the port of Sanya which is located on the south coast of Hainan island. This island is situated off the south coast of China in the eastern part of the Gulf of Beibu and has been designated as an economic free zone in which development of production and tourism facilities will proceed at an accelerated pace relative to mainland China. The port of Sanya city has been identified for possible significant expansion that will permit its use by large commercial vessels. Presently, these types of ships must anchor outside the mouth of the harbour where they are loaded and offloaded by smaller barges. The engineering designs for an expanded Sanya harbour must consider the nature of local sedimentation patterns as well as particle transport mechanisms and rates so that ambient siltation effects in key user areas of the port can be minimized. The nature and rate of some of the hydrodynamic and depositional processes controlling these effects in Sanya harbour have been evaluated by

    * Ying Wang, Dakui Zhu, Charles T. Schafer, John N. Smith: Ying Wang, Charles T. Schafer eds., *Island Environment and Coast Development*, pp. 125 – 156. 1992, Nanjing University Press.

participants of an international Chinese-Canadian research project that surveyed some of these phenomena during a November, 1988 field study. Geological investigations of coastal dynamic geomorphology and sedimentary processes on this part of the Hainan island coast were carried out by Nanjing University's Department of Geo & Ocean Sciences during 1966, 1976, 1984. Virtually all of the geological, climate-hydrological and sedimentary information outlined in the following sections of this paper is a product of these earlier efforts.

1. Field Methods

In 1964 and 1985, Nanjing University carried out a regional coastal survey from Yulin bay to Baoping Bay. The work involved a marine hydrological survey in Sanya Bay including whole tidal process survey simultaneously of tidal currents, SPM measurements, tidal level and wave monitoring (Endeco 949) along critical sections of five to control the hydrological dynamic in the harbour area; coastal geomorphology & sedimentology study including coastal dynamic geomorphological survey and mapping, beach and submarine coastal slope, echo sounding survey monitoring along 20 profiles reached to the water depth of 20m, shore & submarine sediment grab and tug sampling, fluorescent sand tracing and submarine topography mapping, and also collected the previous data on climate, typhoon, river discharges, sea level records etc.

In 1988, surveys in Sanya Bay were conducted from a 30 m long boat. A support frame was welded onto the stern of the ship to permit the towing of gear, and for core and grab sampling operations. A Del Norte Trisponder system with two remote stations was installed to facilitate the location of sampling stations, sounding line and reflection seismic transects. Coring was done with a Lehigh gravity corer that used 10.3 cm×3 m long PVC core barrels and a core head weighing about 37 kg. A Shipek sampler was used to collect surficial sediments. The vessel was large enough to allow the setup of a laboratory thus eliminating the need for a land-based facility and allowing immediate splitting and description of sediment cores. A light weight Ekman dredge grab sampler was deployed from a 3 m inflatable boat in shallow nearshore areas to obtain sediment samples of these environments. In these instances, the location of the sample was determined using a navigation chart and water depth information. Data on sediment structures and thicknesses were collected with an ORE Geopulse reflection seismic system. Water depths were measure from the research vessel using a 12 kHz Raytheon recording echosounder. Water samples were collected for SPM and salinity analysis by three boats anchored for a 15 hour period in the inner harbour, the eastern part of the outer harbour and at the western end of the outer harbour near the mouth of the embayment. Samples were collected from two or three depths every hour for 15 hours i. e., from low water to high water. On the research vessel, the water column was profiled at the same hourly interval using a portable Guildline Instruments CTD (Model 8770). CTD data were recorded on cassette tapes and processed at the Bedford Institute of Oceanography. The CTD sensor was coupled to an Oregon Red SPM profiler that was used to study the distribution of SPM in the water column under various tidal stage conditions.

SCUBA divers were used in 1988 to deploy sediment traps and to collect short cores. The sediment traps were manufactured from 6.6 cm diameter piston core liner. One end of a one metre length of liner was sealed with a core cap using standard electrical tape. A 40 cm long × one cm diameter steel rod was attached to the base of the liner so that the trap could be anchored to the sediment in a vertical position. The traps were deployed near navigation buoys to facilitate their recovery. 1.5 m lengths of the core liner were used by divers to recover 50~60 cm long sediment cores. The liners were initially pressed into the sediment and then hammered to the appropriate depth using a sliding weight that was fitted to the top of the core liner. When the appropriate penetration was achieved, the hammer device was removed and the core was plugged with a rubber stopper. It was then rotated through a 10 cm radius, pulled out of the bottom, and capped at its lower end before being transported in a vertical orientation to the surface.

2. Laboratory Methods

Marine hydrology: All marine hydrological surveying data collected by Ekman and electronic current metres were corrected for water depth and then processed by computers to produce diagrams of tidal flow characteristics, Residual flow vectors, and suspended load transport curves for Sanya Harbour.

Sediments: The sieve method was used for sizing sands and micro photo sizer (SKC2000) for finer parts of silt and clay. Heavy mineral analysis was carried using a polarizing microscope. For the core samples from the harbour area, analyses of pollen & spores, foraminifera, organic content, chemical composition and trace elements were completed.

Dating: All sediment cores were dated by the method of ^{210}Pb isotopic dating in the Chinese-Canadian collaborative isotopic laboratory of Nanjing University. Shells, peat, and carbonate materials were dated by ^{14}C method, both of the labs are in Nanjing University.

Climate

The climate of the Sanya Harbour coastline reflects its location adjacent to a tropical ocean and falls within the sphere of influence of a monsoonal atmospheric circulation system. The trajectory of the "mainland" monsoon is mainly from the northeast; the "ocean" monsoon moves toward Sanya mainly from the southwest. The modal climate of the Sanya area is wet and hot. The rainy season occurs from May to October when high rainfall and warm temperatures are the norm. Relatively dry and cool conditions dominate during the dry season between November and April. Sanya's annual mean temperature is 25.4 ℃ and monthly mean temperature throughout the year is above 20 ℃. June mean temperature is 29.5 ℃ and the January mean is 21 ℃. The annual maximum temperature can reach 36 ℃ and the minimum may be as low as 20 ℃. Fog is virtually unknown along this part of the coast.

Rainfall

During the autumn and summer seasons, the southwest monsoon controls the rainfall in this part of the island. The annual mean rainfall is 1 279.3 mm with a maximum and minimum of 1 870.5 mm and 746.5 mm respectively. 91% of the total annual rainfall occurs between May and October. Intense rainfall events are typical during typhoons which frequent this area in August and September. The results of 35 years of observation (1958 to 1983) show that the maximum one day rainfall (287.5 mm) occurred on May 29, 1971 and that the maximum continuous rainfall was on May 28 - 29, 1971 (545.8 mm).

Monsoon

The winter monsoon (October-March) is controlled by a high pressure cyclonic atmospheric circulation; the summer monsoon (April-September) originates from the Indian low pressure system and has an anticyclonic pattern during July-August period but high winds can also occur during the winter months. The winter wind usually blows from the northwest to northeast and is controlled by the Mongolian high pressure system. On average, wind intensity in Sanya is above Beaufort 6 for about 7.3 days each year.

Typhoon

The Sanya area experiences two types of typhoon, one of which originates in the Northwest Pacific ocean. During the past 35 years, the Pacific typhoon has occurred 122 times. The other typhoon type approaches Sanya from the South China Sea. This type has occurred in Sanya 61 times over the past 35 years (1959—1981). During a typhoon, the wind velocity in Sanya can reach 40 m/s with concurrent rainfall. If the centre of the typhoon comes within 2.8 degrees latitude of Sanya, the city will typically experience wind velocities of greater than 8 on the Beaufort scale. If the typhoon centre approaches to within 1.5 degrees, wind gusts can be expected to reach 40 m/s (Beaufort equivalent 8). At these times, the maximum wave height in the vicinity of Sanya harbour would be in the 7~8 m range (Table 1).

Table 1　Wave Height Conditions in the Sanya Bay Area When the Typhoon Centre is at a Distance of 180 Nautical Miles (3° latitude)

Wind Speed (m/s)	Wave Height (m)	Swell Height (m)	Max. Wave Height (m)	Max. Swell Height (m)
<15	2	2.5	4	5
20~25	3	3	5	6
30	3	4	6	7
35~40	3	4	6	7
40~45	4	5	7~8	8~11

COASTAL GEOMORPHOLOGY AND DYNAMICS

The total length of the Sanya Bay coastline is 26.5 km. It begins at Lu Hui Tou peninsula to the southeast and follows the base of Nanbian mountain to Sanya harbour. On the west side of the harbour, the coast has a westerly strike until Tian Ya Hai Jiao and then ends in the Nanshan mountains (Figure 1). Sanya Bay is open to the south and is part of a lagoon embayment coast system. There are six embayments along this part of the coast. Yulin harbour is at the eastern end while Nanshan mountain marks the western end. Most of the embayments were depocenters for grey marine sediments that are presently expressed as salt marshes and coastal plains. Three large sand barriers separated by lagoons are presently located behind the modern coastline of this area. Sanya harbour is located within this barrier/lagoon coastline complex.

The original bedrock coast lies inland and consists mainly of Mesozoic granite. Weathering and marine erosion of this rock provides the quartz-rich sands that formed the sand barriers which lie between headlands. The barrier coast was formed during middle to late Pleistocene time. Thereafter, Sanya bay gradually evolved into a flat sand barrier lagoonal coast that has remained in dynamic equilibrium with the post Pleistocene rise in relative sea level. Rock headlands, offshore islands and coral reefs have sheltered the sandy coast of Sanya Bay from southwesterly waves.

Sediment dynamics in the coastal zone of Sanya Bay are controlled principally by geological and oceanographic elements. This process can be viewed most effectively by subdividing the coast into three compartments based on wave exposure, longshore and tidal current characteristics, and on geological evidence detailed in this paper. The three compartments are i) the east cape of Sanya bay which lies to the south of the harbour (Lu Hui Tou area), ii) the tidal inlet of Sanya harbour, and iii) the sandy barrier shoreline of Sanya bay on the north side of the harbour (see Figure 1).

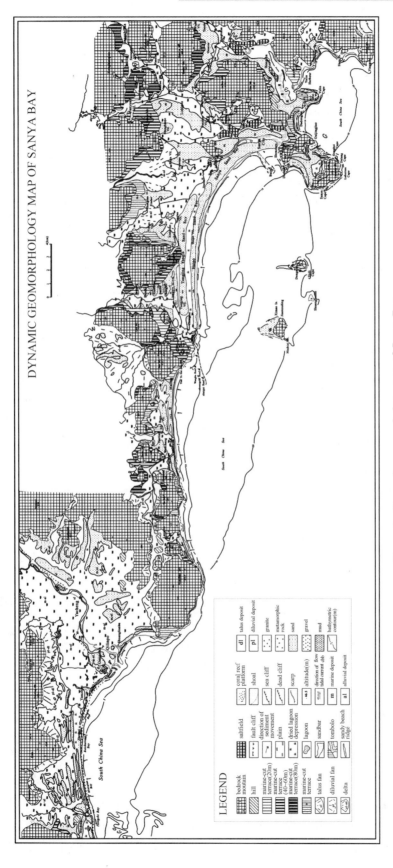

Figure 1 Dynamic geomorphology map of Sanya Bay

1. East Cape Area

This section of coastline is 8 km long and lies between Lu Hui Tou peninsula and Nanbian Mountain. This peninsula protects Sanya harbour from southeast waves. Lu Hui Tou mountain is composed mainly of Mesozoic intrusions of fine to coarse-textured granite that reaches an elevation of 275 m. This bedrock is covered by a dense mountain forest that minimizes soil erosion; this area is a negligible source of sediment for the harbour. The southern part of Lu Hui Tou mountain faces the open sea along a steep cliff coastline. The 20 m isobath at this location lies very close to the shoreline ($<$10 m). Erosion along this steep coastline produces sediment which is transported mostly to deep water areas where it remains isolated from longshore sediment transport processes. Large gravels are deposited in some concave parts of the coastline forming pocket beaches. Residual coarse sediment accumulates along other parts of the coast forming cobble beaches. A coconut forest sand bar connects Lu Hui Tou mountain and Nanbian mountain to form a tie island bar (tombolo) which is elevated about 5 m above sea level. This tombolo is effectively isolated from modern wave action. It consists of a succession of three small sand bars. The material comprising these smaller bars is mainly coral reef fragments and shingles (gravel, small cobbles and coral reef fragments). Excavations and exposures indicate at least a 4 m thickness of reworked coral reef material underlying sandy gravel beach sediments. Excavations for the South China Sea Institute have encountered dolomitized coral limestone below coral reef detritus. The surface of the limestone has a ^{14}C date of 7 900$\pm$145 yBP (sample $\sharp$ ND84089) which suggests that the entire tombolo formed during the Holocene.

The west side of the tombolo consists of a crescent-shaped beach embayment that has a total area of 2 km^2. The bay is about 10 m deep at its mouth. Its landward side is fringed by a coral reef platform that lies in about 1 m of water. The platform is 200 m wide at the south end of the bay and increases to a width of 800 m at the north end. Between the coral platform and the tombolo is a 30 m wide coral sand beach. The beach is protected from SE wind and wave action and is usually a calm water environment. Because it is bordered by the reef platform, there is little or no sediment transport on this beach. In the southern part of the bay, a 0.6 to 1.0 m thick layer of beachrocks situated 0.5 $\sim$ 1.5 m above the high tide line have ^{14}C ages of around 5 485 yBP (ND83090). Xiao Zhou island, a remnant of the west headland of Nanbian mountain, is located at the northern end of the bay. The island is presently connected to the mountain by a rock bench. The peak of the headland is 24.4 m above sea level (ASL) and is surrounded by a 10 m high terrace composed of ferruginous quartzite. Because of intense wind and wave erosion in this area, there is no sediment deposited on the southwest side of this island. On the northeast or leeward side of the island, remnants of coral and gravel beach rocks can be found attached to the cliffs (Figure 2). The northern part of the island faces landward. At the foot of the terrace in this area, exposures of upper beach sediment consist mainly of sand, shingles and coral reef fragments that reach a thickness of 0.5 m. Below the upper beach are reddish weathered deposits formed on the rock surfaces. Coral reef fragments found on the top of the terrace suggest that relative sea level has decreased. Also, since the age of the coral reef complex fringing Hainan island is known to be of Holocene age (based on all ^{14}C date of Nanjing University it appears that the oldest coral reef is about 8 000~9 000 yBP), it may be presumed that terrace formation was a Holocene rather than a modern event.

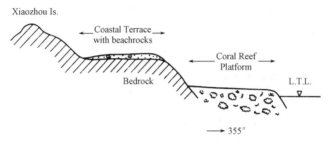

Figure 2　Profile of Xiaozhou Island coast, Sanya Bay

There are several 250 m wide coral reef platforms developed along the northern part of the island that extend northeasterly to connect with the west coral reef platform of Shenzhou Reef (Figure 1). Platforms are composed of solitary coral Shells and coral reef fragments, and of detritus from these sources deposited in depressions. The platform environment supports a large population of brown algae. Water depths over these tabular features are in the 0~4 m range. In 1940, a causeway was constructed to the mainland from the eastern shore of Xiao Zhou island. This causeway was quickly abandoned but has continued to inhibit longshore sediment drift between Coconut Grove bay and Sanya harbour. In 1966, during Nanjing University's initial survey of the area, the foundation rock of the causeway was still exposed. A resurvey in 1986 indicated that all of the foundation rocks were buried by sand to form a sand barrier that now connects the island to the mainland during low tide. This sand bar has two characteristics: it contributes to the protection of the island's foundation from wave erosion and also restricts sediment exchange between Coconut Grove bay and Sanya harbour. Sediment movement in Coconut Grove bay has distinctive characteristics. The coral reef is the main sediment source. Sediments are supplied locally from biogenic sources and from intra-bay sediment exchanges. Small sediment-starved beach cusps occur along the bay's coral sand beach. Thus, there is no sediment supply to Sanya bay from this area despite the evidence of sediment erosion in Coconut Grove bay.

Nanbian mountain is located at the north end of Coconut Grove bay; its headland peak is 176.8 m ASL and its highest peak is 213 m ASL. This mountain consists of

Cambrian and Ordovician dolomitic limestone conglomerates and quartzite. The conglomerates are capped by reddish weathered soil deposits. The upper layer of the conglomerate is weathered and desiccated but the underlying fresh conglomerate can be used as a building stone. This tropical mountain environment has dense ground cover so that very little sediment from this area reaches the harbour. Along the foot of the mountain are ancient sea cliffs that rise to within $5\sim6$ m ASL. The base of the mountain seaward of these cliffs is a rocky bench or wave platform that describes the position of an older coastline. The outer perimeter of the bench presently lies $200\sim300$ m offshore where water depths are typically <0.5 m. Corals have colonized the top of the platform and their distribution marks the extent of the original platform.

Shenzhou Reef lies between Nanbian mountain and the channel leading to the inner harbour. The 1940 Japanese navigation chart shows Shenzhou rock beach extending well to the north with no sediment covering i. e. , the surface is a combination of rocks and pure coral reef areas. The 1961 chart shows that the area from Shenzhou reef to the edge of the Sanya harbour tidal channel is one of sandy shoals that have covered the upper part of the rocky bench. On the outer part of the bench, one of the sandy shoals extends seaward to a water depth of 1 m. This shoal is about 300 m wide from land to sea and 400 m long from west to east. Most of the shoal consists of sandy silt. In 1966, Nanjing University drilled the nearshore side of the shoal. Its upper 0.25 m consists of yellowish coloured shingles made up of shells and coral reef fragments (MGSL, 1986). Some $0.2\sim0.4$ cm diameter pebbles and some rock fragments composed of quartzite and siliceous limestone are also present. The $0.25\sim1.2$ m interval is a grey coarse sand that contains a small percentage of mud and coral shell fragments with long diameters of <2 mm. The next section, between 1.2 and 2.1 m, consists of brownish yellow sandy clay. Most of the sand grains are quartz mixed with some shell and coral fragments. Below 2 m, the sediments are yellowish-brown sticky clays with quartz pebbles. Between 2.0 and 2.4 m, the sediments change to a greyish blue clay with yellowish-coloured limestone fragments. Below 2.4 m is weathered bedrock. According to the morphological evidence, Shenzhou shoal has developed on a wave cut platform and the sediments deposited on the shoal belong to an early stage bay head bar. Both shoals i. e. , the one located under the present Sanya dock and the adjacent one on the south side of the harbour, controlled the outflow from Sanya lagoon and helped to shape the modern Sanya harbour tidal channel.

2. Sanya Harbour

Sanya harbour is, in effect, a small bay located at the southern end of Sanya bay. Bai Pai reef, and its associated northeast projecting sandy tail, constitutes the north side of outer Sanya harbour. The total area of this small bay is 1.4 km². Both the north and south headlands of the outer harbour have been eroded by wave action. As a result, sediments are carried from the headlands to the head of the outer harbour by longshore currents. Sediment from the Bai Pai reef headland is deposited mainly on the leeward (east) side of this feature to form a tombolo that is growing from the reef toward the mainland. On the south side of the outer harbour, sediments derived from Xiao Zhou island are carried primarily towards the outer harbour head to form the juvenile bay head bar that is evident in this area. The hard rock of Xiao Zhou island is resistant to erosion so that waves have generated only a small volume of sediment; a small proportion of this sediment originates from the erosion of the associated coral reef. Because of the small volume of long shore drift on the south side of the harbour, sediments only begin to be deposited when they reach the bay head. At that point, longshore current velocities and wave turbulence are significantly reduced; sediment concentration reaches the saturation level and is deposited. This process has continued up to the present time at a very slow rate.

Evolution of Shenzhou shoal begins with the formation of a bay head bar made up of coral fragments that were deposited on a wave cut bench. The total thickness of Shenzhou shoal sediments is 2. 0~3. 5 m. Most of the sediment is dark-coloured and is rich in H_2S gas which suggests a very low level of sediment transport i. e. , the sediment is probably in a reduced undisturbed state for long periods of time. Based on these observations, it has been concluded that a navigation channel could be excavated through the sandy shoal, that it would not suffer any serious siltation, and that its design depth of 5 m would be maintained (MGSL, 1986). Between 1968 and July, 1985, the siltation in this channel has only amounted to 40 cm. Most of this accumulation probably represents natural readjustments of the channel slopes following the dredging operation in 1966. According to a succession of harbour surveys (MGSL, 1986), maximum siltation reached 60 cm in a small part of the channel. In 1978, the navigation channel was expanded and deepened to 6 m with no apparent increase in siltation over the following 10 years (MGSL, 1986). On the other hand, all of the rocky beach coastline in the southern part of the harbour was developed during early Holocene time. The coral reef platform that surrounds Shenzhou reef has evolved since 7 000 yBP and has reached a thickness of 2~3 m. This implies that the net deposition rate on the platform is between 0. 3~0. 4 m/1 000 yr. The 3. 5 m thick shoal deposit on top of the coral reef platform raises this average rate to 0. 5 m/1 000 yr.

Bai Pai reef forms the northwest side of outer Sanya harbour. It consists of a bedrock high capped by coral that strikes N85°E parallel to Shenzhou reef and Nanbian mountain (Figure 1). It is separated from these adjacent features by the 700~1 000 m wide outer harbour basin. The rock component of the reef is quartzite; it is 300 m long and 4. 2 m above sea level. Water and wave erosion have produced mini-inlets along N23°

W-striking joints. On the west side of the reef, water depth rapidly reaches 7 m; depths average about 5 m on the harbour side. A 1 000 m long N35°E trending coral reef tail is developed on the shore-facing side of the reef. The reef itself has developed on a bedrock high and in a wave shadow environment. Water depth on the reef is <0. 5 m and it is almost completely exposed during spring low tides. There is a 500 m wide pass between the end of the coral reef tail and the mainland coast. Water depth in this pass is < 1 m. In 1966, a study recommended that the pass be filled with stone to decrease current velocities through its opening which would lower the long shore sediment drift from the west (MGSL, 1986). A stone submarine barrier was placed there in 1979 which effectively reduced water depths there to zero.

The inner harbour area is situated along the southeastern end of Sanya barrier bar on which much of the city has been built. It is surrounded by Nanbian, Hu Bao and Hai Luo mountains. The inner harbour area was originally a large wide lagoon. However, with the passage of time, natural siltation and land reclamation have produced swamps and marshes. The area can be sub-divided into three morphological entities:

(1) Mountain river; this environment occurs along the periphery of the harbour. It is characterized by abundant runoff, steep slopes and high river gradients, well developed river channels, and by bottom sediments that are mostly sand and gravels.

(2) Lagoon depression and river channel; this part of the harbour is developed on a gently sloping coastal plain that is dry and exposed for most of the year except during the rainy season. It is an environment that is marked by sudden high velocity river flows (floods) that are responsible for river bank erosion processes that transport coarse material (sands) to the distal edge of the coastal plain. Near this distal edge, mangroves and salt water marsh vegetation trap the sand and inhibit its transport to the lower reaches of the harbour. In the lagoon depression-river channel environment, channels are maintained primarily by river runoff with minimal influence from tidal currents. Freshwater capacity of the mountain and coastal plain environments are estimated from numeric models at $< 2 \times 10^8$ m^3 (MGSL, 1986). Multi-year records of water resources indicate that the average value of total annual runoff from these environments is $1. 7 \times 10^8$ m^3. Each year a total of about 2 420 tonnes of sediment is transported from these two environments to the sea (MGSL, 1986).

(3) Tidal inlet; south of Jin Ji and Hu Bao mountains, there is a second sand bar (Government sand bar) which divides the tidal inlet into two distinctive elongate basins. Dapo channel forms the basin on the west side of the harbour and Yue Chuan channel defines a basin on the eastern side. They join at the southern tip of Government bar to form the lower reaches of Sanya estuary. The estuary is a large water area when compared to the river channel environment and its waters are regulated by the tidal prism. Dynamic morphologies of the estuary differ from those noted coast and river

channels. Because of the critical nature of the tidal dynamic process, it will be necessary to consider this feature of the environment in future city and port development.

Inner Sanya harbour is located at the southern end of Sanya barrier bar adjacent to the mouth of Sanya estuary. Here the tidal channel is 90 m wide. Because of tropical thunderstorms and typhoons during the summer and autumn, runoff is relatively intense and large volumes of water discharge through this channel to the coast. The one in 50 year flood discharge is 2 375 m^3/s. Dapo channel, the larger of the two inner harbour channels, has an estimated discharge under low water conditions of about 0. 3 m^3/s. This low discharge level implies that Sanya estuary is controlled mainly by tidal currents and is, in fact, an arm of the sea. During the flood part of the tidal cycle, the tidal current passes closer to the southern bank of the narrow tidal inlet channel where it has eroded and maintained a deeply incised trough.

The flood tide current moves 7 km upstream, its effect being felt up to the base of Jin Ji and Hu Bao mountains. The tidal circulation of Sanya harbour is of the mixed diurnal type. The semi diurnal component has an average period of 11 days. The flood tide period varies between $8 \sim 17$ hours; the ebb tide duration is $4 \sim 8$ hours. The average tide range is $0.75 \sim 0.93$ m and the maximum tide range reaches 2. 13 m. As a result of the >10 hour flood period and the >2 m tidal range, the entire estuary basin is filled with seawater especially near the confluence of the two tidal channels. The tidal prism of the flood has an estimated volume of 1. 5×10^8 m^3. The ebb tide period is relatively short so that, given the volume of water transported into the tidal channels during the flood stage, the ebb currents flowing through the 90 m wide tidal channel outlet can reach velocities of about two knots (1 ms^{-1}). These ebb currents have scoured a deep channel on the north side of the ebb tidal channel. The two deeply eroded channels formed by the ebb and flood currents join at the narrow tidal channel inlet in the inner harbour where water depths reach 6 m. The maximum water depth in the main tidal channel is about 6. 8 m over a distance of 700 m.

Sediments at the one km wide junction of Dapo and Yue Chuan channels (i. e. the Sanya estuary) is mostly fine quartz sand and clayey silt. Near the eroded part of the river bank, on the western side of Government bar, there are deposits of coarse sand. Upstream, near the Sanya highway bridge, water depths in Dapo channel during flood tide reach 2. 5 m and the bottom sediment is sand. Further upstream from the bridge, water depths decrease rapidly to about 1. 5 m during the flood stage. During this stage, the channel width reaches 400 m. The sediment in this part of the channel is a mixture of fine quartz sand and mud. Near Ronggen village, the channel changes direction from north to northwest. Here there are many sandy shoals and the channel is only 100 m wide. This part of the channel is also characterized by meander features and water depths are still about 1. 5 m during the flood stage of the tide. At Jin Ji mountain, near the upstream end of the tidal inlet environment, the water surface is $100\sim150$ m wide during high tide. Some parts of the channel are 1.2 to 1.5 m deep at this time but are suitable only as a turning area for shallow draft san pans. The bottom sediment along this part of the channel is sandy in texture. The area of the inlet lying to the north of the East Highway bridge is part of the lagoon depression-river channel environment. It is marked by numerous river distributaries and by the absence of a clearly definable main channel. Further to the north, i. e., on the north side of the railway bridge, there are only small creeks of up to 10 m in width. The associated bottom sediment is a dark organic mud that supports a large population of mangrove bushes suggesting that tide water reaches this location during the high spring tides.

The cross-sectional area of Yue Chuan channel is maintained by tidal currents and is relatively wide. The flood tide that courses through this channel can reach up to Hai Luo village on the northside of Hu Bao mountain. Major commercial salt evaporating ponds are distributed along this channel. The success of the pond operations indicates that this section of the channel is not influenced by the freshwater table. There is a plan to develop rice fields between the edge of the salt ponds and the village. No distinctive channels are evident in this area suggesting that there is no possibility of freshwater sediment transport to the harbour.

3. Sanya Bay Sand Barrier

The modern coastline in the vicinity of Sanya city consists of a series of west to east oriented sand barriers that begin in the west near Tian Ya Hai Jiao, across the whole of Sanya bay to the east and that end in Sanya harbour. Their length is about 18 km. These barriers have enclosed a series of former embayments to form the present-day straight sandy coastline. A large portion of the coast of the lowest bar consists of a yellowish, medium to fine sand beach that has a wide surf zone. Most of this part of the coast undergoes a small degree of erosion which results in the development of some sand scarps. The eastern part of the Sanya bay sand barrier starts at Jiucun-airport. Its total length is 6 km and much of this length consists of low, flat sand bars that have an average width of about $600\sim700$ m. The barrier has a maximum width of about one km. The elevation of its associated sand bars is about 5 m; Sanya city is located mainly on these sand bars.

The sediments of Sanya barrier consist mainly of quartz and feldspar minerals. Their texture is predominantly fine sand with a small proportion of medium sand and shell fragments. According to sediment core data, Sanya barrier formed during the Holocene. The upper 5 m layer of the barrier is coarse and fine feldspathic quartz sand. Between $5\sim6$ m, these sands give way to an interval of comparatively large coral fragments. Both of these layers are sand barrier deposits that reflect a former relatively high energy environment. The barrier seems to have formed during the Holocene since the coral reefs that surround the coast of Hainan Island have only colonized this are since about 10 000 yBP following the return to interglacial climatic conditions (as all ^{14}C data from coral reef). The 6～18 m interval is characterized by a greenish muddy clay that reflects marine sediment deposited in comparatively deep water before 10 000 yBP, and which is indicative of a relatively high sea level stand in that time. Between 18 m to a depth of 25 m, the sediments consist of coarse sand that appears to reflect a relatively low sea level stand of late Pleistocene age.

MARINE HYDROLOGY

1. Tidal amplitudes and currents

Along the coast near Sanya city, the tides are of the regular diurnal type. The dominant tidal cycle is 14 days long but this primary cycle is modulated by an irregular semi-diurnal tide having a variable 5～14 day cycle (average = 11 days). Maximum high tide level in this area is 2. 13 m above datum; the mean high tide level is 1. 43 m and the mean low level is 0. 64 m. The maximum tide range is about 1. 9 m and the average range is 0. 79 m. During the November 1988 field operations; the average tide range observed at three monitoring stations in Sanya harbour (No. 37, No. 38 and No. 39; Figure 3) was 0. 82, 0. 84 and 0. 86 m.

Flood tide velocity averaged over many stations in the harbour and bay is 11. 2 cm/s. The maximum current speed (57 cm/s) measured during the flood stage occurred at station 12 at the mouth of the harbour on August 17 - 18, 1985. At this time, the average ebb tide current velocity was 19. 4 cm/s and a maximum ebb tide velocity of 75 cm/s was measured at the outlet of the inner harbour. A maximum ebb tide velocity of 95 cm/s was measured in 1964 (♯2 station). All data collected to date show that ebb current velocity is larger than flood velocity. During the November 1988 field operation, flood stage current velocity averaged 9～10 cm/s at each of three stations (No. 37, 38 and 39) and maximum flood velocities ranged between 22～28 cm/s at these locations. At this time, the average suspended particulate matter concentration (SPM) was 4 mg/L. During the ebb part of the cycle, average flows out of the harbour ranged between 8 ～19 cm/s and maximum ebb velocity varied from 21 to 28 cm/s; average SPM concentration decreased to 3 mg/L.

The duration of the flood stage in the surface water layer is about 12. 4 hours. At mid-water depths, the flood stage lasts for 14 hours and reaches 15. 4 hours in the bottom water layer. The ebb tide is 11. 8 hours long in the surface water layer and respectively 9. 6 and 8. 6 hours in the middle water and near bottom water layers.

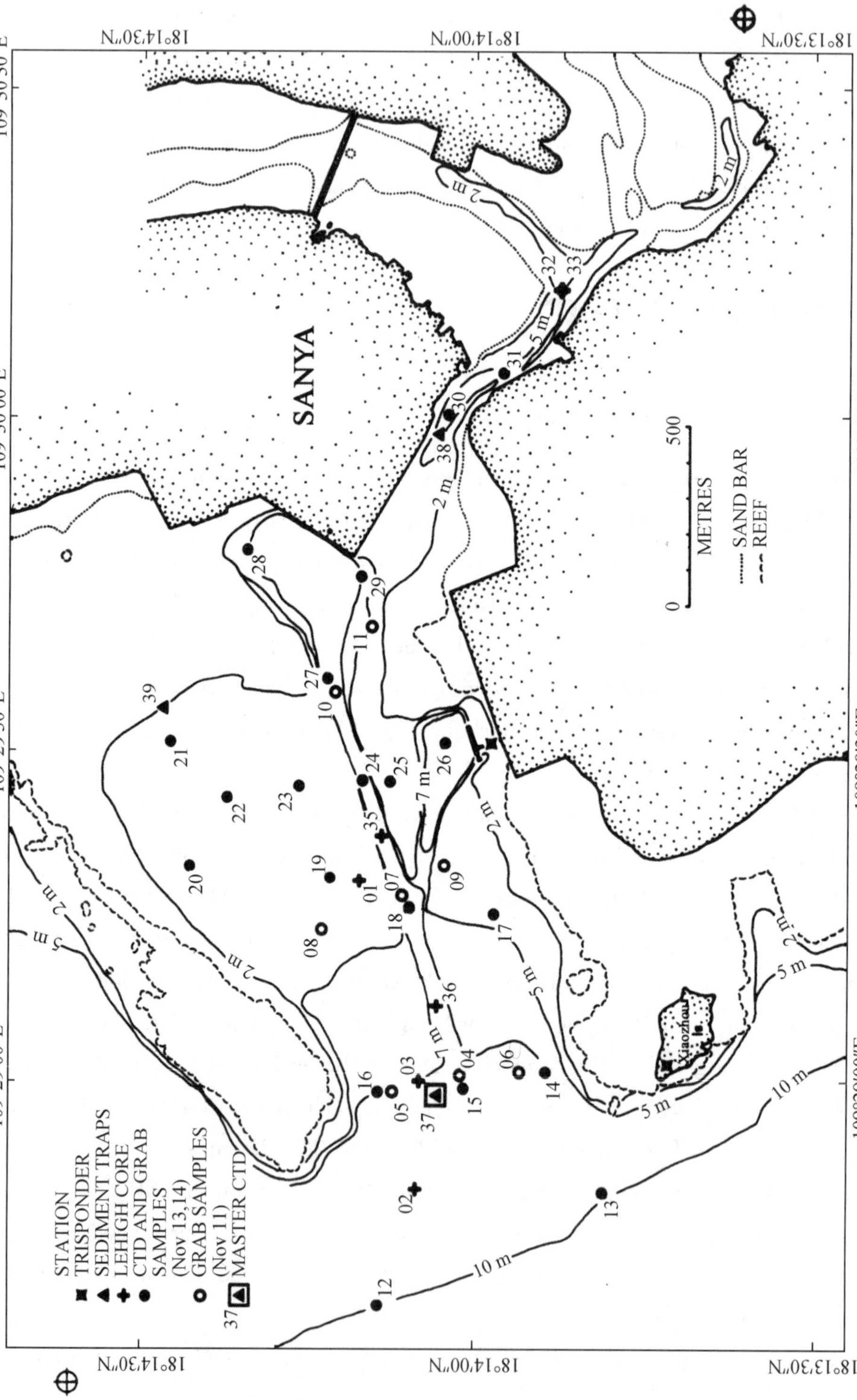

Figure 3　Grab-sample, Lehigh core and water monitoring stations in Sanya Harbour

Differences in tidal stage duration between the top and bottom water layers produces shear current conditions. The volume of the tidal prism measured on August 20, 1985 was 3.9×10^6 m³.

The primary flood current enters the outer harbour on a northeast trajectory; a sub branch of this primary current comprising about 30% of the total flow moves toward the north (Figure 4). A subordinate flood tide current enters Sanya bay from the west passing to the north of Bai Pai reef and then curving around to the south parallel to the outer harbour side of the reef. On this leeward side, ebb and flood surface current flows are always moving from northeast to southwest. In the middle and bottom water layers, flood currents flow in a northeasterly direction (towards the inner harbour) and ebb flow follows a southwest trajectory i. e., in the same direction as outer harbour ebb flow (Figure 5). The inner harbour surface ebb current always moves to the southwest. On passing the inner harbour mouth, it divides into a main component that follows along the south shore of the bay while a smaller component moves northwesterly into the eastern end of the outer harbour (Table 2).

Table 2　Tidal current patterns in Sanya Harbour during November 1988 field operations.

Location	Flood Stage	Ebb Stage
mouth of Sanya inlet(Station 38)	N20°E	S220°W
middle harbour(Station 39)	E110°S	W300°N
outer harbour(Station 37)	S230°W	E115°S~S230°W

The residual current system flows in a seaward direction to a depth of about 10 m. Below this level it divides into two components; one component follows a south-southwest direction and a smaller branch moves towards the northwest around the southern tip of Bai Pai reef.

2. Wave Climate

The wave climate of Sanya harbour is governed by the monsoon wind. Waves generated by northeast prevailing winds are characteristic during the winter season. They produce waves with an average height of about 1. 3 m; maximum heights of 6 m have been recorded. During the summer season, waves approach the harbour from the southwest. Their average and maximum heights are 0. 7 and 2. 0 m respectively. Between September and November of 1985, wave characteristics were measured in Sanya harbour using an ENDECO instrument moored at a depth of 2. 9 m. During this interval, the average and maximum wave period was 5. 5 and 25 seconds respectively. Three typhoons passed through the Sanya area during this survey. The wave climate on these three occasions was characterized by wave heights of <0.5 m (69. 9%) and >1.0 m (12. 4%). Wave periods in the range of 1. 5~7 seconds occurred 85. 6% of the

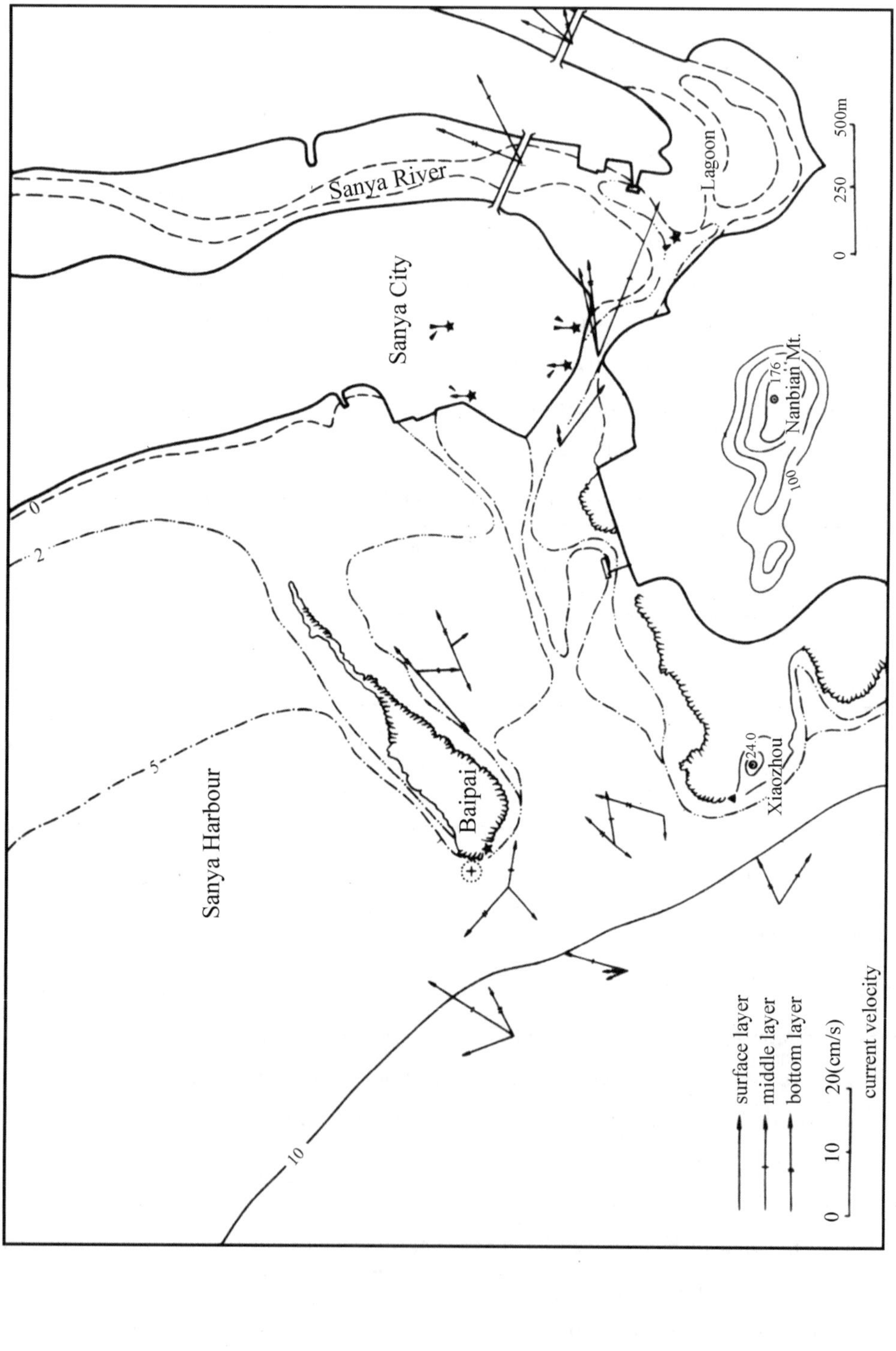

Figure 4 Flood tidal current pattern of Sanya Harbour showing inter-relationships between watermass layers

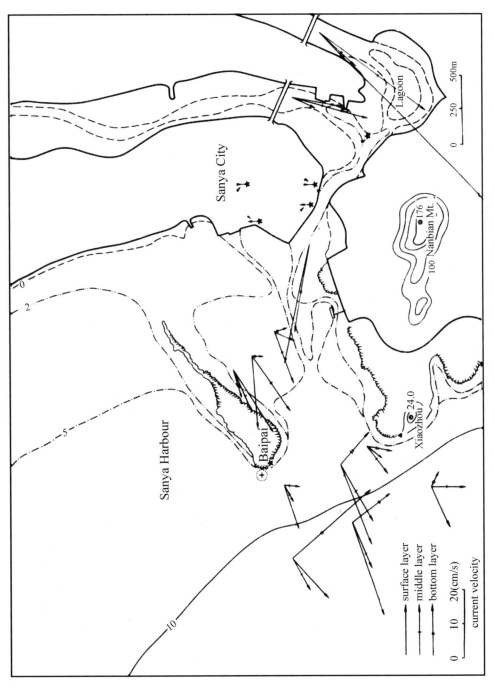

Figure 5　Ebb tidal current pattern of Sanya Harbour showing inter-relationships between watermass layers

time. Swells accounted for 11% of the total wave action; their maximum height was about 0. 5 m and they averaged about 0. 1 m.

SEDIMENT DISCHARGE

The Sanya river is the major fresh water source discharging into the estuary. It is about a 28. 5 km long feature with an average gradient of 0. 029 m/km. It drains an area of about 353 km² comprising mountain and hill environments (75%) and plains (24%). The mountain soil of the river's drainage basin is muddy and comparatively indurated with a neutral pH. It contains a high proportion of clay-size particles and has a porosity of between $42\% \sim 52\%$. Plains sediment is comparatively sandy, unconsolidated and slightly acidic. It has a loose structure and a porosity of between $43\% \sim 53\%$. The river's drainage basin is primarily mountain rainforest (90%) between 800 m and mid-river foothill elevations of $200 \sim 400$ m. The rainforest is colonized by tropical species that provide a vegetative cover of between $80\% \sim 90\%$. This vegetative cover is relatively less dense on the sandy coastal plain below an elevation of 200 m.

The average runoff (R) for the Sanya river can be estimated from measured data (Table 3) using the relationship (according to MGSL, 1986):

$$R=P-E$$

where P is the mean annual precipitation and E is mean annual evapotranspiration.

Table 3　Water discharge data for the Sanya river drainage basin.

Variable	Mountains	Plains
Rainfall	1 478 mm/yr	1 373 mm/yr
Evaporative capacity	850 mm/yr	892 mm/yr
Water excess	628 mm	421 mm
Total water runoff (water excess× area of drainage basin)		
Ground water	0. 534 8×10⁸ m³	
Surface water	1. 679 5× 10⁸ m³	

Sediment discharge into Sanya bay is a function of key parameters such as drainage basin area, vegetative cover and amount of precipitation. On average, the drainage basin has a sediment delivery capacity of 6. 9 tonnes km²/yr which gives an average annual sediment discharge figure of about 2 417 tonnes/yr.

SEDIMENT TRANSPORT THROUGH SANYA HARBOUR AND BAY

The eastern part of the Sanya bay sand barrier was formed during Holocene time

(Wang and Cheng, 1982). The large size of the bar suggests that it was built up under abundant coast sediment supply conditions. One sediment source was from the erosion and redeposition of old sand bar material i. e., sediments derived from the western part of Sanya bar near Shao Qi River. A second source involved sediment from offshore i. e., from the bottom of Sanya bay, and particularly from remnant coastal sediments that are presently located between the 10 and 20 m isobaths. A small volume (10%) of material originated from coastal river discharge (according to Nanjing University Survey data since 1966).

All sediment transport in the Sanya bay area occurs under conditions of southwest and southerly wave action that gradually moves material towards the shoreline to produce the bar. The Sanya sand barrier consists of two parts; the old western part and the more recently constructed eastern part. Based on the spatial configuration of this sand barrier, it is judged to have been relatively stable since the early Holocene because both the Middle Holocene and modern coastlines are still in place (Wang and Cheng, 1982).

Profiles of Sanya bay barrier beaches have been collected since 1965. They show that the coastline at this location is in a state of dynamic equilibrium without significant rates of erosion and deposition and only minimal net erosion over the long term. This dynamic equilibrium state seems to reflect the current existence of an offshore sand source in addition to sediments that are being supplied by rivers. The status of the equilibrium does not appear to have changed significantly since middle Holocene time.

Beach profiles show a small amount of erosion in upper beach environments and net accumulation in lower beach areas. This condition is reversible on a seasonal basis. The entire coastline of Sanya bay is undergoing adjustment at this time. A slight rise in relative sea level is promoting some coastal erosion. Between the Foreign Trade Company offices and Sijiayuan village, immediately west of Sanya city, the coastline has retreated 30 m since 1980. Owing to this retreat, a more clayey sediment of former lagoon deposits has been exposed underneath the bar. This lagoon deposit has provided a municipal freshwater source because precipitation is trapped in the sands which overly impermeable clayey deposits. Coastal retreat has also occurred on the southern side lee wards of Baipai pass. In 1940, the Japanese army set up fortifications along this part of the coastline. By 1965, the foundations of these structures were located at the low tide line (17 m from the high tide line). By 1985, they were under 1 m of water. Erosion has occurred in this area because of the placement of stones on the tail of the Baipai pass sine 1976 thereby preventing sediment resupply of this part of the shoreline by long shore drifting from the western coast of the harbour. However, the artificial submarine groin has only reach 0 m altitude which is still low enough to allow sea water to pass through during high tide.

Coastal measurements indicate that gradients of the submerged slope of Sanya bay is between 1 : 150 and 1 : 400. The steepest part of the bay slope is at its western end near "End of the Earth" cape (1 : 150). Near the Shaoqi river mouth, the slope decreases to 1 : 400 owing to fluvial sediment deposition. The slope in the middle section of the bay ranges between 1 : 250 to 1 : 360. In the eastern part near the end of Sanya bar it varies from 1 : 280 to 1 : 300. All of these slope gradients are consistent with the angle of exposure of the bay slopes to the prevailing wave climate.

Landward of Sanya bay, there is a series of lagoonal depressions that may be the remnants of an older bay. The evolution of the present bay includes a decrease in size and its disconnection from the inner shelf which has caused it to silt up. Only the southern end has remained connected to the sea through the Sanya estuary tidal channel. Most of the lagoon is now filled in; it is presently about 500 m wide with silty surficial sediments overlying older clayey deposits. Li minority people live on the sand bar and use its clay-based depressions for agricultural applications. A 3 m long core drilled through a section of these lagoonal sediments near Hai Po village showed that the upper 80 cm of the deposit consists of brownish-green fine sand and silt that was deposited by fluvial, overwash or aeolian agents. Between the 0.8 and 1.8 m levels, the sediment is fine silt with laminae that alternates with anoxic deposits. From 1.8 to 3.0 m the deposit consists mostly of fine sand of marine origin (Nanjing University, 1976). This sedimentary sequence indicates that the coast in this area was open and exposed to changing lagoonal conditions. The grey anoxic sediment reflects a stagnant environment that may have developed as a consequence of a reduction in tidal exchange between the lagoon and the inner continental shelf. The iron concretions observed in the core between the 1.8 and 3.0 m levels indicate the timing of this closure.

SEDIMENT DISTRIBUTION

Size analysis results suggest that the sediment in Sanya harbour is from local sources (Nanjing University, 1986). The main textural types observed in the harbour environment are clay and silt. These textures stand in contrast to those observed on the east side of the outer harbour which are mainly sand and coral fragments. In the southwest part of the harbour, the sediment is mostly fine sand. This general sediment distribution pattern is consistent with the distribution of wave energy (turbulence) and the pattern of tidal currents.

Heavy minerals included in Sanya harbour sediments comprise 37 species that mainly show the terrigenous source of these materials from the surrounding granite and granodiorite bedrock (Wan & Cheng, 1982, Nanjing University Report, 1986). Minor constituents (e. g. , rutile and garnet) are derived from metamorphic rock that occurs in hinterland areas.

The 1988 field survey provided further information on the spatial distribution of sediment types in the inner and outer harbour (Figure 6). The observed distribution patterns are explainable in terms of hydrodynamics and sediment sources. A sand-silt-clay type of sediment is distributed over most of the outer harbour bottom. This pattern is related to the tidal current pattern which, in turn, can be correlated to current velocity fields. Sandy (fine and coarse silty sands) material is distributed along the pathways of relatively high speed tidal currents e. g., in the pass on the north side of Bai Pai reef and along the leeward side of the reef where there is continual overwash by flood and ebb tidal currents. Textural distribution patterns of sediments in the navigation channel is consistent with the changing tidal current velocity regime in this area of the harbour. Fine sand is distributed over the deeper parts of the inlet channel bottom where current velocities are between $1 \sim 2$ knots ($0.5 \sim 1.0$ m/s). Silty sand is deposited at both ends of the narrow tidal channel where its cross-section increases, and where local bottom current values decrease to <1 knot (< 0.5 m/s). Coarse sand is found near coral reef, offshore island, cliff, and rocky bench environments which are all potentially erodible sources of sediment.

Results to date suggest that the inner harbour end of the navigation channel is becoming shorter and shallower because the original sediment was fine sand while the modern sediment is silty sand that reflects a decrease in hydrodynamic energy. The outer part of the navigation channel has been shortened and straightened by dredging to focus ebb tide currents so that their natural path now follows the channel axis. Originally, most of the channel bottom was covered by silty clay. Today, these areas are blanketed by deposits of silty sand and sandy silty clay. This comparatively coarse sediment appears to reflect enhanced current velocities that have resulted from the dredging activity. Other parts of the harbour lying outside the navigation channel are characterized by increasingly finer-grained sediments that may be indicative of more quiescent conditions, or of a change in the flux of sediments reaching this part of the harbour.

WATER TEMPERATURE AND SALINITY DISTRIBUTION

During the first quarter of the flood stage, surface water temperatures in the tidal inlet channel and in the inner harbour ranged between 24.4 ℃ and 24.8 ℃; the bottom water range was 24.2 ℃ to 24.3 ℃. The base of the thermocline was at a depth range of 2.5 to 4.5 m.

In the central part of the outer harbour, surface water temperatures were in the 24.4 ℃ to 24.5 ℃ range between low water and first quarter flood tidal stages; bottom

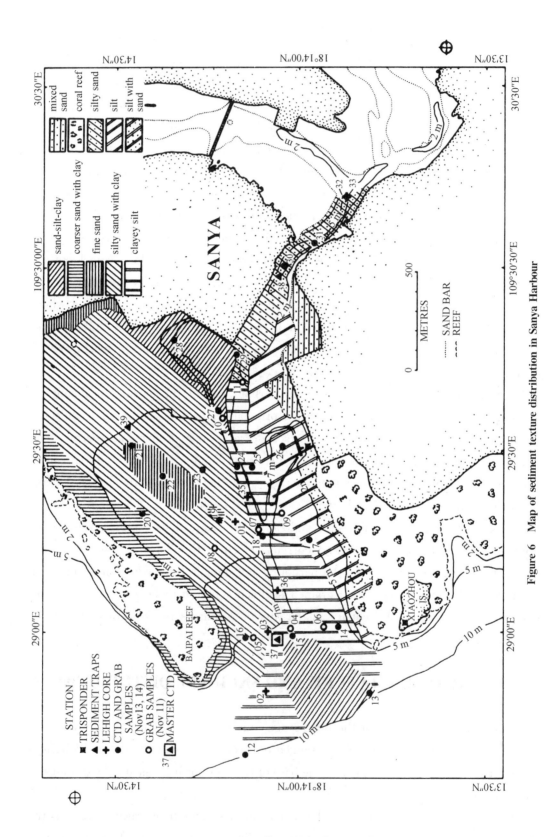

Figure 6　Map of sediment texture distribution in Sanya Harbour

water temperatures varied from 24. 30 ℃ to 24. 35 ℃. Intermediate depth water was the coldest and ranged between 24. 24 ℃ and 24. 30 ℃ (Figure 7) showing representative temperature profiles from inner, middle, outer and inner shelf areas. Outside the mouth of the inner harbour, surface water temperatures ranged from 23. 7 ℃ to 24. 1 ℃. Relatively cold surface water occurred on the south side of the harbour mouth (station 13). Bottom water temperatures in this area averaged about 24. 34 ℃. At low water, the base of the thermocline occurred at about 8 m at station 12 and was at about 6. 5 m at station 13 on the north side of the harbour mouth.

In the inner harbour and tidal channel, the surface water salinity ranged between 31. 1‰ and 32. 0‰ during the first quarter of the flood; bottom water salinities averaged 33. 1‰ on the north side of the channel (station 30) and 32. 9 to 33. 0‰ on the south side of the tidal channel and inner harbour (stations 31 and 32). In this area of the harbour, the base of the halocline ranged between 2. 7 and 4. 5 m deep i. e. , it sloped towards the inner harbour in a classic salt wedge configuration. Near station 30, on the north side of the tidal channel, the base of the halocline was at about 4. 3 m suggesting that the salt wedge was moving into the inner harbour along the south side of the tidal channel and that fresh water was discharging from the inner harbour as a surface layer flow on the north side of the channel. This ebb tide phase lag effect is also suggested by station 37 SPM time series observations (see below).

In the middle part of the outer harbour, surface water salinity range from 31. 3‰ to 33. 0‰. Bottom water salinities between low water and the end of the first quarter of the flood stage varied between 32. 25‰ and 33. 45‰. During this stage of the flood, the halocline had a comparatively steep slope near station 19 on the north side of the navigation channel indicating relatively mixed conditions. At low water, stations on the south side of the navigation channel show a distinctive low salinity surface layer that extends to a depth of about 2. 5 m. The station 19 salinity profile reflects a first quarter flood stage condition and indicates that the low water stratification observed in south side profiles breaks down during the flood as relatively salty water moves into the outer harbour. The low salinity surficial water layer observed on the south side of the outer harbour during low water appears to be consistent with the ebb tide current pattern.

During low water, the surface water salinity just outside the south side of the outer harbour mouth (station 13) is about 0. 6‰ higher than noted adjacent to the north side of the mouth (station 12). Bottom water salinities in this inner shelf environment averaged about 33. 45‰. Major differences in the low water salinity profiles at these two stations are: (a) the base of the halocline is about 3 m deeper at station 12 and (b) the profile of station 12 indicates a distinctive decrease in subsurface (3. 0∼3. 5 m) salinity of < 33. 05‰. These differences appear to mark the cyclonic pattern of the ebb tide circulation near the mouth of the outer harbour.

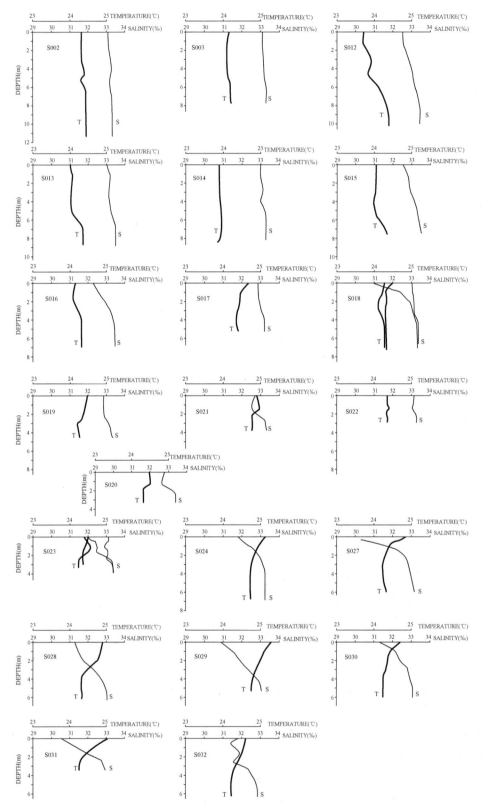

Figure 7　Profiles of salinity and temperature at various locations in Sanya Harbour representing conditions during November，1988（see Figure 3 for station locations）

TOTAL SUSPENDED PARTICULATE MATTER CONCENTRATION

Total suspended particulate matter (SPM) concentrations show a comparatively homogeneous distribution in Sanya bay and inlet. Average SPM in the surface and middle layers ranges between 97 and 99 mg/L. Bottom water values are slightly higher averaging about 101 mg/L (Table 4). In all three layers, maximum SPM concentration occurs during the flood stage about 8~11 hours after the time of lowest low tide. During the ebb, maximum SPM concentrations occur 3.9~4.4 hours before lowest low tide. In general, the maximum SPM concentration lags the maximum tidal current velocity by about 1~2 hours. During the flood stage, the average maximum SPM concentration is about 250 mg/L; during the ebb, this values increases to 345 mg/L. The comparatively high SPM concentrations that mark the ebb component of the tidal cycle reflect "flushing" of suspended particles through the navigation channel. The total annual maximum and minimum SPM discharge budgets for Sanya harbour are respectively 6.33×10^6 and 2.0×10^6 tonnes.

Table 4　SPM Concentrations at Two Stations in Sanya Harbour that were Monitored During 1988 Field Operations.

Station	Tidal Stage	
	Ebb	Flood
Station 1 (located along navigation channel in front of harbour dock)	Avg. 40 mg/L	34 mg/L
	Max. 147 mg/L	
Station 3 (outer harbour, "mouth")	Avg. 47 mg/L	55 mg/L
	Max. 146 mg/L	

In 1988, the SPM distribution in harbour water was measured and monitored in Sanya harbour using an Oregon red transmissometer. This device gives a measure of the attenuation of a 660 mm light source over a 25 cm path. Attenuation is expressed as a percentage where 100% represents the "in air" value. The transmissometer reading (T) is therefore a composite value that reflects, in addition to water colour, both organic and inorganic particle concentrations.

Transmissometer profiles collected between low water and the first quarter of the flood stage show a general decrease in surface water (upper 1~2 m) T values that are indicative of relatively increased concentrations of SPM between the outer and inner harbour (Figure 8). Inner harbour T profiles in tidal channel environments during this stage of the flood tide suggest relatively high and uniform SPM concentrations of between 75% and 85%. In the outer harbour, T values of the surface water layer show a general increase from north to south and have a range of 88% to 92% during the early

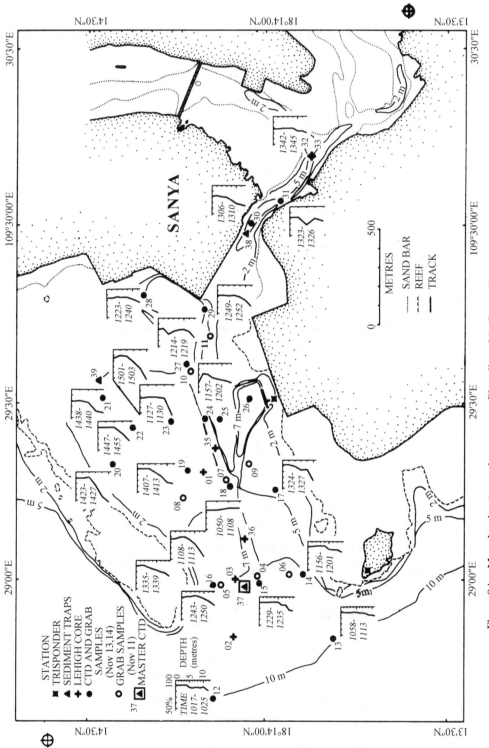

Figure 8A　Map showing transmissometer profile collected in Sanya Harbour during November, 1988.

These data correspond to T/S profiles shown in Figure 7.

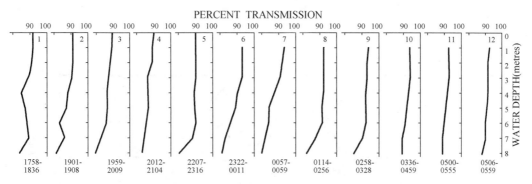

Figure 8B Suite of transmissometer profiles collected over the greater part of the flood cycle of the tide at master station 37 in the outer part of Sanya Harbour.

stages of the flood. Surface water T values between the main dock and the life saving station dock in the outer part of the harbour vary between 66% and 82%. West of the life saving station, T values of the surface water layer are comparatively high and at Station 18, near the western end of the dredged navigation channel, they show about a 6% variation over a 3.5 hour period. Outside of the mouth of the outer harbour, T values of the surface water layer are typically in the 87%~93% range during the first quarter of the flood stage.

T profiles collected in Sanya harbour appear to show a general direct association between SPM concentration and water depth. In the tidal channel environment of the inner harbour, bottom water T values ranged between 60% and 75%. This comparatively high SPM concentration watermass is restricted generally to the lowest 2 m interval of the water column in this area. Relatively high bottom water SPM concentrations also occur at some locations within the northern half of the outer harbour. The general pattern of bottom water T values in the outer harbour suggests a decreasing trend of bottom water SPM concentration from north to south. There was no distinctive pattern of high SPM concentration at mid-water depths during this stage of the tide.

A 12 hour time series of SPM profiles was completed at station S 37 near the mouth of the harbour. This series is representative of the SPM distribution between the first quarter of the flood stage and high water. Data suggest that both the surface water and bottom water layers are characterized by a significant range of temporal variation during this part of the tidal cycle (Figure 9). The surface layer shows a slight decrease in T values between 1 900 and 2 300 hours during the second quarter of the flood stage. Surface layer SPM concentrations are lowest between 0300 and 0500 or about two to three hours before high water. T values of the bottom water layer show a direct association to top layer values with minimum SPM concentrations occurring between 0500 and 0600 just before high water. The decreasing SPM concentration trend of the surface water layer appears to lead that of the bottom water layer by about three hours.

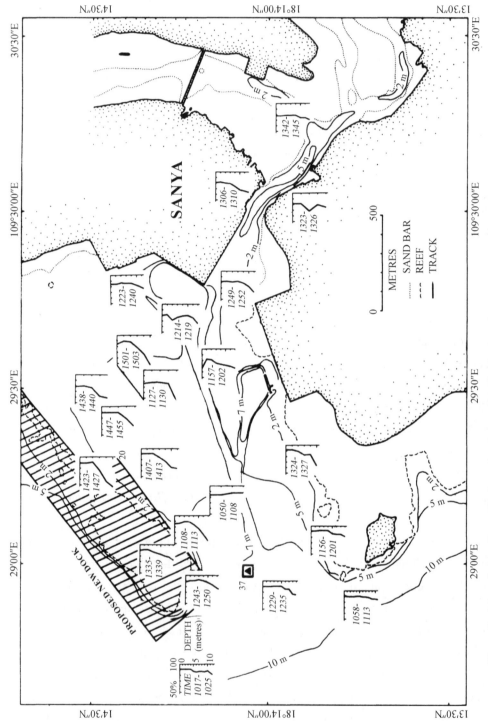

Figure 9 Map of outer Sanya Harbour showing spatial distribution of SPM data in relation to proposed new dick facility. Dredging on the southeastern side of the new dock is expected to alter bottom water SPM distribution.

RESIDUAL CURRENT PATTERNS

In the Sanya harbour, there is a residual flow system with a weak residual current velocity (max. velocity 14 cm/s) in the ebb tide. It flows from the Sanya River mouth to the NW. After entering the harbour area, it turns to the SW, approximately along the south side of Baipai reef is rung out of the Baipai-Xiaozhou section. During the flood tide, it starts from −10 m water depth to N or NWE, goes around the end of Baipai reef to the north side of the reef, and then enters the harbour area from the cross section between the tail of baipai and dock.

SEDIMENT CORE RESULTS

Six Lehigh cores were collected in Sanya harbour during the 1988 field survey. Three of these were chosen to illustrate the shallow stratigraphy of the outer, tidal channel and inner harbour areas. Core S002 is located at the mouth of the outer harbour in a water depth of 5.8 m (Figure 3). It was collected along with the other grab and core samples just after the end of the 1988 typhoon season. The upper 7 cm section is a mixture of brownish-yellow silt and green coarse sand that contains coral fragments. The yellow silts overlie the sands in an unconformable relationship, which was deposited just after typhoon period of Nov. 7 − 8th 1989. Between the 7 and 21 cm levels, the sediment is a muddy greenish grey sandy silt to silty sand. The texture of this layer differs from the subjacent deposit but the boundary between them is gradational. In the 21~100 cm interval, the sediments are greenish grey silty sands. The sand component is coarse with shell fragments. Within this interval, there is a coarse silt layer (68~95 cm) with upper and lower gradational boundaries. Because silty sand is the main textural type near the mouth of the outer harbour, the sediment reflect a dynamic field of wave turbulence action in a semi-closed environment which is sheltered by offshore islands lying to the west, and by Lu Hui Tou tombolo in the south. The silt and mud component of this deposit indicate a nearshore accretionary environment because only a small proportion of this particle size can be transported offshore during heavy rainfalls. Relatively high organic matter concentrations in the upper part of the core are suggestive of sediment transport to the deeper parts of the outer harbour (east to west) probably as a consequence of river runoff processes operating during the typhoon storm period.

Core S035 is located in the middle part of outer Sanya harbour on the axis of the navigation channel and in a water depth of 6.7 m. The upper 13 cm of this core consists of a homogeneous green silt sand with no distinct laminae. Between 13 and 32 cm, there is a homogeneous greyish yellow silt and dark grey clay with an overall silty clay texture. The 32~56 cm interval is dark grey silt clay mixed with a small proportion of sand. The colour of this layer is lighter than that of the superajacent section and includes several distinctive thin and dark-coloured organic-rich layers. Lens structures observed in this interval are predominantly sands. From 56 to 145 cm the core is a greenish grey to grey clayey silt or silty clay. Field tests indicate that this interval has a greater proportion of silt particles than is contained in the superajacent interval. It is a homogeneous layer with several worm tubes throughout that are filled with sandy sediment. However, it was a relatively physically active benthic environment while depositing the sandy layer. Silt is the main textural component in all of the core S035 intervals.

The sediment textures and structures of core S035 suggest the following conditions for the outer harbour sedimentary environment. Firstly, the dynamic condition in the middle part of the outer harbour are different than observed near its mouth. Wave turbulence decreases because of wave refraction and silt is the predominant size of the particles being deposited. Tidal currents are concentrated along the axis of the tidal inlet channel and along certain paths that flow around the bay (Figure 3). Under the circumstance of stronger currents, only sands or silty sand can be deposited. The middle part of the outer harbour is an area of sediment accumulation for material transported out of the inner harbour during the ebb tide. Consequently, the middle part of the outer harbour basin is characterized by sediments having transitional textural features that reflect both inner harbour and inner shelf influences. The comparatively higher clay and organic matter content of the upper part of the core appears to be related to anthropogenic activity. The lower part of the core reflects semi-closed bay conditions with more benthic biological activity.

Core S033 is located in the inner harbour at the lagoon end of the tidal channel in a water depth of 6.4 m (Figure 3). From the top to the 2 cm level, the sediment in this core is a brownish yellow fine sand with soft clay or soft fluid mud showing a recent typhoon storm influence that deposited fine sand in the inner harbour. The yellowish fluid mud moved around during the storm period and deposited after the typhoon passes. In the 2~3 cm interval, thin clay bands with uneven convex surfaces show the effect of periods of erosion during typhoon conditions. Dark brownish-yellow fine sands and very fine sand with a black mottled appearance and H_2S smell comprise the 3~6 cm interval (Figure 10). There is an unconformable surface on both the top and bottom of this interval. Between 6 and 16 cm, there is a greyish-yellow medium to fine sand that contains some coarse shell fragments. From 16 to 19 cm, the core consists of a homogeneous greyish-yellow fine sand layer with comparatively thin laminae suggestive of sediment transport during each tidal cycle under high SPM concentration conditions. The fine sand reflects the dynamic environment of the tidal channel i. e., smaller silt and

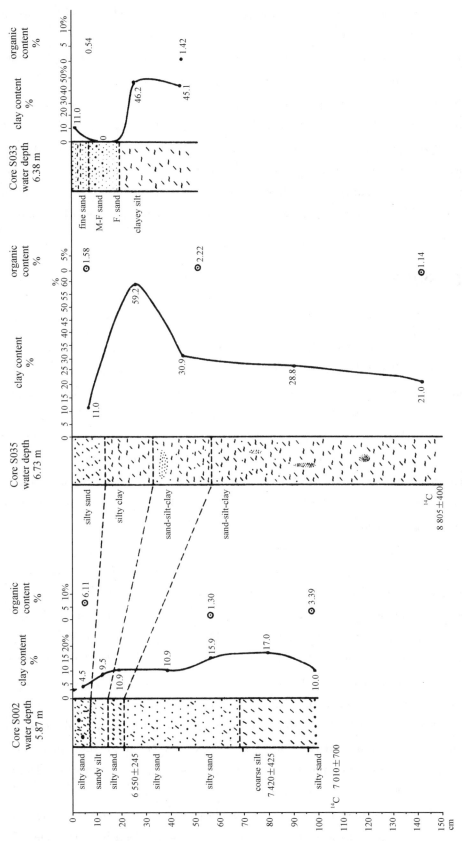

Figure 10 Textural description of sediment sequences observed in Lehigh cores collected at three locations in Sanya Harbour（see Figure 3 for core locations）

clay-size particles cannot be deposited in the deeper parts of the tidal channel because of comparatively high bottom current velocities that are developed throughout this region during each tide. The $20 \sim 50$ cm interval is a grey clayey silt with higher clay percentages in its upper part. Throughout this layer are thin shell fragments having elongate shapes. This interval appears to be illustrative of the dominant type of lagoonal facies before the onset of anthropogenic activity. This paleoenvironment would also have been characterized by relatively low energy hydrodynamic conditions compared to the present day outer harbour which implies a gradual swallowing of water depths in this area. Clayey silt is the main type of sediment in the inner harbour. Gradually decreasing water depths in this area are clearly seen in the core record by the upward change to relatively coarser textures.

DISCUSSION

The intensive flushing effect of the ebb current is the main mechanism for accounting the deep water conditions adjacent to Sanya harbour's main dock and in the navigation channel. The tidal channel/sand bar system is therefore a significant element in maintaining the navigable water conditions of the harbour and should be developed with a view to preserving the effectiveness of this natural process.

The inner harbour serves both as a shelter for the local fishing fleet and for recreational vessels and its shoreline has future potential for the development of seaside recreational areas. However, this resource is presently jeopardized by the unrestricted discharge of human waste and urban pollutants, and by land reclamation that is gradually reducing the volume of the lagoon. About 33% of the lagoon area has been reclaimed to date for salt production from evaporation ponds.

Current recommendations (Sanya Port Bureau, 1986) call for the construction of a large finger pier on Bai Pai reef. This structure would have several advantages for local and international shipping. The most significant benefit would be a doubling of Sanya's harbour capacity. Shelter within the harbour would be enhanced because wave energy from the west would be significantly diminished. On the harbour side of the proposed pier, dredging could be carried out to accommodate 20 000 ton vessels i. e., $9 \sim 10$ m water depths could be created further along the leeward side of the reef to provide additional berths for commercial vessels. The present harbour dock could then be reserved for 5 000 ton ships and the inner harbour would cater to smaller fishing and recreational craft. These recommendations would result in a more cost-effective utilization of the harbour's natural potential.

The three cores described above provide some basic data on the type of sedimentary sequences and structures that were observed in harbour deposits. Particle size distribution patterns reveal a gradual fining from the outer harbour to the inner harbour. Core data also illustrate the nature of the local hydrodynamic gradient i. e. , small to medium wave energy levels and medium to strong current velocities. Evidence of extensive benthic activity is observed in the deeper core sections. This activity decreases in younger deposits perhaps as a consequence of anthropogenic inputs and alteration of the natural sediment transport regime. Anthropogenic inputs to the harbour are manifested primarily by the relatively high organic matter concentrations of the sediments and by the high SPM concentration of harbour waters.

If a coastal profile does not reach a state of dynamic equilibrium, the coastline will remain unstable and erode. Erosion of sediment from foreshore areas leads to further instability which promotes slumping and transport of sediments in a seaward direction. Slumping has not been observed by Chinese scientists who have been working in the Sanya area for over 20 years. This stable situation is also suggested indirectly by the fact that new buildings have been constructed successfully in this area for more than 10 years.

SUMMARY

At present, the general configuration of Sanya harbour can be viewed as an elongate and relatively well-flushed system. Any future developments should be planned with an aim to preserve these characteristics. For example, the construction of docks exposed to ambient tidal currents should be favoured over closed-end channels. Closed end channels are natural sediment traps that maximize sediment transport by density currents that result from SPM concentration gradients between the mouth of the closed-end channel and the main tidal inlet channel (e. g. , Chung and Mehta, 1989).

REFERENCES

Marine Geomorphology and Sedimentology Laboratory, Centre of Marine Sciences, Nanjing University. 1986. *Study on Coastal Environment and Sanya Harbour Development of Baoping-Sanya Bay, Hainan Island.*

Chung, P. L. and Mehta. A. J. 1989. Turbidity-Induced Sedimentation in Closed-End Channels. *Journal of Coastal Research*, 5: 391 – 401.

Wang, Ying and Cheng, Wanli. 1982. Some Problems of Coastal Geomorphology of Sanya Bay. *Marine Science Bulletin*, 1(3): 37 – 45.

Sediment Transport Processes
in Yangpu Bay, Hainan Island[*]

INTRODUCTION

Yangpu Harbour is located on the west coast of Hainan Island(19°43′N, 109°12′
E), 200 km southwest of Hong Kong (Figure 1). The northern part of Hainan Island
borders on a large east-west trending graben feature and Yangpu Harbour lies within the
southern part of an associated trough feature. Between Pliocene and Quaternary time,
3 000 m thick of sediment was deposited in the graben as a shallow nearshore facies
called the Zhanjiang Formation. Since Pliocene time a substantial amount of volcanic
activity has been associated with the tectonic movements that have occurred in this part
of the island. This is evidenced by the basaltic nature of the rocks exposed on the north
side of the harbour. The rocks on the south side of the harbour are part of the Zhanjiang
Formation and consist primarily of gravel, sand, and clay in variable proportions. The
harbour is the topographic expression of the fault that separates the basalt from the
Zhanjiang Formation sediments. The area north of the harbour has been subject to
neotectonic movements from early Holocene time up to the present (Wang and Zhou,
1990).

In 1974, The South China Sea Oceanographic Institute, Zhongshan University and
the Hainan Harbour Administration carried out a coastal geomorphological survey in
Yangpu Harbour which provided many useful data on the harbour environment. Since
1983, the Marine Geomorphology and Sedimentology Laboratory (MGSL) of Nanjing
University was the recipient of a project from Department of Transportation of China for
a siltation study in connection with a deep water harbour development program for
Yangpu inlet. A coastal dynamic survey and a coastal geomorphologic study had been
carried out by Nanjing University between December 1983 and January 1984, September
to November 1984, between January and February 1985, and during the summer of
1987. The work included a regional and coastal zone geologic-geomorphological
investigation, surveys and mapping; coastal and submarine sampling totally 700 samples
including surficial sediment and core samples, marine oceanographic simultaneous

* Dakui Zhu, Ying Wang, John N. Smith, Charles T. Schafer; Ying Wang, Charles T. Schafer
eds., *Island Environment and Coast Development*, pp. 157 – 182. 1992, Nanjing University Press.

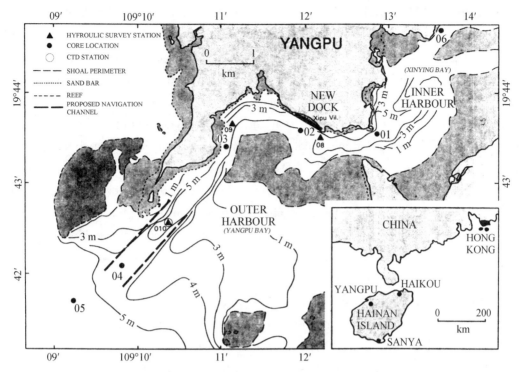

Figure 1　Location map and bathymetry of Yangpu Harbour

section surveys of tidal level currents, suspended particulate matter (SPM), wind and wave action for 40 stations in 15 sections, climate and river data collection and processing, airphoto and satellite image analyses, laboratory analyses on sediment grain size, heavy mineral, chemical elements, trace elements, surface textures of quartz grains, and ^{14}C dating (MGSL). Based on results of these research, Yangpu Harbour has been planned as the largest deep water harbour in Hainan Island. This paper is a product of MGSL's earlier efforts and of recent work carried out under a joint project by Nanjing University and Bedford Institute of Oceanography during 1989—1991. This collaboration was supported by the International Development Research Centre (Canada) and the State Commission of Science and Technology (China).

A joint Canadian-Chinese cruise was undertaken in Yangpu Harbour as part of the Hainan harbours project in November, 1988. A 30 m ferry boat was used for the offshore survey field operation. A support frame was welded onto the stern of the ship to permit coring and water sampling and for towing geophysical gear. Wet and dry laboratory facilities were constructed in the passenger quarters of the ship for core sampling, computer operations and for some preliminary data processing. This eliminated the necessity for a land-based laboratory and permitted the immediate subsampling, inspection and processing of samples at sea. The ship operated in Yangpu Harbour between November 23 - 30, 1988.

Sediment cores were collected using a Lehigh gravity corer. This devices utilized a

10 cm diameter PVC core barrel, which was coupled to a 36 kg core head. A sphincter core catcher was installed in each core barrel to retain the sediment. The cores were raised using a hand-operated 3 ton winch and stored upright on the ship until they were subsampled. In each instance, the PVC barrel was removed from the core head, excess water was drained from the top of the core and the empty space near the top was packed with styrofoam to minimize core distortion during the subsampling operation. The cores were then split along their longitudinal axis using a circular saw and the two halves were laid open on a bench for photography, description and subsampling. The working half of the core was subsampled at 1 cm intervals for the upper 40 cm and at 2 cm intervals below 40 cm using plastic dividers and spatulas. The sampling was performed immediately upon splitting of the core in order to retain as much of the ambient sediment moisture as possible because the downcore porosity profile is critical to an accurate estimate of the sediment accumulation rate. The core subsamples were returned to the laboratory at Nanjing University in sealed containers.

Benthos core liners (1 metre long, 6.6 cm i. d.) were initially forced into the sediment by a diver with minimal disturbance of the sediment-water interface. The core was then hammered farther into the sediment with a "core hammer" to obtain a sample of approximately 60 cm in length. Next, the core was capped and then rotated through a 10 cm radius to facilitate its extraction from the seafloor. A cap was placed on the bottom of the core by the diver immediately upon its extraction and the core was transported to the surface in a vertical orientation.

Three types of grab samplers were used to collect surficial sediment—Shipek, Van Veen and Eckman. Bulk samples were stored in plastic bags and a small subsample of surface sediment was preserved in potassium dichromat for foraminifera analysis.

Sediment traps were constructed by securing a core cap with tape to the bottom of a benthos core liner (6.6 cm i. d., 1 metre long) and attaching a heavy metal rod to the bottom of the trap so that it extended below the trap base. The trap was deployed by a diver by inserting the rod into the sediment until the bottom of the trap was level with the seafloor. The location of the trap was marked with a "high flyer" marker buoy. The traps were deployed for periods of 3 to 5 days. Upon recovery, the trap opening was covered with a core cap by the diver prior to removal from the sediment.

A portable Guildline Instrument CTD (Model No. 8770) was used to measure profiles of depth, salinity and temperature. These data were recorded on cassette tapes as well as manually at 1 m depth intervals for each station. Surface salinity samples were collected using a bucket for calibration of the salinity cell of the CTD. A portable Oregon red turbidity profiler was used to measure SPM at each station. A Raytheon 12 kHz recording echosounder was used to measure water depths at each station prior to each profiling or operation.

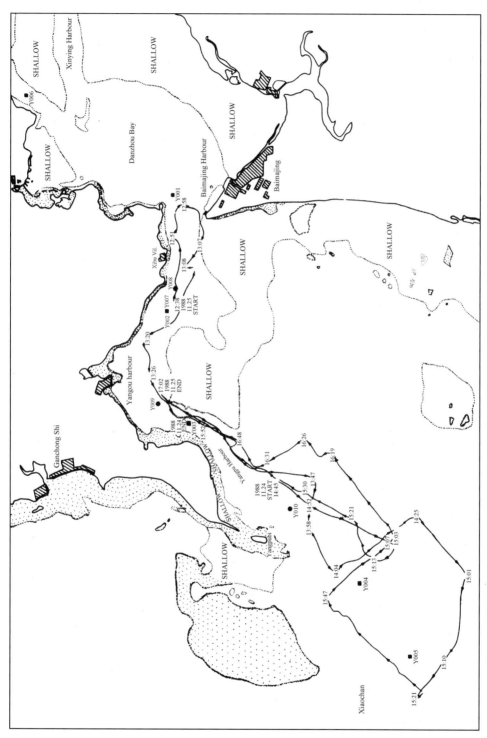

Figure 2　Location of reflection seismic survey lines in Yangpu Harbour

Two litre seawater samples were collected using 5 litre Niskin water sampling bottles. Samples were preserved by adding 0. 25 ml of Dettol and were subsequently filtered at the Nanjing University laboratory.

A Del Norte Trisponder System was employed for navigation in Yangpu Harbour. Two remote stations were established at Baimajing and Yangpubi.

An ORE Geopulse Boomer reflection seismic system was set up on the research vessel and tested by a representative of China-ORES. The system was configured for optimum data recording in the shallow waters of the survey area. A series of lines were run through the harbour area (Figure 2), to define the geometry of sediment deposits, especially in the ebb tidal delta region. A portable EG&G 100 kHz sidescan sonar was also used in the harbour survey.

The locations of stations for the various cruise operations in Yangpu Harbour are given in Figure 1. Eight Lehigh gravity cores were collected at 7 stations in Yangpu Harbour (a duplicate core of Y007 was collected at Station Y002) in an effort to sample the three key types of depositional regimes (inner harbour, navigational channel and "block gate Shoal" (ebb tidal delta)). A core was also collected by divers in the "block gate shoal" (ebb tidal delta) area. Grab samples were collected at stations identified in Figure 1.

Tidal current survey and water samples collection were carried out every hour during 10:00 am. Nov. 27th to 10:00 am. Nov. 28, 1988 at three stations indicated in Figure 1. Salinity, temperature and transisometer water depth measurements were carried out synoptically during the same tidal cycle but only at the station of Y010 (Figure 1). Water samples were returned to the laboratory for SPM analyses. Current meter measurements were conducted in conjunction with the water column temperature, salinity and transisometer measurements. In the laboratory, SPM determinations were carried out by filtering samples through 0. 4 micron filter paper.

A seismic reflection profiles were collected in the harbour along a series of lines illustrated in Figure 2.

Sediment core samples were returned to a ^{210}Pb dating facility which has been established at the Geo and Ocean Sciences Department of Nanjing University. A ^{208}Po tracer was added to the oven dried sediment samples which were then digested in HF and HNO$_3$ in teflon bombs and subsequently redissolved in dilute (0. 5 N) HCl solution. Ascorbic acid was added to the solution and the polonium isotopes were plated out on nickel discs at ca. 90 ℃ over a period of approximately 4 hours. The discs were dried and placed in an alpha spectrometer (Canberra Series 35 multichannel analyzer; Ortec 576A spectrometer with 450 mm^2 surface barrier detectors) and the ^{208}Po activity were determined by alpha counting. Procedures and laboratory equipment were virtually identical to those outlined in Smith and Walton (1980).

ENVIRONMENTAL SETTING

Yangpu Harbour is composed of an outer harbour, called Yangpu Bay, and an inner harbour, called Xinying Bay (Figure 1). Xinying Bay is a tidal embayment with an area of 50 km². The mouth of the embayment, referred to as Baimajing Strait is 550 m wide and has a cross-sectional area of about 5 000 m². Two rivers enter the embayment from the east (Dashui Jiang) and from the southeast (Chun Jiang), respectively. Their combined drainage basin area is 1 419 km². Fluvial sediment, transported into the eastern side of the embayment by the rivers, has formed two distinctive river deltas. During each tidal cycle, the fine fraction of the delta sediment is eroded and transported seaward on the ebb tide. In the vicinity of the deltas, more than 90% of the residual sediment in the tidal channel and lower beach areas is coarser than 0.1 mm. Muddy sediments are carried onto the tidal flats during the flood tide where they tend to be trapped by the mangrove vegetation. The net result of these erosional and depositional processes is a seaward migration of the coastline. Evidence of this effect is found in historical records which indicate that the towns of Zhonghe and Xinzhou were the sites of the harbour 700~1 000 years ago during the Tong Dynasty, but are currently landlocked. Zhonghe is now 5 km from the coastline and Xinzhou is about 4 km from the present shoreline.

The northern part of Yangpu bay has been formed on a basalt platform with the north shore consisting of a series of basalt cliffs. On the south side of the bay, cliffs of the Zhanjiang Formation are presently undergoing a process of net erosion. The main tidal channel (Figure 3) extends from Baimajing Strait to the ebb tidal delta located at the mouth of the bay. It is 10 km long, 400~500 m wide and 5~25 m deep. The bottom sediments of the main channel consist of three distinctive, stratigraphic units. The upper layer is muddy marine sediment. The underlying middle layer is composed of sand and gravel deposited by river currents and the lowermost layer consists of silt and clay size particles typical of Zhanjiang Formation sediments. The main channel is an ancient river valley that was eroded along two major faults in pre-Holocene time. During the Holocene, a rise in relative sea level drowned the ancient river valley. The upper and lower reaches of the channel became the sites of sediment deposition that eventually infilled those parts of the ancient river valley. However, the channel area between Baimajing and the ebb tidal delta was kept open because of daily scouring of deposited sediment by powerful tidal currents.

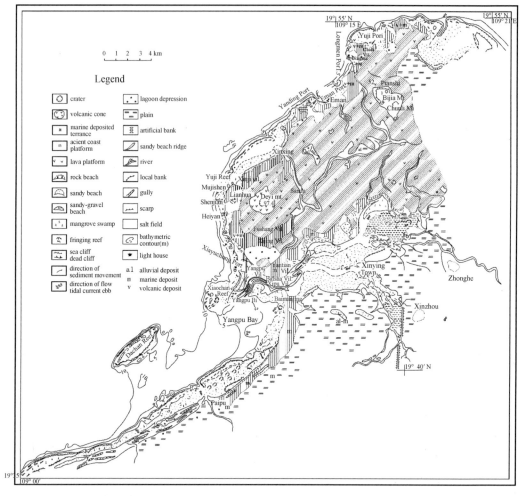

Figure 3　Map of dynamic geomorphology feature in Yangpu Bay

An ebb tidal delta at the western end of the tidal channel is characterized by shoals having a maximum water depth of about 5 m. This shallow area is about 400 m long and $80\sim150$ m wide. Its stratigraphy consists of the same three layers observed in the channel sediments; an upper layer of primarily marine mud ($0\sim3$ m); a middle layer of about 7 m thick consisting of sand and gravel deposited by river currents and a lowermost layer consisting of Zhanjiang Formation sediments (indurated silt and clay).

Block gate shoal (ebb tidal delta) has developed in three stages. Initially, prior to 8 500 y BP, the channel was cut by river currents which eroded the Zhanjiang Formation and deposited the 7 m thick layer of sand and gravel. Secondly, during the Holocene relative sea level rise, the entire embayment (inner and outer harbour area) was flooded. Finally, a tidal inlet, which developed in the bay, deposited marine mud resulting in a gradual shoaling of the ebb tidal delta until it attained its present depth of about 5 m. Core data suggest that this approximate water depth has been maintained over the past 3 000 years with net deposition of only about 70 cm since that time.

A large sand bank area forms the south side of Yangpu Bay (Figure 1). The area of the bank lying above the zero m contour is about 2.5 km². The bank has developed on an erosional platform of Zhanjiang Formation sediment. It forms a thin veneer of about $0\sim2$ m in thickness on the platform surface. Bathymetric records indicate that, between 1955 and 1980, this veneer area increased by about 0.7 km². An area delineated by water depths less than 2 m represents about 40% of the total area of the bay. This area also shows an enlargement of about 0.7 km² between 1955—1980 i. e., an increase of approximately 5%.

The tidal current regime in Yangpu Harbour was characterized through synchronous hydrological measurements made at three water depths at 30 stations and along 10 profiles covering the entire harbour area (MGSL, 1985). The measurements were made three times (3×25 hours) by using Eckman current meters fitted with automatic recorders and water samplers.

Tides in Yangpu Harbour are of the irregular diurnal type and have a mean range of 2 m. During the November, 1988 survey, the measured range was 3.8 m. Observations over several tidal cycles show that the ebb period is shorter (12 hours) than the flood period (13 hours). This difference is reflected in ebb current velocities which tend to be higher than those observed during the flood. The mean flood velocity (V_f) is about 22 cm/s and the mean ebb velocity (V_e) is 27 cm/s. However, the flood and ebb velocities exhibit considerable variation with water depth. During the flood tide, the mean surface water velocity (V_f, s) is about 23 cm/s while middle and bottom water velocities (V_f, m and V_f, b, respectively) attain a speed of 24 cm/s. During the ebb tide, surface water currents (V_e, s) have a mean velocity of 31 m/s, while middle (V_e, m) and bottom water velocities (V_e, b) are 27 cm/s and 24 cm/s, respectively.

During the ebb, tidal current velocities exhibit a westward decreasing trend from the mouth of the inner harbour near Baimajing. At Baimajing, maximum surface current velocities during the flood reach 51 cm/s (Table 1). Middle and bottom water velocities are considerably higher, reaching values of 88 and 84 cm/s respectively. During the ebb, the surface water flows have a maximum velocity of 84 cm/s while middle and bottom water velocities attain values of 86 cm/s and 97 cm/s, respectively. At Yangpu village, maximum surface water flow during the flood stage of the tide attains a value of 42 cm/s. Middle and bottom water flows are 67 and 56 cm/s. During the ebb cycle, surface, middle and bottom water maximum velocities are 68, 53 and 45 cm/s. At the mouth of the outer harbour, in the area of block gate shoal, the maximum velocity of surface water during the flood stage is 57 cm/s. Middle and bottom water velocities during this stage are 49 and 22 cm/s. During the ebb stage, the surface water maximum flow is 54 cm/s. Middle water flow is relatively high reaching a value of 75 cm/s. Bottom water maximum velocity is comparable to surface water values (56 cm/s). All bottom water velocities are above the threshold value necessary for the erosion and transport of silt and fine sand particles. Maximum ebb velocities are generally higher than flood values so that the net transport direction of sediment is from the inner to the outer bay.

Table 1　Maximum tidal velocities (cm/s): Yangpu Harbour

Station	Tidal Period	Surface	Middle	Bottom
Baimajing	flood	51	88	84
	ebb	84	86	97
Yangpu Vil.	flood	42	67	56
	ebb	68	53	45
Ebb Tidal Delta	flood	57	49	22
	ebb	54	75	56

Residual currents represent net current velocities following removal of the tidal component. These currents exercise an important control over the sediment distribution patterns of Yangpu Harbour. At Baimajing, surface water residual currents are 6 cm/s and have a southerly direction (Table 2). Middle and bottom water residual currents at this location are diminished by a factor of 2 compared to surface water values and their direction is easterly and southeasterly, respectively. At Yangpu village, the surface residual flow is 12 cm/s in a northwesterly direction. Middle water and bottom water residual velocities are comparable to surface water values, but their direction is southeasterly. At the ebb tidal delta, the surface water residual flow is 7 cm/s in a southwesterly direction. Middle and bottom water residual flows are 5 and 8 cm/s in southeasterly and southwesterly directions, respectively.

Table 2　Residual current velocities (cm/s)/Direction (deg)

Station	Surface	Middle	Bottom
Baimajing	6/187	3/95	3/102
Yangpu Village	12/301	11/116	12/107
Ebb Tidal Delta	7/230	5/164	8/215

The residual flow of surface water has a large river component and tends to follow a direction parallel to the axis of the tidal channel. During the winter season, there is less coherency of the residual flows between surface and bottom waters.

Wind speed and direction patterns exhibit two dominant modes. The first mode consists of a NNE-ENE component (in a direction off the shore). This is a relatively strong wind component, but it is not associated with large wave generation because of the shelter effect of the land, i. e. there is a short wave fetch with respect to the outer harbour. The second mode is composed of a wind component which has a SSW-WSW direction. Because of the exposed nature of the outer harbour, this wind mode produces the largest waves and is responsible for both wind-driven sediment transport and most of the coastal erosion in the harbour. This wind mode also produces the greatest interference to ships that use the outer harbour as an anchorage. The components of the total wave climate include wind-generated waves (78%) and sea swell waves (12%). The highest frequency of waves entering the harbour is from a southwesterly direction. Waves propagating from this direction are also among the largest that enter the harbour. The average height of SW propagating waves is 0.83 m with an average period of 4.5 s. Because the breaking depth of the 4.5 s waves is only 0.8 m, their greatest erosional impact is on the southwestern edge and on the top of the sand bank. Under SW wave conditions, sediments from the outer strait are resuspended, transported in a northeasterly direction and deposited on the top of the bank.

SPM distributions were estimated through synoptic sampling at several stations and water depths. The concentration of SPM in the water column of Yangpu Harbour has a mean value of about 100 mg/L. Near Baimajing, the integrated (surface, middle and bottom water) SPM concentration during the flood stage averaged 88 mg/L (Table 3). On the ebb cycle, this value increased to 103 mg/L. At Yangpu village, the SPM level during the flood was somewhat higher than at Baimajing and was similar to the ebb value. SPM levels at the ebb tidal delta station were slightly lower than those observed at Yangpu village and were comparable to the ebb values recorded for the Baimajing station.

Table 3 Suspended particulate matter concentrations (mg/L)

Station	Flood	Ebb
Baimajing	88	103
Yangpu Village	106	119
Ebb Tidal Delta	102	102

Salinity and temperature profiles measured during a tidal cycle at a station in the navigation channel are illustrated as a function of time in Figure 4. Transmissometer data appear in Figure 5. These data are summarized below for each component of the tidal cycle.

1. High Tide

During high tidal conditions the current regime is reduced in magnitude and the temperature, salinity and SPM distributions are relatively uniform with water depth. These results indicate that there is minimal stratification of the water column at high tide.

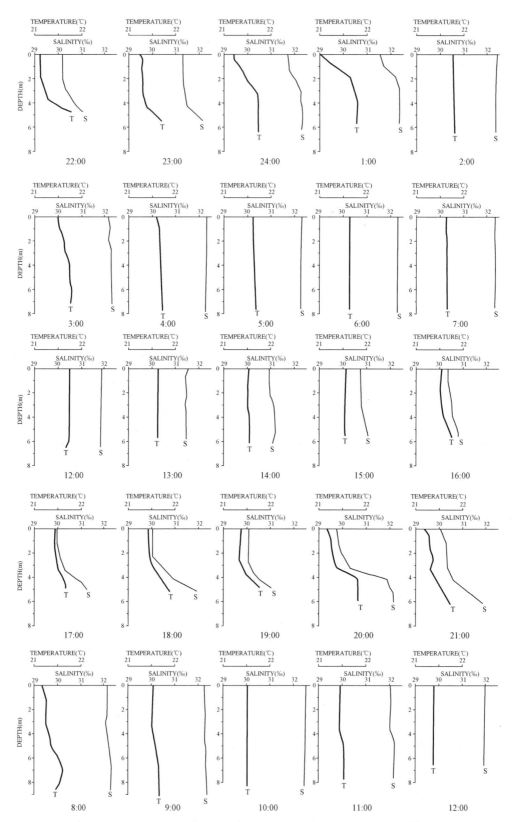

Figure 4 24 hours temperature and salinity data for Station Y010

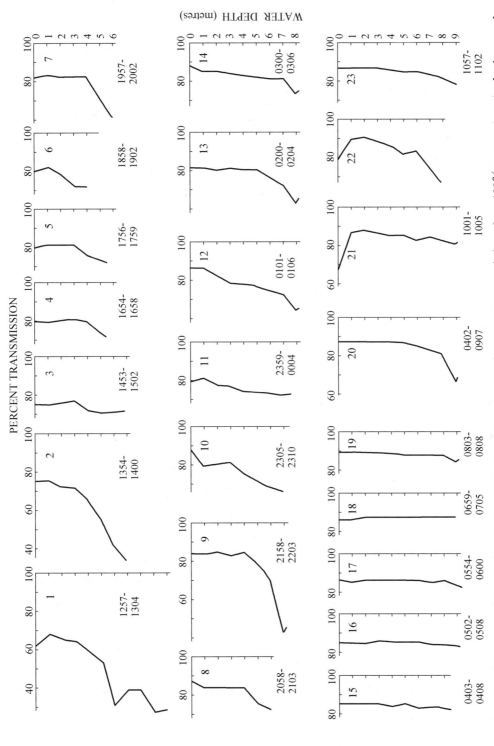

Figure 5 24 hours turbidity data for Station Y010. Results are expressed as percent transmission where 100% represents water having very low concentrations of suspended particulate matter (SPM)

2. Ebb Tide

Water is flowing seaward of the navigation channel and the current is increasing with time. Current direction and velocity are approximately the same at each depth. The salinity and temperature values are also uniform with depth indicating that the water column is well mixed under these conditions. An increase in the bottom SPM levels is evident during this component of the tidal cycle. An increase in surface SPM is also evident which is not reflected by changes in water temperature and salinity (TS).

At mid ebb tide, current velocities increase at all depths to maximum values. This tidal stage generates maximum in SPM concentrations in bottom water (30% transmission). After the mid ebb, bottom currents begin to decrease and are reflected by a corresponding decrease in bottom SPM levels. The surface salinity decreases from 31.8% at mid falling tide to 29.9% at low tide. The bottom salinity only decreases moderately during this tidal stage.

3. Low Tide

During low tidal conditions, current velocities are diminished at each water depth and the current direction is offshore. Salinity and temperature profiles are characterized by a surface mixed layer having reduced TS values. These increase with increasing water depth. SPM values are low and uniform in the surface mixed layer and increase with increasing salinity and water depth.

4. Flood Tide

Time lags were observed in the current velocity field according to depth during this tidal stage. Initially, the bottom current velocity begins to increase in magnitude in an onshore direction while the surface current velocity remains low. Bottom salinities and temperature begin to increase as warmer, saltier water moves in from the bay. The elevated salinity values in bottom water are accompanied by elevated levels of SPM.

Mid-water depth current velocities begin to increase moving water in an onshore direction with no corresponding, significant change in T, S or SPM. Finally, the surface current velocity begins to increase in magnitude but in a direction at right angles to the tidal movement; salinity remains at a value of about 29‰. Current velocities at each depth eventually attain similar values, relatively large in magnitude and moving in an onshore direction. Profiles of salinity, temperature and SPM also become uniform with water depth during the final stage of the flood tide.

SEISMIC SURVEY RESULT

Seismic profiles from Yangpu Harbour have improved our understanding of the geological evolution of this basin. The main tidal channel is superimposed on an ancient river valley. Three valley shaped sections can be recognized at subsurface depths of 9 m, 16 m, and 22 m below the present sea floor (based on a sound velocity of 1 500 m/s). The ancient valley is filled with stratified sediments (Figure 6). In the area adjacent to the Xipu newdock, the modern tidal channel is also superimposed on ancient valley morphologies at subsurface depths of 3 m, 12 m and 18 m (Figure 7). Both profiles suggest that the 16 m and 12 m depths valley were relatively V-shaped compared to their younger and older counter parts. Sediment strata of Block Gate Shoal (the ebb tidal delta) are at least sediments comprising the large sand bank on the south side of the outer harbour are only 2 m thick. These sandy sediment are essentially a veneer that covers the Zhanjiang Formation. The Zhanjiang Formation is about 4 to 10 m thick and is superimposed on an erosional surface.

Sediment Sources: the grain size distributions of surface sediment in Yangpu Harbour indicate that sand-size material is concentrated in offshore and inner harbour regions and a large, sandy tidal flat has formed east of the navigation channel. Clay-size sediment is found primarily on the floor of the navigation channel near the ebb tidal delta. Clay is also deposited in a basin on the inner harbour side of Baimajing Strait.

The sediment sources for Yangpu Harbour have been identified on the basis of mineralogical, particle size and major ion analyses of sediment grab samples (Wang et al., 1987). The three major sources of sediment are, in order of importance, river transport, coastline erosion and coral reef erosion. The Dashui Jiang River enters Yangpu Harbour from the northeast and has a drainage basin of 648. 3 km². The annual discharge of sediment (Ws) from this basin into the harbour is about 7.4×10^4 tonnes/yr. The Chun Jiang River enters the harbour from the southeast and has a drainage basin of 577. 8 km². The annual sediment discharge of the Chun Jiang River is about 6.7×10^4 tonnes/yr. The remaining, smaller rivers which drain into the harbour have a total drainage basin area of about 81. 3 km² and account for an annual sediment discharge of 1.5×10^4 tonnes/yr. The total discharge from all river sources is 15.6×10^4 tonnes/yr which corresponds to a sediment volume of about 6.2×10^4 m³/yr.

Most of the sediment derived from coastal erosion originates from the Zhanjiang Formation along the southern coast of the outer harbour. Each year this source contributes approximately 2.2×10^4 m³ of sediment. The northern basaltic coastline is also being eroded, but at a significantly slower rate compared to the southern coast. Its annual contribution of sediment is about 0.2×10^4 m³. The total contribution of

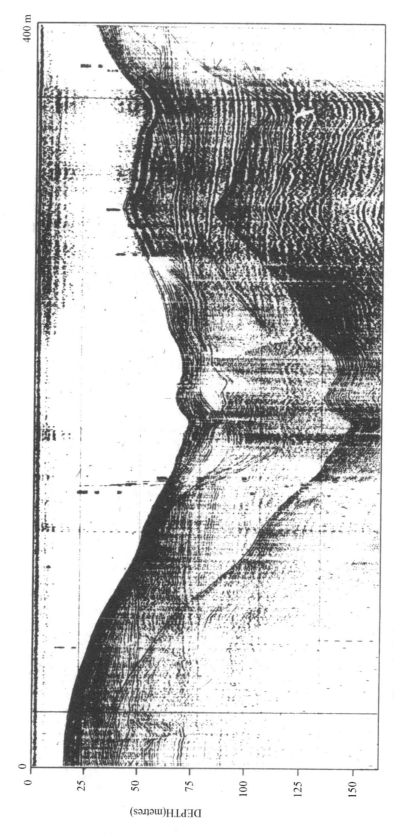

Figure 6　Reflection seismic profile (transverse section) showing sediment infilling of ancient Yangpu River valley

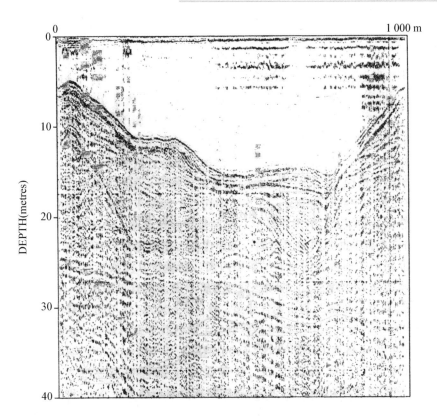

**Figure 7　Reflection seismic profile near the new deep water dock at Xipu
showing channel infilling of the ancient valley floor**

sediment from coastal erosion processes is 2.4×10^4 m³/yr.

　　Coral reefs that have been established on offshore islands, at the northwestern end of the harbour, and at various locations along the south coast are a third source of sediments (Figure 3). Of the two main coral reefs, the closer one to the harbour, Xiaochan which has an exposed area of 2 km², provides a considerably larger quantity of sediment to the harbour basin compared to the larger coral reef (Dachan, area of 4 km²). Sediments in the block gate shoal area contain about 33% $CaCO_3$ while those along the subtidal areas of the south shore contain about 22%~23% $CaCO_3$. The distribution of $CaCO_3$ in outer harbour sediments reflects the distribution of reefs along the coastline. In the inner harbour, the sediments contain only about 0.6%~0.9% $CaCO_3$. The low $CaCO_3$ concentrations in inner harbour sediments indicate that minimal transport of material occurs into this region from the outer harbour. The total amount of sediment eroded from the coral reef is 1 400~4 000 m³/yr. The total transported to the other harbour may be only 1 400 m³/yr. Because the living coral is limited and the growth rate is much more less than 7 m every thousand years. The total amount of sediment supplied to the harbour area from all sources is about 9.1×10^4 m³/yr.

　　Drilled Cores: In 1984, engineering cores of 20~30 m in length were collected at several locations by drill ships from the Guangdong Province Harbour Board. These cores（described below）have been used to reconstruct the geological history of the harbour described in the introduction. The textural features and ^{14}C dates for various core strata are outlined in Tables 4 - 7.

Table 4　Core A - 10

Core A - 10　was collected in a water depth of 3. 43 m on
the southern bank of the channel near Baimajing Strait.

Depth Interval	Description/Interpretation
−3. 43～−20. 79 m	sand-fluvial deposition during low sea level.
−20. 8～−36. 12 m	clay-Quaternary deposit more than 100 000 years old，part of the Zhanjing Formation.

Table 5　Core A - 3

Core A - 3　was collected in a water depth
of 5. 02 m on Block Gate Shoal.

^{14}C Results for Core A - 3

Depth Interval	Description/Interpretation
−5. 02～−9. 41 m	soft mud
−9. 41～−12. 37 m	medium-fine sand
−12. 37～−21. 17 m	sandy mud all of the above units are marine deposits.
−21. 17～−25. 00 m	sand from fluvial sources deposited during low sea level conditions
−25. 00～	clay derived from marine deposits of the Zhanjiang Formation（basically，bedrock）

Depth	^{14}C Date
0. 7 m	1 880±85 YBP
2. 5 m	3 220±100 YBP
6. 1 m	6 350±150 YBP
13. 6 m	8 500±230 YBP

Table 6　Core A - 2

Core A - 2　was collected near Block
Gate Shoal in a water depth of 5. 75 m.

^{14}C Results for Core A - 2

Depth Interval	Description/Interpretation
−5. 75～−10. 45 m	sandy mud—recent marine deposition. The absence of river deposits in this interval indicates that the channel may have changed position and missed this location.
−10. 45～−15. 8 m	cohesive clay derived from Zhanjiang Formation.

Depth	^{14}C Date
1. 3 m	2 930±100 YBP
2. 5 m	4 780±120 YBP
3. 4 m	4 980±120 YBP

Table 7   Core A - 4

Core A - 4   was collected in a water depth of
5. 41 m near the Block Gate Shoal.

Depth Interval	Description/Interpretation
−5. 41～−10. 06 m	mud
−10. 06～−13. 16 m	soft sandy mud
−13. 16～−14. 91 m	Zhanjiang Formation sediments

^{14}C Results for Core A - 4

Depth	^{14}C Date
1. 3 m	2 310±90YBP
3. 0 m	5 010±140YBP
5. 8 m	5 840±140YBP 5 990±150YBP (duplicate analyses)

SEDIMENT CORE RESULTS

Sediment structures in seven cores collected during the 1988 field survey reflect the historical record of circulation patterns and sedimentary processes in this embayment (Figure 8).

Core Y001 was collected from the deep channel of the inner bay; the water depth there is 20. 4 m with rapid current. The core is 35 cm long and consists of three layers.

1) 0～6 cm   dark grey sandy mud with shell fragments

2) 6～15 cm   dark grey sandy mud with coarser sands

3) 15～35 cm   dark grey sandy mud with a coarser sands lenses with mainly shell fragments; continuing deposited layers.

Core Y006 was collected from the main channel of the inner harbour in a water depth of 11. 4 m. The boundaries between adjacent layers are distinguished by very faint unconformities. Sands and shingles mixed with clay particles are typical of sediments deposited in the inner bay channel. The relatively coarse particle composition of these deposits reflects the dominant process of deltaic deposition in this area.

1) 0～2 cm   yellowish silty clay containing fine sand particles. This layer is a mobile fluid mud that is the result of typhoon storm processes.

2) 2～17 cm   shell fragments and quartz sands consisting of fine to medium dark grey sand particles that are interspersed with shingles.

3) 17～32 cm   dark grey medium sand mixed with clay. The clay is distributed as mottles or as thin bands that include a certain quantity of shell fragments.

Core Y002 was collected in the deep channel off Xipu Village; water depth is 10. 22 m and total length of the Core is 3 m.

1) 0～19 cm   yellowish grey "fluid mud" with silt and a dark organic band in the lower part. The fluid mud may have been deposited as the result of a typhoon storm.

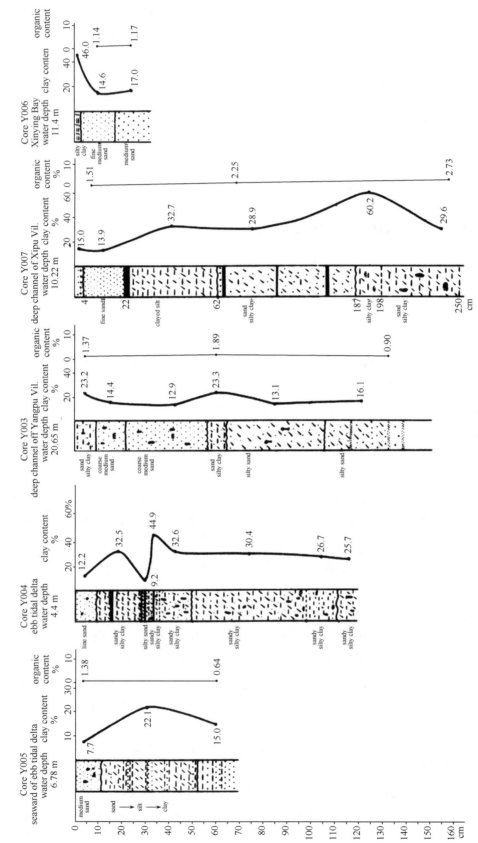

Figure 8　Core logs showing textural facies observed in Lehigh gravity cores collected along the primary axis of Yangpu Bay

2) 19~45 cm　dark grey sandy mud mixed with coarse sands and sea urchin fragments. This layer is unconformable with respect to the layers above and below it.

3) 45~63 cm　dark brownish grey sandy mud, a few coarse sands and shell fragments; the layer is unconformable with respect to the underlying layer.

4) 63~140 cm　generally brownish grey sandy mud; there is a lenticular of coarse sand and shell fragments and a coarse sand layer in the 100~110 cm interval. It may be the worm burrow deposits mosaiced in the mud layer.

5) 140~260 cm　grey mud, pure clayey silt, well sorted, with worm burrow infill deposits of sands and shell fragments. The layer shows clear boundaries with respect to overlying and underlying layers.

Core Y007 was collected in the main channel just west of Xipu Village in a water depth of 11.00 m, and in an environment similar to that of Core Y002. Sarid and clay are the main components of the sediment. These textural modes reflect the dominant bi-directional character of the hydrodynamic processes operating in this part of the inlet. Clay size particles are carried to this location primarily by flood tide currents. Sands are deposited mainly by fluvial and ebb currents emanating from the inner bay. Grading upward from the bottom of the core is a weak textural transition from silty sand to clayey sand. This trend is indicative of a gradual decrease with time in hydrodynamic energies associated with transport from the fluvial processes there, and the possibility of increased flood tide transport in relation to rising relative sea level.

1) 0~4 cm　sediment consists of yellowish grey shingles, sand and silt, some of the coarse fragments are biogenic.

2) 4~22 cm　grey fine sand mixed with silt and shell fragments, there is an unconformable surface between the upper two layers.

3) 22~24 cm　dark reduced band of clay.

4) 24~62 cm　grey clayey silt; the entire layer is a homogeneous deposit.

5) 62~187 cm　grey coloured sand-silt-clay; a homogeneous deposit with laminae. There are three prominent dark bands at the 62~63 cm, 92~93 cm, and 117~119 cm levels. The dark bands represent reduced sediments having a characteristic H_2S smell. Relatively thin fine sand lenses (<1 mm) are distributed in the 178 cm~187 cm interval. These may reflect bioactivity processes. The upper and lower boundaries of this layer are unconformities.

6) 187~198 cm　yellowish grey silt clay with numerous sand lenses formed by bioturbating process.

7) 198～250 cm　grey sand-silt-clay with a large proportion of sandy mottles and lenses reflecting bioturbation activity.

Y003 is in the deep channel off Yangpu villge. Water depth of this location is 20. 65 m. The textural features of the core indicate that both sands and clays are deposited in the channel under different tidal and hydrodynamic energy conditions. Sands (especially fine sands) are deposited under high energy—(ebb tidal) conditions while clays are deposited under lower energy conditions associated with tidal highs and lows. The simultaneous appearance of both sands and clays indicates that the hydrodynamic conditions generally prevailing in the channel are characteristically bidirectional.

The sand component of the sediment becomes coarser from the bottom to the top of the core indicating that the current regime has increased in magnitude in the recent past. Textural results also suggest that current speeds in the deep channel have been greater in the past compared to those prevailing today at this location. The influence of storms is reduced at this location (compared to the Y007 location) because of the enhanced shelter afforded by the coastline to the prevailing and strong NE winds.

1) 0～9 cm　yellowish grey, sand-silt-clay material containing shell fragments and coarse sand grains.

2) 9～22 cm　grey coarse-medium sands mixed with clay characterized by mottled features.

3) 22～58 cm　grey coarse-medium sands with shell fragments.

4) 58～66 cm　grey sand-silt-clay. Homogeneous deposits of silt/clay with a small quantity of coarse and fine sands. Unconformities are evident between each of these two layers.

5) 66～108 cm　grey silty-sands with clay mottling.

6)108～148 cm　grey clayey-sands and coarse sands with intermittent shell fragments. This layer has distinctive, alternating rhythmic bands of coarser and fine sands of about 1 cm in thickness. These bands may reflect alternating times of storms and more quiescent regimes.

Core Y004 is located on the Block Gate Shoal near the seaward end of the tidal channel; the water depth is only 4. 11 m.

1) 0～10 cm　dark grey sandy mud with small shells. Sand grains are fine, and with more clay particles.

2) 10～29 cm　brownish grey sandy-silty-clayey mud with 1 mm thick laminae. The layer has lots of shells and an erosional surface at the top.

3) 29～32 cm　dark grey fine silt deposits with 1～2 mm thick layers.

4) 32～37 cm　yellowish grey sandy-silty-clayey mud with shells and thin laminae

(1 mm or less).

5) 37~51 cm　yellowish grey clayey mud, contains sands and mud; quartz sand shows relatively good roundness, some worm burrows are filled with silty grains.

6) 51~100 cm　yellowish grey clayey silt, with many shell fragments in the lower layer. Some worm burrows are filled with clayey silt.

7) 100~114 cm　dark grey sandy mud with lots of shell fragments and coarse sands. This layer is unconformable with respect to the underlying layer.

8) 114~121 cm　dark grey mud with worm burrows filled with sand and shell fragments.

Core Y005 is located outside of the harbour seaward of Block Gate Shoal; water depth there is 6. 78 m. The core exhibits a coarse to fine grain textural change grading upwards from the bottom which reflects a reduction in tidal current velocity there. An alternating pattern of coarse sand and silt layers reflects the impact of storm events on deposition at this location. The coarse sandy material was deposited primarily during storm events while the finer sands represent material reworked by wave action. Both the silts and clays have been transported to this location from the inner harbour during ebb tidal conditions.

1) 0~12 cm　dark grey sandy mud, poorly-sorted sediment, coarse sands have a random distribution without clear stratification.

2) 12~53 cm　dark grey mixed deposits of sand and clay. Clayey bands are 3~4 cm thick with irregular boundaries. Sand layers are 1 cm thick and contain larger numbers of more shell fragments. Sand layers may be storm deposits.

3) 53~70 cm　clayey coarse sands dark grey in colour. There are small gravels and shell fragments in the deposits. All the material in this core is coarser than in cores collected inside of Block Gate Shoal.

In summary, Yangpu Harbour sediments represent natural tidal inlet deposits and have a bi-modal textural distribution. There is minimal evidence of recent human (anthropogenic) influences evident in these deposits. Textural variations indicate that only minor changes in the depositional regime have occurred since late Holocene time. Sediment textures and stratigraphy of the Yangpu cores reflect the following specific features of the hydrodynamic regime that has prevailed in Yangpu harbour since late Holocene time:

1. Inner Xinying Bay: These sediments are composed mainly of river delta deposits of sand and gravel interspersed with fine silts transported on the flood tide. This is a bi-directional depositional regime characterized by the dominance of fluvial inputs.

2. Tidal Channel: This region is characterized by sand/silt and silt/clay deposition and can also be considered to be a bi-directional, tidal current-dominated depositional regime. Different locations in the channel, having different current patterns, give rise to different textural compositions. Core Y003 reflects a comparatively stronger current speeds resulting in the deposition of coarser sand. Y007 is characterized by its silt component appears to be indicative of a reduced current velocity regime during historic time. Y007 has smaller quantities of sand, but a number of sandy intervals reflecting the greater impact of storm events at this location. In general, the channel deposits can be characterized as tidal current deposits that are occasionally influenced by storm events.

3. Block Gate Shoal: These deposits are composed principally of fine sand indicative of the dynamic balance maintained between wave action and ebb tide current dynamics. They are characterized by silty sand in the sheltered regions and by medium sand, silt and clay in the outer, offshore regions which are affected to a greater degree by storm events.

^{210}Pb analyses were performed at Nanjing University laboratory on sediment samples from these cores. The sedimentation rates at the various core is detailed below:

Inner Harbour

Core Y006 was collected in the inner harbour near the delta of the Dashui River. The sandy deltaic sediments ($>60\%$ sand) in Core Y006 are reflected generally by low ^{210}Pb activities throughout most of this short (36 cm) core. One exception is a 2 cm thick, yellowish silty-clay layer at the sediment surface that has elevated levels of ^{210}Pb (5 dpm/g). This layer represents a mobile fluid mud that has been sorted from the sediments by typhoon storms. The ^{210}Pb profile of this core is consistent with a sedimentation rate in excess of 0.8 cm/y, although there is a large ($>50\%$) uncertainty associated with this estimate.

Core Y001 is also a relatively short (35 cm). Although there was considerable variability in the ^{210}Pb results for this core, they were consistent with a sedimentation rate in excess 1 cm/y. This core also contained high concentrations ($>50\%$) of sand and exhibited several unconformities in sediment texture which could have been caused by high energy storm events.

Core Y007 exhibited alternating layers of sands and clays having distinct boundaries or unconformities. The upper 22 cm of the core contained 55% sand by weight while below this level the sand content decreased to 20%. The excess ^{210}Pb activity decreases exponentially through the clay sediment layers below the 22 cm level of this core. A least squares fit to the log of the excess ^{210}Pb activity profile below the surficial sandy layer in Core Y007 gives a sedimentation rate of 1.47 cm/y.

Core Y002 was collected at the same location (one day later) as core Y007 and exhibits a similar ^{210}Pb distribution, although it had a less pronounced sand layer at the sediment surface. The sedimentation rate determined from the excess ^{210}Pb distribution for core Y002 was 0. 87 cm/y.

Core Y003 was collected from the deepest part of the tidal channel. ^{210}Pb concentrations were lower in this core, compared to cores Y002 and Y007, and the sedimentation rate was also significantly reduced 0. 52 cm/y (Table 8). The reduced ^{210}Pb concentration in this core is consistent with its lower clay content (<20%) compared to that of cores Y002 and Y007 (>30%). Because its location is characterized by rapid currents, fine material bipasses this part of the channel.

Table 8　^{210}Pb sedimentation rates and ^{210}Pb inventories for Yangpu Harbour sediments.

Sample Number	Location	Water Depth (m)	Mean Density (g/cc)	Sedimentary Rate (cm/y)	^{210}Pb Inventory (Dpm/cm^2)
Y006	inner bay channel of Dashui River	11. 4	1. 0	>0. 8	
Y001	inner bay channel near entrance	20. 4	1. 2	>1. 00	
Y007 Y002	tidal channel near new dock of Xichun Village	11. 0 10. 22	1. 0 1. 0	1. 47 0. 87	256 184
Y003	deepest part of tidal channel	20. 65	1. 1	0. 52	29. 9
Y004	Block Gate Shoal	4. 11	1. 1	1. 06	73. 1
Y005	seaward of Block Gate Shoal	6. 78	1. 2	>2. 00	

Core Y004 is distinguished by numerous uniformities and sharp textural changes along its length. Elevated ^{210}Pb activities are associated with the clay layers in the core while significantly reduced ^{210}Pb activities characterize the sandy layers. A sedimentation rate of 1. 06 cm/y can be determined for the clay layers using a linear regression curve-fitting method ($r=0. 873$; $n=6$). Since this sedimentation rate was estimated using only the data in the clay layers, it should be applied only to the deposition of the clay material. To estimate the sedimentation rate in the sand layers, the excess ^{210}Pb activity has been divided by the fractional concentration of clay for that sediment interval (shown as open square data points in Figure 4). A linear regression fit to these data ($r=0. 915$, $n=8$) is represented by the dashed line in Figure 4 gives an excellent fit to the total data set. The resulting sedimentation rate (1. 22 cm/y) is very close to the value 1. 06 cm/y previously determined by omitting the low ^{210}Pb values in the sand layers. The excellent linearity of the upper dashed regression curve in Figure 4 indicates that the sedimentation rate for core Y004, including the net deposition of both sands and clays, has been relatively constant during the past 100 years, i. e. , that the intervals of sand deposition are spaced in time in a manner that results in similar sedimentation rates.

Even in the sand layers of core Y004 that have been normalized for excess ^{210}Pb, the ^{210}Pb values are lower than those in adjacent clay layers. This result suggests that the inputs of sands result in decreases in ^{210}Pb for two reasons: (1) a smaller quantity of ^{210}Pb is adsorbed onto the less particle-reactive surfaces of the sand-size material, and (2) there is, in fact, some very short term increase in the sedimentation rate during the deposition of the sands. Although relatively constant on time scales of a hundred years, sand deposition clearly has a "pulsed" component on the seaward side of the tidal channel.

Core Y005 was collected on the sea floor outside of the harbour. This core has a large sand content (82% sand) in the upper 12 cm and a mean sand content of 50% in the 12～52 cm interval. The core exhibits a coarse to fine grain textural change grading upwards from the bottom of the core which suggests a recent reduction in the current regime.

The high sand content has resulted in low ^{210}Pb levels of approximately 1.0 dpm/g, approaching background ^{226}Ra supported levels throughout the core. A lower limit on the sedimentation rate of approximately 2 cm/y can be established based on the small decrease in the ^{210}Pb activity throughout the upper portion of the core; the uncertainty in this estimate is at least ±50%.

In conclusion, it appears that the sedimentation rate increases gradually from the inner harbour to channel to the block gate shoal and to the offshore sea bottom (Table 8). However, even the largest sedimentation rate of 2 cm/y is still a negligible quantity with respect to navigation channel siltation.

CONCLUSION

1. Sediment Dynamics

(1) The total area of Xinying Bay is 50 km^2. The entrance cross-section of Baimajing is 5 000 m^2 with an average tidal range of 1.81 m. Maximum tidal range from our survey is 3.8 m. Thus, there is 100～200 million m^3 tidal prism. The mean surface water tidal current speed is＞1 m/s and the bottom water mean current speed is＞0.25 m/s at the narrow entrance of Baimajing. The large volume of the tidal prism and the level of tidal current speed are important in avoiding siltation of the navigation channel.

(2) The speed of the flood current is smaller than the ebb current; flood tide period is longer than the ebb tide by two or three hours. The flood tidal current approaches Yangpu Harbour through the outer harbour sand bank and then enters the Xinying Bay. In the Bay, the current divides into two branches that follow two channels. The main flood current is in the south branch and the main ebb flow is in the north branch. The ebb current (3/4 of water quantity) flows out mainly along the deep channel, and the remainder passes through the outer harbour sand bank to return back to sea.

The velocity of the ebb current decreases from the inner harbour to the outer bay, but it is still powerful enough to transport sediment out of the harbour.

(3) The concentration of SPM in Yangpu Harbour is less than 0.1 kg/m^3. It consists of 67% of quartz in flakes; 15.2% of the SPM consist of microbes, biofragments and flocculent carbonate, which forms a micro-conglomerate of particles having diameters less than 0.008 mm, the other components of SPM are limonite, basalt, pyroxene and zircon fragments.

2. Sediment grain size distributions indicate that river discharges are the main sediment source of the inner harbour deposits, and of the coarse sediment deposits in the inner bay, the channels (sands are 80%), and the shoals between the channels (50%~70% sands). Fine sediment composed of silt and clay-size materials, are transported out of the inner bay. Sediments deposited in the deep tidal channel are mainly clayey silt. Clay content is about 40%~50% in the channel on both sides of the shoal and gradually decreases towards the top of the shoal. The sand on Block Gate Shoal are 33% of CaCO$_3$ which indicates that their source is Xiaochan coral reef.

3. According to geological historic data, including ^{14}C, ^{210}Pb and bathymetric comparative studies, the average sedimentation rate of Block Gate Shoal is 0.16 cm/y since 8 500 YBP, and has been about 0.1 cm/y since 3 000 YBP. These data fit the rate of world sea level rise. Consequently, the natural evolution of Block Gate Shoal is a slow, steady-state process. In the past 50 years, both the deep channel and Block Gate Shoal have been in a state of dynamic balance as several years suffered from erosion such as 1.2~5 cm/y in the deep channel and 2~10 cm/y on the Block Gate Shoal during 1947—1974, but with accumulated process during 1974—1983, the depositional rate is 5 cm/y in the deep channel and 2.2~4.5 cm/y in the Block Gate Shoal. The changes are little without significant influence to the harbour. In conclusion, the natural environment of Yangpu harbour is excellent for setting up a deep water harbour to handle ships in the 50 000 to 100 000 ton Class.

References

[1] Wang, Ying and Zhou, Lufu. 1990. The volcanic coast in the area of Northwest Hainan Island. *Acta Geographica Sinica*, 45(3): 321 - 330.

[2] Marine Geomorphology and Sedimentology Laboratory (MGSL), Nanjing University. 1984. *Survey Report on Geomorphology of Yangpu Harbour, Hainan Island*.

[3] MGSL, Nanjing University. 1985. *Research Report on Sediment Provenance and Siltation of Yangpu Harbour, Hainan Island*.

[4] Smith, J. N. and Walton, A. 1980. Sediment accumulation rates and geochronologies measured in the Saguenay Fjord using the ^{210}Pb dating method. *Geochim. Cosmochim. Acta*, 44: 225 - 240.

A Comparative Study on Harbour Siltation and Harbour Development, Hainan Island, China*

INTRODUCTION

Hainan Island is the second largest island in the South China Sea. It extends from latitude 18°9′ to 20°11′ North and longitude 108°36′ to 111°3′ East with an area of 33 920 square kilometers. The distance of the island from north to south is 180 km, and 258 km from east to west. The total length of the coastline is 1 528 km that includes about 70 embayments. For this size of island, the major transportation for contacting the outside world is by sea (cargo and passenger) and by air (passenger and fresh seasonal products). Of all the embayments, tidal inlets are the best for setting up deep water harbours (Fig. 1).

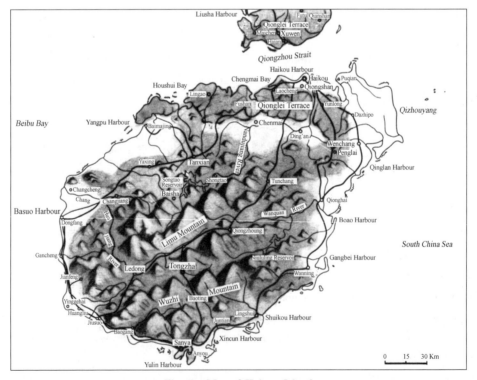

Fig. 1　Map of Hainan Island

---

* Ying Wang: Ying Wang and Charles T. Schafer eds, *Island Environment and Coast Development*, pp. 373 - 391, Nanjing: Nanjing University Press, 1992.

Three harbours, Sanya(三亚), Yangpu(洋浦), and Haikou(海口), have been chosen for comparative studies, as these three are of economic importance to the island and are undergoing development that can benefit from new understandings on how these inlets work. Haikou Harbour is the functional gateway to the capital city (Haikou), and plays an important role in linking the mainland and as the oldest harbour of the island. Haikou as literally "sea mouth" is so named because it is located in the delta of the Nandu Jiang River. The harbour itself is located in a bayhead, leeward of the modern river mouth, and faces the Qiongzhou Strait(琼州海峡). Sanya and Yangpu are located in natural tidal inlets. Sanya, in the south, is a medium rank commercial harbour for landing 1 000 to 5 000 tonne ships. It has great natural potential as a deepwater berth for 10 000 to 50 000 tonne ships in its outer harbour, and can be combined with a modernized marina in its lagoon. Yangpu is located on the northwest coast of the island. It is the largest deep water harbour and can accommodate 20 000 to 50 000 tonne ships; it is still under construction. Yangpu can be developed for multipurpose use by combining the large outer harbour for commercial traffic and for fishing vessels working in the Beibu Gulf area of the South China Sea.

A tidal inlet is primarily an arm of the sea that extends inland. There is usually only a small amount of fresh river discharges into the inlet. Thus, the tide represents the major dynamic process of the inlet. Three types of tidal inlet have been recognized in Hainan Island according to their genetic pattern:

(1) Tidal inlets developed along structural fault zones, where there is also a boundary zone of weak volcanic rock and old sandy deposits. Yangpu Inlet and Dong Zhai-Pu Qian Inlet are developed along a modern earthquake fault zone and subsidenced consequently.

(2) Tidal inlets formed by sand barrier and lagoon systems which have enclosed former bedrock embayments. This is the major type of the tidal inlet developed along the southern and eastern coast of Hainan (e. g. , Sanya, Xincun, Gangbei(港北港) and TieluGang(铁炉港)).

(3) The Holocene transgression flooded mountain and river valleys. Yulin and Qinglan bays belong to this type of inlet.

All of Hainan Island's tidal inlets are good for setting up multipurpose harbours. Because of the large tidal prism of the inlet, which can produce powerful currents capable of flushing out sediments and maintaining adequate water depth, maintenance dredging costs are typically low. Tidal prism size is controlled by tidal range and by the duration ratio of flood and ebb tides. A large tidal range or a longer period of flood tide brings huge quantities of sea water which produces powerful currents during the relatively short duration of the ebb tide. For minimum siltation, it is desirable to keep or extend the area and volume of the inlet for maintaining the flow of large quantities of sea water.

COMPARISON

1. Sanya Harbour

Sanya Harbour is located at the southern end of Hainan Island near 18°14′06″N and 109°29′58″E; it is the nearest harbour to international navigation routes of all the Chinese harbours in the South China Sea (Fig. 2).

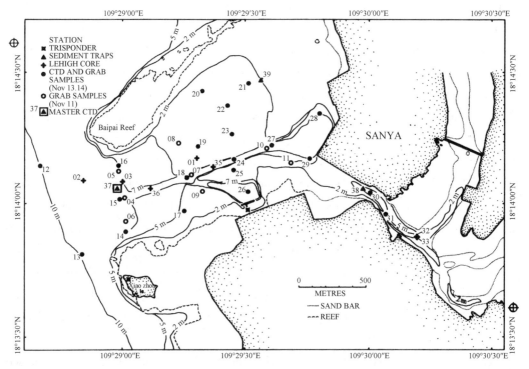

Fig. 2　Map of Sanya Harbour

Main docks are set up on the southern end of a flat sand barrier, where the mouth of Sanya Inlet is located. The harbour is well protected from wind and wave attack. A mountain range, which is 400~500 m above sea level, obstructs the northeasterly gales of winter and spring cold waves, or of typhoons during summer and autumn. Luhuitou Peninsula(鹿回头半岛), on the east side of the harbour, stretches 4 km out to the sea. The peninsula consists of a coral reef tombolo tied between the southern mountain of Luhuitou (275 m above sea level) and Nanbian Mountain(南边岭)(176.8 m) on the north. The peninsula completely protects the harbour from wind, wave, or sea swell approaching from the open sea to the east and south. Two islands, Dongmao(东瑁) and Ximao(西瑁) are located offshore in a water depth of 20 m on the west side of the harbour. Two reefs, Baipai(白排) and Da Zhou(大洲, also called Xiao Zhou 小洲) located along the west side of the harbour form two ranks of obstruction to waves approaching from the south and southwest, i. e., from the prevalent wind/wave direction. Thus, Sanya Harbour has the larger water area characterized by low wave energy. The average wave height is less than 0.5 m and the wave period is shorter than 4 seconds. An instantaneous maximum wave height reached 3 m during a typhoon that passed the local coast in 1985, and which had an influence over both the outer harbour and the anchorage. However, the cumulative frequency of 1 m high waves is only 12.4%, and the frequency of wave height less than 0.5 m is 69.9% (MGSL, 1986).

The tides in Sanya Harbour are a pattern of mixed diurnal types with an average of 11 days of semi-diurnal tides per month. The duration of flood is 8~17 hours, but the ebb tide lasts only 4~8 hours. The average tidal range is 0.79 m and the maximum tidal range is 1.89 m. Maximum velocity of flood currents is 57 cm/sec, and reaches 75~95 cm/sec for ebb currents; the average velocity of the flood current is 11.2 cm/sec, and 19.4 cm/sec for the ebb. The maximum flood current appears 8 to 10 hours after low tide, and the maximum ebb current appears 2 to 5 hours before low tide level. The longer period of flood tides and the characteristic tidal ranges causes a tidal prism reaching 1.5×10^6 to $44 \times 10^6 \, \mathrm{m}^3$ per tide. (Fig. 3)

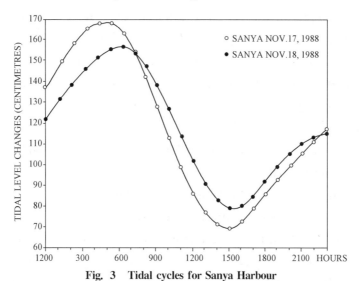

Fig. 3　Tidal cycles for Sanya Harbour

During the relatively short period of the ebb tide, current speed is 100 m/sec, which is twice than that of the flood current speed. The powerful ebb current is the main factor in forming and maintaining a 700 m long and 6~9 m deep channel in the narrow pass at the southern end of the sand bars. By following the natural ebb current pattern, a dredged navigation channel has been naturally maintained in this harbour since 1967 without any serious siltation problems.

Suspended particulate matter (SPM) is lower in the tidal current water mass of Sanya Harbour. According to measurement made in 1985, the average concentration of SPM in the tidal current water is only 0.99 g/L. Maximum SPM concentration of flood water is 0. 250 g/L appearing at 8. 8 to 11 hours after low tide. Maximum SPM concentration of ebb tide water is 0. 345 g/L appearing 3. 9 to 4. 7 hours before low tide. Maximum SPM concentration is associated with maximum current velocity concurrently or up to 2 hours later. Because the concentration of SPM is determined by tidal current velocity, the SPM volume of ebb currents is larger than that of flood currents indicating that sediment is carried completely out of the inner harbour by the fast ebb current. This process is also evidenced by the SPM concentration distribution which shows a high concentration trend of SPM toward the outer harbour along the main navigation channel. Survey data indicates that the maximum daily SPM discharge is 8 328 tonnes and the total annual SPM discharge ranges between 6. 33 million and 20 000 tonnes (MGSL, 1986).

Sediment supply to the outer harbour is minimal because of minimal coastal erosion in the vicinity of the harbour; fringing reefs have blocked long shore sediment movement from south. A sandy coast in the northern part of the outer harbour suffers slightly from erosion by southwesterly waves from the open sea. However, the long shore sediment supply has been prevented from entering the harbour from the west by a submarine jetty built up in the 1970's to connect Baipai Reef with the shoreline; the height of the submarine jetty is $\pm$ 0 m above local sea level. Therefore, the outer harbour is well located and without serious sediment inputs and wave turbulence.

The inner part of Sanya Harbour consists of a narrow inlet and lagoon depression. River discharges from inland mountain ranges are minimal ($1. 7 \times 10^8$ m³ runoff annually). The annual sediment discharge is about $2. 4 \times 10^3$ tonnes which is deposited mainly in the inner bay head. Fresh water input is actually only significant during the summer season; Sanya Harbour is maintained by sea water.

Sedimentation rates in Sanya Harbour have been estimated from core sample analyses and from bathymetric monitoring. Core samples from six cores were collected in the harbour, mainly along the navigation channel from the outer harbour to the inner lagoon (Fig. 4, cores S002, S035, S033). They have been dated using both by ^{210}Pb and ^{14}C. The results show the same trend of sedimentation rate increases from the outer harbour to the inner harbour but the net value is small, only 1. 0 cm more in the inner harbour. According to ^{210}Pb data, rates range from $0. 26 \sim 1. 24$ cm/yr in the outer harbour, and less than 0. 2 cm/yr outside of the outer harbour (anchorage area) according to ^{210}Pb data (Table 1). Sedimentation rates calculation from ^{14}C data are even smaller than the ^{210}Pb results; $0. 016 \sim 0. 018$ cm/yr inside of the outer harbour, and $0. 003 \sim 0. 014$ cm/yr outside of the harbour. Because sediments have been compacted since their accumulation, and because there was no influence from human beings as the harbour did not exist 4 000 years ago. The inner harbour has more sedimentation indicating that wave and current energy decreases there.

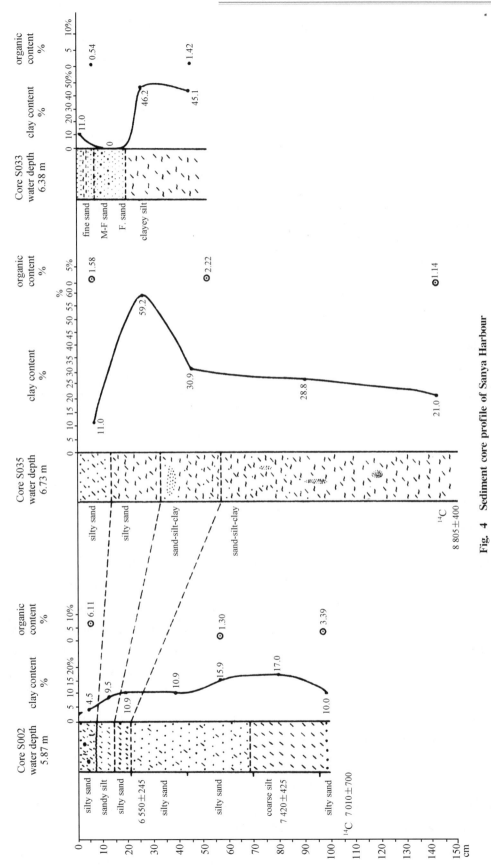

Fig. 4　Sediment core profile of Sanya Harbour

Table 1　Sedimentation Rates: Sanya Harbour

Core #	Location	²¹⁰Pb results		¹⁴C results		
		Sedimentation Rate cm/yr	Sediment Accumulation Rate g/cm²/yr	Interval of the core (m)	¹⁴C date yrBP	Sedimentation Rate cm/yr
S002	Anchorage-outside of harbour	0. 18	0. 22	0. 25 0. 85 0. 96	6 550±246 7 420±425 7 010±700	0. 003 0. 011 0. 014
S003	Harbour entrance	0. 38	0. 46			
S036	Outer harbour-navigation channel	1. 24	1. 24			
S001	Middle harbour-outside of navigation channel	0. 26	0. 31	0. 37～0. 57	3 395±235	0. 016
S035	Middle harbour-navigation channel	0. 91	0. 91	1. 46	8 305±400	0. 018
S033	Inner harbour-end of the deep channel	>1. 0				

Bathymetric Monitoring Result (cm/yr)

Section	Period of Survey	
	1987—1990	1990—1991
Outer harbour channel from No. 2 light buoy to the west	30	30～40
Navigation channel between No. 2 and No. 3 light buoy	30	20～30
Navigation channel from No. 3 light buoy to the east	<10	10

Hainan Province was set up in 1988, and Sanya County expects to be one of the three major urban areas of the island. Many buildings including hotels, fishing docks, an ice-making factory and cold storage facilities have been constructed along the shores of the inner lagoon and coastline around Sanya Harbour. These developments have caused an increasing amount of soil erosion because forests have been cut for the land needed for new buildings. There has also been a 51. 6% decrease in the original total tidal prism volume because large-scale land reclamation decreased the water area of the lagoon by 2/3 (Table 2). The reclamation changed the natural flushing pattern of the ebb current such that sediments are now carried only halfway to the open sea. Siltation rates have been approaching annual levels of 10 cm (inner section of navigation channel) to 30 cm (outer section of navigation channel). This process may form an ebb tidal delta in the future. Sanya Harbour has to be dredged every two to three years. Even though the annual sedimentation rate of 10～30 cm is not a serious problem, new trends of ever increasing siltation that reflect increased human activity should be taken seriously. Even the low rate of sedimentation determined from ^{210}Pb and ^{14}C sediment cores also reflect the changes in sediment characteristics attributable to human activity in the harbour area. Cores S002, S035 and S033 have a textural background of pure silty sand. This older natural sediment characteristic gives way to more clay, organic matter and less worm activity that marks a change to a less oxidative environment after the harbour was first setup (1617—1728).

Table 2　Change of Tidal Prism Volume in Sanya Harbour

	Original Lagoon	Lagoon after salt pan construction	After reclamation (1990)
Tide flooded area (m³)	4 140 000	1 560 000	1 393 300
Percent of original lagoon area (%)	100	37. 7	32. 4
Tidal prism (m³)	4 906 800	2 842 800	2 436 320
Percent of original tidal prism	100	57. 8	49. 7

It is important to keep the largest possible basin area of the inner bay to maintain a large tidal prism volume for enhancing natural flushing processes. Land reclamation, evaporating ponds for sea salt, the cutting of mangrove forests in the inner bay, or of forests along coastline, are all measures that will lead to an increase in harbour siltation.

The tidal inlet of Sanya can serve as a multipurpose harbour. The inner bay is suitable for small fishing craft and pleasure boats. The outer bay of the present harbour is a medium rank harbour suitable for 1 000 to 5 000 tonne commercial ships. Additionally, it can be developed as a base harbour for servicing offshore oil and natural gas fields. The berths and the navigation channel of Sanya Harbour can be dredged to 9 m water depth for landing 10 000 tonne ships. 30 000~50 000 tonne ships can be accommodated at the entrance to the present outer harbour by constructing a pier to connect Baipai reef with the mainland coast just to the north of the modern harbour. The leeward side of the structure can be dredged to 13 m and used as a deep water berth.

2. Yangpu Harbour

Yangpu Harbour is located on the northwest coast of Hainan Island. This tidal inlet was formed under the control of a NE trending fault and associated volcanic activity. An old river incised a deep channel during the Holocene that connects the estuary of Xinying Bay(新英湾) inside, the outer bay of Yangpu, and the shallow sea of Beibu Gulf.

Xinying Bay is a 50 km² embayment which is under the influence of dynamic processes of river flow and tidal currents. Annual fluvial sediment discharge is 15. 6× 10^4 tonnes which is supplied mainly by two rivers: Dashui Jiang River in the northeast side discharges 7. 39 × 10^4 tonnes/yr of sediment; Chunjiang River (春江) on the southwest side has a sediment discharge of 6. 72×10⁴ tonnes/yr. 40% of the bottom sediment in the estuary is fine sand and shingles which has been developed as a fluvial fan. The other 60% of the sediment consists of silt and clay that is carried out of the inner bay by tidal currents (MGSL, 1985).

The Northern part of Yangpu Bay is developed on a basalt platform, which consists of a series of cliff shorelines. On the south side of the bay, cliffs are developed in the Zhanjiang Formation, an unconsolidated sedimentary strata of an elevated 10 m high old beach that is presently being eroded (Fig. 5).

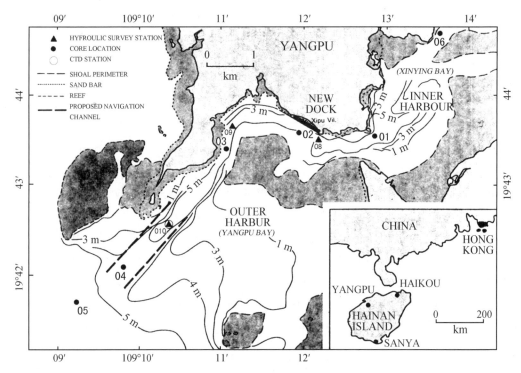

Fig. 5 Map of Yangpu Harbour

There is a deep channel linking the two bays. The entrance section is 550 m wide and has a cross-section area of about 5 000 m². The main tidal channel, which extends from the entrance of inner bay to the outer bay at "Block Gate shoal" (ebb tidal delta), is 10 km long, 400~500 m wide and 5~25 m deep. The main channel is an ancient river valley formed by erosion along two major faults in pre-Holocene time. During the Holocene, a rising relative sea level drowned the ancient river valley, and sedimentation eventually filled some parts of it. However, the area between the main channel entrance and Block Gate shoal has kept its original form because of daily flushing and scoring of sediments by tidal currents.

Block Gate shoal is about 400 m long and with a maximum water depth of about 5 m. Shoal stratigraphy consists of three layers. The top layer is primarily marine mud (0~3 m); the middle layer is about 7 m thick and consists of fluvial-deposited sand and gravel. The bottom layer consists of Zhanjiang Formation rocks (indurated silt and clay).

Block Gate shoal has experienced three stages in its development. Before 8 500 YBP, the channel was cut by river currents which eroded the Zhanjiang Formation and deposited a 7 m thick layer of sand and gravel. During the Holocene sea level rise, the entire embayment was flooded and lastly, tidal currents deposited marine mud that has caused a gradual shoaling of the ebb tidal delta to a present water depth of about 5 m. Core data suggest that this water depth has been maintained over the past 2 000 years with net deposition of only about 70 cm since that time.

The south side of Yangpu bay is a sand bank developed on an erosional platform of the Zhanjiang Formation. This formation has provided a sediments for the accumulation of a veneer of sand (about $0\sim2$ m thick) on the surface of the platform. The area of the bank lying above the 0 m contour is about 2.5 km^2.

There are three major sources of sediment in the Yangpu Harbour area. The most important source is material derived from the local river. The river's annual sediment discharge is about 6.3$\times10^4$ m^3/yr. The second sediment source is coastal erosion. Most coastal erosion occurs in the Zhanjiang Formation along the southern coast and contributes 2.2$\times10^4$ m^3 of sediment each year. The northern basaltic coastline of the harbour is also being eroded providing an annual contribution of about 0.2$\times10^4$ m^3 of sediment. The total contribution of sediment from coastal erosion processes is 2.4$\times10^4$ m^3/yr. The third source of material is coral reefs. There are fringing reefs along the basalt coastline of Yangpu Bay, and a barrier reef lying just outside of the Yangpu Bay along the main coastline. The total amount of sediment from this biogenic source is estimated to be no greater than about 4 000 m^3/yr. The total amount of sediment supplied to the harbour area from all sources is about 9.1$\times10^4$ m^3/yr (Wang and Zhu, 1990). Dominant southwest waves stir up fine sediment and then deposited in the estuary by flood tidal currents. The storm disturbing is often here in Yangpu than in Sanya. According to statistics, whenever the typhoon centre enters to the west area of 113°E and $17°\sim22°$N, it will influence the Yangpu Harbour. There were 144 times of typhoon during 1949—1986 influenced Yangpu Harbour (Zhu and Deng, 1990). Because of the coastal outline of Yangpu Bay stretch inland for quite a distance and the bay is well protected so that the maximum rise of local sea level caused by a typhoon was 0.89 m (1980, No. 7 typhoon). Concurrently, the local sea level rise in Xiuying Harbour and Haikou Harbour was 2.42 m and 2.49 m. The rise time of sea level during that typhoon period in Yangpu was $3\sim5$ hours and was attributable to strong northwest winds. After the northwest wind speed slowed for $5\sim6$ hours, sea level returned to normal (Zhu and Deng, 1990). The storm events are reflected in the sediment cores by an increase in the frequency of sand layers (Fig. 6, core Y003, Y004, Y005, and description).

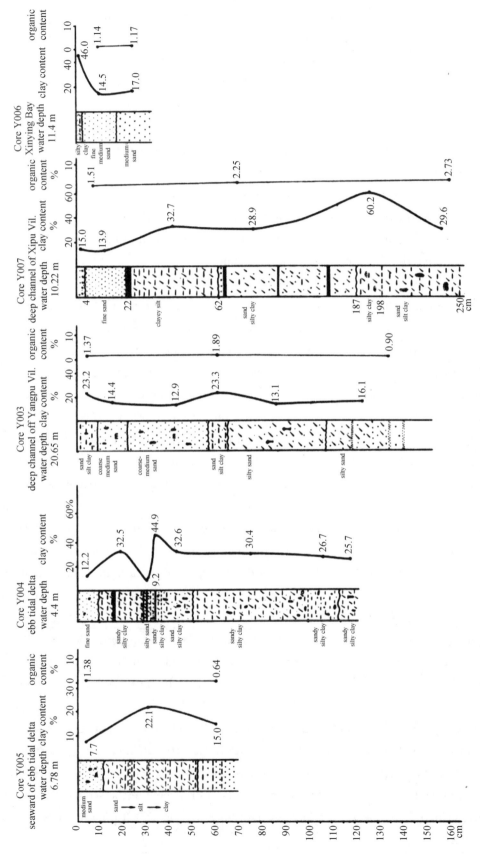

Fig. 6　Sediment core of Yangpu Harbour

Tides in Yangpu are of the irregular diurnal type. The mean tidal range is 2 m and the tidal prism is 2×10^8 m³. Observations over several tidal cycles show that the ebb current velocities are greater than flood velocities even through the ebb period is longer than that of flood. This reflects the fact that the ebb current always combines with runoff (MGSL, 1985). During the ebb tide, tidal current velocities show a decreasing trend from east (the inlet entrance) to west (ebb tidal delta). At the entrance, the maximum tidal velocity is 97 cm/sec while at the ebb tidal delta, maximum tidal velocity is 56 cm/sec (MGSL, 1985). All bottom water velocities are above the threshold value necessary for eroding and transporting silt and fine sand particles. Maximum ebb tidal velocities are generally greater than flood values, so that the net transport direction of sediment is from the inner bay to the outer bay. The deep channel maintains its depth through this tidal current transport mechanism. Thus, the other important factor after a large tidal prism volume is a strong ebb current. (Fig. 7)

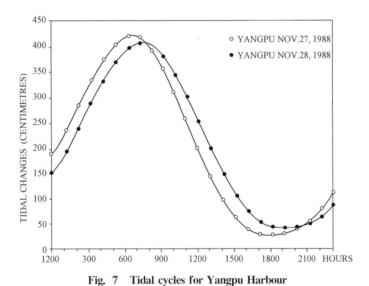

Fig. 7 Tidal cycles for Yangpu Harbour

The modern suspended sediment concentration shown by monitoring data are the same for both the sand shoal and for the deep channel, 0.05 kg/m³. Tidal current speeds have shown a decreasing trend from the channel entrance to Block Gate shoal. However, the maximum ebb current speed of 80 cm/sec observed on the shoal is strong enough to transport the silt sediment in suspension to areas beyond the shoal.

The analysis of sedimentation rate data for Yangpu Harbour shows that the rate increases from the inner harbour towards the sea, and that it also has been increasing from historic times to the present. Nevertheless, the net sedimentation is low, i.e., the maximum value is only 7 cm per year (Table 3). These results indicate that Yangpu Harbour is a natural deep water harbour with minimal siltation.

Table 3　Sedimentation Rate：Yangpu Bay

Location		Inner Harbour		Deep Channel			Block Gate Shoal			Outer Sea
Sample ♯		Y006	Y001	Y007	Y002	Y003	Y004	2	4	Y005
Method Result (cm/yr)	^{210}Pb	>0.8	>1.0	1.47	0.87	0.52	1.06			2.0
	^{14}C			0.045 ↑ 0.044		0.028 ↑ 0.034	0.028 ↑ 0.034	0.05 ↑ 0.42	0.6 ↑ 0.32	
Bathymetric Monitoring	1947—1974 period			−1.2~1.5			−2~10			
	1974—1983 period			5.0			2.2~4.5			
	1983—1989			7.0			6.0			

Yangpu inlet can serve as multipurpose harbour for fishing boats and larger ships. There are quite a number of fishing boats from Hongkong, Macao, and local sources that anchor in the outer bay during storms. Two 20 000 tonne docks have been built along the basalt coastline between Xipu Village and the inlet entrance. The new dock area can be used for 50 000 tonne ships, the only problem being the sandy shoal on the outside end of the channel. Dynamic geomorphology analyses have confirmed the advantageous natural conditions for construction of a harbour in Yangpu inlet. These analyses suggest that the optimum direction for a navigation channel is through the ebb tidal delta along 45°~225° direction following the natural route of tidal currents. These results also predict that the siltation rate following dredging of the channel will be about 0.5 m per year.

3. Haikou Harbour

Haikou Harbour is located on the northern coast of Hainan Island on a plain coast between two deltas of the Nandujiang River(南渡江). The harbour area coastline is 57 km long from the present river mouth in the east to the Cape Xijia of Macun port(马村港) in the west. The whole coastline lies on Qiongzhou Strait(琼州海峡) and 20 km is the average distance of Leizhou Peninsula(雷州半岛) which forms a southerly extension of the mainland. Haikou is the oldest harbour and is also the main gateway to Hainan Island. The harbour presently consists of three ports (Fig. 8). The inner port handles passenger and ferry boats that link Haikou to the mainland. It also services small cargo ships of under 500 tonnes. Macun port is a dock for 20 000 tonne ships which mainly supply coal to the Macun power plant. Xiuying port(秀英港) is the largest and oldest port of Haikou Harbour. Fourteen berths can accommodate ships from 1 000~5 000

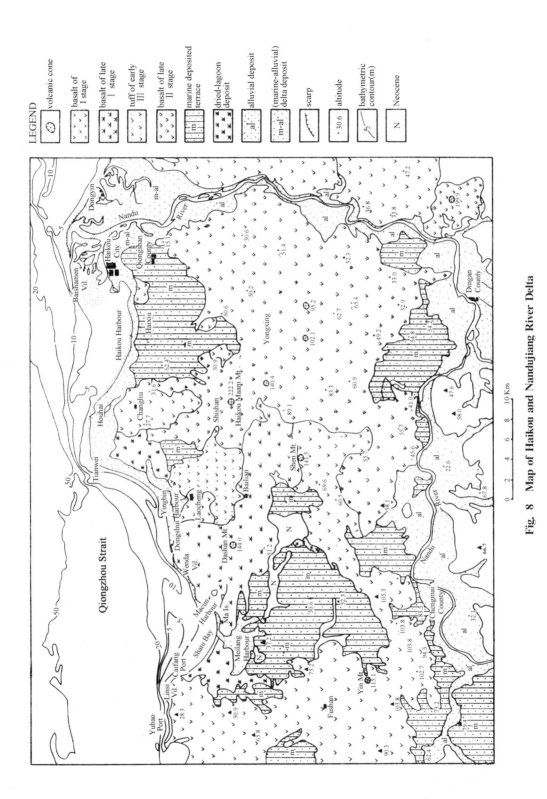

Fig. 8 Map of Haikou and Nandujiang River Delta

tonnes that are used for passengers, cargo and harbour maintenance.

The harbour is built up at the head of a shallow bay in the pattern of an artificial barrier bar parallel to the sandy bay head. There is insufficient natural flushing processes in the harbour. It depends instead on a wave shadow area as a berth. It has always suffered from siltation. The total quantity of siltation including berth and channel areas is $26 \times 10^4 \sim 31 \times 10^4$ m^3 (Wang Baocan et al., 1985). The bay was formed by the coalescing of two lobes of the Nandujiang River delta. Presently, the main channel of the river is 311 km long and has an annual total discharge of approximately 6.0×10^9 m^3. Its annual sediment discharge, before the construction of Songtao reservoir and the dam at Longtang in 1959, was 48.6×10^5 tonnes and later is 37.7×10^5 tonnes. Longshore drift, strengthened by strong northeast wind waves, which are related to the narrow channel affect of Qiongzhou Strait, and the onshore movement of sediment are causing siltation in the bay. Thus, the harbour is always having siltation problems, a feature which has limited its expansion.

Survey data indicate that the sediment rate in Haikou Harbour is as high as $50 \sim 75$ cm/yr in the berth area, and reaches $75 \sim 200$ cm/yr in the navigation channel. This means that the harbour needs to be dredged often.

The old lobe of the river delta is on the west side of the bay (Fig. 8). It was formed by sedimentation processes but is also characterized by several bedrock crops. Seismic profiles clearly show the level layers of deltaic deposits and with basalt rock intrusions (Fig. 9). The delta lobes have similar outlines. They are protruding lobes that have been longshore drift from northeast to southwest under the control of the prevailing northern-northeasterly winds that dominate Qiongzhou Strait. The old lobe was abandoned as the river shifted during the late Pleistocene and early Holocene because of the Mt. Maanling(马鞍岭) (horse saddle shaped hell above sea level of 224.2 m) volcanic eruption. Eruption-related processes forced the river to migrate from a northern direction toward the east until it reached the east side edge of the volcano in the Dingan County. At that point, the river turned to the north and formed the modern course of Nandujiang River. The eruptions elevated the original river bed forming terraces that are elevated $20 \sim 40$ m above the modern river bed on the south side and north side of Maanling Volcano. These terraces are characterized by flat surfaces that are underlain by exposed 3 metres of brownish yellow silty clay layer on the top and brownish clayey silt on the bottom of Dachang Cun in the north. The terraces deposit of fluvial sand and gravel with laminae and cross-bedding evident in the section have been baked by basalt lava on the top. On the southern side of Mt. Maanling, and an abandoned river channel section and meander spur and scarp preserve the original river features (near Junniao Cun and Chuangcao Cun). This abandoned channel is elevated $20 \sim 40$ m above modern river bed. From the meander of the abandoned channel, it can

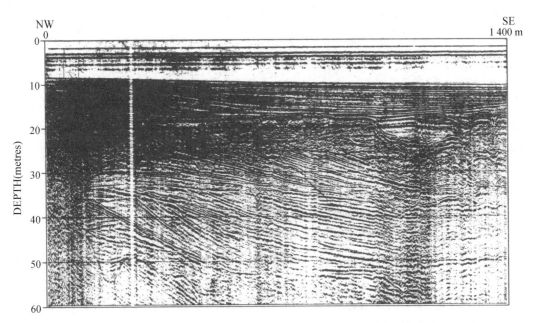

Fig. 9　Seismic profile of basalt and delta deposits in the old lobe of Nandujiang River delta

be deduced that the river tried to keep its original course in the early stage of the crustal upheaval caused by the volcanic eruption, but at last, the river flow changed to the lower land on the south side of the volcano. On the northern side of Mt. Maanling Volcano, the abandoned river mouth and river mouth bar have been reformed by prevailing longshore drift processes as a narrow lagoon behind a sandy bar-a new system of Dongshui Inlet. Understandings about the evolution of the Nandujiang River delta can be used to deduce that, (1) the river delta was developed in the same pattern since late Pleistocene time, and longshore drift always started at the river mouth and then flowed in a southwest direction; (2) Dongshui Inlet was a river mouth channel in late Pleistocene to early Holocene time. Consequently, the inlet is much more stable than the lagoon. The water area of the inlet can be dredged to form a deep water berth without risk of serious siltation problems which is much better for re-creation than the shallow water of the lagoon.

To meet the requirements of developing a 10 000 tonne harbour for the capital city of Haikou, present harbour capacity can be improved by setting up a group of multipurpose harbours along the river delta coastline including the present delta and the old delta. For example, one of these could be built in the deep channel of Baishamen(白沙门) located leeward of a modern river mouth sand bar. A second could be constructed on the shores of Dongshui(东水) Inlet abandoned river mouth channel on the southwest side of the old delta. A third could be developed along the deep channel of Houshui Bay. These future harbour locations are near Haikou City, and can serve different purposes depending on future economic development trends. To combine this group of harbours together can satisfy the city's current needs for greater transportation capacity.

CONCLUSIONS

(1) Sanya, Yangpu and Haikou are among the larger commercial harbours of Hainan Island. Among these, Yangpu Harbour is the best one with the largest protected water area and the deepest navigation channel. There is little sediment supply to the harbour; deposited sediments are mainly coarse, which makes them much more stable than fine-grained material. The current sedimentation rate is minimal and with little human impact, the harbour is easy to maintain in a good condition without annual dredging or an unusually large amount of engineering work. The natural features of the harbour can be developed for various ships from fishing boats to 50 000 tonne ships in the one area.

Sanya Harbour also has good natural characteristics for harbour development with little siltation even though the sedimentation rate has increased during historical time especially since the 1980's. The harbour needs to be planned and managed well, with definite engineering projects such as a pier to connect Baipai Reef with the main coastline to accommodate 50 000 tonne ships. It is important to develop the inner lagoon properly to avoid a decrease of the tidal prism so as to preserve the natural flashing pattern of navigational waters.

Haikou Harbour is oriented facing the open sea and in a leeward position relative to river sediment drifting and to coastal zone deposition as is seen in the bay head area. Compared to the others, this harbour has more natural problems both in terms of wind wave disturbance and sediment siltation. The harbour structure must include a breakwater and sediment dredging. However, because the harbour serves directly for the capital city, it needs to be expanded. The best way to increase its capacity is to set up a group of small to medium multipurpose harbours nearby.

(2) According to the results of the comparative study, it is clear that tidal inlet embayment environments are better for setting deep water harbours than the sandy plain coast or delta/sandy bay systems of Hainan Island. Among the tidal inlets, the tectonic subsidence type is better for harbour construction purposes than the ones formed by sand barrier and lagoon systems.

(3) The net sedimentation rates of all three harbours are not enormous. However, given the trend of increasing siltation from human activity, impacts have been increasing in the coastal zone especially since the last decade.

References

[1] MGSL (Marine Geomorphology and Sedimentology Laboratory), Nanjing University. 1985. *Research Report on Sediment Pravenance and Siltation of Yangpu Harbour, Hainan Island*.

[2] MGSL, Nanjing University. 1986. Study on Coastal Environment and Sanya Harbour Development of Boping-Sanya Bay, Hainan Island.

[3] Wang, Baocan; Lao, Zhisheng; Lu, Jinrong *et al*. 1985. The Report on natural condition analyses of harbour construction in the area of Baishamen, Haikou. China's Harbour, 67 – 79.

[4] Wang, Ying and Zhu, Dakui. 1990. Study on the Coastal Geomorphology and Engineering of Yangpu Harbour. *Journal of Nanjing University (Geography)*, 11: 1 – 13.

[5] Zhu, Suofeng and Deng, Weidong. 1992. Basic Studies on Tropical Storm Surges in Yangpu Harbour, Hainan Province. *In: Island Environment and Coast Development*. Nanjing University Press. 31 – 38.

Dynamic Geomorphology and Coastal Engineering of Yangpu Harbour, Hainan Island, China*

COASTAL GEOMORPHOLOGY

The area of Yangpu Harbour includes three structural systems. Consequently, this location has been subject to a large number of tectonic movements since Pliocene. The major structural system trends toward the northeast. The northern part of Hainan Island borders on a large east-west trending graben feature and Yangpu Bay lies within the southern part of the associated trough feature. During Pliocene and Quaternary, 3 000 m of sediment have deposited in the graben as a shallow nearshore facies called the Zhanjiang Formation (湛江系). There is a substantial amount of volcanic activity associated with the tectonic movements that have occurred in this part of the island. This is evidenced by the rocks, primarily basaltic, exposed on the north side of the bay. The rocks on the south side of the bay are part of the Zhanjiang Formation and consist mostly of gravel, sand and clay in various proportion. The embayment is the topographic expression of the fault that separates the basalt from the Zhanjiang Formation sediments. The area north of the embayment has been subject to neotectonic movements since early Holocene that have continued to the present (Ding, 1964).

Yangpu Harbour consists of two embayments-Yangpu and Xinying. Xinying Bay is a tidal embayment with an area of 50 km². Two rivers enter in the embayment, the Dashui River (大水河)* from the east, and the Spring River (春江) from the southeast. Their combined drainage basin area is 1 419. 44 km² (Figure 1).

The mouth of the embayment is 550 m wide and has a cross-section area of about 5 000 m². The mouth area is referred to as the Baimajing Strait. Fluvial sediment carried into the eastern side of the embayment by the two rivers has formed two distinct river deltas. During each tidal cycle, the fine fractions of the delta sediment is eroded and transported seaward during the ebb tide. Near the deltas, more than 90% of the residual sediment in tidal channel and lower beach environments is coarser than 0. 1 mm. Muddy sediments are carried onto the tidal flats during flood where they are trapped by

* Ying Wang, Dakui Zhu: Narendra K. Saxena ed. , *Recent Advances in Marine Science and Technology*, *92*, pp. 355 - 362, May 1993, Honolulu, Hawaii, the U. S. A. : PACON International.

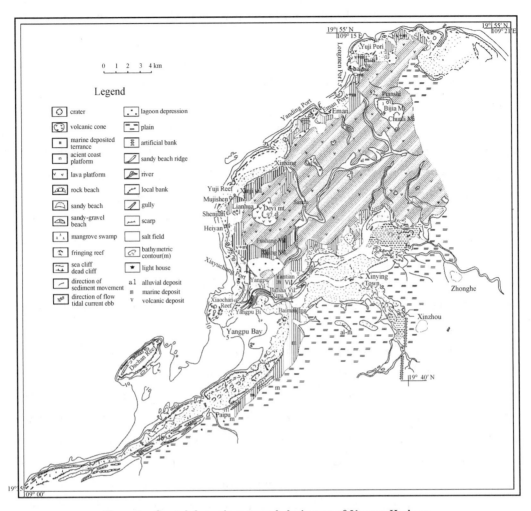

Figure 1　Coastal dynamic geomorphologic map of Yangpu Harbour

the mangrove vegetation. Consequently, the coastline is migrating seaward. According to historical records, the towns of Zhonghe(中和) and Xinzhou(新洲) were the sites of seaport 1 000 years ago, during the Tang Dynasty. Currently, however, they are landlocked. In Qing Dynasty, seaport moved to Xinying Town. But now, even a small 20-ton boat can only navigate to Xinying only during hightide time.

The northern part of Yangpu Bay is developed on a basalt platform. There were two periods of volcanism. The first one occurred in the late Pleistocene, dating back to 52 000 YBP. It formed a lava field which covered a sand deposit. Another eruption took place 25 000 YBP. And the latest eruption occurred in the Holocene (Wang, 1990). The lava which emitted at that time consists of olivine basalt and tholeiite basalt. It is the olivine basalt of oceanic crust, and the tholeiite of mid-ocean ridge. So it may be an interesting phenomenon if there was/is sea floor spreading in the Beibu (Tonkin) Bay (北部湾)(Nicholls and Ringwood, 1973; Wang, 1990).

The north shore of Yangpu Bay consists of a series of basalt cliff. In the inner part deposits muddy sediment and grows up mangrove while the outer part is presently in erosion.

On the south side of the bay, cliffs that were developed in the Zhanjiang Formation are presently eroded. The Zhanjiang Formation supplies sand and silt for the embayment.

The main tidal channel from Baimajing strait (白马井海峡) to the ebb tidal delta (block gate shoal) is 10 km long, 400~500 m wide and about 5~25 m deep. The bottom sediment of the main channel consists of three distinctive layers. The upper layer is muddy marine sediment. The middle layer is composed of sand and gravel deposited by river currents. The lower layer consisted of Zhanjiang Formation sediments (silt and clay). The main channel is an ancient river valley that was eroded along two major faults in pre-Holocene river valley. The upper and lower reaches of the channel became the sites of sediment deposition that eventually infilled those parts of the ancient river valley. However, the channel was kept open because of daily erosion of sediment by tidal currents.

The ebb tidal delta at the western end of the tidal channel is currently an area of shoals with maximum water depths of about 5 m. This shallow area is about 400 m long and 80~150 m wide. The ebb tidal delta stratum consists of three layers. The top layer is primarily marine mud (0~3 m). The middle layer is about 7 m thick and consists of sand and gravel deposited by river currents. The lower layer consists of Zhanjiang Formation sediments (indurated silt and clay). The ebb tidal delta developed in three stages. Firstly, before 8 500 YBP, the channel was cut by river currents which eroded the Zhanjiang Formation and deposited a 7 m thick layer of sand and gravel. Secondly, during the Holocene sea level rise, the entire embayment was flooded. Finally, tidal currents which developed in the bay deposited marine mud resulting in a gradual shoaling of the ebb tidal delta to its present depth of about 5 m. Core ^{14}C data suggest that this water depth has been maintained over the past 2 000 years with net deposition of only about 70 cm since that time.

The sand bank area forms the south side of the Yangpu Bay. The area of the bank lying about the zero contour is about 2. 5 km². Between 1955—1980, this area has increased by about 0. 7 km². An area of less than 2 m water depth represents about 40% of the total area of the bay. This area also shows an enlargement of about 0. 7 km² between 1955—1980, i. e. an increase of <5%. The sand bank is developed on an erosion platform of Zhanjiang Formation. Its sediment forms a thin veneer about 0~2 m thick on the platform surface.

COASTAL DYNAMICS

Wind and Wave

Wind variation shows two dominant modes. One of these blows NNE-ENE off the land. This wind is relatively strong but does not produce large waves because of the shelter effect of land, i. e., there is a short wave fetch with respect to the outer bay. The other major component blows from the sea in a SSW-WSW direction. Because of the exposed nature of the outer harbour, this mode of wind produces the largest waves and is responsible for most sediment transport in the harbour, and of coastal erosion. It can also have a disturbing influence on ships that use the harbour as an anchorage.

The components of the tidal wave climate include wind-generated waves (78%) and sea swell (12%). The most frequent waves entering the harbour are from the southwest. Waves propagating from this direction are also among the largest that enter the harbour. The average height of SW propagating wave is 0. 83 m with an average period of 4. 5 second. Because the breaking depth of the 4. 5 second waves is only 0. 8 m, their effect on the erosion of sediment is felt primarily on the southwestern edge and top of the sand bank.

Tide

Tides in Yangpu Bay are of the irregular diurnal type. The mean tidal range is 1. 81 m. During the survey of November, 1988, the measured range was 3. 8 m. Observations over several tidal cycles show that the ebb portion is shorter (12 hours) than the flood portion (13 hours). This difference is reflected in ebb current velocities with tend to be higher than those observed during the flood. Mean flood velocity (V_f) is above 22 cm/s, and mean ebb velocity (V_e) is 27 cm/s (Table 1). There are differences in flood and ebb velocity depending water depth. During flood, mean surface water velocity (V_f, s) is above 23 cm/s. Middle water velocity (V_f, m) and bottom water velocity (V_f, b) reach 24 cm/s. During the ebb, surface water currents (V_e, s) have a mean velocity of 31 cm/s. Middle water (V_e, m) and bottom water (V_e, b) flows are respectively 27 cm/s and 24 cm/s. During the ebb, tidal current velocities show a decreasing trend from east to west starting at the mouth of the inner harbour near Baimajing. At Baimajing, surface velocities during the flood reach 51 cm/s. Middle and bottom water velocities are considerably high reaching values of 88 and 84 cm/s, respectively. During the ebb, the surface water flows have a maximum velocity of 84 cm/s while middle and bottom water velocities flow at a rate of 86 and 97 cm/s, respectively.

Table 1　Maximum tidal velocity (cm/s)

Station		Surface	Middle	Bottom
Baimajing	flood	51	88	84
	ebb	84	86	97
Main Channel (Yangpu Vil.)	flood	42	67	56
	ebb	68	53	45
Ebb Tidal Delta	flood	57	49	22
	ebb	54	75	56

At the main channel (Yangpu Village), maximum surface water flow during the flood stage of the tide is 42 cm/s. Middle water and bottom water flows are 67 and 56 cm/s. During the ebb cycle, surface, middle and bottom water maximum velocities are 68, 53 and 45 cm/s. At the mouth of the outer harbour, in the ebb tidal delta, the maximum velocity of surface water during the flood stage is 57 cm/s. And middle and bottom water velocities are 49 and 22 cm/s. And during the ebb stage, the surface water maximum flow is 54 cm/s. Middle is 54 cm/s. Bottom is comparable to surface values 56 cm/s.

All of the bottom water velocities are above the threshold value necessary for the erosion and transport of silt and fine sand particles. Maximum ebb velocities are generally higher than flood values so that the modal transport direction of sediment is from the inner to the outer bay.

Residual flow presents water velocities and directions in the absence of tidal flow. These flow are important in understanding sediment distribution patterns (Table 2). The residual flow of surface water has a large river component and tends to follow a direction parallel to the axis of the tidal channel. During the wet season there is less coherency of the residual flows between surface and bottom water (Wang and Aubrey, 1987).

Table 2　Residual flow velocities (cm/s) and direction

Station	Surface	Middle	Bottom
Baimajing	6/187	3/95	3/102
Main Channel	12/301	11/116	12/107
Ebb Tidal Delta	7/230	5/164	8/125

Suspended Particulate Matter

In general, the concentration of suspended particulate matter (SPM) in the water column of Yangpu Harbour is about 0.1 kg/m³. Near Baimajing the integrated (surface, middle and bottom water) SPM concentration during the flood stage average

0.088 kg/m³ (Table 3). During the ebb cycle, this value increased to 0.103 kg/m³. At Yangpu Village, SPM during the flood is somewhat higher than at Baimajing and is similar to the ebb value. SPM measurements at the swash platform station are slightly lower than those observed at Yangpu Village and are comparable to the ebb value noted for the Baimajing station.

Table 3　Suspended particulate matter concentration (kg/m³)

Station	Flood	Ebb
Baimajing	0.088	0.103
Yangpu Village	0.116	0.119
Ebb Tidal Delta	0.102	0.102

SEDIMENT SOURCES

There are three major sources of sediment in the Yangpu Bay area. The most important source is material derived from the Dashui River. The drainage basin of this river is 648.3 km². The annual discharge of sediment from the basin into the inner bay (Xinying Bay) is about 74 000 tonnes per year. The Spring River enters the bay from the southeast. Its drainage basin has an area of 577.8 km². The river's annual sediment discharge is about 67 000 tonnes per year. All of other small rivers draining into the bay have a total drainage basin area of about 81.3 km² and account for an annual sediment discharge of 15 000 tonnes per year. So the total discharge from all river sources is 156 000 tonnes per year. The weight equates to a sediment volume of about 62 000 m³ per year.

The second most important sediment source is from coastal erosion. Most of the coastal erosion occurs in the Zhanjiang Formation along the southern coast of the outer harbour. Each year this source contributes 22 000 m³ of sediment. The northern basaltic coastline is also being eroded but at a significantly slower rate than the southern coast. Its annual contribution of sediment is about 2 000 m³ of sediment. The total contribution of sediment from coastal erosion processes is 24 000 m³ each year.

The third source of material is from coral reefs that are established on offshore islands, at the northwestern end of the bay, and at various locations along the south coast (Figure 1). Sediments in the ebb tidal delta area contain about 33% $CaCO_3$, those along the subtidal areas of the south shore contain about 22%~23% $CaCO_3$. In the inner bay, the sediments contain only about 0.6%~0.9% $CaCO_3$. The distribution of $CaCO_3$ in outer bay sediments is consistent with the distribution of reefs along the coastline. It also shows that the sediment from the outer bay is not transported too far into the inner bay. The total amount of sediment from this biogenic source is estimated to be not greater than about 4 000 m³ per year. The total amount of sediment supplied to the bay area from all sources is about 91 000 m³ per year.

COASTAL ENGINEERING

Yangpu Harbour is a natural harbour. The main channel is $10 \sim 23$ m deep and $500 \sim 800$ m wide. The huge water and large currents are major factors for keeping channel open. The Northern, Eastern and Southern parts of the harbour are 10 m-deep marine terraces that stop the wind wave from north, northeast, east and south. The frequency of strong winds is a little, >6 wind (force) scale only 0. 42%. Wave heights $H_1/10 < 0. 5$ m are 99%, $H_1/10 \leqslant 0. 8$ m are 99. 88%, but those $H_1/10 \geqslant 1$ m are only 0. 06%. The frequency of wave height$\geqslant 1. 2$ m is zero. So, Yangpu Harbour is a good harbour. The harbour area, geologically, is a platform of basalt and the Zhanjiang Formation. The geologic base is quite good. And an abundant rocks material, broad land area. But now, the problem is only a blocked gate shoal and shallow waters, requiring dredging for a navigation channel.

Yangpu Harbour is a tidal inlet, in which Xinying Bay is a tidal water trap: the main channel is the tidal water pass; the block gate shoal is the ebb tidal delta. Core data and regional geomorphological analysis show: the main channel and block gate shoal were originally a fossil river valley 8 500 YBP. The valley's elevation is minus $25 \sim 30$ m. The top surface of the river deposits has a minus 20 m elevation. The ancient river's gradient was 0. 07%. The ancient Dashui River and Spring River joined the ancient river valley in the area of today's Xinying Bay, which finally entered the open sea at the -20 m contour through the main channel and block gate shoal. With the post-glacial sea level rise, the coastal plain was drowned as Yangpu Harbour and ancient river valley were filled with marine sediment while the ancient river valley was kept as today's deep water channel because of the tidal flushing and scoring of tidal prism. But, the section of the block gate shoal disappeared due to siltation.

Dating data (according to ¹⁴C) from the block gate shoal show: the average sedimentation rate was 0. 16 cm/a during the last 8 500 years. And during the last 3 000 years the rate was 0. 10 cm/a, which is consistent with the world wide sea level change processes and now (according to ²¹⁰Pb inventory analysis) is 0. 52~1. 06 cm/a. Hence, the natural formation process of the block gate shoal is shown.

We have worked out the erosion and siltation budget for the last decades of the area by comparing the six 1947—1983 navigation maps and the bathymetric maps. In this work, we transferred the different depth bases of various maps into a same theoretical depth base, and then calculated the volumes of the block gate shoal, the main channel at different periods. We find the block gate shoal with an erosion rate $2 \sim 10$ cm/a during

1947—1974，a siltation rate 7. 2～4. 5 cm/a during 1974—1983；and the main channel with an erosion rate of 1. 2～5 cm/a during 1947—1974，a siltation rate of 5 cm/a during 1974—1983. In general，however，the block gate shoal and main channel were both subjected to slight erosion during last fifty years. The total erosion volume of the block gate shoal is one million cubic meters and that of the main channel is 0. 7 million cubic meters.

Therefore，for the past 8 000 years，the sedimentation rates of the area have been very small and the submarine reliefs are nearly stable although sometimes there was slight erosion or siltation，which does not interfere with constructing a harbour and digging a navigation channel in the area.

Tides in Yangpu are irregular diurnal type. The mean tidal range is 2 m and tidal prism is 200 million cubic. Tidal cycles show that the ebb current velocities are above the threshold value necessary for the eroding and transporting silt and fine sand particles (Figure 2).

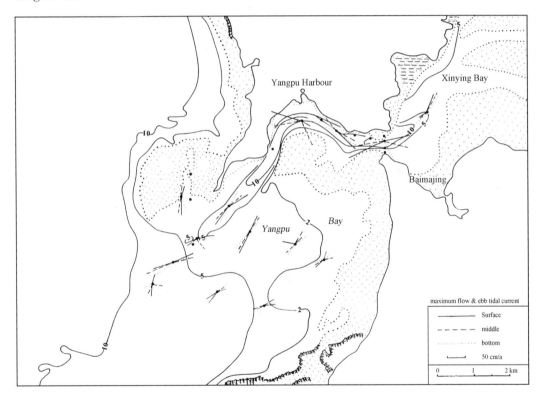

Figure 2　Tidal current in Yangpu Harbour

According to dynamic geomorphology analysis，the paper verified the advantageous conditions for construction of Yangpu Harbour，suggested the optimum direction as 45°～225° of navigation channel passing through the block gate shoal (ebb tidal delta)，and predicted that the siltation rate following digging the channel will be about 0. 5 m/y.

References

［1］Ding，G. 1964. On some problems about the Quaternary geology of Hainan Island. Quaternary Geology of China. Beijing：Science Press. 207－233 （in Chinese）.

［2］Nicholls，L. A. and A. E. Ringwood. 1973. Effects of water on olivine stability in tholeiites and production of silioasatureted magmas in the island-arc environment. *Jour. of Geol.*，81：285－300.

［3］Wang，Y. 1990. The volcanic coast in the area of northwest Hainan Island. *Acta Geographica Sinica*，45：321－330 （in Chinese with English abstract）.

［4］Wang，Y.，and D. G. Aubrey. 1987. The characteristics of the China coastline. *Continental Shelf Research*，7：329－349.

The Dynamic Geomorphology of Yangpu Bay, Hainan Island, China [*]

Ⅰ. COASTAL GEOMORPHOLOGY

The area of Yangpu Bay includes three structural systems. Consequently, this location has been subject to a large number of tectonic movements since the Pliocene period. The major structural system trends northeast. The northern part of Hainan Island boarders on a large east-west trending graben feature and Yangpu Day lies within the southern part of the associated trough feature (Wang Ying, 1977). During the Pliocene and Quaternary periods, 3 000 m of sediment have deposited in the graben as a shallow nearshore facies called the Zhanjiang Formation. There is a substantial amount of volcanic activity associated with the tectonic movements that have occurred in this part of the island. This is evidenced by the rocks primarily basaltic, exposed on the north side of the bay. The rocks on the south side of the bay are part of the Zhanjiang Formation and consist mostly of gravel, sand and clay in various proportions. The embayment is the topographic expression of the fault that separates the basalts from the Zhanjiang Formation sediments. The area north of the bayment has been subject to neotectonic movement from the early Holocene period continuing on to the present (Ding Guoyu, 1964).

Yangpu Bay consists of two embayments—Yangpu and Xinying bay is a tidal embayment with an area of 50 km². There are two rivers entering the embayment, the Dashui River from east and the Chun Jiang River from the southeast. Their combined drainage basin area is 1 419. 44 km². The mouth of the embayment is 550 m wide and has a cross section area of about 5 000 m². The mouth area is referred to as the Baimajing Strait. Fluvial sediment carried into the eastern side of the embayment by the two rivers has formed two distinct river deltas. During each tidal cycle, the fine fraction of the deltas sediment is eroded and transported seaward during the ebb tide. Near the deltas, more than 90% of the residual sediment in tidal channel and lower beach environments is coarser than 0. 1 mm. Muddy sediments are carried onto the tidal flats during flood where they are trapped by the mangrove vegetation. Consequently, the coastline is migrating seaward. According to historical records, the towns of Zhonghe and Xinzhou were the sites of seaports

---

* Dakui Zhu, Ying Wang: *COASTAL ZONE'93*, Vol. 3: pp. 3296 - 3304.

1 000 years ago, during the Tang Dynasty. Currently, however, they are landlocked. In the Qing Dynasty, the seaport moved to Xinying Town, but now even a 20 ton boat can navigate to Xinying only during high tide time.

The northern part of Yangpu Bay is developed on a basalt platform. There were two major periods of volcanic activity. The first one occurred in the late Pleistocene period, dating back to 52 000 YBP. It formed a lava field which covered a sand deposit. Another eruption took place 25 000 YBP, and the latest eruption occurred in the Holocene period. The lava which emitted at that time consists of olivine basalt and tholeiite basalt. It is the olivine basalt of oceanic crust, and the tholeiite of mid-ocean ridge, so it may be an interesting phenomenon if there was/is sea floor spreading in the Beibu (Tonkin) Bay.

The north shore of Yangpu Bay consists of a series of basalt cliffs. In the inner part consists of deposits of muddy sediment and mature stands of mangrove while the outer part is being eroded away.

On the south side of the bay, cliffs that were developed in the Zhanjiang Formation are being eroded, supplying sand and silt for the embayment.

The main tidal channel from Baimajing strait to the ebb tidal delta (block gate shoal) is 10 km long, 400~500 m wide and about 5~25 m deep. The bottom sediment of the main channel consists of three distinctive layers. The upper layer is muddy marine sediment. The middle layer is composed of sand and gravel deposited by river currents. The lower layer consists of Zhanjiang Formation sediments (silt and clay). The main channel is an ancient river valley that was eroded along two major faultlines in a pre-Holocene river valley. The upper and lower reaches of the channel became the sites of sediment deposition that eventually infilled those parts of the ancient river valley. However, the channel was kept open because of daily erosion of sediment by tidal currents.

The ebb tidal delta at the western end of the tidal channel is currently an area of shoals with maximum water depths of about 5 m. This shallow area is about 400 m long and 80~150 m wide. The ebb tidal delta stratum consists of three layers. The top layer is primarily marine mud (0~3 m). The middle layer is about 7 m thick and consists of sand and gravel deposited by river currents. The lower layer consists of Zhanjiang Formation sediments (indurated silt and clay). The ebb tidal delta developed in three stages. Firstly, before 8 500 YBP, the channel was cut by river currents which eroded the Zhanjiang Formation and deposited the 7 m thick layer of sand and gravel. Secondly, during the Holocene sea level rise, the entire embayment was flooded. Finally, tidal currents which developed in the bay deposited marine mud resulting in a gradual shoaling of the ebb tidal delta to its present depth of about 5 m. Core 14-C data suggest that this water depth has been maintained over the past 2 000 yrs with net deposition of only about 70 cm since that time.

The sand bank area forms the south side of the Yangpu Bay. The area of the bank lying around the zero contour is about 2.5 km². Between 1955—1980, this area has increased by about 0.7 km². An area with less than a 2 m water depth represents about 40% of the total area of the bay. This area also shows an enlargement of about 0.7 sq km between 1955—1980, i. e. an increase of<5%. The sand bank is developed on an erosion platform of Zhanjiang Formation. Its sediment forms a thin veneer about 0~2 m thick on the platform surface.

Ⅱ. COASTAL DYNAMICS

Wind and Wave

Wind variation shows two dominant modes. One of these blows NNE-ENE off the land. This wind is relatively strong but does not produce large waves because of the shelter effect of land, i. e. there is a short wave fetch with respect to the outer bay. The other major component blows from the sea in a SSW-WSW direction. This type of wind produces the largest waves, is responsible for most of the sediment transport in the harbour, and coastal erosion because of the exposed nature of the outer harbour. It can also have a disturbing influence on ships that use the harbour as an anchorage.

The components of the tidal wave climate include wind-generated waves (78%) and sea swell (12%). The most frequent waves entering the harbour are from the southwest. Waves propagating from this direction are also among the largest. The average height of SW propagating wave is 0.83 m with an average period of 4.5 second. Because the breaking depth of the 4.5 second waves is only 0.8 m, their effect on the erosion of sediment is felt primarily on the southwestern edge and top of the sand bank.

Tide

Tides in Yangpu Bay are of the irregular diurnal type. The mean tidal range is 1.81 m. During the survey of November, 1988, the measured range was 3.8 m. Observations over several tidal cycles show that the ebb portion is shorter (12 hours) than the flood portion (13 hours). This difference is reflected in ebb current velocities which tend to be higher than those observed during the flood. Mean flood velocity (V_f) is above 22 cm/s, and mean ebb velocity (V_e) is 27 cm/s (Table 1). There are differences in flood and ebb velocity depending on water depth. During flood, the mean surface water velocity (V_f, s) is above 23 cm/s; middle water velocity (V_f, m) and bottom water velocity (V_f, b) reach 24 cm/s. Dring the ebb, surface water currents (V_e, s) have a mean velocity 31 cm/s; middle water (V_e, m) and bottom water (V_e, b) flows are respectively 27 cm/s and 24 cm/s. During the ebb, tidal current velocities show a decreasing trend from east to west starting at the mouth of the inner harbour near Baimajing. At Baimajing, surface velocities during the flood reach 51 cm/s. Middle and bottom water velocities are considerably high reaching values of 88 and 84 cm/s, respectively. During the ebb, the surface waterflows have a maximum velocity of 84 cm/s while middle and bottom velocities flow at a rate of 86 and 97 cm/s, respectively.

Table 1　Maximum Tidal Velocity（cm/s）

Station		Surface	Middle	Bottom
Baimajing	flood	51	88	84
	ebb	84	86	97
Main Channel（Yangpu Vil.）	flood	42	67	56
	ebb	68	53	45
Ebb Tidal Delta	flood	57	49	22
	ebb	54	75	56

At the main channel (Yangpu Village), maximum surface water flow during the flood stage of the tide is 42 cm/s. Middle water and bottom water flows are 67 and 56 cm/s. During the ebb cycle, surface, middle and bottom water maximum velocities are 68, 53 and 45 cm/s. At the mouth of the outer harbour, in the ebb tidal delta, the maximum velocity of surface water during the flood stage is 57 cm/s. And middle and bottom water velocities are 49 and 22 cm/s. And during the ebb stage, the surface water maximum flow is 54 cm/s. Middle is 54 cm/s. Bottom is comparable to surface values 56 cm/s.

All of the bottom water velocities are above the threshold value necessary for the erosion and transport of silt and fine sand particles. Maximum ebb velocities are generally higher than flood values so that the modal transport direction of sediment is from the inner to the outer bay.

Residual flow represents water velocities and directions in the absence of tidal flow. These flows are important in understanding sediment distribution patterns (Table 2). The residual flow ofsurface water has a large river component and tends to follow a direction parallel to the axis of the tidal channel. During the wet season there is less coherency of the residual flows between surface and bottom water.

Table 2　Residual flow velocities（cm/s）and direction

Station	Surface	Middle	Bottom
Baimajing	6/187	3/95	3/102
Main Channel	12/301	11/116	12/107
Ebb Tidal Delta	7/230	5/164	8/215

Suspended Particulate Matter

In general, the concentration of suspended particulate matter in the water column of Yangpu harbour is about 0.1 kg/m^3. Near Baimajing the integrated (surface, middle and bottom water) SPM concentration during the flood stage averages 0.088 kg/m^3 (Table 3). During the ebb cycle, this value increased to 0.103 kg/m^3. At Yangpu village, SPM during the flood is somewhat higher than at Baimajing and is similar to the ebb value. SPM measurements at the ebb tidal delta station are slightly lower than those observed at Yangpu Village and are comparable to the ebb value noted for the Baimajing station.

Table 3 Suspended Particulate Matter Concentration (kg/m^3)

Station	Flood	Ebb
Baimajing	0.088	0.103
Yangpu Village	0.116	0.119
Ebb Tidal Delta	0.102	0.102

Ⅲ. SEDIMENT SOURCES

There are three major sources of sediment in the Yangpu Bay area. The most important source is material derived from the Dashui River. The drainage basin of this river is 648.3 km^2. The annual discharge of sediment from the basin into the inner bay (Xinying Bay) is about 74 000 tonnes per year. The Chun Jiang River enters the bay from the southeast. Its drainage basin has an area of 577.8 km^2. The river's annual sediment discharge is about 67 000 tonnes per year. All other small rivers draining into the bay have a total drainage basin area of about 81.3 km^2 and account for an annual sediment discharge of 15 000 tonnes per year. So the total discharge from all river sources is 156 000 tonnes per year. The height equates to a sediment volume of about 62 000 m^3 per year.

The second most important sediment source is from coastal erosion. Most of the coastal erosion occurs in the Zhanjiang Formation along the southern coast of the outer harbour. Each year his source contributes 22 000 m^3 of sediment. The northern basaltic coastline is also being eroded but at a significantly slower rate than the southern coast. Its annual contribution of sediment is about 2 000 m^3 of sediment. The total contribution of sediment from coastal erosion processes is 24 000 m^3 each year.

The third source of material is from coral reefs that are established on offshore islands, at the northwestern end of bay, and at various locations along the south coast (Fig. 1). Sediments in the ebb tidal delta area contain about 33% $CaCO_3$, those along the subtidal areas of the south shore contain about 22%~23% $CaCO_3$. In the inner bay, the sediments contain only about 0.6%~0.9% CaCO₃. The distribution of CaCO₃ in outer bay sediments is consistent with the distribution of reefs along the coastline. It also shows that the sediment from the outer bay is not transported too far into the inner bay. The total amount of sediment from this biogenic source is estimated to be not greater than about 4 000 cubic m per year. The total amount of sediment supplied to the bay area from all sources is about 91 000 cubic m per year.

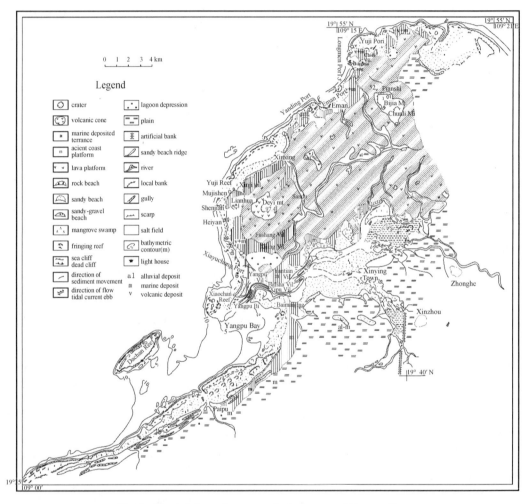

Fig 1　Coastal dynamic geomorphologic map of Yangpu Bay

Ⅳ. CONCLUSIONS AND DISCUSSIONS

1. The area of Xinying Bay is 50 sq km, the mouth of the bay (Baimajing Strait) has a cross-section area of 5 000 sq m. The mean tidal range is 1.81 m and the maximum tidal range is 3.8 m (Nov. 1988). The tidal prism is 100~200 million cubic m. The surface velocity of the mouth is above 1 m/s. The bottom velocity is above 0.25 cm/s. The huge area of water and large currents are the major dynamics for keeping the

channel open. In the Yangpu bay, ebb tidal velocity is larger than that of floods. Between the mouth of Xinying Bay and the ebb tidal delta, the current velocity decreases. But in general, the both velocities of surface and bottom are quite large. They can transport sediment particles of various grain sizes.

During the flood, the current overflows into the inner bay from the sand bank and the main channel. During the ebb, tidal currents are divided into two flows: the major flow, 3/4 of the total amount of waters, along the main channel, while the rest of the water passes through the sand bank. So the velocity is less than that of the channel.

The suspended particulate matter of Yangpu Bay is about 0.1 kg/m^3, of which 67% consists of easily transported quartz and tabular illite. Micro fossils, organism fragments and floccular carbonate take up 15.2%. Minerals include ferohydrite, basalt, pyroxene, acorite etc. In addition, there are fragmental and floccular calcium carbonate in the form of impalpable aggregate with a less than 0.008 mm grain size.

2. On the substrate map based on the grain size analysis, the coarse sand and gravel deposits in the mouth and along the large shoal; medium and fine sand mainly occupy the shoal of the Xinying Bay, the main channel, Yangpu shoal and along the coast of the Yangpu Bay; and clay appears only on the mangrove tidal flat at the east end of the Xinying Bay and in the main channel. From the main channel to the block gate shoal the sediment grain size displays a transition: sandy clay—clayey fine sand—fine sandy clay. Such a distribution suggests: (1) The developed estuarine deposit changes from coarse to fine from mouth to outside the bay, which indicates the sediment in the Xinying Bay comes from the rivers. (2) Sand takes up 80% in the main channel; 50%~70% on the shoal of the Xinying Bay. Hence, Xinying Bay's major sediment discharge is silt and clay. (3) The dominant sediment of the main channel and the block gate shoal is of sand which grain size and mineral constitution are consistent with that of the Zhanjiang Formation strata, which indicates the sand comes from the erosion of the Zhanjiang Formation. There is an area of fine sediment (clay) between the large shoal and the block gate shoal. The block gate shoal contains 40%~50% clay, similar to the main tidal channel, transported through the main tidal channel. The block gate shoal contains 20%~30% sandy sediment, which is not continuous to the shoal and is a little higher than that of the main channel. 33% of the block gate shoal sediment is CaCO$_3$, relating with Xiaochan (a small coral reef). Therefore, the sediment of the block gate shoal is mainly from Xiaochan.

3. Yangpu Bay is a tidal inlet, in which Xinying is a tidal water trap; the main channel is the tidal water pass; the block gate shoal is the ebb tidal delta. Core data and regional geomorphological analysis show that the main channel and the block gate shoal were originally a fossil river valley 8 500YBP. The valley's elevation is minus 25~30 m. The top surface of the river deposits has a minus 20 m elevation. The ancient river's

gradient was 0. 07%. The Dashui River and the Chun Jiang River joined the ancient river valley in the area of today's Xinying Bay, which finally entered the Daipu Day at the −20 m contour through the main channel and the block gate shoal. With the post-glacial sea level rise, the coastal plain was drowned as Yangpu Bay, and Xinying Bay. The ancient river valley was filled with marine sediment but was kept open as today's deep water channel because of the daily flushing and scouring of the tidal prism of Xinying Bay and Yangpu Bay. However, the ancient river valley section in the Xinying Bay and at the block gate shoal disappeared due to siltation.

Dating data from the block gate shoal shows that the average sedimentation rate was 0. 16 cm/a during the last 8 500 years, and during the last 3 000 years the rate was 0. 1 cm/a, which is consistent with the world wide sea level change processes. Hence, the natural formation process of the block gate shoal is slow.

We have worked out the erosion and siltation budget for the last decades of the area by comparing six 1947—1983 navigation maps and the bathymetric maps. In the work, we transferred the different depth bases of various maps into a same theoretical depth base, and then calculated the volumes of the block gate, the main channel and the large shoal at different periods. We find that the block gate shoal had an erosion rate 2∼10 cm/a during 1947—1974, and a siltation rate 7. 2∼4. 5 cm/a during 1974—1983; the main channel had an erosion rate of 1. 2∼5 cm/a during 1947—1974, and a siltation rate of 5 cm/a during 1974—1983. In general, however, the block gate shoal and the main channel are both subject to slight erosion during the last fifty years. The total erosion volume of the block gate shoal is one million cubic m and that of the main channel is 0. 66 million cubic m.

Therefore, we can see from the 8 000-year history, that the sedimentation rates of the block gate shoal and the main Chanel are very small. The topographic changes during the last 50 years suggest that the block gate shoal and the main channel are nearly stable with only intermittent slight erosion or siltation, which would not interfere in the construction of a harbour and deep water navigation channel in the area.

References

[1] Ding, Guoyu. 1964. On Some Problems About The Quaternary Geology of Hainan Island. In: Quaternary Geology of China, Science Press, Beijing. 207 – 233.

[2] Wang, Ying. 1977. The Submarine Geomorphology Of The South China Sea. In: Marine Geography, Science Press, Beijing. 33 – 35.

海南岛洋浦湾沉积作用研究[*]

　　洋浦湾位于海南岛西北部儋县,由新英湾及洋浦湾组成,是正在加速开发的天然良港。1983年,南京大学应交通部水运规划设计院、海南港务局邀请,进行洋浦港选址初步可行性研究。1984—1985年,为洋浦港一期工程,作者等赴现场做了全潮水文泥沙测验、大面积海岸地貌和港湾沉积调查,分析了海底表层及钻孔样。1988—1991年,中国—加拿大海洋科技合作计划组对洋浦海域做了地层剖面、柱状取样、海流、温度、盐度、密度、悬浮体等测验研究。本文以上述工作为基础对港湾沉积作用做一综合分析研究。

一、沉积环境

　　洋浦湾在地质构造上属于雷琼断陷南部,上新世至第四纪沉积了厚度超过3 000 m的滨海-浅海相碎屑岩建造(湛江组)。洋浦湾沿着东西向大断裂产生。在北岸,底部是湛江组,上部是由大面积的玄武岩流(52 000aB. P.)构成的台地[1],该处自全新世初期至今仍有抬升。南岸是湛江组砂砾层台地,其上部是未胶结的海相砂砾层。

　　新英湾是纳潮水域,面积50 km²,出口处(白马井峡)宽仅550 m。新英湾主要是堆积岸,有两条主要的河流(大水江和春江)在湾顶注入,其流域面积合计为1 419 km²。河流泥沙在河口堆积,岸线随三角洲发展而推进。据历史记载,中和镇和新州镇在唐朝时均为海港,后逐渐淤高成陆,至清朝时海港移至新英镇,而今海船(渔轮)只能至湾口(白马井)。目前,大水江、春江三角洲的面积占新英湾面积的50%。潮流作用将湾底侵蚀成一条条水道,同时将河口堆积中<0.01 mm的细粒泥沙带走,使河口沉积物中砂粒占90%以上。当河口浅滩逐渐增高时,外海潮流带来的悬移质泥沙沉积其上。由于浅滩上成片密集的红树林捕捞了悬浮体,促进了浅滩发育,从而形成了下部冲积相砂砾层和顶部薄层海相淤泥层的二元相河口潮滩沉积。

　　洋浦深槽从新英湾口内(白马井)延伸至拦门沙浅滩,长10 km,宽400～500 m,水深5～25 m,呈明显的河谷形态。深槽沉积层可分成三部分:底部为湛江组亚黏土,中层为河流相砂砾层,上部为海相砂质淤泥或淤泥。若把湛江组地层的顶面作深槽的原始谷底,则白马井为-24.34 m,至洋浦灯标为-28.03 m,拦门沙为-30.13 m,其平均坡降为0.58‰。深槽切在湛江组中,其走向与断裂构造走向(NE,NW)一致,是沿断裂谷发育的河谷。在全新世海面上升过程中,河谷上游段(新英湾内)及下游段(拦门沙向西海域)逐渐被海相沉积所充填。目前的深槽是由潮流作用,特别是新英湾内蓄纳的潮量落潮时冲刷所维持的。

　　拦门沙浅滩是深槽通向海域的浅水段,水深5 m,长400 m,向海坡度3%,与洋浦向

　　* 王颖,朱大奎:《第四纪研究》,1996年第2期,第159-167页。

海的自然坡度一致,拦门沙的沉积层与深槽相同,以湛江组砂质黏土为基底,上覆河流相砂砾层及海相淤泥层。拦门沙基底的原始地形也具河谷形态,由古河谷淤积而成。其地貌发育经历三个阶段:① 古河谷阶段,河谷中砂砾堆积厚 7 m,其年龄为 8 000aB. P. ;② 海侵-古河谷淤积阶段,洋浦湾平原成为海湾,古河谷成为潮汐通道,水道中沉积了新英湾带出的细粒沉积(砂质淤泥),其年龄为 8 000~6 000aB. P. ;③ 拦门沙形成阶段,6 000aB. P. 以来,拦门沙逐渐形成,出现一粗砂层(粗砂含细砾),表明南部湛江组地层有过强烈侵蚀,以后物质逐渐变细。3 000aB. P. 以来,沉积厚度是 70 cm。

洋浦湾南部为一片大浅滩,高出海图 0 m 的面积为 2.5 km²,是侵蚀湛江组地层形成的侵蚀-堆积阶地,在湛江组侵蚀面上覆盖 0~2 m 现代海相砂砾。根据 1955—1980 年水深资料对比,0 m 线以上的浅滩面积增加 0.7 km²,而水深 5 m 的面积减少 1 km²,即浅滩在扩大,水深较大的面积在缩小。

自 1984 年 1 月至 1985 年 1 月,及 1988 年 11 月,分 4 次对洋浦湾做了 15 个断面近 40 船次的全潮水文泥沙测验,获得了洋浦湾潮流的基本特征。洋浦湾为不规则日潮,平均潮差 1.81 m,1988 年 11 月曾测得最大潮差 3.8 m。落潮历时(12 h)小于涨潮历时(13 h)。平均涨潮流速约为 22 cm/s,而平均落潮流速是 27 cm/s,但在不同水层中有所差别。涨潮时,表层平均流速是 23 cm/s,而中层和底层均是 24 cm/s;落潮时,表层为 31 cm/s,而中层和底层分别力 27 cm/s 和 24 cm/s。在落潮时,从口门向海流速减小,其最大值为 97 cm/s。总的看,底层及上部各层流速均大,能搬运该海区粉砂和细砂颗粒。落潮的最大流速一般大于涨潮,故沉积物净的输送方向是从内港(新英湾)向外海。余流是控制沉积物运动的重要因素,洋浦湾的余流主要受径流影响,其方向受水下地形(深槽)控制,表 1 所列大体代表冬季,而夏半年随着径流的增大,余流亦将增强。

表 1　洋浦港最大潮流速和余流流速及方向

测站	最大潮流速/cm·s⁻¹						余流的流速/cm·s⁻¹　　余流的方向/°					
	表层		中层		底层		表层		中层		底层	
	涨潮	落潮	涨潮	落潮	涨潮	落潮	流速	方向	流速	方向	流速	方向
白马井	51	84	88	86	84	97	6	187	3	95	3	102
洋浦村	42	68	67	53	56	45	12	301	11	116	12	107
拦门沙	57	54	49	75	22	56	7	230	5	164	8	215

影响该海区的风主要有 NNE—ENE 及 SSW—WSW。前者是强风,但因系离岸方向,故吹程小而不能形成大浪。SW 向风来自开敞海区,形成大浪,促使洋浦湾外港海岸侵蚀,同时这方向的风浪在湾内受阻,使洋浦湾成为船只的避风锚地。该区风浪占 78%,涌浪 12%,而 SW 向波浪频率最大,强度最大,其平均波高 0.83 m,平均周期 4.5 s。由于破波水深为 0.8 m,故其最大侵蚀作用产生在大浅滩的西南边缘,而沉积物在西南向风浪作用下再悬浮并向东北搬运,沉积在大浅滩顶部。

悬浮沉积物的含量平均值为 100 mg/L。在白马井测站将表、中、底各层综合,其涨潮时为 88 mg/L,落潮时稍大,为 103 mg/L,洋浦村涨潮 106 mg/L,落潮 119 mg/L,而至拦门沙略有降低,涨潮 102 mg/L,落潮 102 mg/L。可见,洋浦湾海域悬移沉积物含量是相当低的。

温度和盐度的实测剖面表明,它们在一个潮周期中随时间而有变化。当高潮时,整个水体流速降至很小,其温度、盐度及悬浮体随水深的分布是一致的,即水柱内无分层的特性。落潮时,水流向海而流速逐渐增加,其流向与流速在各深度大致相近,温度和盐度在各水层中也是一致的,这表明落潮时整个水柱内是混合的。在底部,悬浮体是增加的,同样表层悬浮体也增加,它们并不反映水温和盐度的变化。当落潮至一半时,整个水层流速都最大,这时悬浮体集中在水底层(占 30%)。以后,底流速开始降低,而底层悬浮体也相应地降低。这时表层的盐度为 31.8‰,而底层的盐度只是稍有降低。

低潮时,流速在各水层中均已减弱,流向是朝外海。温度和盐度在表层是低的,而随深度增大而增加。悬浮体在表层是均一的,含量较低,亦随水深而增大。当从低潮进入涨潮,流速场随着水深产生延迟现象,底流速开始增加,流向朝岸,而表流速仍是很低。这时底层温度、盐度亦开始增加,即较咸的海水向湾内运动。在底层,盐度的增加与悬浮体含量的增加是同步的。而后中层水流速增加并向岸运动,但与温度、盐度及悬浮体的变化并不同步。更后,表层流速增加,但其流向与潮流运动方向成直角,其盐度约为 25‰。最后,流速增大到整个水层一致,流向朝岸,同时温度、盐度及悬浮体含量在整个水层达到均一状态。

二、地层剖面测量与 ^{14}C 年龄测定

为了解洋浦湾地层结构及港湾的演化,工作中使用了 ORE 地脉冲地震剖面仪做地层结构的测量,同时也用 EG&G 100 kHz 旁侧声呐做了同步测量。洋浦深槽是一古河谷,可从剖面记录中分辨出 3 个埋藏的谷地形态,它们分别在现代海底以下 9 m、15 m 及22 m,埋藏谷地中充填了成层的沉积层。在西浦(新建码头处),其剖面同样记录到老谷形态,其分界面分别在水底以下 3 m、12 m 和 18 m。这些剖面是用声速为 1 500 m/s 测得的。同样,剖面记录了拦门沙浅滩亦有埋藏古谷地形态,其表面覆盖了现代砂质沉积,厚仅 2 m。

洋浦湾表层沉积物的粒度分布表明,粗砂、砾石主要分布于河口区及洋浦大浅滩的边缘,砂质沉积物主要分布于新英湾内浅滩、洋浦大浅滩及洋浦湾沿岸,黏土分布于湾顶红树林滩地及洋浦深槽的深泓。从白马井口门经深槽至拦门沙的粒度是砂质黏土—黏土质砂—细砂质黏土。

沉积物来源主要有 3 类:河流的、海岸侵蚀的及珊瑚礁生物的。大水河从东北面注入洋浦湾内,其流域面积 648.3 km²,年输沙量 7.4×10⁴ t。春江从东南注入湾内,流域面积577.8 km²,年输沙量 6.7×10⁴ t。沿岸一些小河其流域面积总计为 81.3 km²,年供沙1.5×10⁴ t。所以河流供沙总计为 15.6×10⁴ t/a,相当于 6.2×10⁴ m³/a。

海岸侵蚀提供的沉积物亦很重要,南岸湛江组地层海蚀岸的侵蚀量,据地貌调查,2.5 km 岸线平均后退 0.5 m/a,据 3 年测波资料和《海洋水文》规范求得沿岸自南向北输沙量为 2.2×10⁴ m³/a。北岸玄武岩侵蚀岸,以海蚀崖后退速率求得为 0.2×10⁴ m³/a。故

总的侵蚀成因的沉积物供应量为每年 $2.4 \times 10^4 \mathrm{m}^3$。

生物成因的沉积物主要是珊瑚碎屑,湾外有小铲($2 \mathrm{km}^2$)、大铲($4 \mathrm{km}^2$)两个珊瑚岛,其四周均为珊瑚礁沉积。拦门沙沉积物中 $CaCO_3$ 含量 33%,其南岸为 22%～23%,而湾内大水河口区仅 0.6%～0.9%,显然湾外部分沉积物中 $CaCO_3$ 物质来自邻近的珊瑚岛礁,据此可算出供应量为 $0.4 \times 10^4 \mathrm{m}^3/\mathrm{a}$。所以,洋浦湾沉积物供应量每年为 $9.0 \times 10^4 \mathrm{m}^3$。

在拦门沙,有 3 个钻孔做了 ^{14}C 年龄测定,其结果如表 2 所示。

表 2　洋浦湾拦门沙钻孔资料及其 ^{14}C 年龄测定结果

孔号	水深/m	深度/m	岩性	^{14}C 年龄/aB. P.
A-2	5.75	0～4.7	粉砂质黏土	2 930±100(1.3)* 4 780±120(2.5) 4 980±120(3.4)
		4.7～10.05	致密的黏土 (起源于湛江组)	
A-3	5.02	0～4.39	粉砂质黏土	1 880±85(0.7) 3 220±100(2.5)
		4.39～7.35	中细砂	6 350±150(6.1)
		7.35～16.15	砂、黏土	8 500±230(13.6)
		16.15～20.00	砂(河流相)	
		>20.00	湛江组地层	
A-4	5.41	0～4.55	粉砂质黏土	2 310±90(1.3) 5 010±140(3.0)
		4.55～7.55	砂质黏土	5 840±140(5.8)
		7.55～9.50	湛江组地层	

* 括号内为样品深度(m)。

三、现代沉积柱状样分析

工作中用 Lehigh 重力取样管采取不扰动海底柱样,以研究港湾的沉积过程。

Y001 取自新英湾的深槽中,水深 20.4 m(该处有强的流速)。0～6 cm 为深灰色砂质软泥,带有贝壳碎片;6～15 cm 为深灰色砂质泥带有粗砂;15～35 cm 为深灰色砂质泥,有粗砂透镜体。

Y006 取自新英湾内深槽,水深 11.4 m。0～2 cm 为灰黄色粉砂质黏土,含有细砂颗粒,是一种呈流动状态的砂质泥,是风暴作用的沉积;2～17 cm 为浅灰色的砂和贝壳碎屑,混有细砾;17～32 cm 为深灰色中砂混有黏土,黏土呈斑点状或呈泥球状,泥球中包含有贝壳碎屑。

Y002 取自西浦村外深槽,水深 10.22 m。0～19 cm 为灰黄色流动状软泥,亦为新近的风暴天气沉积,含有粉砂,底部有深色有机质泥球;19～45 cm 为深灰色砂质泥混有粗砂及海胆碎片,与上、下层之间均为不连续沉积;45～63 cm 为深褐灰色砂质泥,有少量粗砂及贝壳碎屑,与其下层亦不连续;63～140 cm为褐灰色砂质泥,在100～110 cm处有一粗砂贝壳碎屑的扁豆体及粗砂的夹层,可能是虫孔沉积镶嵌在泥层中;140～260 cm为灰色泥及分选好的黏土质粉砂,有充填了砂及贝壳碎屑的虫穴,与上覆盖层有明显的分界。

　　Y007取自西浦村西的深槽中,水深11.0 m。0～4 cm为灰黄褐色细砂和粉砂,有生物碎屑;4～22 cm为灰色细砂混有粉砂贝壳碎屑,与上、下层均不连续;22～24 cm为深色的黏土带条;24～62 cm为灰色黏土质粉砂,是一整体的均质沉积层;62～187 cm为灰色砂、粉砂和黏土,具均质的沉积带条,在62～63 cm,92～93 cm及117～119 cm处有很明显的颜色特别深的条带,是具H_2S臭味还原的沉积物,在178～187 cm间有一些细砂的透镜体,亦反映了生物的作用,整个层与上、下层均不连续。187～198 cm为黄灰色粉砂、黏土,有众多的生物扰动形成的砂质透镜体;198～250 cm为灰色砂、粉砂和黏土,有许多生物作用形成的砂质扁豆体和透镜体,占了大部分的面积。

　　Y003取自洋浦村外深槽,水深20.65 m。0～9 cm为黄灰色砂、粉砂和黏土,并含有贝壳碎片及粗砂颗粒;9～22 cm为灰色粗-中砂混杂着黏土斑块;22～58 cm为灰色粗砂、中砂夹贝壳碎片;58～66 cm为灰色砂、粉砂和黏土。这些层明显与上、下层不连续;66～108 cm为灰色粉砂质砂,带有黏土斑点;108～148 cm为灰色黏土质砂和粗砂,构成韵律的条带,带有间断的贝壳碎屑条带。

　　Y004位于拦门沙靠近深槽一侧,水深仅4.11 m。0～10 cm为深灰色砂质泥夹小贝屑,含较多黏粒;10～29 cm为褐灰色砂、粉砂和黏土,构成1 mm纹层,层内生物碎屑较多,顶部为侵蚀的表面;29～32 cm为深灰色细粉砂,呈1～2 mm厚的纹层;32～37 cm为灰黄色砂、粉砂和淤泥,含贝壳,呈1 mm或更薄些的纹层;37～51 cm为灰黄色黏土质泥,石英砂粒磨圆好,有充填了粉砂颗粒的虫穴;51～100 cm为灰黄色黏土质粉砂,底层含较多的贝壳碎屑,有一些充填了黏土质粉砂的虫穴;100～114 cm为深灰色砂质泥,含有大量贝壳碎屑及粗砂,与下部沉积层不连续;114～121 cm为深灰色泥,具有砂及贝壳碎屑充填的虫穴。

　　Y005位于拦门沙向海一侧,水深6.78 m。0～12 cm为深灰色砂质泥,分选性差,粗砂杂乱分布不成层;12～53 cm为深灰色砂和泥的混杂沉积,黏土呈厚约4 cm的条带状,层的界面不规则,砂层厚约1 cm,含有大量贝壳碎片,可能是风暴沉积;53～70 cm为深灰色黏土质粗砂,含细砾及贝壳碎屑。该柱整个沉积物都比拦门沙内侧的粗。

　　总之,洋浦湾的沉积物代表了潮汐汊道的沉积,具有双向的特征,受现代人类活动的影响极小。自晚全新世以来,沉积体系只有很微弱的变化,反映了主要的水动力的影响。

　　在新英湾,主要是由砂及细砾组成的河流三角洲沉积,间隔着潮流搬运的细粉砂,但此双向沉积体系中以河流作用沉积为主。

　　在洋浦深槽,其特征为砂(或粉砂)和粉砂(或黏土)的沉积,是以潮流作用为主的双向结构。在深槽的不同位置流态不同,因而具有不同的结构:Y003沉积柱反映了强的流速,沉积了较粗的颗粒;Y007以粉砂为主,反映了沉积时期流速减弱的过程,大量的砂质夹层反映在该部位风暴作用起了重要影响。因此,洋浦深槽中以潮流沉积作用为主导,同时有风暴的影响。

　　在拦门沙浅滩,主要是细砂沉积,反映了波浪作用及落潮流的动力平衡。其特征是遮掩区域为粉砂质砂,而外围是中砂、粉砂及黏土,外海受风暴作用而颗粒较粗。

这些样品经^{210}Pb同位素测定,获得洋浦湾各区域的现代沉积速率(表3)。自湾内向深槽、拦门沙浅滩至外海,沉积速率逐渐增加,但即使是外海也只有每年2 cm,所以整个洋浦湾海域是稳定的。

表3　洋浦湾沉积速率及^{210}Pb残余量

孔号	位置	水深/m	平均密度/g·cm^{-3}	沉积速率/cm·a^{-1}	^{210}Pb残余量/DPM·cm^{-2}
Y006	湾内,大水江三角洲水道	11.4	1.0	>0.8	
Y001	湾内,近口门处	20.4	1.2	>1.0	
Y007	西浦附近深槽	11	1.0	1.47	256
Y002	西浦附近深槽	10.22	1.0	0.87	184
Y003	深槽中最深处	20.65	1.1	0.52	29.9
Y004	拦门沙浅滩	4.11	1.1	1.06	73.1
Y005	拦门沙向海侧	6.78	1.2	>2.0	

四、结论

(1) 洋浦湾为一潮汐汊道港湾,新英湾是汊道体系的纳潮水域,深槽是潮流通道,拦门沙浅滩是落潮流堆积体。据钻探及地貌分析,深槽及拦门沙原为一古河谷,谷底标高在−25～−30 m,河流沉积层的顶面为−20 m,古河谷纵比降0.7‰。海面上升过程中淹没为洋浦湾和新英湾,古河谷为海相沉积所充填。洋浦湾和新英湾由于涨落潮流的冲刷使古河谷保留了深槽的水深,而新英湾内及拦门沙段的古河谷被淤积而消失。

(2) 新英湾面积50 km^2,口门断面仅5 000 m^2,平均潮差1.81 m,实测最大潮差为3.8 m,所以纳潮量为$(1\sim2)\times10^8$ m^3,平均流速约1 m/s,巨大的潮流量及大的流速使深槽中沉积速率很低,不被充填。

洋浦湾落潮流速大于涨潮流速,从新英湾口门经深槽至拦门沙外海,潮流流速稍有减少,但总的看,表层及底层流速均较大,能搬运该区域各粒级沉积物输向外海,使洋浦湾沉积过程非常缓慢。

(3) 拦门沙8 500aB. P. 以来的平均沉积速率是0.16 cm/a,3 000aB. P. 以来的平均沉积速率是0.1 cm/a,此数值同世界海面上升相适应。因此,洋浦湾自然演化是缓慢的。近百年来的沉积速率,在新英湾内为0.8～1 cm/a,深槽中为0.5～1.5 cm/a,拦门沙浅滩大于2 cm/a,这些变化仍然相当微小。

参考文献

[1] 王颖,周旅复. 1990. 海南岛西北部火山海岸. 地理学报,45(3):321-330.

海南岛潮汐汊道的现代沉积特征研究[*]

1　引　言

海南岛有许多潮汐汊道(也称深槽或通道),可归纳为两种类型:港湾溺谷型和沙坝潟湖型。潮汐汊道体系中的汊道通常被用作进出港口的深水航道[1]。因此,潮汐汊道的研究,特别是稳定性研究,具有重要的现实意义。潮汐汊道现代沉积特征和沉积模式的研究可以为判别未知沉积、恢复古地理环境提供识别标志,为潮汐汊道的稳定性提供判别依据[2,3]。

2　两类潮汐汊道的地貌发育体系

港湾溺谷型洋浦潮汐汊道(图 1)是低海面时流水沿玄武岩与湛江组接触带冲刷形成的河谷,冰后期高海面时被淹没后,经涨落潮流的再造所形成。汊道南岸是由"湛江组"沉积物组成的松散堆积体,北岸为玄武岩基岩海岸。地貌组成单元有:深槽、拦门沙(落潮流三角洲)、涨潮流三角洲。深槽紧挨着北岸发育,呈反 S 型大致沿北东方向延伸;其起点为白马井角,终点为小铲附近拦门沙;全长 8 km,0 m 等深线之间宽度 400～500 m,最窄处 300 m;水深 10～20 m,最深处 24.5 m;深槽口门宽 600 m,水深 22 m。涨潮流三角洲发育在新英湾内东浦村东南侧,面积约 1 km²,水深 2～5 m。拦门沙发育在深槽末端,小铲附近,最浅处水深 5 m,水深小于 8.5 m 的长度达 3.5 km。洋浦大浅滩是一个特殊的地貌单元,紧挨着深槽南侧发育,是波浪、潮流和沿岸流带来的泥沙在湛江组浪蚀平台上堆积形成。新英湾是洋浦潮汐汊道的纳潮水域,总面积 50 km²,纳潮量 4.55×10^7 m³,有 87% 的面积位于零米等深线以上,纳潮水域主要有大水江与春江水沙汇入,年径流量分别为 24.1 亿 m³ 和 2.7 亿 m³,供沙分别为 7.39 万 t/a 和 4.38 万 t/a。

沙坝潟湖型三亚潮汐汊道(图 2)是由多条水下沙坝在不同时期因地壳抬升或海面下降拦封海湾所形成。汊道南岸为珊瑚礁海岸,北岸为沙坝。地貌单元有:深槽(汊道)、拦门沙(落潮流三角洲)(在 20 世纪 60 年代建三亚港时被挖掉)、涨潮流三角洲。通道走向近东西向,全长 300 m 左右,最大水深 6 m,口门处水面宽 80 m,平均水深 3.6 m,最大水深 6.3 m。拦门沙在开挖前长约 850 m,水深不足 4 m,最浅处 2.7 m。涨潮流三角洲发育在口门内侧,水深不足 2 m。三亚潮汐汊道的纳潮水域为两条南北走向的狭长潟湖,分别有月川水与大坡水两条河的水沙注入。纳潮水域面积为 1.1 km²,其中滩地占 81.8%,大

　　* 邵全琴、王颖、赵振家:《地理研究》,1996 年第 15 卷第 2 期,第 84 - 91 页。南京大学海岸与海岛开发国家试点实验室资助(SCIEL21196122)。

潮纳潮量为 $0.9 \times 10^6 \text{ m}^3$，水深小于 5 m，其中东潟湖只有 2 m 左右的水深。

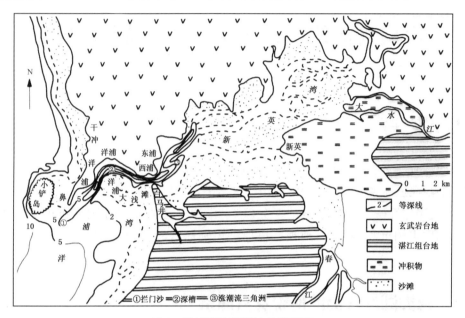

图 1 洋浦潮汐汊道地貌类型图

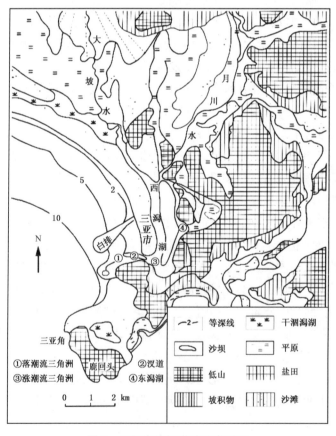

图 2 三亚潮汐汊道地貌类型图

从表1可以看出,两类潮汐汊道的涨潮流三角洲均不如落潮流三角洲发育:洋浦潮汐汊道拦门沙(落潮流三角洲)的发育系数(拦门沙长度与深槽长度之比)比0.44,涨潮流三角洲的发育系数(涨潮流三角洲长度与深槽长度之比)为0.12,这主要是因为拦门沙是在古河谷的水下三角洲基础上发展起来的,而且落潮流因为有河流径流的加入,携带泥沙量因此也大于涨潮流;三亚潮汐汊道的拦门沙(落潮流三角洲)和涨潮流三角洲的发育系数分别为2.8和1左右,这是因为拦门沙不仅有落潮流带来的沉积物,还有沿岸流带来的沉积物。

表1　两类潮汐汊道的地貌体系

地貌单元	深槽(汊道)			拦门沙(落潮流三角洲)		涨潮流三角洲		纳潮水域	
	长 km	宽 m	深 m	长 km	深 m	长 km	深 m	面积 km²	纳潮量 m³
洋浦(港湾溺谷型)	8	300～600	10～24.4	3.5	5～8.5	1±	2～5	50 (87％在0 m以上)	4.55×10⁷
三亚(沙坝潟湖型)	0.3	80±	<6	0.85	2.7～4	0.3	2～5	1.1 (81.1％为滩地)	0.9×10⁶

港湾溺谷型潮汐汊道的发育规模大于沙坝潟湖型,但涨落潮流三角洲的发育系数小于沙坝潟湖型潮汐汊道,这主要是两者成因不同所造成。这反映了港湾溺谷型潮汐汊道比沙坝潟湖型稳定,前者适合于做大型海港,后者适于做中小型海港。

3　两类潮汐汊道的现代沉积特征

3.1　粒度特征

沉积物的粒度性质主要受物源及沉积环境两个方面因素控制。

从表2可以看出,洋浦港湾型潮汐汊道:深槽(汊道)沉积物相对较细,除口门外均是黏土质粉砂和粉砂质黏土;拦门沙(落潮流三角洲)和涨潮流三角洲沉积物相对较粗,为砂质黏土和黏土质粉砂。三亚沙坝潟湖型潮汐汊道则相反:汊道中沉积物相对较粗,为中细砂;拦门沙和涨潮流三角洲的沉积物相对较细,分别为泥质砂和细砂。造成这一差异的主要原因是:洋浦深槽中有8.6 m左右厚的粉砂质黏土沉积物,而深槽正处于冲刷状态。

表2　两类潮汐汊道的粒度特征

粒度特征		深槽	拦门沙	涨潮流三角洲	纳潮水域
洋浦(港湾溺谷型)	组成物质	粉砂质黏土、黏土(口门为砾质砂)	粉砂质黏土、黏土质粉砂	黏土质粉砂	粉砂质砂、粗中砂、砾质砂
	中值粒径与分选系数(一元回归)	$Q_{d\phi}=7.9-0.83M_{d\phi}$ $N=10$ $R=-0.923\,4$ $R_{0.05}=0.576$	$Q_{d\phi}=0.69+0.025M_{d\phi}$ $N=10$ $R=-0.599\,3$ $R_{0.05}=0.576$		$Q_{d\phi}=-0.267\,6+0.631\,3M_{d\phi}$ $N=12$ $R=0.895\,3$ $R_{0.05}=0.576$

（续表）

粒度特征		深　槽	拦门沙	涨潮流三角洲	纳潮水域
三亚 （沙坝 潟湖型）	组成物质	中细砂	泥质砂	细　砂	细砂、黏土质粗 砂中砂、砂质黏土
	中值粒径 与分选系 数 （一元回归）	分选系数约0.85		分选系数约 0.65	$Q_{d\phi}=1.23-$ $0.203M_{d\phi}$ $N=8$ $R=-0.877$ $R_{0.05}=0.86$

3.2　重矿物特征

沉积物中重矿物含量的高低取决于动力、沉积物粒径和物源三要素的综合作用。重矿物一般富集于细砂，在粒度和物源相同的情况下，动力越大，重矿物含量越大。ZTR 指数（成熟度）是由锆石、电气石、金红石（全是稳定矿物）组成的透明矿物组成的百分含量。根据成熟度和重矿组合以及特征矿物可以追踪沉积物的泥沙来源。

从表3可以看出，洋浦深槽沉积物的重矿含量高于拦门沙和涨潮流三角洲；深槽沉积物的成熟度也高于拦门沙和涨潮流三角洲；拦门沙和涨潮流三角洲沉积物中的重矿组合不同，反映三者泥沙来源不同。

表3　两类潮汐汊道的重矿特征

地区	重矿特征	深　槽			拦门沙	涨潮流 三角洲	纳潮水域	
		口门内	口门	口门外				
洋浦 （港湾溺 谷型）	含量(g)	1.2	0.6	3.0	0.6	0.3	3.0	
	ZTR	30	40	12	8	18	16	
	重矿组合 （10%以上）	电气石(26%) 帘石类(20%) 辉石(14%)	锆石(25%) 帘石类(17%) 辉石(14%)	钛铁矿 (48%)	辉石(31%) 角闪石(19%) 白钛石(13%)	帘石类 (31%) 角闪石 (11%)	钛铁矿 (59%) 锆石(12%)	
三亚 （沙坝潟 湖型）	含量(g)	0.49			0.28	0.3 （东潟湖）	0.26 （西潟湖）	
	ZTR	10			32	25 （东潟湖）	27 （西潟湖）	
	重矿组合 （10% 以上）	钛铁矿(37%) 磁铁矿(24%)			磁铁矿 (28%) 钛铁矿 (27%) 白钛石 (16%)	磁铁矿 (52%) 锆石 (21%) 钛铁矿 (18%)	钛铁矿 (44%) 锆石 (20%) 磁铁矿 (19%)	

三亚潮汐汊道的汊道沉积物中重矿含量高于涨潮流三角洲；汊道沉积物的成熟度则低于涨潮流三角洲；各地貌单元沉积物的重矿组合一致，反映沉积物来源相同。

三亚沙坝潟湖型潮汐汊道沉积物中重矿含量普遍低于洋浦港湾溺谷型潮汐汊道，反映了两类潮汐汊道的物源与动力的差异。三亚沙坝潟湖潮汐汊道重矿组合单一,沉积物的重矿组合均为:钛铁矿、磁铁矿和锆石;洋浦港湾溺谷型潮汐汊道的重矿组合复杂,深槽、拦门沙与涨潮流三角洲沉积物的重矿组合分别为:钛铁矿+电气石+锆石+帘石类+辉石、辉石+角闪石+白钛石、帘石类+角闪石。这是因为三亚地区母岩是花岗岩,洋浦地区母岩有玄武岩、湛江组沉积物以及热液蚀变沉积物。

3.3　地球化学特征

Al+K+Fe 通常被认为是沉积物中泥质含量的指标。一般地,淡水沉积物的 Sr/Ba<1,海相沉积物的 Sr/Ba>1。

从表 4 可以看出洋浦深槽的 Al+Fe+K 高于拦门沙,反映深槽沉积物含泥量高、沉积物细;深槽沉积物的 Sr/Ba 小于拦门沙,反映拦门沙的海相性高于深槽。

表 4　两类潮汐汊道的地球化学特征

地区	地球化学特征	深槽		拦门沙	涨潮流三角洲	纳潮水域	
		口门内	口门				
洋浦(港湾溺谷型)	Sr/Ba	1.04	0.51	3.52		0.27	
	Al+Fe+K(ppm)	9.4×10^4	20.2×10^4	8.4×10^4		13.8×10^4	
	Ca(ppm)	3.04×10^4	1.5×10^4	6.3×10^4		0.40×10^4	
三亚(沙坝潟湖型)	Sr/Ba				0.332	0.277(W)	0.310(E)
	Al+Fe+K(ppm)				5.3×10^4	3.4×10^4(W)	4.5×10^4(E)
	Ca(ppm)				0.50×10^4	0.4×10^4(W)	0.7×10^4(E)

注:E 指东潟湖,W 指西潟湖。

3.4　有孔虫特征

单位重量(或体积)样品中有孔虫数量的多寡是沉积物沉积速度的标志[4,5]。一般地,有孔虫含量高反映陆源碎屑沉积速度慢。反之,陆源碎屑沉积速度快。复合分异度 $H(s)$ 采用熵函数公式 $H(s)=\sum\limits_{i=1}^{s}P_i\ln P_i$ 计算,其中,S 为样品中某一门类生物种(属)数,P_i 为第 i 个种(属)的个数 N_i 在全群总个数 N 中所占的比例 $\left(P_i=\dfrac{N_i}{N}\right)$,$\ln P_i$ 为 P_i 的自然对数。$H(s)$ 越高,反映沉积物的海相性程度越高。多变度(v)为将壳体或壳瓣累计占壳体或壳瓣总数 5% 以下的罕见种除去,其余占总数 95% 之内的种数。优势度(dm5)指每个样品中前五个优势种的百分比总和。多变度(v)越大,优势度越小,反映沉积物海相性程度越高。

从表 5 中可以看出,洋浦港湾型潮汐汊道,拦门沙沉积物中的底栖有孔虫的含量、种类、复合分异度、多变度均高于深槽;优势度和低盐种类(Ammonia)的含量低于深槽。反映拦门沙沉积物的海相性高于深槽,即拦门沙处受河流径流影响小。

表5　两类潮汐汊道的有孔虫沉积特征

地区	有孔虫沉积特征	深槽（汊道）	拦门沙（落潮流三角洲）	涨潮流三角洲	纳潮水域
洋浦（港湾溺谷型）	含量	较高	高		低
	种类（种）	34	54		26
	$H(s)$	3.3	3.7		1.6
	dm5（%）	49	37		95
	v（种）	21	33		7
	组合特征（含量在10%以上）	Ammonia（28%）Elphidium（26%）Quinoueloculina（20%）Spiroloculina（12%）	Quinqueloculina（22%）Elphidium（16%）Triloculina（12%）Ammonia（10%）		Ammonia（89%）
三亚（沙坝潟湖型）	含量（枚/克）	86		68	7
	种类（种）	62		28	14
	$H(s)$	3.6		2.2	2
	dm5（%）	50		76	92
	v（种）	42		17	5
	组合特征（含量在10%以上）	Ammonia（38%）Elphidium（20%）Quinqueloculina（12%）		Ammonia（62%）Elphidium（25%）	Ammonia（80%）

　　三亚沙坝潟湖型潮汐汊道,汊道沉积物中的底栖有孔虫含量、种类、复合分异度、多变度均高于涨潮流三角洲;优势度和低盐种类(Ammonia)低于涨潮流三角洲。这反映深槽(汊道)沉积物的海相性高于涨潮流三角洲,即汊道受河流径流的影响小于涨潮流三角洲。

　　三亚潮汐汊道和洋浦潮汐汊道沉积物中的有孔虫均为底栖有孔虫,而且以低盐种类卷转虫(Ammonia)和希望虫(Elphidium)为主;三亚月川水河口沉积物中有大量的货币虫(Nummulites)和花篮虫(Cellanthus),而洋浦则没有。

4　结　论

　　潮汐汊道是海岸带特殊的地貌体系,海陆交互作用最为强烈,沉积环境复杂。潮汐汊道的成因决定了其发育规模,而物源条件、动力条件、气候条件等因素的综合影响决定了其沉积特征。通过对洋浦港湾溺谷型潮汐汊道和三亚沙坝潟湖型潮汐汊道的研究[6,7],对两类潮汐汊道的现代沉积特征可以得出以下几点结论:

　　(1)港湾溺谷型潮汐汊道和沙坝潟湖型潮汐汊道的发育模式异同点:港湾溺谷型潮汐汊道发育规模大,沙坝潟湖型小;港湾溺谷型潮汐汊道的涨落潮流三角洲的发育系数小于沙坝潟湖型,前者为 0.12 和 0.44,后者为 1 和 2.8。两类潮汐汊道体系均有通道(或称汊道、深槽)、拦门沙(落潮流三角洲)和涨潮流三角洲组成,而且拦门沙比涨潮流三角洲发育。

(2) 港湾溺谷型潮汐汊道和沙坝潟湖型潮汐汊道的沉积模式异同点:港湾溺谷型潮汐汊道沉积物相对较细,涨、落潮流三角洲沉积物相对较粗,沙坝潟湖型潮汐汊道则相反;沙坝潟湖型潮汐汊道内的沉积物普遍粗于港湾溺谷型潮汐汊道;港湾溺谷型潮汐汊道沉积物中重矿物含量和成熟度(ZTR 指数)均高于沙坝潟湖型潮汐汊道;港湾溺谷型潮汐汊道、涨潮流三角洲和拦门沙沉积物中的重矿组合不一,沙坝潟湖型则一致;两类潮汐汊道沉积物中有孔虫均为底栖有孔虫,且均以卷转虫(*Ammonia*)和希望虫(*Elphidium*)为主;拟日货币虫(*Nummultes*)和花篮虫(*Cellanthus*)在三亚月川水河口沉积物中大量出现,洋浦则没有。

(3) 港湾溺谷型潮汐汊道的发育规模大于沙坝潟湖型潮汐汊道,而且港湾溺谷型潮汐汊道比沙坝潟湖型潮汐汊道稳定。因此,前者适合于做大型海港,后者适于做中小型海港。

“现代沉积是打开古环境的钥匙”,潮汐汊道的现代沉积特征研究能为识别古海岸沉积环境提供判别依据[8~11]。潮汐汊道通常被人类建设为海港。因此,汊道的稳定性及拦门沙的开挖问题最受关注,沉积特征是动力、物源、气候等因素综合作用的静态反映,只能作为判别稳定性及拦门沙可否开挖的依据之一,还需对潮汐汊道体系的动力特征进行深入的研究。其实,潮汐汊道体系是由若干个相互关联的子系统构成的一个有机整体,处于动态平衡之中,一个子系统发生变化,则其他子系统也发生相应的变化,使整个系统达到一个新的平衡。纳潮水域是一个重要的子系统,潮棱柱大小决定汊道水深。因此,有关部门应注意保护纳潮水域,减少围垦工程。

参考文献

[1] 朱大奎,邵全琴.1987.沉积学在海港选址研究中的应用.海洋工程,(2):69 - 76.

[2] 王文介.1984.华南沿海潮汐通道类型特征的初步研究.南海海洋科学集刊,5:19 - 29.

[3] 任美锷,张忍顺.1984.潮汐汊道若干问题.海洋学报,6(3):352 - 360.

[4] 汪品先等.1980.第四纪地层微体化石的研究方法及其应用.见:海洋微体古生物学论文集.北京:海洋出版社.172 - 191.

[5] 汪品先等.1980.北部湾第四纪晚期的微体化石群与海面升降的初步探讨.见:海洋微体古生物学论文集.北京:海洋出版社.140 - 145.

[6] 宋朝景.1984.海南岛东南岸地貌特征与潮汐通道.南海海洋科学集刊,5:31 - 50.

[7] 王文介,李春初,杨干然.1978.洋浦港港湾地貌的形成和发育.见:南海海岸地貌学论文集,3:80 - 100.

[8] Johnson, W J. 1973. Characteristics and Behavior of Pacific Coasts Tidal Inlets. Proceedings of ASCE. *Journal of the Water ways*, *Harbors and Coastal Engineering Division*, 99(3): 325 - 339.

[9] Bruun, P. 1978. Stability of Tidal Inlets, Theory and Engineering. Elsevier Scientific Pub. Co.

[10] Escoffier, E F. 1977. Hydaulics and Stability of tidal inlets. GITI 13.

[11] Richard A. and Delvis Jr. 1978. Coastal Sedimentary Environments. Springer-verlag, U. S. A.

海南岛洋浦深槽与拦门沙沉积
特征和沉积环境分析[*]

1　前　言

1.1　潮汐汊道的国内外研究概况

自 L. J. Leconte 和 M. P. O'Brien 于 20 世纪 30 年代初提出潮汐汊道理论以来,由于潮汐汊道在海岸航运上的重要性,以及在海岸演变过程、海洋和陆地之间物质和能量交换、口内海湾或潟湖生态和环境保护等方面意义重大,促使人们长期以来十分重视对这一特殊海岸动力地貌类型的研究[1],潮汐汊道的研究取得了很大进展。国外对于潮汐汊道的研究基本上可以分为两个方向[2]:① 从动力学的角度研究潮汐汊道(主要是堆积海岸的潟湖型潮汐汊道)(M. P. O'Brien,P. Bruun,J. T. Jarrett,T. Shigemura,G. H. Keulegan,E. F. Escoffier,T. L. Walton 等),通过对汊道各种动力条件、地形参数和沿岸泥沙输移数量的大量数据的统计分析,寻找某些参数之间的相互关系,分析泥沙越过口门的方式,确立潮汐汊道稳定性判据,中心问题是研究潮汐汊道的稳定性;② 从沉积学角度研究潮汐汊道(M. O. Hayes Jonc,J. D. Boothroyd,D. M. Fitz Gerald 等),研究潮汐汊道的沉积环境(包括动力条件)和沉积特征(沉积物水平分布和垂直层序)以及地形的发育演变,主要用作"将今论古",为寻找矿产资源服务。我国研究潮汐汊道的历史相对较短,但由于潮汐汊道的重要性,研究潮汐汊道的学者越来越多,不仅交通部门为了工程建设需要而支持学者们对潮汐汊道进行调查研究,而且国家基金委也支持对潮汐汊道的基础研究[2,3]。任美锷、张忍顺、张乔民、王文介等从动力学的角度对华南沿海和黄渤海沿岸潮汐汊道的稳定性进行了分析研究[1~6]。但任美锷[4]认为风暴时的波浪强度等参数难以准确测定,而且我国许多潮汐汊道并非都有长期的(包括风暴时期)泥沙流、波浪和潮汐的实测资料,单纯用公式计算来判断汊道是否能够利用或提出整治方案是有困难的;他认为海岸动力地貌的研究在潮汐汊道整治中具有重要的意义,建议把沉积学与海岸水文资料结合起来综合分析。宋朝景[7]从地貌形态出发,结合动力因素对海南岛东南岸的几个潮汐汊道的发育和稳定性进行过讨论,我国学者从沉积学角度对潮汐汊道的研究不仅是为寻找矿产资源服务,而更主要是为海港建设服务,如王颖[①]等在三亚港和洋浦港建设实践中应用。有些学者则同时从动力学和沉积学角度对潮汐汊道进行研究[8,9]。

*　邵全琴,王颖,王益峋:《海洋学报(中文版)》,1997 年 19 卷 6 期,第 134－145 页。
①　南京大学地理系海洋室,三亚港泥沙来源与海岸发展趋势研究报告,1973;洋浦港泥沙来源与回淤问题研究报告,1984。

1.2　洋浦潮汐汊道的区域自然环境

在地质上，洋浦地区位于雷琼凹陷南部的邻昌凹陷内，海南岛王五—文教大断裂之北，海南岛西北岸。火山和断裂活动活跃，晚更新世和全新世共发生 5 次火山喷发[10]。第一次喷发是沿 NEE 向断裂多中心溢流喷发，形成整个火山区的玄武岩台地和丘陵，玄武岩流向南流至洋浦—白马井—新英市—中和镇一线即中止，第一次火山喷发的玄武岩烘烤层热释光定年为 52 000±400aB. P. 。其余四次火山喷发的范围均比第一次小，且一次比一次小。其中第四次喷发玄武岩下伏潮滩沉积中的贝壳^{14}C 定年为 26 100±960aB. P. ，第五次喷发热释光定年为 4 000±300aB. P. 。新英湾—洋浦湾北侧，由于火山喷发，地层被掀开隆起，形成以火山口为中心的盾状高地。新英湾—洋浦湾东侧和南侧，由于新构造运动影响，均有明显的上升现象：南部地层发生了由南向北的倾斜；东部蚂蝗岭隆升抬高达161. 1 m①。因此，新英湾—洋浦湾在构造上处于东、南、北三面隆升而其本身则为相对下沉的洼陷地带，而洼陷地带中又以东西方向玄武岩与湛江组地层两种岩性的接触带最脆弱。流水沿着玄武岩与湛江组地层的接触带发生冲刷侵蚀，形成沟谷。18 000aB. P. 前后为玉木冰期鼎盛时期，南海北部最低海平面在−80 m 以下[11]，为距今海平面−150 m[12]。海平面的下降，使冲刷沟谷强烈切割加深形成河谷。今洋浦深槽发育的部位，基本上与低海面时的古河谷位置一致，今深槽的发育是在低海面位置时侵蚀切割河谷的基础上遭海水浸没后进一步发展起来的，而古河谷在"湛江组"地层中呈嵌入状态（见图1）。

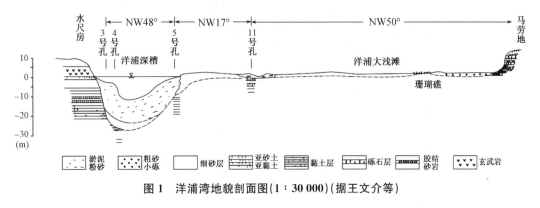

图 1　洋浦湾地貌剖面图(1∶30 000)(据王文介等)

洋浦地区属热带海洋性气候，年平均降水量为 1 257. 4 mm，年平均气温 24. 2 ℃。海水平均温度 2 月为 22 ℃，8 月为 29 ℃。洋浦深槽全长 8 km，0 m 等深线之间宽度为400～500 m，最窄处 300 m，水深 10～20 m，最深处为 24. 4 m；拦门沙（落潮流三角洲）最浅处水深 5 m，水深小于 8. 5 m 的长度达 3. 5 km；涨潮流三角洲长 1 km 左右，水深 2～5 m；纳潮水域总面积为 50 km²，纳潮量为 4. 55×10⁷ m³，有 87% 的面积位于 0 m 等深线以上，湾内浅滩、潮沟极为发育。洋浦湾内深槽南侧发育有洋浦大浅滩，低潮时可出露，大浅滩中部发育有宽浅的潮流通道。

---

① 王文介，李春初，杨干然. 洋浦港港湾地貌的发育. 南海海岸地貌论文集，第三集，1977，80～100。

2　洋浦深槽与拦门沙的现代沉积特征与沉积环境分析

深槽及拦门沙的现代沉积主要取决于其周围物源及水动力(正常天气下的波浪和潮流、极端天气下的风暴潮、河川径流、沿岸流)。不同物源的物质都有可能被波浪和风暴潮掀起或侵蚀或被潮流冲刷,然后由潮流或沿岸流携带在适当动力条件下沉积于深槽或拦门沙处。

2.1　物源环境分析

深槽北岸为玄武岩侵蚀海岸,由上而下地层为:① 沙坝沉积:含有较多生物碎屑的灰黄色中细砂,年代为 4 900±200aB. P. ;② 玄武岩:厚 1. 59~8. 59 m,层顶标高为＋4. 01~−1. 81 m,层底标高为−2. 45~−5. 21 m;③ 沙坝沉积:灰黄色中粗砂,局部存在;④ 湛江组沉积:深灰色亚黏土,层顶标高为−3. 4~−9. 08 m,钻深−40. 46 m,未穿。深槽南岸为湛江组沉积,新英湾纳潮水域主要有大水江与春江汇入年供沙量分别为 7. 39×10⁴t/a 和4. 38×10⁴t/a,河流上游地区为古老变质岩层,下游地区为湛江组地层洋浦湾湾口发育有离岸珊瑚堡礁大铲岛、小铲岛,在排浦发育有沿岸珊瑚礁。本区动力地貌类型示于图 2。

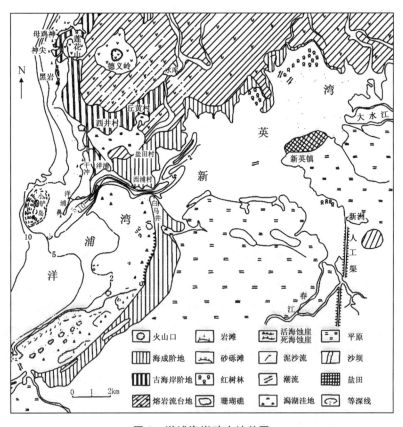

图 2　洋浦海岸动力地貌图

本区物源可归为四类:玄武岩、湛江组地层、珊瑚礁和古老变质岩,此外,不同沉积单元的沉积物也互为物源或自为物源(图 3),发生再沉积。各种物源的沉积特征见表 1。

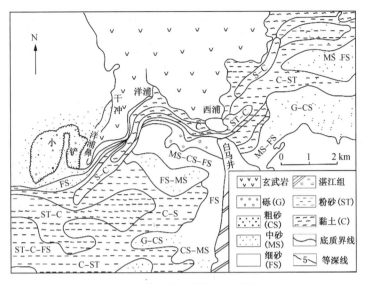

图3　洋浦湾—新英湾底质图

表1　不同地貌沉积单元的沉积特征

地区	重矿特征		地球化学元素特征			有孔虫特征			
	含量	重矿组合	Ca	M*	Sr/Ba	$H(S)$	V	dm_5	有孔虫组合
春江河口	7.6	钛铁矿＋紫苏辉石＋金红石	0.1	5.2	0.21				
大水江河口	2.8	钛铁矿＋锆石＋角闪石	<0.1	3	0.25				
新英湾浅滩	3.1	钛铁矿＋锆石＋电气石＋绿帘石＋辉石＋重晶石	0.2	2.8	0.24	1.6	6	94	Ammonia（88%）＋Elphidium（6%）
洋浦大浅滩	0.3～3.5	辉石＋褐铁矿＋钛铁矿＋电气石＋锆石	7.2	3.4	1.05	2.2	1.7	79	Ammonia（58%）＋Elphidium（16%）＋Quinqueloculina（14%）＋Cellathus（8%）
湛江组	<1.0	褐铁矿＋钛铁矿＋锆石＋电气石＋重晶石	<0.5	1.2～11	<0.25				
深槽口门	1.4	电气石＋帘石类＋辉石＋角闪石＋白钛石＋褐铁矿＋钛铁矿	3.2	9.6	0.95	2.4	23	75	Ammonia（55%）＋Elphidium（19%）＋Spiroloculina（10%）＋Quinqueloculina（7%）
深槽口门外	1.2	钛铁矿＋辉石＋角闪石＋白钛石＋锆石＋黄铁矿＋电气石	1.5	20.2	0.51	3.3	21	48	Ammonia（27%）＋Elphidium（26%）＋Quinqueloculina（21%）＋Spiroloculina（11%）＋Cibicides（6%）＋Triloculina（6%）

（续表）

地区	重矿特征		地球化学元素特征			有孔虫特征			
	含量	重矿组合	Ca	M*	Sr/Ba	H(S)	V	dm_5	有孔虫组合
拦门沙	0.6	辉石＋角闪石＋白钛石＋钛铁矿＋电气石							*Quinqueloculina*（24%）＋*Elphidium*（19%）＋*Triloculina*（11%）＋*Ammonia*（10%）＋*Spiroloculina*（7%）＋*Cibicides*（6%）＋*Tretomphalu*(6%)

注：重矿含量为 100 g 干样中的重矿质量(g)；Ca 和 M(即 Al＋K＋Fe)的单位为 $10\ 000 \times 10^{-6}$.

＊化学元素分析数据及部分重矿分析数据引自南京大学大地海洋科学系的"洋浦港泥沙来源调查报告"(1984).

2.2 动力环境分析

2.2.1 沿岸流

沿岸流是海水在盛行风吹动下,沿海岸的一种长期稳定的流动有沿岸流存在的海岸,其沉积物的成熟度向沿岸流下游方向增加,并在岬角或河口出现沙嘴地貌。据分析,本区存在干冲→洋浦鼻、排浦→白马井两支强度较弱的沿岸流。

2.2.2 波浪及风暴潮

本区出现的波浪主要是风浪,1976 年全年出现风浪 985 次、涌浪 72 次(测点位于深槽)。水下岸波水深 10 m 外,拦门沙、深槽处的最大 $H^{1/10}$ 波高 1976 年实测值分别为 2.4 m、1.0 m、0.5 m,最大平均周期分别为 5.5 s、3.2 s、2.6 s,最大波级分别为 5、3、2。由此可知,拦门沙处波浪作用强,而深槽波浪作用则相对较弱。本区主要波向冬季为 ENE 和 SE,夏季为 SW。洋浦湾内波浪作用对海岸侵蚀作用极为明显,洋浦大浅滩则是海岸遭侵蚀后退留下的浪蚀平台。影响本区较大的台风约平均每年 2 次,而极端天气下的风暴潮对深槽与拦门沙是有影响的。

2.2.3 径流

本区较大的河流即为注入纳潮水域新英湾的春江和大水江,年径流量分别为 2.7×10^8 m³ 和 4.1×10^8 m³。这两条河流的注入,增加了纳潮水域的潮棱柱,顶托了涨潮流,加强了落潮流,淡化了海水,使生态环境有所变化。由表 1 中有孔虫组合可看出,口门处受河流径流影响较大,洋浦大浅滩受河流径流影响大于口门外的深槽,拦门沙处径流影响最小,这也可由表 1 的 H(S)(底栖有孔虫复合分异度)和 Sr/Ba(微量元素 Sr 和 Ba 的比值)得到印证。这是因为淡水密度小,落潮时口门处淡水主要在上层,出口门后上层淡水主要从洋浦大浅滩上流出。

2.2.4 潮流及含沙量

洋浦港区潮汐类型属正规日潮,潮位特征值:最高高潮位为 4.06 m,最低低潮位为

0.33 m;平均高潮位为 2.96 m,平均低潮位为 1.09 m;最大潮差为 3.60 m,平均潮差为 1.82 m;平均大潮差为 3.23 m,平均小潮差为 0.13 m。潮流类型属于不正规日潮流,具有往复流性质,涨潮流方向与落潮流方向基本相反(据实测资料)。深槽内涨落潮流速均大于湾内其他测点(图 4),说明洋浦湾的涨落潮主流线位于深槽内,由于地形及径流影响,涨落潮流的方向和速度随时都发生变化,表 2 给出的是根据每小时实测值(共 25 h)计算的平均流速。由表 2 可知,洋浦深槽中涨落潮流流速均大于拦门沙,表明深槽内潮流作用较强,而拦门沙处潮流作用则相对较弱;洋浦深槽内绝大部分测点的落潮流大于涨潮流流速,"凡是落潮流流速大于涨潮流流速的地方,汊道一般能维持水深,或甚至被继续冲刷"[4],因此可认为,深槽中冲大于淤,这也可由表 3 中得到印证,拦门沙上北部涨潮流流速大于落潮流流速,南部落潮流流速大于涨潮流流速,说明拦门沙上既发生沉积作用,也发生冲刷作用,综合表 3 可知,拦门沙上淤大于冲,深槽与拦门沙的这一冲淤规律也可由表 4 看出,深槽中含沙量从口门至拦门沙逐渐增加,而从拦门沙向外,含沙量逐渐减少,说明泥沙在深槽中的沉积量少于冲刷量,泥沙沿着深槽被搬运到外海过程中,在拦门沙处有沉积,潮流流速相对较高,而且含沙量却相对较少,则其不饱和程度相对较高,其侵蚀冲刷能力也就较强,综合表 3 和表 4 可知,潮流对深槽的冲刷能力从口门至拦门沙逐渐减弱。

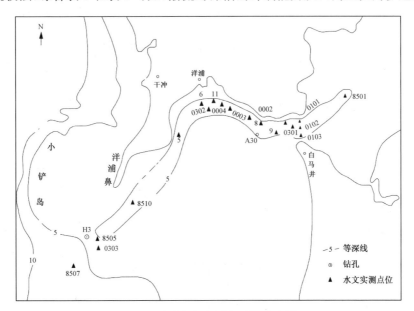

图 4　洋浦湾潮流测点及钻孔位置

表 2　洋浦深槽和拦门沙涨落潮流平均流速(cm/s)/历时(h)

站位		深槽								拦门沙			
潮流速		0101	0102	0103	0301	0001	0003	0004	0302	8510	8505	0303	8507	0005
涨潮流	表层	48/8	51/5	15/7	33/14	19/12	23/11	22/11	11/7	33/10	25	13/8	20/11	12/15
	中层	46/9	27/16	15/11	34/12	30/14	43/15	37/15	30/15	23/15	25	16/11	22/13	9/12
	底层	36/9	40/15	15/13	41/13	52/11	35/15	33/18	29/18	27/15	28	7/12	14/11	8/13
	平均	43/9	39/12	15/10	36/13	34/12	34/14	31/15	23/13	28/13	26	12/10	19/12	10/13

（续表）

潮流速		深　槽									拦门沙			
站位		0101	0102	0103	0301	0001	0003	0004	0302	8510	8505	0303	8507	0005
落潮流	表层	38/17	51/20	34/18	55/11	41/13	32/14	40/14	40/18	33/15	23		28/14	
	中层	26/16	56/9	27/14	32/13	46/11	27/10	27/10	31/10	33/10	18		29/12	
	底层	20/16	42/10	25/12	31/12	49/14	21/10	21/7	21/7	29/10	20		26/14	
	平均	29/16	50/13	29/15	39/12	45/13	24/12	29/10	31/12	32/12	20		28/13	

表 3　洋浦湾各地貌单元冲淤变化地图对比

地貌单元 年代	新英湾浅滩		拦门沙		深　槽		洋浦大浅滩	
	厚度(m)	体积(m³)	厚度(m)	体积(m³)	厚度(m)	体积(m³)	厚度(m)	体积(m³)
1947—1955			−0.2	−4.1×10⁵				
1951—1963	约0	约0	−0.8	−2.0×10⁶	−0.1	−2.0×10⁵	−0.3	−2.5×10⁶
1963—1974	约0	约0	+0.2	+4.9×10⁵	−0.3	−5.6×10⁵	+0.1	+4.0×10⁵
1974—1980	+0.6		+0.4	+9.0×10⁵	+0.4	+1.0×10⁶	+0.2	+1.7×10⁶
1980—1983					−0.4	−9.0×10⁵		

注："+"表示淤积，"−"表示冲刷；洋浦大浅滩指 0 m 线以下，"约0"指海图上难以看出变化情况。

表 4　洋浦深槽与拦门沙实测平均含沙量(kg/m³)

水流		深　槽						拦门沙			
站位		0101	0102	0103	0301	0302	8501	8505	0303	8507	0005
涨潮流		0.082 8	0.077 0	0.064 4	0.096 8	0.103 4	0.119 7	0.167 2	0.100 4	0.120 2	0.076 1
落潮流		0.099 4	0.075 2	0.146 0	0.093 4	0.114 3	0.124 3	0.137 7	0.086 9	0.107 4	0.094 1
平均		0.099 1	0.076 1	0.105 2	0.095 1	0.108 8	0.122 0	0.152 5	0.093 7	0.113 8	0.085 1

综上所述，拦门沙上的波浪作用强于深槽内的波浪作用，深槽内的潮流作用强于拦门沙上的潮流作用；拦门沙上的沉积作用强于冲刷作用，总体处于淤积状态，而深槽内冲刷作用强于沉积作用，总体处于冲刷状态。

2.3　洋浦深槽拦门沙表层沉积特征与泥沙来源分析

深槽沉积物除口门（含砾质砂）外，均为黏土、砂质黏土和粉砂质黏土，拦门沙沉积物为黏土质粉砂和砂质黏土，这与 J. C. Boothroyd 提出的潮汐通道由较粗物质所组成的发育模式[①]相反。深槽沉积物的中值粒径与分选系数呈负相关（口门处除外），即颗粒越细，分选越好，相关极显著，而拦门沙沉积物的中值粒径与分选系数呈正相关，即颗粒越粗，分选越好，相关较显著。这可能是拦门沙处波浪作用较强，而深槽内潮流作用较强之故，洋浦深槽与拦门沙表层沉积物的其他沉积特征示于表1。由表1可知，湛江组地层中未出现辉石与角闪石，而辉石与角闪石则是玄武岩的主要成分，因此可根据辉石含量分布来探索玄武岩侵蚀物质或者江河物质运移规律；钛铁矿在河流沉积及湛江组沉积中含量

---

① J. C. Boothrogd. 海岸潮汐汊道沉积环境、海洋地貌与沉积学问题. 地理科技资料，1981，(29)。

较高,而且本区出现的钛铁矿特征与湛江组地层及河流沉积中产出的一致(周旅复,1984),而钛铁矿在玄武岩中的含量较低,占重矿含量的百分比小于14.59%,因此可根据钛铁矿含量分布来探索湛江组物质及河流物质运移方向。绿帘石的生成与热液作用有关,广泛见于中低级变质岩和硅卡岩以及遭受热液蚀变的岩浆岩和沉积岩中,是由斜长石分解形成的。本区母岩无硅卡岩,河流上游有古老变质岩,但河流沉积中并未出现绿帘石。由此可知,本区出现的绿帘石,是玄武岩喷发时,其下伏物质遭受热液蚀变而形成。口门处沉积物中主要重矿物为绿帘石、角闪石、辉石、电气石、重晶石,并有大量玄武岩屑,表明其物质主要来自玄武岩及其下伏湛江组沉积物,口门外深槽表层沉积物中,钛铁矿含量高,并有较高含量的电气石和锆石,而辉石、角闪石和绿帘石含量较低,表明深槽物质主要来源于河流及湛江组沉积,拦门沙表层沉积物中,辉石、角闪石含量很高,表明其物质主要来源于玄武岩侵蚀物质,拦门沙沉积物中Ca元素含量很高,表明其沉积物中生物碎屑含量较高,可能来源于小铲珊瑚礁。

3　洋浦深槽和拦门沙的古沉积特征与沉积环境分析

对比钻孔A30和H3揭示的沉积序列和沉积特征(表5和表6),可得知:

表5　深槽A30钻孔沉积特征(孔口标高:−9.30 m,孔底标高:−34.65 m)

取样编号	取样深度(m)	钻孔剖面	钻孔剖面描述	粒度特征	H(S)	P/B	有孔虫组合
A30−1	−10.15~−12.15		青灰色淤泥夹虫孔砂				Quinqueloculina(26%)—Trilquculina(14.5%)—Ammonia(13%)—Cibicides(11%)—Elphidium(10%)—Rasalina(10%)—Spiriloculina(6%)
A30−2	−13.15		灰黑色淤泥夹虫孔砂				
A30−3	−14.80~−15.00		淤泥				
A30−4	−16.25~16.45		淤泥				
A30−5	−17.70~−17.90		灰色淤泥,无层理		2.32	0	Cibicides(43%)—Globretella(31%)Brizalina(8%)—Rasalina(8%)Globigerina(64.4%)—Cibicides(9.5%)—Brisalina(6%)—Ammania(6%)
A30−6 ①	−21.45~−22.45		粗砾夹"巨砾",有一颗长5.5cm,夹贝壳屑				
A30−7	−24.65~−25.45		粗砾砂		2.52	0.46	
A30−8 ②	−26.15~26.35		灰色亚黏土		3.18	1.52	
A30−9	−28.2~−28.35		灰色亚黏土				
A30−10 ③	−29.90~−30.10		灰色亚黏土				
A30−11	−32.10~−33.00		浅黄色亚黏土				
A30−12	−33.40~−33.55		浅黄色扰动亚黏土				

① 无列式壳有孔虫和浮游有孔虫；② 列式壳有孔虫 9.7%，浮游有孔虫占 60.36%，有孔虫表层有一层黑色氧化锰；③ 列式壳有孔虫 9.46%，有孔虫表层有一层黑色氧化锰；——分选系数；－－－－中值粒径。

表6　拦门沙3号钻孔沉积特征(孔口标高：－5.02 m，孔底标高：－28.57 m)

取样编号	取样深度(m)	钻孔剖面	钻孔剖面描述	粒度特征	重矿特征	有孔虫特征	地球化学特征	14C年代(aBP)
H3-3 ①	－5.52～-5.72		无层理，表层为淤泥，3 cm下为粉砂夹淤泥，有贝壳及碎屑					1 880±85
H3-5 ②	－6.27～-6.47		无层理，有细砂薄层、生物扰动孔完整贝壳，孔内充满贝壳碎屑					
H3-6	－6.47～-6.52		无层理，灰色淤泥，粉砂夹中砂					
H3-11	－7.77～-7.97		粉砂淤泥夹砂块，砂块为呈层理状(粗中砂，夹贝壳屑)					3 220±100
H3-12	－7.97～-8.02		无层理，粉砂淤泥，上细下较粗					
H3-13 ③	－9.97～-10.17		粗砂细砾，含贝壳屑，夹中砾、粗砾，下部含块状泥，上粗下较细					6 350±150
H3-17	－11.42～-11.62		粉砂淤泥，有虫孔，孔径2.5 cm，虫孔中充填中粗砂，有一黄色细砂不连续夹层					
H3-18	－12.52～-12.72							
H3-19	－13.92～-14.12		淤泥粉砂，波状层理，干燥，8 cm厚有6层波状黄色粘土夹层。一段为2 mm厚					
H3-20	－15.42～-15.62		粉砂淤泥，波状层理明显，层理厚度比19样的大					
H3-21	－16.97～-17.27		淤泥夹黄色泥薄层和黄色砂透镜体，有玉螺					
H3-22 ④	－18.42～-18.62		黑色粉砂淤泥，有水平层理，干燥					8 500±230
H3-23	－19.92～-20.12		灰色粉砂夹细砂，有水平层理					
H3-24	－21.17～-21.37		灰黄色中细砂，颜色不均匀					
H3-25	－22.27～-23.32		粗中砂夹细砾，中砾，灰黄色					
H3-26	－23.37～-25.37		灰黄色中粗砂，无层理					
H3-27 ⑤	－28.02～-28.57		褐色黏土夹中粗砂，微细水平层理					
	－28.57～?		灰色亚黏土					

① 出现黄铁矿和重晶石；有孔虫含量很高，并有列式壳有孔虫，有孔虫组合为 *Elphidium*(18%)—*Triloculina*(16%)—*Ammonia*(14%)—*Cibicides*(9%)—*Rosalina*(6%)；② 黄铁矿含量较高，有孔虫中充填有黄铁矿，有孔虫个体较小，有孔虫组合为 *Elphidium*(27%)—*Quinqueloculina*(20%)—*Ammonia*(18%)—*Triloculina*(13%)—*Cibicides*(6%)；③ 黄铁矿含量较高，玄武岩碎屑占40%，有孔虫含量较

低,但种类多,列式壳有孔虫占 16%,有孔虫中充填黄铁矿,个体小,有孔虫组合为 *Ammonia*(37%)—*Brizalina*(12%)—*Bulimina*(10%)—*Elphidium*(8%)—*Rosalina*(8%);④ 有 60% 以上的酸盐物质,有孔虫含量较低,没有列式壳有孔虫,有孔虫表面沉积有碳酸盐物质,有孔虫组合 *Ammonia*(98%);⑤ 有孔虫个体较大,外壳洁净,有孔虫含量较低,没有列式壳有孔虫,有孔虫组合为 *Elphidium*(41%)—*Ammonia*(34%)—*Triloculina*(17%)

　　▭分选系数、辉石、H(s)、Ca×10^{-2};▤中值粒径、钛铁矿、dm5、Sr/Ba;▨电气石;▩角闪石;
　　▥磁铁矿;▭玄武岩

　　(1) 深槽−26.15 m 和拦门沙−28.57 m 以下的沉积为"湛江组"沉积,这与脚注(王文介等,1977)中钻孔数据揭露的"湛江组"地层出现深度相吻合:深槽中 4 号钻孔于−26.2 m 以下见到"湛江组"地层,而深槽南部边缘 5 号钻孔于−10.2 m 以下见到"湛江组"地层,深槽南侧大浅滩 11 号钻孔−2.5 m 以下即见到"湛江组"地层(图 1)。该层沉积物上部为灰色亚黏土,下部为浅黄色黏土。有孔虫含量很高,深槽钻孔−26.3 m 处达 4 233 枚/100 g,−31 m 和 33.5 m 处 5 408 枚/100 g,远高于表层沉积物,且浮游有孔虫和列式壳有孔虫含量分别高达 60.36% 和 9.46%,而本区表层沉积的 5 个样品中仅出现了个别的浮游有孔虫和列式壳有孔虫,长江、钱塘江、西江及滦河河口沉积物中浮游有孔虫含量分别仅为 12.4%、12%、0、0[13]。因此,可以判断该层沉积物为正常浅海相沉积。该层沉积物中的有孔虫外壳有一层黑色氧化锰沉积,这是老沉积物的标志,表明该层沉积物形成时代较老,文献[14]将"湛江组"地层定为早更新世,"湛江组"地层形成后,由于新构造运动被抬升,实际上,"湛江组"地层在洋浦区分布高程达+10 m 左右(见图 1):洋浦大浅滩是南岸"湛江组"地层侵蚀后退残留的浪蚀平台,深槽是玉木冰期时,流水沿玄武岩(覆盖于北岸的"湛江组"地层形上)和"湛江组"地层接触带侵蚀切割形成的古河谷(详见本文第一部分)。

　　(2) 深槽钻孔−21.45～−26.15 m 和拦门沙钻孔−21.7～−28.57 m 处沉积为古河谷侵蚀切割形成过程中的堆积物。该层沉积物颗粒粗大,分选较好。沉积物中有孔虫含量少,为 318 枚/100 g,且以广盐性瓷质壳有孔虫占优势,无浮游有孔虫和列式壳有孔虫等窄盐性有孔虫。有孔虫外壳洁净,个体较大。虽然深槽钻孔该层沉积物上部的有孔虫组合类似于拦门沙处表层沉积物中有孔虫组合,而且有孔虫的复合分异度 H(S)达 2.32(深槽钻孔)和 2.9(拦门沙钻孔),表明沉积物的海相性比较高,但是,Sr/Ba 仅有 0.3,表明沉积物的海相性很低。这说明有孔虫的复合分异度反映沉积物的海相性比较高,是因为古河谷是在"湛江组"地层上发育的,有些有孔虫是被从"湛江组"地层中冲刷出来后的再沉积。该层沉积物的形成年代介于 26 100±960aB. P. 和 8 500±230aB. P. 之间。因为第一次火山喷发形成洋浦湾北岸玄武岩台地的时间是 26 100±960aB. P. [10],而覆盖在该层沉积物上的沉积物形成年代为 8 500±230aB. P. (见表 6)。此期间经历了玉木亚间冰期和玉木冰期后期(玉木Ⅱ期),南海北部的海平面变化为[11]:约 30 kaB. P. ,海平面在−30 m 以下;28ka～25kaB. P. ,海平面上升到−28 m 或−23 m 左右;24kaB. P. ,相对海平面上升到了−12 m 左右;24kaB. P. ,海平面开始迅速下降;18 kaBP,为玉木冰期的鼎盛时期,是全球显著的低海平面时期,南海海平面为−150 m 左右;末次冰期至早全新世,南海北部海平面迅速上升,12ka～11kaB. P. ,海平面上升到−50 m 左右,8kaB. P. ,海平面上升到−20 m左右。26ka～24kaB. P. 仅2 000年左右的时间,且这期间海平面是上升的,流水能冲刷出20 m余深的沟谷是不可能的,因此玉木亚冰期−12 m的高海平面对该层的形成没有影响。

(3) 深槽钻孔−9.30～−21.45 m和拦门沙钻孔−5.20～−21.70 m处沉积为现代沉积。拦门沙处的沉积根据拦门沙钻孔的沉积特征又可以分为:潮滩沉积(−12.17～−21.70 m)、水下沙坝或拦门沙早期沉积(−9.41～−12.17 m)、落潮流三角洲(拦门沙后期)沉积(−5.02～9.41 m)。

由沉积构造可看出,拦门沙钻孔−21.70～−12.17 m为高潮滩沉积—中潮滩沉积—低潮滩沉积,是一个海进沉积序列[15]。沉积物较细,为粉砂质淤泥,分选较差。有水平层理、波状层理和透镜状层理,−18.42～−18.62 m处的样品沉积物中,有60%左右为碳酸盐物质,有孔虫外壳上也沉积有碳酸盐物质,表明沉积时蒸发较强,经常出露;有孔虫含量很少,种属也少,且卷转虫(Ammonia)占98%;形成年代为8 500±230aB. P. 。再考虑地壳抬升因素,第五次火山喷发时间为4 000±300aB. P. 喷发中心位于母鸡神、龙门一带,这次喷发使中心附近的小港湾抬升成10 m的海积阶地,沿岸的海相沉积层抬高到20 m高[10],喷发中心距拦门沙不过10 km左右,且洋浦湾北岸玄武岩平台上覆盖海相沙坝沉积,年代为4 900±200aB. P. ,已构成5 m阶地,说明洋浦港区抬升量为5 m左右。当时海平面应为−23.5 m左右,这与8kaB. P. 南海北部古海平面在−20 m左右的位置[11]是一致的,这时,古河谷慢慢开始被海水淹没,潮流开始对古河谷产生作用。

拦门沙钻孔−9.41～−12.17 m处沉积为水下沉积或拦门沙早期沉积。沉积物上粗下细:上部为粗砂质细砾,夹中粗砾和块状泥,含贝壳碎屑;下部为粉砂质淤泥,虫孔中充填粗中砂,−9.97～−10.17 m样品为粗粒沉积物:有孔虫含量低,但种类多,且列式壳有孔虫占16%,说明该处陆源碎屑沉积速度较快,受海水影响较大,受径流影响较小;有孔虫中充填有黄铁矿,且重矿物中黄铁矿含量也较高,表明沉积环境为还原环境,水深较大,动力作用小;玄武岩碎屑占40%,重矿物组合为钛铁矿、褐铁矿、角闪石,表明这个时期拦门沙处物质主要来源于玄武岩之侵蚀物质,洋浦鼻侵蚀是最严重的时期;形成年代为6 350±150aB. P. 。7kaB. P. ,古海平面已接近现代海平面[11],据此该层沉积物形成时的水深应为10 m左右,再考虑地壳抬升因素,该层沉积物形成时的实际水深应为15 m左右。这一时期,古河谷已完全被淹没,形成洋浦深槽,洋浦潮汐汊道体系开始形成。拦门沙钻孔的该层沉积物不是落潮流三角洲沉积,因为沉积物并非主要由落潮流带来,而主要由沿岸流和波浪带来。物质主要来源于洋浦鼻侵蚀物质,原因是水深大,涨潮主流并非位于深槽,而可以从大浅滩上通过,因此该层沉积只能称为水下沙坝沉积或拦门沙早期沉积。也因为水深的缘故,这一时期洋浦深槽以沉积为主。

拦门沙钻孔−9.41～−5.52 m为落潮流三角洲(拦门沙后期)沉积,沉积物较细,为粉砂质淤泥、淤泥,分选中等至差,无层理。该层沉积物的上部(下部未进行有孔虫、重矿、地球化学分析)与表层沉积物相比:有孔虫组合相似,底栖有孔虫的复合分异度$H(S)$、多变度V和优势度的值以及Ca含量基本相等,说明该层沉积物上部的沉积环境与拦门沙现代沉积环境相同,该层沉积物中不同深处的重矿物组合不同,反映出物质来源以玄武岩侵蚀物质为主→以湛江组物质为主→以玄武岩侵蚀物质为主→以湛江组物质为主的交替变化规律。有孔虫含量很高,并有列式壳有孔虫,有孔虫个体较小,并充填有黄铁矿。Ca 和 Fe＋Al＋K 含量以及 Sr/Ba 比值明显高于其他各层,拦门沙钻孔的该层沉积物是 6 350±150aB. P. 以来形成的,其中:－7.77～－7.97 m 处样品沉积物年代为 3 220± 100aB. P. ;－5.52～－5.72 m 处样品沉积物年代为 1 880±85aB. P. ,文献[11]对南海北 部中全新世界是否存在高海面进行了讨论,在分析各种资料以后,初步认定为 5kaB. P. 前 后,南海北部的古海面比现今海面还要低。文献[16]认为,5kaB. P. 的海平面高度为 －7.3 m 左右,4kaB. P. 海平面高度为－2.5 m 左右。据此可知,6ka～4kaB. P. 前后,海平 面是下降的。因此,这一时期洋浦港区逐渐变浅,涨落潮流主线逐步归向深槽。4 000± 300aB. P. 时,第五次火山喷发使本港区抬升 5 m,大浅滩出露,涨落潮流主线完全归入深 槽,洋浦潮汐汊道系统完全形成。此后,本港区的地壳和海平面基本是稳定的,因为洋浦 湾顶部高潮线附近有一厚 40 cm 的海滩岩,年代为 3kaB. P. 前后,而海滩岩形成于潮间带 和海浪带[17]。6 350～4 000aB. P. 期间,拦门沙和深槽以沉积为主,4kaB. P. 以来有冲 有淤。

　　根据拦门沙钻孔特征(见表 6)可知,拦门沙的平均沉积速率自 6 350aB. P. 以来为 0.088 1 cm/a,其中 6 350～3 200aB. P. 期间为 0.073 cm/a,3 200～1 880aB. P. 期间为 0.16 cm/a,1 880aB. P. 以来为 0.032 cm/a。

参考文献

[1] 王文介. 1984. 华南沿海潮汐通道类型特征的初步研究. 见:南海海洋科学集刊,第 5 集. 北京:科学 出版社,19 - 29.

[2] 张忍顺,李坤平. 1994. 黄渤海沿岸海湾—溺谷型潮汐汊道的地貌结构. 黄渤海海洋,12(4):1 - 10.

[3] 张乔民,郑德廷. 1992. 潮汐汊道沉积动力与现代地貌过程国外研究进展. 海洋通报,11(1):84 - 92.

[4] 任美锷,张忍顺. 1984. 潮汐汊道的若干问题. 海洋学报,6(3):352 - 360.

[5] 张忍顺. 1994. 中国潮汐汊道研究的进展. 地球科学进展,9(4):45 - 49.

[6] 张乔民等. 1995. 湛江港潮汐汊道落潮流三角洲沉积动力过程. 地理学报,50(5):422 - 429.

[7] 宋朝景. 1984. 海南岛东南岸地貌特征与潮汐通道. 南海海洋科学集刊,第 5 集. 北京:科学出版 社,31 - 40.

[8] 王文介,李绍宁. 1988. 清澜潟湖—沙坝—潮汐通道体系的沉积环境和沉积作用. 热带海洋,7 (3):27 - 35.

[9] 张忍顺. 1995. 渤海湾淤泥质海岸潮汐汊道的发育过程. 地理学报,50(6):506 - 513.

[10] 王颖,周旅复. 1990. 海南岛西北部火山海岸的研究. 地理学报,45(3):321 - 329.

[11] 冯文科,薛万俊,杨达源. 1988. 南海北部晚第四纪地质环境. 广州:广东科技出版社,178 - 191.

[12] 李国胜. 1996. 对晚更新世以来南海海平面变化的成因探讨. 南海研究与开发,(2):23 - 25.

[13] 汪品先. 1980. 我国若干河口有孔虫、介形虫埋葬地点及其地质意义. 见:海洋微体古生物论文集. 北京:海洋出版社,22 - 25.

[14] 中国科学院南海海洋研究所海洋地质研究室. 1978. 华南沿海第四纪地质. 北京:科学出版社,57 - 61.

[15] 沉积构造与环境编著组. 1984. 沉积构造与环境解释. 北京:科学出版社,64 - 66.

[16] 黄镇国,李平日,张仲英,宗永强. 1986. 华南晚更新世以来的海平面变化. 见:中国海平面变化. 北 京:海洋出版社,178 - 194.

［17］赵希涛,沙庆安,冯文科. 1978. 海南岛全新世海滩岩. 地质科学,13(2):163－173.

［18］汪品先,夏伦煜,郑范. 1980. 北部湾第四纪晚期的微体化石群与海面升降的初步探讨. 见:海洋微体古生物学论文集. 北京:海洋出版社,140－145.

［19］邵全琴,王颖,赵振家. 1996. 海南岛潮汐汊道现代沉积特征. 地理研究,15(2):84－91.

［20］Brunn P. 1978. Stabillty of tidal inlets-theory and engineering. Elsevier Scientific Pub. Co. 560－570.

［21］Hayes Mo. 1980. General morphology and sediment patters intidal inlets. *Sedimentary Geology*,26:139－156.

［22］Sha I. P. 1990. Surface sediment and sequence model in the ebb-tidal delta of Texel Inlet,Wadden Sea. *The Netherlands Sedimentary Geology*,68:125－141.

Coastal Plain Evolution in Southern Hainan Island, China[*]

The study area is located in Hainan Island, China between $18°10'4''\sim20°9'40''$N and $108°36'43''\sim111°2'31''$E. The island is separated from mainland China to the north by the 18 km wide Qiongzhou Strait. The island is about 33 920 km^2 with a total coastline length of 1 528.4 km. It has a central mountainous area with peaks reaching 1 876 m asl, surrounded by hills, terraced highlands and coastal lowlands[1].

Hainan Island is located on a passive continental margin setting in the northern part of the South China Sea. The island has a complex geological history. It is composed of several terranes accreted during the Paleozoic as they migrated northward from the Australian zone to the Chinese zone. The accretion led to several orogenies, remnants of which constitute the predominantly metamorphic and granitic central mountains and hills. Since Mesozoic times the geological history of the island has developed in unison with mainland China. This has led to repeat Mesozoic emplacement of granite and some rhyolite volcanism, and to the development of the E-W oriented extensional/transtensional basin centered in the Qiongzhou Strait. The northern part of the island has experienced extensive Cenozoic to Holocene basaltic volcanism[2]. During more recent times, the coastal areas of the island have experienced major extensional and transtensional stresses indirectly related to the Himalayan orogeny (transcurrent movements associated with the Red River strike slip fault system) and the distant Philippine plate subduction[1]. Similar to other parts of the world, the island has been affected by major sea level variations due to Pleistocene glaciations[3,4]. The contacts of the various terrenes and other geological zones are recorded in morphotectonic alignments that follow major ancient faults, in part reactivated through the ages. The major ones are oriented NE-SW across the island, E-W oriented ones in the north and south and a NW-SE fault sets that dissect the other trends. The major faults determine, among other things, the overall parallelogram-shape of the island, the major valley direction and the drainage of some of the major rivers, and the localization of promontories and embayments.

---

　*　Ying Wang, I. Peter Martini, Dakui Zhu, Yongzhan Zhang, Wenwu Tang: *Chinese Science Bulletin*, 2001, Vol. 46 (supp. 1): pp. 90 - 96, plate 1.

英文摘要 2000 年载于: *IAG 2000 Thematic Conference*: *Monsoon Climate, Geomorphologic Processes and Human Activities, Schedule and Abstracts*, pp. 106. Nanjing, 2000. 8. 25 - 29.

Hainan Island has monsoon tropical to subtropical climate with an annual average temperature between 22 ℃ and 26 ℃. The annual average rainfall is between 1 500~2 000 mm, but variously distributed with the dryer areas to the west due to orographic effect of the high central mountains (Fig. 1)[1]. Oceanographic conditions are characterized by micro- to meso-tides (1~3 m tidal excursions), mostly of the diurnal type except along the southern and southeastern coasts where irregular diurnal tide balances over one month period[5,6]. NE wave prevail during winter, SE and SW waves prevail during summer (Fig. 1). Storm waves are usually on the order of 2~3 m, but they can reach up to more than 5 m during typhoons. The strongest waves develop along the east coast and part of the SW coasts. The southern coast and sheltered bays receive the lowest waves. Typhoons are frequent, particularly during summer, and they derive both from the west Pacific Ocean and from the South China Sea. During the past 35 years the southern largest town of Sanya has experienced a total of about 180 typhoons, with wind velocities as high as 40 m/s (Beaufort equivalent 8)[7].

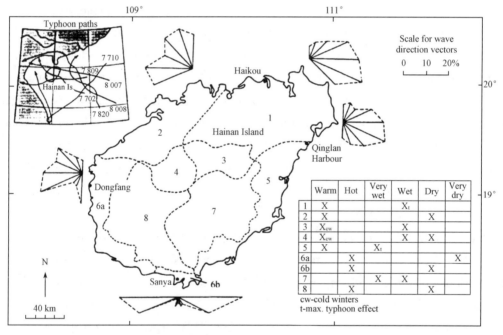

	Warm	Hot	Very wet	Wet	Dry	Very dry
1	X			X_t		
2	X				X	
3	X_{cw}			X		
4	X_{cw}			X	X	
5	X			X_t		
6a		X				X
6b		X			X	
7			X	X		
8		X			X	

cw-cold winters
t-max. typhoon effect

Fig 1 Climatic zone and rose diagrams showing frequencies of occurrence of maximum wave height during year. Inset indicates most common paths of typhoons affecting the island.

1　Study design

Nanjing University has studied the coastal zone of Hainan Islands for approximately half a century. Considerable amount of information has been obtained both offshore and inshore, through mapping of coastal morphology and sediments, surveying of seismic profiles, analyzing of cores and dating of reefs[8,9]. Recently Ground Penetrating Radar (GPR) has been used to establish the internal structures of the coastal deposits, particularly of coastal sand ridges.

The GPR utilized was a pulse EKKO IV radar system equipped with a 1 000 V transmitter and 100 MHz antennas. The instrument transmits to the ground short pulses of high frequency electromagnetic (EM) energy that is refracted and reflected at material discontinuities underground. The receiver monitors the portion of the energy reflected back to the surface against delay time. The delay time of the pulse from transmission to reception is a function of the EM propagation velocity though the substrate material and the depth of the reflectors. By moving the transmitter and receiver along a transect, a profile of the substrate can be recorded showing horizontal distance in meters and vertical two-way time of the EM wave in nanoseconds. The latter can be transformed into depth-meters knowing the velocity of transmission of the EM though the sediments present at the site. Assuming a velocity of 0. 1 m/ns, the resolution of the GPR with the 100 MHz antennas is of about 0.25~0.50 m. The depth of penetration of the EM wave depends on the lithological composition of the deposits and their interstitial fluids. It ranges from maximum penetration obtainable in dry sand to no penetration in clay layers or porous material with salty interstrial fluids. The data are recorded and stored digitally, and therefore they can be processed similarly to seismic data by sophisticated processing software. A GPR survey consists of two parts:

(i) To calculate the near surface average velocities needed to estimate depth of reflectors, a Common Mid Point (CMP) technique was used at a few characteristic sites. This consists of relatively short (up to 30~50 m) transects where the antennas and transmitter and receiver are progressively moved away from each other in relation to central point in incremental short (5 cm) steps. In this way the central reflectors are the same and their distance from the instrument and the arrival delay time of the reflected wave increase.

(ii) Routinely, a reflection survey was used along pre-established, topographically surveyed transects, keeping the distance between antennas (thus transmitter and receiver) constant (1 m) and shifting their position in 25 cm steps. A standard setting was used with a time window of 512, sampling interval of 800, stacks of 64. The field data were corrected for topographic elevation and processes with automatic gain control (AGC) and various other settings to optimize resolution and lower noise.

2　South coastal zone of Hainan Island

The southern coastal zone of Hainan Island is characterized by terraced bedrock hills backed by high mountains, a relatively narrow sandy coastal plain, a highly indented promontory and embayment coastline, and a shallow offshore area.

The mountains and hills are mostly composed of Palaeozoic metamorphic and sedimentary rocks, intruded by Paleozoic and Mesozoic granites. They are dissected by major NE-SW faults, locally indented by E-W and NW-SE trending faults. The lower hills show several well-developed terraces. The highest and smallest terrace is just a notch found at several locations at about 80 m asl. The second terrace is a sloping one that grades from 60 m to about 40 m elevation. A well-defined terrace, locally with preserved marine deposits occurs at 20 m asl[8,9] (Plate I).

The promontories are an extension of the central hills. Their position is dictated by the NE-SW trending fault system, but they are also indented by NW-SE and E-W trending faults. The interplay between these faults and the marine reworking of sediments along the coasts has led to local complex structures. The Luhuitou peninsula is an example of this (Plate I). The peninsula consists of the present promontory of the Luhuitou Mt. composed of Mesozoic granite, steep sided, with the 20 m isobath hugging close to land (less than 10 m from the shoreline). This is joined to the main island (Nanbian Mt.) by a lowland formed by a series of sandy bars (tombolos), lagoons and reefal platforms. The bedrock of the Nanbian Mt. is bedded Paleozoic sandstone, conglomerates and carbonates. Characteristic erosional coastal alcoves are found at about 15~20 m elevation along the eastern side of the Luhuitou cape and other parts of the Nanbian Mt. and adjacent hills.

The quasi-regular spacing of the promontories, jetting out to sea for considerable distance all along the southern coast of Hainan Island has the effect of breaking longshore sediment transport into non-communicating cells. They also protect some embayments, such as Sanya Bay, from large swells associated with typhoons.

The embayments consist of (i) narrow drowned valleys bounded by steep bedrock hills, (ii) embayments characterized by surficial alluvial-deltaic deposits, and (iii) embayments of various dimensions filled by sands of coastal ridges/barriers and associated elongated lagoons.

Drowned valleys. Very narrow, relatively deep embayments are preferentially formed along NE-SW faults. They receive little sediment from inland and for the most part they remain submerged, constituting valley drowned during the postglacial transgression. Yulin Bay and harbor area is a typical example. The embayment is surrounded by hills composed of Paleozoic sedimentary and metamorphic rocks and Mesozoic granite. The embayment consists of a narrow, long inland northern branch, in part silted up and in part reaching depth of about 10 m, and an inverted funnel-shaped, relatively deep (10~20 m) southern bay 1~2 km wide at the mouth and 25 km long. Its overall trend and shores have been indented and segmented by relatively close-spaced NW-SE trending faults.

Alluvial-coastal plain. Extensive alluvial-coastal plains have developed to the

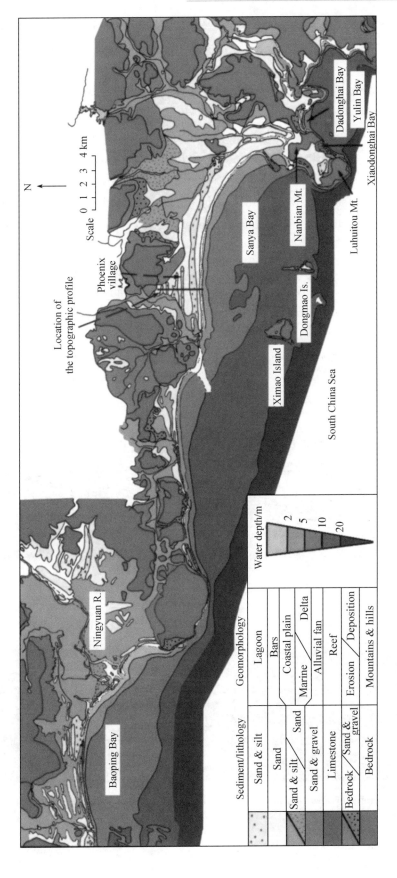

Plate Ⅰ　Map showing the principal geomorphic features of part of the south coast of Hainan Island（附彩图）

southwest such as those of the Loloxi River (west to Baoping Bay) and of the Ningyuan River (Plate I).

The Loloxi River originates from the Changmao Reservoir, flows southwest, and finally enters into Wanglougang Bay. The Loloxi River alluvial-coastal plain is relatively wide (about 20 km at the shore), dominated by the irregular meandering river and its lobate delta. The Ningyuan alluvial-coastal plain is, instead, typical of a relatively narrow embayment (about 10 km wide at the shore). It is partially filled with fine-grained alluvial and estuarine deltaic-coastal sediments. The structure of the alluvial-coastal plain is characterized by remnant fluvial bars and channels due to switching and migration of the river channel, and by residual coastal bars exposed along the margins of the embayment.

Sandy ridge/lagoon coastal plain. Coastal plains dominated by well-developed coastal sand ridges are well developed in parts of southern Hainan Island. Some of the largest ones have developed in YalongBay (Fig. 2), Sanya Bay, Xiaodonghai Bay (Fig. 3), southeastern part of Lingshui Bay, and the southwestern of Yinggehai area. The sand ridges have formed on various substrata ranging from reef platforms and pre-Pleistocene bedrock.

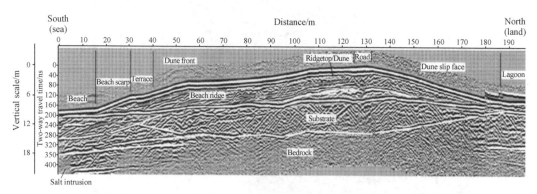

Fig 2　GPR profile of Yalong Bay beach ridge (east to Yulin Bay)

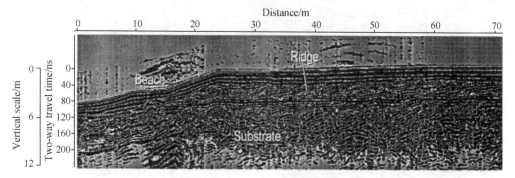

Fig 3　GPR profile of Xiaodonghai beach ridge (in the east coast of Luhuitou Peninsula)

A good succession of sand ridges occurs from inland to the shore in the narrow embayment north of the Phoenix village in the Sanya Bay area (Plate Ⅰ; Fig. 4). The succession consists of five sand ridges/baymouth bars, plus a higher (40 m asl) marine terrace (Baodao) rimmed by marine sand[10]. The ridges are generally composed of coarse-grained sand, becoming increasingly finer grained in successive younger ones and they are highly weathered (brick red coloration). The internal structure of older ridges is difficult to detect because of deep weathering, some homogenization and slight cementation. The Liangqin bar is the most seaward of the baymouth bars fully enclosed in the narrow Phoenix embayment. It has an approximate elevation of 20 m. The Tongjin-Yanglan bar that has an elevation of 10～15 m follows it seaward. This is a very large (about 9 km long, 600 m wide and 13 m high), composite sand ridge that closes the Phoenix embayment and spills over onto the large Sanya Bay inlet. It is composed of very coarse-grained quartzose sand, highly weathered with some interstitial neogenic clay matrix. The surficial sediments have a reddish color and massive appearance with few ghosts of former bedding. The GPR penetrates the sediments of the ridge and detect few lensing layers alternating with quasi-chaotic reflectors and the water table at about 8 m depth.

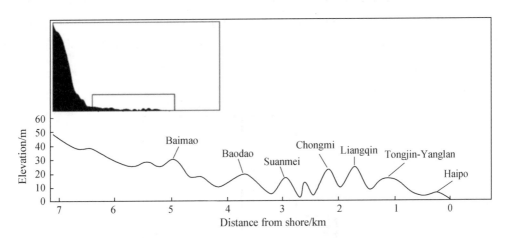

Fig 4　Topographic profile of sand ridges developed along the Phoenix narrow embayment to the present coast (location shown in Plate Ⅰ)

Farther seaward from the Tongjin-Yanglan bar there are two large, composite ridges/barriers that differ greatly from the previous ones. They are the Haipo bar and the Sanya bar/barrier.

The Haipo bar is a large structure about 10 km long, 450 m wide and 10 m high. Where surveyed, the bar is composed of a 90 m wide frontal beach system and a main body. It is composed of coarse-to medium-grained, light gray-yellowish/brownish, slightly weathered, feldspathic quartz sand with occasional comminuted, abraded shell fragment. Good GPR profile shows that the internal structures of the sand ridge have a composite nature being formed by several superimposed secondary ridges. In most cases, the cross-beds of the component secondary ridges dip seaward indicating progradation in that direction. Occasional landward dipping cross-beds in the back part of the ridge record washover events. At the time the Haipo ridge was forming, most of the wide inland area of the present Sanya Bay was inundated, and an arm of this water formed the lagoon of the Haipo bar.

The Sanya bar/barrier is a large composite structure rimming and enclosing the semi-lunate apex of the large bay (Plate I). It is mainly composed of medium to light gray fine quartzose feldspathic sand. Several cores taken in the bar indicate an upper succession of alternating strata (each 2~3 m thick) of gray sand and dark gray clayish sand, which overlays a prevalently fossiliferous, dark gray clay succession (Fig. 5). In places some gravels are presented. The basal unit is probably an open marine bay deposits. The upper succession indicates that the sand of the ridge has repeatedly migrated landward onto and interfingers with lagoon silty deposits. Silty lagoon deposits are also exposed at the seaward margin of the barrier. No GPR profiles were taken on the Sanya bar because it is mostly built up (urbanized). Dates of shells from cores of the Sanya bar and associated fine lagoonal/tidal inlet sediments indicate a range from 8 305 ±80 a (a＝yrs BP)for the deepest samples (1. 46 m) to about 7 420±425 a at depth of 0. 84 m to 3 395±235 a at depth of 0. 37 m. It is during the formation of this ridge/ barrier that elongated, narrow tidal inlets developed and are still active today carrying salt water several kilometers inland.

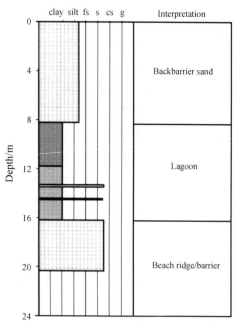

Fig 5　Core from the Sanya bar/barrier showing ridge sand over lagoon material

At Dadonghai Bay on the eastern side of the Luhuitou peninsula there is development of another large coastal sand ridge enclosing a small lagoon, which has been accurately dated (Plate I). Shells at the base of this coastal sand ridge in contact with the underlying lagoon deposits have been dated at $4\,640 \pm 165$ a, and they are now found at 2 m asl. The top of the ridge has been dated at $3\,110 \pm 725$ a and it is now found at 5 m asl.

Reef development and Holocene relative sea level change. Whereas dating of the coastal ridges is generally difficult, coral reefs and shells cemented in beachrock can be readily dated and this allows reconstructing a rather precise history of the relative sea level change during the Holocene. Reefs have developed around promontories and offshore islands. In the Sanya Bay area, reefs have been studied in some details on the Luhuitou peninsula and Ximao Island (Plate I). The oldest recorded date ($(7\,680 \pm 145)$ a) is of a reef head in living position from the base of the carbonate platform on the western side of the Luhuitou peninsula. It now lays at about 1 m asl. At the top of the same platform about 4 m higher, another coral has been dated at $(5\,160 \pm 139)$ a, now at about 5 m asl. On the eastern side of the peninsula, near the southeastern edge of the promontory at Maanling cape, a coral head-stack has been dated at $(6\,820 \pm 154)$ a and is now found at about 2 m asl. Beachrock has been dated from $(4\,890 \pm 120)$ a to $(4\,750 \pm 115)$ a and it is now found at about high-tidal level. The coralline sandy tombolos that join the southernmost cape of the peninsula to the main island rise to about 3 to 5 m asl.

Similar reef conditions occur on the western side of the Ximao Island in the middle of Sanya Bay. Coral heads have been dated at $(4\,290 \pm 160)$ a, which are now found at about 2 m asl. They are topographically higher than the erosion surface of the reef platform dated at $(5\,425 \pm 130)$ a, which is at approximately present sea level.

Other reef material has been dated along the south shore of Hainan Island ranging down to 300 a, and corals are still living in the area. Beachrock associated with coralline areas usually show a slight development time lag in respect to the reef.

3　Disscussion

(i) Terraces and coastal sand ridges. No numerical dates are available for the terraces found on the hills and promontories. However, their position in the landscape (80, $40 \sim 60$ and 20 m asl) and deep brick red weathering suggest their relative antiquity. It is believed that they developed since late Tertiary—Early Pleistocene for the highest ones, to the Middle Pleistocene for the 20 m high terrace* [9,11].

---

* Study group of marine geomorphology and sedimentology of Nanjing University, Research Report on Coastal Environment along Baopin Bay.

Similarly, no numerical dates are available from the older sand ridges systems, all calcareous shells having been dissolved and no organic matter having been found. In any case, the older ridges would be beyond the datable age with simple methods. For the most recent ridges, dates from shells indicate a mid-Holocene age younger than 8~7000 a.

The relative time sequence and the approximate antiquity of the various sand ridges are suggested by their position in the landscape, by their component materials, and by the degree of weathering. It is believed that the older sand ridges probably developed since Early Pleistocene[9,11]. During the Middle Pleistocene embayments like the Phoenix one got filled with sediments and were fully enclosed by large, composite ridges such as the Tongjin-Yanglan bar. The fact that the sand ridges/baymouth bars in the narrow Phoenix embayment are well preserved indicate that the sand ridges were for the most part formed during an overall regression, perhaps punctuated by temporary transgressions. At that time, the short mountain streams carried relatively coarse material to these sites, mostly sand from the granite hills, which was reworked onto the coastal structures.

From the Tongjin-Yanglan bar to the Haipo bar a drastic change occurs, from a highly weathered, partially homogenized system of the former to a slightly weathered to fresh system with internal structures well preserved in the latter. The Haipo bar was probably formed during the penultimate interglacial highstand, showing seaward progradation associated with abundant sedimentation of material in part derived from rivers and, for the most part, from reworking of the shallow shoreface.

During the subsequent glacial-maximum regression much material was distributed onto the shallow shelf. These sediments were later partly reworked onshore to form the Sanya bar and similar other coastal ridges along the south shore of Hainan Island, probably during the Holocene optimum. Differently from other ridges of the area, the Sanya bar shows marked landward migration.

At the present, under tropical climatic conditions and luxuriant vegetation cover, the coast receives little sediment from inland rivers (less than about 10%)[12], and the constituting material mostly derives from offshore relict deposits. The present rise in sea level measured by tidal gauges is small (0. 64 mm/a, measurement from 1954 to 1992) and a dynamic coastal equilibrium has developed[9]. Some erosion occurs on the more exposed parts of the Haipo bar and the material is transported along shore and deposited together with what is derived from the shoreface on the Sanya bar/barrier in the inner part of the bay.

(ii) Reefs and related features. During the Holocene transgression, water conditions became propitious for the development of reefs at approximately 8 000 a. The reef grew vigorously until approximately 5~4 000 a during the transgression and high sea level stand at the Holocene temperature optimum. At its highest, the sea level is thought to have been about 2 m higher than the present[4]. At that time and just subsequent to the high sea level, conditions were right for the formation of extensive beachrock development in tropical settings. During the transgression, the Luhuitou cape became for a time an island separated from the Nanbian Mt. to the north by a narrow isthmus. As reef developed and material eroded from them was redistributed along the isthmus tombolos formed separated by a middle lagoon. These tied the Luhuitou cape with the main island forming the Luhuitou peninsula.

Subsequent to the 8 000~4 500 a initial reef growth, reef development continued probably at varying rates through time. Periods of apparently more intense reef development have occurred between 8 000 a and 4 500 a, at around 3 400 a and 2 500 a. Almost invariably, rapid reef growth seems to have been accompanied and/or followed with a slight time gap by the formation of beachrock.

There are clear indications that Holocene shoreline deposits are now found about 1 to 2 m above sea level. This may be partly due to eustatic changes associated with the variable Holocene climate, partly to slight isostatic movements that have been reported but not yet satisfactorily documented from various parts of the island.

4　Conclusions

Although Hainan Island is located on a cratonic area, its coastal architecture is dictated by the bedrock lithology and the fault trends (NE-SW, E-W, and NW-SE), the latter governs the major geomorphologic features there including the overall shape of the island, the orientation of the central high mountains, the river drainage patterns, and the localization of the coastal promontories and embayments.

The island contains also a good record of interaction between eustatic and isostatic movements during the Quaternary and their effect on sedimentation and geomorphic features. Utilizing relative dating, the late Tertiary-Pleistocene evolution of the coastline is based primarily on succession of highstand coastal features. This is indicated by the formation of terraces and coastal sand bars/baymouth bars that are progressively uplifted. More accurate date control allows reconstruction of the onset of reef development during the early Holocene transgression, and suggests periodic rate fluctuations in reef and beachrock development.

Acknowledgements　Concerned research works have got the powerful support from State Pilot Laboratory of Coast & Island Exploitation, Nanjing University. Most staffs from this lab have joined the series of fieldwork, and also done many kinds of indoor analyses. Recent studies have gained the financial support from CIDA (project number 282/19736), also, benefit from the local government (Bureau of Land, Resources and Environment of Hainan Province).

References

[1] Zeng, Z. X. and Zeng, X. Z. 1989. Physical Geography of Hainan Island (in Chinese). Beijing: Science Press, 3.

[2] Wang, Y. and Zhou, L. F. 1990. The volcanic coast in the area of northwest Hainan Island. *Acta Geographica Sinica (in Chinese)*, 50(2): 118.

[3] Zhao, X. T. 1979. Dating of coral reef and coastline migration in Luhuitou in Hainan Island. *Science Bulletin (in Chinese)*, 24(21): 995.

[4] Zhao, X. T., Peng, G., Zhang, J. W. 1979. Studies on Holocene strata and sea level change along the coasts of Hainan Island. *Geologic Science (in Chinese)*, (4): 350.

[5] Wang, Y. 1980. The coast of China. *Geoscience Canada*, 7(2): 109.

[6] Wang, Y. and Aubrey, D. G. 1987. The characteristics of China coastline. *Continental Shelf Research*, 7(4): 339.

[7] Zhu, S. F. and Deng, W. D. 1992. Basic studies on tropical storm surges in Yangpu harbor, in Island Environment and Coastal Development (ed. Wang, Y.). Nanjing: Nanjing University Press, 31.

[8] Wang, Y. (ed.) 1992. Island Environment and Coastal Development. Nanjing: Nanjing University Press, 2.

[9] Wang, Y. (ed.) 1998. Tidal Inlet-Embayment Coasts of Hainan Island, China (in Chinese). Beijing: Environmental Science Press of China, 3.

[10] Wang, Y. and Chen, W. L. 1982. Studies on coastal geomorphology of Sanya Bay, Hainan Island, China (in Chinese). *Marine Bulletin*, 1(3): 37.

[11] Wang, Y. 1986. Sedimentary characteristics of Hainan Island coast: effect of tropical climate and active tectonism. 12th International Sedimentological Congress, Canberra, Australia.

[12] Wang, Y., Zhu, D. K., Schafer, C. 1992. Marine geology and environment of Sanya Bay, Hainan Island, China. In: Wang Ying ed., Island Environment and Coastal Development. Nanjing: Nanjing University Press, 125.

海南岛海岸环境特征[*]

1　概况

海南岛位于南海西北部,界于 $18°10'04''N \sim 20°9'40''N$ 以及 $108°36'43''E \sim 112°2'31''E$ 之间。全岛面积 33 920 km²,人口 750 万,其中汉族 540 万,黎族 110 万,回族 6.75 万,苗族 5.55 万,其他民族 2.23 万。约有 70 万海南岛人移居东南亚各国。历史上海南由广东省管辖,于 1988 年 8 月 23 日正式成立海南省,辖 3 市 16 县 3 群岛及海域。

海南岛属热带季风岛屿型气候[1],气候变化呈半年交替。冬季盛行偏北季风(NE 为主),天气干旱;夏季盛行偏南季风(SW,SE),多降水。终年温暖无冬,日照数多(1 750~2 700 h/a),年平均气温 22.5 ℃~26 ℃,最冷月平均气温为 15.3 ℃(表1)。中部山区气温较低,北部冬日受寒潮影响有一周气候低于 12 ℃,局部盆地有霜冻,寒潮过后气温复升。按月温 18 ℃以上为夏季,北部有 10 个月夏季,五指山以南全年为夏。海南岛年降水量为 1 000~2 600 mm,雨季始于 5 月止于次年 10 月,夏季与夏秋之交降雨量占 80% 以上,台风雨热雷雨占降雨之 50%,春季少雨。岛屿东部雨量大,西部雨量少(东方县 1969 年降雨量为 275 mm),北部干旱,南部高温多雨。干湿季节受季风影响,岛屿东西气候不同系山地影响所致。

表 1　海南岛各地气候要素(曾昭璇,1986 年)

序号	地名海拔 (m)	日照 (h)	年总辐射量 (kPa/cm²)	年平均气温(℃) (括弧内最低温)	雨量 (mm)	蒸发量 (mm)
1	海口(14.1)	2 238.8	129.4	23.8(3.2)	1 643.6	1 210.8
2	文昌(21.7)	2 068.1	125.0	23.9(4.7)	1 740.5	1 115.7
3	琼海(23.5)	2 162.1	125.3	24.0(4.3)	2 005.4	1 160.9
4	琼中(251)	1 778.5	111.5	22.4(0.1)	2 406.5	983.3
5	屯昌(118)	2 032.0	123.4	23.4(3.4)	2 008.7	1 017.0
6	定安(30)	1 927.7	114.4	23.8(2.7)	1 960.6	1 092.8
7	澄迈(31.7)	2 125.7	121.4	23.7(1.1)	1 764.3	1 117.5
8	万宁(6.2)	2 188.5	124.9	24.3(6.2)	2 151.0	1 181.5

＊ 王颖:《海洋地质动态》,2002 年第 18 卷第 3 期,第 1~9,Ⅰ 页.

2001 年载于:《人类与海岸海洋资源环境学术研讨会会议议程与论文、论文摘要汇编》,1 - 8 页. 海南,2001.11 - 17.

2003 年载于:《第四届海峡两岸地形与环境学术研讨会》,1 - 8 页. 海南岛,2003.2.8 - 13.

（续表）

序号	地名海拔 （m）	日照 （h）	年总辐射量 （kPa/cm²）	年平均气温（℃） （括弧内最低温）	雨量 （mm）	蒸发量 （mm）
9	临高(30.6)	2 172.9	124.2	23.4(2.8)	1 446.5	1 162.1
10	儋县(168)	2 040.0	122.7	23.1(0.3)	1 775.2	1 130.0
11	通什(329)	1 959.6	113.4	22.4(0.1)	1 688.9	1 046.2
12	白沙(217)	1 991.5	121.3	22.7(−1.4)	1 905.1	1 057.7
13	东方(8)	2 737.4	147.5	24.5(1.4)	989.9	2 400.0
14	崖县(6)	2 490.4	141.0	25.4(5.1)	1 246.5	2 080.0
15	乐东(159.9)	2 139.0	126.8	23.8(1.1)	1 584.0	2 043.2
16	保亭(148)	1 927.0	116.4	24.1(2.2)	1 918.6	1 042.1
17	陵水(9)	2 400.0	126.0	24.8(5.6)	1 576.1	1 800.0
18	昌江(100)	2 300.0	132.6	24.2(4.2)	1 677.1	1 313.1
19	天池(760)	1 931.0		19.6(−3)	2 651.6	1 893.6
20	莺歌海(10.6)	2 607.7	109.0	25.2(5.6)	1 078.3	2 397.3

海南岛为南海大陆架岛屿,中央为山地,被丘陵环绕,外围为玄武岩台地与海积阶地,临海为多种类型海岸[2]。海南岛原与陆地相连,晚第四纪时,北部火山活动频繁,琼州海峡隔断,全新世海面上升,海南岛与大陆分隔[1]。

海南岛基底为前泥盆纪砂页岩系与变质岩系经加里东运动成为坚硬陆块,嗣后沉积了晚泥盆世、石炭纪至二叠纪的海陆交替相的砂页岩及石灰岩系,碎屑岩与碳酸盐岩层,均覆盖于原褶皱基底之上。中生代印支运动与燕山运动使基底断裂并伴以花岗岩入侵。花岗岩沿 NE-SW 向断裂入侵,并沿东西向发生断裂。中生代亦沿断裂带沉积红色砂砾岩(如乐东、白沙、安定等)。新生代喜马拉雅运动,使海南岛发生断块升降,一些中生代红色盆地沉积在第三纪被抬升为高山(如鹦哥岭,海拔 1 811.6 m)。中央山地是上升中心,花岗岩体被剥落为高山,如五指山东北西南走向,主峰达 1 876 m,花岗岩约 $1.4 \times 10^8 \sim 1.7 \times 10^8$ a,沿垂直节理侵蚀为五指型山峰;黎母岭(1 441 m)是由粗粒花岗岩组成的东西向山岭及尖峰岭(1 412 m),吊罗山(1 499.2 m)、坝王岭等尚保存着数级抬升高度不等的侵蚀面。断块抬升亦使沿岸浅海抬升为海岸阶地,高度从 5 m 到 80 m,甚至 100 m 不等[2]。构造抬升使岛屿面积逐渐扩大。除海南岛北部有沉降断裂外,断块上升活动一直持续到第四纪,晚更新世与全新世有频繁的火山活动,玄武岩流充填了北部低洼谷地,形成 20 m、40 m 及 60 m 高的熔岩台地,沿玄武岩流喷出孔与溢流带发育了火山海岸及火山喷发所抬升的海成阶地[2]。

海南岛河流受降水补给,发源于中部山地向四周海岸呈放射状。独流入海河流 1 510 条,山溪性河流比降大,水利资源丰富,但季节性分布不均匀。雨季与台风季节为洪峰期,洪峰高,历时短,最大洪峰流量可达年平均流量的 25～45 倍。干季时为枯水期,部分河川断流。集水面积在 3 000 km² 的有三条:南渡江(年径流量 296.7 亿 m³,年输沙量 60 万 t)、昌化江(径流量变化大,保桥站测得最大 20 000 m³/s(1969),最枯 3.58 m³/s(1963),年输沙量 88 万～298 万 t)、万泉河(17 亿～83 亿 m^3 年径流量,60 万 t 年输沙量),其次为陵水河、珠碧河及宁远河(表 2)。

表 2　海南岛河川水文特性(曾昭璇,曾宪中,1989)

河名	源地	流域(km^2)	长度(km)	落差(m)	水量(m^3/s)	水力(10^4kW)
南渡江	白沙南峰山	7 176.5	311	703	209	19.6
昌化江	五指山北麓	5 070.0	230	1 270	122	28.8
万泉河	五指山东坡	3 683.0	163	523	166	26.24
陵水河	保亭峨隆岭	1 120.7	75.7	1 059	47.4	6.77
宁远河	保亭马咀岭南坡	1 082.0	90.2	1 100.6	18.4	3.62
珠碧河	白沙南高岭	1 101.2	85.5	605	19.2	2.04
望楼河	乐东尖峰岭南坡	827.0	87.0	800	13.6	1.81
文澜江	儋县马鞍岭	795.0	70.5	270	14.5	0.91
北门江	儋县马排岭	653.3	62.1	301	11.2	0.75
太阳河	琼中长沙岭	576.3	82.5	876	23.8	1.15
藤桥河	保亭峨天岭	705.5	57.7	1 284	18.3	2.5
春　江	儋县高石岭	550.0	54.1	334.6	8.7	0.46
文教河	文昌坡口村	522.0	56.0	65.7	11.6	0.2
文昌江	文昌斗牛	380.9	37.1	15.5	9.09	0.16
感恩河	东方朦瞳岭	373.8	62.6	1 192.4	4.8	0.65
安仁渡河	文昌排山良	362.7	48.0	29.9	7.14	0.08
三亚河	崖县炭板岭	337.0	32.7	959.2	5.86	0.36
演州河	琼山四元村	263.0	53.1	148.3	5.7	0.09
九曲江	琼海望天朗	279.8	50.0	68.0	10.2	0.14
八所河	东方西方岭	248.0	47.9	290.0	2.43	0.27
石壁河	文昌昌城	243.2	33.5	101.2	5.4	0.13
通天河	东方瞎牛岭	231.0	33.5	315.3	2.91	0.26
南罗河	昌江干村	215.5	26.0	45.0	2.74	0.06
龙滚河	万宁香根岭	197.0	61.6	641.0	7.5	0.2
龙尾河	万宁六连岭	145.7	38.6	652.0	6.0	0.19
北黎河	东方茅安西岭	214.6	47.4	200.0	2.54	0.14
佛罗河	乐东铁色岭	118.0	23.0	259.0	1.5	0.03
光村溪	儋县老童地	176.2	32.2	120.8	2.62	0.09
白沙溪	乐江尖峰岭西南	166.6	26.3	101.4	2.16	0.14
龙头河	万宁黄竹岭	140.1	42.2	328.0	6.1	0.13

(续表)

河名	源地	流域(km²)	长度(km)	落差(m)	水量(m³/s)	水力(10⁴ kW)
北溪水	文昌唐教	143.4	26.9	28.0	2.73	0.02
排浦水	儋县打表村	130.5	22.0	94.0	1.39	0.04
花场河	澄迈鹧鸪岭	117.5	29.0	371.2	1.37	0.14
南港河	东方独岭	117.7	29.0	371.2	1.37	0.14
马袅河	临高多文岭	100.6	19.8	242.9	1.67	0.03
英州河	陵水天岭	129.0	28.0	399.0	3.61	0.36
大茅水	崖县甘什岭	121.3	19.8	411.0	2.16	0.92
山鸡江	儋县南确岭	115.2	25.6	111.8	0.04	0.04

海南岛自然资源主要有 5 类[3]：① 农业与热带经济作物：稻米、番薯、玉米；橡胶林、甘蔗、椰子、香蕉、咖啡、可可、腰果、油棕、茶叶、槟榔、益智、砂仁、西沙尔麻等。② 矿产资源：已发现的 88 种矿物中 59 中具开采价值。著名的有：铁矿（石碌铁矿的蕴藏量达 8 000 万 t，占全国富铁矿储量的 71%）、钴矿（13 000 t，占全国钴矿储量的 1/2）、铝土矿（25 000 t）、水晶矿、花岗石矿、砂矿（石英砂、重晶石、钛铁矿）及蓝刚玉（蓝宝石）与红皓石共生。③ 石油（1.54×10^{10} t 储量）及天然气（1.117×10^{11} m³ 储量）。已有海底管道向外输气，年输气 2.9×10^{10} m³ 至中国香港，另有 5×10^8 m³ 供海南本岛使用，估计可供气 20 a。④ 旅游资源：山地、海岸、林木、民俗土风及众多的河流，胜于夏威夷。⑤ 水产与港湾资源。

2　海岸特点

海南岛主要是基岩岬湾海岸，海岸线长 1 528.4 km，有 64 个港湾，周围岛屿 300 多个。海岸带 5 m 水深的海域面积有 1 116 km²，5～10 m 之间的面积 1 215 km²。南部海岸水深大，北部海岸水深小，东部海岸水深较西部大。

2.1　海南岛的潮汐及风浪

海南岛潮汐属日潮型，东部与南部为不规则日潮混合潮，日潮 15～18 d，半日潮平均为 11 d；西岸与北岸属规则型全日潮，潮差约 1～3 m，潮流作用显著。

季风波浪对海岸作用显著。冬季与台风季节多东北向与偏北向波浪，风力强。在北部与东部海岸亦为常向波浪[4]，但对南部海岸影响小。海南自早春二月有西南季风作用，4—5 月盛行东南季风与偏东风，至 9 月仍频繁。夏季多东南与东向波浪，尤其东岸，因面向开阔外海，东向波浪袭击频繁，大浪甚至高达 6 m。南岸多偏南向风浪，1983 年 10 月 26 日爪哇海钻探平台翻倾，系由突发性瞬时大浪造成。近海岸由于港湾曲折及岛屿环绕，故风浪减小，但在三亚港外，测记到瞬时波高约 3 m（1986 年）[2]。通常波高均在 1 m 以下，周期 4～6 s，均属近岸风浪。本岛海岸临开阔外海，风浪作用显著。但是，由于海岸多港湾，湾内水流与潮流作用仍强，风浪与潮流均为海岸主要动力。河流作用对此热带海岛具特殊性。短小山区河流坡降大,加上季节性暴流,故而河流携沙力量大,海岸发育过程中,积聚了河流带自上游的花岗岩质泥沙。但是,热带山地植被茂密,故而水土流失量仍较大陆为小。由于海岸带已发育了较宽的海岸平原或沙坝潟湖带,现代河流沙多堆积于河流出山处或河口带以内,除三大河流外,其他河流直接输向海岸的泥沙减少,这是目前河流对海岸作用之特点[5]。海岸带泥沙供给也有海蚀岸段供沙,但量皆不大,因为海蚀岸段多为堆积岸所取代,而海滩砂部分来自海底泥沙的"补偿性"供应,其特点是多细砂。可以认为海南岛海岸属于海蚀-海积型,而且处于动态平衡的发展阶段[6]。海南岛北部多玄武熔岩台地与熔岩喷溢带[6],发育为海蚀型基岩岸或火山海岸,但不少岸段镶嵌有珊瑚礁平台及残留的沙坝海岸,故仍为海蚀-海积型。南渡江口为堆积型三角洲海岸,目前由于河流泥沙供应减少,海滩亦遭受侵蚀后退。东部多为海积阶地、海积平原与沙丘沙坝-潟湖海岸,反映出海岸带的风浪作用、海蚀作用强,河流向海岸带供应泥沙亦丰,泥沙运动活跃,海积地貌发育宽广,但海积地貌发展的结果减缓了海岸坡度并阻隔了河流泥沙向海输送,进而使海积作用减缓。在海面上升的环境背景下,东部海积岸微微受蚀,泥沙再运移或加高为沙丘掩覆陆地或延长沙坝,发生局部变化改造,所以海岸亦属海蚀-海积型,尤其是部分岛岬仍临海。南部海岸仍保留较多岬角海湾,湾内多海积平原。就沙坝与潟湖体系而言,沙丘不及东岸普遍,反映出其环境特色,南岸是典型的海蚀-海积岸。西岸具有较宽的海积阶地与沙丘岸段,河流供沙较南部丰富,却不及东部多,风浪掀带泥沙作用亦不及东岸强,但由于干旱,不少海岸段落具有钙质海滩岩或钙质盖层保护的沙丘,部分岬角亦临海,仍属海蚀-海积型。总之,海南岛具有多种海岸地貌,海岸以处于动态平衡的海蚀-海积型海岸为主[7]。

2.2　海南岛海岸的构造运动与气候变化[2~10]

更新世初期,海南岛海岸属基岩港湾型,众多的港湾,湾阔水深,南部、东部与西部沿岸仍能清晰地分辨出当初港湾海岸的原始轮廓。这些古海岸线目前多位于山麓地带,有些地区仍保留着海蚀与海积遗迹,伴随着中央山地的上升,海岸亦有轻微上升。

晚更新世构造运动频繁,北部与西北部火山频频喷发,一直延续至全新世。熔岩流充填着低洼谷地与海湾,覆盖了部分高地,因此原始海岸轮廓变化较大[8]。构造运动,使不少海湾底部抬升为阶地,如西部母鸡神海岸段 40 m 高海岸阶地(其下沉积层中贝壳测年为 27 300±1 200aB. P. ,ZK - 1506)[8]。三亚南边岭南坡、北坡 40 m 高阶地及鹿回头岭东部之马鞍岭 20 m 高阶地等,均保留着原始海盆形态的地貌,开口面向大海,阶地上仍残存扁平状海成砾石。马鞍岭阶地下有死海蚀崖及海蚀穴(已高出现代海面 10 m)[2]。落笔洞内的石化贝壳堆积层,1996 年 4 月 4 日 ^{14}C 测年为 15 745±102(HL86027)至 833±195aB. P. (ND91008),目前其高度在 60 m 左右,洞口外邻近沉积层中贝壳不及洞内丰富,主要为笔螺,测年为 18 030±860aB. P. (ND91009),洞外潟湖相黏土沉积与贝壳测年为 19 685±890 aBP(HL86028)。这些反映了落笔洞外在 20 000~18 000aB. P. 时为海岸线位置,岩洞内在 15 000~8 000 多年前仍可为潮侵,全新世初期可能仍为海滨。嗣后,岸线向海推展,发育了一系列沙坝与潟湖组合的地貌,至今老沙坝表现为棕红色砂质陇岗,而潟湖洼地呈青灰色或灰白色淤泥质低地(已辟为稻田),落笔洞已高出现代海面 60 m,

溶洞底已露干而停止发育。

构造抬升运动可能至全新世,使一些珊瑚礁抬升为高出海面5m的阶地,如鹿回头连岛坝已为陆堤,其基底珊瑚礁中之滨珊瑚经[14]C测年为7 680±145aB. P.(ND84089)至5 160±130aB. P.(ND90001),而海滩岩[14]C测年为4 750±115aB. P.(ND83090)或4 810±105aB. P.(ND90002);鹿回头岭东侧马鞍岭下礁平台高潮线上圆盘礁定年为6 820±154aB. P.(HL86021);西瑁岛西岸圆盘形礁出现于礁平台内陆侧与海滩高潮线上,采样经[14]C测年为4 290±160aB. P.(ND90004)至2 520±135aB. P.(ND91010);礁平台上圆盘礁测年为5 425±130aB. P.(ND91012)[2~9];西海岸洋浦村玄武岩台地上附有薄层海相沉积,其贝壳测年为5 500±210aB. P.(ND84060),目前已为5m高的海岸阶地,海相沙延伸至村内并渐增厚,因受海浪冲蚀,岸边沙层薄[11~12]。这些现象反映出全新世的海岸变化,5 000aB. P.前,海岛有抬升,局部的抬升甚至发生于2 000a前。

气候影响在海岸带表现亦很明显。中晚更新世的沉积砂砾层已经红土化,晚更新世喷发的玄武岸已具红土型风化壳。沿岸珊瑚礁均于全新世形成,自8 000a至5 000a前再至1 000a前左右,是气候变暖的明显标志。海滩钙质砂岩是干湿变化明显的热带海岸气候反映。所采样品测年为7 090±380aB. P.(南岸红塘村ND85004)至4 180±205aB. P.(莺歌海,ND85005)及近代700±104aB. P.(西瑁岛,HL8623),比珊瑚礁发育时间略微延迟些,可能在7 000、4 000及700aB. P.左右为海滩岩发育兴盛阶段。海南岛东岸、西岸沙丘发育广泛,东岸沙丘尤为高大,反映东西海岸段落风力强盛。珊瑚礁在西岸与北部湾分布位置较北,反映海域温度值差异,可能与北部湾内部地幔物质强烈上涌引起海域地热条件的不同有关[8~12]。验潮记录反映出近数十年来海平面之变化[13](表3)。

表3　海南岛沿岸现代海平面上升速率(据王颖,吴小根,1995)

岸段	代表站位	站位地理坐标		海平面相对上升速率(mm/a)	资料年限
南岸	榆林	18°13′N	109°32′E	0.64	1954—1992
西岸	东方	19°07′N	108°37′E	1.21	1960—1992
北岸	秀英	20°01′N	110°16′E	1.83	1954—1957 1960—1970 1976—1990
东岸	港北	18°53′N	110°31′E	0.92	1977—1978 1980—1993

2.3　海南岛海岸类型

按其成因与形态可分为两大类:基岩港湾海岸与砂砾质平原海岸。

(1)基岩港湾海岸

分布于山地与丘陵临海处,如北部澄迈、临高、儋县等地海岸,南部乐东、崖县、三亚及陵水等地海岸。东部与西部也间断分布着基岩港湾海岸,如文昌、琼海、万宁、东方等丘陵临海处。

(2)砂砾质平原海岸或三角洲平原海岸

其表现形式多为沙坝与潟湖海岸,沙坝上可叠置发育海滩、沙堤及沙丘,随着海岸加积展宽或地壳上升,老的沙坝可抬高成为海积阶地。这类海岸岸线平坦浅缓,与前类海岸截然不同,如北部南渡江三角洲平原,东部文昌、琼海与万宁等地大部分海岸,西部昌江与东方等地大部分海岸。南部海岸的一些开阔海岸亦发育了海积平原与沙坝潟湖海岸,如三亚湾有4～6条沙坝及坝后潟湖,它们逐渐改变着基岩港湾海岸,从海蚀型→海蚀-海积型→海积型的平原海岸,保平湾亦类似。

海南岛原始海岸大多为山地丘陵组成的基岩港湾海岸,而后随着海岸的侵蚀堆积过程,有些地区发育了众多的沙坝堆积岸,使全岛海岸具有海蚀-海积型的特点,并且两类岸线交错分布。在基本海岸类型上又可叠加发育次生的海岸。如玄武岩流喷溢于原基岩港湾岸(如临高)或砂砾质海岸带(如洋浦湾)可形成火山海岸或基岩海岸;珊瑚礁沿基岩岬角两侧或岛屿波影区分布可形成珊瑚礁海岸,如三亚大东海、鹿回头、东西瑁岛等处;红树林沿潟湖或基岩港湾内部分布又会形成红树林沼泽岸,如北部铺前港的东寨红树林自然保护区、西部的洋浦新英湾、东部的青澜港等处。

由于全新世气候变暖海面上升,海水入侵河口、海湾与山谷洼地,甚至近期海面上升,潮水侵袭潟湖与湿地,使海岸港湾与沙坝潟湖具有潮汐汊道港湾特征。潮水维持着内湾水域,提供了比较稳定的半封闭的海洋环境。潮流作用对港湾众多的海南岛的发育演变以及开发利用均有重要作用。从潮汐汊道港湾体系与港口开发角度,海南岛潮汐汊道港湾按其成因类型可分成三种。

① 沉溺的谷地　全新世海侵淹没了山地或丘陵内的谷地与河谷,形成潮汐汊道,港湾大多狭长,水域开阔,水深条件良好,其内侧可能有中、小型河流汇入,但水、沙影响小,仍为有潮汐作用控制的海洋环境。其口门可能有岛屿夹峙,可能有沙坝围封,或兼而有之,如榆林港、青澜港等。这类潮汐汊道港湾规模较大,水域面积达数十平方千米。

② 沿构造断裂带或软硬岩层交接地带发育的港湾　这类港湾的规模较大,如东寨—铺前港是1605年(明朝万历三十三年)琼北八级大地震时海岸大片陆地沉入海中时形成的,水域面积达 $58.0 \ km^2$,但水深不大,至今在铺前港水下还有村庄遗址可见。洋浦港水域面积为 $50.2 \ km^2$,系沿王五断裂带以及玄武岩流边缘发育而成,湾内部分汊道原为古河道,但规模不大,主要为潮流冲刷而成的汊道港湾。港湾形态受岩性控制,而火山岩流的分布受地质构造的控制。

③ 沙坝潟湖体系　由于海面上升使潟湖面积得以维持或延缓淤填,这类潮汐汊道港湾发育普遍,或发育于冲积平原外缘或发育于港湾海道被泥沙充填后的后期阶段,不乏良港,而且多半无河流泥沙汇入之虞,如三亚港、铁炉港、新村港、黎安港、坡头港、港北港、博鳌港等。

有些潮汐汊道湾的成因是复合的,如东水港,可列入沙坝潟湖体系,因为外有沙坝内为潟湖,也可列入沉溺的河谷或受构造岩性控制的汊道。该潟湖由于火山喷发,玄武岩流堵塞了原河道,残留的河口与口门外沙嘴形成潟湖-沙坝形态。北部的马袅港亦兼有潟湖沙坝与岩性沉溺河谷的双重特性。因此,上述之分类只是相对的,因为各类汊道皆具海侵、潮流作用为主这一共同特征。

2.4 潮汐汊道港湾的地貌演化与动态平衡

潮汐汊道港湾是海洋伸向陆地的支汊。狭义的汊道(Inlet)是指连接海洋与潟湖、海湾或河口湾的水道。"潮汐汊道港湾"是一个体系,这个体系包括连接港湾的狭窄水道,其所连接的内部水域与外部浅海或外海湾。整个体系是由潮汐涨落携运的海水所维持,外湾、狭窄通道,以及内湾由一潮流流路系统相连为一整体。内湾可能有淡水径流输入,但其径流量与含沙量皆小,不能与潮流相比。潮型、延时与潮流特征、外湾沿岸输沙及潮流通路状况、内湾面积形状与容纳水量状况、水道的过水断面等特点,决定着潮汐汊道港湾的动力特征与发展进程。这个相互关联体系的关键是汊道[14]。

汊道是一个狭窄的潮流通道,或为海湾狭窄的出口,或为口门岛屿分割与约束的水道,或为沙坝、沙嘴等沙体或人工堤所限制的狭窄水道,形似一咽喉,连接着内外水域,控制着潮水出入。潮汐汊道具有如下特征:

(1) 具有中心深槽,当潮流通过通道时,由于水道狭窄,流速增大,冲刷汊道成为深槽,水深可超过 $20\sim30$ m,但在汊道区,涨落潮流流路复杂,涨潮流流路往往与落潮流流路分开。主深槽方向常与海岸垂直,主要为落潮流所流经,而涨潮流则成为片状,有时分为几条涨潮流深槽。在两条深槽交汇的地方,潮流冲刷力强,往往形成局部深潭(Hole),水深可超过 $40\sim50$ m。

(2) 深槽两端有时有潮流三角洲如广东汕头港,口门有两座岛屿形成狭窄通道,潮流经汊道口门时束水流急,但进入湾内后,水流扩散,流速减小,泥沙逐渐堆积起来,成为涨潮流浅滩,颗粒较细,主要为悬移质泥沙(粉砂与黏土)。落潮时,有一部分泥沙被落潮流带向外海至一定深度,潮流力减卸下沉积物,形成落潮流浅滩,落潮流较涨潮流强,故浅滩物质较粗,主要为粗砂或砂级物质。这类浅滩系潮流形成,故称为"潮流三角洲"。潮流三角洲口常有些深槽,成为潮流进出的通道,在航道上具有重要意义。门内浅滩物质细,由于堆积于海湾内,风浪作用小,故而堆积体规模大,常成为指状、鸟足状或扇状的三角洲形态。口门以外的潮流三角洲受浪、流影响较大,其生长受到限制,沙体较小,往往形成新月形水下沙坝。潮流三角洲的大小与泥沙供应量以及纳潮量(潮棱柱)的规模有关,视通过汊道的潮棱柱(Ω)与年泥沙流量(M)间的对比关系而定。经验证明: $\Omega/M \leqslant 150$ 的汊道其两端有较大的潮流三角洲[14~17]。

(3) 潮汐汊道中因潮流流速较大,细粒物质多被冲走,只留下粗砂、贝壳碎块、石砾等粗粒物质,是所谓的蚀余堆积或滞留沉积物。汊道底部粗粒沉积物受水流运移,常成为巨大的沙坡(高达数米,长达数十米)。汊道中若为单向水流(如涨落潮流速不等,其中之一较大),则沙坡的剖面不对称,陡坡向水流下方。如汊道中有双向水流(即涨落潮流速大致相同),则沙坡的剖面也大致对称。

(4) 潮汐汊道一般处于动力均衡状态,但由于潮流、波浪基岩海岸泥沙流的状况经常变化,在发育过程中汊道会改变其位置和形态,尤其是在沙坝海岸段落,汊道可向泥沙下游方向偏移。此在海南岛南部、砂质海岸颇多例证。

平均纳潮量(P)和狭窄通道口门在平均海平面下的过水断面(A)是反映潮汐汊道特征的主要因素,一个稳定的潮汐汊道口门断面反映着通过口门的泥沙流(波浪携带泥沙)

与潮流冲刷口门维持畅通之间的动力均衡。早在1886年,Stevenson即指出"如果海水的进入量减少,则水流通过的口门断面亦减小,如海水的进入量减少巨大,则水道会淤塞而不能通航[18]"(张乔民,1987)。1931年,O'Brien提出纳潮量(P)和口门段平均海平面下的均衡过水面积(A)之间的关系为[19]:

$$A = CP^n$$

式中:A为口门过水面积平方英尺;P为纳潮量立方英尺。

此关系式至今仍被广泛应用,只是在各种不同的自然和人工情况下,系数和幂数有差异[14]。

张乔民对华南40多个潮汐汊道中的32个较大汊道进行了研究,发现既有良好的P-A相关,但也有明显的不符合,系数C的变化显著[18]。

海南岛的潮汐汊道港湾体系属于有泥沙流作用的砂质海岸的潮汐汊道,其自然发展趋势是逐渐淤浅,原因在于:① 汊道长度因海岸沙嘴的伸长而增长。汊道增长,则汊道中的潮流逐渐减弱,不能将沿岸流带来的泥沙从汊道中冲刷出去。② 风暴时大浪与岸流将泥沙堵积于汊道口门,使纳潮量减小,因而使汊道很快淤浅并最后封闭,但强浪或风暴潮也可冲出新的汊道。③主汊道分裂为两条或两条以上的分汊道。各分汊道的纳潮量总数往往与主汊道的潮棱柱相同,因此,通过各条分汊道的潮棱柱都较小,汊道易淤浅。④与汊道联通的海湾或潟湖,水体面积缩小。

实际上,汊道在其发展历史中总是不断地改变位置与形态,但在航运上却要求汊道在位置与断面上相对的稳定。应充分考虑影响汊道稳定性的因素:最大潮流量、汊道形状、悬移质泥沙、波浪作用、泥沙流数量、河流流量,其中以潮流量与泥沙流数量最为重要。利用汊道自然环境的有利方面,改造不利条件,以达到汊道的稳定、维持或增加航道水深[20]。

参考文献

[1] 曾昭璇,曾宪中. 1989. 海南岛自然地理[M]. 北京:科学出版社.

[2] 王颖等. 1998. 海南潮汐汊道港湾海岸[M]. 北京:中国环境科学出版社.

[3] MENG Qing-ping. 1992. Natural resources of Hainan Province[M]. In: WANG Ying ed. Island Environment and Coastal Development. Nanjing University Press. 9–12.

[4] ZHU Suo-feng and DENG Wei-dong. 1992. Basic studies of tropical storm surges in Yangpu Harbour. In: WANG Ying ed. Island Environment and Coastal Development. Nanjing University Press. 31–38.

[5] 王颖,陈万里. 1982. 三亚湾海岸地貌的几个问题[J]. 海洋通报,1(3):37–45.

[6] 丁国瑜. 1964. 海南岛第四纪地质的几个问题. 中国第四纪地质问题[M]. 北京:科学出版社,207–233.

[7] WANG Ying and Aubrey D G. 1987. The Characteristics of China[J]. *Coastline Shelf Research*, 7(4): 329–349.

[8] 王颖,周旋复. 1990. 海南岛西北部火山海岸研究[J]. 地理学报,45(3):321–330.

[9] WANG Ying and ZHU Da-kui. 1992. Schafte C. Marine geology and environment of Sanya Bay,

Hainan Island, China. In: WANG Ying ed. Island Environment and Coastal Development. Nanjing University Press. 125 - 156.

[10] WANG Ying. 1986. Sedimentary Characteristics of Hainan Island Coast: Effect of Tropical Climate and active Tectonism [J]. 12th Inter Sedimentological Congress Camberra. Australia.

[11] WANG Ying, Schafer C, Smith J N. 1987. Characteristics of tidal inlet designated for deep water harbor development, Hainan Island [J]. In: China Proceeding of Coastal & Port Engineering in Developing Countries 1(1). Beijing: China Ocean Press, 363 - 369.

[12] Smith J N and PAN Shao-ming. 1992. Sedimentation rates in Sanya and Yanpu Harbour based on ^{210}Pb dating [M]. In: WANG Ying ed. Island Environment and Coastal Development, 199 - 214.

[13] 王颖,吴小根. 1995. 海平面上升与海岸侵蚀[J]. 地理学报,50(2):118 - 127.

[14] 任美锷,张忍顺. 1984. 潮汐汊道若干问题[J]. 海洋学报,6(3):352 - 360.

[15] Keulegan G H. 1967. Tidal flow in entrances, water-level fluctuations of basins in communication with the sea [C]. In: Tech. Bulletin, 14. Committee on Tidal Hydraulics. Corps of Engineers.

[16] Bruun P. 1960. Stability of Coastal Inlets [M]. North Holland Publishing Company.

[17] Bruun P. 1978. Stabilty of Tidal Inlet: The Theory and Engineering [M]. Elsevier Scientific Publishing Company.

[18] ZHANG Qiao-min. 1987. Analyses of P-A correlationship of tidal inlet along the coast of South China [C]. In: Yin kai ed. Proceedings of Coastal Port Engineering. China Ocean Press, 414 - 422.

[19] O' Brien M P. 1931. *Civil Engineering*, 1(8): 738 - 739.

[20] ZHU Da-kui, WANG Ying, Smith J N, Schafter C. 1992. Sediment transport processes in Yangpu Bay, HainanIsland, China [M]. In: WANG Ying ed. Island Environment and Coastal Development, 157 - 182.

Coastal Sandy Ridges and Reefs of Southern Hainan Island (China) Developed During Quaternary Sea-level Variations[*]

1 INTRODUCTION

1.1 Objectives

The objective of this paper is to report on some of the most important coastal features of the southeastern Hainan Island, China, and to establish the time of their formation during Pleistocene, relative sea-level changes (Fig. 1).

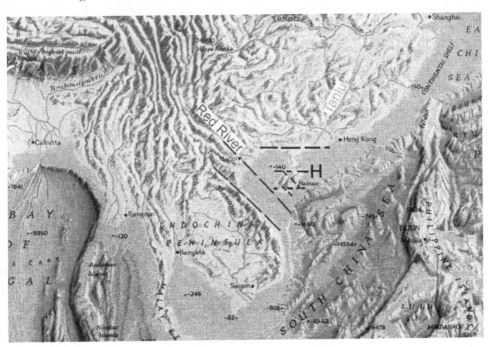

Fig 1　Map of South Asia with major fault trends and location of Hainan Island (H) (after National Geographic, Internet)

* I. P. Martini, Y. Wang, D. Zhu, Y. Zhang, W. Tang: *QUATERNARIA NOVA* 2004, Ⅷ, pp. 277 – 296.

The study consisted in collecting information on the geomorphologic and climatic conditions of a portion of the southern coast of Hainan Island, determining the internal stratigraphy of siliciclastic and carbonate coastal ridges, and sampling shells and coral material for ^{14}C dating. The mapping of the coastal features was greatly aided by detailed 1 : 250 000 topographic maps and remote sensing images. Available pits, shallow cores and Ground Penetrating Radar(GPR) were used to determine the internal structure of the coastal ridges, primarily to establish their internal structures and therefore their origin; that is, whether they were developed primarily by wave or by wind action. The GPR technology utilizes an instrument that transmits electromagnetic (EM) waves that are refracted and reflected at material discontinuities (www. groundpenetratingradar. com). The time taken by the EM waves to travel from the transmitter to the reflecting surface and back to the receiver antenna is measured in nanoseconds (TWT = Two Ways Travel Time). By moving along topographically surveyed transects and transmitting the EM waves at regular intervals, it is possible to map underground reflecting surfaces, similarly to seismic profiles (Fig. 2). Differently from seismic, the EM waves used by the GPR are strongly influenced by the ground fluids and by the particle size of the materials: EM waves cannot penetrate saline waters or clays. This technology is useful in establishing stratigraphy down to approximately 12～18 m depth (in dry sandy terrains), with a resolution of about 30 cm. The energy becomes too attenuated farther down to provide useful information. Surface obstacles, such as buildings, large trees, aqueducts and similar structures, and other anomalies, such as radio transmitters and high-tension lines, generate noise, making interpretation of the results difficult. The instrument used in this study was a pulse EKKO IV radar system with a 1000V transmitter and 100 Mhz antennas, built by Sensor and Software (Toronto, Canada) (www. sensoft. on. ca). The 100 Mhz antennas were selected for this study because they provided the best penetration and bedding resolution. The 50 Mhz antennas provided only slightly deeper penetration in this material but with considerable loss of resolution; the 250 Mhz antennas did not provide sufficient penetration and unnecessary high resolution of the topmost layers of only the younger coastal ridges. The station interval along the profile lines was 0. 25 m for the 100MHz antennas. Post-acquisition data processing involved zero time adjustments, corrections for elevation, low pass time filtering, horizontal trace to trace averaging (1 to 3 traces) and the application of various automatic gain control (AGC) functions with max gain of 500 (Davis, Annan 1988). The profiles are here presented as a black and white version of the screen color-trace plots, rather than wiggle-trace plots to emphasize the general trends.

1. 2　Geographic and Geological settings

The subtropical Hainan Island is located 18 km off mainland China, on the northern

Fig 2　Set-up for Ground Penetrating Radar survey. Standing person moves the transmitter and receiver antennas along a topographically surveyed transect; the sitting person monitors recording of reflected wave arrival to the console and computer.

continental shelf of the South China Sea (Fig. 1). The island is affected by monsoons and numerous typhoons (3~4 per year) primarily derived from SE. During typhoons, high waves (5+m) can develop along the eastern exposed shores (Zhu, Deng 1992). The island is affected generally by microtides. Tidal gauge data at the nearby Sanya harbor indicate that the high high tidal level(HHTL) is 2.13 m, low low tidal level (LLTL) is 0.06 m, mean high tidal level(MHTL) is 1.43 m, mean low tidal level (MLTL) is 0.64 m, mean tidal level (MTL) is 1.02 m, the maximum tidal range is 1.89 m, and mean tidal range is 0.79 m.

The island is located on a passive continental margin setting. It formed by accretion of various Paleozoic terranes, Mesozoic granite intrusions and, to the north, Cenozoic-Holocene basaltic volcanism (Fig. 3A).

During Cenozoic-Quaternary times, Hainan Island has experienced major extensional and transtensional stresses indirectly related to the Himalayan orogeny (transcurrent movements associated with the Red River strike slip fault system) and the distant Philippine plate subduction zone (Fig. 1; Zeng, Zeng 1989). It is dissected by three major fault systems oriented NE-SW, E-W and NW-SE, which determine the overall shape of the island and the local coastal morphology (Fig. 3B). Some of these faults are still active, as indicated by triangular-shaped steep hill flanks and recent earthquakes in parts of the island (Martini et al. 2003, 2004). During the late Tertiary-Pleistocene, the area has experienced is ostatic and eustatic movements associated with neotectonics and climatic changes. Such history is recorded in terraces at various altitudes (80, 40, 20 m asl; elevation from topographic map at 1 : 25 000) and sequences of coastal sand ridges/bay mouth bars (all locally called "bars") (Yao Qingyi

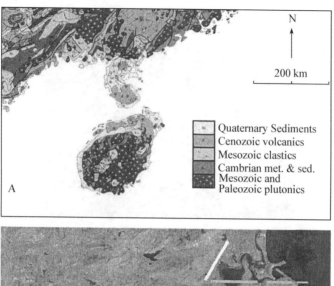

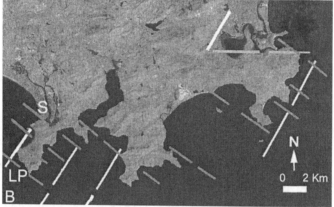

Fig 3　Maps of Hainan Island: A. Geology of Hainan Island and adjacent mainland China (after Chinese Academy of Geological Sciences 1980); B. Southeastern corner of Hainan Island dissected by three sets of normal faults. (S＝Sanya town; LP＝Luhuitou Peninsula).

et al. 1981; Wang Ying, Chen Wanli 1982; Wang Ying et al. 1992). The Holocene variations in relative sea level are in part documented in the dated coastal ridges, coral reefs and beach rocks of the southern coast (Zhang Zhongying et al. 1987; Wang Ying et al. 2001). Ground Penetrating Radar profiles show the internal structures of the sand ridges to have composite nature, being formed by several superimposed sediment successions.

2　RESULTS

2.1　Dating

^{14}C dating was done using the Liquid Scintillation counting method at the Key Laboratory of Coast & Island Development of Nanjing University, following standard chemical and microscopic techniques to remove and check for any calcite contamination.

[14]C-dates are reported here calibrated utilizing the algorithms presented by Bard (1988),
after having subtracted 400 years from the measured dates of marine samples to
compensate for the reservoir effect (F. Antonioli, pers. communication) (Table 1).
The dated materials consisted of either mollusk shells in siliciclastic sandy ridges and
beach rock (consisting of either coarse-grained, highly fossiliferous, siliciclastic
sandstone, or carbonate, gravely sandstone local with minor, disseminated siliciclastic
pebbles and sand), and coral fragments and coral heads in living position in emerged reef
platforms (Table 1; Fig 4). In siliciclastic deposits sample were collected from the
surface, from shallow pits, and some from cores (samples 19～23) from the Sanya
harbor in the back of the Sanya barrier/coastal bar. Surficial samples form beach rocks
and reef platforms were taken from the intertidal zone up to supratidal settings,
including along two intertidal transects in the NW (sample 15～18) and SE (samples
3～5 and 6) bays of the Luhuitou peninsula (Table 1; Fig. 4). Samples affected by
recognizable diagenetic alteration were discarded. Mean high tidal level (present day sea
level-psl) is taken as the datum of the elevations (heights) of the samples, as measured
in the field.

Table 1 Calibrated ^{14}C dates from the Sanya area, southern coast of Hainan Island
(dated in bold have been referred to in text). Calibration was done utilizing the algorithms
indicated by Bard 1998. The datum level is MHWL (mean high water (tidal) level).

No.	Lab No.	Location	Height above MHWL (p sl in m) (D=depth below surface in core)	[14]C Dating (calibrated)	Material	Annotations
1	HL86024	Dadonghai Bay	2	4 471±165	Shell	Bottom of bar
2	HL86025		5	2 826±725	Shell	Top of bar
3	ND90005	Luhuitou peninsula SE Bay (Xiaodonghai)	0～0.5	6 569±110	Beachrock	Shoreline
4	ND90006		−0.5	5 298±240	Coral	Intertidal reef platform
5	ND90007		−1	4 295±150	Coral	..
6	ND90008		−1.5	Modern	Coral	Shallow subtidal
7	ND85008	Luhuitou peninsula SE cape (Maanling)	1.5	2 244±145	Coral debris	Sediment at tht foot of old cliff
8	HL86021		1	7 272±154	Coral head	Raised reef platform
9	ND83090	Luhuitou peninsula NW cape	0～0.5	4 907±115	Beachrock	Shoreline
10	ND83091		0～0.5	5 079±120		
11	ND90002		0～0.5	4 981±105		

（续表）

No.	Lab No.	Location		Height above MHWL (p sl in m) (*D=depth below surface in core*)	^{14}C Dating (calibrated)	Material	Annotations
12	ND84089	Luhuitou peninsula		1	8 168±145	Coral	Base of raised platform
13	ND90001			5	5 406±130	Coral	Higher in raised platform
14	ND90003			5	4 584±120	Coral debris	Bar
15	HL86005	Luhuitou NW Bay		0	3 093±110	Coral	Intertidal reef platform
16	HL86006			−0.5	3 061±100	Coral	..
17	HL86007			−1	120±50	Coral	..
18	HL86008			−2	Modern	Coral	Shallow subtidal
19	HL86010	Sanya harbour	Core S002	D 0.25	6 984±246	Shell	Bay
20	HL86011			D 0.84	8 312±425	Shell	Bay
21	HL86012			D 0.96	7 472±700	Shell	Bay
22	HL86015		S001	D 0.37~0.57	3 189±235	Shell	Bay
23	HL86018		S035	D 1.46	8 818±400	Shell	Bay
24	ND90004	Western coast of Ximao Island		0.5	4 333±160	Coral head	Landwards of reef platform
25	ND91010			1	2 092±135	Coral head	Landwards of reef platform
26	ND91011			1	1 713±175	Coral head	Landwards of reef platform
27	ND91012			0.5	5 720±130	Coral head	Reef platform

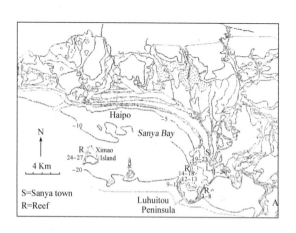
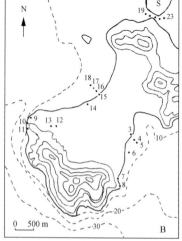

Fig 4　Schematic maps showing location of samples collected for dating. A. Sanya area（see figure 5A for better definition of geomorphic features）; B. Details of the Luhuitou Peninsula（−10＝water depth in meters）. Elevations are derived from topographic and bathymetric maps.

2. 2　Coastal sand ridges (bars)

Coastal sand ridges are well developed in coastal plains of southern Hainan Island (Fig. 5A). They are mostly formed of siliciclastic material, and locally by reefal carbonate clasts. A succession of five large, raised, siliciclastic sand ridges (HB, TY, LI, CB, SH) and a high (40 m asl) inland marine terrace (BA) occurs inland from Sanya Bay at Haipo (Figs. 5A, B). Farther inland, along the sea-ward flank of high hills, narrow terraces cut into bedrock occur at 60 and 80 masl (BA in Fig. 5B, and black profile in insert). (Note that the elevations of terraces and other coastal features derived from published maps are still referred to as "above sea level" (asl); that is, above average sea level). The coastal ridges show a landward overall grain size increase and progressively more intense weathering. The farther inland ones (LI, CB, SH) developed as bay mouth bars within a relatively narrow embayment. The more recent ones evolved as barriers in a more open coastal zone. The most seaward of the bay mouth bars (LI) has an approximate elevation of 20 m. The next seaward bar (locally called Tongjin-Yanglan bar) (TY in Fig. 5B) is a composite structure that closes the narrow embayment and spills over onto the large Sanya Bay inlet (Fig. 5A). It is about 9 km long, 600 m wide, and 13 m high. It is composed of very coarse-grained, quartzose sand, highly weathered with some interstitial neogenic clay matrix. The surficial sediments have a reddish color and massive appearance with just ghosts of former bedding (Fig. 6A). The GPR can detect only few lenticular layers alternating with quasi-chaotic reflectors, and the water table at about 8 m depth.

Farther seaward there is the composite coastal barrier inclusive of the so-called Sanya barrier (SB, Fig. 5A) and the Haipo bar (HB; Figs. 5, 7A). This barrier differs significantly in composition and structure from the more inland raised coastal ridges. It is tens of kilometers long, approximately 8~10 m high, and varies in width from 450 m to the west (Haipo bar) to several kilometers at the eastern end (Sanya barrier). The barrier is generally composed of coarse-to medium-grained, light gray-yellowish to brownish, slightly weathered, felds-pathic-quartzose sand with occasional comminuted, abraded shell fragment (Fig. 6B). Cores in the eastern part of SB indicate an upper succession (more than 8 m thick) of alternating strata (each 2~3 m thick) of gray sand and dark gray silty sand, which overlays a 6~8 m thick, fossiliferous, dark gray, clay-silt succession. The basal unit is probably an open marine bay deposits. The upper succession indicates that the sand of the ridge has repeatedly migrated landward onto, and interfingers with, lagoon silty deposits. Silty lagoon deposits are also exposed at the seaward margin of the barrier. Dated of shells taken from cores from the bay near the Sanya barrier indicate a range from 8 818±400 a (a＝calibrated [14]Cyrs BP) for the deepest samples, to about 8 312±425 a at intermediate depth, to 3 189±235 a near the surface. GPR profiles obtained from the western part of the Haipo bar show the internal structures of the sand ridge to have a composite nature, being formed by several superimposed secondary ridges (Fig. 7B). In most cases, the cross-beds of the component secondary ridges dip seaward indicating beach progradation. Occasional landward dipping cross-beds in the back part of the ridge record washover events. The barrier has typical transgressive components, but it also shows considerable seaward enlargement during middle and upper Holocene sea-level highstand.

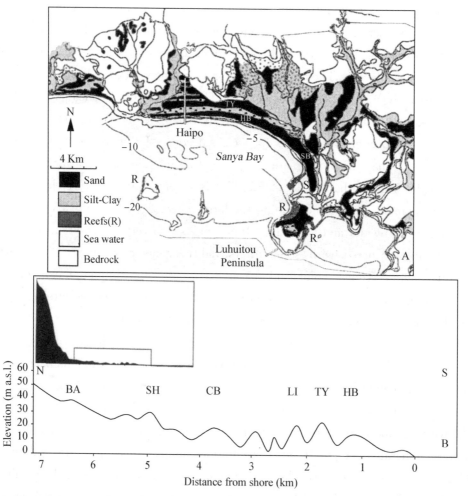

Fig 5　Geomorphic features of the Sanya area: A. Geomorphic map: sand is deposited mostly in coastal ridges (bars, barrier, baymouth bars, and tombolo) (S＝Sanya town; R＝reef; −10＝ water depth in meters; TY＝Tongjin-Yanglan bar; HB＝Haipo bar; SB＝Sanya barrier); B. Schematic topographic profile along a transect inland from Haipo showing the various emerged coastal ridges and 40 m terrace. In the insert the transect is extended to the inland high hill showing notches (terraces) at 60 and 80 m asl (after Wang et al. 2001; Martini et al. 2003, 2004). Elevations are derived from a 1：25 000 topographic map.

Fig 6　Internal characteristics of siliciclastic coastal ridges. A. Pliocene Tongjin-Yanglan bar(TY): reddish weathered sand with poorly defined layers; B. Holocene Haipo bar(HB): gray slightly fossiliferous sand (with well defined layers, visible on fresh cuts and GPR profiles).

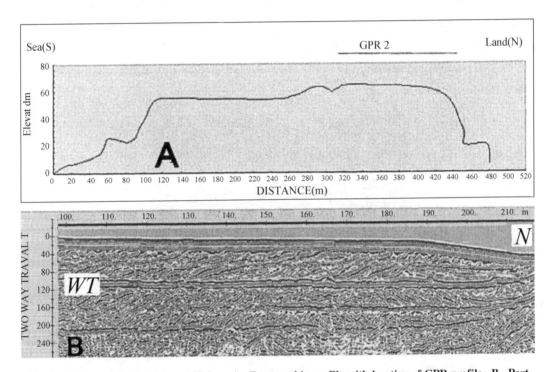

Fig 7　Most recent coastal bar at Haipo: A. Topographic profile with location of GPR profile; B. Part of GPR 2 profile positioned between 340 m and 450 m on top of the Haipo bar. (WT = water table; N = North direction; depth is in Two Way Travel Time of electromagnetic waves) (after Martini et al. 2003, 2004).

Another Holocene coastal barrier in a small embayment (Dadonghai bay) east of the town of Sanya has been dated. A sample taken at the contact between the lagoonal clay and the barriers sand has been dated at 4 471±165 a. This contact is now exposed at about 2 m psl (datum mean high tidal level). A second sample taken within the barriers and now at about 5 m psl has been dated at 2 826±725 a.

2.3 Reefs

Whereas dating of coastal ridges is generally difficult, coral reef and shells cemented in beachrock can be readily dated. They have been studied in some detail on the Luhuitou Peninsula and Ximao Island in Sanya Bay (Fig. 4A; Lu Bingquan et al. 1984). The peninsula is composed of a bedrock promontory (former small island) joined to the main island through two tomboli enclosing a now drained-out lagoon (Figs. 4A, 5A). The tombolo rise to about 5 m psl and are composed of coralline sandy gravels. Isolated pits and GPR profiles show them to be characterized by regular seaward-dipping cross-beds, indicating that they were formed primarily by beach accretion (Fig. 8). The oldest recorded date (8 168±145 a) is of a coral head in living position, from the base of the carbonate platform on the western side of the Luhuitou Peninsula. It now lies at about 1 m psl. At the top of the same platform about 4 m higher, another coral has been dated at 5 406±130 a, now at about 5 m psl. Near the southeastern edge of the promontory, a coral head-stack has been dated at 7 272±154 a, and is now found at about 1 m psl (sample 8, Figs. 4B, 9). Beachrocks dated from 5 079±120 a to 4 907± 115 a are now found at about high tidal level (0~0.5 m psl).

Fig 8 Tombolo of the Luhuitou Peninsula: A. Gravel (fossil fragments) pit showing well developed cross-beds; B. GPR profile from the top of the eastern tombolo to sea level, showing well development seaward-dipping layers.

Fig 9　Eastern shore of the Luhuitou Peninsula:
A.　In-situ coral head used for dating; B.　Raised reef platform

Other reef material has been dated along the intertidal transect of the north-western bay of the Luhuitou Peninsula, to as recent as 120±50 a, and corals are still living in the subaqueous part from 0.5 to 9 m water depth with optimum growth between 2~6 m depth.

Similar conditions occur on the northwestern side of the Ximao Island in the middle of Sanya Bay (Figs. 4A, 5A). Coral heads have been dated at 4 333±160 a and 5 720±130, which are now found at about high tidal level (0.5~0 m psl).

The available information indicates that during the Holocene transgression, water conditions became propitious for the development of reefs at approximately 8 000 a. The reef grew vigorously until approximately 5~4 000 a during the transgression and high sea-level stand at the Holocene temperature optimum, and again later at about 3 300 a and 2 200 a (Lu Bingquan et al. 1984; Zhao Xitao 1979; Wang Ying et al. 2001). Almost invariably, rapid reef growth seems to have been accompanied and/or follow, with a short time lag, by the formation of beachrock.

3　DISCUSSION

The information obtainable from the elevated, inland terraces and coastal ridges, in the absence of either numerical dates for the older units, allows only approximate inferences to be made. The older, highly weathered, elevated coastal ridges and terraces indicate not only that they may have formed during higher sea-level stands, as probably is the case for the Tongjin-Yanglan bar (TY) that probably formed during the penultimate interglacial period, but also that they were affected by tectonic uplift. The latter is indicated by the older ridges that are now up to 40 m asl and terraces that are up to 80 m asl. The top of the Holocene coastal ridges such as Haipo and Sanya is about 8 m psl, but it has been in part modified by man and it may have some wind-blown component. This 8 m value conforms to the thickness of the sandy portion of the ridge over lagoonal clay and silt. The sandy component of the ridge was not necessarily all deposited subaerially; that is, the contact between lagoonal clay and ridge sand does not indicate shoreline, only proximity to it, perhaps 1~2 m below. Furthermore, the internal structures of these Holocene ridges consistently indicate formation by beach accretion. The present elevation of the highest of the surfaces over which accretion took place is now at about 5 m psl, over which an approximately 2~3 m thick wave-generated bar developed. Construction of the beach may have occurred during storm conditions. Allowing for average 3 m storm wave height in the area, the relative average fair weather beach level during Holocene may have been about 5 m higher than the present time. The dating of samples between lagoonal clay to ridge sand in Dadonghai Bay coastal ridge, now exposed at 2 m psl, further indicates that the relative sea level of about 4 600 a may have been more than 3 m higher than at the present time. These estimated elevations compare favorably with the 5 m elevation of the wave-constructed, coralline gravely tomboli of the Luhuitou Peninsula.

The *in-situ* coralline material of reef platforms provides more accurate dates, but some uncertainty still exists on the precise water depth at which some of the corals lived. Nevertheless, based on modern analogies, it is safe to infer that the corals lived near the surface to perhaps to 2~3 m below it. The older(~8 000 a), lower part of the coralline platform is now at about 1 m psl; progressively younger parts of the platform are found at progressively higher elevations from about 2 to 5 m above present sea level. One of the highest samples gives a date of approximately 5 400 a. Depending on the living water depth of some of the corals dated, these data suggest that the relative sea-level elevation reached approximately 5 m higher than the present.

It remains now to estimate the contributions of eustatic and isostatic movements to the change in relative sea level. Several sea-level curves have been drawn for southwest Asia (Peltier 1998; Pizzaroli 1991; Tanabe. et al. 2003). Peltier (1998) analyzed the global isostatic adjustment in coastal areas of Pacific islands. His data derive from islands from Japan in the north and New Zealand to the south, and from Tahiti in the east to the Philippines in the west. Peltier used data mostly derived from tectonically active areas (Fig. 10) Hainan Island is located, instead, on a relatively more stable area. Along a transect from Japan to New Zealand, Peltier confirmed the existence of a relative sea level high-stand at about 5 ka (sideral time from ^{14}C dates) (Fig. 11). Such a highst and is subdued at high latitudes, but it is about 2 m above present sea level in the equatorial zone, such as in Rota (Roti) Island and in northern Australia. Except for areas where anomalously high rates may have occurred, data corrected for a Pleistocene average tectonic uplift of less than 1 mm/year fit the sea-level curves predicted by theoretical models (model ICE-4G (VM2); Peltier 1998), and the mid-Holocene, 2 m high, sea-level stand persists at low latitudes. In a few places where the data have been corrected for the living depth of corals, such as at Tahiti, there is a slight shift in the paleo-sea-level depth, but the overall trend of the sea-level curve does not change significantly (Peltier, 1998). Tanabe et al. (2003) obtained similar results along the coasts of mainland Asia in the Red River area (their figures 8, 9), Yongqiang Zong (2004), instead, in reviewing the available dates from the southern coast of mainland China found discrepancies in the age and elevation of the mid-Holocene high stand in various settings, them being 1 000 years older than in other settings and a few meters below the present sea level elevation in large delta areas, at about present sea level elevation in stable areas, and 1~2 m higher in tectonically active areas. Considering the regional information and our local data, we retain it reasonable to ascribe 2 m of the estimated mid-Holocene, higher-than-present, relative sea-level stand in Hainan Island to the higher-than-present, relative sea level reported from the region.

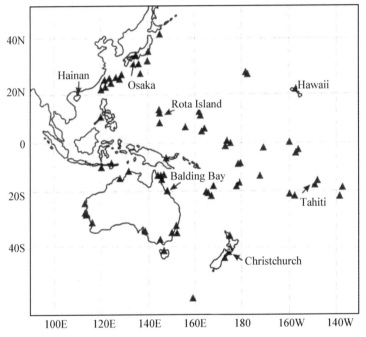

Fig 10　Map showing location of sample (triangles) used by Peltier(1998) in the Southern Pacific.

The remainder of the even higher, relative stand suggested for Hainan Island, estimated to an additional 2 m (Dadonghai Bay) to 3 m (Luhuitou Peninsula), may be associated with a tectonic uplift of approximately 0.2 to 0.4 mm/year. These rates of uplift are lower than, but comparable with the rates of uplift estimated by Peltier (1998). They are still an order of magnitude too high for the average rate of uplift of less than 0.01 mm/year obtained if the highest marine coastal bar/terrace in the Haipo area, now located at about 40 m asl, is assumed to have formed during the mid-Pleistocene. This discrepancy in Pleistocene-Holocene rates of uplift and the slight difference in elevation of the dated samples at Dadonghai Bay and Luhuitou Peninsula may be explained by local differential land movements through time and space. The area studied is, in fact, cross-cut by still active, normal faults, as indicated by geomorphic features such as hills with steep triangular-shaped flanks.

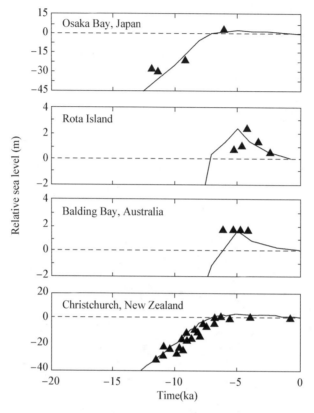

Fig 11 Sea-level curves in the Southern Pacific Ocean area (from Peltier, 1998). Time in thousand of years (ka) based on the sidereal transformation of ^{14}C dates (triangles show approximate elevation-age of sample).

4 CONCLUSIONS

A detailed study of the coastal morphology of southeastern Hainan Island, China, has provided evidence on the mode of formation of Pleistocene-Holocene coastal siliciclastic coastal sand ridges and carbonate gravelly tomboli, and on the distribution of raised reef platforms. The older, Pleistocene coastal ridges that are now found at various elevations up to 40 m asl, are intensely weathered and lack significant internal structures and any material datable (numerical dating) with conventional means. The internal structures of the younger Holocene ridges and carbonate tomboli are instead well defined showing formation by beach accretion through wave action. The elevation of such paleo-beaches is on the order of 5 m above present sea level. ^{14}C dating of shells collected in the siliciclastic deposits indicates that~4. 5 ka-old material from the contact between lagoon day and bar (barrier) sand is now found at 2 m psl. Such a contact does not pinpoint the ancient shoreline, rather a level approximately 1~2 m below the paleo-sea surface. Dating of in situ coral heads and other coral material indicates that the development of the reef platform started about 8 ka. The base of such a platform is now found at about 1 m psl. Other coral heads dated~5 ka are found at elevation 4. 5 m above present sea level. The corals lived from near the water surface to about 2~4 m below. Although information varies from location to location in southern Hainan Island, it is safe to estimate that the mid-Holocene relative sea level was at least 3. 5 m higher than at the present time. About 2 m of such mid-Holocene highstand reflect the over an situation through the tropical to sub-tropical western Pacific zone, as latterly reiterated by Peltier (1998). The remaining 1~3 m excess elevation may be associated with tectonic uplift. The variations in estimated elevations of the paleo-shorelines of local land blocks may be partly associated with differential movements along the numerous, still active, normal faults that cross-cut the area.

Acknowledgements

This study started as part of the Canada-China Higher Education Program project entitled "Environmental Training for Integrated Monitoring and Management in the Coastal Zone of Hainan Province, China" financed by the Canadian International Development Agency (CIDA Project No. 27159). The Natural Science and Engineering Research Council of Canada (NSERC Grant A7371to IPM), University of Guelph, State Pilot Laboratory of Coast and Island Exploitation in Nanjing University, provided additional financial and logistic support for completing the scientific analysis of the data and final write-up. The Government of Hainan Province provided some field transportation and some local logistic support. Many people helped in the study. Among them we like to thank Dr. R. Bullock, Dr. Xinqing Zou, Dr. Chendong Ge and Mrs. ShaoqunHe. Thanks are given to the numerous field assistants and to the technicians of the State Pilot Laboratory of Coast and Island Exploitation in Nanjing University for the various sediment analyses, including ^{14}C dating. Dr S. Kershaw helped in focusing the final version of the paper with a critical review. Dr. F. Antonioli provided encouragement during preparation of the manuscript and the algorithm for calibrating the measured ^{14}C dates.

References

[1] DAVIS J. L. , ANNAN A. P. 1988. Ground-penetrating radar for high resolution mapping of soil and rock stratigraphy. *Geophysical Prospecting*, 37: 531 - 551.

[2] CHINESE ACADEMY OF GEOLOGICAL SCIENCES. 1980. Geological map of Asia (1:5,000,000). Map Publishing House, Beijing.

[3] LU BINGQUAN, WANG GUOZHONG, QUAN SONGQING. 1984. The characteristics of fringing

reefs of Hainan Island. *Geographical Research* 3: 1~15 (in Chinese).

[4] MARTINI I. P. , ZHU DAKUI, GAO XUETIAN, YIN YONG. 2003. Coastal landscape and landuse Hainan Island, China. Environmental Studies Monograph, University of Waterloo, Canada. 258pp+ hyperlinked CDrom with original, color images.

[5] MARTINI I. P. , ZHU DAKUI, GAO XUETIAN, YIN YONG. 2004. Coastal landscape and landuse-Hainan Island, China (http://wxvw. uoguelph. ca/geology/hainan: website to be released in 2004).

[6] PELTIER W. R. 1998. Global glacial isostatic adjustment and coastal tectonics. In: I. S. STEWART, C. VITA-FINZI (eds.), *Coastal Tectonics*. Geological Society, London Special Publications 146: 1 - 29.

[7] PIRAZZOLI P. A. 1991. World atlas of Holocene sea-level changes. Elsevier, Amsterdam, 300 pp.

[8] TANABE SUSUMU, HORI KAZUAKI, SAITO YOSHIKI, HARUYAMA SHIGEKO, Vu VAN PHAI, KITAMURA AKIHISA. 2003. Song Hong (Red River) delta evolution related to millennium-scale Holocene sea-level changes. *Quaternary Science Reviews*, 22: 2345 - 2361.

[9] WANG YING, CHEN WANLI. 1982. Basic studies on coastal geomorphology in Sanya Bay. *Marine Bulletin*, 1 (3): 37~45 (in Chinese).

[10] WANG YING, ZHU DAKUI, SCHAFER C. T. , SMITH J. N. 1992. Marine geology and environments of Sanya bay, Hainan Island, China. In: YING WANG, C. T. SCHAFER (eds.), Island environment and coastal development. Nanjing University Press, Nanjing. 125 - 156.

[11] WANG YING, MARTINI I. P. , ZHU DAKUI, ZHANG YONGZHAN, TANG WENWU. 2001. Coastal plain evolution in southern Hainan Island, China. *Chinese Science Bulletin*, 46 Supp: 91 - 96.

[12] YAO QINGYI, CHEN HUATANG, LU GUOQI, TAN PIXIAN. 1981. A study on geomorphic types of Qionglei region. *Tropical Geography*, 1: 13 - 20 (in Chinese).

[13] YONGQIANG ZONG. 2004. Mid-Holocene sea-level highstand along the Southeast Coast of China. *Quaternary International*, 117: 55 - 67.

[14] ZHANG ZHONGYING, LIU RUIHUA, HAN ZHONGYUAN. 1987. Quaternary stratigraphy along the coastal area of Hainan Island. *Tropical Geography*, 7: 54 - 63 (in Chinese).

[15] ZHAO XITAO. 1979. Dating of coral reef and coastline migration in Lubuitou in Hainan Island. *Science Bulletin*, 24: 995 - 998 (in China).

[16] ZHU SUOFENG, DENG WEIDONG. 1992. Basic studies on tropical storm surges in Yangpu Harbor. In: YING WANG, C. T. SCHAFER(eds), Island Enviroment and Coastal Development. Nanjing University Press, Nanjing. 31 - 38.

[17] ZENG ZAOXUAN, ZENG XIANZONG. 1989. Physical Geography of Hainan Island. Beijing: Science Press (in Chinese).

Quaternary Evolution of the Rivers of Northeast Hainan Island, China: Tracking the History of Avulsion from Mineralogy and Geochemistry of River and Delta Sands [*]

1　Introduction

Bulk geochemistry of sediments has been widely used in ancient rocks as an indicator of the tectonic setting of the sediment source (Bhatia, 1983; Ryan and Williams, 2007). It has the advantage of being a relatively inexpensive and rapid method of assessing provenance (von Eynatten et al., 2003). Its utility in discriminating precise sources from similar geological terranes is less clear (Pe-Piper et al., 2008). Provenance studies in such situations rely on more expensive and time consuming mineralogical and geochronological data.

In northern Hainan Island, southern China, neotectonic tilting and basaltic volcanism influenced the course of the largest river, the Nandu River, in the Holocene (Wang, 1998). The northeastern coast of Hainan has reworked prograded beach ridges, rich in heavy minerals, yet has insignificant Holocene river supply of sediment. Potential river supply in the past would have drained the mountainous interior of the island, comprising principally granites and metamorphic rocks (Fig. 1).

This study had two interlinked objectives. First, to investigate the history of avulsion of the Nandu River using bulk geochemical and mineralogical composition of river and deltaic sands. Second, to assess the success in using solely geochemical tracers of provenance in modern sediments to discriminate between drainage basins with rather similar geology. We hope to be able to apply the results of the second objective to our studies of Cretaceous sedimentary rocks in the Scotian Basin (Gould et al., 2010; Zhang et al., 2014), which had similar paleoclimate and source-area geology in the early Cretaceous to Hainan Island in the Quaternary.

[*]　Georgia Pe-Piper, David J. W. Piper, Ying Wang, Yongzhan Zhang, Corwin Trottier, Chendong Ge, Yong Yin: *Sedimentary Geology*, 2016, Vol. 333: pp. 84 – 99.

Supplementary data 请参见 http://dx. doi. org/10. 1016/j. sedgeo. 2015. 12. 008.

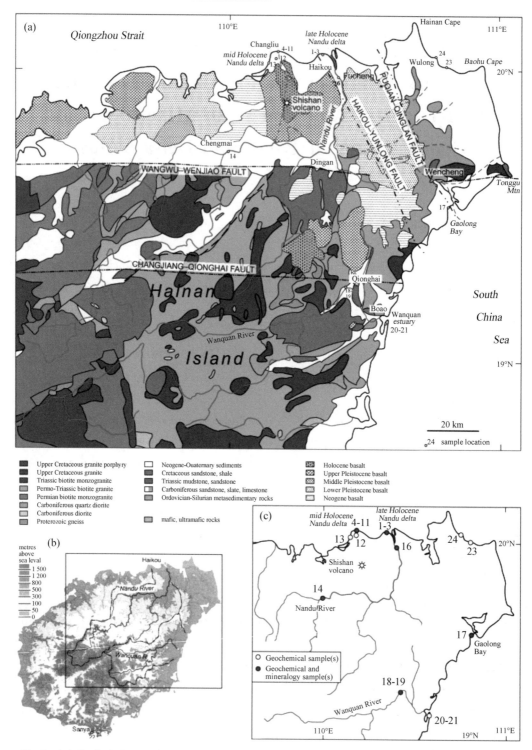

Fig 1 (a) Geological map of northeastern Hainan Island. Geology from Guangdong BGMR(1988) with ages of volcanic rocks from Ho et al. (2000). Dashed lines show major dry valleys on the northeastern peninsula. (b) Topography of Hainan Island and the drainage basins of the Nandu and Wanquan rivers. (c) Map showing sample locations. (附彩图)

2　Geological setting

2.1　Geomorphology

Hainan Island (Fig. 1) is a mountainous (reaching 1867 m asl), tropical island, with bedrock dominated by granite and orthogneiss (37% of the land area) in the centre and south, and widespread Quaternary basalts in the north. Active neotectonic faults follow E-W and NNW-SSE trends (Ma, 1989; Lei et al., 2009). The island receives over 1.6 m of rain annually and lies in one of the main cyclone tracks in the western Pacific Ocean (Hu et al., 2014). The island is drained by a generally radial pattern of sandy, mountainous rivers mostly <100 km in length. The longest is the Nandu River, in the north (170 km). The study area includes the Nandu River basin and the smaller Wanquan River basin to the southeast.

The Nandu River drains much of the northeastern part of the island (Fig. 1b). The abandoned delta lobe at Changliu represents an earlier path of the Nandu River (Wang, 1998), which was dammed by lava from the Shishan volcano and was thus forced to flow eastwards and then northwards to its present route along the edge of the Holocene lava field (Fig. 1a). Before construction of dams in its upper reaches in 1959, the Nandu River had a water discharge of $6\ 800 \times 10^6$ m³/year and carried 385 000 tonnes of sediment to the sea annually (Wang,1992). The Wanquan River drains the eastern part of the island immediately south of the Changjiang-Qionghai fault; the area of its drainage basin is about 70% of that of the Nandu River (Fig. 1b).

The northeastern peninsula of Hainan, east of the active Puqian-Qinglan tectonic lineament (Ma, 1989), consists principally of Quaternary sands in fluvial deposits, raised marine terraces that suggest Quaternary uplift (Wu and Wu, 1987), and coastal aggradational features. Headlands and offshore shoals are pinned on granitic bedrock. At present, no rivers drain from the mountainous interior to this area and Ti-rich placer sands are abundant along the coast (Zeng et al.,2011). The area could have received sediment in the past via a paleo-Nandu River or possibly from drainage from the present Wanquan drainage basin or from a mainland river during times of lowered sea level (Li et al., 2015).

Previous heavy mineral studies include a regional assessment of provenance of Neogene sediment in the Qiongdongnan Basin, to the south of Hainan Island (Cao et al., 2015; Liu et al., 2015). Li et al. (2015) studied the provenance of heavy mineral deposits on the northwestern shelf of the South China Sea using mineral chemistry of selected heavy minerals, principally tourmaline and amphibole. They characterised the minerals at the mouths of the Wanquan and Nandu rivers, and inferred that during the

Late Pleistocene lowstand the Nandu River discharged westwards through Qiongzhou Strait into Beibu Gulf (Gulf of Tonkin). Ma et al. (2010) presented REE analyses from the mouth of the Nandu River and from prodeltaic areas out to the 20 m isobath.

2. 2　Bedrock geology

Bedrock geology controls the composition of sediment supplied to the rivers. Basement rocks are the Mesoproterozoic high greenschistto amphibolite-grade metamorphic rocks and granitoid plutons (Zhang et al., 2011). Overlying Paleozoic shallow marine siliciclastic strata have experienced low grade metamorphism. Late Paleozoic metamafic rocks of ophiolitic origin outcrop locally south of the Wangwu-Wenjiao fault (Li et al., 2002; Wang et al., 2013). Permian orthogneiss in the centre of the island is intruded by granites of Triassic, Jurassic and Cretaceous ages, which are overlain by Cretaceous sandstones and shales. Molybdenumtungsten mineralisation is associated with the Late Cretaceous granites in southern China (Mao et al., 2007) including those in southwestern Hainan (Fu et al., 2013).

Neogene-Quaternary basalts are widespread in the northern part of the island (Wang and Zhou, 1990; Ho et al., 2000). They are related to ongoing regional extension and mantle upwelling associated with the Hainan mantle plume (Lei et al., 2009). Abundant volcanism began in the Early Pliocene (Ho et al., 2000). Holocene volcanism is centred around Shishan volcano, 20 km SW of Haikou (Fig. 1a). Most basalt is tholeiitic, with abundant ilmenite, whereas titanomagnetite is dominant in the less common, principally Late Pleistocene, alkalic basalt (Ho et al., 2000).

3　Materials and methods

Bulk sediment samples, ~1. 5 kg in weight, were collected from 10~20 cm deep sections on river beds, beaches and old beach ridges (Table 1, Fig. 1c, Supplementary data Table S1). River sediment samples characterise the three main headwater areas: the SW Nandu River(sample 14), the SE Nandu River headwaters (sample 16, which also includes sediments from the SW headwaters), and the Wanquan River (samples 18 and 19). Beach and beach-ridge samples characterise former and modern river deltas. From NW to SE, these are the mid-Holocene Nandu delta at Changliu (samples 4~13), a mid-Quaternary Nandu delta at Haikou (sample 1), the modern Nandu delta at Haikou(samples 2, 3), probable mid-Pleistocene reworked deltaic deposits on the northeast peninsula (samples 23, 24 and 17) and a beach at the modern Wanquan River mouth (samples 20, 21). In addition, one beach sand sample from southern Hainan Island was studied (sample 22).

Table 1　Samples analysed.

Sample	Material [a]	Geological unit
1	Paleosol	Mid-Quaternary Nandu delta
2~3	Beach	Late Holocene Nandu delta
4~6, 8, 10, 12~13	Beach	Mid-Holocene Nandu delta
7, 11A, B	Paleosol	Mid-Holocene Nandu delta
9	Beach [b]	Mid-Holocene Nandu delta
14	River	Lower Nandu River
16	River	Upper Nandu River
17	Beach	Gaolong Bay
18	River	Wanquan River
19	River [b]	Wanquan River
20~21	Beach	Wanquan estuary
22	Beach	Sanya
23~24	Beach	Baohu Cape

Samples (in bold-faced type) have heavy mineral analysis.

See Supplementary dataTable S1 for further details including location coordinates.

[a] Note: beach includes inland beach ridges.

[b] Sample with heavy mineral concentration.

Bulk sediment geochemical analyses were determined from 24 sand samples (two of which were duplicated to check reproducibility). The more time consuming detailed heavy mineral counts were made on six samples, three (samples 14, 16, 19) representing the major drainage basins and three (samples 1, 9, and 17) representing the location of former deltaic sediments (Fig. 1C). These samples provide mineralogical control that can be used to interpret the larger number of geochemical analyses. Grain size analysis by sieving was carried out on all samples (Supplementary data Table S2).

Samples for bulk geochemical analysis were crushed to <60 μm. Major elements and trace elements were determined by Activation Laboratories Limited according to their Code 4 Lithoresearch and Code 4B1 packages (Activation Laboratories Ltd, 2012), which combine lithium metaborate/tetraborate fusion ICP analyses with a trace element ICP-MS package. Abundance of major elements was recalculated on a volatile-free basis and for those beach samples with abundant biogenic carbonate (samples 17, 22, 23, 24) the abundance of all elements was recalculated on a carbonate-free basis.

Heavy minerals were separated from the $63\sim177$ μm fraction by centrifuge (gravity) separation in aqueous solution of sodium polytungstate with a specific gravity of 2.9. We are aware of the strong arguments for not using a restricted size range (Garzanti et al., 2009), but our original motivation to make comparison with the Cretaceous Scotian Basin led us to use a size fraction that in the lithified rocks of the Scotian Basin excludes coarser sizes which comprise large amounts of insufficiently disaggregated rock fragments and diagenetic minerals, and also silt sizes that have a large proportion of crushed mineral fragments. Polished thin sections of grain mounts (Fig. 2) were examined by scanning electron microscope (SEM). The SEM used is a LEO 1450 VP with a maximum resolution of 3. 5 nm at 30 kV. The SEM uses a tungsten filament to supply electrons to produce a back-scattered electron (BSE) image of the grains in the polished thin section. The SEM was also used to identify minerals through the use of energy dispersive spectroscopy (EDS), with an analytical spot size of ~10 μm. Element detection limit is > 0. 1%. More than 1800 grains were identified by their geochemistry and counted in each sample, except for sample 01 where there was insufficient separate to count more than 623 identifiable grains. The chemical criteria used for identifying minerals are documented in Table 2. It is difficult to determine whether some heavy "minerals", particularly in fluvial and soil samples, are altered detrital magnetite or are ferruginous crusts formed by pedogenesis (Fig. 3b, e). Likewise, some ilmenite grains show complex alteration to a titania mineral (probably rutile, Fig. 3a), with textures similar to those documented by Pe-Piper et al. (2005), and some ilmenite is present in lithic clasts (Fig. 3c).

Fig 2　Examples of heavy minerals from (a) river (upper Nandu River, sample 14) and (b) beach on mid-Holocene Nandu delta (sample 09). And = andalusite (possibly sillimanite or kyanite); Bio = biotite; C= ferruginous cement and mostly quartz and clays; Cpx= clinopyroxene; Ep= epidote; Fe-Chl= septochlorite; Glt = glauconite; Grs = grossular garnet; Hbl = hornblende; Ilm = ilmenite; Mag = magnetite; Mnz=monazite; Pl=Plagioclase; Qz=quartz; Rut=rutile (possibly anatase); Ti=altered Ti-rich mineral; Timag=titanomagnetite; Tur=tourmaline; Zr=zircon. (附彩图)

4　Results

4. 1　Detrital mineralogy—abundance of minerals

Relative abundance of heavy minerals is shown in Table 2. Abundance of most heavy minerals is illustrated in summary pie diagrams in Fig. 4, which exclude the

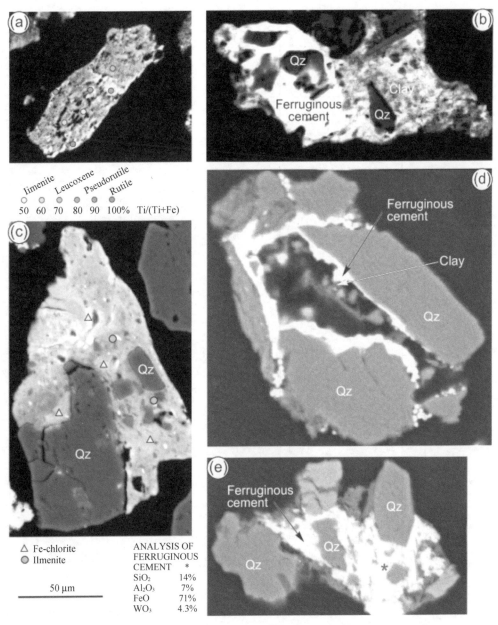

Fig 3 Backscattered electron images of grains with irregular Fe-Ti oxides of various origins. (a-b) from paleosol (sample 01); (c-e) from upper Nandu River (sample 14). (a) Original ilmenite grain with patchy alteration to leucoxene, pseudorutile and rutile. Colours of analysed spots indicate atomic ratio of Ti/(Ti + Fe). (b) Ferruginous aggregate of quartz grains (Qz), clays (fine grained, low backscatter) and iron (hydr)oxides (ferruginous cement, high backscatter). (c) Altered lithic clast of quartz-chlorite schist including minor ilmenite; analysed spots of ilmenite and Fe-chlorite are shown. (d) Loose aggregate of quartz grains with ferruginous cement (high backscatter), with cement largely removed by abrasion on outer part of aggregate. Some clays (fine grained, low backscatter) appear trapped within the aggregate. (e) Loose aggregate of quartz grains and lesser clay, with ferruginous cement (chemical analysis of cement shown). (附彩图)

Table 2　Summary of mineral counts and geochemical criteria for mineral identification by EDS

Sample descriptions — CH–01: Paleosol, mid-Q Nandu delta; CH–09: Beach, mid Holocene Nandu delta; CH–14: Upper Nandu River; CH–16: Lower Nandu River; CH–17: Beach, Gaolong Bay; CH–19: Lower Wanquan River. Sample columns give *percentage in heavy mineral fraction*. Columns Other–REE are *Chemical criteria (%) for mineral recognition*.

Mineral	Group	CH–01	CH–09	CH–14	CH–16	CH–17	CH–19	Other	SiO_2	TiO_2	Al_2O_3	$FeO(T)$	MnO	MgO	CaO	Na_2O	K_2O	P_2O_5	Cr_2O_3	REE
Magnetite		0.6	3.9	1.9	4.1	0.2	5.9					**>90**	<3							
Ti–Magnetite		1.1	13.1	3.9	14.7	0.8	1.9			>3		**>63**								
Hematite		2.2	0.1	0.4	0.2	0.1	0.1					**86–90**	<3							
Goethite/limonite		0.7	0.0	1.1	0.4	0.0	0.2					**80–83**	<3							
Ilmenite		10.1	55.5	7.0	36.3	32.3	30.3			~55		~42	~3							
Leucoxene/pseudorutile		0.3	1.1	2.5	1.7	2.5	0.6			70–90										
Rutile		1.3	3.5	4.9	2.9	11.0	1.1			>90										
Hornblende	Amphibole	0.1	0.2	3.3	4.1	2.0	24.5		39–54		**4–10**	0–28		2–14	9–13					
Actinolite	Amphibole	0.0	0.2	1.5	1.6	0.9	8.1		47–60		**0–4**	0–30		0–26	10–13					
Pargasite	Amphibole	0.0	0.0	0.3	0.5	0.4	6.1		36–48		**10–17**	2–32		2–20	9–14					
Almandine	Garnet	0.0	0.0	0.1	0.0	0.1	0.4		34–42		19–23	21–39	<1	**1–14**	<1					
Andradite	Garnet	0.0	0.0	0.0	0.0	0.0	0.1		35–42		**0–10**	16–30			22–33					
Grossular	Garnet	0.0	0.1	0.5	0.1	0.1	0.5		39–43		**9–24**	**0–15**	<1	<1	31–34					
Spessartine	Garnet	0.0	0.0	0.0	0.0	0.0	0.1		35–40		20–22	2–14	22–40	<3	1–5					
Chromite	Spinel	0.1	0.1	0.1	0.0	0.3	0.0				7–21	16–33		4–14					**35–59**	
Spinel	Spinel	0.0	0.2	0.0	0.2	0.0	0.0				**23–70**	11–46		5–20					26–41	
Clinopyroxene	Pyroxene	0.1	0.1	0.9	0.2	0.1	0.4		50–60		1–8	6–14		11–18	14–22					
Orthopyroxene	Pyroxene	0.0	0.1	0.6	0.7	0.0	0.8		50–60		0–4	12–28		14–30	0–9					
Andalusite etc.		1.5	0.2	21.1	2.3	2.1	0.1		~38		~62									
Epidote	Epidote	0.1	0.1	4.3	3.7	1.8	4.1		37–47		**17–27**	9–17			14–25					
Clinozoisite	Epidote	0.0	0.2	1.0	0.9	0.6	1.9		41–48		**27–34**	<9			20–24					
Titanite	Epidote	0.1	0.0	0.1	0.4	0.6	2.1		28–44	**30–40**	<6	<4			24–29					
Tourmaline		5.3	0.7	11.9	2.8	11.5	0.3		41–47		**28–50**	1–20		1–12			**1–3**			
Zircon		7.0	16.0	1.9	5.8	11.8	2.5	68% ZrO_2	~32											
Monazite		0.1	2.4	0.4	1.4	0.3	0.5											~35		~65
Allanite		0.0	0.0	0.0	0.0	0.0	0.2		36–38		14–15	13–15		1–2	11–13					18–22
Al–phosphate mineral		0.0	0.1	0.3	0.0	0.2	0.1				~35							~35		~30
Apatite		0.0	0.1	0.0	0.0	1.1	0.1	~7% F							**~46**			**~41**		
Phosphate mineral		0.1	0.1	0.1	0.0	0.0	0.0								None			High		None
Staurolite		0.0	0.0	0.3	0.0	0.4	0.0		30–41		**50–56**	2–15								
Calcite		0.0	0.0	0.0	0.0	2.0	0.0								**>95**					
Mn–(hydr)oxides	Other	0.6	0.0	0.1	0.0	0.1	0.1	Some BaO, WO_3					Some	**~75**						
Pyrite	Other	0.0	0.1	0.1	0.1	0.0	0.0	~53% SO_3				**~47**								
Olivine	Other	1.4	0.0	0.2	0.2	0.0	0.2		37–60			12–49		0–46						
Corundum	Other	0.0	0.0	0.0	0.0	0.1	0.0				>95									
Thorite	Other	0.0	0.1	0.0	0.1	0.0	0.1	~75% ThO_2	Some									Some		
W–minerals	Other	0.0	0.0	0.0	0.1	0.8	0.0	>90% WO_3					Some		Some					
Nb–mineral	Other	0.0	0.0	0.0	0.0	0.1	0.0	High $(Nb,Ta)_2O_5$												
Cassiterite	Other	0.0	0.0	0.0	0.0	0.1	0.0	>95% SnO_2												
Ti–rich altered grains		1.4	1.3	7.6	3.0	10.1	1.7	TiO_2 > FeO, considerable SiO_2, Al_2O_3												
Fe–rich altered grains		20.4	0.4	9.8	5.2	0.6	0.9	FeO > TiO_2, considerable SiO_2, Al_2O_3												
Unidentified		9.1	0.2	2.5	2.2	2.5	3.1													
Biotite		0.6	0.1	1.9	0.6	0.8	0.9		34–39	<4	12–17	15–29		7–13			6–9			
Mg–Chlorite		2.2	0.0	1.7	1.2	1.0	0.2		20–35		14–27	**0–35**		0–22			<3			
Fe–Chlorite		33.3	0.1	5.4	2.7	0.6	0.2		20–35		15–25	**35–70**		0–22			<3			
Total counts		715	1804	2034	1611	1422	1724	Bolded values are most useful in discriminating the mineral												
Mean grain size (φ)			1.9	2.2	0.5	2.5	1.5													

abundant Fe-(hydr)oxides and Fe-Ti oxides, platy minerals with borderline density (chlorite and biotite), and unidentified grains, many of which may be lithic clasts.

The principal heavy minerals are Fe-(hydr)oxides (magnetite and its alteration products) and Fe-Ti oxides (ilmenite and its alteration products) (Table 2, Figs. 2, 3). Magnetite and titanomagnetite are common in both the lower Nandu River (sample 16: 4.1% and 14.7% respectively of total heavy minerals) and the mid-Holocene Nandu River delta beach (sample 09, 3.9% and 13.1% respectively; Fig. 2B), but only magnetite is common in the Wanquan River (sample 19; 5.9%, but only 1.9% titanomagnetite). Ilmenite is particularly abundant in the lower Nandu River (sample 16; 36.3%), the mid-Holocene Nandu River delta (sample 09; 55.5%), the Wanquan River (sample 19; 30.3%), and Gaolong beach (sample 17; 32.3%). Rutile is least

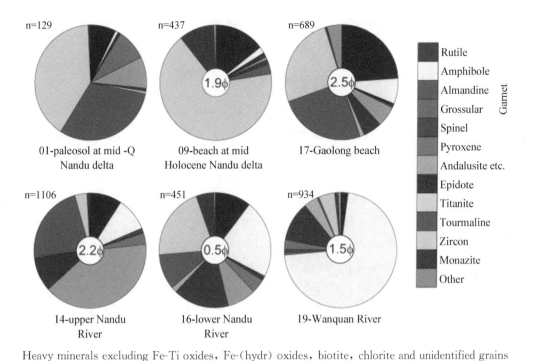

Heavy minerals excluding Fe-Ti oxides, Fe-(hydr) oxides, biotite, chlorite and unidentified grains

Fig 4 Pie diagrams showing abundance (counts) of heavy minerals in six representative samples. Diagrams exclude the abundant Fe-(hydr)oxides and Fe-Ti oxides, platy minerals with borderline density (chlorite and biotite), and unidentified grains (many of which are lithic clasts). Mineral identification based on chemical composition by EDS and on morphology. Andalusite may include some sillimanite and kyanite. Mean grain size of the sand sample (in φ units) is shown. For relative abundance of all heavy minerals, see Table 2. (附彩图)

abundant in the Wanquan River (sample 19; 1.1%) and most abundant in the Gaolong beach (sample 17;11.0%).

Heavy minerals in the Wanquan River (sample 19) are dominated by amphibole (38.7%), which is of lesser importance in the Nandu River (samples 14, 16; 5.1%, 6.2%) and almost absent in beach samples (samples 09, 17; 0.4%, 3.3%) (Table 2, Fig. 4). Likewise, epidote is moderately common in all river samples (4.6%~6.0%), but almost absent in beach samples (0.3%~2.4%). Conversely, zircon is concentrated on beaches (11.8%~16.0%) compared to rivers (1.9%~5.8%; Fig. 2). Tourmaline is abundant in the upper Nandu River (sample 14; 11.9%), less common in the lower Nandu River (sample 16; 2.8%), and uncommon in the mid-Holocene Nandu delta (sample 09; 0.7%). It is, however, abundant at Gaolong beach (sample 17; 11.5%). Andalusite (based on mineral morphology, perhaps including a few grains of its polymorphs sillimanite and kyanite) is abundant in the upper Nandu River (sample 14; 21.1%; Fig. 2a), presumably from Paleozoic metasedimentary rocks upstream(Fig. 1a). It is common in the lower Nandu River (sample 16; 2.3%) and at Gaolong beach (sample 17; 2.1%), but very rare in the Wanquan River (sample 19; 0.1%).

Staurolite, also sourced from metasedimentary rocks, is found only in samples 14 (upper Nandu River) and 17 (Gaolong beach) (Table 2).

4.2 Detrital mineralogy—chemistry of minerals

Only a few minerals show systematic chemical variation in major elements detectable by EDS analysis that may be diagnostic of provenance (Supplementary data Table S3). Amphiboles in the Wanquan River (sample 19) have a wide range of compositions, principally ferroedenite, actinolite and magnesiohornblende (Fig. 5a-c). The least silicic amphiboles have Ti contents of >0.2 atoms per formula unit (a. p. f. u.). All other analysed samples have a smaller range of composition, principally magnesiohornblende and lesser actinolite, and lack silicic amphiboles with Ti>0.2 a. p. f. u. In the discrimination diagram of Fleet and Barnett (1978) that is based on the ratio of ^{IV}Al to ^{VI}Al (Fig. 5d, e), all the amphiboles are of metamorphic origin using the calibration of Li et al. (2015), and almost all are from high-grade metamorphic rocks. Amphibole grains from high pressure high grade metamorphic rocks make up $81\% \sim 83\%$ of amphiboles at Gaolong beach (sample 17) and the upper Nandu River (sample 14), but only 61% in the Wanquan River (sample 19). Low pressure amphibole is found in the Wanquan River (sample 19; 29%) and upper Nandu River (sample 14; 9%), and grains having ^{IV}Al/^{VI}Al <0.9, which Li et al. (2015) regarded as unclassifiable, make up $10\% \sim 20\%$ of all samples.

Tourmaline is particularly abundant in the upper Nandu River (sample 14) and at Gaolong beach (sample 17) (Fig. 6a). In general, there are no large differences in the tourmaline assemblages between samples, except that the proportion of low-Mg tourmaline, likely of granitic origin using the criteria of Kassoli-Fournaraki and Michailidis (1994), is higher in the upper Nandu River (sample 14; 50%), mid-Holocene Nandu delta (sample 09; 58%) and Gaolong beach (sample 17; 48%) compared with other samples ($37\% \sim 40\%$). Only five grains could be analysed from sample 19 from the Wanquan River, but most have low Al content suggesting an origin from metapelite orcalc-silicate rock.

Two types of garnet are present in more than trace amounts and geochemical variation is summarised in Fig. 6b. Grossular, showing some solid solution with andradite, is common in the upper Nandu River (sample 14) and Wanquan River (sample 19). Almandine with some solid solution with pyrope is common in the Wanquan River (sample 19) with one grain identified at Gaolong beach (sample 17) and two in the upper Nandu River (sample 14). The few chrome spinels analysed (Fig. 5f) have a wide range of compositions, typical of complex ophiolitic sources (e. g., Tsikouras et al., 2011), and show no systematic variation between samples.

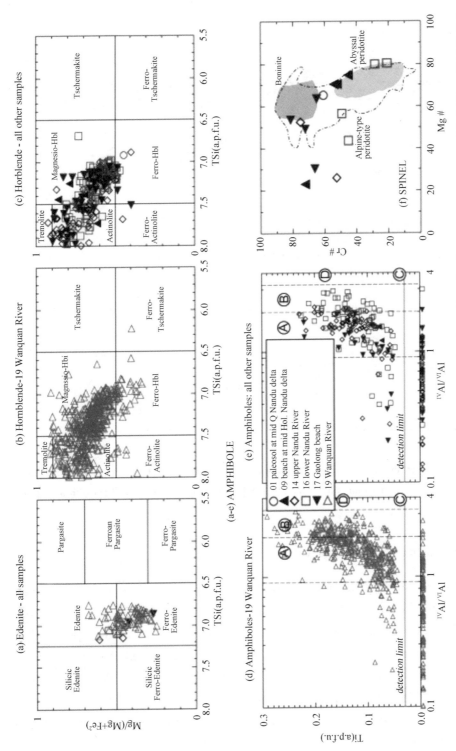

Fig 5 Plots showing chemical composition of analysed amphibole and spinel: (a–c) Mg/(Mg+Fe) *vs.* Si for amphibole, (d–e) Ti *vs.* ^{IV}Al/^{VI}Al for amphibole, (a) = high grade, high pressure metamorphic amphibole, (b) = high grade, low grade metamorphic amphibole, (c) = low pressure metamorphic amphibole, (d) = igneous amphibole (from Li et al., 2015). Percentages quoted in text based on grains with Ti above detection limit. (f) Cr *vs.* Mg(after Dick and Bullen, 1984) for spinel. Caution: analyses by EDS with a 10 μm analytical spot.

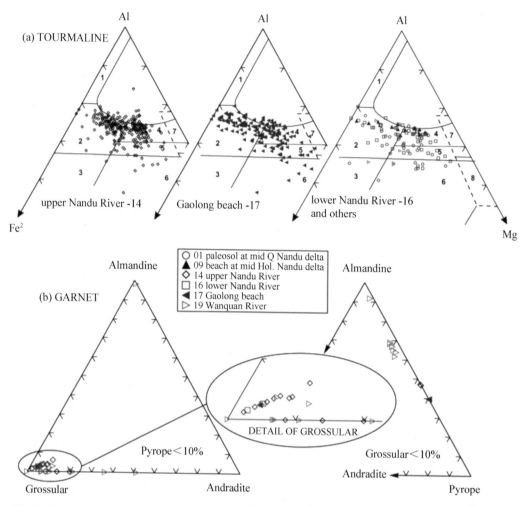

Fig 6 Plots showing chemical composition of analysed (a) tourmaline and (b) garnet. Fields 1～8 for tourmaline from Kassoli-Fournaraki and Michailidis (1994). Caution: analyses by EDS with a 10 μm analytical spot.

4.3 Bulk sediment geochemistry

Geochemistry data for the bulk samples with heavy mineral counts are given in Table 3; fulldata are in Supplementary data Table S4. Element variation between samples is illustrated in Fig. 7. A plot of Al_2O_3 vs. SiO_2 (Fig. 7a) shows that most of the samples are principally terrigenous sands, with some muds (higher Al_2O_3) in river sands, as confirmed by grain size data (Supplementary data Table S2). Four beach samples with high biogenic CaO plot off the trend line (samples 17, 22, 23, 24; small symbols in Fig. 7a), as do two samples (samples 8 and 9) with high TiO_2 content due to concentration of ilmenite.

Table 3　Geochemistry of samples with detailed mineralogy.

Sample	01	09	14	16	17	19	Detection limit
Major elements，wt.%，recalculated volatile free							
SiO_2	80.98	73.52	89.67	91.89	74.22	80.32	0.01
TiO_2	0.78	11.02	0.29	0.12	0.26	2.89	0.001
Al_2O_3	9.39	2.44	5.48	4.08	2.05	6.63	0.01
Fe_2O_{3T}	7.53	11.24	1.31	0.85	1.51	4.73	0.01
MnO	0.04	0.48	0.02	0.01	0.03	0.16	0.001
MgO	0.27	0.22	0.14	0.05	1.03	0.46	0.01
CaO	0.07	0.19	0.18	0.09	19.51	0.85	0.01
Na_2O	0.16	0.15	0.30	0.29	0.47	0.74	0.01
K_2O	0.68	0.58	2.51	2.55	0.83	3.08	0.01
P_2O_5	0.06	0.14	0.06	0.04	0.08	0.07	0.01
LOI	5.32	0.14	1.46	0.41	14.12	0.51	0.01
S	0.04	0.007	0.005	0.001	0.09	b.d.	0.001
Total	98.77	98.97	99.95	99.42	98.94	100.4	0.01
Trace elements，ppm							
Be	1	b.d.	1	b.d.	b.d.	1	1
V	101	188	22	9	12	74	5
Cr	130	210	20	30	20	40	20
Co	6	9	3	2	b.d.	4	1
Ni	23	9	7	4	5	7	1
Cu	17	16	5	4	2	3	1
Zn	29	141	22	12	13	43	1
Sc	9	19	2	b.d.	2	6	1
Ga	12	12	5	4	3	8	1
Ge	1.8	2.4	1.5	1.1	0.8	1.7	0.5
As	17	9	b.d.	b.d.	8	b.d.	5
Rb	43	23	83	77	29	96	1
Ba	133	82	335	340	84	444	3
Cs	4.3	0.8	2.7	2.5	0.8	1.8	0.1
Sr	22	36	68	69	1509	154	2
Y	17.1	203	8.7	3.6	13	25.8	0.5

（continued）

Sample	01	09	14	16	17	19	Detection limit
Zr	348	15 400	186	61	286	622	1
Nb	19. 2	242	7. 5	3. 3	12. 7	50. 1	0. 2
Mo	b. d.	2	b. d.	b. d.	b. d.	b. d.	2
Ag	b. d.	0. 3	b. d.	b. d.	b. d.	b. d.	0. 3
Sn	5	41	2	1	5	5	1
Hf	8. 6	382	4. 9	1. 6	6. 9	15. 8	0. 1
Ta	1. 82	32. 1	0. 88	0. 41	1. 48	5. 21	0. 01
W	3. 2	14. 9	1. 7	1. 5	1. 6	4	0. 5
Tl	0. 08	b. d.	0. 48	0. 5	0. 06	0. 48	0. 05
Pb	19	54	17	15	9	20	3
Bi	b. d.	0. 7	b. d.	0. 5	0. 1	1	0. 1
Th	12. 1	274	6. 35	3. 25	4. 48	94. 4	0. 05
U	2. 94	58. 5	2. 06	1. 07	1. 86	9. 98	0. 01
La	18	455	12. 8	6. 97	15. 9	108	0. 05
Ce	43. 2	884	24. 8	13. 8	30. 6	224	0. 05
Pr	4	95. 6	2. 54	1. 41	3. 79	23. 8	0. 01
Nd	13. 8	315	8. 94	4. 86	14. 3	83. 9	0. 05
Sm	2. 88	57. 9	1. 7	0. 97	3. 23	15. 1	0. 01
Eu	0. 545	1. 71	0. 342	0. 234	0. 407	0. 941	0. 005
Gd	2. 34	42. 6	1. 41	0. 74	2. 52	10. 2	0. 01
Tb	0. 46	6. 08	0. 25	0. 12	0. 46	1. 2	0. 01
Dy	2. 88	32. 6	1. 45	0. 67	2. 45	5. 41	0. 01
Ho	0. 62	6. 61	0. 31	0. 13	0. 45	0. 93	0. 01
Er	1. 87	20. 3	0. 9	0. 36	1. 22	2. 46	0. 01
Tm	0. 302	3. 65	0. 142	0. 055	0. 187	0. 372	0. 005
Yb	2. 01	24. 9	1. 01	0. 37	1. 22	2. 46	0. 01
Lu	0. 3	4. 19	0. 153	0. 056	0. 177	0. 354	0. 002

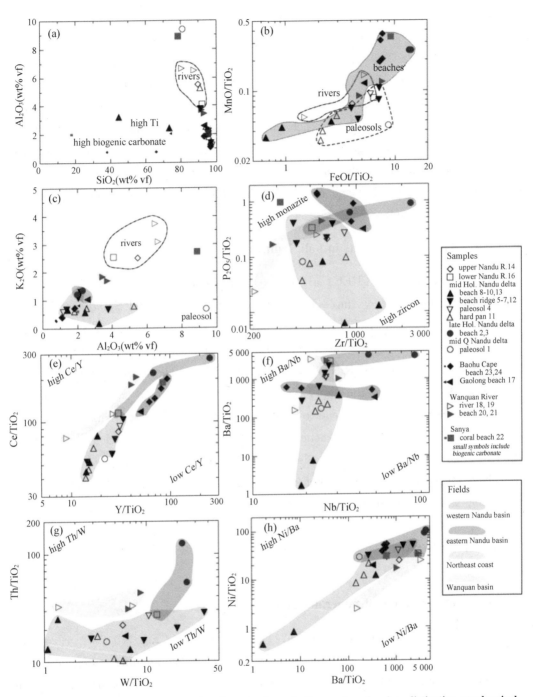

Fig 7 Elemental biplots for various elements from bulk samples showing distinctive geochemical features, as discussed in text. Seven samples with abundant biogenic carbonate have been recalculated on a calcium carbonate free basis. Most trace elements are normalised to TiO_2. (附彩图)

In general, beach and soil samples can be discriminated using a plot of MnO/TiO_2 vs. FeO_t/TiO_2 (Fig. 7b), with the soils tending to have higher FeO_t and lower MnO. River samples consistently show high K_2O (Fig. 7c), reflecting the abundance of K-feldspar and muscovite in sediment transported in the rivers, that is destroyed by chemical weathering in soils and by physical abrasion on beaches.

Geographic variation in particular elements has been studied, normalising the chemical analyses to TiO_2 to minimise the effects of heavy mineral concentration on beaches. For example, samples from the river and beaches of the Wanquan River system (samples 18~21) have high Ce for low Y (Fig. 7e), high Th for low W (Fig. 7g) and high Ba for low Ni (Fig. 7h). Beach and river samples of the lower Nandu River (samples 2, 3, 16) have high W/TiO_2, as do two beach ridge samples (samples 12, 13) from the mid-Holocene Nandu delta (Fig. 7g). In contrast, W was low (1.6 ppm) at Gaolong beach (sample 17) and below detection (0.5 ppm) near Baohu Cape on the northeast coast (samples 23, 24).

4.4 Rare earth elements in bulk sediments

Rare earth elements (REE) are plotted against average upper continental crust (UCC, Fig. 8)and selected samples are also shown normalized to C1 chondrite (Fig. 8f). Most patterns are close to UCC, but there are a few exceptions. Sample 19 (Wanquan River; Fig. 8d) has strong relative enrichment in light (L) REE and corresponding beach samples also show LREE > heavy (H) REE when normalised to UCC. In contrast, sample 14 from the upper Nandu River (Fig. 8a) and some Late Holocene Nandu delta samples (Fig. 8c) tend to have LREE<HREE when normalised to UCC, as do samples 23 and 24 from the beaches of Baohu Cape. The ilmenite-rich beach sand at the mid-Holocene Nandu delta (sample 09; Fig. 8b) and the Wanquan River (sample 19) both have strong negative Eu anomalies. One beach sample (sample 02) from the Late Holocene Nandu delta shows strong HREE enrichment (Fig. 8c).

5 Discussion

5.1 Interpretation of mineral sources

The modal abundance of different heavy minerals (Fig. 4) has a complex relationship to bedrock sources. The availability of a particular mineral depends on its abundance in source rocks, and heavy minerals are commonly more abundant in basalt and many metamorphic source rocks than in granite. Hydraulic sorting during transport may influence relative abundance. The heavy mineral fraction was analysed from the 63~177 μm (4~2.5 ϕ) size range, but that size range is noticeably enriched in higher density grains (specific gravity>4) such as zircon and Fe-Ti oxides in coarse-grained river sand (e. g., sample 16, mean grain size 0.5 ϕ; Fig. 4), whereas fine-grained sands such as sample 14 (mean 2.2 ϕ) have abundant minerals such as andalusite and tourmaline with specific gravity of<3.2. Finally, abrasion and chemical weathering

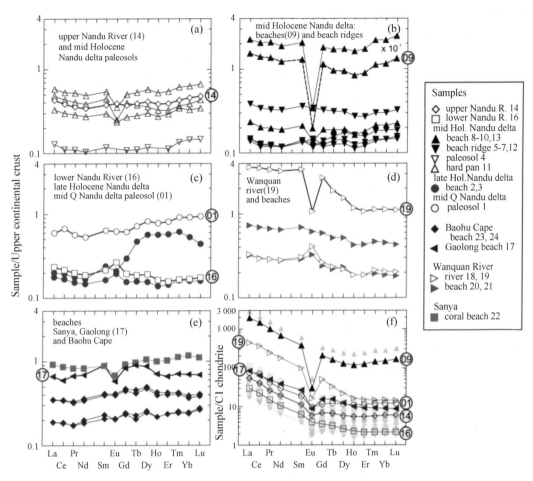

Fig 8 (a-e) REE plots normalised to Upper Continental Crust (UCC). Symbols as in Fig. 7. (f) REE plots of samples studied in petrographic detail, normalised to C1 chondrite.

during river transport and particularly on beaches will tend to concentrate ultrastable minerals such as zircon and rutile (samples 9 and 17, Fig. 4). All this environmental bias (Garzanti et al., 2009) complicates the interpretation of heavy mineral abundance and assessment of their relationship to bulk geochemical properties.

The distribution of ilmenite suggests that there is a major source in the east central part of the island, drained by the Wanquan River (sample 19) and the eastern headwaters of the Nandu River, that contributes to sample 16, but with lesser supply to the southwestern Nandu River at Chengmai (sample 14) (Fig. 4). This distribution suggests that much of the ilmenite is derived from the Pliocene-Quaternary basalts (Fig. 1A), which contain microphenocrystal ilmenite and titanomagnetite (Ho et al., 2000), although some ilmenite may be derived from granites or schists. Ilmenite is most abundant in tholeiites, whereas titanomagnetite is dominant in the less common alkalic basalts that are principally of Late Pleistocene age (Ho et al., 2000). Basalts of this age are mapped southwest of Haikou and south of Dingan (Fig. 1A). This distribution of source rocks is consistent with the lowest titanomagnetite abundance being present in sample 01, which predates most of the alkali basalt, and in samples 14, 17, and 19 which are distant from the Upper Pleistocene basalt. However, chemical weathering may also have played an important role in destroying magnetite in samples 01 and 17, as discussed below. The ratio of magnetite to titanomagnetite in the Wanquan River (sample 19) is much higher than in any other sample, suggesting an additional source of magnetite within the Wanquan River basin.

Metamorphic rocks are interpreted as the source for several heavy minerals, such as staurolite, andalusite and garnet, but in the absence of detailed mineralogical studies of bedrock geology, the precise sources are unknown. Some may be polycyclic from Cretaceous sandstones, which are relatively abundant in both the Nandu and Wanquan basins, and Carboniferous sandstones, which are most abundant in the SW Nandu river basin. Low-grade Lower Paleozoic metamorphic rocks are common in the SW Nandu and Wanquan basins. Both staurolite, notably in the upper Nandu River (sample 14) and at Gaolong beach (sample 17), and andalusite, which is most abundant in the upper Nandu River (sample 14), and common in the lower Nandu River (sample 16) and at Gaolong beach (sample 17), suggest that Gaolong beach sediment may be derived from the Nandu basin. Grossular garnet is most abundant in the upper Nandu and Wanquan rivers (samples 14, 19). It is normally sourced from metacarbonate skarns, potentially found where plutons have intruded Carboniferous limestone in the areas southwest of Chengmai and southwest of Boao (Fig. 1A). High-grade Mesoproterozoic metamorphic rocks outcrop in small areas of the Wanquan basin west of Qionghai and may be the source of almandine garnet, which is common only in the Wanquan River (sample 19).

Relatively few heavy minerals are derived from granite, despite the abundance of granite in Hainan Island. The principal such mineral is zircon, which is common in the lower Nandu River (sample 16) and strongly concentrated on beaches (samples 09, 17) because of its resistance to abrasion. Several minerals may be sourced from either metamorphic rocks or granite, including monazite and tourmaline. Monazite is most abundant in the lower Nandu River (sample 16) and the mid-Holocene Nandu delta (sample 09). The highest proportion of tourmaline sourced from graniteis found in the upper Nandu River (sample 14), mid-Holocene Nandu delta (sample 09) and Gaolong beach (sample 17), with higher proportions of tourmaline from metapelites elsewhere (Fig. 6). Some magnetite and ilmenite may also be derived from granite or metamorphic rocks. Small quantities of REE-bearing minerals including allanite, xenotime and titanite are either predominantly or exclusively from granites.

Chrome spinels are sparse (Fig. 5f) and are most abundant in sample 16 (lower Nandu River), sample 09 (the beach at the mid-Holocene Nandu delta) and sample 17 (Gaolong beach). They are likely derived from gabbro in the eastern headwaters of the

Nandu River south of Dingan, although metamafic rocks of ophiolite affinity are also reported (Li et al., 2002) south of Chengmai (Fig. 1).

5.2 Changes from river to beach sands: impact on provenance interpretation

Differences between river and corresponding beach sands are most clearly shown by differences between samples 14 (upper Nandu River) and 09 (beach at the mid-Holocene Nandu delta). On the beach, resistant heavy minerals, particularly zircon and ilmenite, predominate. Because many of the beaches are undergoing erosion, there is also a tendency for heavier heavy minerals to accumulate (Garzanti et al., 2009). In general, the zircon to monazite ratio is higher on beaches than from rivers (Figs. 4, 7d), which cannot be ascribed to density sorting as monazite is a little denser than zircon. Grains are subrounded, compared with subangular in the river (Fig. 2). Ferruginous crusts have broken up on the beaches, and heavy minerals such as epidote and amphibole that are common in the rivers are sparse on beaches.

Heavy minerals contain a large part of the trace element signature of the bulk sediment samples, so that care is needed in interpreting the provenance significance of bulk geochemistry. Nevertheless, some trace elements normalised to TiO_2 (to reduce the effect of bulk concentration of heavy minerals) show geographically coherent patterns (Fig. 7). Compared with the Nandu River basin (rivers and beaches), the Wanquan River basin has higher Ce/Y (Fig. 7e), reflecting more LREE fractionation in the source rocks that include abundant Cretaceous granite porphyry. Tungsten (normalised to TiO_2) is more abundant in the Wanquan River and lower Nandu River than in the upper Nandu River (Fig. 7g), reflecting the greater abundance of Cretaceous granites, potentially with high W, in the east (Fig. 1). The Ni/Ba ratio is lower in the Wanquan River basin (Fig. 7h), probably because of lesser supply of olivine from basalt (as shown by the low titanomagnetite in the Wanquan River).

In general, the beaches on the northeast coast, at Baohu Cape (samples 23, 24) and Gaolong beach (sample 17), have elemental ratios that overlap those of the Nandu River basin. The Ce/Y ratio is lower (Fig. 7e), which may reflect concentration of HREE in zircon in the beach sands. The zircon/monazite ratio is very high (~50) and the proportion of magnetite and other Fe (hydr) oxides is very low (~1% of heavy minerals). Both monazite and magnetite are destroyed by chemical weathering, suggesting that the sand was originally supplied by rivers in the Pleistocene.

In summary, individual rivers and their major tributaries in adjacent drainage basins (represented by samples 14, 16 and 19) have significant differences in detrital mineralogy (Fig. 4) and geochemistry (Fig. 7) despite relatively similar hinterland geology (Fig. 1).

5.3 The origins of REE distributions

Rare-earth element patterns have been widely used in provenance studies (e. g., Bhatia, 1985; Cao et al., 2015). REE abundance in most river and beach sands from this study is 0.1~0.3 times the abundance in average upper continental crust (UCC). This probably reflects the deep weathering of the source rocks, so that ferromagnesian minerals and feldspars that carry much of the REE signature break down to clays, which are winnowed out of river and beach sands. Mid-Quaternary soil sample 01, by contrast, with a high proportion of finegrained material (Supplementary data Table S2), has a REE pattern close to average UCC (Fig. 8c) and REE were detected by EDS analysis in a few ferruginous crusts. Significant enrichment in REE to values above UCC, and strong Eu depletion relative to UCC, are a consequence of a small number of accessory minerals that are strongly enriched in REE. Among heavy minerals identified in this study, monazite, allanite and titanite show relative enrichment in light LREE, whereas xenotime and zircon have heavy HREE concentrations almost as high asmonazite, but LREE concentrations are 2~3 orders of magnitude lower (Bea, 1996). These REE-bearing minerals are principally derived from granites and thus might be of value in discriminating different granite sources.

The relative contribution of different REE minerals to the LREE, middle(M) REE, HREE and Eu has been calculated (Fig. 9) based on measured abundance of REE minerals in the 63~177 μm fraction(Supplementary data Table S5), REE abundance in minerals from the literature (Bea, 1996; Rasmussen et al., 1998; Emsbo et al., 2015) and assumptions noted in the figure caption. Sample 19 (Wanquan River) has the lowest ratio of zircon tomonazite plus allanite. It is the sample with the highest LREE to HREE ratio. The LREE distribution (Figs. 8f, 9a) appears to be dominated by monazite, with some contribution from allanite and titanite. The HREE values are five times more abundant than predicted for monazite and allanite, but similar to that predicted for the observed zircon, thorite and titanite. Monazite and titanite characteristically show little change in normalised abundance of the LREE from La to Nd, whereas allanite shows at least a 3-fold decrease from La to Nd. The 2-fold decrease in sample 19 thus reflects a strong influence of allanite. The Eu anomaly (Eu/ Eu *)~4 is probably influenced by a strong negative Eu anomaly in allanite, thorite and zircon; the minerals shown in Fig. 9c are those that contribute to high Eu values.

Sample 09 from the beach on the mid-Holocene Nandu delta shows enrichment in HREE (Fig. 8b), as does the mid-Quaternary paleosol sample 01 (but less strongly), related to their particularly abundant zircon contents (Fig. 9d). The strong Eu anomaly in the beach sample, absent in the paleosol sample, is related tomonazite and thorite, which are both present in the beach sample but rare or absent in the paleosol.

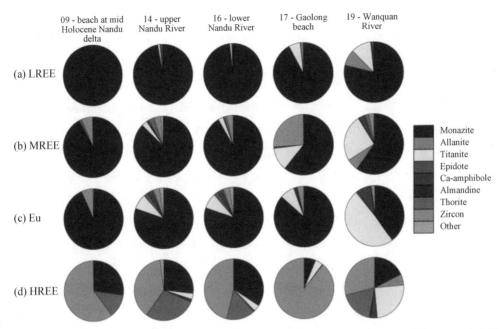

Fig 9 Estimated contribution of different minerals to (a) LREE; (b) MREE; (c) Eu; and (d) HREE, in five samples with detailed mineralogy. Abundance of REE in various minerals is mostly from Bea (1996); also Emsbo et al. (2015) and Rasmussen et al. (1998). Mineral abundances were counted for the 63~177 μm fraction and calculated using the total weight of heavy minerals in that fraction. Content of heavy minerals in other fractions was estimated: the 177~2 000 μm fraction as 25% and for the<63 μm as 50% of the abundance in the 63~177 μm fraction for river samples and as 20% for both fractions in beach samples. See Supplementary data Table S5.（附彩图）

River samples 14 and 16 from the upper and lower Nandu River differ in absolute abundance of REE (Fig. 8a, c): sample 14 is twice as abundant in total REE and has a negative Eu anomaly, yet has lower thorite, titanite and zircon and similar monazite compared with sample 16 (Fig. 4). Sample 16 is coarser grained than 14, and the proportion of REE heavy minerals in the > 177 μm fraction may have been overestimated. More likely, the high proportion of ferruginous crusts in sample 14 is an important carrier of REE. Sample 16 is also noteworthy for its lack of a Eu anomaly, probably because Eu depletion due to monazite is balanced by Eu enrichment due to a higher proportion of epidote (Fig. 4). In sample 17 (Gaolong beach), zircon is about 50 times more abundant than monazite, but shows no strong enrichment in HREE (Fig. 8e). This may indicate an important role for small amounts of thorite, xenotime or titanite. The strong negative Eu anomaly reflects the presence of monazite and tourmaline.

The influence of zircon on REE abundance can be examined from bulk chemical analyses of sediments, because zirconium is predominantly present in zircon. The suite of samples from the mid-Holocene Nandu delta (blue triangle symbols in Fig. 10) shows a better correlation between Zrand La，Eu/Eu * and Lu（Fig. 10）than between Zr and other elements concentrated in different resistant minerals along with zircon，such as Cr in chromite（Fig. 10c）and Ti in ilmenite and rutile（Fig. 10b）. Plots of selected REE elements against P_2O_5 also show a good correlation（Fig. 10g，h），confirming that monazite，together with other aluminophosphates and possibly xenotime，also have influenced REE content. In contrast，samples from the lower Nandu River（sample 16）and Wanquan River（sample 19）have higher relative abundance of Cr（from ophiolite sources）（Fig. 10c）and LREE（Fig. 10f），and a smaller Eu anomaly（Fig. 10e）. The northeastern beaches show little increase in REE with increasing P_2O_5（Fig. 10g，h），suggesting that REE-phosphates are unimportant. As noted for Gaolong beach，this is a consequence of sand being supplied in the Pleistocene and subsequent chemical weathering at marine lowstands.

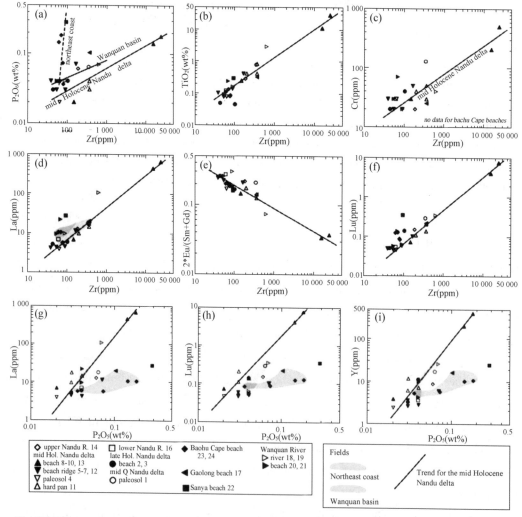

Fig 10　Biplots of selected elements that may be indicators of REE sources. For further explanation, see text. (For interpretation of the references to colour in this figure legend, the reader is referred to the web version of this article.)

We conclude that while REE patterns should be used as provenance indicators in sands and sandstones with caution (as noted by Cullers, 2000), but when coupled with some knowledge of REE mineral abundance they can be used in provenance studies. It takes a very different heavy mineral assemblage, such as that in the Wanquan River (sample 19), to produce a distinctive REE pattern (Fig. 8d). Rather similar REE distributions, such as in samples 14 (upper Nandu River) and 17 (Gaolong beach) (Fig. 8a, e) may result from sands with substantially different accessory heavy mineral components (Fig. 4), with different minerals controlling the REE patterns (Fig. 9). Thus minor amounts of REE minerals, which may only be reliably quantified if thousands of grains are identified and counted, may significantly influence bulk sediment REE patterns.

5.4　Paleogeographic changes in river patterns

The active neotectonics and basaltic volcanism in northern Hainan Island have the potential to produce river avulsion, both by basement block tilting and by lava flows damming rivers, illustrated by the mid-Holocene Shishan volcano. The bulk geochemistry data from beaches of the northeast coast, informed by the general relationships between minerals and geochemistry determined in this study, are used to assess the provenance of the beach ridges and placer minerals of the northeast coast. Strong tidal currents in Qiongzhou Strait have built a submarine tidal delta east of Hainan Cape (Ni et al., 2014) and waves may transport some of this sediment to the coastline. There is no evidence for significant longshore drift from the modern Nandu or Wanquan rivers. The wide beach ridge complex north of Tonggu Mountain (Ji and Zhang, 2012) suggests a more voluminous sediment supply from rivers at some time in the past. Sediment supply since the mid-Holocene sealevel rise at Gaolong Bay has been insufficient to infill the large estuary, suggesting that sediment supply took place in the Pleistocene. Such Pleistocene supply is also indicated by low abundance of monazite and magnetite, susceptible to chemical weathering. At least three raised marine terraceson the northeast peninsula (Wu and Wu, 1987) suggest late Quaternary uplift. Possible dry valleys are interpreted from satellite imagery to cross parts of the northeast peninsula (Fig. 1a).

The beach samples from the northeast coast (samples 17, 23, 24) have the closest geochemical matches for many elements with the beaches of the modern Nandu delta (Fig. 7b, d, e, h). Ba/TiO_2 is lower (Fig. 7f), probably because of loss of feldspars by abrasion or weathering. Mineralogy of the Gaolong beach sample (sample 17) confirms this finding, with the common occurrence of chrome spinel, staurolite and andalusite, the latter two being characteristic of the SW headwaters of the Nandu River (sample 14, Table 2) and chrome spinel likely from the SE headwaters. There is no evidence for significant contribution of sediment from the Wanquan river, for example by longshore drift or reworking of Pleistocene lowstand deposits: many geochemical markers are quite distinct (Fig. 7b-e, g, h) and mineralogy is also quite different (e. g. , titanomagnetite/ magnetite ratio, more abundant almandine and allanite, different amphibole assemblage: Table 2, Figs. 4, 5). Neither is it reasonable to invoke tidal current transport so far from Qiongzhou Strait, particularly since Li et al. (2015) found no evidence of a Nandu River supply to their station F5, 50 km northeast of Hainan Cape. The mineralogical and geochemical evidence thus suggest that at times in the Quaternary, the Nandu River flowed across the northeast peninsula. The beach ridges close to present sea level north of Tonggu Mountain suggest that the most recent eastwards flow of the Nandu River could have dated from the last interglacial period around 125 ka, with the river flowing along the continuation of the Wangwu-Wenjiao fault, either to Gaolong Bay or to the north of Tonggu Mountain (Fig. 11d).

By the mid-Holocene, the Nandu River had avulsed to a northwards course to the old delta at Changliu (Fig. 11b). This drainage route may have been the stable drainage system throughout the last sea-level lowstand (Fig. 11c), draining westwards to the Beibu Gulf (Gulf of Tonkin), as there is no evidence for lowstand deposits of the Nandu River on the continental shelf east of Hainan Island (Li et al. , 2015). The low relief basalt flows east of the modern Nandu River are of early Pleistocene age, and are cut by several W-E dry valleys, so that active basalt volcanism is unlikely to have caused river avulsion. Rather, tectonic tilting related to dip-slip on the Haikou-Yunlong fault could have blocked original eastwards flow of the Nandu River at some time in the Late Pleistocene.

The presence of well developed beach ridges at Changliu indicates that the Nandu paleo-delta was still active after the sea rose to close to modern sea level at about 7 ka. The presence of chrome spinel, probably from gabbro in the eastern headwaters of the Nandu River south of Dingan (Fig. 1), suggests that the southeastern headwaters of the Nandu River drained westwards along the Wangwu-Wenjiao fault and then northwards to Chengmai, entering the Qiongzhou Strait (Fig. 11b). High W in some samples (Fig. 7g) indicates that the southwestern headwaters of the Nandu River, above Chengmai, also drained to the delta at Changliu.

The most recent major avulsion was the blocking of the Nandu River in the Late Holocene by basalt flows around Shishan volcano (Wang, 1998), resulting in the modern course being deflected eastwards along the Wangwu-Wenjiao fault and then northwards along the Haikou-Yunlong fault to Haikou (Fig. 11d). This was apparently also the route of the Nandu River at some time in the mid-Quaternary, as the paleosol at Haikou (sample 01) contains indicators of the Nandu River drainage, with abundant andalusite, some tourmaline (Table 2, Fig. 4), and a characteristic Th/W ratio (Fig. 7g).

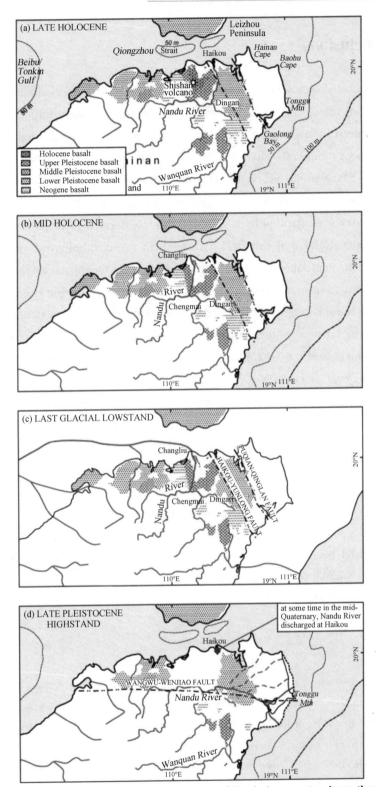

Fig 11 Maps illustrating the inferred evolution of the drainage system in northeastern Hainan Island. (a) Present geography; (b) mid-Holocene; (c) last glacial lowstand; (d) Late Pleistocene interglacial highstand, probably the last interglacial at～125 ka.

6　Conclusions

1. Although the headwaters of the Nandu and Wanquan rivers drain mountainous, deeply weathered terrain principally of granite, variation in heavy mineral abundance is strongly influenced by metamorphic and mafic igneous source rocks that have a higher abundance of heavy minerals than granite. Variable distribution of such source rocks means that the SW Nandu, SE Nandu and Wanquan basins have different heavy mineral assemblages. These differences are reflected in variations in trace element geochemistry. REE geochemistry is controlled by accessory minerals derived principally from granites.

2. Geochemical and mineralogical data confirm geomorphological evidence of river avulsion due to neotectonics and lava flows. In the mid-Quaternary, the Nandu River supplied sediment to near its present mouth. Later, probably during the last interglacial highstand, the Nandu River supplied sediment to the northeast coast north of Tonggu Mountain. In the latest Pleistocene and Holocene the Nandu River discharged again to Qiongzhou Strait, first at Changliu and then, following damming by the Shishan volcano, farther east at Haikou.

3. Detrital geochemistry alone shows too much variability to interpret changes in river supply. The heavy mineral study was necessary to provide an understanding of the geological origins of geochemical variation. REE geochemistry of sand samples is particularly influenced by trace quantities of REE-bearing accessory minerals.

Acknowledgements

Geochemical and heavy mineral analyses were funded by a Natural Sciences and Engineering Research Council (NSERC) Discovery Grant to GP-P. We thank the editor and two anonymous reviewers for their critically constructive reviews.

References

[1] Activation Laboratories Ltd. 2012. Lithogeochemistry Available at: http://www. actlabs. com/ page. aspx? menu=74&app=244&cat1=595&tp=2&lk=no Accessed 27 August 2012.

[2] Bea, F. 1996. Residence of REE, Y, Th and U in granites and crustal protoliths: implications for the chemistry of crustal melts. *Journal of Petrology*, 37: 521 - 552.

[3] Bhatia, M. R. 1983. Plate tectonics and geochemical composition of sandstones. *Journal of Geology*, 91: 611 - 627.

[4] Bhatia, M. R. 1985. Rare earth element geochemistry of Australian Paleozoic graywackes and mudrocks: provenance and tectonic control. *Sedimentary Geology*, 45: 97 - 113.

[5] Cao, L. , Jiang, T. , Wang, Z. , Zhang, Y. , Sun, H. 2015. Provenance of Upper Miocene

sediments in the Yinggehai and Qiongdongnan basins, northwestern South China Sea: evidence from REE, heavy minerals and zircon U-Pb ages. *Marine Geology*, 361: 136 - 146.

[6] Cullers, R. L. 2000. The geochemistry of shales, siltstones and sandstones of Pennsylvanian-Permian age, Colorado, USA: implications for provenance and metamorphic studies. *Lithos*, 51: 181 - 203.

[7] Dick, H. J. B., Bullen, T. 1984. Chromium spinel as a petrogenetic indicator in abyssal and alpine-type peridotites and spatially associated lavas. *Contributions to Mineralogy and Petrology*, 86: 54 - 76.

[8] Emsbo, P., McLaughlin, P. I., Breit, P. N., du Bray, E. A., Koenig, A. E. 2015. Rare earth elementsin sedimentary phosphate deposits: solution to the global REE crisis? *Gondwana Research*, 27: 776 - 785.

[9] Fleet, M. E., Barnett, R. L. 1978. Al^{IV}/Al^{VI} partioning in calciferous amphiboles from the Frood Mine, Sudbury, Ontario. *Canadian Mineralogist*, 16: 527 - 532.

[10] Fu, W., Xu, D., Wu, C., Yang, C., Zhou, Y., Wang, Z. 2013. Molybdenite Re-Os isotopic dating of Hongmenling Mo-W deposit in Hainan province and its geological implications. *Journal of East China Institute of Technology (Natural Science)*, 36: 135 - 142.

[11] Garzanti, E., Andó, S., Vezzoli, G. 2009. Grain-size dependence of sediment composition and environmental bias in provenance studies. *Earth and Planetary Science Letters*, 277: 422 - 432.

[12] Gould, K., Pe-Piper, G., Piper, D. J. W. 2010. Relationship of diagenetic chlorite rims to depositional facies in Lower Cretaceous reservoir sandstones of the Scotian Basin. *Sedimentology*, 57: 587 - 610.

[13] Guangdong BGMR (Bureau of Geology Mineral Resources of Guangdong Province), 1988. Regional Geology of Guangdong Province. Geological Publishing House, Beijing, pp. 1～602 (in Chinese).

[14] Ho, K. S., Chen, J. C., Juang, W. S. 2000. Geochronology and geochemistry of late Cenozoic basalts from the Leiqiong area, southern China. *Journal of Asian Earth Sciences*, 18: 307 - 324.

[15] Hu, B., Li, J., Cui, R., Wei, H., Zhao, J., Li, G., Fang, X., Ding, X., Zou, L., Bai, F. 2014. Clay mineralogy of the riverine sediments of Hainan Island, South China Sea: implications for weathering and provenance. *Journal of Asian Earth Sciences*, 96: 84 - 92.

[16] Ji, X., Zhang, Y. 2012. Observational study of morphodynamics of sandy beaches in Yueliang Bay, northeastern Hainan Island. 6th Chinese-German Joint Symposium on Hydraulic and Ocean Engineering, CGJOINT 2012, pp. 633 - 638.

[17] Kassoli-Fournaraki, A., Michailidis, K. 1994. Chemical composition of tourmaline in quartz veins from Nea Roda and Thasos areas in Macedonia, Northern Greece. *Canadian Mineralogist*, 32: 607 - 615.

[18] Lei, J., Zhao, D., Steinberger, B., Wu, B., Shen, F., Li, Z. 2009. New seismic constraints on the upper mantle structure of the Hainan plume. *Physics of the Earth and Planetary Interiors*, 173: 33 - 50.

[19] Li, X. H., Zhou, H. W., Chung, S. L., Ding, S. Z., Liu, Y. 2002. Geochemical and Sm-Nd isotopic characteristics of metabasites from Central Hainan Island, South China and their tectonic significance. *Island Arc*, 11: 193 - 205.

[20] Li, G. , Yan, W. , Zhong, L. , Xia, Z. , Wang, S. 2015. Provenance of heavy mineral deposits on the northwestern shelf of the South China Sea, evidence from single-mineral chemistry. *Marine Geology*, 363: 112 - 124.

[21] Liu, X. , Zhang, D. , Zhai, S. , Liu, X. , Chen, H. , Luo, W. , Li, N. , Xiu, C. 2015. A heavy mineral viewpoint on sediment provenance and environment in the Qiongdongnan Basin. *Acta Oceanologica Sinica*, 34: 41 - 55.

[22] Ma, X. 1989. Lithospheric Dynamics Atlas of China. China Cartographic Publishing House, Beijing, p. 70.

[23] Ma, R. , Yang, Y. , He, Y. 2010. Geochemistry of rare earth elements in coastal and estuarial areas of Hainan's Nandu River. *Zhongguo Xitu Xuebao*, 28: 110 - 114.

[24] Mao, J. W. , Wang, Y. T. , Lehmann, B. , Yu, J. J. , Du, A. D. , Mei, Y. X. , Li, Y. F. , Zhang, W. S. 2007. Large-scale tungsten-tin mineralization in the Nanling region, South China: metallogenic ages and corresponding geodynamic processes. *Acta Petrologica Sinica*, 23: 2329 - 2338.

[25] Ni, Y. , Endler, R. , Xia, Z. , Endler, M. , Harff, J. , Gan, H. , Schulz-Bull, D. E. , Waniek, J. J. 2014. The "butterfly delta" system of Qiongzhou Strait: morphology, seismic stratigraphy and sedimentation. *Marine Geology*, 355: 361 - 368.

[26] Pe-Piper, G. , Piper, D. J. W. , Dolansky, L. M. 2005. Alteration of ilmenite in the Cretaceous sands of Nova Scotia, southeastern Canada. *Clays and Clay Minerals*, 53: 490 - 510.

[27] Pe-Piper, G. , Triantafyllidis, S. , Piper, D. J. W. 2008. Geochemical identification of clastic sediment provenance from known sources of similar geology: the Cretaceous Scotian Basin, Canada. *Journal of Sedimentary Research*, 78: 595 - 607.

[28] Rasmussen, B. , Buick, R. , Taylor, W. R. 1998. Removal of oceanic REE by authigenic precipitation of phosphatic minerals. *Earth and Planetary Science Letters*, 164: 135 - 149.

[29] Ryan, K. M. , Williams, D. M. 2007. Testing the reliability of discrimination diagrams for determining the tectonic depositional environment of ancient sedimentary basins. *Chemical Geology*, 242: 103 - 125.

[30] Tsikouras, V. , Pe-Piper, G. , Piper, D. J. W. , Schaffer, M. 2011. Varietal heavy mineral analysis of sediment provenance, Lower Cretaceous Scotian Basin, eastern Canada. *Sedimentary Geology*, 237: 150 - 165.

[31] von Eynatten, H. , Barceló-Vidal, C. , Pawlowsky-Glahn, V. 2003. Composition and discrimination of sandstones: a statistical evaluation of different analytical methods. *Journal of Sedimentary Research*, 73: 47 - 57.

[32] Wang, Y. 1992. A comparative study on harbour siltation and harbour development, Hainan Island, China. In: Wang, Y. , Schafer, C. T. (Eds.), Island Environment and Coastal Development. Nanjing University Press, Nanjing, pp. 373 - 391.

[33] Wang, Y. 1998. Tidal Inlet, Embayment Coast of Hainan. China Environmental Science Publishing House, Beijing (282 pp).

[34] Wang, Y. , Zhou, L. 1990. The volcanic coast in the area of northwest Hainan Island. *Acta Geographica Sinica*, 45: 321 - 330.

[35] Wang, Z. L. , Xu, D. R. , Wu, C. , Fu, W. W. , Wang, L. , Wu, J. 2013. Discovery of the Late

Paleozoic ocean island basalts (OIB) in Hainan Island and their geodynamic implications. *Acta Petrologica Sinica*, 29: 875 - 886.

[36] Wu, Z. ,Wu, K. 1987. Sedimentary characteristics and developing model of coastal dunes in north coast of Hainan Island. *Acta Geographica Sinica*, 42: 129 - 141.

[37] Zeng, X. , Yang, N. , Jin, X. , Zhang, M. 2011. Exploitation and utilization scheme of Zr-Ti placers in Baoding Sea. *Zhongnan Daxue Xuebao (Ziran Kexue Ban)/Journal of Central South University (Science and Technology)*, 42: 277 - 284.

[38] Zhang, F. , Wang, Y. , Chen, X. , Fan, W. , Zhang, Y. , Zhang, G. , Zhang, A. 2011. Triassic highstrain shear zones in Hainan Island (South China) and their implications on the amalgamation of the Indochina and South China Blocks: kinematic and ^{40}Ar/ ^{39}Ar geochronological constraints. *Gondwana Research*, 19: 910 - 925.

[39] Zhang, Y. Y. , Pe-Piper, G. , Piper, D. J. W. 2014. Sediment geochemistry as a provenance indicator: unravelling the cryptic signatures of polycyclic sources, climate change, tectonism and volcanism. *Sedimentology*, 61: 383 - 410.

The Volcanic Coast and Tectonic Activities of Northwestern Hainan Island, China[*]

Three periods of volcanism are documented in the area of northwestern Hainan Island. The first eruption occurred during the middle Pleistocene (5.2×10^4 Y. B. P.), forming a basaltic hinterland, with some lava covering the lower sandy deposit and forming a veneer of contact-metamorphic alteration. The second eruption was in the late Pleistocene (2.5×10^4 Y. B. P.) depositing volcaniclastics with well-preserved ripple bedding, olivine basalt and olivine tholeite basalt. The latest eruptions were in the Holocene with many small breached craters forming volcano coasts and islands in the area of Beibu Bay. The last eruption elevated an adjacent shingle deposit and shallow sediment to form a 20 m high terrace.

The volcanic belts with a northeasterly trend may result from a tectonic event along a deep tension fault zone which developed in the Beibu Bay and the Yingge Sea area, especially considering that the thickness of the earth's crust there is only 30 km. Present day barrier coral reefs have only developed along the Holocene volcanic coast of China, perhaps reflecting the longshore eruptive basalt as the basement of the barrier reefs, and also reflecting the geothermal gradient situation of the sea bottom.

---

* Ying Wang, Lu-fu Zhou: DETLEF BUSCHE, WÜRZBURG, eds. *Second International Conference on Geomorphology "Geomorphology & Geoecology", Abstracts of Posters and Papers*, pp. 310, 1989. 9. 3~9, FRANKFURT.

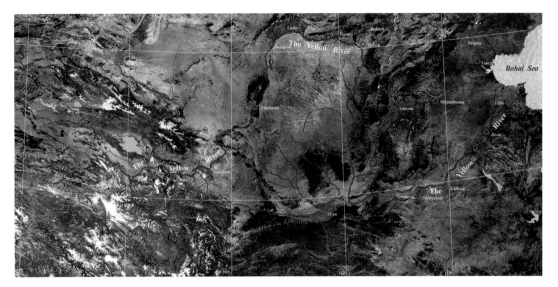

The Yellow River drainage basin（363页 Fig.1）

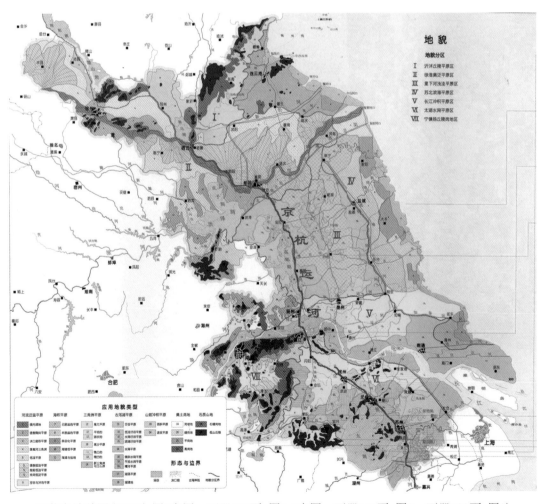

江苏省地貌图（390页 图3底图，下册110页 图1-1底图，下册167页 图2，下册182页 图3）

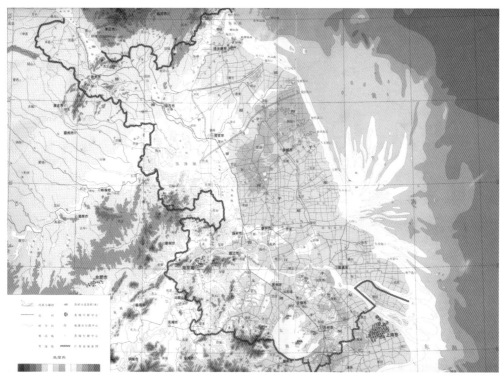

江苏海岸图（389页 图2底图，下册93页 图5，下册104页 Fig.7左，下册171页 图10底图）

苏北平原地貌遗迹（391-392页 图4）

Modern cheniers（100页 Fig.6，172页 Fig.4b）

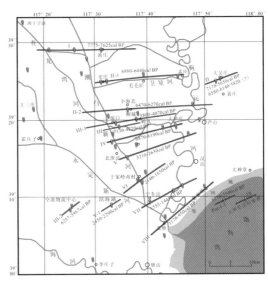

渤海湾牡蛎礁平原礁群时空分布图(67页 图3)

塔克拉玛干沙漠卫星照片(479页 图5)

塔克拉玛干沙漠边缘之山麓冲积扇(479页 图6)

塔克拉玛干沙漠具有挖掘坑、无V
痕或撞击点石英颗粒(490页 图16)

TK-10-7   TK-10-10

TK-10-11   TK-10-13

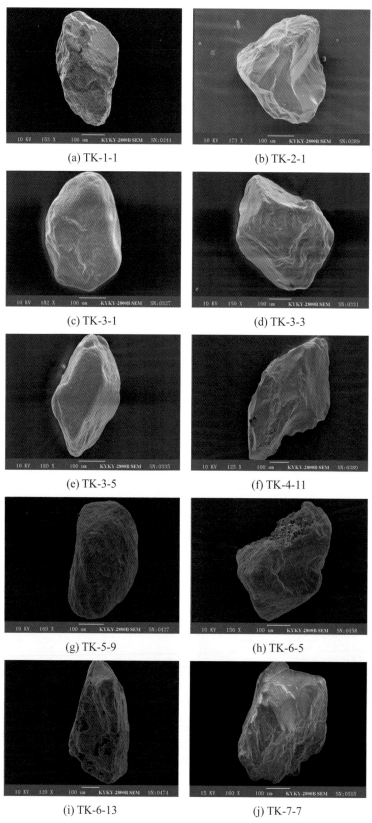

(a) TK-1-1

(b) TK-2-1

(c) TK-3-1

(d) TK-3-3

(e) TK-3-5

(f) TK-4-11

(g) TK-5-9

(h) TK-6-5

(i) TK-6-13

(j) TK-7-7

塔克拉玛干沙漠石英砂表面结构(488-489页 图15)

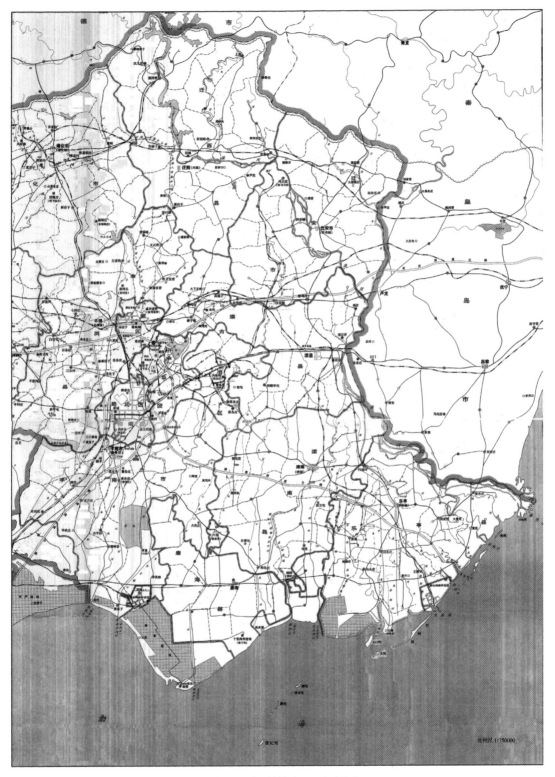

古滦河三角洲体(477页 图9)

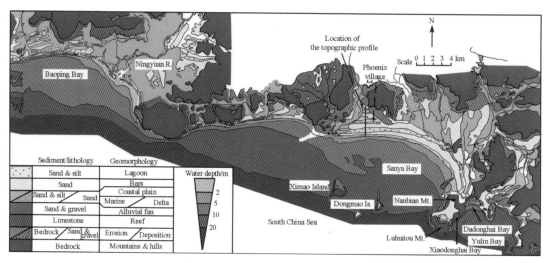

海南岛南部充填港湾的沙坝潟湖海岸（500页 图8，923页 Plate I）

三亚亚龙湾地区遥感影像（503页 图10）

海口西海岸海滩遭受侵蚀（496页 照片1）　　沿三亚湾发育的潟湖沙坝体系（501页 照片2）

海南三亚湾海滩侵蚀，碉堡相继入海
（501页 照片3）

三亚市金鸡岭路口外，在人工补沙施工
前的被侵蚀海滩（501页 照片4）

三亚市金鸡岭路口至光明路口间恢复的
海滩（501页 照片5）

沙坝顶端宾馆建筑阻流翻越
（502页 照片6）

某宾馆的沙坝前坡，回流冲刷移沙外流
（502页 照片7）

亚龙湾湾顶沙坝西段受蚀，沙坝顶上林
木坍坠入海（503页 照片8）

潮滩上部草滩
（551页 照片4-2，下册110页 图2）

潮滩外淤积带-粉砂波痕带
（554页 照片10）

现代黄河三角洲遥感影像(639页 图5)

黄河口偏移与扇形三角形(639页 图6)

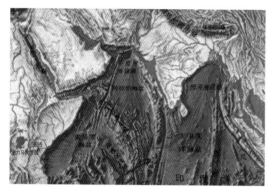

分布于印度洋海盆的恒河海底扇与阿拉
伯海盆的印度河海底扇(640页 图8)

亚马逊河口外侧陆架(641页 图9)

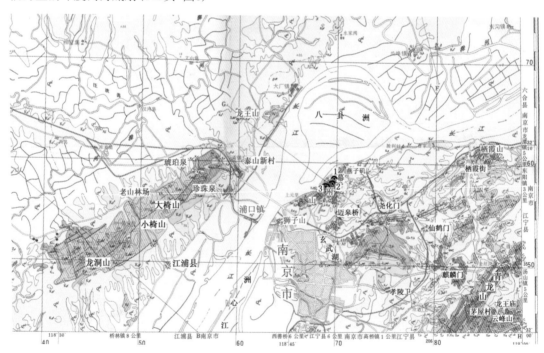

南京地质图和幕府山地理位置示意图(1—头台洞、2—二台洞、3—三台洞)
(697页 图4,710页 图1)

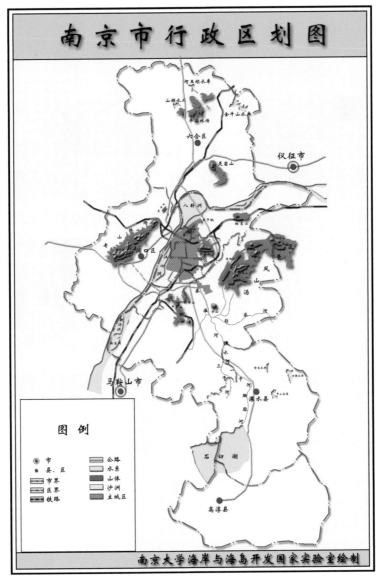

南京绿肺图(706页 图11，下册631页 图8)

港口与沿江风光带考查组全体成员
(706页 照片1)

长江上考察指挥(706页 照片2)

沿长江采取水样（707页 照片3）

新生圩深水港区（707页 照片5）

建设中的龙潭深水港集装箱码头（707页 照片6）

冲蚀的长江岸与堆放的石材（707页 照片7）

长江夹江北河口取水（708页 照片9）

沿长江采沙堆场（708页 照片10）

烟尘笼罩的梅山段沿江(708页 照片14)　　　长江江心洲洲头(708页 照片15)

三台洞崖壁及洞穴内洪水位遗迹图(711页 图2)

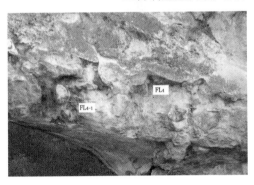

头台洞洪水位遗迹图(712页 图3)　　　燕子矶崖壁洪水位遗迹图(712页 图4)

采石矶崖壁水位遗迹图(713页 图5)

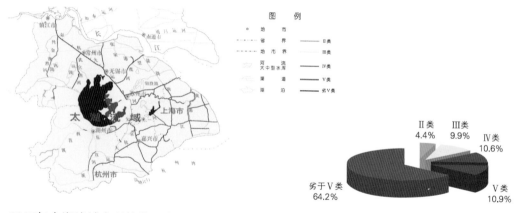

2007年太湖流域水环境状况(748页 图5)

2007年太湖流域河流全年期水质类别比例
(748页 图6)

三台洞3号洞、洞顶悬钙板及定年样品取样位置(756页 图4)

采石矶饮用水源一级保护区取水站水位遗迹
(713页 图6)

三台洞3号洞附近岩壁的岩相(756页 图5)

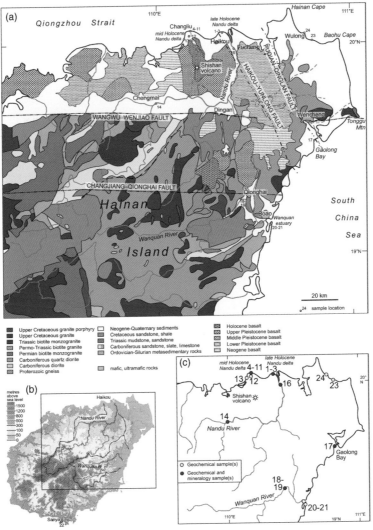

(a) Geological map of northeastern Hainan Island. (b) Topography of Hainan Island and the drainage basins of the Nandu and Wanquan rivers. (c) Map showing sample locations.（958页 Fig.1）

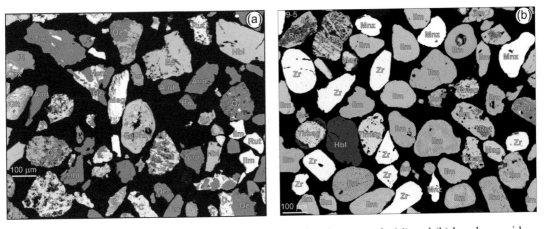

Examples of heavy minerals from (a) river (upper Nandu River, sample 14) and (b) beach on mid-Holocene Nandu delta (sample 09).（962页 Fig.2）

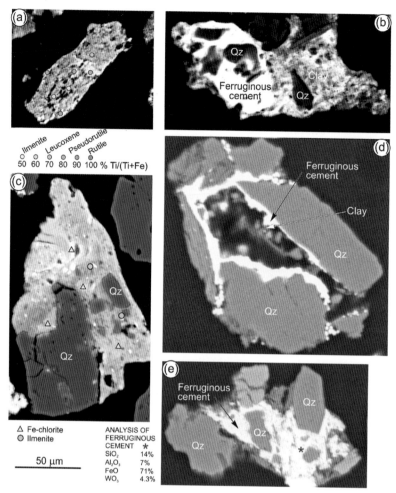

Backscattered electron images of grains with irregular Fe-Ti oxides of various origins.（963页 Fig.3）

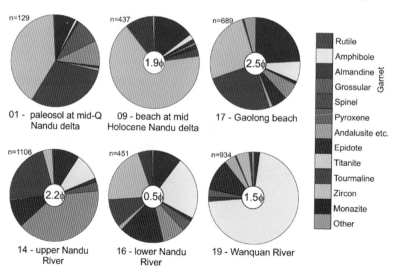

Heavy minerals excluding Fe-Ti oxides, Fe-(hydr)oxides, biotite, chlorite and unidentified grains

Pie diagrams showing abundance (counts) of heavy minerals in six representative samples.
（965页 Fig.4）

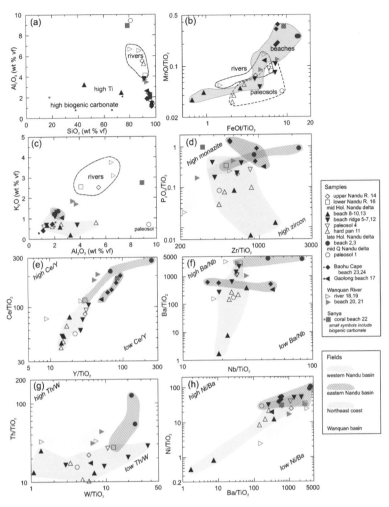

Elemental biplots for various elements from bulk samples showing distinctive geochemical features.
(971页 Fig.7)

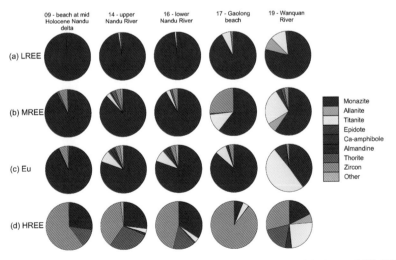

Estimated contribution of different minerals to (a) LREE; (b)MREE; (c) Eu; and (d) HREE, in five
samples with detailed mineralogy.(977页 Fig.9)